KU-300-941

Biochemistry for the
Medical Sciences

Biochemistry for the Medical Sciences

E. A. Newsholme

Department of Biochemistry,
University of Oxford

and

A. R. Leech

Gresham's School, Holt, Norfolk

JOHN WILEY & SONS

Chichester · New York · Brisbane · Toronto · Singapore

Copyright © 1983 by John Wiley & Sons Ltd.

Reprinted with corrections December 1984

All rights reserved.

No part of this book may be reproduced by any means, nor transmitted, nor translated into a machine language without the written permission of the publisher.

Library of Congress Cataloging in Publication Data:
Newsholme, E. A.
  Biochemistry for the medical sciences.

  Includes index.
  1. Metabolism. I. Leech, A. R. II. Title.
[DNLM: 1. Metabolism. QU 120 N558m]
QP171.N43 1983   612'.39   82-13643

ISBN 0 471 90056 7
ISBN 0 471 90058 3 (pbk.)

British Library Cataloguing in Publication Data:
Newsholme, E. A.
  Biochemistry for the medical sciences
  1. Metabolism.
  I. Title   II. Leech, A. R.
  612'.39   QP171

ISBN 0 471 90056 7 (Cloth)
ISBN 0 471 90058 3 (Paper)

Filmset by Speedlith Photo Litho Ltd., Manchester.
Printed by Page Brothers, Norwich.

CONTENTS

FOREWORD

There are basically two kinds of textbooks of biochemistry. One is concerned primarily with molecules and reactions, the other with biochemical explanations of physiological and clinical phenomena. The former is intended for biochemists and would-be biochemists, while the latter is most profitably read by clinicians, medical students and other students of the life sciences.

This new book by Eric Newsholme and Tony Leech is clearly of the second category. Basic biochemistry is not ignored: Readers will find all the essential information about biomolecular structure and the mechanism of biochemical reactions. However, the emphasis throughout is on the biological purposes of biochemical phenomena—on metabolic pathways and their control, and on the physiological significance and clinical relevance of the topics being discussed.

Much of this book is based on a lecture course in metabolic biochemistry that Dr. Newsholme has been giving to medical students at Oxford for many years. I had the privilege of spending a sabbatical year in the Department of Biochemistry at Oxford seven years ago, and I well remember the pleasure afforded by those lectures. They were models of clarity and conciseness, and they conveyed the beauty and excitement of modern biochemistry in a way that I had never seen before. This book captures the flavor of those lectures, but of course contains far more detail and substance.

As a clinician and medical teacher, I particularly appreciate the accessibility and logical simplicity of this book. The writing is clear, direct, and economical.

Biochemistry is chemistry with a purpose, and Dr. Newsholme explains the biological purpose of every molecule and every reaction he discusses. Students of medicine will like the way he deals with the functions and regulation of metabolic pathways and their disturbance by disease and physiological stress. This is an area in which the author himself has made notable original contributions.

There are many competent introductory textbooks of biochemistry, but none more suited to the special interests of physicians and physiologists. For those seeking a lucid and readable account of the metabolic machinery of the body, this new book should be just what the doctor ordered.

ARNOLD S. RELMAN, M.D.
Editor, *New England Journal of Medicine*
Professor of Medicine, Harvard
Medical School

PREFACE

Biochemistry is not always the most popular subject in the medical school curriculum and the reason is not too hard to see. Now that most of the reactions of the degradative and synthetic processes occurring in man have been described it has become a requirement for medical students to demonstrate their knowledge of these pathways. The tacit assumption is that this knowledge will automatically benefit the practising physician when he comes to deal with disturbances of these processes. In fact a knowledge of pathways is no more and no less useful than a knowledge of anatomy. Both are essential but in themselves quite inadequate. What the student needs is a great deal of help in appreciating the connection between metabolic pathways and metabolic disturbances.

Twenty years ago one of us was told by his Professor that no comprehensive biochemistry text would ever again be written—the subject was too vast. This statement underestimated the organizational ability of many biochemists; at regular intervals since then many excellent biochemistry textbooks have appeared. Paradoxically, these have increased the problems of the medical student because they have made the understanding of biochemistry easier without helping the student appreciate the medical implications. In fairness, many of these implications have only emerged as a result of the spate of quantitative metabolic research taking place in the 1970s. We feel that medical students deserve a more helpful and relevant textbook and we have sought to provide it. It is our intention that it should stand alone and provide all the metabolic biochemistry needed by the first- or second-year medical student, although we would never discourage a student from undertaking wider reading!

The book falls into five parts. In the first (Chapters 1–3) those fundamental principles on which all of metabolism is based are explained as clearly as we can manage. This is followed (in Chapters 4–9) by a description of the oxidative and degradative pathways of carbohydrates and lipid metabolism, how these are controlled and integrated in health and how failures of this integration explain a number of diseases and hence how they can be treated. In the third part (Chapters 10–15) the metabolism of amino acids is introduced into the metabolic scheme and again examined from the standpoint of control and integration in health and disease. In the penultimate part (Chapters 16–19) synthetic pathways are described together with a range of clinical topics including obesity and non-insulin-dependent diabetes. Finally, the metabolism and disturbance thereof, of the molecules that integrate other processes, namely hormones and neurotransmitters, is described.

Anyone writing a textbook is faced with decisions that hitherto could always be avoided. One of the most troublesome has been nomenclature; in its present state of change we are bound to have offended someone. We have done our best to be correct but have used as the final criterion utility and clarity for the student, even if that did mean an anticipation of future trends. For the names of enzymes we have used throughout the book the trivial names recommended by the Enzyme Commission. Another contentious decision will be the inclusion of references and a bibliography. We decided that, despite our inability to offer a comprehensive bibliography, we would include a substantial number of references both to demonstrate the relevance of biochemistry to clinical work and to help those with deeper interest in any area to begin a literature search.

Few of the ideas in the book have claim to originality on our part but while we hope we have not committed plagiarism we are grateful to all those whose writings have been clear enough to enable us to make use of them.

We wish to express our gratitude to our colleagues who have, either wittingly or unwittingly, helped with the preparation of this text, in particular: Dr. B. Crabtree, Dr. H. R. Fatania, Professor M. Gelder, Dr. G. M. Hall, Dr. L. Hermansen, Professor N. L. Jones, Dr. P. Lund, Dr. J. Mellanby, Dr. B. D. Ross, Professor K. W. Taylor, Dr. D. H. Williamson.

We would also like to thank the typists who at various times were involved in the preparation of the manuscript, Clare Bass, Shirley Greenslade, Patricia Stallard and especially Andrea Bates. Finally, we thank our families for their patience and support. Many times they must have thought the writing would never end but didn't say so.

We have tried, throughout the text, to provide a balanced account of controversial areas and consider that the presentation of different views is important to reinforce the point that theories, both of biochemistry and of the nature of disease, must always be open to reinterpretation with the advancement of knowledge. It is our hope that indications of doubt that are raised in this text will provide impetus for further research by the interested student, teacher and practising physician. We have tried to write a text that will be of use and interest not only to the student but to the doctor who wishes to be brought up-to-date in this area of medicine. We hope that the enthusiasm we find for the subject is communicated to our readers and the book will be read in order, as the editorial in the *Lancet* on 'Medical Education on Trial' (*Lancet i*, 837–838, 1982) requested, to enable doctors to meet 'the challenge imposed by the rapidly changing medical scene'.

CHAPTER 1

METHODS AND APPROACHES IN METABOLISM

Progress in science is made either from discovering new facts or from re-interpreting well-established ones. Theories are necessarily transient; they last only until replaced by ones more consistent with the accumulated facts. Now that medical practice is based firmly on the theories of biological science, it too must change with the theories; in many aspects of metabolism, these changes have been radical. The elucidation of the pathways of chemical conversion in living organisms began about 60 years ago and led, over the following 40–50 years, to an accepted set of theories describing metabolism. This acceptance is demonstrated by the proliferation of impressive metabolic 'maps' that adorn the walls of many laboratories and teaching institutions, simultaneously giving to the student a justifiable sense of complexity and frightening him lest he be required to recall even a small fraction of it in the examination room. The realization has slowly dawned that such maps provide a very limited description of metabolism. They turn out to be no more useful to the doctor needing to understand physiological processes than a map of routes is to the rail traveller. What the traveller needs in addition is a timetable, but the doctor needs more than a quantitative description of fluxes through pathways if he is to predict the consequences of his treatment; he needs an understanding of how the pathways are controlled (that is, the train controller's manual). To acquire this information, a great variety of methods must be used and some of the most important are outlined in this chapter. A more specific reason for discussing methods stems from the fact that the information on human metabolism and its changes in disease, which is provided in this book, has been gathered from most areas of biochemical research. Not surprisingly, this has frequently involved considerable extrapolation from unphysiological experiments and from results on animals other than man. These interpretations must always be made with caution and this chapter may help the reader understand why this should be.

Studying the intact organism has the advantage that it should be functioning normally but the disadvantage that few of the processes occurring within it are accessible to study. Conversely, while total disintegration may lead to identification of components, the way in which they interact may not be as in the intact organism. By analogy, consider the difficulties involved in gaining information about the working of an internal combustion engine if the mechanic was only able to observe it working or grind it to a fine powder and determine its composition. To obtain a complete understanding of physiological processes, the scientist may have to investigate at many levels including, in increasing order of complexity, purified enzymes, subcellular organelles, tissue homogenates, isolated cells, tissue slices, isolated organs, and the intact animal. (Because of the complexity of modern investigations the same scientist is not always able to do this hence the need for efficient communication in science.) Facts obtained at one level are used to make predictions at another. These are checked by experiment and if found valid used to enlarge the theory.

The levels of investigation listed above fall into two categories, cell-free systems and those involving intact cells. Both are described below together with their advantages and disadvantages, together with references to their specific use in later chapters. Investigations on the intact animal are referred to as *in vivo*; those on tissues and more fragmented systems as *in vitro* (literally, in glass).

A. CELL-FREE SYSTEMS

In order to investigate what goes on inside a cell it is usually necessary to break it open and so study that particular process in a cell-free system. To do this the cell membrane must be ruptured. This is frequently achieved by mechanical means in an homogenizer which consists of a strong glass tube with a glass or plastic plunger (Figure 1.1). There is a very small clearance between the plunger and the tube so that, as the plunger is depressed and twisted, the cells passing between the plunger and tube are exposed to considerable shearing forces. This damages the cell membrane so that the contents of the cell leak into the medium, usually a buffer, in which the cells are suspended. This technique can be used for separate cells or pieces of tissue. Frequently, it is necessary only to cut the tissue with scissors into small pieces before homogenization but with tissues containing a large proportion of connective tissue (e.g. heart, skeletal muscle) a power-driven homogenizer with rotating blades is often used in a preliminary stage. An alternative means of breaking membranes is to use ultrasonic vibrations which induce cavitation in the medium and hence large local pressure differences, which also damage the cell membrane. This action may also damage the membranes of subcellular organelles but in some experiments this may be desirable since it exposes enzymes located within them to the experimental conditions. Osmotic shock and detergents may also be used to aid disruption of membranes (for further details, see Lloyd and Coakley, 1979).

1. Isolated enzymes

Unequivocal information about the catalytic and regulatory properties of enzymes can only be obtained after their purification. The first stage is usually tissue homogenization (above) and this is followed by procedures designed to separate the proteins present. Preliminary purification is often achieved by precipitation with particular concentrations of salts or organic solvents. This is generally followed by more selective chromatographic procedures.

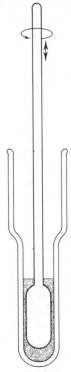

Figure 1.1 Simple hand-held tissue homogenizer. The arrows indicate the direction of movement of the plunger during homogenization

Despite the precision of data obtained from purified enzymes, doubts always remain about their validity in the intact cell where conditions may be quite different (but largely unknown).

2. Cell-free homogenates

The whole homogenate, prepared as described above, can be usefully used for a variety of investigations. It lacks membranes to restrict the access of substrates and regulators to the enzymes under study but conditions will be closer to those pertaining in the cell than is the case with purified enzymes. Provided that sufficiently specific assays can be devised for the enzymes under study (in the presence of hundreds, if not thousands, of other enzymes) the maximum activities of key enzymes in these homogenates can provide quantitative indications of the maximum fluxes through pathways *in vivo* (see Chapter 5; Appendix 2; Newsholme *et al.*, 1980). In addition, provided that the initial homogenization has not disrupted the internal membranes of the cell, the homogenate can be used as the starting point for the preparation of subcellular organelles. Some of the evidence for the identification of the individual reactions that constitute the metabolic pathways described in this text has been obtained from studies with homogenates; for a discussion of the experimental techniques leading to the concept of the tricarboxylic acid cycle (Chapter 4; Section B) see Krebs, 1981. This is also the level at which the maximum catalytic activity of an enzyme is most easily measured since purification always results in loss of some activity.

Cells also have an extensive network of internal membranes, known as the endoplasmic reticulum, which divides the cytosol into interconnected chambers. Such membrane systems may serve as an internal communication system, store calcium ions (especially the sarcoplasmic reticulum in muscle), store proteins prior to excretion (Golgi apparatus) or provide a surface to which certain enzymes are attached. When cells are disrupted these membrane systems are broken into small pieces which re-seal to form spherical 'organelles' known generally as microsomes. Further details of subcellular organisation are given in Appendix 1.1.

Differential centrifugation is used to separate organelles and depends on the fact that the larger organelles are sedimented rapidly at relatively low speeds (that is, at low centrifugal forces) while the smaller and lighter organelles are sedimented more slowly and at higher speeds (e.g. at 100 000 times the force due to gravity). The supernatant remaining after the highest speed centrifugation is called the soluble fraction and is assumed to contain those components originally in the cytosol. A typical fractionation scheme is presented in Figure 1.2. It is important to note that although enzymes usually remain within, or attached to their respective organelles during centrifugation, smaller molecules such as most metabolic intermediates do not.

Once organelles have been isolated they can be used to determine which aspects of metabolism are restricted to what organelles. For example, experiments with isolated mitochondria demonstrate that they are the main site for oxidation of lipids, carbohydrates, and amino acids and for the generation of ATP (Chapter 4; Section A.1). Indeed the abundance of mitochondria in a cell (as visualized in electron micrographs) provides a useful qualitative measure of the importance of aerobic energy production in a cell.

B. CELLULAR SYSTEMS

The cell is a functional unit and investigations carried out on intact cells can be expected to provide more physiological models of processes occurring in intact organisms. However, cell membranes are selectively permeable barriers and the experimental problem of getting materials into and out of cells limits the kinds of experiments that can be carried out on them. With isolated cells a very large surface area is exposed so that entry of substances is usually limited only by the permeability properties of the membrane. In aggregates of cells, e.g. tissue slices, the rate of entry may be limited by the diffusion distance between the cells and the

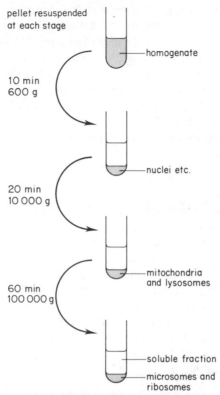

pellet resuspended
at each stage

homogenate

10 min
600 g

nuclei etc.

20 min
10 000 g

mitochondria
and lysosomes

60 min
100 000 g

soluble fraction

microsomes and
ribosomes

Figure 1.2 Scheme for differential centrifugation to separate organelles. The exact times and forces will vary according to the cell type and the kinds of particles being prepared. For details see Lloyd and Poole, 1979

medium. To some extent this problem can be overcome by utilizing the natural circulation pathway by perfusion of the whole organ (for discussion of the use of these various systems, see Ross, 1979).

1. Isolated cells

Erythrocytes have been used for many years as an experimental intact cell preparation for metabolic studies since they are easy to prepare and are resistant to mechanical damage during incubation. Unfortunately, their metabolic capacity is very limited. Renewed interest in isolated cell preparations has arisen since the discovery that isolated cells of a variety of tissues can be produced by incubation (or perfusion) of small pieces of the tissue with enzymes that break down the intercellular matrix (e.g. collagenase and proteolytic enzymes; for description of technique see Elliot, 1979). Cells from the liver, adipose tissue, heart, small intestine, and brain have been prepared in this way and are viable during several hours of incubation so that metabolic processes can be readily studied. Such cells have the particular advantage of genetic homogeneity so that control experiments can be carried out more easily and statistical validation is simpler. The ability to culture single types of a cell has further extended the possibility of metabolic studies at this level. The major disadvantages are that few cells can be incubated so that the absolute rate of metabolism or the concentration of metabolites might be too small to measure; in addition, the digestive preparation can produce changes in the cell membrane (e.g. loss of hormone receptor, changes in fluidity) that could seriously modify intracellular metabolism.

2. Tissue slices

A slice is prepared as the name implies, by slicing the tissue (either by hand with a razor blade or with a mechanical device). The slices can then be incubated in a 'physiological' medium. The advantage of the slice (compared to whole tissue preparation) is its thinness so that diffusion of gases and fuels is not a limitation (Umbreit *et al.*, 1964). The disadvantage is the damage to the cells of tissues by slicing can be large. In addition, the amount of tissue that can be incubated is small and many types of cell may be present. Nonetheless, it was one of the first intact tissue preparations to be used in metabolic studies and is still used in investigations of tissues such as the brain.

3. Isolated tissues and organs

The isolated tissue preparation has the advantage over cells and slices that it retains its structural integrity; interactions between cells can occur but interactions with other tissues and the ability of the whole animal to adapt to the experimental conditions are removed. It is therefore easier to control the environmental conditions. The major problem is to ensure that the preparation is physiological, that is, that it receives an adequate supply of fuels and oxygen; and that toxic end products can be satisfactorily removed. This can best be achieved if the tissue is thin and metabolism is not very active. For example, part of the adipose tissue that surrounds the epididymis of the rat (the epididymal fat pad) can be dissected away from other tissues and incubated in a physiological medium. Similarly the diaphragm or the epitrochlearis muscle of a young rat (in which the muscles are thin) can also be incubated intact. However, the most useful preparations at this level are isolated perfused organs in which the organ, either in or out of the body, is perfused by an artificial medium through its normal blood vessels so that adequate oxygenation is ensured. Organs that have been perfused in this way include the heart, liver, kidney, adipose tissue, skeletal muscle, endocrine organs and the intestine of the rat (see Ross, 1972, 1979).

One major problem which may be overlooked by the metabolic biochemist is that these preparations have usually been denervated and control of metabolism by nervous activity cannot be readily studied. Another problem is that, despite the physiological nature of the preparation, its physiological control processes may not be in a normal state and the maximum response to hormones or other physiological stimuli may not be achieved. (For example, even the best perfused rat heart preparations may not be able to perform mechanically as well as they can *in vivo*; Cooney *et al.*, 1981.) Nonetheless, a considerable amount of the information presented in this text has been obtained from such preparations.

4. Whole animals

To be really meaningful, a physiological experiment must be carried out on the whole animal. However, there are several major problems with experiments on whole animals. The system is complex so that there are a large number of interacting sub-systems and any change in one may be rapidly balanced by an adaptive change in another. Thus the usual experimental approach of perturbing the system and measuring and interpreting changes occurring may not be possible since the changes may not be large enough to measure. This latter problem is compounded by variation in the magnitude of any parameter from one animal to another, so that statistical validation of any changes may require large numbers of experimental animals. Finally, experimental animals are usually expensive and experiments upon them quite properly controlled (in many countries) by vivisection laws. Experiments on man must also take into account ethical considerations. Despite these problems, a considerable amount of information is available from intact animals including man. For example, the use of fuels by various tissues in man can be most readily measured by inserting catheters into the artery and vein supplying and draining a given organ or tissue and measuring arterio-venous differences (see Chapter 9; Section B.2; Chapter 14; Section A). Information about metabolic pathways and their control can be obtained with whole animals by the use of isotopes and following changes in labelling patterns of metabolic substrates and intermediates in the accessible body fluids (blood, urine, cerebrospinal fluid) or tissue samples, which can be taken from tissues of man with a biopsy needle (Chapter 5; Appendix 2.1; see Katz, 1979). Finally, use is being made of nuclear magnetic resonance (NMR) for the study of metabolism in either isolated tissues or in tissues and organs in the whole animal. This involves the use of ^{31}P, ^{13}C, or ^{1}H NMR and has the advantage that it is non-invasive. This technique is still at an experimental stage but may prove to be of considerable value to clinical practice in the future (Gadian and Radda, 1981). The effects of feeding various diets must of necessity be carried out on whole animals, and human diseases, for which there is no corresponding animal model, can only be investigated in the patient. Even so, it is possible to investigate the diseased state at other levels since small tissue samples can be taken from man with a biopsy needle and, from this piece of tissue, isolated cells can be prepared and their metabolism investigated. Alternatively the tissue can be homogenized and enzymes, receptors or organelles prepared and their properties or capacities investigated before and after treatment.

APPENDIX 1.1   CELL ULTRASTRUCTURE

All cells are divided internally by membranes into compartments. Some of the more important in the context of this book are outlined below. In addition to those described, many other kinds of organelle are restricted to certain cells, for example myofibrils (in muscle cells); storage vesicles (in neurons); fat droplets (in adipocytes).

1. Nucleus

The site of DNA synthesis and storage. Limited by a double membrane penetrated by pores through which mRNA can pass. Contains a small denser body, the nucleolus, which is the site of ribosomal RNA synthesis.

2. Mitochondria

The basic structure of a mitochondrion is shown in Figure 1.3 but they are somewhat variable in size, shape and number of cristae. Muscle mitochondria are typically sausage-shaped and several micrometres long while liver mitochondria are more nearly spherical, $1-2 \mu m$ across. (They are comparable in size to bacteria from which it is widely believed they arose originally as symbionts.) Enzymes of the TCA cycle and β-oxidation are in the matrix; those of oxidative phosphorylation and electron transfer are attached to the inner mitochondrial membrane (see Chapter 4; Section C.2b). The outer membrane is freely permeable to many small molecules but the inner membrane is much less so.

3. Endoplasmic reticulum

A network of double membranes extends through much of the cell and encloses cavities called cisternae (see Figure 1.3). The surface of some endoplasmic reticulum is covered with ribosomes which synthesize proteins destined for release from the cell. These give it a granular appearance and the name of rough endoplasmic reticulum. Also attached to the endoplasmic reticulum are a number of enzymes, including many catalysing hydroxylation reactions. In muscle, the cisternae of a similar membranous structure, the sarcoplasmic reticulum, sequester calcium ions, the release of which initiates contraction of the myofibril (see Chapter 7; Section C.1.c). During homogenization, the endoplasmic reticulum is disrupted and reforms into small spherical particles called microsomes.

4. Golgi complex

Cells which synthesize proteins and mucopolysaccharides for secretion possess a region in which layers of smooth interconnected double membranes are regularly stacked. This is the Golgi complex. Proteins are transported through the

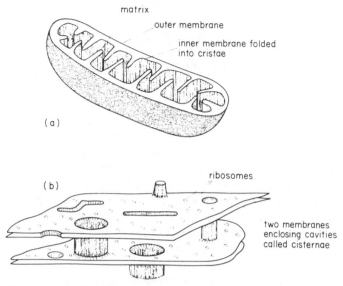

matrix

outer membrane

inner membrane folded into cristae

(a)

(b)

ribosomes

two membranes enclosing cavities called cisternae

Figure 1.3 Stereograms depicting the structure of (a) a mitochondrion and (b) a small piece of rough endoplasmic reticulum

endoplasmic reticulum to the Golgi apparatus where they are glycosylated and 'packaged' into small vesicles which discharge their contents through the plasma membrane by exocytosis (Chapter 21; Section C.2; Chapter 22; Section D.2.b).

5. Lysosomes

Membrane-limited structures (approximately $1\,\mu$m in diameter) containing hydrolytic enzymes. These vesicles are produced by the Golgi complex but their contents are not secreted by the cell. Lysosomes fuse with vesicles containing either external material which has entered the cell by endocytosis or cell contents around which a membrane has formed (Chapter 10; Section B.3.b).

6. Soluble fraction

As prepared by centrifugation, the soluble fraction contains not only the cytosol in which so many reactions take place but the remains of a variety of structures which electron microscopy shows to be present* but which are not limited by membranes. These include microtubules (which impart local rigidity to the cell contents), microfilaments (contractile strands), and the microtrabecular lattice (an all-pervasive system of strands which imparts some of the properties of a gel to the cell contents, see Porter and Tucker (1981)).

* There always remains some doubt as to whether images in electron micrographs represent real structure within a cell or artefacts of the fixation procedure.

CHAPTER 2

THERMODYNAMICS IN METABOLISM

Thermodynamics is the branch of science dealing with energy and its conversion from one form to another. In a chemical reaction, for example, chemical energy is redistributed and some may be converted into other forms of energy. Thermodynamics also describes what can and cannot happen spontaneously. The thermodynamic principles underlying energy interconversions (the first law of thermodynamics) and the directionality of reactions (the second law of thermodynamics) are very important in understanding the basis of metabolism, metabolic inter-relationships and control. Living organisms can in no way violate the laws of thermodynamics; reactions occurring in the cell are subject to the same thermodynamic constraints as those occurring in the test-tube. This remains true despite the fact that the rates of some metabolic reactions change enormously from one second to the next, while those of other reactions remain constant for very long periods of time. Only with some knowledge of thermodynamic principles can the mechanisms regulating the rates of metabolic reactions be fully understood.

This chapter attempts to provide an account of the ways in which the laws of thermodynamics can lead to an improved understanding of metabolism. Unfortunately, there are a number of problems to be overcome even in a relatively simple approach to metabolic thermodynamics. First, there is the problem of terminology. In any branch of science it is convenient to reduce elaborate ideas to single words or phrases and it is inevitable that, initially, such terms will be unfamiliar. Paradoxically, some thermodynamic terms confuse just because they are familiar; for example, energy, work, equilibrium, coupling, and state are commonly used in everyday language, but, in thermodynamics, they have much more restricted meanings. Although this chapter is intended to convey general ideas and principles, it has not proved possible to avoid specialized terms: nor is it desirable to do so, since it is important that the subject matter in the chapter should relate easily to that in conventional thermodynamic texts. In an attempt to reduce this problem, definitions and explanations have been given wherever possible. Secondly, since the ideas of thermodynamics are essentially quantitative, they are conveniently represented mathematically. As this chapter is not concerned with thermodynamic ideas *per se*, no attempt is made to explain the development of thermodynamic principles or the derivation of many of the equations used. For these, the reader is referred to physical chemistry texts, particularly those describing the application of physical chemistry to biology (for example, Spanner, 1964; Linford, 1966; Wyatt, 1967) and, for an excellent detailed account of the application of thermodynamic principles to cellular energy metabolism, to Atkinson (1977). Since we appreciate that many medical students find mathematical manipulations difficult to follow, we have, in those cases where a derivation is necessary to an understanding of the application of thermodynamic principles to metabolism, sought to make it as comprehensible as possible.

The first section of this chapter aims to show, in a straightforward non-mathematical manner, how the first and second laws of thermodynamics can be applied to metabolic processes (and indeed to medicine). In the second section, the thermodynamic principles involved in the use of ATP by living organisms are discussed. Finally, what may be termed the thermodynamic structure of a metabolic pathway is considered in the last section. Although this may seem esoteric, it will allow the reader to appreciate the principles of metabolic regulation which will be utilized frequently in subsequent chapters. It will also explain how metabolic pathways can remain out of equilibrium for prolonged periods (indeed, for the lifetime of the animal) and yet the flux through the pathway can adapt rapidly to changes in environmental conditions.

A. APPLICATION OF THE LAWS OF THERMODYNAMICS TO METABOLISM

The early study of thermodynamics was related to physics and engineering and the laws were formulated to describe the properties of devices such as heat engines. The laws, of course, apply equally well to metabolic reactions but the emphasis is somewhat different. Metabolic thermodynamics is concerned primarily with the ways in which concentrations of substances affect the direction of a reaction. Consequently, there is less emphasis on factors such as temperature and pressure unless they affect the concentrations of substances under consideration.

For thermodynamic purposes it is not necessary to consider the nature of the reaction itself, so that a variety of metabolic processes can be considered in the same way. These can all be represented by the general equation:

$$A \underset{v_r}{\overset{v_f}{\rightleftarrows}} B$$

where v_f and v_r represent the rates in the forward and reverse directions respectively. There are three types of metabolic processes that need to be considered; these are as follows.

1. A single chemical reaction. It is uncommon in metabolism for a single reaction to be isolated, since the substrates and products of such a reaction usually participate in other reactions.
2. A series of consecutive enzyme-catalysed reactions. The conversion of A to B may involve a number of separate and individual intermediate reactions. In this situation, the over-all rate of conversion of A to B is termed the flux

through the pathway. The net forward flux is given by $v_f - v_r$. Thermodynamic principles apply to the over-all pathway as well as to the individual reactions.

3. The transport of a substrate across a membrane. Although transport is not a chemical reaction because A and B are chemically identical, the process can occur by mechanisms similar to those involved in enzyme-catalysed reactions. Thus thermodynamic principles apply equally to transport processes.

Thermodynamics is essentially the study of energy. Energy is the capacity for doing work and the same unit, the joule,[1] is used for both energy and work. This rather abstract definition of energy is clarified by consideration of the different forms that energy can take. Some of these forms are familiar in everyday life and include heat energy, electrical energy, light energy, and mechanical energy. Machines exist that can convert energy from one form to another. For example, a dynamo converts mechanical energy into electrical energy and a light bulb converts electrical energy into light energy. In both cases some of the energy is turned into heat energy. The relationship between these forms of energy is described by the first law of thermodynamics (see below).

Less immediately obvious as a form of energy, but of more importance in the present context, is chemical energy. In a chemical reaction, bonds are formed and broken and the distribution of electrons is changed; the chemical energy is being redistributed between the reactants. The equivalence of chemical energy with the other forms of energy is more clearly seen when interconversions are considered. If the products of a chemical reaction contain less energy than the reactants, the balance is released in some other form. In many familiar chemical reactions this energy is released in the form of heat as, for example, in the burning of ethanol:

$$CH_3CH_2OH + 3O_2 \longrightarrow 2CO_2 + 3H_2O + heat$$

In a battery, the energy is released mainly as electrical energy and in some other situations directly as light energy. Conversely heat, electricity, and light can all be used to drive chemical reactions, in which case these forms of energy are converted into chemical energy. Life is based on energy conversions: in photosynthesis, light energy is converted into chemical energy; in respiration, chemical energy is converted, first into other forms of chemical energy (e.g. ATP) and then into heat energy, mechanical energy (for movement), electrical energy (for communication in nerves), etc.

The chemical energy (or internal energy as it is also known) in a substance depends on two things: the nature of the molecule and the number of molecules, that is, the amount of the substance. Therefore chemical energy must always be expressed per mole.

1. The first law of thermodynamics

The first law of thermodynamics is the law of conservation of energy; it states that energy can neither be created nor destroyed, it can only be converted from one form into another. From the above comments, it should be clear that the first law can be restated in the following terms; the change in chemical (internal) energy when one mole of A is converted into B is equal to the heat produced (or consumed) plus the work done on (or by) the reaction. Using the sign convention that heat (q) absorbed and work (w) done by the reaction are positive:

$$\Delta H = q - w$$

ΔH is known as the change in enthalpy of the reaction.[2] If no work is done by the reaction then the whole of the enthalpy change appears as heat; that is, when $w = 0$, $\Delta H = q$. Therefore, ΔH can be measured by the heat produced (or absorbed) in a reaction when no work is done and, for this reason, it is also known as the heat of the reaction. If heat is produced, ΔH is negative and the reaction is said to be exothermic; if heat is absorbed, ΔH is positive and the reaction is said to be endothermic. When one mole of a substance reacts in the standard state (a pressure of one atmosphere for a gas or, in most cases, a concentration of one molar for a solute), the standard enthalpy change (ΔH^0) occurs. For a given reaction at a particular temperature, this is a constant. The values of ΔH^0 for some metabolic reactions are given in Table 2.1.

Table 2.1. Enthalpy changes for some metabolic reactions

Reaction	ΔH^0 (kJ.mole^{-1} of organic substrate)
palmitate $+ 23O_2 \rightarrow 16CO_2 + 16H_2O$	$-10\,024$
glucose $+ 6O_2 \rightarrow 6CO_2 + 6H_2O$	$- 2\,813$
alanine $+ 3O_2 \rightarrow 2\frac{1}{4}CO_2 + 2\frac{1}{2}H_2O + \frac{1}{2}$ urea	$- 1\,304$
$ATP + H_2O \rightarrow ADP + P_i$	$- \quad 21$

It is most important to note that the value of ΔH is independent of the chemical path followed by the reaction, that is, it is independent of the mechanism of the reaction. Whether glucose is oxidized by burning in oxygen in the laboratory or by the concerted action of the enzymes of glycolysis and the tricarboxylic acid cycle, ΔH is the same. ΔH is, therefore, said to be a thermodynamic or state function. This is not true for the work or the heat involved in the reaction. For example, although the enthalpy change when one mole of ATP is hydrolysed to ADP and phosphate will be the same whatever the mechanism for conversion, the amount of work done will depend on the mechanism. If the hydrolysis is catalysed by acid in a test-tube, no work is done and the entire enthalpy change appears as heat but, if the reaction is catalysed by the myofibrillar ATPase within the muscle, mechanical work is performed and less heat is released.

For the conversion of one mole of substrate into product, the maximum available energy from that reaction is ΔH. This will not be changed by modifying the mechanism of the reaction or changing the concentration of the substrate or product. However, the value of the *total* ΔH will depend upon the *number* of moles transferred in the reaction. This is of considerable importance in metabolism, since an increase in the rate of conversion of substrate into product (i.e. an increase in flux through the pathway) is the means by which the cell responds to an increased demand for energy. In other words, when the rate of ATP utilization by a cell is increased, the concentrations of most metabolic intermediates remain almost constant but the flux of substrate through the pathway must increase to provide more ATP.

Two related applications of the first law must be considered, if less than consciously, by anyone concerned with the planning of diets. These are the calculation of the energy value of food and the heat produced by metabolism. Since they provide interesting examples of the applications of the first law to medical matters, they are briefly discussed below.

(a) *Weight reduction and the first law*

The only energy input into a human being (assuming an environmental temperature below that of the body and the absence of a source of thermal radiation) is in the form of food. The energy output is in the form of mechanical work and heat. If the subject maintains a constant body weight, then the first law of thermodynamics states that the chemical energy consumed must equal the mechanical and heat energy produced. Conversely, if these energy terms do not balance, a change in body weight must occur, since the body weight represents the chemical energy content of the subject. It follows that, from a knowledge of energy expenditure, it is possible to calculate the amount of food necessary to maintain constant body weight or to change this in a particular direction.

Weight control, therefore, requires knowledge of both energy output and energy input. Although the average daily expenditure of energy of an individual can be measured experimentally (see Durnin and Passmore, 1967; Davidson *et al.*, 1975; Garrow, 1978) this is an elaborate undertaking and, in practice, sufficiently reliable information is now available that it is possible to estimate this expenditure from a knowledge of the subject's daily activities. For example, assuming the major energy expenditure occurs during the working day of the individual, a secretary may expend about 8000 kJ.day^{-1} while a manual worker expends over 16 000 kJ.day^{-1} (see Table 2.2). These values could be increased by, for example, participation in sporting activities after work. Moreover, these values assume 'normal' environmental temperatures and therefore 'normal' levels of heat production. In cold climates a greater production of energy in the form of heat must occur to maintain body temperature. In hot climates the converse is true. A second source of variation is body size. Published tables generally refer to a standard man of 65 kg and a standard woman of 55 kg but, clearly, a larger person will expend more energy in, say, climbing stairs.

Turning now to the quantity of energy taken in, this can be calculated precisely

Table 2.2. Energy expenditure by men and women during various activities*

General description of work intensity	Activity	Sex	Approximate mean energy expenditure during activity $(kJ.min^{-1})$
Basal metabolic rate	Resting	M	4.7
		F	4.1
Very light	Typing and clerical work	F	6
	Lorry driving	M	6
Light	Scrubbing floors	F	17
Moderate	Swimming, dancing	M	21
		F	17
Heavy	Squash playing	M	31
		F	25
	Road digging with pick and shovel	M	30
Very heavy	Carrying logs	M	51
Exceedingly heavy	Marathon running (see Chapter 9; Section B)	M	84

* Data taken from Durnin and Passmore (1967).

The Food and Agricultural Organization of the United Nations (FAO) Committee on Calorie Requirement met in 1957 and recommended that the reference man requires 13 400 kJ (3200 kcal) daily and the reference woman requires 9600 kJ (2300 kcal) daily. The reference men and women are healthy, live in a temperate climate and work in a non-sedentary occupation for eight hours each day. The reference man weighs 65 kg and the reference woman weighs 55 kg (see also Davidson *et al.*, 1979).

The basal metabolic rate is determined when the subject is lying down at complete physical and mental rest wearing light clothing at a comfortable temperature and at least 12 hr after the last meal. It is usually expressed in $kJ.m^{-2}$ since the rate varies with surface area. The value given assumes a surface area of $1.85 \, m^2$ for males and $1.66 \, m^2$ for females. Since the basal metabolic rate also varies with age it is assumed that both subjects are aged 30 years (Davidson *et al.*, 1979). The metabolic rate during sleep is very similar to the basal metabolic rate, it is slightly higher in the early part of sleep (probably due to the effect of the last meal) and slightly slower in the later part of sleep. It is also likely that the matabolic rate is lower during paradoxical sleep than slow-wave sleep (Roussel and Bittel 1979).

from a knowledge of the composition of the food ingested. (This is generally known as 'calorie counting' but perhaps it should now be known as 'joule counting' or more simply as 'energy counting'. Note that the 'calorie' used in calculations of energy content of food is usually a kilocalorie, often written Calorie.) For this, the amount of the different foods ingested and the values of the ΔH for their oxidation must be known. The calculation is simplified by the fact that, despite the diversity of substances present in the food, the major components fall into three groups—carbohydrate, lipid, and protein. Although there are differences in values of ΔH for oxidation between the many difference substances in any one group (e.g. between glucose and glycogen) these are sufficiently small to be ignored for the purpose of weight control. Note that this approach only works because ΔH is a state function, so that the values of ΔH for the oxidation of foodstuffs are unaffected by the nature of the metabolic reactions involved, provided that the end-products are the same. Thus it does not matter, for example, whether ingested glucose is oxidized directly or first converted to triacylglycerol (triglyceride). Furthermore, the value is unaffected by concentrations of the substrates, intermediates, or products, so that the *in vivo* conditions do not have to be taken into account in these calculations.

The enthalpy changes are calculated from the heat produced when the foodstuff is combusted with excess oxygen in a bomb calorimeter. In combustion, all the carbon is oxidized to carbon dioxide and all the hydrogen to water so that for fats and carbohydrates the over-all reaction is identical to that occurring in the body. (It is of no consequence to the calculation that some of the carbohydrate may be metabolized in the tissue to lactate, since this compound does not leave the body but is oxidized to carbon dioxide and water in other tissues.) However, before enthalpy changes measured in this way can be used to calculate the energy input, two corrections are necessary. First, digestion or absorption may not be complete. In man, under normal conditions, these processes are very efficient, so that, for a mixed diet, 99 % of the carbohydrate, 95 % of the fat, and 93 % of the protein are typically absorbed. However, less of these compounds is absorbed if the diet contains a large amount of 'unavailable carbohydrate' or dietary fibre (see Chapter 5; Section A.3). Secondly, the end-products of protein metabolism in the living organism differ from those produced in the bomb calorimeter. Thus oxidation of protein in the calorimeter produces, in addition to carbon dioxide and water, oxides of nitrogen, whereas in the body the nitrogen from protein is excreted in the form of urea. This represents a loss of energy equivalent to about $6.25 \, kJ.g^{-1}$ protein. These two factors mean that in man, on a typical mixed diet, the available energy is about 85 % of the gross energy in the ingested food. These losses of energy were investigated in detail by Atwater and coworkers in the early part of this century and they are known as 'Atwater' factors. (The reader is referred to Southgate and Durnin (1970) for further reading.)

The biological values of ΔH for protein, fat, carbohydrate and ethanol are given in Table 2.3. Ethanol has been included in this list since in many Western

Table 2.3. Energy content of major constituents of food*

Constituent	Value of ΔH $(kJ.g^{-1})$
Protein	17
Fat	37
Carbohydrate	16
Ethanol	29

* Data are taken from Paul and Southgate, 1978.

† Values given are for biological oxidations, corrected for average extent of absorption.

countries its consumption accounts for about 10 % of the energy value of ingested food. The energy intake of any subject can be calculated, therefore, by adding together the products of mass and ΔH for each of these major food constituents. Tables of food composition are widely available and include McCance and Widdowson (1960), Paul and Southgate (1978) ((British foods), Watt and Merrill (1963) (foods in U.S.A.) and those published by the Food and Agriculture Organization of the United Nations (1949) for international use.

It should be obvious that, in order to increase body weight, the energy intake must exceed the energy expenditure over a period of time and, conversely, to lose weight energy expenditure must exceed energy intake. There is no doubt that if energy intake exceeds energy expenditure in adult subjects over a long period, obesity results. However, there is now considerable evidence that normal subjects possess mechanisms that will increase energy expenditure in order to maintain body weight. This energy appears, of course, as heat which is lost from the body. A metabolic discussion of these processes, together with a detailed discussion of obesity, its causes, consequences and treatment, will be given in Chapter 19; Section C.3. However, a brief discussion of some of the more thermodynamic aspects of heat production is given below.

(b) *Heat production in living organisms*

Many organisms are able to exert some degree of control over their body temperature. This ability is best developed in birds and mammals which are able to maintain their body temperature to within a degree or so of a particular temperature despite large fluctuations in that of the environment. These animals are known as homoiotherms. The importance of this maintenance of body temperature is that the effect of variation in environmental temperature on the rate of metabolic reactions is removed. This ability has enabled mammals and birds to colonize areas of the world not available to amphibians and reptiles. Maintenance of body temperatures in homoiotherms is achieved in two ways— by altering the rate of heat loss (e.g. by insulation with fur, feathers or subcutaneous adipose tissue, see Irving, 1966) and by increasing heat production.

Metabolism involves the redistribution of chemical energy between compounds. In this process, some of the energy is converted into heat. Thus the normal processes of metabolism produce heat and the greater the rate of metabolism the greater the heat production. For this reason, exercise, which increases the rate of metabolism markedly, increases heat production. The heat production of a man at rest is about $0.07 \, kJ.kg^{-1}.min^{-1}$ whereas, for example, during a marathon run it rises to about $1.1 \, kJ.kg^{-1}.min^{-1}$. This 15-fold increase in the rate of heat production presents a major problem for the marathon runner who must achieve adequate heat loss in order to prevent serious, and even fatal, hyperthermia. (If a marathon runner had no means of losing heat produced by the exercise, the body temperature would rise to about $45 \, °C$ in 30 min, Wyndham, 1977; Nicholson and Somerville, 1978.)

Apart from heat production as a consequence of normal metabolism, there is evidence for the existence of specific metabolic heat generating processes in homoiotherms which are controlled either hormonally or by nervous activity.[3] In addition to the maintenance of body temperature when the ambient temperature is low, these processes might also be involved in the increase in the rate of heat production that occurs after a meal (thermic response of food) or during an infection (fever, which may increase the mortality of the pathogen—see Kluger, 1979; Weinberg, 1980; Atkins, 1983).

Can the first law be used to calculate the heat produced during metabolism? The answer is 'yes' provided that no external work is being done by the subject and its mass remains constant. Under these conditions, all the chemical energy entering the organism leaves it as heat, that is, $\Delta H = q$. Following the arguments set out in the previous section, the heat (q) produced by the oxidation of each fuel can be calculated by multiplying ΔH (determined by bomb calorimetry) by the weight of fuel used. If more than one fuel is used, the heats can be summed to give the total heat produced. This calculation requires a knowledge of the amounts of fuel used (see Chapter 5; Appendix 2, for methods) but not of the metabolic pathways followed. Conversely, the amount of substrate necessary to support a particular rate of heat production can also be calculated. A 'standard' 65 kg man has a basal metabolic rate (when no work is being done) of about $7000 \, kJ.day^{-1}$. From Table 2.3 it can be seen that glucose oxidation has an enthalpy of $16 \, kJ.g^{-1}$, so that such a man would need to ingest about 450 g of glucose per day to support this basal metabolic rate.

It is tempting to calculate the heat produced from a knowledge of the enthalpy change of ATP hydrolysis ($\Delta H = -21 \, kJ.mole^{-1}$) but this ignores the fact that the hydrolysis of ATP demands the simultaneous production of ATP to maintain its steady state concentration. Hence an increased rate of ATP hydrolysis increases the rate of metabolism, which results in further heat generation. In other words, since the concentration of ATP does not change in the tissue (under steady state conditions) heat is produced not only due to hydrolysis of ATP but due to the increased rate of the oxidation of the fuel that produces the ATP. Thus, provided that no work is done, the total ΔH for the oxidation of the fuel is released as heat, as described above. Nonetheless, in order to compare the quantity of heat that can be produced by different metabolic mechanisms, the *total* amount of heat produced as a consequence of the hydrolysis of one mole of ATP is important. This value is obtained by dividing the ΔH for the oxidation of the fuel by the number of moles of ATP produced in that oxidation process (and, therefore, utilized in the steady state).

Thus in the bomb calorimeter, the following reaction:

$$glucose + 6O_2 \rightarrow 6H_2O + 6CO_2$$

releases $2813 \, kJ.mole^{-1}$. However, in the cell, glucose oxidation is coupled to ATP formation as follows:

$$\text{glucose} + 6\,O_2 \longrightarrow 6\,H_2O + 6\,CO_2$$
$$38\,\text{ADP} + 38\,P_i \qquad\qquad 38\,\text{ATP} + 38\,H_2O$$

Consequently, for the hydrolysis of one mole of ATP *in vivo*, the total amount of heat produced is 2813/38 or 74 kJ. This value for the heat production from ATP hydrolysis will be made use of when the role of substrate cycles in heat generation and control of body weight is discussed in Chapter 19; Section C.3.b.

2. The second law of thermodynamics

One of the most important facts that needs to be known about a metabolic pathway is its direction. Will glucose be split to form lactate or will lactate be converted to glucose? All chemical reactions are reversible so why does a metabolic pathway proceed in a particular direction and under what circumstances can this direction be changed? Furthermore, even when the direction has been established, knowing how near an individual reaction is to equilibrium is important; it provides information about the ease of reversibility of the reaction and about the nature of control of flux through that reaction. The key to understanding direction and equilibrium in chemical reactions is the second law of thermodynamics, which introduces the concept of entropy.

(a) *Entropy*

It was once thought that only changes in enthalpy need to be considered to determine the direction of a reaction. It was assumed that the reaction would go in the direction associated with a decrease in enthalpy (i.e. a negative ΔH) so that heat would always be produced. However, this was shown to be incorrect when endothermic reactions were discovered, that is, reactions which proceed spontaneously with the uptake of heat. A dramatic demonstration of an endothermic reaction is provided by stirring together solid ammonium thiocyanate and solid barium hydroxide $8H_2O$ in a vessel standing in a pool of water:

$$Ba(OH)_2 \cdot 8H_2O + 2NH_4SCN \rightarrow Ba(SCN)_2 + 2NH_3 + 10H_2O$$

Sufficient heat is absorbed from the environment as these substances react to cause the water to freeze. Clearly, knowledge of some other factor is required to predict whether a reaction will occur; this factor is the change in entropy (ΔS). Spontaneous reactions always proceed with an increase in total entropy (i.e. $\Delta S > 0$). Since entropy is a thermodynamic (or state) function (like enthalpy) its value depends only on the state of a system and not on the route by which it arrived at that state. This leads us to one way of stating the second law of thermodynamics; all processes proceed in a direction that increases the total entropy.

In formal terms, entropy is the heat (q) absorbed in a thermodynamically reversible reaction (at $T\,°K$) divided by the absolute temperature, T, thus $\Delta S = q/T$. However, for the present purpose, a more qualitative representation of entropy as the degree of disorder will suffice. The more disordered or random a system becomes the more entropy it has, so that, in a spontaneous reaction, disorder must increase.

As an example of the importance of entropy, consider a straight line of ball bearings equidistant from each other on a flat tray. Very little energy is required to disturb this regular arrangement, for example, by tilting the tray, but it is highly unlikely that the ordered arrangement will be re-established simply by tilting the tray back again. The ball bearings gained a great deal of entropy when the tray was tilted. But the process *can* be reversed if care is taken to place each ball bearing in position. Has entropy thus been lost in defiance of the second law? No, because the second law does not specify where the entropy increase must occur and in this case the loss of entropy in the ball bearings is more than compensated for by the increase in entropy occurring (as a result of metabolism) in the tissues of the person replacing the ball bearings. Considering a chemical reaction, the increase in entropy can occur in the reactants or the environment (defined as everything other than the reactants), or both, so that for a reaction to occur:

$$\Delta S_{\text{reactants}} + \Delta S_{\text{environment}} > 0$$

The degree of order can take a number of forms. Gases are less ordered than liquids and liquids less ordered than solids (because of the arrangement of molecules in the substance). In the example of an endothermic reaction given above there is a considerable increase in entropy because two solids are reacting to form a liquid (water) and a gas (ammonia). In addition, some molecules have a greater degree of internal order than others, and so they have an inherently lower entropy. For example, proteins are highly ordered and possess a constrained conformation, but this changes to a much more random structure upon denaturation (see Chapter 3; Section B.2) and hence the increase in entropy during denaturation is considerable.

(b) *Gibbs' free energy*

Although the total entropy change is a sufficient criterion upon which to establish which processes can and cannot occur, it is not always easy to measure the changes in entropy of both the reactants and the environment. However, a chemical reaction can be carried out so that the only effect of an entropy change in the environment is seen as heat produced or taken up (i.e. ΔH). The change in enthalpy can therefore be combined with a term involving the entropy change of the reactants to give the required criterion. At constant temperature, a reaction will occur spontaneously if

$$\Delta H - T\,\Delta S_{\text{(reactants)}} < 0$$

The quantity, $\Delta H - T \Delta S$, is known as the change in Gibbs' free energy (ΔG).[4]

Gibbs' free energy is a thermodynamic (or state) function which describes a system's maximum potential for doing work, so that the second law can also be stated in these terms; spontaneous reactions are those which, when carried out under suitable conditions, can be made to perform work. Since ΔG is a thermodynamic function, it is possible to obtain a value for an over-all ΔG by adding the values of ΔG for each component reaction. However, ΔG is not a form of energy in the conventional sense and is not conserved in the way described by the first law. This arises because, although enthalpy is a true form of energy, the term $T \Delta S$ is not; there is not a fixed amount of entropy in the universe, it is always increasing.

(c) *Gibbs' free energy and equilibrium*

If ΔG for a reaction is negative, the reaction can occur spontaneously; if ΔG is positive, it cannot. In the latter case, the reverse reaction will have a negative ΔG so that the reaction will occur in that direction. If ΔG is zero, the reaction proceeds in neither direction and is said to be in a state of equilibrium. Consider the reaction in which substance A is converted to substance B and the equilibrium is represented as follows:

$$A \rightleftarrows B$$

Very early on in any chemistry course it is learned that such an equilibrium may be displaced (that is, the reaction can be made to have a direction) by changing the concentration of one of the reactants. (This is sometimes stated as one aspect of Le Chatelier's principle.) Here, for example, increasing the concentration of A will cause the reaction to move to the right until a new equilibrium is established. Since ΔG has been shown to be the quantity which determines the direction in which a reaction proceeds, it follows that the value ΔG must depend on reactant concentrations. It does so according to the following equation (see Crabtree and Taylor, 1979):

$$\Delta G = \Delta G^0 + RT \ln \frac{[B]}{[A]}$$

where R is the gas constant and T the absolute temperature, so that at a fixed temperature the value of ΔG depends on the constant ΔG^0 (see below) and the concentrations of the substrates and products (designated by square brackets). (Note that if there is more than one substrate or product the concentrations of each are multiplied together.) In order to obtain the value of ΔG for a reaction in a living tissue, it is necessary to know the concentrations of substrate and product of that reaction in that tissue (i.e. A and B in the above example). The ratio of the concentrations of product and substrate (i.e. [B]/[A]) for a reaction in a living organism is known as the mass action ratio, which is given the symbol Γ.

The constant, ΔG^0, is the standard free energy change and is a constant for a particular reaction at a given temperature. The nature of the constant can be most clearly explained by establishing its relationship with a more familiar constant, namely the equilibrium constant. It is well known that the position of equilibrium of a reaction depends on the value of the equilibrium constant (K_{eq}). This is the ratio of concentrations of product and substrate when the reaction is at equilibrium $\left(K_{eq} = \dfrac{[B_{eq}]}{[A_{eq}]}\right)$. The relationship of the two constants is expressed in the equation:

$$\Delta G^0 = -RT\ln K_{eq}$$

The value of ΔG^0 is therefore determined solely by the equilibrium constant (at a given temperature). It can be seen from the earlier equation that ΔG^0 is the free energy change that would occur if all substrates and products were in their standard state (i.e. a concentration of one molar[5] for all solutes) when the term, $RT\ln\dfrac{[B]}{[A]}$, becomes zero.

In summary, therefore, the sign and magnitude of ΔG (and hence the direction of a reaction) depends on two factors; the actual concentrations of the substrates and products of the reaction and the value of the constant ΔG^0 (itself determined by the value of K_{eq}).

It should be clear that values of ΔG^0 alone are insufficient to determine the direction of a reaction and there is little point in providing only these values, although this is widely done in textbooks of biochemistry. What are important are the values of ΔG which, as will be shown, can provide information on the thermodynamic structure of metabolic pathways and thus increase our understanding of the control of the flux through the pathways. Values of ΔG for the reactions in the metabolic pathways described in this book will be provided wherever possible.

(d) *Role of Gibbs' free energy in metabolism*

The role of the free energy change in determining the direction of a reaction has been indicated above; ΔG must be negative for a reaction to occur. It should be stressed that not only must ΔG for the over-all metabolic pathway be negative but this must be so for all the constituent reactions. Consequently, there is no point in discussing whether ΔG for a reaction is positive or negative, it must be negative if the flux through the pathway proceeds in the expected direction. However, knowledge of the magnitude, in contrast to the sign, of ΔG is very important since it enables differentiation to be made between near- and non-equilibrium reactions (see Section C.2).

To determine ΔG for a reaction *in vivo* it is necessary to know both the equilibrium constant and the concentrations of substrates and products.

Measurement of the former usually presents little difficulty for it can be measured *in vitro*. The substrates and products (in the presence of the appropriate enzyme) are allowed to come to equilibrium before their concentrations are measured and K_{eq} calculated. The measurements of actual reactant concentrations in the tissue present more difficulty since they must be made on the tissue itself without perturbing the concentrations of the substrates and products of that reaction. Some of the methods used and problems involved are discussed by Newsholme and Start (1973) (see also Chapter 7; Note 1). As examples, the values of ΔG, together with those of ΔG^0, K_{eq}, and Γ, for most of the reactions of glycolysis are given in Table 2.4.

At equilibrium, when $\Delta G = 0$, the mass action ratio is equal to the equilibrium constant, that is $K_{eq}/\Gamma = 1$. As the reaction departs further from equilibrium, this ratio rises. Since a major application of thermodynamic theory to metabolism involves the assessment of how close a reaction is to equilibrium (see Section C.2) this ratio, obtained directly from experimental data, is often presented in preference to the value of ΔG. However, it is very easy to calculate ΔG from the ratio, K_{eq}/Γ, since:[6]

$$\Delta G = -RT\ln K_{eq}/\Gamma$$

These calculations have been carried out, where possible, since we consider that it is more useful to present values of ΔG, if only to emphasize the irrelevance of ΔG^0.

The data in Table 2.4 show that while values of ΔG for the glycolytic reactions are inevitably negative, those of ΔG^0 can be either negative or positive. Textbooks which emphasize values of ΔG^0 for metabolic reactions make statements of the kind that 'despite the positive ΔG^0, the reaction proceeds because the products are removed by the subsequent reaction in the pathway', thereby admitting the importance of the second factor (substrate and product concentrations) which determine ΔG. This influence of one reaction on another is termed coupling and is often taken for granted. However, a more detailed consideration of the thermodynamics of coupling explains just how a metabolic pathway can exist despite containing reactions with widely differing ΔG^0 values.

The word 'coupling' has been used to describe two different situations in metabolism. Reactions are said to be coupled when the product of one is the substrate of the next, as indicated by:

$$A \rightarrow B \rightarrow C \rightarrow D$$

This will be called coupling-in-series. A slightly different situation pertains when one chemical reaction is inseparably accompanied by a second as indicated by:

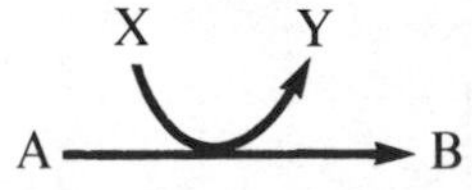

The linking of the two reactions is described as coupling-in-parallel, although it must be emphasized that they cannot occur separately since a single mechanism

Table 2.4. Standard free energy change, apparent equilibrium constant, mass action ratio, free energy change and equilibrium nature of some glycolytic reactions in the myocardium of the rat*

Reaction catalysed by	ΔG^{0} (kJ.mole^{-1})	Equilibrium constant	Mass action ratio	Equilibrium constant/ mass action ratio	ΔG† (kJ.mole^{-1})	Suggested or assumed equilibrium nature
Hexokinase	−20.9	4700	0.08	59 000	−27.9	non-equilibrium
Glucosephosphate isomerase	+ 2.1	0.4	0.24	1.7	− 1.3	near-equilibrium
6-Phosphofructokinase	−17.1	1050	0.03	35 000	−26.5	non-equilibrium
Aldolase	+23.0	0.000 1	0.000 009	11	− 6.1	near-equilibrium
Glyceraldehyde-phosphate dehydrogenase plus phosphoglycerate kinase	+ 7.9	850	9	94	−11.5	near-equilibrium
Phosphoglyceromutase	+ 4.6	0.14	0.12	1.2	− 0.5	near-equilibrium
Enolase	− 3.3	3.6	1.4	2.6	− 2.4	near-equilibrium
Pyruvate kinase	−24.7	2000	40	500	−15.8	non-equilibrium

* Data taken from Crabtree and Taylor (1979) and Newsholme and Start (1973); p. 97.

† ΔG is calculated from equation $\Delta G = -RT\ln K_{eq}/\Gamma$. For most reactions, the values of the equilibrium constant and mass action ratios are dimensionless. However, for aldolase the units are mol per l and for the complex, glyceraldehyde 3-phosphate dehydrogenase-phosphoglycerate kinase, the units are (mol per l)$^{-1}$.

causes both reactions. In metabolism, X may be regenerated (from Y) in a separate reaction, e.g.:

Hence, by means of reactions coupled-in-parallel, pathways are linked together. A rather small number of pairs of compounds with the functions of X and Y play a major role in metabolism (for example, ADP/ATP, NAD^+/NADH, $NADP^+$/NADPH, coenzyme A/acetyl-coenzyme A, all of which will be discussed in detail in later chapters).

(i) *Coupling-in-series* The ΔG^0 for the hypothetical reaction A $\rightleftarrows$ B, which is catalysed by enzyme E_2 and is part of the metabolic pathway,

$$S \overset{E_1}{\rightleftarrows} A \overset{E_2}{\rightleftarrows} B \overset{E_3}{\rightleftarrows} P$$

could be either negative or positive, whereas ΔG must be negative. Consider the situation if ΔG^0 is positive. For ΔG to be negative, the value of the term, $RT\ln\frac{[B]}{[A]}$, (in the equation $\Delta G = \Delta G^0 + RT\ln\frac{[B]}{[A]}$) must be negative and larger than the value of ΔG^0 (i.e. the value of Γ must be less than that of K_{eq}). As indicated above, the concentrations of A and B play a very important role in governing the direction of this reaction but the important question here is how are these concentrations determined? They will depend upon a number of factors, one of which is whether the reactions catalysed by enzymes E_1 and E_3 are non- or near-equilibrium. For example, if K_{eq} for reaction E_1 is large and if the reaction is near-equilibrium, this would result in a high concentration of A. Similar conditions for reaction E_3 would produce a low concentration of B. A high concentration of A and a low concentration of B are the conditions necessary to ensure that the value of the term, $RT\ln\frac{[B]}{[A]}$, is large and negative, so that ΔG for reaction E_2 would be negative. This is one way in which it is possible for reaction E_2 to proceed with a negative ΔG despite the fact that ΔG^0 is large and positive (see Crabtree and Taylor, 1979).

(ii) *Coupling-in-parallel* Coupling-in-parallel is most simply explained by reference to a specific reaction, for example the phosphorylation of glucose. This is an early reaction of glycolysis (Table 2.4) and is necessary for metabolism of glucose in all tissues. It is known that this reaction, which is catalysed by the enzyme hexokinase, occurs as follows:

$$\text{ATP} \qquad \text{ADP}$$

glucose $\longrightarrow$ glucose 6-phosphate

However, it is pertinent to ask why a simpler and more direct reaction cannot take place, as follows:

$$\text{P}_i{}^* \qquad \text{H}_2\text{O}$$

glucose $\longrightarrow$ glucose 6-phosphate

This reaction does occur in some cells (e.g. liver) but the reaction always occurs in the opposite direction (i.e. glucose formation); it is catalysed by the enzyme glucose 6-phosphatase. The free energy change for the glucose 6-phosphatase reaction (i.e. hydrolysis of glucose 6-phosphate) is given by:

$$\Delta G = \Delta G^0 + RT \ln \frac{[\text{glucose}]\,[\text{P}_i]}{[\text{glucose 6-phosphate}]\,[\text{H}_2\text{O}]}$$

from which it is possible to calculate the concentrations of substrates and products that would be necessary to produce a negative value of ΔG for the *synthesis* of glucose 6-phosphate by this reaction. (The value of ΔG^0 is 13.8 kJ.mol^{-1} and the concentration of water is assumed to be unity.) If the physiological concentrations (i.e. those that occur in the living cell) of glucose 6-phosphate and phosphate are inserted in this equation, the glucose concentration would need to be greater than 1.6 M for ΔG to be less than zero (see Section B.1.b for details of the calculation). This concentration is far in excess (by more than 300-fold) of the blood glucose concentration and probably 3000-fold greater than the intracellular glucose concentration. Such a glucose concentration would be physiologically unacceptable; if present in the bloodstream, it would be exceedingly difficult to absorb glucose from the lumen of the intestine and to prevent its loss through the kidneys. In addition, such glucose concentrations would cause massive osmotic effects and side reactions might occur which could cause directly damage to the cells or the accumulation of unwanted and even toxic side products (e.g. fructose, sorbitol – see Chapter 13; Section E.2; Chapter 15; Section G).

On the other hand, a negative ΔG for this reaction could be achieved by decreasing the concentration of glucose 6-phosphate. Why is this not feasible? Since glucose 6-phosphate is an important metabolic intermediate, which is involved in several metabolic pathways (e.g. glycogen synthesis, glycolysis,

---

* The symbols P_i and PP_i will be used in equations to denote phosphate and pyrophosphate respectively. These widely used abbreviations originated from the designations inorganic phosphate and pyrophosphate. Under physiological conditions the ionic forms HPO_4^{2-} and $HP_2O_7^{3-}$ will predominate.

pentose phosphate pathway) lowering its concentration by an order of magnitude or more could have serious effects on the direction and the rates of reactions of these other pathways. Hence, the involvement of ATP in the phosphorylation of glucose can be seen as a means of achieving this reaction without the need for excessively high concentrations of substrate or low concentrations of product. In other words, the phosphorylation of glucose is coupled, in parallel, to the conversion of ATP to ADP.

From the thermodynamic point of view, it is possible to separate glucose phosphorylation into two reactions as follows:

$$P_i + \text{glucose} \longrightarrow \text{glucose 6-phosphate} + H_2O$$

ΔG^0 for this reaction is $+13.8 \, \text{kJ.mole}^{-1}$, and:

$$ATP + H_2O \longrightarrow ADP + P_i$$

ΔG° for this reaction is $-30.5 \, \text{kJ.mole}^{-1}$.

These two reactions and their ΔG^0 values can be algebraically summed to give:

$$ATP + \text{glucose} \longrightarrow \text{glucose 6-phosphate} + ADP$$

ΔG° for this reaction is $-16.7 \, \text{kJ.mole}^{-1}$.

From this it can be seen that the involvement of ATP in the coupled reaction changes the value of ΔG^0 from large and positive to large and negative. Consequently, phosphorylation of glucose can occur in the cell at concentrations of glucose and glucose 6-phosphate that are physiologically acceptable (i.e. 10^{-5} to 10^{-4} M).

A further example of the use of ATP in a 'facilitating' role is the initial reaction involved in metabolism of acetate and fatty acids:

$$\text{acetate} + \text{CoASH} + ATP \longrightarrow \text{acetyl-CoA} + AMP + PP_i$$

This is slightly different from the reaction considered above since ATP is hydrolysed to AMP and pyrophosphate. The advantage of this may be two-fold. First, the concentrations of AMP and pyrophosphate are maintained lower than those of ADP and phosphate so that ΔG will be more negative. (This is due, in part, to the coupling-in-series of the hydrolysis of pyrophosphate to phosphate catalysed by inorganic pyrophosphatase.) Secondly, the ΔG^0 for the hydrolysis of ATP to AMP and pyrophosphate ($-35.9 \, \text{kJ.mole}^{-1}$) is more negative than the hydrolysis of ATP to ADP and P_i ($-30.5 \, \text{kJ.mole}^{-1}$). Thus the change in the values of both terms (ΔG^0 and $RT\ln$ [products]/[substrates]) in the equation will be instrumental in providing a more negative ΔG.

It is important to note that coupling-in-parallel is not achieved by the simultaneous occurrence of two separate reactions (although they can be treated in this way for thermodynamic considerations as above); the two reactions must be mechanistically linked (i.e. a new reaction is required with an appropriate enzyme). A detailed discussion of this role of ATP has been given by Atkinson (1977).

(iii) *Energy transfer in coupled systems* If a reaction that involves coupling-in-parallel is notionally separated into two reactions for thermodynamic analysis, chemical energy can be said to be transferred from one reaction to another. It is possible, and of some interest, to calculate the efficiency of this transfer for metabolic reactions. That is, to calculate how much of the chemical energy is released as heat. However, it is important to know that only the first law of thermodynamics can be applied to this manipulation.

Applying the first law of thermodynamics to the hypothetical example used on p. 23.

$$\Delta H_{(X+A \rightarrow Y+B)} = \Delta H_{(A \rightarrow B)} + \Delta H_{(X \rightarrow Y)}$$

(The reactions in parentheses indicate those to which the ΔH applies.)

If the reaction, $X \rightarrow Y$, proceeds with a negative enthalpy change and reaction $A \rightarrow B$ proceeds with a positive enthalpy change, the former reaction has transferred chemical energy to the latter. (Note that we can only speak of the transfer of chemical energy between reactions and not between metabolites since the absolute value of H are not known.) The maximum amount of chemical energy that can be transferred from the reaction $X \rightarrow Y$ is equal to $-\Delta H_{(X \rightarrow Y)}$. Consequently, a percentage efficiency for the transfer can be calculated as:

$$\frac{100 \times \text{energy transferred}}{\text{energy available}}$$

or

$$\frac{100 \times \Delta H_{(A \rightarrow B)}}{-\Delta H_{(X \rightarrow Y)}}$$

In a sense, the reaction $X \rightarrow Y$ is doing chemical work. From the first law of thermodynamics it follows that, since $\Delta H = q - w$, the energy not transferred to the second reaction is released as heat.

As an example, we can consider the process of glucose oxidation and ATP formation. In the oxidation of glucose, a proportion of $-\Delta H$ for this process is transferred to the synthesis of ATP (the specific reactions of which are given in Chapters 4 and 5). The remainder is lost as heat. The oxidation of one mole of glucose is accompanied by the synthesis of 38 moles of ATP from ADP and phosphate. This can be represented as follows:

$$\text{glucose} + 6O_2 \longrightarrow 6CO_2 + 6H_2O$$

$$38P_i + 38ADP \qquad 38ATP + 38H_2O$$

The ΔH for glucose oxidation is $-2813 \, \text{kJ.mole}^{-1}$ and for ATP synthesis is $+21 \, \text{kJ.mole}^{-1}$. Thus the efficiency of transfer is

$$\frac{100 \times 21 \times 38}{2813} = 28.4\%$$

This means that during the oxidation of glucose only about 30% of the available chemical energy is conserved in the formation of ATP. The remainder will be released as heat.

It is also possible to calculate an efficiency of transfer of Gibbs' free energy. Like ΔH, ΔG is a thermodynamic function and so, for fixed concentrations of A, B, X, and Y in the example given above, the following equation applies:

$$\Delta G_{(X+A \to Y+B)} = \Delta G_{(A \to B)} + \Delta G_{(X \to Y)}$$

From this, an efficiency of transfer of free energy may be calculated in the same way as for ΔH above, indeed such efficiencies are frequently quoted. However, unlike the enthalpy efficiency, this efficiency of transfer of free energy cannot be usefully applied to metabolic situations. This is because Gibbs' free energy is not a conventional form of energy so that it is not conserved according to the first law. It is not, for example, correct to speak quantitatively of ΔG being converted into heat and work (Crabtree and Taylor, 1979). Therefore, in any calculation or any discussion of the quantitative nature of the transfer or transduction of energy, values of ΔH and not ΔG must be used.

B. THE THERMODYNAMICS OF THE ROLE OF ATP IN METABOLISM

The metabolic processes involving ATP are outlined in the introduction to Chapter 4 and described in detail in subsequent chapters. ATP is synthesised (from ADP and P_i) during the catabolism of fuels such as glucose, glycogen, or lipids; it is used in such processes as muscle contraction, active transport and biosynthesis. The ATP/ADP + P_i system, therefore, serves to transfer energy between the producing processes and the utilizing processes in living organisms (see Figure 2.1). ATP is the only form of chemical energy that can be converted into all other forms of energy used by living organisms. It therefore fills a role analogous to that of money in the economy and has been described as the 'energy currency' of the cell. The main weakness of the analogy is that money, unlike ATP, can be accumulated for later use.

Although it is not possible to use Gibbs' free energy changes as a measure of the amount of energy transferred from one such reaction involving ATP to another

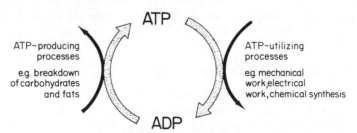

Figure 2.1 The ATP–ADP cycle

(see above), it is known that, within the cell, the concentrations of ATP, ADP and P_i are such that ΔG of the reactions,

$$ATP \rightarrow ADP + P_i$$

and

$$ATP \rightarrow AMP + PP_i$$

are large and negative. The fact that these reactions are far displaced from equilibrium has a number of important consequences.

1. In muscle, some of the change in chemical energy that occurs when ATP is hydrolysed is converted into mechanical energy to produce movement. It has been shown that muscles can perform up to 24 kJ of work for each mole of ATP hydrolysed (Wilkie, 1970). Since ΔG is a measure of the maximum work that can be done by a reaction (Section A.2.b) it follows that the ΔG for ATP hydrolysis at the site of the myofibrillar ATPase must be *greater* than $-24\,\text{kJ.mole}^{-1}$.

2. The hydrolyses, ATP to ADP or ATP to AMP, are used in initiating the metabolism of glucose and fatty acids. Thus glucose is phosphorylated to glucose-6-phosphate and fatty acids are acylated to produce fatty acyl-CoA. In these reactions the coupling to the hydrolysis of ATP ensures that the formation of the metabolic intermediate occurs at physiologically acceptable concentrations of substrates (glucose or fatty acids) and products (glucose 6-phosphate or fatty acyl-CoA) (see Section A.2.d). In a similar way, the hydrolysis of ATP, or one of its nucleotide analogues (e.g. GTP, CTP, UTP), occurs in the biosynthesis of carbohydrate, lipids, and proteins. Again this ensures that the biosynthetic processes occur at physiologically acceptable concentrations of the precursors of the biosynthetic pathways.

Although it is accepted that the value of ΔG for the hydrolysis of ATP is important in these reactions, the reader should be aware of the differences of opinion as to the importance of the value of ΔG^0 for ATP hydrolysis. It has been seen both as central to the understanding of the energetics of metabolism (Lehninger, 1975) and as totally irrelevant (Banks and Vernon, 1970). It is the authors' belief that these extreme views are the consequence of inadequate understanding of the role of ΔG^0, so that the remainder of this section attempts to provide a simplified account of the origin and importance in metabolism of ΔG^0 for the hydrolysis of ATP.

1. Standard free energy of hydrolysis of ATP

Since ΔG^0 is one of the two factors that determine the thermodynamic feasibility of any reaction, a great deal of effort has gone into measuring the value of ΔG^0 for the hydrolysis of ATP (Krebs and Kornberg, 1957). A value of $\Delta G^{0\prime}$ of $-30.5\,\text{kJ.mole}^{-1}$, at 37 °C and saturating Mg^{2+} concentration, is now widely accepted. Attention has been focussed on ATP hydrolysis for several reasons. First, when kinetically coupled to another reaction, it plays a very important role in metabolic biochemistry (see Section A.2.d). Secondly, its reversal is part of the over-all reaction by which ATP is synthesized. Thirdly, if the ΔG^0 values for hydrolysis of a series of organic phosphates is known, it is a simple matter to calculate the ΔG^0 values for phosphate transfer between any pair of them. It avoids the need to produce long lists of ΔG^0 values for all possible transfers. In other words, the ability of water to accept phosphate groups (which is what is happening in hydrolysis) is used as a reference point against which to compare the tendency for different organic compounds to accept phosphate groups. An example should make this clear. The transfer of a phosphate group to creatine from ATP (or, conversely, from phosphocreatine to ADP) is an important metabolic reaction, catalysed by creatine kinase. If the ΔG^0 for the hydrolysis of ATP and for the hydrolysis of phosphocreatine are known, then the ΔG^0 for the creatine kinase reaction can be calculated:

$$\text{phosphocreatine} + H_2O \longrightarrow \text{creatine} + P_i \qquad \Delta G^{0\prime} = -43.1 \text{ kJ mole}^{-1}$$

$$\text{ATP} + H_2O \longrightarrow \text{ADP} + P_i \qquad \Delta G^{0\prime} = -30.5 \text{ kJ.mole}^{-1}$$

Summation of these reactions gives the creatine kinase reaction and summation of their $\Delta G^{0\prime}$ values, the $\Delta G^{0\prime}$ for the creatine kinase reaction:

$$\text{creatine} + \text{ATP} \longrightarrow \text{phosphocreatine} + \text{ADP}$$

$$\Delta G^{0\prime} = -30.5 + 43.1 = +12.6 \text{ kJ.mole}^{-1}$$

Standard free energy values for the hydrolysis of other phosphates of biochemical interest are presented in Table 2.5. Values of ΔG^0 for the hydrolysis of organic phosphates, which were first obtained in the 1930s, suggested that such compounds fell into two groups; those with a relatively low ΔG^0 for hydrolysis and those, including ATP, with a high ΔG^0 for hydrolysis. This division supported the concept of the 'high energy phosphate bond' which is discussed in more detail below. In fact, improved data show that ΔG^0 values for the hydrolysis of organic phosphates of biochemical interest cover a more or less continuous range with ATP possessing an intermediate value (Table 2.5). Despite this, ATP is still classified as a 'high energy compound', a term which, although having some biological usefulness (see below) has given rise to much misunderstanding concerning the thermodynamics of ATP. However, before considering its significance, the chemical basis for the ΔG^0 of ATP hydrolysis being $-30.5 \text{ kJ.mole}^{-1}$ deserves explanation.

(a) *Chemical basis for magnitude of $\Delta G^{0\prime}$ of ATP hydrolysis*

The reason for the relatively large change in free energy when ATP is hydrolysed under standard conditions is that the products of the reaction (ADP and P_i) are much more stable than ATP, at or around pH 7.0. At least three factors

Table 2.5. Values of $\Delta G^{0\prime}$ of hydrolysis for phosphates of biochemical interest*

Phosphates	Hydrolysis product	$\Delta G^{0\prime}$ (kJ.mole^{-1})
Phosphoenolpyruvate	Pyruvate	-61.9
3-phosphoglyceroyl phosphate	3-phosphoglycerate	-49.3
Phosphocreatine	Creatine	-43.1
ATP	ADP	-30.5
ADP	AMP	-27.6
Pyrophosphate	Phosphate	-27.6
Glucose 1-phosphate	Glucose	-20.9
Glucose 6-phosphate	Glucose	-13.8
AMP	Adenosine	-14.2
Glycerol 3-phosphate	Glycerol	-9.2

* Data taken from Krebs and Kornberg (1957).

contribute to this increased stability. They will be understood more readily if reference is made to the structure of ATP (Figure 2.2).

(i) *Resonance stability* Compounds are more stable when their bonding electrons are distributed over more than one covalent bond (delocalization). This can be indicated by drawing a number of extreme structures (resonance or

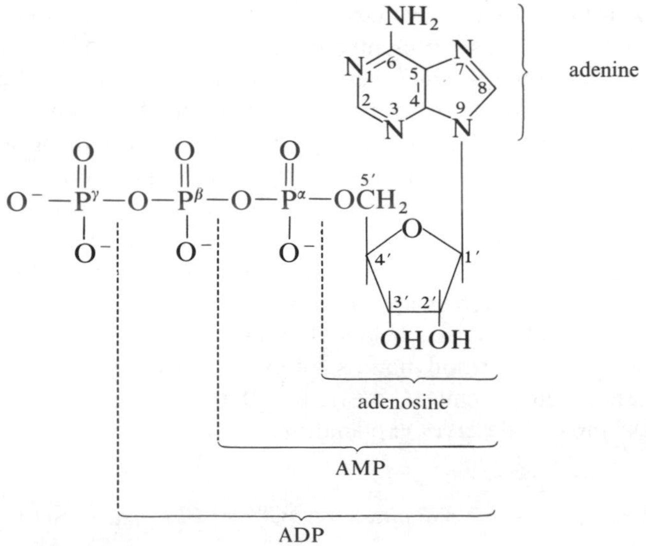

Figure 2.2 Structural formula of ATP. The molecule is shown in its fully ionized form. Primes (′) are used to distinguish the atoms of the ribose ring from those of the purine ring. Greek letters are used to identify the phosphorus atoms

canonical forms) in which the electrons are delocalized; the real structure is then considered to be intermediate between these forms. In the phosphate ion, all the oxygen atoms are equivalent so that the real structure is intermediate between the following resonance forms:

$$
\begin{array}{cccc}
\text{O} & \text{O}^- & \text{O}^- & \text{O}^- \\
\| & | & | & | \\
\text{O}^-\!-\!\text{P}\!-\!\text{O}^- \rightleftharpoons & \text{O}^-\!-\!\text{P}\!=\!\text{O} \rightleftharpoons & \text{O}^-\!-\!\text{P}\!-\!\text{O}^- \rightleftharpoons & \text{O}\!=\!\text{P}\!-\!\text{O}^- \\
| & | & \| & | \\
\text{O}^- & \text{O}^- & \text{O} & \text{O}^-
\end{array}
$$

However, when the phosphate is part of the ATP molecule there are fewer resonance possibilities, so that phosphate and ADP are more stable than ATP.

(ii) *Charge repulsion* At physiological hydrogen ion concentrations the phosphorus atoms of ATP (see Figure 2.2) bear negatively charged oxygen atoms. Repulsion between these charges causes strain in the molecule which is relieved on hydrolysis when the charges can separate.

(iii) *Ionization* On hydrolysis of ATP, two new acid groups make an appearance:

$$
\text{H}_2\text{O} + \text{adenosine}\!-\!\text{O}\!-\!\overset{\displaystyle \text{O} \atop \|}{\underset{| \atop \text{O}^-}{\text{P}}}\!-\!\text{O}\!-\!\overset{\displaystyle \text{O} \atop \|}{\underset{| \atop \text{O}^-}{\text{P}}}\!-\!\text{O}\!-\!\overset{\displaystyle \text{O} \atop \|}{\underset{| \atop \text{O}^-}{\text{P}}}\!-\!\text{O}^- \rightarrow
$$

$$
\text{adenosine}\!-\!\text{O}\!-\!\overset{\displaystyle \text{O} \atop \|}{\underset{| \atop \text{O}^-}{\text{P}}}\!-\!\text{O}\!-\!\overset{\displaystyle \text{O} \atop \|}{\underset{| \atop \text{O}^-}{\text{P}}}\!-\!\text{O} + \text{HO}\!-\!\overset{\displaystyle \text{O} \atop \|}{\underset{| \atop \text{O}^-}{\text{P}}}\!-\!\text{OH}
$$

At pH 7.0 one of these groups is partly ionized according to the equation:

$$
\text{HO}\!-\!\overset{\displaystyle \text{O} \atop \|}{\underset{| \atop \text{O}^-}{\text{P}}}\!-\!\text{OH} \rightleftharpoons \text{HO}\!-\!\overset{\displaystyle \text{O} \atop \|}{\underset{| \atop \text{O}^-}{\text{P}}}\!-\!\text{O}^- + \text{H}^+
$$

This ionization has a large negative ΔG^0 which contributes to the over-all ΔG^0 for ATP hydrolysis.

Factors (ii) and (iii) suggest that the ΔG^0 of ATP hydrolysis will be highly pH dependent. That this is the case is seen by comparing the value of ΔG^0 at pH $= 0$ ($-1.25\,\text{kJ.mole}^{-1}$) with that at pH $= 7.0$ ($-30.5\,\text{kJ.mole}^{-1}$).

(b) *Metabolic importance of* $\Delta G^{0\prime}$ *of ATP hydrolysis*

The significance of the fact that the value of ΔG for ATP hydrolysis in the cell is moderately large and negative has been described in the introduction to Section B. The value of ΔG^0 may appear to have no direct significance because, for any value of ΔG^0, a large negative ΔG can arise if the reactant concentrations are appropriate (see equation in Section A.2.c). Conversely, however, the reactant concentrations pertaining when ΔG is large and negative are influenced by the value of ΔG^0, so that it does become of considerable metabolic importance.

The influence of the value of ΔG^0 can best be seen by considering an actual reaction in metabolism and calculating the consequences of halving the ΔG^0 for ATP hydrolysis. The example chosen is the reaction catalysed by hexokinase so that essentially the same reasoning is used as in Section A.2.d. The reaction is:

$$\text{glucose} + \text{ATP} \rightarrow \text{glucose 6-phosphate} + \text{ADP}$$

The value of $\Delta G^{0\prime}$ for this reaction is $-16.7\,\text{kJ.mole}^{-1}$ and the value of ΔG in the isolated perfused heart[7] can be calculated as $-23.1\,\text{kJ.mole}^{-1}$. If however the value of $\Delta G^{0\prime}$ for ATP hydrolysis happened to be just half of its actual value (a purely hypothetical situation) this would change the value of ΔG^0 for the hexokinase reaction to $-1.5\,\text{kJ.mole}^{-1}$. In order to maintain the same ΔG for the reaction, the mass action ratio would need to change from 0.08 to 0.0002. If the concentrations of ATP, ADP, and glucose 6-phosphate were maintained at normal physiological concentrations, this would result in a 400-fold increase in glucose concentration within the cell. The problems associated with such an increase have already been discussed (see Section A.2.d).

2. The concept of the 'high energy phosphate bond'

In the 1930s, work on glycolysis, the adenine nucleotides, and related compounds led to the identification of a large number of phosphorylated intermediates. K. Lohmann and O. Meyerhof, in 1934, postulated that in glycolysis the phosphate ester bond of the hexose phosphates was changed into a novel 'energy-rich' phosphate bond in ATP. The available values for the ΔG^0 of hydrolysis of these phosphorylated intermediates (although less accurate than the data in Table 2.5) led Lipmann (1941) to propose that the organic phosphates could be divided into a 'high energy' class (which included ATP) and a 'low energy' class (which included glucose 6-phosphate). Although the basis for this classification has been revised (see Table 2.5) and ATP is now seen as occupying an intermediate position in the range of ΔG^0 values for hydrolysis, the term 'high-energy phosphate bond' has persisted to describe bonds in such compounds as ATP, ADP, phosphocreatine, phosphoenolpyruvate, etc. (Phosphate groups which are hydrolysed with large standard free energy changes are often represented by the symbol $\sim$ P.) Before considering the utility of the term, the validity of some of the objections to its use will be considered.

(a) *Problems associated with the 'high energy bond' terminology*

(i) *Confusion with bond energy* Bond energy is a term which applies specifically to the enthalpy change when a single chemical bond is broken. The free energy change of ATP hydrolysis is the net result of breaking several bonds and forming some others (see Section B.1.a) so that the 'high energy' is not associated with a single bond. This looseness of terminology has, perhaps justifiably, irritated many chemists. This objection does not stand if the terms 'high energy compound' or 'energy-rich phosphate' rather than 'high energy bond' are used.

(ii) *Irrelevance of hydrolysis* Since the use of ATP in many reactions does not involve hydrolysis, it may be asked whether the ΔG^0 for the hydrolysis reaction has any biological relevance. The answer to this question has already been given in the affirmative in the introduction to Section B.1.

(iii) *Relevance of* ΔG^0 A more fundamental objection to the 'high energy bond' concept is that it is based on values of free energy changes under standard conditions which do not apply in the cell. Since ΔG^0 is only one of two terms that determine ΔG (and hence the feasibility of reaction) it alone cannot determine whether a reaction can occur. This matter has been considered in detail in Section A.2.

(b) *Usefulness of the 'high energy' terminology*

Despite these objections, the terms 'high energy phosphate bond', 'energy-rich bond', 'high energy compound', and 'energy-rich phosphate'* have persisted. Why is this? The answer is that in biology it is sometimes important to know the amount of available energy in the cell, the rate at which energy can be produced and utilized and the amounts of energy that can be obtained from various reserves under different conditions. These aspects of energy metabolism can be more simply quantified using energy-rich phosphate as a unit of available energy. For example, it can be calculated that the rate of turnover of ATP in sprinting man is approximately $160\ \mu mol.min^{-1}.g^{-1}$ muscle (see Chapter 9; Section A.2) whereas the concentration of ATP is only $5\ \mu mol.g^{-1}$ and that of phosphocreatine is only about $20\ \mu mol.g^{-1}$ muscle. (The concentrations of other 'energy-rich phosphate' compounds, e.g. phosphoenolpyruvate, are too low to be considered.) Thus the total concentrations of energy-rich phosphate (i.e. ATP plus phosphocreatine) could last only a few seconds, which emphasizes that the rate of glycolysis must be increased very rapidly during sprinting. Furthermore, the total concentration of ATP must be turned over once every 2–3 sec during such activity. Knowledge of which compounds contain 'energy-rich phosphate' groups

---

* This plethora of synonyms, and the widespread use of inverted commas, probably signify the discomfort shown by biochemists at their continued use of such a chemically sloppy but biologically useful term.

allows such calculations to be made. Another example is provided by consideration of the important metabolic problems that arise if the blood supply to the heart or a portion of the myocardium is suddenly occluded. The survival of that tissue, and indeed of the whole animal, will depend upon the availability of compounds such as ATP, phosphocreatine and phosphoenolpyruvate, which are conveniently described as energy-rich phosphates, together with the compounds that can produce ATP under anaerobic conditions (e.g. glycogen). If the concentrations of these compounds and the rate of utilization of ATP by the contracting myocardium are known, the time of survival of that part of the myocardium affected by the occlusion can be calculated. This enables various possible clinical interventions to be discussed quantitatively in relation to the production and utilization of ATP in the heart (see Chapter 5; Section G.2).

In such examples, the term 'energy-rich phosphate' describes a known and accepted biological function (i.e. the ability to provide, directly or by phosphate transfer, energy, in the form of ATP, for the vital process of contraction). The use of the term in this manner should imply no specific physicochemical nature of the energy source or the mechanism of energy transfer.

3. Kinetic factors and the metabolic role of ATP

The central position of ATP in metabolism can be viewed in thermodynamic terms as discussed in the above sections. Of equal importance, however, are mechanistic (kinetic) considerations. It is likely that an important factor that permits the widespread use of the ATP/ADP couple in energy transfer within the cell is the kinetic stability of ATP. If the hydrolysis of ATP actually approached equilibrium, the concentration of ATP in the cell would be very low indeed. Attainment of equilibrium is prevented by the high activation energy (see Chapter 3; Section A.2.a) for hydrolysis of ATP and the remarkable control exerted over enzymes (e.g. myofibrillar ATPase) capable of lowering the activation energy of hydrolysis. Mechanisms of control that apply to these enzymes as well as the more 'metabolic' enzymes are described in Chapter 7.

C. THE THERMODYNAMIC STRUCTURE OF A METABOLIC PATHWAY

Thus far in the chapter, thermodynamic principles have been applied to two extremes, metabolism in the whole animal (Section A) and the individual biochemical reactions (Section B). The intermediate level at which metabolism can be viewed is the metabolic pathway which comprises individual reactions arranged in sequence to achieve a particular chemical change. During the last 25 years, the biochemical details of all the major metabolic pathways, and probably most of those that actually exist in mammals, have been elucidated. This information has enabled very complex metabolic wall charts to be produced which provide details of the individual reactions of a large number of such pathways. Although such charts can be useful as a ready reference and to impress one's friends of the complexity of metabolism, they have done a certain disservice to metabolic biochemistry. They imply that pathways are no more than a sequence of enzyme-catalysed reactions and they provide a seemingly authoritative stamp of approval as to the existence of such pathways and therefore their physiological importance. However, they give no indication of the complex inter-relationship between different tissues and between different organs or the means by which the flux through pathways can be regulated in the whole animal. This section attempts to demonstrate that a metabolic pathway is not just a series of linked reactions but that it has an important thermodynamic (and kinetic) structure. It also questions what is meant by the term 'metabolic pathway' and provides a physiologically useful definition. The function of metabolic pathways, the inter-relationships between different pathways, their integration and regulation provide the basis of much of the remainder of the book.

In the ensuing analysis, the thermodynamic principle most in evidence is the extent to which an individual reaction *in vivo* approaches equilibrium. To appreciate the significance of this it is necessary to examine the equilibrium state, and its alternatives, in some detail.

1. Closed and open systems

Consider the following hypothetical enzyme-catalysed reaction:

$$A \underset{v_r}{\overset{v_f}{\rightleftharpoons}} B$$

where the reactants are A and B and where v_f and v_r represent the rates of the forward and reverse components. If such a reaction is isolated from its surroundings, so that A and B neither enter nor leave the system, that is, it is a closed system, the concentrations of these substances will approach values that will equalize v_f and v_r. Such a state, in which the rates of the forward and reverse reactions are equal, is referred to as an equilibrium. For a closed system, equilibrium is the only state in which the concentrations of A and B do not vary with time. There is no *net* interconversion of the reactions (since $v_f = v_r$) and such systems can do no useful work (since $\Delta G = 0$). Hence a closed system does *not* provide a model of the over-all metabolic situation in living cells, which does result in net conversions of substances at rates which can be varied and in which useful work is done. (Nonetheless, it should be emphasized that, in the cell, some individual reactions may be at equilibrium and, in metabolic pathways, some reactions may be very close to equilibrium (see below).)

Metabolic pathways are examples of open thermodynamic systems, which are characterized by a continuous exchange of both matter and energy with their surroundings. Such an open system can be illustrated as follows:

$$E_1 \quad E_2 \quad E_3$$

$$S \rightarrow A \underset{v_r}{\overset{v_f}{\rightleftarrows}} B \rightleftarrows P$$

This system contains the same hypothetical reaction as in the closed system described above but, in this case, there is a reaction (E_1) which continuously supplies A from the surroundings and a reaction (E_3) which continuously removes B from the system. In this hypothetical pathway, conditions can be such that reaction E_1 generates a constant flux (i.e. a constant rate of flow of molecules of A) which is transmitted through reactions E_2 and E_3 by the concentrations of the substrates for these reactions. In other words, the flux through all the reactions is the same and the flux through reactions E_2 and E_3 is a function of the concentrations of A and B respectively. In this condition, in which there is a constant rate of conversion of S to P and the concentrations of the intermediates A and B remain constant, the metabolic pathway is said to be in a steady state.

Three types of reactions play an important role not only in the maintenance of the steady state but also in permitting the flux to change from one steady state to another (i.e. they provide the basis for regulation). These are near-equilibrium, non-equilibrium, and flux-generating reactions; the flux-generating reaction is a special case of a non-equilibrium reaction (see below). Each has a role to play in a metabolic pathway.

2. Near-equilibrium, non-equilibrium, and flux-generating reactions

(a) *Near- and non-equilibrium reactions*

Reactions in a metabolic pathway can be divided into two classes: those that are very close to equilibrium (near-equilibrium) and those that are far removed from equilibrium (non-equilibrium). For a near-equilibrium reaction, the concentration ratio of products and substrates is similar to the value of the equilibrium concentrations. The value of the mass action ratio thus approaches that of the equilibrium constant (i.e. the value of the ratio K_{eq}/Γ is close to unity) so that ΔG is close to zero. For a non-equilibrium reaction, the concentration ratio of products and substrates is much smaller than the value of the equilibrium constant (i.e. the value of the ratio K_{eq}/Γ is much larger than unity) so that the value of ΔG is large and negative.

This description of two types of reactions is no more than an extension of what was discussed in Section B, but it re-emphasizes the meaning of near- and non-equilibrium reactions. However, it is also possible to explain the equilibrium nature of a reaction by reference to kinetics and, in particular, to the rates of the forward (v_f) and reverse (v_r) components of the reaction. Although it preempts some of the discussion in Chapter 3, a kinetic explanation of near- and non-equilibrium reactions is given here for reasons of clarity.

A reaction in a metabolic pathway is non-equilibrium because the activity of the enzyme which catalyses the reaction is low in comparison to the activities of other enzymes in the pathway. Consequently, the concentration of substrate of the reaction is maintained high, whereas that of the product is maintained low (since it will be removed by the activity of the next enzyme). The rate of the reverse component of the reaction is thus very much less than the rate of the forward component. Conversely, a reaction is near-equilibrium if the catalytic activity of the enzyme is high in relation to the activities of other enzymes in the pathway. Rates of the forward and the reverse components of the reaction will be much greater than the over-all flux in this situation.

The situation in a simple hypothetical pathway containing both types of reaction might be indicated as follows:

$$\text{S} \underset{0.01}{\overset{\overset{\text{E}_1}{10.01}}{\rightleftarrows}} \text{A} \underset{90}{\overset{\overset{\text{E}_2}{100}}{\rightleftarrows}} \text{B} \underset{0.1}{\overset{\overset{\text{E}_3}{10.1}}{\rightleftarrows}} \text{P}$$

where the numbers represent the rates (not rate constants) of the forward and reverse components of the reactions. The flux through the pathway is 10 and, since it is in the steady state, each reaction must proceed at this rate. Thus for each reaction, $v_f - v_r$ equals 10. The enzyme converting S to A has a low activity (possibly because the concentration of the enzyme in the tissue is small), so that the rate of the forward component of the reaction is very similar to the flux through the pathway. Since the reverse component has a rate that is 1000-fold lower, it is quantitatively insignificant and the reaction is described as non-equilibrium. In the conversion, $\text{A} \rightarrow \text{B}$, the catalytic activity of the enzyme is much greater than that catalysing the first reaction. Hence the rates of the forward and reverse components are high, but the net flux is still only 10 units. Since the rates of both components are similar (differing by only 10%) the reaction is described as near-equilibrium. The final reaction, $\text{B} \rightarrow \text{P}$, is also described as non-equilibrium but it is not as far removed from equilibrium as the first reaction. The relationship between the thermodynamic and kinetic treatments of near- and non-equilibrium reactions is explained in mathematical terms in Appendix 2.1.

(b) *Metabolic significance of near-equilibrium reactions*

There are at least two advantages of near-equilibrium reactions (often loosely called equilibrium reactions) in a metabolic pathway: they allow easy reversal of the pathway, and small changes in the concentrations of substrates or products can produce large changes in flux. This latter point is important in metabolic control (see Chapter 7; Section A.2.c).

Some pathways need to function in opposite directions at different times. For example, under some conditions, the liver converts glucose to pyruvate, whereas under other conditions it converts pyruvate to glucose (i.e. gluconeogenesis). Although it is possible to reverse a non-equilibrium reaction by markedly reducing the concentration of the substrates or markedly increasing those of the products (i.e. by converting a large negative ΔG into a positive ΔG) such large changes in metabolite concentration would be physiologically unacceptable (see Section A.2.d). Consequently, if all the reactions of glycolysis were non-equilibrium, at least ten separate reactions would be required in the synthesis of glucose. In fact, seven of the glycolytic reactions are near-equilibrium so that they are readily reversed by *small* changes in the concentrations of their substrates and products. (Since the values of ΔG for these reactions are close to zero (Table 2.4) small concentration changes can change the value of ΔG from negative to positive.) The remaining three reactions in the glycolytic pathway are non-equilibrium (hexokinase, 6-phosphofructokinase and pyruvate kinase) so that separate non-equilibrium reactions are required at these positions in the reverse pathway. These are the reactions specific to gluconeogenesis which are described in Chapter 11.

A major advantage of near-equilibrium reactions in all metabolic pathways is that the flux through the reaction is very sensitive to changes in the concentrations of substrates or products. A very small increase in substrate concentration (or a very small decrease in that of the product) can produce a very large increase in flux through the reaction. This also means that a large change in flux can be accommodated by a near-equilibrium reaction without a large change in substrate or product concentration.

One disadvantage of a near-equilibrium reaction is its lack of sensitivity to allosteric regulation (i.e. regulation by factors other than changes in substrate or product concentrations—see Chapter 7; Note 3) when compared with non-equilibrium reactions. A very large change in the concentration of the allosteric regulator would be needed to change the flux through a near-equilibrium reaction (Crabtree, 1976). (This is not solely an academic point; it indicates to research workers in the pharmaceutical industry that they should be searching for compounds which inhibit the activities of enzymes that catalyse non-equilibrium reactions. Such compounds should have a greater chance of success as metabolic drugs than those which inhibit 'near-equilibrium' enzymes.)

(c) *Metabolic significance of non-equilibrium reactions*

Non-equilibrium reactions play an important role in providing directionality in a metabolic pathway. Their existence provides the answer to the question posed in Section A.2, namely, how is it that, in muscle, glucose is converted to lactate but not *vice versa*. The major advantage of the non-equilibrium reaction in metabolic control is that the flux through the reaction can be controlled by allosteric factors, so that such reactions are potential sites for feedback and hormonal control of flux through metabolic pathways (see Chapter 7; Section A.1).

This last statement reveals two problems, hitherto put to one side. First, what constitutes a metabolic pathway and, secondly, what determines the flux through it. The answer to both lies in the identification of a special kind of non-equilibrium reaction—the flux-generating reaction.

(d) *The flux-generating reaction*

That allosteric effectors can modify the flux through a non-equilibrium reaction allows such reactions to be regulated by factors external to the pathway. The question then arises, which if any non-equilibrium reaction in a pathway actually determines the flux through the pathway? Only one non-equilibrium reaction in the pathway can do this at any one time. Consider an enzyme (catalysing a non-equilibrium reaction) whose pathway-substrate* concentration is much less than that required to saturate the enzyme. The catalytic activity of this enzyme will result in a decrease in the concentration of its substrate, and hence in its activity, if the substrate concentration is not maintained. Such a reaction could not maintain a constant flux through a pathway so that a steady state could never be achieved. However, if such an enzyme is saturated with its pathway-substrate, a decrease in the concentration of this substrate will not affect the activity of the enzyme so that a constant flux through the reaction and hence through the pathway can be maintained at least for a period of time. A non-equilibrium reaction that is saturated with pathway-substrate is termed a flux-generating reaction. The identification of flux-generating reactions depends on kinetic criteria and is considered in Chapter 3; Section C.2.d. Examples of flux-generating reactions will be given when specific pathways are discussed in later chapters.

It is reasonable to suppose that each metabolic pathway possesses a flux-generating step—a reaction that generates the flux through the pathway and to which the rates of all other reactions in the pathway conform (as a consequence of coupling-in-series discussed in Section A.2.d). A change in the activity of this enzyme would cause a change in the flux through the pathway, whereas a change in the activity of any other enzyme in the pathway would not necessarily affect the flux.[8]

---

* Pathway-substrate is defined as that substrate which represents the flow of matter through the pathway under consideration. Care must be taken to identify correctly the pathway-substrate; the problem is illustrated by the following hypothetical example:

$$X \quad Y$$

$$S \longrightarrow A \rightleftarrows B \rightarrow P$$

In this pathway, the first reaction is flux-generating since it is saturated with S. However, it need not necessarily be saturated with X which can usefully be termed a co-substrate. In this case the latter is continuously regenerated, by the conversion of Y to X that may occur in another pathway, and has been termed a coenzyme (see Chapter 3; Section A.3).

The concept of the flux-generating step provides something hitherto impossible—a physiologically useful definition of a metabolic pathway. It has been tacitly assumed that a metabolic pathway is defined by its designation as such in a textbook or on a wallchart. Series of linked reactions have become accepted as metabolic pathways if the series is amenable to biochemical study. For example, glycolysis-from-glucose is usually accepted as a pathway since it is possible to supply a tissue with glucose and detect the appearance of lactate (or pyruvate) and the enzymes, which form the coupling-in-series system known as glycolysis, have been isolated and identified. This subjective definition of convenience can, with the appreciation of the nature of flux-generating steps, be replaced with a more objective definition. Namely, that a metabolic pathway is a series of enzyme-catalysed reactions, initiated by a flux-generating step and ending with either the loss of products to the environment, to a stored product (a metabolic 'sink') or in a reaction that precedes another flux-generating step (that is, the beginning of the next pathway). Such a pathway may be short or long. One interesting consequence of this functional definition is that the pathway need not be contained in a single tissue but can span several. Thus glycolysis-from-glucose is not a complete metabolic pathway. In this case, the physiological pathway begins at glycogen in the liver with glycogen phosphorylase as the flux-generating step or absorption in the gut with glucose transport as the flux-generating step (see Chapter 5; Section B.1). Glucose is then transported, by the blood, to muscle, where the pathway continues with the conversion of glucose to pyruvate. Other examples of metabolic pathways defined in this way will be given in subsequent chapters. The usefulness of this definition in understanding control of the flux through pathways should become clear in Chapter 7.

3. Determination of the equilibrium status of reactions *in vivo*

As implied in the discussion above, a complete understanding of a metabolic pathway requires knowledge of the equilibrium status of the constituent reactions. Although there is, as yet, no ideal experimental means of acquiring this knowledge, two approaches have been used with reasonable success; the comparison of equilibrium constants with mass action ratios and comparison of maximum enzyme activities for each of the reactions in the pathway. The details of how such experiments are carried out have been described elsewhere (see Newsholme and Start, 1973).

In the following chapters, where possible, the equilibrium nature of the various reactions is indicated. This is necessary in order to appreciate the mechanisms of control of such pathways and why, for example, inadequate amounts or impaired control of certain enzymes can result in diseased states. The metabolic pathways will be described according to the usual 'wall chart' interpretations of pathways since this will ease the job of cross-reference from this text to others. However, since it is important from the point of view of metabolic control in the whole animal, pathways will also be described according to the definition given in this chapter. For those who wish to read more deeply into this subject, the following references are recommended: Newsholme and Crabtree (1979, 1981a, b); Crabtree and Taylor, (1979).

APPENDIX 2.1 RELATIONSHIP BETWEEN THERMODYNAMIC AND KINETIC INTERPRETATIONS OF EQUILIBRIUM NATURE OF REACTIONS

Reaction $A \rightleftarrows B$ is considered from the kinetic viewpoint:

$$S \rightarrow A \underset{v_r}{\overset{v_f}{\rightleftarrows}} B \rightarrow P$$

where v_f and v_r are the rates in the forward and reverse directions, respectively. These are given by:

$$v_f = k_1 [A] \quad \text{and} \quad v_r = k_{-1}[B]$$

where k_1 and k_{-1} are the rate constants of the forward and reverse directions (see Chapter 3; Section C.2.a).

Combining the above two equations gives:

$$\frac{v_f}{v_r} = \frac{k_1 [A]}{k_{-1}[B]}$$

But, at equilibrium $\dfrac{k_1}{k_{-1}} = K_{eq}$

Therefore $\dfrac{v_f}{v_r} = K_{eq} \dfrac{[A]}{[B]}$

But $\dfrac{[B]}{[A]}$ in the living cell is the mass action ratio, Γ, therefore

$$\frac{v_f}{v_r} = K_{eq} \left(\frac{1}{\Gamma}\right) = \frac{K_{eq}}{\Gamma}$$

Thus the ratio of the rate in the forward direction to that in the reverse direction is equal to the ratio; equilibrium constant/mass action ratio.

Since $\Delta G = -RT\ln \dfrac{K_{eq}}{\Gamma}$

it follows that $\Delta G = -RT\ln \dfrac{v_f}{v_r}$

In other words, if we know the rates in the forward and reverse directions of a reaction in the *living* cell, we can calculate the free energy change.

Conversely, if we know ΔG, (or K_{eq}/Γ) the ratio, v_f/v_r, can be calculated. Thus for a ratio v_f/v_r of 1000, the value of ΔG is approximately $-17.5\,\text{kJ}$ (i.e. a non-equilibrium reaction) whereas for a ratio v_f/v_r of 2, the value of ΔG is $-3.0\,\text{kJ}$. For non-equilibrium reactions in glycolysis (e.g. 6-phosphofructokinase) the ratio v_f/v_r is approximately 10 000.

NOTES

1. The use of the joule (J) rather than the calorie as the unit of energy is necessitated by adoption of SI (Système Internationale d'Unités). Medical school staff are slowly adapting to the new units. Since one calorie equals 4.18 joules, readers can convert from calories to joules during lectures or when reading older literature by multiplying by 4.18. The abbreviations kJ for kilojoule (J $\times 10^3$), and MJ for megajoule (J $\times 10^6$) are frequently used.

2. The symbol Δ placed before a quantity signifies a change in that quantity. ΔH is the change in internal energy when pressure is constant, under which conditions changes in volume can occur. These are likely to be the conditions applying in metabolic reactions. Under conditions of constant volume, another internal energy function, ΔU, is of greater applicability. In reactions involving large increases in volume, ΔH will be significantly greater than ΔU because some of the work that could otherwise be done by the system will be used in expansion ($\Delta H = \Delta U + P\Delta V$).

3. An example of such a mechanism is substrate cycling which will be discussed in detail in Chapter 7; Section C.3. Another mechanism occurs in brown adipose tissue which is an important heat generating tissue in hibernating animals and in the newborn of some mammals including man (see Chapter 4; Section E.4).

4. The symbol ΔF is used to denote the change in Gibbs' free energy in the U.S.A. Unfortunately, ΔF is also used, in Britain, to denote the change in free energy which pertains at constant volume (i.e. $\Delta U - T\Delta S$).

5. In biochemical reactions it is frequently more convenient to define standard free energy changes at a proton concentration of 10^{-7} M (pH = 7.0) rather than at the standard state of 1 M (pH = 0). In such cases the standard free energy change is designated $\Delta G^{0\prime}$. Differences between ΔG^0 and $\Delta G^{0\prime}$ may be appreciable if substrates and products have different ionization constants.

6.
$$\Delta G = \Delta G^0 + RT\ln\frac{[\text{B}]}{[\text{A}]}$$

where [A] and [B] are the concentrations of the substrate and product of the reaction respectively.

But
$$\Delta G^0 = -RT\ln K_{eq}$$

and
$$[\text{B}]/[\text{A}] = \Gamma$$

So that,
$$\Delta G = -RT\ln K_{eq} + RT\ln\Gamma$$

$$\therefore \quad \Delta G = -RT\ln K_{eq}\left(\frac{1}{\Gamma}\right)$$

$$\Delta G = -RT\ln\frac{K_{eq}}{\Gamma}$$

7.
$$\Delta G = \Delta G^0 + RT\ln\Gamma$$
$$= -16.7 + 2.3RT\log\Gamma$$

Since the mass action ratio for the hexokinase reaction in heart is 0.08 (see Table 2.4) and since R is 0.0082 kJ.mole^{-1}, at 37 °C (310 °K),

$$\Delta G = -16.7 + (2.3 \times 0.0082 \times 310 \times \log 0.08)$$
$$= -23.1 \text{ kJ.mole}^{-1}$$

8. The flux-generating step is important in producing a steady-state flux in any pathway but the regulation of that flux must depend upon the structure of the remainder of the pathway. For example, if the pathway which transmits a flux J branches into two separate fluxes, J_1 and J_2, such that $J = J_1 + J_2$, then the enzyme activities at the branch point can determine the relative values of J_1 and J_2. In this case, it is possible to vary the flux (i.e. either J_1 or J_2) without affecting the flux-generating reaction (Newsholme and Crabtree, 1981a, b); see Chapter 8; Section D.1 for discussion of control of the branched pathway of fatty acid oxidation in liver.

CHAPTER 3

KINETICS AND METABOLISM

The application of thermodynamic principles provides information about whether a chemical reaction can proceed spontaneously (see Chapter 2) but not about the rate at which it will proceed. This is determined by kinetic factors. Hence a knowledge of kinetics is essential to an understanding of how the flux through metabolic pathways is governed. Almost all of the reactions of metabolism occur so slowly in the absence of catalysts that their rates are not measurable. Consequently, organisms possess catalysts, known as enzymes, that are capable of enormously accelerating these rates. The study of the kinetics of metabolism is, therefore, the study of enzymes. Nonetheless, it is important to emphasize, even at this early stage, that the significance of enzymes in living organisms is not just that of catalysis. The existence of enzymes not only increases the rates of metabolic processes but it enables them to be regulated. As a consequence of this regulation, individual reactions and metabolic pathways can be integrated into the over-all metabolic system which functions so effectively in the whole organism. If metabolic reactions could occur sufficiently rapidly in the absence of enzymes, regulation and integration would not be possible so that life, as we know it, could not exist. It follows that studies of the properties of enzymes will provide considerable information about metabolic processes. Indeed, knowledge of enzymology takes us much further towards an understanding of metabolism than does a knowledge of thermodynamics.

The aim of this chapter is to provide a comprehensible account of those aspects of enzymology relevant to an understanding of metabolism. First, a brief account of enzyme classification is given which is followed by an outline of the mechanisms by which catalysis occurs and a discussion of some of the properties of enzymes. Enzyme kinetics, and in particular their application to metabolic studies, are described in more detail. Finally, a brief outline is given of those aspects of enzymology which are most relevant to clinical practice.

A. THE NATURE OF ENZYMES

1. Nomenclature and classification

The early work on the elucidation of metabolic pathways resulted in the discovery of a large number of enzymes which were generally named according to their function. Many of these enzymes catalysed degradation reactions and were simply named by addition of the suffix, -ase, to the name of the substrate (for example, arginase, which hydrolyses arginine to urea and ornithine; urease, which hydrolyses urea to ammonia and carbon dioxide). This practice was extended to non-degradative enzymes by the addition of -ase to a term descriptive of the catalysed reaction to give, for example, kinases, isomerases, oxidases, and synthetases. These terms were generally prefixed by the name of the substrate or, in the case of the synthetases, the name of the product. More etymologically inventive biochemists have coined such names as cocoonase (for an enzyme secreted by some emerging silk moths to hydrolyse the silk of their cocoon) and nickase (for an enzyme catalysing the hydrolysis of a single phosphodiester bond on one strand of a double-stranded DNA molecule). The natural growth of this unsystematic terminology has resulted in single enzymes possessing more than one name and, worse still, in more than one enzyme bearing the same name. To overcome these difficulties, the International Union of Biochemistry adopted, in 1961, recommendations for a systematic nomenclature and classification of enzymes prepared by its Enzyme Commission (EC) (see Enzyme Nomenclature, 1979). For each well-characterized enzyme, the EC proposed a systematic name and a unique numerical designation. Since the systematic names were frequently cumbersome and the numbers difficult to memorize, the EC also proposed that a single recommended (trivial) name should be retained (or invented) for each enzyme. For example, the enzyme catalysing the reaction:

$$ATP + AMP \rightleftarrows 2ADP$$

bears the systematic name ATP:AMP phosphotransferase, the number EC 2.7.4.3. and the recommended name adenylate kinase. (The previously used name of myokinase, which erroneously suggested that the enzyme was restricted to muscle, has been abandoned.)

The EC system of classification and numbering is based on the division of enzymes into six major classes according to the reaction they catalyse. These classes are as follows:

1. *Oxidoreductases*, which catalyse oxidation-reduction reactions.
2. *Transferases*, which catalyse group transfer reactions.
3. *Hydrolases*, which catalyse hydrolytic reactions.
4. *Lyases*, which catalyse the non-hydrolytic removal of groups to form double bonds.
5. *Isomerases*, which catalyse isomerizations.
6. *Ligases*, which catalyse bond formation with the concomitant breakdown of nucleoside triphosphate.

Each class is subdivided. In class 1, for example, this is done according to the nature of the electron donor: 1.1 with CHOH group as donor; 1.2 with aldehyde or keto group as donor; 1.3 with CH_2-CH_2 group as donor, etc. A further subdivision is achieved by consideration of the electron acceptor: 1.1.1 with NAD^+ as acceptor; 1.1.2 with cytochrome as acceptor; 1.1.3 with O_2 as acceptor, etc. Finally, each individual enzyme is given a fourth number so that, for example, enzyme 1.1.1.1 is alcohol dehydrogenase (or, to give it its systematic name, alcohol:NAD^+ oxidoreductase) which catalyses the reaction:

$$\text{alcohol} + NAD^+ \rightleftarrows \text{aldehyde (or ketone)} + NADH + H^+$$

However, in order to define an enzyme uniquely, its source must be stated since the properties of an enzyme may differ from one species to another. For example, horse liver alcohol dehydrogenase has a molecular weight which is half that of yeast alcohol dehydrogenase.

There is one further problem in classification. A single organism and even a single cell may contain more than one enzyme catalysing the same reaction. Such enzymes are termed isoenzymes (or isozymes) and, not surprisingly, they generally exhibit considerable similarity in structure and properties. Since they catalyse the same reaction, isoenzymes must be distinguished in other ways; this is usually done by differences in electrophoretic mobility and occasionally by a difference in kinetic properties. The various isoenzymes are indicated by letters or numerals; for example, lactate dehydrogenase occurs as five isoenzymes, designated H_4, H_3M, H_2M_2, HM_3 and M_4 (see Chapter 5; Section D.4a) and hexokinase as four isoenzymes designated I, II, III, and IV. (The more recent recommendation of the Enzyme Commission is that isoenzymes should be designated by Arabic numerals (1, 2, 3, etc.) with the lowest number being given to the form migrating most rapidly towards the anode.)

2. Mechanism of enzyme action

One important and often quoted property of enzymes is their very high catalytic activity; for example, hexokinase can accelerate the rate of phosphorylation of glucose by a factor of about 10^{10}. It is now accepted that this remarkable property is explicable in physicochemical terms. Despite the complexity of enzymes, considerable progress has been made in elucidating the mechanisms by which they increase the rate of chemical reaction and it is the purpose of this section to indicate the chemical principles underlying these mechanisms. This involves describing what is meant by the terms, transition state, enzyme-substrate complex, and the active site of an enzyme.

(a) The transition state

In a chemical reaction, one stable arrangement of atoms (i.e. the substrate) is changed to another (i.e. the product). As this change proceeds, the atoms will pass through unstable arrangements, the least stable of which is termed the transition state (or activated complex), which can be thought of as the 'half-way house' between the substrates and the products. For example, in the alkaline hydrolysis of methyl iodide, the transition state is a carbanion in which five atoms, including the oxygen of the hydroxyl group and the iodine, are linked to the central carbon atom—a highly unstable arrangement:

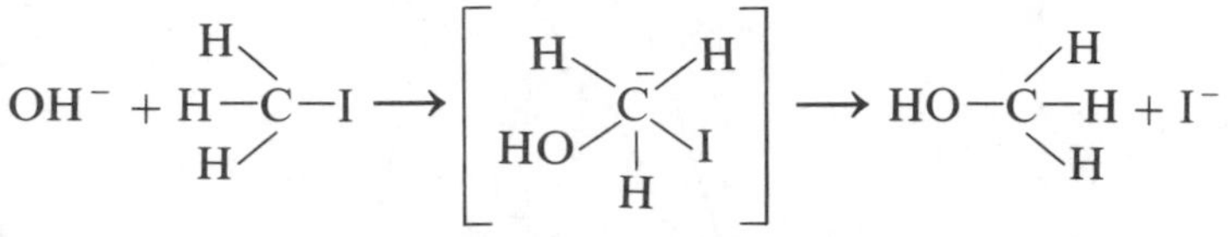

$$OH^- + H-\underset{H}{\overset{H}{C}}-I \longrightarrow \left[\underset{H}{\overset{H}{HO-\overset{\cdots}{\underset{|}{C}}}} \cdots I \right] \longrightarrow HO-\underset{H}{\overset{H}{C}}-H + I^-$$

The importance of the transition state is that the rate of the over-all reaction is limited by the number of molecules in this state. If the formation of the transition state $AB^{\ddagger}$ is represented by the equation:

$$A + B \rightleftarrows AB^{\ddagger}$$

then the equation describing the relationship between the rate of the over-all reaction (V) and the concentration of the transition state ($[AB^{\ddagger}]$) is:

$$V = \frac{RT}{Nh} [AB^{\ddagger}]$$

(R is the gas constant, T is the absolute temperature, N is Avogadro's number, and h is Planck's constant.) Anything that increases the concentration of $AB^{\ddagger}$ will therefore speed up the reaction. One of the factors that determine $[AB^{\ddagger}]$ is the equilibrium constant for the reaction forming it, designated $K^{\ddagger}_{eq}$. This is related to the standard free energy change for the formation of the transition state ($\Delta G^{0\ddagger}$) by the equation:

$$\Delta G^{0\ddagger} = -RT\ln K^{\ddagger}_{eq}$$

which is similar to that encountered previously in Chapter 2; Section A.2.c. Furthermore, the free energy change for the reaction forming the transition state will depend upon the concentrations of the substrate and products of this reaction, that is:

$$\Delta G^{\ddagger} = \Delta G^{0\ddagger} + RT\ln \frac{[AB^{\ddagger}]}{[A][B]}$$

$\Delta G^{\ddagger}$ is known as the activation energy for the reaction. This is relevant to catalysis since the effect of any catalyst (including an enzyme) is to lower the activation energy for a given reaction. In other words, catalysts lower the energy barrier for the reaction and this is represented graphically in Figure 3.1. In chemical terms this is achieved by making possible a novel reaction mechanism for the formation of a transition state which has a lower value of $\Delta G^{0\ddagger}$ than the non-catalysed reaction mechanism.

It should be stressed that the mechanism of most reactions is more complex than the profile in Figure 3.1 indicates. Most enzymic and non-enzymic reactions involve the formation of at least one intermediate which has a structure that is

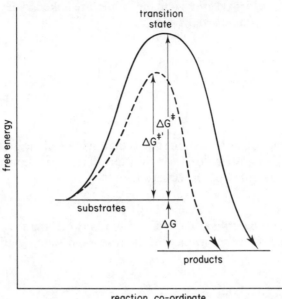

Figure 3.1 Simple reaction profile for catalysed and uncatalysed reactions. Solid line indicates the uncatalysed reaction; broken line indicates the catalysed reaction. ΔG is the free energy change for the over-all reaction. $\Delta G^{\ddagger}$ is the free energy change of activation for the uncatalysed reaction, i.e. the activation energy. $\Delta G^{\ddagger\prime}$ is the free energy change of activation for the catalysed reaction

more stable than that of the transition state, but less stable than those of substrate and product (Figure 3.2). The elucidation of the catalytic mechanisms employed by enzymes is an area of intense research activity and many have now been described (see Fersht, 1977). Such investigations involve the identification of the intermediates, mostly by indirect means since they are rarely stable enough to permit isolation. Apart from providing further information about the mechanism of the enzymic reaction, a knowledge of the structure of these intermediates also provides evidence of the chemical nature of the transition state. This is important since, with a knowledge of the chemistry of the transition state, it should be possible to synthesize compounds that so closely resemble this state that they will bind very tightly to the enzyme. Such compounds are called transition state analogues and because they are *not* true transition states, but are nevertheless bound tightly to the enzyme, they are usually very potent inhibitors of enzyme activity. Such analogues have now been synthesized for several enzymes (see Lienhard, 1973, and Figure 3.3) and may prove to be of considerable practical importance, since they should provide the basis for the preparation of very specific inhibitors of individual metabolic reactions. This field is of particular interest to the pharmaceutical industry which is striving to develop drugs that will inhibit only one reaction in a single metabolic pathway.

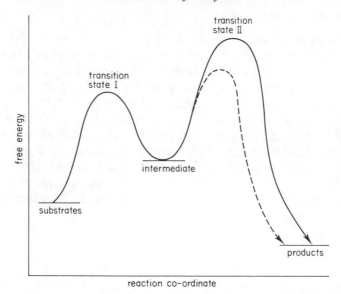

Figure 3.2 Reaction profile showing formation of intermediate. The broken line indicates the course followed by a catalysed reaction

adenylate kinase

Proposed transition state:

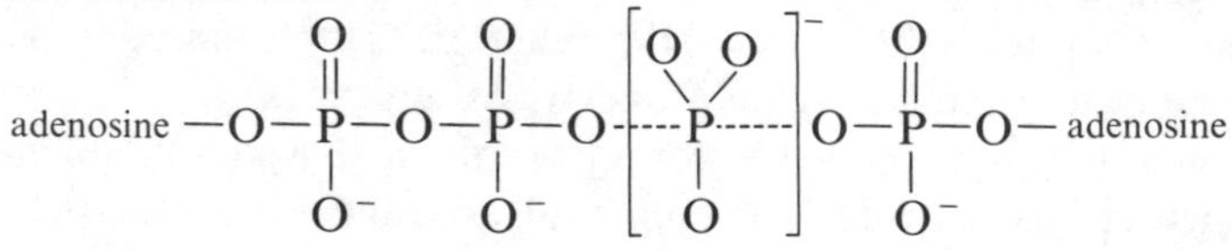

$$\text{adenosine} -O-\overset{\overset{\displaystyle O}{\|}}{\underset{\underset{\displaystyle O^-}{|}}{P}}-O-\overset{\overset{\displaystyle O}{\|}}{\underset{\underset{\displaystyle O^-}{|}}{P}}-O \cdots \left[\overset{\overset{\displaystyle O\ \ O}{\diagdown\diagup}}{\underset{\underset{\displaystyle O}{|}}{P}}\right]^- \cdots O-\overset{\overset{\displaystyle O}{\|}}{\underset{\underset{\displaystyle O^-}{|}}{P}}-O- \text{adenosine}$$

Inhibiting analogue:

$$\text{adenosine} -O-\overset{\overset{\displaystyle O}{\|}}{\underset{\underset{\displaystyle O^-}{|}}{P}}-O-\overset{\overset{\displaystyle O}{\|}}{\underset{\underset{\displaystyle O^-}{|}}{P}}-O-\overset{\overset{\displaystyle O}{\|}}{\underset{\underset{\displaystyle O^-}{|}}{P}}-O-\overset{\overset{\displaystyle O}{\|}}{\underset{\underset{\displaystyle O^-}{|}}{P}}-O-\overset{\overset{\displaystyle O}{\|}}{\underset{\underset{\displaystyle O^-}{|}}{P}}-O- \text{adenosine}$$

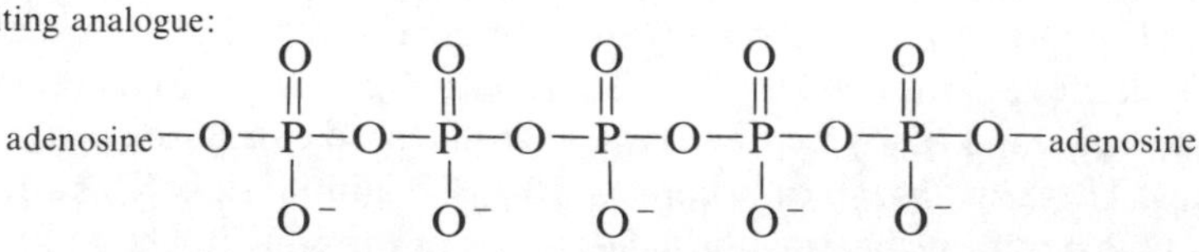

P^1,P^5-di(adenosine-5'-)pentaphosphate

triosephosphate isomerase

Proposed transition state:

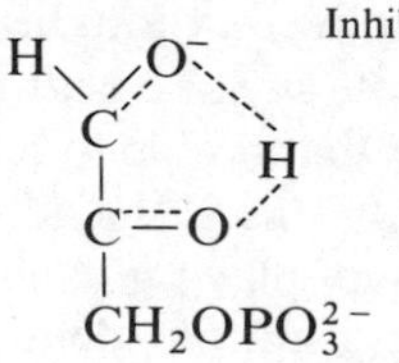

Inhibiting analogue:

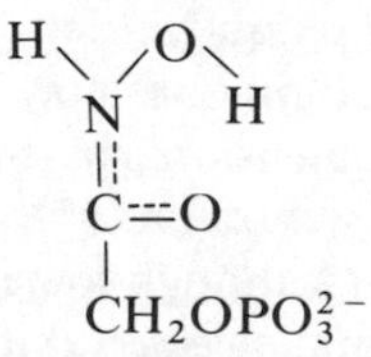

hydroxamate of phosphoglycolic acid

Figure 3.3 Examples of transition state complexes and inhibiting analogues

(b) *The enzyme-substrate complex*

An enzyme must interact with its substrates to form an enzyme-substrate complex in order to modify the mechanism of the reaction and thus achieve catalysis. That is, at least one intermediate must involve the enzyme molecule. The interaction between enzyme and substrates usually involves non-covalent bonds although in some cases covalent bonds are formed, but they must be sufficiently weak to permit the eventual dissociation of the enzyme-product complex. For a few enzymes, the enzyme-substrate complexes are sufficiently stable to be isolated as such, but in most cases the elucidation of their structure depends on spectroscopic methods. Recently, X-ray crystallography has been used to establish more precisely the three-dimensional structure of the enzyme-substrate complex. Such information enables the structure of possible transition states to be proposed with greater certainty.

(c) *The active site of an enzyme*

Enzymes composed of a single polypeptide chain usually combine with only one molecule of each substrate at a time and, indeed, this may be true for some enzymes that contain more than one polypeptide chain. The fact that substrate molecules are generally small whereas the enzyme molecule is large suggests that only a small part of the enzyme is involved in the interaction. Such observations lead to the concept of the active site, that is, a limited region on the enzyme that binds the substrates and catalyses the reaction. This concept is of particular importance in understanding the properties of enzymes.

In order to fully appreciate the concept of the active site, some knowledge of enzyme structure is essential. However, a full account of the structure of enzymes would require a whole text to itself so that only the barest outline can be given here. (For further information on the structure of enzymes, the reader is referrred to Ferdinand, 1976; Fersht, 1977.) Enzymes are proteins and therefore consist of amino acids linked by peptide links into linear (i.e. unbranched) polypeptide chains. These chains typically contain 100–300 amino acids. This structure gives a regular backbone bearing the side-groups of the constituent amino acids. The 20 amino acids commonly found in proteins possess great diversity of side-group structure as indicated in Figure 3.4. Interactions between these groups cause the polypeptide chain to fold up and adopt a precise three-dimensional structure that is unique to that enzyme. This folding serves to bring into proximity amino acid groups that may have been remote in the linear sequence to produce a three-dimensional assemblage of atoms that is able to bind the substrates, catalyse the reaction and release the products. This is the active site.

Although some amino acid groups play a greater part in binding than catalysis (and *vice versa*) it is not always possible to identify which side-groups do what and, indeed, some may play a part in all three processes indicated above. The number of amino acids in the active site is obviously not fixed and will vary

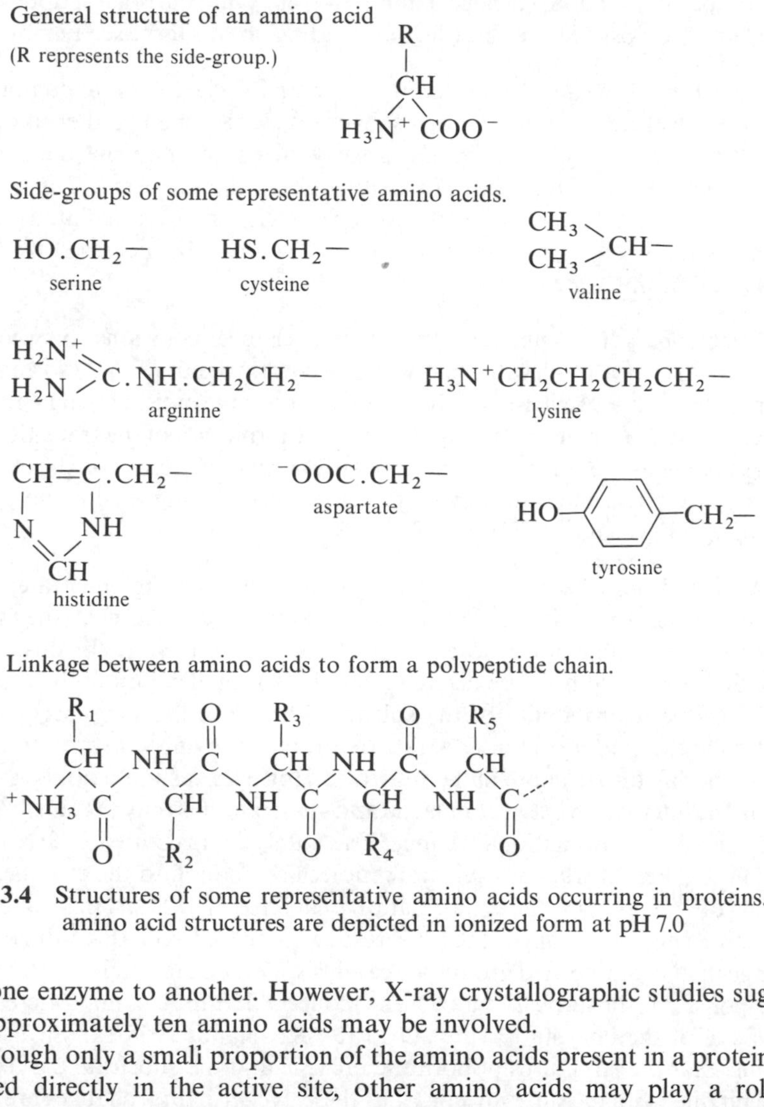

General structure of an amino acid

(R represents the side-group.)

$$\underset{H_3N^+ \quad COO^-}{\overset{\displaystyle R}{\overset{\displaystyle |}{\overset{\displaystyle CH}{\diagup \diagdown}}}}$$

Side-groups of some representative amino acids.

$HO.CH_2-$ $HS.CH_2-$ $\underset{CH_3}{\overset{CH_3}{\diagdown}}\!\!\diagup CH-$

  serine   cysteine   valine

$\underset{H_2N}{\overset{H_2N^+}{\diagdown}}\!\!\diagup C.NH.CH_2CH_2-$ $H_3N^+CH_2CH_2CH_2CH_2-$

      arginine   lysine

$$\underset{\underset{\displaystyle CH}{\underset{\diagup}{N \quad NH}}}{\overset{\displaystyle CH=C.CH_2-}{| \quad |}}$$
     $^-OOC.CH_2-$ aspartate $HO-\!\!\langle\bigcirc\rangle\!\!-CH_2-$

   histidine   tyrosine

Linkage between amino acids to form a polypeptide chain.

Figure 3.4 Structures of some representative amino acids occurring in proteins. The amino acid structures are depicted in ionized form at pH 7.0

from one enzyme to another. However, X-ray crystallographic studies suggest that approximately ten amino acids may be involved.

Although only a small proportion of the amino acids present in a protein are involved directly in the active site, other amino acids may play a role in maintenance of the three-dimensional structure of the enzyme, in attaching the enzyme molecule to intracellular structures (e.g. membranes) or in binding molecules that regulate the activity of the enzyme.

(d) *Rate enhancement*

At least four mechanisms can contribute to rate enhancement caused by enzymes (i.e. catalysis) but their relative importance is not always clear and may vary from enzyme to enzyme. These mechanisms are proximity, orientation, strain and the effect of specific functional groups in the active site. Only the briefest discussion of these factors is possible here. For further reading on this topic see Fersht (1977).

(i) *Proximity*   The effects of proximity are, perhaps, the easiest to comprehend. In a two-substrate reaction, the two substrates must come together in order to react. The chance of this occurring is greatly increased in an enzyme-catalyzed reaction since the enzyme binds both substrates at the active site. This, in effect, increases the concentrations of the substrates. It can be calculated that the concentration of the substrate at the active site could be increased by 100- to 10 000-fold in this way.

(ii) *Orientation*   It is unlikely that many chance collisions between two molecules in free solution will occur with the correct orientation for formation of the transition state. However, on the enzyme surface the molecules may be bound in such a way that their orientation favours the formation of the transition state. This is known as the orientation effect or orbital steering since it is the electronic orbitals of the two molecules that are being correctly aligned for the reaction (Storm and Koshland, 1972).

(iii) *Strain*   It has been proposed that the binding of the substrates to the enzyme causes strain in the substrates so that the structure of the latter is modified to resemble more closely that of the transition state. In this case the value of $\Delta G^{0\ddagger}$ would be more negative, not due to stabilization of the transition state but due to instability of the substrate. Some of the free energy change involved in the binding of the substrate to form the enzyme-substrate complex is used to modify the structure of the substrate. However, it should not be assumed that, in the interaction between enzyme and substrate, it is only the substrate that can undergo a conformational change. The 'catalytic' structure of the active site may not be present when no substrate molecule is bound to the enzyme. Upon binding of the substrate, the conformation of the enzyme, at least in the vicinity of the active site, can change and this results in the correct three-dimensional arrangement of amino acid groups for catalysis. This change in the conformation in response to substrate is known as 'induced fit' since it may increase the specificity of the enzyme for the substrate (Koshland, 1958). It is, of course, possible that the structures of both the enzyme and the substrate are modified upon binding with the substrate and that these two structural changes are linked, so that the conformational change in the enzyme molecule causes strain in the substrate molecule.

(iv) *Functional groups*   Catalysis by protons or hydroxyl ions is common in organic chemistry. This is known as *specific* acid or *specific* base catalysis, and it is proportional to the concentration of these ions. However, except for a few situations (e.g. in the stomach) the concentrations of these ions are very low in living organisms. Nonetheless, general acid or general base catalysis, which can be achieved via proton donors and proton acceptors respectively, is important in many enzyme-catalysed reactions. Examples of amino acid groups that function as proton donors are the $-NH_3^+$ group in lysine and the $=NH^+-$ group in histidine, while proton-acceptor properties are shown by COO^- in aspartate and glutamate and by $=N-$ in histidine (Figure 3.4). The individual effects of these groups is markedly increased if they can act in a concerted manner, so that a proton in one part of the substrate is removed by the enzyme whereas in another part of the molecule a proton is donated by the enzyme. This is sometimes known as a 'push-pull' mechanism and it illustrates the importance of the precise positioning of catalytic groups in the active site of the enzyme. One of the most important amino acids in general acid-base catalysis is histidine, since the pK of the ionizable group ($=NH^+-$) is about 6.0 which enables it to act as both an acceptor and donor of protons at intracellular pH (about 7.0). For this reason histidine is present in the active site of many different enzymes.

3. Cofactors and prosthetic groups

Despite the diverse chemical nature of side groups of amino acids, at the active site of enzymes compounds other than amino acids may be 'co-opted' to provide additional reactive groups. These compounds play a role in the catalytic mechanism and hence they remain unchanged at the end of the catalytic process. Some compounds may be tightly bound to the enzyme in which case they are known as prosthetic groups. Examples of prosthetic groups include haem in the cytochromes and flavin adenine dinucleotide in some dehydrogenase enzymes (Chapter 4; Section C.1). Other compounds are bound less tightly and these are best described as cofactors. Metal ions provide a good example of cofactors; many enzymes require such ions to provide strong electrophilic centres in the active site. Since it depends upon the strength of binding to the enzyme, the distinction between prosthetic group and cofactor is not a very precise one and it is of little theoretical importance.[1] In contrast, it is important to be aware of the difference between cofactors and coenzymes; uncritical use of these terms has led to confusion. Coenzymes (such as NAD^+, coenzyme A, and ATP) do not function as part of the enzyme but fulfil the role of substrate (and product) since they are chemically modified at the end of a catalytic process. Although there is no fundamental distinction between a coenzyme and any other substrate, the term coenzyme is used to describe substrates which are readily regenerated by other linked reactions in metabolism (see Chapter 2; Section A.2.d). In this way, coenzymes function as group-transferring agents between metabolic pathways. The best example of such group transference is that of 'energy-rich phosphate' in the ATP/ADP coupled system that links energy-producing with energy-utilizing processes (see Figure 2.1). The term co-substrate (Chapter 2; Section C.2.d) has the same meaning as coenzyme used in this (correct) sense but has been introduced to provide a complementary term to 'pathway-substrate' and to avoid the implication that coenzymes have anything to do with the mechanism of the enzyme itself.

B. PROPERTIES OF ENZYMES

The most important property of an enzyme is its catalytic activity. As a consequence of the protein nature of enzymes, this catalytic activity is influenced by a large number of factors. These include the nature of the substrate, temperature, pH, conditions that alter protein structure and the presence of other molecules. These are discussed qualitatively in this section. A more quantitative discussion is reserved until Section C for the very important effects of the concentrations of substrates and inhibitors.

1. Specificity

Formation of the enzyme-substrate complex can occur only if the substrate possesses groups which are in the correct three-dimensional arrangement to interact with the binding groups of the active site. A 'lock and key' analogy has been widely used to explain specificity but this is inadequate because the formation of the enzyme-substrate complex involves more than a steric complementarity between enzyme and substrate. It involves chemical interaction between enzyme and substrate that have no counterpart in the mechanical analogy. Moreover, binding of the substrate can modify the structure of the enzyme within the vicinity of the active site so that additional groups interact with the substrate at the active site and this 'induced fit' further increases the specificity.

Varying degrees of specificity can be distinguished in enzymes, ranging from absolute specificity, when only a single type of substrate will react (for example, urease will hydrolyse only urea), to group specificity when catalysis occurs with a family of related compounds (for example, hexokinase will phosphorylate a number of hexoses). In addition to this chemical specificity, the vast majority of enzymes exhibit stereospecificity; thus hexokinase phosphorylates D-glucose but not L-glucose (see Appendix 4.1 for an account of the terminology of stereoisomerism). Such distinction between enantiomers may seem remarkable, but it must be remembered that the three-dimensional structures of L- and D-glucose (which differ in configuration at all optically active centres) are much less similar than are the structures of, for example, D-glucose and D-galactose (which differ in configuration at only one carbon atom). Stereospecificity of enzymes is anticipated from an appreciation of the three-dimensional nature of an active site and the likelihood that an enzyme interacts with its substrate at more than two points.

2. Denaturation

A number of factors destroy enzyme activity as a consequence of their ability to denature proteins, that is, to produce an extensive disturbance of the three-dimensional structure of the enzyme. Denaturing agents include heat, extremes of pH, organic solvents, detergents and high concentrations of urea. The effects of heat and pH are considered in more detail in subsequent sections. All of these agents are able to disrupt the non-covalent forces between amino acid residues in the enzyme that determine its over-all three-dimensional structure and hence the precise geometry of the groups in the active site. The forces that are affected by denaturing agents are primarily hydrophobic interactions[2] and hydrogen bonds.

Organic solvents (for example, ethanol and propanone) and detergents disrupt enzyme structure because their presence around the enzyme molecule makes it less favourable for non-polar side chains to remain in the centre of the molecule. The non-polar side chains move towards the surface of the enzyme since they can now interact with solvent or detergent molecules. This causes a marked change in the three-dimensional structure and sufficiently disturbs the geometrical arrangement of the amino acid groups in the active site so that catalytic activity is lost.

Strong solutions of urea, $CO(NH_2)_2$, or guanidine, $NH{=}C(NH_2)_2$, denature enzymes primarily because these substances are particularly good at forming hydrogen bonds. Hydrogen bonds between side chains and within the backbone of the protein are important in maintaining the native conformation of enzymes. At high concentrations (e.g. 8.0 M), urea or guanidine molecules displace some of these intramolecular hydrogen bonds and form intermolecular hydrogen bonds with the enzyme. If a sufficient number of errant hydrogen bonds are formed, denaturation results.

The extent to which denaturation can be reversed is variable but, in many cases, the native protein structure is restored only with difficulty. Success is more likely if the denaturing agent is removed very slowly. However, in many cases, the molecules of the denatured enzyme interact strongly with each other so causing the protein to precipitate. Reversal is then particularly difficult or impossible. It is probable that enzymes are denatured at a fixed rate *in vivo* and that it is an important physiological process. It is suggested that before proteins are broken down to their constituent amino acids within the cell (for each protein is being continuously synthesized and broken down) they must first suffer denaturation and, indeed, this process may control the rate of degradation of the protein (see Chapter 10; Section B.3).

It is important to realize that denaturation can occur under conditions much milder than indicated above, and for some enzymes, rates of denaturation are high even at room temperature. Since measurements of enzyme activities are used routinely in hospitals as a diagnostic aid, knowledge of the factors affecting the process of denaturation is of some importance to all personnel involved in the measurement of enzyme activity. This includes the physician who takes the sample of blood (or tissue) and the technician who performs the assay of the enzyme activity. A low activity of an enzyme could be real or due to denaturation of the enzyme in the sample after collection.

3. Hydrogen ion concentration

The hydrogen ion concentration can have a marked effect on enzyme activity because many of the amino acids in the enzyme bear ionizable groups. Changes in pH will modify the degree of ionization of some or all of these groups and this will affect the ionization of the enzyme molecule as a whole. This can modify enzyme activity in at least three ways. First, extremes of pH (typically below 4.0 or above 10.0) will change the degree of ionization of many groups in the enzyme so that the three-dimensional structure may be disturbed. This could be sufficiently large to cause denaturation and irreversible loss of enzyme activity.[3] (Even the fall in pH which can occur in samples of body fluid or tissue after damage or death can cause denaturation of enzymes.) Secondly, smaller changes of pH (within the range of approximately pH = 4 to 10) will affect a smaller number of ionizable groups but if these happen to be present in the active site they will cause a marked change in enzyme activity. This effect, however, is completely reversible. Thirdly, changes in pH may also cause changes in the ionization of the substrate and this could modify the binding of the substrate to the enzyme. For many enzymes, a plot of activity against pH gives a bell-shaped curve with a well-defined pH optimum (see Figure 3.5) but for some enzymes the shape of this curve is very different (if, for example, several ionizable groups are involved in enzyme-substrate interaction). (The points of inflection of the pH/activity curve should correspond to the pK_a values of the groups responsible for the pH dependence. This information has been used to identify amino acids in the active site.)

Some enzymes are clearly adapted to function at the particular pH of their normal environment. For example, pepsin has a pH optimum of 2.0 which corresponds approximately to the pH of the stomach contents. However, for a number of intracellular enzymes, their pH optima are far removed from the intracellular pH (for example, glycerol kinase has a pH optimum of 9.8 and phosphoglyceromutase has a pH optimum of 5.9). This may be due to the enzymes having different properties *in vitro* to those *in vivo*, where interaction with cellular components may modify pH dependence.

4. Temperature

A rise in temperature will increase the rate of enzyme-catalysed reactions, as it will for all chemical reactions. This is due to an increase in the number of

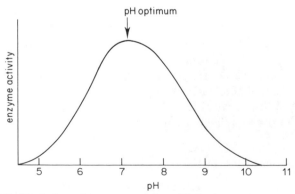

Figure 3.5   Variation of enzyme activity with hydrogen ion concentration

molecules possessing sufficient kinetic energy to overcome the energy barrier for the reaction (Section A.2.a). However, enzymes are denatured by heat so that the activity will tend to decrease as the temperature rises. The consequence of these opposing effects of an increase in temperature is that graphs of enzyme activity against temperature exhibit a maximum (Figure 3.6). However, since thermal denaturation is a time-dependent process, the shape of the graph and the position of this 'optimal temperature' will depend on the length of time the enzyme has remained at that temperature. It is expected that all enzymes will be at least moderately stable *in vivo* at temperatures normally experienced by the organism. However, extraction of the enzyme from its *in vivo* environment may increase its susceptibility to thermal denaturation. Most enzymes are very rapidly denatured at 100 °C and show a decreased activity after quite short exposures to temperature above 60 °C. (The exceptional ability of a few enzymes of higher animals to withstand boiling, for example, adenylate kinase and ribonuclease, greatly facilitates their purification since most other enzymes can be destroyed by a preliminary heat treatment.) Exceptional heat resistance is found in the enzymes of bacteria capable of living in hot springs at a temperature of 60–80 °C. More surprisingly perhaps, a small number of enzymes can be denatured *in vitro* by a fall in temperature. This is explained by postulating that the free energy of

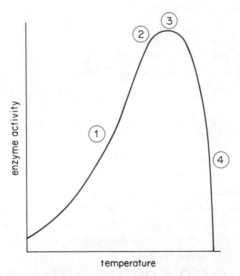

temperature

Figure 3.6 The effect of temperature on enzyme activity. In this hypothetical example, enzyme activity was measured while the enzyme was held at the given temperature for a constant short period.

1. The effect of temperature on reaction rate predominates.
2. Denaturation reduces the effect of temperature rise on enzyme activity.
3. The so-called optimum temperature. This will be displaced to the left as the period for which the enzyme is held at the given temperature is increased.
4. Denaturation predominates

formation of the conformation-determining bonds contains a large contribution from the entropic term, which is temperature dependent.

The relationship between the rate of a chemical reaction and temperature is described by the Arrhenius equation:

$$k = Ae^{-E/RT}$$

where k is the rate constant for the reaction (see Section C.2.a), T is the absolute temperature, R is the gas constant, A is a constant, and E is a second constant known as the activation energy of the reaction. The Arrhenius equation explains the remarkable sensitivity of the rate of a reaction to a change in temperature, since the relationship between k and T is exponential; a small increase in temperature causes a large increase in k and hence in the reaction velocity. This explains the importance of a homoiothermic animal being able to maintain its body temperature constant at about $310\,°K$ ($37\,°C$). If the temperature of the animal was to fall with the ambient temperature to, say, $273\,°K$ ($0\,°C$) this would change the rate of metabolism approximately 16-fold despite the closeness of the temperatures on the absolute scale.

For practical purposes a more convenient expression of the relationship between the rate of a reaction and temperature is the Q_{10} value. This is the ratio of the velocity of a reaction at temperature $T + 10\,°C$ to its velocity at $T°C$. The Q_{10} values for enzyme-catalysed reactions usually fall within the narrow range of 1.5–2.5 while those for non-enzymic reactions are usually in the range 2.0–4.0. It is sometimes important to know the Q_{10} value for a given enzyme-catalysed reaction if the activity is measured at a temperature other than the normal physiological temperature.

5. Mechanisms of inhibition

Any substance able to reduce the activity of an enzyme is called an inhibitor. It is convenient to discuss the mechanisms of enzyme inhibition in this section although a detailed discussion of the kinetics of inhibition is deferred to Section C.3. The inhibition of enzyme activity that results from the general irreversible modification of three-dimensional structure has been dealt with under the heading of denaturation; more subtle changes in structure that lead to loss of enzyme activity are discussed below.

Inhibition is likely to occur if the active site binds molecules other than the substrate (or cofactor). If such inhibitors can be readily displaced by substrate molecules, this results in competitive inhibition (Section C.4.a). Structural analogues of the substrate frequently inhibit in this way. If, however, an inhibitor binds very strongly at the active site it may not be displaced by the substrate. Inhibition of this kind can involve the formation of a covalent bond between the inhibitor and a specific group in the active site which is virtually irreversible[4] and the compounds are known as 'irreversible inhibitors'. For example, diisopropylfluorophosphate (DFP)* inhibits enzymes that possess a serine residue at the active site (Figure 3.7). A number of proteolytic enzymes involved in digestion (e.g. trypsin, chymotrypsin—see Chapter 10; Section B.2.a) contain such serine residues and are consequently inhibited by DFP. Another enzyme inhibited by organophosphorus compounds of this type is acetylcholinesterase. This enzyme, which also contains serine in the active site, is found at synapses and neuromuscular junctions where it destroys acetylcholine after it has performed its neurotransmitter role (see Chapter 21; Section B.1.b). The inhibition of this enzyme prevents the breakdown of acetylcholine so that, for example, innervation of motor nerves produces a continuous stimulation of the muscle resulting in tetanic contraction. This inhibition can have some macabre implications. DFP is a volatile liquid so that one of the first effects of the vapour is on the respiratory tract. It finds its way into the smooth muscle of the bronchioles which contract strongly and remain contracted so that death by asphyxiation occurs rapidly. Similar compounds to DFP have found uses as nerve gases in warfare (with a lethal dose an order of magnitude less than for hydrogen cyanide) as insecticides, as agents for the treatment of myasthenia gravis[5] and a restricted

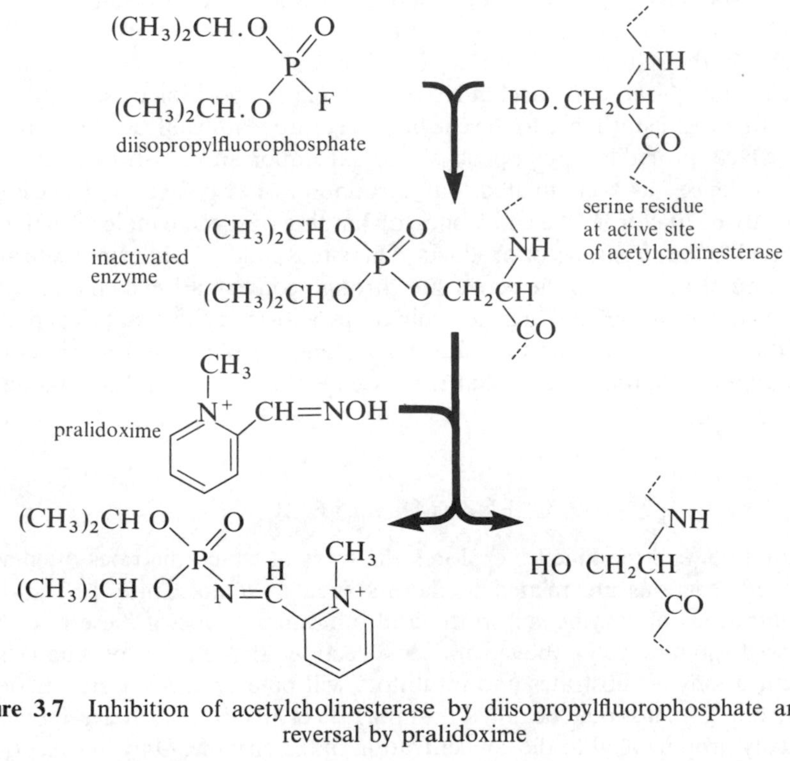

Figure 3.7 Inhibition of acetylcholinesterase by diisopropylfluorophosphate and its reversal by pralidoxime

* Also known as diisopropylphosphofluoridate.

use as long-acting myotic agents (to constrict pupils) in ophthalmology. Indeed one of the first effects of the vapour of organophosphorus compounds is on sight as graphically described by B. C. Saunders (1957).

'We entered the testing chamber, in which DFP was sprayed so as to give a concentration of 1 part in 1,000,000 ... and remained in the chamber for five minutes. No effects were detected while we were in the chamber nor until some five minutes afterwards. Intense myosis (pupil constriction, etc.) then set in and often persisted for as long as seven days, and there was usually little relaxation of symptoms until after 72 hr. This myotic or eye effect may be summarized as follows:

(a) pupil constriction, often down to pin-point size. The amount of light entering the eye was greatly reduced. Incapacitation was naturally greater in a poor light;
(b) powers of accommodation were reduced;
(c) photophobia and headaches, and pain experienced when changing from a bright to a dull light ...

At higher concentrations the toxicity was such as to cause a quick 'knock-out' action ... The symptoms were muscular weakness, gasping and finally cessation of respiration.'

Other inhibitors can bind reversibly to enzymes at a site distinct from the substrate-binding site and are not therefore displaced by the substrate. Such inhibitors do not cause competitive inhibition, that is, their effect is not reversed by raising the substrate concentration but only by reducing the inhibitor concentration. Some may well bind to the enzyme at a position close to the active site so that the disturbance to the enzyme structure readily affects the active site. Such effects probably have no physiological importance. However, in recent years evidence has accumulated that certain enzymes possess a site, which is physically distinct from the active site, for binding of specific molecules that are unrelated to the substrates or products. This site is called the allosteric site and is in 'communication' with the active site through conformational changes of the protein so that the binding of a molecule to the allosteric site results in a reduced catalytic activity. Such inhibitions are considered to be physiologically important in regulation of enzyme activity *in vivo* (see Section C.5 for full discussion).

C. ENZYME KINETICS

The study of enzyme kinetics explores the ways in which the rates of enzyme-catalysed reactions are related to factors such as temperature, pH, and the concentrations of enzyme, substrate, and inhibitors. Some of these have been discussed qualitatively above; in this section the effects of changes in concentrations of substrates and inhibitors will be examined in greater detail. Before doing so, it should be noted that the rate of an enzyme-catalysed reaction is usually proportional to the concentration of the enzyme. Only in exceptional circumstances (for example, when the concentration of enzyme is comparable to that of the substrate) does this simple relationship not hold.

1. The measurement of enzyme activity

In order to make precise statements about enzyme kinetics, it is necessary to have a satisfactory means of measuring the rate of the enzyme-catalysed reaction, that is, of assaying enzyme activity. Enzyme activity is conventionally reported in units of μmoles substrate converted to product per minute. The activity of pure enzymes is sometimes given as μmol.min^{-1}.mg^{-1} enzyme, when it is known as the specific activity. The activity of an enzyme also provides a measure of the amount of enzyme in a biological system (body fluid or tissue extract, for example). More direct measurements of enzyme concentration (in contrast to its activity) are difficult, because any one enzyme is present at very low concentrations in biological systems together with many other enzymes.[6] Variations in enzyme activity from one tissue to another are usually taken to represent differences in concentration of enzyme.

Enzyme activity is measured by determining the rate of disappearance of a substrate or the rate of appearance of a product. In general, it is more accurate to measure the latter.[7] Experience shows that the rate of an enzyme-catalysed reaction usually decreases as the reaction progresses. This may be due to such factors as substrate depletion, increase in product concentration (which increases the rate of the reverse reaction), inhibition by products, pH changes or enzyme denaturation. For these reasons, enzyme activities should always be measured as initial rates. In principle, the initial rate is measured by plotting a reaction-progress curve of product appearance against time and drawing a tangent to this curve at the origin. In practice, most enzyme rates are found to remain constant for a small percentage conversion of substrate to product and for short periods of time (several minutes). Therefore, the activity is usually measured from the concentration of product that has accumulated at the end of such a short period.

Although there is virtually no limit to the variety of analytical methods that can be used to measure rates of enzyme-catalysed reactions, they can be divided into two classes: sampling methods, in which the product (or substrate) concentration is measured after the reaction has been stopped by inactivation of the enzyme, and continuous methods, in which the change in product (or substrate) concentration is measured whilst the reaction proceeds. Whatever the method used, it must be specific and sufficiently sensitive. For information on the techniques of enzyme activity determinations see Bergmeyer (1978) and Crabtree *et al.* (1979).

(a) *Sampling methods*

Of the two classes, sampling methods are the more versatile since any technique can be used to measure product concentration once the reaction has been stopped.[8] If the product possesses distinctive physical properties which are proportional to its concentration (e.g. absorption of light) the change in this parameter may be measured. More frequently, the product does not possess a suitable property and must be quantitatively converted (by chemical reaction) into one that does.

An important class of sampling assays is the radiochemical (or radiometric) assay in which the enzyme is incubated with radioactively-labelled substrate. This leads to the formation of radioactively-labelled product. After the reaction has been stopped, the product is separated from the substrate and the amount of radioactivity now present in the product provides a measure of the activity of the enzyme. Radiochemical methods are especially useful when no suitable chemical methods of product estimation are available. They may be more sensitive and more specific than other techniques and ingenious methods for the rapid separation of substrate and product have been devised (Oldham, 1968; Crabtree *et al.* 1979). Major disadvantages of sampling methods, and of radiochemical methods in particular, are that they are tedious and that information is not obtained until after completion of all the procedures involved in the assay.

(b) *Continuous methods*

Most continuous methods of enzyme assay rely on differences in light absorption between substrate and product. The reaction is carried out in a small transparent cell in a spectrophotometer so that changes in light absorption at a specific wavelength can be followed continuously. In metabolic and clinical biochemistry, enormous use is made of the large difference in absorption between the oxidized (NAD^+ and $NADP^+$) and reduced pyridine nucleotides (NADH and NADPH) at 340 nm (Figure 3.8). Any of the many enzymes catalysing reactions involving these nucleotides can be readily assayed in this way. For example, the activity of

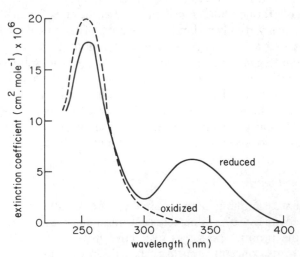

Figure 3.8   Spectra of reduced and oxidized pyridine nucleotides. The spectra of NAD^+ and $NADP^+$ and of NADH and NADPH show no significant differences over this range of wavelengths

lactate dehydrogenase can be measured by the rate of decrease in absorption at 340 nm when NADH is incubated with pyruvate[9] according to the reaction:

$$NADH + H^+ + CH_3COCOO^- \longrightarrow NAD^+ + CH_3CH(OH)COO^-$$

However, the usefulness of this method is not restricted to the assay of dehydrogenase reactions. It can be extended to reactions in which no spectral change occurs, provided that the reaction to be measured can be linked (coupled) to a dehydrogenase reaction. For example, pyruvate kinase precedes lactate dehydrogenase in the metabolic pathway of glycolysis (Chapter 5; Section B) and this can be used to assay the activity of pyruvate kinase. These reactions are coupled together as follows:

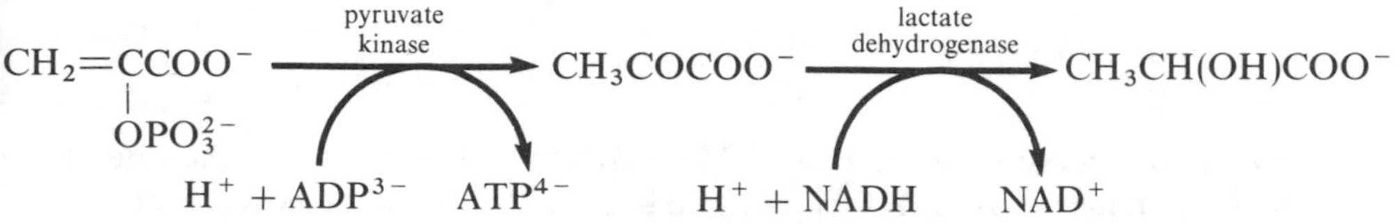

Sufficient ADP, PEP, NADH and lactate dehydrogenase are included in the assay so that the rate of NADH oxidation is limited by the activity of pyruvate kinase.

Reactions which involve the production or uptake of a gas (or which can be coupled to such reactions) can be followed in a manometer. In the Warburg manometer, which has in the past been used extensively for enzyme assays, the volume changes occurring in the gas phase during a reaction are converted to pressure changes which are readily measured (see Umbreit *et al.*, 1964).

Specific electrodes can also be used to follow concentration changes continuously and thus measure enzyme activity. The measurement of changes in hydrogen ion concentration (using a glass electrode) can be used to assay reactions in which substrate and product differ markedly in degree of ionization. If a suitable pH is chosen, the production (or consumption) of protons can be made equal to the number of substrate molecules reacting. For example, the creatine kinase reaction can be followed at pH 8.0 by the production of protons according to the equation:

$$creatine + ATP \longrightarrow phosphocreatine + ADP + H^+$$

Since enzyme activities change markedly with pH, the glass electrode is normally employed together with an autotitrator and pH-stat which continually adds alkali (or acid) to the reaction in order to maintain constant pH. The rate of H^+ production is then equal to the rate of alkali addition. Another specific electrode, the oxygen electrode, has largely replaced the manometer as a means of following enzyme reactions in which oxygen is produced or consumed (see Schuler, 1978).

2. Dependence of catalytic activity on substrate concentration

(a) *Preliminary principles in chemical kinetics*

In order to follow the derivation of the equations described below, it is necessary to consider a few relevant principles of chemical kinetics. For further information, a textbook of physical chemistry should be consulted.

(i) *Rate expression and rate constant*   A rate expression relates the rate of a reaction to the concentration of its reactants. For the following reaction

$$A + B \rightarrow AB$$

the rate expression is

$$\frac{-d[A]}{dt} = k[A][B]$$

This is an expression of the Law of Mass Action which states that the rate of chemical change varies directly with the active concentrations of the reactants. Of course, this law applies to enzyme-catalysed as well as non-catalysed reactions.

The proportionality constant, k, in the rate expression is known as the rate constant for that reaction. When the reactants (i.e. A and B) are at a concentration of unity (usually one molar), the rate of reaction equals the rate constant. The value of the rate constant for an enzyme-catalysed rection will be very much larger than that for the same reaction which is not enzyme-catalysed. However, because the mechanism of the enzyme-catalysed reaction is much more complex, measurement of the rate constants for the individual processes that constitute an enzyme-catalysed reaction is usually difficult.

(ii) *Order of reaction*   The order of a reaction is the sum of the powers to which the reactant concentrations are raised in the rate expression. In the above equation the reaction is second order since it depends on the concentration of A and B each raised to the power of one.

It is possible to have a non-integral order of reaction and indeed this is the case for enzyme-catalysed reactions at most concentrations of substrate (see below). Although the order of reaction may appear to be a physicochemical parameter of little relevance to physiology, this is not the case. Knowledge of the order of reaction is important in understanding some aspects of metabolic control (see Chapter 7; Section A.1).

(b) *The hyperbolic response of enzyme activity to substrate concentration*

The rate of an enzyme-catalysed reaction varies with its substrate concentration. In many cases, this relationship is found experimentally to be hyperbolic (Figure 3.9). This means that, at very low substrate concentrations, the reaction is

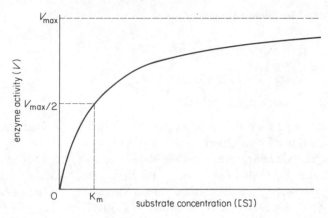

Figure 3.9 The hyperbolic response of enzyme activity to substrate concentration

approximately first order, at high substrate concentrations the reaction approaches zero order and at intermediate substrate concentrations the order is intermediate between zero and unity. For simplicity, enzymes with single substrates are considered at this stage. The same principles apply to reactions with more than one substrate (see Section C.2.g). All rate measurements refer to initial velocities (so that there is no significant change in substrate concentration during the reaction).

A hyperbolic curve can be described mathematically by an equation of the form:

$$x = \frac{C_1 y}{C_2 + y}$$

Two questions can now be posed. First, how does the mechanism of enzyme catalysis account for this hyperbolic relationship? Secondly, what is the metabolic significance of the constants C_1 and C_2 in the above equation?

(c) *The Michaelis–Menten equation*

The answers to the two questions posed above are credited to L. Michaelis and M. L. Menten. By assuming a simple model for enzyme action they were able to derive, in 1913, a hyperbolic rate equation. In 1925, G. E. Briggs and J. B. S. Haldane derived a similar equation using a more general model of enzyme catalysis. The derivation of Briggs and Haldane is given below but the names of Michaelis and Menten have always been associated with the general form of the equation.

The model of enzyme action, upon which the derivation is based, is that the enzyme (E) binds substrate (S) to form a complex (ES) which breaks down to give product (P) and to release the enzyme:

$$E + S \underset{k_{-1}}{\overset{k_1}{\rightleftharpoons}} ES \overset{k_2}{\longrightarrow} E + P$$

where k_1, k_{-1}, and k_2 are rate constants.

The assumptions made are as follows:

1. The over-all reaction is in a steady state so that the rate of formation of ES is equal to its rate of breakdown.
2. The substrate concentration, S, is considerably greater than the total enzyme concentration, E_t, so that formation of the ES complex does not significantly alter [S]. However, ES formation does reduce the concentration of free enzyme. At steady state, this concentration is given by ($[E_t] - [ES]$).
3. The rate of reverse reaction is negligible (that is, only initial rates are considered).

Assumption 1 states that the rate of change of concentration of ES $\left(\text{i.e. } \dfrac{d[ES]}{dt}\right)$ is zero. This means that the algebraic sum of the rates of the two reactions removing ES and the one reaction forming it are zero. Since the rate of a chemical reaction is given by the product of a rate constant and a reactant concentration this leads to:

$$\frac{d[ES]}{dt} = k_1([E_t] - [ES])[S] - k_{-1}[ES] - k_2[ES] = 0 \qquad (1)$$

Equation (1) can be rearranged in a number of steps. The term with brackets is first multiplied out:

$$k_1[E_t][S] - k_1[ES][S] - k_{-1}[ES] - k_2[ES] = 0 \qquad (2)$$

Rearranging (2) by taking all but the first term to the other side:

$$k_1[E_t][S] = k_1[ES][S] + k_{-1}[ES] + k_2[ES]$$
$$= [ES](k_1[S] + k_{-1} + k_2) \qquad (3)$$

Equation (3) is rearranged:

$$[ES] = \frac{k_1[E_t][S]}{k_1[S] + k_{-1} + k_2} \qquad (4)$$

Division of both numerator and denominator by k_1 gives:

$$[ES] = \frac{[E_t][S]}{[S] + \dfrac{k_{-1} + k_2}{k_1}} \qquad (5)$$

Since the initial rate of the over-all reaction (v) is given by k_2[ES] equation (5) gives:

$$v = k_2 [\text{ES}] = \frac{k_2 [\text{E}_t] [\text{S}]}{[\text{S}] + \dfrac{k_{-1} + k_2}{k_1}} \tag{6}$$

The maximum catalytic activity of the enzyme is reached when all the enzyme is involved in the ES complex (that is, ES = E_t) so that $V_{max} = k_2 [\text{E}_t]$. Substituting in equation (6) gives:

$$v = \frac{V_{max} [\text{S}]}{[\text{S}] + \dfrac{k_{-1} + k_2}{k_1}} \tag{7}$$

If the term $\dfrac{k_{-1} + k_2}{k_1}$ is replaced by K_m (the Michaelis constant) equation (7) becomes:

$$v = \frac{V_{max} [\text{S}]}{[\text{S}] + K_m} \tag{8}$$

Equation (8)—the Michaelis–Menten equation—describes a hyperbolic relationship between initial rate (i.e. catalytic activity) and substrate concentration which, as indicated above, is found experimentally (Figure 3.9). Thus the simple model of enzyme mechanism described above predicts a hyperbolic relationship between enzyme activity and substrate concentration. It should, however, be emphasised that this agreement between theory and experiment supports, but does not prove, the model. Other models also predict a hyperbolic relationship between v and [S]; for example, the substrate could be absorbed onto a large, but finite, number of independent sites on the enzyme with catalysis occurring at each site.

The two constants of the model, V_{max} and K_m, can be measured relatively easily for any enzyme (see Section C.2.f). Their metabolic significance is discussed below.

(d) *Significance of* K_m

The Michaelis constant is defined as the substrate concentration at which enzyme velocity is half-maximal. This follows from the Michaelis–Menten equation:

$$v = \frac{V_{max} [\text{S}]}{[\text{S}] + K_m} \tag{8}$$

but if

$$v = \frac{V_{max}}{2}$$

then

$$\frac{V_{max}}{2} = \frac{V_{max}[S]}{[S] + K_m} \tag{9}$$

Therefore, dividing each side by the V_{max}

$$\frac{[S]}{[S] + K_m} = \frac{1}{2} \tag{10}$$

Rearrangement leads to

$$[S] + K_m = 2[S] \tag{11}$$

so that $K_m = [S]$.

However, K_m can also be interpreted in terms of rate constants for the component reactions in the enzyme mechanism model. In the original model of Michaelis and Menten (1913) it was assumed that $K_m = k_{-1}/k_1$; in other words, K_m was the dissociation constant of the ES complex. For this to be the case, the magnitude of k_2 must be so small that it does not disturb the equilibrium between E, S, and the ES complex (i.e. $k_2 \ll k_{-1}$). In the Briggs–Haldane treatment (above) no such restriction is made and it can be shown that $K_m = \dfrac{k_{-1} + k_2}{k_1}$. Only direct measurement of rate constants k_1, k_{-1}, and k_2 can indicate whether the equilibrium (Michaelis–Menten) or steady state (Briggs–Haldane) assumptions are correct. In the case of the relatively few enzymes for which rate constant determinations have been made, the dissociation constant for the ES complex has not been found equal to the K_m^{10}. The equilibrium assumption of Michaelis–Menten is therefore not valid in these cases. One consequence of this is that an inhibitor causing a change in the value of K_m may not necessarily be interfering with the binding of substrates to the enzyme, but may be affecting the value of k_2.

Although there may be problems concerning the physical interpretations of K_m, it must be emphasized that this is of little physiological importance. The operational definition of K_m as the substrate concentration of half-maximal velocity is quite sufficient for most metabolic purposes.

There are at least two important uses to which the value of the K_m can be put.

1. In the elucidation of a metabolic pathway it is frequently necessary to establish whether a particular enzyme plays a part. The activity of this enzyme is usually measured at saturating substrate concentrations to give V_{max}. If the enzyme is to play a role in the postulated pathway it must be demonstrated that its catalytic activity is adequate to account for the maximum flux through the pathway in the tissue *in vivo*. Although the maximum activity of the enzyme can be measured (*in vitro*, of course) the substrate concentration *in vivo* may be much lower than that required to give the maximum activity. However, if the concentration of substrate *in vivo* is known, together with the value of V_{max} and

K_m, the 'physiological' activity can be calculated from the Michaelis–Menten equation. For example, hexokinase from brain catalyses the phosphorylation of both glucose and fructose and the question arises whether both substrates are physiologically important; that is, whether hexokinase plays a physiologically significant role in the phosphorylation of fructose in the brain in comparison to the phosphorylation of glucose. The V_{max} for fructose phosphorylation is in fact greater than that for glucose phosphorylation (Table 3.1). However, comparison of the K_m values for the two substrates with the physiological concentration of these substrates in brain (Table 3.1) shows that glucose will be phosphorylated at approximately 50% of the maximal activity whereas fructose will be phosphorylated at less than 1% of this activity (the concentration of fructose in the brain is very low). Consequently, hexokinase is not involved in fructose phosphorylation in the brain to any significant extent.

Table 3.1. Estimated rates* of phosphorylation of glucose and fructose in brain

Sugar	Properties of brain hexokinase V_{max} (μmol.min^{-1}.g^{-1})	K_m (M)	Sugar concentration in brain cell (M)	Calculated rate of phosphorylation *in vivo* (μmol.min^{-1}.g^{-1})
Glucose	17	10^{-5}	10^{-5}	8.5
Fructose	25	10^{-3}	10^{-6}	2.5×10^{-2}

*The rate of phosphorylation is calculated from the values given in the table using the Michaelis–Menten equation.

2. Another important use of the K_m is to indicate whether an enzyme-catalysed reaction *in vivo* is zero, first, or intermediate order, a matter of some consequence in the identification of the flux-generating reaction (see Chapter 2; Section C.2.d). If the substrate concentration is considerably lower than the K_m, the reaction will approximate to first order so that a change in substrate concentration will cause a proportional change in catalytic activity. If the *in vivo* substrate concentration is greater than approximately ten times the K_m, then the reaction will approach zero order so that large changes in substrate concentration will have little effect on the rate of catalysis (i.e. a flux-generating step).

(e) *Significance of V_{max}*

When the substrate concentration is considerably greater than K_m (so that $K_m + [S] \simeq [S]$), the Michaelis–Menten equation simplifies to:

$$v = \frac{V_{max}[S]}{[S]} = V_{max} \qquad (12)$$

V_{max} is therefore the rate of the reaction at very high concentrations of substrate. Such concentrations are said to be saturating.

It should be clear that the value of V_{max} will depend upon the concentration of the enzyme. For purified enzymes, values of V_{max} are usually presented on the basis of milligrams of protein (this is known as the specific activity). In crude extracts of tissues, values of V_{max} are usually referred to grams of the tissue. Indeed, if the specific activity of the purified enzyme is known, the concentration of the enzyme can be estimated from the activity in a crude extract.

There are a number of physiological uses to which values of the V_{max} can be put.

1. Values of V_{max} can be used to compare the activity of one enzyme with that of another or to compare the rate of catalysis of an enzyme with one substrate to that with another substrate (see Table 3.1).
2. Comparison of maximum activities of enzymes in a metabolic pathway can be used as one means of indicating whether a reaction is near- or non-equilibrium (Chapter 2; Section C.2.a). A high value of the V_{max}, in comparison with other enzymes, indicates that the reaction is near-equilibrium whereas a low V_{max} indicates a non-equilibrium reaction (see Newsholme and Start, 1973).
3. The V_{max} of an enzyme that catalyses a non-equilibrium reaction in a pathway may provide an indication of the maximum flux through that pathway (see Appendix 5.2 and Newsholme *et al.*, 1978, 1980). For example, the V_{max} of glycogen phosphorylase or of 6-phosphofructokinase gives an indication of the maximal capacity of anaerobic glycolysis from glycogen in the intact tissue. Similarly, the V_{max} of oxoglutarate dehydrogenase can provide a quantitative indication of the aerobic capacity of an intact muscle (see Tables 4.6 and 5.4).

(f) *Determination of kinetic constants*

Plots of initial velocity against substrate concentration (Figure 3.9) do not provide accurate means of determining the constants K_m and V_{max}. This is because the V_{max} is approached asymptotically and, furthermore, it is difficult to fit experimental points to a hyperbola. By algebraic manipulation of the Michaelis–Menten equation (8) a number of linear relationships can be obtained from which the values of the constants can be determined more accurately. One of the more widely used is attributed to H. Lineweaver and D. Burk. The basic Michaelis–Menten equation is inverted to give:

$$\frac{1}{v} = \frac{K_m + [S]}{V_{max}[S]} = \frac{K_m}{V_{max}[S]} + \frac{[S]}{V_{max}[S]} \tag{13}$$

Separating the terms, this becomes:

$$\frac{1}{v} = \frac{K_m}{V_{max}} \cdot \frac{1}{[S]} + \frac{1}{V_{max}} \tag{14}$$

Therefore, if $\dfrac{1}{v}$ is plotted against $\dfrac{1}{[S]}$, a straight line is obtained with a slope of $\dfrac{K_m}{V_{max}}$ and an intercept on the ordinate of $\dfrac{1}{V_{max}}$ (Figure 3.10(a)). This plot has been criticized, however, because it gives greater weight to the points obtained at very low substrate concentrations which, because of the low enzyme activity, are least accurate. An alternative treatment due to G. S. Eadie and G. H. J. Hofstee, overcomes this problem. The Michaelis–Menten equation (8) is divided by v to give:

$$\frac{v}{v} = \frac{V_{max}\,[S]}{v(K_m + [S])} = 1 \tag{15}$$

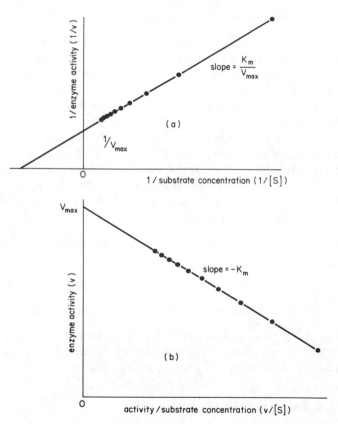

Figure 3.10 Linear plots for the determination of K_m and V_{max}. The same data have been used for both plots.

(a) Lineweaver–Burk plot.
(b) Eadie–Hofstee plot

Therefore

$$V_{max}[S] = vK_m + v[S] \tag{16}$$

or

$$v[S] = V_{max}[S] - vK_m \tag{17}$$

Dividing both sides of (17) by $[S]$:

$$v = V_{max} - K_m \frac{v}{[S]} \tag{18}$$

Thus a plot of v against $\dfrac{v}{[S]}$ gives a straight line with a slope of $-K_m$ and an intercept on the ordinate of V_{max} (Figure 3.10(b)).

(g) Kinetic constants in two-substrate reactions

In the discussion so far, only reactions involving single substrates have been considered, although most enzyme-catalysed reactions involve two substrates. The corresponding equations for two substrate reactions can be very much more complex (Engel, 1977) but V_{max} and K_m remain metabolically useful quantities. The V_{max} in a two-substrate reaction is defined as the activity of the enzyme when both substrate concentrations are saturating. Similarly, the K_m for a given substrate is defined as the concentration of that substrate when the activity is half V_{max} in the presence of a saturating concentration of the other substrate. Thus the K_m can be measured as for a single substrate reaction provided that the concentration of the second substrate is maintained at a saturating concentration. Alternatively, and in preference, the initial activities can be measured with varying amounts of one substrate (S_1) at a series of fixed concentrations of the second substrate (S_2). Plots of $\dfrac{1}{v}$ against $\dfrac{1}{[S_1]}$ at the various concentrations of S_2 provide a series of slopes and intercepts. These are known as primary or double reciprocal plots. The slopes (or intercepts on the ordinate) of the primary plots are then plotted against $\dfrac{1}{[S_2]}$ to give a secondary plot from which values of K_m for S_2 and V_{max} can be readily obtained (Figure 3.11(a) and (b)). In a similar way, the concentration of S_2 can be varied at fixed concentrations of S_1, so that a K_m for S_1 is obtained.

3. Deviations from hyperbolic kinetics

Although many enzyme-catalysed reactions do exhibit hyperbolic kinetics, a significant number do not. These deviations were first appreciated in the mid-1950s and since that time many enzymes have been re-examined and shown to possess sigmoid kinetics, that is, a plot of activity against substrate concentration

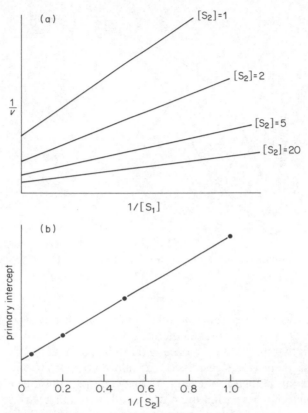

Figure 3.11 Determination of kinetic constants for two-substrate reactions. The intercepts on the ordinate of the double reciprocal plot (a) are given by $\dfrac{1}{V_{max}} + \dfrac{K_m}{V_{max}[S_2]}$ (for S_2).

When the intercept is plotted against $\dfrac{1}{[S_2]}$, secondary plot (b), gives a straight line with an intercept of $\dfrac{1}{V_{max}}$ and a slope of $\dfrac{K_m \text{ (for } S_2)}{V_{max}}$

is S-shaped or sigmoid (Figure 3.12). In this section, the causes of sigmoid behaviour are briefly described, leaving the full consideration of its significance in the regulation of metabolism to Chapter 7.

It has already been shown that acceptance of the Michaelis–Menten or Briggs–Haldane model for enzyme action leads to a prediction of hyperbolic kinetics. In both of these models one substrate molecule binds with one enzyme molecule. However, some enzyme molecules bind more than one molecule of the same substrate at a time and this can increase the complexity of the kinetic response. Nonetheless, even in this case, provided that each binding site on the

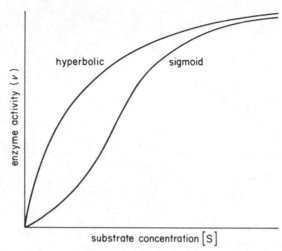

Figure 3.12   Comparison of sigmoid and hyperbolic responses to changes in the concentration of substrate

protein is identical, the response should be hyperbolic. If, however, the binding of one molecule of substrate causes an increase in affinity of the other binding sites it can be shown that a sigmoid response will result. This is called the co-operative effect because there is a transmission of information from one active site to another. Such an influence of events at one site upon another is mediated by conformational changes within the protein molecule, and a variety of molecular models have been proposed to explain the mechanism. Their relative merit is of little metabolic significance and the interested reader is recommended to consult Chapter 2 of Newsholme and Start (1973).

The co-operative effect between binding sites is not restricted to enzymes and indeed its physiological significance is most clearly seen in the case of haemoglobin. The sigmoid nature of the oxyhaemoglobin dissociation curve was first reported by Bohr *et al.* (1904). Since that time not only has its physiological significance become evident (see Chapter 5; Section F.3.c) but the conformational changes involved have been determined by X-ray crystallography (Perutz, 1970). How far the knowledge of the mechanism of co-operativity in haemoglobin will be of value in understanding the mechanism of co-operativity in enzymes remains to be seen.

The major physiological importance of co-operative effects in enzymes is to increase the relative change in enzyme activity for a given relative change in substrate or regulator concentration (that is, the sensitivity is improved). In an allosteric enzyme, the activity is affected by the binding of a regulator molecule to a site spatially distinct from the active site (Section C.5). In many allosteric enzymes, co-operative effects occur not only between the binding sites for the substrate molecules but also for the regulator molecules so that the response of enzyme activity to either substrate concentration or regulator concentration is sigmoid. Enzymes possessing such properties play a central role in the regulation of metabolism as discussed in Chapter 7; Section A.1.

4. Kinetics of reversible inhibition

Anyone who has carried out practical work in the field of enzymology will be aware of the ease with which enzymes can be inhibited. In addition to the agents causing non-specific denaturation considered in Section B, a large number of compounds exist which reversibly inhibit specific enzymes. Since an increasing number of drugs are inhibitors of specific enzymes there is considerable clinical interest in this field. These drugs are administered to patients in order to decrease the flux through a particular metabolic pathway. While it is not necessary to have a full understanding of the mechanism of inhibition to appreciate the metabolic importance of such effects of drugs, it is important to have some knowledge of the way in which changes in inhibitor concentration affect activity of enzymes. This section describes the major types of inhibitor kinetics, the parameters which describe them (inhibitor constants) and the way in which these parameters are determined.

From a study of inhibitor kinetics (that is, how the concentration of inhibitor affects the K_m and V_{max} of the enzyme under study) simple models of the mechanisms of action of inhibitors are developed. This approach is used here to describe and distinguish three different classes of inhibition. It should be noted from the outset, however, that although a kinetic analysis of inhibition can suggest a mechanism by which an inhibitor is acting, it cannot *prove* the mechanism, since other mechanisms may give rise to identical kinetics.

The simplest general model of inhibition allows the inhibitor, I, to react reversibly with either E or ES to produce complexes EI or EIS which lack catalytic activity:

$$
\begin{array}{ccc}
& K'_M & \\
E + S & \rightleftharpoons ES & \rightarrow E + \text{products} \\
+ & + & \\
I & I & \\
\Big\updownarrow K_{EI} & \Big\updownarrow K_{EIS} & \\
EI & EIS &
\end{array}
$$

The general equation derived from this mechanism is (in reciprocal form):

$$
\frac{1}{v} = \frac{1}{V_{max}}\left(1 + \frac{[I]}{K_{EIS}}\right) + \frac{K'_m}{V_{max}}\left(1 + \frac{[I]}{K_{EI}}\right)\frac{1}{[S]} \tag{19}
$$

where $[I]$ is the inhibitor concentration, K_{EI} is the dissociation constant for complex EI, and K_{EIS} is the dissociation constant for the complex EIS. (In most cases the dissociation constant, $K'_m = \dfrac{k_{-1}}{k_1}$ will not be equal to K_m (see Section

C.2.d). An alternative symbol for this dissociation constant is K_S. (Here, K'_m has been used to emphasize the similarity in form between the Michaelis–Menten equation and the inhibitor equations.) The three main classes of inhibition are known as competitive, non-competitive, and uncompetitive. Each can be seen to arise from restrictions placed on the general equation given above.

(a) *Competitive inhibition*

For competitive inhibition, the value of K_{EIS} approaches infinity so that the inhibitor reacts only with the uncomplexed enzyme (i.e. E). Thus the general inhibitor equation (19) simplifies to:

$$\frac{1}{v} = \frac{1}{V_{max}} + \frac{K'_m}{V_{max}} \left(1 + \frac{[I]}{K_{EI}}\right) \frac{1}{[S]} \tag{20}$$

Since V_{max} depends on the concentration of ES, its value remains unaltered by the presence of a competitive inhibitor. However, to attain the maximum concentration of ES (so that $[ES] = [E_t]$) in the presence of inhibitor, a much higher concentration of substrate is required. The K_m is therefore increased. Double reciprocal plots of $\frac{1}{v}$ against $\frac{1}{[S]}$ in the presence of different concentrations of inhibitor show that the slope, but not the intercept on the ordinate, is affected by a competitive inhibitor (Figure 3.13(a)). (The ratio, slope in the presence of inhibitor/slope in its absence, equals $1 + \frac{[I]}{K_{EI}}$, allowing K_{EI} to be calculated. Both K_{EI} and K_{EIS} are sometimes known as inhibitor constants and given the symbol K_i.) However, more accurate values of K_{EI} can be obtained from secondary plots. These are obtained by plotting the slopes of double reciprocal (primary) plots against the concentration of inhibitor. This should yield a straight line with the intercept on the ordinate equal to $\frac{K'_m}{V_{max}}$ and the slope equal to $\frac{K_m}{V_{max}K_{EI}}$ (Figure 3.13(b)). Consequently the ratio, slope/intercept, for the secondary plot, equals $\frac{1}{K_{EI}}$.

(b) *Non-competitive inhibition*

In simple non-competitive inhibition, the constants K_{EI} and K_{EIS} are identical, i.e. the inhibitor reacts equally well with the free enzyme and the ES complex. Most examples of non-competitive inhibition occur in two-substrate reactions, when a product often inhibits competitively with respect to the substrate which it resembles structurally, but inhibits non-competitively with respect to the other substrate. In non-competitive inhibition, the K_m remains unaltered but V_{max} is

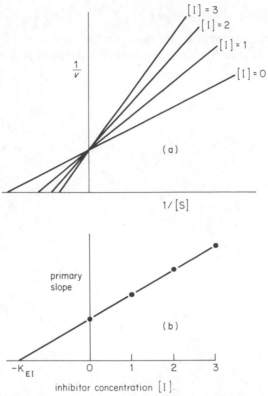

Figure 3.13 Determination of inhibitor constants for competitive inhibition. The slope of the double reciprocal plot (a) is $\dfrac{K'_m}{V_{max}}\left(1 + \dfrac{[I]}{K_{EI}}\right)$ (see text). When the slope is plotted against [I] this gives a secondary plot (b) with a slope of $\dfrac{K'_m}{V_{max}K_{EI}}$, an intercept on the ordinate of $\dfrac{K'_m}{V_{max}}$ and an intercept on the abscissa of $-K_{EI}$

decreased because, at all concentrations of inhibitor, some of the ES complex will be in the inactive EIS form. The general inhibitor equation (19) can be simplified only to the extent of noting that K_{EI} equals K_{EIS}:

$$\frac{1}{v} = \frac{1}{V_{max}}\left(1 + \frac{[I]}{K_{EIS}}\right) + \frac{K'_m}{V_{max}}\left(1 + \frac{[I]}{K_{EI}}\right)\frac{1}{[S]} \tag{21}$$

The slope and intercept (on the ordinate) of the double reciprocal plot are altered to the same extent with non-competitive inhibition (Figure 3.14(a)). The value of K_{EI} can be calculated from the ratio, slope in the presence of inhibitor/slope in the absence of inhibitor, which equals $1 + \dfrac{[I]}{K_{EI}}$. Secondary plots of slope against inhibitor concentration (or intercept against inhibitor concentration) can be used to obtain more accurate values of the inhibitor constant (Figure 3.14(b)).

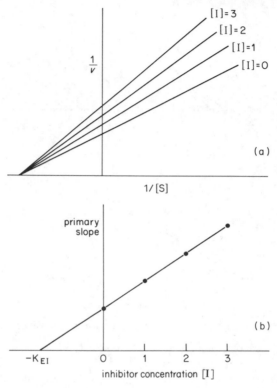

Figure 3.14 Determination of inhibitor constant for non-competitive inhibition. The slope of the double reciprocal plot (a) is $\dfrac{K'_m}{V_{max}}\left(1 + \dfrac{[I]}{K_{EI}}\right)$. When the slope is plotted against [I], this gives a secondary plot (b) with a slope of $\dfrac{K'_m}{V_{max}K_{EI}}$, an intercept of the ordinate of $\dfrac{K'_m}{V_{max}}$ and an intercept on the abscissa of $-K_{EI}$

(c) *Uncompetitive inhibition*

In the less common situation of the inhibitor reacting only with the ES complex, K_{EI} is infinity and uncompetitive inhibition is observed. In this case, the general inhibitor equation (19) simplifies to:

$$\frac{1}{v} = \frac{1}{V_{max}}\left(1 + \frac{[I]}{K_{EIS}}\right) + \frac{K}{V_{max}} \cdot \frac{1}{[S]} \tag{22}$$

A double reciprocal plot for uncompetitive inhibition is presented in Figure 3.15(a). The slope is the same in the presence or absence of the inhibitor; values of K_{EIS} may be obtained from the ratio, intercept in the presence of inhibitor/intercept in the absence of inhibitor, which equals $\left(1 + \dfrac{[I]}{K_{EIS}}\right)$. In a secondary plot of intercept against inhibitor concentration, the ratio, slope/intercept equals $\dfrac{1}{K_{EIS}}$ (Figure 3.15(b)).

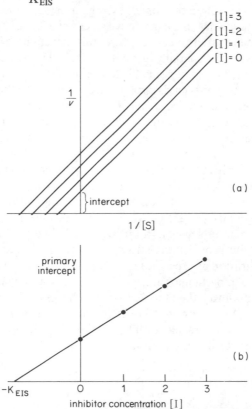

Figure 3.15 Determination of the inhibitor constant for uncompetitive inhibition. The intercept on the ordinate of the double reciprocal plot (a) is $\dfrac{1}{V_{max}}\left(1 + \dfrac{'I]}{K_{EIS}}\right)$. When the intercept is plotted against [I] this gives a secondary plot (b) with a slope of $\dfrac{1}{V_{max}K_{EIS}}$, an intercept on the ordinate of $\dfrac{1}{V_{max}}$ and an intercept on the abscissa of $-K_{EIS}$

Several factors complicate this simple classification of types of inhibition. First, in two substrate reactions it is possible for an inhibitor to act, for example, non-competitively towards one substrate and competitively towards a second substrate. Secondly, if K_{EI} is not equal to K_{EIS} and neither is infinite, mixed kinetics are obtained. Finally, it is possible that the EIS complex retains some catalytic activity. These situations give rise to more complex kinetics and those with a special interest in this subject are referred to Dixon and Webb (1979).

5. Allosteric regulation

The inhibitors described above all bind to the active site of the enzyme but allosteric inhibition is different. Some key regulatory enzymes possess a specific regulator site that is distinct from the active site but which is able to communicate with the active site in order to change the catalytic activity of the enzyme. Thus the enzyme possesses two spatially distinct binding sites on the protein molecule; the active site and the regulator site. Since the regulator site is distinct from the active site, it can bind compounds structurally unrelated to the active site so that it has been termed the allosteric site. It is proposed that when a metabolic regulator molecule binds at the allosteric site it produces a change in the three-dimensional structure of the enzyme, so that the geometrical relationship of the amino acid groups in the active site is modified. An allosteric inhibitor decreases the enzyme activity whereas an allosteric activator increases it. It is important to realise that the kinetics of such inhibition may be competitive, uncompetitive, non-competitive or mixed. The kinetics will depend upon the modifications produced in the geometry of the active site—but this is due to a conformational change in the protein molecule and not direct binding to the active site.

The allosteric site provides the structural basis for metabolic regulation. A change in concentration of a specific regulator molecule in the cell can change the catalytic activity of a particular enzyme. If this is a key enzyme determining the flux through the pathway, then the flux will be changed. For example, when it is necessary to maintain a constant concentration of a product of a metabolic pathway this is most simply achieved if the product inhibits an early reaction (e.g., the flux-generating step) in the pathway:

$$A \rightarrow B \rightarrow C \rightarrow D \rightarrow product$$

If the concentration of product rises, this inhibits its rate of formation so that the concentration will fall. Conversely, if the concentration of product falls, its rate of formation is increased. Note that this negative feedback depends on the inhibition of reaction $A \rightarrow B$ by a molecule structurally unlike A or B, that is, an allosteric inhibitor. Because of the widespread occurrence of feedback inhibition, the term allosteric regulation has become almost synonymous with inhibition but it should be noted that allosteric regulators can be either activators or inhibitors.

The biological importance of allosteric regulation lies in the fact that the structural independence of the catalytic and regulator sites has allowed each to evolve separately. Thus no restriction is placed on the regulatory mechanism by the nature of the catalytic process and selection of regulators for the metabolic pathway would depend on advantages for the metabolism of the cell as a whole rather than on the individual reaction of the pathway. This has permitted great flexibility in metabolic control since the flux through one pathway may be controlled by the concentration of an intermediate in a different pathway or indeed a compound that is produced from outside of the cell (e.g. hormone). Appreciation of these possibilities and the challenge of unravelling them has brought new life to the study of metabolic biochemistry in recent years.

Most enzymes that are regulated allosterically also exhibit sigmoid substrate kinetics (Section C.3). In addition the response of the enzyme activity to changes in regulator concentration is sigmoid rather than the usual response which is hyperbolic. The metabolic importance of this sigmoid response to the regulator concentration is the same as that for the substrate; it serves to increase the sensitivity of the response of the enzyme activity to a given change in concentration of regulator (Figure 3.16). This point is elaborated in Chapter 7; Section A.2.b.

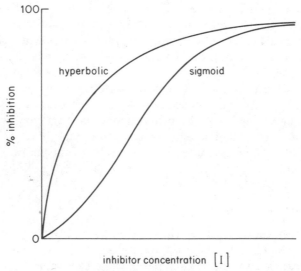

Figure 3.16 Comparison of sigmoid and hyperbolic responses to changes in the concentration of an inhibitor

D. MEDICAL ASPECTS OF ENZYMOLOGY

It has been emphasized in the Preface that one of the aims of this book is to relate the basic biochemistry, wherever possible, to the clinical field and, in particular, to clinical practice in which physicians will be involved almost every day of their working life. Although enzymology, like thermodynamics, appears to be almost totally removed from the work of the physician, this is no longer the case. The remainder of this chapter is devoted to a summary, under six headings, of the ways in which an understanding of enzymes and their properties is aiding the physician in clinical practice.

1. Enzymes as diagnostic agents

Differences in enzyme activities between normal subjects and patients are now used to aid diagnosis in a large number of conditions. Since many articles and several books (see, for example, Schmidt and Schmidt, 1974) have been written on this subject, only the general principles underlying clinical use of enzymes will be described although they will be illustrated by reference to specific conditions.

(a) *Source of the enzyme*

The source of the enzyme to be measured will depend upon the suspected condition and the normal localization of the enzyme. Body fluids such as blood (or plasma), lymph, cerebrospinal fluid, amniotic fluid, and urine are relatively easy to sample as are the red or white blood cells and faeces. However, in some situations, particularly when an inborn error of metabolism is suspected, it is necessary to measure the activity of the enzyme in a tissue (e.g. skeletal muscle, liver) so that a small sample of the tissue must be removed. This is usually done with a biopsy needle (see Appendix 5.2; Section 2) and approximately 50–100 mg of tissue can be removed without causing serious damage.

Enzymes in blood can be divided into two categories: those that are normally present and have a specific role in that body fluid (e.g. enzymes of the blood clotting mechanism; cholinesterase) and those that have no physiological function in blood but are released into the bloodstream from damaged (or diseased) tissues.

(b) *Differential diagnosis*

A problem in diagnosis can arise when the symptoms presented could result from one of several conditions. Since not all tissues contain the same complement of enzymes, those that are specific to a particular tissue can be important in aiding diagnosis. Enzyme activity measurements on plasma may be able to identify the tissue that is damaged and this could be of obvious importance in making a final diagnosis. For example, a patient may present with nausea and pain of short duration (i.e. $<24\,hr$) in the lower chest. Clinical examination and chest X-ray can rule out respiratory chest infection, gastric disease, and pancreatitis but it may not be able to distinguish between myocardial infarction, pulmonary embolism, viral hepatitis, or cholestatic jaundice (bile duct obstruction). The activities of five metabolic enzymes (aspartate aminotransferase, alanine aminotransferase, lactate dehydrogenase, 3-hydroxybutyrate dehydrogenase, and creatine kinase) when measured in samples of the patient's blood provide a

Table 3.2. Relative changes in activities of some enzymes in blood in diseased states

Enzyme	Magnitude of change in enzyme activities			
	Myocardial infarction (heart)	Pulmonary embolism (lung)	Viral hepatitis (hepatocellular tissue)	Cholestatic jaundice (bile duct obstruction)
Aspartate aminotransferase	+++++	0	+++	++
Alanine aminotransferase	0	0	+++++	++
Lactate dehydrogenase	+++	++	++	0
3-Hydroxybutyrate dehydrogenase	+++	+	0	0
Creatine kinase	++++	0	0	0

+ Indicates the relative increase in activity of the enzyme; 0 indicates no change in activity.

basis for differential diagnosis between these diseases (see Table 3.2). Even in cases where the ECG is negative, a distinctive pattern of blood enzyme activities can indicate the occurrence of a myocardial infarction.

To take a second example, the diagnosis of acute viral hepatitis can be aided by measuring activities in the blood of the enzymes, isocitrate dehydrogenase ($NADP^+$), alcohol dehydrogenase, and glycerol-3-phosphate dehydrogenase (NAD^+) which are elevated about 30-fold in this condition. (These are all enzymes that have a high activity in the cytosol of the liver, so that upon tissue damage the enzymes escape readily into the bloodstream.) Further information may be gained from a measurement of ratios of activities. The ratio, aspartate aminotransferase/alanine aminotransferase activities, is of particular importance since in serum from normal patients it is 3.0 but it decreases to 1.0 in 85 % of cases of viral hepatitis.

(c) *Monitoring therapy*

The half-life of an enzyme in the bloodstream is short, typically several hours. Since damage to a tissue results in increased membrane permeability and hence increased leakage of some enzymes, improvement in the condition of the patient would be expected to result in reduced activities in the bloodstream. Consequently, measurement of the activities of enzymes in the blood can be used to assess therapy and indeed to control the dose of therapeutic agents given to the patients. For example, the efficacy of steroid or immunosuppressive therapy for active chronic hepatitis can be followed by measuring changes in the blood levels of the aminotransferase enzymes. If the dose of the drug is not sufficient, the enzyme activities will not decrease or, alternatively, if an effective dose is withdrawn too rapidly, this will result in a rapid increase in the enzyme activities in the blood (Schmidt and Schmidt, 1974).

(d) *Diagnosis of inborn errors of metabolism*

The presence of an abnormally low activity of an enzyme in one or more tissues produces a class of diseases known as 'inborn errors of metabolism'. Since the disease is caused by a deficiency of a specific enzyme, precise diagnosis depends upon the demonstration of a low activity of the enzyme either in a body fluid or in a biopsy sample from the relevant tissue. One example is the low activity in muscle of the enzyme glycogen phosphorylase, which initiates the degradation of glycogen (see Chapter 5; Section C.2). This disease is known as McArdle's syndrome (first reported by B. McArdle in 1951) in which patients suffer pain and exhaustion during the initial stages of exercise (see Chapter 5; Section F.1.c for full discussion). The activity of glycogen phosphorylase in muscle of patients suffering from McArdle's syndrome is so low as to be non-detectable by the usual spectrophotometric methods.

(e) *Factors affecting use of enzyme activity data in diagnosis*

If a diagnosis is to depend upon the value of an enzyme activity in a given body fluid or tissue, it is essential that any change in activity is due to the diseased state and not to the preparation of the sample or assay of the enzyme activity. From the properties of enzymes discussed in the earlier part of this chapter, it should be evident that a large number of factors can modify enzyme activity so that care must be taken to avoid loss of enzyme activity during the sample collection, during preparation of the sample for enzyme assay or during the assay itself. Thus the effects of the method of sample collection, storage of the sample, temperature, pH, as well as the effect of possible activators, inhibitors, and the presence of drugs should be taken into account in assessing the importance of any difference from the normal range of enzyme activity. Furthermore, the activity of the enzyme could differ from the accepted normal range due to age, sex or race of the patient.

2. Enzymes as laboratory reagents

Many metabolic diseases are characterized by changes in the concentrations of particular metabolites in the blood or other body fluids. For example, the changes in the blood glucose concentration after ingestion of 100 g of glucose in the morning before breakfast are very different in patients suffering from diabetes mellitus compared with normal subjects (see Chapter 15; Section B.1). The measurement of the concentration of many such metabolites involves the use of enzymes (enzymatic assays). The enzymes are in fact used as the reagents in the assay and, for this purpose, many purified or partially purified preparations of enzymes are commercially available. The main advantage of an enzymatic assay is that the concentration of individual metabolites can be assayed specifically and accurately without any purification of the metabolite from the body fluid. The assay takes advantage of one biological property of enzymes—their ability to react specifically and rapidly with one compound in the presence of a large number of chemically similar compounds.

The measurement of glucose concentrations, essential in the diagnosis and monitoring of diabetes mellitus, provides an example of the importance of enzymatic assays in clinical practice. Several methods are available; one method uses the enzyme glucose oxidase (of fungal or bacterial origin) which catalyses the oxidation of glucose by oxygen:

$$\text{glucose} + O_2 \xrightarrow{\text{glucose oxidase}} \text{gluconolactone} + H_2O_2$$

In the presence of another enzyme, peroxidase, the hydrogen peroxide produced in this reaction oxidizes an organic compound, here represented by AH_2, to a second compound of different colour (A):

$$H_2O_2 + AH_2 \xrightarrow{\text{peroxidase}} 2H_2O + A$$

Conditions of the assay are arranged such that all the glucose is oxidized and all the hydrogen peroxide reacts with the organic compound, so that the change in colour, measured in the spectrophotometer, is proportional to the amount of glucose in the sample. It has even proved possible to impregnate strips with these reagents so that they may be dipped into a urine sample to give a semi-quantitative indication of glucose concentration without the requirement of laboratory facilities. A detailed description of many enzymatic assays is provided in Bergmeyer (1974).

3. Enzymes and drug sensitivity

Many drugs are metabolized by enzyme-catalysed reactions in the tissues of the body (particularly the liver). This metabolism usually reduces the effectiveness of the drug and the recommended dose takes this into account. However, if a patient has low activities of the drug-metabolizing enzymes, the drug can have a much greater effect and its effect may be maintained for a much longer period of time. This can lead to a life-threatening situation. For example, succinylcholine is used as a muscle relaxant by anaesthetists. (The surgery is made very much easier if the muscles are relaxed.) Succinylcholine is structurally similar to acetylcholine so that it binds to the same receptors in the neuromuscular junctions but fails to elicit a response. Consequently, it competes effectively with acetylcholine for these receptors (i.e. it is a competitive inhibitor) and thus causes muscle relaxation. Succinylcholine is degraded to inactive products (succinate and choline) by the cholinesterase which is present in the blood, so that the effect of a normal dose of the drug persists for only 5–10 min. However, in patients with a low activity of the cholinesterase, inactivation of the drug occurs only very slowly so that its effects persists for long periods. During such periods, prolonged apnoea can occur and death may result. An indication of sensitivity to succinylcholine can be obtained in advance by measurement of cholinesterase activity in serum obtained from the patient.

Since many drugs are metabolized by the liver, the effectiveness of the drug-metabolizing enzymes is reduced in patients with reduced liver function. In normal subjects, the effectiveness of these enzymes in the metabolism of a drug can be temporarily reduced by the ingestion of alcohol, since this is metabolized by the same enzymes. The importance of these conditions in relation to the use of the oral hypoglycaemic agent phenformin (phenethylbiguanide) is discussed in Chapter 13; Section D.3.b.

4. Enzymes and poisons

Poisons can damage health and may even result in death due to inhibition of one or more enzymes that are important in metabolism. The classic example of cyanide toxicity, due to inhibition of cytochrome c oxidase (see Chapter 4; Section E.2) is known to almost every biology student. Although the awareness of the dangers of poisoning by chemicals that are accidentally or deliberately released into the environment has increased over recent years, the use of toxic chemicals will continue to be a major industrial and even domestic hazard. Their widespread use requires clinicians in both general practice and in industry to be aware of the symptoms that can develop from exposure to such poisons since, in many cases, the victims will be unaware of the toxicity of the chemicals with which they work.

Although drugs that are prescribed by the physicians are not always considered as poisons, many drugs can cause tissue damage even when their usage is carefully controlled. The liver is particularly susceptible to damage by drugs and the physician should be aware of this when investigating the cause of liver disease.

The obvious effects of cyanide and the description of the effects of diisopropylfluorophosphate given in Section B.5, might suggest that poisons always produce pronounced and straightforward symptoms. This is not so. In some cases the symptoms are such that poisoning may not be the obvious cause. For example, delayed neuropathy can develop (Johnson, 1975) and several farm workers who have been exposed to organophosphorus insecticides can develop psychoses, possibly as a result of inhibition of acetylcholinesterase in the brain (Conyers and Goldsmith, 1971). Redhead (1968) reported the case of a farmworker who, while suffering from organophosphorus poisoning, lost control of his tractor which ran into a dyke.

5. Enzymes in chemotherapy

In some situations, it is chemotherapeutically desirable to increase the activity of an enzyme. The obvious examples are in the diseases known as inborn errors of metabolism. Although in theory it should be possible to increase the activity of an enzyme with a drug, in practice it is known that it is easier to find drugs that inhibit enzymes. Consequently, consideration has been given to the use of enzymes themselves as therapeutic agents; this has become known as enzyme-replacement therapy (Beutler, 1981). Two problems have severely limited this approach: first, enzymes do not readily cross cell membranes so that they do not enter cells; secondly, since enzymes are proteins, they usually induce an immune response. Nonetheless, work is being carried out in this field and there is cautious optimism that progress is being made in the treatment of certain lipid storage diseases, especially those in which the enzyme deficiency occurs in the liver since liver cells may be more permeable to proteins than cells of other tissues (see Brady *et al.*, 1975; Kakkar and Scully, 1978). In addition it is possible to trap an active enzyme within a small lipoprotein vesicle, known as a liposome, in which the enzyme is protected from degradation and which may be prepared in such a way as to favour transport into the affected cells (Ryman and Tyrrell, 1980).

Some success has also been achieved using enzymes that have a beneficial catalytic effect in the bloodstream or the extracellular fluid of the patient. In this case, enzymes do not have to enter the cells and they can be administered by intravenous or subcutaneous injection. For example, thrombosis can occur in veins of the legs and pelvis (particularly during immobilization due to accident, illness, or surgery when the slow circulation through the extremities favours thrombosis) and a clot can travel in the blood to the lung where it can block an arteriole or even an artery and result in pulmonary embolism. Treatment involves the use of thrombolytic (also known as fibronolytic) agents[11] to dissolve the clot (which is composed mainly of fibrin) followed by anticoagulant therapy (Fratantoni *et al.*, 1975). Thrombolytic agents can be classified into two groups according to their mechanism of action: plasminogen activators and proteolytic enzymes. Both groups are relevant to this brief discussion since they involve enzymes. Plasminogen is a protein present in the plasma, which is itself inactive but can be cleaved by partial proteolytic action to yield plasmin. The latter is a proteolytic enzyme which digests fibrin and hence it causes the dissolution of blood clots. Two enzymes, streptokinase and urokinase, have been used to activate plasminogen artificially. Streptokinase is an extracellular protein produced by the β-haemolytic streptococci and it is injected into patients suffering from a life-threatening pulmonary embolism in order to activate maximally the natural fibrinolytic system in the bloodstream. Streptokinase is the most widely used thrombolytic agent, since it is very potent, easy to prepare and hence relatively inexpensive. However, it has several disadvantages: it is antigenic, it can cause pyrogenic or toxic reactions, and it does not act preferentially towards plasminogen within the clot. An enzyme that has a similar effect to streptokinase is urokinase which has been isolated from the urine of normal human subjects. Urokinase is a β-globulin produced by the kidney. Its use in the treatment of such patients has the advantage that it is non-antigenic and it does not cause allergic reactions. Furthermore, it may have a preference for plasminogen that is present within the clot, so that there is less danger of activation of circulating plasminogen in the blood and production of a potent proteolytic enzyme. (Since many of the enzymes in the blood clotting mechanism are activated by proteolytic digestion (e.g. prothrombin, factors V and VIII), release of an active proteolytic enzyme in the bloodstream can actually induce clotting.)

It is also possible to use proteolytic enzymes directly in order to digest the clot: these include trypsin, plasmin and brinase. Although these enzymes have a direct effect on the fibrin of the clot, they may also activate the clotting mechanism as indicated above. Brinase, which is obtained from *Aspergillus oryzae*, appears to have advantages over the other two enzymes in that it is more specific for fibrin.

6. Enzymes as sites for drug action

It is one of the aims of the pharmaceutical industry to design drugs that will produce specific metabolic changes to counteract the metabolic lesions that cause, or are a consequence of, a disease. At the present time, the precise metabolic abnormalities of many, if not most, diseases are unknown so that the rational approach to drug design is limited. Nonetheless, attempts are being made, if not to treat the basic cause of the disease with such drugs, at least to alleviate the symptoms of the disease. For example, endogenous depression is considered to be due to a reduced concentration of monoamines (e.g. noradrenaline, dopamine, 5-hydroxytryptamine) in specific areas of the brain (see Chapter 21; Section E.1) although the biochemical basis for the reduced concentration of monoamines is not known. Monoamines are degraded by the enzyme amine oxidase (usually known as monoamine oxidase), which catalyses the following type of reaction:

$$RCH_2NH_2 + \tfrac{1}{2}O_2 \rightarrow RCHO + NH_3$$

A number of useful drugs that have been used in the treatment of depression (anti-depressant drugs) are inhibitors of this enzyme (the monoamine oxidase inhibitors). It is presumed that these drugs inhibit the activity of this enzyme in the neurones of the brain that are involved in the control of mood, so that the amine concentration is increased towards normal levels. The use of drugs in the field of neurological disorders has been particularly fruitful. Some of these drugs and their effects will be discussed in more detail in Chapter 21.

In the more clinical parts of this book, the use of other drugs which are, or may be, potent enzyme inhibitors will be discussed. Only when metabolic effects of both the disease and the drugs are understood can treatment be put on a rational basis.

NOTES

1. The distinction between prosthetic group and cofactor may be of considerable practical importance in the assay of enzyme activities because a cofactor, but not a prosthetic group, can be easily lost from the enzyme by dilution, during extraction or purification or removed by agents that will bind the cofactor. For these reasons an excess of cofactor is routinely added to the assay medium for the measurement of maximal enzyme activity.
2. A close association between water and hydrophobic substances is energetically unfavourable. For example, oil and water do not mix. Hydrophobic substances, in the presence of water, associate with each other thus preventing their association with water. For proteins in an aqueous environment, this means that the more non-polar amino acid groups (for example, those possessed by valine, alanine, leucine, and isoleucine) will accumulate towards the centre of the protein where few water molecules penetrate. This tendency places a significant constraint on the conformation of globular proteins. (See Kauzmann, 1959, for a detailed discussion of hydrophobic interactions in protein structure.)
3. A practical use is made of the denaturation by extremes of pH, since acidification or alkalination of a solution containing an enzyme will immediately stop catalysis. During the assay of an enzyme by a sampling technique (Section C.1.a) it is necessary to arrest the reaction after a specified time so that the amount of product that has been formed during that period can be measured. A rapid change in pH is usually very effective.

4. No reaction, strictly speaking, is irreversible, so that in this context the term means that the reaction is very difficult to reverse in practice. It is sometimes possible to reverse the effects of such inhibitors by the action of 'reactivators' which contain even more highly reactive groups than the enzymes themselves. For example, many oximes can remove diisopropylfluorophosphoryl groups from enzymes and one, pralidoxime, (Figure 3.7) is used as an antidote to poisoning by organophosphorus compounds of the DFP type.

5. In myasthenia gravis the postsynaptic receptors for acetylcholine are deficient. DFP is used to raise the concentration of acetylcholine in the synaptic cleft and so partly overcomes the problem.

6. In a few cases, enzyme concentration has been measured by immunological techniques in which the enzyme in a crude extract of the tissue is specifically separated from the large number of other proteins by interaction with the antibody to the purified protein (van Weeman and Schuurs, 1978). This method is similar in principle to the immunological assay of hormones (see Chapter 22; Section E.3).

7. The necessity of measuring initial, or near to initial rates, means that only small changes in substrate or product concentration can be permitted during an enzyme assay. It is less accurate to measure small changes in the amount of a given substance in the presence of large quantities of that substance than it is to measure small amounts of the substance. Since the initial concentration of the products is zero, while that of the substrate is large compared with the changes to be measured, it is clearly better to measure the former.

8. The reaction can be stopped by the use of heat, acids, alkalis, or more specific inhibitors of the enzyme activity. Perchloric acid and trichloracetic acid are widely used for this purpose. They not only stop the reaction, but precipitate proteins (facilitating their removal by centrifugation and thereby preventing their possible interference with subsequent chemical estimations). In addition, perchlorate can be readily removed by precipitation as the potassium salt and trichloracetate by extraction with ether (Hess and Brand, 1974).

9. It is conventional in biochemistry to refer to acids that are ionized at normal physiological hydrogen ion concentrations as their anions.

10. Peroxidase is one enzyme for which data are available The catalysed reaction can be represented:

$$\text{peroxidase} + \text{H}_2\text{O}_2 \underset{k_{-1}}{\overset{k_1}{\rightleftarrows}} [\text{peroxidase}-\text{H}_2\text{O}_2] \underset{\text{AH}_2 \quad \text{A}}{\overset{k_2}{\longrightarrow}} \text{peroxidase} + 2\text{H}_2\text{O}$$

where AH_2 represents a reduced dye (colourless) and A is the oxidized dye (coloured). The K_m is approximately 4×10^{-7} M. The three rate constants were determined by Chance (1943) as follows, $k_1 = 10^7 \text{l.mole}^{-1}.\text{sec}^{-1}$, $k_{-1} = 0.2 \text{sec}^{-1}$ and $k_2 = 4.2 \text{sec}^{-1}$ so that the dissociation constant for the ES complex is 2×10^{-8} M and the ratio, $\dfrac{k_{-1} + k_2}{k_1}$ is 4.4×10^{-7} M.

11. There is some evidence that acutely occluded coronary arteries can be re-canalized by the intracoronary injection of streptokinase and there appears to be some benefit in the treatment of acute myocardial infarction with this enzyme (see Muller *et al.*, 1981).

CHAPTER 4

THE OXIDATION OF ACETYL-COENZYME A

A. INTRODUCTION: GENERAL ASPECTS OF CATABOLISM

The enzyme-catalysed chemical reactions constituting metabolism include some in which large molecules are broken into smaller ones and others in which small molecules are used for the synthesis of larger and more complex molecules. The former can be grouped under the heading of catabolism and the latter under the heading of anabolism. Although the distinction is not a rigorous one, it is maintained here and consideration of anabolic processes is deferred until later.

Catabolism involves two major phases. First there is digestion in the alimentary canal which serves to break the large organic molecules present in food into smaller molecules which can be absorbed, distributed throughout the organism, and enter cells in the various tissues. The digestion of carbohydrate, lipid, and protein and the absorption and utilization of the products are described in Chapters 5, 6, and 10, respectively. The absorbed molecules are themselves further degraded in the individual cells of the body. Some of the chemical energy present in these compounds is transferred to other molecules, notably ATP, which can be used directly to provide energy for processes that require it. In addition, the same products of digestion are used to synthesize new molecules in anabolic processes. These are required for a variety of purposes: to achieve growth and repair of the organism, to replace molecules of the organism which have broken down and to replenish the stores of fuel which can subsequently undergo catabolism (for, although energy expenditure is continuous, eating is intermittent).

These metabolic generalizations cannot be elaborated without considering the chemical nature of the molecules involved. It is a useful simplification to consider food as a mixture of carbohydrates, lipids and proteins. Although other substances are vital constituents of the diet, their *quantitative* role in metabolism is less significant. Traditionally, carbohydrates and lipids have been considered as the fuels destined for energy supply through catabolism and, in the form of glycogen and triacylglycerol, they form the major stores of fuel held in the body. When required, these storage compounds are mobilized to form the true fuels of respiration, glucose, and long-chain fatty acids respectively. Until recently, the role of amino acids (the products of protein digestion) as respiratory fuels has been neglected. It had been considered that they served only to generate glucose and ketone bodies when consumed in excess or when derived from endogenous protein during starvation. Appreciation that oxidation of amino acids occurs in

93

the small intestine, in some muscles and in other tissues under normal situations is changing our understanding of their role in metabolism. It remains convenient, however, to delay consideration of this role until Chapters 10 and 11 when their deamination and functions as respiratory fuels and gluconeogenic precursors can be considered together.

The important energy-yielding (i.e. ATP producing) processes in the degradation of carbohydrates and lipids are oxidative reactions with molecular oxygen acting as the final oxidizing agent. The oxidative processes leading most directly to ATP synthesis, namely the tricarboxylic acid cycle, electron transfer, and oxidative phosphorylation, are common to both carbohydrate and lipid catabolism and are considered in detail in the present chapter. The process of glycolysis which serves to feed carbon atoms from carbohydrates into the central oxidative processes is described in Chapter 5 while those processes which feed carbon atoms from lipids into this central machinery are considered in Chapter 6.

1. An outline of the major pathways of catabolism

Before fragmenting the processes of ATP generation and assigning them to different chapters (a matter of descriptive convenience rather than metabolic reality as, we hope, will be appreciated in Chapters 8 and 14 when the integration of these pathways is considered) it may be useful to summarize them. Their general inter-relationship is shown in Figure 4.1.

(a) *Glycolysis*

Under many conditions, glucose is the important respiratory fuel in the body. It is degraded by oxidation to pyruvate (with the reduction of NAD^+ to NADH—see Figure 5.8). The fate of the pyruvate produced depends on the tissue involved and the availability of oxygen. Under anaerobic conditions it is reduced to lactate while, under aerobic conditions, oxidation continues forming first acetyl-coenzyme A (acetyl-CoA) and then, through the tricarboxylic acid cycle and the electron transfer pathway, carbon dioxide and water.

(b) *Oxidation of fatty acids*

Long-chain fatty acids are made available from stored triacylglycerol when glucose is unable to provide sufficient energy for the needs of the tissues of the body. This can be a frequent occurrence; fatty acids become the major respiratory fuel during starvation or prolonged exercise. The oxidation of fatty acids, by the β-oxidation pathway, produces acetyl-CoA (with the concomitant reduction of NAD^+ to NADH and FAD to $FADH_2$—see Figure 6.9). The acetyl-CoA then enters the tricarboxylic acid cycle for complete oxidation.

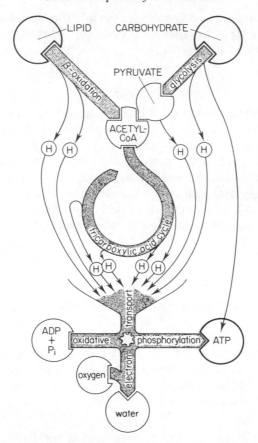

Figure 4.1 Major pathways for ATP synthesis from lipid and carbohydrate

(c) *Tricarboxylic acid cycle*

The major fate of the acetyl-CoA produced from both glycolysis and the β-oxidation of fatty acids (and also the acetyl-CoA produced from amino acid oxidation) is oxidation via the tricarboxylic acid cycle. This cyclic sequence of reactions (see Section B) oxidizes the acetyl group of acetyl-CoA to carbon dioxide (with the reduction of more NAD^+ to NADH and FAD to $FADH_2$).

(d) *Amino acid metabolism*

Since there are 20 different amino acids their metabolism is considerably more diverse than that of glucose or fatty acids. In general, the α-amino group is removed from the amino acid to produce an oxoacid which may be a simple metabolic intermediate itself (e.g. pyruvate) or be converted to a simple metabolic intermediate via a specific metabolic pathway. The intermediates are either oxidized via the tricarboxylic acid cycle or converted to glucose via gluconeogenesis (see Chapter 10).

(e) *Electron transfer and oxidative phosphorylation*

The reduced coenzymes, NADH and $FADH_2$, generated in the above pathways, do not react directly with oxygen but donate electrons to other electron carriers (compounds able to undergo reversible oxidation). Electrons are transferred along a series of such carriers which constitute the electron transfer chain until, finally, oxygen is reduced to water. A significant proportion of the chemical energy change associated with these oxidoreduction reactions is not released as heat but is utilized for the synthesis of ATP from ADP and phosphate. This process of ATP formation is known as oxidative phosphorylation (see Section D).

2. Reactions of metabolism

The pathways outlined above comprise many reactions, each catalysed by a different enzyme, and at first sight such pathways appear dauntingly complex. However, some simplification is possible when it is realized that many of the reactions encountered are chemically similar and involve only slightly different substrates. Indeed, as many as three-quarters of the reactions involved in glucose and fatty acid catabolism can be described by as few as six general types of reaction which are outlined below.

(i) *The oxidation of alcohols and reduction of carbonyl compounds*   Primary alcohols are oxidized to aldehydes and secondary alcohols to ketones. In metabolism these reactions usually involve NAD^+ (or $NADP^+$) as oxidizing agent which becomes reduced to NADH (or NADPH). (See Section C.1.e for details of the structure of these compounds.)

$$\begin{array}{c} R_1 \\ \diagdown \\ R_2 \end{array} C \begin{array}{c} H \\ \diagup \\ \diagdown OH \end{array} + NAD^+ \rightleftarrows \begin{array}{c} R_1 \\ \diagdown \\ R_2 \end{array} C{=}O + NADH + H^+$$

secondary alcohol   ketone

(ii) *The oxidation of* $-CH_2-CH_2-$ *and reduction of* $-CH{=}CH-$ *groups*   Reactions of this type most frequently utilize flavins (FAD or FMN); see Section C.1.b for details.

$$R_1{-}CH_2{-}CH_2{-}R_2 + FAD \rightleftarrows R_1{-}CH{=}CH{-}R_2 + FADH_2$$

(saturated)   (unsaturated)

(iii) *Hydrolysis reactions and reactions in which compounds are joined with the elimination of water* Hydrolytic reactions of biochemical importance include the hydrolysis of esters (to a carboxylic acid and an alcohol), phosphoesters (to a phosphoric acid and an alcohol) and peptides (to a carboxylic acid and an amine). Acylglycerols, nucleotides and proteins are all synthesized by the reverse of reactions of this kind.

$$
\underset{\text{phosphodiester}}{R_1-O\overset{\displaystyle O}{\underset{\displaystyle O^-}{\overset{\|}{P}}}O-R_2} + H_2O \rightleftarrows \underset{\substack{\text{organic} \\ \text{phosphate}}}{R_1-O\overset{\displaystyle O}{\underset{\displaystyle O^-}{\overset{\|}{P}}}OH} + \underset{\text{alcohol}}{R_2-OH}
$$

$$
\underset{\text{dipeptide}}{R_1-CONH-R_2} + H_2O \rightleftarrows \underset{\text{amino acids}}{R_1-COO^- + R_2-NH_3^+}
$$

(iv) *Addition to unsaturated* $(-CH{=}CH-)$ *compounds and the formation of such double bonds by elimination* (*most frequently of water*)

$$
R_1-CH{=}CH-R_2 + H_2O \rightleftarrows R_1-CH_2CH(OH)-R_2
$$

(v) *Addition to a carbonyl group and its reverse reaction*

$$
\underset{R_2}{\overset{R_1}{\diagdown}}C{=}O + R_3-H \rightleftarrows \underset{R_3}{\overset{R_1}{\diagdown}}R_2-\overset{\displaystyle}{\underset{\displaystyle}{C}}-OH
$$

(vi) *Decarboxylation of carboxylic acids* (*especially 2- and 3-oxocarboxylic acids*)

$$
R-CO.CH_2COOH \longrightarrow R-CO.CH_3 + CO_2
$$

Decarboxylation reactions involving 2-oxocarboxylic acids are generally more complex than indicated above since the aldehyde product is often further oxidized. Carboxylation reactions, although they are of importance in metabolism, usually proceed by a more complex mechanism than simple reversal of the example given above.

Of the six types of reactions given above, all except (vi) can be readily reversed. Whether they are near-equilibrium *in vivo* depends upon the concentrations of substrates and products in the cell (Chapter 2; Section A.2.c).

3. Acetyl-coenzyme A

Many carboxylic acids, in particular the fatty acids, undergo their biochemical transformations as derivatives of coenzyme A (CoA). Acetyl-CoA is the best known of the coenzyme A derivatives since it is the standard two-carbon unit of metabolism; it is produced in catabolic processes and used in many biosynthetic pathways. This central role is the reason that the oxidation of acetyl-CoA in the tricarboxylic acid cycle is considered first in the ensuing discussion of catabolic processes.

Coenzyme A can be considered as a derivative of the nucleotide ADP (see Figure 2.2). It is notionally (but not biosynthetically)[1] derived from ADP by the esterification of an additional phosphate group at the 3'-position of the ribose moiety and the esterification of the 5'-diphosphate group with a long-chain alcohol (see Figure 4.2). The latter consists of pantothenic acid (a vitamin—see Appendix A) linked to 2-amino ethanthiol. For all its structural complexity, it is the single thiol (sulphydryl or —SH) group at the distal end of the side-chain which confers upon coenzyme A its special biochemical properties. For this reason, the molecule is often represented by the abbreviation CoASH in biochemical equations. It is this thiol group which esterifies with carboxylic acids to form acyl-CoA compounds which are, therefore, thioesters:

$$CoASH + R\text{-}COOH \longrightarrow R\text{-}CO.SCoA + H_2O$$

thioester

Acyl-CoA compounds are used metabolically as acylating agents in an analogous way to that in which acyl chlorides are used by organic chemists. Both acetyl-CoA and acetyl chloride donate their acetyl groups very much more readily than do acetyl derivatives of alcohols (i.e. esters[2]). This can be seen if the

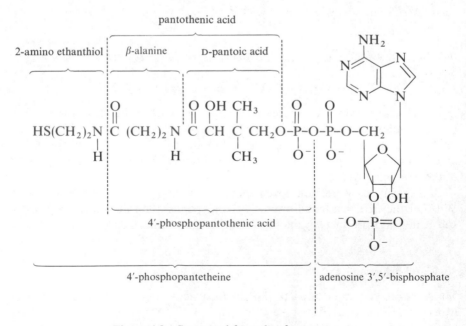

Figure 4.2 Structural formula of coenzyme A

$\Delta G^{0\prime}$ values for the hydrolysis (i.e. donation of the acetyl group to a hydroxyl ion) of corresponding esters and thioesters are compared:

$$CH_3CH_2OC\overset{\displaystyle O}{\underset{\displaystyle CH_3}{\diagdown}} + H_2O \longrightarrow CH_3CH_2OH + CH_3COOH$$

ethyl ethanoate (ethyl acetate) $\qquad \Delta G^{0\prime} = -21.3\,\text{kJ.mole}^{-1}$

$$CoAS.C\overset{\displaystyle O}{\underset{\displaystyle CH_3}{\diagdown}} + H_2O \longrightarrow CoASH + CH_3COOH$$

$$\Delta G^{0\prime} = -32.2\,\text{kJ.mole}^{-1}$$

acetyl-CoA

The corollary of the large negative $\Delta G^{0\prime}$ for acetyl-CoA hydrolysis is that $\Delta G^{0\prime}$ for synthesis of acetyl-CoA from acetate and coenzyme A is large and positive. Hence, for ΔG for this reaction to be negative, the concentrations of acetate and coenzyme A would need to be unphysiologically high or that of acetyl-CoA would need to be unphysiologically low or both. This problem is overcome by coupling the acetylation of coenzyme A to the hydrolysis of ATP in a single reaction catalysed by a synthetase enzyme. This makes use of the large and negative value of $\Delta G^{0\prime}$ of ATP hydrolysis (see Chapter 2; Section A.2.d):

$$CoASH + ATP + R\text{-}COOH \longrightarrow R\text{-}CO\ SCoA + AMP + PP_i$$

B. TRICARBOXYLIC ACID CYCLE

Although tricarboxylic acid (TCA) cycle is the name by which the series of reactions oxidizing acetyl-CoA will be referred to in this book, the pathway has been given several other names. Tricarboxylate cycle is a modern version gaining acceptance (see Chapter 3; Note 8) and it is widely known as Krebs cycle thus honouring the major contribution made to its elucidation by H. A. Krebs in 1937 (Krebs and Johnson, 1937). The name citric acid cycle has also been used but is rejected here because it suggests an unwarranted importance of one particular intermediate.

The function of the TCA cycle is to oxidize the acetyl group of acetyl-CoA* to

---

* The amount of acetyl-CoA in tissues is very small (e.g. about $0.01\,\mu\text{mol.g}^{-1}$ fresh wt) and the maximum rate of the cycle in heart is about $10\,\mu\text{mol.min}^{-1}\,\text{g}^{-1}$, so that acetyl-CoA must turn over very rapidly in aerobic tissues. However, acetylcarnitine (see Chapter 6; Section D.1.c for structure of carnitine) is found in such tissues at a concentration approximately 100-fold greater than that of acetyl-CoA. These two compounds are in equilibrium via the reaction catalysed by carnitine acetyltransferase as follows:

acetylcarnitine + CoASH $\rightleftarrows$ acetyl-CoA + carnitine

Hence, it is suggested that acetylcarnitine acts as a buffer system to maintain constant concentration of acetyl-CoA whenever the flux through the cycle changes in a similar way to which phosphocreatine buffers the ATP concentration (Chapter 5; Section F.1.b). It may also be important in the control of the rate of long-chain fatty acid oxidation in muscle (see Chapter 7; Section D.2).

carbon dioxide. This occurs with the concomitant reduction of the electron carriers, NAD^+ and FAD, whose subsequent oxidation is accompanied by the synthesis of ATP. The reactions of the cycle may be summarized by the general equation:

$$CH_3CO.SCoA + 3H_2O \longrightarrow 2CO_2 + 4[2H] + CoASH$$

where [2H] represents a pair of hydrogen atoms removed to reduce an electron carrier. Since the TCA cycle plays a central role in the process of making chemical energy available for biological use, it is no surprise that its highest activity is found in aerobic muscles. Indeed, most of the early work on the cycle was carried out on preparations of pigeon pectoral muscle since, during long-distance flight, this muscle generates all its ATP by aerobic metabolism. The importance of the cycle in energy production in various tissues is discussed in Section E.1.

All the constituent enzymes of the cycle are located within the matrix of the mitochondria, with the exception of succinate dehydrogenase which is located on the inner surface of the inner mitochondrial membrane. The reactions of the cycle are considered below and summarized in Figure 4.3.

1. Reactions of the cycle

(a) *Citrate synthase reaction*

Citrate synthase catalyses the reaction between acetyl-CoA and oxaloacetate to form citrate; this is an aldol condensation of acetate on a 2-oxocarboxylic acid:

$$CH_3CO.SCoA \; + \; \begin{matrix} CO.COO^- \\ | \\ CH_2COO^- \end{matrix} \; + \; H_2O \longrightarrow \begin{matrix} CH_2COO^- \\ | \\ HO.CCOO^- \\ | \\ CH_2COO^- \end{matrix} \; + \; CoASH + H^+$$

acetyl-CoA   oxaloacetate   citrate

Subsequent reactions of the cycle regenerate the oxaloacetate thereby allowing the entry of further molecules of acetyl-CoA into the cycle. Citrate synthase catalyses a non-equilibrium reaction (Table 4.1) and may, therefore, be regulated by allosteric effectors. In addition, the evidence suggests that it catalyses a flux-generating step (see Section B.2).

(b) *Aconitate hydratase reaction*

Aconitate hydratase (aconitase) catalyses the conversion of citrate to isocitrate. For a long time, it was presumed that this conversion proceeded by the removal of water from citrate with the intermediate formation of *cis*-aconitate but there is some evidence that this is not an obligatory pathway and that the enzyme can

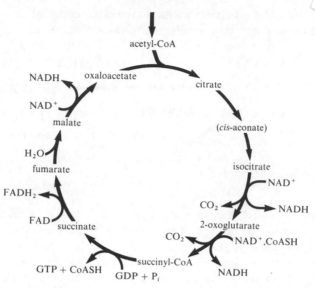

Figure 4.3   Pathway of tricarboxylic acid cycle

Table 4.1. Equilibrium nature of reactions of the TCA cycle

Reaction catalysed by	ΔG (kJ.mole^{-1})*	Suggested or assumed equilibrium nature
Citrate synthase	-53.9	non-equilibrium
Aconitate hydratase	$+0.8$	near-equilibrium
Isocitrate dehydrogenase (NAD$^+$)	-17.5	non-equilibrium
Isocitrate dehydrogenase (NADP$^+$)	-17.5†	non-equilibrium
Oxoglutarate dehydrogenase	-43.9	non-equilibrium
Succinyl-CoA synthetase	—	near-equilibrium‡
Succinate dehydrogenase	—	non-equilibrium‡
Fumarate hydratase	—	near-equilibrium‡
Malate dehydrogenase	—	near-equilibrium‡

* Values for ΔG are calculated as described in Table 2.4. The data for calculations are taken from Rowan and Newsholme, 1979; Paul, 1979.

† It is assumed that mitochondrial NADP$^+$/NADPH concentration ratio is the same as that of NAD$^+$/NADH—see Krebs (1973).

‡ Equilibrium nature indicated from maximum enzyme activities.

catalyse the isomerization of citrate without the involvement of *cis*-aconitate. The reaction is therefore best represented as follows:

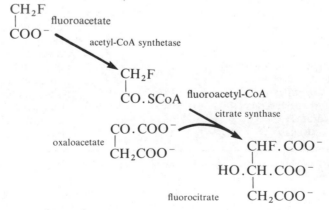

It is generally assumed that aconitate hydratase catalyses a near-equilibrium reaction *in vivo* with the ratio of citrate:isocitrate:*cis*-aconitate close to 89:4:7.

Aconitate hydratase is strongly, but competitively, inhibited by the citrate analogue, fluorocitrate. Since the cycle is of fundamental importance in energy metabolism, inhibition of aconitate hydratase has serious consequences and fluorocitrate is very toxic. Even more toxic, however, is fluoroacetate $(CH_2F\,COO^-)$, the oral lethal dose of which for man is estimated to be $2-5\,mg.kg^{-1}$. This compound is the toxic principle in the South African plant known as gifblaar (*Dichapetalum cymosum*). Fluoroacetate itself has no direct inhibitory action on aconitate hydratase but it traverses the cell and mitochondrial membranes more readily than fluorocitrate, thus accounting for its greater toxicity. Once within the mitochondrion, fluoroacetate is converted to fluoroacetyl-CoA by acetyl-CoA synthetase. This is a substrate for citrate synthase, which condenses it with oxaloacetate to form the inhibitory fluorocitrate (Figure 4.4). This conversion of a biologically inactive compound to one that is dangerous was termed 'lethal synthesis' by Peters (1951).

Figure 4.4 The lethal synthesis of fluorocitrate

The ability of enzymes to modify compounds *in vivo* can be utilized by the pharmaceutical industry to produce active drugs from inactive precursors. For example, enzymes that degrade fibrin clots can also degrade other proteins in the bloodstream, but acyl-enzymes can be produced that are catalytically inactive until they bind to a clot. Here deacylation occurs and the enzymes remain bound to the clot to catalyse fibrinolyses. Such derivatives represent a new approach to thrombolytic therapy (Smith *et al.*, 1981).

(c) *Isocitrate dehydrogenase reaction*

Isocitrate dehydrogenase catalyses the oxidative decarboxylation of isocitrate to 2-oxoglutarate. This compound was formerly known as α-ketoglutarate and both names appear in biochemical literature at the present time. (Biochemists have been understandably reluctant to relinquish older names for both enzymes and intermediates merely for the sake of uniformity.) In this two-stage reaction, which proceeds via the intermediate formation of oxalosuccinate, both essential functions of the TCA cycle are realized, namely the production of carbon dioxide and the reduction of a coenzyme:

$$
\begin{array}{l}
\text{HO.CH.COO}^- \\
\mid \\
\text{CH.COO}^- \\
\mid \\
\text{CH}_2\text{COO}^- \\
\text{isocitrate}
\end{array}
+ \text{NAD(P)}^+ \longrightarrow
\begin{array}{l}
\text{CO.COO}^- \\
\mid \\
\text{CH.COO}^- \\
\mid \\
\text{CH}_2\text{COO}^- \\
\text{oxalosuccinate}
\end{array}
+ \text{NAD(P)H} + \text{H}^+
$$

$$
\begin{array}{l}
\text{CO.COO}^- \\
\mid \\
\text{CH.COO}^- \\
\mid \\
\text{CH}_2\text{COO}^-
\end{array}
+ \text{H}^+ \longrightarrow
\begin{array}{l}
\text{CO.COO}^- \\
\mid \\
\text{CH}_2 \\
\mid \\
\text{CH}_2\text{COO}^- \\
\text{2-oxoglutarate}
\end{array}
+ \text{CO}_2
$$

The involvement of oxalosuccinate in this way has only been established for the NADP^+-linked enzyme (see below). In this enzyme, both oxidation and decarboxylation reactions occur at the same active site and the oxalosuccinate remains bound to the enzyme.

Most tissues contain two isocitrate dehydrogenase enzymes: one specific for NAD^+ and the other for NADP^+. Although it had been believed that the NAD^+-linked enzyme alone was involved in the TCA cycle, current evidence suggests that both enzymes must play a part in the conversion of isocitrate to oxoglutarate, but it is unclear why two enzymes are necessary (Alp *et al.*, 1976; Fatania and Dalziel, 1980). It is likely that both isocitrate dehydrogenases catalyse non-equilibrium reactions (Table 4.1). Allosteric effectors of possible physiological significance have been reported for the NAD^+-linked enzyme only (see Chapter 7; Section B.3.b).

(d) Oxoglutarate dehydrogenase reaction

Oxoglutarate dehydrogenase catalyses the oxidative decarboxylation of 2-oxoglutarate, with the formation of succinyl-CoA. The overall reaction can be represented as follows:

$$\text{CoASH} + \begin{array}{c} \text{CO.COO}^- \\ | \\ \text{CH}_2 \\ | \\ \text{CH}_2\text{COO}^- \end{array} + \text{NAD}^+ \longrightarrow \begin{array}{c} \text{COO}^- \\ | \\ \text{CH}_2 \\ | \\ \text{CH}_2\text{CO.SCoA} \end{array} + \text{NADH} + \text{CO}_2$$

2-oxoglutarate   succinyl-CoA

The reaction is, however, more complex than the previous oxidative decarboxylation (i.e. isocitrate oxidation) and oxoglutarate dehydrogenase is a complex of three enzymes so that the constituent reactions occur at separate active sites (see Figure 4.5(a)). In such multienzyme complexes,[3] intermediate reactants are not released from the enzyme but are transferred from one active site to another. The way that this is generally achieved is by the intermediate becoming covalently linked to one of the proteins of the complex by a long, flexible arm (composed of an amino acid side-group together with a prosthetic group). In this way the intermediate is prevented from 'escaping' from the complex and its chance of encountering the next active site is much enhanced.

Each of the three components of the oxoglutarate dehydrogenase complex contains a prosthetic group, two of which (thiamine pyrophosphate and FAD) are derived from B vitamins (see Appendix A). The third prosthetic group, lipoate, forms part of the flexible arm (with the amino acid lysine, that is part of the enzyme, forming the other half) which transfers the intermediate between active sites. The mechanism of the oxoglutarate dehydrogenase reaction will be described in some detail since it illustrates the functioning of multienzyme complexes and the role of prosthetic groups. Its mechanism is precisely analogous to that of pyruvate dehydrogenase, an important enzyme in pyruvate metabolism discussed in Chapter 5; Section B.13.

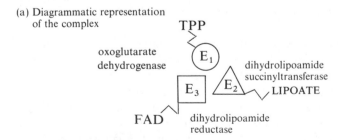

(a) Diagrammatic representation of the complex

Figure 4.5 Mechanism of the oxoglutarate dehydrogenase reaction. See text for explanation

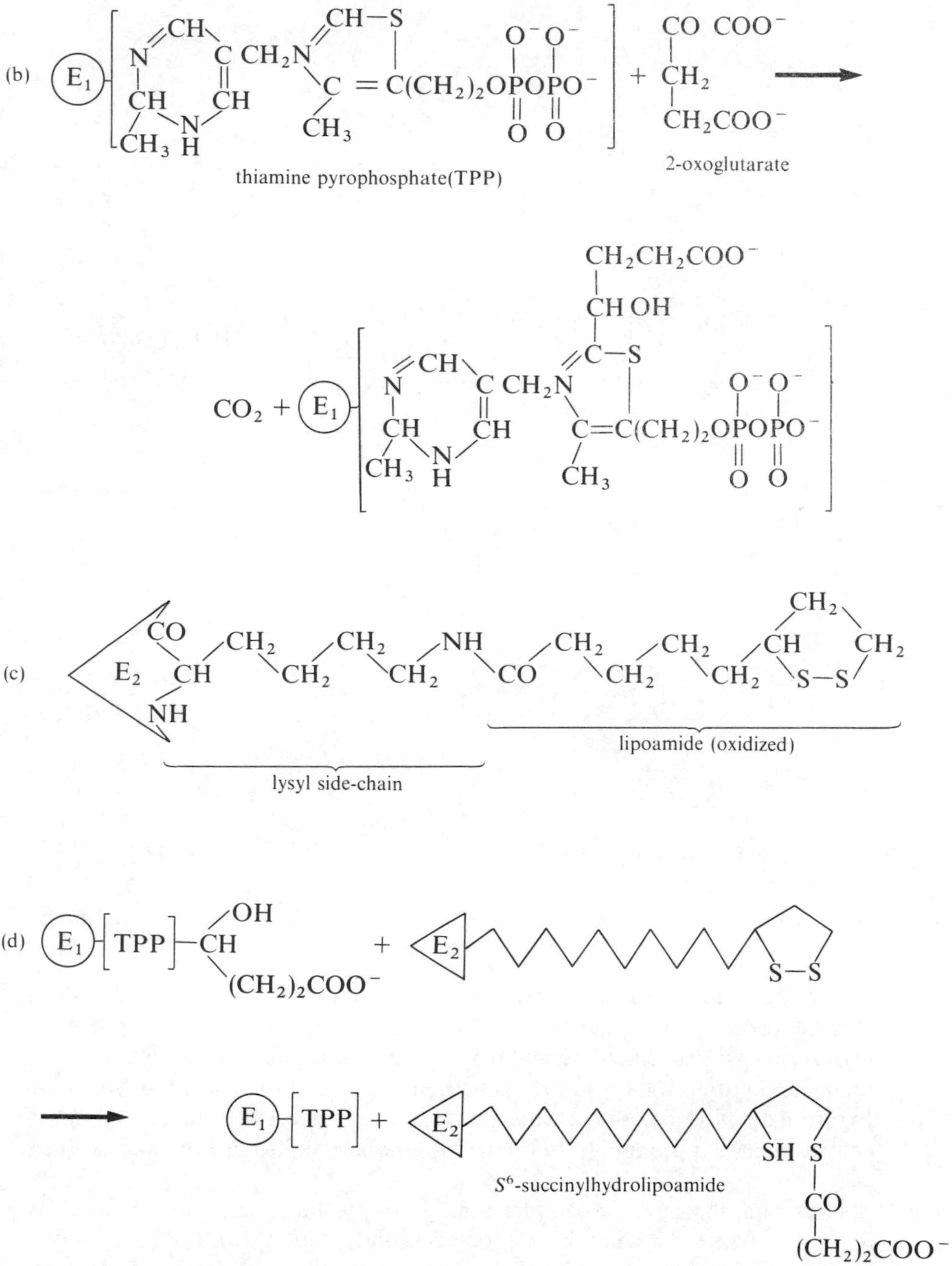

Figure 4.5 (*cont.*) Mechanism of the oxoglutarate dehydrogenase reaction. See text for explanation

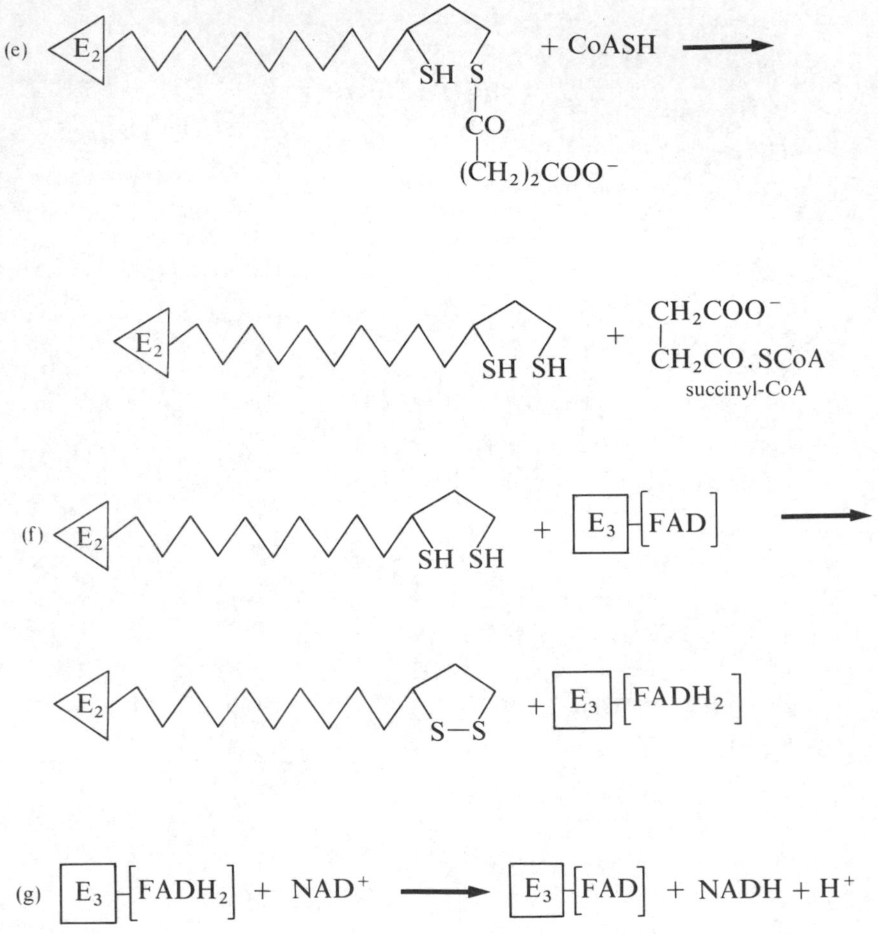

Figure 4.5 (*cont.*) Mechanism of the oxoglutarate dehydrogenase reaction. See text for explanation

The first reaction is catalysed by oxoglutarate dehydrogenase (E_1) which removes carbon dioxide from 2-oxoglutarate, leaving the residual 1-hydroxy-3-carboxypropyl (succinate semialdehyde) group covalently attached to the thiamine pyrophosphate (TPP) prosthetic group of the decarboxylase (see Figure 4.5(b)). (Note that confusion can arise since the name oxoglutarate dehydrogenase is given both to the whole complex and to one of its constituent enzymes.)

The second enzyme of the complex is dihydrolipoamide succinyltransferase (E_2) in which a lysine side chain is extended by condensation with lipoate to form a lipoyllysine (lipoamide) arm (see Figure 4.5(c)). Lipoate is an oxidizing agent and as the succinate semialdehyde group is transferred to the lipoyl group the former becomes oxidized to a succinyl group and a thioester is formed (see Figure 4.5(d)).

The same succinyltransferase then transfers the succinyl group from one thiol (dihydrolipoate) to another (coenzyme A). This generates one of the products of the over-all reaction (succinyl-CoA) and leaves the lipoyl group in the fully reduced (dihydrolipoyl) state (see Figure 4.5(e)). The third enzyme of the complex, dihydrolipoamide reductase (E_3) then catalyses the reoxidation of the dihydrolipoyl group by NAD^+ in a series of steps. The reductase possesses an FAD prosthetic group which is first reduced by the dihydrolipoyl moiety of the succinyltransferase to produce $FADH_2$ (see Figure 4.5(f)). While still bound, this flavin is oxidized by NAD^+ in a reaction also catalysed by the reductase (see Figure 4.5(g)). This last reaction is one of the few examples encountered in which NAD^+ oxidizes a reduced flavin rather than a flavin oxidizing NADH. The over-all reaction catalysed by the complex is non-equilibrium and the evidence suggests that it catalyses a flux-generating step (see Section B.2).

(e) *Succinyl-CoA synthetase reaction*

Succinyl-CoA synthetase catalyses the formation of succinate from succinyl-CoA and the formation of GTP from GDP and phosphate:

$$
\begin{array}{l}
CH_2COO^- \\
| \\
CH_2CO.SCoA
\end{array}
+ GDP^{3-} + P_i^{2-} \longrightarrow
\begin{array}{l}
CH_2COO^- \\
| \\
CH_2COO^-
\end{array}
+ GTP^{4-} + CoASH
$$

$$\text{succinyl-CoA} \qquad\qquad\qquad \text{succinate}$$

The reaction involves the intermediate phosphorylation of the enzyme (on a histidine residue) and proceeds in the following stages where E represents the enzyme:

$$\text{succinyl-CoA} + E \longrightarrow \text{E-SCoA} + \text{succinate}$$
$$\text{E-SCoA} + P_i \longrightarrow \text{E-P} + \text{CoASH}$$
$$\text{E-P} + \text{GDP} \longrightarrow \text{GTP} + E$$

The formation of succinyl-CoA, rather than succinate, as an endproduct of the oxoglutarate dehydrogenase reaction may be seen as a means of conserving a larger part of the chemical energy associated with 2-oxoglutarate oxidation. The succinyl-CoA synthetase reaction enables this energy to be conserved in the high energy phosphate compound, GTP. However, the reason why guanine rather than adenine nucleotides are used in this reaction is not clear. Although GTP has a number of specific roles in the cell (e.g. in protein synthesis) it is readily generated for these purposes from ATP by the action of nucleosidediphosphate kinase in the mitochondria. The same enzyme, which catalyses a near-equilibrium reaction, is also responsible for the generation of ATP from GTP.

$$ADP + GTP \rightleftarrows ATP + GDP$$

Many tissues contain a second enzyme capable of catalysing the conversion of succinyl-CoA to succinate. This enzyme, 3-oxoacid CoA-transferase, catalyses the transfer of coenzyme A from succinyl-CoA to acetoacetate to form acetoacetyl-CoA:

$$
\begin{array}{cccc}
\text{CH}_2\text{COO}^- & \text{CO.CH}_3 & \text{CH}_2\text{COO}^- & \text{CO.CH}_3 \\
| \quad + & | & \longrightarrow \quad | \quad + & | \\
\text{CH}_2\text{CO.SCoA} & \text{CH}_2\text{COO}^- & \text{CH}_2\text{COO}^- & \text{CH}_2\text{CO.SCoA} \\
\text{succinyl-CoA} & \text{acetoacetate} & \text{succinate} & \text{acetoacetyl-CoA}
\end{array}
$$

The importance of this reaction is in the 'activation' of ketone bodies so that they can be converted to acetyl-CoA for oxidation (see Chapter 6; Section D.2.6). The succinyl-CoA synthetase reaction is near-equilibrium.

(f) *Succinate dehydrogenase reaction*

Succinate dehydrogenase catalyses the oxidation of succinate to fumarate. In common with many other desaturation reactions, the oxidizing agent is a flavin, in this case FAD (see Section C.1.b) which is firmly bound to the enzyme:

$$
\begin{array}{ccc}
\text{CH}_2\text{COO}^- & & \text{HC.COO}^- \\
| \quad + \text{E.FAD} & \longrightarrow & \| \qquad\qquad + \text{E.FADH}_2 \\
\text{CH}_2\text{COO}^- & & ^-\text{OOC.CH} \\
\text{succinate} & & \text{fumarate}
\end{array}
$$

Succinate dehydrogenase, unlike other enzymes of the TCA cycle, is located in the inner mitochondrial membrane. This permits electrons to be transferred directly into the electron transfer chain (most probably at the level of ubiquinone-10, see Section C.1.c) when the flavin is reoxidized. Since the enzyme is relatively easy to assay, and because it can be visualized histochemically, it has long been used as a marker enzyme to identify mitochondria and mitochondria-rich tissues in cytological studies and subcellular fractionations. Its unique location for an enzyme of the TCA cycle, however, renders it of doubtful value as an indicator of intact functional mitochondria. Furthermore, since it has not been un-equivocably established as a non-equilibrium enzyme, its activity cannot be regarded as a good quantitative indicator of the maximum capacity of the TCA cycle (see Chapter 5; Appendix 5.2).

(g) *Fumarate hydratase reaction*

Fumarate hydratase (fumarase) catalyses the hydration of fumarate to malate in a reaction that appears to be near-equilibrium (Table 4.1):

$$\begin{array}{c} HC.COO^- \\ \parallel \\ {}^-OOC.CH \end{array} + H_2O \longrightarrow \begin{array}{c} HO.CH.COO^- \\ | \\ CH_2COO^- \end{array}$$

$$\text{fumarate} \qquad\qquad\qquad \text{malate}$$

This reaction is analogous to that catalysed by aconitate hydratase (see Section B.1.b) and probably proceeds by a similar mechanism.

(h) *Malate dehydrogenase reaction*

Malate dehydrogenase catalyses the oxidation of the secondary alcohol, malate, by NAD^+ in a near-equilibrium reaction. Oxaloacetate is the ketone produced, so that this final reaction of the TCA cycle regenerates the means by which a further acetyl-CoA molecule can enter the cycle for oxidation. The reaction is near-equilibrium (Table 4.1):

$$\begin{array}{c} HO\ CH.COO^- \\ | \\ CH_2COO^- \end{array} + NAD^+ \longrightarrow \begin{array}{c} CO.COO^- \\ | \\ CH_2COO^- \end{array} + NADH + H^+$$

$$\text{malate} \qquad\qquad\qquad \text{oxaloacetate}$$

2. Overview of the TCA cycle as a metabolic pathway

By summing the individual reactions described above it is possible to produce a more detailed over-all equation for the oxidation of acetyl-CoA than that given in the introduction to Section B:

$$CH_3COSCoA + 3NAD^+ + E.FAD + GDP^{3-} + P_i^{2-} + 2H_2O$$

$$\longrightarrow CoASH + 2CO_2 + 3NADH + E.FADH_2 + GTP^{4-} + 2H^+$$

The immediate yield of 'high energy phosphate' is minimal but the cycle will only operate when electron transfer (and thus normally oxidative phosphorylation) occurs to convert the chemical energy in the reduced coenzymes into ATP.

It should be noted that none of the intermediates of the TCA cycle enters into the over-all equation since they are neither formed nor destroyed in the net operation of the cycle. This emphasizes the function of the TCA cycle which is to oxidize acetate (in the form of acetyl-CoA) and concomitantly to produce reduced coenzymes (from which ATP may be generated). Indeed, in a real sense, this is the *only* function of the cycle for if intermediates were removed (e.g. for synthetic purposes) the cycle could not continue to function since the amount of oxaloacetate regenerated would be less than that consumed. However, it is well established that some intermediates of the pathway are also members of other metabolic pathways, as conventionally defined. For example, in liver, oxaloacetate can be converted (via phosphoenolpyruvate) to glucose; but this can only occur if carbon atoms are fed into the cycle in addition to those from acetyl-CoA (e.g. as 2-oxoglutarate formed by the deamination of glutamate). In other words, some of the reactions of the cycle, but not the cycle itself, can be used synthetically. This 'multiple use' of reactions of the TCA cycle in some tissues will influence the way in which flux through the cycle is regulated. The situation is clarified by applying the definition of a metabolic pathway developed in Chapter 2; Section C.2.d to the cycle; this analysis indicates that the tricarboxylic acid cycle is not a single metabolic pathway.

Table 4.2. Substrate concentrations* and K_m values for some enzymes of the TCA cycle

Reaction catalysed by	Substrate	Range of substrate concentration (μM)	Range of K_m values (μM)
Citrate synthase	{ acetyl-CoA	100–600	5–10
	oxaloacetate	1–10	5–10
Isocitrate dehydrogenase (NAD$^+$)	isocitrate	150–700	50–200
Oxoglutarate dehydrogenase	2-oxoglutarate	600–5900	60–200

* Concentrations of substrate are those calculated to be in the matrix of the mitochondrion. Since there is a large difference in values presented in the literature, a range is given. (Data collected by Paul, 1979.)

There is no doubt that, in muscle at least, citrate synthase is saturated with acetyl-CoA, which is the pathway-substrate for the cycle (although it is *not* saturated with oxaloacetate—Table 4.2) and, in addition, 2-oxoglutarate dehydrogenase is saturated with 2-oxoglutarate (see Table 4.2). Thus the TCA cycle has at least two flux-generating steps so that it can be considered to consist of at least two physiological pathways: the span from acetyl-CoA to 2-oxoglutarate and the span from 2-oxoglutarate to oxaloacetate (Figure 4.6). In some tissues, under some conditions (e.g. muscle during sustained exercise), this division of the cycle into two pathways is only academic, since the flux through the two pathways must be identical and must be regulated in a concerted manner. However, the division does explain how it is possible for intermediates to feed into the cycle at, or after, the level of 2-oxoglutarate and be withdrawn from the cycle at the level of malate or oxaloacetate (Figure 4.6). Since the span from 2-oxoglutarate to oxaloacetate is a separate pathway, intermediates can enter and leave it without necessarily interfering in the normal operation of energy generation via the conventional cycle. The division of the cycle into two pathways is particularly relevant when amino acid metabolism is discussed in Chapter 10 and when the regulation of the flux through the cycle is discussed in Chapter 7; Section B.3.

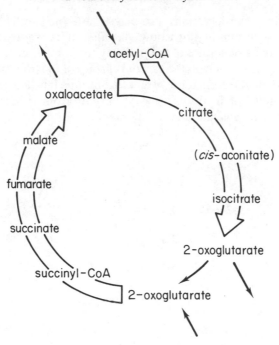

Figure 4.6 Interpretation of the TCA cycle emphasizing its construction from two physiological pathways. The first pathway commences with acetyl-CoA and terminates at 2-oxoglutarate, which can leave the cycle to form glutamate. The second pathway starts with 2-oxoglutarate, which can feed into the cycle from glutamate, and ends with the formation of oxaloacetate which may be further metabolized

Although the TCA cycle functions within the mitochondria, several of the reactions also occur outside the mitochondria. In some cases, such extramitochondrial reactions are known to play a specific role in metabolism. For example, extramitochondrial malate dehydrogenase plays a role in the malate/aspartate shuttle for the re-oxidation of cytosolic NADH to NAD^+ (see Chapter 5; Section D.1). However, the function of other TCA cycle enzymes (e.g. aconitate hydratase) in the cytosolic compartment of the cell is not known.

3. **Stereochemical considerations and the elucidation of the TCA cycle pathway**

Four of the intermediates of the TCA cycle exhibit stereoisomerism, namely *cis*-aconitate, *threo*-D_s-isocitrate, fumarate (*trans*) and L-malate. (An account of stereoisomerism and its nomenclature is given in Appendix 4.1.) The enzymes catalysing the reactions of these stereoisomers are, like most other enzymes, stereospecific and in most cases will neither bind nor react with 'unnatural' stereoisomers.

Stereochemical considerations also played an important role in establishing the pathway by accounting for isotope labelling data whose misinterpretation had obscured the true nature of the pathway for some time. Administration of [^{14}C]acetate, labelled only in the carboxyl group of the molecule, to a respiring tissue resulted in the formation of 2-oxoglutarate labelled in the 5-carboxyl carbon atom but not in the 1-carboxyl carbon atom (Figure 4.7). This was originally taken to preclude the involvement of citrate in the cycle, since this

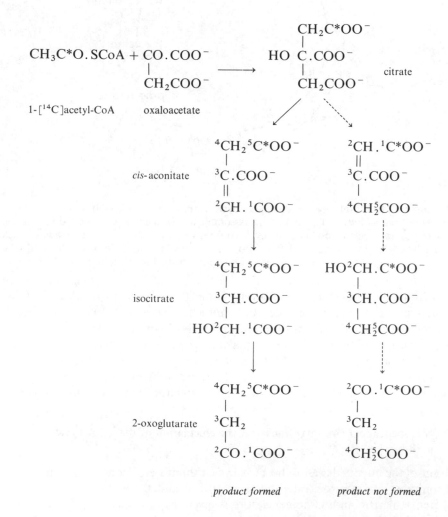

Figure 4.7 Fate of a ^{14}C atom introduced via acetyl-CoA into the tricarboxylic acid cycle. The labelled carbon atom is indicated by an asterisk

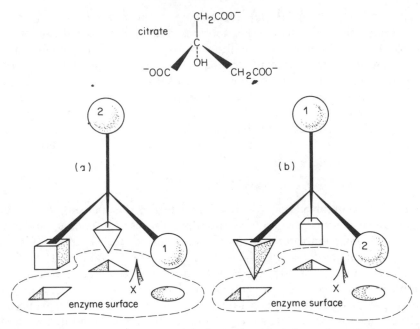

Figure 4.8 Interaction of citrate with the active site of aconitate hydratase. The tetrahedral model represents a molecule of citrate. The spheres represent the $-CH_2 COO^-$ groups which are identical.

In (a), the model is in the correct orientation for attack (at X) on the bond to the $-CH_2 COO^-$ group arbitrarily labelled (1).

If the model is rorated (b) to bring the other $-CH_2 COO^-$ group (labelled (2)) into the correct position for reaction, it can be seen that the substrate no longer fits the enzyme

compound is a symmetrical molecule whose terminal carboxyl groups are apparently indistinguishable. It was argued that if citrate participated in the cycle it should lead to the appearance of ^{14}C in both carboxyl groups of 2-oxoglutarate. Accordingly, a non-symmetrical seven-carbon condensation product of oxaloacetate and pyruvate was sought. It was, however, never found. Some time later, Ogston (1948) restored citrate as an accepted intermediate of the cycle by pointing out that the observed labelling pattern is explained if (as now seems very likely) aconitate hydratase interacts with citrate at more than two points. Although citrate is a symmetrical molecule, the enzyme is not and Figure 4.8 shows how a hypothetical active site could distinguish between the two terminal carboxyl groups of citrate and ensure that the elements of water are always removed from the same pair of carbon atoms.

C. ELECTRON TRANSFER CHAIN

The processes of glycolysis, β-oxidation, and the TCA cycle generate reducing agents in the form of NADH and reduced flavins. These are oxidized by oxygen, not directly, but via a series of oxidations and reductions (oxidoreduction reactions) that terminate in a reaction with oxygen. The intermediates in this series constitute the electron transfer chain, so named because each oxidoreduction reaction involves the transfer of electrons from a reducing agent (which becomes oxidized) to an oxidizing agent (which becomes reduced). Some, but not all, of these oxidoreduction steps also involve the transfer of hydrogen atoms (often represented [H]) from one intermediate to the next. For this purpose, a hydrogen atom can be considered equivalent to a proton plus an electron:

$$[H] = H^+ + e^-$$

so that an electron can be seen to be involved here too. The electron transfer chain is also known as the electron transport pathway or respiratory chain and forms part of the inner mitochondrial membrane. Indeed, the components of the respiratory chain together with the enzyme that generates ATP from ADP and phosphate represent 30–40% of the total protein of the inner mitochondrial membrane. Thus the electron transfer chain consists of a series of enzymes bearing prosthetic groups which become alternately reduced and oxidized as electrons (and in some cases, hydrogen atoms) are passed from one to the next within the inner membrane. In some ways, therefore, the electron transfer chain is a linear metabolic pathway but the very high degree of spatial organization and the lack of free intermediates makes it more like a large multienzyme complex (see Section B.1.d). Most of the electron carriers (i.e. prosthetic groups) are also found in oxidoreduction enzymes which are not part of the electron transfer chain. Before considering the way in which these carriers are organized in the chain, their chemical nature will be described.

1. The electron carriers

Five main types of electron carrier take part in mitochondrial electron transfer: haems, flavins, ubiquinones, non-haem iron compounds, and nicotinamide nucleotides. These will be discussed separately below.

(a) *Haems*

Haems consist of a planar porphyrin ring system composed of four pyrrole rings surrounding a central iron atom (Figure 4.9). In the electron transfer chain, haems are the prosthetic groups of a series of proteins called cytochromes; outside the chain, haems form part of the oxygen-carrying proteins haemoglobin and myoglobin, and the enzymes catalase and peroxidase. The iron atom in haem can form co-ordinate bonds with six other atoms. (Groups containing atoms which bond with a transition metal in this way are known as ligands.) Four of these are the nitrogen atoms of the pyrrole rings in the haem and these four bonds

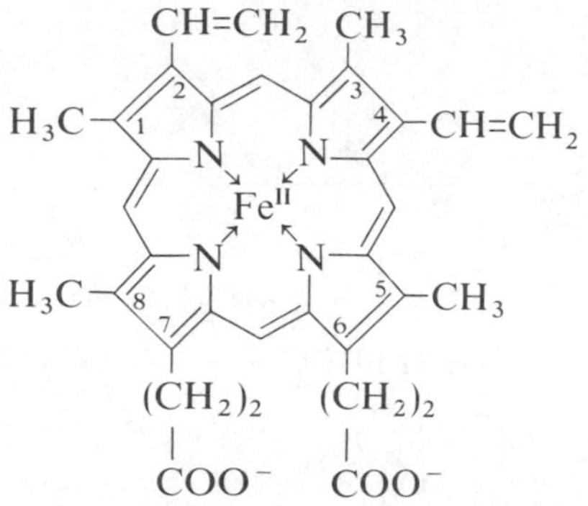

Figure 4.9 Structural formula of a typical haem. Haem B (ferroprotoporphyrin XI), the haem of cytochrome *b*, is shown

hold the iron atom in the plane of the porphyrin ring. The remaining two co-ordinate bonds form with atoms of amino acid groups (e.g. histidine) in the cytochrome. In those haemoproteins which bind oxygen, namely haemoglobin, myoglobin and cytochrome *c* oxidase (the last enzyme in the electron transfer chain), only one of these two ligands is provided by the protein so that the sixth co-ordination position can be occupied by an oxygen molecule. At least four different cytochromes, *b*, c_1, *c*, and aa_3, occur in the electron transfer chain, differing both in protein structure and in the nature of the substituent groups on the haems. In the case of cytochrome *c*, there is an additional covalent link between the haem and the protein.

In reduced cytochromes the iron atom is in the Fe^{II} (i.e. ferrous or Fe^{2+}) oxidation state. When the cytochrome is oxidized, a single electron is lost and the iron atom assumes the Fe^{III} (i.e. ferric or Fe^{3+}) oxidation state. This reversible change occurs during electron transfer along the respiratory chain so that, for example, the reduction of cytochrome c_1 by cytochrome *b* can be represented:

$$\text{cyt } c_1.Fe^{3+} + \text{cyt } b.Fe^{2+} \rightarrow \text{cyt } c_1.Fe^{2+} + \text{cyt } b.Fe^{3+}$$

(b) *Flavins*

Flavins are the prosthetic group of enzymes known as the flavoproteins, several of which, for example, dihydrolipoamide reductase (Section B.1.d) and succinate dehydrogenase (Section B.1.f) have been referred to. Two types of flavin are found in flavoproteins: flavin mononucleotide (FMN) and flavin adenine dinucleotide[4] (FAD). FMN is the simpler molecule (Figure 4.10(a)) and is the 5'-phosphate of the B vitamin, riboflavin (see Appendix A). FAD can be considered as a molecule of FMN linked to AMP by the elimination of a molecule of water between their respective phosphate groups. The flavin nucleotides are prosthetic groups and remain bound (covalently or non-covalently, depending on the enzyme) to the protein during oxidation and reduction. Unlike NADH, for example, flavins do not transport hydrogen atoms from one part of the cell to another.

(a)

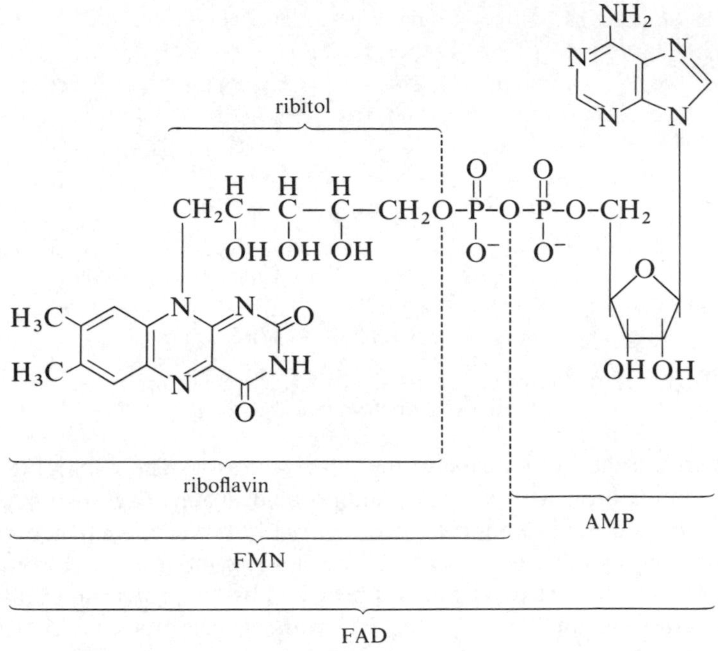

(b)

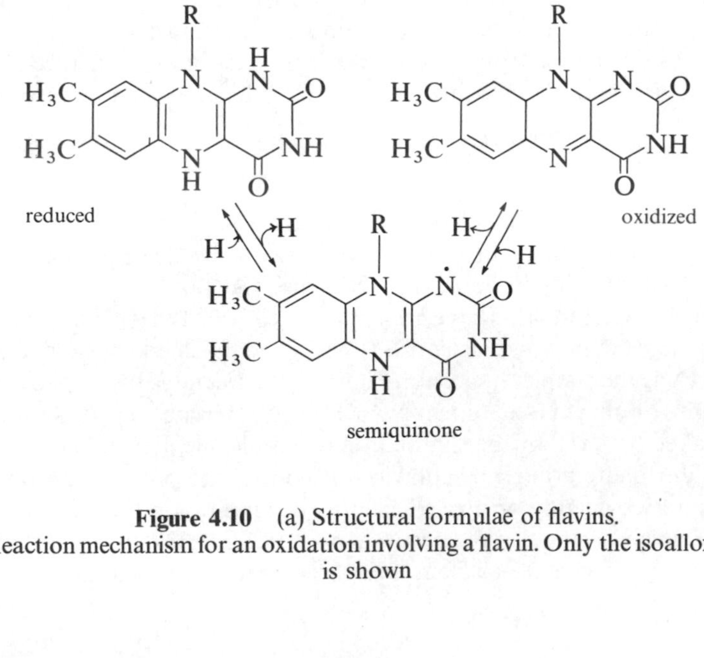

Figure 4.10 (a) Structural formulae of flavins.
(b) Reaction mechanism for an oxidation involving a flavin. Only the isoalloxazine ring is shown

Flavins can act as oxidizing agents because of the ability of the isoalloxazine ring system (Figure 4.10(b)) to accept a pair of hydrogen atoms. This reaction proceeds by two successive one-electron transfers with the intermediate formation of a semiquinone free radical (see Note 7) as shown in Figure 4.10(b).

Because flavoproteins exhibit a wide range of standard redox potentials[5] (depending on the environment of the flavin created by the protein) they play a wide variety of roles in metabolism. Some function entirely outside the electron transfer chain as intermediate electron carriers in oxidoreduction reactions. The enzyme dihydrolipoamide reductase serves such a function in the oxidative decarboxylation of pyruvate and 2-oxoglutarate in the pyruvate and oxoglutarate dehydrogenase complexes respectively. In these cases, NADH is the final reduced product. In the case of flavoprotein oxidases, the electrons are transferred to oxygen with the formation of hydrogen peroxide (see xanthine oxidase—Section E.5.a). Other flavoprotein enzymes react with their reduced substrate and pass the electrons directly into the electron transfer chain with which they are physically associated. Examples include succinate dehydrogenase (Section B.1.f) and acyl-CoA dehydrogenase (in β-oxidation of fatty acids, Chapter 6; Section D.1.d), both of which transfer electrons to ubiquinones (below). Finally, flavoproteins within the electron transfer chain itself accept electrons from the NADH generated in reactions of glycolysis and the TCA cycle and pass them on down the chain. Such flavoproteins are among the few containing FMN rather than FAD.

(c) *Ubiquinones*

Ubiquinones (also known as coenzyme Q) consist of a substituted 1,4-benzoquinone structure with a lengthy polyprenoid side-chain (see Chapter 20) at position 2. The ubiquinone occurring in mammalian mitochondria bears a side chain of ten isoprene units (ubiquinone-10). This constitutes a hydrophobic group of considerable size and probably fits the molecule for association with lipids and hydrophobic membrane proteins. Ubiquinones do not appear to be strongly bound to specific proteins.

The reduction of a ubiquinone involves the addition of two hydrogen atoms (and hence two electrons) to form the 1,4-dihydroxybenzene derivative:

ubiquinone-10

Ubiquinones are the only electron carriers described in this section which function solely in electron transfer chains where their role is to pass electrons from flavoproteins to cytochromes.

(d) *Iron-sulphur proteins*

In a number of iron-containing proteins which take part in oxidoreduction reactions the iron atom is not part of a haem but is linked to sulphur atoms in the protein. Such proteins have been called iron-sulphur proteins or non-haem iron proteins. Some of the sulphur atoms involved are part of the amino acid cysteine in the protein, while others are not. The latter may be released as hydrogen sulphide when the structure is destroyed by acid and are consequently known as 'labile sulphide'. A number of arrangements of iron and sulphur atoms are found in iron-sulphur proteins and a typical one is shown in Figure 4.11. Iron-sulphur proteins take part in oxidoreduction reactions by virtue of their ability to undergo single electron transitions from the Fe^{II} state to the Fe^{III} state and *vice versa*.

Proteins with iron-sulphur centres are found in the inner mitochondrial membrane and are involved in electron transfer but, like haems and flavins, they are also found in enzymes outside the chain. For example, mammalian xanthine oxidase (see Section E.5.a) contains iron-sulphur centres as well as flavins.

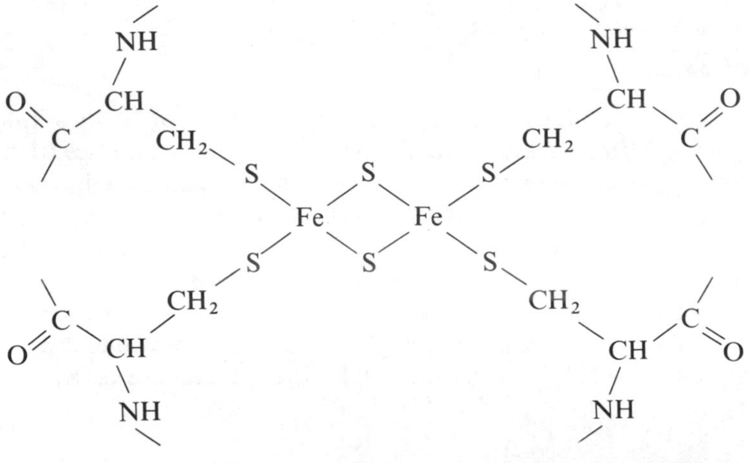

Figure 4.11   Arrangement of iron and sulphur atoms in a typical iron-sulphur protein. This iron-sulphur centre involves four cysteinyl residues

(e) *Nicotinamide adenine dinucleotides*

Nicotinamide adenine dinucleotides are involved in a very large number of oxidoreduction reactions both in the cytosol and in mitochondria. In general,

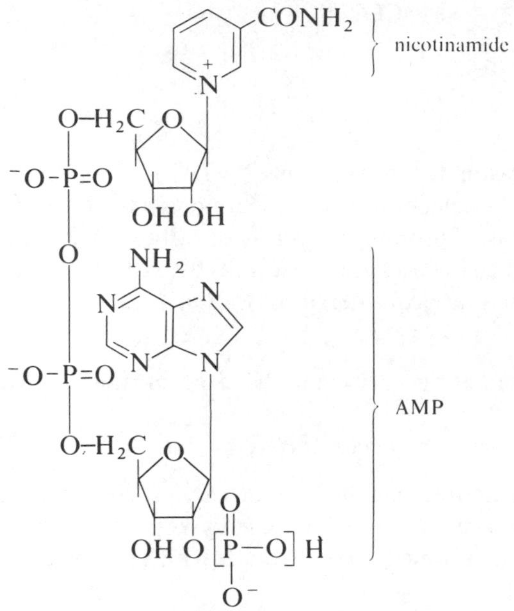

Figure 4.12 Structural formula of oxidized nicotinamide adenine dinucleotides. In $NADP^+$ the phosphate group in the 2′-position of the lower ribose moiety is present

they are not tightly bound to enzymes and are best considered as true substrates although they are often referred to as coenzymes (see Chapter 3; Section A.3). Nicotinamide adenine dinucleotide (NAD^+) itself (Figure 4.12) consists of a nicotinamide nucleotide linked to AMP by an anhydride link between the phosphate groups of the two nucleotides. A second nicotinamide-containing nucleotide, nicotinamide adenine dinucleotide phosphate ($NADP^+$) possesses an additional phosphate group in the 2′-position of the adenine nucleotide. The nicotinamide moiety is derived from nicotinic acid, a vitamin of the B group (Appendix A). Although NAD^+ and $NADP^+$ are very similar and undergo reversible reduction to NADH and NADPH (respectively) in the same way, they fulfil very different roles in the cell. The major role of NADH is to transfer electrons from metabolic intermediates into the electron transfer chain, whilst NADPH acts as a reducing agent in a large number of biosynthetic processes (see Chapter 17; Section A.8 for a fuller discussion of the metabolic significance of this difference).

Although NAD^+ and $NADP^+$ are involved in two-electron transfers, only one hydrogen atom is accepted by the dinucleotide. This changes the character of the nicotinamide ring from aromatic to quinonoid with resultant changes in the absorption spectra (see Figure 3.8). The second hydrogen atom appears as a proton, having effectively donated its electron to neutralize the positive charge on the nitrogen atom in the nicotinamide ring:

$$\text{[structure]} \quad +2[H] \longrightarrow \quad \text{[structure]} \quad +H^+$$

For this reason, the designations $NAD(P)^+$ and $NAD(P)H$ for the oxidized and reduced coenzymes respectively are preferred to $NAD(P)$ and $NAD(P)H_2$. The significance of proton production or utilization in oxidoreduction reactions will be seen when theories about the links between electron transfer and oxidative phosphorylation are considered in Section D.2.

2. Organization of the carriers in the electron transfer chain

(a) *Evidence for the sequence of electron carriers*

In the above description of the electron carriers, little indication has been given of the way in which they are functionally organized within the mitochondrial membrane, that is, which carrier reduces which other carrier. The current view of

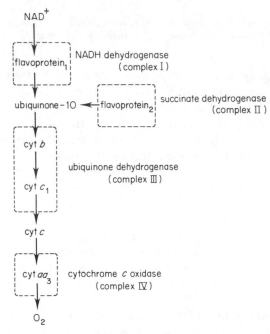

Figure 4.13   Sequence of carriers in the electron transfer pathway. All the carriers are shown in their oxidized form. The regions enclosed by broken lines indicate the association of carriers in the complex. Their constituents are listed in Table 4.3. Carriers outside these complexes also bind but less strongly. Note that electrons may enter at the level of ubiquinone-10 from sources other than succinate

the linear sequence of carriers is given in Figure 4.13. It is not possible in the space available in this text to present in any detail the evidence that has led to the currently accepted sequence of electron carriers but some of the general methods that have been used are indicated briefly below.

(i) *Mitochondria* These have been disrupted (using ultrasonic vibration, detergents and organic solvents) and subjected to separation techniques to yield four complexes each containing only part of the electron transfer chain; namely, NADH dehydrogenase or complex I; succinate dehydrogenase or complex II; ubiquinone dehydrogenase or complex III; and cytochrome *c* oxidase or complex IV (see Hatefi *et al.*, 1962; Table 4.3). Cytochrome *c* and ubiquinone remain free and provide the links between the complexes. From a knowledge of which complex transfers electrons to which other complex (gained from approaches described in Sections C.2.a.ii and C.2.a.iii below) it has proved possible to piece together a functional link between the various components. Furthermore, recombination of all the complexes produces a functional electron transfer chain in which the hydrogen atoms from NADH can reduce oxygen.

(ii) *Inhibitors of electron transfer* A number of inhibitors of electron transfer, including rotenone, amytal, antimycin A, and cyanide (see Figure 4.20) act at specific points in the chain (Figure 4.14). The addition of one of these compounds to respiring isolated mitochondria causes a 'cross-over', that is, an increase in the state of reduction of the carrier prior to the site of inhibition and a decrease in the state of reduction of the carrier subsequent to the inhibition. Since changes in the

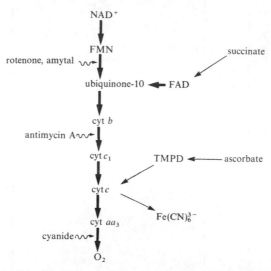

Figure 4.14 Sites of action of inhibitors of electron transfer and of artificial electron donors and acceptors. See Figure 4.20 for molecular structures of inhibitors. Abbreviation: TMPD (tetramethyl-*p*-phenylenediamine)

Table 4.3. Composition of complexes in the electron transfer chain

Number	Complex	Molecular wt. (approximate)	Number of subunits	Major constituents	'Substrates'	'Products'
I	NADH dehydrogenase	0.5×10^6	16	FMN, iron-sulphur centre	NADH, oxidized ubiquinone	NAD$^+$ reduced ubiquinone
II	Succinate dehydrogenase	—	6	FAD, iron-sulphur centre	succinate, oxidized ubiquinone	fumarate, reduced ubiquinone
III	Ubiquinone dehydrogenase	—	6–8	cytochrome b, cytochrome c_1, iron-sulphur centre	reduced ubiquinone, cytochrome c-Fe^{3+}	oxidized ubiquinone, cytochrome c-Fe^{2+}
IV	Cytochrome c oxidase	0.2×10^6	6	cytochrome aa_3, Cu^{2+}	cytochrome c-Fe^{2+}, oxygen	cytochrome c-Fe^{3+}, water

oxidation state of electron carriers are associated with changes in their visible or ultraviolet spectra, which can be followed in intact mitochondria, the relative order of the two carriers involved in the 'cross-over' can be deduced.

(iii) *Artificial electron acceptors and donors* These have been used to functionally isolate sections of the electron transfer chain so that they can be more easily studied. For example, ascorbate, in the presence of tetramethyl-*p*-phenylenediamine (TMPD), will provide electrons to specifically reduce cytochrome *c* in intact mitochondria thus bypassing the earlier part of the system (Figure 4.14). Conversely, $Fe(CN)_6^{3-}$ ions will readily accept electrons from cytochrome *c* in intact mitochondria so bypassing the cytochrome oxidase complex (see Figure 4.14). Note that, although succinate is a natural electron donor for the mitochondrial electron transfer chain, it provides electrons which feed in at the ubiquinone level, thereby bypassing the NADH dehydrogenase complex.

(iv) *Value of standard redox potential* Considerable emphasis has been placed on the value of the standard redox potentials (E_h^0) of the electron carriers as indicators of their position in the electron transfer chain. (These values are analogous to standard free energy changes so that they indicate the direction of electron flow under standard conditions—see Note 5.) It has been assumed that the sequence of reduction of the electron carriers in the mitochondrion will correspond to the order of increasing value of E_h^0. However, two difficulties arise. First, in an analogous manner to the difference between ΔG and ΔG^0 (see Chapter 2; Section A.2.c) the direction of electron transfer depends on the change in electrode potential (ΔE), given by the equation:

$$\Delta E = \Delta E^0 + \frac{RT}{nF} \ln \left[\frac{\text{oxidized components}}{\text{reduced components}} \right]$$

(where R is the gas constant, T is the absolute temperature, n is the number of electrons transferred and F is Faraday's constant). It is thus apparent that ΔE is determined not only by the value of ΔE^0 but by the concentrations of the oxidized and reduced components in the reaction. Some indication of these concentrations can be obtained from the spectroscopic studies on intact mitochondria outlined above. Secondly, the values of ΔE^0 are measured with the isolated carriers in an aqueous environment and it is possible that they are very different in the non-aqueous environment of the intramitochondrial membrane. Nonetheless, despite these limitations, the position of the electron carriers of the chain as judged by the values of standard redox potentials for the carriers (Table 4.4) is very similar to that obtained from the other lines of experimental evidence outlined above.

(b) *Structure of inner mitochondrial membrane*

Comparatively little is known of the structural, as opposed to the functional,

Table 4.4. Standard redox potentials of some of the electron transfer components*

Electron transfer component	Approximate standard redox potential (volts)
NAD$^+$	−0.30
ubiquinone	0.00
cytochrome b	+0.05
cytochrome c	+0.30
cytochrome a	+0.35
oxygen	+0.80

* Data from Lehninger (1973).

organization of the respiratory chain in the inner mitochondrial membrane. The chemiosmotic theory of oxidative phosphorylation (see Section D.2.c) has provided a strong impetus for the elucidation of mitochondrial membrane structure. However, like so many problems challenging molecular biologists at present, it falls beyond the useful resolution of the electron microscope but lacks the regularity necessary for the successful application of X-ray crystallography. The ingenious application of a variety of spectroscopic techniques to this problem has provided some useful information despite the bluntness of the tools. It is possible that the organization of soluble multienzyme complexes (e.g. oxoglutarate dehydrogenase, Section B.1.d) is analogous to that in the electron transfer chain. However, the membrane location of the respiratory chain means that its components are in close association with lipids and the extent to which these are an important structural element has yet to be established (for discussion, see Racker, 1976).

D. OXIDATIVE PHOSPHORYLATION

The basic chemistry of oxidative phosphorylation occurring in the multienzyme complex known as ATP synthase can be represented by the equation:

$$ADP^{3-} + P_i^{2-} + H^+ \rightarrow ATP^{4-} + H_2O$$

What has proved difficult to establish is the mechanism by which this reaction is coupled to the reactions of the electron transfer chain. The ΔG^0 for these latter reactions must be sufficiently large and negative so that when coupled to the above ATP synthase reaction (for which ΔG^0 is large and positive) ΔG for the over-all reaction is negative at cellular nucleotide concentrations.

1. ATP formation in isolated mitochondria

Before presenting a brief description of what is known of the mechanism of oxidative phosphorylation it is useful to describe some of the studies carried out on isolated mitochondria which have been prepared from intact tissues by gentle homogenization and differential centrifugation (see Chance and Williams, 1956, for details). It is indeed fortunate that carefully isolated mitochondria (most studies have been carried out with liver mitochondria) appear to function as they would in their natural environment. With such a preparation it is possible to control the concentration of substrates for electron transfer and oxidative phosphorylation and to measure the yield of ATP without interference from competing reactions. The results of such experiments are presented here for two reasons: first, they provide some of the evidence used to establish the mechanisms discussed below and, secondly, they have a greater relevance to clinical practice than a knowledge of the exact mechanisms of the process.

(a) *Coupling of oxidative phosphorylation to electron transfer*

With a source of intact mitochondria a number of relatively simple (they can be done by any medical student) but highly informative experiments become possible. First, however, means of measuring the rates of oxidative phosphorylation and electron transfer are needed. Oxidative phosphorylation can be measured by determining the rate of ATP production using a relatively straightforward enzymatic assay. Electron transfer is most easily measured by following the rate of disappearance of oxygen from the medium using an oxygen electrode (which can be made to plot a continuous graph of oxygen concentration against time as shown in the upper trace in Figure 4.21). One more problem needs to be overcome—that of getting oxidizable substrate into the mitochondrion because NADH dehydrogenase can only use intramitochondrial NADH and the inner mitochondrial membrane is impermeable to this substance (see Chapter 5; Section D). However, pyruvate is readily taken up and will generate (through the action of pyruvate dehydrogenase) NADH within the mitochondrion. Fortunately, the process that transfers ADP and ATP in and out of the mitochondrion (the adenine nucleotide translocator—Section D.4) remains functional in isolated mitochondria.

Thus equipped, it should be a simple matter to demonstrate that the oxidation of pyruvate is accompanied by ATP production. If electron transfer is prevented (by, for example, lack of an oxidizable substrate or oxygen) then no ATP is produced. This is as expected from our knowledge of the thermodynamics of these processes but what is more surprising, and significant, is that in the absence of oxidative phosphorylation (occasioned by lack of ADP or P_i) no electron transfer occurs. This is what is meant by the two processes being 'tightly coupled'. The situation is analogous to that described in Chapter 2, Section A.2.d, under the heading of coupling-in-parallel. Electron transfer and oxidative phosphorylation can be considered as a single reaction (albeit an exceedingly complex one) that will not proceed unless all substrates are present:

$$NADH + \tfrac{1}{2}O_2 + 3ADP^{3-} + 3P_i^{2-} + 4H^+ \longrightarrow NAD^+ + 3ATP^{4-} + H_2O$$

However, using the mitochondrial preparation described above it is possible to uncouple the two processes so that electron transfer occurs (and oxygen is consumed) in the absence of ADP. In fact, any agent causing damage to the mitochondria tends to uncouple the processes, suggesting that coupling depends on the structural integrity of the mitochondrion. In addition, a wide variety of compounds serve to uncouple *in vitro* and some are also effective *in vivo*. Uncoupling agents are described in more detail in Section D.3.c. (The ratio of oxygen uptake in the presence of ADP to that in its absence is called the respiratory control (receptor control) ratio. It is used as an index of the functional integrity of prepared mitochondria since it falls from above ten to unity in aged or damaged mitochondria.)

In the intact tissue, coupling provides the means by which the oxygen uptake of a tissue can be regulated by its metabolic activity. When the demand for energy is high, the ATP/ADP concentration is decreased and the raised concentration of ADP stimulates the rate of phosphorylation and, through coupling, the rates of electron transfer and oxygen uptake (see Chapter 7; Section B.4 for more details).

(b) *Quantitative relationship between oxidative phosphorylation and electron transfer*

The next question that can be tackled (equally simply by the student) is how much ATP is synthesized for a given amount of electron transfer. Traditionally, the results of such experiments have been presented as a P/O ratio, that is, the number of molecules of ATP synthesized per atom of oxygen consumed (equivalent to the transfer of two electrons down the chain). With the best preparations and using an NADH-generating substrate (e.g. pyruvate) P/O ratios approaching 3.0 are obtained. Substrates which feed electrons into the main chain at ubiquinone (e.g. succinate) give P/O ratios of approximately 2.0. By using artificial electron donors (e.g. ascorbate in the presence of tetramethyl-*p*-phenylenediamine) which introduce their electrons further down the chain, ratios of about 1.0 are obtained. Data of this kind are extremely useful for the calculation of the potential ATP yield from pathways such as the TCA cycle (twelve from each molecule of acetyl-CoA), glucose oxidation (38 for each glucose molecule) and fatty acid oxidation (129 for each molecule of palmitate) (see Table 5.3 for details). The fact that measured P/O ratios were rarely integral values was assumed to be due to experimental error. However, the current view of the mechanism of oxidative phosphorylation (Section D.2) is such that it is unlikely that P/O ratios would be integral.

(c) *Positions of ATP formation along the electron transfer chain*

Quite naturally the next challenge is to locate the sites at which each of the three ATP molecules is synthesized along the chain. A great deal of effort was put into investigations of this kind because they were thought likely to lead the way to the elucidation of the mechanism of coupling. As events have turned out, this expectation was misguided because it was based on a false assumption concerning the kind of mechanism involved. Results were obtained, however: site I for ATP generation lies between NADH and ubiquinone; site II between ubiquinone and cytochrome c and site III between cytochrome c and oxygen (see Figure 4.13). The student need not spend time repeating these experiments (indeed some are far from simple) but the methods involved are of some interest. Some indication of the location of the 'sites' has already emerged from studies of P/O ratios described above. By isolating different sections of the electron transfer pathway using different electron donors and acceptors a more detailed picture of the sites can be built up. A second approach takes advantage of the phenomenon of coupling and the reasoning behind the 'cross-over' experiments described in Section C.2. In a tightly coupled system, omission of ADP will have the same effect on electron transfer as the presence of an inhibitor of electron transfer acting at that site. Spectral changes were therefore used to identify 'cross-overs' and hence sites of coupling in stretches of the electron transfer chain (isolated by use of artificial electron donors and acceptors) when electron transfer and oxidative phosphorylation were stimulated by the addition of ADP (Chance and Williams, 1956).

(d) *Near-equilibrium nature of electron transfer and oxidative phosphorylation*

In Section D.1.a above, the metabolic advantage of coupling was explained. In fact, the rate of oxygen consumption is regulated not by ADP concentration alone but by changes in the ATP/ADP concentration ratio. Control by the product as well as the substrate can be explained if oxidative phosphorylation and at least part of electron transfer are close to equilibrium. Indeed, there is now considerable evidence that the reactions of the electron transfer chain are near-equilibrium up to, but not including, the terminal reaction with oxygen. Three lines of evidence support this contention. First, for ATP synthesis associated with electron transfer from NADH to cytochrome c, the ΔG value calculated from mass-action ratio data in isolated mitochondria and the equilibrium constant is close to zero indicating a near-equilibrium reaction (Erecińska and Wilson, 1978). Secondly, the maximal activities of the isolated components of the respiratory chain are considerably greater than the maximum flux of electrons in either isolated mitochondria or the intact tissue. Thirdly, it is possible to demonstrate reversed electron flow through the chain (Klingenberg, 1964). The addition of succinate to isolated liver mitochondria causes the reduction of acetoacetate to 3-hydroxybutyrate. Since this reaction is catalysed by 3-hydroxybutyrate dehydrogenase, which uses only NADH as the reducing agent, it implies that the concentration of NADH has risen. But succinate, through the activity of succinate dehydrogenase feeds electrons into the chain at the level of ubiquinone. The only satisfactory explanation for the reduction of acetoacetate

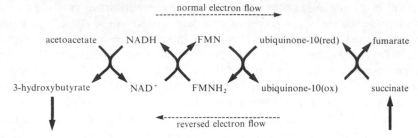

Figure 4.15   Demonstration of reversed electron transfer. The reversed electron flow requires hydrolysis of ATP

by succinate is that reduced ubiquinone reduces FMN, in the NADH dehydrogenase complex, and this reduces NAD^+ to NADH; that is, reversed electron transfer (Figure 4.15). Furthermore, the reduction of acetoacetate, but not the oxidation of succinate in isolated mitochondria, is inhibited by rotenone which inhibits the NADH dehydrogenase. The importance of the near-equilibrium status of the electron transfer and its functional relationship with the TCA cycle will be discussed in Section E.1.

2. Mechanism of ATP synthesis

At the beginning of Section D, the basic equation of oxidative phosphorylation was presented thus:

$$ADP^{3-} + P_i^{2-} + H^+ \rightarrow ATP^{4-} + H_2O$$

The $\Delta G^{0'}$ for this reaction is $+31.5$ kJ.mole^{-1} so, in order to achieve the negative ΔG necessary for it to occur, the concentrations of ADP, P_i, or H^+ must increase greatly or the concentration of water or ATP must fall greatly. There are two general ways in which this could be achieved, coupling-in-series or coupling-in-parallel (explained in Chapter 2; Section A.2.d). The previous section has left no room for doubt that electron transfer is coupled to the ATP synthase reaction. In order for the transfer of electrons from NADH to oxygen to be coupled to the synthesis of ATP, the two processes must share a common intermediate. Elucidation of the nature of this intermediate has presented one of the major challenges to biochemists. Not surprisingly, by analogy with other reactions (e.g. glyceraldehyde-phosphate dehydrogenase in glycolysis) it was assumed for many years that the intermediate would be a chemical compound participating in both electron transfer and ATP synthesis. This is the covalent-intermediate hypothesis, described only briefly below as it no longer seems a very promising candidate. The search has widened in recent years with the realization that the 'intermediate' might not be a compound formed by covalent bonding but could be some other means of storing chemical energy, such as a concentration gradient, a membrane potential or a conformational change in a macromolecule.

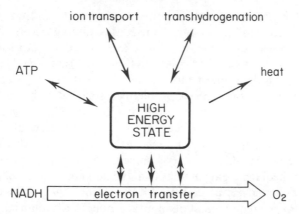

Figure 4.16 The relationship between the energy-linked functions of the mitochondrion. The high energy-state produced by electron transfer can be used to synthesize ATP, to transport ions into the mitochondrion, to transfer hydrogen from NADH to $NADP^+$ to form NADPH, or it can be dissipated as heat

To avoid the need to specify its nature, the 'intermediate' has received the designation of a 'high energy state'. Electron transfer is now seen as generating this 'high energy state' which must be utilised if electron transfer is to continue. This 'high energy state' is used to form ATP in the ATP synthase reaction but it can also be used directly to drive other energy-requiring processes, such as mitochondrial ion transport or heat production, without the involvement of ATP (see Figure 4.16). Only one hypothesis for the nature of this 'high energy state' has acquired a significant amount of evidence, namely the chemiosmotic hypothesis due to P. Mitchell and described in Section D.2 below. Although many important details remain to be established, it would appear that the fundamental nature of the link has been established. However, for completeness, a brief account of the covalent intermediate and conformational hypotheses are included below. (For a review of the hypotheses, see Dawson and Selwyn, 1974.)

(a) *Covalent intermediate hypothesis*

In 1953, E. C. Slater proposed that discrete chemical intermediates link electron transfer and oxidative phosphorylation. The simplest scheme that could be supported by the available evidence involved two steps linking the processes and can be represented:[6]

$$AH_2 + B + I \rightarrow A \sim I + BH_2$$
$$A \sim I + X \quad\rightarrow X \sim I + A$$
$$X \sim I + P_i \quad\rightarrow X \sim P$$
$$X \sim P + ADP \rightarrow ATP + X$$

Here, A and B are components of the electron transfer chain that undergo alternate oxidations and reductions; I and X are coupling factors and $\sim$ indicates the 'high energy' status of the hypothetical intermediates. In such schemes, a number of the electron transfer carriers (in this case A) are directly involved in the formation of a covalent intermediate.

There were, however, a number of problems associated with this hypothesis. In particular, despite an enormous amount of effort there has been no satisfactory identification of intermediates such as $A \sim I$, or indeed any of the other intermediates. (Note that it is identification of $A \sim I$ etc., rather than $X \sim I$ or $X \sim P$ that is crucial to this hypothesis, since the other hypotheses do not exclude covalent intermediates being involved at some stage in the phosphorylation process.) The second factor casting doubt on the reality of such a coupling mechanism was the requirement of an intact inner mitochondrial membrane for oxidative phosphorylation. Fragments of the inner mitochondrial membrane will carry out oxidative phosphorylation but only if the edges of the fragment fuse together to form a sealed vesicle. In contrast, phosphorylation of ADP by the 'high energy' intermediates of glycolysis can take place with purified enzymes and does not require the presence of membranes.

(b) *Conformational change hypothesis*

The possibility that the coupling might not involve covalent intermediates was recognized by P. D. Boyer in 1964. He proposed that the changes occurring during the transfer of chemical energy from the electron transfer chain to the ATP synthesising reactions were conformational. That is, the altered three-dimensional structure of a coupling protein, determined by non-covalent forces, could transmit chemical energy from one process to another. A mechanical analogy would be that of a spring compressed by one process being subsequently used to provide the mechanical energy for a second process. At its simplest, a conformational link of this kind could be represented by the following scheme in which Pr* represents a coupling intermediate in an altered conformational state:

$$AH_2 + B + Pr \rightarrow A + BH_2 + Pr*$$
$$Pr* + P_i \rightarrow Pr \sim P$$
$$Pr \sim P + ADP \rightarrow Pr + ATP$$

This hypothesis, although virtually impossible to prove experimentally, did at least have the virtue of widening possibilities beyond the confines of covalent intermediates.

(c) *Chemiosmotic hypothesis*

In 1961, P. Mitchell proposed the revolutionary hypothesis that electron transfer produces a concentration gradient of protons (i.e. a pH difference) across the inner mitochondrial membrane. This gradient constitutes the 'high energy state' and can be used to 'drive' ATP synthesis. After a lengthy period of scepticism by others, Dr. Mitchell was awarded the Nobel prize in chemistry in 1978 for his work in elucidating the mechanism of ATP synthesis in mitochondria. This indicates the current widespread (although not universal) acceptance of this hypothesis which should perhaps, therefore, be elevated to the status of a theory (see Nicholls, 1982, for a review).

The treatment of this mechanism given here is admittedly brief. It is hoped that it will be sufficient for the reader to understand the principles involved without becoming bogged down with physiochemical intricacies. It is convenient to consider the mechanism in two parts; first, the way in which the proton concentration gradient is established and, secondly, how this can be used to achieve ATP synthesis.

(i) *Establishment of the proton gradient* Mitchell pointed out that many of the electron transfer steps in the respiratory chain involve protons as substrates or products. For example:

$$H^+ + NADH + FMN \rightarrow NAD^+ + FMNH_2$$
$$ubiQH_2 + 2cyt\, b.Fe^{3+} \rightarrow ubiQ + 2cyt\, b.Fe^{2+} + 2H^+$$
$$2H^+ + 2cyt\, a_3.Fe^{2+} + \tfrac{1}{2}O_2 \rightarrow 2cyt\, a_3.Fe^{3+} + H_2O$$

If these reactions are arranged in the inner mitochondrial membrane in such a way that in each case the protons are taken from inside and released on the outside of the membrane, electron flow through the chain will lead to an efflux of protons from the mitochondrial matrix and the establishment of a pH difference across the membrane. This vectorial movement of protons (sometimes known as a 'proton pump') is at the centre of Mitchell's theory and would explain the requirement for intact vesicles of inner mitochondrial membrane (to provide the two necessary compartments). An increase in external pH has indeed been detected during electron transfer, and the membrane has been shown to be normally impermeable to protons. Significantly, uncoupling agents (see Section D.3.c) render the membrane permeable to protons. This would dissipate the 'high energy state' and neatly account for the action of uncouplers. Since its inception, however, this fundamental mechanism has had to be modified in detail. One major problem was the magnitude of the pH difference; it can be calculated that to synthesize ATP (see below) a pH difference across the membrane of some 3.5 units would be required. Such large gradients have not been observed. Hence, the first modification is the suggestion that proton movement leads not only to a concentration difference but also to a potential difference across the membrane with the inside negative with respect to the outside. This potential would oppose the further release of free protons on the outside but would provide part of the store of the energy required for ATP synthesis. Both the energy stored in the potential difference plus the concentration gradient of protons would then be involved in the generation of ATP. The term, proton motive force (Δp) has been used to describe the combined effect of the pH gradient (ΔpH) plus membrane potential ($\Delta\psi$). The relationship is as follows:

$$\Delta p = \Delta\psi - 2.303 \left(\frac{RT}{F}\right) \Delta pH$$

(R is the gas constant; T is the absolute temperature and F the Faraday constant). At 30 °C, the factor $2.303\dfrac{RT}{F}$ is equal to approximately 60 mV so that $\Delta p = \Delta\psi - 60\,\Delta$pH. A further modification is that a concentration gradient of other cations, set up by exchange with protons, could also contribute to the 'high energy state'.

(ii) *Coupling of the proton motive force to ATP synthesis*   The first graphic demonstration that a pH-gradient could be used to bring about ATP synthesis came from the so-called 'acid-bath' phosphorylation experiments with plant chloroplasts (Jagendorf and Uribe, 1966). These structures harness light energy for ATP synthesis (photophosphorylation) in a mechanism which also involves the transfer of electrons between carriers. Isolated chloroplasts suspended in a

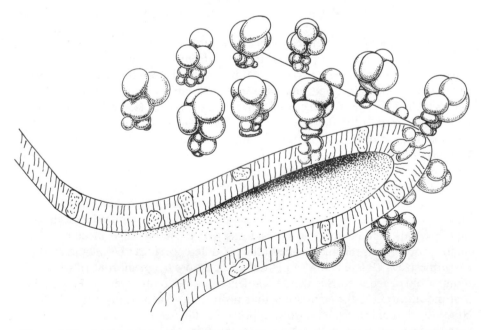

Figure 4.17   Artist's impression of ATPase synthase complex projecting from a fold of the inner mitochondrial membrane. The figure is based on electron micrographs and biochemical studies of the structure of this complex (see Racker, 1976)

buffered medium of pH 4.0 acquire that pH internally. When transferred to a medium of pH 8.5, in the presence of ADP and P_i, there is a burst of ATP synthesis despite the whole operation being carried out in the dark.

The ATP synthase complex is situated in close proximity to the complexes constituting the electron transfer pathway. In all probability, the small spheres (or particles) that can be seen in high-resolution electron micrographs, covering the inner surface of the inner mitochondrial membrane are the ATP synthase complexes (Figure 4.17). In 1966, Kagawa and Racker succeeded in removing these particles from mitochondrial vesicles and showed that the 'stripped' vesicles could not carry out oxidative phosphorylation. However, this ability was restored by adding back the particles. The particles contain a number of proteins, but of primary importance is the F_1-ATPase which is considered to be involved in the formation of ATP, particularly since this activity is inhibited by oligomycin, which is known to be a potent inhibitor of oxidative phosphorylation in isolated mitochondria. (It is called an ATPase because experimentally it is characterized by its ability to hydrolyse ATP. Enzymes cannot modify the equilibrium position of reactions they catalyse and in this case the substrate concentrations in the disrupted particles are such that hydrolysis rather than synthesis occurs.)

The precise mechanism by which the proton motive force is used to bring about ATP synthesis is not yet understood. One theory, proposed by Mitchell, is that water, one of the products of the synthase reaction, is not produced by the ATPase complex as water itself but as hydrogen and hydroxyl ions which are released separately on different sides of the inner mitochondrial membrane. This is feasible even if difficult to prove. If the proton (hydrogen ion) is released on the inner side of the membrane where the pH is highest, it would be effectively removed (i.e. maintained at very low concentration) by combination with hydroxyl ions to form water. Conversely, hydroxyl ions released on the outside side of the membrane would be removed by the protons accumulated there as a result of electron transfer, again forming water. In this way, ATP synthesis would be coupled-in-series (twice in fact) to the reaction:

$$H^+ + OH^- \rightarrow H_2O$$

under conditions when the latter reaction is displaced to the right. This hypothesis is summarized in Figure 4.18.

Unfortunately, knowledge of the spatial organization of ATP synthesis (above) is not easily reconciled with this mass action hypothesis of ATP formation and other mechanisms have been suggested. Each F_1-ATPase complex contains five different proteins and it is connected to the membrane by another protein (F_0) which can also be obtained in a soluble form (Figure 4.19). It is possible that the proton motive force causes a conformational change in the membrane which is transmitted to the active site of the synthase. Such a change could conceivably affect the binding of a reactant in the ATP synthesis reaction so that its local concentration would be altered in such a way as to make the ΔG for ATP

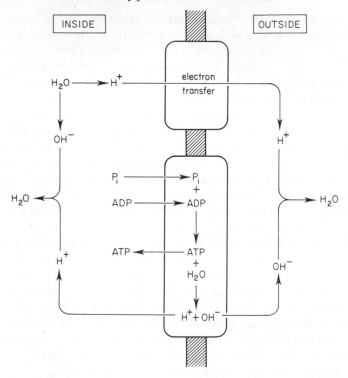

Figure 4.18 Simplified summary of Mitchell's original chemiosmotic mechanism for coupling electron transfer to ATP synthesis. Although the general nature of the relationship is now widely accepted, this particular model is not fully supported by experimental data

synthesis negative. There are several binding sites for ATP and ADP on the F_1-ATPase and changes in their affinity constants could play a role in ATP formation (Harris, 1982). If conformational changes were found to be involved in such a way, Boyer's conformational hypothesis would be seen to be essentially correct, but only half the story.

(iii) *Other functions of the proton motive force*   Mitochondria are able to use the proton motive force generated by electron transfer for purposes other than ATP synthesis, such as the transport of substances against concentration gradients into and out of mitochondria. For example, the accumulation of phosphate or calcium ions by mitochondria against a concentration gradient requires electron transfer but not oxidative phosphorylation.

3. Inhibition of electron transfer and oxidative phosphorylation

A large number of compounds have now been found which interfere with electron transfer or oxidative phosphorylation. Many have played important parts in the

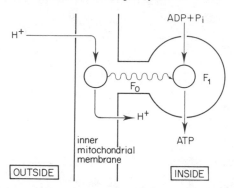

Figure 4.19 An interpretation of the chemiosmotic hypothesis for ATP synthesis which is consistent with current views on the structure of the inner mitochondrial membrane (see Figure 4.18). The nature of the connection between the proton flux and ATP synthesis is unknown but could involve conformational changes or even covalent intermediates

elucidation of the mechanisms described above. This section is designed to summarize their modes of action and gives, as examples, some of those most likely to be encountered. In addition, a number have found a limited use in medical practice and these are discussed in Section E.2.

Four classes of inhibitory effect on electron transfer and oxidative phosphorylation are distinguished below. Structural formulae of representative examples are shown in Figure 4.20 and their effects on electron flow and ATP synthesis summarized in Figure 4.21.

(a) *Inhibition of electron transfer*

Compounds such as amytal, rotenone, antimycin A, 2-heptyl-4-hydroxyquinoline N-oxide (HHQ), and cyanide inhibit electron transfer itself and therefore also prevent phosphorylation. Since they inhibit at specific sites in the electron transfer process, they have been particularly useful in studies to establish the sequence of carriers in the chain.

(b) *Inhibition of oxidative phosphorylation*

Some compounds can inhibit oxidative phosphorylation but they act directly, that is, they have no effect on electron transfer except through the inhibition of phosphorylation. It is possible to distinguish such inhibitors from those that affect electron transfer since, in the presence of an uncoupling agent, they do not affect the rate of electron transfer (Figure 4.21). Oligomycin, an antibiotic, is perhaps the best known of this class; it binds to the protein that links the F_1-ATPase to the membrane and presumably prevents the normal inward movement of protons that result in ATP formation.

(a)

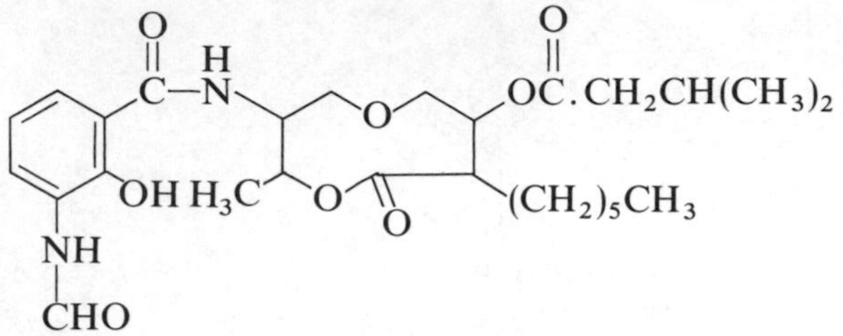

antimycin A

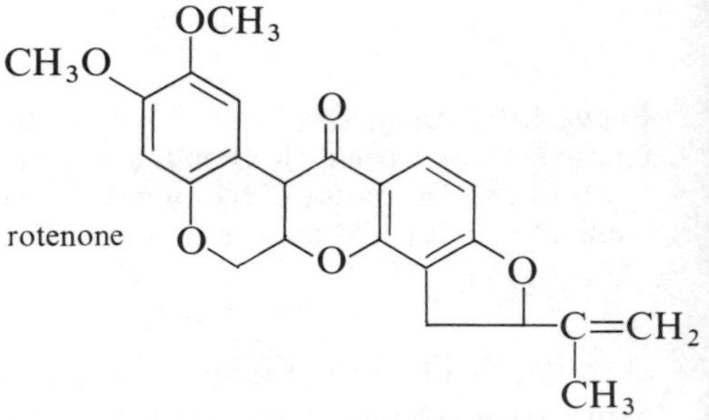

rotenone

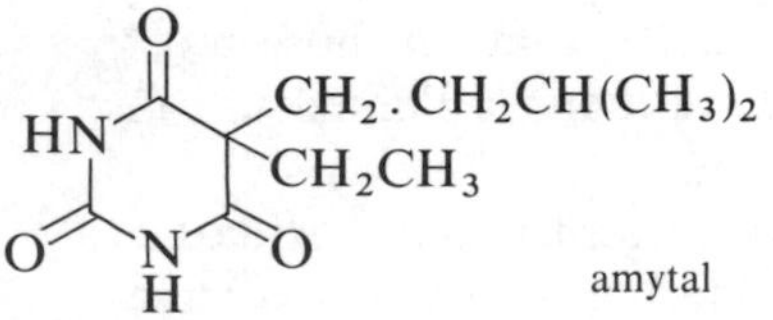

amytal

CN⁻
cyanide

N₃⁻
azide

(b)

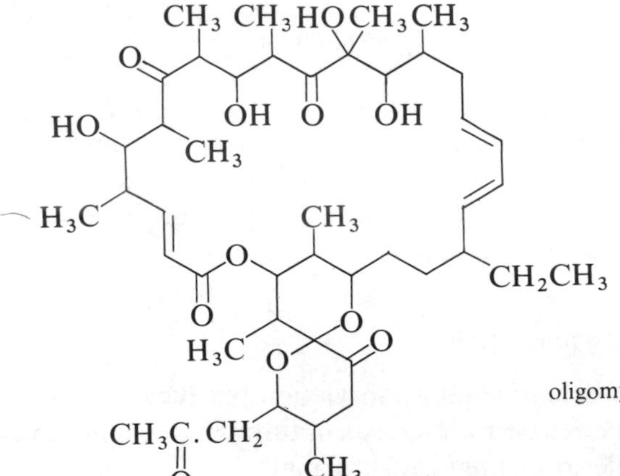

oligomycin B

Figure 4.20 Structural formulae of some compounds which interfere with mitochondrial ATP synthesis.

(a) Inhibitors of electron transfer.
(b) Inhibitor of oxidative phosphorylation.

(c)

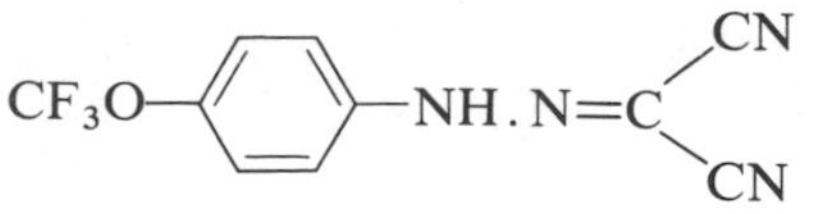

carbonylcyanide *p*-trifluoro-
methoxyphenylhydrazone (FCCP)

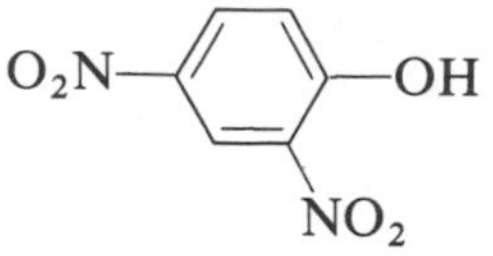

2,4-dinitrophenol⁻

(d)

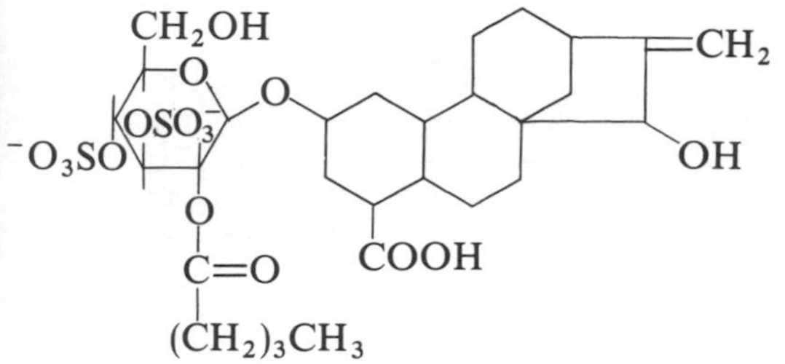

atractyloside

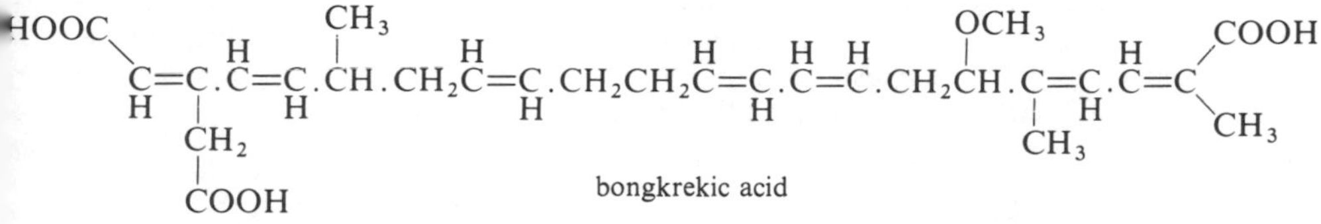

bongkrekic acid

(c) Uncoupling agents.
(d) Inhibitors of adenine nucleotide translocation

(c) *Uncoupling*

Uncoupling agents have the effect of separating the processes of oxidative phosphorylation and electron transfer so that the latter process can occur without concomitant ATP synthesis. In this situation, the energy normally conserved in ATP formation is lost as heat. It is now known that uncoupling agents can act through a variety of mechanisms. Menadione, for example, provides an alternative route of electron transfer, bypassing much of the electron transfer chain, so that protons are not translocated and the 'high energy state' is not produced. Many uncouplers, like 2,4-dinitrophenol (DNP) and carbonylcyanide *p*-trifluoromethoxyphenylhydrazone (FCCP) appear to uncouple by increasing the permeability of the mitochondrial membrane to protons. This allows the proton gradient (i.e. the 'high energy state') to dissipate without generating ATP. The fact that so many compounds that act as uncoupling agents are chemically

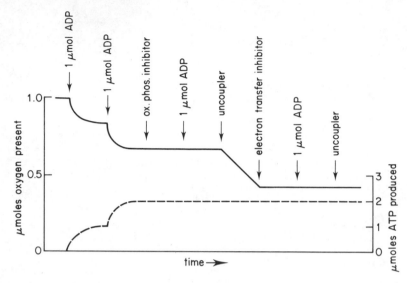

Figure 4.21 The effect of inhibitors and uncouplers on oxygen uptake and ATP production in mitochondria. The upper line represents the changes in the concentration of oxygen present and is a stylized representation of the tracing that would be obtained if a recording oxygen electrode were used for the experiment. The lower line represents the synthesis of ATP. Excess phosphate and sufficient oxidizable substrate (e.g. pyruvate) to reduce all the oxygen are initially present in the medium

unrelated but have in common the ability to increase proton permeability of membranes is evidence in favour of the existence of a proton gradient that can be used to synthesise ATP (i.e. the chemiosmotic theory).

The antibiotic valinomycin is a member of a class of uncoupling agents that require monovalent cations for their actions. These compounds cause the inner mitochondrial membrane to become permeable to specific cations (potassium ions in the case of valinomycin). The 'high energy state' is then used to transport potassium ions across the membrane into the mitochondria rather than for ATP synthesis.

Since in most cells, almost all of the ATP is synthesized in the oxidative phosphorylation process described above, the existence of a *physiological* uncoupling mechanism would be unexpected, to say the least. Indeed, no such mechanism exists in any tissue, with one very important exception; brown adipose tissue. The main, if not the only, function of this tissue is heat generation and this is achieved by uncoupling oxidative phosphorylation from electron transfer so that fuel oxidation occurs without concomitant ATP formation and the energy is released in the form of heat. The mechanism and the physiological importance of this uncoupling are discussed in Section E.4.

(d) *Inhibition of adenine nucleotide transport*

A small group of inhibitors exert their effect on ATP synthesis by inhibiting the translocation of ATP out of, and ADP into, the mitochondria across the inner mitochondrial membrane (see below). Two natural products have proved particularly useful in the experimental study of this process, atractyloside and bongkrekic acid (Figure 4.20). Atractyloside is a glycoside found in the rhizomes of a thistle (*Atractylis gummifera*) occurring in the Southern Mediterranean region. The plant became of interest when animals that had eaten the rhizomes died and, in particular, when a class of school children were poisoned, some fatally. Several derivatives of atractyloside have also been used to inhibit translocation and one, carboxyatractyloside, which crosses the cell membrane, has been particularly useful in investigating the importance of the translocation process *in vivo* (Stubbs, 1979). Bongkrekic acid is a complex branched-chain, cyclic fatty acid produced as an antibiotic by a mould which causes decomposition of coconut meal.

4. Transport of adenine nucleotides across mitochondrial membranes

(a) *General features of mitochondrial transport*

Mitochondria possess an inner and an outer membrane. The outer membrane appears to be more of a protective sheath rather than an effective permeability barrier and the passage of molecules with molecular weight less than 5000 is not restricted by this membrane. However, the inner membrane is very different, only a few small neutral molecules can diffuse rapidly across it; these include oxygen, water, carbon dioxide, ammonia, acetate, and ethanol, which thus equilibrate rapidly with the matrix space. Nonetheless, it is obvious that a number of other compounds must cross this membrane. The adenine nucleotides and phosphate are immediate examples, in addition, various fuels for the mitochondrial pathways, such as pyruvate and fatty acids, must cross the membrane as must a number of other key metabolic intermediates.

For the transport of such compounds, the inner mitochondria membrane possesses specific carriers (which are probably proteins). The existence of such carriers (or translocators) has been demonstrated by similar experimental approaches to those described for glucose transport across the muscle cell membrane (i.e. saturation kinetics, specificity, inhibition by specific compounds, and competition by analogues that are also transported—see Chapter 5; Section B.2). Many of the compounds indicated above are anions and there is no doubt that the transport of an ion poses a problem which does not arise in the case of the transport of an uncharged molecule; the movement of the ion across the membrane takes with it an electric charge and this will set up a potential that will inhibit further transport. In addition, since the membrane is already charged (due to proton pumping) the transport of any ion will be affected by this membrane potential. The problem can be overcome either by the carrier carrying out an equimolar exchange of anions (that is, one anion is transported into, while another is transported out of, the mitochondria—known as an antiport system) or by the anions being transported along with protons so that, effectively, a neutral metabolite crosses the membrane (known as a symport system) (see Figure 4.22). These two transport systems will therefore be electroneutral. Two examples of the former process are the dicarboxylate and the 2-oxoglutarate carriers which catalyse the electroneutral exchange of divalent anions. The substrates for the dicarboxylate carrier include malate, succinate, and phosphate, and for the 2-oxoglutarate carrier include 2-oxoglutarate, malate and succinate. Two examples of the symport system are the phosphate (each ion of which is transported into the mitochondrion with about 1.5 protons) and pyruvate carriers. The inward movements of these important anions are thus favoured by the proton gradient across the mitochondrial membrane. There are, however, two carriers that are electrogenic, that is, they transfer a net electric charge (Figure 4.22); these are the adenine nucleotide and the glutamate/aspartate translocators. It should be appreciated that, at steady-state, the electrogenic carriers are unlikely to produce a net change in charge since any change produced will be balanced by a compensatory movement of another ion. However, the ion

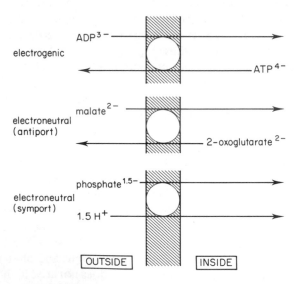

Figure 4.22 Representative examples of transport systems across the inner mitochondrial membrane

movements may not be simultaneous and the operation of the electrogenic carrier is not restricted by the movement of a specific 'balance' ion as in the case of the electroneutral carriers.

(b) *Adenine nucleotide translocator*

Since almost all ATP is synthesized within mitochondria in aerobic tissues and almost all of it is used outside, the adenine nucleotide translocator plays a vital role in metabolism. It serves to transport one molecule of ADP into the mitochondrion in exchange for one molecule of ATP transported out. The translocator is sometimes referred to as the adenine nucleotide translocase, thus acknowledging the similarity between membrane carriers and enzymes. The essential properties of the translocator are outlined below but for further details the reader is referred to reviews by Klingenberg (1976), Stubbs (1979) and Vignais and Lauguin (1979).

(i) *Specificity* The translocator is absolutely specific for adenine nucleotides (as is oxidative phosphorylation). ADP is predominantly transported inwards and ATP outwards.

(ii) *Membrane potential* As discussed in Section D.4.a, the adenine nucleotide translocator is electrogenic and its operation in isolation would tend to make the inside of the mitochondrion positive with respect to the outside. It is suggested that this is balanced by the proton pumping associated with electron transfer which generates a potential that is positive on the *outside* of the membrane; that is to say, that a proportion, possibly as much as 25 %, of the 'high energy state' is used for transport of nucleotides across the mitochondrial membrane and is not therefore used for ATP formation.

(iii) *Equilibrium nature of the translocator* There is evidence that adenine nucleotide transport is near-equilibrium (Stubbs, 1979). The K_m for inward ADP translocation as determined with isolated mitochondria is approximately 5 μM. This is much lower than the estimated concentration of free ADP in the cytosol (20–50 μM) but competition with ATP raises the effective K_m to these values so that *in vivo* the carrier can respond readily to changes in ADP concentration.

(iv) *Control of the ATP/ADP concentration ratio* Measurements of the ATP/ADP concentration ratios in the cytosol and mitochondria of liver cells (involving the very rapid separation of the two fractions) or in isolated, incubated mitochondria indicate that both *in vivo* and *in vitro* the external ratio is much higher than the internal ratio (Seiss and Wieland, 1976; see Table 4.5). The higher

Table 4.5. The ATP/ADP concentration ratios inside and outside incubated mitochondria* and in the intact isolated hepatocyte†

| System | Conditions | ATP/ADP concentration ratio | | |
		Cytosolic compartment	Mitochondrial matrix	External Internal ratio
Liver mitochondria	control	1.2	0.05	24.0
	uncoupled mitochondria	1.1	1.8	0.6
Isolated hepatocytes	hepatocytes incubated with no substrate	6.0	2.0	3.0

* Isolated mitochondria were isolated, incubated and separated from the medium by centrifugation (Heldt *et al.*, 1972).

† Isolated hepatocytes were incubated, the cell membrane broken by digitonin and mitochondria separated by centrifugation (Seiss and Wieland, 1976).

ATP/ADP concentration ratio in the cytosol is due to the fact that the equilibrium established by the translocator involves the proton motive force. This is analogous to, but more complex than, the situation pertaining in neurones where the Nernst equation relates the membrane potential to the uneven distribution of monovalent cations across the membrane. If the proton motive force is dissipated (for example by an uncoupling agent) the cytosolic ATP/ADP concentration ratio is decreased (Table 4.5).

5. Overview of electron transfer and oxidative phosphorylation

Electron transport and oxidative phosphorylation have been discussed in the same chapter as the TCA cycle since they are functionally part of that pathway; that is, the cycle removes hydrogen atoms (or electrons) and these react with molecular oxygen after transfer along the electron transfer system during which ATP is synthesized. Electron transfer is, however, even more dependent upon the cycle since the flux-generating step(s) for the hydrogen (or electron) transfer occur in the TCA cycle (at citrate synthase and oxoglutarate dehydrogenase). As indicated above, most of the electron transfer process is close to equilibrium; the non-equilibrium reactions that provide directionality are those of the TCA cycle and the terminal cytochrome *c* oxidase reaction. The flux-generating step for oxidative phosphorylation is the hydrolysis of ATP in the cytosolic compartment of the cell. The translocation of ADP into and ATP out of the mitochondria and the oxidative phosphorylation reactions are near-equilibrium (Figure 4.23).

The TCA cycle is usually considered in terms of carbon flux; the electron transfer chain in terms of hydrogen or electron flux and oxidative phosphorylation as $\sim P$ flux. Obviously, in the coupled state, these fluxes are stoichiometrically related but the dovetailing of the three different fluxes poses

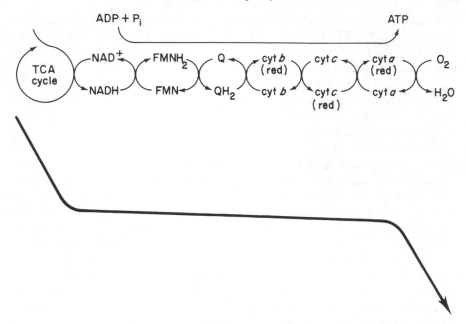

Figure 4.23 Diagrammatic representation of the equilibrium status of the reactions described in Chapter 4. The greater the slope, the further the reactions are from equilibrium

interesting questions concerning the precision by which the rate of respiration is controlled in order to satisfy the energy demands of the tissue. These are pursued in Chapter 7; Section B.4.

E. PHYSIOLOGICAL AND MEDICAL ASPECTS

1. Importance of electron transfer and oxidative phosphorylation in the whole organism

Because most biochemical investigations involve fragmenting a system (be it an animal, a tissue, a cell or a metabolic pathway) into smaller units for convenient study, it is easy to lose sight of the role of the system in the whole organism. The importance of aerobic metabolism is apparent from the oxygen uptake of organisms and tissues. Although molecular oxygen does have other parts to play in metabolism, its role as a terminal electron acceptor for the oxidation of glucose, fatty acids, and amino acids is quantitatively by far the most important. The complete oxidation of glucose, in contrast to anaerobic conversion to lactate, produces 20 times the number of ATP molecules for each molecule of glucose catabolized, thus reducing the amount of fuel that must be stored. This is important in terrestrial animals in order to maintain mobility, but it is particularly important in animals that fly long distances since not only do they have to produce ATP at a high rate for a long period of time but they *must* also maintain a low body weight. An extreme example is provided by migrating birds, some of which can fly more than 1000 miles non-stop, using only stored fuel in the form of triacylglycerol (Odum, 1965).

For similar reasons, muscles that are in continuous or almost continuous use depend on aerobic metabolism; the continuous contractile activity requires a large expenditure of energy over a period of time which can only be provided by aerobic metabolism. The heart is an obvious example of such a muscle; over the lifetime of an average man, the muscles will contract 2.5×10^9 times. Hence this muscle has a large capacity of the TCA cycle (Table 4.6) and a high rate of oxygen consumption (Table 4.7). Another very different muscle that illustrates the same principle is the radula retractor muscle of gastropods (e.g. whelk, limpet, periwinkle). This muscle moves the radula which, in these animals, is used for rasping food (e.g. carcass of an animal) and making it available for ingestion. To 'prepare' the food, when it is available, this muscle may have to operate continuously for prolonged periods of time—up to 24 hr in the case of the whelk (Zammit and Newsholme, 1976). From the activity of oxoglutarate dehydrogenase, the key enzyme indicating the maximum capacity of the TCA cycle given in Table 4.6, it can be seen that the aerobic capacity of the radula retractor muscle of the humble whelk is almost four times higher than that in the quadriceps muscle of man! The aerobic nature of the radula muscle is readily apparent from its beautiful red appearance, a consequence of its content of myoglobin and cytochromes; this can be appreciated from a simple dissection performed with a penknife on a holiday beach.

However, the highest power output for a prolonged period is achieved by insect flight muscles. Indeed, Weis–Fogh (1961) has calculated that, on a weight basis, locust flight muscle has a power output similar to an engine in a Volkswagen car. All the ATP generated in this muscle involves the operation of the aerobic pathways described in this chapter. The maximum rates of the TCA cycle and the activities of relevant enzymes from some insect flight muscles are given in Table 4.6. Perhaps the best way of appreciating the importance of aerobic metabolism is to compare the physiological function of those muscles that depend almost exclusively on anaerobic metabolism with those that depend upon the TCA cycle (see Table 5.4).

The importance of aerobic metabolism is graphically underlined by electron micrographs of insect or bird flight muscle (Figure 4.24). Not only do they show large numbers of mitochondria but the regular arrangement of these mitochondria, in very close association with the ATP-demanding myofibrils, especially in insect flight muscle, emphasizes the importance of electron transfer and oxidative phosphorylation in these tissues.

The importance of oxidative metabolism in man can be seen by reference to the oxygen consumption for the various tissues given in Table 4.7. The high rates of oxygen consumption in brain, heart, kidney, and liver reflect the continuous

Table 4.6. Maximal catalytic activities of enzymes of the TCA cycle and calculated maximal flux in different animals*

Animal	Muscle	Calculated maximum rate of the TCA cycle *in vivo* (μmol.min^{-1}.g^{-1} at 25°C)	Enzyme activities (μmol.min^{-1}.g^{-1} at 25°C)			
			Citrate synthase	NAD$^+$- plus NADP$^+$-linked isocitrate dehydrogenase	Oxoglutarate dehydrogenase	Succinate dehydrogenase
Common whelk (*Buccinum undatum*)	radula retractor	—	25	29	4.7	—
Locust (*Schistocerca gregaria*)	flight	28.0	242	62	24	—
Cockroach (*Periplaneta americana*)	flight	30.0	185	77	54	57
Honey bee (*Apis mellifera*)	flight	64.0	346	98	46	—
Silver-Y moth (*Plusia gamma*)	flight	26.0	352	129	26	—
Dogfish (*Scylliorhinus canicula*)	red	0.6	35	31	—	4.5
Mackerel (*Scombrus scombrus*)	red	0.4	—	—	—	—
Trout (*Salmo gairdneri*)	red	2.4	50	125	2.0	2.7
Pigeon (*Columba livia*)	pectoral	7.6	115	55	4.2	9.0
Rat (Wistar strain)	heart	7.4	96	77	11.0	37
Human (*Homo sapiens*)	quadriceps	1.0	—	—	1.2	—

* Data obtained in the laboratory of E. A. Newsholme; details available in Paul (1979). The rate of the TCA cycle is calculated from oxygen uptake of the insects and pigeon during flight, mackerel and trout during swimming, man during marathon running and the rat heart while perfused and working *in vitro* (see Paul, 1979 for references).

Table 4.7. Oxygen consumption, equivalent glucose* consumption and ATP turnover rates by human tissues

	Rates					
	Approximate oxygen consumption		Equivalent glucose uptake		Equivalent ATP turnover	
Tissue	mol.day^{-1}	μmol.g^{-1} tissue.min^{-1}	g.day^{-1}	μmol.g^{-1} tissue.min^{-1}	mol.day^{-1}	μmol.g^{-1} tissue.min^{-1}
Brain	3.4	1.7	103	0.3	20.4	10.2
Heart	1.9	4.5	57	0.7	11.4	27.0
Kidneys	2.9	7.1	88	1.1	17.4	42.6
Liver	3.6	1.6	108	0.3	21.6	9.6
Skeletal muscle (rest)	3.3	0.08	98	0.01	19.8	0.5
Skeletal muscle (marathon running)	257	6.4	7710	1.1	1542	—
Lactating mammary gland	—	—	86†	—	38.4	—

* The equivalent glucose uptake is the amount of glucose required to account for the oxygen consumption if it was the sole fuel. The equivalent ATP turnover is the ATP that would be produced under those circumstances. Summation of the values for the oxygen consumption of tissues at rest gives the total consumption of oxygen per day as 14.7 moles or 329 l. Since 1 l of oxygen is equivalent to 21 kJ of energy when glucose is oxidized, this represents 6909 kJ of energy expenditure each day, which is only slightly less than that measured for a sedentary human subject (Table 2.2). This indicates that other tissues contribute little to the magnitude of the metabolic rate. Not all of the tissues listed will use glucose; for example, under normal dietary conditions, liver and muscle (at rest) will obtain considerable energy from oxidation of amino acids, and during starvation most tissues will obtain energy from oxidation of fatty acids or ketone bodies.

† Calculated on the basis that all the lactose and 50 % of the lipid secreted by the gland are produced from glucose. If some of the energy for biosynthesis and secretion is obtained from glucose, the rate of glucose utilization will be higher than shown in the table.

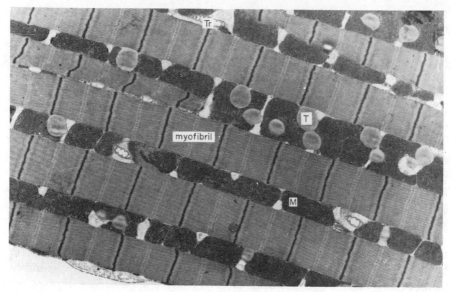

Figure 4.24 L.S. dorso-longitudinal muscle (flight muscle) of the bumble bee showing the abundance of mitochondria (M). T, -droplets of triacylglycerol, an intracellular store of fuel. Tr, tracheole which supplies oxygen directly to mitochondria. The electron micrograph was kindly supplied by Mrs. B. Luke, Department of Zoology, Oxford University.

requirement for ATP for ion transport in the brain, for tubular re-absorption in the kidney, for contraction in heart and for biosynthesis in the liver. The rate of oxygen consumption for skeletal muscle is markedly increased in sustained exercise (Table 4.7).

2. Physiological effects of inhibitors of electron transfer and oxidative phosphorylation

Since inhibitors and uncouplers (described in Section D.4) interfere with the major pathway of ATP generation, it is to be expected that they will be highly toxic. The degree of toxicity depends on the rate at which they are absorbed into an organism and their access to mitochondria. As little as 50 mg of hydrogen cyanide constitutes a lethal dose for a human and its ingestion is rapidly followed by a few convulsive respiratory movements, collapse, convulsions, and death. The metal cyanides take somewhat longer to act and may permit the effective use of an antidote designed to react with the cyanide before it can enter the tissues and inactivate cytochrome *c* oxidation. Rotenone, which is obtained from the root of a plant, inhibits electron transfer in the NADH dehydrogenase complex but, in man, is much less toxic than cyanide since it is poorly absorbed. Fish, which absorb rotenone readily, are very sensitive to its toxic effects.

Workers handling the explosive picric acid (2,4,6-trinitrophenol) during the First World War suffered from fever and loss of body weight. These symptoms occur because picric acid is an uncoupling agent, permitting high rates of electron transfer for prolonged periods. More detailed studies were subsequently carried out on the related compound, 2,4-dinitrophenol, which has a similar physiological effect (Cutting *et al.*, 1933). The most obvious immediate effect of such uncoupling agents is to increase the metabolic rate; oxygen uptake is increased by 20–30% and there is a similar increase in heat production. Indeed, with high doses of these agents, the rate of heat production exceeds the physiological capacity for heat dissipation and fatal hyperthermia develops. The increased respiration results in the rapid utilization of stored fuel reserves and dinitrophenol was in fact introduced as a slimming agent in 1933, long before the mode of action of the compound was understood. Unfortunately, its use for this purpose had serious consequences (skin lesions, peripheral neuritis, anaemia, generalized pain, anorexia, vomiting, cataracts, liver, and kidney damage) and was rapidly discontinued (see Goodman and Gilman, 1955).

The lethal effects of inhibition of mitochondrial processes have been exploited by numerous species of fungus which produce antibiotics to kill competing bacteria. The list of such antibiotics is long and includes antimycin A, oligomycin, valinomycin, and bongkrekic acid, but relatively few have been pressed into therapeutic service.

3. Disorders due to deficiencies of mitochondrial proteins

Since aerobic metabolism is so important for energy provision in vital organs such as the brain, heart, kidney, and liver, it is not surprising that there are no reports of genetically determined deficiencies of the mitochondrial proteins in these organs. Skeletal muscles, however, provided they are only used for short periods, can generate sufficient ATP anaerobically (Chapter 5; Section F.1) and there are a few reports of patients with deficiencies of such proteins in their muscle. Two such genetically determined primary myopathies are described below. (See Morgan-Hughes, 1978, for a general discussion of myopathies.)

(a) *Luft's syndrome*

Patients with this syndrome (two have been reported to date) appear to suffer from partially uncoupled mitochondria in their muscles (Luft *et al.*, 1962). Mitochondria isolated from such muscle have a respiratory control ratio of 1.2–2.4 whereas mitochondria from normal subjects show a ratio of 4–16 (DiMauro *et al.*, 1976). The clinical characteristics of the myopathy are severe heat intolerance, fever, profuse sweating, and dyspnoea. The basal metabolic rate is raised (almost twofold) and there is generalized muscular weakness. The precise biochemical lesion in these patients is not known, but it seems likely that the inner mitochondrial membrane is much more permeable to protons (or other cations) so that the 'high energy' state is dissipated without concomitant ATP production.

(c) *Deficiencies of electron transfer proteins*

As indicated above, it is not surprising that reports of total deficiencies of key mitochondrial proteins are rare. However, there is a report of one child who was totally deficient in cytochrome c oxidase in skeletal muscle (van Biervliet *et al.*, 1977). Muscular weakness was very apparent and, since this included the respiratory muscles, the patient suffered severe respiratory distress and artificial ventilation was essential.

A reduction in the concentration of cytochromes, as opposed to their total absence, is more likely to be consistent with life. Patients deficient in cytochrome b or cytochrome c oxidase show susceptibility to muscle fatigue, general muscle weakness, and, at times, lactic acidosis (see Chapter 13; Section D.3). One case of a patient suffering from severe muscular weakness whose mitochondria had a deficiency of the F_1-ATPase has also been reported (Land and Clark, 1978).

4. Brown adipose tissue and thermogenesis

Brown adipose tissue is highly specialized and is found in some animals, particularly hibernating animals and the new-born of certain species. This tissue has presumably been termed 'adipose tissue' on account of its high lipid content but it is structurally, metabolically and functionally distinct from white adipose tissue. In adult humans most of the brown adipose tissue is localized in the dorsal part of the thorax under the shoulder blades, although there are often smaller amounts around the adrenals and kidneys and in the peri-anal and inguinal regions.

The particular association of brown adipose tissue with hibernating animals, with animals whose habitat is a cold region and with neonatal infants, has given rise to the idea that this tissue might be important in heat generation (Joel, 1965). Hibernating animals will warm up slowly as the ambient temperature increases in spring but many are capable of warming up rapidly despite a low ambient temperature (for example, for short periods in winter). Such animals must possess a special mechanism for the production of heat and it has been found that brown adipose tissue develops extensively prior to hibernation. In some very young animals (for example, human babies and newly born rabbits) the ability to produce heat by shivering (involuntary muscular activity which results in heat production rather than movement) is not developed, but brown adipose tissue is present. The proximity of brown adipose tissue to the vital organs (such as heart and kidneys) is consistent with the theory that the tissue is responsible for heat generation. More direct evidence for the involvement of this tissue in heat production has been obtained from experiments on hibernating animals in which thermocouples have been placed in brown adipose tissue and in several other tissues. During the arousal period, the temperature of brown adipose tissue increases before that of any other tissue and this temperature difference is maintained for some time (Smith and Hock, 1963). It is now generally accepted that the function of brown adipose tissue is to generate heat. Indeed it has been calculated that at least 50 % of the heat produced during arousal in hibernating animals is derived from this tissue (Smith and Horwitz, 1969).

(a) *Mechanism of thermogenesis*

Some indication of the metabolic nature of brown adipose tissue has been obtained from investigation into its ultrastructure. The tissue is highly vascular and consists of small polygonal cells, each containing many separate lipid droplets (in contrast to white adipose tissue which contains large cells with single lipid droplets (see Figure 6.13). The most striking feature of brown adipose tissue cells is the high content of mitochondria (see Figure 4.25). From this it can be deduced that oxidative metabolism plays a vital role in heat production and that

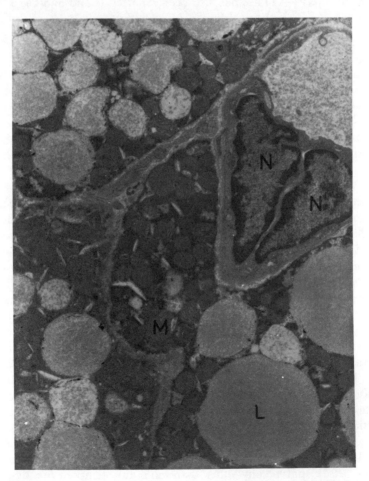

Figure 4.25   An electron micrograph of a brown adipose tissue cell. × 6000

triacylglycerol provides the fuel. This is hydrolysed to fatty acids which provide the substrate for mitochondrial oxidation. Therefore, an increase in the catalytic activity of the triacylglycerol lipase in this tissue is a prerequisite for increased heat production. It has been shown that an increase in cyclic AMP content in brown adipose tissue is associated with increased lipolysis. The content of cyclic AMP in brown adipose tissue is raised in response to noradrenaline and there is evidence that during arousal in hibernating animals this hormone is released from sympathetic nerve endings in this tissue. Therefore, the sequence of events initiating heat production during arousal may be as follows: stimulation of the sympathetic nervous system raises the local concentration of noradrenaline in the brown adipose tissue and this increases the intracellular level of cyclic AMP which activates protein kinase; the latter phosphorylates, and hence activates, triacylglycerol lipase (see Chapter 8; Section C.1 for details of this mechanism of activation). The intracellular concentration of fatty acids is thus increased and these are oxidized by the mitochondria (Himms-Hagen, 1979; Nicholls, 1979).

Obviously, the next question is why should an increase in the rate of fatty acid oxidation lead to such a considerable increase in heat production. The inclusion of the discussion on brown adipose tissue in this chapter indicates that the mechanism of heat generation is related to mitochondrial metabolism and indeed there is considerable evidence that it is due to uncoupling of oxidative phosphorylation from electron transfer. Mitochondria from brown adipose tissue (especially if obtained from hibernating or perinatal animals) have poor respiratory control and P/O ratios are usually low. In addition, the rate of proton movement across the mitochondrial membrane (proton conductance) is high (about 100 times greater than that of liver mitochondria) but the activity of the F_1-ATPase is low. Mitochondria from brown adipose tissue carry out electron transfer and proton pumping in a similar manner to other mitochondria but they possess a specific carrier that transfers protons from outside to inside without concomitant production of ATP (that is, they possess an energy-dissipating proton carrier). That this carrier is important in thermogenesis is supported by the fact that it is present only in mitochondria isolated from brown adipose tissue of neonatal animals, hibernating animals or in animals that have been adapted to the cold. The proton carrier is a protein which has been given the name thermogenin.

The activity of this energy dissipating proton carrier can, however, be regulated (Nicholls, 1976). It is inhibited by several purine nucleotides (including ADP and ATP) but most effectively by GDP. A specific protein (molecular weight 32 000), which has a high affinity for GDP, has been isolated from brown adipose tissue mitochondria and is considered to be either the specific proton carrier or a regulatory portion of the carrier. It is suggested that, when thermogenesis is not required, GDP binds to this protein and reduces the activity of the carrier so that the proton motive force is established and, since the activity of F_1-ATPase is very low, the rate of respiration falls to a low level. However, when thermogenesis is required, this GDP-binding protein may be modified in such a way that the GDP no longer causes inhibition of the proton carrier so that the proton motive force is dissipated leading to heat generation and an increased rate of respiration. This GDP-binding protein may provide the link between increased lipolytic activity and increased respiration; it is suggested that the increased concentration of either fatty acyl-CoA or even fatty acid, which results from the lipolytic activity, decreases the extent of purine nucleotide binding to thermogenin.

(b) *Diet-induced thermogenesis*

There has been a renewed interest in metabolism in brown adipose tissue with the suggestion that it is present in adult man and that it might be important in generation of heat after a meal—a phenomenon known for over 70 years and originally termed the specific dynamic action of food but more recently known as diet-induced thermogenesis or the thermic effect of food. It has been suggested that controlled rates of thermogenesis in brown adipose tissue could be important in 'burning off' excess energy consumed in the diet so that an optimal body weight is maintained. Obese subjects would, according to this hypothesis, suffer from a malfunctioning brown adipose tissue. This hypothesis and the problem of obesity are discussed in detail in Chapter 19; Section 3.b.

5. **Oxygen toxicity and the superoxide radical**

The admonition 'moderation in all things' is in no way better demonstrated than by the toxicity of pure oxygen. As is obvious from the above discussions, oxygen is necessary for life but, at a pressure of five atmospheres, it causes rapid death through deleterious effects on the central nervous system. Even at atmospheric pressure, breathing 100% oxygen for a long period (longer than 48 hr) causes respiratory distress and eventually death. The toxicity of oxygen is not due to the oxygen molecule itself but to the production of highly reactive reduced products from oxygen. Even under normal conditions of oxygen availability, these toxic agents are produced by the body (indeed they are part of normal metabolism) but their rate of production does not exceed the capacity of the tissues to remove these agents.

(a) *The cause of oxygen toxicity*

Molecular oxygen (dioxygen) readily gives rise to partial reduction compounds that are very reactive oxidants and which are consequently very dangerous. These compounds are the superoxide anion (a free radical[7]), hydrogen peroxide and the hydroxyl radical. The origin of each is described below, before a discussion of their harmful effects. For further details, see McCord and Fridovich (1978); Fridovich (1978); Hill (1979); Chance *et al.* (1979); Deneke and Fanburg (1980); Bannister and Hill (1982).

(i) *Origin of superoxide anions* The superoxide anion ($O_2^-\cdot$) is a charged free radical with a very short half-life (milliseconds). It arises from the acquisition of a single electron by a molecule of oxygen:[7]

$$O_2 + e^- \rightarrow O_2^-\cdot$$

It has long been known to be generated by the action of ionizing radiation on oxygen and for many years its only biological interest lay in its role (with other radicals) in causing radiation damage to tissues. More recently it has been identified as a normal intermediate in certain biological processes. For example, xanthine oxidase, an enzyme which catalyses the oxidation of xanthine to uric acid (see Figure 4.26) generates superoxide anions as a normal product (McCord and Fridovich, 1968). Normally, the superoxide anion is a short-lived intermediate which is converted to hydrogen peroxide in a reaction catalysed by the enzyme superoxide dismutase (see below) which maintains the steady state concentration at a very low level ($<10^{-11}$ M). Several other flavoprotein oxidases (e.g. aldehyde oxidase) also generate superoxide anions in this way. The anion is also produced by the auto-oxidation (that is, oxidation not involving an enzyme) of, for example, reduced quinones (e.g. ubiquinone, see Section C.1.c), catecholamines and thiols. Traces are also produced when oxygen combines with haemoglobin or myoglobin.

(ii) *Origin of hydrogen peroxide* In contrast to the recent appreciation of the role of superoxide anions, the generation of hydrogen peroxide in cellular oxidations (and the need to reduce the concentration of this toxic metabolite) has been known for 50 years. Among those enzymes which generate hydrogen peroxide are D-amino acid oxidase (see Chapter 10; Section C.2.a) and amine oxidase (see Chapter 12; Section A.3) as well as superoxide dismutase (see below).

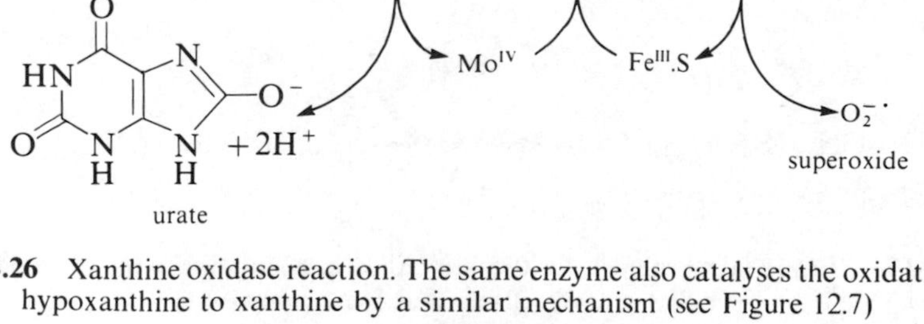

Figure 4.26 Xanthine oxidase reaction. The same enzyme also catalyses the oxidation of hypoxanthine to xanthine by a similar mechanism (see Figure 12.7)

(iii) *Origin of hydroxyl radicals* Hydroxyl radicals (OH·) probably arise from the interaction of hydrogen peroxide and the superoxide anion in the Haber–Weiss reaction, the rate of which is increased by ions of iron or copper (Bannister and Hill, 1982):

$$H_2O_2 + O_2^- \cdot \longrightarrow O_2 + HO^- + HO\cdot$$

Hydroxyl radicals are considerably more reactive, and therefore more toxic, than either hydrogen peroxide or superoxide. Indeed it is possible that the toxicity of the superoxide radical and hydrogen peroxide is due to the formation of the hydroxyl radical.

(iv) *Mechanism of oxygen toxicity* It appears that unsaturated lipids, proteins and DNA are the cell components most sensitive to the effects of these oxidants. Although hydrogen peroxide and superoxide anions may have deleterious effects of their own, the major problem may be the generation of hydroxyl radicals (see above) which promote a chain reaction. The chemical basis for the effect of radicals on DNA is possibly due to peroxidation and chemical modification of the bases; but it is likely that damage to proteins is caused by the oxidation of sulphydryl groups (i.e. —SH groups are converted to —S—S—groups). A plausible scheme put forward to explain the effect on lipids is that hydroxyl radicals attack unsaturated fatty acids in phospholipids and other lipid components of membranes and this results in the formation of lipid hydroperoxides (Slater and Benedetto, 1981) (Figure 4.27). Hydroperoxidation of membrane lipids has two consequences. First, it increases the hydrophilic nature of the lipid and this changes the structure of membranes so that their normal function is disturbed. Secondly, the lipid hydroperoxides (and their immediate breakdown products) are powerful inhibitors of some enzymes, so

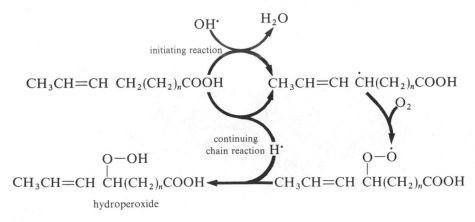

Figure 4.27 Proposed mechanism for the free radical peroxidation of unsaturated lipids. (see Slater, 1982; Slater and Benedetto, 1981)

that processes within the membrane and within the cell may be seriously inhibited. For example, the red cell membrane is sufficiently damaged by hydroperoxidation that the cells are more rapidly degraded and the result is anaemia. A further problem in the red cell is that the iron atoms in haemoglobin are readily oxidized (even by hydrogen peroxide) and the resulting methaemoglobin[8] is unable to transport oxygen. This exacerbates the tissue hypoxia due to the anaemia. Two other tissues in which cell membranes are easily damaged by oxidation are those of the lung and brain. Breathing pure oxygen damages the alveoli[9] and this causes respiratory distress. However, at very high oxygen concentrations, cell membranes in the brain may be affected leading to problems with electrical activity and interference in brain function. Unfortunately, many premature babies suffer from respiratory distress syndrome (see Chapter 17; Section C.2.a) and to alleviate some of the problems, pure oxygen has been administered. However this results in damage to the lung[9] and to the retina (growth of fibrous tissue in front of the retina—retrolental fibroplasia and in consequence, blindness). At one time this was the greatest single cause of blindness in children; to avoid this problem, such babies are now given air with no more than 40% oxygen (Lanman *et al.*, 1954; Nichols and Lambertsen, 1969; Weiter, 1981).

(b) *Cellular protection against oxidants*

At least four mechanisms appear to play some part in reducing the harmful effects of these oxidants in the cell. Indeed, at normal oxygen tensions they eliminate the problem entirely.

(i) *Superoxide dismutase reaction*   The enzyme, superoxide dismutase, catalyses the destruction of the superoxide radical according to the equation:

$$2O_2^{-\cdot} + 2H^+ \longrightarrow H_2O_2 + O_2$$

This enzyme is present in most if not all tissues and serves to lower the normal intracellular superoxide concentration to extremely low levels ($< 10^{-11}$ M). The hydrogen peroxide produced is removed by the action of catalase.

(ii) *Catalase reaction*   A protective role has long been assigned to the ubiquitous and highly active enzyme catalase, which catalyses the reaction:

$$2H_2O_2 \longrightarrow O_2 + 2H_2O$$

However, some patients with the rare inherited disorder, acatalasaemia, in which the activity of catalase in all tissues is very low, show few, if any, symptoms. Presumably, in such cases, the other enzymes, particularly glutathione peroxidase, compensate for the low activity of catalase.

(iii) *Glutathione peroxidase reaction*   The enzyme glutathione peroxidase*
catalyses the reduction of organic hydroperoxides and hydrogen peroxide in a
reaction that involves reduced glutathione:[10]

$$ROOH + 2GSH \longrightarrow GSSG + ROH + H_2O$$
$$H_2O_2 + 2GSH \longrightarrow GSSG + 2H_2O$$

This enzyme may have three roles: first, to convert hydrogen peroxide to water so
that its concentration is maintained very low; secondly, to convert peroxidized
fatty acids to hydroxy fatty acids, and thirdly to reverse the oxidation of protein
sulphydryl groups. Oxidized glutathione is reduced by NADPH in a reaction
catalysed by glutathione reductase:

$$GSSG + NADPH + H^+ \longrightarrow 2GSH + NADP^+$$

The $NADP^+$ is reduced by the oxidation of glucose 6-phosphate, catalysed by
glucose-6-phosphate dehydrogenase and phosphogluconate dehydrogenase
acting in sequence (Figure 4.28). This is part of the pentose phosphate pathway
described in detail in Chapter 17; Section A.8. The consequences of glucose-6-
phosphate dehydrogenase deficiency in the erythrocyte are described in Note 8.

(iv) *Antioxidants*   A number of naturally occurring compounds have the ability
of reacting with free radicals without generating further radicals. Such scavengers
will therefore quench chain reactions.[7] Both tocopherol (vitamin E) and β-
carotene have this property and there is some evidence that they can provide
protection from free radicals *in vivo* (Oski, 1980). The problems associated with
red cell glucose-6-phosphate dehydrogenase deficiency (see above) markedly
improve after vitamin E is administered to the patient (Corasch *et al.*, 1980) and
the bronchopulmonary dysplasia in infants given oxygen-enriched air to breath
is ameliorated by administration of vitamin E (Ehrenkranz *et al.*, 1979).

(v) *Localization of protective enzymes*   The three enzymes described above are
found in different intracellular locations and appear to have complementary
roles. Catalase is found predominantly in peroxisomes, small organelles
($0.5-1.0\,\mu m$ in diameter in liver and rather smaller in some other tissues where
they are called microperoxisomes or microbodies) which, in addition to catalase,
contain hydrogen peroxide-producing oxidases. Glutathione peroxidase is more
widely distributed in mitochondria and cytosol but is absent from peroxisomes.
Its main function may be to eliminate any organic peroxides once they have been
formed. Superoxide dismutase occurs predominantly in the cytosol (where, for
example, xanthine oxidase occurs) but a second form of the enzyme does occur in
mitochondria.

---

* The enzyme contains selenium which may be an essential mineral in man and other animals.[11]

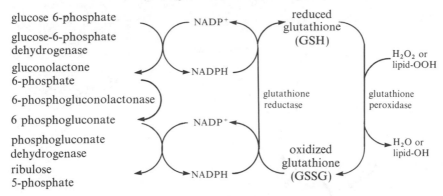

Figure 4.28 The action of glutathione reductase in erythrocytes to prevent the accumulation of hydrogen peroxide and peroxidized lipids

(c) *Features of biological interest*

(i) *Role of cytochrome* c *oxidase* At first sight it seems surprising that, despite the importance of oxidation in the generation of ATP from fuels, only one reaction involves molecular oxygen directly. This is, of course, the one catalysed by cytochrome *c* oxidase. This restriction is probably explained by the fact that any reduction of oxygen to water is likely to involve the formation of the superoxide radical, but any release of this radical must be avoided or reduced to the minimal rate. Indeed the work of B. Chance has indicated that the catalytic reaction proceeds via a superoxide intermediate but that the enzyme is embedded in the inner membrane and the active site is in a hydrophobic area so that there is little chance of the radical escaping. Perhaps these structural requirements have so constrained the evolution of a suitable enzyme that only one has evolved; all fuel oxidations are therefore channelled through this 'safe' system. There is some indirect evidence that structural integrity of the membrane is important in preventing 'leakage' and hence production of the superoxide radical. After death, when this organization is destroyed, oxygen consumption increases (and may actually exceed that in life) and this has been attributed to the oxidation of the large quantities of unsaturated lipids in biological membranes by such radicals (Dormandy, 1978).

The complete reduction of a molecule of oxygen by cytochrome *c* oxidase requires four electrons:

$$4\text{cyt } c.\text{Fe}^{2+} + \text{O}_2 + 4\text{H}^+ \longrightarrow 4\text{cyt } c.\text{Fe}^{3+} + 2\text{H}_2\text{O}$$

Cytochrome *c* oxidase has two haem groups, one in cytochrome *a* and one in cytochrome a_3, together with two copper atoms, in close association. However, only cytochrome a_3 appears to be obligatorily involved in the reaction sequence in which both iron and copper atoms change their oxidation state (Chance *et al.*, 1979; Chance, 1981).

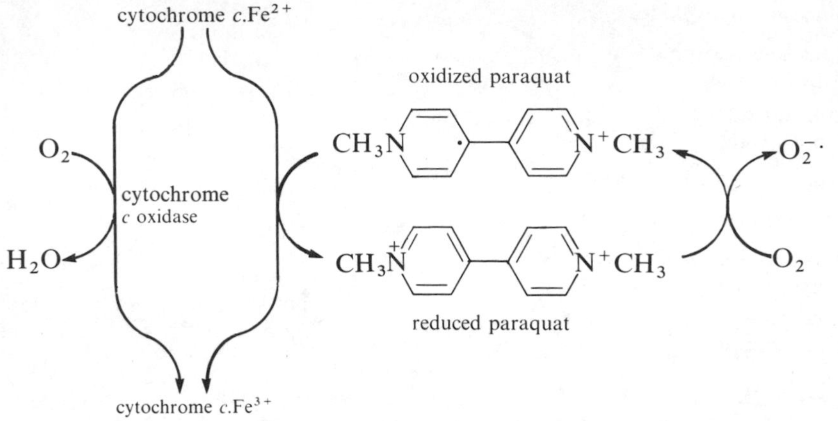

Figure 4.29 The provision of an alternative route for cytochrome c oxidation by the herbicide paraquat. Note the superoxide anion generated by this pathway

(ii) *Toxicity of paraquat* The widely used herbicide paraquat (Figure 4.29) has been responsible for a large number of deaths. The liquid is usually kept in a garden shed or garage and is therefore easily available, particularly to children. Upon drinking this liquid, the person soon suffers respiratory distress from which death can rapidly occur. Its toxicity is probably due to formation of the superoxide radical.

Paraquat readily accepts electrons from cytochrome c in the electron transfer chain. Reduced paraquat is oxidized by molecular oxygen (which bypasses the cytochrome c oxidase complex—see Figure 4.29) and releases superoxide anions in such large quantities that they swamp the destruction processes. Death is probably due to damage to the membranes of the epithelial cells lining the bronchioles and alveoli in the lung (Fridovich and Hassan, 1979).

(iii) *Role of oxidants in granulocyte phagocytosis and autoimmunity* The possession of highly toxic oxidants appears to have been turned to good purpose by phagocytic cells. There is evidence that granulocytes (polymorphonuclear leucocytes) increase their rate of production of these oxidants some ten-fold when engaged in engulfing bacteria and are, in fact, using these metabolites to kill the bacteria (Babior, 1978; Badwey *et al.*, 1979). Lymphocytes may use a similar mechanism so that these oxidants may be partially responsible for the inflammatory reaction. It is further possible that part of the deleterious action of lymphocytes and macrophages on the body's own tissues in autoimmune diseases (see Chapter 17; Section B.2.d) could be caused by the production of such oxidants. Accordingly, attempts have been made to treat patients suffering from autoimmune diseases with purified superoxide dismutase in the hope that this will reduce the local concentrations of the superoxide radical and hence reduce the inflammation (Menander-Haber and Haber, 1977).

APPENDIX 4.1   STEREOISOMERISM

A covalent compound will have stereoisomers if its atoms can be arranged in more than one way in space so that the structures formed cannot be interconverted without breaking covalent bonds. (This latter proviso is needed to exclude conformations which are the shapes taken up by compounds as a result of any freedom of movement possessed by the atoms within them.) Stereoisomerism is important in metabolism because it determines the shape of molecules and consequently their interaction with enzymes. In practice, the problems with stereochemistry are mainly the technical ones of representing their structures unambiguously on paper and devising names to distinguish them.

Two factors give rise to stereoisomerism; restricted rotation about double bonds (geometrical isomerism) and molecular asymmetry (optical isomerism). For the following account it is entirely adequate to consider covalent compounds as made of balls (representing atoms) and rigid sticks (representing bonds). Indeed, the construction of such models will greatly clarify the points made.

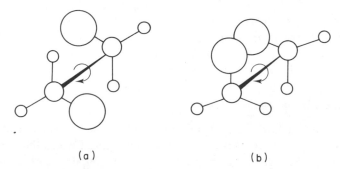

(a)   (b)

Figure 4.30   Conformational change arising from free rotation of a single covalent bond. The spheres represent atoms and the rods joining them, bonds. This model corresponds to compounds of the type $CH_2Cl.CH_2Cl$

1. Geometrical isomerism

Free rotation of other bonds can occur around the axis of a single bond. This means that the two structures depicted in Figure 4.30 are not stereoisomers: they are conformers (i.e. differ in conformation) since one can be converted to the other without breaking any bonds. (Note, however, that some conformers are more stable than others and in the example given, fewer molecules would be in conformation (b) in which the bulky groups come closest.) When double bonds occur between carbon atoms, all atoms connected to the carbon atoms lie in a plane and free rotation is not possible. When each of the carbon atoms is attached to two different groups, two geometrical isomers can exist, as illustrated by aconitate, an intermediate in the TCA cycle:

$$\begin{array}{ccc} H & & COO^- \\ & C=C & \\ {}^-OOC & & CH_2COO^- \end{array}$$

trans-aconitate

$$\begin{array}{ccc} H & & CH_2COO^- \\ & C=C & \\ {}^-OOC & & COO^- \end{array}$$

cis-aconitate

These compounds cannot be interconverted without breaking covalent bonds and have substantially different physical and chemical properties.

2. Optical isomerism

The second type of stereoisomerism, optical isomerism, occurs in compounds whose structure lacks a plane of symmetry. Since the valence bonds of a saturated carbon atom are directed towards the corners of a tetrahedron, the commonest cause of asymmetry is a carbon atom linked to four different groups, as in malic acid (Figure 4.31). Pairs of stereoisomers (enantiomers) form mirror images, behave identically in symmetrical environments and for the most part, therefore, show identical physical and chemical properties. Only in interaction with asymmetrical agents do their properties differ so that, for example, enantiomers rotate the plane of polarized light in opposite directions. Since the active site of an enzyme provides an asymmetrical chemical environment, only one of a pair of enantiomers will generally fit and can react.

(a) *Assignment of configuration*

In order to assign the configuration of an optical isomer (that is to establish which of the two possible structures rotates polarized light in which direction) it is first necessary to establish a means of depicting on paper the three-dimensional structure of a molecule. This is widely done by use of Fischer projection formulae. The longest chain of carbon atoms is drawn vertically with C_1 at the top and with the four bonds of each carbon atom directed towards the four points of the compass; bonds to the east and west are envisaged as projecting out of the plane of the paper and those to the north and south as projecting below this plane. Use is made of X-ray crystallography to assign configuration absolutely. Thus for glyceraldehyde, the dextrorotary ($+$) enantiomer has been shown to have the structure depicted in (a) rather than (b), and is designated D-glyceraldehyde. (Note that the prefixes *d*- (dextrorotatory) and *l*- (laevorotatory) are no longer used.)

$$\begin{array}{c} CHO \\ | \\ H-C-OH \\ | \\ CH_2OH \end{array}$$

(a)

D-glyceraldehyde

$$\begin{array}{c} CHO \\ | \\ HO-C-H \\ | \\ CH_2OH \end{array}$$

(b)

L-glyceraldehyde

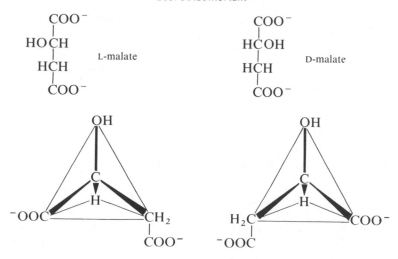

Figure 4.31 Representation of the two optical isomers of malic acid superimposed onto tetrahedra

(The prefix DL- indicates a racemic modification (or racemate) with both D- and L- forms present in equal amounts.)

The configuration of relatively few compounds has been assigned absolutely by X-ray crystallography and in most cases assignment is relative. For example, since D-glyceraldehyde can be oxidized to glyceric acid without breaking any of the bonds attached to the asymmetric carbon atom, the configuration at this atom must be retained.

$$
\begin{array}{ccc}
\text{CHO} & & \text{COOH} \\
| & & | \\
\text{H}-\text{C}-\text{OH} & \xrightarrow{[\text{O}]} & \text{H}-\text{C}-\text{OH} \\
| & & | \\
\text{CH}_2\text{OH} & & \text{CH}_2\text{OH} \quad \text{D-glyceric acid}
\end{array}
$$

The product of this reaction is, therefore, known as D-glyceric acid. This compound rotates the plane of polarized light anticlockwise and its full designation is therefore D-(−)-glyceric acid. Note that there is no simple relationship between structure and the direction of rotation of polarized light.

The allocation of sugars to the L- or D- series depends on the configuration of the highest-numbered asymmetric carbon atom (i.e. the penultimate carbon atom). The carbon atom is assigned relative to glyceraldehyde so that the D-series of sugars are depicted with the hydroxyl group on the penultimate carbon atom directed to the right (when C_1 is at the top of the projection formula). Not all compounds, however, can be readily related to glyceraldehyde and other reference compounds are used, for example, serine:

$$
\begin{array}{cc}
\text{COOH} & \text{COOH} \\
| & | \\
\text{H}-\text{C}-\text{NH}_2 & \text{H}_2\text{N}-\text{C}-\text{H} \\
| & | \\
\text{CH}_2\text{OH} & \text{CH}_2\text{OH}
\end{array}
$$

D-serine   L-serine

To avoid confusion in more complex molecules, the reference compound is sometimes indicated with a subscript so that D_g and L_s mean configuration relative to D-glyceraldehyde and L-serine respectively.

(b) *Multiple asymmetric centres*

The problems of nomenclature increase when there is more than one optically active centre in a molecule. With two such centres, four stereoisomers exist and correspond to D_1D_2, D_1L_2, L_1D_2, and L_1L_2. Of these, the pair D_1D_2 and L_1L_2 have mirror image structures, as do D_1L_2 and L_1D_2. There are, therefore, two pairs of enantiomers. However, the other possible pairs are not mirror image structures and do not have identical physical and chemical properties. Optical isomer pairs of this kind, in which the configuration at some of the centres only is reversed, are known as diastereoisomers. In the sugars, diastereoisomers are given different trivial names, so that, for example, glucose and galactose are diastereoisomers, each of which consists of a pair (D- and L-) of enantiomers. The stereochemistry of sugar ring structures is generally indicated by the Haworth convention in which the bottom edge of the ring is considered to project out of the paper and groups depicted as pointing upwards and downwards extend above and below the plane of the ring, respectively:

D-glucose   D-galactose

(Note that for hexose sugars, in the six-membered ring structure only, there are five optically active centres hence 32 optical isomers—16 pairs of enantiomers.) When only two asymmetric centres are present, the prefixes *erythro-* and *threo-* are sometimes used to distinguish diastereoisomers, by analogy with the aldo-tetrose sugars, erythrose and threose. In *erythro*-compounds, the configuration is the same at both carbon atoms while in *threo*-compounds it is different. The prefix *allo-* is also sometimes used to distinguish between diastereoisomers, as with *allo*threonine and threonine.

This nomenclature is not only cumbersome but can be ambiguous and virtually impossible to apply to compounds with large numbers of asymmetric centres. The latter two criticisms are overcome by the 'S/R' notation which is widely used by chemists. For each asymmetric carbon atom in turn, the four substituents are ranked according to a set of rules. One then imagines a three-dimensional model of the molecule in which the asymmetric centre is viewed from a point opposite the substituent of lowest priority. If the priority of the three remaining groups increases clockwise, then the centre is designated R (preceded by the number of the carbon atom involved). If the priority increases anticlockwise then the centre is designated S. The rules for establishing priority, although arbitrary, are universally applicable. The four atoms attached to the asymmetric carbon atom are arranged in order of decreasing atomic number. If more than one atom has the same atomic number then the second atoms (with the highest atomic number) are compared and so on.

The results of applying these notations to isocitrate, the only intermediate of the TCA cycle with more than one asymmetric centre, are shown in Figure 4.32.

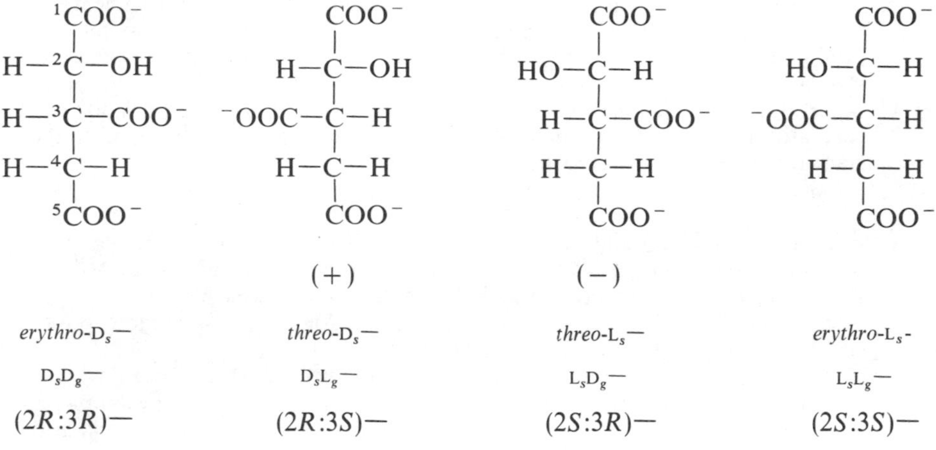

$$
\begin{array}{cccc}
^1COO^- & COO^- & COO^- & COO^- \\
| & | & | & | \\
H-^2C-OH & H-C-OH & HO-C-H & HO-C-H \\
| & | & | & | \\
H-^3C-COO^- & ^-OOC-C-H & H-C-COO^- & ^-OOC-C-H \\
| & | & | & | \\
H-^4C-H & H-C-H & H-C-H & H-C-H \\
| & | & | & | \\
^5COO^- & COO^- & COO^- & COO^- \\
\end{array}
$$

$$(+) \qquad\qquad (-)$$

erythro-D$_s$—	*threo*-D$_s$—	*threo*-L$_s$—	*erythro*-L$_s$-
D$_s$D$_g$—	D$_s$L$_g$—	L$_s$D$_g$—	L$_s$L$_g$—
(2R:3R)—	(2R:3S)—	(2S:3R)—	(2S:3S)—

Figure 4.32 Structural formulae of the four optical isomers of isocitrate according to the Fischer convention, with alternative designations. The subscripts s and g refer to serine and glyceraldehyde respectively. The configuration of C_2 is assigned relative to serine and of C_3 relative to glyceraldehyde. The naturally occurring isomer is *threo*-D$_s$-isocitrate

NOTES

1. The biosynthesis of coenzyme A in man involves phosphorylation of pantothenic acid, peptide bond formation with cysteine (followed by decarboxylation) to give 4′-phosphopantetheine, condensation with ATP (eliminating pyrophosphate) to give dephospho-CoA and finally phosphorylation in the 3′-position of the ribose.
2. The high reactivity of thioesters, compared with oxygen esters, is explicable in terms of the increased partial positive charge (δ^+) on the carbonyl carbon atom. This δ^+ arises

as a result of the polarization of the carbonyl atom. In oxygen esters, the δ^+ is reduced in magnitude by the existence of a resonance form in which the bridging oxygen atom bears a full positive charge:

$$\underset{R_1-O-\overset{\displaystyle \overset{O^{\delta-}}{\underset{\displaystyle \|\, \delta+}{}}}{C}-R_2}{} \longleftrightarrow \underset{R_1-O^+=\overset{\displaystyle O^-}{\underset{\displaystyle |}{C}}-R_2}{}$$

The formation of this $-C{=}O^+\!-$ bond involves the overlap of two p-orbitals but to form the corresponding $-C{=}S^+\!-$ bond would involve overlap between a p-orbital and a d-orbital. This is much less favourable so that the second resonance structure contributes less and delocalization is reduced. The increased δ^+ of thioesters accounts for their susceptibility to attack by hydroxyl ions with consequent hydrolysis.

3. Multienzyme complexes differ greatly in the extent to which the dissociated subunits can function independently of one another and to which recombination restores function. In the case of the oxoglutarate dehydrogenase complex the three constituent enzymes have been separated and recombined with complete restoration of activity (Koike *et al.*, 1980).

4. Despite their generally accepted names, flavin mononucleotide and flavin adenine dinucleotide are not, strictly speaking, nucleotides at all since the carbohydrate moiety is derived from an alcohol (ribitol) not a sugar.

5. The standard redox potential (E_h^0) of a redox couple is a measure of the tendency of a reducing agent to part with its electron. In principle it is measured as the e.m.f. produced when a half-cell containing the redox couple is suitably connected to a standard hydrogen electrode (a half-cell in which the reaction, $H^+ + e^- \rightleftarrows \frac{1}{2}H_2$, is occurring). In a redox reaction, the standard electrode potential, ΔE^0, is the e.m.f. generated in a cell composed of half-cells, each containing one redox couple in their standard state. It is related to ΔG^0 by the equation:

$$\Delta G^0 = -nF\,\Delta E^0$$

where F is the Faraday and n is the number of electrons transferred. When combined with a term relating to reactant concentrations it provides, therefore, like ΔG^0 (and K_{eq}), an indication of direction in which the reaction will proceed. A redox couple with a more negative redox potential will reduce a redox couple with more positive standard redox potential *under standard conditions*.

6. This scheme is one example of several similar schemes, which were devised to accommodate observed properties of the phosphorylation process. For example, 2,4-dinitrophenol uncouples oxidative phosphorylation in isolated mitochondria in the absence of phosphate or ADP which suggests the existence of a high energy intermediate that is formed before phosphate is involved in the reaction; that is, the existence of I $\sim$ X. If this were hydrolysed to I and X by an uncoupling agent it would provide a mechanism for uncoupling.

7. Free radicals contain unpaired electrons represented by a superscript dot placed after the atom. Most are extremely short-lived and in consequence highly reactive, but a few are quite stable and others, such as the superoxide radical assume an intermediate position on this scale. To understand the nature and properties of free radicals some knowledge of structure and chemical bonding is required. Fortunately a very simplistic model will suffice and is given here using oxygen as the example.

Atoms consist of positively-charged nuclei surrounded by negatively-charged electrons. These electrons occupy discrete energy levels, each of which can accommodate two electrons. With this information, and the fact that it is particularly favourable for a nucleus to be surrounded by eight electrons, the structure of the oxygen atom can be examined in more detail. The nucleus of a single oxygen atom is surrounded by six electrons (Figure 4.33(a)) but there are a number of other stable configurations available to the oxygen atom.

(a) A single oxygen atom can acquire two electrons (making a total of eight) and become an oxide ion (Figure 4.33(b)). Out of contact with water, in a crystal lattice, this is very stable.

(b) Two oxygen atoms can unite in such a way as to share a pair of electrons with its neighbour so each has a share in eight (Figure 4.33(c)). This produces the diatomic oxygen molecule but the electron-pairing rule breaks down and two of the electrons occupy different energy levels so that molecular oxygen is in fact a diradical, $O_2\cdot\cdot$, albeit a very stable and unreactive one. (This structure is not easily represented by diagrams based on the valence bond theory.)

(c) If an extra electron is gained by an oxygen molecule in the course of a reaction it becomes a negatively charged monoradical, namely superoxide, $O_2^-\cdot$ (Figure 4.33(d)).

Reactive free radicals are able to break covalent bonds by homolytic fission, that is, with one of the two electrons from the covalent bond going with each atom. This generates two more free radicals, one of which typically combines with the attacking radical to form a new covalent bond and the second is available to attack a further bond and thus continue what becomes a chain reaction. Such a sequence will end if a radical too unreactive to generate further radicals is formed. Compounds able to quench radical reactions in this way are known as inhibitors. A hypothetical reaction sequence in which $HO\cdot$ is the initiating radical and IH the inhibitor can be represented:

$$HO\cdot + X{-}X \longrightarrow X{-}OH + X\cdot \qquad \text{initiating reaction}$$

$$\left.\begin{array}{l} X\cdot + R{-}H \longrightarrow X{-}H \;+ R\cdot \\ R\cdot + X{-}X \longrightarrow R{-}X \;+ X\cdot \end{array}\right\} \text{chain reaction}$$

$$X\cdot + I{-}H \longrightarrow X{-}H \;+ I\cdot \qquad \text{inhibiting (or quenching) reaction}$$

The predominant over-all reaction occurring in this case is:

$$X{-}X + R{-}H \longrightarrow X{-}H + R{-}X$$

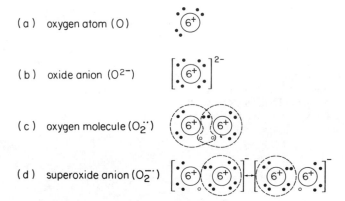

(a) oxygen atom (O)

(b) oxide anion (O^{2-})

(c) oxygen molecule ($O_2\cdot\cdot$)

(d) superoxide anion ($O_2^-\cdot$)

Figure 4.33 Representations of the atomic structures of oxygen atoms, molecules, and ions. Note that in the diatomic forms, the electrons are shared equally by both atoms. Electrons in pairs are indicated by small solid circles; single electrons by open circles

8. Methaemoglobin is haemoglobin in which the Fe^{II} atom has become oxidized to Fe^{III}. It is incapable of binding oxygen and if it accumulates causes the erythrocyte to be destroyed. The oxidation of a significant amount of haemoglobin to methaemoglobin thus causes anaemia. This may occur either if the rate of haemoglobin oxidation is increased or the mechanism for methaemoglobin reduction impaired. The former situation occurs in the presence of compounds which will accept electrons from molecular oxygen and a frequent culprit is the nitrite ion. This is particularly so in infants in which micro-organisms growing in the anaerobic and only slightly acidic conditions of the stomach can carry out the reduction of nitrate (a common pollutant of the environment) to the toxic nitrite.

Several of the pathways which reduce methaemoglobin back to haemoglobin (or destroy oxidants before they can oxidize haemoglobin in the first place) require NADPH. This is largely generated through operation of the pentose phosphate pathway (see Chapter 17; Section A.8) the first enzyme of which is glucose-6-phosphate dehydrogenase. Individuals with a reduced activity of glucose-6-phosphate dehydrogenase (the most common disease-producing enzyme deficiency) in their erythrocytes are very much more sensitive to agents which increase methaemoglobin formation (Beutler, 1972). Those which have caused problems of this kind include primaquine, a widely used antimalarial drug, and fava beans, consumed as a vegetable in the Mediterranean region.

9. Damage to the lung caused by oxygen is non-specific. It consists of atelectasis (lungs do not expand completely), oedema, alveolar haemorrhage, inflammation, fibrin deposition and thickening and hyalinization of alveolar membranes (Clark and Lambertsen, 1971; Denecke and Fanburg, 1980). This may be the reason for the development of bronchopulmonary dysplasia in infants suffering from respiratory distress syndrome and given oxygen-enriched air (Northway, 1978).

10. Reduced glutathione (GSH) is a tripeptide, γ-glutamylcysteinylglycine which has the following structure:

$$H_3N^+CHCH_2CH_2CO \cdot NH \cdot CH \cdot CO \cdot NH \cdot CH_2COO^-$$
$$\underset{\textstyle COO^-}{|} \qquad\qquad \underset{\textstyle CH_2SH}{|}$$

On oxidation, two molecules link via a disulphide bridge.

11. It has been shown that selenium is an essential mineral in a number of animals; deficiency results in muscular dystrophy in ruminants, pancreatic degeneration and exudative diathesis in poultry, liver necrosis in rats, and it may be one of the principal factors in Keshan disease (a particular form of cardiomyopathy) (Johnson *et al.*, 1981). The only known role so far for selenium in higher animals is as a component of glutathione peroxidase and hence in protection against the effects of oxygen radicals (Young, 1981). Deficiencies of selenium, and other trace elements, can occur in man with long-term parenteral nutrition if these are not included in the feeding solutions (Lancet editorial, 1983a).

CHAPTER 5

CATABOLISM OF CARBOHYDRATES

The previous chapter has emphasized the central role of acetyl-CoA and the importance of the tricarboxylic acid cycle in the metabolism of many cells. But how does this acetyl-CoA arise? The answer depends on the tissue involved, the availability of fuels within the tissue or in the bloodstream and the metabolic activity of the tissue. A precise quantitative answer will only be possible when the regulation and integration of catabolism has been discussed in Chapters 7 and 8. In general, it can be said that most of the major tissues (e.g. muscle, liver, kidney) are able to convert glucose, fatty acids, and amino acids to acetyl-CoA. However, brain and nervous tissue, in the fed state and in the early stages of starvation, depend almost solely on glucose. The pathway for the formation of acetyl-CoA from glucose is described in this chapter; that from fatty acids and other lipid-fuels in Chapter 6, and that from amino acids in Chapter 10.

Not all tissues, however, obtain the major part of their ATP requirements from the TCA cycle. Red blood cells, tissues of the eye and the kidney medulla, for example, gain most, if not all, of their energy from the anaerobic conversion of glucose to lactate. The physiological significance of this is discussed in Section F. In both anaerobic and aerobic tissues, glucose is first converted to pyruvate. However, in anaerobic tissues this is converted to lactate, whereas in aerobic tissues it is converted to acetyl-CoA. The pathway as far as pyruvate has become known as anaerobic glycolysis but this terminology is misleading since the same reactions are involved in both aerobic and anaerobic glycolysis. The fate of pyruvate will depend upon the tissue and the environmental conditions at that time, so that it is important to define the conditions, not the pathway, as aerobic or anaerobic.

Although the pathway of glycolysis is usually assumed by biochemists to start at the level of glucose in a particular cell, the physiological pathway really begins with the digestion of dietary carbohydrates in the fed state (Figure 5.1). For this reason the digestion of carbohydrate is discussed prior to the other reactions of glycolysis.

A. DIGESTION OF CARBOHYDRATES

The manner in which we eat, the kind of things we eat, and indeed whether we eat or not, can have profound effects on the metabolism of the body, so that it is necessary to have some knowledge of the processes of digestion and absorption.

167

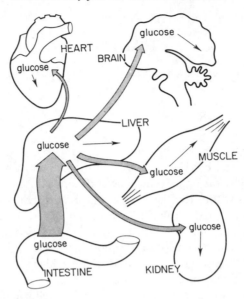

Figure 5.1 Pathway of glucose metabolism in the whole body in the fed, sedentary state.
A number of other tissues also utilize glucose but their quantitative contribution is small.
See Table 4.7 for quantitative information

Sufficient description is given here to enable the reader to understand the over-all processes and to relate them to metabolism. For further reading in this area the reader is referred to Gray (1975) and Crane (1977).

Approximately 50% of the calories ingested by the average Western man is in the form of carbohydrate. This comprises about 160 g of starch, 120 g of sucrose, 30 g of lactose, and 10 g of glucose plus fructose each day, with traces of maltose and trehalose. It should be noted that only three monosaccharides result from the digestion of these carbohydrates, namely glucose, fructose, and galactose. Appendix 5.1 should be consulted for information on the structure and nomenclature of carbohydrates.

1. Enzymes of carbohydrate digestion

Since monosaccharides are the only carbohydrate that can be absorbed from the intestine, carbohydrate digestion is essentially a process of glycoside hydrolysis achieved through the action of a series of hydrolytic enzymes known collectively as glycosidases. The effect of their action is best illustrated by considering the hydrolysis of starch, the predominant carbohydrate in most human diets.

Starch is a mixture of two polysaccharides: amylose, in which glucose units are joined together by α-1,4-linkages to form very long chains, and amylopectin, in which shorter chains of α-1,4-linked glucose units are connected by α-1,6-links to give a branched structure. (Amylopectin has a structure similar to that of glycogen—see Section C.1.) Two general classes of hydrolytic enzymes are involved in the complete hydrolysis of starch in the intestine: α-amylases and the oligosaccharidases (here used to include the disaccharidases).

(a) *α-Amylases*[1]

α-Amylases are present in saliva and in the exocrine pancreatic secretion. Enzymes of this class hydrolyse α-1,4-links in the middle of chains but will not hydrolyse links immediately before, or two units after, branch points. Thus the action of α-amylase converts amylose into maltose (with some maltotriose) and amylopectin into oligosaccharides containing one or more α-1,6-branch points and more than five glucose units (average eight). Digestion of starch to these compounds begins in the mouth and is completed in the lumen of the duodenum.

(b) *Oligosaccharidases*

These enzymes are attached to the surface of the brush border of the intestinal columnar epithelial cells in the jejunum and most of the ileum. They are large glycoproteins (Chapter 16; Section D.3) with a pH optimum about 6.0. There is one β-galactosidase, for which the only substrate is lactose and at least three α-glycosidases. The first, *exo*-1,4-α-D-glucosidase (also known as glucoamylase and α-amylase) catalyses the sequential hydrolysis of terminal glucose units from the non-reducing (i.e. the free C_4 of the pyranose ring rather than C_1) end of linear oligosaccharides so that long chains are progressively shortened to produce glucose. The second, sucrose α-D-glucohydrolase (also known as sucrase) catalyses the hydrolysis of α-1,6-linkages of the fragments arising from the action of α-amylase on amylopectin and also the hydrolysis of the α-1,2-linkage in sucrose. The third, trehalase, catalyses the hydrolysis of the disaccharide trehalose (found in insects and fungi) to glucose.

Both amylose and amylopectin can be completely hydrolysed to monosaccharides by these two classes of hydrolytic enzymes (Figure 5.2). The rate of production of the monosaccharides by the action of the brush border oligosaccharidases is such that it may exceed the capacity of the monosaccharide transport mechanism. The excess monosaccharides diffuse back into the mixed intestinal contents to be absorbed subsequently at more distal sites in the jejunum or ileum. Thus much of the monosaccharide transport system in the columnar epithelial (absorptive) cells is saturated with substrate during the absorptive period (i.e. it constitutes a flux-generating step). However, this probably does not occur when lactose is the predominant sugar in the diet since this is hydrolysed much more slowly than the other sugars.

2. **Monosaccharide absorption**

The absorption of monosaccharides involves transport across the luminal

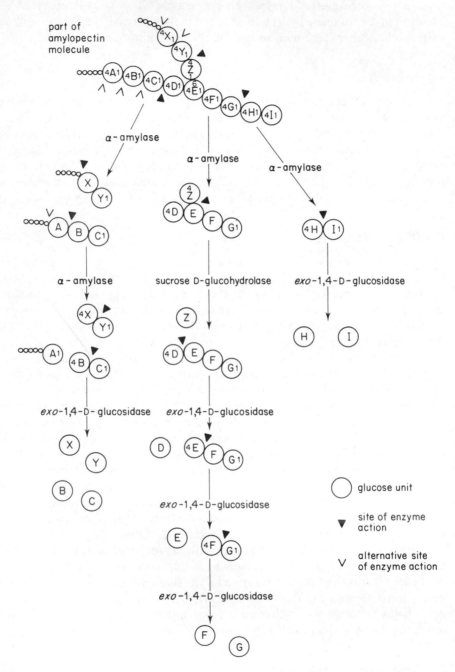

Figure 5.2 Hypothetical sequence of enzyme action in the digestion of amylopectin. See text for description of enzyme specificities

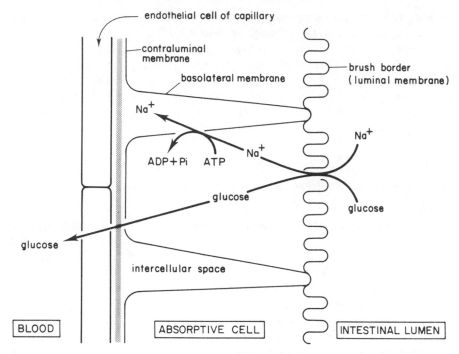

Figure 5.3 Diagrammatic representation of sodium-linked glucose transport from the lumen of the intestine into the blood. Note that the exact location of the sodium pump has not been established

membrane of the columnar epithelial (absorptive) cell and further transport, from this cell, across the contraluminal or basolateral membrane into the intercellular space prior to entry into the capillaries (Figure 5.3). There are at least three mechanisms involved in this process and, surprisingly, it is not clear what proportion of the sugar is absorbed by each of these mechanisms. This current dilemma has considerable metabolic implications for understanding the control of carbohydrate metabolism in the whole animal.

(a) *Transport and absorption of glucose by passive diffusion*

As emphasized above, the hydrolysis of oligosaccharides at the brush border could produce very high *local* concentrations of glucose adjacent to the epithelial cells, particularly during the middle period of the digestion of a meal that contained a large amount of carbohydrate. The concentration of glucose on the luminal side of the membrane could become greater than the intracellular glucose concentration, so that glucose could enter the cell via a non-energy requiring transport process. This would, of course, require a specific glucose carrier in the membrane which may be similar to that required to transport glucose across the muscle and other cell membranes (see Section B.2). The concentration of glucose in the epithelial cell will be higher than that in the intercellular space on the vascular side of the cell so that glucose will be transported passively across the contraluminal membrane (presumably via another carrier mechanism) into this space, from where it will enter the bloodstream. In the early and late periods of digestion, however, the glucose concentration in the lumen will be low and hence an energy-requiring process must operate.

(b) *Transport and absorption of glucose linked to the transport of sodium ions*

At least two monosaccharide transporting systems may come under this heading: one is specific for glucose or galactose and the other for fructose. The former has been more extensively studied, but it seems likely that both function in the same complex manner described below.

The complexities are introduced by the fact that the system that transports glucose across the luminal membrane also transports sodium ions. Thus, both glucose and sodium ions are transported into the cell. The latter are transported down a concentration gradient. Thermodynamically, the energy to transport glucose molecules up their concentration gradient is provided by the energy made available through the transport of sodium ions down their concentration gradient. (That is, a process with a large and positive ΔG is coupled to a process with a large and negative ΔG—see Chapter 2; Section A.2.d.) The current biochemical model for the coupling is illustrated in Figure 5.3 and is described as follows. On the luminal surface of the epithelial cell membrane, glucose (or galactose) can bind to the carrier protein only if sodium ions are also bound. Within the cell, the concentration of sodium ions is much lower than in the lumen so that they, and consequently the glucose molecules, are released from the carriers. The low intracellular sodium ion concentration, and hence the sodium ion concentration gradient, is maintained by the activity of a second carrier system which specifically transports sodium ions outwards against a concentration gradient. This latter part of the transport process is directly linked to the hydrolysis of ATP, which has led to it being termed a 'sodium pump'. Current evidence suggests that this pump is located in the basolateral membrane. This transport process ensures that the intracellular glucose concentration is maintained higher than that in the intercellular space on the vascular side of the cell so that glucose will be transported across the contraluminal membrane, and hence into the bloodstream, passively.

The elegant biochemical demonstration of the complex link between the transport of sodium ions and that of glucose has led to the view that this is the only quantitatively important mechanism for transport and absorption of glucose. This may not always be so.

(c) *The conversion of glucose to lactate in the absorptive cells of the intestine*

There is no doubt that glucose is absorbed as such into the columnar epithelial cell either by passive or active transport. However, this cell contains the usual complement of glycolytic enzymes with which to metabolize this glucose. The interesting question as to whether glucose absorbed from the lumen is metabolized when there is a normal availability of glucose in the bloodstream has been answered recently. Experiments with an *in vitro* intestinal preparation, which is perfused both through the lumen and through the normal vasculature, have shown that, when glucose is present on both sides of the cell, less than 50 % of the glucose absorbed from the lumen is translocated to the bloodstream unchanged. Almost all of the remainder of this glucose is converted to lactate which diffuses into the bloodstream (Hanson and Parsons, 1976). Hence, on the basis of such *in vitro* experiments, it has been suggested that, of the glucose transported into the epithelial cells of the intestine from the lumen *in vivo*, a high proportion (at least 50 %) may be converted into lactate prior to entry into the bloodstream (see Porteous, 1978). In order to complete the absorption of glucose, lactate must be re-converted into glucose. This is achieved in the liver via the pathway known as gluconeogenesis (see Chapter 11; Section B.2). The glucose produced by the liver is released into the general circulation via the inferior vena cava. Thus, the absorptive cells of the intestine, the hepatic portal vein and the liver can all be considered to be involved in glucose absorption (Figure 5.4). The conversion of glucose to lactate can be seen as a means of maintaining a concentration gradient for the passive entry of glucose into the epithelial cell. The process can be considered to be active overall (i.e. energy requiring) since more ATP is used in gluconeogenesis than produced in glycolysis. It is also possible that the absorption of other monosaccharides (especially fructose) may be achieved, in part, by conversion to lactate in the epithelial cells of the intestine.

There is evidence both for and against the quantitative importance of glycolysis in the absorption of carbohydrate in the intact rat. In an anaesthetized rat, the hepatic portal vein was surgically connected to the inferior vena cava, thus bypassing the liver (a portocaval shunt). Glucose was injected into a segment of the small intestine and blood was sampled from the systemic circulation (aorta). In the control animals with an intact circulation to the liver, the increment in systemic blood glucose concentration was about 1.6 mM; however, it was almost non-detectable (0.1 mM) in the rats with a portocaval shunt, but there was a large increase in the concentration of lactate. It was calculated that only 15 % of the glucose removed from the intestine in the rats with a portocaval shunt entered the bloodstream as glucose and at least 60 % could be accounted for as lactate (Shapiro and Shapiro, 1979). In contrast, in experiments carried out by Windmueller and Spaeth (1980), a simulated meal (containing 70 mM glucose and 20 different amino acids) was administered to the rat and only 3 % of the glucose transported across the intestine was metabolized. The quantitative significance of glycolysis in glucose absorption is therefore unclear.

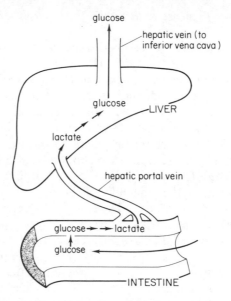

Figure 5.4 Absorption of glucose by intestinal glycolysis and hepatic gluconeogenesis. Glucose is converted to lactate in the absorptive cells of the intestine. Lactate is transported to the liver, via the hepatic portal vein, where it is converted to glucose which is released into the general circulation

3. Dietary fibre or unavailable carbohydrate

A typical human diet contains plant polysaccharides and other polymers that undergo little or no hydrolysis in the intestine[2]. These are described by the general term, *fibre*, which includes a range of different substances. It is possible to divide them into five general classes as follows: cellulose, hemicellulose, lignin, pectic substances, and gums and mucilages.

(i) *Cellulose* This is a linear polymer of glucose with β-1,4-linkages (Figure 5.5) and is the major component of the cell walls of higher plants. Consequently, it is quantitatively the most important component of fibre. No known vertebrate produces enzymes capable of hydrolysing the β-linkages in cellulose. Some degradation can occur in the large intestine due to the action of the micro-organisms that do contain cellulases but there is considerable variation in the reports of the extent of cellulose digestion in man; this may depend upon the microflora and the length of time food residues remain in the colon (Bryant, 1978; Cummings, 1981). Microbial degradation is particularly important in herbivores since it results in cellulose being converted to smaller molecules that are absorbed into the animal and used as a fuel. This process has developed most markedly in the ruminants in which a part of the alimentary canal (the reticulorumen) is effectively a continuous culture system in which cellulose-producing micro-

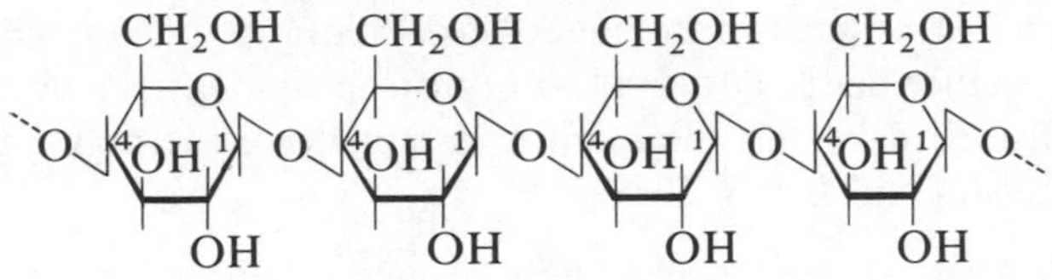

Figure 5.5 Molecular structure of part of a molecule of cellulose, consisting of glucose units linked β-1,4. For comparison with an α-1,4 linked polymer of glucose, see Figure 5.13

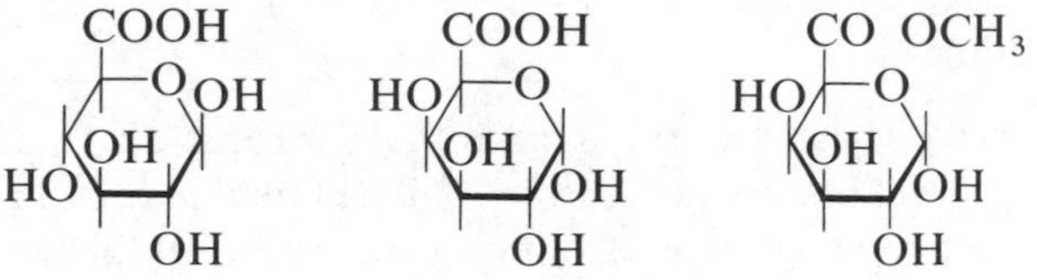

β-D-glucuronic acid

α-D-galacturonic acid

methyl-α-D-galacturonate

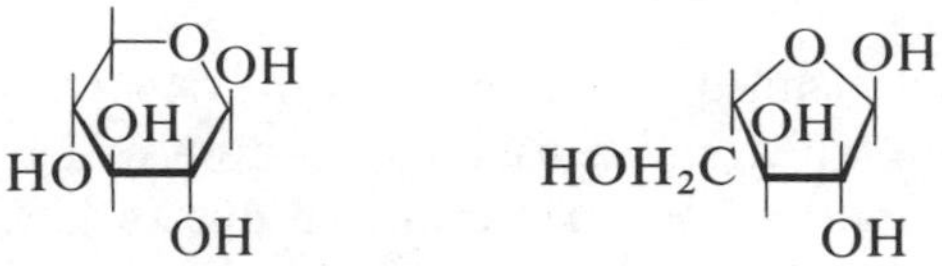

β-D-xylose

α-L-arabinose

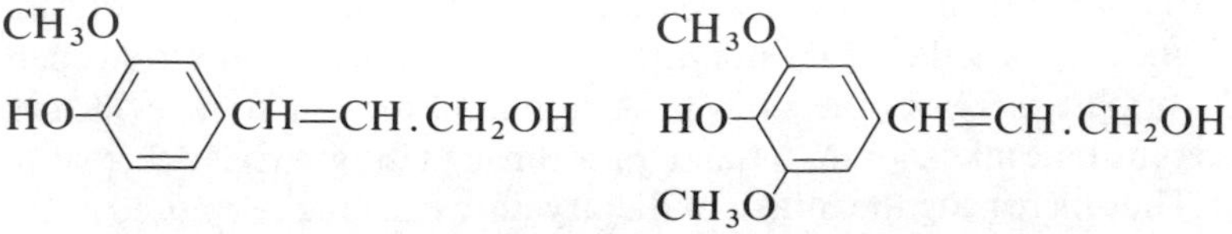

coniferyl alcohol

sinapyl alcohol

Figure 5.6 Molecular structures of monomer units in plant polysaccharides and lignin. The monosaccharides are shown in the ring form and as the anomer most frequently occurring; see Appendix 5.1

organisms hydrolyse most of the ingested cellulose to glucose which is further metabolized to the volatile fatty acids (butyrate, propionate, lactate, and acetate) which are absorbed by the ruminant and constitute the major fuel for these animals (Annison and Lewis, 1959).

(ii) *Hemicelluloses* These are a heterogeneous group of polysaccharides containing mainly the pentose sugars, xylose (in xylans) and arabinose (in arabinoxylans). They also contain a smaller component of polymerized acidic sugars, particularly galacturonic or glucuronic acid (Figure 5.6). The hemicelluloses suffer from degradation in the colon of man.

(iii) *Lignin* This is present in woody tissue. It is not, however, a polysaccharide but an aromatic polymer based on coniferyl and sinapyl alcohols (Figure 5.6) and it is totally indigestible.

(iv) *Pectic substances* These are polymers of galacturonic acid and varying amounts of other sugars. They are not fibrous and are found in the soft tissues of fruits. In pectins, which are components of this group, many of the galacturonic acid residues are methylated (see Figure 5.6).

(v) *Gums and mucilages* These are mixed polymers of arabinose, xylose, mannose and glucuronic and galacturonic acids. An example of this class is guar gum, a galactomannan, from the cluster bean (a leguminous seed).

Dietary fibre is a term which now includes all the fibrous material (cellulose, hemicellulose, and lignin) as well as the non-fibrous pectic substances and gums.

The term fibre has been defined by Trowell *et al.* (1976) as 'the plant polysaccharides and lignin which are resistant to hydrolysis by the digestive enzymes of man'. This definition has been criticized due to lack of chemical precision and the fact that some of the above components can be digested by man (Cummings 1981). Fibre is most abundant in foods such as cereals and flours made from whole grain, starchy roots (e.g. potatoes), leafy vegetables (brussels sprouts, cabbage), mature leguminous seeds, nuts, and fruits. The average individual crude fibre intake in Britain and the U.S.A. is about 20 g.day^{-1} but, for example, this can increase to 35 g.day^{-1} on diets rich in wholemeal cereal and vegetarians may consume even larger amounts. The intake in parts of Africa may be as much as 100 g.day^{-1} (Southgate, 1978). There is no doubt that such a high intake of fibre increases the size of the stool and there is some evidence that it reduces the time taken for material to pass through the gut (i.e. it decreases transit time). Thus increasing the intake of dietary fibre has long been used as a remedy for constipation (see Cleave, 1974). Fibre has also been used in the treatment of some diseases of the colon such as diverticulitis and irritable colon syndrome (Connell, 1978; Almy and Howell, 1980; Trowell and Burkitt, 1981).

Epidemiological studies have suggested that diets low in fibre may be responsible for other diseases of the colon, including cancer. Furthermore, similar studies have been used to support the claim that many of the 'metabolic' diseases now prevalent in industrial societies (e.g. diabetes, obesity, cardiovascular disease, cancer) are due to a low intake of fibre as a consequence of the over-refining of foods in such societies (Cleave, 1974; Jenkins, 1979; Kromhout *et al.*, 1982). The importance of an increased fibre content of the diet in the control of the blood glucose concentration in diabetic patients is discussed in Chapter 15; Section F.3.

4. Clinical disorders of digestion

A deficiency of pancreatic α-amylase occasionally occurs in infants so that for the first few months of their life they are unable to digest starch. In adults, there is such a large excess of this enzyme that, even in patients with severe exocrine pancreatic insufficiency, enough enzyme is secreted into the intestine to hydrolyse normal amounts of starch.* Thus maldigestion of starch rarely, if ever, occurs in man. Primary deficiencies (i.e. congenital deficiency) of the oligosaccharidases (see above) occasionally occur. Secondary deficiencies can also occur in diseases such as coeliac sprue. In this case, there is a marked reduction in the activities of all the oligosaccharidases. (The symptoms of oligosaccharidase deficiency can be simply produced by ingestion of a complex polysaccharide inhibitor of the enzyme known as acarbose—see Caspary, 1978.) The consequence of these deficiencies is that the oligosaccharides cannot be digested and hence they pass into the large intestine. Here the sugars increase the osmotic pressure which prevents the re-absorption of water and this causes diarrhoea. Additional discomfort results from bacterial fermentation of this carbohydrate in the large intestine which produces carbon dioxide, methane and hydrogen. (Some of the hydrogen appears in the breath and this can be measured and used as an index of the quantity of carbohydrate entering the colon.) Most of these conditions are rare but the inability to digest lactose, present in milk, is more widely encountered in adults. Before weaning, children have high activities of the intestinal β-galactosidase but these may fall after weaning. Patients present with a history of milk intolerance (nausea, diarrhoea, and abdominal cramps after intake of milk) although the symptoms may be mild. However, their tolerance to milk would have been normal during infancy. Symptoms disappear completely if milk is excluded from the diet. Alternatively, a fungal lactase can be added to the milk prior to ingestion. An interesting feature of this condition is that its incidence is much higher in some races (notably negro and oriental groups) than in others (see Bayless and Rosensweig, 1966; Menzies, 1980). Advantage has been taken of this intolerance to provide a simple means for lowering the pH of the large intestine in the treatment of hepatic coma. It is important that circulating ammonia levels are maintained low when the liver is not functioning. Since a high intestinal pH favours ammonia absorption from the intestine, lactose tends to reduce this by supporting the growth of bacteria which produce acidic metabolic end products.

The fact that the transport of sodium ions and glucose into epithelial cells is

---

* An inhibitor of α-amylase obtained from the red kidney bean, has been marketed as a 'starch-blocker' for the treatment of obesity. There is, however, considerable controversy as to the efficacy of this inhibition as an anti-obesity agent (Lancet editorial, 1983d).

coupled (see Section A.2.b) has been exploited in the treatment of patients suffering from cholera. In this disease, the toxin produced by the cholera bacteria induces a massive loss of fluid and electrolytes into the intestinal lumen (van Heyningen, 1976). Indeed, the clinical manifestations of cholera are solely the consequence of these losses which in severe cases must be replaced by intravenous infusion. As the patient recovers, fluids and electrolytes can be administered orally and it has been found that the inclusion of glucose facilitates the absorption of sodium ions (Hendrix, 1975; Molla *et al.*, 1982).

B. THE REACTIONS OF GLYCOLYSIS

1. The glycolytic pathway

In most textbooks it is assumed that glycolysis starts with intracellular glucose, thus ignoring the important process of glucose transport across the cell membrane. However, even if this process is included, glycolysis-from-glucose, in contrast to glycolysis-from-glycogen, is not a physiological metabolic pathway as defined in Chapter 2. The non-equilibrium reactions in glycolysis are catalysed by the glucose transport system, hexokinase, 6-phosphofructokinase, and pyruvate kinase (see Table 2.4) and none of these reactions is saturated with substrate (Table 5.1). The flux-generating step for this process is either the absorption of glucose from the intestine (see Section A.2) or, when glucose arises endogenously, the breakdown of glycogen in the liver (Figure 5.7). Thus, in these cases, a metabolic pathway spans more than one tissue: it includes the intestine or liver, the blood, and the given tissue, for example, muscle. The metabolic importance of

Table 5.1. K_m values of enzymes that catalyse non-equilibrium reactions in glycolysis and *in vivo* pathway-substrate concentrations.‡

Reaction	Pathway-substrate	K_m for enzyme (mM)	Substrate concentration *in vivo* (mM)
Glycogen phosphorylase	glycogen	1.0	20–30
Glucose transport	extracellular glucose	10.0	5.0
Hexokinase	intracellular glucose	0.1	<0.1†
6-Phosphofructokinase	fructose 6-phosphate	about 0.1*	0.1
Pyruvate kinase	PEP	0.1	0.05

* The kinetics are such that if the enzyme is inhibited by ATP, which is considered to be the case *in vivo* (see Chapter 7; Section B.1.b), the enzyme cannot be saturated with fructose 6-phosphate.
† Intracellular glocuse is difficult to measure precisely.
‡ Data taken from New?holme, 1976; Beis and Newsholme, 1975.

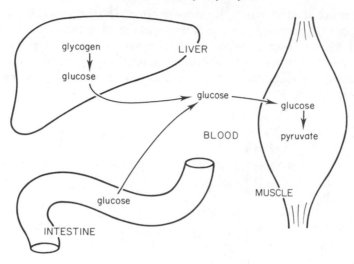

Figure 5.7 Source of glucose for glycolysis in muscle. Other glucose-utilizing tissues (e.g. brain, kidney) receive glucose from the same sources. Release of glucose from the intestine is probably important as a flux-generating step in the absorptive state whereas the degradation of liver glycogen is important in the post-absorptive state

this interpretation of a metabolic pathway is related to the control of the blood glucose concentration and the rate of glucose utilization by various tissues. Thus, in order for muscle to utilize more carbohydrate, for example during exercise, liver glycogen must be degraded more rapidly to provide the blood glucose that will be utilized by the muscle. Furthermore, since the blood glucose concentration remains remarkably constant during exercise, the rate of degradation of glycogen and hence the rate of glucose release by the liver must be the same as the rate of glucose utilization by the muscle. How does the liver assess how much glucose is being used by muscle (or indeed any other tissue, such as the brain, that can use large amounts of glucose)? Similarly when a high carbohydrate meal has been digested, the rate of absorption of glucose by the intestine will be high, and, in order to prevent a very large increase in the blood glucose concentration, the rate of glucose utilization by muscle and other tissues must be increased. These problems of control and integration of metabolism between tissues will be discussed in Chapters 8, 9, and 14.

Despite the above comments, the 'pathway' of glycolysis described here is the conventional one, that is, it starts at the level of extracellular glucose. The incomplete pathway is described since it provides the essential biochemical details for an understanding of energy production under aerobic and anaerobic conditions. The *complete* pathway will be discussed in Chapter 7; Section B.1.c, where it is appropriate for the understanding of the control of these processes.

All the reactions of glycolysis occur in the cytosolic compartment of the cell, where it is generally considered that the enzymes involved are in free solution and that the transfer of intermediates from one enzyme to the next occurs by diffusion. The reactions are given in Figure 5.8; they appear to be common to all vertebrate cells (and, indeed, to most, if not all, invertebrate cells). Other reactions that supply glycolytic intermediates or affect the utilization of pyruvate and which do show some variation in different tissues are considered in subsequent sections of this chapter.

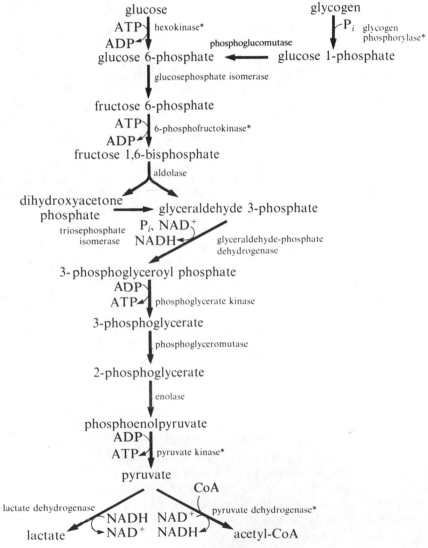

Figure 5.8 The conventional metabolic pathway of glycolysis. Note that the physiological pathway begins with glucose absorption from the intestine or breakdown of glycogen in the liver. All enzymes can function physiologically in the reverse direction except those marked with an asterisk

2. Glucose transport across the cell membrane

The first reaction of the glycolytic pathway in any tissue is the transpo
glucose across the cell membrane. Transport involves the transient combination
of a glucose molecule with a carrier (probably a protein) at the outer surface of the
membrane and the subsequent release of the glucose at the inner surface. In most
tissues, this process of glucose transport is down a concentration gradient and is
therefore a passive process. The active transport of glucose is seen only into the
epithelial cells of the intestine (see Section A.2), into the epithelial cells of the
kidney tubule and into the choroid plexus (where the cerebrospinal fluid is
formed) (Lund-Andersen, 1979). Evidence for the involvement of a carrier in the
passive transport of glucose has accumulated from a wide variety of experiments
which are summarized below:

(i) *Saturation kinetics* The relationship between the rate of glucose transport
and glucose concentration is hyperbolic (Figure 5.9). The plateau in the curve
suggests that the transporting mechanism has a limited number of glucose-
binding sites. The K_m for glucose transport in a number of tissues including heart,
skeletal muscle, brain, and erythrocytes is approximately 10 mM.

(ii) *Specificity* Although the transport system has a relatively broad specificity
(for example, in addition to glucose, L-arabinose, D-xylose, 3-O-methyl-D-glucose,
and 2-deoxy-D-glucose are transported, see Figure 5.10) by no means all sugars
can enter cells. L-glucose or D-sorbitol, for example, are not transported.

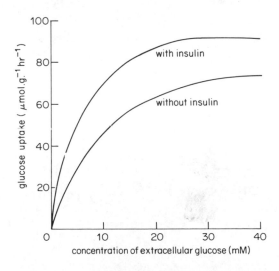

Figure 5.9 Dependence of rate of glucose transport into a perfused rat heart on the
concentration of extracellular glucose. Based on data of Morgan *et al.* (1961)

Transported

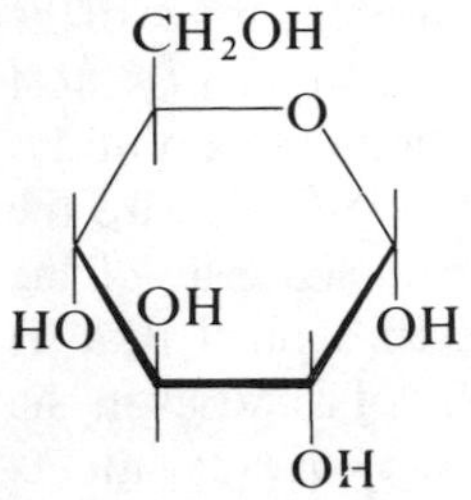

D-glucose

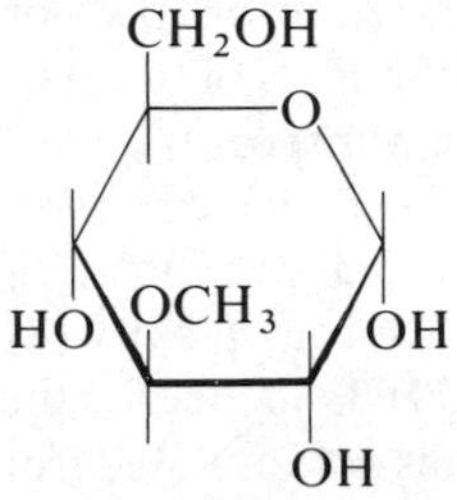

3-*O*-methyl-D-glucose

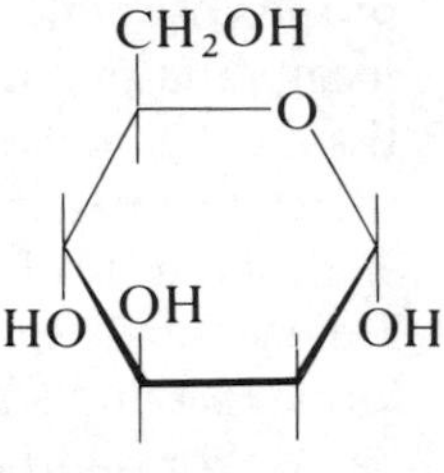

2- deoxyglucose

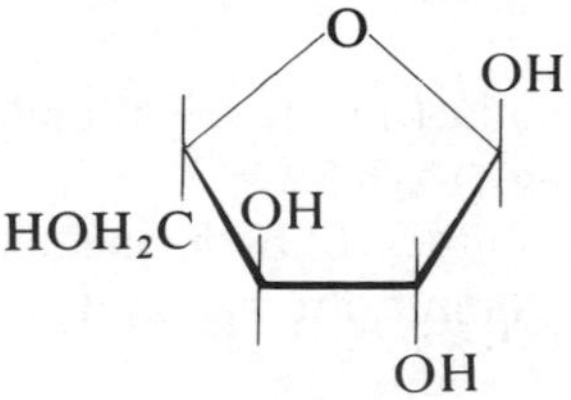

L-arabinose

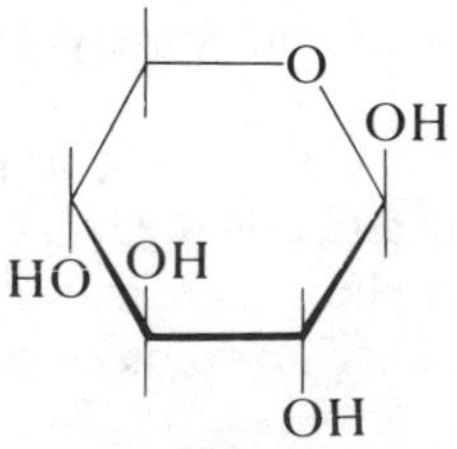

D-xylose

Not Transported

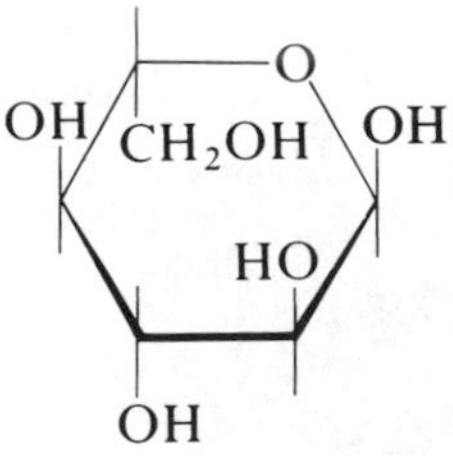

L-glucose

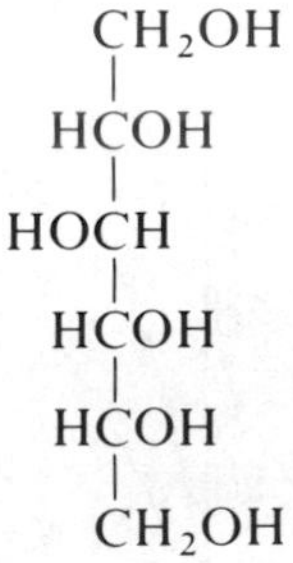

D- sorbitol

Figure 5.10 Structure of some molecules which can, and some which cannot, be transported into cells by the glucose transport mechanism

(iii) *Competition*   The rate of transport of one sugar is reduced by the transport of another sugar; for example, the transport of L-arabinose reduces the rate of glucose transport. This is because one sugar will compete with the second for binding to the carrier. The kinetics are identical to that of competitive inhibition of enzymes (see Chapter 3; Section C.4.a).

(iv) *Countercurrent transport*   The phenomenon of countercurrent transport is explained on the basis of such competition. Cells can be 'loaded' with a high concentration of a sugar that is not metabolized (e.g. L-arabinose) by incubating them in a medium containing a high concentration of that sugar. If glucose is then added to this medium, L-arabinose will move out of the cell *against* its concentration gradient. This is most readily explained by the existence of a carrier for glucose and arabinose. Outside the cell, glucose and arabinose compete for the carrier and are transported inwards at rates dependent on their concentrations and on the affinity of the carrier for these sugars. When glucose is added to the medium it will be transported inwards but inside the cell, where glucose is metabolized so that its concentration is low, arabinose will compete more successfully for sites on the carrier and will be transported out (Figure 5.11).

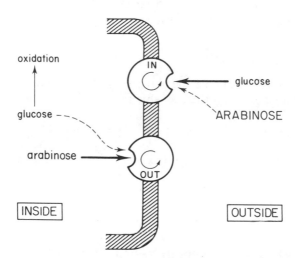

Figure 5.11   Model to explain countercurrent transport. The carriers are identical but are shown at different phases of activity. Under these conditions the arabinose concentration outside the cell is greater than the arabinose concentration inside

(v) *Specific activation and inhibition*   A number of compounds are known which alter the activity of the glucose transporting system without themselves becoming transported. Thus the hormone insulin increases the activity of the transporting system (Figure 5.9), while the plant glycoside, phloridzin, and its aglycone phloretin, inhibit glucose transport:

OH

HO—⟨⟩—CO.CH$_2$CH$_2$—⟨⟩—OH

OH

phloretin

These properties are reminiscent of those exhibited by enzymes and strongly support the existence of a protein carrier for glucose. In the majority of tissues (e.g. muscle, brain, adipose tissue, erythrocytes) the glucose transport process is far from equilibrium.[3] Consequently, it is a potential site for allosteric and hormonal control. In muscle and adipose tissue, glucose transport is stimulated by insulin, thus providing one mechanism for the control of the blood glucose level by this hormone (see Chapter 15; Section D.1).

The glucose transport system in the liver, in contrast to those other tissues, is a near-equilibrium process, so that the intracellular concentration of glucose approaches that in the extracellular fluids (i.e. ΔG is close to zero). From the discussion of the control of near-equilibrium reactions (Chapter 2; Section C.2.b) it is predicted that this transport process in liver will be controlled only by changes in concentrations of intra- and extracellular glucose. In accord with this, insulin does not affect the glucose transport process in liver cells.

3. Hexokinase reaction

The first intracellular reaction is the phosphorylation of glucose by ATP to form glucose 6-phosphate. In most tissues this reaction is catalysed by hexokinase:

$$\text{CH}_2\text{OH} \qquad\qquad +\text{ATP}^{4-} \longrightarrow \qquad \text{CH}_2\text{OPO}_3^{2-} \qquad\qquad +\text{ADP}^{3-}+\text{H}^{+}$$

α-D-glucose   α-D-glucose 6-phosphate

As in very nearly all reactions involving nucleoside triphosphates, Mg^{2+} is required since ATP reacts as its Mg^{2+} complex ($MgATP^{2-}$). Hexokinases show a relatively broad specificity and will phosphorylate several sugars (including fructose, mannose, 2-deoxy-D-glucose, and allose) but in most cases at a lower maximal rate and with a higher K_m (see Chapter 3; Section C.2.d) than with glucose. The hexokinase reaction is non-equilibrium (Table 2.4) so that it is a potential site for allosteric inhibition. The inhibition of the enzyme by its product, glucose 6-phosphate, provides a mechanism for regulation of this enzyme in concert with the regulation of the activity of 6-phosphofructokinase (see Chapter 7; Section B.1.b).

In liver there is a second enzyme which catalyses the phosphorylation of glucose, glucokinase. As the name implies, this enzyme is more specific for glucose than is hexokinase, but the physiologically important properties of the enzyme are that it has a high K_m for glucose (about 10 mM) and it is not inhibited by glucose 6-phosphate. These properties of glucokinase (together with the near-equilibrium nature of the glucose transport system in liver) fit it for a role in the regulation of the blood glucose concentration as described in Chapter 11; Section E.1.a.

The investment of a molecule of ATP at this stage in glycolysis is necessary for two reasons. First, it enables glucose to be phosphorylated at physiological concentration (i.e. at <0.1 mM), see Chapter 2; Section A.2.d for discussion. Secondly, phosphorylated intermediates cannot usually traverse the cell or mitochondrial membranes, so that the formation of glucose 6-phosphate restricts this compound to the cytosol where the glycolytic enzymes reside.

Glucose 6-phosphate is not only an intermediate in glycolysis but it is also a substrate for two other important pathways; glycogen synthesis and the pentose phosphate pathway. In this latter pathway, glucose is oxidized not for the generation of ATP but for the production of NADPH, which is an important reducing agent in a number of biosynthetic pathways. This pathway will be discussed in detail in Chapter 17; Section A.8.

4. Glucosephosphate isomerase (phosphoglucoisomerase) reaction

In preparation for the second phosphorylation of glycolysis, glucose 6-phosphate isomerizes to fructose 6-phosphate in a near-equilibrium reaction (Table 2.4) catalysed by glucosephosphate isomerase. Since the reaction involves the interconversion of the open-chain forms of the sugars, the same enzyme also catalyses ring opening of α-D-glucopyranose 6-phosphate:

α-D-glucose 6-phosphate   α-D-fructose 6-phosphate

The fructose 6-phosphate formed cyclises to a furanose ring.

5. 6-Phosphofructokinase reaction

6-Phosphofructokinase catalyses the phosphorylation of fructose 6-phosphate by ATP to form fructose bisphosphate:

$$CH_2OPO_3^{2-} \quad O \quad CH_2OH \quad HO \quad OH \quad + ATP^{4-} \longrightarrow CH_2OPO_3^{2-} \quad O \quad CH_2OPO_3^{2-} \quad HO \quad OH \quad + ADP^{3-} + H^+$$

α-D-fructose 6-phosphate α-D-fructose 1,6-bisphosphate

This second phosphorylation ensures that when the hexose is split each resultant triose will bear a phosphate group.

This enzyme catalyses a non-equilibrium reaction (Table 2.4) and there is now considerable evidence that the enzyme is subject to allosteric control by a number of metabolites that are indicators of energy demand in the cell. In this way, the rate of glycolysis is adjusted to the rate of ATP utilization so that, for example, when the latter increases in a muscle cell, due to an increase in mechanical activity, the activity of 6-phosphofructokinase is increased. The mechanisms involved in this regulation are described in Chapter 7; Section B.1.b. The enzyme may also be subject to hormonal control in both liver and muscle (see Chapters 7 and 11).

6. Fructose-bisphosphate aldolase reaction

Fructose bisphosphate is cleaved, to yield the two triose phosphates glyceraldehyde 3-phosphate and dihydroxyacetone phosphate, by fructose-bisphosphate aldolase (generally known simply as aldolase) in a reaction which is the reverse of an aldol condensation:

$$CH_2OPO_3^{2-} \quad O \quad CH_2OPO_3^{2-} \quad HO \quad OH \longrightarrow \begin{array}{c} CHO \\ | \\ CHOH \\ | \\ CH_2OPO_3^{2-} \end{array} + \begin{array}{c} CH_2OH \\ | \\ C{=}O \\ | \\ CH_2OPO_3^{2-} \end{array}$$

α-D-fructose 1,6-bisphosphate   D-glyceraldehyde 3-phosphate   dihydroxyacetone phosphate

Aldolase will also catalyse the splitting of other ketohexose 1-phosphates, including fructose 1-phosphate, and thereby plays an important role in fructose metabolism (see Chapter 11; Section F.2.a). However, conclusions on the breadth of specificity of this enzyme are complicated by the existence of isoenzymes which can be readily established by electrophoretic techniques. The isoenzymes occurring in muscle and liver have different catalytic properties; muscle aldolase has a V_{max} for fructose 1,6-bisphosphate some 50 times greater than that for fructose 1-phosphate whereas the liver enzyme shows equal activity towards the two substrates. This probably reflects the quantitative importance of the liver in fructose metabolism. Despite the contraindication of the data in Table 2.4, aldolase is usually considered to catalyse a near-equilibrium reaction in the cell.

7. Triosephosphate isomerase reaction

Dihydroxyacetone phosphate and glyceraldehyde 3-phosphate are isomers related to each other in the same way that glucose 6-phosphate is related to fructose 6-phosphate. The interconconversion of the triose phosphates is similarly catalysed by an isomerase (triosephosphate isomerase) which has a similar mechanism of action to that of glucosephosphate isomerase:

$$
\begin{array}{ccc}
CH_2OH & & CHO \\
| & & | \\
C{=}O & \longrightarrow & CHOH \\
| & & | \\
CH_2OPO_3^{2-} & & CH_2OPO_3^{2-}
\end{array}
$$

dihydroxyacetone phosphate   D-glyceraldehyde 3-phosphate

The enzyme catalyses a near-equilibrium reaction in the cell. At equilibrium, the concentration of dihydroxyacetone phosphate is 22 times that of glyceraldehyde 3-phosphate, but assessment of equilibrium is complicated by the existence of hydrated forms (gem-diols) of both trioses. For example, the hydrated form of glyceraldehyde 3-phosphate contributes 97 % to the total although the aldehyde form is the substrate of the isomerase.

8. Glyceraldehyde-phosphate dehydrogenase reaction

The formation of triose phosphates, and their isomerization, brings to a conclusion the preliminary phase of glycolysis, in which ATP is consumed. In the second part of glycolysis, ATP is produced.

The glyceraldehyde-phosphate dehydrogenase reaction is the central reaction of glycolysis. In this reaction, NAD^+ functions as an oxidizing agent and a molecule of inorganic phosphate is incorporated so that the product of the reaction is 3-phosphoglyceroyl phosphate, a mixed acid anhydride:

$$
\begin{array}{ccc}
CHO & & COOPO_3^{2-} \\
| & & | \\
CHOH \; + NAD^+ + P_i^{2-} & \longrightarrow & CHOH \quad + NADH + H^+ \\
| & & | \\
CH_2OPO_3^{2-} & & CH_2OPO_3^{2-}
\end{array}
$$

D-glyceraldehyde 3-phosphate   3-phospho-D-glyceroyl phosphate

Not surprisingly, the reaction proceeds in several steps. The structures of some of the intermediate states have been elucidated. They are described here because of the intrinsic importance of the reaction and because it has provided one possible model for the mechanism of oxidative phosphorylation in the mitochondria (Chapter 4; Section D.2.a).

The active site of the enzyme contains a binding site for NAD^+ and a thiol group which forms a covalent link with glyceraldehyde 3-phosphate. The binding of NAD^+ is unusually strong compared with most other dehydrogenases but, probably, does not involve the formation of a covalent bond:

$$E.SH + NAD^+ \longrightarrow E\begin{smallmatrix} \diagup NAD^+ \\ \diagdown SH \end{smallmatrix}$$

The binding of NAD^+ occurs first, and this affects the reactivity of the nearby thiol group which then forms a thiohemiacetal with the glyceraldehyde 3-phosphate:

$$E\begin{smallmatrix} \diagup NAD^+ \\ \diagdown SH \end{smallmatrix} \quad + \quad \begin{smallmatrix} CHO \\ | \\ CHOH \\ | \\ CH_2OPO_3^{2-} \end{smallmatrix} \quad \longrightarrow \quad E\begin{smallmatrix} \diagup NAD^+ \\ \diagdown S-CHOH \\ \quad | \\ \quad CHOH \\ \quad | \\ \quad CH_2OPO_3^{2-} \end{smallmatrix}$$

Electron transfer now occurs, so that the NAD^+ becomes reduced and the thiohemiacetal is oxidized to a thioester:

$$E\begin{smallmatrix} \diagup NAD^+ \\ \diagdown S-CHOH \\ \quad | \\ \quad CHOH \\ \quad | \\ \quad CH_2OPO_3^{2-} \end{smallmatrix} \quad \longrightarrow \quad E\begin{smallmatrix} \diagup NADH \\ \diagdown S-CO \\ \quad | \\ \quad CHOH \\ \quad | \\ \quad CH_2OPO_3^{2-} \end{smallmatrix} \quad +H^+$$

The NADH on the enzyme-substrate complex exchanges with free NAD^+. The reaction is completed by attack on the thioester by inorganic phosphate with the formation of 3-phosphoglyceroyl phosphate and regeneration of the free enzyme:

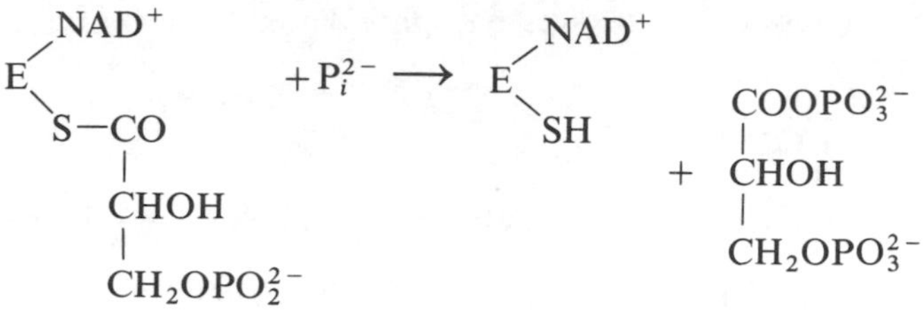

$$
E\overset{\displaystyle NAD^+}{\underset{\displaystyle \underset{\displaystyle \underset{\displaystyle CH_2OPO_2^{2-}}{|}}{\underset{\displaystyle CHOH}{|}}}{S-CO}} \quad + P_i^{2-} \longrightarrow \quad E\overset{\displaystyle NAD^+}{\underset{\displaystyle SH}{}} \quad + \quad \overset{\displaystyle COOPO_3^{2-}}{\underset{\displaystyle \underset{\displaystyle CH_2OPO_3^{2-}}{|}}{\underset{\displaystyle CHOH}{|}}}
$$

It is likely that the glyceraldehyde-phosphate dehydrogenase reaction is near-equilibrium in the cell.

9. Phosphoglycerate kinase reaction

The two phosphate groups in 3-phosphoglyceroyl phosphate are quite dissimilar so that this name is much better than the name formerly used for the compound, namely 1,3-diphosphoglycerate. The newly introduced phosphate group is involved in an acid anhydride linkage which is highly susceptible to hydrolysis and puts 3-phosphoglyceroyl phosphate into the category of a 'high-energy phosphate' compound from which the phosphate group can be readily transferred to ADP to form ATP. This reaction is catalysed by phosphoglycerate kinase and it is considered to be near-equilibrium:

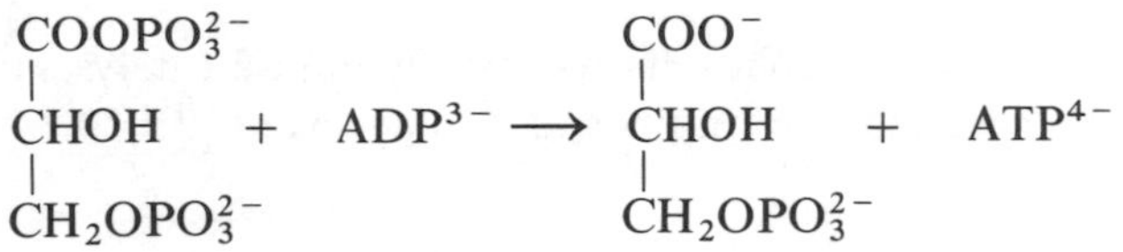

$$
\overset{\displaystyle COOPO_3^{2-}}{\underset{\displaystyle \underset{\displaystyle CH_2OPO_3^{2-}}{|}}{\underset{\displaystyle CHOH}{|}}} \quad + \quad ADP^{3-} \longrightarrow \quad \overset{\displaystyle COO^-}{\underset{\displaystyle \underset{\displaystyle CH_2OPO_3^{2-}}{|}}{\underset{\displaystyle CHOH}{|}}} \quad + \quad ATP^{4-}
$$

3-phospho-D-glyceroyl phosphate   3-phospho-D- glycerate

It should be noted that between them the two enzymes, glyceraldehyde-phosphate dehydrogenase and phosphoglycerate kinase, have catalysed the synthesis of ATP from ADP and inorganic phosphate. They have converted the chemical energy made available through the oxidation of glyceraldehyde 3-phosphate into a more immediately useful form of chemical energy for the cell. This synthesis of ATP is known as a substrate-level phosphorylation in order to distinguish it from the oxidative phosphorylation which occurs in association with the mitochondrial electron transfer chain.

10. Phosphoglyceromutase reaction

The next reaction in glycolysis, the transfer of the phosphate group from the 3-position to the 2-position, can be seen as preparation for its transfer to ADP. Phosphoglyceromutase catalyses this isomerization in a reaction that is near-equilibrium (Table 2.4). The reaction proceeds via the transient formation of 2,3-

bisphosphoglycerate and involves phosphorylation of the enzyme (at a serine residue):

$$\begin{array}{c} COO^- \\ | \\ CHOH \\ | \\ CH_2OPO_3^{2-} \end{array} + E.OPO_3^{2-} \longrightarrow \begin{array}{c} COO^- \\ | \\ CHOPO_3^{2-} \\ | \\ CH_2OPO_3^{2-} \end{array} + E.OH$$

3-bisphospho-D-glycerate   2,3-bisphospho-D-glycerate

$$\begin{array}{c} COO^- \\ | \\ CHOPO_3^{2-} \\ | \\ CH_2OPO_3^{2-} \end{array} + E.OH \longrightarrow \begin{array}{c} COO^- \\ | \\ CHOPO_3^{2-} \\ | \\ CH_2OH \end{array} + E.OPO_3^{2-}$$

2-phospho-D-glycerate

To prime this reaction, a small quantity of the diphosphorylated intermediate must be present to phosphorylate the enzyme. This 2,3-bisphosphoglycerate[4] is independently formed from 3-phosphoglyceroyl phosphate under the influence of bisphosphoglyceromutase.

11. Enolase reaction

Enolase, also known as phosphopyruvate hydratase, catalyses the dehydration of 2-phosphoglycerate to form phosphoenolpyruvate (PEP):

$$\begin{array}{c} COO^- \\ | \\ CHOPO_3^{2-} \\ | \\ CH_2OH \end{array} \longrightarrow \begin{array}{c} COO^- \\ | \\ COPO_3^{2-} \\ || \\ CH_2 \end{array} + H_2O$$

2-phospho-D-glycerate   phosphoenolpyruvate

This enzyme requires Mg^{2+} for activity. The removal of water causes such a modification in the structure that phosphoenolpyruvate can be described as a 'high energy phosphate' compound so that the phosphate group can be readily transferred to ADP to form ATP. The reaction is near-equilibrium (Table 2.4).

12. Pyruvate kinase reaction

Pyruvate kinase catalyses this transfer of the phosphate group from phosphoenolpyruvate to ADP with the formation of pyruvate. With this reaction, the part of glycolysis that is common to both aerobic and anaerobic metabolism is complete:

$$
\begin{array}{c}
COO^- \\
| \\
COPO_3^{2-} \\
|| \\
CH_2
\end{array}
+ ADP^{3-} + H^+ \longrightarrow
\begin{array}{c}
COO^- \\
| \\
CO \\
| \\
CH_3
\end{array}
+ ATP^{4-}
$$

phosphoenolpyruvate   pyruvate

The equilibrium constant for the reaction is very large and much larger than the mass action ratio so that, unlike the other ATP-forming reaction in glycolysis, it is a non-equilibrium reaction (Table 2.4). One factor contributing to this large equilibrium constant is the instability of the *enol*-pyruvate that is initially formed in the reaction. Tautomerism to the favoured *keto*-form (and its hydration to the *gem*-diol) pulls the pyruvate kinase reaction over by removing the immediate product:

$$
\begin{array}{c}
COO^- \\
| \\
COH \\
|| \\
CH_2
\end{array}
\longrightarrow
\begin{array}{c}
COO^- \\
| \\
CO \\
| \\
CH_3
\end{array}
\xrightarrow{+H_2O}
\begin{array}{c}
COO^- \\
| \\
HOCOH \\
| \\
CH_3
\end{array}
$$

enol-   keto-   gem-diol

At equilibrium, only a trace of the *enol*-form is present whereas the amount of the *keto*-form is approximately 94% and that of the *gem*-diol is about 6%.

Isoenzymes of pyruvate kinase which differ in catalytic properties are found in differing proportions in different tissues. The isoenzyme found predominantly in mammalian liver shows a sigmoid relationship between activity and the concentration of its substrate, phosphoenolpyruvate; it is activated by fructose bisphosphate and inhibited by alanine and ATP. These properties may ensure that the activity of pyruvate kinase in liver is reduced under conditions when glucose is synthesized rather than degraded (see Chapter 11; Section B.4.b for a fuller discussion). In contrast, the isoenzyme predominating in muscle from birds and mammals shows a hyperbolic response to increasing phosphoenolpyruvate concentration, is not affected by fructose bisphosphate or alanine and is only inhibited slightly by ATP. The regulation of this reaction is discussed in Chapter 7; Section B.1.b.

Under anaerobic or hypoxic conditions, vertebrate tissues reduce pyruvate to lactate by means of NADH in a reaction catalysed by lactate dehydrogenase; this reaction regenerates the NAD^+ required for the continued oxidation of glyceraldehyde 3-phosphate (see Section D.4 for further discussion). In aerobic tissues under aerobic conditions, the pyruvate is oxidatively decarboxylated to form acetyl-CoA in a reaction catalysed by the pyruvate dehydrogenase complex of enzymes. Although the reaction does not constitute part of the glycolytic pathway, it is convenient to consider it in this section since it links glycolysis with the TCA cycle which has already been discussed in Chapter 4.

However, an additional process is required before the conversion of glycolytically produced pyruvate to acetyl-CoA; this is the transport of pyruvate across the mitochondrial membrane. The inner mitochondrial membrane presents a barrier to the transfer of most molecules between the cytosol and the mitochondrial matrix so that specific transporter mechanisms must exist (see Chapter 4; Section D.4.a).

13. Conversion of pyruvate to acetyl-CoA

(a) *Mitochondrial transport of pyruvate*

Pyruvate is transported into the mitochondria via a specific carrier. In order to preserve electrical neutrality, the pyruvate anion is transported in with a proton (Halestrap, 1975).

(b) *Pyruvate dehydrogenase*

Pyruvate dehydrogenase catalyses the conversion of pyruvate into acetyl-CoA in a complex series of reactions summarized by the equation:

$$CoASH + CH_3CO\,COO^- + NAD^+ \longrightarrow CH_3CO\,SCoA + NADH + CO_2$$

Pyruvate dehydrogenase consists of three enzymes in a multienzyme complex which is attached to the inner side of the inner mitochondrial membrane. The three enzymes are pyruvate dehydrogenase, dihydrolipoamide acetyltransferase, and dihydrolipoamide reductase. Together these three enzymes catalyse reactions depicted in Figure 5.12, which are entirely analogous to those involved in the oxidative decarboxylation of 2-oxoglutarate catalysed by the oxoglutarate dehydrogenase complex (see Chapter 4; Section B.1.d): indeed the dihydrolipoamide reductase appears to be identical in both enzymes (Koike *et al.*, 1980).

The pyruvate dehydrogenase complex catalyses a non-equilibrium reaction. Consequently, in the energy-demanding tissues (e.g. muscle, brain, kidney), it commits pyruvate to oxidation via acetyl-CoA and the TCA cycle since there is no biochemical means by which acetyl-CoA can be converted back to pyruvate. In other tissues (e.g. adipose tissue, liver) this acetyl-CoA will be used for the biosynthesis of lipids, but once again, the pyruvate dehydrogenase reaction is an important *committing* reaction for pyruvate in these tissues. [Acetyl-CoA may also be used for the synthesis of such compounds as acetylcholine (in brain), cholesterol and other steroids (in liver, adrenal cortex, placenta, testes and ovaries; see Chapter 20; Section B.1]. It is not surprising, therefore, to find that it is controlled by a variety of allosteric effectors and by hormones via an interconversion cycle (see Chapter 7; Section B.2 and Chapter 8; Section B.1.b). Indeed, it is likely that the enzyme is saturated with its pathway substrate, pyruvate, so that it catalyses a flux-generating step. Hence, under aerobic conditions, glycolysis ends at the mitochondrial transport of pyruvate.

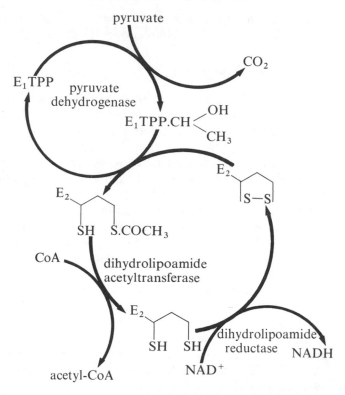

Figure 5.12 Reactions of the pyruvate dehydrogenase complex. The reactions are entirely analogous to those presented in more detail for 2-oxoglutarate dehydrogenase in Figure 4.4

Furthermore, readers will be aware that citrate synthase also catalyses a flux-generating step, so that the conversion of pyruvate to acetyl-CoA may constitute a metabolic pathway as defined in Chapter 2; Section C.2.d).

Pyruvate dehydrogenase provides an important metabolic link between glycolysis and the complete oxidation of glucose. Although the oxidation of the various lipid-fuels requires the TCA cycle, they do not involve pyruvate dehydrogenase. Consequently a deficiency of this enzyme could have serious consequences in those tissues that depend largely on glucose oxidation for ATP production. This is the case for brain and nervous tissue, and there is some evidence that some cases of spino-cerebellar ataxia, especially Friedrichs ataxia, may be due to pyruvate dehydrogenase deficiency (Blass, 1978). Friedrichs ataxia is usually familial and the condition typically begins in childhood when there is difficulty in walking. The legs are unco-ordinated, and the gait develops a reeling character. *Post mortem* studies reveal degeneration of the thoracic nucleus, the spino-cerebellar tracts, the posterior columns, and the pyramidal tracts, presumably due to deficiency of this important enzyme. Patients have been treated with a diet high in lipid so that ketone bodies become available in the blood for oxidation by the brain (see Chapter 6; Section D.2).

C. CATABOLISM OF GLYCOGEN

Glucose is not the only carbohydrate to be catabolized through the reactions of glycolysis. In some muscles during exercise and in many tissues during energy stress, glycogen is quantitatively much more important than glucose as the substrate for glycolysis. Unlike glycolysis-from-glucose, glycolysis-from-glycogen is saturated with its substrate, so that glycogen phosphorylase catalyses a flux-generating step (see Table 5.1). Indeed from a functional point of view, glycolysis-from-glucose (i.e. glucose provided from the bloodstream) and glycolysis-from-glycogen, at least in muscle, can be regarded as two separate pathways which, although they have many enzymes in common, rarely if every occur at the same time. Glycogen is usually degraded to lactate under hypoxic conditions whereas glucose is usually completely oxidized via pyruvate dehydrogenase and the TCA cycle. Even during sustained exercise when glucose or glycogen can be completely oxidized glucose utilization may inhibit that of glycogen (Jansson, 1980) and muscle glycogen degradation may only occur when liver glycogen has become exhausted (see Chapter 9; Section B.2.b).

In addition to glycogen, fructose (mainly from sucrose), and galactose (from lactose) are also metabolized to glycolytic intermediates. Since their metabolism occurs primarily in the liver it will be considered in connection with gluconeogenesis (Chapter 11; Section F).

1. **Structure of glycogen**

Glycogen is an amylopectin-like (see Section A.1) polysaccharide which functions as a storage form of glucose in many if not all tissues. It is composed entirely of glucose units but differs from amylopectin in having a larger number of α-1,6-branch points with 10–12 glucose residues between each of the inner branches. Three kinds of chain can be distinguished in glycogen (Figure 5.13).

1. There is a single C chain which terminates in the only glucose residue with a free C_1-hydroxyl group in the whole molecule (i.e. the reducing end).
2. There are many B chains, the C_1-hydroxyl ends of which form α-1,6-linkages with residues on the C chain. Some C_6-hydroxyl groups on B chain residues also participate in α-1,6-linkages.
3. There are many A chains, the C_1-hydroxyl ends of which form α-1,6-linkages with residues on B chains but which form no other α-1,6-linkages.

About 93 % of all glycoside linkages are α-1,4 and 7 % are α-1,6. Although a molecular weight can be determined for glycogen (a wide range of values up to about 2×10^8 are found) it does not exist in the cell as discrete molecules. Rather, it forms glycogen particles, visible under the electron microscope (Figure 5.14)

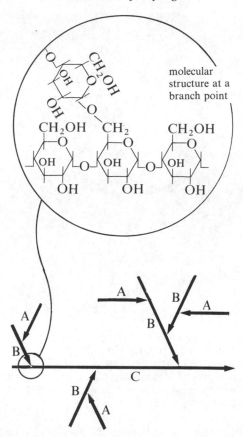

Figure 5.13 Diagrammatic representation of a small part of a glycogen molecule. Each line represents a chain of α-1,4-linked glucose units. The arrowhead represents the C_1-hydroxyl end of each chain. See text for significance of letters

which contain not only the polysaccharide but also the protein which provides the backbone for the initial synthesis of glycogen (Chapter 16; Section A.1) and the enzymes responsible for the synthesis, degradation and the control of these processes (Drockmans and Dantan 1968; Huijing, 1975). The branched structure of glycogen is important for the rapid breakdown and for the control of the rate of the degradation. It is probable that the formation of glycogen particles depends upon the branched nature of the glycogen molecule, since in one type of glycogen storage disease, where the glycogen is improperly branched, glycogen particles are absent and glycogen breakdown does not occur (see Chapter 16; Section E.4).

2. Glycogen breakdown (glycogenolysis)

Glycogen stores are mobilized by the action of the enzyme glycogen

Figure 5.14   Electron micrograph of muscle tissue from a baby with type III glycogen storage disease (see Chapter 16; Section E). Glycogen particles are the small dark objects between the myofibrils associated particularly with the Z-disc (courtesy of Professor Brenda Ryman, Department of Biochemistry, Charing Cross Hospital Medical School, London with permission from Royal College of Pathologists)

phosphorylase (more widely known simply as phosphorylase). This enzyme catalyses the phosphorolysis (splitting by inorganic phosphate) of terminal glucose units from the non-reducing end of glycogen to yield glucose 1-phosphate:

$$(\text{glucose})_n + P_i \longrightarrow (\text{glucose})_{n-1} + \text{glucose 1-phosphate}$$

Phosphorylase removes glucose units sequentially from the outer branches (Figure 5.13) but ceases to act some four units before a branch point. This leaves what is known as a limit dextrin, which consists of the core of the molecule bearing short branches of about four glucose units in length. As only 50% of glucose units are in the outer branches, this leaves about half of the molecule intact. The remaining units are made available to the phosphorylase through the activity of debranching enzyme. This enzyme is unusual in possessing two separate catalytic activities apparently in the same polypeptide chain. The first activity of this 'double-headed' enzyme is that of a glucosyltransferase which catalyses the transfer of a trisaccharide consisting of three glucose units, with α-1,4-linkages, from the end of a branch to form an α-1,4 linkage with another chain. This results in an extended chain which is now subject to the further action of glycogen phosphorylase while the single glucose 'stub' (linked α-1,6 to the main chain) is hydrolysed and produces glucose. This latter reaction is catalysed by the second activity of debranching enzyme, namely amylo-1,6-glucosidase activity. This results in some 7% of the glucose residues in glycogen being released as glucose and not glucose 1-phosphate and probably accounts for the increase in glucose concentration that occurs in muscle during periods of rapid glycogen breakdown. The debranching process is illustrated in Figure 5.15. The gradual

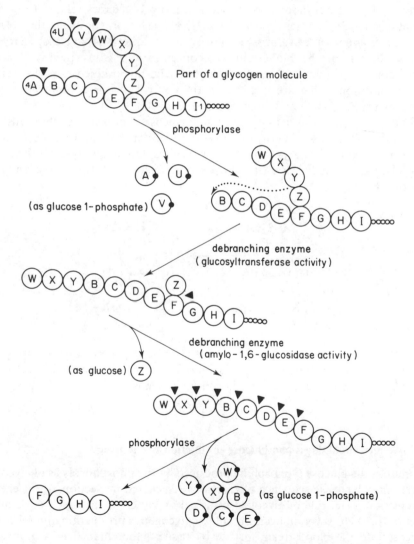

Figure 5.15 Progressive degradation of glycogen by the combined action of glycogen phosphorylase and the debranching enzyme. Arrowheads indicate sites of enzyme action

degradation of glycogen molecules by phosphorylase and debranching enzyme will occur at the same time *in vivo*, so that a 'limit dextrin' is not produced under normal physiological conditions.

It has already been stressed that glycogen phosphorylase catalyses a flux-generating step and, since the rate of glycogenolysis (and hence glycolysis) varies enormously under different conditions, the enzyme is subject to a variety of allosteric and hormonal controls via an interconversion cycle. The main role of glycogen in muscle is to provide energy for the contractile process so that the activity of phosphorylase is controlled primarily by factors related to energy metabolism (see Chapter 7; Section C.1). However, in the liver, the role of glycogen is as a reserve of glucose for use by other tissues, so that it plays a role in the regulation of the blood glucose concentration. Consequently, hepatic phosphorylase is regulated by the changes in the blood glucose concentration and by hormones that are involved in the control of this concentration (see Chapter 11; Section D and Chapter 16; Section C).

Since phosphorylase activity produces glucose 1-phosphate, this must be converted into glucose 6-phosphate in order to enter the pathway of glycolysis. The isomerization is catalysed by the enzyme phosphoglucomutase which acts by a mechanism analogous to that already described for phosphoglyceromutase (Section B.10).

α-D-glucose 1-phosphate   α-D-glucose 6-phosphate

A very small quantity of glucose 1,6-bisphosphate is necessary to prime the reaction; this is generated by the phosphorylation of glucose 1-phosphate (by ATP, in a reaction catalysed by phosphoglucokinase).

3. Glycogen breakdown and glucose production in the liver

In muscle, the glucose 6-phosphate produced by the phosphorolysis of glycogen enters the glycolytic sequence of reactions (discussed in Section B) for energy production. (The role of glycogen as a fuel in muscle is discussed in detail in Section E.2.) However, in liver, the role of glycogen is production and release of glucose into the bloodstream for use by tissues other than liver (e.g. muscle, brain). Thus, in the liver, glucose 6-phosphate is converted to glucose in a hydrolytic reaction catalysed by glucose-6-phosphatase:

$$CH_2OPO_3^{2-} \qquad\qquad CH_2OH$$

α-D-glucose 6-phosphate $\quad+H_2O \rightarrow\quad$ D-glucose $\quad+P_i^{2-}$

The glucose is transported across the cell membrane by the glucose carrier system (which is close to equilibrium in the liver) into the interstitial space and hence into the bloodstream.

A variety of diseases are caused by inherited deficiencies of enzymes concerned with glycogen metabolism. None are common and several are very rare. Nevertheless they are of interest for the light they shed on glucose and glycogen metabolism and their inter-relationships; some of these deficiencies are considered in Chapter 16; Section E, by which stage such related processes as gluconeogenesis, glycogen synthesis and the complex process of the regulation of blood glucose concentration will have been considered.

D. MECHANISMS FOR THE REOXIDATION OF CYTOSOLIC NADH

The conversion of one molecule of glucose to two of pyruvate in glycolysis generates two molecules of NADH. Since the amount of NAD^+ in the cell is limited (approximately 0.8 $\mu mol.g^{-1}$) and is small compared with its maximum rate of turnover in glycolysis (for example, about 60 $\mu mol.min^{-1}.g^{-1}$ in the leg muscle of man during sprinting) it is essential for the continuation of glycolysis that NAD^+ is rapidly reformed from the NADH produced. In aerobic tissues there would appear to be no problem since the mitochondrial electron transfer process has a large capacity for the oxidation of NADH. Thus glycolytically-produced NADH would enter the mitochondrion and suffer oxidation. Nevertheless, a potential problem remains because the inner mitochondrial membrane is not permeable to NADH or NAD^+ and there is no mitochondrial transport system for these nucleotides similar to that for ADP and ATP. This means that the NADH formed in the cytosol by glycolysis does not have direct access to the electron transfer chain. There is, however, no need for transport of the nucleotides themselves provided that electrons can traverse the mitochondrial membrane. This is achieved by substrate shuttles which, in effect, transport hydrogen atoms (known as reducing equivalents) across the membrane. In principle, these shuttles involve a reaction between NADH and an oxidized substrate in the cytosol, followed by transport of the reduced substrate into the mitochondrion and its subsequent oxidation by the electron transfer chain (Dawson, 1979). Two of these shuttles have been well-established and are described below.

1. Malate/aspartate shuttle

One of the first shuttles to be proposed was the malate/aspartate shuttle in which oxaloacetate is reduced to malate in the cytosol and the reverse reactions occurs in the mitochondrion. Both reactions are catalysed by the enzyme malate dehydrogenase, which is present in both the cytosolic and the mitochondrial compartments of the cell.

$$\underset{\text{oxaloacetate}}{\begin{array}{l}CO.COO^- \\ | \\ CH_2COO^-\end{array}} + NADH + H^+ \longrightarrow \underset{\text{malate}}{\begin{array}{l}HO.CH.COO^- \\ | \\ CH_2COO^-\end{array}} + NAD^+$$

The NADH produced within the mitochondrion will be oxidized by the electron transfer process and produce three molecules of ATP. The shuttle requires a transport system for the translocation across the mitochondrial membrane of both oxaloacetate and malate. However, although such a transporter exists for malate (see Chapter 4; Section D.4.a) there is no transporter for oxaloacetate. The problem of the translocation of oxaloacetate is overcome by the transport of the carbon skeleton of oxaloacetate in the form of aspartate. The aspartate is produced by a process known as transamination (see Chapter 10; Section C.2.b) in which the amino group of glutamate is transferred to oxaloacetate thereby forming aspartate and leaving 2-oxoglutarate; this reaction is catalysed by aspartate aminotransferase:

$$\underset{\text{oxaloacetate}}{\begin{array}{l}CO.COO^- \\ | \\ CH\,_2COO^-\end{array}} + \underset{\text{glutamate}}{\begin{array}{l}{}^+H_3N.CH.COO^- \\ | \\ CH_2 \\ | \\ CH_2COO^-\end{array}} \longrightarrow \underset{\text{aspartate}}{\begin{array}{l}{}^+H_3N.CH.COO^- \\ | \\ CH_2COO^-\end{array}} + \underset{\text{2-oxoglutarate}}{\begin{array}{l}CO.COO^- \\ | \\ CH_2 \\ | \\ CH_2COO^-\end{array}}$$

Both aspartate and 2-oxoglutarate can be transported across the inner mitochondrial membrane to the cytosol where oxaloacetate is regenerated by reversal of the same aminotransferase reaction. The complete scheme is depicted in Figure 5.16. This shuttle appears to be quantitatively the most important for oxidation of cytosolic NADH in all vertebrate tissues (Crabtree and Newsholme, 1972a; Williamson *et al.*, 1973).

2. Glycerol phosphate shuttle

An alternative mechanism for the transport of reducing equivalents into mitochondria is provided by the glycerol phosphate shuttle (Figure 5.17). In the cytosol, NADH reduces dihydroxyacetone phosphate to glycerol 3-phosphate[5] in a reaction catalysed by glycerol 3-phosphate dehydrogenase (NAD^+):

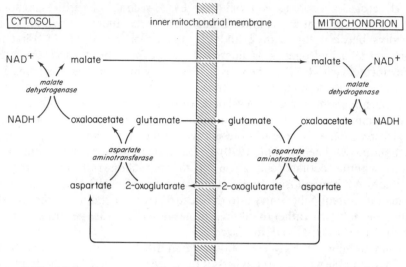

Figure 5.16 Malate/aspartate shuttle for the transport of reducing equivalents from cytosol into the mitochondrion

$$
\begin{array}{l}
\text{CH}_2\text{OH} \\
| \\
\text{CO} \qquad + \text{NADH} + \text{H}^+ \longrightarrow \\
| \\
\text{CH}_2\text{OPO}_3^{2-}
\end{array}
\qquad
\begin{array}{l}
\text{CH}_2\text{OH} \\
| \\
\text{CH.OH} \qquad + \text{NAD}^+ \\
| \\
\text{CH}_2\text{OPO}_3^{2-}
\end{array}
$$

dihydroxyacetone phosphate   L-glycerol 3-phosphate

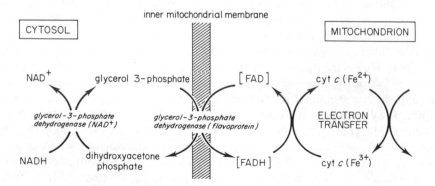

Figure 5.17 Glycerol phosphate shuttle for the transport of reducing equivalents from cytosol into the mitochondrion. It is likely that the FAD and cyt *c* are components of the inner mitochondrial membrane (see text)

The glycerol 3-phosphate is reoxidized by a second glycerol-3-phosphate dehydrogenase present in the mitochondria. However, there are two important differences between this shuttle and the malate/aspartate shuttle. First, the mitochondrial dehydrogenase is different in nature from the cytosolic enzyme; it is a flavoprotein with FAD as the prosthetic group which removes electrons from glycerol 3-phosphate and donates them directly to the electron transfer chain within the mitochondrion at the level of ubiquinone. Secondly, the mitochondrial dehydrogenase is located on the outside of the inner membrane so that it reacts with glycerol 3-phosphate that is present within the intermembrane space of the mitochondria (Donnellan *et al.*, 1970). In this way, mitochondrial ubiquinone is reduced without actual penetration of the mitochondrion by glycerol 3-phosphate. (A mitochondrial transporter system for glycerol 3-phosphate has not been demonstrated.) Since entry into the electron transfer chain is at the level of a ubiquinone, only two, rather than three, molecules of ATP are generated for each molecule of cytosolic NADH oxidized.

The functioning of the glycerol phosphate shuttle has been well established in insect flight muscle where the activities of the enzymes involved are high. In most mammalian tissues these activities are not only low but considerably lower than the maximum rate of oxidation of glycolytically-produced pyruvate, so that this shuttle cannot account for much of the cytosolic NADH oxidation (see Table 5.2). The activities of the mitochondrial glycerol-3-phosphate dehydrogenase are higher in mammalian brain and liver than in muscle but the quantitative importance of the shuttle in these tissues is not known.

3. Significance of the reducing equivalent shuttles

It is interesting and metabolically instructive to consider the reasons for mitochondria being impermeable to nicotinamide adenine nucleotides. The inner mitochondrial membrane is impermeable to charged molecules, especially those as large as nucleotides. Nonetheless, for the large number of compounds that must traverse this membrane, carriers have evolved. It is unlikely that there is anything inherent in the structures of NAD^+ and NADH that prevents their transport. However, rapid transport across the mitochondrial membrane would tend to equalise the concentrations in the two compartments (i.e. the cytosol and the mitochondrial matrix) and this appears to be highly undesirable. Although it is not possible to measure directly the local concentrations of metabolites in different cell compartments, it is possible to estimate the $NAD^+/NADH$ concentration ratios in the cytosol and in the mitochondrion (see Chapter 7; Note 1 and Krebs, 1973). The available data indicate that the values of the ratios are markedly different in the two compartments; approximately 1000 in the cytosol and approximately eight in the mitochondrion. In other words, the cytosol is considerably more oxidized than the mitochondrial matrix. A high $NAD^+/NADH$ ratio is necessary in the cytosol to ensure that the value of ΔG for the glyceraldehyde-phosphate dehydrogenase reaction remains negative without

Table 5.2. Maximum activities of lactate dehydrogenase and the enzymes of the glycerol phosphate shuttle together with glycolytic flux* in muscles from different species

| Animal | Muscle | Enzyme activities (μmol.min^{-1}.g^{-1}) at 25 °C | | | Glycolytic flux |
| | | Glycerol-3-phosphate dehydrogenase | | Lactate dehydrogenase | |
		Cytosolic	Mitochondrial		
Locust (*Schistocerca gregaria*)	flight	141	43	3	34
Waterbug (*Lethocerus cordofanus*)	flight	51	8	1	12
Honey bee (*Apis mellifera*)	flight	257	44	1	40
Bumble bee (*Bombus hortorum*)	flight	513	90	2	66
Polar hawk moth (*Laothoe populi*)	flight	36	13	3	18
Blowfly (*Phormia terranova*)	flight	300	110	2	92
Dogfish (*Scyliorhinus canicula*)	red	14	0.1	110	28
Frog (*Rana temporaria*)	sartorius	24	0.4	398	44
Domestic pigeon (*Columba livia*)	pectoral	33	1.2	314	48
Domestic fowl (*Gallus gallus*)	pectoral	103	2.8	542	246
Laboratory rat (Wistar strain)	heart	6	0.3	310	20
	quadriceps	48	1.2	448	94
Man (*Homo sapiens*)	quadriceps	9	—	121	58

*Glycolytic flux estimated by multiplying maximum activity of 6-phosphofructokinase by two and equivalent to μmol and pyruvate produced min^{-1}·g^{-1} at 25°C. Data taken from Crabtree and Newsholme, 1972a.

a large 3-phosphoglyceroyl phosphate/glyceraldehyde 3-phosphate concentration ratio (see Chapter 2; Section A.2.c). However, in the mitochondria,
NADH is the substrate for the electron transfer chain, so that the NAD^+/NADH
concentration ratio must be sufficiently small to ensure reduction of the flavin
carriers (i.e. the flow of electrons from NADH to flavin carriers). Consequently,
the two compartments must be totally separate from the point of view of the
NAD^+/NADH ratio, although a functional link to transport reducing
equivalents is necessary. This is achieved via one or more shuttles as described
above, which must of necessity be non-equilibrium processes (i.e. at least *one*
reaction in the shuttle must be non-equilibrium). In the glycerol phosphate
shuttle, the mitochondrial dehydrogenase catalyses a non-equilibrium reaction
but, for the malate/aspartate shuttle, it is probably one of the transport processes
(Williamson, *et al.*, 1973).

The possibility remains that not all the reducing power generated in the cytosol
is transported into mitochondria; some may be used for reductive syntheses in the
cytosol. This would require the conversion of NADH to NADPH, possibly via
the pyruvate/malate cycle (see Chapter 17; Section A.3.b).

4. Formation of lactate

Under anaerobic conditions, oxygen is not available to accept electrons in the
cytochrome *c* oxidase reaction and the electron carriers in the electron transfer
chain become almost totally reduced. Consequently, cytosolic NADH cannot be
oxidized in the mitochondria but, in order for glycolysis to proceed, which it must
since it is the only pathway capable of generating energy under these conditions,
the NAD^+/NADH ratio must be maintained at the high values characteristic of
the cytosol. In vertebrates, quantitatively the most important process for re-
oxidation of NADH is the reduction of pyruvate to lactate in the reaction
catalysed by lactate dehydrogenase:

$$CH_3CO.COO^- + NADH + H^+ \longrightarrow CH_3CH(OH)COO^- + NAD^+$$

pyruvate $\qquad\qquad\qquad\qquad\qquad$ L-lactate

Another reaction also plays a role in this re-oxidation namely the cytosolic
glycerol-3-phosphate dehydrogenase reaction. Although the activity of the
mitochondrial dehydrogenase is low in vertebrate muscle (see Table 5.2), the
activity of the cytosolic dehydrogenase is high. It is likely that this reaction plays
an important role in the re-oxidation of NADH in the initial stages of
anaerobiosis (Klingenberg and Bücher, 1960).

$$
\begin{array}{lll}
\mathrm{CH_2OH} & & \mathrm{CH_2OH} \\
| & & | \\
\mathrm{CO} & +\,\mathrm{NADH} + \mathrm{H^+} \longrightarrow & \mathrm{CH.OH} \qquad +\,\mathrm{NAD^+} \\
| & & | \\
\mathrm{CH_2OPO_3^{2-}} & & \mathrm{CH_2OPO_3^{2-}}
\end{array}
$$

dihydroxyacetone phosphate $\qquad\qquad$ glycerol 3-phosphate

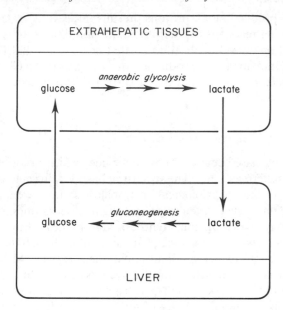

Figure 5.18   Cori cycle

Thus the concentration of both glycerol 3-phosphate and lactate increase in muscle during anaerobic conditions. Lactate can be produced continuously, since it is released from the muscle into the bloodstream, but glycerol 3-phosphate cannot pass across the cell membrane so that it increases rapidly to a high concentration which remains within the muscle. The glycerol 3-phosphate concentration decreases when the muscle becomes aerobic once again, probably due to oxidation via conversion to pyruvate. The lactate that diffuses into the bloodstream is carried to the liver where it is taken up and reconverted to glucose in the process known as gluconeogenesis (Chapter 11; Section B.2.a). The production of lactate by extrahepatic tissues, its transport to the liver and its resynthesis into glucose to resupply the tissues is known as the Cori cycle (Figure 5.18). It is of interest to note that the removal of lactate by the blood and its subsequent uptake by the liver to form glucose extend the 'physiological pathway' of glycolysis. It can be considered to finish with the lactate dehydrogenase reaction in the liver or other tissue which oxidizes the lactate, but it may be extended further if lactate is converted to glucose (see Chapter 11; Section B.2.c for discussion of physiological pathway of gluconeogenesis in the liver).

The over-all process of anaerobic glycolysis, from either glucose or glycogen, results in the formation of lactate plus protons, so that glycolysis can be written as follows:

$$\text{glucose} \longrightarrow 2 \text{ lactate}^- + 2\text{H}^+$$

Normally the production of protons in this process is balanced by their utilization in other reactions (e.g. aerobic oxidation) but in some conditions the rate of production exceeds utilization so that acidosis results. This problem is discussed in relation to proton production during sprinting in Chapter 9; Section A.4.a and in relation to the clinical disturbance known as lactic acidosis in Chapter 13; Section D.

(a) *Isoenzymes of lactate dehydrogenase*

Lactate dehydrogenase occurs as five isoenzymes which are found in differing proportions in differing tissues. The structural basis for this heterogeneity is well established but the physiological rationale is less clear. The lactate dehydrogenase molecule is a tetramer and the five isoenzymes arise due to the existence of two types of monomer (synthesized from the information encoded in two different genes) which can associate in any combination. If the monomers are designated H subunits (which predominate in the heart) and M subunits (which predominate in skeletal muscle) then the five possible tetramers are H_4, H_3M, H_2M_2, HM_3, and M_4.[6] In skeletal muscle from a wide variety of species there is a good correlation between the ability of muscles to work anaerobically and the predominance of M forms of lactate dehydrogenase. In tissues other than skeletal muscle, the correlation is less good since in liver, an aerobic tissue, the M_4 isoenzyme predominates and in erythrocytes, which are anaerobic, the H_4 isoenzyme is most abundant. A physiological interpretation of some aspects of these findings has been put forward by N. O. Kaplan based on the differing kinetic properties of the lactate dehydrogenase isoenzymes (Kaplan and Everse, 1972). The H_4 isoenzyme has a lower K_m for pyruvate but is strongly inhibited by high concentrations of that substrate, the M_4 isoenzyme has a higher K_m but is less strongly inhibited by pyruvate; the other isoenzymes have intermediate properties. It is claimed that these properties make the M_4 isoenzyme well adapted to convert pyruvate to lactate in anaerobic skeletal muscle whereas, in heart, this is not favoured so that pyruvate is directed into the mitochondrion for oxidation. In this tissue, it is claimed, lactate dehydrogenase is likely to work in the direction of pyruvate formation since the latter can be completely oxidized by the heart. This interpretation has been given credence by inclusion in several textbooks. However, there are several important difficulties associated with the hypothesis. It has been doubted whether the properties described above pertain in the cell. For example, the differences in kinetics between M and H forms are less marked at 37 °C than at 25 °C, and the inhibition by pyruvate may not occur at the high concentrations of the enzyme encountered in cells. (Note that enzyme activities and kinetic properties are normally measured in greatly diluted cell extracts.) Furthermore, the inhibition may be caused, not by the *keto* form of pyruvate (which is the substrate of the enzyme) but by traces of the *enol* form (see Section B.12). Since tautomerization is slow, the proportion of the *enol* form could be much lower in the cell than in samples of pyruvate used in the laboratory. Readers who have fully comprehended the contents of Chapter 2 should be aware of a second argument against Kaplan's hypothesis; lactate dehydrogenase catalyses a near-equilibrium reaction so that pyruvate inhibition (which is effectively an allosteric type of inhibition) would have a negligible effect on the flux through the reaction. Finally, there is strong experimental evidence that in heart, both *in vivo* and *in vitro*, the rate of conversion of pyruvate to lactate is increased under anaerobic conditions even though the concentration of pyruvate is increased. The arguments concerning the possible roles of the isoenzymes of lactate dehydrogenase have been discussed above in some detail since it is important for the reader to be aware of the dangers inherent in an extrapolation of the properties of an enzyme *in vitro* to the situation that exists *in vivo* (Newsholme and Crabtree, 1981a). Furthermore, it is salutary to note that, although the three-dimensional structure of lactate dehydrogenase has been elucidated (Holbrook *et al.*, 1975), we are still not clear of the precise physiological role of the isoenzymes of lactate dehydrogenase.

Notwithstanding this ignorance, the existence of multiple forms of lactate dehydrogenase is of considerable diagnostic significance. Tissues damaged by disease or injury release enzymes into the blood, where their detection and the measurement of their activities can provide information about which tissue is affected by the disease (see Chapter 3; Section D.1.b). The pattern of lactate dehydrogenase enzymes present in the blood is particularly useful in distinguishing between myocardial infarction and liver diseases (such as infective hepatitis), since liver contains the M-type of the dehydrogenase and heart, of course, contains the H-type.

E. ATP PRODUCTION FROM AEROBIC AND ANAEROBIC METABOLISM

1. Aerobic metabolism

The basic metabolic information provided in this chapter and in Chapter 4 enables the number of ATP molecules produced in the metabolism of glucose to be calculated. For this calculation the pathway of glucose oxidation can be arbitrarily divided into separate 'processes' as follows.

1. The conversion of one molecule of glucose to pyruvate is associated with the net production of two molecules of ATP.
2. Two molecules of NADH are produced during the conversion of one molecule of glucose to pyruvate. If it is assumed that the NADH is reoxidized via the malate/aspartate shuttle (Section D.1) this will result in the formation of six molecules of ATP.
3. In the conversion of two molecules of pyruvate to acetyl-CoA, two molecules of NADH are produced and the oxidation of these molecules of NADH will produce six molecules of ATP.

4. Glucose produces 2 molecules of acetyl-CoA, via glycolysis and the pyruvate dehydrogenase reaction, and the oxidation of these via the TCA cycle produces 24 molecules of ATP (Chapter 4; Section B).

Table 5.3. Yield of ATP from various fuels under aerobic and anaerobic conditions. It is assumed that fuels are utilized by pathways described in Chapters 4, 5, and 6

Fuel	Conditions	ATP yield (mol) per mol of fuel utilized
Glucose	aerobic, complete oxidation	38
Glucose	anaerobic, conversion to lactate	2
Glycogen	aerobic, complete oxidation	39
Glycogen	anaerobic, conversion to lactate	3
Palmitate	aerobic, complete oxidation	129
Acetoacetate	aerobic, complete oxidation	24

Hence the complete oxidation of one molecule of glucose produces 38 molecules of ATP (see also Table 5.3). However, if the glycerol phosphate shuttle is used to reoxidize glycolytically-produced NADH, only 36 molecules of ATP will be produced (since glycerol 3-phosphate is oxidized via a flavin-linked rather than an NAD^+-linked dehydrogenase, see Section D.3).

If glycogen is used as the substrate for glycolysis, 39 molecules of ATP will be produced for the complete oxidation of every glucose equivalent (since the first phosphorylation is achieved using P_i rather than ATP).

Thus, from a knowledge of the metabolic pathways, the number of ATP molcules that can be generated from the oxidation of a given number of molecules of substrate can be calculated. Although the process of β-oxidation of fatty acids has not yet been described, it will be shown that the oxidation of one molecule of palmitate $(CH_3(CH_2)_{14}COO^-)$ produces 129 molecules of ATP (see Table 5.3). This must not be taken to mean necessarily that fatty acid is a better fuel than glucose; one reason for the larger number of ATP molecules produced from fatty acid oxidation is that palmitate is a bigger molecule than glucose (i.e. it contains more atoms of carbon and hydrogen). This statement provides the opportunity to re-emphasize two important points concerning fuel oxidation and energy production. First, the actual number of ATP molecules that are produced in the living cell will depend upon the value of the P/O ratio, which is measured with isolated mitochondria *in vitro* (see Chapter 4; Section D.1.b, for discussion). If the *in vivo* ratio is different from 3.0, the number of ATP molecules given in Table 5.3 will need to be reappraised. (However, we do not know the value of the P/O ratio *in vivo*.) Secondly, one *biologically* important difference between fuels is the energy released upon oxidation of a given weight of fuel. Since we eat intermittently, fuel must be stored and lipid is a much more efficient store of energy than is carbohydrate (see Chapter 6; Section E.1).

2. Anaerobic metabolism

The conversion of glucose to lactate, which produces 2 molecules of ATP, is important in a number of tissues in higher animals including erythrocytes, white blood cells, the kidney medulla and tissues of the eye (see Section F). However, in muscle, it is the conversion of *glycogen*, rather than glucose, to lactate that is the quantitatively important process for energy production under anaerobic or hypoxic conditions. This has the energetic advantage that three ATP molecules are produced per glucose residue converted to lactate, which represents an increase in 'efficiency' of 50% over the conversion of glucose to lactate. It should be noted that anaerobic metabolism of glucose (even if the glucose residues are derived from glycogen) produces less than 10% of the energy available from oxidative metabolism. This biochemical fact has some interesting physiological and clinical consequences, some of which are discussed in the following sections.

F. PHYSIOLOGICAL IMPORTANCE OF ANAEROBIC GLYCOLYSIS

Most tissues that receive an adequate supply of oxygen will derive their ATP requirements from the oxidation of carbohydrates, lipids or even amino acids. The inter-relationships between these various oxidizable fuels will be discussed further in subsequent chapters. However, in the absence of oxygen or in tissues that do not contain mitochondria, glycolysis is the only pathway capable of generating ATP. Similarly, the extra energy required when the demand for energy exceeds that which can be supplied from aerobic metabolism is also obtained from glycolysis.

In most animals (with the notable exception of the insects) oxygen is carried to the tissues by the blood, so that an inadequate blood supply, in relation to metabolic need, is an obvious reason why tissues might possess a significant capacity for anaerobic metabolism. On the other hand, some tissues in higher animals have a ready supply of oxygen and yet rely mainly or totally on anaerobic glycolysis for energy production (e.g. erythrocytes). The reasons for this will be explored below.

Much of the argument presented in this section will hinge on measurements of the relative contribution of different fuels and different pathways to ATP generation. These measurements are not always easy to make and frequently doubt is expressed as to their reliability. It is therefore important to have some knowledge of the methods used; these are given in Appendix 5.2. An indication of the relative importance of anaerobic versus aerobic metabolism in muscles can be obtained by comparing the activity of 6-phosphofructokinase, a quantitative indicator of anaerobic glycolysis, with that of oxoglutarate dehydrogenase, a quantitative indicator of the TCA cycle (see Table 5.4).

1. Skeletal muscle

The chemical energy made available by anaerobic glycolysis (i.e. glycogen to lactate) is less than 10 % of that arising from the complete oxidation of glycogen (i.e. three compared to 39 ATP molecules per glucose molecule, see Section E). Nonetheless, anaerobic glycolysis makes a vital contribution under at least three conditions: it is always important for the provision of energy for mechanical activity in the more anaerobic type of fibre (Type IIB or white fibres—see below); it is important in all muscles during the initial period of exercise before the exercise-stimulated increase in blood supply occurs which increases the oxygen supply to the muscles; and it is important when the rate of ATP demand exceeds the maximum rate of aerobic ATP production. In order to understand more clearly the role of anaerobic glycolysis in human muscle, it is necessary to appreciate the metabolic and physiological differences between different types of muscles.

(a) *Classification of skeletal muscles*

Muscles differ widely in both their physiological properties (e.g. speed of contraction) and metabolic properties (e.g. oxidative capacity). Attempts have been made to classify muscles according to metabolic criteria but at least two kinds of problem arise. First, single anatomical muscles are composed of many fibres and there are several types of fibre. These fibres differ both physiologically and metabolically. Furthermore, different muscles possess variable proportions of these types of fibre and this proportion varies from one individual to another, so that it is more useful to classify fibres within a muscle than muscles *per se*. Secondly, classification schemes that apply to fibres from one species are not always applicable to those from another species. In particular, it appears that fibres from human muscles are less metabolically distinct than those in certain other well-studied species (e.g. guinea-pig, see Peter *et al.*, 1972).

Originally, muscle fibres (and in many animals whole muscles—see below) were distinguished by their colour, red or white (see Needham, 1926), which is dependent upon the amounts of myoglobin and cytochromes in the muscle.[7] However, it was found that the redness of a muscle was not necessarily a good indication of its physiological or metabolic characteristics (Denny-Brown, 1929) although the terms red and white are still frequently used in the literature. The better means of distinguishing between different fibre types is by making a section of a muscle and measuring the activity of a particular enzyme by histochemical techniques (see Figure 5.19). Although a number of different enzymes have been used, the sensitivity of myosin ATPase to extremes of pH is the most widely used method and on this basis human muscle is classified into at least three types of fibre, I, IIA, and IIB (Saltin *et al.*, 1977).[8]

Type I fibres are also known as slow twitch fibres since the time course of the maximal twitch is longer than type II fibres, which are known as fast twitch fibres; type IIA fibres are known as fast twitch-oxidative and type IIB are known as fast twitch-glycolytic (see Chapter 9; Section A.1).

The metabolic differences between fibre types are relevant to the present

Table 5.4. Maximum activities of hexokinase, phosphorylase, 6-phosphofructokinase, arginine or creatine kinase, and oxoglutarate dehydrogenase†

Animal	Muscle	Enzyme activities (μmol.min^{-1}.g^{-1} fresh tissue at 25 °C)				
		Hexokinase	Glycogen phosphorylase	6-Phospho-fructokinase	Arginine or creatine kinase	Oxoglutarate dehydrogenase
Great scallop (*Pecten maximus*)	phasic adductor	0.1	8.5	9.4	930	—
Lobster (*Homarus vulgaris*)	deep abdominal flexor	0.6	8.6	9.9	1754	—
Locust (*Schistocerca gregaria*)	flight	11.5	7.5	17.0	205	21
Bumble bee (*Bombus* sp.)	flight	114	8.0	33	138	47
Yellow underwing moth (*Noctua pronuba*)	flight	21.4	—	—	218	40
Dogfish (*Scyliorhinus canicula*)	red	1.9	12.0	14.0	100	2.2
	white	0.1	62.0	50.0	258	<0.1
Trout (*Salmo gairdneri*)	red	2.6	14.0	12.2	609	1.8
	white	1.5	48.0	58.3	1098	<0.1
Domestic pigeon (*Columbia livia*)	pectoral	3.0	18.0	24.0	456	3.2
Domestic fowl (*Gallus gallus*)	pectoral	1.1	83.0	105.0	391	0.1
Laboratory rat (*Wistar strain*)	heart	6.1	12.0	10.0	123	6.8
	quadriceps	1.9	50.0	47.0	851	—
Man (*Homo sapiens*)	quadriceps	0.8	—	57	446	1.2

* Note that all activities are given at 25 °C and not at the physiological temperature of the organism.

† Data taken from Crabtree and Newsholme, 1972a; Zammit and Newsholme, 1976; Newsholme et al., 1978; Paul, 1979. Arginine kinase is present in the invertebrates and creatine kinase in the vertebrates. Both enzymes are measured in the direction of phosphagen formation. Activities are four-fold higher in the opposite direction.

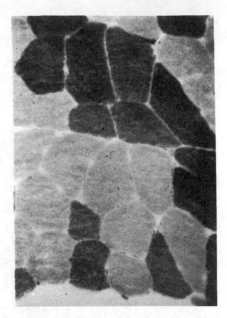

Figure 5.19   Staining for type I and II fibres of man. On the left-hand side, the ATPase stain has been used in which type II fibres stain dark (see note 8). On the right-hand side the succinate dehydrogenase stain has been used. (Courtesy of Dr. Archie Young, Nuffield Department of Surgery, Oxford)

discussion. Type I fibres have a high oxidative capacity (i.e. high activities of enzymes of the TCA cycle, of fatty acid oxidation and of the electron transfer chain), a high content of triacylglycerol and low glycolytic capacity. They are also known as fatigue resistant, a property that depends upon their capacity for oxidation of fatty acids (see Chapter 9; Section B.1). Type IIA fibres have a high oxidative and high glycolytic capacity and an intermediate content of triacylglycerol. Type IIB fibres have a low oxidative and a high glycolytic capacity and a low content of triacylglycerol (see Table 5.5.).

(b) *ATP production in type II fibres*

Although all human muscles contain both type I and type II fibres, some animals possess muscles which consist almost exclusively of a single type of fibre. It is from comparative studies of such muscles that a picture of the metabolism of the different fibre types and their physiological role has emerged. Examples of 'white' muscles containing virtually only type IIB fibres are lobster abdominal muscle, fish white abdominal muscles, game bird pectoral muscles (including the domestic fowl), and the psoas muscles of the rabbit. All these muscles can contract very rapidly and vigorously but for very short periods of time; not surprisingly, they power escape reactions (e.g. the tail flick of the lobster, the flight of the

Table 5.5. Characterization of human muscle fibre types*

Property	Type I (Slow twitch)	Type II A (fast twitch-oxidative)	Type II B (fast twitch-glycolytic)
ATPase activity after pre-incubation at pH 10.3	—	+ + +	+ + +
ATPase activity after pre-incubation at pH 4.6–4.8 and 10.3	—	—	+ + +
Speed of contraction	slow	fast	fast
Glycolytic capacity	low	moderate	high
Oxidative capacity	high	moderate	low
Glycogen store	moderate-high	moderate-high	moderate-high
Triacylglycerol store	high	moderate	low
Capillary supply	good	moderate	poor

* Information from Saltin *et al.* (1977).

pheasant). They possess a poor blood supply so that the provision of glucose as a fuel for mechanical activity is totally inadequate; they contain very few mitochondria (Figure 5.20), very low activities of the enzymes of the TCA cycle but very high activities of the glycolytic enzymes, except for hexokinase (Table 5.4). The fuel used by these muscles is endogenous glycogen, which is present in large amounts (in man, about $80 \mu mol.g^{-1}$ fresh wt.) and, since the glycolytic capacity is large, the degradation of glycogen to lactate could occur in seconds rather than minutes. Since the power output of short-term violent exercise (e.g. flight for escape in birds, the running for cover by the rabbit, sprinting in man) is higher than the power output of continuous exercise (e.g. sustained flight of the pigeon, marathon running in man) it is perhaps surprising that the energy is obtained from the inefficient process of glycolysis. What is the advantage of forsaking the high ATP yield of aerobic metabolism for the lower yield of glycolysis? One answer is that it may be difficult to get oxygen to the fibres sufficiently rapidly; vasodilation and the change in distribution of blood within the body take several minutes to establish. A second point to be borne in mind is an architectural one; more myofibrils can be located within a fibre in the absence of mitochondria. Since many of these muscles are used in escape reactions, presumably the evolution of a muscle able to perform a greater amount of work in a given time is highly advantageous. The price paid is that mechanical activity can only be maintained for short periods of time either because of the limited store of glycogen within the muscle or the accumulation of protons during the exercise (see Chapter 9; Section A.4.a). If the glycogen phosphorylase were fully activated,

Figure 5.20   Anaerobic muscle. L.S. pectoral muscle (white muscle) of the pheasant showing the low frequency of mitochondria (M)

the glycogen would be depleted in less than twenty seconds in many of these white muscles.[9] Indeed, the limited duration of the escape reaction has been exploited by the hunter for many centuries. In *Anabasis*, Xenophon told the story of the expedition which Cyrus the Younger led against his brother Artaxerxes II, king of Persia, in 401 B.C. in the hope of gaining for himself the Persian throne. In order to obtain food for the army, the Greeks had to be aware of the behaviour of their quarry. Thus Xenophon wrote, in about 394 B.C.,

'The bustards[10] on the other hand can be caught if one is quick in starting them up, for they fly only short distances, like partridges, and soon tire; and their flesh was delicious.'

In most fish, 95 % or more of the muscle mass is composed of white (type IIB) fibres (see Figure 5.21). The red muscle is located just under the skin along the lateral line, especially towards the posterior of the fish and is used for normal cruising swimming, whereas the white muscle is used in vigorous swimming such as required for escape from danger (Bone, 1966). Presumably when a fisherman 'plays' his fish, after hooking it, he is allowing the fish to deplete the glycogen content of its white muscle so that, when the fish is landed, mechanical activity is restricted to the very small amount of red muscle. If all the muscle of the fish comprised type I fibres, the fisherman would wait all day (and perhaps all night) to land his catch!

Glycolysis is not the only process for the generation of ATP in muscle under anaerobic conditions, although quantitatively it is the most important. Both

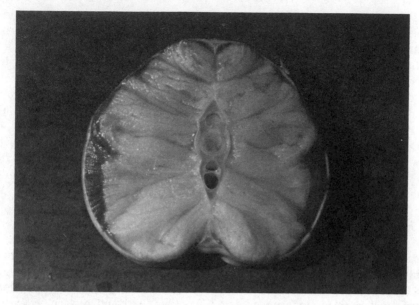

Figure 5.21 Transverse section through tail of dogfish showing bulk of white muscle flanked by small amount of red muscle close to the lateral line of the fish

muscle and nervous tissue contain phosphagens, that is phosphorylated guanidine compounds which can be described as 'energy-rich compounds' since the phosphate group can be transferred directly to ADP to form ATP. The phosphagen in vertebrate tisues is invariably phosphocreatine (also known, less correctly, as creatine phosphate) whilst that in invertebrates is generally phosphoarginine. The transfer of the phosphate group is catalysed by creatine (or arginine) kinase in a reaction with an equilibrium constant not far from unity and the reaction is very near to equilibrium *in vivo*:

$$^-OOC.CH_2N\ C\overset{\displaystyle /NH.PO_3^{2-}}{\underset{\displaystyle \diagdown NH}{\big|}} \quad +ADP^{3-}+H^+ \longrightarrow\ ^-OOC.CH_2N\ C\overset{\displaystyle /NH_2}{\underset{\displaystyle \diagdown NH}{\big|}} \quad +ATP^{4-}$$

$$\underset{\text{phosphocreatine}}{CH_3} \qquad\qquad\qquad\qquad\qquad \underset{\text{creatine}}{CH_3}$$

The concentration of phosphocreatine in human quadriceps muscle is $18\,\mu\text{mol.g}^{-1}$. The phosphorylation of ADP by phosphocreatine is particularly important in the initial stages of the exercise (the first 4–5 sec in sprinting in human subjects) before the rate of glycolysis has been increased to a new steady state (see Chapter 9; Section A.2). It is possible that the ATP required in a very short explosive burst of activity (e.g. in high jumping, shot putting) is obtained almost exclusively from phosphocreatine. If this is so, then the concentration of the phosphocreatine and the activity of creatine kinase should be very high in the muscles of top class high jumpers and shot putters. There is some indirect evidence that virtually all the ATP required for the tail-flicks of the lobster arises from phosphoarginine (Zammit and Newsholme, 1976). It is, however, wrong to consider that phosphocreatine plays an important role only in white muscle; it is important in red (i.e. type I fibres) and in heart muscle, since it can help to prevent large changes in the ATP concentration during a short lasting energy-stress condition (e.g. at the onset of a sudden demand for energy) prior to the stimulation of glycolysis and the TCA cycle (see Chapter 9 for full discussion). Since creatine kinase catalyses a near-equilibrium reaction, its maximum activity in any muscle cannot provide quantitative information on flux of energy-rich phosphate from phosphocreatine to ADP. Only qualitative information can be obtained (Table 5.4).

(c) *ATP production in the initial period of exercise*

During rest, it is likely that most of the ATP required in all types of muscle fibres is obtained from aerobic metabolism (oxidation of glucose, fatty acids, etc.). Under this condition, the demand for ATP is very low so that even a very poor blood supply will provide sufficient glucose (or other fuel) and oxygen. However, initiation of exercise changes the situation dramatically. The ATP demand will increase enormously (perhaps several hundred-fold) as soon as the exercise begins and, since the store of phosphocreatine is small compared to the rate of ATP utilization, for most types of exercise, metabolism of fuels must occur. The major fuel-supply problem at the initiation of exercise is that, although demand for energy arises immediately, an increased blood supply to the muscle (vasodilation) takes several minutes. (Increased blood flow to the muscle is dependent upon increased cardiac output and reduction of blood flow to other organs, Rowell, 1974.) During this initial period, ATP must be generated from endogenous fuels, phosphocreatine and glycogen. As indicated above, ATP synthesis from phosphagens will occur first but the stores are sufficient for only a few seconds of vigorous contraction. It is probable that some of the glycogen in the muscle can be completely oxidized to carbon dioxide and water since there will be a small quantity of oxygen in the blood within the muscles and, furthermore, type I fibres contain myoglobin,[11] an oxygen-binding molecule similar in structure to haemoglobin. However, it is unlikely that this oxygen store is sufficient to provide all the energy required prior to vasodilation at the beginning of exercise. Hence the conversion of glycogen to lactate must provide a considerable proportion of the ATP required.

Apart from the metabolic arguments advanced above, there are two other pieces of evidence in support of the idea that anaerobic glycolysis is the major process providing ATP at the onset of exercise. First, lactate release from working muscle reaches a peak during the first few minutes after the beginning of prolonged exercise in human subjects and then decreases (Figure 5.22). Secondly,

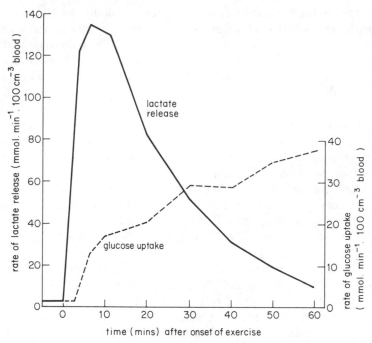

Figure 5.22 Lactate release and glucose uptake by human forearm muscle during exercise. Exercise carried out with a hand ergometer. Based on data from Wahren (1979)

there is a small number of subjects who suffer from a congenital deficiency of glycogen phosphorylase in their muscles; this is known as McArdle's syndrome. During the first few minutes of exercise these patients develop severe pain and cramps in their muscles so that they stop exercising. However, if they begin exercise at a low power output (i.e. very mild exercise) and withstand the initial pain, they find that they can continue to exercise without difficulty at a power output which would otherwise have caused severe pain and fatigue. Presumably the blood supply to the muscle increases so that glucose and fatty acids are provided for oxidation and glycogen phosphorylase is not required (Pernow *et al.*, 1967).

(d) *When the demand for ATP exceeds the aerobic capacity*

After the initial period, the ATP required by an exercising muscle will be generated by the oxidation of glucose or fatty acids (see Chapter 9; Section B.2). Presumably the rate at which ATP can be generated by these oxidative processes is limited by the availability of oxygen in the muscle fibres. There are two ways in which this limitation may be overcome; the first is applicable both in the physiology laboratory and in athletic training, while the second is extensively used on the sports field and in competitive athletics. First, if after vasodilation has occurred, short bursts of intense exercise (e.g. 10–15 sec) alternate with similar periods of rest, the overall rate of ATP turnover can be 25 % greater than that which is possible in continuous aerobic exercise (Essén, 1978). This increase is achieved without anaerobic glycolysis. The extra ATP is obtained from two sources: during the rest periods the oxygen stored in the muscle (i.e. that bound to myoglobin) is increased to normal levels, and phosphocreatine stores are replenished. Hence, during the next period of exercise, phosphocreatine and extra oxygen are available. It is considered that this 'periodic' activity is important in athletic training, when it is known as interval training. It is possible that the higher repetitive rates of oxygen consumption by the muscle lead to an increase in the amount of myoglobin in muscle and an improvement in the blood supply (i.e. the capillary circulation) to muscle. It may also lead to an increase in the capacity of the TCA cycle and the β-oxidation process for oxidation of fatty acids (see Chapter 9; Section B.5.a).

The second means by which the rate of ATP production can exceed the aerobic capacity is by anaerobic glycolysis. Thus the muscle may be oxidizing both glucose and fatty acids and the rate of ATP generation will be limited by the oxygen supply to the electron transfer chain. If now the rate of glycogen conversion to lactate is increased markedly, the rate of ATP generation will increase and the muscle will be able to achieve a greater power output. This will only be possible for a limited period, until the glycogen in the muscle is exhausted or until the decrease in pH causes inhibition of 6-phosphofructokinase. In athletic races longer than perhaps 400 m, the athlete may increase his speed as he approaches the finish of the race (i.e. 'sprinting to the tape'). Since this 'sprint-finish' depends in part on anaerobic glycolysis, lactate will accumulate in the muscle and the blood. The athletic reporter's comment that an athlete has completed his run 'in a sea of lactic acid' probably has some metabolic basis, but it is the accumulation of protons rather than lactate which accounts for fatigue and exhaustion (see Chapter 9; Section B.4 for a discussion of the metabolic basis of fatigue).

2. Tissues of the eye

The function of the eye is to collect, transmit and focus light and to respond electrically to changes in its intensity. The main tissues of the eye are cornea, lens, and retina (Figure 5.23). Light enters the eye through the cornea which causes some refraction. It then passes through the fluid-filled anterior chamber to the lens; the shape of the lens can be altered by the ciliary muscles so that the light is focussed on the retina. The retina absorbs the light and transduces it into electrical energy which appears in the optic nerve as a series of impulses.

It is obviously essential for certain tissues of the eye to transmit light with a high degree of efficiency. This precludes the presence in these tissues of large numbers

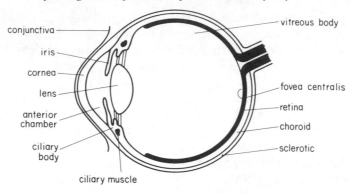

Figure 5.23 Horizontal section through left eye showing features referred to in the text

of optically dense structures such as mitochondria and capillaries which would scatter and absorb the light. For this reason, the tissues of the eye rely extensively, but not totally, on anaerobic glycolysis.

(a) *Cornea*

The cornea is not a homogeneous tissue, so that estimates of its metabolic activity based on total weight are misleading; the bulk of the cornea is the metabolically inactive connective tissue which forms the stroma. Surrounding this are actively metabolizing tissues: the epithelium on the exterior (accounting for 10% of the total weight of the cornea) and the endothelium on the inner surface (accounting for 1% of the weight). Studies on incubated corneas, freely supplied with oxygen, indicate that 84% of the glucose used is metabolized anaerobically and the remainder aerobically. Despite this high glycolytic rate, some 75% of the ATP required by the tissue is produced aerobically by the few mitochondria present (because of the efficiency of aerobic metabolism). However, no blood vessels are required as the epithelium can obtain oxygen directly from the air (a factor which must be considered when designing contact lenses). The endothelium obtains oxygen and glucose from the aqueous humour, which also receives the lactate formed during glycolysis. The aqueous humour is in contact with the bloodstream via the blood vessels that supply the ciliary body. This is the thickened anterior portion of the choroid (the layer of the eyeball behind the retina), the blood vessels of which provide fuels for the retina and the lens, either directly or via the vitreous body. It is generally assumed that lipids make no contribution to corneal metabolism.

One feature of glucose metabolism in the cornea is the high activity of the pentose phosphate pathway. This pathway oxidizes glucose to carbon dioxide in the cytosol and does not require either the TCA cycle or the electron transfer system. Its role is generally the production of NADPH which is involved in reductive synthesis of, for example, lipids, and the pathway is considered in detail in that context in Chapter 17; Section A.8. Some lipid synthesis occurs in the cornea, but it is probable that the NADPH is used to reduce oxidized glutathione (GSSG) in a reaction catalysed by glutathione reductase:

$$\text{GSSG} + \text{NADPH} + \text{H}^+ \longrightarrow 2\text{GSH} + \text{NADP}^+$$

Reduced glutathione may be important in the cornea for the maintenance of reduced sulphydryl groups in proteins and for removal of hydrogen peroxide and 'repair' of lipids which have been peroxidized by superoxide and hydroxyl radicals (see Chapter 4; Section E.5). Metabolism in the cornea has been reviewed by Maurice and Riley (1970).

(b) *Lens*

The lens is primarily a 'bag of proteins' together with the machinery for their synthesis. Most of the ATP produced in the tissue is utilized for ion and water transport in order to maintain osmotic balance in the epithelial layers of the lens. Failure to do this results in swelling of the lens and loss of transparency. Mitochondria are almost totally absent from the lens, so that virtually all the ATP required is· produced by anaerobic glycolysis. The glucose required is supplied from the vitreous body and aqueous humour which also remove the lactate produced. The lactate concentration in the vitreous body is high ($>15\,\text{mM}$) but the significance of this is not known.

  The lens, like the cornea, has an appreciable activity of the enzymes of the pentose phosphate pathway which probably plays a similar role in the lens. In addition, this tissue possesses the sorbitol or polyol pathway, in which glucose is converted to fructose with the intermediate formation of the polyol known as sorbitol:

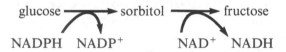

These reactions are catalysed by aldose reductase and iditol dehydrogenase (also known as sorbitol dehydrogenase), respectively (see Chapter 11; Section F.3.a). The function of this pathway has not been established since the fructose formed does not seem to be further metabolized in the lens. Unfortunately, the pathway has pathological implications in conditions where blood glucose concentration is chronically elevated since the rate of formation, and hence the concentrations, of sorbitol and fructose are increased (for example, in diabetes mellitus—see Chapter 13; Section E.2). This leads to the ingress of water with consequent swelling and eventually the formation of a cataract (clouding of the liquid contents of the lens, probably due to a change in solubility of the protein). A related polyol, galactitol, accumulates (to concentrations exceeding $100\,\text{mM}$) in

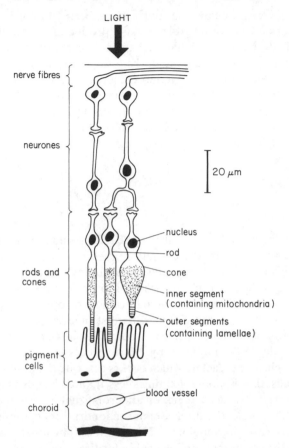

Figure 5.24 Diagrammatic section through retina. The light enters the retina from the top of the diagram and must pass through the neurones and much of the rod and cone layer before reaching the light-sensitive outer segments

the lens in cases of defective galactose metabolism in which blood galactose concentration is raised (galactosaemia, see Chapter 11; Section F.1.b). This is even more serious than sorbitol formation since the galactitol cannot be further metabolized and cataracts invariably form. Metabolism in the lens has been reviewed by Kuck (1970).

(c) Retina

There is considerable evidence that anaerobic glycolysis is an important pathway for ATP production in avian and mammalian retinas, but the extent of its contribution varies from species to species. The retina contains the rods and cones which are the visual receptors. Although these cells contain mitochondria, the outer segment, which contains the photosensitive pigment (Figure 5.24) appears devoid of mitochondria and it may be in this part of the cell that glycolysis plays an important role in ATP production. Although some blood vessels overlie the retina in the mammalian eye, this is not the case in that part of the retina where acuity is the greatest, the *fovea centralis* (which is being used at this moment by the reader). Furthermore, there are very few, if any, blood vessels overlying the retina in birds: their legendary visual acuity correlates with a dependence of the tissue on anaerobic metabolism (Krebs, 1972).[12] In these cases, glucose and any oxygen that is used must diffuse from the choroid. Metabolism of the retina has been reviewed by Glaymore (1970).

3. Erythrocytes

(a) *ATP generation*

Compared with tissues discussed above, the mature mammalian erythrocyte has a fairly simple metabolism. It lacks all organelles including mitochondria and, not surprisingly therefore, many pathways including those of fatty acid oxidation and the tricarboxylic acid cycle are absent. However, it is far from being an inert bag filled with haemoglobin. It must maintain its biconcave shape (by a process similar to contraction in muscle) and it must maintain its ionic environment against the entry of unwanted ions and water. Despite the highly aerobic internal environment, the ATP required by these processes is generated from the conversion of glucose to lactate which occurs at a rate of about 0.1 μmol glucose $min^{-1}.g^{-1}$ cells, that is, about 400 times less than that consumed by exercising anaerobic muscles. The dependence on anaerobic metabolism is explained in the same way as for muscle; the absence of mitochondria allows more room for other things, in this case, haemoglobin. The pentose phosphate pathway is also important in the erythrocyte and probably for the same reason as in the cornea, that is, reduction of oxidized glutathione which is required for the glutathione peroxidase reaction. Erythrocytes that have a low activity of glucose-6-phosphate dehydrogenase are very prone to haemolysis (Chapter 4; Section E.5.b). These cells are unable to reduce oxidized glutathione rapidly and hence are unable to withstand an increased production of superoxide radical. Administration of the free radical scavenger, vitamin E, improves the life-span of the erythrocytes in these patients (Corasch *et al.*, 1980).

(b) *The 2,3-bisphosphoglycerate shunt*

Although, in general, erythrocyte glycolysis shows no marked qualitative differences to the process in other tissues, it is distinguished by high activities of the two enzymes of the 2,3-bisphosphoglycerate shunt. These are bisphospho-glyceromutase, which isomerizes 3-phosphoglyceroyl phosphate to 2,3-bisphosphoglycerate (2,3-BPG)[13] and bisphosphoglycerate phosphatase which hydrolyses phosphate from the 2-position of the 2,3-BPG to form 3-

phosphoglycerate (Figure 5.25). There is evidence that, in fact, these two activities reside on the same protein molecule and, indeed, the enzyme may possess a common active site for the two reactions (see Chiba and Sasaki, 1978, for review). The role of the shunt is to produce 2,3-BPG. Very small quantities of 2,3-BPG are required as a cofactor for phosphoglyceromutase (Section B.9) but in the erythrocyte the concentration of 2,3-BPG is very high (about $5\,\mu\mathrm{mol.g^{-1}}$). The high activities of the two enzymes of the shunt are necessary to produce and maintain this high concentration, the reason for which is described below.

(c) *The function of 2,3-bisphosphoglycerate*

The importance of 2,3-BPG in the erythrocyte lies in its ability to alter the extent to which haemoglobin binds with oxygen. As described in Chapter 3; Section C.3, C. Bohr reported in 1904 that the dissociation curve obtained by plotting the percentage saturation of haemoglobin against the partial pressure of oxygen is sigmoid (Figure 5.26). If haemoglobin solutions are dialysed (to give what was known as 'stripped haemoglobin') the curve becomes less sigmoid (Figure 5.26). In 1921, Adair *et al.* speculated that some additional factor was involved in the interaction between haemoglobin and oxygen. However, it was not until 1967 that 2,3-BPG was identified as the factor involved (Chanutin and Curnish, 1967; Benesch and Benesch, 1967). It occurs in approximately the same concentration ($5\,\mu\mathrm{mol.g^{-1}}$) as haemoglobin and X-ray studies have indicated precisely where on the protein molecule the 2,3-BPG binds (Arnone and Perutz, 1974).

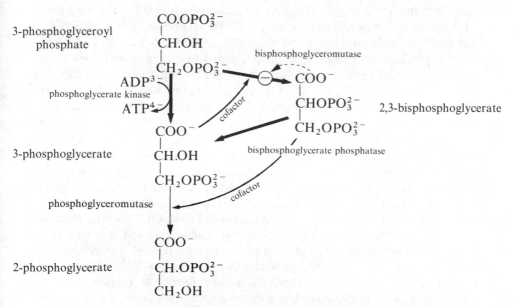

Figure 5.25 The 2,3-bisphosphoglycerate shunt

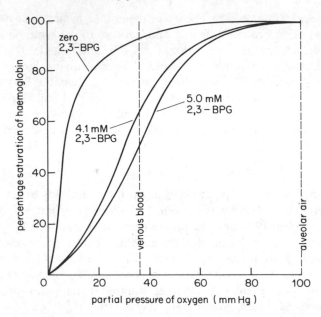

Figure 5.26   The effect of 2,3-bisphosphoglycerate on the oxyhaemoglobin dissociation curve

A quantitative appreciation of the physiological role of 2,3-BPG requires a careful examination of the two curves given in Figure 5.26. Stripped haemoglobin is half-saturated with oxygen at a partial pressure of 6 mmHg (i.e. P_{50} is 6 mmHg) while haemoglobin in the presence of normal cellular concentration of 2,3-BPG is half-saturated at 26 mmHg (P_{50} is 26 mmHg). Thus, in the presence of 2,3-BPG, erythrocytes give up a much higher proportion of their oxygen load at the partial pressure of oxygen present in the capillaries of the tissues (about 36 mmHg).[14] In the absence of 2,3-BPG in the erythrocyte, tissues would suffer hypoxia. Furthermore, quite small changes in 2,3-BPG concentration have a marked effect on oxygen release; a 24% increase in the concentration of 2,3-BPG above the normal shifts the half-saturation point for haemoglobin upwards by 3 mmHg so that there is a 22% increase in the amount of oxygen released at the partial pressure of oxygen in capillaries. Consequently, changes in the concentration of 2,3-BPG permit adjustment of the dissociation of oxygen from haemoglobin according to the physiological conditions. The steady state concentration of 2,3-BPG can be controlled by allosteric regulation of the activities of both enzymes of the shunt but the presence of both activities on one protein and even within one active site hinders the interpretation of *in vitro* properties. Nonetheless, the inhibition of the mutase activity by the free (i.e. the 2,3-BPG not bound to haemoglobin) concentration permits a simple means of regulation of the dissociation curve of haemoglobin according to the conditions. The 2,3-BPG binds strongly to deoxygenated haemoglobin but much less strongly to the oxygenated form. Hence, high concentrations of 2,3-BPG favour unloading of oxygen as follows:

$$HbO_2 \longrightarrow O_2 + Hb \longrightarrow Hb\text{-}2,3\text{-}BPG$$
$$2,3\text{-}BPG$$

If the partial pressure of oxygen in the blood is chronically reduced, the concentration of deoxygenated haemoglobin increases, which binds more 2,3-BPG. Hence the free concentration of the latter will decrease with the consequence that the activity of the mutase will increase in order to maintain the free 2,3-BPG concentration. Hence the dissociation curve will move to the right and more oxygen will be unloaded in the tissues. This is the likely explanation for the change in the dissociation of haemoglobin during acclimatization to high altitude. There is an increase in the concentration (both free and bound) of 2,3-BPG which increases the amount of oxygen unloaded in the tissues so compensating for the poorer loading of oxygen onto haemoglobin in the lungs. Similarly, in the chronic hypoxic conditions associated with such conditions as congenital heart disease, chronic obstructive pulmonary disease, and anaemia, there is an increase in the concentration of 2,3-BPG so that more oxygen is released in the tissues to compensate for the reduced amount of oxygen carried.

Not surprisingly, the concentration of 2,3-BPG is decreased during storage of red blood cells in a blood bank, so that there is an increase in the affinity of the haemoglobin for oxygen.

The oxygen affinity of haemoglobin in foetal blood (containing a molecular variant of normal adult haemoglobin) is higher than that of adult blood. This facilitates oxygen diffusion from mother to foetus. Part of the difference is due to the fact that foetal haemoglobin (deoxyhaemoglobin-F) binds 2,3-BPG less well than adult haemoglobin (deoxyhaemoglobin-A).

There are at least two inborn errors of metabolism that affect the glycolytic process in the red cell and hence the concentrations of 2,3-BPG. In one patient suffering from pyruvate kinase deficiency, the concentration of 2,3-BPG increased from 4.1 μmol.cm^{-3} to 10.3 μmol.cm^{-3} red cells and the P_{50} increased from 24.4 mmHg to 38 mmHg. This resulted in larger amounts of oxygen being released to the tissues, but, in turn, this decreased the rate of synthesis in the kidney of erythropoietin, the hormone that stimulates the rate of erythropoiesis in the bone marrow. Hence an increase in the 2,3-BPG concentration in the red cells resulted, surprisingly, in anaemia. In a patient with hexokinase deficiency, the concentration of 2,3-BPG decreased from 4.1 μmol.cm^{-3} to 2.7 μmol.cm^{-3} red cells and the P_{50} decreased from 24.4 mmHg to 19 mmHg so that the affinity of haemoglobin increased with consequent tissue hypoxia (Delivoria-Papadopoulos *et al.*, 1969). This stimulated erythropoietin synthesis so that the number of red cells was increased.

4. Leucocytes (white blood cells)

The white cells of the blood, which are concerned in the body's response to injury and infection, consist of the granulocytes (40–75 %), lymphocytes (20–50 %) and monocytes (2–10 %). Of the 40 g of glucose that is used each day by the cells of the blood (see Chapter 14) about 4 g is used by the white cells. The granulocytes are amoeboid and phagocytic and are the cells first mobilised to deal with injury and infection. The monocytes can develop into the large phagocytic cells known as the macrophages. The lymphocytes are involved both cellularly and chemically in the immune response to infection. Perhaps because of their action in the immune response and the production of antibodies they have received most attention from the metabolic biochemist. Thus it is well established that the lymphocytes have a high rate of glycolysis and that this can be increased by the stimulus for growth and cell division (even when this is stimulated artificially by the plant lectins such as concanavalin A, see Hume *et al.*, 1978). The requirement for a high rate of glycolysis in these cells is still not understood but it may be related to their ability to grow and divide rapidly. This is a property that the lymphocytes have in common with the cancer cell and it is discussed in Section G.1. It is possible that the rate of glucose utilization by the white cells increases markedly when the immune response is activated (e.g. due to infection).

5. Kidney medulla

From the metabolic point of view, the kidney is virtually two organs, the cortex and the medulla. The cortex contains the glomeruli, through which the blood is filtered, the proximal tubules and part of the distal tubules, from which small ions and molecules are reabsorbed (Figure 5.27). Large amounts of ATP must be generated in order to provide energy for this reabsorption (the kidneys of man would need to oxidize about 0.34 mmol of glucose every minute or about 88 g each day to provide energy for the reabsorption processes). The cortex is well supplied with blood so that ATP can be generated by the oxidation of glucose, fatty acids, ketone bodies or glutamine.

The medulla is metabolically quite different. Here the ATP is required for the reabsorption of ions from the loops of Henle, which penetrate the medulla (Figure 5.27) and it is generated predominantly by anaerobic glycolysis. The supply of blood, and therefore of oxygen, to the medulla is much poorer than to the cortex. There is also much more glycogen stored in the medulla than the cortex.

The loops of Henle are organized in such a way that their two arms lie in close apposition so that a countercurrent multiplier system is formed which produces a salt concentration gradient with the highest concentration at the papillary tip. This causes water to leave the collecting duct into the hypertonic interstitial fluid of the medulla thus leading to concentration of the urine. Since this tonicity gradient parallels the gradient of decreasing oxygen tension (and therefore of

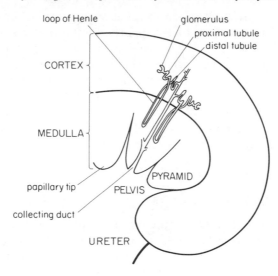

Figure 5.27 Diagrammatic section through the kidney showing major regions and two renal tubules (nephrons)

increasing dependence on glycolysis) it is tempting to propose a functional connection between these properties. It is possible that vascularization of the inner medulla is minimized to avoid the movement of water from the blood vessels into the hypertonic surroundings. Whatever the reason for this arrangement, the question arises as to how the medullary cells obtain their glucose in the absence of a good blood supply. One likely possibility is that the glucose is obtained from the lumen of the tubules since significant reabsorption of glucose occurs from the distal tubule. It has also been speculated that the glucose consumed by the medulla arises via gluconeogenesis in the cortex, but there is no evidence for this.

6. Skin

Skin is yet another tissue to rely extensively on anaerobic glycolysis for reasons which are not understood. Although the TCA cycle exists in this tissue and some glucose is oxidized, the rate of glucose conversion to lactate is about ten-fold higher than that of oxidation. One puzzling feature of carbohydrate metabolism in skin is the exceptionally high glucose content of the tissue (about $1.0 \, \mu\text{mol.g}^{-1}$); it may function as an intracellular fuel store for the tissue. The lactate concentration in the skin is also high (about $10 \, \mu\text{mol.g}^{-1}$) and this may be explained by the high concentration of lactate in sweat (about $15 \, \text{mM}$) which is secreted by the sweat glands. The secretion ensures that the surface of the skin has a high concentration of lactic acid which can act as an antibacterial agent (Kealey, 1983). This poses a problem in the assay of the lactate concentration in biological fluids, since incautious handling of pipettes, etc. may cause lactate contamination of these fluids. The small amount of information available suggests that the metabolism of the skin has been neglected, which is surprising because the quantity of this tissue in man is high; the average 70 kg man possesses about 5 kg of skin. The information that is available raises several important questions. For example, calculations based on glucose uptake or oxygen uptake of slices or on enzyme activities suggests that skin could utilize at least 100 g of glucose each day and this is not taken into account in the usual calculations on glucose requirements in man (see Chapter 8; Section A). We do not know if skin can use other fuels during starvation.

7. Absorptive cells of the intestine

It has been calculated that the intestinal mucosa must consume about 5 % of the chemical energy it transports in order to bring about that transport but it is only very recently that attention has been paid to the nature of the fuels used by this tissue (see Parsons, 1979, for review; Windmueller and Spaeth, 1980). The issue is complicated by the fact that for a proportion of the time (perhaps 30 %) fuels are available to this tissue via absorption from the lumen as well as from the vasculature. It is possible that the conversion of glucose to lactate in the absorptive cells is quantitatively important in the absorption of glucose from the lumen (Section A.2.c) so that some ATP required by these cells would be obtained from anaerobic glycolysis. A major source of ATP for this tissue is obtained from the oxidation of glutamine (derived both from the bloodstream and the lumen of the intestine). The importance of glutamine metabolism in the intestine will be discussed in Chapter 10; Section F.4.

8. Neonatal tissue

The mammalian foetus is supplied with oxygen and fuels via the placenta. The placenta is impermeable to fatty acids and triacylglycerol but it is permeable to glucose and ketone bodies, the oxidation of which provide most of the energy needs of the foetus. However, during parturition, the placenta separates from the mother and, occasionally, this can happen before the baby is fully delivered, especially if the birth is slow. Under such circumstances, the baby will become hypoxic. It is well established that neonatal tissues can withstand such conditions for longer periods than corresponding adult tissues. It has been reported that a human foetus can survive 30 min without oxygen (Bullough, 1958) although, in the adult, brain damage results from very short periods of anoxia, perhaps less than 3 min.

The mechanism of this resistance to anoxia involves a number of factors, but it is clear that anaerobic glycolysis must provide most if not all of the ATP used during this period. The marked increase in glycogen content of nearly all foetal tissues, including the heart, immediately prior to birth suggests that tissues will utilize this fuel if required. In addition, the content of glycogen in the liver is very high, so that this could maintain the concentration of glucose in the blood, so that tissues could also convert glucose to lactate for energy production. Thus survival in the absence of oxygen would require the maintenance of circulation, which is thought to be the case (Dawes and Shelley, 1968). These considerations suggest that the foetal heart is able to function for perhaps a prolonged period of anaerobiosis. Since this is in contrast to the situation in the adult heart, it would be of interest to know more of the metabolic adaptation shown by the neonatal heart; it might be of value in the treatment of patients suffering from acute myocardial infarction (see Section G.2).

G. CLINICAL ASPECTS

In Section F, the involvement of anaerobic glycolysis in a number of tissues was discussed in relation to the physiological characteristics of these tissues. There are, however, two clinical conditions in which anaerobic glycolysis plays a particularly important role: in the tumour cell and in the myocardium affected by an occlusion in the coronary blood supply.

1. Anaerobic glycolysis and the tumour cell

The exceptionally high rate of anaerobic glycolysis occurring in some tumour cells has been recognized since the work of O. Warburg in the 1920s (see Warburg, 1956a). It was he who showed that extreme forms of rapidly growing ascites tumour cells are able to convert an amount of glucose equal to about 30% of their dry weight to lactate in an hour (Warburg, 1956b). This represents a rate of glycolysis that cannot be maintained in any other mammalian tissue. For comparison, human skeletal muscle performing strenuous anaerobic exercise consumes an amount of glycogen equivalent to only about 6% of its dry weight per hour (i.e. 200 μmol.min^{-1}.g^{-1} dry wt). In a series of experimentally induced hepatomas in the rat it has been shown that the potential for anaerobic glycolysis (as measured with tissue slices under nitrogen) is proportional to their rate of growth. The importance of anaerobic glycolysis has been confirmed *in vivo* by arterio-venous differences across these hepatomas. Even the most slowly growing tumours possessed a glycolytic capacity in excess of that of the cells from which they had arisen and tumours of the brain, an organ which one suspects is never anaerobic under physiological conditions, also show an enhanced capacity for anaerobic glycolysis. This high rate of glycolysis may account for the hypoglycaemia that is observed in many patients suffering from a malignant tumour (see Chapter 11; Section C.2.f).

Despite this increased glycolytic rate in tumour cells, they possess the full complement of TCA cycle enzymes and the carriers of the respiratory chain. Indeed, many tumours, especially the slow-growing well-differentiated types, rely heavily on aerobic metabolism and utilize both carbohydrate and lipid fuels. This ability of the oxidative machinery to function in cancer cells refutes Warburg's original suggestion that a defect in oxidative metabolism is a primary lesion and closely related to the cause of malignancy. Consequently, this view is no longer held but the advantage of anerobic glycolysis to the tumour cell is not readily apparent. Nonetheless, the observations that other cells in which the rate of cell division is high also possess a high capacity for glycolysis adds credence to the view that the process is important in such cells. Thus the capacity of glycolysis is high in the lymphocyte, especially when transformation has been stimulated (Hume *et al.*, 1978; Ardawi and Newsholme, 1982); it is high in embryonic tissues and it is high in the chorion during the first days of embryonic development (see Warburg, 1956b). There are a number of suggestions as to the importance of glycolysis in rapidly dividing tissues including the following.

1. The rapid proliferation of the cells may disorganize their blood supply to such an extent that sufficient oxygen cannot reach all the cells and anaerobic glycolysis is necessary to provide the energy for cell growth and division. Nonetheless, it should be borne in mind that some blood supply will be essential to provide glucose for the glycolytic process, and that discrete cells such as ascites tumour cells and lymphocytes also have high glycolytic capacities.
2. In rapidly growing and dividing cells, several essential macromolecules need to be produced (e.g. DNA, RNA, and protein, polysaccharides, lipids). It has been suggested that a high capacity of glycolysis is necessary to maintain high concentrations of metabolic intermediates that can be used as precursors for macromolecular synthesis (Hume and Weidemann, 1979).
3. The change towards a greater glycolytic potential is also accompanied by changes in isoenzyme patterns that are similar to those found in foetal tissue. It is possible that the genetic changes that are ultimately responsible for the cause of the tumour may coincidentally cause a change to a cell with a greater glycolytic capacity.

One obvious feature of the cancer cell is that the normal regulation of cell growth and division is impaired or even lost. Consequently, most antitumour agents are directed at interfering with the processes that are involved in growth and/or cell division. If a number of cancers are dependent upon glycolysis for energy production to maintain their rates of growth, it may seem surprising that this dependence has not been exploited chemotherapeutically. The reason is that many other tissues have a requirement for anaerobic glycolysis (see Section F) and a number of tissues including the brain oxidize pyruvate obtained from glycolysis. Nonetheless, it is now established that, at least in the aerobic tissues of the body, there are alternative fuels to replace glucose (e.g. ketone bodies—see Chapter 8; Section B.2) so that for short periods of time it may be possible to lower the blood glucose level sufficiently to deprive the tumour cells of their main fuel, but yet enable tissues like the blood cells and kidney medulla to survive. A full knowledge of metabolic biochemistry of carbohydrate and fat metabolism does at least allow these possibilities to be discussed in a rational manner (see Hume and Weidemann, 1979).

2. Anaerobic glycolysis and the heart attack

The mammalian heart obtains its energy for contractile activity from the oxidation of glucose, lactate, or lipid fuels. In the fed state, glucose and lactate are the predominant fuels but these are largely replaced by the lipid-fuels during starvation or sustained exercise (see Chapter 8 for discussion of regulation by alternative fuel oxidation). Although the heart is normally aerobic, anaerobic glycolysis may be important for the provision of extra ATP during extreme exercise when the heart rate is very high (e.g. 180 beats or more each minute). This, however, may be extremely rare for modern civilized man and it may only be achieved by athletes in the 'heat' of competition. It is perhaps likely that anaerobic glycolysis in heart muscle will only be important in civilized man when there is a serious decrease in blood flow to the heart due, for example, to atherosclerosis, during cardiac surgery, or in the event of a heart attack. This is discussed below (see Braunwald, 1978a; Hearse, 1980; Hearse *et al.*, 1981, for a more detailed discussion).

The myocardium is supplied with blood through the coronary arteries. The diameters of the arteries in many, if not all, adult subjects in the advanced societies of the world are reduced due to atherosclerosis (this is discussed in Chapter 20; Section G.1). A severe reduction in diameter due to atherosclerosis together with the occurrence of one or more blood clots within that artery or arteriole can produce complete occlusion. This will result in the failure of the blood supply to part of the myocardium, which, in turn, leads to a failure of normal contractile activity in that part of the myocardium. If the artery which is occluded is large, the proportion of the myocardium without any blood supply is likely to be large. When a tissue, which is normally supplied with blood, has this supply removed, the tissue is said to be ischaemic. Irreversible damage results after about 20 min of ischaemia (Sobel, 1974).[15] Such damage means that this portion of the myocardium will die. It will, eventually, be replaced by fibrous tissue provided that the patient survives the initial trauma due to the occlusion and this is known as the infarction. However, some of the myocardium may only be partially affected because it may receive some of its blood supply from arteries not affected by the occlusion (known as the collateral circulation). This region of the muscle will not be ischaemic but hypoxic; in other words, blood will flow through the muscle but not at a sufficient rate to provide all the oxygen required. It has been termed the border or grey zone (Figure 5.28). There is clinical interest in attempts to protect this zone in the early period after the arterial occlusion, since, if the patient survives and if the myocardial cells in the border zone can be kept alive, collateral circulation to this zone will gradually develop so that the blood supply and hence the function will return towards normal. If this zone can be protected, the amount of tissue that eventually dies (that is the size of the

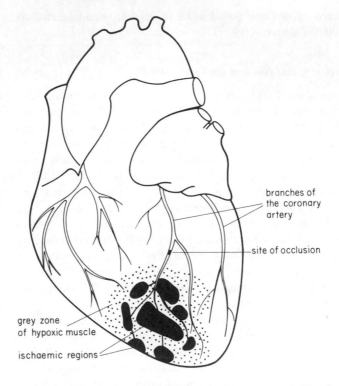

Figure 5.28   Diagram of heart showing regions of ischaemia surrounded by the grey zone
after occlusion of a branch of the coronary artery

infarction) will be reduced and this is considered to be an important determinant
of morbidity as well as mortality in post-infarction patients. (The importance of
this problem is emphasized by the fact that there are more than 7 million patients
in the U.S.A. alone who have suffered and survived at least one heart attack.)

Knowledge of the metabolic changes that occur in hypoxic or anoxic heart
muscle have led to a number of suggestions for increasing the chance of survival of
the tissue. The decreased flow of blood to the border zone means that aerobic
metabolism cannot generate sufficient ATP to maintain normal contractile
activity. Consequently, the concentrations of both ATP and phosphocreatine
decrease whereas those of AMP, IMP, NH_4^+ and P_i increase. These changes cause
a stimulation of glycogenolysis and glycolysis so that the rate of ATP production
from anaerobic glycolysis will increase (see Chapter 7; Section C). Oxidation of
pyruvate and mitochondrial respiration will undoubtedly occur in the hypoxic
state but will be limited by oxygen availability rather than availability of ADP,
phosphate, and NADH (see Chapter 7; Section B.4).

Experiments with the isolated perfused rat heart provide some information on
the maximum rates of ATP production and utilization in the aerobic and

Table 5.6. Rates of ATP production by the perfused isolated rat heart under various conditions

Conditions	Rate of ATP production (and utilization) (μmol.min^{-1}.g^{-1} at 37 °C)
Maximal capacity under aerobic conditions	260*
Maximal rate under anaerobic conditions	60†
Working heart under aerobic conditions (steady state)	78‡
Non-contracting heart under anaerobic conditions (steady state)	11†
Non-contracting heart (arrested by K$^+$)	10‡

* Calculated from maximum activity of oxoglutarate dehydrogenase (Table 5.4).
† Calculated from rates of lactate production in perfused heart.
‡ Calculated from rate of oxygen consumption.

anaerobic tissue, but it is not clear how applicable these are to the human heart. From the maximal rate of lactate production in the isolated perfused rat heart, under anaerobic conditions, it can be calculated that the maximum rate of ATP production is 60 whereas that in norml *aerobic* heart is only 78 μmol.min^{-1}.g^{-1} (Table 5.6). This suggests that the maximal rate of anaerobic glycolysis could almost satisfy the energy requirement of the *normal* heart, although this rate of anaerobic glycolysis cannot be continued for longer than a few seconds since either the store of glycogen will be depleted or proton accumulation will inhibit the rate of glycolysis. However, the activity of hexokinase is high in the heart so that a substantial rate of anaerobic glycolysis from glucose is possible; at physiological levels of blood glucose this process can provide about 14% of the ATP requirement of the normal isolated perfused rat heart. This is a substantial contribution to the energy requirement of the heart and, provided that glucose is available and lactate and protons can be removed by the blood that flows through that portion of the myocardium in the border zone, it could play an important role in survival of the tissue. These and other considerations have led to a number of suggestions for treatment of the patient who has recently suffered an acute myocardial infarction. These fall into two classes: those to decrease its energy demand and those to increase the availability of oxygen and fuels to the myocardium.

(a) *Reduction of energy demand*

1. The requirement for ATP can be lowered by reducing the work-load of the heart. In the immediate period after the heart attack, total rest in the supine position is particularly important. However, the fall in blood pressure which often accompanies a heart attack raises the catecholamine level in the blood which increases the heart rate and the force of contraction (a phenomenon

known as sympathetic drive). Consequently, it has been suggested that drugs that block the action of adrenaline on the heart, compounds known as the cardioselective β-blockers,[16] should be beneficial to such patients since they will reduce the work load of the heart and there is some evidence in favour of this (the beneficial effect could be due to a decrease in the incidence of arrhythmias—see below; Multicentre International Study, 1977; Anderson *et al.*, 1979; Sleight, 1981). However, a number of patients may rely on sympathetic drive after a heart attack in order to maintain cardiac output. Use of β-blockers in such patients could result in heart failure.

2. The body temperature could be reduced (hypothermia) which would lower the energy demand of all the tissues and, therefore, reduce the work-load of the heart. This procedure has been used in heart surgery but there are no reports of its use after infarction.

3. It is well established that the link between electrical excitation of the cell membrane of the muscle (the action potential) and the activation of the contraction process is provided by Ca^{2+}. The action potential causes an increase in the cytosolic concentration of Ca^{2+} in the muscle fibre, due to an increase in the rate of Ca^{2+} release from the sarcoplasmic reticulum. The increased concentration of Ca^{2+} in the cytosolic compartment stimulates the myofibrillar ATPase resulting in contraction (see Chapter 7; Section C.1.c). Drugs (e.g. verapamil, nifedipine) which reduced the rate of Ca^{2+} transport either across the cell membrane or the sarcoplasmic reticulum membrane can lead to a reduction in contractile activity of the heart and they have been used after a heart attack (Opie, 1980).

The damage to part of the myocardium as a result of an arterial occlusion may interfere in the normal propagation of electrical activity over the whole of the myocardium, which could result in cardiac arrhythmias and ventricular fibrillation (see Chapter 6; Note 14). Indeed ventricular fibrillation, rather than simple lack of oxygen and fuel, is considered to be one of the major causes of death immediately after the occlusion has occurred (sudden death). Hence, there is considerable interest in anti-arrhythmic drugs. Both β-blockers and the drugs that interfere with Ca^{2+} movement (see above) have anti-arrhythmic properties so that these drugs may be beneficial in more than one way in the treatment of such patients (Zipes, 1981). Finally, it is possible that the elevated fatty acid concentration (particularly within the myocardial cell) that accompanies a heart attack may also play a role in causing arrhythmias (Oliver *et al.*, 1968). Specific antilipolytic agents (e.g. β-pyridylcarbinol) may be beneficial if given as soon as possible after the infarction (see Chapter 6; Section F.2.b for discussion of blood fatty acid levels and the heart attack).

Recently clinical trials have suggested that the drug sulfinpyrazone reduces the incidence of sudden death in patients after a heart attack but its mechanism of action is unclear although there is an indication that it reduces platelet aggregation (Braunwald, 1980; Hood, 1982).

(b) *Increasing energy production*

1. A possible metabolic means for increasing the rate of ATP production is the infusion of a solution containing glucose, insulin, and potassium (known as GIK therapy) which has been promoted by Sodi-Pallares and colleagues (1969). The rationale behind this treatment is that raising the blood glucose level should increase the rate of glucose utilization by the heart. (This is only possible because there is no flux-generating step in glycolysis-from-glucose in muscle (see Section B.1).) Furthermore, an elevated insulin level, which stimulates transport of glucose, should also lead to an increase in the rate of glycolysis. The addition of K^+ to the infusion medium compensates for the loss of this ion from muscle during hypoxia or ischaemia. This loss reduces the rate of glycolysis because of the dependence of the kinase enzymes on K^+ ions for maximal activity. Despite the metabolic basis for this treatment and some experimental evidence of its efficacy (Opie and Owen, 1976; Opie and Stubbs, 1976) it has not been accepted by most clinicians. Indeed a clinical trial, organized by the British Medical Research Council indicated that there was no improvement of the patients' condition, although the trial has been criticized for using low doses of glucose, insulin and potassium (Opie, 1970).

2. It has been suggested that hyperbaric oxygen would be valuable for such patients since this could increase the amount of oxygen transported in the blood. However, this carries the danger of oxygen toxicity discussed in Chapter 4; Section E.5.

3. An increase in flow of blood through the coronary circulation could result in more blood reaching the damaged myocardium so that use of coronary artery vasodilators has been suggested (e.g. nitroglycerin, dihydropyridine derivatives such as dipyridamole—see Chapter 7 for discussion of the mechanism of action of dipyridamole). The use of coronary vasodilators in such circumstances has recently received encouragement from the suggestion that acute vasoconstriction of the coronary arteries (coronary spasm) may be an important factor inducing the occlusion and the heart attack; coronary vasoconstriction may cause reduced blood flow resulting in the formation of blood clots. In this case, treatment with coronary vasodilators may help not only to reduce the damage once the occlusions have occurred but to prevent the heart attack in patients who are known to be susceptible, that is, patients with angina pectoris (see Braunwald, 1978b; Maseri *et al.*, 1978).

APPENDIX 5.1 STRUCTURE AND NOMENCLATURE OF CARBOHYDRATES

1. Monosaccharides

Monosaccharides are polyhydric alcohols bearing either an aldehyde group (aldoses) or a ketone group (ketoses) and thus having general structures of the type:

$$
\begin{array}{cc}
& CH_2OH \\
H\!\!\diagdown_{C}\!\!\diagup\!\!^{O} & | \\
| & C\!=\!O \\
(CH.OH)_n & | \\
| & (CH.OH)_n \\
CH_2OH & | \\
& CH_2OH \\
\text{aldose} & \text{ketose}
\end{array}
$$

Monosaccharides are further classified according to the number of carbon atoms present; trioses (three carbon atoms), pentoses (five carbon atoms), and hexoses (six carbon atoms) are the monosaccharides most frequently encountered in metabolism. Because of the large number of asymmetric centres present in many of the monosaccharides, their stereochemistry is complex. Some aspects of stereochemical nomenclature are discussed in the Appendix to Chapter 4. If written according to the Fischer convention, monosaccharides of the D-series have the hydroxyl group on the penultimate carbon atom directed to the right.

Since monosaccharides possess both carboxyl and hydroxyl groups, many are able to undergo intramolecular cyclization reactions involving aldol condensations to produce five-membered (furanose) and six-membered (pyranose) rings. For most hexoses the pyranose ring predominates. Cyclisation introduces a new asymmetric carbon atom (the so-called anomeric carbon) into the molecule (C_1 in aldoses; C_2 in ketoses) so that each ring structure exists as two diasterioisomers which are distinguished by the prefixes α and β.[17] The

Table 5.7. Some derivatives of monosaccharides

Type of compound	Modification of monosaccharide	Example
Deoxy sugars	$-CHOH$-group reduced to $-CH_2-$	2-deoxyribose
Amino sugars	Hydroxy group replaced by amino group	2-aminoglucose (glucosamine)
-uronic acids	terminal $-CH_2OH$ group oxidized to COOH	glucuronic acid
-onic acids	aldehyde group oxidized to COOH	gluconic acid
Esters	condensation of any hydroxyl group with an acid. Esters with phosphoric acid are particularly important in metabolism	glucose 6-phosphate
Glycosides	condensation of C_1 hydroxyl (C_2 in ketose) group with a second alcohol	methylglucoside

cyclization reactions of D-glucose are depicted in Figure 5.29 using the Haworth convention for the representation of cyclic monosaccharides in two dimensions. Note that the reactions forming these cyclic hemiacetals are relatively easily reversed in aqueous solution at normal temperatures so that equilibrium is rapidly established between all forms. This explains mutarotation which is the change in optical activity as pure α- or β-monosaccharides are converted into an equilibrium mixture of both in aqueous solution.

Relatively few of the wide variety of possible monosaccharide structures are encountered in metabolism. Diversity is, however, increased by the existence of a

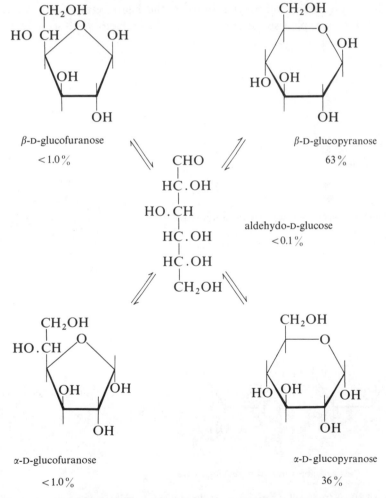

Figure 5.29 Equilibrium between straight-chain and cyclic forms of D-glucose. Percentage figures refer to fraction in that form under physiological conditions

variety of derivatives of the simple monosaccharides, some of which are listed in Table 5.7. Of these, the glycosides are of particular significance for it is through glycosidic linkages that monosaccharides polymerize to form oligosaccharides and polysaccharides (see Chapter 16; Section A).

2. Oligosaccharides and polysaccharides

In disaccharide formation, a molecule of water is eliminated between the hydroxyl group on the hemiacetal carbon atom of one monosaccharide and any hydroxyl group of a second monosaccharide. In naming oligosaccharides and polysaccharides it is necessary to specify both the position of linkage and the configuration of the hydroxyl group on the hemiacetal carbon atom. Thus lactose, for example, (see Figure 5.30) is β-D-galactosyl-1,4-D-glucose and is

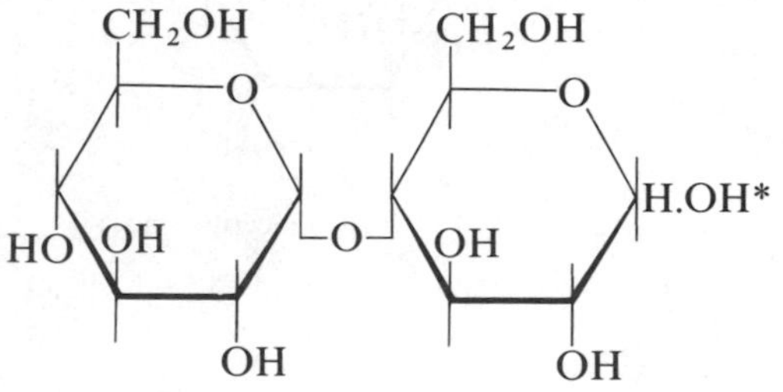

maltose (4-(α-D-glucopyranosyl)-D-glucopyranose)

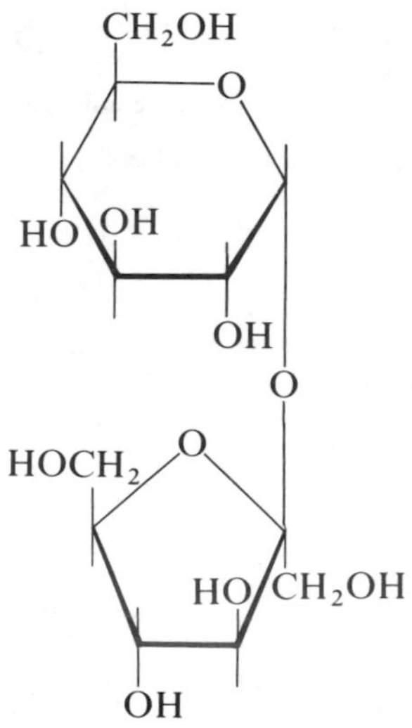

sucrose (α-D-glucopyranosyl -β-D- fructofuranose)

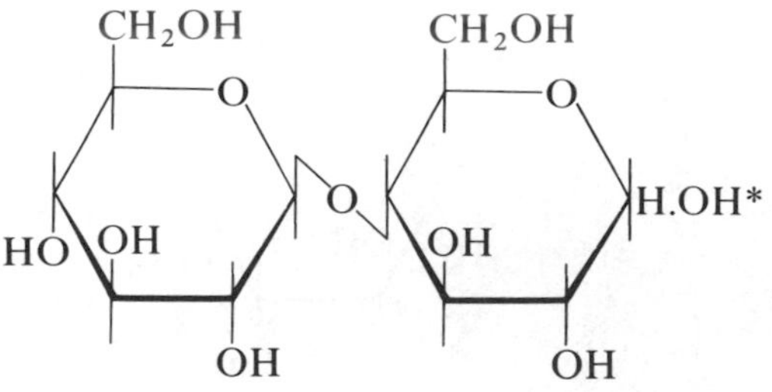

lactose (4-(β-D-galactopyranosyl)-D-glucopyranose)

Figure 5.30 Structural formulae of the three disaccharides occurring most commonly in food

* Mutarotation possible (to give α and β forms).

formed by the elimination of water between the C_1 hydroxyl group of β-D-galactose and the C_4 hydroxyl group of D-glucose. Note that this glycoside formation prevents ring-opening of the galactosyl residue but not of the glucose residue so that lactose also equilibrates in α and β forms and therefore exhibits mutarotation. If the hemiacetal hydroxyl group of both monosaccharides is involved in glycoside formation (as, for example, in sucrose—see Figure 5.29) a non-reducing disaccharide is formed since ring opening is prevented and no free carbonyl groups are available. Mutarotation cannot occur either.

Except in the case of non-reducing disaccharides, the second monosaccharide unit may form a further glycoside linkage with a third and so on to form a linear polysaccharide. A branch can originate if three hydroxyl groups on any one monosaccharide unit are involved in glycoside linkages (see, for example, the structure of glycogen in Figure 5.13).

APPENDIX 5.2 METHODS OF INVESTIGATING FUEL UTILIZATION IN A TISSUE

The first part of this appendix describes the more widely used quantitative methods for determining which fuel is used by a particular tissue. In the second part, a consideration is given to how these may be applied to the study of human metabolism bearing in mind the experimental restrictions necessarily pertaining.

1, Methods of investigating fuel utilization

(a) *Respiratory exchange ratio*

The respiratory exchange ratio (R) (originally known as respiratory quotient or R.Q.) is the ratio of carbon dioxide produced to oxygen consumed and can be measured for a whole organism, an isolated organ (*in vivo* or *in vitro*) or a tissue preparation. Aerobic carbohydrate metabolism results in a ratio of 1.00 whereas a value of 0.71 indicates total dependence on lipid metabolism (Christensen and Hansen, 1939). However, the interpretation of the ratio is complicated by the contribution of amino acid oxidation (which would give an R value of approximately 0.8 if it was the only fuel being oxidized). A serious limitation is that R values cannot distinguish between different carbohydrate fuels (e.g. glucose and glycogen) or different lipid fuels (e.g. fatty acid and ketone bodies). If breath oxygen and carbon dioxide are measured it may give only a poor estimate of fuel utilization by muscle since gas exchange across the lungs represents a mean for the whole body. In addition, any alterations of acid/base balance which would influence the release of carbon dioxide from the hydrogencarbonate pool can change the exchange ratio.

(b) *Changes in fuel concentration*

In a closed system, such as an incubated tissue or perfused organ, it is often relatively easy to measure the concentration of fuels before and after a period of time and thus determine the fuel used. Further information can be obtained from analysis of the end-products of metabolism; for example, lactate formation will indicate the extent of anaerobic metabolism if the possibility of further metabolism of the lactate is excluded. Whole animals may be treated in the same way, although the information is less useful because there is no way of assigning the changes to the metabolism of a particular tissue.

Intact organs *in situ* may receive all or part of their fuel from their blood supply. By measuring the concentration of fuels in the arterial supply and comparing them with concentrations in the venous blood draining the organ (arterio-venous difference), the amount of fuels taken from the blood can be calculated (provided the blood flow through that organ is known—see Ekelund, 1969). When these are added to the changes in the concentration of endogenous fuels, a complete picture is obtained. Problems arise, however, if the amount of fuel removed is small but the blood flow is high since, in this case, the arterio-venous concentration difference is too small to measure accurately.

(c) *Enzyme activities*

A rather less direct approach to the problem is to measure the maximum activities (V_{max}, see Chapter 3; Section C.2.e) of key enzymes in different pathways, each of which characterizes the use of a particular fuel. The quantitative value of the approach depends upon the care with which the enzymes are chosen. They must function only in the pathway being assessed and must catalyse non-equilibrium reactions (see Chapter 2; Section C.2.c). Furthermore, that the maximum activity can indicate the maximum flow through the pathway must be tested for tissues in which the maximum flux is either known or can be calculated (see Newsholme *et al.*, 1980 for further details of the approach). For example, the maximum activity of 6-phosphofructokinase provides a quantitative indication of the maximum capacity of anaerobic glycolysis whereas oxoglutarate dehydrogenase provides a quantitative indication of the maximum rate of aerobic metabolism (see Table 5.4). It should be emphasized that maximum enzyme activities indicate only the maximum potential for the utilization of a particular fuel and cannot determine whether a fuel is actually used in a given situation.

2. Fuel utilization in human tissues

The simplest measurement to make on humans is that of the respiratory exchange ratio (R). Unfortunately, it is also the least informative for reasons already stated and because the contribution of different organs is not always easy to assess. During exercise, however, muscle metabolism predominates and changes in R

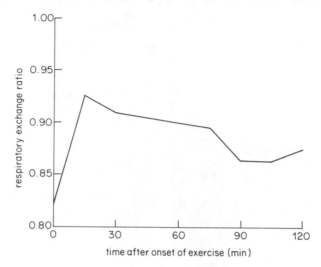

Figure 5.31 Changes in respiratory exchange ratio during severe exercise. Data from Bergström *et al.* 1967

have provided indications of major shifts in fuel utilization (Figure 5.31). In laboratory animals, much use has been made of isolated cells and perfused organ studies. The latter are useful in that they allow close simulation of physiological conditions but suffer from the disadvantange that the contribution from differing kinds of cells in the organ cannot be assessed. For ethical reasons, experiments of this kind are not normally possible with humans, although cell cultures may provide useful information.

In recent years, a number of techniques involving minor surgical operations have provided considerable information about human metabolism. Although these investigations are often considered harmless, the ethics of their use for purely experimental investigations must always be carefully scrutinised. The first of these techniques is the arterio-venous difference which is widely applicable provided that catheters can be placed in suitable blood vessels. Such important discoveries as the major dependence of the human brain on oxidation of ketone bodies during prolonged starvation (see Chapter 8; Section B.2) and the release of alanine and glutamine from skeletal muscle, also during starvation (see Chapter 10; Section F.2) could only have been made through such measurements.

The second technique is the removal of small samples of tissue from human subjects by the biopsy technique. (Bergström, 1962; Edwards *et al.*, 1980). It has been used for diagnostic purposes and for the assay of enzyme activity or metabolite concentration measurement. For use in diagnosis this technique provides a 'static' picture of metabolism in a resting, and usually ill, subject. However, a major extension of this technique has provided the Scandinavian physiologists in particular with the opportunity of studying the changes that occur in muscle under varying physiological conditions, for example, long and

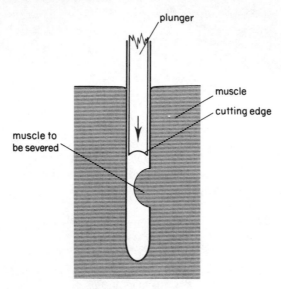

Figure 5.32 Biopsy needle being used to take a sample of muscle tissue

short duration exercise. Before beginning the experiment, the surface pain receptors in the skin overlying the muscle are anaesthetized and a small incision made. The subject then begins to exercise (on a bicycle ergometer, for example) and at an appropriate time a biopsy needle is inserted through the incision. This needle, a few millimeters wide, is hollow with a rounded tip (Figure 5.32). Just above the tip is a small window through which a piece of muscle intrudes. This is rapidly severed by a plunger which is pushed down the needle. The whole needle is then removed and, if metabolite assays are to be performed, plunged into liquid nitrogen. In this way, approximately 50 mg of muscle can be taken with minimal (15–30 sec) interruption of exercise. Several samples can be taken from the same subject, with, it is claimed, no more after effect than a mildly aching muscle the next day.

The ideal diagnostic tool would be a means for measuring quantitatively the flux through metabolic pathways in a given tissue by a non-invasive technique and without altering the function of the tissue. The flux through glycolysis under anaerobic conditions can be measured in such a way by nuclear magnetic resonance. This technique depends upon the measurements of specific radio-frequency emissions produced in a strong magnetic field from molecules containing nuclei with magnetic properties (e.g. ^{31}P, the most prevalent isotope of phosphorus). Phosphorus nuclear magnetic resonance (^{31}P NMR) has been used to measure the concentrations of phosphorylated metabolic intermediates in living tissue *in vitro* (Hoult *et al.*, 1974) and more recently in human muscle *in vivo* (Cresshull, *et al.*, 1981). These intermediates include ATP, phosphocreatine, and phosphate. Furthermore, there is a shift in the position of the peak of the signal emitted from inorganic phosphate as HPO_4^{2-} is converted to $H_2PO_4^-$, or *vice versa*, with a change in pH. Hence, it is possible to follow the increase in proton accumulation in muscle by ^{31}P NMR and, since protons are one of the end-products of glycolysis, this indicates the rate of formation of lactic acid. The technique has been used to study the rate of glycolysis during muscular work under anaerobic conditions; in normal subjects, the pH in the muscle decreased rapidly, from 7.0 at rest to 6.4 within 5 min, but in a patient suffering from McArdle's syndrome (Section F.1.c) the pH did not change (Ross *et al.*, 1981). Whether this technique can be used to measure the flux in other pathways remains to be ascertained. Perhaps NMR spectroscopy using carbon-13 (^{13}C NMR) may be more applicable to metabolic pathways (Shulman, 1982).

NOTES

1. The Greek letters used here do *not* indicate the type of glucoside linkage attacked. The β-amylases, found only in plants, also attack α-1,4-linkages, removing maltose molecules from the non-reducing ends of glucose polymers.
2. Some oligosaccharides such as raffinose and stachyose, which are present in small amounts in legumes, cannot be hydrolysed by enzymes in the small intestine. They are acted upon by bacteria in the colon which produce 2- or 3-carbon compounds, hydrogen and carbon dioxide.
3. The concentration of extracellular glucose is about 4 mM whereas that of intracellular glucose is 0.1 mM. Hence the value of the mass action ratio (0.025) is much lower than that of the equilibrium constant (1.0).
4. 2,3-Bisphosphoglycerate is of particular importance in erythrocytes in which it modifies the affinity of haemoglobin for oxygen (see Section F.3).
5. Glycerol phosphate was formerly known as α-glycerophosphate which should become glycerol 1-phosphate in the nomenclatural system that uses numbers to indicate the position of substituents. However, the compound is widely known as L-glycerol 3-phosphate because it is notionally derived from L-glyceraldehyde 3-phosphate (correctly named because the aldehyde carbon atom bears the lowest number). Because of the conventions used in naming stereoisomers the correct name becomes D-glycerol 1-phosphate but this is rarely used.
6. These forms are also known, respectively, as LD_1, LD_2, LD_3, LD_4, and LD_5. Isoenzyme LD_1 bears the highest negative charge.
7. The division of muscle into red and white types was known many centuries ago and physiological differences were observed in the last century; for example, Ranvier (1873) observed that the red muscles of the rabbit contracted more slowly than the white muscles.
8. Cross-sections of muscle are incubated for a short period of time with ATP and then heated with a solution of a lead salt. Any phosphate liberated from the ATP combines with lead ions to form insoluble lead phosphate. Treatment with hydrogen sulphide converts the lead phosphate to lead sulphide which is black and can be seen under the microscope (Figure 5.19). This is an example of a sampling enzyme assay (see Chapter 3; Section C.1.a). Both type I and II fibres stain with a similar intensity when the myosin ATPase is stained in this way. However, if the muscle sections are preincubated at pH 10.3 prior to assay, the myosin ATPase in type I fibres is denatured. Hence type I fibres do not show positive staining for ATPase activity (i.e. they are light) after pre-incubation at pH 10.3, whereas type II fibres show strong positive staining (i.e. they are dark) (see Figure 5.19). Two further methods have been used to subdivide the type II fibres into groups A and B (see Table 5.5). First, some of

the type II fibres show positive staining (i.e. dark) after preincubation at both pH 10.3 and pH 4.6 (type IIB) whereas the others do not respond (i.e. light stain) after preincubation at pH 4.6 (type IIA) (see Saltin *et al.*, 1977). The precise biochemical basis for the difference in pH sensitivity of the ATPase in the different fibres is not clear but it presumably relates to differences in structure of the myosin molecule (Weeds, 1980). Secondly, a histochemical technique which depends upon the activity of succinate dehydrogenase can also distinguish between type II fibres. The fibres staining darkly in the procedure (type IIA) possess a higher activity of succinate dehydrogenase and thus have a higher oxidative capacity than the type IIB fibres that stain only lightly.

Fibre typing is necessary in human muscles since, as indicated above, the individual muscles comprise different proportions of the main fibre types, and these proportions can differ from one individual to another. Thus the muscles of man are mixed with the fibres existing in a mosaic pattern within a muscle (Figure 5.19). For example, in one study of 16-year old human subjects (70 male and 45 female) the vastus lateralis contained 52% type I, 37% type IIA, and 14% type IIB fibres, the soleus 75–90% type I fibres and the triceps 60–80% type II fibres (Saltin *et al.*, 1977). There was no evidence of a sex difference. Furthermore, in some muscles the proportion of the fibres may differ in different parts of the muscle. Usually the more superficial portions of the muscle contain more of the type II fibres. In addition to these differences, marked variations in the proportions of the fibre in a given muscle can occur in different subjects. The muscle of top class sprinters is found to contain mainly type II fibres whereas those of top class long-distance runners contain mainly type I fibres (see Chapter 9; Section A.1 and B.1 for full discussion).

9. Pectoral muscle of the pheasant contains about 50 μmol.g^{-1} fresh muscle of glycogen. Phosphorylase activity at physiological temperatures for pheasant is about 3 μmol.sec^{-1}.g^{-1} fresh muscle. However, in any single bout of activity, proton accumulation in the muscle rather than glycogen depletion is likely to inhibit the glycolytic process; in such anaerobic exercise the pH of the muscle rapidly falls well below 7.0 and this inhibits phosphorylase and 6-phosphofructokinase activities—see Chapter 9; Section A.4.a.

10. A large bird related to the cranes, once common in open country in Europe but now restricted to a few localities in central Europe and southern Spain.

11. At the oxygen tensions pertaining in muscle, myoglobin has a greater affinity for oxygen than does haemoglobin and so serves to 'pull' oxygen into the cell (Cole, 1982). The myoglobin also serves as a small store of intracellular oxygen which can be used briefly to support aerobic metabolism when the demand for oxygen within the fibre exceeds its supply from outside, as will be the case at the onset of exercise. (In diving mammals this storage function of myoglobin is paramount and large amounts are found in the muscles of seals and whales.)

12. Isolated pigeon retinas generate 88% of their ATP requirements by anaerobic glycolysis, even in the presence of oxygen and, apart from certain tumour cells and working muscles, show the highest rate of glycolysis (up to 15 μmol.min^{-1}.g^{-1}) reported for any vertebrate tissue (Krebs, 1972).

13. Since 2,3-bisphosphoglycerate was previously known as 2,3-diphosphoglycerate, it is also known as the 2,3-DPG shunt.

14. A high partial pressure of oxygen is required in the capillaries in order to provide a sufficient pressure gradient for rapid diffusion of oxygen across the interstitial space between capillaries and cells of the tissue. If myoglobin was the oxygen carrier in the blood (with its hyperbolic kinetics) the partial pressure of oxygen in the capillaries would have to be very low for efficient release of oxygen, so that the rate of diffusion to the cells of the tissue would be severely limited. The biological advantage of the

sigmoid dissociation curve is that it permits a greater rate of supply of oxygen to maintain a high rate of aerobic metabolism in the tissues.

15. The initial biochemical problem in the ischaemic tissue is the accumulation of protons as an end-product of anaerobic glycolysis. Since there is no blood flow, not only is oxygen not available but this end-product cannot diffuse away. Consequently, the pH of the cell decreases and this causes inhibition of 6-phosphofructokinase (see Chapter 9; Section A.4.a). This results in a rapid depletion of ATP which not only stops contraction but also prevents maintenance of ion and water balance. The first organelle to show marked structural damage is the mitochondrion; the changes are enlargement, loss of density in the matrix and appearance of granules. Such mitochondrial damage may reflect the point of no return for the ischaemic heart (see Sobel, 1974).

16. A hormone affects cells by binding to a specific receptor either on the cell membrane or within the cell (see Chapter 22; Section A.3). Agents that can bind to the receptor but elicit no response from the cell are known as antagonists or blockers, whereas those that bind and elicit a response are known as agonists. The pharmaceutical industry has produced a large number of such agonists and antagonists for many hormones but especially the catecholamines (Motulsky and Insel, 1982). This research has demonstrated that receptors may be slightly different from one tissue to another so that it is possible to design receptor blockers that are specific or almost specific for one tissue. Cardioselective β-adrenergic blockers are one example.

17. The prefix α is used if the configuration at the hemiacetal carbon atom is the same as that of the final asymmetric centre (C_5 in a hexose, the one that determines whether the sugar is of the L- or D-series). If the configuration of the hydroxyl groups on these two carbon atoms is different, the sugar is β.

CHAPTER 6

CATABOLISM OF LIPIDS

A. INTRODUCTION

The importance of acetyl-CoA as an intermediate in the generation of ATP has been described in Chapter 4. As well as arising from carbohydrates (via glycolysis and the pyruvate dehydrogenase reaction, described in Chapter 5) acetyl-CoA can also be derived from lipids. Lipids are defined as long-chain fatty acids and their esters, which include acylglycerols, waxes, and phospholipids (see Appendix 6.1).[1] A metabolic disadvantage of this definition is that it excludes ketone bodies which function as an important lipid-fuel (see Section C.3). It also excludes steroids, terpenes, carotenes and other hydrocarbons which are sometimes called the non-saponifiable lipids (i.e. they cannot be converted into glycerol and a fatty acid moiety by boiling with alcoholic potassium hydroxide solution).

A recurring theme in discussion of lipid transport and function is the nature of the relationship between lipids and water. Five categories of lipid-water interaction can be usefully discerned.

1. Lipids that are insoluble and do not form a stable association with water (e.g. cholesterol esters).
2. Lipids that are insoluble but are just polar enough to form stable monolayers at air-water interfaces (e.g. triacylglycerols, diacylglycerols, protonated long-chain fatty acids, cholesterol).
3. Lipids that are insoluble but sufficiently polar to allow association with water in a regular manner (to form liquid crystals) in the bulk phase (e.g. phospholipids, monoacylglycerols).
4. Lipids (e.g. bile salts, salts of long-chain fatty acids) that are soluble at low concentrations but at higher concentrations, that is, above the critical micellar concentration, form a micellar 'solution'. (Micelles are aggregates of molecules which are similar to particles in an emulsion but are considerably smaller (typically 4–6 nm in diameter). They cannot be seen under the light microscope and appear to form a clear 'solution'.)
5. Lipids that are readily soluble (e.g. ketone bodies, very short-chain fatty acids).

A useful functional classification of the lipids is into those that are ultimately oxidized in order to provide ATP (i.e. ketone bodies, long-chain fatty acids, triacylglycerols and, in some animals, waxes) and those which serve a structural function (e.g. phospholipids and other compound lipids (see Appendix 6.1)).

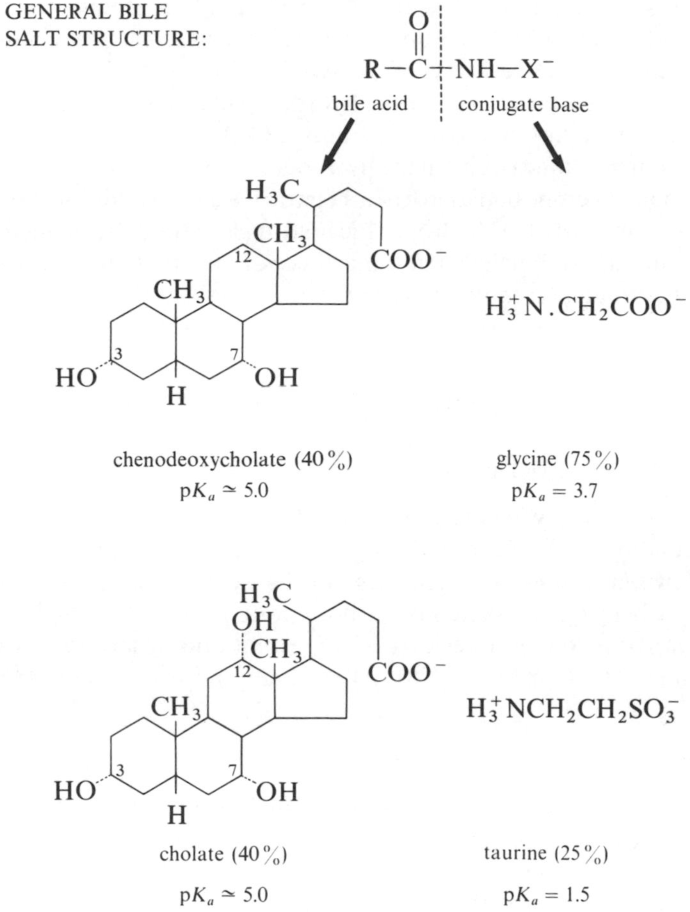

GENERAL BILE
SALT STRUCTURE:

$$R-\overset{\overset{\displaystyle O}{\|}}{C} \mid NH-X^-$$

bile acid $\quad\vdots\quad$ conjugate base

chenodeoxycholate (40%)
$pK_a \simeq 5.0$

$H_3^+N.CH_2COO^-$

glycine (75%)
$pK_a = 3.7$

cholate (40%)

$pK_a \simeq 5.0$

$H_3^+NCH_2CH_2SO_3^-$

taurine (25%)

$pK_a = 1.5$

Figure 6.1 Structural formulae of bile salts occurring in man. Percentages represent the abundance of each group in human bile (see Chapter 20; Section C.1). The remaining 20% of bile acids are secondary products resulting from the action of bacteria on the two primary bile acids. Predominant among these secondary acids is deoxycholic acid which is formed by the loss of the 7-hydroxyl group from cholic acid. Note that the acid groups in the bile acids (for which pK_a values are given) are not free in the bile salt conjugates

However, this distinction cannot be drawn solely on the basis of the lipid chemistry since long-chain fatty acids and acylglycerols are also components of structural lipids.

The functional distinction is underlined by the location of these lipids within the cell and within the animal. The structural lipids are mostly found as components of membranes, whereas oxidizable lipids are found in intracellular lipid stores (especially in the adipose tissue of higher animals) and in the plasma.

In this chapter, the pathways of acetyl-CoA formation from lipid-fuels are described. These pathways include the hydrolysis of triacylglycerol to fatty acids, the β-oxidation of the latter to produce acetyl-CoA and the reactions involved in the formation of ketone bodies in the liver together with the reactions involved in the oxidation of ketone bodies in extra-hepatic tissues. In addition, the problems inherent in transport of the hydrophobic lipid-fuels in the aqueous medium of the blood are discussed. Finally some clinical conditions that are a consequence of abnormal fat metabolism are described.

B. DIGESTION AND ABSORPTION

1. Triacylglycerols (triglyceride)

Western man typically ingests about 100–150 g of triacylglycerol per day. Under normal conditions, this dietary triacylglycerol is the major source of lipid for the body. However, significant quantities can be synthesized from carbohydrate, especially when the proportion of carbohydrate in the diet is high and that of triacylglycerol is low. For reviews on digestion and absorption of lipids, see Thomson (1978); Johnston (1978); Alfin-Slater and Aftergood (1980).

(a) *Digestion*

The digestion of triacylglycerol occurs mainly in the small intestine into which flow both bile and the secretions of the pancreas. In the context of fat digestion, the most important constituents of the bile are the bile salts. These are sterols (i.e. hydroxylated steroids) with side-chains (taurine and glycine) which contain acidic groups (see Figure 6.1). (The metabolism of the bile salts is discussed in Chapter 20; Section C.1.) At the pH in the small intestine (6.0–8.0) most of these groups are ionized (for pK_a values see Figure 6.1). In this form they have a detergent action, especially in the presence of monoacylglycerol, which is produced during triacylglycerol digestion, and cause the emulsification of the triacylglycerol to form particles approximately 1 μm in diameter. This facilitates the enzymatic hydrolysis of the triacylglycerol which is carried out by a triacylglycerol lipase secreted by the pancreas (pancreatic lipase). This enzyme hydrolyses ester links in the 1- and 3-positions of the triacylglycerol to yield the 2-monoacylglycerol and fatty acids:

$$CH_2O.CO(CH_2)_nCH_3$$
$$|$$
$$CH\ \ O.CO(CH_2)_nCH_3 + 3H_2O \longrightarrow$$
$$|$$
$$CH_2O.CO(CH_2)_nCH_3$$

triacylglycerol

$$CH_2OH$$
$$|$$
$$CH_2O.CO(CH_2)_nCH_3 + 2CH_3(CH_2)_nCOOH$$
$$|$$
$$CH_2OH$$

fatty acid

monoacylglycerol

Emulsification facilitates hydrolysis since it greatly increases the area of the lipid–water interface, at which pancreatic lipase acts. In the laboratory, investigation of the kinetics and properties of lipases has been hampered by the difficulty of preparing substrates in a suitable form. One of the problems is that chemical agents that aid the formation of an emulsion usually cause inhibition of the lipases. Indeed, bile salts above their critical micellar concentration inhibit pancreatic lipase, but this is prevented *in vivo* by the presence of a protein, colipase, that binds to the bile salt micelles and reduces their inhibitory action. Colipase is secreted by the pancreas (Brindley, 1974).

Emulsifying agents other than the bile salts, are present in the lumen of the small intestine. They include the monoacylglycerols and lysolecithin. The latter is formed by the hydrolytic removal of a fatty acid from the 2-position of the phospholipid lecithin (present in the diet and also secreted in the bile (see Appendix 6.1 for structure)) by phospholipase A_2 from the pancreas. Other dietary phospholipids are hydrolysed to lysophosphoglycerols (lysophospholipids) by this phospholipase.

Although pancreatic lipase is quantitatively the most important enzyme in catalysing the hydrolysis of triacylglycerol, a gastric lipase, secreted by the gastric glands of the stomach, and a lingual lipase originating in the tongue and soft palate, also exist. Gastric lipase may be particularly important in the hydrolysis of triacylglycerols containing short- and medium-chain fatty acids.

(b) *Absorption*

The products of triacylglycerol digestion, mainly monoacylglycerol and long-chain fatty acids, must form a stable interaction with water before uptake into the epithelium of the intestine can occur. Once again, this stabilization is achieved by the action of the bile salts which are present at concentrations in excess of their critical micellar concentration of 2–5 mM, so that micelles are formed (see Section A). These bile salt micelles incorporate monoacylglycerols, lysophosphoglycerols, and long-chain fatty acids, the products of lipid digestion, to form mixed bile salt/monoacylglycerol/lysophosphoglycerol/fatty acid micelles. Tri- and diacylglycerols, however, are not appreciably soluble in bile salt micelles and so form a separate oily phase. Exactly how mixed micelle formation aids the absorption of monoacylglycerols, lysophosphoglycerols and fatty acids is not known but it is likely that it conveys the non-polar lipid molecules through the aqueous contents of the intestinal lumen to the epithelial cell surface. Here the micelle dissociates to produce locally high concentrations of monoacylglycerols, lysophosphoglycerols and fatty acids which are absorbed, while the bile salts remain in the lumen (Figure 6.2) (to be absorbed in the lower regions of the intestine, see Chapter 20: Section C.1). The absorption of these compounds does not appear to involve expenditure of energy by the epithelial cells. There is no evidence for specific receptors or uptake sites for the absorbed lipids, nor is there evidence for uptake of whole micelles by pinocytosis.

The uptake of lipids into the absorptive cells involves transfer from a favourable micellar environment into an apparently unfavourable aqueous one. This transition is facilitated by the existence of a fatty acid binding protein (FABP) in the cytosol of epithelial cells. FABP is a small protein with a very high affinity for long-chain fatty acids (Ockner and Manning, 1974). It increases in concentration in response to a triacylglycerol-rich diet.

The importance of bile salts is emphasized by the diminished fat absorption and the steatorrhoea which results from an abnormally low concentration of bile

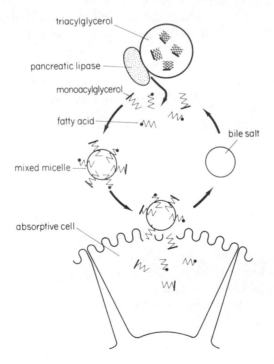

Figure 6.2   Diagrammatic representation of triacylglycerol digestion and absorption. (Not to scale)

salts in the lumen of the small intestine. There are a number of causes of this low luminal concentration; the most obvious is biliary obstruction in which the bile duct is blocked, but it can also occur when the liver is diseased, thus causing decreased bile production. Another cause is abnormal deconjugation of bile salts by bacteria in the large intestine, since this produces altered bile acids which possess higher pK_a values than normal acids so that their solubility and consequently, their rate of reabsorption from the intestine is lowered (see Figure 6.1). As a result, the rate of bile salt excretion in the faeces increases and the luminal concentration of bile salts decreases. In some conditions, the pH of the small intestine is lowered and this also reduces the solubility and, therefore, reabsorption of bile salts. Although fat absorption is reduced in all these situations, 50–70% of the ingested fat can still be absorbed even in the absence of bile. It seems likely in such cases that a greater proportion of lipid is taken up in the form of slightly soluble fatty acids rather than less soluble monoacylglycerol since the former is more readily absorbed than the latter in the absence of bile salts. A large proportion of these fatty acids enters the hepatic portal blood and is metabolized by the liver. Since the fatty acids are esterified within the liver this may result in a large accumulation of triacylglycerol in the liver to such an extent that the functions of the liver may be impaired. Consequently, patients with reduced bile salt concentration are maintained on a low fat diet.

(c) *Lipid metabolism within absorptive cells*

After their entry into the epithelial cell, fatty acids and monoacylglycerols are re-esterified to form triacylglycerols. As with all reactions involving fatty acids, it is preceded by reaction of the fatty acid with coenzyme A to form the acyl-CoA; a reaction catalysed by the enzyme acyl-CoA synthetase (fatty acid thiokinase):

$$R.COO^- + ATP^{4-} + CoASH \longrightarrow R.CO.SCoA + AMP^{2-} + PP_i^{3-}$$

This acyl-CoA synthetase is specific for fatty acids with more than twelve carbon atoms. Shorter-chain fatty acids (which form a small proportion of a normal diet but are found in high concentrations in some foods, for example milk and milk products) are not converted to acyl-CoA derivatives in the epithelial cells since medium-chain acyl-CoA synthetase is not present in this tissue. These fatty acids enter the hepatic portal vein and are transported to the liver for further metabolism.

Two pathways of triacylglycerol synthesis have been identified in the absorptive cells. The quantitatively important pathway is the 'monoacylglycerol path' in which a monoacylglycerol is condensed with two molecules of acyl-CoA to form a triacylglycerol (see Figure 6.3). This pathway accounts for about 85% of intestinal triacylglycerol synthesis and is restricted to the epithelial tissue where it occurs in association with the smooth endoplasmic reticulum. The

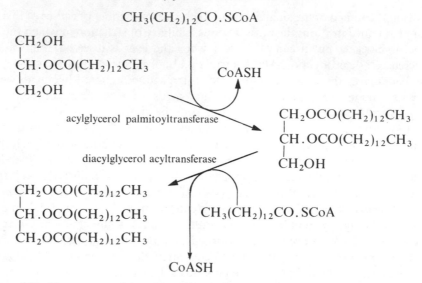

Figure 6.3 The monoacylglycerol pathway for the synthesis of triacylglycerols in intestinal absorptive cells. Although depicted for palmitoyl-CoA, other long-chain fatty acids can participate. Both transferase enzymes, together with acyl-CoA synthetase, associate to form a triacylglycerol synthase complex

remaining triacylglycerol is synthesized from glycerol 3-phosphate and acyl-CoA in association with the rough endoplasmic reticulum via a pathway that occurs in many tissues and is described in detail in Chapter 17; Section A.5.

In the epithelial cells of the small intestine, triacylglycerol, phospholipids, cholesterol, cholesterol esters and a specific protein called apolipoprotein[2] (see Section C.1) combine to form spherical chylomicrons, with a diameter > 75 nm. The processes involved in the synthesis of chylomicrons from dietary triacylglycerol are summarized in Figure 6.4 and take place within the endoplasmic reticulum and the Golgi apparatus. The chylomicrons are released by exocytosis into the lymphatic system via the lacteals and pass through the thoracic duct into the left subclavian vein; subsequently they are taken up by the liver and by adipose tissue. They are sufficiently large to scatter light so that serum that contains chylomicrons is markedly turbid. Their structure and further metabolism is discussed in Section C.1.a.

2. Lipids other than triacylglycerols

Dietary phospholipids are probably digested and absorbed in a similar manner to that of dietary triacylglycerol. Pancreatic lipase itself has some hydrolytic activity towards phospholipids, removing the fatty acid from the 1-position. Phospholipase A_2 is secreted by the pancreas and it removes fatty acid from the 2-position. It is likely that these partially degraded phospholipids play a role in

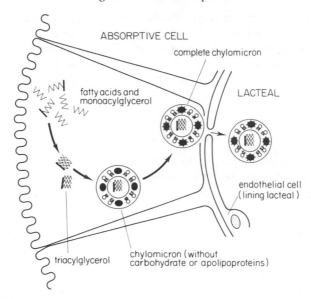

Figure 6.4 Diagrammatic representation of chylomicron synthesis in an intestinal absorptive cell. Not to scale. See Figure 6.5 for key to components of chylomicron

emulsification. (Indeed the phospholipid lecithin is secreted into the bile to provide the necessary substrate.) The partially hydrolysed phospholipids enter the epithelial cells from mixed micelles containing the other digested lipid components. As with triacylglycerols, resynthesis of the phospholipid occurs within the epithelial cells (see Chapter 17; Section C.2 for details of the pathway).

Approximately 1 g of cholesterol is ingested by man per day and an equal amount arises endogenously in the intestinal lumen (from intestinal secretions and sloughed epithelial cells). Uptake of free cholesterol occurs from the mixed micelles, primarily in the jejunum. Most of the dietary cholesterol is in the form of cholesterol esters (Appendix 6.1) and cannot be absorbed as such. The esters are hydrolysed by cholesterol esterase, secreted by the pancreas, within the lumen. Some resynthesis of cholesterol esters occurs in the epithelial cells and these esters, together with some free cholesterol, leave the intestinal cells in the chylomicrons. The possible importance of dietary cholesterol in the development of atherosclerosis is discussed in Chapter 20; Section G.1.

In contrast to cholesterol, the bile salts are absorbed mainly in the ileum. They are returned to the liver through the hepatic portal vein (in association with proteins) and can thence be resecreted into the bile. The transport of bile salts between liver and intestine is known as the enterohepatic cycle of bile salts (see Chapter 20; Section C.1). Some of these salts are lost in the faeces and must be replaced by bile salts synthesized *de novo* in the liver. Certain types of dietary fibre increase the faecal loss of bile salts and this loss can also be increased artificially by the administration of cholestyramine (see Section F.3.b).

A number of other non-saponifiable lipids of importance, for example, vitamins A, D, E, and K, are also absorbed from mixed bile salt micelles. Consequently, patients suffering from inadequate concentrations of bile salts in the intestine (including those patients being treated with cholestyramine) are prone to deficiencies of these vitamins. In such patients, these vitamins can be administered by injection.

C. CIRCULATING LIPID FUELS

Lipid fuels appear in the blood in three forms: triacylglycerols (lipoproteins), free fatty acids, and ketone bodies. Possible reasons for this diversity of lipid-fuels will emerge in subsequent discussions.

1. Lipoproteins

Although triacylglycerols are stored in cells apparently without association with water or protein, they are transported between tissues as lipoprotein complexes. This is necessary because triacylglycerols are not soluble in water and the combination with protein confers stability in an aqueous environment. In addition, varying amounts of phospholipids (mostly lecithin and sphingomyelin) and cholesterol (free and esterified) are also present in lipoprotein complexes. Although there are four main classes of circulating lipoproteins only two are quantitatively important in transport of triacylglycerol, chylomicrons and very low-density lipoproteins. However the other two, low-density and high-density lipoproteins, play an important role in the transport of the lipid in the bloodstream. These four are usually separated and classified according to either their density (a function of the lipid/protein ratio) or their electrophoretic mobility (dependent on the composition of the apolipoprotein).[2] The classification of lipoproteins is summarized in Table 6.1. In order to appreciate the roles of the various lipoprotein classes, a brief account of their metabolism is provided in Section C.1.c. Since elevated concentrations of blood cholesterol and possibly triacylglycerol are important as predisposing factors in coronary heart disease (see Chapter 20 Section G.1) there is an immense amount of literature in the field of lipid transport (see Lewis, 1976; Osborne and Brewer, 1977; Scanu, 1978, for reviews). Despite this, the metabolic picture is far from complete. One of the major problems is that during the release of fatty acids from the chylomicrons and very low-density lipoproteins there is a considerable exchange of components (e.g. cholesterol esters) between the lipoproteins. Furthermore, the metabolism of the lipoproteins in man may be considerably different from that in the rat.

(a) *Chylomicrons*

Chylomicrons are the largest of the lipoprotein particles occurring in blood and

Table 6.1 Properties and composition of plasma lipoproteins†

Lipoprotein	Density (g.cm^{-3})	Electrophoretic mobility classification	Molecular weight ($\times 10^6$)	Diameter (nm)	% composition					Apoproteins	
					Triacylglycerol	Protein	Cholesterol	Cholesterol ester	Phospholipid	Major	Minor or trace
Chylomicrons	<1.006	origin	>400	>75	85–90	1–2	2–3	3–4	6–8	A,B,C,E	—
VLDL	<1.006	prebeta	5–10	25–75	50–55	8–10	6–8	14–16	16–20	B,C,E	A
LDL	1.006–1.063*	beta	2–5	22–24	6–10	18–22	8–12	35–45	20–25	B	C
HDL	1.063–1.12	alpha	0.2–0.4	6–14	3–6	47–52	2–4	12–18	25–30	A	C,D,E

* LDL is frequently separated into two density classes IDL, 1.006–1.019 and LDL$_2$, 1.019–1.063 g.cm^{-3}.

† Data taken from Lewis (1976); Osborne and Brewer (1978) and Simons and Gibson (1980).

contain the highest proportion of triacylglycerol. Chylomicrons contain approximately 85% triacylglycerol, 8% phospholipid, 2% cholesterol, 3% cholesterol ester and 2% protein (Table 6.1). They arise solely in the intestine and contain triacylglycerol of dietary origin only. The major apoprotein component is apolipoprotein B which also occurs in low-density and very low-density lipoproteins (see below). Chylomicrons are released into the lymphatic system which drains into the main circulation. Although it has been known since the work of Claude Bernard that the absorbed fat enters the bloodstream via the lymphatic system, it is not yet clear if this has any functional significance. It is possible that this route ensures that adipose tissue is exposed to a higher concentration of chylomicrons than would be the case if they passed through the liver first. Nonetheless, the liver does remove some of the chylomicrons that enter the blood and eventually most of the triacylglycerol that enters the body may be metabolized within the liver (see below). A speculative proposal that the lymphatic route of entry of triacylglycerol into the body is important as a natural immunosuppressive mechanism is discussed in Chapter 17; Section B.2.d.

Triacylglycerol, whether in the form of chylomicrons or other lipoproteins, is not taken up directly by any tissue. It must be hydrolysed outside the cell to fatty acids and glycerol, which can then enter the cell. This hydrolysis is carried out by lipoprotein lipase[3] (which is also known as clearing-factor lipase). In extrahepatic tissues including adipose tissue, skeletal muscle, heart, lung, and mammary gland, the enzyme is attached to the outer surface of the endothelial cells lining the capillaries. (In the liver, the lipoprotein lipase is attached to the outer surface of the hepatocytes.) At this site, the triacylglycerol is hydrolysed to glycerol and fatty acids and the latter taken up by the cells of the tissue in which the hydrolysis occurs. The glycerol is transported in the blood to the liver and kidney for further metabolism (see Chapter 11; Section B.3.b). Evidence for this extracellular location of lipoprotein lipase is provided by the ability of heparin, injected intravenously, to release the enzyme into the bloodstream. This effect may not be of physiological significance but is useful in diagnosis; the activity of the lipase can be measured in a sample of blood taken from a patient after injection of heparin. Post-heparin lipolytic activities (abbreviated to PHLA) are important in diagnosis of type I hyperlipoproteinaemia (see Section F.3.a). Chylomicrons obtained directly from the lymphatic duct are a poor substrate for lipoprotein lipase. In order to become an effective substrate for extrahepatic lipoprotein lipase, chylomicrons must acquire apolipoprotein C from the high-density lipoproteins in the blood.

Once in the bloodstream the half-life of chylomicrons is quite short; less than one hour in man and only a few minutes in rats (Eisenberg and Levy, 1975). However, the digestion and absorption of lipid and the formation and secretion of chylomicrons takes several hours, so that lipaemia develops several hours after a fat meal.

The activity of the lipoprotein lipase complex depletes chylomicrons of their triacylglycerol, so that their lipid/protein ratio falls. This is a stepwise process

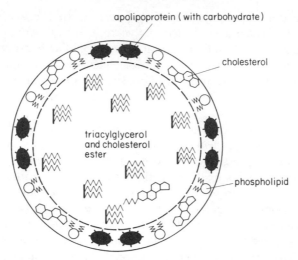

apolipoprotein (with carbohydrate)

cholesterol

triacylglycerol
and cholesterol
ester

phospholipid

Figure 6.5 Diagrammatic structure of chylomicron or very low-density lipoprotein. A polar shell some 2 μm thick surrounds a core of triacylglycerol also containing cholesterol ester. (Not to scale.)

resulting in the production of a remnant chylomicron (or 'remnant particle') which is metabolised in the liver. Further details of this degradation are given in Section C.1.c.

(b) *Very low-density lipoprotein*

The second major source of triacylglycerol in the body is that synthesised endogenously (that is, not arising directly from dietary triacylglycerol). Many tissues are able to esterify fatty acids to form triacylglycerol but (with the exception of the intestinal cells) only the liver secretes them into the blood in significant amounts. The fatty acids esterified in the liver arise from two sources. First, fatty acids can be synthesized *de novo* from acetyl-CoA which is derived mainly from lactate, alanine or glucose (see Chapter 17; Section A.1 and A.2). Secondly, fatty acids can be taken up as such from outside the cell where they arise either from the hydrolysis of triacylglycerol in remnant chylomicrons, or as fatty acids that are present in the blood and that have originated in adipose tissue. Thus there is evidence that fatty acids are released from adipose tissue and taken up by the liver for esterification. Triacylglycerols are secreted from the liver as very low-density lipoprotein particles (VLDL) which are smaller than chylomicrons and have a lower lipid/protein ratio (see Table 6.1). A model of the possible structure of VLDL is illustrated in Figure 6.5. Although high concentrations of VLDL give plasma a turbid appearance, this lipoprotein does not separate on standing to form a creamy layer. However, VLDL is similar to chylomicrons in that it contains apolipoprotein B as its major protein component and it must acquire apolipoprotein C before it is an effective substrate for lipoprotein lipase. The fate of the triacylglycerol in VLDL would appear to be similar to that in the chylomicrons, that is, extracellular hydrolysis by tissues possessing a lipoprotein lipase with consequent uptake into the tissue of the released fatty acid. In the fed state, the uptake will be predominantly into adipose tissue. The half-life of VLDL triacylglycerols (2–4 hr) is somewhat longer than that of the chylomicrons.

For the liver to hydrolyse triacylglycerol from chylomicrons, to reassemble them again and to secrete them in the form of VLDL appears, at first, to be an energetically wasteful process without obvious benefit. The absorption of dietary triacylglycerol into the lymphatic system (as chylomicrons) rather than into the hepatic portal vein, probably prevents a good proportion of dietary triacylglycerol from undergoing this fate. However, it may be important that some triacylglycerol is 'processed' by the liver in this way. The fatty acid composition of stored triacylglycerols in adipose tissue is not necessarily that of the triacylglycerols present in the diet (although dietary composition does have some influence on stored fat composition). In reassembling triacylglycerols, the liver is able to alter their composition.

It has recently been discovered that the intestinal cells can also secrete some VLDL into the bloodstream. However, the triacylglycerol that is found in intestinal VLDL may not have been derived directly from the diet. It is possible that this triacylglycerol is synthesized from fatty acids taken up from the blood or from fatty acids synthesised from acetyl-CoA, as in the liver, and that it is the glycerol 3-phosphate path of triacylglycerol synthesis rather than the monoacylglycerol path (see Chapter 17; Section A.5) that is involved. The quantitative significance of intestinal VLDL is not yet known.

(c) *Metabolism of the triacylglycerol in chylomicrons and VLDL*

A considerable amount of work has been done on the metabolism of lipoproteins and the following discussion can only be a brief summary. It is now clear that lipoprotein metabolism in man may be different from that in the rat and the summary will describe, as far as possible, what is known in man (see Owen & McIntyre, 1982). The initial process in the pathway of triacylglycerol metabolism is hydrolysis of the triacylglycerol in chylomicrons and VLDL by lipoprotein lipase. This decreases their lipid/protein ratio and produces a 'remnant chylomicron' (from chylomicrons) and an intermediate-density lipoprotein (IDL) (from VLDL). These are formed by a progressive process, with triacylglycerol gradually being removed, as the lipoproteins circulate through the tissues that contain active lipoprotein lipase. Under normal conditions most of this hydrolysis occurs in the adipose tissue but the metabolism of the remnant chylomicron is different from the IDL. The remnant is metabolised in the liver probably because this tissue possesses a specific uptake process for the remnant particle; its metabolism in the liver is similar to that of LDL (Chapter 20; Section D.2.a). The intermediate-density lipoprotein is further metabolised by the adipose tissue and by exchange of proteins and cholesterol with HDL to form eventually low-density lipoprotein (LDL). In simple terms, the pathways can be described as follows.

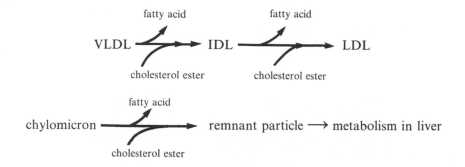

It is possible that lipoprotein is saturated with substrate, so that it catalyses a flux-generating step and the pathway starts with lipoprotein in the bloodstream and ends with the formation of triacylglycerol in adipose and other tissues. However the 'pathway' in the bloodstream is complicated by the need to maintain the stability of the lipoproteins as the non-polar triacylglycerol is lost and the lipid/protein ratio decreases. Hence at least two other enzymes, phospholipase A_2, attached to capillary endothelium, and lecithin-cholesterol acyltransferase (LCAT), and one further lipoprotein, high-density lipoprotein (HDL) are involved in the degradation process. The phospholipase acts on phospholipids on the surface of the lipoproteins, probably in concert with the lipoprotein lipase. The enzyme LCAT is present in the blood, but the cholesterol and phospholipid substrates are present in HDL. A fatty acid is transferred from lecithin to unesterified cholesterol on the HDL to form the non-polar cholesterol ester (Chapter 20; Section D.2.b). The latter is transferred from the HDL to the lipoprotein particle as the triacylglycerol core is removed by lipoprotein lipase; the transfer of the non-polar cholesterol ester is necessary to maintain the stability of the lipoproteins as the triacylglycerol is removed in passage through adipose and other tissues.[4] A summary of lipoprotein metabolism is presented in Figure 6.6 and further detials of HDL and LDL metabolism in man are given in Chapter 20; Section D.2, where cholesterol metabolism is discussed.

2. Free fatty acids

The importance of long-chain fatty acids as a circulating lipid-fuel was overlooked until the mid-1950s, when it was shown that, despite their low concentration in blood (0.3–2.0 mM), they had a very short half-life (less than 2 min, Havel and Fredrickson, 1956). Their importance was further indicated by the observation that their concentration increased during starvation and decreased on feeding (Gordon and Cherkes, 1956). This suggested that they were not involved in lipid transport from the gut but in the mobilization of lipid stores when required. During starvation the flux of long-chain fatty acids through the

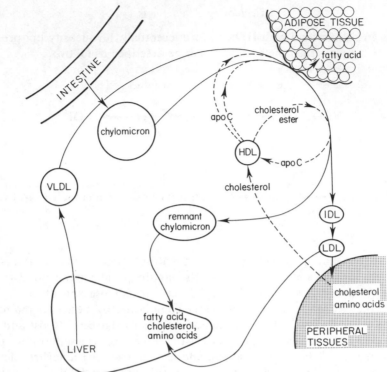

Figure 6.6 Simplified representation of lipoprotein metabolism. For metabolism of HDL and LDL, see Chapter 20; Section D.2

blood is considerable (about 270 g per day in sedentary man) and this accounts for a high proportion of man's total caloric requirement in this condition.[5] The long-chain fatty acids arise from the hydrolysis of triacylglycerol within the adipose tissue and are released into the bloodstream (they are not released by any other tissue.) Fatty acids are oxidized by most tissues; liver, kidney, heart, brown adipose tissue, and the aerobic muscles have a particularly high capacity for fatty acid oxidation but the rate of oxidation is very low in brain, white adipose tissue and type II (anaerobic) muscle fibres (see Table 5.5). The long-chain fatty acids that are present in the blood of man are mainly oleic (43%), palmitic (24%), stearic (13%), linoleic (10%) and palmitoleic (5%) (Havel *et al.*, 1964). They are very nearly insoluble in water; at physiological pH they exist in true aqueous solution only at concentrations below 10^{-6} M. Their transport in blood is facilitated by the existence of albumin, a soluble protein which binds long-chain fatty acids with high affinity. At least ten fatty acid molecules may be bound by each protein molecule although this capacity is never reached under physiological or even pathological conditions. Three of these ten sites have a high affinity for fatty acids and when these three sites are full (at a total concentration of fatty acids of about 2 mM) the concentration of fatty acid *not* bound to the albumin increases markedly (Spector and Fletcher, 1978).

In order to distinguish fatty acids from the fatty acid esterified in triacylglycerol, the former have been variously termed free fatty acids (FFA), non-esterified fatty acids (NEFA) and unesterified fatty acids (UFA). Arguments can be put forward to defend each of these terms but in recent years the term, free fatty acid, seems to have gained wide acceptance and will be used here when it is necessary to make the distinction. Except where otherwise qualified, the term free fatty acid is taken to mean *long-chain* fatty acid, that is, having 14 or more carbon atoms and to include both fatty acids in solution and those bound to albumin. The term fatty acyl-CoA, or simply acyl-CoA, is used to describe the coenzyme A derivatives of these acids. It is possible that, in time, biochemists will accept current chemical terminology in which case they will become known as alkanoyl-CoA.

The formation of complexes with protein may be necessary not only to facilitate the transport of FFA in the blood but also to prevent the formation of fatty acid micelles which would be produced at the concentrations normally present in the blood. Micellar 'solutions' of fatty acids act as detergents disrupting the conformation of proteins and the organization of membranes and can therefore cause considerable damage to tissues (see Section F.2).

The transport of free fatty acids into the cells is a passive process and, as yet, there is no evidence for the existence of a carrier so that it may occur by simple diffusion. Entry into the cell will therefore only occur if the concentration of fatty acid in solution in the cell is less than that in true solution in the extracellular fluid (i.e. $<10^{-6}$ M). It is likely that the very low intracellular concentration necessary for inward transport is maintained by the presence of a fatty acid binding protein which has a high affinity for fatty acids. Such a protein is known to occur in the absorptive cells of the intestine (see Section B.1.b) and in hepatocytes (The binding protein would be expected to play an analogous role to myoglobin, which increases the rate of diffusion of oxygen into muscles by reducing the concentration of free oxygen within the cell. Indeed, there is some evidence that, in heart muscle, myoglobin may also function as the fatty acid binding protein (Gloster and Harris, 1977.)

3. Ketone bodies

The compounds acetoacetate $(CH_3COCH_2COO^-)$ and 3-hydroxybutyrate $(CH_3CH(OH)CH_2COO^-)$ are the only freely soluble circulating lipids. They are known as ketone bodies and arise (from the partial oxidation of fatty acids) in the liver and can be used by most if not all aerobic tissues (e.g. muscle, brain, kidney, mammary gland, small intestine) except the liver. They are known to be an important lipid-fuel in starvation but they may also be important in other conditions (see Chapters 8 and 14). The ketone bodies are misnamed, for they are not bodies and 3-hydroxybutyrate is not a ketone. The appellation ketone was given towards the end of the last century by German physicians who found that the urine of diabetic patients gave a positive reaction with reagents used to detect ketones. In these circumstances, not only acetoacetate but also acetone (propanone) would contribute to the reaction. Acetone is formed by the spontaneous decarboxylation of acetoacetate and is only detectable when the concentration of the latter is abnormally high. The acetone is not further metabolized but excreted through the kidneys and lungs (where it accounts for the characteristic sweet smell on the breath of severely diabetic patients). The pathways of ketone body synthesis and utilization are described in Section D.2. The physiological importance of ketone bodies is considered in Section E.3.c.

D. PATHWAYS OF LIPID CATABOLISM

In the majority of situations in which triacylglycerols are transported into or out of cells, the process involves their hydrolysis. The hydrolyses occurring before absorption of lipids from the gut and during the entry of triacylglycerols from lipoproteins into cells have already been described. Hydrolysis also occurs when triacylglycerols within a cell are to be mobilised, for example, in adipose tissue for release of fatty acids into the blood. These hydrolyses are catalysed by lipases acting consecutively on triacylglycerols, diacylglycerols, and monoacylglycerols. Of these lipases, triacylglycerol lipase has received by far the most attention since its activity is low and, therefore, the over-all rate of lipolysis is regulated by the activity of this enzyme (i.e. it is a flux-generating step—see below). Indeed, its sensitivity to a variety of hormones (discussed in Chapter 8; Section C.1) has engendered its alternative name of 'hormone-sensitive lipase' although this is somewhat ambiguous since lipoprotein lipase activity is also affected by hormones. There appears to be an active monoacylglycerol lipase in many cells but the existence of a specific diacylglycerol lipase is less well established. The distribution of the various lipases is summarized in Table 6.2. The result of the activities of these lipases is that triacylglycerol is hydrolysed to three fatty acid molecules and glycerol as follows:

$$
3H_2O + \begin{matrix} CH_2O.CO(CH_2)_nCH_3 \\ | \\ CH\ O.CO(CH_2)_nCH_3 \\ | \\ CH_2O.CO(CH_2)_nCH_3 \\ \text{triacylglycerol} \end{matrix} \longrightarrow \begin{matrix} CH_2OH \\ | \\ CH.OH \\ | \\ CH_2OH \\ \text{glycerol} \end{matrix} + 3CH_3(CH_2)_nCOO^- + 3H^+
$$

Other tissues (e.g. muscle and liver) contain the same complement of lipase enzymes to convert triacylglycerol to fatty acids and glycerol, but they are quantitatively less important than the adipose tissue enzymes in higher animals.

1. The pathway for oxidation of triacylglycerol

(a) *The physiological pathway*

There is no doubt that the triacylglycerol lipase within the adipose tissue is saturated with substrate, since on a dry weight basis approximately 95 % of the fat cell is triacylglycerol, whereas the concentrations of monoacylglycerol and

Table 6.2. Distribution and properties of lipases

Recommended name	Substrate specificity	pH optimum	Activators	Inhibitors
Pancreatic triacylglycerol lipase	tributyrin > triacylglycerol containing long-chain fatty acids; triacylglycerol > monoacylglycerol	8.0–9.0	taurocholate	NaF (weak)
Adipose tissue triacylglycerol lipase	tributyrin > triacylglycerol containing long-chain fatty acids	6.0–7.5	phosphate	NaF (strong)
Adipose tissue monoacylglycerol lipase	monoacylglycerol > {diacylglycerol and triacylglycerol}	7.5–8.5	albumin, taurocholate	NaF (weak)
Lipoprotein lipase	tributyrin not hydrolysed; diacylglycerol also hydrolysed	8.0–8.5	albumin	protamine (strong) NaCl (strong)

Data taken from Hübscher (1970)

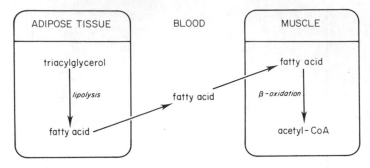

Figure 6.7   The physiological pathway of lipid oxidation. In addition to muscle, other
tissues such as kidney will also oxidize fatty acids

diacylglycerol are very low. Furthermore, none of the reactions that are involved
in fatty acid oxidation, from exogenous fatty acids to the formation of acetyl-CoA,
are saturated, so that the physiological pathway for fat oxidation, in most if not
all tissues, starts at the level of triacylglycerol. Since most of the store of
triacylglycerol in man and other higher animals occurs in the adipose tissue, the
flux-generating step and hence the start of the lipid oxidation pathway is the
triacylglycerol lipase within the adipose tissue* (Figure 6.7). This is another
example of a metabolic pathway that spans more than one tissue, that is, adipose
tissue, blood and the actual tissue that utilises the fatty acid (e.g. muscle). The
mechanism of control that links the rate of fatty acid oxidation in muscle with
triacylglycerol hydrolysis in adipose tissue is discussed in Chapter 8; Section
B.2.b.

(b) *Fatty acid activation*

There is no evidence that long-chain fatty acids enter cells other than by passive
diffusion, so that their rate of uptake is proportional to the difference in their
concentrations inside and outside the cell. The further metabolism of the long-
chain fatty acids is dependent upon the formation of thioesters with coenzyme A
(see Chapter 4; Section A.3). This process is known as 'activation' and serves to
convert the fatty acids into a form more amenable to further biochemical changes.
Activation can be achieved by a number of reactions catalysed by different
enzymes but the most important are the acyl-CoA synthetases,[6] formerly known
as thiokinases. They catalyse reactions of the type:

$$R.COO^- + CoASH + ATP^{4-} \longrightarrow R.CO.SCoA + AMP^{2-} + PP_i^{3-}$$

Three types of acyl-CoA synthetase are widely distributed and distinguished by

---

* Triacylglycerol lipase is probably flux-generating for oxidation of endogenous triacylglycerol in, for
example, muscle and liver, but this fuel is much less important than exogenous fatty acid (see Chapters
9 and 14).

their preferences for fatty acids of different chain lengths and by their intracellular distribution.

1. Acetyl-CoA synthetase activates acetate and several other low molecular weight carboxylic acids. In muscle, its activity is restricted to the mitochondrial matrix but in other tissues it is also present in the cytosol.
2. Medium-chain acyl-CoA synthetase activates fatty acids containing from four to eleven carbon atoms. (The name recommended by the Enzyme Commission for this enzymes is butyryl-CoA synthetase which is somewhat misleading since it has a much wider specificity.) It has been detected only in liver mitochondria. This is consistent with the fact that the medium-chain length fatty acids in man are obtained only from the diet (e.g. from triacylglycerols in dairy products[7]). These medium-chain length fatty acids released in digestion are absorbed and transported to the liver via the hepatic portal vein. They are transported to the mitochondria without further metabolism (unlike long-chain fatty acids, see below) where they become available as substrate for medium-chain acyl-CoA synthetase.
3. Acyl-CoA synthetase activates fatty acids containing from six to 20 carbon atoms and occurs both in microsomes and on the outer mitochondrial surface. Its substrate specificity, location, and activity suggest that it plays the major role in fatty acid activation. In addition to these AMP-forming fatty acyl-CoA synthetases, there occurs in mitochondria an enzyme which catalyses the reaction:

$$R.COO^- + CoASH + GTP^{4-} \longrightarrow R.CO.SCoA + GDP^{3-} + P_i^{2-}$$

Nothing is known of its physiological role but, like other mitochondrial activating enzymes, it may serve to activate carboxylic acids other than fatty acids destined for oxidation (e.g. acetate for biosynthesis).

(c) *Entry of fatty acids into mitochondria*

Although mitochondria contain all the enzymes for β-oxidation (see below), isolated mitochondria are unable to oxidize long-chain fatty acids or their coenzyme A derivatives. Although both of these compounds are anions and might be expected to be transported in an antiport or symport carrier system (see Chapter 4; Section D.4.a) no direct transport of either occurs. Instead, their transport requires the formation of an ester of the fatty acid with carnitine[8] in a reaction catalysed by carnitine palmitoyltransferase:

$$(CH_3)_3N^+CH_2\underset{\underset{\text{carnitine}}{|}}{\overset{|}{CH}}.CH_2COO^- + CH_3(CH_2)_nCO.SCoA \longrightarrow$$

$$CoASH + (CH_3)_3N^+CH_2\underset{\underset{\text{acylcarnitine}}{|}}{\overset{|}{CH}}.CH_2COO^-$$
$$O.CO.(CH_2)_nCH_3$$

This enzyme is probably able to transfer a variety of fatty acyl groups from coenzyme A to carnitine. There is a specific carrier to transport fatty acyl-carnitine across the membrane and, in order to achieve effective transport of acyl-CoA, carnitine palmitoyltransferase is also present on the inner surface of the inner mitochondrial membrane (Figure 6.8). The properties of the enzymes from the two locations are somewhat different and the 'outer' enzyme is known as carnitine palmitoyltransferase-I and the 'inner' enzyme as carnitine palmitoyl-transferase-II (Frenkel and McGarry, 1980). The former enzyme generates acyl-carnitine (from acyl-CoA) which is transported across the inner membrane, after which transferase-II regenerates the acyl-CoA and free carnitine. The latter is transported out of the mitochondrion by the same carrier that transports acyl-carnitine in (LaNoue and Schoolwerth, 1979). Indeed, the entry of acyl-carnitine is obligatorily linked to the exit of carnitine as with the mitochondrial transport of ATP and ADP. There is some evidence that the transferase-I enzyme is an important site for the regulation of fatty acid oxidation in the liver (see Chapter 8, Section D.1.b).

There are a number of reports of cases of the congenital deficiency of muscle carnitine palmitoyltransferase (Bank *et al.*, 1975; Warshaw, 1975). Normal metabolism is maintained except during sustained exercise and starvation, both conditions in which fatty acid oxidation is important. The patient usually presents with a history of cramps, fatigue after exercise and myoglobinuria. The latter is an indication of muscle damage. A more severe disease, which can be fatal, occurs when the enzyme is deficient in the liver: clinical features include an enlarged fatty liver, hypoglycaemia, hyperammonaemia and possibly coma, which occur on fasting normally due to an infection (e.g. upper respiratory tract infection) (Bougneres *et al.*, 1980). A number of cases of carnitine deficiency have

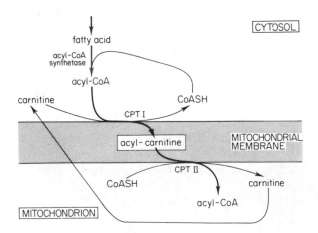

Figure 6.8   Entry of fatty acids as acyl-CoA into the mitochondrion. Abbreviations: CPT, carnitine palmitoyltransferase (see text for role of CPT I and CPT II)

been reported (Chapoy *et al.* 1980) and the patients may be divided into two types: those with a deficiency of muscle carnitine (myopathic carnitine deficiency) probably due to an inability to transport carnitine across the cell membrane, and those with a deficiency in liver, plasma and muscle (systemic carnitine deficiency) probably due to a defect in the synthetic pathway in liver. The former deficiency produces clinical features similar to those of a deficiency of the muscle carnitine palmitoyltransferase whereas the latter deficiency produces features similar to those of a deficiency of the liver enzyme. The deficiency of carnitine can, however, be treated with carnitine either intravenously or orally (Chapoy *et al.*, 1980).

(d) *β-Oxidation*

Fatty acid oxidation involves the progressive removal of two-carbon units from the carboxyl end of the acyl-CoA and occurs entirely within the mitochondria. The process is known as β-oxidation since the β (i.e. 3-) carbon atom is oxidized prior to cleavage between carbon atoms 2 and 3 of the acyl-CoA. The complete oxidation of a saturated acyl-CoA (with an even number of carbon atoms) requires four enzymes acting sequentially and repeatedly. The process, sometimes termed the β-oxidation spiral, is described in more detail below and diagrammatically in Figure 6.9. It has not been possible to detect the intermediates of the pathway although both acyl-CoA and acetyl-CoA are readily detected in tissues that oxidize fatty acids. It seems likely that the product of one reaction is transferred directly to the enzyme catalysing the next, so that β-oxidation can be considered to be catalysed by a multienzyme complex.

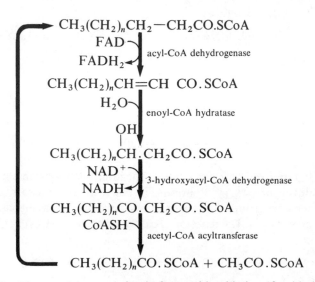

Figure 6.9 The reactions occurring in fatty acid oxidation (β-oxidation spiral)

(i) *Acyl-CoA dehydrogenase reaction*   The first stage in β-oxidation is the oxidation of the acyl-CoA to the corresponding Δ^2-*trans*-enoyl-CoA[9] (2,3-dehydroacyl-CoA). The reaction is catalysed by several acyl-CoA dehydrogenases with varying substrate specificities so that, between them, all fatty acids with more than four carbon atoms can be oxidized. All of these dehydrogenases possess a tightly bound FAD molecule (as a prosthetic group) so that the reaction can be represented by the equation:

$$R.CH_2CH_2CO.SCoA + E.FAD \rightarrow R.CH{=}CH.CO\,SCoA + E.FADH_2$$

   fatty acyl-CoA Δ^2-*trans*-enoyl-CoA

The prosthetic group is re-oxidized through the activity of the electron transferring flavoprotein which transfers the electron into the electron transfer chain at the ubiquinone level. Consequently, the initial dehydrogenation of acyl-CoA leads to the generation of two ATP molecules (in a similar way to succinate oxidation in the TCA cycle).

(ii) *Enoyl-CoA hydratase reaction*   In the next reaction, a molecule of water is added across the double bond of the *trans*-enoyl-CoA to give an L-3-hydroxyacyl-CoA. The reaction is catalysed by enoyl-CoA hydratase (crotonase):

$$R.CH{=}CH.CO.SCoA + H_2O \rightarrow R.CH(OH).CH_2CO.SCoA$$

   Δ^2-*trans*-enoyl-CoA   L-3-hydroxyacyl-CoA

(iii) *3-Hydroxyacyl-CoA dehydrogenase reaction*   The preceding hydration prepares the fatty acid for its second oxidation reaction, catalysed by 3-hydroxyacyl-CoA dehydrogenase:

$$R.CH(OH).CH_2CO.SCoA + NAD^+ \longrightarrow R.CO.CH_2CO.SCoA + NADH + H^+$$

   L-3-hydroxyacyl-CoA   3-oxoacyl-CoA
                                               (β-ketoacyl-CoA)

This oxidation, like most in which a secondary alcohol is being oxidized, involves NAD^+ and results in the formation of NADH each molecule of which can enter the electron transfer pathway and generate three molecules of ATP

(iv) *Acetyl-CoA acyltransferase reaction*   The molecule is now in a suitable form to undergo splitting of the C_2-C_3 bond. The reaction takes the form of a transfer of the shortened acyl group to a molecule of coenzyme A, leaving carbon atoms C_1 and C_2 as acetyl-CoA. The reaction is catalysed by acetyl-CoA acyltransferase (β-ketothiolase):

$$R.CO.CH_2CO.SCoA + CoASH \rightarrow R.CO.SCoA + CH_3CO.SCoA$$

   3-oxoacyl-CoA   acyl-CoA   acetyl-CoA

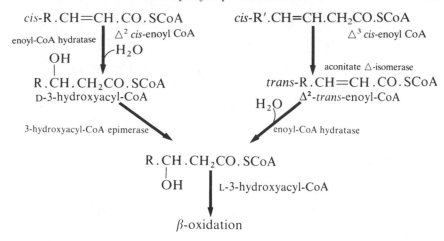

Figure 6.10 Reactions enabling unsaturated fatty acids to enter the β-oxidation spiral

The shortened acyl-CoA is now a substrate for acyl-CoA dehydrogenase and can undergo a further turn of the β-oxidation 'spiral'. A number of fates are possible for the acetyl-CoA; it may be oxidized via the tricarboxylic acid cycle (e.g. in muscle or kidney); it may be used to synthesize other lipids or, in liver, it may be converted into ketone bodies (see Section D.2.a).

(e) *Other reactions in fatty acid oxidation*

Unsaturated fatty acids (see Chapter 17; Section B.1) must enter the β-oxidation spiral as Δ^2-*trans*-enoyl-CoA or L-3-hydroxyacyl-CoA intermediates. Difficulties arise, however, due to the fact that naturally occurring unsaturated acids have a *cis* configuration (see Appendix 4.1) and the position of the double bonds are often such that a Δ^3-enoyl (rather than a Δ^2-enoyl) intermediate results. Such a Δ^3-enoyl-CoA is hydrated to a D-hydroxyacyl-CoA rather than an L-hydroxyacyl-CoA. Two auxiliary enzymes carry out the necessary conversions which allow these intermediates to enter the β-oxidation 'spiral' (Figure 6.10). The first is aconitate Δ-isomerase which converts a Δ^3-*cis*-enoyl-CoA to a Δ^2-*trans*-enoyl-CoA. The second is 3-hydroxyacyl-CoA epimerase which converts a D-3-hydroxyacyl-CoA into an L-3-hydroxyacyl-CoA.

Additional enzymes are also required when fatty acids with odd numbers of carbon atoms are to be oxidized completely. These acids are not common in animal tissue but small amounts are likely to arise from plant material in the diet. The oxidation of these acids proceeds via β-oxidation until the three-carbon fatty acid derivative, propionyl-CoA, is formed. The further metabolism of this compound involves, first, carboxylation to D-methylmalonyl-CoA, secondly, epimerization to L-methylmalonyl-CoA, and thirdly, isomerization to succinyl-CoA (Figure 6.11).

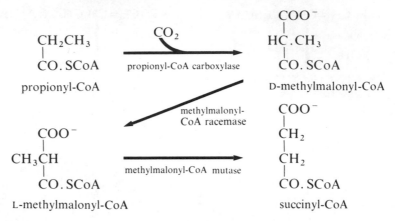

Figure 6.11   Pathway for the metabolism of propionyl-CoA. The conversion of succinyl-CoA to oxaloacetate occurs via the tricarboxylic acid cycle

   Methylmalonyl-CoA mutase, the enzyme catalysing the last reaction in this sequence, is of some interest since it is one of the small number of enzymes requiring deoxyadenosylcobalamin (a vitamin B_{12} derivative) as cofactor. Patients suffering from pernicious anaemia (in which cobalamin cannot be properly absorbed) excrete methylmalonate in their urine. However, quantitatively the most important source of propionyl-CoA in humans is not fatty acids with odd numbers of carbon atoms but the branched-chain amino acids, isoleucine, and valine. The metabolism of these amino acids is discussed in Chapter 10, Section C.3 and Appendix 10.1.

2. Ketone body formation and utilization

(a) *Pathway of ketogenesis*

Under physiological conditions, only the liver is able to synthesise ketone bodies. It is probable that the flux-generating step for ketone body formation is the lipolysis of triacylglycerol in adipose tissue, so that the physiological pathway spans more than one tissue and includes adipose tissue, blood, and liver. The importance of this physiological pathway is in understanding the control of ketogenesis which is considered in Chapter 8, Section D. Usually the pathway is considered to start at the level of long-chain fatty acids in the liver. Thus the first sequence of reactions in ketone body formation involves the activation of fatty acids, the transport of acyl-CoA from the cytosol into the mitochondria and the oxidative cleavage of long-chain acyl-CoA to acetyl-CoA in the process of β-oxidation, described in detail above. The conversion of acetyl-CoA to ketone bodies requires three more stages: the condensation of two molecules of acetyl-CoA to form acetoacetyl-CoA, the deacylation of acetoacetyl-CoA, and the reduction of some of the resulting acetoacetate to 3-hydroxy butyrate. All of these processes take place in the mitochondria of the liver.

The first of these steps is catalysed by acetyl-CoA acetyltransferase (thiolase):

$$CH_3CO.SCoA + CH_3CO.SCoA \longrightarrow CH_3CO.CH_2CO.SCoA + CoASH$$

As acetoacetyl-CoA is also an intermediate in the β-oxidation of fatty acids it could arise from this source, although such a contribution would be of minor importance. The deacylation of acetoacetyl-CoA is achieved in two reactions, catalysed by hydroxymethylglutaryl-CoA synthase (HMG-CoA synthase) and hydroxymethylglutaryl-CoA lyase (HMG-CoA lyase). The former reaction involves the condensation of a third molecule of acetyl-CoA with acetoacetyl-CoA in a reaction analogous to that catalysed by citrate synthase. The product is 3-hydroxy-3-methylglutaryl-CoA (HMG-CoA) which is cleaved by HMG-CoA lyase to form acetoacetate and acetyl-CoA (Figure 6.12). (HMG-CoA is also an intermediate in the biosynthesis of steroids, but for this purpose it is synthesized in the cytosol of the liver (see Chapter 20; Section B.1).) Note that, although the acetyl-CoA released contains different carbon atoms to those incorporated in the previous reaction, there is no over-all consumption of this acetyl-CoA, so that it plays a 'catalytic' role.

There is an enzyme in liver mitochondria which catalyses the direct hydrolysis of acetoacetyl-CoA (acetoacetyl-CoA hydrolase) but its role in ketogenesis is not thought to be important. Its maximum activity is considerable lower than that of the HMG-CoA cycle enzymes (see Table 6.3) and it requires a high concentration

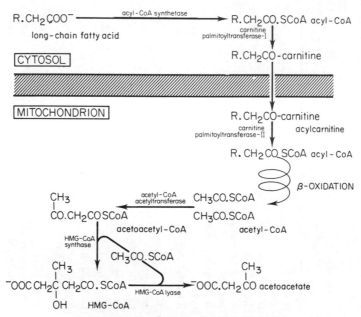

Figure 6.12 Pathway of ketone body synthesis. For details of fatty acid transport into mitochondria see Figure 6.8

Table 6.3. Maximal activities of enzymes involved in formation of ketone bodies in livers from vertebrates*

Animal	Enzyme activities (μmol substrate utilized $.min^{-1}.g^{-1}$ fresh wt. tissue)				
	HMG-CoA synthase	HMG-CoA lyase	Acetoacetyl-CoA hydrolase	3-Hydroxybutyrate dehydrogenase	3-Oxoacid CoA-transferase
Pisces					
Bass (*Dicentrarcus labrax*)	0.25	0.45	—	<0.01	9.15
Plaice (*Pleuronectes platessa*)	0.22	0.47	—	<0.01	4.8
Dogfish (*Scylliorhinus canicula*)	0.41	0.55	<0.01	0.16	0.80
Ray (*Raja clavata*)	0.32	0.71	—	—	0.21
Reptilia					
Green lizard (*Lacerta viridis*)	0.51	3.6	<0.01	1.5	0.68
Aves					
Domestic pigeon (*Columba livia*)	1.7	8.2	0.26	0.15	1.5
Domestic fowl (*Gallus gallus*)	1.7	4.3	0.33	0.22	1.0
Mammalia					
Laboratory mouse	1.1	3.8	0.44	2.3	0.86
Laboratory rat	1.8	9.8	0.21	12.4	0.46
Rabbit (*Oryctolagus cuniculus*)	0.58	2.8	0.05	2.3	0.21

* Data taken from Zammit *et al.* (1979). Enzyme activities were measured at 10°C for enzymes from fish and at 25°C for enzymes from other species.

of substrate (that is, the K_m for acetoacetyl-CoA is high). It is likely that this enzyme plays a general role in the removal of CoA from acyl-CoA and thus may serve to prevent the accumulation of high concentrations of such compounds within the cell.

The final reaction in ketone body production is the conversion of acetoacetate to D-3-hydroxybutyrate[10] catalysed by the enzyme 3-hydroxybutyrate dehydrogenase:

$$CH_3CO.CH_2COO^- + NADH + H^+ \longrightarrow CH_3CH(OH).CH_2COO^- + NAD^+$$

acetoacetate   D-3-hydroxybutyrate

At least in the liver of the rat, the enzyme is sufficiently active to catalyse a near-equilibrium reaction. Since the mitochondrial $NAD^+/NADH$ concentration ratio is normally maintained approximately constant at about 8.0 (see Chapter 17; Section A.8.c) this sets the ratio of 3-hydroxybutyrate/acetoacetate concentration in the liver, and hence that released into the blood, to within the range 3–6. Thus a considerably greater amount of 3-hydroxybutyrate than acetoacetate is produced by the liver. Why two, rather than one, ketone bodies are produced is not entirely clear. Comparative studies indicate that acetoacetate is the more 'primitive' ketone body as it occurs in the absence of 3-hydroxybutyrate in invertebrates and some fish (Zammit and Newsholme, 1979; Table 6.3). However, 3-hydroxybutyrate is the better fuel since it is more reduced, so why has acetoacetate persisted? The answer may be that to produce solely one or the other ketone body in large amounts could render ketone body synthesis dependent on the redox state of the liver. However, if an equilibrium mixture of the two is produced, changes in the redox state of liver mitochondria simply alter the ratio of ketone bodies released.

Of some clinical importance is the fact that, in diabetic coma, the concentration ratio, 3-hydroxybutyrate/acetoacetate, can rise to as high as 15, probably due to disturbances of the $NAD^+/NADH$ concentration ratio and the intracellular pH. Since a frequently used rapid test for ketonuria (using Clinistix or similar material) detects only acetoacetate (and acetone) this can result in a serious underestimate of the extent of the ketonuria.

(b) *Ketone body oxidation*

Ketone bodies are oxidized in the mitochondria of aerobic tissues such as muscle, heart, kidney, intestine and brain. The more reduced ketone body, 3-hydroxybutyrate, is first oxidized to acetoacetate by NAD^+ in a reaction catalysed by 3-hydroxybutyrate dehydrogenase. This is a direct reversal of its synthesis in the liver. Activation of acetoacetate (that is, its conversion to acetoacetyl-CoA) occurs by the transfer of coenzyme A from succinyl-CoA to acetoacetate in a reaction catalysed by the enzyme 3-oxoacid CoA-transferase:

$$
\begin{array}{ccccc}
\underset{|}{CH_2CO.SCoA} & & \underset{|}{CH_2COO^-} & \underset{|}{CH_2COO^-} & \underset{|}{CH_2CO.SCoA} \\
CH_2COO^- & + & CO.CH_3 & \longrightarrow & CH_2COO^- & + & CO.CH_3
\end{array}
$$

succinyl-CoA   acetoacetate   succinate   acetoacetyl-CoA

The major source of succinyl-CoA for this reaction is undoubtedly the tricarboxylic acid cycle. Since succinate is formed from succinyl-CoA, the transferase bypasses the succinyl-CoA synthetase reaction of the cycle (see Chapter 4; Section B.1.e). As this latter reaction would have produced one molecule of GTP for each succinyl-CoA split, the activation of acetoacetate is seen to require the equivalent of one molecule of ATP. This mechanism of activation is quite different from that involved in the activation of fatty acids, and other carboxylic acids (see above). Why should this be so? When other fuels cannot or should not be used by the tissue (e.g. glucose during starvation— Section E.3.c) and to ensure complete oxidation of the acetyl-CoA produced from the ketone bodies, oxaloacetate must be available for citrate synthase and the TCA cycle. The activation of acetoacetate causes the conversion of succinyl-CoA to succinate and the latter is converted to oxaloacetate via the reactions of the TCA cycle. A high concentration of acetoacetate will favour the conversion of some succinyl-CoA to oxaloacetate so that the concentration of the latter will increase at the expense of that of succinyl-CoA and so facilitate the entry of more acetyl-CoA into the cycle.

Acetoacetyl-CoA is split to acetyl-CoA in a reaction involving the same enzyme (acetyl-CoA acetyltransferase) as that used in acetoacetyl-CoA synthesis. The direction of the reaction depends upon the concentration ratio of substrates and products, which must be different in the liver to that in the tissues that utilize ketone bodies.

The physiological pathway for ketone body utilization probably starts at triacylglycerol in adipose tissue, so that ketone body production and utilization (as far as acetyl-CoA) are really *one* physiological pathway, as follows:

$$ \text{triacylglycerol} \rightarrow \text{fatty acid} \rightarrow \rightarrow \text{ketone bodies} \rightarrow \rightarrow \text{acetyl-CoA} $$

This is important in control, since it means that the rate of ketone body utilization by the various tissues depends upon their concentration in the bloodstream (see Section E).

E. PHYSIOLOGICAL IMPORTANCE OF LIPID OXIDATION

1. Energy content of lipid

The major store of chemical energy in terrestrial animals is triacylglycerol. In birds, mammals, and probably reptiles, most of the triacylglycerol is stored in adipose tissue which may be present in considerable amounts; at least one quarter of the weight of an average young woman is due to body lipid (Pond,

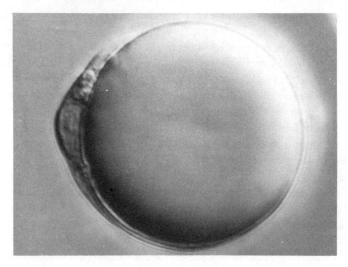

Figure 6.13 Interference contrast photograph of a white adipose tissue cell. Note the spherical lipid droplet which fills most of the cell. The photograph was kindly supplied by Professor C. N. Hales, Department of Chemical Pathology, The Welsh National School of Medicine, Cardiff

1978—see Chapter 19; Section C.3. for further discussion). The predominance of this storage fuel is accounted for by its superior energetic value (the amount of energy liberated on oxidation) when compared with other possible storage fuels (see Table 2.3). Comparison with carbohydrate shows that fat is nearly 2.5 times more efficient than carbohydrate on a weight basis. However, two factors can affect the biological relevance of this difference. First, it must be asked whether the 'efficiency' by which this energy is made available as ATP is comparable between fat and carbohydrate. Since the pathways of ATP generation from both carbohydrate and lipid are largely identical (i.e. oxidative phosphorylation) the 'efficiencies' must be very similar. Secondly, the state of stored material in the cell must be considered. The data in Table 2.3 refer to dry substances. Since lipids are highly hydrophobic they are stored 'dry'—indeed a typical adipocyte contains 90% of its total weight as pure triacylglycerol (see Figure 6.13). Storage carbohydrate, however, presents a very different picture. Glycogen is hydrophilic and exists *in vivo* in highly hydrated glycogen granules. Some 65% of this 'physiological' glycogen is water so that the enthalpy change on oxidation (heat of combustion) of 1 g of this biological glycogen is only about 6kJ.g^{-1}. This makes lipid more than five times better than carbohydrate as a biological store of energy. This difference between lipid and carbohydrate is especially important in terrestrial animals in which an increase in mass would interfere with mobility. If a normal man stored the same number of joules in carbohydrates as he does in triacylglycerol he would be nearly twice as heavy and his mobility severely reduced. A typical 70 kg man stores 7 kg lipid (and a 65 kg woman, twice this amount.) Stores of this magnitude enable man to survive starvation for more than a month, whereas the carbohydrate stores (about 100 g) are virtually depleted in a single day (see Chapters 8 and 14 for detailed discussion). Although the average human subject in Western society may be rarely called upon to starve for prolonged periods, the ability to do so may have been an important survival factor during his evolution and may be important today in some parts of the world. The fuel/weight ratio is of even greater importance in those animals that fly and yet need to store large amounts of energy for long-distance flight (e.g. migrating birds and insects). These animals depend almost exclusively on lipid oxidation during migration (Odum, 1965; Weis-Fogh, 1952).

Another consideration in the choice of fuel-storage compound could be the amount of water released during oxidation. For tripalmitoylglycerol (tripalmitin), 1.09 g water is produced during the complete oxidation of 1 g of the triacylglycerol (Table 6.4). This 'metabolic water' may make a significant contribution to the fluid balance of the animal. It is thus very much more economical from all points of view for the camel's hump to contain adipose tissue and not water.[11]

2. Constraints on the use of lipids as fuels

Although triacylglycerol may appear to be the ideal storage fuel, there are at least three drawbacks to the storage of all fuel reserves in this form. First, tissues that function under anaerobic conditions or lack a sufficient oxidative capacity (e.g. erythrocytes and white blood cells, kidney medulla, tissues of the eye) are

Table 6.4. Formation of water by the oxidation of storage fuels

	M. wt.	moles ATP/mole oxidized	moles water/mole oxidized	g water/g oxidized	moles water/mole ATP
Tripalmitoylglycerol	807	409*	49†	1.09	0.12
Glucose	180	38‡	6‡	0.60	0.16
'Wet' glycogen§	'162'	13‖	7.7¶	0.85	0.59

* One mole palmitate yields 129 moles ATP (Table 5.3) and 1 mole glycerol yields 22 moles ATP. $(129 \times 3) + 22 = 409$ (tripalmitoylglycerol + $3H_2O \rightarrow 3$ palmitate + glycerol.

† Oxidation of each mole of palmitate yields 16 moles water. Oxidation of a mole of glycerol yields 4 moles water. Three moles water are consumed in hydrolysis. $(3 \times 16) + 4 - 3 = 49$.

‡ $C_6H_{12}O_6 + 6O_2 \rightarrow 6H_2O + 6CO_2$

§ Assuming that two-thirds of the weight of physiological glycogen is water.

¶ One 'mole' (162 g) dry glycogen produces 5 moles water on oxidation. Thus a similar weight of 'wet' glycogen produces $\frac{5}{3}$ moles water from glycogen oxidation plus water released ($\frac{2}{3}$ of 162 g = 108 g; $\frac{108}{18} = 6$ mole), that is a total of $7\frac{2}{3}$ moles.

‖ One third of ATP yield (39 moles) of dry glycogen.

dependent on carbohydrate for energy provision. Secondly, brain does not oxidize fatty acids (although it can use ketone bodies when their concentration in blood is sufficiently high—see Chapter 8) so that it too is dependent upon glucose oxidation for its energy provision. Since almost all animals are unable to convert lipid into carbohydrate, they must store some carbohydrate for these tissues. Thirdly, since the normal diet of man consists of a high proportion of carbohydrate, it is perhaps metabolically simpler for tissues to utilize glucose under normal dietary conditions. Indeed, the transport of the insoluble lipids in the aqueous medium of the blood requires specific mechanisms which have already been discussed in this chapter. Although ketone bodies can be transported in simple solution, their synthesis in any quantity requires the transport of fatty acids from adipose tissue to the liver.

3. Significance of multiple forms of lipid-fuel

It is intriguing to speculate on why there are three classes of lipid-fuel that are transported in the blood (triacylglycerol, fatty acid, and ketone bodies) but only one carbohydrate (glucose). Although no definitive answer can emerge to questions of this nature, speculation provides an important driving force in the investigation of physiological function. In this section, the special roles of each of these classes of lipids are considered in turn with the intention of shedding light on this question. A first clue to the problem is provided by looking at the site of origin and the site of uptake of each of the fuels (Table 6.5). This information is summarized in Figure 6.14. Quantitative data on rates of fuel utilization can be

Table 6.5. Sites of origin and uptake for different circulating lipid-fuels

Lipid-fuel	Form	Origin	Major site of uptake
Triacylglycerol	chylomicrons, VLDL	intestine (diet) liver, intestine	adipose tissue, muscle, lactating mammary gland, liver*
Fatty acid	FFA-albumin	adipose tissue	liver, muscle, kidney
Ketone bodies	acetoacetate, 3-hydroxybutyrate	liver	heart, brain, kidney, skeletal muscle, intestine

* Cycling of VLDL by the liver (that is, simultaneous production and utilization within that tissue) is minimized by the attachment of apoliprotein C to the VLDL. This activates the lipoprotein lipase of other tissues. Only when apolipoprotein C leaves the lipoprotein (at a late stage of VLDL degradation) can the liver enzyme compete effectively for the triacylglycerol.

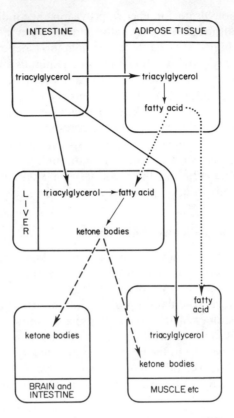

Figure 6.14   The production and utilization of lipid-fuels by different tissues. In the blood, fatty acids are transported as complexes with albumin and triacylglycerol as lipoprotein. Triacylglycerol is normally taken up by tissues for storage, except by muscle in starvation, and uptake requires extracellular hydrolysis by lipoprotein lipase (Section C.1.c).

obtained by measuring the activity of key enzymes in the pathways involved. For fatty acid and endogenous triacylglycerol utilization, the activities of the enzymes carnitine palmitoyltransferase and triacylglycerol lipase, respectively, provide useful indications of pathway flux. The maximal rate at which exogenous triacylglycerols can be oxidized is generally assumed to be equal to the maximum activity of lipoprotein lipase. However, the assessment of the rate of ketone body utilization is more problematical, since both of the enzymes specific to that pathway, namely 3-hydroxybutyrate dehydrogenase and 3-oxoacid CoA-transferase, catalyse near-equilibrium reactions which cannot provide quantitative information (see Appendix 5.2 for an account of how enzyme activities can be used to provide this information). Some of the activities of enzymes involved in lipid metabolism are presented in Table 6.6.

Table 6.6. Activities of enzymes of lipid catabolism in tissues of rat and man*

Animal	Tissue	Enzyme activities (μmol.min^{-1}.g^{-1} at 25 °C)				
		Triacylglycerol lipase	Lipoprotein lipase	Carnitine palmitoyltransferase	3-Hydroxybutyrate dehydrogenase	3-Oxoacid CoA-transferase
Rat	Heart	0.12	3.0	2.2	1.4	48.3
	Diaphragm	—	0.8	—	0.3	17.1
	Skeletal muscle (gastrocnemius)	0.03	—	1.9	0.1	8.0
	Liver	—	—	—	31.0	0.5
	Kidney	—	—	—	26.0	4.0
	Brain	—	—	—	2.1	1.5
	Adipose	—	0.05	—	—	0.3
	Non-lactating mammary gland	—	0.3	—	—	—
	Lactating mammary gland	—	3.0	—	—	139.0
Man	Skeletal muscle (quadriceps)	—	0.03	—	18.6	4.0
	Brain	—	—	—	1.6	0.3
	Adipose (abdominal)	—	0.12	—	—	—

*Data from Crabtree and Newsholme (1972b); Hamosh *et al.* (1970); Krebs *et al.* (1971); Lithell *et al.* (1979); Page and Williamson (1971) and Zammit *et al.* (1979). The activities for the lipases are based on release of fatty acid.

(a) *Triacylglycerol*

In order for exogenous triacylglycerol to be utilized by a tissue, lipoprotein lipase must be present and active in an extracellular site and in the fed animal most of the active lipoprotein lipase is present in adipose tissue. Consequently, the triacylglycerol released into the blood will be hydrolysed in the adipose tissue and the fatty acids will be taken up by that tissue, esterified, and stored without being released into the general circulation. The apparently complicated series of hydrolyses and esterifications that take place in the absorption and utilization of dietary lipid can be understood in this light. If most of the dietary lipid appeared as free fatty acid in the blood, it would be readily oxidised by most tissues of the body in preference to glucose and this would have serious consequences for the integration of fuel metabolism in the fed state. The transport of fatty acid in the form of triacylglycerol, therefore, effectively 'conceals' the fatty acids from tissues other than those which possess an active lipoprotein lipase. In this way, fatty acids are 'directed' to the storage tissues.

In starvation, the activity of lipoprotein lipase in adipose tissue decreases while it increases in heart and skeletal muscle (Lithell *et al.*, 1979). The decline in lipoprotein lipase activity in adipose tissue is expected since there is no point in this tissue hydrolysing extracellular triacylglycerol when it is releasing fatty acids from endogenous triacylglycerol. The triacylglycerol in the blood is, therefore, directed to other tissues such as heart muscle. However, the quantitative importance of exogenous triacylglycerol as a source of energy in aerobic muscles compared with fatty acids and ketone bodies is unclear. The activity of lipoprotein lipase in human muscle (Table 6.6) is such as to suggest that triacylglycerol is much less important in exercise compared to fatty acids (compare lipoprotein lipase activity with that of carnitine palmitoyltransferase).

An important switch in the use of triacylglycerol occurs in the mother in preparation for lactation. Experiments with rats have shown that a few days before parturition, the level of triacylglycerol (as lipoprotein) in the blood increases and this increase continues until parturition. This increase is not due to elevated rates of synthesis or to greater lipid intake but is due to a decrease in the activity of lipoprotein lipase in adipose tissue. After parturition, the level of triacylglycerol in the blood decreases towards the normal due to a marked increase in lipoprotein lipase activity in the mammary gland. These changes in lipase activity direct available blood triacylglycerol away from adipose tissue to the mammary gland. The triacylglycerol is used in the mammary gland for the synthesis of milk lipids which form the major source of energy for the infant before weaning (Otway and Robinson, 1968).

(b) *Long-chain fatty acids*

Fatty acids are released into the bloodstream only by adipose tissue. However, in contrast to triacylglycerol, fatty acids can be used by most tissues of the body since the enzymes necessary are widely distributed (see Table 6.6 and Figure 6.14). Brain and small intestine may be exceptions to this general statement. The inability of long-chain fatty acids to cross the blood-brain barrier may be responsible for the failure of the brain to oxidise this fuel and the activities of the β-oxidation enzymes may be very low in the intestine. In contrast to the situation described above for lipoprotein lipase, the enzymes involved in fatty acid activation and oxidation are present in the cell in an active form even when fatty acids are not being oxidised to any large extent. The rate of fatty acid oxidation is controlled, therefore, by the intracellular concentration of fatty acids, which, in turn, is determined by their concentration in the blood. Thus a rise in the level of fatty acids in the blood ensures that fatty acid oxidation is immediately increased in those tissues capable of using fatty acids (e.g. muscle, kidney cortex, liver). Consequently, increased mobilization of fatty acid from adipose tissue (the flux-generating step for fatty acid oxidation), which raises the concentration of fatty acids in the blood, is an immediate signal for tissues to increase their rate of fatty acid oxidation.

Fatty acids are mobilized from adipose tissue under a number of conditions, which include starvation, stress and prolonged exercise (see Table 6.7). In these conditions, the increased fatty acid concentration in the blood not only provides an additional fuel for tissues but it reduces the rate of glucose utilization and oxidation and thus helps to maintain the blood glucose concentration. This integration between lipid and carbohydrate metabolism will be discussed in detail in Chapter 8. During starvation, fatty acid oxidation provides most of the energy required by the tissues of the body except for the 'anaerobic' tissues and the brain (see Section E.3.c and Chapter 14; Section A.2–4). Fatty acids play an important role in sustained exercise and the longer the exercise proceeds the greater is the contribution to the energy requirement of the muscle by fatty acids. It has been shown that after 4 hr of sustained exercise in man, more than 60 % of the energy requirement of muscle is provided by fatty acid oxidation (see Chapter 9; Section B.2.c).

(c) *Ketone bodies*

Since their discovery in the urine of diabetic patients in the latter part of the nineteenth century, ketone bodies have had a chequered history, being variously considered as indicators of disturbed metabolism and as important fuels. Their association with a diseased state branded them initially as undesirable metabolic products. However, the demonstration in the 1930s that various tissues could oxidise ketone bodies led to the suggestion that they could be important as a lipid fuel. Thus in a review in 1943, E. M. MacKay wrote:

'Ketone bodies may no longer be looked upon as noxious substances which are necessarily deleterious to the organism.'

In 1956, evidence was presented (see Section C.2) that long-chain fatty acids provided the important lipid-fuel in starvation, so that ketone bodies were

Biochemistry for the Medical Sciences

Table 6.7. Plasma concentrations of fatty acids in man under varying conditions†

Conditions	Plasma fatty acid concentration (mM)
Normal fed	0.30–0.60
Stress (racing driver)	1.72
Starvation (8 days)	1.88
Prolonged exercise (240 min)	1.83
Immediate pre-surgery*	0.90
Three hour post-surgery	0.80
Diabetic coma	1.60
Glycogen storage disease (type III) (after 12 hr starvation)	1.90
Severely burned patients	1.30
Severely injured patients	0.75

* The increased fatty acid concentration before surgery could be caused by stress, which would be reduced by the anaesthesia thus lowering the concentration of fatty acids.

† Data taken from Tables 6.8; 8.2; 9.8; 14.1; 15.2 and Hall *et al.* (1978); Carlson (1970). There is usually considerable variation in the concentrations of fatty acids in normal fed controls. This probably reflects the degree of anxiety in the subject when a blood sample is taken, and may also reflect differences in the methods used by different workers for measuring the fatty acid concentration.

neglected as a fuel and were once more associated in the minds of biochemists, physiologists, and clinicians with pathological conditions. Indeed, in a review article in 1958 entitled *Transport of Fatty Acid*, D. S. Fredrickson and R. S. Gordon did not mention ketone bodies. Even in 1968, doubts were present in at least some minds and were expressed in a review by G. D. Greville and P. K. Tubbs (1968) who stated:

'Clearly it is not obvious in what ways ketogenesis in fasting is a good thing for the whole animal; should the liver be regarded as providing manna for the extrahepatic tissues or does it simply leave them to eat up its garbage?'

Research since the latter review was written has provided strong support for the view that ketone bodies have a very important role in the transport of lipid-fuels in the blood. Furthermore, recent developments suggest that, as well as being a fuel, ketone bodies play an important role in integrating fuel mobilization and utilization in the whole animal.

The concentration of ketone bodies in the blood of fed, healthy human subjects is very low, but it is elevated in starvation. The total blood ketone body concentration in normal subjects rises from about 0.1 mM after an overnight fast to 2.9 mM after three days starvation (see Table 8.3). This increase in concentration is important in providing a fuel for use by muscle, kidney and intestine (and perhaps other tissues) during the period of starvation. If the period of starvation is extended, the concentration of ketone bodies increases further and they become an important fuel for brain. The data in Table 14.1 show that no other fuel rises so markedly in concentration during starvation. In Chapter 8, a more detailed discussion of the role of ketone bodies both as a fuel and a metabolic regulator will be presented, and this will be extended in Chapter 14.

Ketone bodies are produced in the liver by partial oxidation of long-chain fatty acids arising from triacylglycerols stored in adipose tissue. Thus the question arises, why should one lipid-fuel be converted into another in the liver? In order to answer this question, it is necessary to look at the limitations of fatty acids as a fuel. The problem of transporting long-chain fatty acids in a non-toxic form in the aqueous medium of the blood has been solved by binding them to albumin. However, this has an important drawback; although the total concentration of fatty acids in the blood can increase to approach 2 mM (see Tables 6.7, 8.3, and 14.2) the free (non-complexed) concentration will be very much lower (possibly as low as 10 μM) since a large proportion is complexed with albumin (Spector and Fletcher, 1978). The rate of diffusion into the cell will consequently be low. This may restrict the fatty acids from fulfilling their role of providing a fuel that will be utilized in preference to glucose. This restriction will be particularly important if fuels have to diffuse over a long distance within a tissue in which there is a high rate of energy demand. If fatty acids are converted into ketone bodies, which are freely soluble and do not require albumin for their transport, the concentration of the lipid-fuel can be maintained at much higher levels in the blood and the interstitial fluid. The ketone body concentration in the blood rises to 2–3 mM after a few days starvation and to 7–8 mM after prolonged starvation. At these levels, ketone bodies can compete very effectively with glucose, which has a normal concentration of approximately 5 mM in blood. This ability of ketone bodies to provide a ready replacement for glucose plays two particularly important roles.

(i) *The provision of a readily diffusable fuel for essential muscles*   There are a number of muscles in the body which carry out vital physiological functions (e.g. heart for the circulation of blood, diaphragm for respiration, some smooth muscles for digestion and absorption of food, myometrium during parturition). If hypoglycaemia coincided with a reduction in the blood supply to these muscles (e.g. during hypotension or hypoperfusion), insufficient fuel (i.e. fatty acids and glucose) might reach these vital muscles resulting in their fatigue and hence precipitate a life-threatening situation. In such situations a high concentration of ketone bodies, which could diffuse rapidly into the tissue, would replace glucose as a fuel. Such vital muscles do indeed possess high activities of the enzymes of the ketone body utilization pathway (Beis *et al.*, 1980).

(ii) *The provision of a fuel to the brain during periods of starvation*   Measurements of arterio-venous differences across the brain in obese human patients undergoing therapeutic starvation have shown that ketone bodies progressively displace glucose as a fuel for this organ. After about three weeks of starvation, the human brain gains most (60%) of its energy from the oxidation of ketone bodies (see Chapter 14; Section A.3). The remainder of the energy is provided by the oxidation of glucose. Further starvation does not appear to alter this degree of dependence on ketone body oxidization. The increased rate of ketone body oxidation appears to depend solely on the increase in the blood concentration of ketone bodies (which rises from about 2 mM at 3 days to 7–8 mM at 24 days starvation; see Owen *et al.* 1967). It has been suggested that fatty acids cannot provide a fuel for brain because, unlike ketone bodies, they cannot diffuse across the blood-brain barrier.[12]

F. PATHOLOGICAL CHANGES IN CONCENTRATIONS OF LIPID-FUELS

The importance of lipid metabolism in the aetiology of a wide variety of disease states, including some very common ones, is being increasingly recognized. In many of the common diseases, however, it is not possible to attribute the pathological condition to a simple lesion in a single pathway. For this reason, detailed discussions of such conditions as diabetes mellitus, obesity, atherosclerosis, and coronary heart disease, although involving lipid metabolism, are considered in later chapters when other metabolic pathways and the ways in which these pathways are regulated and integrated have been described. A lipid storage myopathy due to deficiency of carnitine palmitoyltransferase has been described in Section D.1.c. In this section, discussion is limited to the causes and effects of diseases characterized by abnormal concentrations of circulating lipid-fuels.

1. Ketoacidosis

An elevated ketone body concentration in the blood has been assumed for many years to be indicative of deranged metabolism. However, it is now known that a raised concentration *per se* is not necessarily indicative of a pathological condition. Thus in obese subjects, undergoing therapeutic starvation, the ketone body concentration in the plasma increases to 7–8 mM. In pathological states of ketoacidosis, of which the most common is diabetic coma, the concentration may not be much greater than this, although it can be higher than 20 mM. A ketotic coma may be the first indication, especially in young children of the severe insulin deficiency that is characteristic of the juvenile type diabetes (see Chapter 15 for a discussion of diabetes mellitus). Such patients probably develop the extreme degree of ketosis over a considerable period during which time hyperglycaemia, glycosuria, polyuria, etc. will also gradually develop to indicate diabetes mellitus. However, sometimes another illness, especially gastro-enteritis, can have a precipitating effect. This can result in some patients, especially young children who have not been diagnosed as diabetic, suddenly going into coma and being rushed to hospital. A similar condition can occur in diabetic patients who are well controlled on daily injections of insulin but the acute illness suddenly increases the requirement for insulin and a ketotic coma ensues. The pH of the blood of such patients in coma can fall to 6.9 (the normal value is 7.4) which is due not only to the high concentration of ketone bodies (acetoacetic and 3-hydroxybutyric acids are relatively strong organic acids[13]) but also to high concentrations of fatty acids and lactic acid. The condition is known as ketoacidosis. Further details of ketoacidosis are given in Chapter 13; Section E.

2. Excessive concentrations of fatty acids

The possible dangers of an elevated blood fatty acid concentration have been realised only in recent years. The authors suggest that a plasma fatty acid concentration above 2 mM should be considered as abnormal and potentially dangerous for the reasons indicated below. Perhaps the term hyperfatty-acidaemia should be used to emphasize the existence of a clinical problem not unlike ketoacidosis or hyperlipoproteinaemia. Possible causes of such high levels of fatty acids are considered below.

(a) *Stress, fatty acids, triacylglycerol and atherosclerosis*

The condition of stress, whether it is caused by anxiety or aggression, will elevate the total catecholamine (i.e. both adrenaline and noradrenaline) concentration in the blood. The noradrenaline level is raised more in anxiety-stress (e.g. parachuting, airline flight, dental surgery) whereas that of adrenaline is raised more in aggression-stress (driving in city traffic, committee meetings, competitive sports). Nonetheless, both hormones stimulate the activity of triacylglycerol lipase in adipose tissue and this elevates the fatty acid concentration in the blood. In addition, stress may increase the blood concentrations of cortisol and growth hormone, which can also increase the rate of lipolysis in adipose tissue. Furthermore, adrenaline will stimulate glycogen phosphorylase activity in the liver and this will elevate that blood glucose concentration. Thus the effect of these hormones is to markedly increase the plasma concentrations of fatty acids and glucose thus ensuring the ready availability of fuels for the 'fight or flight' response to the stress situation. At least, this is as it applies (or did apply) to primitive man and animals in the wild. For modern man, in a civilized society, the stress-inducing situations are very different from those of primitive man. Driving a car in crowded traffic conditions, discussion of a contentious issue at a committee meeting, or even watching certain programmes on television can induce stress in some individuals. In these cases, blood glucose and fatty acid concentrations will be raised but, since exercise is unlikely to follow the stress-inducing situation, the fuel concentrations in the blood can remain elevated for prolonged periods. M. E. Carruthers has proposed that this failure to oxidize the fatty acids and glucose, after the stress condition, results in an increased rate of esterification of fatty acids in the liver and subsequent release of these as VLDL. Frequent elevations of the blood triacylglycerol concentration over many years, especially in the obese or unfit individual, could contribute to the development of atherosclerosis (Carruthers, 1969). Taggart and Carruthers (1971) have obtained some evidence in support of this hypothesis. They have measured the concentrations of the catecholamine hormones, fatty acids and triacylglycerol in the plasma of racing drivers before and after a race and compared it to the concentrations in normal subjects. The concentrations of the hormones and fatty acids are elevated both before and immediately after the race, but the triacylglycerol concentration is markedly increased one hour after the race (Table 6.8). Although this evidence is slender, there is no doubt that the enormous increase in deaths from vascular and coronary heart disease has paralleled the increasing stress of modern life. Causes and effects are, however, very difficult to disentangle.

(b) *Stress, fatty acids and myocardial infarction*

The hypothesis of Carruthers and colleagues suggests that the major danger of elevated plasma concentrations of fatty acids is not the elevation *per se* but the increase in blood triacylglycerol to which they eventually give rise. However, there is no doubt that high concentrations of fatty acids *per se* can be dangerous. At concentrations above 2mM the concentration of free (i.e. uncomplexed to protein) fatty acid in the blood will increase markedly and lead to high intracellular concentrations of fatty acids, which will have detergent effects. This detergent action may explain some of the damaging effects of high concentrations of fatty acids; outside the cell they damage cell membranes and cause aggregation of platelets; within the cell they interfere with electrical conduction in nerves and muscle, inhibit enzymes non-specifically and uncouple oxidative phosphorylation (see Newsholme, 1976, for references). It is possible that high concentrations of

Table 6.8. Concentrations of catecholamines, fatty acids and triacylglycerol in plasma of racing car drivers before and after the race*

	Concentrations in plasma			
Time of sample	Noradrenaline $(\mu g.l^{-1})$	Adrenaline $(\mu g.l^{-1})$	Fatty acid (mM)	Triacylglycerol (mM)
1–3 min before race	1.26	0.19	1.72	0.38
1–3 min after race	3.54	0.55	1.37	0.99
15 min after race	1.51	0.17	1.48	0.80
1 hr after race	0.81	0.09	0.61	1.79
$1\frac{1}{2}$–3 hr after race	0.74	0.03	0.67	0.99
Normal subjects	0.69	0.06	0.58	0.55

* The values represent the means of measurement on blood samples from 16 drivers taken before or after international and club races. Data from Taggart and Carruthers (1971).

fatty acids, especially when they are present in the bloodstream for long periods, could play a role in damaging the endothelial lining of arteries and arterioles, which could be one factor enhancing the rate of development of atherosclerosis (see Chapter 20; Section G.1.a. for further discussion of this problem). Furthermore, a stress condition in such patients could elevate the fatty acid concentration sufficiently to cause aggregation of platelets in the atherosclerosed coronary artery and result in a thrombus and subsequent infarction. In addition, there is some evidence that high concentrations of fatty acids especially within the heart under hypoxic conditions may be one cause of arrhythmias[14] in the heart (Oliver *et al.*, 1968; Cowan and Vaughan-Williams, 1977; Oliver, 1978). Even a mild acute infarction, if compounded by fatty acid-induced arrhythmias, could be fatal. We suggest that this may provide some basis for the explanation of deaths that occur in apparently healthy men during stress conditions. One particular occupation in which there are periods of excessive stress is that of the airline pilot. The periods during take-off and landing are particularly stressful and if a pilot suffers a heart attack at these times, disaster could easily result.[15]

If indeed an elevated fatty acid concentration is partially responsible for arrhythmias, then reduction of these levels should be beneficial for the patient who has suffered a heart attack. It has been suggested that antilipolytic agents should be administered to patients as soon as possible after a heart attack.

(c) *Fatty acids and sudden death in sport*

Despite the widely accepted beneficial effects of exercise (see Paffenbarger and Hyde, 1980) there are a number of reports of fatalities either during or immediately after the exercise, especially for hard competitive games such as rugby football, soccer or squash (Opie, 1975; Tunstall-Pedoe, 1979; Haskell, 1982; see Table 6.9). The plasma fatty acid concentration will be increased despite the fact that these sports are characterized by many short bursts of explosive activity which

Table 6.9. Sudden deaths during various sports*

Type of sport	Number of deaths	Average age at time of death
Rugby football	7	26
Soccer	2	43
Golf	1	40
Mountaineering	1	47
Tennis	2	54
Jogging	1	46
Yachting	1	38

* Details were collected during an 18-month period in South Africa (see Opie, 1975).

use energy generated by anaerobic metabolism. Furthermore, the stress of competition, especially in an individual with a type A personality (see Chapter 20; Note 22) could lead to a further elevation in the plasma fatty acid concentration. Since in these sports there is a greater dependence upon 'anaerobic' exercise, there will be less opportunity for oxidation of these fatty acids to reduce their concentration. Hence the blood fatty acid concentration could remain very high for prolonged periods of time thereby increasing the risk of tissue damage, platelet aggregation, and a myocardial infarction. Since high concentrations of fatty acids can affect the electrical activity of the heart particularly during hypoxic conditions (Cowan and Vaughan-Williams, 1977) this could increase the risk of arrhythmias during the short bursts of activity characteristic of some of the sports indicated above. This implies that 'aerobic' exercise (i.e. that which involves exclusively aerobic metabolism) may be preferable for older individuals especially those who are not extremely fit. Jogging, cycling, swimming and rowing are examples of this type of exercise. However, for the unfit, such activities must always be entered into very gradually to give time for the aerobic capacity of the muscle to adapt with the increase of exercise (Chapter 9; Section B.5.a).

3. Hyperlipoproteinaemias

The measurement of concentrations of triacylglycerol and cholesterol in the plasma has long been a standard technique in diagnostic clinical chemistry since the concentrations of these lipids are elevated in a number of diseases (Fredrickson *et al.*, 1967; Simons and Gibson, 1980). More recently, measurement of the concentration of the individual lipoprotein classes in serum has proved to be an even more valuable diagnostic indicator. The separation of lipoproteins is most conveniently achieved by electrophoresis of fresh plasma.

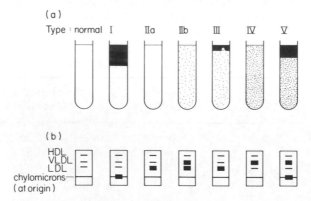

Figure 6.15 Diagnosis of hyperlipoproteinaemias. For sedimentation, the plasma sample is collected 12 hr after the last meal and stored at 4 °C for 12 hr.

(a) Sedimentation of plasma.

(b) Electrophoretic pattern (plasma)

This can be followed by a staining procedure to measure the concentrations of the lipoprotein classes. (The World Health Organization has listed the practical details that should be followed if reliable results are to be obtained—see Beaumont, 1970). An indication of lipoprotein abnormalities may, however, be obtained in some instances from visual observation of plasma obtained after an overnight fast and left standing at 0–4 °C for 18–24 hr (Figure 6.15). Chylomicrons rise to form a creamy layer and high concentrations of other lipoproteins increase the turbidity of the plasma.

Before considering pathological deviations from the normal serum lipoprotein pattern, consideration should be given to non-pathological factors affecting the normal pattern. Paramount among these is the time between sampling and the previous meal. An abundance of chylomicrons appears in the blood 2–4 hr after a meal and persists for a further 6–8 hr. This hyperlipaemia is most marked after a fatty meal but may be present even after meals low in fat. For this reason, samples for diagnostic purposes are generally taken at least ten hours after the last ingestion of food. Other 'normal' factors which affect the levels of lipoprotein are fat composition of the diet (e.g. degree of unsaturation of fatty acids), fibre content of the diet, stress, sex (triacylglycerol content is generally lower in females than in males), and age (triacylglycerol content increases with age).

(a) *Classification of hyperlipoproteinaemias*

The generally accepted classification of hyperlipoproteinaemias is based on the lipoprotein pattern observed after electrophoresis of serum. As expected, this classification aligns broadly with the aetiology of the disease but, as yet, knowledge of the causes of disturbances of triacylglycerol transport is unable to provide an adequate basis for classification. That is to say, hyperlipoproteinaemias are classified according to changes in plasma concentrations of the lipids rather than causes. At least six types (I, IIa, IIb, III, IV and V) of hyperlipoproteinaemia are distinguished (see Table 6.10) and described briefly below.

(i) *Type I: Hyperchylomicronaemia* The chylomicron level in the serum is elevated even 24–48 hr after a meal, whereas other lipoprotein levels (including VLDL) are within the normal range or only slightly increased. Total blood triacylglycerol concentrations are highly elevated. Unlike the other types, the metabolic cause of type I hyperlipoproteinaemia is known; it is due to a low level of lipoprotein lipase activity. This is revealed by the assay of lipase activity in the plasma after administration of heparin. The latter releases some of the lipoprotein lipase from its location in the capillary wall into the bloodstream (Section C.1.a). Type I hyperlipoproteinaemia, which is usually apparent in infancy and childhood, is rare. Clinical features include abdominal pain,[16] enlarged liver, pancreatitis,[16] and xanthomata (pigmented fatty deposits under the skin or in tendons). Patients can utilize lipid in which the chain length of the fatty acids is

Table 6.10. Clinical aspects of hyperlipoproteinaemia

Type	Frequency	Onset	Plasma cholesterol concentration	Plasma triacylglycerol concentration	Clinical features	Treatment usually prescribed	Cause
I	rare	infancy and childhood	normal	marked increase	hepatosplenomegaly; abdominal pain; pancreatitis; xanthomata; post-heparin lipase activity decreased	restrict fat intake (about 30 g.day^{-1}); add medium-chain triacylglycerol	deficiency of lipoprotein lipase
IIa	common	childhood and adult	increased	normal	atherosclerosis and vascular disease; xanthomata; elevated LDL-cholesterol	diet; cholestyramine	defect in removal of low density lipoprotein
IIb	common	childhood and adult	increased	increased	atherosclerosis and vascular disease; xanthomata	diet; cholestyramine	over-production of VLDL
III	rare	adult	marked increase	marked increase	atherosclerosis and vascular disease; xanthomata; frequently abnormal glucose tolerance	clofibrate; diet; nicotinic acid	not known
IV	very common	adult	slight increase	marked increase	atherosclerosis and vascular disease; xanthomata; abnormal glucose tolerance	reduce weight; limit carbohydrate; clofibrate;	not known; can be secondary to hypothyroidism
V	rare	depends on primary condition	slight increase	marked increase	atherosclerosis and vascular disease; xanthomata	diet; clofibrate	diabetes, alcoholism, nephrotic syndrome (see Havel et al., 1980)

For further details see Lewis, (1976)

less than twelve carbon atoms, since these fatty acids will *not* be esterified in the epithelial cells but will pass into the hepatic portal blood to be metabolized by the liver. In this disease there is no increased incidence of cardiovascular complications.

(ii) *Type II: Hyper-β-lipoproteinaemia* Two forms of type II hyperlipo-proteinaemia, type IIa and type IIb are distinguished. Both are characterized by an elevation in the level of LDL in the serum but in type IIb, the VLDL level is also raised. A marked hypercholesterolaemia is apparent in patients with the type IIa disease who are suffering from a genetic abnormality known as familial hyper-β-lipoproteinaemia; this is caused by a defective LDL-receptor on peripheral cells—see Chapter 20; Section D.2.a. Type IIa or IIb disease may be a secondary result of disturbances such as hypothyroidism, nephrotic syndrome, biliary obstruction or autoimmunity towards plasma lipoproteins. Thus it is found in both children and adults. Clinical features (other than those causing the hyperlipoproteinaemia) include premature development of atherosclerotic plaques and xanthomata, especially in tendons. There is often a history of premature death from heart disease in the family of these patients; indeed this high incidence provides evidence of a causal link between high levels of LDL (and VLDL) and atherosclerosis (see Chapter 20).

(iii) *Type III* This disorder is caused by a defect in the processes for conversion of VLDL to LDL so that an abnormal lipoprotein accumulates. It has a density comparable to VLDL but a mobility close to that of LDL. The type III pattern is sometimes referred to as 'broad-β' or 'floating-β'. It is found in adults especially in obese individuals and those suffering from hypothyroidism; it is rare. The disease is characterized by skin xanthomata and deposits of lipids in the heart, arteries and reticulo-endothelial cells.

(iv) *Type IV: Hyperpre-β-lipoproteinaemia* In type IV the level of VLDL is increased but there is no increase in LDL or chylomicrons. This is likely to be due to a marked increase in triacylglycerol output (and therefore possibly increased triacylglycerol synthesis) from the liver. It is also known as endogenous hypertriglyceridaemia and is found in adults. There may be a number of causes of this disease and it may be secondary to hypothyroidism, diabetes, or nephrotic syndrome. It is the commonest of the classes of hyperlipoproteinaemias; the incidence can be as high as 20 % of the adult male population in Western societies. Patients may have the following symptoms and clinical features: xanthomata, premature vascular disease, impaired glucose tolerance, and pancreatitis.

(v) *Type V: Hyperpre-β-lipoproteinaemia and chylomicronaemia* This disorder, known as mixed lipaemia, is characterized by an increased concentration of both chylomicrons and VLDL. The major metabolic impairment is in the clearance of dietary fat from the blood but this state contrasts with that in type I in that the post-heparin lipolytic activity is not depressed. Like type I, however, the degree of hypertriglyceridaemia is severe. Symptoms and clinical features include increased incidence of cardiovascular disease, abdominal pain, enlarged liver, and spleen, xanthomata, increased plasma concentration of insulin and glucose intolerance. The type V pattern may arise as a consequence of an inherited defect or of diseases such as diabetes, nephrotic syndrome, pancreatitis, alcoholism, or hypo-thyroidism and may therefore be found in children or in adults.

(b) *Treatment of hyperlipoproteinaemias*

The treatment of hyperlipoproteinaemia depends both on its type and its cause (see Table 6.10 and Lewis, 1976). Where the hyperlipoproteinaemia is secondary (e.g. due to hypothyroidism), treatment may be best directed at the primary cause. Since hyperlipaemia may significantly increase the chances of premature coronary disease, much attention has been directed at means of reducing the levels of circulating triacylglycerols and cholesterol. Two types of treatment of hyperlipoproteinaemia are practised; dietary control and drug therapy. Clearly, chylomicron formation is dependent on a supply of dietary triacylglycerol so that a reduction in the latter will mollify types I and V. To some extent, normal triacylglycerols in the diet may be replaced by triacylglycerols containing medium-chain and short-chain length fatty acids. These are not esterified in the gut and so are released into the blood as free fatty acids rather than triacylglycerols (see Section B.1.c). To reduce the levels in type II, III, IV, and V hyperlipoproteinaemias, the total amount of fat and cholesterol in the diet must be reduced and the ratio of polyunsaturated to saturated fatty acids increased (see Chapter 20; Section G.1). In addition, in type IV, the carbohydrate intake must

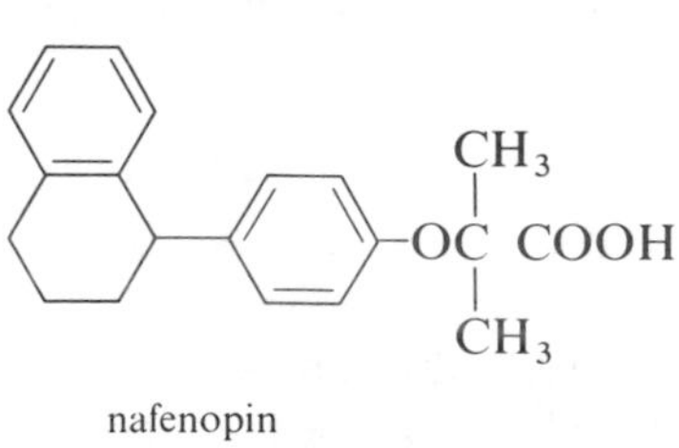

nafenopin 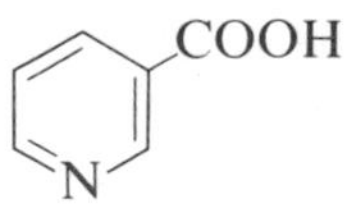   nicotinic acid

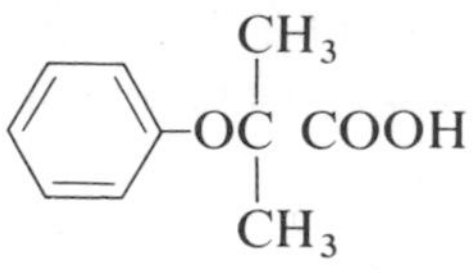

clofibrate

Figure 6.16 The structural formulae of commonly prescribed hypolipidaemic agents

be reduced to decrease triacylglycerol synthesis in the liver. Although strict dietary control can be effective in some of these conditions, it is not satisfactory in most patients suffering from familial hypercholesterolaemia. In most of these patients, coronary atherosclerosis is severe and can often be fatal at an early age (second or third decade). Consequently both dietary control and drug therapy is used, and even surgery has been attempted (e.g. ileal bypass, portocaval shunt).

The drugs that are or have been used include nicotinic acid, nafenopin and clofibrate (Figure 6.16). Nicotinic acid, in large doses ($3–9\,\mathrm{g\,day^{-1}}$), inhibits adipose tissue lipolysis. This reduces concentrations of fatty acids and of VLDL and ultimately, LDL, so that nicotinic acid is particularly useful in the treatment of types III, IV, and V. Clofibrate has been used widely in the treatment of types II, IV, and V, since it reduces the synthesis and increases the rate of utilization of VLDL so that the blood concentration of VLDL is decreased. Clofibrate can also decrease the cholesterol concentration in some patients. Since cholesterol levels are an indication of the likelihood of atherosclerosis, the World Health Organisation sponsored a study into the value of clofibrate treatment in reducing the incidence of coronary heart disease (Chapter 20; Note 22).

Some success in reducing cholesterol levels in the serum has been achieved with cholestyramine,[17] an ion-exchange resin which binds bile salts in the intestine and prevents their reabsorption (i.e. it interferes in the entero-hepatic circulation—see Section B.1.b; Chapter 20; Section C.1). This has the effect of increasing the rate of conversion of cholesterol to bile salts in the liver and hence diminishing serum cholesterol levels (Heaton, 1976). Cholestyramine, and a number of other bile salt sequestering agents, are therefore employed in the treatment of type II hyperlipoproteinaemias.

APPENDIX 6.1   CLASSIFICATION OF LIPIDS (WITH REPRESENTATIVE STRUCTURAL FORMULAE)

1. Simple lipids

(a) *Long-chain fatty acids*

Long-chain fatty acids are linear monocarboxylic acids, mostly with an even number of carbon atoms between 12 and 22. They may be saturated (i) or unsaturated with one or more double bonds (ii).

palmitic acid $CH_3(CH_2)_{14}COOH$ (i)

linoleic acid $CH_3(CH_2)_4CH{=}CH\;CH_2CH{=}CH(CH_2)_7COOH$ (ii)

(b) *Acylglycerols (glycerides)*

Glycerol possesses three hydroxyl groups and so can esterify with one, two, or three molecules of fatty acid to form monoacylglycerols, diacylglycerols, and triacylglycerols (iii) respectively. Triacylglycerols are called fats if they are solid at room temperature and oils if they are liquid. Melting point falls with decreasing molecular weight and increasing unsaturation. Naturally occurring triacylglycerols are liquid at the body temperature of the organism in which they occur. Triacylglycerols were formerly known or triglycerides.

$$
\begin{array}{ll}
\mathrm{CH_2O.CO(CH_2)_{16}CH_3} & \mathrm{(iii)} \\
| & \\
\mathrm{CH\ O.CO(CH_2)_7CH{=}CH(CH_2)_7CH_3} & \\
| & \\
\mathrm{CH_2O.CO(CH_2)_{16}CH_3} &
\end{array}
$$

(c) *Waxes*

Waxes are esters of long-chain fatty acids with long-chain alcohols (iv). Many organisms secrete waxes externally to waterproof their surface and many marine species store waxes as a fuel. The cholesterol esters (v) (in which cholesterol functions as an alcohol) resemble the waxes in general properties.

triacontanyl hexadecanoate $\quad \mathrm{CH_3(CH_2)_{29}OCO(CH_2)_{14}CH_3}$ (iv)

cholesteryl stearate

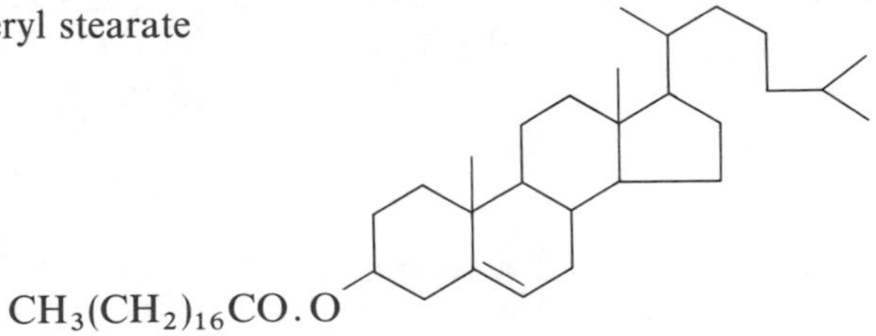

(v)

$\mathrm{CH_3(CH_2)_{16}CO.O}$

2. **Compound lipids**

(a) *Glycerol-based compound lipids*

These are most simply considered as diacylglycerols with the third hydroxyl group linked to a group other than a fatty acid.

(i) *Glycerophospholipids (acylphosphoglycerols, phosphoglycerides)* The third hydroxyl group is esterified with phosphoric acid which forms a second ester link with one of the alcohols: choline (vi), serine, ethanolamine or inositol. They are named as derivatives of phosphatidic acid (that is, a diacylglycerol esterified with phosphoric acid). Together with phosphosphingolipids (below) and other related compounds they constitute the phospholipids, all of which are polar molecules mostly serving structural functions in membranes.

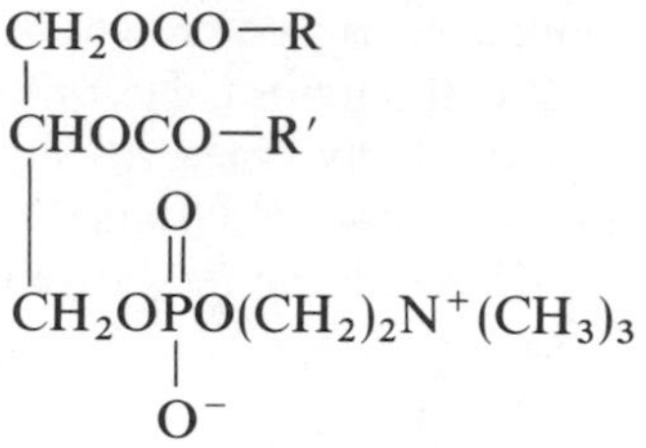

$$CH_2OCO-R \qquad \text{(vi)}$$
$$CHOCO-R'$$

phosphatidylcholine

(Note, lysolecithin
lacks R'.)

$$\underset{O^-}{\overset{O}{\overset{\|}{CH_2OPO(CH_2)_2N^+(CH_3)_3}}}$$

(Note: When the nature of the acyl groups is known, the compound is named as a derivative (e.g. dipalmitoyl-) of glycerophosphocholine etc.)

(ii) *Alkyl ether diacylglycerols*   The third hydroxyl group of the glycerol forms an ether linkage with an alkyl group.

(iii) *Diacylglycerol glycosides* (*glycosylglycerides*)   The third hydroxyl group of the glycerol forms a glycoside linkage with a sugar. Found in plants but not animals.

(b) *Sphingosine-based compound lipids*

Sphingosine (vii) is a long-chain alcohol with two hydroxyl groups and an amino group. The amino group forms an amide link with a long-chain fatty acid. Like glycerol-based compound lipids, the terminal hydroxyl group can link with a variety of groups.

(i) *Phosphosphingolipids* (*sphingophospholipids*)   Here the terminal hydroxyl group is esterified to phosphoric acid which, in turn, esterifies with one of the bases found in glycerophospholipids. An example of a phosphosphingolipid is sphingomyelin (viii) widely found in cell membranes.

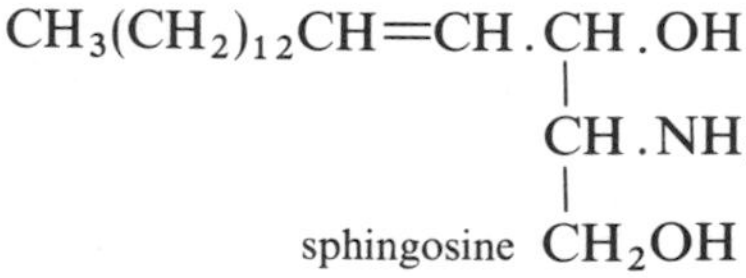

$$CH_3(CH_2)_{12}CH=CH.CH.OH \qquad \text{(vii)}$$
$$CH.NH_2$$

sphingosine CH_2OH

$$CH_3(CH_2)_{12}CH=CH.CH.OH \qquad \text{(viii)}$$
$$CH.NH.CO-R$$

sphingomyelin

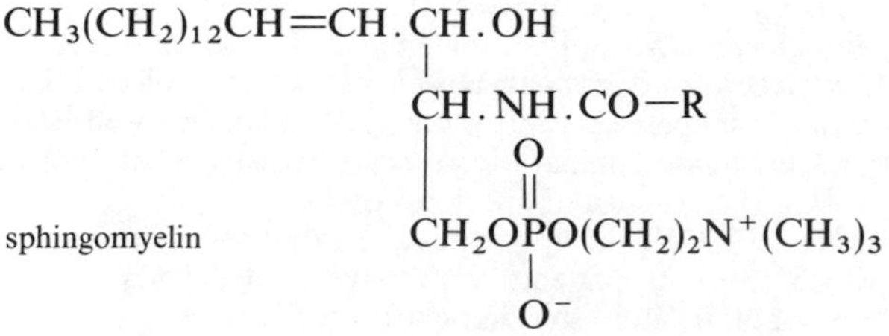

$$\underset{O^-}{\overset{O}{\overset{\|}{CH_2OPO(CH_2)_2N^+(CH_3)_3}}}$$

(ii) *Glycosphingolipids*   Sphingosine is also present in yet more complex lipids; the glycosphingolipids. In these compounds the terminal hydroxyl group forms a glycoside linkage with simple carbohydrates (to form cerebrosides) or with carbohydrate chains containing neuraminic (sialic) acid residues and sulphate groups (to form gangliosides). These compounds are all found in cell membranes.

3. Water-soluble lipids

Short-chain fatty acids (e.g. ethanoate (acetate), propanoate (propionate) and butanoate (butyrate)) have appreciable water solubility as do their derivatives acetoacetate and 3-hydroxybutyrate which are known as ketone bodies (see Section C.3).

4. Non-saponifiable lipids

Compounds which are insoluble in water but soluble in organic solvents and not derived from long-chain fatty acids are sometimes grouped under this heading. Such compounds include steroids (e.g. cholesterol, see (v)), terpenes (e.g. vitamin A (ix)) and ubiquinone (Chapter 4; Section C.1.c).

$$CH=CH.C=CH.CH=CH.C=CH.CH_2OH \qquad (ix)$$

vitamin A

NOTES

1. The term lipid has been used in a broad sense to include all those molecules insoluble in water but soluble in non-polar solvents (e.g. ether) but this is too wide a definition to be biochemically useful.
2. Proteins which associate with non-protein molecules to form a single functional unit are known as apoproteins. Apoproteins occurring in lipoproteins (therefore known as apolipoproteins) are mostly glycoproteins, that is, they have carbohydrate residues attached covalently to the protein backbone. A lipoprotein is therefore an apolipoprotein-lipid complex held together by non-covalent bonds. At least eleven apolipoproteins have now been discovered in man.
3. The lipoprotein lipase complex catalyses the complete hydrolysis of triacylglycerol to fatty acids and glycerol. More than one enzyme is present and there appears to be separate triacylglycerol and monoacylglycerol lipases as well as a phospholipase. The separate existence of a lipase specific for diacylglycerol is less well established although some lipoprotein lipase preparations will hydrolyse diacylglycerol (but not triacylglycerol) in the absence of apolipoprotein.
4. The importance of LCAT in lipoprotein metabolism is indicated by the clinical problems which arise in patients who have a deficiency of the enzyme. The concentrations of HDL and esterified cholesterol are reduced, but the amounts of

unesterified cholesterol and triacylglycerol are usually increased and the triacylglycerol associated with LDL is also increased. In addition, the VLDL particles are much larger than normal and there is accumulation of cholesterol and lipid in the kidney, liver, spleen, cornea (the cornea has a milky appearance so that opthamologists may be first to diagnose the condition) and in the major arteries and arterioles (atherosclerosis). The most serious damage is found in the kidneys and patients usually suffer from kidney failure (Norum and Gjone, 1967; Gjone, 1974; Havel *et al.*, 1980).

5. In starvation, the plasma concentration of long-chain fatty acids increases by at least 1.0 mM. In 3 litre of plasma there are $1.0 \times 3 = 3.0$ mmoles. Since the average molecular weight of a fatty acid is approximately 250 this is equal to 0.75 g fatty acid. The fractional turnover is approximately $0.25 \min^{-1}$. Therefore the flux through the plasma in 24 hr is $0.75 \times 0.25 \times 60 \times 24 = 270$ g fatty acid. The oxidation of 270 g of fatty acid produces about 10 000 kJ, which almost satisfies the normal daily energy requirements of the average man (see Table 2.2).

6. The Enzyme Commission has restricted the term synthetase to the recommended names of enzymes catalysing the joining together of two molecules when the reaction is coupled with the hydrolysis of a pyrophosphate bond in a nucleoside triphosphate. The term synthase is used when nucleoside triphosphates are not involved (e.g. HMG-CoA synthase, Section D.2.a).

   Reactions catalysed by acyl-CoA synthetases proceed by the intermediate formation of an acyl-AMP intermediate containing a mixed acid anhydride link:

$$R.CO.O.\overset{\overset{\displaystyle O}{\|}}{\underset{\underset{\displaystyle O^-}{|}}{P}}.O.CH_2\text{-ribose-adenine}$$

7. Bovine milk is unusual in containing a high proportion of medium-chain fatty acids in its lipids. The fatty acids that comprise triacylglycerol in human milk are long-chain (i.e. they contain more than twelve carbon atoms).

8. Carnitine, $(CH_3)_3N^+CH_2CH(OH).CH_2COO^-$, is synthesized in the liver by hydroxylation of γ-butyrobetaine and is normally abundant in tissues able to oxidise fatty acids. Concentrations of 1.25 mM and 0.75 mM have been reported for adult rat heart and skeletal muscle respectively. The levels are very much lower, however, in the heart of the newborn rat (within two weeks of birth) when the organ is virtually unable to oxidize fatty acids (see Broquist, 1982).

9. The symbol Δ is used to indicate the position of double bonds in unsaturated compounds; for example, Δ^2 indicates a double bond between carbon atoms 2 and 3, while $\Delta^{3,5}$ indicates that there are two double bonds, one between carbon atoms 3 and 4 and the second between carbon atoms 5 and 6.

10. It is of interest to note that the 3-hydroxybutyrate produced by the reduction of acetoacetate is the D-enantiomer, in contrast to the L-3-hydroxyacyl-CoA intermediates in the β-oxidation of saturated fatty acids (Section D.1.d).

11. The oxidation of 1 g of pure glucose yields 0.6 g water and the oxidation of 1 g of wet (physiological) glycogen yields approximately 0.85 g (Table 6.4). However, the point must be made that if water generation were the prime factor determining the nature of the fuel stored then glycogen yields 0.59 moles of water *per mole of ATP* while the triacylglycerol yields only 0.12 moles (Table 6.4).

12. The importance of ketone bodies as a fuel for the essential muscles and the brain is illustrated by a patient with a deficiency of the enzyme glycogen synthase in the liver.

This patient was thus unable to synthesise liver glycogen. During the overnight fast, the patient's blood glucose concentration fell to about 1.0 mM (< 20 mg.dl^{-1}) but rose to 10 mM after feeding. Conversely, the concentration of ketone bodies in the blood rose to almost 8 mM during the overnight fast but fell to about 0.2 mM after feeding (Aynesley-Green *et al.*, 1977). It is suggested that the ketone bodies provided an important fuel not only for the patient's brain during the hypoglycaemia of the overnight fast (see above) but also for the essential muscles. (The patient has been satisfactorily treated with a high protein diet—see Chapter 16; Section E.8.)

13. The pK_a of acetoacetic acid is 3.58 and that of 3-hydroxybutyric acid is 4.70 (at 25 °C). For comparison, the pK_a of acetic acid is 4.75.

14. An arrhythmia is an irregularity of the heart beat. There are a number of reasons for arrhythmias but in the post-infarcted patient they are thought to be due to the interruption in the conduction of electrical activity across the myocardium caused by the damage. It is possible that a high concentration of fatty acids (within the heart muscle) interferes further with the electrical conduction and increases the likelihood of arrhythmias (Cowan and Vaughan-Williams, 1977). The heart rate increases with arrhythmia and, in the case of severe arrhythmia, the function of the heart as a pump may be impaired with possibly fatal consequences.

15. Although there are a number of reports of pilots being incapacitated by a heart attack (Barclay, 1969), perhaps the best known is the crash of the British European Airways Trident soon after take-off on 18 June 1972. (See Civil Aircraft Accident Report 4/73 by Accidents Investigation Branch, Department of Trade and Industry, published by HMSO, London.) A *post mortem* on the pilot, Captain Key, showed severe atherosclerosis in the coronary arteries and evidence of a previous minor infarction. Despite this, as the official report states 'Captain Key presented to those who knew him a picture of robust good health and he was passed fit to fly by a medical examination in November 1971'.

    Of particular relevance to the present discussion is that one and a half hours before take-off Captain Key had been involved in an altercation in the crew room over whether pilots should take strike action in support of a claim for higher salaries. Captain Key was directly involved in a heated argument and he was reported to be very angry indeed. We suggest that this was a stress condition which may have lasted for most of the period prior to take-off and would have been extended by the pre-flight preparation and the take-off itself. The official report concludes that Captain Key had a heart attack during take-off but that this did not cause total incapacitation. It considers, however, that Captain Key was suffering from pain and malaise which would have impaired his judgement and his mental faculties. Thus the aircraft failed to maintain sufficient speed and the droops (retractable leading-edge high-lift devices) were retracted too early during the take-off. This produced a stall and resulted in the crash. Did the stress of the pre-flight argument contribute to the heart attack by raising the fatty acid concentration in the blood above the 'safe' limit of 2 mM?

    A study carried out after the Athens earthquake demonstrated a marked increase in death due to heart attack over a period of 5 days after the quake (Trichopoulos *et al.*, 1983). The psychological stress caused by this sudden disaster could have raised the blood fatty acid levels above the 'safe' level in some subjects and this could have increased the risk of a heart attack.

16. Abdominal pain that is caused by pancreatitis is one of the most serious consequences of hyperlipidaemia. The reason for the association between hyperlipidaemia and pancreatitis is not known; one interesting suggestion is that the clumping of chylomicrons in capillaries within the pancreas causes local ischaemia, damage to cells and release of triacylglycerol lipase. This hydrolyses the triacylglycerol, producing high local concentrations of fatty acids which cause aggregation of platelets and formation of thrombi and further damage to cell membranes (Section F.2.b) and hence release of

more lipase. Consequently, a vicious circle is set up that results in severe damage to the pancreas with inflammation, invasion of the tissue by lymphocytes etc. and the development of pancreatitis (Havel *et al.*, 1980).

17. Cholestyramine is a polystyrene polymer with divinylbenzene cross-linkage and quaternary ammonium groups (in the chloride form) so that it is strongly positively charged. Its molecular weight is 1 million and it probably acts by exchanging chloride ions for the bile salt ions.

CHAPTER 7

REGULATION OF GLUCOSE AND FATTY ACID OXIDATION IN RELATION TO ENERGY DEMAND IN MUSCLE

The pathways of glycolysis, β-oxidation and the TCA cycle plus electron transport and oxidative phosphorylation have been described in the preceding chapters. Since these are quantitatively the most important processes for ATP production, knowledge of how the flux through these pathways is regulated is essential to an understanding of metabolic integration. Although discussion is limited to the regulation of flux in relation to the energy demand in muscle, upon which much of the experimental work has been carried out, it is likely that similar if not identical control mechanisms occur in other tissues. The control of glucose utilization in relation to the oxidation of other fuels is discussed in Chapter 8. This information, together with that in the present chapter, is applied to the regulation of metabolism in exercise in Chapter 9.

A. INTRODUCTION TO METABOLIC REGULATION

One important requirement for all living organisms is to adapt to changes in their external environment. Physiological changes are sometimes very obvious, for example, the increase in respiration and heart rate upon exercise, but such changes are always accompanied by changes in rates of metabolism. The mechanisms by which changes in the rates of metabolic pathways are achieved are known collectively as 'metabolic regulation'.

The flux through a pathway is varied by changes in the activities of enzymes of that pathway. These can be achieved through modification of the catalytic activity or through a change in the concentration of an enzyme. The discussion in this chapter will be restricted to changes in the catalytic activity which can occur very rapidly (seconds or minutes). Changes in enzyme concentration are discussed in Chapters 10 and 18.

Two everyday experiences serve to illustrate the importance of metabolic regulation. First, man eats intermittently, so that, even in modern industrial societies, a period of 8–12 hr of starvation (overnight starvation) occurs regularly. During this period, the body must utilize fuel stores that have been laid down after previous meals. The mechanism of control of the breakdown of these stores in relation to the demands for energy are discussed in Chapters 8 and 14. Secondly, climbing stairs or running to catch a bus or train requires an enormous increase in the rate of ATP utilization and synthesis in the leg muscles, which involves a hundred-fold increase in the rate of anaerobic glycolysis.

Since metabolic control is central to an understanding of metabolism and since it is intimately involved in many clinical disorders, a considerable part of this book is devoted to this subject. This chapter is intended as an introduction and the first part deals with the principles underlying the different ways in which enzyme activity can be regulated since these can be applied to any metabolic pathway. This part has to be very brief but for those who wish to read more deeply into this subject, the following reviews are recommended: Newsholme and Crabtree 1973, 1976, 1979, 1981a, b; Newsholme, 1980. The remainder of this chapter contains a description of the mechanisms controlling glycolysis, glycogenolysis, pyruvate oxidation, the citric acid cycle, electron transfer, and the β-oxidation of fatty acids in muscle; that is, the major processes involved in ATP synthesis.

1. Principles of regulation of flux

(a) *Theoretical*

The three types of reactions that constitute a metabolic pathway, non-equilibrium reactions that approach saturation with pathway-substrate, non-equilibrium that are far from saturated with pathway-substrate and near-equilibrium reactions, together with an outline of their importance in metabolic control, have been discussed in Chapter 2; Section C.2. The different role that each plays in the control of flux is most easily explained by reference to a hypothetical pathway in which S is converted, via a series of intermediates, to P:

$$S \xrightarrow{E_1} A \xrightarrow{E_2} B \xrightarrow{E_3} C \xrightarrow{E_4} P$$

The first reaction, catalysed by enzyme E_1, is flux-generating and, for the present, it is assumed that all reactions are non-equilibrium. At steady state, the flux through all these reactions will be equal to that generated at E_1; but how do these reactions respond precisely to the flux generated at this initial reaction? One simple answer is by means of the changes in substrate concentrations. For example, if the activity of enzyme E_1 increases, the concentration of A increases and this will increase the activity of E_2 until it reaches that of E_1. Similarly, for the reaction catalysed by E_3, the concentration of B will adjust the flux through this reaction to that generated at E_1. In this simple system, the flux generated at the initial reaction is transmitted through the remainder of the reactions of the pathway by the concentrations of the pathway-substrates, A, B, and C. This type of regulation is termed *internal* regulation. It must be emphasized that internal regulation can never change the flux through the pathway. For example, if, with the pathway at steady-state, the activity of enzyme E_2 alone was increased, it would only lower the concentration of A, which would reduce the activity of E_2 so that it was once more equal to the flux generated at E_1. It would not lead to a change in the steady-state flux through this pathway. (The decrease in concentration of A would have a negligible effect on the flux through the preceding reaction, since the latter is non-equilibrium.) Similarly, a decrease in the activity of enzyme E_2 would lead, at steady state, only to an increase in the concentration of A. If, however, the activity of E_2 was decreased such that its maximum activity was below that of E_1, the concentration of A would accumulate and saturate the enzyme, so that a new flux-generating step (i.e. E_2) would be established. Indeed, the concentration of A could increase so much that it might escape from the cell into the bloodstream and appear in the urine. This situation occurs in diseases known as inborn errors of metabolism, in which the activity of an enzyme is abnormally low or even non-detectable. It is interesting to note that, in some of these diseases, it is the accumulation of the metabolic intermediate or a derivative of the intermediate which poses the major danger to the well-being of the patient, rather than the decreased flux through the metabolic pathway.

It should be clear that the enzyme that catalyses the flux-generating step cannot be controlled by internal regulation since it approaches saturation with pathway-substrate. Other factors must regulate the activity of the enzyme and these are termed *external* regulators; they include allosteric regulators (see Chapter 3; Section C.5) and cosubstrates (Chapter 2; Section C.2.d). In the hypothetical system described in Figure 7.1, compound X is an allosteric regulator of the flux-generating step and hence changes in the concentration of X will control the flux through the pathway.

Although an increase in flux may be transmitted through all the reactions of the pathway by internal regulation, it is possible that one or more reactions (especially the non-equilibrium reactions) cannot respond sufficiently to the increase in the concentration of pathway-substrate. Consequently, in order to maintain a steady-state flux, such a reaction must be regulated by additional factors (i.e. external regulation). For example, the external regulator, X, could increase the activity of E_4 as well as that of E_1. The effect of X on E_4 will be to reduce the increase in concentration of C that is necessary to increase the activity of E_4 to transmit the new flux (Figure 7.1). For example, if the stimulatory effects of X on both E_1 and E_4 are quantitatively identical, the concentration of C will remain constant, despite changes in flux through the pathway; if X has a greater effect on E_4 than E_1, the concentration of C will be reduced. Indeed, the external

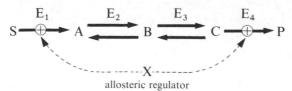

Figure 7.1 Hypothetical reaction scheme in which an allosteric regulator (X) stimulates the flux-generating reaction (E_1) and a non-equilibrium reaction near the end of the pathway (E_4). Both E_2 and E_3 are near-equilibrium. See text for explanation

regulators for E_1 and E_4 could be different compounds. However, it must be emphasized that a specific regulator of E_4 could not influence the flux through the over-all pathway, since it could not change the activity of the flux-generating step.

If, for the pathway described in Figure 7.1, the reactions catalysed by enzyme E_2 and E_3 are near-equilibrium, they would be very sensitive to changes in concentrations of the reactants (see Section A.2.c). For example, only a small increase in the pathway-substrate concentrations (A and B) would be necessary to increase the flux. If the reactions also involved co-substrates or co-products these would be external regulators to which the reactions would also be very sensitive.

(b) *Experimental*

In an investigation of the control of flux through any pathway, the first requirement is to identify the near- and non-equilibrium reactions. The experimental methods for this have been described elsewhere (see Chapter 2; Section C.3; Newsholme and Start, 1973). The second requirement is to identify the flux-generating step. This is usually done by measurement of the pathway-substrate concentration in the intact tissue[1] and comparison with that of the K_m of the isolated enzyme for that substrate (Chapter 3; Section C.2.d). This approach has already been described for the TCA cycle (Chapter 4; Section B.2), glycolysis (Chapter 5; Section B.1), and fatty acid oxidation (Chapter 6; Section D.1.a).

The identification of a flux-generating step permits the physiological metabolic pathway to be identified, within which all the reactions, except for the flux-generating step, will be controlled by internal regulation. In addition, they may be controlled by external regulation. To demonstrate external regulation, experiments are carried out in which the flux through the pathway is varied and the concentrations of all the pathway-substrates are measured. If the concentration of pathway-substrate changes in the opposite direction to the flux, this indicates that external regulation operates at that reaction. The rationale is very simple: if the flux increases but the concentration of the pathway-substrate decreases, the activity of the enzyme must have been stimulated by factors other than the pathway-substrate (e.g. in Figure 7.1, when the flux increases but the concentration of C decreases this indicates external control at reaction E_4). The identity of the external regulator(s) is established by investigating the properties of the enzyme in an extract of the tissue. For a non-equilibrium reaction, cosubstrates, and allosteric effectors[2] are candidates for external regulators, whereas, for a near-equilibrium reaction, only cosubstrates and coproducts need be considered.[3] On the basis of the results of these studies, a theory of control of the enzymes of the pathway in the intact cell can be formulated. The theory is tested by measuring the concentration, and the changes in concentration, of potential external regulators in the intact tissue when the flux through the pathway is modified.

A major problem in metabolic control is that activities of enzymes *in vitro* can be frequently modified by so many compounds that hypotheses of control can be formulated all too easily. Use of the theoretical and experimental principles given above provide a logical basis on which the properties of enzymes can be judged and their physiological relevance considered. This experimental approach will be applied in this and subsequent chapters to the elucidation of control mechanisms.

2. Sensitivity in metabolic regulation

Whether the regulator of an enzyme is substrate, cosubstrate, coproduct, or allosteric effector, an important question is the sensitivity of the enzyme to regulator. Sensitivity[4] is a measure of the quantitative relationship between the relative change in enzyme activity and the relative change in concentration of the regulator. If, for example, the concentration of the regulator increases twofold, how large an increase in enzyme activity will this produce? The greater the response of the enzyme to a given increase in regulator concentration, the greater is the sensitivity.

There are several important mechanisms for increasing sensitivity, but before describing these, it is necessary to consider the basic interaction between enzyme and regulator and how this provides a physicochemical limitation to sensitivity in metabolic control.

(a) *Equilibrium-binding of a regulator to an enzyme*

As far as we know all regulators (internal or external) modify the activity of an enzyme or protein by binding in a reversible manner to the protein; such binding, which is described as equilibrium-binding, will control the activity of the enzyme as follows:

$$E + R \rightleftarrows E^*R$$

where E is the inactive form of the enzyme and E* is the active form. The asterisk indicates that the binding of R has changed the conformation of the catalytic site of the enzyme to the active form. The normal response of enzyme activity to the binding of the regulator is the same as that described for the binding of the substrate in Chapter 3; Section C.2.b; that is, a hyperbolic or a 'Michaelis–Menten' response (Figure 7.2). Unfortunately, this response is relatively 'inefficient' for metabolic regulation; for example, a twofold change in regulator concentration will change the enzyme activity by no more than twofold (i.e. the maximum sensitivity is unity; Figure 7.2). This may be difficult to accept when simply observing the steepness of the initial part of a hyperbolic curve. However, it must be appreciated that sensitivity is *not* the slope of the plot of activity versus concentration of substrate or regulator. Sensitivity is the relationship between the *relative* change in activity to the *relative* change in concentration.[4] This limitation of the hyperbolic response in sensitivity in

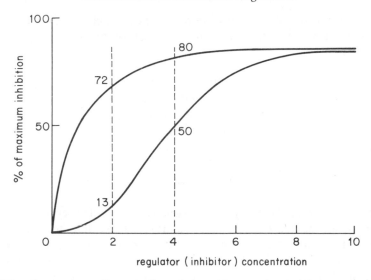

Figure 7.2 Comparison of hyperbolic and sigmoid curves for the allosteric regulation of enzyme activity. In this case the regulator is an inhibitor but similar curves can be drawn for an activator. The numbers show the change in percentage inhibition occurring when the inhibitor concentration is doubled from two to four units

metabolic regulation has been discussed further (Newsholme and Crabtree, 1973). Since the hyperbolic response is the simplest relationship between protein and regulator it can be considered as the basic response with which any mechanism for improving sensitivity can be compared. Four such mechanisms will now be examined.

(b) *Mechanisms for improving sensitivity at non-equilibrium reactions*

(i) *Multiplicity of regulators* It is assumed, in Section 1.a above, that there is only one regulator for the enzyme. However, it is possible for an enzyme to be regulated by several different external regulators which all bind at different sites on the enzyme. In this case, if the concentrations of all the regulators change in the same direction (or in directions to change the activity of the enzyme in the same way) the effect of all external regulators will be cumulative. This can improve the response to a given situation—see control of 6-phosphofructokinase in Section B.1.b.

(ii) *Co-operativity* For many enzymes that play a key role in metabolic regulation, the response of their activity to the substrate or regulator concentration is sigmoid (see Chapter 3; Section C.3). This phenomenon is known as positive *co-operativity*. In this case, the protein is usually polymeric so that it binds more than one molecule of regulator such that the binding of the first molecule of regulator increases the affinity of the other binding sites. Hence the response is no longer hyperbolic and the limitation on sensitivity in regulation is reduced. For part of the concentration range of the regulator, the sensitivity will be greater than that provided by the hyperbolic response, i.e. greater than unity (Figure 7.2).

(iii) *Substrate cycles*[5]   A totally different mechanism for improving sensitivity is known as the substrate cycle. It is possible for a reaction that is non-equilibrium in the forward direction of a pathway (i.e. $A \rightarrow B$, see below) to be opposed by a reaction that is non-equilibrium in the reverse direction of the pathway (i.e. $B \rightarrow A$):

$$S \xrightarrow{E_1} A \underset{E_5}{\overset{E_2}{\rightleftarrows}} B \xrightarrow{E_3} P \xrightarrow{E_4}$$

The reactions must be chemically distinct and consequently they will be catalysed by different enzymes (i.e. E_2 and E_5, above). It is possible that these two opposing reactions are components of two separate pathways which function under different conditions (e.g. glycolysis and gluconeogenesis in the liver—see Chapter 11; Section B.2.a). However, the 'reverse' reaction (E_5 in the above example) may not be part of any pathway and only present in the cell to provide a cycle for metabolic control.

If the two enzymes are simultaneously active, A will be converted to B and the latter will be converted back to A, thus constituting the substrate cycle. There are thus two fluxes, a linear flux converting S to P and a cyclical flux between A and B. Both fluxes are to a large extent independent and calculations show that the improvement in sensitivity is greatest when the cyclical flux is high but the linear flux is low (i.e. the ratio, cycling rate/flux, is high—Newsholme and Crabtree, 1976; Newsholme, 1978a). The means by which a substrate cycle improves sensitivity in metabolic control may not be immediately obvious. After reading the simple account of the basis for the improvement in sensitivity below, the reader is recommended to read Section C.3, in which the role of the fructose 6-phosphate/fructose bisphosphate cycle in the control of the rate of fructose 6-phosphate phosphorylation in muscle is described.

The role of a cycle can best be understood when it is appreciated that, in some conditions, an enzyme activity may have to be reduced to values closely approaching zero. Even with a sigmoid response this would require that the concentration of an activator be reduced to almost zero or that of an inhibitor to an almost infinite level. Such enormous changes in concentration probably never occur in living organisms, since they would cause osmotic and ionic problems and unwanted side reactions (Chapter 2; Section A.2.d). However, the net flux through a reaction can be reduced to very low values (approaching zero) via a substrate cycle. Consider the following argument. As the product of the forward enzyme (E_2) is produced (i.e. B in the above example) it is converted back to substrate A by the reverse enzyme (E_5). This ensures that the *net* flux (i.e. A to B) is very low despite a *finite* activity of the forward enzyme and a moderate concentration of an activator. Now if the concentration of this activator is

Table 7.1. Effect of increase in regulator concentration on net flux through a reaction by a substrate cycle

Concentration of regulator	Enzyme activities* (units.min^{-1})		Net flux A to B	Relative fold increase in flux
	E_2	E_5		
Situation 1†				
Basal	10	9.8	0.2	
Fourfold above basal	90	9.8	81.2	406
Situation 2‡				
Basal	10	9.8	0.2	
Fourfold above basal	90	1.0	89.0	445

* The activities are hypothetical.

† In situation 1, the regulator has no effect on the reverse reaction catalysed by E_5. The ratio, cycling rate/flux is 49.

‡ In situation 2, the regulator not only increases the activity of E_2 but decreases that of E_5. The improvement in the relative change of the net flux is, however, not much greater than that in situation 1.

increased, by only a small amount above that at which the activities of the two enzymes are almost identical (and the flux is almost zero), the activity of E_2 will increase so that the net flux through the reaction will increase from almost zero to a moderate rate (see Table 7.1). Such a cycle therefore provides a large improvement in sensitivity; indeed, it can be seen as a means of producing a threshold response with a simple metabolic system. The sensitivity is further increased if the regulator activating E_2 also inhibits E_5 but this is by no means essential (Figure 7.3).

Since in the substrate cycle both reactions are non-equilibrium, it is not possible to operate even one turn of the cycle without conversion of chemical energy into heat. Usually this comes about by the hydrolysis of ATP to ADP and phosphate since ATP is involved as a substrate in one reaction. This is illustrated by the glucose/glucose 6-phosphate cycle in Figure 7.4; glucose is converted to glucose 6-phosphate in the hexokinase reaction that requires ATP, and glucose 6-phosphate is converted to glucose by the glucose 6-phosphatase reaction. The net result of the cycle is the hydrolysis of ATP. For a considerable number of years it was considered that this loss of energy was too high a price to pay and that metabolic control would ensure that such apparently energetically-wasteful cycles would not occur. Indeed, such cycles are sometimes known as 'futile' cycles but there is now considerable evidence to show that these cycles do exist and that the remarkable improvement in sensitivity provided by the cycles justifies the metabolic cost to the organism (see Newsholme, 1976a for an historical account of the development of these ideas).

In some circumstances, substrate cycles may operate not only to regulate flux

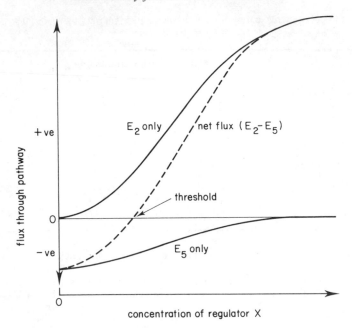

Figure 7.3 Kinetics of a substrate cycle. The upper curve shows the response of enzyme E_2 to change in the concentration of a regulator X, that activates the enzyme. The lower curve represents the activity of E_5 which is inhibited by regulator X. The net effect of the simultaneous activity of E_2 and E_5 is shown by the dotted curve. This will be the net flux through the reaction as the concentration of regulator is changed. Whereas, in the absence of enzyme E_5, the net flux could not be reduced to zero, only a small decrease in the concentration of the regulator, in the presence of active E_5, can now reduce the net flux to zero. Thus the flux through a pathway can be turned off completely when the concentration of regulator is low but not zero; that is, a threshold response is achieved.

through metabolic pathways but to achieve the controlled conversion of chemical energy (i.e. ATP) into heat, either to maintain body temperature (non-shivering thermogenesis), to raise the temperature (pyrexia) or to reduce body mass by burning off fuel (that is, weight control). These topics are discussed further in Chapter 19.

(iv) *Interconversion cycles* A number of enzymes (e.g. glycogen phosphorylase, pyruvate dehydrogenase) are known to be regulated by a different mechanism to that described above. These enzymes exist in two forms, conventionally designated *a* and *b*, one being a covalent modification of the other. The conversion of one form to the other is generally brought about by reaction with ATP and in most cases one form is a phosphorylated modification of the other. In general, only one of the two forms, (*a*), has significant catalytic activity so that the flux can be regulated by altering the amount of enzyme in this form. The interconversions between the forms are carried out by enzymes, E_1 and E_2, one

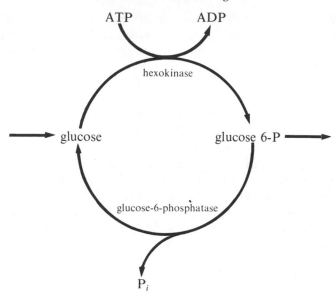

Figure 7.4 The glucose/glucose 6-phosphate substrate cycle. Note that in the liver, where the cycle will be most active, the enzyme glucokinase will be responsible for catalysing glucose phosphorylation

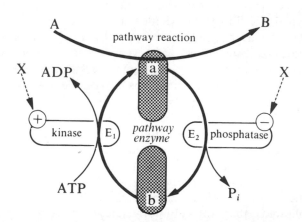

Figure 7.5 A general example of an enzyme interconversion cycle. The enzyme catalysing the pathway reaction exists in two forms, one active (*a*) and the other inactive (*b*). In this case *a* is phosphorylated and is thus formed from *b* by the action of a kinase enzyme. The phosphate is removed, thus inactivating *a*, by the action of a phosphatase enzyme. An activator (X) of the pathway enzyme can exert its effect by activating the kinase, inhibiting the phosphatase, or both. This style of diagram will be maintained for all subsequent interconversion cycles. Note that with some enzymes the phosphorylated form is the inactive one and that covalent modifications other than phosphorylations occur in some interconversion cycles. See Note 6.

for each direction, that catalyse non-equilibrium reactions. The activity of one or both of these enzymes can be altered by external regulators and such a scheme is outlined in Figure 7.5. (To distinguish the enzyme involved in the metabolic pathway from the enzymes catalysing the interconversions, the former enzyme is sometimes known as the pathway-enzyme.) The activity of the pathway-enzyme and hence the flux through that reaction depend upon the balance of the activities E_1 and E_2 and hence on the concentrations or the changes in concentration of the regulators of these two enzymes (see, for example, Section C.1.a for description of the control of phosphorylase in muscle). The mechanism of regulation via the conversion of an inactive form of an enzyme into an active form, or *vice versa*, has similarities to a substrate cycle and, indeed, this 'interconversion cycle' represents a logical extension of the substrate cycle. However, although the improvement in sensitivity provided by such cycles is likely to be large, the precise improvement is difficult to calculate (Newsholme and Crabtree, 1973, 1976).

It is important to point out that these four mechanisms are not mutually exclusive. Indeed, it is probable that at some reactions all four mechanisms play a role in regulation of flux and this combination could provide an enormous increase in sensitivity.

(c) *Sensitivity of near-equilibrium reactions*

Each of the four mechanisms outlined above for improving sensitivity involves some increase in complexity of enzyme mechanism beyond that which gives rise to the basic hyperbolic (Michaelis–Menten) kinetics described in Chapter 3. However, regulation of near-equilibrium enzymes by changes in concentration of pathway substrates or cosubstrates is achieved without additional complexity. The improvement in sensitivity depends upon the fact that the rates of the reaction in the forward and reverse directions are considerably greater than the flux; this is illustrated in Figure 7.6 (for full description see Crabtree and Newsholme, 1978). The sensitivity depends upon the fact that the catalytic activity is large, a consequence of the high concentration of such enzymes. This explains why the cell synthesizes considerably more enzyme-protein than is required to accommodate the maximum flux through the reaction. The reason that most pathways do not consist of all near-equilibrium reactions is that the sensitivity of the near-equilibrium reactions can only be achieved if they are preceded and followed in the pathway by non-equilibrium reactions (Crabtree and Newsholme, 1978) and non-equilibrium reactions serve to generate and transmit the steady-state flux.

B. REGULATION OF GLUCOSE UTILIZATION AND OXIDATION IN MUSCLE

The complete oxidation of glucose in muscles involves at least three separate physiological pathways as defined in Chapter 2: these are glycolysis-from-glucose, pyruvate to acetyl-CoA, and oxidation of the latter via the TCA cycle. It

	INITIAL STATE	INITIAL FLUX	AFTER 10% INCREASE IN V_1	RESULTING FLUX	INCREASE IN FLUX
NEAR-EQUILIBRIUM REACTION	$\rightarrow A \underset{V_2=100}{\overset{V_1=110}{\rightleftarrows}} B \rightarrow$	10	$\rightarrow A \underset{V_2=100}{\overset{V_1=121}{\rightleftarrows}} B \rightarrow$	21	x2.1
NON-EQUILIBRIUM REACTION	$\rightarrow A \underset{V_2=1}{\overset{V_1=11}{\rightleftarrows}} B \rightarrow$	10	$\rightarrow A \underset{V_2=1}{\overset{V_1=12.1}{\rightleftarrows}} B \rightarrow$	11.1	x1.11

Figure 7.6　Comparison of the effects of increasing the rate of the forward reaction by 10% (due to an increase in pathway-substrate concentration) on reactions near to equilibrium and far from equilibrium

will be recalled that glycolysis-from-glucose begins either in the liver (at glycogen phosphorylase) or the intestine (with absorption from the lumen) so that a *complete* understanding of the control of the pathway can only be given when these processes are included (see Section B.1.c). However, it is well established that the rates of conversion of extracellular glucose to pyruvate and its conversion to carbon dioxide can be increased by increasing the work performed by an isolated muscle. Consequently these processes must be controlled by external regulation within the muscle, and the question is raised as to the nature of the external regulators.

The two most important regulators in the control of the rates of carbohydrate utilization and oxidation are changes in concentration of Ca^{2+} and the ATP/ADP concentration ratio. Of these, the changes in concentration of Ca^{2+} are the more fundamental since they regulate the activity of the myofibrillar ATPase, which results in contraction and the hydrolysis of ATP so that the ATP/ADP concentration ratio is decreased. The latter either directly or indirectly (that is via changes in concentrations of phosphocreatine, ammonia, phosphate, AMP, or IMP) regulates the activity of key enzymes in the fuel utilization pathways. In addition, Ca^{2+} also regulates directly the activities of the key enzymes phosphorylase, pyruvate dehydrogenase, NAD^+-linked isocitrate dehydrogenase and oxoglutarate dehydrogenase.

The mechanisms involved, which are considered in some detail in Sections B and C, have been investigated in a variety of animals and muscles including insect flight muscle (Rowan and Newsholme, 1979), isolated perfused skeletal and heart muscle of the rat (Houghton, 1971; Opie and Owen, 1975) and skeletal muscle of man (Hermansen *et al.*, 1983). For reviews see Randle and Tubbs (1979) and Vary *et al.* (1981). Since the external regulation appears to be very similar in all animals, the description will be restricted to man.

1. Regulation of glycolysis-from-glucose

The non-equilibrium reactions in glycolysis are glucose transport, hexokinase, 6-phosphofructokinase and pyruvate kinase (Table 2.4) and there is evidence that all are regulated by external factors (i.e. the concentrations of their pathway-substrates decrease when the flux is increased). The properties of the enzymes that catalyse these reactions have been studied in order to identify possible external regulators which are listed below.

(a) *External regulators of the non-equilibrium reactions*

(i) *Glucose transport*   Since transport cannot be studied *in vitro* it is difficult to investigate properties of the system. However, it is known that the transport rate is stimulated by insulin, exercise and any other condition that imposes an energy stress on the muscle (e.g. hypoxia) but the molecular bases for these effects are not known.

(ii) *Hexokinase*   External control of hexokinase is brought about by *allosteric** inhibition by its product, glucose 6-phosphate which can be relieved by phosphate.

(iii) *6-Phosphofructokinase*   External regulation of 6-phosphofructokinase is brought about by a number of regulators: ATP is an allosteric inhibitor and this inhibition is potentiated by phosphocreatine and citrate, whereas AMP, fructose bisphosphate, fructose 6-phosphate (which is therefore an internal and an external regulator) phosphate and ammonia relieve the inhibition (Sugden and Newsholme, 1975). This is an excellent example of multiplicity of regulators and the regulators influence the activity in a sigmoid manner.

(iv) *Pyruvate kinase*   External regulation of pyruvate kinase is brought about by ATP and phosphocreatine, which are allosteric inhibitors and by the cosubstrate, ADP, which increases the activity.

(b) *Theory of control of glycolysis-from-glucose*

The physiological reason for the increased rate of glycolysis during exercise is increased demand for ATP by the contractile process which can thus be said to control the rate of glycolysis. This raises the question of the nature of the information-link between contractile activity and the glycolytic enzymes. The external regulators of 6-phosphofructokinase and pyruvate kinase given above provide the answer to the question. An increase in activity of myofibrillar ATPase causes a small decrease in the steady state concentration of ATP and increases those of ADP, phosphate and protons according to the following reaction:

$$ATP^{4-} + H_2O \longrightarrow ADP^{3-} + P_i^{2-} + H^+$$

Although the changes in the concentrations of these regulators are very small (Table 7.2) they result in larger changes in the concentrations of other regulators as described below and these probably play the major role in regulation of glycolytic flux. These include AMP, phosphocreatine and ammonia. In addition, the change in concentration of phosphate is far greater than that predicted from the above equation.

1. Adenylate kinase, which is highly active in muscle, maintains the following reaction close to equilibrium:

$$ATP^{4-} + AMP^{2-} \rightleftarrows 2ADP^{3-}$$

The relationship between the concentrations of the three nucleotides in muscle is thus determined by the equilibrium constant for this reaction and the

---

* Since hexokinase catalyses a non-equilibrium reaction, its product cannot have a quantitatively significant effect on the rate of the reverse component of the reaction. Glucose 6-phosphate must therefore reduce the catalytic activity of the enzyme by binding at a site separate from the active site, that is, allosterically.

Table 7.2. Contents of some glycolytic intermediates and regulators of glycolysis in insect flight muscle, perfused rat heart and quadriceps muscle of man at rest and during sustained exercise*

| | Contents (μmol.g^{-1} fresh wt. tissue) | | | | | |
| | Locust flight muscle | | Perfused rat heart | | Quadriceps of man | |
	rest	flight	non-pumping	pumping	rest	exercise
ATP	5.1	4.3	4.6	3.6	5.2	4.7
ADP	0.43	1.1	1.1	1.0	—	—
AMP	0.06	0.12	0.29	0.25	—	—
P_i	9.3	11.9	3.7	6.9	21.3	11.7
Phosphocreatine	—	—	6.1	4.8	—	—
Ammonia	0.77	1.2	—	—	—	—
Glycogen	—	—	—	—	81	30
Glucose 6-phosphate	0.10	0.22	—	—	0.37	0.37
Fructose 6-phosphate	0.03	0.04	—	—	0.07	—
Fructose 1,6-bisphosphate	0.05	0.12	—	—	0.07	0.11
Pyruvate	0.06	0.24	0.06	0.09	—	—
Lactate	0.61	0.35	1.9	1.6	1.0	3.0
Citrate	0.78	0.16	1.1	0.7	0.11	0.25
Cytosolic NAD$^+$/NADH	1084	2173	476	502	—	—
Mitochondrial NAD$^+$/NADH	5.8	8.1	—	—	—	—

*Data taken from Rowan and Newsholme (1979); Opie and Owen (1975) and Essen and Kaijser (1978).

Table 7.3. Theoretical changes in concentrations of adenine nucleotides calculated from adenylate kinase equilibrium*

| Concentration (mM) | | | [ATP]/[ADP] | $\dfrac{\% \text{ change in [AMP]}}{\% \text{ change in [ATP]}}$ |
ATP	ADP	AMP		
4.796	0.2	0.004	24.0	—
4.691	0.3	0.009	15.6	102
4.48	0.5	0.02	8.9	49
3.86	1.0	0.11	3.9	42
3.19	1.5	0.31	2.1	16

* It is assumed that the total adenine nucleotide concentration is 5 mM and that the equilibrium constant for adenylate kinase is 0.44. See Newsholme and Start (1973; p. 112) for further details.

ATP/ADP concentration ratio. The latter is such that the AMP concentration in muscle is very low and consequently small changes in the ATP/ADP concentration ratio cause much larger relative changes in that of AMP (Table 7.3).

2. The increase in concentration of AMP and the decrease in that of ATP have the further consequence of increasing the activity of the enzyme AMP deaminase, which hydrolyses the amino-nitrogen from AMP to produce ammonia and inosine monophosphate (IMP) as follows:

$$AMP^{2-} + H_2O \longrightarrow IMP^{2-} + NH_3$$

This leads to an increase in the concentration of ammonia which may even be greater than that of AMP.

The increase in the concentration of ammonia may additionally be derived from the degradation of adenosine to inosine catalysed by the enzyme adenosine deaminase:

$$\text{adenosine} + H_2O \longrightarrow \text{inosine} + NH_3$$

Adenosine arises from AMP by the action of the enzyme, 5'-nucleotidase. The importance of this latter enzyme and the adenosine deaminase is not so much the production of ammonia as the control of the concentration of adenosine, which is a vasodilator in muscle and other tissues. It is released from muscle cells into the interstitial space where it binds to receptors on the cell surface of the smooth muscle of the arterioles causing relaxation and hence increasing the diameter of the arterioles. This results in an increased flow of blood to the muscle, providing an improved supply of oxygen and fuels (for review of adenosine, see Arch and Newsholme, 1978). The conversion of AMP to IMP or adenosine means that there is a reduction in the total concentration of adenine nucleotides (Table 7.2) but, since the concentrations of ammonia and adenosine are very small, the decrease in the nucleotide concentration is also small. The adenine nucleotides lost during exercise etc. are resynthesized during rest. The reactions involved in the resynthesis are part of the system known as the purine nucleotide cycle and are described in Figure 12.1.

3. A further consequence of the change in the ATP/ADP concentration ratio is to favour phosphocreatine breakdown which leads to an increase in the concentration of inorganic phosphate as follows:

$$\text{phosphocreatine}^{2-} + \text{ADP}^{3-} + \text{H}^+ \longrightarrow \text{ATP}^{4-} + \text{creatine}$$

$$\text{ATP}^{4-} + \text{H}_2\text{O} \longrightarrow \text{ADP}^{3-} + \text{P}_i^{2-} + \text{H}^+$$

If these two reactions are summed, we get:

$$\text{phosphocreatine}^{2-} \longrightarrow \text{creatine} + \text{P}_i^{2-}$$

The above changes in concentrations of external regulators will lead to an increase in the activity of 6-phosphofructokinase (Figure 7.7); this will lower the concentration of its substrate, fructose 6-phosphate, and since the enzyme phosphoglucoisomerase catalyses a near-equilibrium reaction, lower the concentration of glucose 6-phosphate, which will increase hexokinase activity

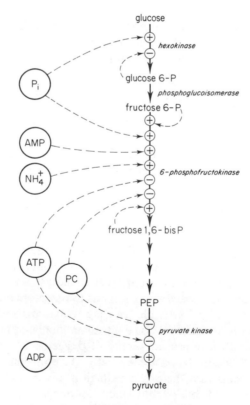

Figure 7.7 The regulation of non-equilibrium enzymes in glycolysis by external regulators

(Figure 7.7). It is well established that the rate of glucose transport is stimulated under conditions of a decrease in the ATP/ADP concentration ratio, but the external regulator(s) responsible is not known. A stimulation of transport will lead to an increase in the concentration of intracellular glucose, an *internal regulator* of hexokinase. (It will also lower the extracellular concentration of glucose, which may act as an information link between muscle and liver to increase the rate of hepatic glycogenolysis (Section B.1.c).)

The stimulation of 6-phosphofructokinase activity will lead to an increase in the concentration of fructose bisphosphate and this will stimulate aldolase activity (internal control) and raise the concentrations of the triose phosphates, which will, in turn, increase the activity of glyceraldehyde 3-phosphate dehydrogenase (internal control). In addition, this enzyme will be stimulated by external regulators; an increase in the concentration of phosphate and an increase in the concentration ratio, $NAD^+/NADH$ (Rowan and Newsholme, 1979). Similarly, phosphoglycerate kinase activity may also be stimulated by the change in the ATP/ADP concentration ratio. The activity of pyruvate kinase is controlled by decreases in the concentration of ATP and phosphocreatine and an increase in that of ADP (Figure 7.7).

It should be noted that the enzymes involved in the glycolytic span from fructose bisphosphate to phosphoenolpyruvate catalyse near-equilibrium reactions, which will be sensitive to small changes in concentrations of substrates, cosubstrates, and coproducts.

(c) *Control of hepatic glycogenolysis in relation to muscle glycolysis*

An increase in the rate of glycolysis-from-glucose in muscle demands a similar increase in the rate of hepatic glycogenolysis, if the overall pathway of glucose mobilization and utilization is to be maintained in the steady state.[7] This is known to be the case in the studies that have been carried out in man (see Chapter 9; Section B.2.a). How is this achieved? There are at least three possibilities. First, the central nervous control of muscle contraction may also control the rate of glycogenolysis in the liver (Edwards, 1972). The disadvantage of this is that the nervous control of glycogenolysis may not be sufficiently precise to satisfy the varying demands for glucose by exercising muscle. Secondly, exercising muscle may release a regulatory compound that stimulates hepatic glycogenolysis. This would provide a feedback link between the two processes that could result in precise regulation, especially if the rate of release of the regulator was porportional to the work performed by the muscle. There is, however, no evidence for such a regulator. Thirdly, small changes in the blood glucose level caused by changes in the rate of glucose utilization might sufficiently regulate the rate of hepatic glycogenolysis. This would have the advantage of being a direct feedback control mechanism but the disadvantage that the glycogenolytic system would need to respond to very small changes in the blood glucose concentration; that is, the mechanism would have to be very sensitive. Such a mechanism, in which glucose regulates both phosphorylase and glycogen synthase activities in liver, has been established, and if this regulatory system is grafted onto a substrate cycle between glycogen and glucose 1-phosphate catalysed by the simultaneous activities of these two enzymes, the system could indeed by very sensitive to small changes in the blood glucose level (Figure 7.8 and Chapter 11; Section E).

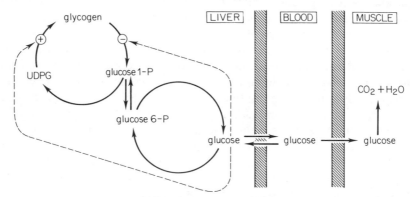

Figure 7.8 The regulation of glycogenolysis in liver by blood glucose. A change in the blood glucose concentration can control rates of glycogenolysis and glycogen synthesis via changes in hepatic glucose concentration

(d) *Quantitative aspects of regulation of glycolysis-from-glucose*

The conversion of glucose to pyruvate and the further oxidation to carbon dioxide occurs under aerobic conditions, that is, when the muscle is at rest or during sustained exercise. Since the aerobic process produces ATP very efficiently (Chapter 5; Section E.1) the increase in glycolytic rate from rest to activity in muscle does not need to be very large (perhaps 30- to 40-fold in muscles of man, Newsholme, 1978a; Wahren, 1979). Consequently, although the changes in the concentrations of the external regulators are not large, the two processes for improving sensitivity—multiplicity of regulators and co-operativity—are probably sufficient to account for the increase in glycolytic flux.

In contrast, sprinting in man (a level of exercise that cannot be sustained) utilizes the less efficient process of glycogen degradation to lactate (Chapter 5; Section E.2). Calculations demonstrate that to satisfy the energy requirement of this type of activity in athletes, the rate of glycolysis must increase about 1000-fold (see Chapter 9; Section A.3). Such a change requires the additional involvement of interconversion and substrate cycles to provide a further improvement in sensitivity (see Section C).

2. **Regulation of pyruvate dehydrogenase activity**

In Chapter 5; Section B.13.b it was emphasized that pyruvate dehydrogenase not only catalyses an important non-equilibrium reaction, which governs the fate of pyruvate, but that it also catalyses a flux-generating step so that it must be regulated by external factors. The main mechanism for regulation of this enzyme is via an interconversion cycle (Denton and Halestrap, 1979; Randle and Tubbs, 1979; Randle, 1981a, b).

Pyruvate dehydrogenase is a multienzyme complex; three enzymes are involved in the catalytic process, namely pyruvate dehydrogenase, dihydro-lipoamide acetyltransferase and dihydrolipoamide reductase. Complexity is further increased by the presence in the same complex of the enzymes of the interconversion cycle, pyruvate dehydrogenase kinase and phosphatase. It is, however, only the dehydrogenase subunit of the dehydrogenase complex that is phosphorylated and dephosphorylated by the interconverting enzymes. In contrast to glycogen phosphorylase, phosphorylation of pyruvate dehydro-genase leads to inactivation (i.e. formation of the *b* form) whereas dephosphory-lation leads to activation (i.e. formation of the *a* form):

$$\text{pyruvate dehydrogenase } a + \text{ATP} \xrightarrow{\text{kinase}} \text{pyruvate dehydrogenase } b + \text{ADP}$$

$$\text{pyruvate dehydrogenase } b + \text{H}_2\text{O} \xrightarrow{\text{phosphatase}} \text{pyruvate dehydrogenase } a + \text{P}_i$$

These interconverting enzymes appear to be specific for pyruvate dehydrogenase and exhibit properties that suggest they are regulated by external regulators. On the basis of these properties it is suggested that four factors may be involved in the regulation of the kinase: the concentration of pyruvate, and the concentration ratios, ATP/ADP, acetyl CoA/coenzyme A, and NAD^+/NADH. High values of the first two ratios and low values of the third ratio activate the kinase and lead, therefore, to inhibition of pyruvate dehydrogenase (Figure 7.9). On the other hand, pyruvate inhibits the kinase, so that an increase in its intramitochondrial concentration leads to activation of the dehydrogenase. Pyruvate dehydrogenase

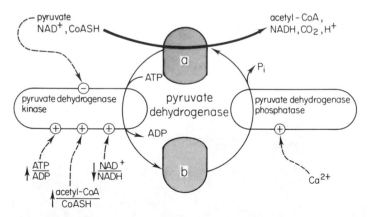

Figure 7.9 Regulation of pyruvate dehydrogenase. Under the influence of the kinase, ATP phosphorylates pyruvate dehydrogenase *a*, thereby inactivating it. Removal of the phosphate, by the phosphatase, activates pyruvate dehydrogenase

phosphatase is activated by very low concentrations (1 μM) of Ca^{2+} so that an increase in Ca^{2+} concentration will activate the phosphatase and hence increase pyruvate dehydrogenase activity. Exercise increases the mitochondrial concentration of Ca^{2+} and the mitochondrial concentration ratios of $NAD^+/NADH$ but decreases that of ATP/ADP (see below). These changes will inhibit the kinase and activate the phosphatase and lead to conversion of inactive to active pyruvate dehydrogenase. Although the changes in these ratios and the concentration of Ca^{2+} may be small, the sensitivity of the interconversion cycle probably ensures that there is a marked increase in the amount of pyruvate dehydrogenase *a*.

The control of kinase activity by the concentration ratio, acetyl-CoA/coenzyme A, may be important in the regulation of the activity of pyruvate dehydrogenase in relation to the oxidation of fuels other than glucose (e.g. fatty acid, ketone bodies) and hence in the integration of carbohydrate and fat metabolism (see Chapter 8). Oxidation of lipid fuels increases this concentration ratio which inhibits pyruvate oxidation and prevents loss of carbohydrate to carbon dioxide during, for example, starvation and sustained exercise and also in pathological states such as diabetes mellitus (Chapter 8; Section B.1.b and Chapter 15; Section E.2). Of some possible clinical importance is the finding that a derivative of acetate, dichloroacetate ($CHCl_2COO^-$) is an inhibitor of the kinase (probably because it mimics the action of pyruvate and binds at the pyruvate-binding site). Dichloroacetate, consequently, activates pyruvate

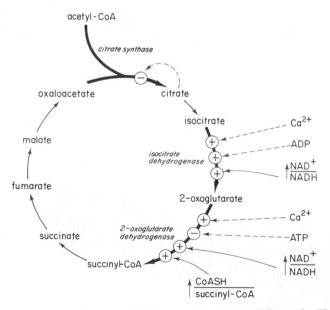

Figure 7.10 Regulation of non-equilibrium reactions in the TCA cycle. The regulators expressed in the form of a ratio are cosubstrate/coproducts (except for succinyl-CoA) and a rise in the concentration ratio as expressed activates the reaction

dehydrogenase and increases the rate of pyruvate oxidation; hence it should favour pyruvate oxidation and lower the concentrations of blood lactate and glucose in diabetes.[8]

3. Regulation of flux through the TCA cycle

In the cycle, the reactions catalysed by citrate synthase, isocitrate dehydrogenase and oxoglutarate dehydrogenase are non-equilibrium and the latter two are flux-generating steps. The regulators of these reactions are summarized in Figure 7.10. In investigations on the control of the cycle, the concentrations of the intermediates have been measured in the isolated perfused working rat heart preparation and in insect flight muscle during rest and flight (Table 7.4). In order to prevent the extraction conditions from changing the concentrations of the metabolic intermediates, the muscle has to be rapidly frozen to the liquid nitrogen temperature.[1] Unfortunately freezing of the muscle destroys the cell structure and mitochondrial and cytosolic intermediates cannot be separated. Since some of the intermediates of the cycle occur both in the cytosol and in the mitochondrial, failure to separate the two compartments interferes in the interpretation of the data (for reviews see Randle and Tubbs, 1979; Williamson and Cooper, 1980).

(a) *Control of the citrate synthase reaction*

The properties of citrate synthase suggest that there are at least two important external regulators: the cosubstrate, oxaloacetate, and the product, citrate, which is an allosteric inhibitor. Because citrate synthase is flux-generating, oxaloacetate can be considered as a cosubstrate for the whole cycle. There is evidence that during exercise the concentration of oxaloacetate increases whereas that of citrate decreases (Table 7.4). The decrease in the concentration of citrate is due to

Table 7.4. Contents of some TCA intermediates in insect flight muscle, perfused rat heart and quadriceps muscle of man at rest and during sustained exercise*

Metabolite	Contents (μmol.g^{-1} fresh wt.)					
	Locust flight muscle		Perfused rat heart		Quadriceps of man	
	rest	flight	low work	high work	rest	exercise
Acetyl-CoA	0.06	0.03	0.04	0.02	—	—
CoA	0.06	0.08	—	—	—	—
Citrate	0.79	0.16	0.56	0.44	0.11	0.25
Isocitrate	0.02	0.04	—	—	—	—
2-Oxoglutarate	0.10	0.07	0.08	0.04	—	—
Malate	0.21	0.40	0.10	0.06	0.08	0.20
Oxaloacetate	0.002	0.004	0.014	0.032	—	—

* Data from Rowan and Newsholme (1979); Randle and Tubbs (1979) and Essen and Kaijser (1978).

external regulation of isocitrate dehydrogenase. Two possible reasons for the rise in oxaloacetate concentration are as follows.

1. Oxaloacetate can be produced by the carboxylation of pyruvate in a reaction catalysed by the enzyme pyruvate carboxylase:

$$\text{pyruvate}^- + CO_2 + ATP^{4-} + H_2O \longrightarrow \text{oxaloacetate}^{2-} + ADP^{3-} + P_i^{2-} + 2H^+$$

This enzyme was considered to be present only in liver and kidney, in which it is a component of the pathway by which lactate and pyruvate are converted to glucose (see Chapter 11; Section B.2). However, it is present, albeit at a low activity, in muscle and the fact that it is present at higher activities in aerobic muscle suggests a role in the cycle (see Crabtree *et al.*, 1972). The activity of this enzyme may increase during exercise due to an increase in the concentration of the substrates, pyruvate and carbon dioxide (Rowan *et al.*, 1978).

2. At rest, the concentrations of metabolites that occur on the 'right-hand side' of the TCA cycle (i.e. citrate, isocitrate and oxoglutarate) are high but they decrease during exercise (see Table 7.4). This is due to a greater stimulation of the activities of the isocitrate and oxoglutarate dehydrogenases than that of citrate synthase. If the total concentration of the cycle intermediates remains constant, the concentrations of intermediates on the 'left-hand side' of the cycle should increase. In particular, since the concentration of oxaloacetate in resting muscle is very low, the relative increase in its concentration could be large.

Finally, it should be emphasized that although citrate synthase catalyses a flux-generating step for the TCA cycle, the rate of oxidation of acetyl units will depend upon their rate of supply from glycolysis and β-oxidation of fatty acids. If the stimulation of the cycle should exceed the rate of production of acetyl units from these sources, the concentration of acetyl-CoA would decrease so that citrate synthase might no longer catalyse a flux-generating step for the cycle. Conversely, if the rate of production of acetyl units exceeded the rate of utilization via the cycle, the concentration of acetyl-CoA would increase and that of coenzyme A would decrease. These changes result in a decreased rate of acetyl-CoA production due to inhibition of pyruvate dehydrogenase (see Section B.2) and inhibition of the rate of β-oxidation (see Section D).

(b) *Control of the isocitrate dehydrogenase reaction*

Isocitrate oxidation is catalysed by two enzymes, the NAD^+- and $NADP^+$-linked dehydrogenases. The relative importance of these two enzymes is still unknown (Chapter 4; Section B.1.c). Both enzymes could be controlled by the availability of the cosubstrates, NAD^+ and $NADP^+$, and, in addition, the NAD^+-linked enzyme exhibits allosteric properties; it is activated by ADP and by Ca^{2+}. The evidence given below suggests that the magnitude of the changes in concentrations of both ADP and NAD$^+$ within the mitochondria will not be large, so that it is unlikely that they will be quantitatively important in the regulation of this enzyme. Consequently, changes in the intramitochondrial concentration of Ca^{2+} may be most important. During the contraction phase of the contraction-relaxation cycle, the sarcoplasmic Ca^{2+} concentration increases and the mitochondria take up Ca^{2+} from the sarcoplasm (Section C.1.c). However, during the relaxation phase of the cycle, the sarcoplasmic Ca^{2+} concentration decreases but, since the release of Ca^{2+} by the mitochondria may be slower than the uptake, the mitochondrial Ca^{2+} concentration should remain elevated* during the entire exercise period, which will lead to an increase in the activity of NAD$^+$-linked isocitrate dehydrogenase. At rest, the intramitochondrial Ca^{2+} concentration will decrease and the activity of the dehydrogenase will be reduced.

(c) *Control of the oxoglutarate dehydrogenase reaction*

Oxoglutarate dehydrogenase activity is inhibited by high concentration ratios of ATP/ADP and succinyl-CoA/coenzyme A and a low concentration ratio of NAD$^+$/NADH. This is similar to pyruvate dehydrogenase, but it is unclear if this is due to the existence of an interconversion cycle. In addition, also similarly to pyruvate dehydrogenase, the activity is increased by Ca^{2+} (McCormack and Denton, 1980).

(d) *Control of the near-equilibrium reactions*

Probably the activities of other enzymes of the cycle are regulated by changes in concentrations of pathway substrates (internal regulation) and changes in the NAD$^+$/NADH concentration ratio (for malate dehydrogenase) and the concentration ratio, oxidized FAD/reduced FAD (for succinate dehydrogenase). If these enzymes catalyse near-equilibrium reactions, (see Table 4.1) the flux will be very sensitive to small changes in the concentrations of internal and external regulators.

4. Regulation of electron transport and oxidative phosphorylation

Ever since the observations of Keilin in 1925 on changes in the spectra of cytochrome c in the flight muscle of insects during wing movement, there has been considerable interest in the regulation of the rate of electron transfer in mitochondria. When it was discovered that oxidative phosphorylation and electron transfer are obligatorily linked, it became obvious that the regulation of the two processes is interdependent. The possibility that increased utilization of

---

* In contrast the sarcoplasmic concentration will increase and decrease with the phases of the contraction-relaxation cycle.

ATP, and hence increased production of ADP, might regulate the rate of electron transfer was raised when it was demonstrated that the rate of oxygen uptake (i.e. electron transfer) by isolated mitochondria could be stimulated by addition of ADP (see Figure 4.21).

From the principles developed in this book, electron transfer and oxidative phosphorylation can be seen to form a sequence of near-equilibrium reactions, the flux through which can be altered by small changes in the concentrations of reactants. Within the mitochondria, the sequence of oxidations and reductions from NADH to cytochrome *a*, together with associated phosphorylations, appear to be near-equilibrium (Chapter 4; Section D.1.d) and can be represented:

$$NADH + 2cyt\ a.Fe^{3+} + 3ADP^{3-} + 3P_i^{2-} + 2H^+ \rightleftharpoons$$
$$NAD^+ + 2cyt\ a.Fe^{2+} + 3ATP^{4-} + 3H_2O$$

The flux-generating step in the pathway is the formation of ADP (and phosphate) by the action of myofibrillar ATPase in the cytosol. A further near-equilibrium process, adenine nucleotide translocation, links events in the cytosol with those in the mitochondrion.

The sequence of events in the control of electron transfer and oxidative phosphorylation can be summarized as follows. The action potential causes an increase in the cytosolic Ca^{2+} concentration, which stimulates the myofibrillar ATPase and results in contraction (see Section C.1.c). This raises the cytoplasmic concentration of ADP and lowers that of ATP. In consequence, the rates of transport of ADP into the mitochondria and hence ATP out of the mitochondria, via the translocase reaction, are increased. This lowers the mitochondrial ATP/ADP concentration ratio and thus stimulates electron transfer so that the rate of ATP synthesis is increased (Figure 7.11). To satisfy the increased rate of NADH oxidation by the electron transfer chain its rate of production must be increased and this is achieved by regulation of the TCA cycle as discussed above.

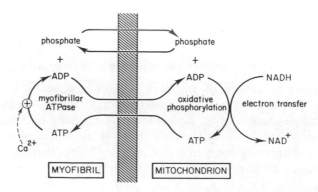

Figure 7.11 The regulation of oxidative phosphorylation by the activity of myofibrillar ATPase (via near-equilibrium reactions except for the ATPase)

C. REGULATION OF RATE OF GLYCOLYSIS-FROM-GLYCOGEN IN MUSCLE

The fact that glycolysis-from-glycogen and glycolysis-from-glucose can be considered, at least from a functional point of view, as two separate pathways was discussed in Chapter 5; Section F.1. This is emphasized in this chapter by discussing the two processes in separate sections. (For review of control of glycogenolysis in muscle, see Newsholme and Start, 1973; Cohen, 1980).

1. Control of glycogen phosphorylase activity

Phosphorylase catalyses the flux-generating step for glycogen conversion to lactate in muscle so that its activity and hence the overall flux through the pathway must be regulated by external factors. The mechanism of control of phosphorylase activity is complex and it involves at least three different control mechanisms. These will be discussed separately, although *in vivo* all these mechanisms will act in concert.

Phosphorylase exists in two forms, *a* and *b*, which are interconvertible via phosphorylation and dephosphorylation reactions catalysed by kinase and phosphatase enzymes, respectively. In resting muscle, most (probably 99%) of the phosphorylase is in the much less active *b* form.

(a) *Control of phosphorylase* b *activity by external regulators*

Phosphorylase *b* can be activated and if maximal activity is achieved it is similar to that of phosphorylase *a*. It is activated by phosphate, AMP and IMP, and is

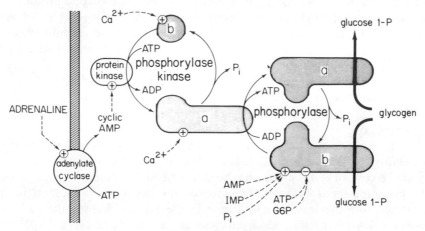

Figure 7.12 Regulation of glycogen phosphorylase activity in muscle. Inactivation of both phosphorylase kinase and phosphorylase *a* is brought about by a phosphatase. The activation of phosphorylase kinase *b* requires a higher concentration of Ca^{2+} than does the activation of phosphorylase kinase *a* (see text)

inhibited by ATP and glucose 6-phosphate. (In this respect, the properties of phosphorylase b are similar to those of 6-phosphofructokinase.) In resting muscle the concentrations of these external regulators are such that the activity of phosphorylase b will be very low. However, when muscle is active, the concentrations of these regulators change in such a way as to activate phosphorylase b (Table 7.2; see also Table 9.3; Figure 7.12). The limitation of this form of control is its low sensitivity. Since the activity of phosphorylase b is so very low in resting muscle, very large changes in the concentrations of the external regulators will be required to produce a marked increase in the activity.

(b) *Control of phosphorylase interconversion by hormones*

The enzyme phosphorylase kinase converts phosphorylase b into phosphorylase a by phosphorylation of a serine residue in phosphorylase b*:

$$\text{phosphorylase } b + 4\text{ATP} \longrightarrow \text{phosphorylase } a + 4\text{ADP}$$

Phosphorylase a is converted back to phosphorylase b via a phosphatase reaction catalysed by phosphorylase phosphatase (more generally known as protein phosphatase-I, since it is involved in the dephosphorylation of other proteins):

$$\text{phosphorylase } a + 4\text{H}_2\text{O} \longrightarrow \text{phosphorylase } b + 4\text{P}_i$$

Phosphorylase a is catalytically active at concentrations of AMP, IMP, phosphate, ATP and glucose 6-phosphate that are present in resting muscle. For full activity, however, it may require an increase in the concentration of phosphate as cosubstrate since this is low in resting muscle (Wilson *et al.*, 1981) but after the first few seconds of exercise, the phosphate concentration will increase due to phosphocreatine breakdown, because of hydrolysis of ATP (Section B.1.b). Hence conversion of b to a, which may take several seconds to occur in muscle, results in a marked activation of phosphorylase activity. Consequently, the activity of this key enzyme is controlled by the relative activities of the interconverting enzymes, phosphorylase kinase and protein phosphatase-I. What controls the activities of these enzymes? The available evidence suggests that hormones, acting indirectly on the kinase, play an important role.

It has been known for many years that adrenaline stimulates glycogen breakdown in muscle and since the late 1950s the sequence of events that provide hormonal control of this process has been identified. It is summarized as follows. The hormone binds to a specific receptor on the outer surface of the cell membrane and this leads to an increase in the activity of a membrane-bound enzyme, adenylate cyclase, which catalyses the formation of cyclic AMP[9] from ATP:

---

* Four ATP and ADP molecules are involved since the enzyme is a tetramer, that is, it is composed of four subunits.

$$\text{ATP}^{4-} + \text{H}_2\text{O} \longrightarrow \text{cyclic AMP}^- + \text{PP}_i^{3-}$$

The immediate effect of adrenaline binding to the receptor is, therefore, to increase the concentration of cyclic AMP (Figure 7.12). This increase in the concentration of cyclic AMP causes activation of the enzyme, protein kinase, which phosphorylates a number of proteins including phosphorylase kinase (see Chapter 22; Section A.3. for full discussion). In an analogous manner to phosphorylase, phosphorylated phosphorylase kinase is more active than the non-phosphorylated enzyme so that, with phosphorylation of the kinase, the rate of conversion of phosphorylase *b* to *a* is increased. If the activity of phosphorylase kinase markedly exceeds that of the phosphatase, the proportion of phosphorylase in the *a* form will rapidly increase resulting in an increased rate of glycogenolysis. This complete sequence, which is shown in Figure 7.12, has become known as an 'enzyme cascade' since several enzymes are sequentially involved in the regulatory process. It can, in fact, be considered as two interconversion cycles in sequence, thereby providing a greater improvement in sensitivity than a single cycle.

(c) *Control of phosphorylase activity by* Ca^{2+}

Despite this cascade providing great sensitivity to changes in hormone concentration, the administration of adrenaline to a muscle does not in fact cause a massive increase in the rate of glycogenolysis. For this, one further factor is required. The clue to its nature was provided by the observation that nervous stimulation of a muscle causes a very rapid and large increase in the rate of glycogenolysis (Helmreich *et al.*, 1965). The link between the electrical activity in muscle and contraction is known to be Ca^{2+}. The action potential travels along the muscle membrane and causes a release of Ca^{2+} from the sarcoplasmic reticulum into the sarcoplasm so that the sarcoplasmic Ca^{2+} concentration increases from about 10^{-8} M to about 10^{-6} M. This change in Ca^{2+} activates the myofibrillar ATPase which results in contraction. Relaxation is brought about by the re-uptake of Ca^{2+} into the sarcoplasmic reticulum (via an ATP-dependent Ca^{2+} uptake process) which reduces the sarcoplasmic Ca^{2+} concentration to 10^{-8} M once again (Figure 7.13). This external regulator of myofibrillar ATPase also activates phosphorylase kinase but the significant point is that the concentration of Ca^{2+} required to activate the enzyme is considerably lower if the kinase is phosphorylated. In fact, the phosphorylated enzyme is fully activated by the same concentration of Ca^{2+} that fully activates the myofibrillar ATPase (Cohen, 1980). Hence hormonal and nervous control function in concert to cause maximal and rapid stimulation of phosphorylase (Figure 7.12). This concerted control between adrenaline and nervous stimulation is predicted on teleological grounds, since adrenaline is an anticipatory hormone (priming the body for 'fight or flight') but, if anticipation led to very high rates of glycogenolysis prior to exercise, a massive and dangerous increase in the concentration of the hexose

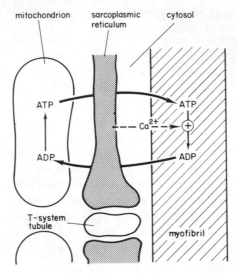

Figure 7.13 Diagrammatic representation of muscle ultrastructure. Depolarization spreading along the T-system tubules causes the sarcoplasmic reticulum to release Ca^{2+} ions and thus stimulate the myofibrillar ATPase. Ca^{2+} ions are rapidly transported back into the sarcoplasmic reticulum (and, to a lesser extent, probably the mitochondria) at the expense of ATP hydrolysis

monophosphates could result. The mechanism outlined above ensures that high rates will only occur if Ca^{2+} is released into the sarcoplasm and the activity of myofibrillar ATPase is also increased, so allowing a rapid increase in the regeneration of ATP. Conversion of phosphorylase kinase to a form sensitive to Ca^{2+} can be seen as a 'priming' action of adrenaline (Figure 7.12).

2. Control of the rate of glycogenolysis via the glycogen/glucose 1-phosphate cycle

Since phosphorylase catalyses a non-equilibrium process, the synthesis of glycogen must occur by another reaction. This is achieved from glucose 1-phosphate in two reactions, catalysed by glucose-1-phosphate uridylyltransferase and glycogen synthase (see Figure 16.1). Since it may not be possible to inhibit glycogen phosphorylase activity completely, even by an interconversion cycle, a low rate of glycogenolysis may occur even in resting muscle. This could be 'opposed' by the activity of the synthetic enzymes which would provide a substrate cycle between glucose 1-phosphate and glycogen (see Figure 9.2). In addition to restriction of the net rate of glycogenolysis, the cycle might improve further the sensitivity of glycogenolysis to changes in the concentrations of regulators of both phosphorylase *b* and the phosphorylase interconversion cycle, especially to changes in the concentration of Ca^{2+}.

3. Role of the fructose 6-phosphate/fructose bisphosphate cycle in the control of fructose 6-phosphate phosphorylation

As explained in Section B.1.d, glycolysis-from-glycogen, as opposed to glycolysis-from-glucose, may involve a 1000-fold increase in glycolytic flux. The changes in concentration of external regulators, reviewed in Section B.1.b, will not be sufficient to cause, by equilibrium binding, such large changes in glycolytic flux. It seems likely that the greater sensitivity required under anaerobic conditions is provided by the operation of a substrate cycle involving the enzymes 6-phosphofructokinase and fructose-bisphosphatase.

The role of fructose-bisphosphatase in muscle may be similar to that described above for glycogen synthase; the activity of 6-phosphofructokinase at rest may be greater than the demand for glycolysis for ATP production so that the net rate of fructose 6-phosphate phosphorylation is reduced by the activity of the reverse reaction. The rate of cycling will also provide an improvement in sensitivity to changes in the concentrations of regulators of 6-phosphofructokinase (AMP, ATP, Pi etc.) as described in Section A.2.b. This will be enhanced since AMP is an inhibitor of fructose-bisphosphatase.

One problem with this cycle is that it could cause a considerable rate of hydrolysis of ATP. Although this might be important in heat production, especially in small animals, in other animals the heat production might be an embarrassment with the danger of hyperthermia. We speculate that this problem could be overcome by a combination of interconversion and substrate cycles, as follows. When there is no likelihood of exercise so that the muscles are completely at rest (e.g. during sleep) the activities of both 6-phosphofructokinase and fructose-bisphosphatase could be very low so that the cycling rate would be very low. However, when exercise is anticipated, the rate of cycling could be raised and the ratio, rate of cycling/glycolytic flux, (which determines the sensitivity of the control system) increased. This would require the activities of both 6-phosphofructokinase and fructose-bisphosphatase to be increased simultaneously prior to exercise and it is suggested that these increased enzyme activities could be brought about by the anticipatory or stress hormones, (e.g. adrenaline, noradrenaline, glucocorticoids) via interconversion cycles. In other words, during anticipation of exercise, 6-phosphofructokinase might be phosphorylated by protein kinase so the activity would increase from a very low resting level to one that is higher than required to provide sufficient ATP from glycolysis at rest. At the same time the activity of fructose-bisphosphatase would also have been increased by an interconversion cycle to oppose the fructose 6-phosphate phosphorylation. The net result is that the rate of cycling between fructose 6-phosphate and fructose bisphosphate increases markedly thereby increasing the sensitivity to changes in the concentrations of ATP, AMP, phosphate, phosphocreatine, etc. and that the muscle is now primed for a large increase in flux (see Chapter 9; Section A.3.b).

In broad terms, the regulation of ATP synthesis from carbohydrate in muscle is

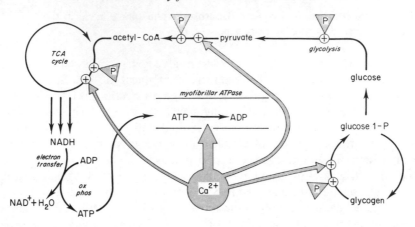

Figure 7.14 Summary of concerted control of ATP synthesis in muscle by Ca^{2+} ions and adenine nucleotides. The effect of adenine nucleotides is indicated by the symbol $\nabla\!\!\!P$. In each case a fall in concentration of ATP or a rise in the concentration of ADP, P_i or AMP activates the process indicated by this symbol

seen to depend on the parallel activation of muscle contraction and ATP synthesis by Ca^{2+} and feedback regulation by changes in the concentrations of adenine nucleotides (Figure 7.14).

D. CONTROL OF ENDOGENOUS LIPOLYSIS AND FATTY ACID OXIDATION IN MUSCLE

1. Endogenous lipolysis

The major store of triacylglycerol in the body is present in adipose tissue but small quantities are present in other tissues including muscle. Hence muscle can utilize fatty acids derived from the bloodstream or those derived from endogenous triacylglycerol. The evidence suggests that exogenous fatty acids are always used in preference. This may be because fatty acyl-CoA inhibits the intracellular lipase and consequently endogenous triacylglycerol will only be degraded when the fatty acyl-CoA concentration is low. Muscle triacylglycerol lipase can be activated by adrenaline, probably in a similar manner to activation of the enzyme in adipose tissue (Chapter 8; Section C.1). The processes of esterification and lipolysis probably occur simultaneously in muscle as in other tissues (i.e. a substrate cycle operates between triacylglycerol and fatty acids) and this may play a role in preventing dangerous accumulation of fatty acids within the muscle (see Chapter 6; Section F.2 and Newsholme and Crabtree, 1976).

2. Fatty acid oxidation

The pathway for fatty acid oxidation in muscle has been given in Chapter 6. The

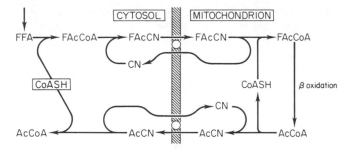

Figure 7.15 Diagram to illustrate how the availability of coenzyme A for fatty acid activation is limited by the rate of acetyl-CoA synthesis.
Abbreviations: AcCN, acetylcarnitine; AcCoA, acetyl-CoA; CN, carnitine; FAcCN, long-chain fatty acylcarnitine; FAcCoA, long-chain fatty acyl-CoA

flux-generating step for oxidation is the triacylglycerol lipase in adipose tissue, so that increasing the blood concentration of fatty acids increases the rate of fatty acid oxidation by muscle (and other tissues) (Neely and Morgan, 1974; Vary *et al.*, 1981; Randle and Tubbs, 1979). However, increasing the amount of work performed by the muscle (most of these experiments have been done on perfused rat heart—see Randle and Tubbs, 1979—or on human subjects by catheterization studies—see Hagenfeldt, 1979) increases the rate of oxidation of fatty acids without change in their extracellular concentration. In other words, the pathway in muscle is controlled externally, but by what regulator? The available evidence suggests that the main external regulator is the cosubstrate, coenzyme A, the concentration of which is changed by variations in the rate of β-oxidation. The sequence of events, which is best followed by reference to Figure 7.15, may occur as follows.

1. If the rate of β-oxidation exceeds that of the TCA cycle (e.g. when the demand for energy in muscle is low), the mitochondrial concentration of acetyl-CoA increases. Since there is a fixed concentration of coenzyme A in the mitochondria, the concentration must fall and this decreases the activity of carnitine palmitoyltransferase-II, so that less fatty acyl-CoA is produced for β-oxidation. This inhibition of the transferase-II might be expected to cause a massive accumulation of fatty acylcarnitine in the cytosol. However, this is prevented by communication between mitochondrial and cytosolic compartments achieved by the enzyme carnitine acetyltransferase. This enzyme is present in both compartments where it catalyses near-equilibrium reactions, and a mitochondrial transporter exchanges acetylcarnitine for carnitine across the inner mitochondrial membrane.[10] Thus an increased concentration of acetyl-CoA in the mitochondria increases that of acetylcarnitine which is then transported into the cytosol (via the mitochondrial exchanger) so that the acetylcarnitine concentration in this compartment is increased. This results in increased formation of acetyl-CoA in the cytosol and a decreased

concentration of coenzyme A which inhibits the activity of fatty acyl-CoA-synthetase (i.e. external regulation). This leads to an accumulation of long-chain fatty acid in the cytosol and, by decreasing the concentration gradient, this reduces the rate of long-chain fatty acid entry into the cell. If the rate of the TCA cycle in the muscle is increased, the concentration of acetyl-CoA will decrease and the above changes will be rapidly reversed, leading to enhanced rates of fatty acid oxidation.

NOTES

1. In order to measure the concentration of metabolic intermediates in a tissue, the metabolic reactions have to be inhibited very rapidly. This is achieved by freezing the tissue between aluminium tongs cooled in liquid nitrogen (i.e. at a temperature of about $-190\,°C$) (Figure 7.16). This lowers the temperature of the tissue to about $-80\,°C$ in less than 0.1 sec (Wollenberger et al., 1960). The frozen tissue is powdered in a percussion mortar at about $-70\,°C$ (see Ross, 1972) and the frozen powder extracted at $0\,°C$ or lower in perchloric acid. This rapidly inactivates and precipitates protein which is removed by centrifugation. Finally, this extract is neutralized and the metabolic intermediate measured by spectrophotometric or fluorimetric techniques (see Bergmeyer, 1974, for methods.) This technique provides concentrations within the whole tissue. However, this can be very misleading if the metabolite in question is restricted to a particular subcellular compartment (e.g. mitochondrion). Furthermore, the extraction in perchloric acid disturbs any non-covalent bonding of metabolites to proteins and other cell constituents, so that total rather than free concentrations are measured. Direct measurements of free metabolite concentration are difficult because disturbances that occur during the normal separation methods for subcellular components will alter the concentration of the metabolite. An indirect approach is to make use of reactions known to be near-equilibrium in the tissue under study. Knowledge of the equilibrium constant (determined *in vitro*) will permit the free concentration of reactants in the vicinity of the enzyme to be calculated. This approach has been particularly useful for determining the $NAD^+/NADH$ concentration ratio. For example, lactate dehydrogenase catalyses the following near-equilibrium reaction:

$$\text{pyruvate} + \text{NADH} + \text{H}^+ \rightleftarrows \text{lactate} + \text{NAD}^+$$

The equilibrium constant (K_{eq}) for this reaction is

$$K_{eq} = \frac{[\text{pyruvate}]\,[\text{NADH}]\,[\text{H}^+]}{[\text{lactate}]\,[\text{NAD}^+]}$$

so that

$$\frac{[\text{NAD}^+]}{[\text{NADH}]} = \frac{[\text{pyruvate}]\,[\text{H}^+]}{[\text{lactate}]} \cdot \frac{1}{K_{eq}}$$

The concentrations of lactate and pyruvate can be measured and substitution in the above equation together with the value of K_{eq} for lactate dehydrogenase enables the $NAD^+/NADH$ concentration ratio to be calculated. Since lactate dehydrogenase is localized in the cytosolic compartment of the cell, this $NAD^+/NADH$ concentration ratio represents that in the cytosolic compartment of the cell. Other dehydrogenases can be used to provide an indication of the $NAD^+/NADH$ concentration ratios in the mitochondrion (see Krebs, 1973). This approach depends on the fact (often only an

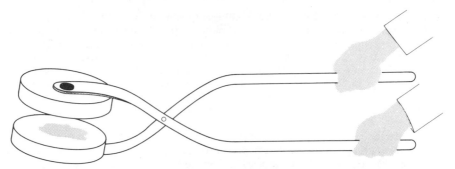

Figure 7.16 Freeze-clamping tongs (after Wöllenberger *et al.*, 1960)

assumption) that the pathway-reactants (in this case lactate and pyruvate) are neither extensively bound nor significantly compartmented. In practice, the validity of these measurements is increased by obtaining similar values using several dehydrogenases.

Similar approaches have been used to establish the concentration of free ADP in the muscle cell which may not be the same as the total ADP since some ADP is bound strongly to actin. In this case the creatine kinase reaction is utilised as follows:

$$\frac{[\text{ATP}]}{[\text{ADP}]} = \frac{[\text{phosphocreatine}]}{[\text{creatine}]} \cdot \frac{1}{K_{\text{eq}}}$$

There is some evidence that most of the total ATP in the cell is not bound (Hoult *et al.*, 1974) so that the free concentration of ADP can be calculated as follows:

$$\text{free}\,[\text{ADP}] = \frac{[\text{creatine}]\,[\text{ATP}]\,[K_{\text{eq}}]}{[\text{phosphocreatine}]}$$

(see Veech *et al.*, 1979).

2. Allosteric effectors are those which bind to enzymes at sites remote from the catalytic site but which influence the binding between substrate and enzyme through conformational changes transmitted through the enzyme. Most enzymes subject to allosteric regulation also exhibit co-operativity between binding sites (for both substrate and regulator) leading to sigmoid kinetics (Chapter 3; Sections C.3 and C.5).

3. The flux through near-equilibrium reactions is insensitive to allosteric regulators. The reason for this is that, for example, inhibition of an enzyme that catalyses such a reaction will result in an increase in the concentration of substrate and a decrease in that of the product, and *both* of these changes will oppose the inhibition of the enzyme by allosteric inhibition. In a non-equilibrium reaction, only the change in substrate will oppose the inhibitory effect (the change in product concentration will have a negligible effect since the rate of the reverse reaction is so small). Moreover, the near-equilibrium reaction is very sensitive to small changes in substrate and product concentrations so that they readily overcome the inhibition (see Crabtree, 1976, for full discussion).

4. Sensitivity is defined mathematically as the ratio of the *relative* change in the activity to the *relative* change in regulator concentration. It is important to appreciate what is meant by the relative change. If the concentration increases from S_1 to S_2, the relative change is $(S_2 - S_1)/S_1$, not $S_2 - S_1$: similarly, the relative change in activity is $(V_2 - V_1)/V_1$, not $(V_2 - V_1)$. If the concentration of the regulator X changes by $\Delta[\text{X}]$ and this causes the flux, J, to change by ΔJ the sensitivity (sometimes known as s) of J to the change in X is given by the ratio $(\Delta J/J)/(\Delta[\text{X}]/[\text{X}])$. The factor can be used to

compare the sensitivities provided by different metabolic mechanisms of control (see Crabtree and Newsholme, 1978; Newsholme, 1978a). One advantage of the relative change is that is is dimensionless, so that the principles apply whether the concentration is μM, mM or even M. The importance of the relative change can be seen when a concentration changes from zero to 0.1 μM. Although, at first sight, this may appear to be a small change, the *relative* increase is infinity.

5. The term 'cycle' has been widely used (and possibly misused) in biochemistry. The following definitions of cycles and related sequences may assist bewildered readers.

   (a) Metabolic cycle: A sequence of metabolic reactions in which one of the starting materials is regenerated. Such cycles provide a pathway for metabolically useful conversions (e.g. TCA cycle, urea cycle) (See Baldwin and Krebs, 1981, for recent discussion of the biochemical principles of such cycles).

   (b) Substrate cycle: (Sometimes misleadingly known as futile cycles). Two or more simultaneously active enzymes which operate to convert one intermediate to another and back again at the expense of chemical energy. Such cycles serve to improve sensitivity of metabolic control or to generate heat from chemical energy (e.g. fructose 6-phosphate/fructose bisophosphate cycle, Section C.3).

   (c) Enzyme interconversion cycle: Such cycles occur when an enzyme can exist in two covalent forms with additional enzymes simultaneously catalysing the interconversion between these forms (e.g. glycogen phosphorylase, Section C.1.b).

   (d) Shuttle: A cyclic sequence of reactions and translocations to achieve the net transport of chemical groups across a membrane (e.g. malate/aspartate shuttle for the transport of reducing equivalents, Chapter 5; Section D.1).

   (e) Shunt: An alternative sequence of reactions in a metabolic pathway (e.g. 2,3-bisphosphoglycerate shunt, Chapter 5; Section F.3b).

6. Although most of the interconversion cycles so far discovered involve phosphorylation of the protein (either at a serine or threonine residue) other covalent modifications are known. For example, using ATP as substrate, proteins can be adenylated. In the adenylation of bacterial glutamine synthetase, AMP is covalently linked to the hydroxyl group of a tyrosine residue (Stadtman and Chock, 1978). In ADP-ribosylation, the unit ADP-ribose is transferred from NAD^+ to the protein. ADP-ribosylation of an elongation factor catalysed by diphtheria toxin is responsible for inhibition of protein synthesis, and ADP-ribosylation of the GTP-binding protein of the intestinal adenylate cyclase complex catalysed by cholera toxin is responsible for reduced rate of water re-absorption and diarrhoea (for reviews see Hilz and Stone, 1976; Purnell *et al.*, 1980; Ueda *et al.*, 1982; see also Chapter 22; Section A.3.b).

7. If the rate of release of glucose is greater than that required by the muscle, the blood glucose concentration will rise and eventually glucose will be lost in the urine. On the other hand, if the rate of release is lower than required, the blood glucose concentration will fall. This would reduce the rate of glycolysis in muscle, reduce ATP formation and result in fatigue.

8. In experimental diabetes, there is evidence that the muscle pyruvate dehydrogenase subunit is phosphorylated in a second position (known as multisite phosphorylation). Although this second site phosphorylation does not further change the activity of the enzyme it makes it a less effective substrate for the phosphatase so that the phosphorylation that causes inactivation is more difficult to reverse and the enzyme is thus more difficult to activate. It is unclear if this happens in clinical diabetes (Randle and Tubbs, 1979; Denton and Halestrap, 1979).

9. In the endocrine control of metabolism, the hormone is known as the first messenger which transmits physiological information to the various tissues in the animal; compounds such as cyclic AMP are known as secondary messengers. They act as the

intracellular messenger for the hormone. (This is discussed in more detail in Chapter 22).

10. Most (95%) of the total coenzyme A concentration is present in the mitochondria whereas most (95%) of the carnitine is cytosolic. Coenzyme A is synthesized from pantothenic acid and the rate of this process is controlled; the rate of synthesis is increased in muscle from fasted or diabetic animals so that the concentration of coenzyme A is increased, thus favouring fatty acid oxidation; on the other hand, the rate is decreased by insulin, which should decrease the concentration of coenzyme A and hence the rate of fatty acid oxidation, thus favouring glucose utilization. These effects would therefore complement control by the glucose/fatty acid cycle (Chapter 8; Section B.1) although their quantitative importance is unclear.

CHAPTER 8

INTEGRATION OF CARBOHYDRATE
AND LIPID METABOLISM

Now that the regulation of carbohydrate and lipid metabolism in muscle has been described it becomes possible for the first time to discuss the integration of metabolic pathways between several tissues; specifically the integration of carbohydrate and lipid metabolism in brain, muscle, adipose tissue and liver. A more complete description of metabolic integration must necessarily include consideration of amino acid metabolism and extend the discussion to other tissues including the kidney and the intestine and this is attempted in Chapter 14.

A. FUEL RESERVES OF THE BODY AND THE
NEED TO UTILIZE FATTY ACIDS

The fuel reserves in the average human subject are given in Table 8.1. These data provide part of the answer to the question as to why glucose and lipid metabolism must be integrated. The amount of carbohydrate stored is very small; it represents only about 2% of that of lipid. The reason for this marked preference for lipid as a reserve fuel, which is not unique to man but is found in most animals, is that it is approximately five times more efficient than carbohydrate as a storage fuel (see Chapter 6; Section E.1).

Liver contains the only store of glycogen that can be broken down into glucose and released into the bloodstream for use by other tissues (muscle glycogen is used solely within the muscle). However, this quantity of stored carbohydrate (80–100 g) is very small in relation to the glucose requirement of the tissues (Table 4.7). At rest, the total glucose requirement of the major carbohydrate-utilizing tissues of the body (brain, kidney, heart, muscle) is over 300 g per day and this is normally met from the dietary intake of carbohydrate (Chapter 5; Section A). In the early period of starvation, liver glycogen is broken down to provide glucose for the tissues; measurement of the glycogen content in small samples of liver of man, removed by biopsy needle, have shown that this store of glycogen is largely depleted after 24 hr starvation (Table 14.2). Since some tissues can oxidize fatty acids as well as glucose, the mobilization of fatty acids from the adipose tissue triacylglycerol store will probably begin during the overnight fast and will increase particularly if breakfast is missed. In this way, some of the glucose derived from liver glycogen will be preserved for tissues that must oxidize glucose (e.g. the cells of the brain). It is well established, in both experimental animals and man, that even short periods of starvation raise the plasma fatty acid

336

Table 8.1. Fuel stores* in average man

Tissue fuel store	Approximate total fuel reserve		Estimated period for which fuel store would provide energy		
			Days of starvation†	Days of walking‡	Minutes of marathon running§
	g	kJ			
Adipose tissue triacylglycerol	9000	337 000	34	10.8	4018
Liver glycogen	90	1500	0.15	0.05	18
Muscle glycogen	350	6000	0.6	0.20	71
Blood and extracellular glucose	20	320	0.03	0.01	4
Body protein	8800	150 000	15	4.8	1800

* Normal man possesses 12% of the body weight as triacylglycerol and normal woman about 26% (see Chapter 19, Section C.3). Davidson *et al.* (1979) state that normal man (65 kg) possesses 9 kg of triacylglycerol and this value is used in this table. Higher amounts of stored triacylglycerol have been given by Cahill (1970) and Wahren (1979). Periods for which the fuel will last are calculated as below.

† Assuming that energy expenditure during starvation is $10\,050\,kJ.day^{-1}$ (i.e. normal energy expenditure of $13\,400\,kJ.day^{-1}$ is reduced by 25% on starvation—Chapter 14; Section B.2).

‡ Assuming that energy expenditure during walking ($4\,miles.hr^{-1}$) for 65 kg man is $31\,248\,kJ.day^{-1}$ (Durnin and Passmore, 1967).

§ Assuming energy expenditure of $84\,kJ.min^{-1}$ (Chapter 9; Section B.1).

The data illustrate the time for which the fuel stores would provide energy, provided this was the only fuel utilized by the body.

concentration (Table 8.2). Calculations based on the increase in the concentration of fatty acids in starvation and their turnover rate indicate that fatty acid oxidation can account for most of the energy requirements of the tissues after 24 hr starvation (see Chapter 6; Note 5).

Sustained exercise is another condition for which there is good evidence that skeletal muscle utilizes fatty acids and calculations demonstrate that not enough carbohydrate can be stored in the body to satisfy the energy demands of the marathon runner (Chapter 9; Section E.3.b).

In the early stages of both starvation and exercise, skeletal muscle will use primarily glucose to satisfy its energy demands. As the carbohydrate stores become depleted, fatty acids are mobilized and their rate of oxidation increases. This occurs despite the fact that the blood glucose concentration falls very little (by no more than about 25% of the normal) and while the absolute blood glucose concentration remains considerably higher than that of fatty acid (see Table 8.2). Since both fuels are available in the bloodstream at the same time, the question arises how does muscle, either at rest or during exercise, utilize fatty acids in preference to glucose? The answer to this question is, in principle, very simple. The elevated concentration of fatty acid in the bloodstream increases the rate of fatty acid oxidation in the muscle and this specifically reduces glucose utilization and oxidation. The details of the control mechanism are given below.

Table 8.2. Concentrations of glucose, fatty acids and ketone bodies during starvation in man and rat

Animal	Fuel or hormone in blood	Concentrations in serum (or plasma) (mM)							
		Fed	1	2	3	4	5	6	8
									(days of starvation)
Man	glucose	5.5	4.7	4.1	3.8	3.6	3.6	3.5	3.5
	fatty acids	0.30	0.42	0.82	1.04	1.15	1.27	1.18	1.88
	ketone bodies	0.01	0.03	0.55	2.15	2.89	3.64	3.98	5.34
	insulin*	>40	15.2	9.2	8.0	7.7	8.6	7.7	8.3
Rat	glucose	6.3	—	4.8	4.4	4.3	—	—	—
	fatty acids	0.66	—	1.30	—	—	—	—	—
	ketone bodies	0.22	—	2.8	3.0	3.3	—	—	—
	insulin*	28.7	—	4.2	—	—	—	—	—

* The units of insulin are $\mu U.cm^{-3}$.

For references from which the data are taken, see Newsholme (1976b).

B. REGULATION OF RATE OF GLUCOSE UTILIZATION AND OXIDATION BY ALTERNATIVE FUELS

1. The glucose/fatty acid cycle

The concept of the glucose/fatty acid cycle was put forward in 1963 (Randle *et al.* 1963; Randle, 1981b) to explain the reciprocal relationship between the rates of oxidation of glucose and fatty acids by muscle.[1] Although some features of the cycle have been modified since that time, there is now considerable evidence to support the important proposal that, under conditions of 'carbohydrate stress' (defined here as when the glycogen store in the liver is depleted) fatty acids are mobilized from adipose tissue so that their rate of oxidation by muscle increases and this, in turn, decreases the rate of glucose utilization (see Newsholme, 1976b). Conversely, when the carbohydrate stress is removed (e.g. by refeeding a starved subject) the rate of fatty acid release by adipose tissue is reduced, decreasing their rate of oxidation, so that the rate of glucose utilization by the muscle increases. These responses serve to stabilize the blood glucose concentration. This regulatory effect of fatty acid mobilization can be appreciated by reference to Figure 8.1.

The regulatory effect of fatty acid on glucose utilization can be seen as a logical necessity when the small reserves of carbohydrate are taken into account together with the fact that some tissues have an obligatory requirement for glucose. The cycle is of such importance in understanding metabolic fuel integration in the whole animal that some of the experimental evidence is presented below. The

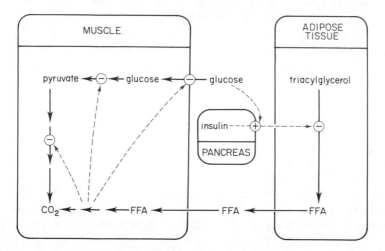

Figure 8.1 The glucose/fatty acid cycle. Note that a change in the peripheral blood glucose concentration is less important than changes in the glucose concentration in hepatic portal blood in eliciting insulin release from the pancreas since glucose absorption from the intestine is accompanied by secretion of duodenal hormones. (See Chapter 14; Section C.1)

biochemical mechanisms by which fatty acid oxidation reduces glucose utilization and oxidation in muscle and other tissues are given in Section B.1.b.

(a) *Experimental support for the glucose/fatty acid cycle*

There are four lines of support for the proposal that fatty acid oxidation reduces glucose utilization and oxidation *in vivo*. Only the key points are given here.

1. The oxidation of fatty acids by *in vitro* perfused preparations of both heart and skeletal muscle of the rat causes inhibition of glucose utilization and oxidation (Randle *et al.*, 1964; Rennie and Holloszy, 1977).
2. A biochemical mechanism to explain the inhibitory effect of fatty acid oxidation on glucose utilization has been established in muscle (see Section B.1.b). This mechanism appears to be very general: it accounts for the inhibitory effects of other fuels (e.g. ketone bodies, pyruvate, leucine) on glucose utilization in muscle; it accounts for the inhibitory effects of fatty acids and ketone bodies on glucose utilization in tissues other than muscle (e.g. brain, kidney, mammary gland—Newsholme and Start, 1973; Newsholme, 1976b; Robinson and Williamson, 1980) and may be present in tissues from animals from many other phyla (Newsholme *et al.*, 1977).
3. In both man and experimental animals, the artificial raising of the blood fatty acid concentration markedly reduces glucose utilization and oxidation (Schalch and Kipnis, 1964; Balasse and Neef, 1974). Similarly, administration of nicotinic acid to man lowers the blood fatty acid concentration but increases the rate of glucose utilization and oxidation (Balasse and Neef, 1973). Furthermore, if the blood fatty acid concentration is raised in man, the rates of glucose utilization and oxidation by the heart and the rate of pyruvate oxidation (and hence glucose oxidation) by skeletal muscle are reduced (as shown using arterio-venous difference measurements—see Appendix 5.2) (Lassers *et al.*, 1971; Hagenfeldt, 1979).
4. Under conditions in which glucose utilization is known to be reduced in the intact animal, the blood fatty acid concentration is elevated and the lipid oxidation (presumably fatty acid) is enhanced. For example, during prolonged starvation, the respiratory exchange ratio (see Appendix 5.2) of human subjects approaches 0.7 (Owen *et al.*, 1979) indicating that lipid is the major fuel of the body, yet the blood glucose concentration is at least 3.5 mM (i.e. higher than that of fatty acids—see Table 8.2). A similar situation exists in sustained exercise when the respiratory exchange ratio indicates some lipid oxidation, yet the blood glucose concentration remains close to normal.

(b) *Mechanism of control of glucose utilization by fatty acids*

The mechanism of control has been established through the approach to metabolic control discussed in Chapter 7. Studies with isolated *in vitro* muscle

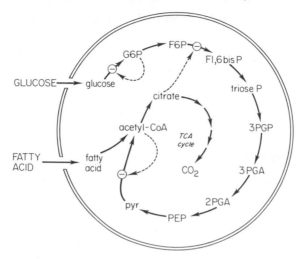

Figure 8.2 Diagram showing the mechanisms by which increased fatty acid oxidation inhibits glucose utilization and oxidation

preparations have shown that fatty acid oxidation results in a decrease in the rate of glucose utilization through effects of external regulators on glucose transport, hexokinase, 6-phosphofructokinase, and pyruvate dehydrogenase. That is, as the flux through glycolysis is decreased, the concentrations of the pathway-substrates for these reactions (namely, extracellular glucose, intracellular glucose, fructose 6-phosphate, and pyruvate) are increased. The theory of metabolic control of these enzymes has been proposed on the basis of their *in vitro* properties. These are as follows.[2] (i) Hexokinase is inhibited allosterically by its product glucose 6-phosphate. (ii) The inhibition of 6-phosphofructokinase by ATP is potentiated by citrate. (iii) Pyruvate dehydrogenase activity is controlled via an interconversion cycle; the active form of the enzyme is converted to the inactive form when the concentration ratio acetyl-CoA/CoA is raised, since this activates the interconverting enzyme, pyruvate dehydrogenase kinase (see Chapter 7; Section B.2). The required changes in concentration of these inhibitors have been demonstrated during fatty acid oxidation in muscle. The raised concentration of citrate* causes inhibition of 6-phosphofructokinase which, in turn, raises the concentration of glucose 6-phosphate, which inhibits hexokinase. This concerted mechanism of control of the non-equilibrium glycolytic reactions and pyruvate dehydrogenase is illustrated in Figure 8.2.

In Chapter 6, it was pointed out that the flux-generating step for fatty acid

---

* The oxidation of fatty acids or ketone bodies raises the intramitochondrial concentration of citrate. However, for this to inhibit 6-phosphofructokinase it must be transported out of the mitochondria to the cytosol. This is achieved by a carrier in which citrate efflux may be 'balanced' by malate influx (i.e. a citrate-dicarboxylate exchange—see Chapter 4; Section D.4.a.).

oxidation in muscle is triacylglycerol lipase in adipose tissue. This point is particularly important in the present discussions since it is, in large part, the increased plasma concentration of fatty acids that increases their rate of oxidation by muscle to cause the changes in concentrations of metabolic regulators indicated above. Therefore, in order to appreciate how carbohydrate metabolism *in vivo* is regulated, it is necessary to understand the control of triacylglycerol lipase activity in adipose tissue (Section C).

2. The glucose/fatty acid/ketone body cycle

A shortcoming of the glucose/fatty acid cycle is that it restricts the consideration of glucose and fatty acid metabolism to muscle and adipose tissue. Although muscle has the highest rate of fuel utilization during exercise, at rest the rate of glucose utilization is greater in tissues such as brain, kidney, intestine and, during lactation, mammary gland (Table 4.7). There is now considerable evidence that, expecially in starvation, the use of ketone bodies by these tissues reduces glucose utilization in an analogous manner to the effect of fatty acids on muscle.

(a) *Effect of ketone bodies on glucose utilization by the brain and other tissues*

An average human brain oxidises about 100 g of glucose each day to carbon dioxide and water. Since hepatic carbohydrate stores could satisfy this requirement for only about 24 hr of starvation, either the brain uses an alternative fuel or glucose must be produced by gluconeogenesis. Although the latter process does supply glucose for the brain during starvation there are serious consequences inherent in a prolonged period of gluconeogenesis sufficient to provide about 100 g of glucose each day. Body protein is the major precursor for glucose formation under these conditions but in order to produce sufficient

Table 8.3. Arterio-venous differences of fuels across the brain of man during starvation

	Arterio-venous concentration difference (mM)					
	Oxygen	Glucose	Lactate	Acetoacetate	3-Hydroxy-butyrate	Free fatty acid
Fed	−3.27	−0.51	—	—	—	—
5–6 weeks starvation	−2.96	−0.26*	+0.20	−0.06	−0.34	−0.02

* When lactate and pyruvate formation is taken into account this value falls to 0.145. No significant rate of lactate production occurs in the fed state.

In the fed state, the glucose uptake (0.32 mmol.min^{-1}), if completely oxidized, would account for 94% of the oxygen consumption (2.1 mmol.min^{-1}) (Reimuth *et al.*, 1965). In the fasted state, glucose oxidation (0.09 mmol.min^{-1}) accounts for 30% of oxygen uptake (1.9 μmol.min^{-1}) (Owen *et al.*, 1967). The negative sign indicates uptake of the fuel by the brain.

glucose, about 50% of the total body protein would be consumed within 17 days of starvation and this would be fatal (see Chapter 14; Section A.3). Since it is known that man can survive starvation for one-two months, the brain must utilize a fuel other than glucose. Studies on the changes in the blood concentrations of ketone bodies (Table 8.2) and arterio-venous differences across the brain of fasting human subjects (Table 8.3) demonstrates that as the rate of glucose utilization is depressed that of the ketone body, 3-hydroxybutyrate, is increased. Fatty acids are, however, not taken up. After 5–6 weeks of starvation of obese human subjects, the oxidation of ketone bodies provided about 70% of the energy requirements of the brain, so that the rate of glucose utilization was reduced to about 30 g each day (Chapter 14; Section A.4).

The mechanism of control of glucose utilization by ketone body oxidation in the brain has been studied in experimental animals (Ruderman *et al.*, 1974). The mechanism is similar, if not identical, to that of the inhibition by fatty acids of glucose utilization in muscle described above. Thus 6-phosphofructokinase activity is decreased by a raised concentration of citrate; hexokinase is inhibited by the resultant increase in the concentration of glucose 6-phosphate and pyruvate dehydrogenase is inhibited, probably by an elevation in the concentration ratio, acetyl-CoA/CoA. Unfortunately, no studies have been carried out on the effects of ketone bodies on the transport of glucose into the brain cell but, since it is an important non-equilibrium step, it is also likely to be inhibited by ketone body oxidation.

The use of ketone bodies during starvation, and the consequent restriction of glucose utilization, also occurs in muscle, kidney cortex, the lactating mammary

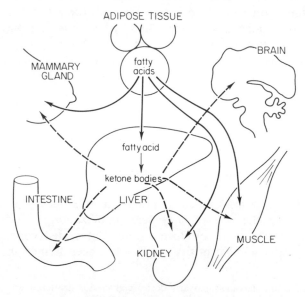

Figure 8.3   Use of fatty acids and ketone bodies by different tissues

gland and the small intestine* (Figure 8.3). These effects are of considerable quantitative importance in the conservation of glucose (and hence body protein) during prolonged starvation.

(b) *Role of ketone bodies in the regulation of fatty acid mobilization*

Originally the glucose/fatty acid cycle was termed a cycle because it was considered that not only could fatty acids modify glucose utilization but that glucose could modify the rate of fatty acid mobilization from adipose tissue. A low blood glucose concentration was considered to increase the rate of fatty acid release whereas a high glucose concentration would decrease the rate of release (through changes in the blood concentration of insulin—Figure 8.1). However, the remarkable constancy of the blood glucose concentration under different conditions (see Tables 8.2; 9.8; 14.1) suggests that the glucose level *per se* is unlikely to be quantitatively important in the regulation of the rate of fatty acid mobilization, so that the dominant regulatory effect of the cycle is directed towards a control of glucose utilization (as discussed above). Nonetheless, it is very important that the rate of fatty acid mobilization from adipose tissue should be precisely related to the energy needs of the tissues, especially of muscle, although it is unclear at the present time how this precision is achieved. Undoubtedly, changes in concentrations of hormones play an important role in the regulation of fatty acid mobilization (Section C) but it is unlikely that they can provide sufficient precision in control to meet rapid changes in energy demand. Because of the dangerous effects of high concentrations of fatty acids in the blood (Chapter 6; Section F.2) it is particularly important that excessive rates of fatty acid mobilization (in relation to energy demand) are prevented. It now seems likely that ketone bodies, and in particular 3-hydroxybutyrate, play an important role in preventing this. A high concentration of 3-hydroxybutyrate has at least three effects which could reduce the blood concentration of fatty acids; it reduces the rate of lipolysis in adipose tissue; it increases the sensitivity of adipose tissue to the effect of insulin, (an antilipolytic hormone (see Section C.1)); and it stimulates insulin secretion (see Green and Newsholme, 1979, for review). Although all three effects could simultaneously reduce the rate of lipolysis, perhaps the increase in insulin sensitivity is quantitatively the most important. The inhibitory role of 3-hydroxybutyrate can be viewed as a feedback inhibition to prevent the concentration of fatty acids increasing to toxic levels (>2 mM) (see Figure 8.4).

The use of ketone bodies as a fuel for brain and other tissues, together with their role in regulating the rate of fatty acid mobilization extends the original glucose/fatty acid cycle to produce a more integrated control mechanism involving a greater number of tissues. This has been termed the glucose/fatty

---

* In the small intestine, starvation may also decrease glucose utilization due to a decrease in the concentration of hexokinase (Windmueller and Spaeth, 1980).

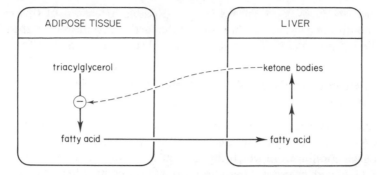

Figure 8.4 Feedback inhibition by ketone bodies on lipolysis in adipose tissue. This is achieved in at least three ways: direct inhibition of lipolysis; increase in the sensitivity of adipose tissue to antilipolytic effect of insulin; stimulation of insulin secretion by pancreas (see Green and Newsholme, 1979)

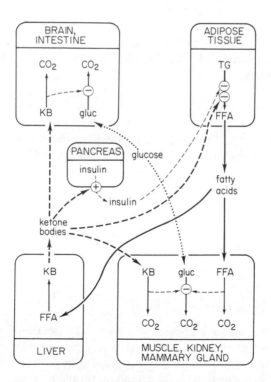

Figure 8.5 Controlling steps in the glucose/fatty acid/ketone body cycle

acid/ketone body cycle (Figure 8.5). The possible extension of this cycle to include control of protein metabolism and the further significance of the cycle in integration of fuel supply during starvation is discussed in Chapter 14.

3. Situations in which the blood glucose concentration is maintained by the control cycles

The glucose/fatty acid and glucose/fatty acid/ketone body cycles described above explain how the use of the three fuels is integrated. Physiological and pathological conditions in which this integration is of fundamental importance include sustained exercise, starvation, stress, refeeding after starvation and diabetes mellitus. They are summarized below but the importance of these conditions is emphasized by the fact that separate chapters are devoted to some of them.

(a) *Sustained exercise*

Since the liver glycogen reserves are partially depleted by the overnight fast even moderate exercise, for example jogging before breakfast, will cause considerable mobilization of fatty acids. (Chapter 9; Section B.3.b.) Readers who are early morning joggers should be aware that it is the operation of the glucose/fatty acid cycle which prevents hypoglycaemia. One of the authors (E.A.N.) frequently fasts for 18 hr prior to running 15–20 miles without obvious signs of hypoglycaemia (Chapter 11; Note 1).

(b) *Starvation*

In the absence of exercise, the overnight fast will result in a small increase in the rate of fatty acid mobilization. (Chapter 14; Section A.) Glycogen degradation will provide most of the energy requirements until a carbohydrate breakfast restores the supply of glucose. However lack of breakfast, so that the fast is extended to 12–18 hr, will cause a marked increase in the rate of mobilization of fatty acids and operation of the glucose/fatty acid cycle to prevent a serious fall in glucose concentration. After 24–48 hr of starvation, the concentration of ketone bodies rise and they become a significant fuel for muscle, kidney, intestine and brain.

(c) *Refeeding after starvation*

During starvation, glucose is conserved (Chapter 14; Section C.). If starvation is terminated with a high carbohydrate meal, failure to reverse this restriction could result in a massive elevation in the blood glucose concentration. This could result in loss of glucose in the urine with the attendant problems of dehydration and loss of ions from the blood as described for the diabetic patient in Chapter 15.

Table 8.4. Blood concentrations of fatty acids and ketone bodies after refeeding of rat*

Time after refeeding	Concentrations (mM)	
	fatty acids	ketone bodies
0	0.8	2.3
10	0.7	1.8
15	0.5	1.5
30	0.3	1.2

* Rats were fed glucose by stomach tube.

However, the response to the absorption of glucose is insulin secretion so that the peripheral blood concentration of this hormone is increased. This will inhibit lipolysis and cause a rapid reduction in the blood fatty acid and ketone body concentrations (Table 8.4). The latter, via the cycle, will enhance glucose utilization and oxidation in muscle and the other tissues. (Of course, uptake of glucose by the liver will occur and this may be of major quantitative importance in removing absorbed glucose, but it is not regulated by fatty acid oxidation—see Chapter 11; Section A).

(d) Stress

Increasing the circulating fatty acid concentration in stress would, through the operation of the glucose/fatty acid cycle, reduce the likelihood of serious hypoglycaemia if 'fight or flight' took place. (Chapter 6; Section F.2.)

(e) Hypoglycaemia

There are several pathological conditions in which the blood glucose concentration is below normal (hypoglycaemia; Chapter 11; Section C) and, in consequence, the concentrations of fatty acids and ketone bodies are elevated. In these conditions, the operation of the cycle prevents a catastrophic fall in the blood glucose concentration.

(f) Diabetes mellitus

In the uncontrolled insulin-dependent diabetic patient, the elevated blood glucose concentration is usually associated with higher than normal concentrations of fatty acids and ketone bodies. Oxidation of fatty acids and ketone bodies by muscle and other tissues will reduce glucose utilization through operation of the cycle and this could be partially responsible for the hyperglycaemia. (Chapter 15; Section E.2.)

C. CONTROL OF THE RATE OF FATTY ACID RELEASE FROM ADIPOSE TISSUE

An essential feature of the glucose/fatty acid/ketone body cycle is the control of the reaction catalysed by triacylglycerol lipase in adipose tissue. This reaction, described in Chapter 6; Section D, is the flux-generating step in the release of fatty acids from adipose tissue and their oxidation in other tissues. The activity of the lipase is increased by the lipolytic hormones and decreased by the antilipolytic hormones (Table 8.5). Although a large number of hormones are known to modify the activity of this enzyme, care must be taken in extrapolation from effects in experimental animals to man since the response of the enzyme depends on the species. The enzyme from the rat has been studied in detail and provides the information given in this section but it may not apply to human adipose tissue (Hales *et al.*, 1978). There is evidence that fatty acid mobilization from adipose tissue is also under nervous control since denervation reduces the rate of mobilization whereas it is increased by stimulation of the sympathetic system. It is probable that nervous control is mediated via the release of local stores of noradrenaline. In addition to acute effects, there is also long-term endocrine control of lipolysis. In particular, glucocorticoids and thyroxine have a 'permissive' effect, that is they are necessary to maintain the response of adipose tissue to other lipolytic hormones.

1. Control of triacylglycerol lipase activity in adipose tissue

It is known that triacylglycerol lipase exists in two forms, an active phosphorylated form and an inactive (or less active) non-phosphorylated form. The proportion of

Table 8.5. Lipolytic and antilipolytic hormones affecting fatty acid mobilization from adipose tissue*

Lipolytic hormones		Antilipolytic hormones	
Rat	Man	Rat	Man
Adrenaline	Adrenaline	Insulin	Insulin
Noradrenaline	Noradrenaline	(Prostaglandin	(Prostanglandin
Glucagon	TSH	E$_1$ and E$_2$)	E$_1$ and E$_2$)
Growth hormone ⎫	Parathyroid		
Glucocorticoids ⎭	hormone		
Thyroxine			
TSH			
ACTH			
Vasoactive intestinal hormone (VIP)			

* Data from Hales *et al.* (1978).

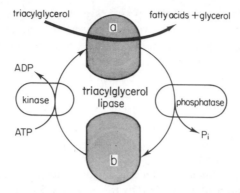

Figure 8.6 Interconversion cycle regulating triacylglycerol lipase activity. The non-phosphorylated form may not be totally inactive

enzyme in either form is controlled by the activities of the interconverting enzymes, a protein kinase and a protein phosphatase (Figure 8.6). Technical difficulties have prevented purification of either the lipase or the interconverting enzymes from adipose tissue but it is known that the protein kinase is activated by cyclic AMP.[3] Since adrenaline, noradrenaline, and glucagon increase the concentration of cyclic AMP in adipose tissue, they probably activate the protein kinase. Indeed, there is very good evidence for adrenaline activation of protein kinase in adipose tissue (Soderling *et al.*, 1975; Steinberg, 1976). The proposed mechanism of activation of the lipase is given in Figure 8.7 and has similarities to the control of glycogen degradation described in Chapter 7. The hormones bind to specific receptors on the outer surface of the adipose tissue cell and either activate (lipolytic hormones) or inhibit (antilipolytic hormones) adenylate cyclase thereby changing the intracellular concentration of cyclic AMP. The most important antilipolytic hormone, insulin, in addition to lowering the cyclic AMP concentration on adipose tissue, may inhibit the activation of the lipase by another mechanism, possibly involving changes in Ca^{2+} concentration (Hales *et al.*, 1978).

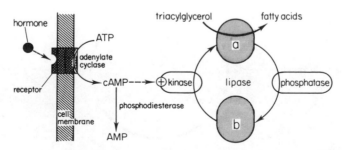

Figure 8.7 Mechanism by which hormones can affect the activity of triacylglycerol lipase

Two antilipolytic compounds which are not hormones and may be of physiological importance, are 3-hydroxybutyrate and adenosine. The inhibitory effect of 3-hydroxybutyrate on lipolysis and its effect on insulin sensitivity, which provide a feedback mechanism to prevent excessive rates of fatty acid mobilization, have already been described in Section B.2.b. Similarly adenosine inhibits lipolysis *per se* and increases the sensitivity of adipose tissue to the action of insulin (Green and Newsholme, 1979). Since insulin is always present in the bloodstream (even in diabetic patients) it is possible that the quantitatively important effect of these compounds in inhibiting lipolysis is the modification of insulin sensitivity rather than any direct effect. Adenosine exerts its effects by binding to a specific adenosine receptor on the external surface of the adipose tissue cell. It is produced from AMP by the action of 5'-nucleotidase, as described for muscle in Chapter 7; Section B.1.b but it is unclear whether the substrate is intra- or extracellular AMP (Arch and Newsholme, 1978). It is also unclear how adenosine modifies insulin sensitivity or indeed how the adenosine concentration is controlled in adipose tissue.

2. The triacylglycerol/fatty acid substrate cycle in the control of the lipolytic rate

The difference between the energy requirement of muscle between rest and intense sustained exercise may be almost 80-fold (Table 4.7). An even greater change in fatty acid mobilization may occur upon refeeding after 12–18 hr of starvation or as a result of taking breakfast immediately after 40 min jogging. In these situations, a period of fatty acid mobilization is immediately followed by fatty acid removal from the blood for esterification. The changes in hormones affecting lipolysis are probably not sufficient to bring about such large changes in rates and direction of fatty acid metabolism unless sensitivity in control is increased by the operation of a substrate cycle. This is provided since lipolysis occurs simultaneously with esterification so that triacylglycerol is broken down

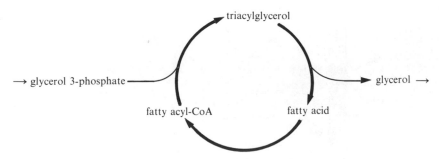

Figure 8.8   Substrate cycle to provide increased sensitivity to changes in concentration of hormones affecting lipolysis

Table 8.6. Effect of noradrenaline and glucagon on rates of triglyceride/fatty acid cycling in isolated rat adipocytes*

Condition of incubation	Cycling rate (μmol.h^{-1}.g^{-1} fresh wt. adipocyte)
Control	3.8
Noradrenaline	8.4
Control	3.4
Glucagon	11.0

* Data from Brooks *et al.* (1982). The control condition is 10 mM glucose plus insulin. Cycling rate is measured by the rate of release of glycerol and fatty acids from the adipose tissue. Cycling is calculated as follows: (3 × rate of glycerol release) − rate of fatty acid release.

to fatty acids, some of which are reactivated and re-esterified to form triacylglycerol (Figure 8.8). The resulting improvement in sensitivity (as explained in Chapter 7; Section A.2.b) would enable small changes in hormone concentration to produce large changes in the rate of fatty acid mobilization. Evidence has been obtained that catecholamines and glucagon, which increase the rate of lipolysis, also increase the rate of the triacylglycerol/fatty acid substrate cycle in rat adipose tissue incubated *in vitro* (Table 8.6; Brooks *et al.*, 1982; see also Chapter 14; Section C.1.b). It is predicted that during sleep the rate of this cycle would be markedly reduced, whereas anticipation of exercise or food would increase the levels of the catecholamines and increase the rate of this cycle. Subsequent changes in other hormones (e.g. insulin) could then have a marked effect on the rate of fatty acid mobilization.

D. REGULATION OF THE RATE OF KETOGENESIS

Ketone bodies are produced exclusively in the liver by the partial oxidation of fatty acids (Figure 6.12). Precise control of the rate of ketogenesis in the liver is of considerable physiological importance to the animal. The flux-generating step, at least for fatty acid oxidation in the liver, is likely to be lipolysis in adipose tissue, but it is also possible that the HMG-CoA cycle does not possess a reaction that approaches saturation with pathway-substrate so that triacylglycerol lipase in adipose tissue and the transport of fatty acids to the liver via the bloodstream may also be part of the physiological pathway of ketogenesis.

1. The branched physiological pathway and control of ketogenesis

If adipose tissue triacylglycerol lipase is the flux-generating step for ketogenesis in liver, it should follow that, whenever fatty acid mobilization from adipose tissue is

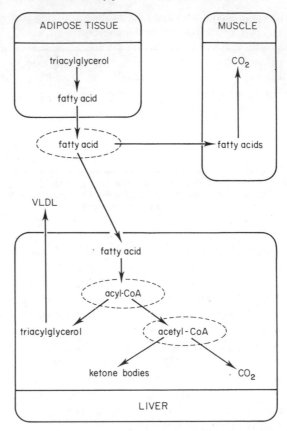

Figure 8.9   Branch points of regulatory significance in the pathway of ketone body
formation. The branch points are encircled

increased, this should increase the rate of ketogenesis in the liver. However, this is
not the case. For example, in sustained exercise, although there is a marked
increase in the rate of fatty acid mobilization, there is little or no increase in the
rate of ketogenesis (Felig and Wahren, 1975). This is a consequence of the
branched nature of the ketogenic pathway (Figure 8.9) and control of the
direction of flux at these branch points is therefore of considerable importance.

(a) *Control of ketogenesis by the blood supply to the liver*

When fatty acids are mobilized from adipose tissues to raise the plasma fatty acid
concentration (Figure 8.9) the rate of utilization by any tissue will depend upon
the metabolic activity of that tissue and the blood flow through it. In sustained
exercise, the blood flow to the active muscles is substantially increased whereas

Table 8.7. The ketone body concentrations in the blood of marathon runners after 90 minutes severe exercise

Time after exercise (hr)	Blood ketone body concentration (mM)	
	Normal diet*	Low carbohydrate diet*
Before exercise	0.1	0.3
0	0.2	0.8
1	0.2	1.9
2	0.2	2.2
3	0.2	2.2
4	0.2	2.5
5	0.3	2.2
8	0.3	2.8

* The diet refers to that eaten several days prior to exercise. The low carbohydrate diet means that the mobilization of fatty acids must be considerably greater than that in subjects on the normal diet. Data from Koeslag *et al.*, (1980).

that to the liver (and the intestine) is decreased. This ensures that most of the fatty acids mobilized from adipose tissue are directed to the muscle, so that little or no ketogenesis takes place in exercise. However, there is an increase in the rate of ketogenesis after exercise and the blood concentration of ketone bodies can be increased for several hours. This phenomenon is known as 'post-exercise ketosis' and was discovered by Forssner (1909). A detailed study of this phenomenon has been carried out by Koeslag *et al.* (1980). They showed that the highest 'level' of post-exercise ketosis occurred when the highest rate of fatty acid mobilization had occurred in the exercise period (i.e. running performed by trained marathon runners who had depleted their carbohydrate stores by diet; Table 8.7). The resumption of a normal flow of blood to the liver soon after cessation of exercise would provide the liver with a high fatty acid concentration (dependent upon the rate of mobilization from adipose tissue) which would result in ketone body formation. Post-exercise ketosis might be considered as part of a mechanism to ensure feedback inhibition of fatty acid release from adipose tissue after exercise and prevent the occurrence of dangerously high concentrations of fatty acids in the bloodstream.

(b) *Esterification and oxidation of fatty acids in the control of ketogenesis*

The second branch point occurs within the liver. Once the fatty acids are activated to fatty acyl-CoA, this metabolic intermediate can either be transported into the mitochondrion for oxidation or esterified with glycerol 3-phosphate to form triacylglycerol (see Figure 8.9). Since the liver can secrete the synthesized triacylglycerol into the bloodstream as VLDL and the capacity of the esterification process is large, a high proportion of fatty acids released from adipose tissue may be re-esterified in the liver. It is likely that both esterification and oxidation of fatty acids will be regulated in concert by external regulators including hormones (thus starvation increases the rate of ketogenesis but decreases that of esterification), but more information is currently available about the mechanism of the control of oxidation (McGarry and Foster, 1980; Zammit, 1981). Oxidation takes place in the mitochondria and transport of the acyl residues across the mitochondrial membrane is catalysed by carnitine palmitoyltransferases I and II (Chapter 6; Section D.1.c). The transferase residing on the outer surface of the inner mitochondrial membrane may be inhibited by insulin and is inhibited by malonyl-CoA which are external regulators of considerable physiological significance. It is suggested that when fatty acid synthesis, and therefore esterification, is occurring in the liver, the malonyl-CoA concentration is elevated and this causes a reduction in the rate of fatty acid oxidation via allosteric inhibition of CPT-I (McGarry and Foster, 1979). This could be part of the mechanism of concerted control over both esterification and oxidation alluded to above. It seems likely that insulin regulates CPT-I directly (that is, by the secondary messenger of insulin) in addition to increasing the concentration of malonyl-CoA through effects on the fatty acid synthesis pathway.

In summary, it should be noted that insulin can reduce the rate of ketone body formation in at least three ways;[4] reduction of the availability of fatty acids (antilipolytic effect), stimulation of esterification and inhibition of hepatic fatty acid oxidation.

(c) *The branch point at the tricarboxylic acid and HMG-CoA cycles*

The third branch point occurs in the mitochondria. The β-oxidation of fatty acids produces acetyl-CoA which can enter either the TCA cycle for complete oxidation or the HMG-CoA cycle for the formation of ketone bodies (Figure 8.9). This branch point has received much attention from biochemists in the recent past since it was considered that the formation of ketone bodies represented an 'overspill' due to the inability of the TCA cycle to satisfactorily oxidize all the acetyl-CoA produced from β-oxidation. However, knowledge of the role of ketone bodies in the whole animal indicates that their formation in the liver is physiologically important and the HMG-CoA cycle is part of a physiological pathway so that ketogenesis is not a mere 'waste disposal system' (Greville and Tubbs, 1968). There is no doubt that the TCA cycle is inhibited during ketogenesis but this is not surprising since the β-oxidation pathway produces reducing equivalents that generate ATP via the electron transfer chain (see Chaper 6; Section D.1.e). Hence the flux through the TCA cycle will need to be reduced. This consideration leads to the conclusion that the TCA cycle is inhibited as a result of ketosis rather than the inhibition being the cause of ketosis. Nonetheless, if the rate of the TCA cycle could be stimulated, this would divert acetyl-CoA residues to oxidation and reduce the rate of ketogenesis. Hence any condition that increases the intramitochondrial concentration of oxaloacetate

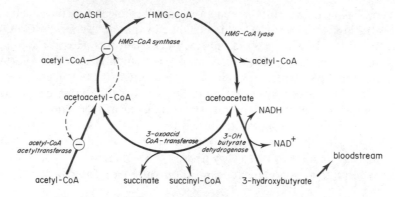

Figure 8.10   The role of 3-oxoacid CoA-transferase in providing a regulatory link for the control of ketogenesis

(an important external regulator of the cycle—see Chapter 7; Section B.3.a) and increases the ATP demand by the liver should reduce the rate of ketogenesis. For example, alanine administration to a ketotic animal reduces the extent of ketosis probably because it increases the concentration of oxaloacetate and increases the rate of ATP hydrolysis by stimulating gluconeogenesis (Alberti *et al.*, 1981).

2. Control of the HMG-CoA cycle

Detailed knowledge of the properties of the enzymes of the HMG-CoA cycle is lacking so that only a very tentative theory of control is proposed (Zammit, 1981). The conversion of acetyl-CoA to acetoacetyl-CoA catalysed by acetyl-CoA acetyltransferase is the first reaction that is specific for formation of ketone bodies so that it would be a suitable candidate for a flux-generating step but there is no evidence for this. It is, however, considered that this reaction is non-equilibrium and that it is inhibited by its two products, acetoacetyl-CoA and CoA competitively with its substrate acetyl-CoA (Figure 8.10). The concentration of acetoacetyl-CoA may play an important role in regulation of ketone body formation since not only does it inhibit the transferase but it also inhibits the next enzyme in the pathway, HMG-CoA synthase. Consequently, the factors that control the concentration of acetoacetyl-CoA could play an important role in the regulation of the HMG-CoA cycle. At least one further enzyme catalyses a reaction involving acetoacetyl-CoA, this is 3-oxoacid CoA-transferase (Figure 8.10). In other tissues the latter enzyme is involved in the utilization of ketone bodies (see Chapter 6; Section D.2.b) and was originally thought to be absent from liver. It is, in fact, present in the livers of many different vertebrates but is *not* thought to function as part of a system for the net utilization of ketone bodies (Table 6.3). Rather it provides a substrate cycle between acetoacetyl-CoA and acetoacetate and in so doing it functions as a regulatory link between the concentrations of acetoacetate and acetoacetyl-CoA. Since the enzyme probably catalyses a near-equilibrium reaction, any changes in the acetoacetate concentration would be reflected in a change in that of acetoacetyl-CoA, which would provide a simple feedback link between acetoacetate and the important regulatory enzyme, acetyl-CoA acetyltransferase (Figure 8.10). Of importance is the fact that the concentration of acetoacetate is dependent upon the $NAD^+/NADH$ concentration ratio via the 3-hydroxybutyrate dehydrogenase reaction. Hence a decrease in this ratio (i.e. an increase in the reduced state of the liver) would favour ketogenesis (Figure 8.10). It is therefore interesting to note that, in general, the higher the blood ketone body concentration, probably reflecting a higher rate of synthesis, the higher is the blood 3-hydroxybutyrate/acetoacetate concentration ratio (see Table 8.8).

Table 8.8. Total ketone body concentration and 3-hydroxybutyrate/acetoacetate concentration ratio in blood of comatosed diabetic patients after treatment with insulin*

Time after insulin (min)	Total ketone concentration (mM)	Concentration ratio 3-hydroxybutyrate/acetoacetate
0	10.7	4.0
1	9.5	3.6
2	7.8	2.9
3	6.0	3.0
4	5.0	2.6
5	2.7	1.0
6	2.3	0.8

* Data from Page *et al.* (1974).

NOTES

1. The term cycle is used here to describe a control relationship between glucose and fatty acids and should not be confused with a metabolic cycle (e.g. TCA cycle) or a substrate cycle (Chapter 7; Section A.2.b). The glucose/fatty acid cycle was so named because of the reciprocal control relationship between the two fuels.
2. Unfortunately, glucose transport can only be studied in intact systems so that the properties of this system which may be relevant to control are not known.
3. The protein kinase in adipose tissue is likely to be very similar, if not identical, to the cyclic AMP-dependent protein kinase in muscle. The role of this kinase in the regulation of phosphorylase activity was discussed in Chapter 7; Section C.1.b.
4. The involvement of insulin in the control of ketogenesis at perhaps all branch points in the physiological pathway is not surprising since a major problem in insulin-dependent diabetic patients is the rapid increase in the blood ketone body concentration that can occur if sufficient insulin is not administered. The experimental withdrawal of insulin from insulin-dependent diabetic patients produces a 100-fold increase in the blood ketone body concentration within 12 hr (Table 15.3).

CHAPTER 9

METABOLISM IN EXERCISE

Exercise is one situation in which knowledge of biochemical control mechanisms provides an explanation of much of the well-established physiology. It is, for example, possible to attempt to answer such questions as: what are the metabolic differences between sprinters and marathon runners or between elite (Olympic champion) and non-elite athletes, and how it is possible that some men can run 100 km whereas others have difficulty running 1 km? The discussion of the metabolism of exercise is important not only because it provides some physiological reference to the contents of previous chapters but because it has considerable implications in medicine. For example, there are a number of conditions characterized by fatigue upon mild exercise (e.g. ageing, post-surgery recovery, peripheral vascular disease) which markedly reduce the enjoyment of life. It is possible that knowledge of the metabolic basis of fatigue could be used to develop training regimes, diets or drugs to help these patients. In addition, the enormous increase in the number of people throughout the world who participate in sporting activities raises the question of the health benefit which can only be considered with a knowledge of the metabolic changes that occur during exercise. The benefits of exercise in overcoming the dangerous effects of stress, for diabetic patients, in prevention of obesity and atherosclerosis are discussed in other chapters (Chapter 6; Section F.2.c, Chapter 15; Section F.2, Chapter 19; Section C.3.c, Chapter 20; Section G.1.b respectively).

Since the metabolism of exercise is a large subject, the treatment is here limited to two extremes, sprinting and endurance (marathon) running. Wherever possible the treatment will be quantitative (for further practical and theoretical information see Newsholme and Leech, 1983).

A. SPRINTING

The energy expenditure in the 100 m sprint (that is, exercise to exhaustion for just over 10 sec) is approximately 200 kJ.min^{-1} as measured by the staircase method of Margaria et al. (1966).[1] It is common knowledge that such an activity can only be maintained for a very short period, whereas exercise at a lower power output can be performed for much longer. To provide a metabolic explanation of exhaustion in both sprinting and endurance exercise it is necessary to consider in some detail muscle fibre types, the fuels utilized by different fibres and the over-all metabolic changes consequent upon exercise.

357

1. Fibre composition in muscle of sprinters

An account of the division of human muscle fibres into types I, IIA, and IIB on the basis of biochemical characteristics of biopsy samples has been given in Chapter 5 (Section F.1.a; Note 8). Sprinters are understandably loath to lose even a small sample of their muscle and very few elite sprinters have so far come forward to provide muscle for such measurements. However, it has been shown that sprinters have a high proportion ($>70\%$) of type IIB (fast twitch) fibres, whereas long distance runners have a high proportion ($>70\%$) of type I fibres (Table 9.1). It is possible that, for world-class sprinters, the percentage of type II fibres is even higher. A similar trend is seen in domestic animals. Horses which are bred for sprinting have a very high proportion of type II fibres. The American Racing Quarterhorse is a breed developed to race over 400 m distances and has 93% type II fibres (Snow and Guy, 1980). On the other hand, the heavy hunters, which are known for their stamina rather than speed, have a relatively high percentage of type I fibres (Table 9.1). In another interesting study, D. H. Snow has shown that muscles of the greyhound, which have been bred for several hundred years for performance over a short distance, contain a very high percentage of type II fibres (in some muscles the percentage is greater than 95%) whereas in the mongrel it is only 69%.

In addition to the metabolic difference distinguishing the two types of fibre (Chapter 5; Section F.1) there are important physiological differences. First, type II fibres are innervated by large motor neurones which transmit impulses very

Table 9.1. Fibre composition* of muscle from horses, dogs and man

	Percentage fibre composition	
	Type I	Type II
Horse		
Quarterhorse	7	93
Thoroughbred	12	88
Heavy hunter	31	69
Dog		
Greyhound	3	97
Mongrel	31	69
Man		
Elite distance runners	79	21
Middle distance runners	62	38
Sprinters	24	76
Untrained	53	47

Data from Guy and Snow (1977) for dogs; Snow and Guy (1980) for horses; Costill *et al.* (1976), and Fink *et al.* (1977) for human subjects.

*The fibre composition of the middle gluteal muscle of horses, dogs, and human subjects was examined using the ATPase staining techniques described in Chapter 5; Note 8.

Table 9.2. Some maximum enzyme activities in fibres from human quadriceps muscle*

Enzyme	Enzyme activities in muscles of man (μmol.min^{-1}.g^{-1} at 37 °C)		
	Whole muscle	Type I fibre	Type II fibre
Myofibrillar ATPase	53	29	86
Creatine kinase	—	5640	7140
Phosphofructokinase	20†	15	26
Hexokinase	1.0	—	—
Succinate dehydrogenase	9.6	11.5	8.0
2-oxoglutarate dehydrogenase	2.0	—	—
Cytochrome oxidase	38	—	—

* Data from Rennie and Edwards (1981).
† This activity is considerably lower than that from Table 4.4.

rapidly (rapid conduction nerves) and the transmission of the action potential along the muscle membrane of the type II fibre is also rapid. Secondly, the innervation of type II fibres is such that the simultaneous contraction and relaxation of all, or most, of the fibres in a muscle is readily achieved. Thirdly, the interaction between the actin and myosin filaments (forming the cross-bridges) in the myofibrils of type II fibres is such that the time for the completion of the contraction-relaxation cycle is short. These physiological characteristics provide the speed necessary for the sprinter; the high power output is achieved by the possession of a large number of fibres. Furthermore, although the actual number of fibres in any one muscle may be fixed at an early stage of life, the size of the individual fibres can be increased by training, especially weight training. The latter is usually considered essential in the training of sprinters (see Shephard, 1978).

2. Fuels for the sprinter

The characteristics of type IIB fibres (Table 5.5) include a poor blood supply,* little myoglobin, few mitochondria, very low activities of the enzymes involved in aerobic metabolism but high activities of glycolytic enzymes and creatine kinase (Table 9.2). The poor blood supply to these fibres means that almost all of the energy for exercise must be obtained from the breakdown of endogenous fuels that can produce ATP under anaerobic conditions; these are phosphocreatine and glycogen (see Chapter 5; Section F.1.b). The contents of these fuels in human muscle at rest and after sprinting are given in Table 9.3.

* The mosaic pattern of fibre distribution in normal human muscle means that some type IIB fibres are surrounded by type I fibres, so that the blood supply to these latter fibres provides blood for type IIB fibres. Hence the above comments may apply mainly to elite sprinters who possess a high percentage of type IIB fibres.

Table 9.3. Contents of some important intermediates in muscle after exhaustion* due to sprinting

| | Contents (μmol.g^{-1} fresh muscle) | | |
| | | After exercise | |
Compound	Rest	15 sec	30 min
Glycogen	88	58	70
Glucose 6-phosphate	0.35	2.6	0.84
Lactate	1.1	30.5	6.5
Glycerol phosphate	0.27	2.23	0.74
Phosphocreatine	17.0	3.7	18.8
ATP	4.6	3.4	4.0
ADP	0.95	1.00	1.00
AMP	0.105	0.103	0.100
IMP	<0.1	0.9	—
P$_i$	9.7	22.0	—
Total adenine nucleotide	5.7	4.5	5.1
pH	7.1	6.3	7.0

* Total exhaustion was produced by three periods of sprinting for one minute separated by rest for only 1 min. Data from Sahlin *et al.* (1978), Hermansen, (1981), Hermansen *et al.* (1983).

There is evidence from the production of lactate during sprinting that only phosphocreatine is used during the first 4 sec after which anaerobic glycolysis provides most if not all of the energy (Margaria, 1976). Support for this is obtained from calculations based on knowledge of the metabolism in human muscle. From the assumption that maximum activity of the glycolytic enzyme, 6-phosphofructokinase, provides a quantitative index of the maximum anaerobic glycolytic flux, it can be calculated[2] that the maximum rate of ATP utilization in human muscle is about 3 μmol.sec^{-1}.g^{-1}. The concentration of phosphocreatine in human muscle is about 17 μmol.g^{-1} (this appears to be the case even for sprinters) and in severe short-lasting exhaustive exercise about 13 μmol of phosphocreatine is actually broken down (Table 9.3), so this would provide energy for about 4 sec, a period identical to that calculated by Margaria. The use of phosphocreatine as the initial source of energy (sometimes known as the 'alactate' period of energy production) is consistent with the near-equilibrium nature of the creatine kinase reaction, since the reaction would respond immediately to the increase in concentrations of ADP and protons (see Section A.3). Furthermore, the use of phosphocreatine during the initial period would give time for metabolic control to increase the flux through the non-equilibrium reactions catalysed by phosphorylase, 6-phosphofructokinase and pyruvate kinase. In this way glycolysis will take over ATP production smoothly from phosphocreatine.

The maximum rate of glycolysis is sufficient to provide energy for sprinting in man for a short period of time. How long is this period? The simple answer might be until the muscle glycogen stores are depleted. (The maximum rate of glycolysis-from-glucose is too low to be quantitatively important in energy provision for sprinters—see Chapter 5; Section F.1.b.) Since 1 g of muscle contains about 80 μmol of glycogen, this could produce a maximum of 240 μmol of ATP which should provide sufficient ATP for 80 seconds of sprinting. However, maximum sprinting cannot be maintained for much longer than 20 sec; the time for 400 m sprint is greater than four-fold that for the 100 m sprint. (The world records for the 100, 200, and 400 m races in 1982 were as follows: 9.95, 19.72 and 43.86 seconds, respectively.) In laboratory experiments sprinting to exhaustion degrades less than half of the muscle glycogen (Table 9.3) showing that fatigue is not due to the limited glycogen store in the muscle. The cause of fatigue is considered in Section A.4.

Other calculations in man have suggested that about 17% of the energy required for the 100 m sprint is obtained from aerobic metabolism (Gollnick and Hermansen, 1973). Estimates of the maximum capacity of the TCA cycle in muscle that contains almost all type IIB fibres (e.g. domestic fowl pectoral muscle) suggest that the capacity of the TCA cycle is very low and could not provide more than 1% of that from anaerobic glycolysis.[3]

Although the amount of ATP produced from aerobic metabolism in the elite sprinter will probably be insignificant for the 100 m race, the contribution may be more substantial in the 400 m race.

3. Control of rate of utilization of phosphocreatine and glycogen

(a) *Phosphocreatine utilization*

The stimulation of the myofibrillar ATPase by Ca^{2+}, which initiates the process of contraction, leads to a decrease in the concentration of ATP and an increase in those of ADP, phosphate and protons (see Table 9.3) as follows:

$$ATP^{4-} + H_2O \longrightarrow ADP^{3-} + P_i^{2-} + H^+$$

During the first seconds of the sprint such changes will favour ATP synthesis from phosphocreatine by displacing the creatine kinase reaction as follows:

$$H^+ + ADP^{3-} + \text{phosphocreatine}^{2-} \longrightarrow ATP^{4-} + \text{creatine}$$

The near-equilibrium nature of this reaction, in which, the rates of the forward and reverse components are very large, renders it sensitive to small changes in reactant concentration (see Chapter 2; Section C.2.b) and permits the regeneration of ATP to occur very rapidly.

(b) *Glycogen utilization*

A major metabolic problem in the control of glycolysis in sprinting is that it must increase from the very low rate, which satisfies the energy requirement of the muscle at rest on the starting blocks, to the very high rate necessary during the race. This increase is at least 1000-fold.* If this massive increase in flux through, for example, the 6-phosphofructokinase reaction, was brought about solely by known metabolic regulators of this enzyme, the concentration changes would have to be several hundred fold (assuming that the only means for increasing sensitivity was a co-operative response between regulator and enzyme activity— see Chapter 7; Section A.2).

The largest change in the concentration of known regulators of glycolysis (Chapter 7; Section B.1.a) occurring in muscle during sprinting is that of phosphocreatine which decreases by about four-fold (Table 9.3). Consequently, for these regulators to be involved in the control of the flux through this reaction, the regulatory mechanism must be very sensitive. It is suggested that this is achieved by co-operativity and the operation of a substrate cycle between fructose 6-phosphate and fructose bisphosphate, catalysed by the simultaneous activities of 6-phosphofructokinase, and fructose-bisphosphatase (see Chapter 7; Section C.3). The maximum activities of these two enzymes in human and rat muscle are given in Table 9.4. The hypothesis for control suggests the following sequence of events in the athlete's muscle (see Figure 9.1). When the athlete is resting in the dressing room some time before the event, the glycolytic flux will be low and the activities of both 6-phosphofructokinase and fructose-bisphosphatase will be very low, so that the ratio, cycling rate/flux, is very low. As the time for the race approaches and particularly when he is on the blocks waiting for the gun, anticipation of the race and competition would be expected to increase the blood concentrations of the catecholamines[4] which, it is predicted, will stimulate the activities of both enzymes. Hence while on the blocks the flux will be low but the cycling rate and hence the ratio, cycling rate/flux, will be very high (Figure 9.1). The initiation of the sprint (i.e. the increase in myofibrillar ATPase activity in the muscle) results in decreases in the concentrations of ATP and phosphocreatine and increases in those of AMP, phosphate and possibly fructose bisphosphate (Table 9.3). Since the ratio, cycling rate/flux is high, small changes in these concentrations can now produce a very large increase in the rate of fructose 6-phosphate phosphorylation. During the latter part of the sprint, when the maximum power output is achieved, the activity of fructose-bisphosphatase may be inhibited (although it does not have to be to provide the sensitivity) so that the cycling rate will be reduced. However, when the race is over, the activity of fructose-bisphosphatase and the cycling rate are probably increased again to improve sensitivity to the opposite changes in the concentrations of metabolic regulators. It is possible that the cycling ratios may be elevated above normal for several hours after cessation of exercise so that the rate of ATP utilization could be greater than that at rest in this recovery period. This could contribute to the phenomenon known as oxygen debt (see Newsholme 1978a, and Chapter 19; Section C.3.b).

* The rate of glycolysis in resting human muscle is approximately $0.05 \, \mu\text{mol.min}^{-1}.\text{g}^{-1}$ fresh wt. but during sprinting the maximum rate increases to almost $60 \, \mu\text{mol.min}^{-1}.\text{g}^{-1}$ fresh wt.

The sensitivity of the other non-equilibrium reactions, those catalysed by phosphorylase and pyruvate kinase, may also be improved by substrate cycles. The activity of some of the enzymes that catalyse the 'reverse' reactions in human muscle are given in Table 9.4. The three possible substrate cycles in the conversion of glycogen to lactate are shown in Figure 9.2. In contrast, the near-equilibrium reactions of glycolysis are very sensitive to changes in concentrations of substrates, cosubstrates and coproducts, so that no special mechanisms are required for improving the sensitivity of these reactions.

Table 9.4. Activities of substrate cycling enzymes in glycolysis of muscle of man and the rat*

	Enzyme activities (μmol.min^{-1}.g^{-1} wet weight at 25 °C)				
	6-Phospho-fructokinase	Fructose-bisphosphatase	Pyruvate kinase	Phosphoenol-pyruvate carboxykinase	Pyruvate carboxylase
Man, quadriceps	28	1.0	—	0.5	0.6†
Rat, quadriceps	35	0.2	800	0.3	—

* Data are taken from Newsholme (1978a), Newsholme and Williams (1978), Zammit et al. (1978).
† Based on a limited number of muscle samples analysed in the laboratory of one of the authors (E.A.N.).

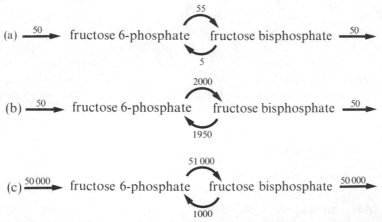

Figure 9.1 Substrate cycling at the 6-phosphofructokinase (upper arrow) and fructose-bisphosphatase reactions (lower arrow) immediately before and during a sprint race. The rates of cycling indicated are speculative. In (b) the rate of cycling is raised by the stress hormones, from which it can be seen that a 25½-fold increase in the rate of 6-phosphofructokinose and a 50% decrease in that of fructose bisphosphatase causes a 1000-fold increase in flux
(a) Resting before race.
(b) On starting blocks.
(c) After 6 sec sprint

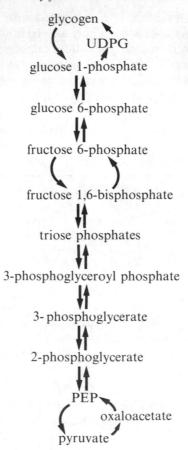

Figure 9.2   Pathway of glycolysis from glycogen, showing sites at which substrate-cycling occurs

4. Metabolic basis of fatigue in sprinting

Muscle fatigue can be defined as an inability of muscle to maintain a given power output (Edwards, 1981). It is an everyday experience that exercise causes exhaustion and that some individuals become exhausted more readily than others. The molecular basis for this exhaustion has taxed the minds of exercise biochemists and physiologists for many years. Furthermore, such considerations are of importance to the clinician who is interested in exhaustion experienced by patients and to the industrial physician interested in fatigue at work. One of the problems in investigating fatigue is its complexity, since both psychological and biochemical factors may be involved. There is evidence that, although the power output may be the same during a given exercise, the effort perceived by a subject may be different under different conditions. For example, the effort perceived by older subjects is greater than that perceived by the young and in response to this an older person will reduce the rate at which exercise is performed (i.e. reduce the power output) so that the perceived effort is about the same. Thus the elderly climb stairs more slowly than the young, although the same effort is perceived by both groups. The physiological explanation for this is probably a phenomenon known as central fatigue which is discussed below.

(a) *Peripheral fatigue*

Peripheral fatigue indicates that the fatigue is due to a failure or limitation of one or more processes within the motor unit (i.e. the motor neurones, the peripheral nerves, the neuromuscular junction or the muscle fibres; Edwards, 1981). Although each of these processes may be limiting under different conditions, the present discussion is limited to fatigue due to metabolic processes within the muscle. Fatigue can be demonstrated in isolated muscle preparations when there is no impairment of the function of motor nerves, neuromuscular junction or propagation of the action potential along the muscle and so is likely to be a consequence of muscle metabolism. In sprinting the accumulation of protons and the consequent decrease in pH within the muscle is probably responsible but the biochemical mechanisms by which protons reduce the power output is not known.

The time taken for fatigue to set in during sprinting is measured in seconds rather than minutes. A similar time course is observed for exhaustion in isometric contractions, that is when tension is maintained without the muscle shortening, for example, when holding a heavy suitcase. In both these types of activity, energy is obtained anaerobically via phosphocreatine degradation and glycogen conversion to lactate. (Isometric contraction depends upon anaerobic metabolism because a force of contraction greater than about 30% of the maximum causes compression of the blood vessels in the muscle and hence the flow of blood is markedly reduced, i.e. the muscle becomes ischaemic (Edwards *et al.*, 1972). In both sprinting and severe isometric contraction, the concentration of lactate in the muscle increases and may reach 30 mM (Table 9.3). Since glycolysis results in the production of protons as well as lactate (Chapter 13; Section D.1) there is an accumulation of protons during such exercise. The intramuscular pH can decrease from a resting value of 7.0 to about 6.5 to 6.4 and the blood pH changes from 7.4 to about 6.8 to 6.9 after laboratory sprinting[5] (Sahlin, 1978). These changes in pH occur despite the fact that some of the protons that are produced will be buffered both intra- and extracellularly by various buffers including intracellular protein and the carbonic acid/hydrogencarbonate system (Chapter 13; Section B.1). It is widely assumed that the lactate ion is responsible for fatigue but this is not the case; muscle continues to contract with a high power output in the presence of a high concentration of lactate, provided the pH remains near 7.0. It is the intracellular accumulation of protons that is responsible for the fatigue, since it can be shown that a decrease in pH below 7.0 reduces the power output in an isolated skeletal muscle preparation (Hermansen, 1979). How the increase in proton concentration leads to fatigue is unclear but there are a number of different suggestions. A decrease in pH has been shown to increase the Ca^{2+}-binding capacity of the sarcoplasmic reticulum so that less Ca^{2+} would be released into the sarcoplasm upon electrical stimulation of the muscle, thus reducing stimulation of the contractile process (Nakamura and Schwartz, 1972). Alternatively, or in addition, a fall in pH could interfere with the interaction between myosin and actin (i.e. reduce myofibrillar ATPase activity). There is evidence for a direct effect of pH on the contractile apparatus since, with isolated muscle fibres that have had their plasma membrane removed ('skinned fibres'), the maximum tension that could be developed upon addition of Ca^{2+} was lower at pH 6.5 than that at pH 7.0 (Donaldson *et al.*, 1978).

A different explanation is based on the properties of the glycolytic enzyme, 6-phosphofructokinase. The inhibition of activity of this enzyme by ATP *in vitro*, a crucial property for its regulation *in vivo*, is markedly increased by a decrease in pH below 7.0. Hence it is proposed that the *in vivo* fall in pH causes inhibition of 6-phosphofructokinase and this results in a severe decrease in glycolytic flux and a marked reduction in the rate of formation of ATP. Consequently, it has been suggested that fatigue is due to a depletion of ATP in the muscle so that the myofibrillar ATPase cannot function due to lack of substrate. There are two problems with this explanation: first, the ATP concentration in fatigued muscle is only decreased to about 3 mM (see Table 9.3), which is still sufficient to saturate myofibrillar ATPase. (The K_m for myofibrillar ATPase for ATP is about 0.1 mM.) Secondly, other ATP-requiring processes (e.g. ion transport) continue to function in a fatigued muscle. The authors speculate that a mechanism exists for an allosteric control of myofibrillar ATPase activity that depends upon changes in the ATP/ADP concentration ratio (in a similar manner to the control of 6-phosphofructokinase activity—see Chapter 7; Section B.1.a). In this way, small changes in the ATP concentration would lead to a reduction in contractile activity and hence cause fatigue. Such a mechanism would serve as a safety device to prevent total exhaustion of ATP and consequent cell death.

It seems unlikely, however, that in short sprints (e.g. 100 m) metabolic fatigue is a factor limiting performance. The major limitations here are probably the quantity of muscle, the co-ordination of the contraction-relaxation cycle between the fibres within a given muscle and a sufficient rate of substrate cycling to provide adequate sensitivity for the massive increase in the rate of glycolysis during the acceleration period. However, in the 200 m, and especially in the 400 m sprint, the process of glycolysis will proceed for a sufficient time to cause a serious decrease in pH within the muscle. This leads to the prediction that the buffering capacity within muscles of elite 400 m sprinters will be especially high.

(b) *Central fatigue*

Central fatigue is assumed to occur proximal to the motor neurones, that is within the brain. One way of distinguishing between central and peripheral fatigue is to fatigue a muscle group through maximal voluntary contractions and then, when fatigue occurs, stimulate the muscles electrically. If the fatigued muscles respond, this indicates central fatigue and this has, in fact, been demonstrated (Asmussen, 1979; Bigland-Ritchie, 1981). It is likely that one cause of central fatigue is due to central inhibition produced by sensory signals from the fatigued muscle. It would be of great interest to know what factors in the fatigued muscle were used by the sensory system to 'detect' fatigue. Not surprisingly, other factors (e.g. mood) can modify the response of the central nervous system to sensory input and result in more or less central inhibition. The possible levels at which fatigue could intervene in the control for voluntary muscle contraction are indicated in Figure 9.3.

5. Training for sprinting

The above discussions would suggest three areas of training that may reduce metabolic and physiological limitations. First, weight-training, which is known to increase the bulk of muscle (i.e. the size of the fibre), should also increase the maximum capacity of both glycolysis and the activity of creatine kinase.

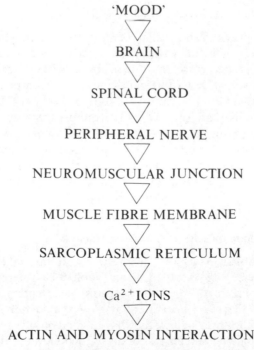

'MOOD'

▽

BRAIN

▽

SPINAL CORD

▽

PERIPHERAL NERVE

▽

NEUROMUSCULAR JUNCTION

▽

MUSCLE FIBRE MEMBRANE

▽

SARCOPLASMIC RETICULUM

▽

Ca^{2+} IONS

▽

ACTIN AND MYOSIN INTERACTION

Figure 9.3 Control sequence for activation of voluntary muscle. Fatigue could operate at any step (based on Edwards, 1981)

Secondly, sprinting from a slow running start may develop the coordination of the contraction-relaxation cycle in all the muscle fibres in a given muscle. Thirdly, start-training may increase the capacity of the substrate cycles in their ability to respond to the catecholamines, so that high rates of cycling and hence a high sensitivity in metabolic control will be possible. How far training can improve or develop these desired characteristics remains to be investigated.

One area of controversy is how far different training regimes can change the proportion of the two main fibre types in muscle. It has generally been accepted that the fibre pattern was genetically determined and could not be changed. Elite sprinters, therefore, had to be born with a high percentage of type II fibres. However, Jansson *et al.* (1978) have shown that anaerobic training can convert some type I fibres to type II whereas aerobic training could produce the opposite changes. In addition, in studies on one long distance skier, Eriksson (1981; 1982) has shown that six weeks immobilization of one leg (due to a knee injury) reduced the proportion of type I fibres from 81 % to 57 % whereas training after recovery from injury reversed the change. This problem is perhaps complicated by the finding that there may be more than three types of fibres (I, IIA, and IIB), possibly up to six types and a pool of undifferentiated fibres may be present in some muscles that can be converted into types I or II by training.

B. MARATHON RUNNING

Marathon running provides an excellent contrast to sprinting. In this endurance event, the 42.2 km (26 miles 385 yards) are completed by elite runners in a little over 2 hr at an average energy expenditure of about 84 kJ each minute[6] or a total expenditure of almost 12 000 kJ. Many non-elite runners complete the distance in 150–170 min at a similar total energy expenditure. P. Milvy (1977) has stated that 'physical activity of this quantity and quality requires an energy expenditure that is both intense and prolonged, and consequently an enormous stress is placed upon the body and its organ systems'. The fuel requirements and the control of the rates of utilization of these fuels have been mentioned in several parts of this book but they will be brought together in this chapter to indicate the problems that face the elite and non-elite endurance runner.

1. Fibre composition in muscles of marathon runners

Analysis of muscle obtained by biopsy from elite marathon runners indicates that they have a high percentage of type I fibres (Table 9.1). The physiological and biochemical characteristics of type I fibres are to some extent opposite to those of the type II fibres: they are innervated by slow conduction nerves and the contraction time of the fibres is slow; they have a good blood supply; they have a high content of myoglobin and possess many mitochondria so that the activities of the enzymes of the TCA cycle and β-oxidation of fatty acids are much higher than in type II fibres. It is usual for physiologists to state that type I fibres possess the property of fatigue resistance; the metabolic basis for this is given in Section B.4.

2. Fuels for the marathon runner

Unfortunately, most of the studies that have been carried out on marathon runners have been physiological in nature and have thrown light only indirectly on metabolic aspects. Other conclusions are based on metabolic studies of muscle of non-athletes. Extrapolations from these studies might not always be valid since the ability to run the marathon in approximately 140 min is restricted to a very few individuals.

It has been established, however, that anaerobic metabolism does not provide any energy for the marathon runner (Gollnick and Hermansen, 1973). From studies on sustained exercise in non-marathon runners, it is likely that the major fuels for muscle during the marathon are glucose plus fatty acids obtained from the bloodstream and glycogen plus triacylglycerol obtained from within the muscle. It is important that these fuels are not used indiscriminately but that the rate of utilization of any one fuel should be controlled not only in relation to the energy demand by the muscle but also in relation to the rates of utilization of the other fuels.

(a) *Hepatic glycogen and blood glucose*

Any glucose taken by the muscles from the blood in the post-absorptive state must be replaced by degradation of liver glycogen or by gluconeogenesis. Experiments with both man and other animals demonstrate that liver glycogen is depleted during sustained exercise (Hultman, 1978). The most extensive studies on glucose metabolism in exercising man have been carried out by Felig and Wahren (1975). They measured arterio-venous differences across both the liver and leg muscles of volunteer Stockholm firemen from which the rate of glucose release from the liver and the rate of glucose uptake by the muscles could be calculated.[7] The rate of glucose uptake by the muscles, approximately 0.2 g.min^{-1}, was similar to the rate of glucose release by the liver, 0.25 g.min^{-1}, demonstrating the integrated control of hepatic release and muscle uptake. However, this rate of glucose uptake accounted for only about 30 % of the oxygen uptake; more than 60 % being accounted for by the oxidation of fatty acids in the later stages of exercise (see Table 9.5 and below). It can be calculated that the elite marathon runner, if he relied solely on blood glucose for energy production, would utilize about 5 g of glucose each minute[8] (or about $1 \text{ } \mu\text{mol.min}^{-1}.\text{g}^{-1}$ muscle). Although the maximum activity of hexokinase has not been measured in elite marathon runners, it is about $1.0 \text{ } \mu\text{mol.min}^{-1}.\text{g}^{-1}$ in muscle of fit normal subjects (Table 5.4) which suggests that the marathon runner could support his mechanical activity using blood glucose generated in the liver. However, the total hepatic store of glucose is only 90 g which would provide the runner with fuel for

Table 9.5. Contribution of glucose, fatty acids and glycogen to oxygen consumption of leg muscles of man during mild prolonged exercise*

Period of exercise (min)	Percentage contribution to oxygen uptake		
	Blood-borne fuels		Muscle glycogen
	Glucose	Fatty acid	
40	27	37	36
90	41	37	22
180	36	50	14
240	30	62	8

* Data taken from Felig and Wahren (1975).

only 20 min (see Table 8.1) so that muscle glycogen and fatty acids from the blood become essential fuels for the marathon runner (see below).

b) *Muscle glycogen*

In sustained exercise the depletion of muscle glycogen is very gradual, in contrast to sprinting. In one experiment, volunteers were exercised on a bicycle ergometer at an intensity such that they became exhausted after about 100 min. Glycogen was assayed in biopsy samples of muscle taken during the exercise period; glycogen depletion occurred gradually over the entire exercise period. Of importance for subsequent discussion is the fact that exhaustion occurred when the muscle glycogen stores were depleted (Hermansen *et al.*, 1967; see Table 9.6). Similar studies showed that, in various individuals, the time taken to the point of fatigue was directly proportional to the initial (i.e. pre-exercise) level of glycogen in the muscle (Ahlborg *et al.*, 1967). Consequently it was argued that, if the

Table 9.6. Depletion of muscle glycogen during sustained exercise*

Duration of exercise (min)	Content of glycogen (μmol.g^{-1} fresh muscle)†	
	Untrained	Trained
0	94	100
20	39	55
40	22	39
60	11	14
80	0.6 (exhaustion)	11
90	—	0.16 (exhaustion)

* Data from Hermansen *et al.* (1967).
† Glycogen measured in quadriceps muscle after removal of biopsy sample.

concentration of glycogen in muscle could be increased prior to exercise, this could delay the point of exhaustion or provide a greater power output during the exercise period. This would be of obvious importance to the competitive long-distance runner. A diet was accordingly designed that could increase the glycogen level up to two-fold in the muscle of some subjects (Table 9.7). The dietary regime was as follows: six days prior to the competition, muscle glycogen levels were decreased by an exhaustive run; the subjects then ate a low carbohydrate diet for three days and once again ran to exhaustion to deplete the glycogen; for the final three days up to the competition the runners ate a high carbohydrate diet. This dietary regime (known as 'glycogen stripping' or the 'unloading–loading' regime) was probably first used by the British marathon runner Ron Hill in preparation for the 1969 European Championship in Athens: Hill won the marathon. Although there is no doubt that many marathon runners believe that it is important to elevate the glycogen levels prior to the race, the need for the three-day period on the low carbohydrate diet has been questioned. Since many marathon runners cover a very large mileage in training every week (100 miles or more), they have to consume a large quantity of carbohydrate to provide the energy needed. A short period on a low carbohydrate diet interferes with the training regime. Costill and Miller (1980) consider that a high carbohydrate diet for a few days before the race, without the previous low carbohydrate period, is sufficient for 'supercompensation'. Unfortunately, direct experimental evidence on this point or on the metabolic mechanism of 'supercompensation' is not yet available.

(c) Lipid–fuels

The plasma concentrations of fuels at rest and during sustained exercise are shown for dogs, horses, and man in Table 9.8. Although there is no doubt that the concentration of fatty acids in the blood is increased during sustained exercise the

Table 9.7. Effect of diet on muscle glycogen content and duration of exercise*

Diet and conditions	Muscle glycogen content before exercise (μmol.g^{-1})	Duration of exercise (min)
1. Normal mixed diet	97	116
2. Low carbohydrate diet for three days	36	57
3. High carbohydrate diet for three days	183	166

The regime is described in the text.
* Data are from Bergström *et al.*, 1967.

Table 9.8. Changes in plasma concentrations of fuels and hormones involved in fuel mobilization during sustained exercise in man, horses and dogs

Animal	Time of exercise (min)	Concentrations of fuels in blood or plasma (mM)				Concentrations of hormones in blood or plasma				
		Glucose	Lactate	Fatty acid	Glycerol	Cortisol ($nmol.l^{-1}$)	Adrenaline ($ng.cm^{-3}$)	Noradrenaline ($ng.cm^{-3}$)	Glucagon ($pg.cm^{-3}$)	Insulin ($\mu U.cm^{-3}$)
Human subjects*	0	4.5	1.1	0.66	0.04	—	0.1	0.4	80	14
	40	4.6	1.3	0.78	0.19	—	0.2	—	80	12
	180	3.5	1.4	1.57	0.39	—	0.7	1.5	200	7
	240	3.1	1.8	1.83	0.48	—	—	—	400	6
Horses†	0	4.4	0.7	0.29	0.14	186	—	—	17	36
	approx. 360	2.0	1.6	1.4	0.49	695	—	—	170	6
Dogs‡	0	6.1	1.6	0.6	—	—	—	—	—	—
	60	5.5	2.4	1.1	—	—	—	—	—	—
	120	5.5	1.6	1.8	—	—	—	—	—	—
	180	5.0	1.6	1.9	—	—	—	—	—	—
	240	5.0	1.9	2.1	—	—	—	—	—	—

* Human subjects were exercised on a bicycle ergometer at a workload of approximately 30 % of maximum. Data taken from Ahlborg *et al.* (1974), Felig and Wahren (1975), Galbo *et al.* (1975; 1979).

† Horses were ridden over a standard course in hilly country at average speed of $12.5\,km.h^{-1}$ (Hall *et al.*, 1981).

‡ The workload for the dogs was increased over the 4 hr period (Paul and Holmes, 1975).

extent of the increase is variable; the increase probably depends upon a variety of factors including previous dietary history, type of exercise (e.g. running or cycling), severity of exercise, duration of the exercise and fitness of the subjects. Since the concentration of fatty acids in the blood represents a balance between the rate of their mobilization by lipolysis in adipose tissue and their oxidation by muscle, the actual steady state concentration tells us little about rates of utilization. Furthermore, control of the rates of these two processes may be better integrated in trained subjects so that the increase in the steady state concentration is less than in untrained subjects (Rennie *et al.*, 1974). For all these reasons, the increase in the blood glycerol concentration (released from adipose tissue during lipolysis) probably provides a better index of the rate of fatty acid mobilization and is included in Table 9.8. In general, the concentrations of glucose and lactate remain fairly constant or decrease whereas, in all three animals, the plasma glycerol and fatty acid concentrations are progressively and markedly elevated during the course of the exercise. Only after 30–40 min of exercise in humans is a new steady state concentration of fatty acid reached in the blood. The increased concentration of fatty acids in the blood will be expected to increase the rate of oxidation of fatty acids by the muscle and this has been confirmed in experiments with human volunteers (Hagenfeldt, 1979).

It is still unclear how much of the energy requirement of the marathon runner can be provided by fatty acid oxidation. In the experiments on the Stockholm firemen, about 60 % of the oxygen consumption could be accounted for from fatty acid utilization by leg muscle (Table 9.5). Indirect evidence suggests that it may be similar for elite marathon and ultradistance runners. Elite ultradistance runners, who were running for 24 hr would be expected to deplete all their carbohydrate reserves after several hours (perhaps 12 hr) of running. The oxygen consumption (i.e. power output) of these runners decreased gradually during the run to

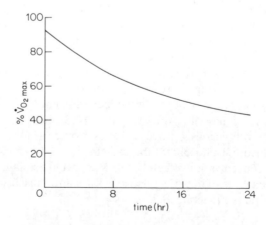

Figure 9.4 The change in aerobic power output with time during ultradistance running. Average of four athletes. Data from Davies and Thompson (1979)

'plateau' at about 50–60% of the maximum (Figure 9.4). At this point the respiratory exchange ratio (Chapter 5; Appendix 5.2) was about 0.7 indicating almost total dependence on a lipid fuel (Davies and Thompson, 1979). Further evidence that fatty acid oxidation alone cannot provide all the energy required for the elite or even non-elite marathon runner is given in note 9.

It is unlikely that the other lipid-fuels (blood triacylglycerol and ketone bodies, muscle triacylglycerol) are quantitatively important as fuels for the marathon runner. The activity of lipoprotein lipase in skeletal muscle is low; from the activities reported in Table 6.6, the maximum contribution to oxygen consumption in the marathon runner would be 10%. Similarly, although some of the endogenous muscle triacylglycerol may be utilized during the marathon, the maximum contribution to the overall rate of energy formation will be small. The blood concentration of ketone bodies is low during sustained exercise suggesting that they too are unimportant, a suggestion confirmed by measurement of arterio-venous differences in the Stockholm firemen (Felig and Wahren, 1975).

3. Regulation of the rates of fuel utilization during marathon running

As outlined in the previous section, blood glucose (from liver glycogen), muscle glycogen, and fatty acids (from adipose tissue triacylglycerol) undoubtedly all serve as fuels during a marathon race. The control mechanisms which operate have largely been discussed in Chapters 7 and 8 but a few additional points may be made in the context of endurance exercise.

(a) *Regulation of individual fuel utilization pathways*

(i) *Glucose from liver glycogen*   The flux-generating step for this process is at glycogen phosphorylase (see Chapter 7; Section C.1). This enzyme will be activated by the increase in concentrations of adrenaline, noradrenaline and glucagon and the decrease in concentration of insulin which occur in the plasma during prolonged exercise (Table 9.8). This hormonally-mediated mechanism may be supplemented by a direct effect of glucose, the concentration of which falls, if only slightly, during exercise. This decrease in glucose concentration would lead to a stimulation of hepatic glycogen phosphorylase and inhibition of glycogen synthase as described in Chapter 11; Section E. Since both these enzymes may be simultaneously active and thus provide a substrate cycle between glycogen and glucose 1-phosphate, the system may be very sensitive to small changes in the concentration of glucose. This could explain the remarkable similarity between rates of hepatic glucose release and muscle glucose uptake in the Stockholm firemen (Section B.2.a).

(ii) *Glucose from muscle glycogen*   Whether the increased rate of glycogenolysis which occurs in muscle during sustained exercise (Table 9.6) is due to activation of glycogen phosphorylase *b* directly by changes in the concentrations of glucose

6-phosphate, ATP, AMP and phosphate or to an increase in the proportion of phosphorylase in the *a* form (Chapter 7; Section C.1) is still unclear. The authors suggest that allosteric control of phosphorylase *b* will be more important.

(iii) *Fatty acid oxidation*   The activity of triacylglycerol lipase in adipose tissue, the flux-generating step for fatty acid oxidation in muscle, will be increased by the hormonal changes shown in Table 9.8. The increase in the blood concentration of fatty acids will lead to an increase in the activities of acyl-CoA synthetase, the carnitine palmitoyltransferases, and the enzymes of the β-oxidation pathway by internal regulation. However, it is likely that the two former enzymes will also be regulated by external factors as discussed in Chapter 7; Section D.2.

(b) *Integration of fuel utilization*

It is possible to calculate for how long each of the fuels used would support the elite marathon runner during the race (Table 8.1). Blood glucose alone would provide energy for about 4 min, hepatic glycogen for about 18 min, muscle glycogen for about 71 min* and triacylglycerol (via fatty acids) for 4018 min. This indicates that both carbohydrate and lipid fuels have limitations as fuels for the marathon runner. Although the rate of glucose oxidation can provide sufficient energy to support the power output of the marathon runner, there is not sufficient reserves for the duration of the race (at least 130 min). On the other hand, although there is an excess of the stored lipid fuel, its rate of oxidation appears unable to provide sufficient energy to maintain the power output.[9] Consequently, both glucose and fatty acids must be oxidized simultaneously to provide sufficient energy for the power output during the entire marathon run. Fortunately, once fatty acids are mobilized from the adipose tissue into the bloodstream they will be used together with glucose, since the raised concentration in the bloodstream will stimulate their oxidation. This, via the control mechanisms discussed in Chapter 8 (Section B.1), will prevent maximal rates of glucose oxidation so that glucose utilization and oxidation will be restricted to that needed to supplement the deficit in energy production that cannot be provided by fatty acids.[10]

4. Metabolic basis of fatigue in the marathon

In sprinting, one possible cause of fatigue is the accumulation of protons within the muscle which inhibit the major, if not the only, ATP-producing system under that condition, glycolysis. Although there is no accumulation of protons during endurance exercise, fatigue may also be caused by a marked decrease in the rate of glycolysis. From considerations outlined above, it seems that the marathon runner (and indeed other track athletes who depend upon aerobic metabolism) must oxidize both carbohydrate and fatty acids to provide energy at a sufficient rate to meet the demands of the muscle. Fatigue will occur when the carbohydrate

---

* This is based on glycogen concentration in normal man; in some individuals training and dietary manipulations can double this concentration (Table 9.7). However, all muscles are not involved in running—so that 71 min is an overestimate.

stores have been depleted, and the last store to be utilized is probably muscle glycogen.[11] When the muscle glycogen store is depleted, so that only fatty acid oxidation can provide energy, the rate of energy production will satisfy about 50% of the maximum power output (perhaps less in non-elite runners). This reduction in power output will be seen as fatigue.* These metabolic considerations lead to the view that the marathon runner should run at such a rate that his glycogen stores in the muscle are depleted just as the race is completed. If the race is completed with glycogen remaining in the muscle, the athlete could have run faster: if all the muscle glycogen was used prior to completion of the race, then the athlete would depend solely on fatty acid oxidation, and the power output would fall by perhaps 50% or more. Thus exhaustion can be delayed either by increasing the store of glycogen in the body—this can be achieved mainly in muscle (Section B.2.b) or by increasing the ability to oxidize fatty acids. Endurance training increases the capacity for fatty acid oxidation and this may be one of the benefits of the high mileage that is covered by marathon runners when training for their event (perhaps up to 200 miles each week). Finally, the studies on the Stockholm firemen have shown that in sustained exercise fatty acids are mobilised slowly so that high blood concentrations are not reached for perhaps 90 minutes. Consequently, more glucose will be used in the earlier part of the run than in the later part. Hence anything that raises the blood fatty acid concentration at the beginning of the race should be beneficial. This accounts for the current practice in many marathon runners of drinking strong black coffee about one hour before the run (Costill and Miller, 1980). The caffeine in the coffee stimulates fatty acid release from adipose tissue possibly due to its ability to inhibit cyclic AMP-phosphodiesterase in adipose tissue and hence raise the cyclic AMP concentration (see Chapter 8; Section C.1).

5. Training for the marathon

There are a number of mechanisms by which training can improve performance in long-distance running. However, the effects of training will differ according to whether an untrained subject is being trained or whether a trained athlete is being brought up to peak performance. Considerably more metabolic information is available about the former situation since experimental animals provide such information, but some speculative suggestions will be made concerning the athlete.

(a) *Effects of training on untrained subjects*

(i) *Physiological effects*   The large number of physiological effects of training can be classified into three subsections: effects on respiration; central circulatory effects and peripheral circulatory effects. The most important respiratory effect of training is the increase in the rate of maximum oxygen uptake by the subject ($\dot{V}_{O_2max}$). A period of two months training increases the $\dot{V}_{O_2max}$ by 15–20% whereas training for a year can increase it by 50–60%. The $\dot{V}_{O_2max}$ is a physiological measurement reflecting improved cardiac output and improved removal of oxygen

---

*Other more physiological factors could also be involved in producing fatigue, for example, dehydration.

by the muscle. It would be of interest to know the improvement in the capacity of the TCA cycle as indicated by changes in the activity of oxoglutarate dehydrogenase.

Training has effects on the heart; for example, the heart rate of a trained subject both at rest and for a given level of exercise is lower than in the untrained subject. The increased cardiac output in the trained subject during exercise is due to an increase in stroke volume so that more blood leaves the heart for each contraction. In addition, training increases the size of the heart (cardiac hypertrophy). Studies on endurance athletes demonstrate that a large proportion have cardiac hypertrophy (see Costill, 1979). For example, the heart of Paavo Nurmi, who was Olympic champion seven times, was three times the normal size.

There is also an important effect on the distribution of the blood after training. Somewhat surprisingly, less of the cardiac output is directed towards the muscles so that more blood goes to the liver and other organs. This improved blood flow to the liver may be particularly important in permitting hepatic gluconeogenesis during sustained exercise. The lower rate of blood supply to the muscle can occur since one of the most important effects of training is an increase in the capillary bed in the muscle. It has been shown that the number of capillaries can increase by 60% with endurance training. This means that there are a greater number of capillaries supplying a given fibre so that the diffusion distance for oxygen and fuels is markedly reduced. This may explain, at least in part, the ability of muscles from trained subjects to extract more oxygen and more fatty acids from the blood than those from untrained subjects.

(ii) *Metabolic adaptations* The physiological adaptations outlined above must occur in concert with the metabolic adaptations. One of the most important metabolic effects of training is the increase in capacity of the oxidative pathways. This is reflected by a large increase in the number of mitochondria and in the activities of the mitochondrial enzymes including those of the TCA cycle, electron transfer and the β-oxidation pathway* (see Table 9.9). The increase in the capacity for β-oxidation is particularly important in delaying the onset of fatigue since the greater the rate of fatty acid oxidation the less dependence there is upon carbohydrate oxidation. In addition, the concentration of myoglobin, the intracellular oxygen-binding protein that improves the rate of diffusion of oxygen into the muscles, is increased. As would be expected, these changes are found in the type I fibres so that the aerobic capacity and, in particular, the capacity to oxidize fatty acids is increased. Perhaps more unexpectedly, these changes also occur in the type II fibres. In other words, these fibres increase their aerobic capacity. In addition, the myofibrillar ATPase also changes as indicated by its altered sensitivity to pH; thus the activity can be detected at both pH 4.6 and 10.4 so that the fibres have become type IIA (see Chapter 5; Section F.1.a). *If* type II fibres *can* be converted into type I fibres by endurance training (see Section A.5) this would be another important effect of such training.

* Quantitative effects of training on the capacity of the aerobic system can only be established by measurement of oxoglutarate dehydrogenase activities; unfortunately such studies have not yet been carried out.

Table 9.9. Effect of training on activities of hexokinase, 6-phosphofructokinase, citrate synthase, and carnitine palmitoyltransferase in different fibres of rat muscles*

Fibre type	Condition of animal	Enzyme activities (μmol.min^{-1}.g^{-1} at 35 °C)			
		Hexokinase	6-Phospho-fructokinase	Citrate† synthase	Carnitine palmitoyl-transferase
Type I	untrained	1.6	20	23	0.6
	trained	2.4	24	41	1.2
Type IIA	untrained	1.5	72	35	0.7
	trained	4.1	58	70	1.2
Type IIB	untrained	0.6	96	10	0.1
	trained	0.7	88	18	0.2

* Data taken from Baldwin *et al.* (1973).

† Unfortunately, citrate synthase activities cannot be used to indicate quantitatively the maximum capacity of the TCA cycle.

(b) *Effects of training on elite athletes*

Information on this subject is difficult to obtain since elite athletes are unwilling to be subjected to physiological and biochemical investigations especially when they are training for peak performance so that discussion must be speculative. There seem to be at least three metabolic consequences of training for the elite athlete.

1. It is likely that the intensive training regimes followed by elite athletes will further increase the capacities of the key enzymes in aerobic metabolism in the mitochondria (e.g. carnitine palmitoyltransferase, oxoglutarate dehydrogenase). Even small increases (e.g. 5 %) in the capacities of these enzymes will be of considerable value in competitive athletics.
2. There is physiological evidence that elite endurance athletes can perform for prolonged periods at a very high percentage of their maximal aerobic capacity (Davies and Thompson, 1979). Indeed, Costill *et al.* (1971) have reported that Derek Clayton was capable of exercising at 86 % $\dot{V}_{O_2max}$ for long periods. This means that the elite athletes are able to utilize glucose at a high rate and that almost all of the pyruvate produced is converted to acetyl-CoA for complete oxidation by the cycle. In other words, elite distance runners are able to maintain a high glycolytic rate but not produce lactate, so that the two substrates for lactate dehydrogenase (i.e. cytosolic pyruvate and NADH) must be maintained at very low concentrations. The training of the elite athlete may ensure a high degree of integration of the control mechanisms that link the key enzymes of glycolysis (hexokinase, glycogen phosphorylase, 6-phosphofructokinase, pyruvate kinase) and pyruvate dehydrogenase as well as the mitochondrial shuttles for transferring hydrogen into the mitochondria (Chapter 5; Section D).

3. It is likely that the capacities and responsiveness of substrate cycles to hormones are increased by the intensive training programmes of elite athletes. Of particular importance for metabolic control is the ratio, cycling rate/flux so that the greater this ratio, the higher is the sensitivity; that is, a smaller stimulus is needed to provide a given flux (Chapter 7; Section A.2). For elite athletes during competition, the fluxes through the metabolic pathway are known to be very close to the maximum so that sensitive control mechanisms are essential.

(c) *Training to prevent sudden death from high fatty acid concentration*

Evidence is gradually accumulating of the beneficial effects of exercise in a number of pathological conditions including obesity, diabetes and coronary heart disease (see Chapters 19, 15 and 20; Sections C.3.c, F.2 and G.1.b, respectively). In this section, an attempt will be made to link the observations on training reported in this chapter with the dangers of high blood concentrations of fatty acids discussed in Chapter 6; Section F.2.c. The reader may remember that it was proposed that sudden death from cardiac arrhythmias and fibrillation could be caused by pathological concentrations of long-chain fatty acids (>2 mM), especially under hypoxic conditions. Since training increases the activities of the enzymes involved in fatty acid oxidation in muscle (especially carnitine palmitoyltransferase; Table 9.9) the muscle of trained subjects will be better equipped to oxidize fatty acids when they occur in the bloodstream. It is known that stress conditions increase the concentration of fatty acids in the blood (Table 6.7). If the stress is severe or several stressful conditions occur simultaneously or sequentially, the fatty acid concentration may be sufficient to increase the likelihood of thromboses and cardiac arrhythmias. Elevated fatty acid concentrations may be especially dangerous if maintained for prolonged periods. The presence of a high capacity for fatty acid oxidation in muscle should enable an elevated blood fatty acid concentration to be reduced more rapidly, even at rest. Since jogging is a form of endurance exercise, it should, if it is sufficiently vigorous, increase the capacity of the fatty acid oxidation system in muscle and hence be beneficial in reducing the raised fatty acid concentration after stress. Unfortunately, up to the present, no studies have been carried out in man to determine the level of exercise required to increase the capacity of this system, or whether a raised plasma fatty acid concentration is more rapidly lowered in trained subjects at rest.[12]

The question was posed in the introduction to this chapter why some individuals find it difficult to run even 1 km. The answer is simple. Lack of fitness means a poor aerobic capacity (i.e. a loss of mitochondria and hence the enzymes of the TCA cycle and β-oxidation pathway), so that energy required for exercise is provided mainly by the anaerobic conversion of glycogen to lactate. Since this is a very inefficient means of producing ATP, it has to occur at a high rate and readily causes fatigue either because of the accumulation of protons (Section A.4.a) or, if the exercise is more gradual (when protons will escape in the bloodstream as lactic acid), depletion of muscle glycogen. Even mild aerobic training (Section B.5.a) can improve the aerobic metabolic capacity and increase the distance that can be run by non-athletes. Many non-athletes who had never run seriously in their life completed the first London marathon (held in March, 1981) after about six months carefully planned training.

NOTES

1. A much higher power output has been measured in alpine skiers during a standing vertical jump off both feet on a force platform (see di Prampero and Magnoni, 1981). The difference between the two methods is probably explained by the fact that the staircase method collects data for only one leg over a period of time during which contractions occur for only part of that time. The power output of the standing jump may reflect the maximum ATPase activity in the muscle of the two legs.

2. The maximum activity of 6-phosphofructokinase in human quadriceps muscle is $57 \mu mol.min^{-1}.g^{-1}$ (Table 5.4) or about $1 \mu mol.sec^{-1}.g^{-1}$. Since anaerobic glycolysis produces three molecules of ATP for each glucose molecule from glycogen converted to lactate and since there is little change in the ATP concentration in the muscle, the maximum rate of ATP utilization in human muscle during sprinting is $3 \mu mol.sec^{-1}.g^{-1}$.

3. The maximum capacity of the TCA cycle in pectoral muscle of the domestic fowl is less than 0.1 whereas that of glycolysis is about $100 \mu mol.min^{-1}.g^{-1}$ (based on the activity of oxoglutarate dehydrogenase—see Table 5.4). Thus the maximum rate of generation of ATP in the cycle could be no greater than $3 \mu mol.min^{-1}.g^{-1}$ whereas that from maximum glycolysis is $300 \mu mol.min^{-1}.g^{-1}$. Unfortunately, this information is not available for muscle from top class sprinters or from pure type IIB human muscle fibres.

4. The blood concentrations of the catecholamines have not in fact been measured in sprinters when they are on their blocks waiting for the gun, since it is very difficult to obtain blood samples at that particular time from athletes engaged in an important competition. A less direct method is therefore used involving the measurement of degradation products of catecholamine metabolism (see Chapter 21; Section B.2.c) in the urine of subjects collected prior to and at the end of a competition. This has been done with a number of different sportsmen under different conditions. The highest urine concentrations of 3-methoxy-4-hydroxymandelic acid (a degradation product of the catecholamines) were observed in the cox and oarsmen of a rowing eight in the anticipation period before an important race but it was also raised after the race (Nowacki *et al.*, 1969).

5. The intracellular pH in human muscle has been measured in a biopsy sample of the muscle taken before and immediately after exercise to exhaustion (see Sahlin, 1978; Hermansen, 1981). Similar changes in pH to those in human skeletal muscle have also been measured in the intact frog and rat heart muscle by direct observation using a nuclear magnetic resonance technique (Dawson *et al.*, 1978; Bailey *et al.*, 1981).

6. The mean oxygen uptake of six nationally ranked (USA) marathon runners during marathon running was $4 l.min^{-1}$ (Costill and Fox, 1969). One litre of oxygen is equivalent to 21 kJ of energy if glucose is oxidized (19.5 kJ if fatty acid is oxidized), so that the energy consumed by the marathon runners is about $84 kJ.min^{-1}$.

7. J. Wahren and P. Felig have carried out most of these studies on Stockholm firemen who although physically fit, did not participate in regular athletic training. Unfortunately, the power output by the firemen during exercise was not very large, 25–30% maximum oxygen consumption, which is almost trivial compared to the

80–90% of elite marathon runners during the race. It is difficult to know if such experimental findings can be extrapolated to the elite athlete.

8. 1 g of glucose produces 16 kJ of energy upon oxidation, so that the elite marathon runner, who expends about 84 kJ.min^{-1}, requires $84 \div 16$ or about 5.2 g of glucose each minute.

9. While the energy requirements of the elite marathon runner could be supported solely by blood glucose (at least for a short period), this is not the case for fatty acids. The evidence that fatty acid oxidation alone cannot support the maximum power output of the elite runner is as follows.

   (a) If carbohydrate stores of the body are depleted by feeding subjects a high fat diet prior to exercise, a given level of exercise produces exhaustion considerably more quickly than for subjects on a normal diet and especially in comparison to those on a high carbohydrate diet (Christensen and Hansen, 1939).

   (b) If the carbohydrate store in the muscle is elevated, a given level of exercise can be maintained for a longer period of time (Table 9.7).

   (c) If the fatty acid concentration in the blood is artificially elevated prior to exercise, a given level of exercise can be maintained for a longer period of time (Hickson *et al.*, 1977; Costill *et al.*, 1977). This manipulation ensures that the plasma fatty acid concentration is elevated at the beginning of exercise rather than 30 min or more later.

   (d) In elite ultradistance runners, running for 24 h, the power output declined towards 50% of $\dot{V}_{O_2max}$ during the race while the respiratory exchange ratio decreased to about 0.7 indicating mainly lipid oxidation (Figure 9.4). This suggests that even in these elite distance runners lipid oxidation could provide perhaps only 50%–60% of the energy required for maximum power output.

   (e) P. Cerretelli (personal communication) has investigated the $\dot{V}_{O_2max}$ of a patient with muscle 6-phosphofructokinase deficiency. The rate of glucose oxidation would be minimal in the patient so that fatty acid oxidation would provide all the energy; Cerretelli found that the $\dot{V}_{O_2max}$ is approximately 60% of what he would expect from other physiological characteristics of the patient.

10. Jansson (1980) has failed to show a reciprocal relationship between glucose and fatty acid utilization by skeletal muscle of man during sustained exercise. Subjects were given a high lipid diet which increased blood fatty acid concentrations during exercise but the rate of glucose uptake was the same in control and lipid-fed subjects. However, muscle glycogen depletion was less in the lipid-fed subjects and muscle concentration of citrate was high, suggesting some inhibition of 6-phosphofructokinase. Unfortunately, exercise was only performed for 25 min.

11. If the rate of fatty acid oxidation was reduced, exercise would demand a greater rate of utilization of the limited carbohydrate stores and result in hypoglycaemia; this has been observed when fatty acid mobilization was reduced by nicotinic acid in one subject investigated by Carlson *et al.*, 1963. Many patients with hypertension are treated with β-adrenergic blockers, which would be expected to inhibit fatty acid mobilization and place a greater emphasis on use of blood glucose as a fuel. It is of interest that such blockers have been reported to lead to fatigue (Simpson, 1977; Pearson *et al.*, 1979).

12. Notwithstanding the fact that muscle may be able to oxidize fatty acids even at rest, the authors recommend that, if possible, jogging or other forms of sustained exercise should be performed as soon as possible after a stressful situation in order to accelerate the removal of fatty acids from the bloodstream and thus reduce the dangers of damage to membranes, thrombosis, and arrhythmias.

CHAPTER 10

AMINO ACID METABOLISM

A. GENERAL INTRODUCTION

The catabolism of amino acids is more complex than that of glucose or fatty acids. The reasons for this are that there are at least 20 amino acids whose metabolism has to be considered and the nitrogen present in the amino acids is metabolized in a different pathway to that of the carbon and hydrogen. However, since the amino acids have a common basic structure their catabolism has certain common features which will be emphasized in this chapter.

The basic structure of the amino acid can be represented as follows:

$$R-\underset{\underset{NH_3^+}{|}}{CH}-\overset{\overset{O}{\|}}{C}-O^-$$

where $-NH_3^+$ and $-COO^-$ are known as the α-amino and α-carboxyl groups respectively. The identity of the amino acid depends upon the chemical nature of the R group, that is, the side-chain (or side-group). Amino acids are the major constituents of proteins which are formed by the condensation of α-amino and α-carboxylic acid groups of adjacent amino acids thus generating a linear amino acid sequence, known as the primary structure of the protein. The biological function of the protein, whether it be an enzyme, hormone, transporter, or structural component, depends upon its correct three-dimensional structure. This results from a folding of the linear chain as a consequence of mostly non-covalent interactions between side-chains (Chapter 3; Section A.2.c). The folding into the three-dimensional structure depends entirely on the primary structure. The latter is generated during the synthesis of proteins which involves the transfer of genetic information from DNA into messenger RNA and translation of the information in the messenger RNA into a sequence of amino acids as described in Chapter 18.

An understanding of at least the general principles involved in protein and amino acid catabolism is necessary for medical students since it is important in many physiological processes and has significant implications in widely differing clinical fields. A problem, from the student's point of view, is that the metabolism of each amino acid is different, so that a sequence of pages describing these pathways is daunting to say the least. To overcome this, the general principles of amino acid degradation are provided in this chapter and these are illustrated by reference to the catabolism of selected examples. For reference purposes, a catalogue of the major routes of catabolism of all the amino acids is provided in Appendix 10.1.

1. Classification of Amino Acids

In order to provide some idea of the diversity of amino acid structures, the amino acids that commonly occur in proteins are listed below. For convenience of reference, the amino acids are classified, according to the chemical nature of their side chains, into seven classes.

(a) *Mono-amino, mono-carboxylic amino acids*

$$\text{glycine} \qquad \begin{array}{c} \text{H}.\text{CH}.\text{COO}^- \\ | \\ \text{NH}_3^+ \end{array}$$

$$\text{alanine} \qquad \begin{array}{c} \text{CH}_3\text{CH}.\text{COO}^- \\ | \\ \text{NH}_3^+ \end{array}$$

$$\text{valine} \qquad \begin{array}{c} \text{CH}_3 \\ \diagdown \\ \text{CH}.\text{CH}.\text{COO}^- \\ \diagup \qquad | \\ \text{CH}_3 \qquad \text{NH}_3^+ \end{array}$$

$$\text{leucine} \qquad \begin{array}{c} \text{CH}_3 \\ \diagdown \\ \text{CH}.\text{CH}_2\text{CH}.\text{COO}^- \\ \diagup \qquad\qquad | \\ \text{CH}_3 \qquad\qquad \text{NH}_3^+ \end{array}$$

$$\text{isoleucine} \qquad \begin{array}{c} \text{CH}_3\text{CH}_2 \\ \diagdown \\ \text{CH}.\text{CH}.\text{COO}^- \\ \diagup \qquad | \\ \text{CH}_3 \qquad \text{NH}_3^+ \end{array}$$

Glycine, the simplest amino acid, is the only one that is not optically active. All other amino acids occur in proteins as the L-enantiomer (see Appendix 4.1). (The D-enantiomers of some amino acids do occur in nature but much less frequently and not in proteins. For example, they occur in the peptides[1] that are found in some bacterial cell walls.) Valine, isoleucine, and leucine are collectively known as the branched-chain amino acids. The side-chains of the amino acids in this section are non-polar.

(b) *Hydroxy-amino acids*

serine

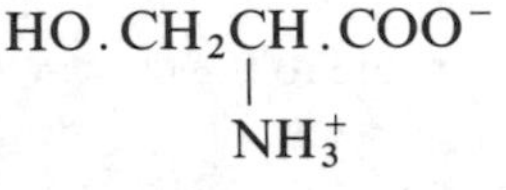

$$HO.CH_2CH.COO^-$$
$$|$$
$$NH_3^+$$

threonine

$$CH_3CH.CH.COO^-$$
$$| \quad |$$
$$OH \quad NH_3^+$$

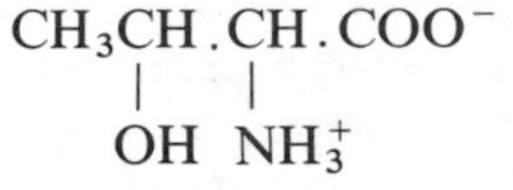

Possession of a hydroxyl group confers polarity on the side-chain but the group remains unionized at pH 7.0. In phosphoproteins, the phosphoryl groups are frequently attached to serine residues.* The aromatic amino acid tyrosine also contains a hydroxyl group but this is phenolic in character. Both lysine and proline (below) occasionally occur as their hydroxylated derivatives in proteins but the hydroxylation takes place after the protein has been synthesized (i.e. post-translationally).

(c) *Basic amino acids*

lysine

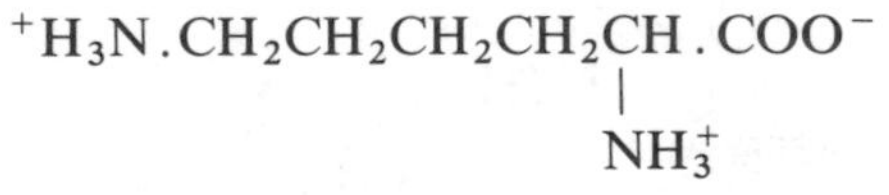

$$^+H_3N.CH_2CH_2CH_2CH_2CH.COO^-$$
$$|$$
$$NH_3^+$$

arginine

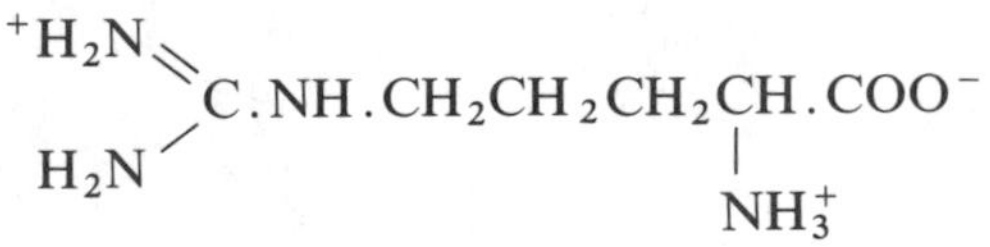

$$^+H_2N$$
$$\diagdown$$
$$C.NH.CH_2CH_2CH_2CH.COO^-$$
$$H_2N \diagup$$
$$|$$
$$NH_3^+$$

histidine

$$HC=C.CH_2CH.COO^-$$
$$| \quad | \quad |$$
$$^+HN \quad NH \quad NH_3^+$$
$$\diagdown \diagup$$
$$C$$
$$H$$

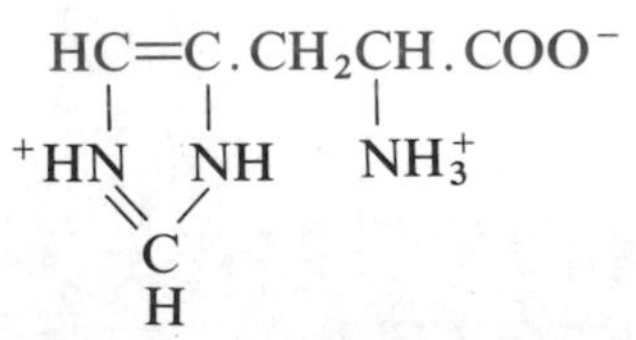

These amino acids, although structurally somewhat diverse, have the common feature of possessing a positively charged group at pH 7.0 in addition to the α-amino group. Both lysine and histidine are occasionally found in proteins in an *N*-methylated form and the lysine side chains can covalently cross-link in proteins. Methylation occurs post-translationally.

---

* For reasons that are not entirely clear, amino acids incorporated into a protein are often referred to as 'residues' in this way.

(d) *Acidic amino acids* (*and amides*)

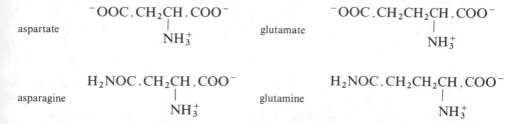

aspartate $^-OOC.CH_2CH.COO^-$ (with NH_3^+ below)

glutamate $^-OOC.CH_2CH_2CH.COO^-$ (with NH_3^+ below)

asparagine $H_2NOC.CH_2CH.COO^-$ (with NH_3^+ below)

glutamine $H_2NOC.CH_2CH_2CH.COO^-$ (with NH_3^+ below)

These amino acids possess a net negative charge at pH 7.0, which is lacking in the amides, asparagine, and glutamine, although the latter do possess a polar side-chain.

(e) *Sulphur-containing amino acids*

cysteine $HS.CH_2CH.COO^-$ (with NH_3^+ below)

methionine $CH_3S.CH_2\ CH_2CH.COO^-$ (with NH_3^+ below)

Cysteine is polar and methionine non-polar but neither bear a net charge at pH 7.0. Pairs of cysteine residues frequently cross-link in proteins with formation of a disulphide link. The double amino acid thus formed is known as cystine.

(f) *Aromatic amino acids*

phenylalanine   (benzene ring)$-CH_2CH.COO^-$ (with NH_3^+ below)

tyrosine $HO-$(benzene ring)$-CH_2CH.COO^-$ (with NH_3^+ below)

tryptophan   (indole ring, N–H)$-CH_2CH.COO^-$ (with NH_3^+ below)

None of these amino acids bear a net charge at pH 7.0, although tyrosine is polar. Tyrosine is another amino acid that can cross-link in proteins.

(g) *Imino acid*

$$\begin{array}{cc} H_2C\!\!-\!\!-\!\!CH_2 & \\ \text{proline} \quad H_2C_{\diagdown N \diagup}CH.COO^- \\ \quad\quad\quad\quad\quad H \end{array}$$

Proline is an imino acid since the amino group attached to the α-carbon atom is a secondary amino group. Its metabolism is closely related to that of the true amino acids and its structure has important implications in protein conformation since it reduces the flexibility of the peptide chain in which it occurs.[1] The 4-hydroxy derivative of proline occurs primarily in the protein collagen.

The 20 amino acids (including glutamine and asparagine) described above occur in proteins but, in addition to these, there are a number of amino acids that occur only in the free state. These include ornithine and citrulline, which are α-amino acids that occur in the urea cycle (see Chapter 12; Section B.2), taurine ($^+H_3NCH_2CH_2SO_3^-$), which is derived from cysteine, and may be important as a neurotransmitter in the brain (Chapter 21; Section B.8), and β-alanine ($^+H_3NCH_2CH_2COO^-$), a β-amino acid produced in the degradation of pyrimidines.

B. SOURCES OF AMINO ACIDS

The majority of amino acids are ingested in combined form as dietary protein. These proteins must be hydrolysed in the process of digestion to produce the constituent amino acids that enter the pathways of metabolism described in this and subsequent chapters. Most humans ingest protein of both animal and plant origin. The amino acid composition of proteins from these two sources is somewhat different and this has nutritional implications which are discussed before the processes of protein digestion and amino acid absorption are described in detail.

1. Nutritional aspects

(a) *Protein deficiency*

Protein (or at least a source of certain amino acids) is an essential nutritional requirement for man. Even when no growth is occurring, nitrogen-containing compounds are lost from the body, indicating a continuous breakdown of protein. This must be replaced if health is to be maintained. (Indeed the loss of protein from certain skeletal muscles may be one cause of death in starvation—see Chapter 14; Section B.) Even when carbohydrates and fats are available in adequate amounts, failure to ingest sufficient protein leads, especially in children, to a disease known as kwashiorkor. This is one of a group or spectrum of diseases known as 'protein-energy malnutrition' which includes kwashiorkor, marasmic kwashiorkor, and marasmus, diseases that are widespread in many parts of the world (Garrow, 1970; Alleyne *et al.*, 1977; Coward and Lunn, 1981). Kwashiorkor results from an inadequate intake of protein, whereas marasmus results from a more general dietary insufficiency (i.e. starvation) but the distinction between them is not always clear. Kwashiorkor is characterized by generalized oedema, cessation of growth, severe body wasting, diarrhoea, mental apathy, and lack of pigmentation of the skin and hair. Peeling of the skin leaving raw areas that resemble the lesions of pellagra (deficiency of the B vitamin, nicotinic acid, see Appendix A) is also common. Marasmus is similarly characterized except that there is more severe body wasting and no oedema.[2] The diseases affect children soon after they have been weaned off their mother's milk onto a diet low in protein or after a period of illness. So many children in the poorer parts of the world are affected by kwashiorkor that it has been claimed to be the most severe and widespread nutritional disorder (see Trowell, 1973). (It seems likely that for adults even the poorest diet, provided it contains enough energy, will contain enough protein for the minimal requirements; hence kwashiorkor primarily affects young children (Bender, 1975).)

Kwashiorkor was not reported in the medical literature until 1933; this was when Cicely Williams, who studied the condition while she was working among the tribes of West Africa, gave it the name used by the African Ga tribe. Since proteins are required for the normal functioning of all metabolic processes, there are probably many sites of metabolic dysfunction in kwashiorkor. The diarrhoea is probably caused by impaired function of the small intestine and the exocrine pancreas. The epithelial cells of the small intestine have a very short life (most cells are replaced every 24 hr) but with protein deficiency they will not be renewed sufficiently rapidly to provide for normal digestion. In addition, the synthesis of the digestive enzymes, in particular pancreatic lipase, is impaired. The failure in digestive function will cause undigested and partially digested food to enter the large intestine where it leads to diarrhoea. The oedema is probably caused by a decreased concentration of plasma proteins, especially albumin, and perhaps an imbalance of sodium concentration between the plasma and interstitial fluid. The decreased solute concentrations in the plasma causing a decrease in the colloid osmotic pressure and possibly increased concentrations in the interstitial fluid will allow movement of water into the interstitial space by osmosis. Inadequate rates of protein synthesis will account for the inadequate replacement of skin and hair and an inability to synthesize essential proteins of the brain at a satisfactory rate could also explain inadequate mental development. The problems caused by the deficiency of protein are often exacerbated by a concomitant deficiency of a number of vitamins.

(b) *Essential and non-essential amino acids*

Nutritional diseases result when the amount of nitrogen containing compounds ingested is less than the amount of nitrogen lost over long periods. However, not all proteins are of equal nutritional value; this reflects their differing amino acid content. Although most proteins contain most of the 20 amino acids described above, these are present in widely differing proportions. For protein synthesis to take place, all the amino acids required must be available within the cell. Studies on the effects of feeding proteins of known amino acid composition, or diets containing specified pure amino acids, on the rate of growth or nitrogen balance of an experimental animal have led to the concept of essential and non-essential amino acids. In biochemical terms, essential amino acids are those which cannot be synthesized at a sufficient rate to supply the normal requirements for protein biosynthesis. These amino acids must therefore be present in the diet. The non-essential amino acids can be synthesized at a sufficient rate (provided of course that the supplies of amino nitrogen and carbon precursors are adequate). The pathways by which these are synthesized are outlined in Section E. There are several means of determining whether an amino acid is essential or not, and the amino acids included under this heading vary slightly according to the criteria used. In man, nitrogen balance is used. In a normal healthy adult, the rate of intake of nitrogen in the diet will be equal to its rate of loss from the body (in faeces, urine, and skin). The situation is described as being in nitrogen balance. However, during periods of active growth, or tissue repair, more nitrogen is ingested than excreted (positive nitrogen balance). Conversely, in malnutrition or starvation, less nitrogen is ingested than excreted (negative nitrogen balance). To determine whether an amino acid is essential, it is omitted from the diet while all the other amino acids are included. If the omission results in negative nitrogen balance, the amino acid is deemed essential. In the absence of this single amino acid, the body has been unable to synthesize certain proteins so that the nitrogen that would have been used in this synthesis is excreted. By this criterion, the following amino acids are essential in man: isoleucine, leucine, lysine, methionine, phenylalanine, threonine, tryptophan, and valine.

Rather more direct experiments are possible with experimental animals and these show that, in addition to the above amino acids, histidine is necessary for proper growth in the rat (as measured by increase in body weight or extension of the long bones). It seems likely that histidine is also essential in humans but is not so indicated in relatively short-term nitrogen balance experiments; it is known to be required for normal growth in a child. It is now possible to carry out more biochemically-orientated experiments to detect essentiality of the amino acids in man. For example, maintenance of the concentration of haemoglobin or plasma proteins can be used (Evered, 1981).

The situation is actually more complex than indicated since, although man can synthesize most of the non-essential amino acids from glucose and ammonia, tyrosine synthesis requires the availability of phenylalanine, and cysteine requires the availability of methionine. Both phenylalanine and methionine are essential amino acids, and if they are available in the diet at or below minimal requirement levels, tyrosine and cysteine can become essential amino acids.

The requirements for the essential amino acids in adults and infants are given in

Table 10.1. Daily requirements of essential amino acids to maintain nitrogen balance in adults and growth in infants under six months*

Amino acid	Daily requirement of amino acids $(mg.kg^{-1})$	
	Adults	Infants
Histidine	—	28
Isoleucine	10	70
Leucine	14	161
Lysine	12	103
Methionine plus cysteine	13	58
Phenylalanine plus tyrosine	14	125
Threonine	7	87
Tryptophan	3.5	17
Valine	10	93

* Data taken from Davidson *et al.* (1979). See also Munro and Crim (1980).

Table 10.1. The additional nine non-essential amino acids are alanine, arginine, aspartic acid, asparagine, glutamic acid, glutamine, glycine, proline, and serine.

(c) *Nutritional value of proteins*

By now it should be clear why some proteins are nutritionally better than others; they contain a more balanced range of the essential amino acids. In general, proteins of animal origin contain adequate amounts of the essential amino acids and hence they are known as first class proteins. However, many proteins of vegetable origin are relatively deficient in certain amino acids, notably lysine and the sulphur-containing amino acids.* Hence vegetarians, in order to receive adequate amounts of the essential amino acids, must eat a varied diet so that the amino acid deficiencies in one plant are balanced by those present in different proteins in other plants. In some of the poorer parts of the world, diets are based predominantly on a single plant (e.g. corn) and they frequently lead to malnutrition. Some of the problems have been alleviated in recent years by breeding varieties of cereals that contain proteins with higher proportions than normal of the essential amino acids such as lysine, (e.g. high-lysine corn, see Nelson, 1969).

Estimates of the protein requirement for man have varied markedly over the years (e.g. from 30 g to 100 g per day). In 1973 the World Health Organization (WHO/FAO report, 1973) reported that the minimum quantity of 'ideal' protein required to maintain a 65 kg man in nitrogen balance was 27 g per day. However, this does not allow for individual variation and a daily intake of 53 g real protein

* Legumes are low in methionine but adequate in lysine whereas cereals are low in lysine but adequate in methionine.

(i.e. protein present in food; $0.8\,g.kg^{-1}$ body wt) is recommended for an adult (Munro and Crim, 1980), and children need considerably more than this on a weight-for-weight basis ($2.4\,g.kg^{-1}$ body wt during the first few months, $1.5\,g.kg^{-1}$ at six months and $1.1\,g.kg^{-1}$ at one year). Pregnant women also require more protein ($1.3\,g.kg^{-1}$). It should also be emphasized that a deficiency of only $1\,g.day^{-1}$ has a cumulative effect and can lead to severe malnutrition.

Various means of calculating the nutritional value of real protein as opposed to ideal protein have been used. Experimentally this can be done by first establishing the rate of nitrogen excretion for an individual on a protein-free diet. Small quantities of the protein to be tested are then administered and the increase in the amount of nitrogen excreted is measured. The biological value of the protein is calculated as follows:

$$\frac{\text{retained nitrogen}}{\text{absorbed nitrogen}} \times 100$$

If the protein is perfectly balanced this should not lead to an increase in nitrogen excretion since all the amino acids could be used for synthesis of protein, and its biological value would be 100. If half the nitrogen ingested was excreted, the biological value would be 50, so that twice as much of the protein would have to be ingested to achieve the same nutritional benefit. Cows' milk protein has a biological value of 95 while that of whole corn has a biological value of 60. Similar values can be determined from a knowledge of the amino acid composition of the protein. Whole chicken egg protein is taken as a standard and the proportion of each essential amino acid in the test protein is expressed as a fraction of that in the egg protein. The lowest fraction is taken as a protein's chemical value. The disadvantage of this approach is that not all the amino acids that are estimated by chemical analysis are fully biologically available. Lysine, for example, may cross-link with other residues in a protein, especially during cooking, and consequently may not be liberated by the processes of intestinal digestion (that is, the amino acid is unavailable). This is taken into account if the net protein utilization is considered since this is defined as:

$$\frac{\text{retained nitrogen}}{\text{intake of nitrogen}} \times 100$$

Consequently, the net protein utilization is equal to the biological value of that protein multiplied by the availability.

2. Amino acids from dietary protein

(a) Digestion of protein

The average daily intake of protein by man in the industrialized countries of the West is approximately 100 g (Table 10.2). Since these proteins cannot enter the

Table 10.2. Over-all protein metabolism in man*

	Amount (g.70 kg^{-1})
Dietary protein intake	100
Amino acids absorbed† (protein equivalents)	160
Loss of nitrogen from body‡ (protein equivalents)	92
Body protein synthesis§	300

* Data taken from Munro (1974), Young and Munro (1978), Munro and Crim (1980) and Waterlow and Jackson (1981). The values are all approximate and general agreement is sometimes lacking; for example the value of 300 g for body protein synthesis is taken from Munro but Waterlow and Jackson consider it is more than this.

† Of 100 g of protein ingested, 10 g is lost in faeces and about 70 g is added from digestive juices and shed mucosal cells.

‡ 80 g is lost in urine, 10 g in faeces, and 2 g through the skin.

§ This includes 70 g for digestive requirements, at least 20 g for plasma proteins, 8 g for haemoglobin, 20 g for white blood cells, and perhaps 100 g (Munro suggests 75 g) each day for total body muscle. The nitrogen balance is 160 g absorbed and 162 g lost (including 70 g for intestinal activity) so that in this calculation there is a 2 g negative nitrogen balance.

epithelial cells of the intestine,[3] they must be digested to amino acids or di- and tripeptides before absorption can occur. The processes of protein digestion and peptide and amino acid absorption have been reviewed by Sleisenger and Young (1979). Digestion involves several stages including extraction of the protein from the food, denaturation of the protein (see Chapter 3; Section B.2) and hydrolysis of the peptide bonds. Protein is extracted from the food in the process of mastication and by the mechanical activity of the stomach. The low pH of the stomach plays a role in denaturation of the extracted protein, thus making it more accessible to the proteolytic enzymes of the gut. There is evidence that gastric emptying is delayed by the presence of proteinaceous food in the stomach and this may be necessary to ensure effective extraction and denaturation of the protein prior to its entry into the duodenum. For the early stages of the digestion of protein, four types of enzymes are important, pepsins (secreted by serous cells in the gastric gland of the stomach), trypsin, elastase, and the chymotrypsins (all secreted by the acinar cells of the pancreas). As with all the enzymes involved in digestion of protein, these are synthesized in the form of the inactive pro-enzymes (also termed zymogens) namely pepsinogen, trypsinogen, pro-elastase, and the chymotrypsinogens. The pro-enzymes contain one or more extra peptide segments which prevent the formation of the three-dimensional structure that possesses catalytic activity. The extra peptide of pepsinogen (containing 42 amino acids) is removed by the hydrolysis of one peptide bond, catalysed by the acidic conditions of the stomach:

$$\text{pepsinogen} \xrightarrow{\text{H}^+} \text{pepsin} + \text{peptide}$$

The brush border cells of the small intestine secrete small amounts of a proteolytic enzyme known as enteropeptidase (formerly enterokinase). This enzyme initiates pro-enzyme activation in the small intestine by catalyzing the conversion of trypsinogen into trypsin. Trypsin is able to achieve further activation of trypsinogen (i.e. an auto-catalytic process) and also activation of chymotrypsinogens and pro-elastase, by the selective hydrolysis of a small number of peptide linkages. (The important initiating role of enteropeptidase is indicated by the disease caused by lack of this enzyme in some human subjects.[4]) The activating processes[5] can be summarized by the equations:

$$\text{trypsinogen} \xrightarrow[\text{trypsin}]{\text{enteropeptidase}} \text{trypsin} + \text{peptide}$$

$$\text{chymotrypsinogen} \xrightarrow{\hspace{2cm}} \text{chymotrypsin} + \text{peptides}$$

$$\text{pro-elastase} \xrightarrow{\hspace{2cm}} \text{elastase} + \text{peptide}$$

These proteolytic enzymes are all proteinases (endopeptidases)[6] that is, they hydrolyse links in the middle of polypeptides but they differ in the peptide links that they hydrolyse. Pepsin has a fairly broad specificity but is most active on links in which the carbonyl group is provided by phenylalanine or leucine; chymotrypsin hydrolyses peptide links in which the carbonyl group is provided by an aromatic amino acid and the amino group by any other amino acid except aspartate or glutamate; trypsin hydrolyses peptide bonds in which the carbonyl group is provided by lysine or arginine but it is unspecific for the amino acid providing the amino group.*

The product of the action of these proteolytic enzymes is a series of peptides of various sizes. These are degraded further by the action of several peptidases (exopeptidases) that remove terminal amino acids, including carboxypeptidases A and B which hydrolyse amino acids sequentially from the carboxyl end of peptides. These enzymes are secreted by the acinar cells of the pancreas in pro-enzyme form (pro-carboxypeptidases A and B) and are each activated by the hydrolysis of one peptide bond, which is catalysed by trypsin. Aminopeptidases, which are secreted by the absorptive cells of the small intestine, hydrolyse amino acids sequentially from the amino end of peptides. In addition, dipeptidases, which are structurally associated with the brush border of the absorptive cells, hydrolyse dipeptides into their component amino acids. The extent to which these act on substrates in the lumen, within the membrane of the cells or within the cell itself is not known. However, there is evidence that at least some dipeptides and tripeptides pass through the brush border membrane of the absorptive cells and are hydrolysed within the cell, (Adibi *et al.*, 1975; Burston *et al.*, 1977). The transport of these peptides is an active process (and depends upon a Na^+ concentration gradient) but it is distinct from that for the amino acids (see

---

* There appears to be a feedback regulatory mechanism for the secretion of these enzymes by the pancreas. Any protein present in the lumen of the duodenum will bind trypsin, which lowers the concentration of free trypsin and so encourages more synthesis and release from the pancreas. When the free concentration increases, this inhibits further secretion.

below). The di- and tripeptides are probably hydrolysed to their constituent amino acids by the dipeptidases located within the absorptive cell.*

Protein digestion in the intestine results in the hydrolysis not only of ingested protein but also endogenous protein, provided in the form of digestive enzymes, other secreted proteins and desquamated epithelial cells. The precise quantity of such protein is difficult to estimate but it is likely to be between 50 g and 70 g each day (Munro, 1969; Munro, 1974). In other words, an amount of *endogenous* protein, which is not much less than that ingested on a normal mixed diet, is broken down during digestion.

There is, in addition, protein degradation occurring continuously in all tissues of the body which may amount to 300 g each day and is discussed in Section B.3.

(b) *Absorption of amino acids and small peptides*

The final stages of the hydrolysis of peptides to dipeptides and amino acids, and their absorption, occur in the jejunum and ileum. The transport of amino acids and di- and tripeptides from the lumen into the cell is an active process and the mechanisms appear to be remarkably similar to that of glucose transport (see Chapter 5; Section A.2.b and Figure 5.3). A specific carrier in the brush border of the absorptive cell combines with both the amino acid and a sodium ion and conveys them to the inner face of the membrane where they are released. Although the amino acid may be moving up a concentration gradient, this is more than compensated for by the fact that the sodium ions are moving down a concentration gradient. The latter is maintained by the active extrusion of sodium ions by the sodium ion 'pump'. This pump exchanges sodium ions for potassium ions in a process that involves the hydrolysis of ATP. What remains in some doubt is the precise location of these transport processes within the absorptive cell. It is possible that the amino acids are not released into the cytosol of the epithelial cell but into a membrane system within the cell. This would communicate both with the lumen and with the interstitial space on the outer surface of the epithelial cell. A further area of amino acid transport which is still not completely clarified is the number and specificities of different carriers that are involved in intestinal amino acid transport. At last seven carrier systems with different specificities are involved in the transport of amino acids into cells in non-intestinal tissues and it seems likely that they are also involved in the intestine (see Lerner, 1978; Morgan *et al.*, 1979; Kilberg *et al.*, 1980); these are described in Section C.1. In addition, there may be specific intestinal carrier systems for cystine plus cysteine and methionine plus tryptophan. It is possible that the less specific amino acid carriers show a preference for transport of the essential amino

* The transport of peptides is likely to be a quantitatively important process in amino acid absorption since patients with an inability to transport certain neutral amino acids into the absorptive cells (Hartnup's disease) grow almost normally. Their requirement for these amino acids is presumably supplied from small peptides absorbed in this way (Wellner and Meister, 1981).

acids since, at equimolar concentrations, the intestine transports essential amino acids more rapidly than the non-essential ones.

There is evidence that a totally different transport system is involved in the tubular re-absorption of some amino acids in the kidney. This is transport via the γ-glutamyl cycle, which is described in detail in Section C.1. However, if the γ-glutamyl cycle does occur in intestinal cells, it probably plays only a minor role in amino acid absorption.

(c) *Amino acid metabolism in the intestinal cells*

Like glucose, some amino acids can be metabolized by the absorptive cells (Chapter 5; Section A.2.c). In particular, aspartate, glutamate, asparagine, and glutamine undergo considerable metabolism so that, for example, very little of the aspartate or glutamate present in the intestinal lumen actually enters the bloodstream. Experiments with a simulated meal indicate that about 40% of the energy requirements of the intestine were met by the oxidation of glutamate, aspartate, and glutamine that had been absorbed from the lumen. In addition, glutamine was removed from the blood and this accounted for a further 40% of the energy required by the absorptive cells. This subject is discussed further in Section F.4.

3. **Amino acids from endogenous protein**

In normal healthy well-fed subjects about 300 g of tissue protein is hydrolysed daily (approximately 5 g.kg^{-1}) and replaced by newly synthesized protein. About 50 g is required for production of the digestive juices and 20 g for the cells of the small intestine that are lost during normal digestion. In recent years, the measurement of the excretion of 3-methylhistidine[7] in the urine has provided a more accurate measurement of the myofibrillar protein degradation of muscle; such measurements indicate that 100 g of protein turns over each day in this tissue

Table 10.3. The half-lives of some enzymes from liver and muscle*

Enzyme	Tissue	Approximate half-life (hr)
Ornithine decarboxylase	liver	0.2
Tyrosine aminotransferase	liver	2.0
Phosphoenolpyruvate carboxykinase	liver	5.0
Glucokinase	liver	12.0
Glucose-6-phosphate dehydrogenase	liver	15.0
Fructose-bisphosphate aldolase	muscle	113
Lactate dehydrogenase	muscle	144
Cytochrome *c*	muscle	150

* Data taken from Dice and Goldberg (1975), Goldberg and St. John (1976).

(Young and Munro, 1978; Waterlow and Jackson, 1981; Table 10.2). Since the addition of each amino acid to a growing polypeptide chain involves the expenditure of six molecules of ATP (Table 18.2), this turnover imposes significant metabolic demands. It has been estimated for example that as much as 15–20 % of the basal metabolic rate (see Table 2.2) is due to protein turnover (Waterlow and Jackson, 1981). Even more surprising is the fact that the turnover of individual proteins varies enormously (Table 10.3). These facts raise several interesting questions: why should this turnover occur; why should it be so variable; what is the pathway of intracellular protein degradation and how is it controlled?

(a) *Explanations for protein turnover*

Several explanations have been put forward to account for protein turnover and the variability in its rate (see Goldberg and St. John, 1976). First, protein degradation is necessary to prevent the accumulation of abnormal and potentially harmful proteins and peptides. Abnormal proteins could be produced from errors in synthesis, failure to complete peptide chains on the ribosome and spontaneous denaturation of normal proteins. For these reasons alone, it would be necessary for the cell to possess a mechanism for degradation of at least abnormal proteins (Indeed, it is known that denatured proteins are very rapidly degraded in mammalian or bacterial cells.) Secondly, for a protein that has a rapid turnover rate, its concentration can be rapidly changed by modifications in the rate of synthesis or degradation. It is not surprising, therefore, that enzymes which play an important role in the regulation of flux through pathways have especially short half-lives (e.g. phosphoenolpyruvate carboxykinase in liver— Table 10.3). Changes in the rate of synthesis of these enzymes will rapidly alter the enzyme concentration and hence the flux through the pathway. This is particularly important in the liver. The greatest sensitivity of control will be achieved when there is rapid cycling between the intact protein and its amino acids (see Chapter 7; Section A.2.b).

(b) *Pathways of protein degradation*

The intellectual challenge of understanding protein synthesis has probably been responsible for reducing the research effort into the mechanism of protein degradation. Nonetheless, since the concentration of any protein depends upon the balance between synthesis and degradation, the mechanism of degradation, together with factors which control the rate and specificity of breakdown are of obvious interest. The limited information available suggests that there are at least two pathways for protein degradation, one involving proteolytic digestion within the lysosomes (Dean, 1980) and the other involving digestion outside the lysosomes (Goldberg and St. John, 1976; Ballard, 1977). The importance of these pathways may vary from one tissue to another. There are a number of features of protein structure that can influence the rate of degradation. Increasing size or increasing acidic nature favour degradation so that histones, which are small basic proteins found associated with DNA in the nucleus of the cell, are very stable. Since denaturation may be the first step in protein degradation, it is possible that large acidic proteins are more readily denatured than smaller basic proteins. Once denatured, the protein can be further degraded by the lysosomal or the non-lysosomal systems.

(i) *Lysosomal pathway* Lysosomes contain at last four proteinases[3] (including cathepsins B, D, and E) and several peptidases (including dipeptidyl peptidases that remove terminal dipeptides) so that proteins could be degraded completely within this cell organelle.* Lysosomes can hydrolyse the constituent proteins of large subcellular structures such as mitochondria and this they appear to do by the lysosome wrapping itself around the structure and engulfing it to form vacuole (an autophagic vacuole, Waterlow *et al.*, 1978) within the lysosome into which digestive enzymes are secreted. (Because of this mechanism, the turnover rates of proteins within the mitochondria are very similar.) It is, however, unclear how simple, soluble enzymes of the cytosol enter the lysosome. Denaturation of a protein would be expected to increase the hydrophobic nature of the protein and this together with a net positive charge may facilitate adsorption onto the lysosomal membrane followed by uptake, perhaps by endocytosis (Dean and Barrett, 1976). Alternatively, the denatured protein may bind to the membrane of the endoplasmic reticulum, which then splits off and is engulfed by the lysosome (Ballard, 1977). In liver, the number of lysosomes increase and their membrane structure becomes more fragile under conditions that favour proteolysis (e.g. starvation—see Section B.3.b above). This supports their involvement in protein degradation in liver, in which lysosomes may be the most important agents for protein degradation.

(ii) *Non-lysosomal pathway* Muscle contains many fewer lysosomes than tissues such as liver but it does possess several soluble proteinases. One in particular is activated by millimolar concentrations of calcium ions and this may be important in the digestion of Z disc proteins of the myofibrils. Two alkaline proteinases, one of which is soluble while the other is attached to the myofibrils may also play a part since the activity of the latter is increased in catabolic states (e.g. starvation, see Chapter 14; Morgan *et al.*, 1979). There are also a number of peptidases in muscle, so that protein hydrolysis can be completed (Goldberg and St. John, 1976). Similarly in reticulocytes, which contain few or no lysosomes, there is evidence that a soluble proteinase, which has an alkaline pH optimum, is responsible for protein hydrolysis.

* The proteinase and peptidases of the lysosomes have acidic pH optima which is consistent with the pH within the lysosome being much lower than that of the cytosol (about pH 5.5). The suggestion has been made that the energy requirement for protein degradation represents that involved in maintenance of a low pH within the lysosome.

Table 10.4. Some factors that modify the rates of synthesis and degradation of protein in muscle

Compound etc.	Direction of change in rate†	
	Protein synthesis	Protein degradation
Insulin	↑↑	↓↓
Growth hormone (in hypophysectomized animals)	↑	—
Glucocorticoids	—	↑
Glucose	—	↓
Ketone bodies	—	↓
Leucine	↑	↓
Contraction (or tension) of muscle	—	↓
Normal level of triiodothyronine*	↑↑	↑
High (catabolic) levels of triiodothyronine	↑	↑↑↑

* At a normal level in a hypothyroid animal, triiodothyronine increases synthesis more than degradation so that net protein synthesis occurs, but at higher levels ('catabolic') the effect on degradation is considerably greater than that on synthesis and net protein degradation occurs.

† The arrows indicate the direction and magnitude of change; ↑ increased rate, ↓ decreased rate (Morgan *et al.* 1979; Goldberg, 1980).

(c) *Control of protein degradation*

The net rate of degradation of body protein will depend upon the rates of both synthesis and degradation of protein. Details of the control of synthesis will be given in Chapter 18; Section C. Unfortunately, there is little information concerning the biochemical mechanism of the control of protein degradation. A number of factors that affect the rate of synthesis or degradation in muscle are given in Table 10.4. The most important factors that affect the rates of synthesis or degradation of muscle protein are insulin, growth hormone, glucocorticoids, glucose, ketone bodies, leucine (it appears to be a metabolite of leucine rather than leucine itself), contraction, and triiodothyronine (Goldberg *et al.*, 1978; Morgan *et al.*, 1979; Goldberg, 1980). Insulin is one of the most important anabolic hormones and consistent with this is the fact that it increases the rate of synthesis and decreases that of degradation. Growth hormone stimulates protein synthesis, as does leucine (see Chapter 19; Section B.2). Glucocorticoids increase the rate of protein degradation but the mechanism is not known. Triiodothyronine increases both the rate of synthesis and that of degradation, but higher concentrations result in a marked increase in the rate of degradation. The effects of hormones and non-hormones to increase rates of protein degradation may be achieved by increasing the number or activity of the lysosomes and possibly increasing the activities of the non-lysosomal proteolytic enzymes (Chua *et al.,* 1978; Mortimore and Schworer, 1980). There is also evidence that protein degradation is an ATP-requiring process since the rate of degradation is decreased when the tissue is subjected to an energy stress (Goldberg *et al.,* 1980a). This supports the view that the rate of degradation is a controlled process but it is unclear how the enzymes or enzyme systems are controlled at a molecular level and it is unclear how glucose or ketone bodies reduce the rate of protein degradation.

There is particular interest in the fact that triidothyronine increases the rate of degradation of muscle protein, since there is evidence that the concentration of triidothyronine is reduced in starvation in man and that this might be one important factor in reducing the rate of protein degradation in prolonged starvation, an effect that has considerable survival value (see Chapter 14; Section B.2 for further discussion). Similarly it may explain the body wasting that is observed in hyperthyroidism (Goldberg *et al.,* 1980b).

C. PATHWAYS OF AMINO ACID DEGRADATION

This section describes the general principles involved in the degradation of the amino acids, treating in greatest detail those reactions that are common to the metabolism of several amino acids.

The final stage in the oxidation of most of the amino acids occurs through the reactions of the TCA cycle but many amino acids are also precursors for gluconeogenesis. Hence their metabolism is inter-related with a number of other metabolic processes. Since there are a large number of amino acids, there are a large number of processes by which they are converted into common metabolic intermediates (i.e. compounds that are components of the main metabolic pathways). Fortunately, knowledge of the details of these pathways is not necessary for an understanding of the over-all aspects of amino acid metabolism. For that reason, the details are relegated to Appendix 10.1, which can be used for reference purposes. In addition, there have been many studies on disorders of amino acid metabolism in man. Most of these are rare so that they will not be discussed here but, when they occur, they can provide details of amino acid metabolism in man (see Rosenberg and Scriver, 1980; Rosenberg, 1981).

1. Transport of amino acids into the cell

Prior to intracellular metabolism, amino acids must be transported from the interstitial space across the cell membrane. As with glucose transport, amino acid transport requires the presence of a carrier system in the cell membrane but there are two major differences. First, the intracellular concentrations of amino acids may be considerably greater than those in the bloodstream so that transport of amino acids into most, if not all, cells is an active process, usually but not always associated with the operation of a sodium ion pump as described for the absorption from the intestinal lumen. Secondly, there are at least seven different

Table 10.5. Amino acid transport systems in the cell membranes of mammalian tissues*

Amino acid carrier	Some preferred amino acids for transport
System A	alanine, glycine, proline, serine, methionine
System ASCP	alanine, serine, cysteine, proline
System L	leucine, isoleucine, valine, phenylalanine, methionine, tyrosine, tryptophan
System Ly	lysine, arginine, ornithine, histidine
Dicarboxylate system	glutamate, aspartate
β system	taurine, β-alanine
N system	glutamine, asparagine, histidine

* See Guidotti *et al.* (1978), Morgan *et al.* (1979), Kilberg *et al.* (1980).

carriers, which have overlapping specificity for the different amino acids. These are known as the A, ASCP, L, Ly, dicarboxylate, N, and β systems. System A transports alanine, glycine, and other neutral amino acids with short side-chains. System ASCP transports alanine, serine, cysteine, and proline and has a higher specificity than system A. System L transports leucine and also neutral amino acids with branches or aromatic side-chains. The Ly system transports basic amino acids such as lysine, arginine and ornithine. Dicarboxylic amino acids, glutamic and aspartic acids are transported by a low activity carrier that is also not very specific. The N system transports the amides glutamine and asparagine together with histidine. The β system transports β-alanine and taurine (see Table 10.5).

It should be emphasized that the specificities of the systems are not absolute and will depend upon several factors including the concentrations of the amino acids.

A totally different mechanism for the translocation of amino acids into cells has been proposed by A. Meister and is known as the γ-glutamyl cycle (Figure 10.1). Six enzymes are involved in this pathway, the central one being γ-glutamyltransferase which is membrane bound. It catalyses the transfer of a glutamyl residue from the tripeptide glutathione (γ-glutamylcysteinylglycine) to the incoming amino acid to form a γ-glutamyl-amino acid and cysteinylglycine:

amino acid + γ-glutamylcysteinylglycine $\longrightarrow$

γ-glutamyl-amino acid + cysteinylglycine

The γ-glutamyl-amino acid is transported across the membrane and in the cytosolic compartment is cleaved by γ-glutamylcyclotransferase to release the amino acid. The other product of cleavage is 5-oxoproline, which is converted to glutamate and in turn back to glutathione in the cytosol. Three molecules of ATP are utilized in this resynthesis which serves to keep the γ-glutamyltransferase reaction out of equilibrium so that the amino acid is 'pulled' into the cell despite its higher concentration inside. There is evidence that this process is important in

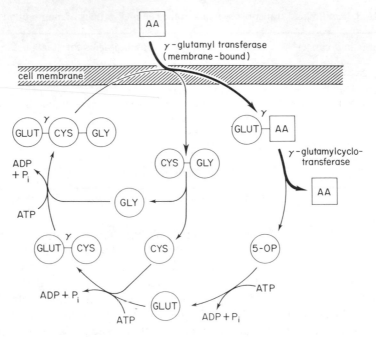

Figure 10.1 The γ-glutamyl cycle for the transport of amino acids across cell membranes. Each reaction is catalysed by a separate enzyme.

Abbreviations: AA, amino acid being transported; GLUT, glutamate; GLY, glycine; CYS, cysteine; 5-OP, 5-oxoproline

the re-absorption of at least some amino acids in the brush border of the kidney tubule cells. It may also be important in amino acid uptake in the erythrocyte and possibly the brain (Rosenberg and Tanaka, 1978; Meister, 1981).

2. Deamination of amino acids

The loss of the α-amino group is an early reaction in the pathway of degradation of most amino acids. In some cases it is lost as ammonia in an oxidation reaction (deamination) while in others it is transferred to an oxoacid (transamination). In both cases the product is usually an oxoacid. In a small number of amino acids, the loss of ammonia is accompanied by the elimination of water or hydrogen sulphide.

(a) Oxidative deamination

Both NAD^+ and flavin nucleotides can act as oxidizing agents in reactions which form oxoacids and release ammonia from amino acids. Both of the reactions can be represented as follows:

$$R-\underset{\underset{NH_3^+}{|}}{CH}.COO^- \xrightarrow{\underset{}{A \quad AH_2}} R-\underset{\underset{NH_2^+}{||}}{C}.COO^- \xrightarrow{\underset{}{H_2O \quad NH_4^+}} R-\underset{\underset{O}{||}}{C}.COO^-$$

amino acid   imino acid   2-oxoacid
            intermediate

However, the reactions by which the two oxidizing agents are regenerated are quite different. A number of enzymes catalysing amino acid oxidation are known; those whose action gives rise to ammonia are described below.

1. The first enzyme discovered that was capable of oxidatively deaminating an amino acid was a D-amino acid oxidase of broad specificity for amino acids, which utilizes FAD (as a prosthetic group on the enzyme, see Chapter 4; Section C.1.b) as the oxidizing agent. The $FADH_2$ does not enter the electron transfer chain but reacts with oxygen in a reaction catalysed by the same enzyme to produce hydrogen peroxide:

$$E.FADH_2 + O_2 \longrightarrow E.FAD + H_2O_2$$

The hydrogen peroxide is decomposed by the enzyme, catalase (Chapter 4; Section E.5.b).

$$H_2O_2 \longrightarrow H_2O + \tfrac{1}{2}O_2$$

Although high activities of D-amino acid oxidase are found in both liver and kidney, its physiological function is not immediately clear since D-amino acids are not common in nature. They do, however, occur particularly in the peptidoglycans of bacterial cell walls, so that the enzyme may be present in liver and kidney to ensure the rapid and irreversible degradation of any D-amino acids absorbed into the body. (The large bacterial population in the colon may provide a continuous source of D-amino acids which can enter the bloodstream.) Many D-amino acids are toxic since they inhibit enzymes that use the L-isomers. A detoxification role for the enzyme in liver is supported by the finding that germ-free rats (which lack bacteria in the large intestine) have extremely low levels of D-amino acid oxidase.

Also widely distributed, but at much lower activity, is a broad specificity L-amino acid oxidase. Although originally implicated in amino acid deamination, its low activity precludes a significant contribution to ammonia release. (The enzyme has a higher activity towards α-hydroxy acids, such as lactic acid, which may represent the physiological substrates for this enzyme.) This enzyme has a similar mechanism to that of D-amino acid oxidase but uses FMN (flavin mononucleotide) in place of FAD. Both the D-amino acid and L-amino acid oxidases are found, together with catalase and some other oxidases, in peroxisomes (known also as microbodies) in liver and kidney. It is possible that within these peroxisomes the enzymes are in close proximity so that little if any hydrogen peroxide accumulates thereby reducing the risk of formation of hydroxyl radicals in the cell (see Chapter 4; Section E.5.a).

2. The most important oxidative deamination reaction is undoubtedly that catalysed by the enzyme glutamate dehydrogenase:

$$^-OOC(CH_2)_2CH.COO^- + NAD(P)^+ + H_2O \longrightarrow$$
$$\underset{NH_3^+}{|}$$
$$^-OOC(CH_2)_2CO.COO^- + NAD(P)H + NH_4^+ + H^+$$

L-glutamate   2-oxoglutarate

This reaction proceeds via formation of the amino acid (see above) so that the oxidation can be seen as formally analogous to the oxidation of a secondary alcohol to a ketone but with nitrogen replacing oxygen. The imino acid is then hydrolysed to release ammonia.

Glutamate dehydrogenase is of central importance in deamination since the amino groups of many amino acids can be transferred to 2-oxoglutarate in the process of transamination, thus forming glutamate (see below). The subsequent deamination of glutamate via glutamate dehydrogenase maintains a supply of 2-oxoglutarate and permits transamination and thus deamination to continue. The combination of transamination and deamination via glutamate dehydrogenase is known as transdeamination. Since the glutamate dehydrogenase reaction is near-equilibrium in the liver, it can catalyse either the deamination of glutamate or the amination of 2-oxoglutarate by ammonia. The direction of the reaction will depend solely on the concentrations of participants in the reaction (see Krebs, 1973).* This has important clinical implications (see Section G).

Glutamate dehydrogenase is unusual amongst nicotinamide-linked dehydrogenases in that it can use either NAD^+ or $NADP^+$ as oxidizing agent. The significance of this is not known but the enzyme may serve to bring both nucleotides to the same redox state in the mitochondrion where the enzyme is located.

(b) *Transamination and transdeamination*

The enzymes known as aminotransferases (formerly as transaminases) are of particular importance in amino acid metabolism. Most amino acids can be converted to their respective oxoacids by aminotransferase reactions in which the α-amino group from one amino acid is transferred to the oxoacid of another, usually 2-oxoglutarate (or less frequently oxaloacetate) as follows:

$$R_1-\underset{\underset{NH_3^+}{|}}{CH}.COO^- + R_2-\underset{\underset{O}{||}}{C}.COO^- \longrightarrow R_1-\underset{\underset{O}{||}}{C}.COO^- + R_2-\underset{\underset{NH_3^+}{|}}{CH}.COO^-$$

---

* Its near-equilibrium status also makes it unlikely that the allosteric effectors of glutamate dehydrogenase, which are studied *in vitro* (see Frieden, 1976) are of any physiological significance in controlling the flux through this reaction *in vivo* (see Chaper 2; Section C.2).

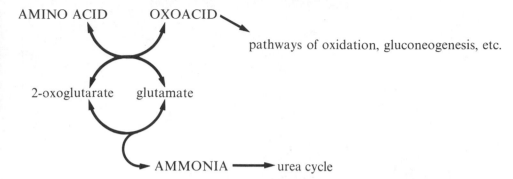

Figure 10.2   Transdeamination of amino acids. This pathway is used in reverse for the synthesis of amino acids

Although all but two amino acids appear able to enter into transamination reactions, it is not always clear how large a part these reactions play in the normal degradation of amino acids in the human liver. Further difficulties arise because the specificity of the aminotransferases is not always known and one aminotransferase may react with more than one amino acid (e.g. there is probably a single aminotransferase for reaction with isoleucine and leucine). The α-amino group of the amides, glutamine, and asparagine, can also be removed by transamination although deamidation (i.e. loss of the amide group as ammonia) is probably the major route of metabolism. A list of amino acids which can enter into aminotransferase reactions (together with the oxoacid formed) is given in Table 10.6. The existence of an aminotransferase reaction does not necessarily mean that transamination is a major pathway *in vivo*. There are no aminotransferase reactions for lysine and threonine but, in addition, histidine, serine, phenylalanine, and methionine are not metabolized to any significant extent by this reaction *in vivo*. This was indicated by early studies using ^{15}N, administered either as ammonia or an amino acid; the ^{15}N was not incorporated into these amino acids although all other amino acids were labelled. This finding is of importance in the clinical use of oxoacids of the essential amino acids (see Section G).

Braunstein and Bychkov (1939) first suggested that the coupled action of an aminotransferase and glutamate dehydrogenase might effect the oxidative deamination of a number of amino acids as follows:

$$\text{amino acid} + \text{2-oxoglutarate}^{2-} \rightleftarrows \text{glutamate}^- + \text{oxoacid}^-$$

$$\text{glutamate}^- + \text{NAD(P)}^+ + \text{H}_2\text{O} \rightleftarrows \text{2-oxoglutarate}^{2-} + \text{NAD(P)H} + \text{NH}_4^+ + \text{H}^+$$

over-all:

$$\text{amino acid} + \text{H}_2\text{O} + \text{NAD(P)}^+ \longrightarrow \text{oxoacid}^- + \text{NAD(P)H} + \text{NH}_4^+ + \text{H}^+$$

Table 10.6. Aminotransferase enzymes, oxoacids, and eventual common metabolic intermediates produced from amino acids

Amino acid	Aminotransferase	Oxoacid	Eventual common intermediate
Alanine	alanine aminotransferase	pyruvate	pyruvate
Arginine	ornithine-oxoacid aminotransferase*	glutamate-semialdehyde	2-oxoglutarate
Aspartic acid	aspartate aminotransferase	oxaloacetate	oxaloacetate
Asparagine	asparagine-oxoacid aminotransferase	2-oxosuccinamate†	oxaloacetate
Cysteine	cysteine aminotransferase	3-mercaptopyruvate†	pyruvate
	aspartate aminotransferase*	3-sulphinylpyruvate	
Glutamic acid	most aminotransferases	2-oxoglutarate	2-oxoglutarate
Glutamine	glutamine-oxoacid aminotransferase	2-oxoglutaramate†	2-oxoglutarate
Glycine	glycine aminotransferase	glyoxylate†	none
Histidine	histidine aminotransferase	imidazol-5-yl pyruvate†	2-oxoglutarate
Isoleucine	leucine aminotransferase	2-oxo-3-methylvalerate	acetyl-CoA / succinyl-CoA
Leucine		2-oxo-4-methylvalerate	acetyl-CoA
Lysine	2-aminoadipate aminotransferase*	2-oxoadipate	acetyl-CoA
Methionine	none normally	2-oxobutyrate	succinyl-CoA
	glutamine-oxoacid aminotransferase)	4-methylthio-2-oxobutyrate†	
Phenylalanine	tyrosine aminotransferase*	p-hydroxyphenylpyruvate	acetyl-CoA
	(phenylalanine aminotransferase)	phenylpyruvate†	fumarate
Proline	none	none	2-oxoglutarate
Serine	serine-pyruvate aminotransferase	3-hydroxypyruvate†	pyruvate
Threonine	none	2-oxybutyrate	succinyl-CoA
Tryptophan	none	indole pyruvate†	acetyl-CoA, alanine
Tyrosine	tyrosine aminotransferase	p-hydroxyphenylpyruvate	acetyl-CoA / fumarate
Valine	valine aminotransferase	3-oxo-3-methylbutyrate	succinyl-CoA

* Not acting on amino acid itself. † Not formed in major degradative pathway.

These combined reactions (depicted in Figure 10.2) may be involved in deamination of perhaps the majority of the amino acids in higher animals but, as indicated above, the quantitative importance in mammals is still unclear. This lack of quantitative information is compounded by the fact that the aminotransferase enzymes are widely distributed and amino acid metabolism occurs in tissues other than liver (see Section F). Indeed, the metabolism of aspartate, asparagine, glutamate, and glutamine from protein digestion occurs almost exclusively in the absorptive cells of the small intestine producing a high proportion of ammonia required for the urea cycle. Despite this caution, the concept of the central role of transdeamination in the overall process of amino acid metabolism within the whole animal is essential to any understanding of amino acid catabolism and its relationship to other processes and is discussed in Section D.

The reactions catalysed by the aminotransferases and by glutamate dehydrogenase are close to equilibrium so that, provided oxoacids are available, the over-all process can be readily reversed and amino acids can be synthesized as well as degraded (see Sections E and G). Nonetheless, after a normal mixed meal, the usual direction in most animals is likely to be transdeamination with the production of ammonia. In order to maintain the reaction in this direction, the oxoacid and the ammonia must be further metabolized and $NAD(P)^+$ must be regenerated from the NADP(H). The latter occurs via the electron transfer chain; the ammonia is converted to urea as described in detail in Chapter 12 and the metabolism of the oxoacids is described below.

One feature that the aminotransferases share (together with many other enzymes involved in amino acid metabolism) is a dependence on pyridoxal phosphate (from vitamin B_6, see Appendix A), which is covalently bound to a lysine residue in the enzyme. During transamination it forms a transient covalent complex (Shiff's base) with the amino acid, rearrangement then occurs and the oxoacid is split off leaving the amino group attached to the pyridoxal phosphate (that is, forming pyridoxamine phosphate). This is then able to aminate the oxoacid involved in transamination (Figure 10.3).

Aminotransferases are found in both mitochondria and the cytosol but, since the highest activities of these enzymes are found in the latter compartment, and glutamate dehydrogenase is localized exclusively in the mitochondria, it is likely that amino groups from the various amino acids are collected onto glutamate within the cytosol and the glutamate is then transported into the mitochondrion by the glutamate carrier which, therefore, plays a central role in amino acid metabolism.

Both aspartate and alanine aminotransferases[8] are released into the blood when tissues are damaged and their levels in serum are widely used in diagnosis (see Chapter 3; Section D. 1.b). In particular, the high activity of aspartate aminotransferase in the cytosol of the liver means that any damage to the liver cells results in the leakage of this enzyme into the bloodstream where it is readily detected.

Over-all reaction: $R_1CH.COO^- + R_2CO.COO^- \longrightarrow R_2CH.COO^- + R_1CO.COO^-$

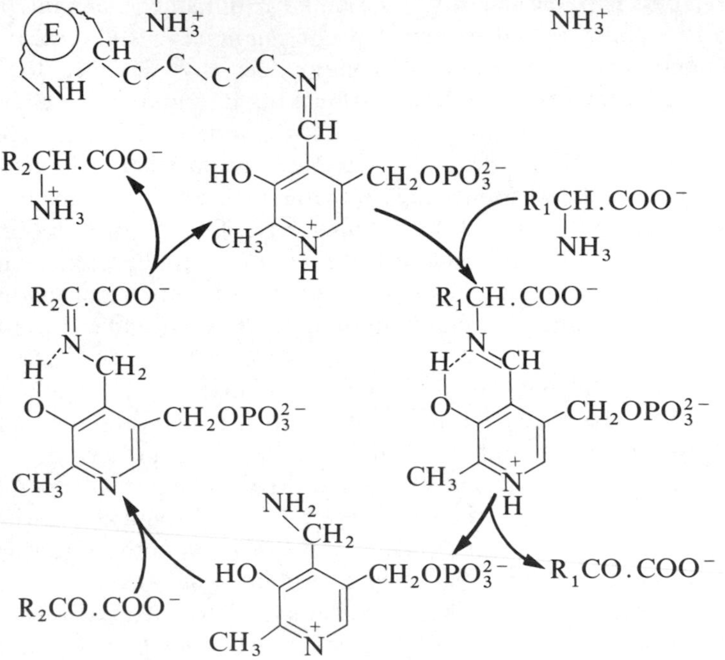

Figure 10.3 Mechanism of the transamination reaction involving pyridoxal phosphate. The pyridoxal phosphate is attached to the aminotransferase (E) non-covalently in addition to the covalent link through a lysyl side-chain. Schiff's base I is shown in aldimine form and Schiff's base II in ketimine form. The forms are interconvertible (tautomerism). Other reactions of amino acids also involve Schiff's base formation with a pyridoxal phosphate prosthetic group. This can undergo decarboxylation, elimination of water or racemization, depending upon the enzyme and amino acid involved

(c) *Dehydration*

The two hydroxy-amino acids (serine and threonine) can be deaminated by a dehydratase enzyme. As with oxidative deamination (above) an imino intermediate is formed. The water produced by the dehydration is then used to hydrolyse the imino intermediate to produce the oxoacid. The same enzyme deaminates both serine and threonine producing pyruvate and 2-oxobutyrate respectively:

$$HO.CH_2CH.COO^- \xrightarrow{H_2O} CH_2{=}C.COO^- \longrightarrow CH_3C.COO^- \xrightarrow{H_2O}$$

serine $\underset{NH_3^+}{|}$ $\underset{NH_3^+}{|}$ $\underset{NH_2^+}{\|}$

$$CH_3CO.COO^- + NH_4^+$$

$$\underset{\text{threonine}}{\underset{\underset{CH_3\ NH_3^+}{|\quad|}}{HO.CH\ CH.COO^-}} \xrightarrow{\ \overset{H_2O}{\nearrow}\ } \underset{\underset{CH_3\ NH_3^+}{|\quad|}}{CH{=}C.COO^-} \longrightarrow \underset{\underset{CH_3\ NH_2^+}{|\quad\ |}}{CH_2.C.COO^-} \xrightarrow{\ \overset{H_2O}{\nearrow}\ }$$

$$\underset{\underset{CH_3}{|}}{CH_2CO.COO^-} + NH_4^+$$

Other pathways of degradation are available for both amino acids (Appendix 10.1).

(d) *Desulphydration*

Cysteine, a sulphur-containing amino acid can be deaminated by a reaction that appears analogous to that described above, in that the products are pyruvate, ammonia, and hydrogen sulphide. However, the mechanism is different as cysteine and thiocysteine are involved as intermediates in a cyclic sequence. The reactions are catalysed by the enzyme cystathionine-γ-lyase:

$$\underset{\underset{NH_3^+}{|}}{HS.CH_2CH.COO^-} + H_2O \longrightarrow CH_3CO.COO^- + NH_4^+ + H_2S$$

Other pathways for cysteine degradation are described in Appendix 10.1.

3. **Pathways of oxoacid degradation**

It is possible to divide amino acids, somewhat arbitrarily, into three classes according to their degradative pathway: those amino acids that are converted via a simple degradation pathway to a common metabolic intermediate (see Section C.4); those amino acids that are converted via a more complex degradation pathway; and those amino acids that are converted to another amino acid prior to degradation.

(a) *Simple degradation pathways*

The transamination of some amino acids gives rise directly to a common metabolic intermediate. For example, alanine gives rise to pyruvate, glutamate to 2-oxoglutarate and aspartate to oxaloacetate.

(b) *Complex degradative pathways*

Amino acids whose oxoacids are metabolized via a more complex pathway include the following: cysteine, tyrosine, lysine, tryptophan, leucine, methionine,

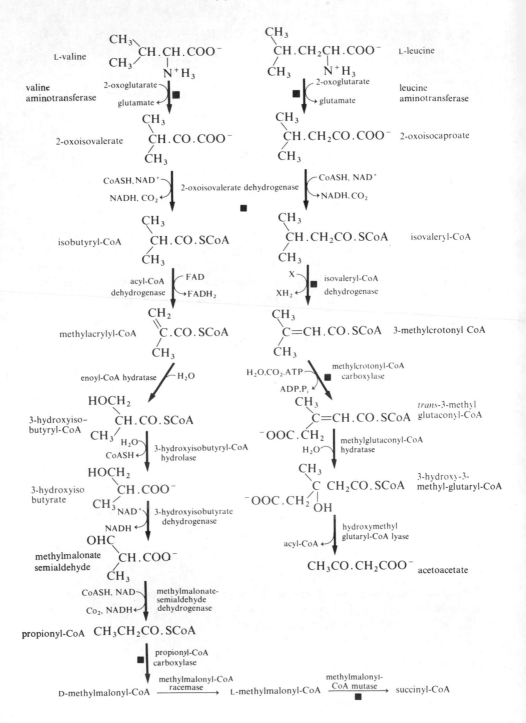

Table 10.7. Common metabolic intermediates derived from amino acids

Amino acid source	Common intermediate
Alanine, glycine, serine, cysteine, tryptophan	pyruvate
Arginine, histidine, proline, glutamine, glutamate	2-oxoglutarate
Valine, isoleucine, methionine, threonine	succinyl-CoA
Phenylalanine, tyrosine	fumarate
Asparagine, aspartate	oxaloacetate
Leucine, phenylalanine, tyrosine, lysine, tryptophan, isoleucine	acetyl-CoA*

* Acetyl-CoA is formed directly from these amino acids and not via pyruvate.

valine, and isoleucine (see Appendix 10.1 for full details). Some of these share common pathways. The pathways of catabolism of leucine and valine are given in Figure 10.4 to illustrate this class. These amino acids were selected since their metabolism (both the early and the late stages of the process) is discussed further in Section F. In a number of cases, different pathways are found in different animals and in some cases more than one pathway may function in the same animal.

(c) *Conversion of one amino acid into another*

Phenylalanine is converted into tyrosine prior to further degradation. The amino acids histidine, proline, and arginine are converted into glutamate which can then be deaminated in the glutamate dehydrogenase reaction (see Appendix 10.1).

4. Metabolism of 'common intermediates' derived from amino acids

The 'common intermediates' (i.e. intermediates of major catabolic pathways) arising from the metabolism of amino acids are acetyl-CoA, pyruvate, 2-oxoglutarate, succinyl-CoA, fumarate, and oxaloacetate (see Table 10.7). Since these intermediates do not accumulate in the cell they must undergo further conversion. The three major fates are as follows: conversion to acetyl-CoA and thence oxidation to carbon dioxide via the TCA cycle; conversion to acetyl-CoA and thence to long-chain acyl-CoA via the fatty acid synthesizing system and, ultimately, to triacylglycerol; and conversion to glucose through gluconeogenesis (except for acetyl-CoA).

Figure 10.4 Metabolic pathways for the degradation of valine and leucine. The pathway from propionyl-CoA to succinyl-CoA is described in more detail in Figure 6.11. Acetoacetate is converted to acetyl-CoA in extrahepatic tissues.
■ Reported enzyme deficiencies

(a) *Oxidation*

It is implicitly assumed in many textbooks that intermediates of the TCA cycle (e.g. 2-oxoglutarate, succinyl-CoA, fumarate and oxaloacetate) can be oxidized directly by the cycle, so that the pathway for amino acid oxidation is described as follows: conversion of the amino acid to oxoacid which, in turn, is converted to one of the cycle intermediates which is then oxidized within the cycle. This cannot be: the only compound of which the carbon-skeleton is oxidized by the cycle is acetate in the form of acetyl-CoA (Chapter 4; Section B.2). Entry into the cycle at positions other than acetyl-CoA will simply increase the concentrations of the intermediates unless they are subsequently removed from the cycle. The pathway for complete oxidation involves the conversion of the cycle intermediates to oxaloacetate, its removal from the cycle and its conversion to acetyl-CoA. This is achieved, in most tissues, as follows. Oxaloacetate is first converted to PEP in the reaction catalysed by phosphoenolpyruvate carboxykinase:

$$^-OOC.CH_2CO.COO^- + GTP^{4-}(IDP^{4-}) \longrightarrow$$

$$CH_2{=}\underset{\underset{\displaystyle OPO_3^{2-}}{|}}{C}{-}COO^- + GDP^{3-}(IDP^{3-}) + CO_2$$

The PEP is then converted to pyruvate by the pyruvate kinase reaction (as in glycolysis) and the pyruvate, after transport into the mitochondrion, is converted to acetyl-CoA by the pyruvate dehydrogenase complex (Figure 10.5). Thus, except for those amino acids that give rise to acetyl-CoA directly (see Table 10.7) all amino acids must be converted to pyruvate prior to complete oxidation. (This reaction sequence for converting oxaloacetate to pyruvate will be referred to later (Section F.2.a) when conversion of amino acids to alanine is discussed.)

(b) *Fatty acid synthesis*

An alternative fate of the acetyl-CoA produced from amino acid catabolism is to provide a precursor for fatty acid synthesis and this occurs in both adipose tissue and liver. Some of the carbon atoms of all the amino acids can be converted to acetyl-CoA either directly or via phosphoenolpyruvate as described above. Thus any excess amino acid in the diet can be used to synthesize lipids for storage so that the chemical energy can be made available by oxidation at a later time (see Chapter 17; Section A).

(c) *Gluconeogenesis*

An important role of the oxoacids derived from amino acids is to provide precursors for the synthesis of glucose. This will be especially important during starvation, when the amino acids arise from the hydrolysis of endogenous

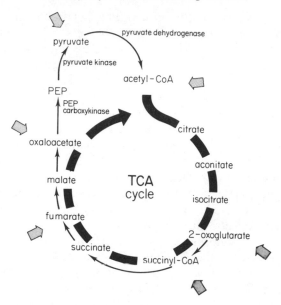

Figure 10.5 General pathway for the oxidation of amino acids. Carbon skeletons of amino acids enter at the positions indicated by stippled arrows and are converted to acetyl-CoA along the pathway indicated by the narrow arrows

proteins, and for carnivores or omnivores eating a meat-rich diet. The reactions of gluconeogenesis are discussed in detail in Chapter 11 (Section B.2.a) but it is important to note, at this stage, that glucose can be synthesized from all the 'common intermediates' except acetyl-CoA. The TCA cycle results in the net loss of two carbon atoms so that acetate (in the form of acetyl-CoA) cannot form glucose. It will be seen from Table 10.7 that leucine and lysine are the only amino acids from which no glucose can be synthesized.

D. CENTRAL ROLE OF TRANSDEAMINATION IN AMINO ACID METABOLISM

The D-amino acids are deaminated by an amino acid oxidase which catalyses a non-equilibrium reaction. The question arises why the natural amino acids are not deaminated by a similar and apparently simple non-equilibrium system? There are several possible answers to this question.

First, the transdeamination process provides metabolic economy. By means of a small number of near-equilibrium reactions, most amino acids can be maintained in equilibrium with their corresponding oxoacids and with each other (Figure 10.2). Thus an amino acid can be deaminated or an oxoacid aminated according to the prevailing concentrations of other amino acids and oxoacids. Provided the oxoacid is available, the corresponding amino acid will be maintained at a given concentration dependent upon concentrations of other amino acids. Such an inter-relationship would not be possible with a non-equilibrium deamination system. Furthermore, having a common amino acid (glutamate), to which most others are related by transamination, requires many fewer specific enzymes for interconversion than if separate enzymes were required for each transamination.[9] Further economy is achieved if the central intermediate is also deaminated. Secondly, the near-equilibrium transdeamination system provides a simple mechanism for maintaining the concentrations of both amino acids and oxoacids fairly constant despite variations in the magnitude and direction of the flux through this system. In other words, the role of near-equilibrium reactions in providing sensitivity for the control of flux, which has been discussed in Chapter 7, can be viewed conversely; large changes in both magnitude and direction of flux can be accommodated with only small changes in the concentrations of substrates and products.

With such a near-equilibrium system, the direction and the magnitude of the flux must be imposed by other reactions. There are at least seven processes that feed reactants into or out of the transdeamination system (Figure 10.6); supply of amino acids from the hydrolysis of dietary protein, supply of amino acids from the hydrolysis of body protein, use of amino acids for protein synthesis, oxidation of the oxoacids, conversion of oxoacids to glucose (or other carbohydrate, e.g. glycogen), conversion of oxoacids to lipid and, finally, conversion of glucose to some of the oxoacids. It is the rates of these various processes that will control the magnitude and direction of the flux through the transdeamination system. Not surprisingly, the control of the rates of these diverse processes and their integration within the whole animal is complex, but some of the mechanisms are discussed in Chapters 11, 14, 17, and 20 and this chapter, Section F.

Some of the important metabolic consequences of the integration achieved through this transdeamination system can be seen by following the changes that occur when the rate of one or more of the related processes is altered. Three examples are given.

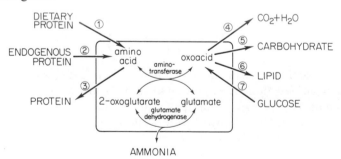

Figure 10.6   The central role of transdeamination reactions. 1. Digestion and absorption. 2. Protein degradation in the tissues. 3. Protein synthesis. 4. Oxidation. 5. Gluconeogenesis or glyconeogenesis. 6. Lipid synthesis (via acetyl-CoA). 7. Glycolysis etc. There is no metabolic significance to the 'box' except to separate the near-equilibrium reactions from the non-equilibrium.

1. During the ingestion of excess protein (in a mixed diet), the rate of amino acid supply from the intestine (process 1 in Figure 10.6) is increased, and any amino acids not required for synthesis of protein (process 3) will produce oxoacids that can be oxidized or converted to carbohydrate and lipid (processes 4, 5, and 6).

2. During starvation, the rate of degradation of body protein (process 2) exceeds the rate of protein synthesis (process 3) and the available amino acids are converted to oxoacids. However, because of inhibition of pyruvate oxidation (process 4) and lipid synthesis (process 6), the oxoacids are converted preferentially to glucose (in the liver) for use by the brain (see Chapter 14).

3. Application of the properties of near-equilibrium systems to transdeamination provides a metabolic explanation for the 'protein-energy malnutrition' disease, such as marasmus and kwashiorkor, described in Section B.1.a. It could be argued that if protein intake became very low, protein turnover, which has to be maintained even at a basic rate to remove unwanted and denatured protein, etc., could be self-sufficient in amino acids. In other words, those amino acids produced from protein degradation could be husbanded solely for the process of protein synthesis (i.e. process 3 would be supplied solely by process 2, in Figure 10.6). In this way, the protein content of the body would be maintained approximately constant.* That this does not happen is explained by the near-equilibrium nature of the transdeamination process. It is not possible to inhibit the near-equilibrium interconversions completely (although the concentrations of the enzymes are reduced on protein-poor diets, Section F.1), so that, if amino acids are available, as they must be for protein synthesis, a proportion will always suffer degradation via the transdeamination system. In marasmus (or starvation) oxoacids will be converted to glucose, whereas in kwashiorkor they will be oxidized to carbon dioxide since these metabolic pathways will not be inhibited (Figure 10.6).

This brief account of the integration of the various aspects of amino acid metabolism is somewhat oversimplified but it serves to illustrate the principles that can be applied to the changes that occur in amino acid metabolism under different conditions. One of the reasons for the further complexity is that amino acid metabolism occurs in various tissues and the extent of metabolism and its function may be different in different tissues. Some of these differences are pursued in Section F.

E. FORMATION OF NON-ESSENTIAL AMINO ACIDS

The non-essential amino acids are alanine, arginine, aspartate, cysteine, glutamate, glycine, proline, and serine (and asparagine and glutamine). The synthesis of

---

* There is a constant loss of protein in the faeces (from digestive enzymes and intestinal cells that are not fully digested and reabsorbed) from the skin in the form of desquamated cells and hair and in sweat but the largest is the obligatory loss in the form of urea in the urine (Munro and Crim, 1980).

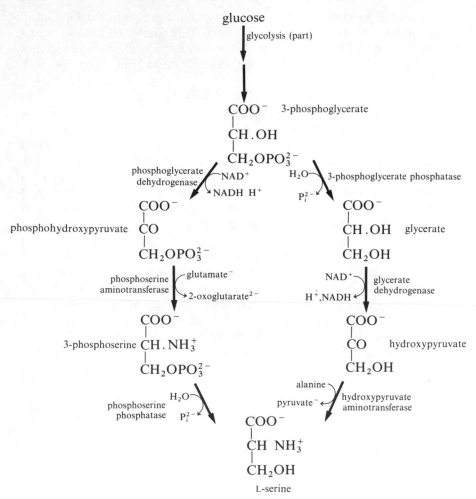

Figure 10.7 Alternative pathways for the synthesis of serine. The pathways differ in the order of reactions. The 'non-phosphorylated' pathway on the right is thought to be the more important

alanine, aspartate and glutamate has already been mentioned above. Since their oxoacids are part of normal metabolism they can, via transamination reactions, be formed directly from the oxoacid (Figure 10.2). In each case the carbon skeleton is obtained from glucose. For alanine, pyruvate is produced from glycolysis; for glutamate, oxoglutarate is produced in the TCA cycle having been formed from acetyl-CoA (via pyruvate and pyruvate dehydrogenase) and oxaloacetate (via pyruvate and pyruvate carboxylase); for aspartate, oxaloacetate is produced from pyruvate via pyruvate carboxylase.

The synthesis of the other amino acids is more complex (see Evered, 1981).

1. Serine

There are two pathways for serine biosynthesis, one termed the 'phosphorylated' pathway, the other the 'non-phosphorylated' pathway. Both pathways start with 3-phosphoglycerate, an intermediate of glycolysis. In the phosphorylated pathway, 3-phosphoglycerate is oxidized to hydroxypyruvate phosphate which is transaminated (serine-pyruvate aminotransferase) to form serine phosphate and the latter is hydrolysed to form serine. The non-phosphorylated pathway is similar except that the first reaction is the hydrolysis of 3-phosphoglycerate to form glycerate (Figure 10.7).

2. Glycine

This amino acid can be produced from serine or from glutamate. The latter is converted to 4-hydroxy-2-oxoglutarate which is cleaved to form pyruvate and glyoxylate. Transamination of glyoxylate (catalysed by alanine-glyoxylate aminotransferase) produces glycine (Figure 10.8).

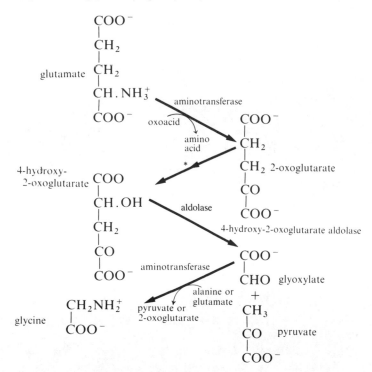

Figure 10.8 Pathway of glycine formation from glutamate. Several aminotransferases capable of catalysing the reactions shown are present in liver; it is possible that glycine aminotransferase catalyses both simultaneously.

* The pathway between 2-oxoglutarate and 4-hydroxy-2-oxoglutarate has not been established

3. Cysteine

This amino acid is produced during the degradation of methionine: methionine is converted to homocysteine, and the latter reacts with serine to produce cystathionine catalysed by the enzyme cystathionine β synthase. Cystathionine is cleaved to form cysteine and 2-oxobutyrate catalysed by cystathionine γ-lyase (see Appendix 10.1). Thus the sulphur in cysteine is obtained from methionine whereas the carbon is obtained from serine.

4. Arginine and Proline

These amino acids are produced from glutamate which is reduced to form glutamic acid γ-semialdehyde which when transaminated gives rise to ornithine (Figure 10.9). Via three reactions of the urea cycle (ornithine transcarbamylase,

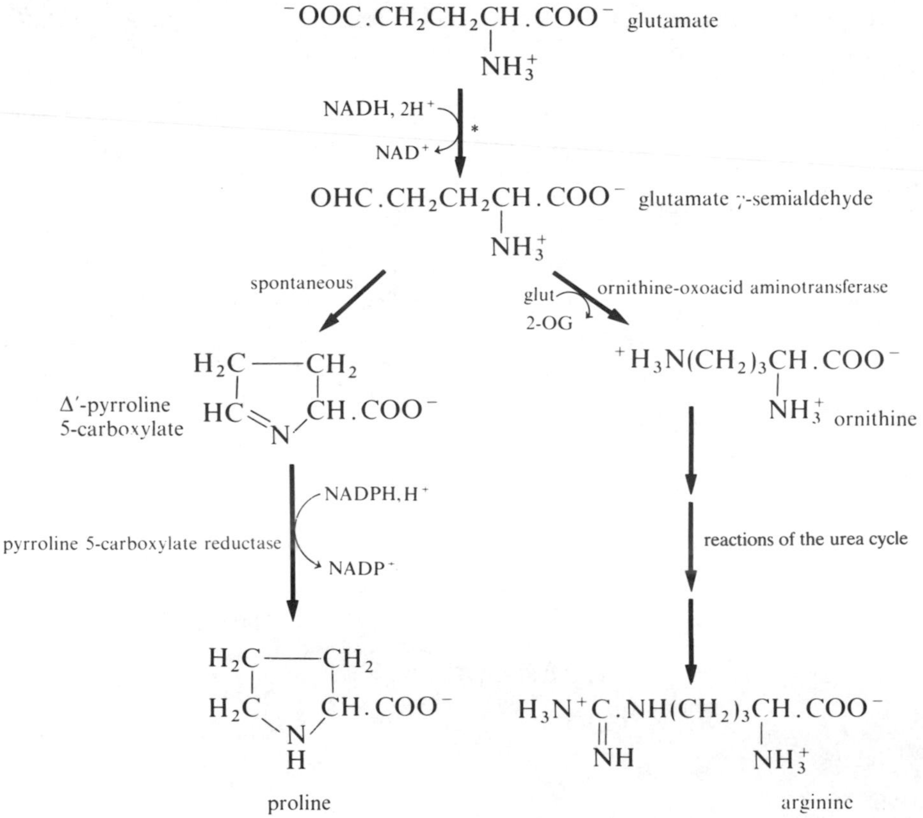

Figure 10.9 Pathway for the synthesis of proline and arginine from glutamate. * The nature of this reaction has not yet been established but may involve a preliminary phosphorylation by ATP. For details of the urea cycle reactions see Chapter 12; Section B.2

argininosuccinate synthetase, and argininosuccinate lyase) ornithine can be converted into arginine. In addition, glutamic acid γ-semialdehyde spontaneously cyclises to produce pyrroline 5-carboxylic acid, which can be reduced to proline (pyrroline 5-carboxylate reductase) (Figure 10.9). These reactions may occur in the liver but experiments in the rat suggest that the small intestine is quantitatively the most important tissue (Windmueller and Spaeth, 1974; Herzfeld and Raper, 1976).

5. Tyrosine

This amino acid is formed on the main pathway of phenylalanine degradation.

F. AMINO ACID METABOLISM IN THE TISSUES

Until relatively recently the reactions outlined in Figure 10.6 have been considered to be the prerogative of the liver. The metabolic picture is now known to be much less simple but definitely more interesting. One of the first indications that tissues other than the liver were involved in amino acid metabolism came from the work of L. L. Miller (1962) who showed that some amino acids could not be oxidized in the perfused liver but could be oxidized (by muscle) in the hepatectomized animal. For other amino acids the opposite was true (Table 10.8). Subsequent work has substantiated and extended this 'division of labour'.

1. Liver

Quantitatively the liver is probably the most important tissue for amino acid metabolism but other tissues such as the intestine, muscle, and adipose tissue are also very important. In general, the essential amino acids (with the exception of the branched-chain amino acids), together with some of the non-essential amino

Table 10.8. Amino acids metabolized by the liver and muscle*

Liver	Liver and muscle	Muscle
Arginine	Aspartate	
Histidine	Glutamate	Isoleucine
Lysine	Glutamine	Leucine
Methionine	Glycine	Valine
Phenylalanine	Proline	
Threonine	Alanine	
Tryptophan		

* Experiments carried out either on the isolated perfused liver and muscle or the hepatectomized animal (Miller, 1962). Some of these amino acids can be used by other tissues, see Section F, and glutamine and alanine are normally released by muscle (Table 10.9). In the fed state aspartate and glutamate are metabolized largely by the intestine.

acids, are degraded (as described in Section D and Appendix 10.1) in the liver. With a minimal protein diet, so that nitrogen balance is just maintained, the activities of the key enzymes for degradation of the essential amino acids (e.g. threonine dehydratase) and the aminotransferase enzymes are low. This ensures that the essential amino acids are protected from high rates of degradation and maintained for protein synthesis. If the intake of dietary protein exceeds the requirement for synthesis, the capacities of these pathways increase (due to increased concentrations of enzymes) and the rate of amino acid degradation rises accordingly.

Indeed it has been shown that the activities of the key enzymes for metabolizing at least some of the essential amino acids increase rapidly only when protein intake exceeds a certain level. This suggests that the liver monitors the demand by the body for the essential amino acids and demonstrates that the liver plays an important role in regulating the blood concentrations of many essential amino acids although the mechanisms by which activities of key enzymes in these pathways are regulated remain largely unknown (Munro and Crim, 1980).

An exception to this role of the liver is provided by the branched-chain amino acids, the blood concentrations of which are regulated by their metabolism in muscle.

2. Muscle

At least six amino acids can be oxidized by muscle, namely alanine, aspartate, glutamate, and the three branched-chain amino acids (leucine, isoleucine, and valine). The physiological significance of amino acid metabolism in muscle is considered to be, not so much complete oxidation, but the ability to convert some of these amino acids to alanine and glutamine which are released from the muscle. Measurements of the rate of amino acid release from the leg or forearm muscles of man first indicated the significance of these conversions. Samples of blood were taken (via in-dwelling catheters) from the femoral artery and the femoral vein for measurement of the concentrations of the amino acids (see Felig, 1975). These studies indicated that starvation caused a net release of amino acids from the muscle and, furthermore, that about 60 % of the amino acids released consisted of alanine plus glutamine (Table 10.9). That during starvation there is a net release of amino acids from skeletal muscle due to the hydrolysis of contractile proteins in order to provide amino acids for gluconeogenesis was no surprise but the preponderance of alanine and glutamine was surprising, since the total content of alanine plus glutamine in contractile proteins is little more than 10 %. It had been assumed that the amino acids would be released in the same proportion in which they were present in the protein being hydrolysed. The question naturally arose as to the origin of the alanine and glutamine and whether their production had any particular physiological importance. These questions have begun to be answered.

Table 10.9. Amino acids released from skeletal muscle of man

Amino acid	Arterio-venous difference*	
	μmol.l^{-1}	Percentage of total
Alanine	−70	30
Glutamine	−70	30
Glycine	−24	10
Lysine	−20	9
Proline	−16	7
Threonine	−10	4
Histidine	−10	4
Leucine	−10	4
Valine	−8	3
Arginine	−5	2
Phenylalanine	−5	2
Tyrosine	−4	2
Methionine	−4	2
Isoleucine	−4	2
Cysteine	+10	—
Serine	+10	—

* The concentration difference presented is that between blood in the femoral artery and in the femoral vein in the postabsorptive period of starvation. The minus sign indicates release from the muscle. Data taken from Felig (1975).

(a) *Origin of alanine*

Of the amino acids known to be oxidized by muscle, isoleucine, valine, glutamate, and aspartate can all give rise eventually to oxaloacetate, whereas leucine gives rise only to acetoacetate and acetyl-CoA. Thus, transdeamination of aspartate produces oxaloacetate directly; transamination of glutamate produces 2-oxoglutarate which can be converted to oxaloacetate by reactions of the TCA cycle; valine and isoleucine are converted, by a more lengthy route, to succinyl-CoA (see Figure 10.4) which is also an intermediate of the TCA cycle and is converted to oxaloacetate. The oxaloacetate is converted to PEP and hence to pyruvate, via the reactions catalyzed by phosphoenolpyruvate carboxykinase and pyruvate kinase (Figure 10.5). Once pyruvate is produced, one of two things can happen to it; it can, after transport into the mitochondria, be converted to acetyl-CoA, and then be completely oxidized via the TCA cycle, or it can be transaminated to alanine in a reaction catalyzed by alanine aminotransferase (Figure 10.10). In the normal fed state, it is likely that much of this pyruvate would be oxidized, thus accounting for the known ability of muscles to oxidize these amino acids. In starvation, or on a low carbohydrate diet, however, the situation is different. Very little pyruvate will be oxidized since the pyruvate dehydrogenase will be present in the inactive form (due to the oxidation of fatty acids and ketone bodies—see Chapter 8; Section B). Pyruvate then follows its alternative fate of transamination to form alanine. The amino group is obtained

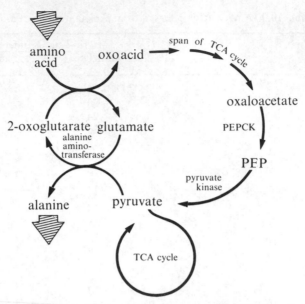

Figure 10.10   Pathway of alanine production from amino acids in muscle. Note that pyruvate can be aminated or oxidized, Abbreviation; PEPCK phosphoenolpyruvate carboxykinase.

indirectly from the original amino acid via transamination (Figure 10.10). Both the carbon and the nitrogen of the original amino acid have thus been used to produce alanine.

In the situation when pyruvate is oxidized, however, it would be necessary to find an alternative fate for the amino group produced in the initial aminotransferase reaction. Pyruvate, derived from glucose via glycolysis, could accept the amino group to form alanine (in which the carbon is not derived from amino acids) or the amino groups could be transferred to 2-oxoglutarate to form glutamate and thence to glutamine (see below).

The possibility that the pyruvate produced from glucose could form alanine led to the proposal of a glucose/alanine cycle operating between muscle and liver (Figure 10.11). In this scheme, it is envisaged that alanine, which was released by the muscle, was taken up by the liver and used to synthesize glucose which, in turn, returned to the muscle to provide pyruvate for continued alanine synthesis (Felig, 1975). The nitrogen atoms required for the synthesis of alanine would be obtained from the amino acids oxidized by the muscle. This cycle could operate, at least in principle, in the fed state although it is unlikely that all the alanine taken up by liver in the fed state would be converted to glucose. However, the cycle cannot be quantitatively important in starvation, since the rate of glucose utilization by muscle under this condition is very small (Chapter 8; Section B.1.b) and almost all the pyruvate used to synthesize alanine is derived from the amino acids (Newsholme, 1976; Snell, 1980). It is possible that the cycle, as depicted in Figure

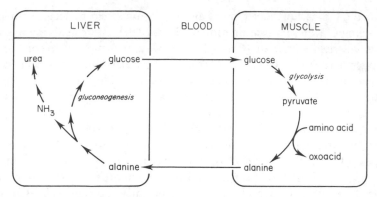

Figure 10.11 The glucose/alanine cycle

10.11, is never quantitatively important in either fed or the starved conditions.

Since leucine degradation gives rise only to acetyl-CoA, this amino acid, although oxidized in muscle, cannot give rise to pyruvate. The fate of the nitrogen atoms derived from leucine in the initial transamination may be to form glutamate from 2-oxoglutarate or to form alanine from pyruvate. In the latter case, the pyruvate would have to be derived from other amino acids or from glucose.

(b) Origin of glutamine

Since glutamine, like alanine, is released from skeletal muscle during starvation in amounts greater than its occurrence in the muscle protein, it too must arise as a result of amino acid metabolism. Glutamine is synthesized from glutamate and ammonia by the enzyme glutamine synthetase in the following reaction:

$$
\begin{array}{c}
\mathrm{COO^-} \\
| \\
\mathrm{(CH_2)_2} \\
| \\
\mathrm{CHNH_3^+} \\
| \\
\mathrm{COO^-} \\
\text{L-glutamate}
\end{array}
\; + \mathrm{ATP^{4-}} + \mathrm{NH_3} \longrightarrow
\begin{array}{c}
\mathrm{CONH_2} \\
| \\
\mathrm{(CH_2)_2} \\
| \\
\mathrm{CHNH_3^+} \\
| \\
\mathrm{COO^-} \\
\text{L-glutamine}
\end{array}
\; + \mathrm{ADP^{3-}} + \mathrm{P_i^{2-}}
$$

To understand the origin of glutamine in muscle, the source of both the ammonia and glutamate must be considered and, since neither appears to be taken up from the bloodstream in appreciable quantities by the muscle, they must be produced endogenously. Glutamate is undoubtedly formed from 2-oxoglutarate and the latter is produced in the first part of the TCA cycle from citrate which arises from the condensation reaction between acetyl-CoA and oxaloacetate, catalyzed by citrate synthase (see Chapter 4; Section B.1.a). Since all the amino acids metabolized by muscle can give rise to acetyl-CoA, and all except leucine to oxaloacetate, the carbon skeletons of any two amino acid molecules could give rise to 2-oxoglutarate. Transamination with one of the amino acids involved in its synthesis will yield glutamate (Figure 10.12). All that remains is to consider how the nitrogen of the second amino acid could be converted into ammonia and thus be incorporated into glutamine. There are two possible routes. The ammonia could arise from the glutamate dehydrogenase reaction in a similar manner to ammonia production in the liver (Section C.2.a). Alternatively, the ammonia could arise from the deamination of AMP to inosine monophosphate (IMP) catalyzed by AMP deaminase. The AMP would be regenerated by the other reactions of the pathway known as the purine nucleotide cycle (Chapter 12; Section A.4) in which the nitrogen is provided from aspartate. Both glutamate dehydrogenase and the enzymes of the purine nucleotide cycle are sufficiently active in skeletal muscle to account for the necessary ammonia production but it is unclear which is the more important pathway (Snell, 1980).

(c) *Significance of amino acid metabolism in muscle*

Although the extent of amino acid metabolism in human muscle has only recently been appreciated, the reactions involved are similar, if not the same, as those occurring in liver which have been known for many years. The fundamental difference is that amino acid catabolism in muscle proceeds only far enough to generate acceptors (i.e. pyruvate and 2-oxoglutarate) which combine with 'ammonia' to form non-toxic compounds (alanine and glutamine) that can be transported safely in the bloodstream. What, then, is the significance of this partial oxidation occurring in muscle rather than liver? There are at least three possible answers. First, the partial oxidation of the amino acids yields a considerable amount of ATP. For example, the conversion of one molecule of valine and one molecule of leucine to glutamine, as depicted in Figure 10.12 produces 16 molecules of ATP. Nonetheless, the absolute rate of branched-chain amino acid catabolism in skeletal muscle is low (about 0.01 μmol.min^{-1}.g^{-1}) and it would only provide about 10% of the ATP requirement for resting muscle (about 1–2 μmol.min^{-1}.g^{-1}), although the complete oxidation of the amino acids to carbon dioxide would double this rate of ATP formation. It might be suggested that the other amino acids that can be completely oxidized by muscle (e.g. aspartate and glutamate) might provide considerably more energy since their rates of oxidation could be higher. However, the evidence indicates that, when aspartate and glutamate arise from dietary protein, they are metabolized in the intestine (Section B.2.c). Secondly, alanine is an important gluconeogenic precursor in the liver, so that its rate of production in muscle may permit the latter tissue to regulate the rate of hepatic gluconeogenesis during prolonged starvation (see Chapter 14; Section A.3). Thirdly, glutamine is required as an obligatory fuel by intestinal cells and other rapidly-dividing cells (Ardawi and Newsholme, 1982) and for ammonia production by the kidney, which is important in acid/base balance. The quantitatively important source of

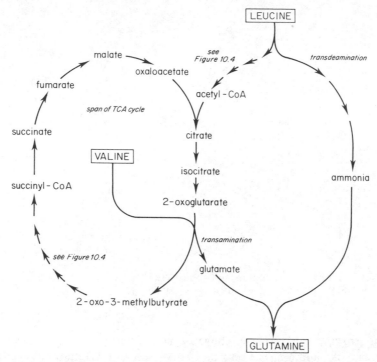

Figure 10.12 Proposed pathway for the synthesis of glutamine in muscle. Any two amino acids can provide all the atoms for glutamine synthesis through established pathways. Amino acids not producing acetyl-CoA directly can do so by reactions indicated in Figure 10.5. Alternatively glucose can provide the acetyl-CoA and/or the oxaloacetate but ammonia and amino-nitrogen must be available to form glutamine

glutamine, other than the lumen of the intestine, is muscle. The ability of muscle to control the rate of conversion of amino acids to either alanine or glutamine may be part of a complex regulatory mechanism that provides alanine for hepatic gluconeogenesis, glutamine for acid/base balance, and glutamine for the small intestine, or other rapidly-dividing cells when required (see Chapter 13; Section C.2).

3. Kidney

Ammonia appearing in the urine is not derived from the glomerular filtrate but from amino acid metabolism in the kidney. The major precursor is glutamine (Chapter 13; Section B.4.c) and current research concerns the mechanism of regulation of this pathway (Chapter 13; Section C).

There is substantial evidence that both nitrogen atoms of glutamine can give rise to ammonia. The pathway for the metabolism of glutamine in the kidney is shown in Figure 10.13. Glutamine is taken up from both the blood and the glomerular filtrate by the proximal and distal tubules of the nephron. Within the

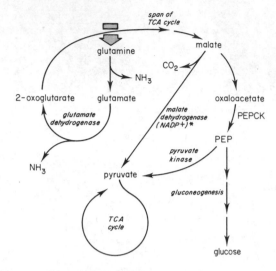

Figure 10.13 Pathway of glutamine metabolism in kidney. Compare with Figure 10.14.
 * This enzyme is also known as the malic enzyme

cell, the enzyme glutaminase converts glutamine to glutamate with the production of ammonia. There are two separate glutaminase enzymes (isoenzymes) in the kidney: one, present in mitochondria, requires phosphate ions for full activity and is known as phosphate-dependent glutaminase; the other is active in the absence of phosphate and is localized on the luminal side of the luminal membrane. The mitochondrial enzyme is considered to play the major role in ammonia production, so that most of the glutamine that is metabolized by the kidney must traverse the mitochondrial membrane. This is achieved by a carrier mechanism linked to the high-energy state of mitochondria (that is, an energy-dependent transport process—see Chapter 4; Section D.2). The glutamate produced by glutaminase is converted to 2-oxoglutarate by glutamate dehydrogenase accounting for the second molecule of ammonia that is produced from the glutamine. The 2-oxoglutarate is converted to malate by reactions that are normally considered part of the TCA cycle. The malate is transported out of the mitochondria and is then either converted to oxaloacetate by the malate dehydrogenase reaction or to pyruvate via the 'malic' enzyme; in the former case, oxaloacetate is converted to PEP by the enzyme phosphoenolpyruvate carboxykinase and PEP is then converted to pyruvate by pyruvate kinase. Both pathways may be involved in the formation of pyruvate in the kidney (Figure 10.13). The pyruvate, after entering the mitochondrion, is converted to acetyl-CoA by pyruvate dehydrogenase for complete oxidation in the TCA cycle. The over-all reaction can be summarized:

$$\text{glutamine} + 4\tfrac{1}{2}O_2 \rightarrow 2NH_3 + 5CO_2 + 2H_2O$$

Since fructose-bisphosphatase and glucose-6-phosphatase are present in kidney, PEP can also be converted to glucose via the gluconeogenic pathway described in Chapter 11. This over-all reaction can be summarized:

$$2H_2O + 2\text{glutamine} + 3O_2 \rightarrow \text{glucose} + 4CO_2 + 4NH_3$$

Whether PEP is converted to acetyl-CoA for oxidation or to fructose bisphosphate for glucose formation will depend upon the conditions. For example, during starvation almost all the PEP produced from glutamine is converted into glucose. Under these conditions, the oxidation of pyruvate in the kidney is severely inhibited since pyruvate dehydrogenase is present in the inactive form (see Chapter 8; Section B.1). The physiological role of the glutamine pathway in maintenance of body pH by the kidney will be discussed in detail in Chapter 13; Section B.4.c.

4. Intestine

It has only recently been appreciated that certain amino acids (particularly glutamine, glutamate, aspartate and asparagine) are metabolized within the cells of the small intestine. Metabolism consists of oxidation of the amino acids to carbon dioxide or conversion to lactate, alanine, or citrulline (Chapter 12; Section B.2). Any nitrogen derived from the metabolism of these amino acids and not retained in alanine or citrulline is released into the bloodstream as ammonia. In this way, a considerable proportion of glutamine and asparagine and perhaps most of the aspartate and glutamate present in the intestinal lumen are metabolized in the absorptive cells of the small intestine (Windmueller and Spaeth, 1980). However, during starvation (and even during the overnight fast) the amount of these amino acids available in the intestinal lumen will be decreased. In this condition, most of the glutamine is obtained from the bloodstream and the oxidation of this glutamine and ketone bodies provide almost all the energy for the small intestine (Table 10.10; Parsons, 1979).

Table 10.10. Percentage contribution of various fuels to carbon dioxide production in the small intestine of the post-absorptive rat*

Fuel	Percentage contribution
Glutamine	35
Glucose	6.8
Lactate	4.6
Acetoacetate	24.1
Hydroxybutyrate	26.1
Fatty acids	3.4

* Data from Windmueller and Spaeth (1978).

Glutamine is required by rapidly-dividing cells (Krebs, 1980); it donates the amide nitrogen in several of the reactions of the pathway of purine nucleotide synthesis (Figures 18.7 and 18.8) and it produces aspartate also used in this pathway and in the pathway for the synthesis of pyrimidine nucleotides (Tate and Meister, 1973). A constant rate of oxidation of this amino acid may ensure that it is always available in this tissue to provide the nitrogen and aspartate whenever cell division is required. This may explain why glutamine is also readily oxidized by other cells capable of rapid division (e.g. lymphocytes—see Ardawi and Newsholme, 1983).

The metabolic pathway for the metabolism of glutamate and aspartate to alanine and carbon dioxide in the intestinal cells is probably similar to that described in Section C.4.a. The metabolism of glutamine by the intestine is, in general, similar to that described for the kidney but there are two specific differences (Hanson and Parsons, 1980); glutamate is converted to 2-oxoglutarate by transamination rather than by glutamate dehydrogenase; and pyruvate is formed from malate.

This occurs in a mitochondrial reaction catalysed by a decarboxylating malate dehydrogenase (known formerly as malic enzyme) which reacts with either NAD^+ or $NADP^+$ as follows:

$$malate^{2-} + NAD(P)^+ \longrightarrow pyruvate^- + NAD(P)H + CO_2$$

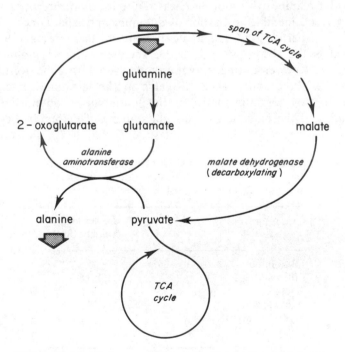

Figure 10.14   Pathway of glutamine metabolism in intestine

The pyruvate produced can either be oxidized via the TCA cycle or converted to alanine via an aminotransferase reaction (see Figure 10.14). At least 50% of the glutamine is completely oxidized in the absorptive cells.

The ammonia that is produced in the conversion of glutamine to glutamate by glutaminase diffuses into the hepatic portal blood from where it is taken up by the liver for urea formation (see Chapter 12; Section A.6).

5. Adipose tissue

Some years ago, it was shown that ^{14}C from $[^{14}C]$leucine or alanine was incorporated into triacylglycerol in isolated adipose tissue (Feller, 1965). This suggested that a capacity for amino acid metabolism existed in adipose tissue. More recent studies have confirmed this and have shown that amino acid metabolism in adipose tissue may be similar to that in muscle (Tischler and Goldberg, 1980). Thus branched-chain amino acids are metabolized by isolated adipose tissue with the release of alanine and glutamine. It is unclear how much of the carbon of the branched-chain amino acids will give rise to TCA cycle intermediates for the formation of these amino acids and how much is converted to acetyl-CoA for lipogenesis.

6. Brain

At least three amino acids are known to act as neurotransmitters in the brain, glutamate, aspartate, and glycine, and several other transmitters are derived from amino acids (e.g. 4-aminobutyrate from glutamate, noradrenaline, adrenaline, and dopamine from tyrosine and 5-hydroxytryptamine from tryptophan; see Chapter 21; Section B; Quastel, 1979). The inactivation of these neurotransmitters, essential for their function, frequently involves deamination. For example, catecholamine degradation by amine oxidase produces ammonia:

$$RCH_2NH_3^+ + \tfrac{1}{2}O_2 \longrightarrow RCHO + NH_4^+$$

and the glutamate dehydrogenase reaction may be involved in the inactivation of glutamate. However, glutamate may be inactivated as a neurotransmitter by being taken up by the glial cells and converted to glutamine. These reactions and others involving amino acids or their derivatives in brain are discussed in Chapter 21; Section B.6.

G. CLINICAL ASPECTS OF TRANSDEAMINATION

Two conditions in which the endproducts of nitrogen metabolism accumulate in the body are hepatic coma (ammonia accumulation) and chronic renal failure (urea accumulation). These nitrogen accumulation diseases are responsible for a large number of deaths every year. The specific problems of ammonia accumulation are discussed in Chapter 12; Section C. Chronic renal failure is usually treated by haemodialysis or kidney transplantation but both treatments are expensive, they involve considerable problems for the patient and neither may be available for treatment of the elderly (Close, 1974). However, simple dietary management of some patients can improve their condition markedly and knowledge of the process of transamination has played a role in the design of such diets. The aim of the management is to maintain adequate energy intake plus nitrogen balance and minimise formation of ammonia and urea (Walser, 1983).

In the early part of this century it was appreciated that low protein diets reduced many of the symptoms of the uraemic patient. Surprisingly, it was not until the 1960s that patients were advised to eat proteins with a high biological value since such proteins reduced the accumulation of ammonia and urea and provided good nutrition (Giovannetti and Maggiore, 1964). (There is less amino acid degradation since the amino acids are available in the correct proportions for protein synthesis—see Section B.1.b.) In the 1970s a further improvement was achieved by supplementing a poor protein diet with essential amino acids (Bergström *et al.*, 1975). The administration of essential amino acids daily, however, results in acidosis (probably from methionine metabolism—see Chapter 13; Section A). Use was therefore made of the fact that it is not the amino acids that are essential but their oxoacids—a fact pointed out by Rose in 1938. With the exception of lysine and threonine, all amino acids can be synthesized if the corresponding oxoacids are supplied in sufficient quantities* (Krebs and Lund, 1977). Uraemic patients were therefore given oxoacids of the essential amino acids (or their hydroxy analogues which can be converted to the oxoacids *in vivo*) together with lysine and threonine and a low protein diet (Heidland *et al.*, 1978). An additional reason for the use of oxoacids rather than amino acids was the evidence that a considerable amount of ammonia could be produced in such patients from the degradation of urea by intestinal bacteria. The oxoacids, via the near-equilibrium of the transdeamination system, would utilize this ammonia and produce the essential amino acids and in so doing reduce the concentration of the toxic ammonia. The treatment has proved to be beneficial, at least in some patients, in that it lowers blood concentrations of urea and ammonia (Walser, 1978; Richards, 1978). The treatment, of course, cannot replace dialysis in patients with chronic renal failure since dialysis removes other end products in addition to urea. Furthermore, it is expensive. However, it may be of value in treatment of patients with acute renal failure (e.g. after injury) and in chronic patients if dialysis, for any reason, is not available or is temporarily contraindicated. It may also be useful in treatment of children with a defective urea cycle (Chapter 12; Section C.2).

Somewhat surprisingly, in view of the benefit of this treatment, the importance of ammonia generation by bacterial metabolism of urea in the large intestine is now under question since it appears to be low, at least in man (Richards, 1978).

---

* The demands for histidine and tryptophan are low (Table 10.1) so that they may not need to be supplied for short term studies. Transamination of these amino acids does occur *in vivo* but the rates are low.

This means that oxoacids must reduce urea formation and favour positive nitrogen balance in some other way and there is some evidence that oxoacids (especially 2-oxo-4-methylvalerate, the oxoacid of leucine) can reduce the rate of endogenous protein breakdown. This would reduce the intracellular concentration of amino acids and hence their rate of deamination so that the rate of urea formation would be decreased (Sapir and Walser, 1977; see also Chapter 14; Section A.4).

A problem in severe energy restriction (e.g. starvation) for treatment of obesity is the negative nitrogen balance and loss of body protein, especially muscle. Infusion of a mixture of essential amino acids (lysine, threonine, histidine, and tryptophan) and oxoacids of isoleucine, leucine, valine, phenylalanine, and methionine in subjects undergoing prolonged starvation, decreased the rate of urea excretion and nitrogen balance was less negative (Sapir *et al.*, 1974). This sparing effect could not be solely explained by the increase in amino acid concentration since the protein sparing effect was observed after the oxoacids and amino acids had been metabolized.

One important feature of this work is that it illustrates the point that even serious clinical problems may be treated successfully by simple nutritional intervention. If recent investigations into the metabolic changes occurring during hibernation of the North American black bear are fruitful it may also illustrate the benefits to medicine of wider-ranging studies of metabolism in the animal kingdom. This animal can survive three to five months of the severe North American winter without eating or drinking and without urinating or defecating. Yet it can arouse itself from hibernation quickly into a mobile reactive state; indeed, the female delivers her cub and suckles it during this hibernation period (Nelson, 1973; Cahill, 1974). The mechanism by which the bear is able to reduce its urea formation without serious accumulation of ammonia is unknown but could have some relevance to the treatment of nitrogen accumulation diseases. The discussions in this chapter would suggest that a marked decrease in the rate of protein degradation (possibly caused by an increase in the concentration of oxoacids) and a very large decrease in the concentration of the aminotransferase enzymes could be responsible for the reduced rate of ammonia formation in the bear.

APPENDIX 10.1 SUMMARY OF PATHWAYS OF AMINO ACID DEGRADATION NOT DESCRIBED IN DETAIL ELSEWHERE IN THE TEXT

For many amino acids, a number of different degradative pathways have been described. Here, wherever information is available, only those pathways known to be quantitatively important in humans are included.

Inborn errors of metabolism, in which one enzyme of the catabolic pathway is deficient, have been particularly useful in providing information about the relative importance of degradative pathways (see Rosenberg and Scriver, 1980;

Wellner and Meister, 1981). Enzymes known to be deficient in a described inborn error of metabolism are marked ■. Some are exceedingly rare.

1. Arginine

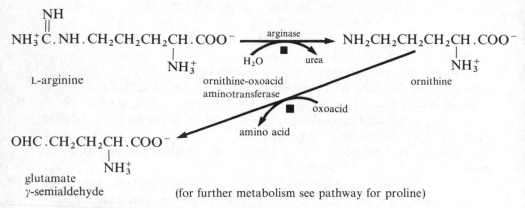

L-arginine

ornithine-oxoacid aminotransferase

ornithine

OHC.CH$_2$CH$_2$CH.COO$^-$

glutamate γ-semialdehyde

(for further metabolism see pathway for proline)

■ Deficiency of arginase gives rise to argininaemia and deficiency of ornithine oxoacid-aminotransferase gives rise to hyperornithinaemia.

2. Cysteine, cystine

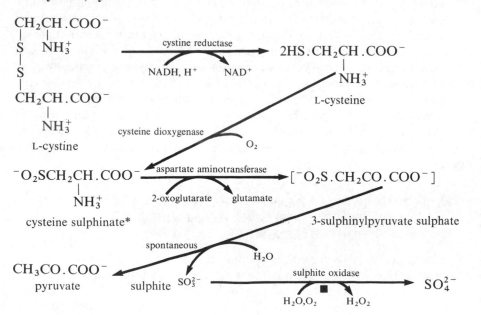

* Some cysteine sulphinate is decarboxylated and oxidized to taurine which is conjugated with bile acids (Chapter 20; Section C.1). Taurine is also synthesized in this way for its role as a neurotransmitter (Chapter 21; Section B.8).

■ There are no known deficiencies of cysteine dioxygenase but sulphite oxidase deficiency is characterized by neurological abnormalities and high urinary excretion rates of sulphite, thiosulphate and sulphocysteine.

Although this is probably the quantitatively important pathway for cysteine degradation in man, a number of other pathways occur:

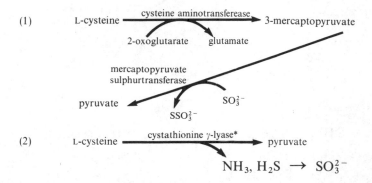

* This enzyme may normally be involved in cysteine synthesis, when cystathionine is the substrate.

3. Glycine, serine

These amino acids are readily interconverted. In addition they enter into a large number of synthetic reactions via the one-carbon pool. The major pathway of glycine degradation is via glycine synthase (the name given by the Enzyme Commission but in view of its importance in degradation perhaps the more frequently used name, glycine cleavage enzyme, is preferable). This reaction also appears to be involved in serine degradation, since serine is converted to glycine via the serine hydroxymethyltransferase reaction. The pathway of serine metabolism via hydroxypyruvate is of minor importance.

Serine is also involved in methionine catabolism (q.v.).

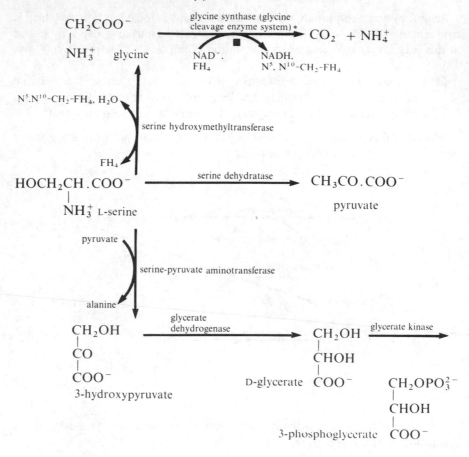

* This enzyme system involves four coenzymes; NAD^+, FAD, FH_4, and pyridoxal phosphate. There are also four protein components; P-protein which contains pyridoxal phosphate, H-protein which possesses a reactive sulphydryl group, T-protein which interacts with FH_4, and L-protein which contains FAD and has lipoyl dehydrogenase activity (Kikuchi, 1973).

■ A deficiency of any of the components of the cleavage enzyme system gives rise to nonketotic hyperglycinaemia, usually with severe cerebral consequences. It is of interest to note that a deficiency of either propionyl-CoA carboxylase or methylmalonyl-CoA mutase (Figure 6.11) gives rise to ketotic hyperglycinaemia in which the elevated blood glycine concentration is accompanied by severe ketoacidosis and hyperammonaemia. Since neither of these enzymes is involved in glycine metabolism the disturbances presumably result from effects on glycine metabolism caused by the accumulation of intermediates (e.g. propionyl-CoA or methyl-malonyl-CoA).

4. Histidine

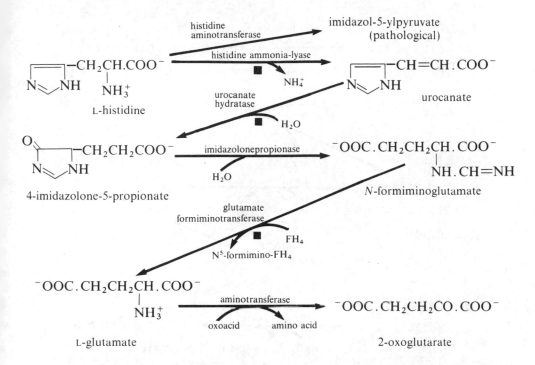

■ Deficiency of histidine ammonia-lyase (histidase) results in histidinaemia; deficiency of urocanate hydratase (urocanase) results in urocanic aciduria; deficiency of glutamate formiminotransferase results in excretion of formiminoglutamate in the urine.

5. Isoleucine

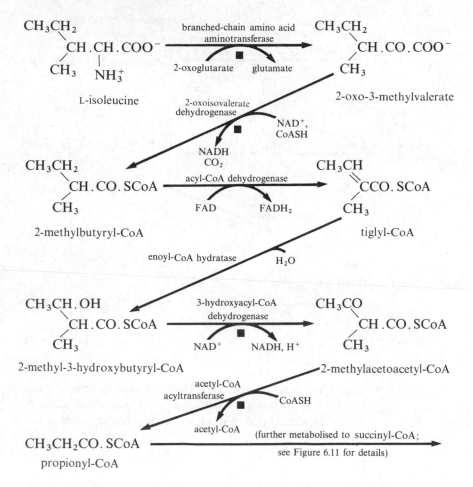

■ For the three branched-chain amino acids there are two aminotransferases; one that reacts with isoleucine and leucine (partial deficiency of which gives rise to leucine-isoleucinaemia) and one that reacts with valine (a deficiency of which gives rise to hypervalinaemia). In deficiency of 2-oxoisovalerate dehydrogenase (which may catalyse the oxidative decarboxylation of all three oxoacids of the branched-chain amino acids), the oxoacids accumulate and as the urine of infants affected has the odour of maple syrup, the disease is known as maple syrup urine disease (branched-chain ketoaciduria).

6. Lysine

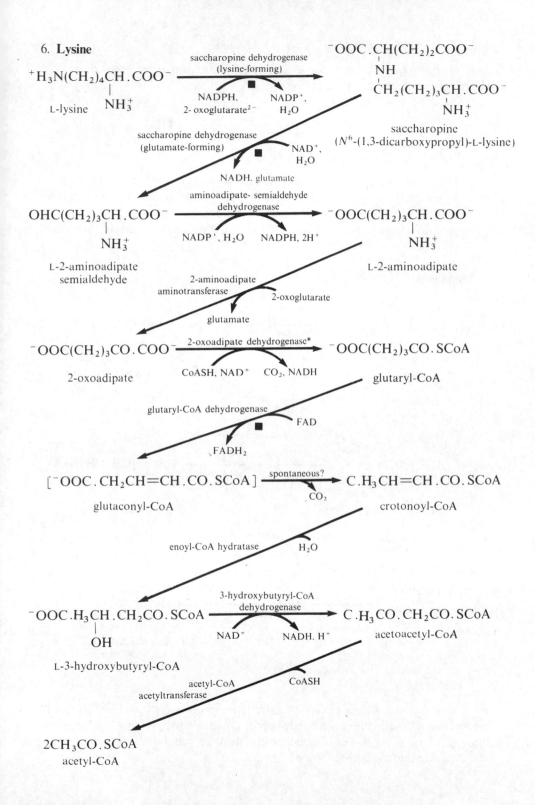

$^+H_3N(CH_2)_4CH.COO^-$

L-lysine NH_3^+

saccharopine dehydrogenase
(lysine-forming)

NADPH,
2-oxoglutarate^{2-}

NADP$^+$,
H_2O

$^-OOC.CH(CH_2)_2COO^-$

NH

$CH_2(CH_2)_3CH.COO^-$

NH_3^+

saccharopine
(N^6-(1,3-dicarboxypropyl)-L-lysine)

saccharopine dehydrogenase
(glutamate-forming)

NAD$^+$,
H_2O

NADH. glutamate

$OHC(CH_2)_3CH.COO^-$

NH_3^+

L-2-aminoadipate
semialdehyde

aminoadipate- semialdehyde
dehydrogenase

NADP$^+$, H_2O NADPH, 2H$^+$

$^-OOC(CH_2)_3CH.COO^-$

NH_3^+

L-2-aminoadipate

2-aminoadipate
aminotransferase

2-oxoglutarate

glutamate

$^-OOC(CH_2)_3CO.COO^-$

2-oxoadipate

2-oxoadipate dehydrogenase*

CoASH, NAD$^+$ CO$_2$, NADH

$^-OOC(CH_2)_3CO.SCoA$

glutaryl-CoA

glutaryl-CoA dehydrogenase

FAD

FADH$_2$

$[^-OOC.CH_2CH=CH.CO.SCoA]$

glutaconyl-CoA

spontaneous?

CO$_2$

$C.H_3CH=CH.CO.SCoA$

crotonoyl-CoA

enoyl-CoA hydratase H_2O

$^-OOC.H_3CH.CH_2CO.SCoA$

OH

L-3-hydroxybutyryl-CoA

3-hydroxybutyryl-CoA
dehydrogenase

NAD$^+$ NADH. H$^+$

$C.H_3CO.CH_2CO.SCoA$

acetoacetyl-CoA

acetyl-CoA
acetyltransferase

CoASH

$2CH_3CO.SCoA$
acetyl-CoA

This is the major pathway for lysine metabolism but there may be minor alternative ones, for example via pipecolate:

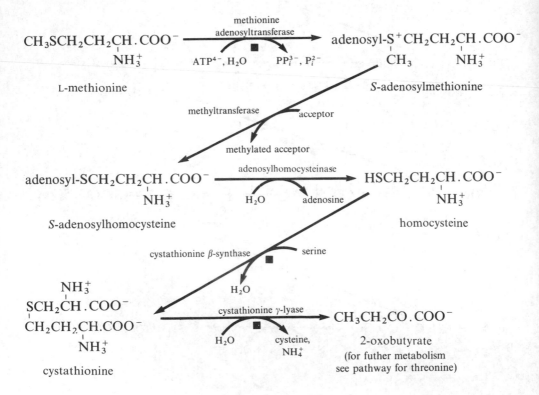

* May be identical to 2-oxoglutarate dehydrogenase.

■ Deficiency of saccharopine dehydrogenase (lysine-forming), also known as lysine-oxoglutaryl reductase, gives rise to hyperlysinaemia; deficiency of saccharopine dehydrogenase (glutamate-forming) gives rise to saccharopinuria and deficiency of glutaryl-CoA dehydrogenase to glutaric aciduria.

7. Methionine

Homocysteine can also be converted back to methionine by transfer of a methyl group from N^5-methyltetrahydropteroyltri-L-glutamate (Appendix A) in a reaction that involves vitamin B_{12} (cobalamin) and is catalysed by tetrahydropteroylglutamate methyltransferase:

$$HS(CH_2)_2CH.COO^- + N^5\text{-}CH_3\text{-}FH_4 \longrightarrow CH_3S(CH_2)_2CH.COO^- + FH_4$$

homocysteine $\overset{|}{NH_3^+}$ N^5-methyltetrahydropteroyl-triglutamate   methionine $\overset{|}{NH_3^+}$ tetrahydropteroyl-triglutamate

■ Deficiency of methionine adenosyltransferase gives rise to primary hypermethioninaemia. 4-Methylthio-2-oxobutyrate may also accumulate due to the action of branched-chain-2-oxoacid dehydrogenase on methionine. Deficiency of cystathionine β-synthase produces elevated concentrations of methionine and homocystine (derived from homocysteine). Deficiency of cystathionine γ-lyase results in asymptomatic cystathioninuria.

8. Phenylalanine, tyrosine

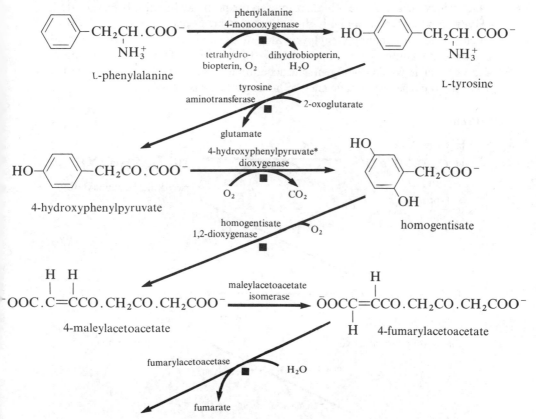

acetoacetate   (further metabolized to acetyl-CoA)

* Ascorbic acid required; activity is depressed in scurvy.

■ A deficiency of phenylalanine 4-monooxygenase (hydroxylase) produces the disease known as phenylketonuria, characterized by hyperphenyl-alaninaemia and mental retardation. However, this may now be described as classical phenylketonuria, since there are two other defects that can give rise to very low rates of phenylalanine metabolism; a deficiency of dihydrofolate reductase, which catalyses the NADH-dependent re-generation of tetrahydrobiopterin, and deficiency of an enzyme involved in the synthesis of dihydrobiopterin from GTP, known as dihydrobiopterin synthetase (see Figure 21.7). Since this cofactor is also required for hydroxylation of tyrosine and tryptophan to produce important neurotransmitters in the central nervous system, deficiency of these enzymes results in severe neurological symptoms that are not reversed or prevented by lowering the blood concentration of phenylalanine.

Deficiency of each of the other enzymes in the pathway has also been reported. 4-Hydroxyphenyllactate and 4-hydroxyphenylpyruvate accumulate in the case of deficiency of 4-hydroxyphenylpyruvate dioxygenase; succinylacetone and succinylacetoacetate accumulate in the case of deficiency of fumarylacetoacetase. Alcaptonuria, in which homogentisate accumulates in the blood and is excreted in the urine, is caused by a deficiency of homogentisate dioxygenase.

9. Proline

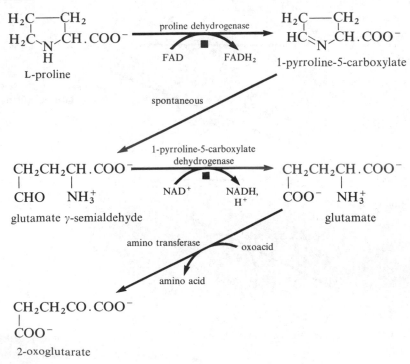

Hydroxyproline may be degraded via this pathway to form hydroxyglutamate, but hydroxyproline dehydrogenase may be a distinct enzyme from proline dehydrogenase.

■ Deficiency of proline dehydrogenase produces type I hyperprolinaemia; deficiency of 1-pyrroline-5-carboxylate dehydrogenase produces type II hyperprolinaemia.

10. **Threonine**

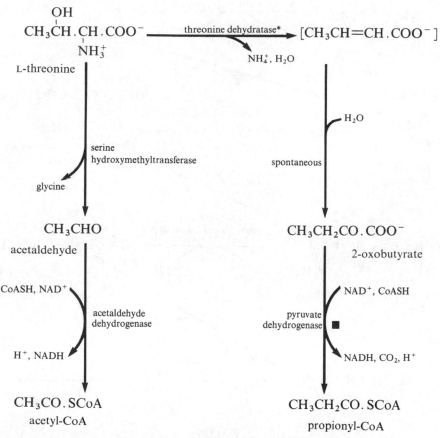

* Probably identical with serine dehydratase.

The relative importance of the two pathways is not known.

■ Hyperthreoninaemia, which is characterized by elevated blood and urine concentrations of threonine, growth retardation and convulsions, has been reported but the cause is not known.

11. Tryptophan

L-tryptophan → (tryptophan-2,3-dioxygenase, O_2) → *N*-formyl-L-kynurenine

N-formyl-L-kynurenine → (formamidase, H_2O, $HCOO^-$) → L-kynurenine

L-kynurenine → (kynurenine 3-monooxygenase, NADPH, H^+, O_2 → $NADP^+$, O_2) → 3-hydroxy-L-kynurenine

3-hydroxy-L-kynurenine → (kynureninase, H_2O, alanine) → 3-hydroxyanthranilate

3-hydroxyanthranilate → (3-hydroxyanthranilate 3,4-dioxygenase, O_2) → 2-amino-3-carboxymuconate semialdehyde

2-amino-3-carboxymuconate semialdehyde → (aminocarboxymuconate-semialdehyde decarboxylase, CO_2) → 2-aminomuconate semialdehyde

2-aminomuconate semialdehyde → (aminomuconate-semialdehyde dehydrogenase, NAD^+, H_2O → NADH, $2H^+$) → 2-aminomuconate

2-aminomuconate → (aminomuconate reductase, NADH, H^+ → NAD^+, NH_4^+) → 2-oxoadipate

2-oxoadipate → see under lysine for remainder of pathway.

NOTES

1. The words peptide and polypeptide are both used to describe chains of amino acids linked by peptide links (i.e. by condensation between α-amino and α-carboxyl groups of adjacent amino acids). In general, peptides are short chains of amino acids which do not form part of a protein; they may be formed by breakdown of proteins but this is not necessarily the case. The number of amino acids involved can be specified by the use of a prefix (e.g. dipeptide, hexapeptide, oligopeptide). Polypeptides (or polypeptide chains) are generally longer sequences which form part of a protein which may thus consist of several polypeptide chains linked covalently (by disulphide bridges) or non-covalently.

2. The crucial diagnostic features for identifying each disease are the degree of body-wasting and oedema: kwashiorkor is diagnosed if children are 60–80% of the expected weight-for-age and oedema is present; marasmic kwashiorkor if children are below 60% of the weight-for-age and oedema is present and marasmus if children are below 60% of this weight without oedema (Coward and Lunn, 1981).

3. In infants, some proteins may be absorbed without hydrolysis. In this way, maternal antibodies, present in the colostrum can enter the bloodstream of the infant to provide a temporary resistance to infection. The proteins may enter the epithelial cells by endocytosis and leave by exocytosis. The ability to absorb intact proteins is probably lost as the infant ages although limited retention of this mechanism could account for allergic responses to some foods shown by some adults (Walker and Isselbacher, 1974).

4. Children lacking enteropeptidase exhibit persistent diarrhoea and fail to gain weight during the first few months of life. Their failure to absorb enough amino acids from the diet prevents the synthesis of proteins including the digestive enzymes so that secondary digestive disorders occur. For example, inadequate production of pancreatic lipase causes partially digested food to enter the large intestine, explaining the diarrhoea. Oral administration of pancreatic extracts restores normal growth rate and removes the symptoms (Tarlow *et al.*, 1970).

5. Partial proteolysis has general importance in the body in several different fields. For example, it is also involved in the secretion of peptide hormones (e.g. insulin is generated from pro-insulin), in the blood-clotting mechanism (e.g. conversion of fibrinogen to fibrin) and in the activation of complement proteins.

6. Current terminology is that enzymes hydrolysing non-terminal peptide links are proteinases and that those removing single amino acids from the end of a peptide chain are peptidases.

7. The contractile proteins, actin and myosin contain histidine and some of these residues are methylated after the proteins have been synthesized (a process known as post-translational methylation) to form 3-methylhistidine:

$$CH_3N \overline{} N \quad \underset{\underset{NH_3^+}{|}}{CH_2CH.COO^-}$$

On hydrolysis of the myofibrillar protein during protein turnover, 3-methylhistidine is released and is not further metabolised by the body but is excreted as such in the urine. Hence the amount of 3-methylhistidine in the urine provides a quantitative index of myofibrillar protein breakdown in the intact animal (Young and Munro, 1978). Although actin is present in tissues other than muscle, the amount is very small.

8. These enzymes have received many names as fashions in nomenclature have changed. Two obsolete names need mentioning here because they have given rise to abbreviations still widely used in clinical practice; glutamate-oxaloacetate trans-aminase (GOT), now aspartate aminotransferase and glutamate-pyruvate trans-aminase (GPT), now alanine aminotransferase. It is unfortunate that the names now sanctioned by the Enzyme Commission do not allow distinction between them by initial letters.

9. To interconvert 20 amino acids by separate and specific transaminations would require 190 enzymes.

CHAPTER 11

CARBOHYDRATE METABOLISM
IN THE LIVER

Thus far, the metabolism of carbohydrate has been discussed mainly from the point of view of glucose utilization for energy production. However, in the liver, the emphasis is elsewhere and carbohydrate metabolism differs in a number of ways from that discussed previously.

1. The liver uses little carbohydrate for ATP production. It has a high capacity for the oxidation of fatty and amino acids and it is likely that, on a normal Western 'diet, most of the ATP requirements of the liver are satisfied by oxidation of short-chain fatty acids, derived mainly from dairy products, and oxoacids derived from dietary amino acids. During starvation, β-oxidation, in which the long-chain fatty acids are converted to acetyl-CoA for ketone body synthesis, will provide most of the ATP. These processes are described in Chapters 6 and 10.
2. Glucose that is removed from the bloodstream by the liver is converted into glucose 6-phosphate which has four major fates: glycogen synthesis; fatty acid and thence triacylglycerol synthesis (which involves the formation of acetyl-CoA via glycolysis); the pentose phosphate pathway (in which the carbon in

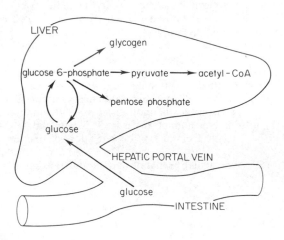

Figure 11.1   Pathways of glucose metabolism in the liver

442

the 1-position of glucose 6-phosphate is oxidized to carbon dioxide with the concomitant production of NADPH for lipogenesis); and, finally, hydrolysis to form glucose again (see Figure 11.1).

3. The liver is the site of metabolism of carbohydrates other than glucose (e.g. galactose, fructose, sorbitol). The pathways involved, together with that for the metabolism of ethanol, are described in Section F.

4. The liver can release glucose into the bloodstream which is important in the maintenance of the blood glucose concentration in most conditions except during the absorption of a carbohydrate-containing meal. This glucose arises either from gluconeogenesis (Section B) or from glycogenolysis (Section D).

A. GLUCOSE UTILIZATION AND GLYCOLYSIS

The complexity of metabolism in the liver is illustrated by reference to glucose uptake and glycolysis. In most other tissues, these two processes are part of the same pathway (i.e. most of the glucose taken up by a tissue is utilized via glycolysis) but in liver, once glucose is converted to glucose 6-phosphate, several quantitatively important pathways are available.

The transport of glucose into the liver cell, unlike many other tissues, is near equilibrium. Liver is also different in that the phosphorylation of glucose is catalysed by glucokinase which has a very high K_m for glucose, approximately 10 mM. Consequently neither process is saturated with glucose, so that the flux-generating step for glucose utilization by the liver is entry of glucose into the bloodstream from the small intestine (see Figure 11.1). However, the pathway of hepatic glucose utilization branches at the level of the hexose monophosphates (i.e. glucose 6-phosphate, fructose 6-phosphate, and glucose 1-phosphate) so that control over the flux through the various branches will be exerted by the activities of the key regulating enzymes of these pathways, glucose-6-phosphate dehydrogenase for the pentose phosphate pathway, glycogen synthase for glycogen synthesis, 6-phosphofructokinase for glycolysis, and glucose-6-phosphatase for glucose formation.

As in other tissues, the pathway of glycolysis-from-glycogen is initiated by glycogen phosphorylase but only under severe hypoxic conditions is glycogen converted to lactate for ATP generation. Although this would provide essential energy for the liver under these conditions, it would also contribute to the development of severe lactic acidosis (Chapter 13; Section D). Glycolysis-from-glycogen may be important, however, in the provision of the precursor, acetyl-CoA, for fatty acid and triacylglycerol synthesis. This suggests that glycogen storage from glucose after a meal is only temporary and that glycolytic intermediates derived from glycogen can be converted into lipid (Walli, 1978). In addition to catalysing flux in the usual glycolytic direction, the enzymes of glycolysis are also involved either directly in, or in the regulation of, the pathway of gluconeogenesis.

B. GLUCONEOGENESIS

The usual definition of gluconeogenesis is the metabolic process by which glucose is formed from non-carbohydrate precursors, which include lactate, pyruvate, glycerol, and the amino acids. Although the process is biosynthetic, since a three-carbon compound is converted into the six-carbon compound, glucose, the pathway is also part of the over-all process by which amino acids are degraded (see Figure 10.6). Gluconeogenesis connects carbohydrate, lipid, and amino acid metabolism in the whole animal; a feature that becomes apparent when metabolism in starvation is treated in Chapter 14.

1. **Occurrence**

In mammals, gluconeogenesis occurs in only two tissues: liver and kidney cortex. An explanation for the process occurring in the kidney will be found in Chapter 14; Section A.3. There is some evidence that glycogen (rather than glucose) can be synthesized from lactate in muscle (i.e. glyconeogenesis—Hermansen and Vaage, 1977) but this is probably of significance only when the lactate concentration within the muscle is very high and the glycogen concentration is low.

2. **The pathway of gluconeogenesis**

(a) *Reactions of the pathway*

Gluconeogenesis can be considered as a *partial* reversal of the glycolytic pathway in that many of the reactions of glycolysis participate in the gluconeogenic pathway. Those that do are the near-equilibrium reactions in which a small increase in the concentrations of the products or a decrease in those of the substrates can reverse the direction of the flux. Conversely, however, reversal of the direction of flux through the non-equilibrium glycolytic reactions would require very large changes in the concentrations of substrates and products and such changes would be physiologically unacceptable (see Chapter 2; Section A.2.d). It is for this reason that these reactions are sometimes described as the 'energy barriers' to gluconeogenesis. The non-equilibrium reactions in glycolysis are those catalysed by hexokinase (or glucokinase in the liver), 6-phosphofructokinase and pyruvate kinase. In order to reverse these steps, separate and distinct reactions must occur in the gluconeogenic direction.

1. Glucose 6-phosphate is converted to glucose in a hydrolytic reaction catalysed by the enzyme glucose-6-phosphatase:

$$\text{glucose 6-phosphate}^{2-} + H_2O \longrightarrow \text{glucose} + P_i^{2-}$$

2. Fructose 1,6-bisphosphate is converted to fructose 6-phosphate in a hydrolytic reaction catalysed by the enzyme fructose-bisphosphatase:

$$\text{fructose 1,6-bisphosphate}^{4-} + H_2O \longrightarrow \text{fructose 6-phosphate}^{2-} + P_i^{2-}$$

3. The conversion of pyruvate into phosphoenolpyruvate in the gluconeogenic pathway requires two separate reactions, catalysed by pyruvate carboxylase and phosphoenolpyruvate carboxykinase. The former enzyme catalyses the carboxylation of pyruvate to oxaloacetate in a reaction involving the hydrolysis of ATP. Phosphoenolpyruvate carboxykinase catalyses the conversion of oxaloacetate to phosphoenolpyruvate in a reaction that involves hydrolysis of ITP (or GTP) to IDP (or GDP):

$$HCO_3^- + CH_3CO.COO^- + ATP^{4-} \longrightarrow {}^-OOC.CH_2CO.COO^- + ADP^{3-} + P_i^{2-} + H^+$$

pyruvate oxaloacetate

$$^-OOC.CH_2CO.COO^- + \begin{cases} ITP^{4-} \\ GTP^{4-} \end{cases} \longrightarrow CH_2{=}\underset{|}{C}.COO^- + CO_2 + \begin{cases} IDP^{3-} \\ GDP^{3-} \end{cases}$$

oxaloacetate

$$OPO_3^{2-}$$

phosphoenolpyruvate

The over-all sequence of reactions from pyruvate to glucose is given in Figure 11.2. An additional complication is that part of the pathway is localized in the mitochondrion and the remainder in the cytosol but for what physiological reason is not clear.

(b) *Intracellular location of the gluconeogenic enzymes*

The gluconeogenic enzymes are located in the cytosol except for pyruvate carboxylase, which is always present within the mitochondrion, and phosphoenolpyruvate carboxykinase, which is mitochondrial in some species. In the rat, however, phosphoenolpyruvate carboxykinase is in the cytosol. This arrangement means that the further metabolism of oxaloacetate produced within the mitochondria must take place in the cytosol. However, as with muscle, neither liver nor kidney possess a mitochondrial carrier for oxaloacetate so that it has to be transported across the mitochondrial membrane in the form of malate or aspartate (see Figure 11.3). One advantage of the transfer of malate into the cytosol is that the oxidation of malate to oxaloacetate results in the reduction of cytosolic NAD^+ to NADH, and the latter is required for the reduction of 3-phosphoglycerate to glyceraldehyde 3-phosphate in gluconeogenesis (see Figure 11.2). In the rat, therefore, the pathway of gluconeogenesis includes both the mitochondrial and cytosolic malate dehydrogenases, together with the mitochondrial transporter for malate.

In human and guinea-pig liver, a proportion of phosphoenolpyruvate carboxykinase (about 50%) occurs inside the mitochondria, so that some oxaloacetate is converted to PEP intramitochondrially. In this case, PEP (as well as malate) must be transported across the mitochondrial membrane and this is achieved by the tricarboxylate mitochondrial carrier. In the birds that have been

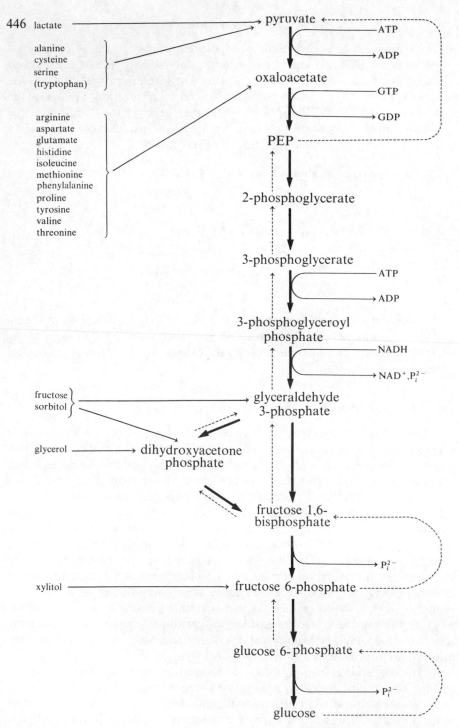

Figure 11.2 Pathway of gluconeogenesis showing points of entry of precursors. Dotted lines indicate reactions of glycolysis

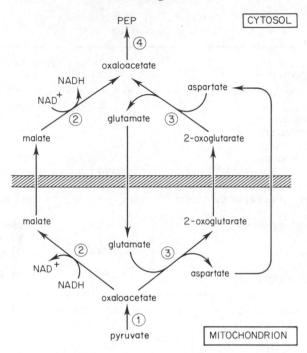

Figure 11.3 Alternative pathways for the transport of oxaloacetate across the inner mitochondrial membrane. 1. Pyruvate carboxylase. 2. Malate dehydrogenase. 3. Aspartate aminotransferase. 4. Phosphoenolpyruvate carboxykinase

investigated (pigeon and chicken) almost all the carboxykinase occurs inside the mitochondria, so that the mitochondrial transport of PEP is an obligatory part of the pathway.

(c) *The physiological pathway*

The ΔG values of some of the reactions of gluconeogenesis are given in Table 11.1, from which it is concluded that those catalysed by pyruvate carboxylase, phosphoenolpyruvate carboxykinase, fructose-bisphosphatase, and glucose-6-phosphatase are non-equilibrium. The concentrations of substrates for these non-equilibrium reactions and the K_m of the enzymes for these substrates indicate that the only enzyme which approaches saturation with pathway substrate is fructose-bisphosphatase. To find the flux-generating step and hence the start of the physiological pathway, it is necessary to consider the precursors of the pathway, namely amino acids (especially alanine), lactate, and glycerol. The gluconeogenic pathway can, in fact, be considered to start with the generation of alanine from muscle, lactate from muscle (and other tissues) and glycerol from adipose (and other) tissues. Gluconeogenesis is thus seen as a complex, branched pathway that spans more than one tissue (Figure 11.4).

Table 11.1. The values of ΔG and the equilibrium status of the reactions of glycolysis and gluconeogenesis in the liver*

Reaction catalysed by	ΔG (kJ.mol^{-1})	Suggested equilibrium nature
Glucokinase	-32.9	non-equilibrium
Glucose 6-phosphatase	-5.1	non-equilibrium
Phosphoglucoisomerase	-1.1	near-equilibrium
6-Phosphofructokinase	-24.5	non-equilibrium
Fructose-bisphosphatase	-8.6	non-equilibrium
Aldolase	-12.1	near-equilibrium
Glyceraldehyde-phosphate dehydrogenase plus phosphoglycerate kinase	-4.1	near-equilibrium
Phosphoglycerate mutase	-1.3	near-equilibrium
Enolase	-1.2	near-equilibrium
Pyruvate kinase	-26.4	non-equilibrium
Pyruvate carboxylase plus phosphoenolpyruvate carboxykinase	-22.6	non-equilibrium

* ΔG values calculated from data in Newsholme and Start, (1973).

3. Precursors for gluconeogenesis

(a) *Lactate*

Something like 120 g of lactate is produced each day by a normally active man (Chapter 13; Section D.1; Krebs *et al.*, 1975). Of this it is estimated that 40 g is produced by tissues which are virtually totally anaerobic in their metabolism (e.g. erythrocytes, kidney medulla, retina—see Chapter 5; Section F). The contribution of the other tissues (including small intestine, brain, skin, and muscle) will vary according to their activity and the prevailing conditions, thus making their individual contribution difficult to assess. That of muscle is particularly variable and vigorous exercise will increase the daily lactate production to well over 120 g.

The normal blood lactate concentration (about 1 mM) represents a steady-state concentration that reflects the balance between the rates of production and utilization. Although the principal tissue for lactate utilization is liver, other tissues, such as heart and red muscles (type I and IIA fibres) can remove lactate from the bloodstream and oxidize it for energy production. Most of the lactate that is removed by the liver is converted to glucose, or glycogen, via gluconeogenesis although a small proportion may be oxidized or converted to triacylglycerol under lipogenic conditions. The conversion of lactate to glucose in the liver and the continuous formation of lactate from glucose in the 'anaerobic' tissues of the body represents a cyclical flow of carbon that has been termed the Cori cycle (after Carl Cori, who originally put forward the idea).

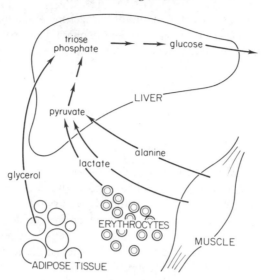

Figure 11.4 The tissues involved in gluconeogenesis. The rate of gluconeogenesis in the kidney is quantitatively much less important than that in the liver except in prolonged starvation

A frequently overlooked aspect of gluconeogenesis from lactate is its role in maintaining acid–base balance, since anaerobic glycolysis results not only in the formation of lactate but also protons (i.e. lactic acid is produced). The conversion of lactate to glucose in the liver utilizes these protons. If gluconeogenesis is unable to remove all the lactic acid produced by peripheral tissues, the dangerous condition of lactic acidosis may result (see Chapter 13; Section D for a fuller discussion of this problem).

(b) *Glycerol*

Glycerol is released into the bloodstream as a result of triacylglycerol hydrolysis from a number of tissues but the most important is adipose tissue. In fasting, resting man, about 19 g of glycerol is released each day and almost all is converted to glucose. This rate will be increased during exercise or stress (see Chapter 6; Section F.2). Even in the fed state, some glycerol is released but most of this will probably be oxidized by the liver.

It is unclear whether glycerol is transported across the cell membrane by a specific transport process or by simple diffusion. Nonetheless, once inside the tissues, glycerol is phosphorylated in a reaction (catalysed by the enzyme glycerol kinase) that involves ATP to form glycerol 3-phosphate:

$$\text{glycerol} + \text{ATP}^{4-} \longrightarrow \text{glycerol 3-phosphate}^{2-} + \text{ADP}^{3-} + \text{H}^+$$

Glycerol 3-phosphate is converted to dihydroxyacetone phosphate in an NAD^+-dependent reaction catalysed by glycerol-3-phosphate dehydrogenase which has been discussed in Chapter 5; Section D.2 in connection with the glycerol phosphate shuttle:

$$\text{glycerol 3-phosphate}^{2-} + NAD^+ \longrightarrow$$
$$\text{dihydroxyacetone phosphate}^{2-} + NADH + H^+$$

(c) *Amino acids*

Experiments in the 1930s demonstrated that amino acids could be divided into three classes; glucogenic, ketogenic, and glucogenic-plus-ketogenic. A glucogenic amino acid was one that, upon administration to a starving animal, increased the blood concentration of glucose; a ketogenic amino acid raised the blood concentration of ketone bodies; and glucogenic-plus-ketogenic amino acids increased the blood concentrations of both. Knowledge of the metabolism of the various amino acids provides an explanation for these observations and this classification. Most amino acids are catabolized to pyruvate or intermediates of the TCA cycle (Table 10.5) all of which can be converted to glucose; these amino acids are glucogenic. On the other hand, there are just two amino acids (leucine and lysine) that give rise solely to acetyl-CoA which cannot be converted to glucose (Chapter 10; Section C.4) but which can, in the starving animal, be converted to ketone bodies. Obviously, amino acids (e.g. phenylalanine) that give rise to both acetyl-CoA and an intermediate of the cycle will be both ketogenic and glucogenic.

Quantitatively, the most important amino acid precursor for gluconeogenesis is alanine, released from muscle and small intestine (see Chapter 10; Section F.2). The flux-generating step for alanine conversion to glucose may be its transport out of the tissue where it arises. For amino acids which are metabolized exclusively by the liver (especially the essential amino acids) the flux-generating steps will be either transport across the intestine or one of the non-equilibrium reactions in the specific degradative pathways in this tissue (see Appendix 10.1). Very little work has been carried out on the equilibrium nature or possible flux generating steps in these pathways.

In a similar manner to the liver, the kidney is able to convert a number of amino acids, of which glutamine and glycine are quantitatively the most important, into glucose. This, however, is probably only of importance during starvation (see Chapter 13; Section C).

4. Control of gluconeogenesis

It should already be apparent from the fact that the gluconeogenic pathway has many substrates and spans many tissues that its control will be complex. In this respect it comes up to expectations and the full details are not yet known (for reviews see Newsholme, 1976b; Pilkis *et al.*, 1978; Exton, 1979). Here, only an outline of some of the factors that control the process will be given but they will be described with a physiological perspective and the story taken up again in Chapter 14.

(a) *Control of precursor supply*

Since the flux-generating steps for gluconeogenesis are the release of alanine, glycerol and lactate from extrahepatic tissues, the rates of production and release of these precursors must be one important factor in determining the rate of gluconeogenesis. Details of factors affecting these rates will be found in other chapters: Chapter 7; Section C.1 for lactate release in muscle glycolysis; Chapter 8; Section C.1 for glycerol release in adipose tissue lipolysis and Chapter 10; Section C.3.c. for amino acid release from muscle. The increased availability of lactate and alanine will raise the concentration of pyruvate in liver mitochondria thereby increasing the activity of pyruvate carboxylase and, in turn, by internal regulation, the activities of the other enzymes of the pathway.

(b) *Control of the pathway from pyruvate*

Internal regulation, consequent upon precursor supply, is certainly not the whole story. For one thing the insensitivity of fructose-bisphosphatase activity to changes in fructose bisphosphate concentration (with which it is saturated), means that some means of external control must operate. In fact, there seem to be many potential regulators which may operate under different circumstances.

One important factor is the role played by the 'glycolytic' enzymes (6-phosphofructokinase and pyruvate kinase) in the regulation of gluconeogenesis. There is now strong evidence that these are active during net gluconeogenesis thereby constituting substrate cycles (Katz and Rognstad, 1976; Katz and Rognstad, 1978). The regulatory significance of such cycles is described in Chapter 7; Section A.2.b. (A substrate-cycle also operates at the level of glucokinase and glucose-6-phosphatase but since it is more directly involved in the regulation of the blood glucose concentration, consideration of this is deferred to Section E.1.)

(i) *Control by hormones*

The main regulators of gluconeogenesis would appear to be hormones. When the concentration of glucose in the blood falls, the secretion of glucagon and glucocorticoids increases while that of insulin decreases. All three hormones affect the rate of gluconeogenesis although full details of the mechanisms involved still await elucidation. Glucagon increases the concentration of cyclic AMP in hepatocytes by stimulation of adenylate cyclase (Chapter 22; Section A.4). Somewhat surprisingly, the action of cyclic AMP appears not to be on gluconeogenic enzymes directly but on 6-phosphofructo-kinase and pyruvate kinase; cyclic AMP inhibits these enzymes by stimulation of the activity of protein kinase, which phosphorylates and hence inhibits the enzymes* (Figure 11.5; Nieto and Castano, 1980). Inhibition of these enzymes will increase the flux through the gluconeogenic pathway through operation of the substrate cycles, and there is now direct evidence that the rate of cycling between fructose 6-phosphate and fructose bisphosphate in rat liver is reduced in starvation (van Schaftingen *et al.*, 1980).

Insulin antagonizes the action of glucagon by lowering the concentration of cyclic AMP within the cell but whether this is achieved by inhibiting the cyclase or activating phosphodiesterase is not known. Glucocorticoids alone have little or no direct effect on gluconeogenesis but show a 'permissive' effect, enhancing the effect of changes in the concentrations of the other hormones.

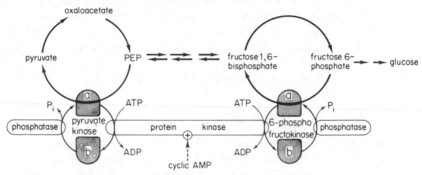

Figure 11.5   Regulation of gluconeogenesis by cyclic AMP

(ii) *Control by the concentrations of acetyl-CoA and citrate*   A rise in the concentration of citrate will decrease the rate of glycolysis by inhibiting 6-phosphofructokinase (Chapter 7; Section B.1.a). By the same token and by virtue of the substrate cycle involving this enzyme, a rise in citrate concentration will increase the rate of gluconeogenesis. By parallel reasoning, a rise in the mitochondrial concentration ratio acetyl-CoA/coenzyme A will inhibit pyruvate dehydrogenase and so increase the availability of pyruvate for gluconeogenesis. In addition, acetyl-CoA is an activator of pyruvate carboxylase (Figure 11.6).

Undoubtedly both citrate and acetyl-CoA concentrations are increased by the oxidation of fatty acids, lactate or alanine, an increase in the availability of which is likely to occur under conditions when gluconeogenesis is required. However, these changes are less important, physiologically, than those involving hormones. Regulation by citrate and acetyl-CoA may represent an evolutionarily more primitive mechanism which is now only important under certain conditions— perhaps after bursts of exercise (e.g. climbing stairs) which raise the blood lactate concentration sharply but are too short to alter blood glucose concentrations sufficiently to cause significant changes in the hormone concentrations. It is also

---

* Recent evidence suggests that glucagon might inhibit the activity of 6-phosphofructokinase not by phosphorylation but by decreasing the concentration of a specific activator, fructose, 2,6-bisphosphate (Hers & van Schaftingen, 1982).

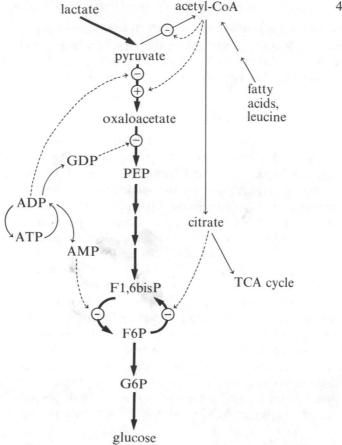

Figure 11.6 Regulation of gluconeogenesis by adenine nucleotides (left) and acetyl-CoA and citrate (right)

possible that this primitive mechanism is more important in the kidney (see Chapter 6 of Newsholme and Start, 1973; Newsholme, 1976b).

(iii) *Control by the ATP/ADP concentration ratio* Changes in the concentration ratio of ATP/ADP are the major means by which the rate of glycolysis is regulated in muscle (Chapter 7; Section B.1.b). The glycolytic enzymes of the liver and kidney possess substantially the same regulatory properties as the muscle enzymes although, at least in the liver, glycolysis is much less important as a means of ATP synthesis. In addition, the key non-equilibrium enzymes of gluconeogenesis possess properties that suggest a decrease in the ATP/ADP concentration ratio would reduce the rate of gluconeogenesis (i.e. fructose-bisphosphatase is inhibited by AMP, phosphoenolpyruvate carboxykinase is inhibited by GDP and pyruvate carboxylase is inhibited by ADP) (Figure 11.6). Indeed, it is well established that anoxia inhibits gluconeogenesis (and increases glycolysis) in both liver and kidney cortex (see Chapter 6 of Newsholme and

Start, 1973) but it is unlikely that the mechanism is important under normal conditions. It may however operate during severe generalized hypoxia (resulting, for example, from haemorrhagic shock) to conserve ATP for processes, such as ion transport, more vital to the life of a liver cell than is gluconeogenesis.

5. Physiological importance of gluconeogenesis

Whenever the carbohydrate content of the diet is low or carbohydrate demand by those tissues that depend exclusively on glucose (that is brain and anaerobic tissues, see Chapter 5; Section F) is high, gluconeogenesis becomes important. If the rate of gluconeogenesis is inadequate under these circumstances, hypoglycaemia results (see Section C). In starvation, the carbohydrate intake is zero and after endogenous carbohydrate reserves have been utilized (which takes about 24 hr) these tissues are totally dependent on gluconeogenesis. Although most human diets are rich in carbohydrate, a few, for example the traditional diet of Eskimos, are not and require that a high level of gluconeogenesis is maintained.

Demand for carbohydrate may well rise in the neonatal period when the danger of hypoxia is increased (see Chapter 5; Section F.8) and during lactation, when glucose is used for both lactose and triacylglycerol synthesis in the mammary gland. The lactating ruminant faces the additional problem of an effectively carbohydrate-free diet since the microorganisms of the rumen, as well as fermenting cellulose, ferment any other carbohydrates present in the food, producing the volatile fatty acids, mainly acetate, propionate, butyrate and lactate. To maintain the blood glucose level in these animals, propionate and lactate are converted to glucose (see Chapter 6 of Newsholme and Start, 1973).

C. GLUCONEOGENESIS AND HYPOGLYCAEMIA

Hypoglycaemia is usually defined as blood sugar concentration less than 2.2 mM ($40\,mg.100\,cm^{-3}$) in the adult. (This represents the concentration in total blood; the concentration in plasma is slightly higher.) The normal blood sugar concentration in the first few days of life may be lower than this; accordingly, hypoglycaemia in babies has been defined as a blood glucose level less than 1.7 mM ($30\,mg.100\,cm^{-3}$). For low birth-weight babies (who are often premature) the concentration may be even lower than this during the first few weeks so that hypoglycaemia for these babies has been defined as a blood glucose concentration below 1.1 mM ($20\,mg.100\,cm^{-3}$). There are several reviews on hypoglycaemia and its causes to which the reader is recommended (Cornblath, 1968; Exton, 1972; Ruderman *et al.*, 1976).

The concentration of glucose in the blood is maintained as a balance between glucose utilization and glucose supply, but hypoglycaemia can occur despite the existence of metabolic mechanisms which are designed to prevent serious changes in the blood glucose concentration.

The current clinical view is that hypoglycaemia is not a disease itself but a clinical feature indicative of impaired control of either glucose utilization or production or both (Yager and Young, 1974; Cahill and Soelder, 1974; Lev-Ran and Anderson, 1981). Nonetheless, there have been attempts to 'upgrade' hypoglycaemia into a disease and, moreover, to a disease that could be responsible for a large number of common symptoms[1] (Steincrohn, 1972).

Although there are a large number of different disturbances that lead to hypoglycaemia, they can, from a metabolic point of view, be divided into two classes (which are not mutually exclusive), increased glucose utilization and decreased glucose production. Once hepatic glycogen stores are depleted, glucose production is dependent solely on gluconeogenesis, so that an abnormality in the process or in its regulation can give rise to hypoglycaemia.

1. Hypoglycaemia due to increased glucose utilization* and decreased hepatic output

This condition applies mainly to the hyperinsulinaemic state, since insulin not only facilitates peripheral utilization but inhibits hepatic glucose output (see Chapter 15; Section D). Infants of diabetic mothers are often hypoglycaemic soon after birth. The cause of this is probably the high blood glucose concentration in the mother, which results in a high blood glucose concentration in the foetus and hence hyperinsulinaemia in both the foetus and the neonate. This can also occur, although in a less severe form, if a normal mother is given an intravenous infusion of glucose during labour.

In both children and adults, islet cell tumour (insulinoma) will produce transient hyperinsulinaemia and corresponding hypoglycaemia. Hypoglycaemia in response to food occurs in some patients with early diabetes mellitus, indicating poor control of the secretion of insulin due to the initial pathological changes in the β-cells. A number of children with hypoglycaemia show an excessive insulin-release in response of the β-cells to leucine for reasons which are not known.

2. Hypoglycaemia due to decreased glucose production

In most of these conditions, the insulin concentration is decreased so that the normal metabolic response to hypoglycaemia can occur—that is increased mobilization of fatty acids and increased rates of ketogenesis. Hence in most, if not all, of the conditions described below the plasma fatty acid and ketone body concentrations are elevated—as would be expected according to the glucose/fatty acid/ketone body cycle (Chapter 8; Section B.2).

(a) *Ketotic hypoglycaemia in children*

The age of onset of this hypoglycaemic disorder varies from nine months to five and a half years with peak incidence between two and three years. Hypoglycaemic

---

* Hypoglycaemia may occur due solely to increased glucose utilisation during marathon or ultramarathon running (Newsholme & Leech, 1983).

episodes recur at intervals for several months or several years but usually disappear once the child has reached ten to twelve years of age. Attacks, which can be severe enough to produce coma, usually occur in the early morning and they usually follow a mild illness on the previous day, particularly a gastrointestinal upset. Thus, in most cases, there has been a short period of starvation during which glycogen stores of the liver could become depleted so that the patient is dependent upon gluconeogenesis. Indeed hypoglycaemia can be induced in these children by a low energy, low carbohydrate ('ketotic') diet. The blood concentrations of fatty acids and ketone bodies are elevated to provide alternative fuels for the tissues. A clue to the possible metabolic defect in these children is provided by the observation that the concentration of alanine in the blood is low during the hypoglycaemic episodes (Pagliara *et al.*, 1972; Sizonenko *et al.*, 1973). It has, therefore, been proposed that the muscle is unable to release alanine sufficiently rapidly to maintain an adequate rate of gluconeogenesis. This may be a particular problem in children since the brain is considerably bigger in relation to muscle mass than in adults.* Hence the rate of protein degradation for a given mass of muscle must be greater in children to provide sufficient glucose for the brain. It is possible that the rate of protein degradation in muscle of these children cannot increase sufficiently to provide for the necessary rate of alanine release. However, the concentrations of other amino acids in the blood of these patients are not reduced during the hypoglycaemic episodes. This suggests that amino acid interconversions within the muscle (Chapter 10; Section F.2) are defective. This might be explained by a low activity of one of the enzymes metabolizing branched-chain amino acids or the enzyme phosphoenolpyruvate carboxykinase, which would limit the rate of alanine formation. As the quantity of muscle increases with age, the total body activity of the enzyme will increase such that it may no longer provide a serious limitation to alanine formation and so explain the disappearance of symptoms.

(b) *Hypoglycaemia of pregnancy*

Starvation during pregnancy is known to reduce the blood glucose concentration in comparison to a non-pregnant control subject (Felig and Lynch, 1970; Felig *et al.*, 1971). The lower concentration of alanine in the blood of the pregnant woman suggests that the hypoglycaemia is due to a lower rate of alanine release from muscle. Whether this represents decreased muscle protein degradation or reduced amino acid interconversion in the muscle is not known. Perhaps, because of the constant glucose demand by the foetus, maternal muscle becomes more resistant to degradation during pregnancy in order to preserve the well-being of the mother and slight hypoglycaemia is tolerated. The blood concentration of ketone bodies is elevated to a greater extent during starvation in the pregnant

---

* At one year of age, the percentage of the weight of the brain to the weight of the body is 11 %, whereas in the adult it is 2 %.

woman and this higher concentration may contribute to a lower rate of protein degradation (see Chapter 14; Section B.1).

(c) *Hypoglycaemia in the small-for-gestational-age infant*

A considerable number of babies are born who are smaller than the average at the same gestational age (small-for-gestational-age infants). A number of maternal factors including malnutrition, infection, ethanol and smoking, are implicated. Such babies may have an unusually small reserve of liver glycogen at birth so that they soon develop hypoglycaemia. Since the blood concentration of the gluconeogenic precursors, alanine and lactate, are elevated in these babies, a deficiency in peripheral mobilization of amino acid precursors is unlikely to be the cause of the hypoglycaemia (Haymond *et al.*, 1974; Stave, 1974). The problem seems to arise from the fact that the synthesis of the enzymes of gluconeogenesis, particularly those that catalyse non-equilibrium reactions, occurs at a relatively late stage during development and they are deficient in these small infants. If the hypoglycaemia becomes marked, and especially if it results in neurological symptoms, treatment (usually infusion of glucose) becomes essential to prevent brain damage.

(d) *Hypoglycaemia in Addison's disease*

Patients with Addison's disease (atrophy of the adrenal cortex probably due in many cases to autoimmunity) can develop severe hypoglycaemia which may be fatal. One of the clinical features is a low blood concentration of cortisol which is necessary for the increased rate of gluconeogenesis and the increased metabolism of fatty acids during starvation (see Chapter 20; Section E.1.c).

(e) *Ethanol-induced hypoglycaemia*

This is caused primarily by inhibition of gluconeogenesis and is discussed in Section F.5.

(f) *Hypoglycaemia and cancer*

Many patients who suffer from cancer have fasting hypoglycaemia, possibly due to the high rate of glucose uptake by the tumour (Shapot, 1972; Chowdhury and Bleicher, 1973; Froesch *et al.*, 1982).

(g) *Inborn errors of metabolism*

Not surprisingly, a congenital deficiency of a gluconeogenic enzyme will give rise to hypoglycaemia during starvation. Deficiencies of each of the specific gluconeogenic enzymes (i.e. those that catalyse the non-equilibrium reactions) have been reported; pyruvate carboxylase, phosphoenolpyruvate carboxykinase, fructose-bisphosphatase, and glucose-6-phosphatase (Howell, 1978). Hereditary fructose intolerance, which results in hypoglycaemia when fructose is ingested, is another inborn error of metabolism that causes hypoglycaemia. However, this is not due to a deficiency of a specific gluconeogenic enzyme but to a low activity of the enzyme fructose-bisphosphate aldolase, which causes accumulation of fructose 1-phosphate. The mechanism by which this accumulation causes hypoglycaemia is discussed in Section F.2. For reviews and other examples of inborn errors that cause hypoglycaemia, see Cornblath (1968), Exton (1972), and Felig (1980).

D. HORMONAL CONTROL OF GLYCOGENOLYSIS

Hepatic glycogen provides the immediately-available reserve of glucose to maintain the blood glucose concentration; this is important in several common situations, for example, during short periods of starvation (the diurnal fast) or sustained exercise, and also in any other conditions that might result in hypoglycaemia (e.g. transient overproduction of insulin or injury). Consequently, the control of hepatic glycogenolysis is of considerable physiological importance and an outline of the mechanisms involved is presented below. Further details can be found in reviews by Stalmans (1976), Hers (1976; 1980), Hue (1979), and Hems and Whitton (1980).

The reactions by which glycogen is degraded to glucose in the liver have been described in Chapter 5; Section C. The enzyme glycogen phosphorylase catalyses the flux-generating step for this conversion. As in muscle (see Chapter 7; Section C.1) two forms of phosphorylase exist in the liver, an active form (phosphorylase *a*) and a considerably less active form (phosphorylase *b*), which are interconvertible via specific enzyme-catalysed reactions. Phosphorylase *a* is formed from phosphorylase *b* by a phosphorylation of the latter enzyme catalysed by phosphorylase kinase. Phosphorylase *b* is formed from phosphorylase *a* by a dephosphorylation catalysed by a protein phosphatase which, unlike that in muscle, may be specific for phosphorylase *a*. Phosphorylase kinase also exists in two forms, active and inactive (or less active); it is activated by a further phosphorylation reaction, this time catalysed by the enzyme protein kinase and is inactivated by a protein phosphatase. As in muscle, both active and inactive forms of hepatic phosphorylase *b* kinase are activated by Ca^{2+}. The activity of protein kinase is controlled by the concentration of cyclic AMP, which itself is controlled by activities of adenylate cyclase and phosphodiesterase. (This control mechanism is the same as that described for muscle—see Figure 7.12).

On the basis of these properties, a theory to explain how various hormones can change the rate of hepatic glycogenolysis can be put forward. Glucagon, which increases the rate of glycogen breakdown in the liver, binds to a specific receptor on the cell membrane and this increases the activity of adenylate cyclase, which increases the intracellular concentration of cyclic AMP. The latter activates

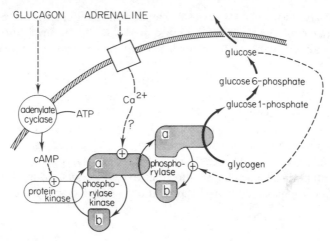

Figure 11.7 Hormonal control of glycogenolysis in liver. The inactivation of both phosphorylase kinase and phosphorylase *a* is catalysed by a protein phosphatase. The activation of phosphorylase *a* phosphatase by glucose is caused by the latter binding to phosphorylase *a* which makes it a better substrate for the phosphatase (see text)

protein kinase, so that inactive phosphorylase kinase is converted to the active form which catalyses the formation of phosphorylase *a* from phosphorylase *b* and hence the rate of glycogenolysis is increased (Figure 11.7). This complex sequence of activating (and inactivating) systems can be more easily understood if it is considered as two interconverting enzyme-systems in sequence (known as an enzyme-cascade).

Insulin is able to reduce the effect of glucagon on the liver and this may be due to its ability to reduce the concentration of cyclic AMP once it has been raised by glucagon. On the other hand adrenaline, which also increases glycogenolysis in the liver, appears not to operate through cyclic AMP. Adrenaline probably has its effect on liver by binding to α-receptors (Chapter 22; Section A.3.a) which influence metabolism through changes in the concentration of Ca^{2+}. The latter increases the activity of phosphorylase kinase (Figure 11.7).

Two questions can be raised concerning the hormonal control of hepatic glycogen phosphorylase activity. First, is the sensitivity of hormonal control sufficient to provide a rate of glucose release that will maintain the blood glucose concentration adequately constant? The answer is probably no, which could explain why, in addition to hormonal control, there is a feedback effect of glucose itself (Figure 11.7). An increase in the concentration of glucose in the liver decreases the proportion of phosphorylase in the active form. This mechanism of control by glucose is discussed in Section E.1.c. Secondly, since the hormonal control of glycogenolysis appears to be very similar if not identical to that of gluconeogenesis—that is regulation via cyclic AMP and protein kinase, why are they not stimulated simultaneously in the early stages of starvation? This second question may be answered if, in the early stages of starvation, the affinity of protein kinase for phosphorylase kinase is greater than that for 6-phosphofructokinase or pyruvate kinase so that, at low levels of activity of protein kinase, phosphorylase is activated but the glycolytic enzymes are not inhibited.

Glycogen is stored with a considerable amount of 'structural' water (Chapter 6; Section E.1). This water is released when the glycogen is degraded but it is usually considered to have no physiological importance. However, the observations that hormones such as vasopressin, oxytocin, and angiotensin II increase the activity of phosphorylase (probably via an increase in Ca^{2+} concentration acting on phosphorylase kinase) and hence the rate of glycogenolysis, suggest that this may be an important mechanism for maintenance of extracellular volume, for example, during haemorrhagic shock, as well as providing blood glucose (Hems and Whitton, 1980).

E. CONTROL OF GLUCOSE UPTAKE AND RELEASE

The liver appears to be the only organ that has the ability to remove glucose from the blood when the concentration is above normal and release it when the concentration is below normal. In this way, the liver plays a unique role in regulation of the blood glucose concentration, which is consistent with its anatomical position. The hepatic portal vein, which provides up to 70% of the blood reaching the liver, drains most of the absorptive area of the intestine so that, after a carbohydrate-containing meal, the largest changes in the blood glucose concentration will be expected to occur in this vein. The rate of response of the liver to a deviation from the normal blood glucose concentration is proportional to the size of this deviation. Perhaps the most remarkable feature of the liver's role in regulating blood glucose concentration is that it achieves this response using a very simple mechanism which does not depend on hormonal or nervous control. Once again, regulation depends on the existence of a substrate cycle.

1. The glucose/glucose 6-phosphate substrate cycle

Evidence has been obtained that the activities of both glucokinase and glucose-6-phosphatase are simultaneously catalytically active in liver cells so that glucose is converted to glucose 6-phosphate and back again at the expense of ATP hydrolysis (Katz and Rognstad, 1978).

(a) *Role of glucokinase*

Glucokinase, which is present only in the hepatocytes and the β-cells of the islets of Langerhans of the pancreas, has a high K_m for glucose (about 10 mM) and is not inhibited by glucose 6-phosphate. The high K_m ensures that the activity of the enzyme varies in approximate proportion to the concentration of glucose in the hepatic portal vein. As the glucose concentration in the hepatic portal vein (and consequently the intracellular concentration within the liver) rises after a meal from about 4 mM (fasting level) to 10–12 mM (after a carbohydrate-rich meal) the activity of glucokinase increases. This results in an increased rate of glucose phosphorylation and hence glucose uptake by the liver. Furthermore, the higher the blood glucose concentration (above normal) the greater will be the rate of glucose phosphorylation and hence glucose uptake (a known property of the liver *in vivo*).

(b) *Role of glucose-6-phosphatase*

The properties of glucokinase do not explain how the direction of glucose metabolism in the liver can change from uptake to release in response to only a small decrease in the blood concentration. The activity of glucose-6-phosphatase is obviously involved in the release of glucose but the enzyme does not appear to possess allosteric properties and may be regulated only by changes in the concentration of its substrate, glucose-6-phosphate. However, if it is assumed that both enzymes are simultaneously active and catalyse a substrate cycle between glucose and glucose 6-phosphate, the properties of glucokinase can be applied to glucose release as well as uptake. As the blood glucose concentration decreases below normal, the intracellular glucose concentration will fall and this will decrease the activity of glucokinase. When the activity of glucokinase falls below that of glucose 6-phosphatase, the rate of hydrolysis of glucose 6-phosphate will exceed that of glucose phosphorylation and net glucose release from the liver will occur. Furthermore, the greater the decrease in the blood glucose concentration, the lower the activity of glucokinase and the greater will be the rate of glucose release (assuming that the activity of glucose 6-phosphatase remains constant) (Figure 11.8). In addition, existence of the cycle will increase sensitivity of the control (Chapter 7; Section A.2.b), especially if the rate of cycling is high, and hence explain how the liver can change both the rate and direction of glucose metabolism in response to small changes in the blood glucose concentration.

(c) *Relationship between the glucose/glucose 6-phosphate cycle and glycogen metabolism*

Operation of the glucose/glucose 6-phosphate substrate cycle as described above would be expected to alter the concentration of glucose 6-phosphate in the liver, increasing the concentration during uptake and decreasing it during release. In fact, just the opposite may occur. To gain a full picture of the control of glucose uptake and release by the liver it is necessary to consider the mechanisms by which the conversion of glucose 6-phosphate to glycogen and the degradation of the latter to glucose 6-phosphate are regulated.

It has been shown that an increase in the concentration of glucose in the liver above the normal decreases the activity of phosphorylase but increases that of glycogen synthase. If the change in glucose concentration is sufficiently large, it

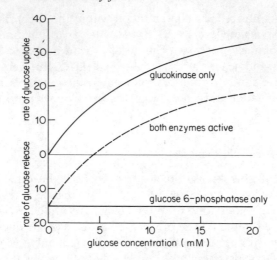

Figure 11.8 The glucose/glucose 6-phosphate cycle. Hypothetical graphs to illustrate how the simultaneous activities of glucose-6-phosphatase and glucokinase can exert fine control over the uptake of glucose into or release from hepatocytes

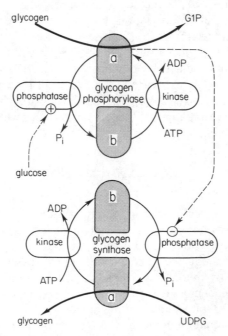

Figure 11.9 The role of glucose in decreasing the activity of phosphorylase and thereby increasing the activity of glycogen synthase

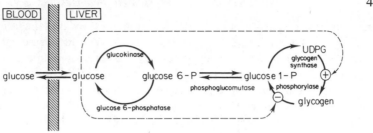

Figure 11.10 The 'push-pull' mechanism for regulation of blood glucose concentration. A raised blood glucose concentration stimulates glucose uptake into hepatocytes by increasing the rate of phosphorylation (see Figure 11.8 for mechanism) and by stimulating the conversion of the resulting glucose 6-phosphate into glycogen (see Figure 11.9 for mechanism)

can result in a change from glycogen degradation to synthesis. Similarly, a decrease in the glucose concentration can change the direction from synthesis to degradation. How is this achieved? Phosphorylase *a* possesses a binding site for glucose and the phosphorylase-glucose complex is a better substrate for phosphorylase phosphatase than non-complexed phosphorylase *a* (Hers, 1976; 1980). Hence an increase in the hepatic glucose concentration will reduce the proportion of phosphorylase in the *a*-form. However the control mechanism is even more complex, since phosphorylase *a* is an inhibitor of glycogen synthase phosphatase, so that a high proportion of phosphorylase in the active form maintains synthase in the inactive form. Therefore, as the glucose concentration is increased above normal, phosphorylase is converted to the inactive form and this causes activation of glycogen synthase (Figure 11.9). Furthermore, the effect of an increase in the blood glucose concentration is greater on glycogen synthase activity than on the activity of glucokinase so that the intracellular concentrations of glucose 6-phosphate (and glucose 1-phosphate and UDP-glucose) actully decrease despite an increased rate of glucose phosphorylation (Hers, 1976; 1980). This has an additional effect on the net rate of phosphorylation of glucose since the decrease in the glucose 6-phosphate concentration decreases the activity of glucose-6-phosphatase thus increasing further the *net* rate of glucose phosphorylation (Figure 11.10). On the other hand, a decrease in the concentration of glucose below normal will reduce the activity of glucokinase and, in addition, increase that of phosphorylase so that the glucose 6-phosphate concentration is maintained despite an increase in the rate of glucose release by the liver (Figure 11.10). The effect on glucokinase and an indirect effect on phosphorylase and glycogen synthase has been described as a 'push–pull' mechanism. It is further discussed in Chapter 16; Section C.2.

(d) *Quantitative problems in glucose metabolism*

Despite strong evidence for this mechanism of regulation of glucose uptake and release in the liver, its quantitative importance has been questioned. Experiments with isolated perfused liver or incubated hepatocytes show that the rate of glucose uptake and glycogen synthesis from glucose is very low, whereas glycogen synthesis from the gluconeogenic precursors, lactate or alanine, is high (see Chaper 16; Section B.1; Boyd *et al.*, 1981). Furthermore, the epithelial cells of the intestine are able to metabolize glucose so that the proportion of glucose that is absorbed as glucose into the hepatic portal blood is still not known with any certainty. Indeed, it has been suggested that glucose is absorbed mainly as lactate and alanine and that glucose is subsequently formed by gluconeogenesis in the liver (see Chapter 5; Section A.2.c for discussion). Although the nature of the problem of the precursors of hepatic glycogen is one of central importance with far-reaching implications, too little time has elapsed since the problem has been recognized for quantitative assessments to have been made. The authors consider that glycogen is synthesized in the liver both from blood glucose and from gluconeogenic substrates but that the quantitative importance of each may depend upon the amount of carbohydrate in the diet; for example, on a low carbohydrate diet, gluconeogenic precursors may provide most of the glycogen whereas, on a high carbohydrate diet, blood glucose may be the most important precursor (see Chapter 16; Section B.1).

F. METABOLISM OF OTHER CARBOHYDRATES AND ETHANOL

Our diet contains a number of compounds that cannot be directly utilized in their absorbed form. Some are potentially toxic, while others, usually present in larger amounts, are chemically modified for utilization by tissues of the body. It is the function of the liver to carry out chemical modifications both to detoxify harmful compounds and to direct potentially useful compounds into the major metabolic pathways. In this section, the latter function will be illustrated by reference to two monosaccharides present in normal diets (galactose and fructose), to two sugar derivatives sometimes administered clinically (sorbitol and xylitol) and to ethanol. The first four compounds are converted to glucose via part of the gluconeogenic pathway. The means by which they enter the gluconeogenic sequence are described together with, in each case, a discussion of points of clinical interest. The hepatic mechanisms involved in detoxification of drugs and similar compounds involve several kinds of reactions including hydroxylations and conjugations which are described in Chapter 20; Sections B.3 and C.2.

1. Galactose

Galactose, together with glucose, is a constituent of the disaccharide lactose which is found in milk (see Figure 5.29). Lactose is hydrolysed to its constituent monosaccharides in the intestine from where they are absorbed and transported to the liver via the hepatic portal vein.

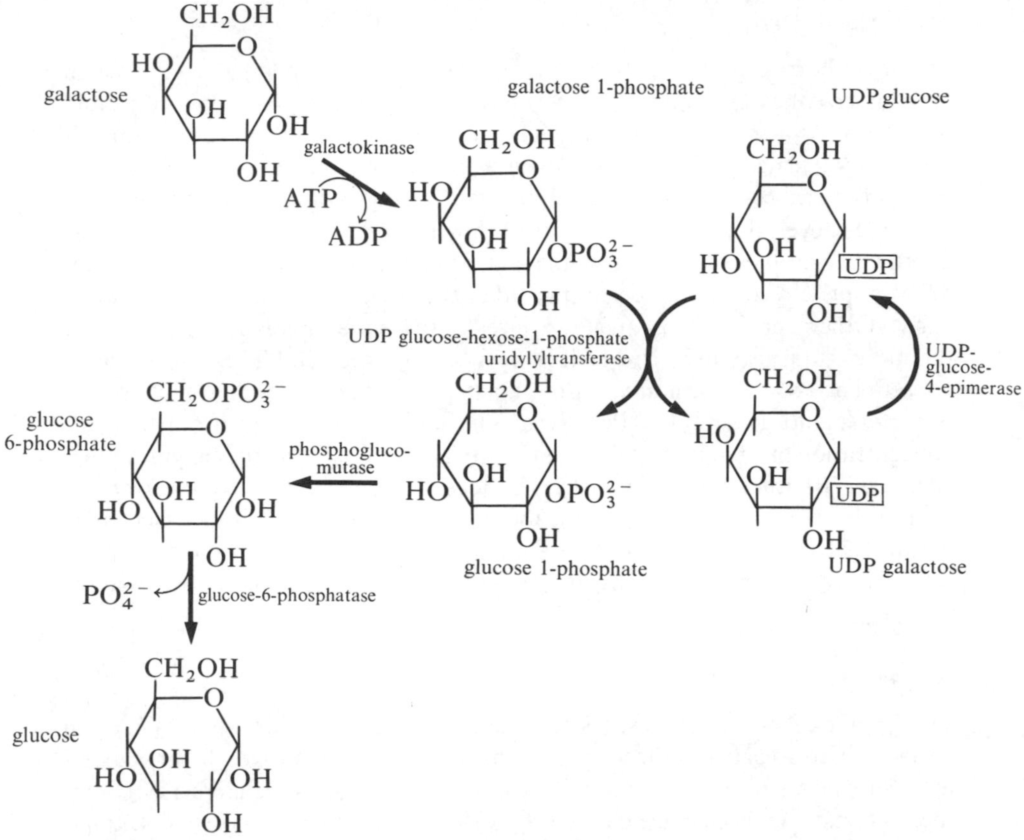

Figure 11.11 Pathway of galactose metabolism

(a) *Metabolic pathway of galactose utilization*

The first reaction in the conversion of galactose to glucose is phosphorylation to galactose 1-phosphate which is catalysed by the enzyme galactokinase. The further metabolism may appear somewhat less usual; galactose 1-phosphate reacts with uridine diphosphate glucose (UDPglucose)[2] to produce UDPgalactose and glucose 1-phosphate, in a reaction catalysed by UDPglucose-hexose-1-phosphate uridylyltransferase. UDPgalactose is then converted into UDPglucose by the enzyme, UDPglucose 4-epimerase. The resulting UDPglucose is reused as a substrate by the uridylyltransferase so that the net conversion is from galactose 1-phosphate to glucose 1-phosphate. The latter is then converted into glucose via the reactions catalysed by phosphoglucomutase and glucose-6-phosphatase (see Figure 11.11).

(b) *Clinical interest*

In rare inborn errors of metabolism, the activity of either galactokinase or UDPglucose-hexose-1-phosphate uridylyltransferase is very low or non-detectable. As a consequence of the lack of either enzyme, galactose accumulates in the blood, causing the disease known as galactosaemia (Segal, 1978). Infants with this disease are usually normal at birth but, once they feed, vomiting and diarrhoea develop. This causes dehydration which can be severe enough to cause death unless milk is removed from the diet. Such removal corrects the symptoms and complete recovery can occur, provided the disease is diagnosed sufficiently early. Other serious consequences of galactosaemia include impaired liver function, cataracts and irreversible mental retardation. These arise as a consequence of the accumulation of galactose or the metabolites galactose 1-phosphate and galactitol in the tissue. The latter is generated (from the high concentration of circulating galactose) in any tissue possessing the enzyme aldose reductase (Chapter 5; Section F.2.b). The actual cause of damage to the liver and brain cells is poorly understood but there is good evidence that cataracts arise because galactitol is produced and accumulates in the lens so that water is drawn into the lens by osmosis. This disrupts the structure of the lens and causes opacity.

2. Fructose

The metabolism of fructose is particularly important in man since, in industrialized societies of the world, a considerable proportion of dietary carbohydrate is in the form of sucrose, which contains 50 % fructose. (Figure 5.29.) Since the suggestion has been made that the ingestion of large quantities of sucrose could be responsible for the development of atherosclerosis, obesity and perhaps other diseases, and since it is sometimes used as an energy source in parenteral feeding, knowledge of its metabolism is important.

(a) *Metabolic pathway of fructose utilization*

Fructose is transported by a specific carrier into the absorptive cells of the intestine where it may either be metabolized to lactate (see Chapter 5; Section A.2.c) or enter the hepatic portal circulation as fructose. From the circulation, fructose is taken up by the liver, in which it is phosphorylated by a specific fructokinase. (The low activity of hexokinase in the liver will catalyse the conversion of some of the fructose to fructose 6-phosphate but the proportion of fructose metabolized by this route is likely to be small ($<10\%$).) Fructokinase activity results in the formation of fructose 1-phosphate which is split by fructose-bisphosphate aldolase[3] into glyceraldehyde and dihydroxyacetone phosphate. Glyceraldehyde is phosphorylated to glyceraldehyde 3-phosphate by the enzyme triokinase, so that the over-all metabolism of fructose in the liver gives rise to the two triose phosphates, glyceraldehyde 3-phosphate, and dihydroxyacetone

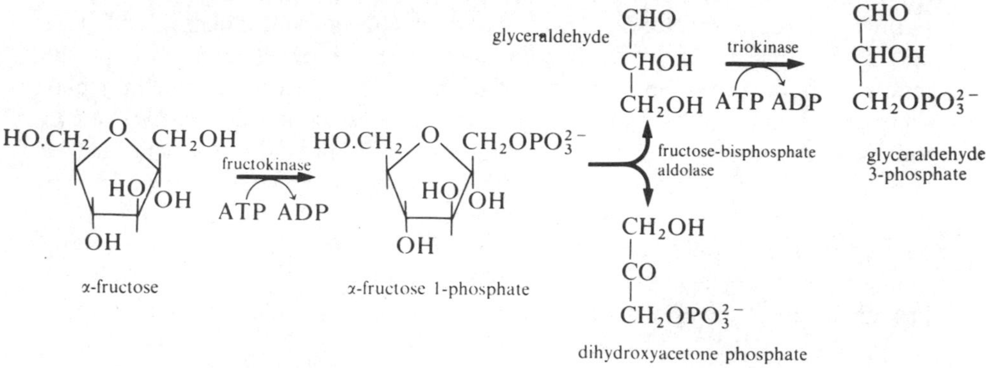

Figure 11.12   Pathway of fructose catabolism

phosphate (Figure 11.12). The triose phosphates will be further metabolized to glucose (gluconeogenesis) or to pyruvate (glycolysis) according to the hormonal control of the pathways (Section B.4.b).

(b) *Clinical interest*

(i) *Inborn errors of metabolism*   There are several rare diseases due to a low activity of enzymes involved in fructose metabolism (Froesch, 1978; Woods, 1980). First, the absence of significant fructokinase activity results in a disease known as essential fructosaemia (EF). Fructose feeding results in higher blood fructose concentrations than normal but this appears to have no serious consequences. Secondly, the absence of a significant activity of hepatic aldolase towards fructose 1-phosphate results in a disease known as hereditary fructose intolerance (HFI) due to accumulation of fructose 1-phosphate in the liver. The major problem is hypoglycaemia after ingestion of fructose which can be sufficiently severe to produce vomiting, convulsions, and coma. Infants fail to thrive when given sucrose as a sweetener and in later life the patient may develop an aversion to sucrose-containing foods. The hypoglycaemia is due to accumulation of fructose 1-phosphate which can cause inhibition of gluconeogenesis (see below) and to the fact that this compound inhibits glycogen phosphorylase.

Similar symptoms and clinical signs are observed after fructose feeding in patients with fructose-bisphosphatase deficiency. In addition, however, lactate may accumulate in the blood in this condition due to the ability to divert fructose to triose phosphate but the inability to form glucose. This deficiency is sometimes classified as a type of glycogen storage disease (see Chapter 16; Section E).

(ii) *Fructose and atherosclerosis*   Comparison of the reactions of the hepatic pathway of fructose metabolism with that of glycolysis-from-glucose (see Figure 5.8) shows that the fructose is converted to triose phosphates without the involvement of the enzyme 6-phosphofructokinase. Since this enzyme may play an important role in the control of the rate of glucose conversion to pyruvate in the liver, the ability of fructose metabolism to bypass this step has led to speculation that fructose will be more readily converted to pyruvate, and hence acetyl-CoA, than glucose. Since lipogenesis can occur from acetyl-CoA in the liver of man, this could account for the finding that fructose or sucrose feeding increases the triacylglycerol level in the bloodstream (Yudkin, 1972). This might provide a causal metabolic link between atherosclerosis and sucrose ingestion.

(iii) *Fructose and intravenous feeding*   Another clinical aspect of fructose metabolism relates to its use for parenteral feeding, a widespread adjunct to the treatment of many conditions. Although the question of what nutrients should be given and under what circumstances would appear to be straightforward, this is not the case as indicated by the extensive medical literature on this subject (and discussed in Chapter 14). One reason for intravenous nutrition is to provide a patient who cannot feed normally with a fuel that can be oxidized by most tissues. Since glucose is the normal blood sugar and can be utilized by all tissues it should be the ideal nutrient for intravenous feeding. However, four objections have been raised to its use.

1. Glucose causes more damage to the vein in which the catheter is placed than other sugars.
2. Other carbohydrates are metabolized more rapidly than glucose.
3. The presence of a reasonable concentration of insulin in the blood is required for glucose utilization.
4. The rate of glucose utilization is reduced in some diseases and following trauma caused by accident, surgery, etc.

For these reasons, fructose, and also xylitol and sorbitol (which are discussed in subsequent sections) have been used as alternatives to glucose.

Since almost all patients who are given intravenous fructose will have fasted for some hours, it is likely that hepatic gluconeogenesis will have been stimulated. Thus a large proportion of the triose phosphates, produced from the metabolism of fructose, will be converted to glucose. On this basis, fructose is a sound alternative to glucose. However, the metabolism of fructose in the liver may be so rapid as to lead to metabolic and clinical problems (see Woods and Alberti, 1972; Woods, 1974; Krebs *et al.*, 1975). Intravenous infusion leads to a high blood and hence a high hepatic concentration of fructose which increase the activity of fructokinase towards its maximum. However, the maximum fructokinase activity is greater than that of the aldolase so that fructose 1-phosphate may reach very high concentrations in the liver (see Table 11.2). Under these conditions aldolase becomes a second flux-generating step in this pathway. This accumulation of a

Table 11.2. Effect of fructose on metabolites in the perfused liver*

Time of perfusion with fructose (min)	Hepatic contents (μmol.g^{-1} fresh wt.)			
	Fructose 1-phosphate	ATP	IMP	Phosphate
0	0.2	2.2	0.2	4.2
10	8.7	0.5	1.1	1.7
40	7.7	0.8	0.4	4.0
80	3.3	1.0	0.3	5.6

* See Woods & Alberti, 1972; Krebs *et al.*, 1975.

sugar phosphate reduces the hepatic concentration of *inorganic* phosphate. (There is only a finite concentration of phosphate in cells and an uncontrolled accumulation of an organic phosphate can remove most of this phosphate.) This lack of phosphate will cause a number of metabolic problems, one of which is the reduction in the rate of mitochondrial oxidative phosphorylation which results in a decrease in the hepatic concentration of ATP (phosphate and ADP are substrates for oxidative phosphorylation) (Table 11.2). A decrease in the ATP/ADP concentration ratio results in inhibition of a number of key enzymes that control the rates of several energy-requiring processes,[4] in particular, gluconeogenesis, and it also results in stimulation of glycolysis (Section B.4.b). The decreased rate of gluconeogenesis can lead to severe hypoglycaemia, but perhaps of more immediate clinical importance is that a decrease in hepatic gluconeogenesis together with an increase in the rate of hepatic glycolysis can lead to the accumulation of lactic acid in the bloodstream and the danger of lactic acidosis. This can be a life-threatening condition (see Chapter 13; Section D for further details). Although the extent of the metabolic problems will depend upon the blood concentration of fructose, and low rates of infusion may not cause any problems, their possibility brings into question the clinical wisdom of the use of fructose as an alternative to glucose for intravenous feeding. It has led Krebs *et al.* (1975) to comment:

'As fructose has no essential advantage over glucose as a parenteral nutrient (but involves a dangerous side effect) its inclusion in nutrient solutions should be abandoned.'

This advice has been heeded by the clinicians and it is one field in which the study of metabolism has had a profound effect on clinical practice. Currently, the usual intravenous carbohydrate fuel is concentrated glucose (10–50 %) which is given with insulin into a central vein.

It is of interest to consider why fructose metabolism by the liver should be such as to endanger life, especially since fructose is a normal constituent of the diet of man. (Honey, which contains about 40 % fructose, has been part of the diet of man for many thousands of years.) The important difference between fructose as a natural food and as a parenteral nutrient is the means of entry into the body. It is likely that the flux-generating step for fructose metabolism, which regulates the rate of this pathway, is the absorption of fructose into the epithelial cells of the intestine, so that the activity of fructokinase is limited by substrate (fructose) concentration. The flux-generating step is bypassed by intravenous infusion so that the blood fructose concentration and hence the activity of fructokinase can exceed the normal. In addition, not all the dietary fructose may enter the bloodstream as fructose since the epithelial cells of the intestine could metabolize some of it to lactate or alanine (see Chapter 5; Section A.2.c).

3. Sorbitol

Sorbitol is one of a number of polyols (polyhydric alcohols) of metabolic or clinical importance. Several polyols can be produced in the body; they are derived from monosaccharides by reduction of the carbonyl groups to an alcohol and so are not capable of ring formation (see Appendix 5.1). Sorbitol can be formed from glucose or fructose in a variety of tissues (Chapter 5; Section F.2); xylitol is formed from xylulose or xylose; galactitol is formed from galactose (or tagatose) when its concentration is elevated and can cause damage in tissues such as the lens (Section F.1.b). Ribitol, formed from ribose or ribulose, is a constituent of FMN and FAD, the prosthetic groups of flavoproteins. None of the polyols are likely to be present in significant amounts in natural diets.

(a) *Metabolic pathway of sorbitol utilization*

This sugar is metabolized by the sorbitol or polyol pathway in which iditol dehydrogenase catalyses the conversion of sorbitol to fructose (Figure 11.13). The latter is further metabolized as described above.

(b) *Clinical interest*

Sorbitol has been used in parenteral nutrition as a replacement for glucose. However, since it is metabolized via fructose, it suffers from the same problems as

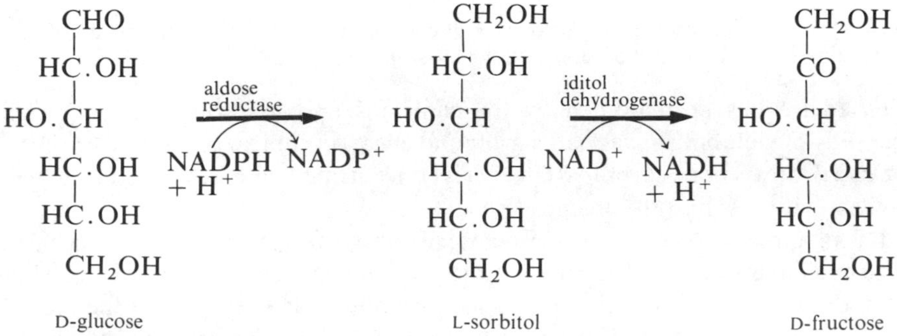

Figure 11.13 The sorbitol pathway

fructose (see above). Furthermore, some synthetic infusion mixtures contain ethanol in addition to sorbitol and ethanol may be an additional factor to inhibit gluconeogenesis (see below; Woods, 1974).

4. Xylitol

(a) *Metabolic pathway of xylitol utilization*

Xylitol is the alcohol derived from the pentose xylulose (or xylose). The initial steps in the hepatic metabolism of xylitol are similar to those of sorbitol. The alcohol is converted to xylulose by xylitol reductase and the resulting ketose is phosphorylated by xylulokinase to xylulose 5-phosphate (Figure 11.14). The latter is an intermediate in the pentose phosphate pathway by which it is eventually converted to fructose 6-phosphate (see Chapter 17; Section A.8) and thence to glucose via the phosphoglucoisomerase and glucose-6-phosphatase reactions.

(b) *Clinical interest*

Xylitol has been, and still is, used as an intravenous nutrient despite the evidence that infusion of relatively large amounts of xylitol can be dangerous (see Woods, 1974).

The lactic acidosis and liver damage that can result from xylitol infusion are probably caused by similar metabolic changes to those produced by fructose. The rapid phosphorylation of xylulose with the accumulation of xylulose 5-phosphate causes a decrease in the hepatic concentration of phosphate and hence a reduction in the ATP/ADP concentration ratio. This leads to inhibition of gluconeogenesis and stimulation of glycolysis. In addition, the first reaction of xylitol metabolism reduces NAD^+ to NADH and this further favours the conversion of pyruvate to lactate.

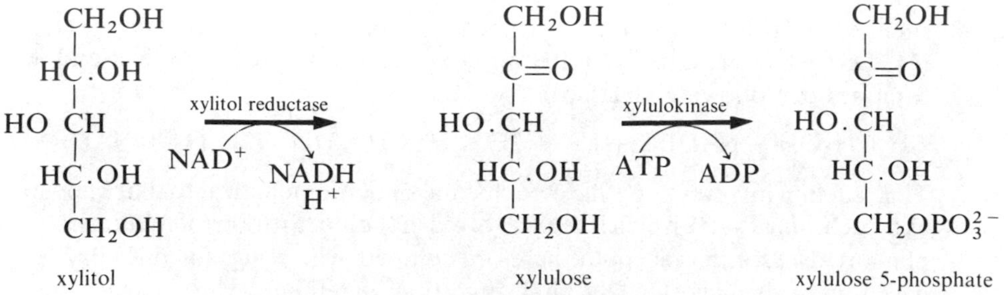

Figure 11.14 Pathway of xylitol metabolism. Xylulose 5-phosphate enters the pentose phosphate pathway

5. Ethanol

Ethanol is not a carbohydrate nor is it a gluconeogenic precursor, but since its metabolism occurs primarily in the liver and since it can inhibit gluconeogenesis, it is discussed in this chapter. (For reviews on ethanol metabolism, see Lieber, 1976, 1977; Badawy, 1978; Kricka and Clark, 1979.)

Ethanol is produced continuously by microbial fermentation in the large intestine of mammals and it is metabolized in the liver in the same way as ingested alcohol. Herbivores, in which fermentation in the large intestine is quantitatively important, produce particularly large amounts of ethanol. (Horse liver contains such a high concentration of alcohol dehydrogenase that this tissue is used as a source of the enzyme for purification.) In man, this source of ethanol amounts to about 3 g each day and gives rise to a *portal* blood concentration of about 0.5 mM; but this can be insignificant when compared with the amount of ethanol ingested from alcoholic beverages.[5] The energy content of the ethanol in a 1 l bottle of table wine (12% v/v ethanol) is about 3000 kJ (see Table 2.3) and in some countries as much as 10% of the daily energy requirement is provided by ethanol.

(a) *Metabolic pathway of ethanol utilization*

As with many other substances, the flux-generating step for the metabolism of ethanol, under normal conditions, is the rate of absorption from the intestine. However, this may not be the case when a large quantity of ethanol is ingested since the blood concentration may approach that necessary to saturate the enzymes utilizing the ethanol in the liver (see below).

Most of the absorbed alcohol is taken up by the liver where it is oxidized to acetaldehyde (ethanal); three separate enzymes or enzyme systems catalyse this reaction.

1. The most widely known is the NAD^+-linked enzyme, alcohol dehydrogenase, which catalyses the reaction:

$$CH_3CH_2OH + NAD^+ \longrightarrow CH_3CHO + NADH + H^+$$

2. A mixed function oxidase, known as the microsomal ethanol-oxidizing system (MEOS), which requires NADPH and oxygen, is also present in liver and it catalyses the following reaction:

$$CH_3CH_2OH + NADPH + H^+ + 2O_2 \longrightarrow CH_3CHO + 2H_2O_2 + NADP^+$$

This reaction involves a specific cytochrome similar or identical to that known as cytochrome P-450, which is also involved in the inactivation (detoxification) of many drugs. Ethanol could therefore compete with drugs for this enzyme which may explain the fact that ingestion of ethanol can increase the circulating concentration of many drugs to such an extent that toxic levels of the drug may be reached (Chakraborty, 1978).

3. In this tissue, catalase is present in the peroxisomes (Chapter 4; Section E.5.b) where, in addition to catalysing the decomposition of hydrogen peroxide, it can also catalyse the oxidation of alcohols, including ethanol and methanol (Chance *et al.*, 1977), according to the equation:

$$CH_3CH_2OH + H_2O_2 \longrightarrow CH_3CHO + 2H_2O$$

All of these three enzymes are involved in conversion of ethanol to acetaldehyde in the liver but quantitative data on their respective roles are difficult to obtain (Brentzel and Thurman, 1977; Peters, 1982). At low hepatic concentrations of ethanol, the dehydrogenase will be more important since the K_m of the dehydrogenase is about 1 mM whereas that for the other two systems is much higher (that for MEOS is about 10 mM). Even with a small intake of alcohol, the portal blood concentration will exceed 1 mM[5] so that the contribution of the oxidase and catalase systems to the oxidation of ethanol will increase. In cases of severe intoxication, the concentration in the blood may approach 100 mM (and the portal blood concentration may well be higher than this) so that the total enzymic capacity in the liver will be saturated. Other tissues which have low activities of alcohol dehydrogenase include the gastric mucosa, kidney, and adipose tissue, but they probably take up only a small proportion of alcohol.

The further oxidation of acetaldehyde to acetate is catalysed by aldehyde dehydrogenase:

$$CH_3CHO + NAD^+ + H_2O \longrightarrow CH_3COO^- + NADH + 2H^+$$

At least two aldehyde dehydrogenases exist in rat liver; one is present in the cytosol and has a K_m of about 1 mM, the other is in the mitochondrial matrix and has a very low K_m for acetaldehyde (0.01 mM); the mitochondrial enzyme is quantitatively more important for oxidation for acetaldehyde, most of which ($>90\%$) is oxidized in the liver so that the hepatic and blood concentrations of acetaldehyde remain very low. This is explained by the high capacity and low K_m of the mitochondrial aldehyde dehydrogenase of the liver. Unlike acetaldehyde, a large proportion of the acetate escapes from the liver and is converted to acetyl-CoA by acetyl-CoA synthetase in other tissues:

$$CH_3COO^- + ATP^{4-} + CoASH \longrightarrow CH_3COSCoA + AMP^{2-} + PP_i^{3-}$$

In most extrahepatic tissues, acetyl-CoA will be oxidized via the TCA cycle (which provides most of the ATP from ethanol oxidation) whereas in liver and in adipose tissue, some of the acetyl-CoA may act as a precursor for the biosynthesis of fatty acids and triacylglycerol.

The final reactions to be considered in the metabolism of ethanol in the liver are those involved in reoxidation of cytosolic NADH and in the reduction of $NADP^+$. The latter is achieved by the pentose phosphate pathway which has a high capacity in the liver (Chapter 17; Section A.8). The cytosolic NADH is reoxidized mainly by the mitochondrial electron transfer system, which means that substrate shuttles must be used to transport the hydrogen atoms into the mitochondria (see Chapter 5; Section D). There is evidence that both the malate/aspartate and the glycerol phosphate shuttles are involved (Lieber, 1976). Under some conditions, the rate of transfer of hydrogen atoms by these shuttles is less than the rate of NADH generation so that the redox state in the cytosolic compartment of the liver becomes highly reduced and the concentration of NAD^+ severely decreased.* This limits the rate of ethanol oxidation by alcohol dehydrogenase.

Chronic consumption of ethanol is associated with increased rates of removal of ethanol from blood (except when the consumption has already caused cirrhosis). An increase in the capacity of the microsomal system may account for this. It is, however, unclear whether the capacity of the other enzymes, aldehyde dehydrogenase and acetyl-CoA synthetase, are also increased.

(b) *Clinical interest*

There is, of course, considerable interest in various effects of ethanol ingestion and metabolism (for review, see Mendelson and Mello, 1979). The available evidence suggests that some of the clinically interesting effect are due not to ethanol *per se* but to its metabolites NADH and acetaldehyde.

The $NAD^+/NADH$ concentration ratio in the cytosol of the liver is maintained at a value of about 1000 (Krebs, 1973). The administration of alcohol can lower this ratio by at least ten-fold (Krebs, 1968). Many dehydrogenase reactions are close to equilibrium so that, for those that react with $NAD^+/NADH$, the concentrations of all the other substrates and products will be affected by a change in the $NAD^+/NADH$ concentration ratio. Hence a decrease in the $NAD^+/NADH$ concentration ratio will lower the concentration of the oxidized reactant and raise that of the reduced reactant of all the dehydrogenation reactions, as follows:

$$NAD^+ + \text{reduced reactant}\uparrow \rightleftarrows NADH + \text{oxidized reactant}\downarrow + H^+$$

where the vertical arrows represent the direction of the change in concentration. If either of these reactants has an important metabolic role in the tissues, marked changes in their concentration could produce abnormal effects (see below, for examples).

Although the toxicity of acetaldehyde has been known for years (Walsh, 1971) it is only recently that evidence for its involvement in the effects of ethanol has been obtained. In man, variations in the blood ethanol concentration from 24 mM to 56 mM were accompanied by acetaldehyde concentrations of between 0.02 mM and 0.03 mM.[6] Despite these very low concentrations, it is the acetaldehyde that is considered to be responsible for many of the effects of ethanol. The concentration

---

* The low concentration of NAD^+ also restricts the rate of conversion of lactate to pyruvate in the liver, which is one factor by which ethanol increases the concentration of lactate in the blood (see below).

of acetaldehyde in the blood is further raised if the activity of aldehyde dehydrogenase is inhibited. Inhibitors of the enzyme include pyrogallol and the higher aliphatic aldehydes which are known to be present in alcoholic beverages* (Lieber, 1976).

The metabolic mechanisms by which changes in the $NAD^+/NADH$ concentration ratio or the acetaldehyde concentration may explain the physiological and clinical effects of ethanol are described below.

(i) *Hypoglycaemia*   After 24 hr starvation, hepatic gluconeogenesis is vital for the production of glucose and maintenance of the blood sugar level (Chapter 14; Section A.2). Ethanol is a potent inhibitor of gluconeogenesis[7] and this effect could contribute to the hypoglycaemia that is common in many chronic alcoholics, particularly after a period of excessive drinking. Under these conditions, the alcoholic usually does not eat, so that gluconeogenesis would be essential to maintain the blood glucose concentration. Severe hypoglycaemia and concomitant lactic acidosis may be the two most important factors that precipitate the coma and collapse of the alcoholic who is 'on the bottle' (see Chapter 13). In addition to inhibition of gluconeogenesis, chronic alcoholics may have reduced ability to secrete some of the hormones involved in control of lipolysis (e.g. cortisol, growth hormone) and this would reduce the rate of fatty acid release in starvation and exacerbate the hypoglycacmia (see Chapter 8; Section C; Marks, 1978).

(ii) *Fatty liver, hepatitis and cirrhosis*   Chronic consumption of ethanol can lead to the development of 'fatty liver', which is the deposition of excess triacylglycerol in the liver.[8] It is still unclear whether this deposition actually causes damage to the liver or if it is a consequence of the damage. In either case, the damage may result in hepatitis and, eventually, cirrhosis. As the 'fatty liver' develops there is a decrease in the amount of endoplasmic reticulum (especially the rough endoplasmic reticulum which is involved in protein synthesis) and, in addition, damage to the mitochondria. This cellular damage may be caused by chronically elevated concentrations of acetaldehyde within the liver cell and is eventually severe enough to cause cell death. Cell damage and death will result in an inflammatory response, that is, infiltration of lymphocytes and activation of the immune system. This condition is then known as hepatitis. If this is not treated it is likely to result in the formation of fibrous tissue with the development of cirrhosis, when the amount of functional liver becomes markedly reduced (Scheuer, 1982). In severe cases, the rates of detoxification become insufficient resulting in hepatic coma and death (see Chapter 12; Section C.3).

(iii) *Gonadal function*   There is evidence of reduced gonadal function in men who suffer from alcoholism. This manifests itself as impotence, testicular atrophy,

---

* It should be noted that even if the aldehyde dehydrogenase catalyses a near-equilibrium reaction so that inhibition should not reduce the flux, it will undoubtedly raise the concentration of pathway-substrate, i.e. acetaldehyde.

gynaecomastia (enlargement of male breasts sufficient to resemble those of a woman), sterility, and changes in distribution of body hair. Indeed, it has been claimed that alcoholism is the most common cause of non-functional impotence and sterility in the U.S.A. (Masters and Johnson, 1970; Van Thiel and Lester, 1974). The metabolic mechanisms by which ethanol produces these changes are likely to be complex. A deficiency of liver function might reduce the rate of metabolism of female sex hormones in man so that the blood concentration of oestrogens could be increased. However, in both experimental animals and man, chronic consumption of ethanol reduces the plasma testosterone concentration. This probably occurs as a result of a diminished rate of testosterone biosynthesis in Leydig cells of the testes (Ellinghoe and Varanelli, 1979). This could be due to a decrease in the $NAD^+/NADH$ concentration ratio in the testes since alcohol dehydrogenase is present in these cells. The oxidation of intermediates in the pathway for testosterone synthesis (see Figures 20.12 and 20.9) could be dependent on the $NAD^+/NADH$ concentration ratio so that the rate of testosterone synthesis might be inhibited by the marked decrease in the $NAD^+/NADH$ concentration ratio.

Shakespeare (Macbeth, Act II, scene III) was clearly aware of the effects of ethanol on gonadal function.

'Macduff: What three things does drink especially provoke?
Porter: Marry, sir, nose painting, sleep, and urine. Lechery, sir, it provokes and unprovokes. It provokes the desire but it takes away the performance.'

Biochemistry provides the explanation for the porter's observations: a decrease in the plasma concentration of testosterone would account for the inability to perform while the reduced extent of feedback inhibition on the pituitary by testosterone would lead to an increased rate of secretion, and hence an increase in the blood concentration, of the gonadotrophic luteinizing hormone (Wright, 1978). There is evidence that a high circulating concentration of luteinizing hormone increases the degree of sexual arousal in men.

(iv) *Neurological effects of ethanol*   Perhaps the most fascinating question concerning ethanol is the biochemical mechanism underlying its neurological effects; the acute effects include stimulation, reduction or removal of inhibition, increased sexuality, increased aggressiveness and an antidepressant action; chronic effects include damage to the brain, tolerance, physical dependence and disturbance of sleep. Current knowledge of the biochemistry of the brain and particularly the biochemistry of mood suggests an involvement of the biogenic amines (dopamine, noradrenaline, 5-hydroxytryptamine, see Chapter 21, Section B) but whether this is caused by ethanol or acetaldehyde is, at present, a controversial subject. There are many hypotheses to explain the acute effects; three are described below.

1. Ethanol is a sufficiently small molecule that it enters the cell membrane and in so doing disturbs the precise positioning of the fatty acyl groups of the phospholipid (Chapter 17; Section C). This could interfere with several processes including the entry of calcium ions into the nerve terminal. These ions are thought to be involved in the control of the rate of release of neurotransmitter into the synapse (Chapter 21; Section C.2). Hence, ethanol would cause a decrease in the rate of neurotransmitter release in certain neurones accounting for the acute effects (Littleton, 1983).

2. The roles of cyclic nucleotides (cyclic AMP and cyclic GMP) and prostaglandins as local and intracellular messengers are described in Chapter 22; Section A.4. In brain, they play an important role as modulators of neurotransmission; prostaglandins increase the concentrations of cyclic AMP probably through stimulation of the activity of adenylate cyclase. Ethanol ingestion decreases the brain concentrations of cyclic AMP and this may be due to a reduced stimulation by prostaglandins (Littleton, 1978). This effect could be due to ethanol entering the cell membrane or acetaldehyde interacting with proteins in the cell, and either effect decreasing the number or affinity of receptors for prostaglandins on the cell membrane.

3. Acetaldehyde could react with biogenic amines to form compounds which resemble the tetrahydroisoquinoline alkaloids found in plants. For example, dopamine and acetaldehyde condense at pH 7.0 at room temperature to form salsolinol (Figure 11.15) (Cohen and Collins, 1970). The alkaloid salsolidine is the bismethoxy derivative of this product, and many more complex alkaloids, including morphine, with profound pharmacological activity share the same ring structure. It has been established that tetrahydroisoquinolines inhibit some enzymes involved in amine metabolism (amine oxidase, catechol

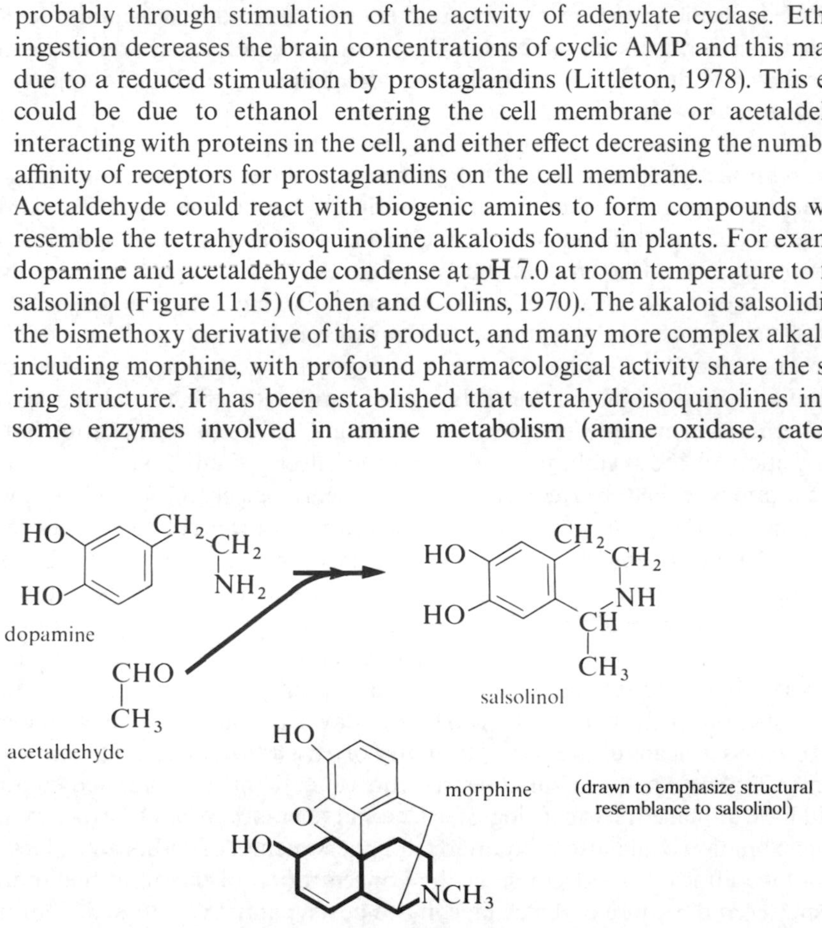

Figure 11.15 The *in vitro* synthesis of tetrahydroisoquinoline compounds from catecholamine and acetaldehyde. The synthesis occurs via the intermediate formation of a Schiff's base

methyltransferase) and could, therefore, influence amine metabolism in the brain (Chapter 21; Section B.2) (*Lancet* editorial, 1982a).

Alternatively, acetaldehyde could inhibit one of the enzymes involved in the normal degradation pathway of the brain amines The latter are normally converted to aldehydes by amine oxidase and these aldehydes are further metabolised to acidic products by aldehyde dehydrogenases. Since acetaldehyde could competitively inhibit the dehydrogenases, this would result in an increase in the concentrations of the aldehydes, which could condense with biogenic amines or their derivatives to produce neurologically active compounds, as indicated above. Evidence to support this suggestion comes from the fact that administration of ethanol to humans and experimental animals decreases the urinary excretion of the acidic derivatives of the biogenic amines (5-hydroxyindoleacetic acid and homovanillic acid— see Chapter 21; Sections 2.c and 3.c) but increases the excretion of the alcoholic drivatives, 5-hydroxytryptophanol and 3-methoxy-4-hydroxyphenylglycol.

The chronic effects of ethanol include brain damage, tolerance and physical dependence. Many alcoholics suffer from brain damage and dementia. This could be due to decreased blood flow to the brain, decreased protein synthesis or vitamin deficiency. Dementia of the Wernicke–Korsakoff type, which involves impairment of memory, is common in alcoholics and may be due to deficiency of thiamine (see Appendix A.1) (Shaw, 1982). Tolerance and physical dependence are more difficult to explain. There are, however, a number of speculative suggestions. First, if ethanol exerts its acute neurological effect by changing the nerve terminal membrane, tolerance could occur if long term modification in the composition of the membrane decreased the effects of ethanol. Thus higher concentrations would be required to elicit neurological effects. Withdrawal symptoms could arise if the membrane rapidly reverted to the normal composition upon removal of ethanol but the neurotransmitter release system maintained a higher sensitivity to calcium ions (see (1) above); the rate of neurotransmitter release would be increased. Secondly, if ethanol increases the concentration of an abnormal amine and if this amine is able to interact with the postsynaptic receptor to elicit a neurological response, the receptors may adapt by a reduction in their number (known as 'down regulation', a phenomenon recognized as a means of changing sensitivity of a tissue to a given hormone—see Chapter 22; Section A.3). Consequently, the same quantity of ingested ethanol would have a smaller pharmacological effect and tolerance would be experienced. This explanation could also account for the phenomenon of withdrawal; lack of ethanol would lead to a decrease in the concentration of the abnormal amine transmitter and because of the change in the postsynaptic receptors, the normal concentration of amine transmitter would be much less effective. Withdrawal symptoms would be experienced until the receptor number or the affinity had increased towards normal.

NOTES

1. The clinical features of hypoglycaemia differ according to the age of the patient. In infants, the presence of convulsions, tremor, and twitching are highly suggestive of hypoglycaemia. Other features include cyanosis, refusal to feed, eye-rolling, and coma. Failure to correct hypoglycaemia in infants can lead to brain damage and consequent mental retardation. Hypoglycaemic children typically show pallor, limpness, inattention, listlessness, and, when severe, stupor, convulsions, and coma. Tachycardia and sweating may also be present. In adults, the features include sweating, malaise, weakness, anxiety, palpitations, paraesthesia, feeling of detachment from the environment, tachycardia, visual disturbances, fine tremor, and sometimes hypothermia. Persistent and chronic hypoglycaemia can produce personality changes, defective memory and extreme depression. Authors such as Steincrohn (1972) see hypoglycaemia as a disease in itself and as a frequent cause of the conditions listed above plus many other symptoms.

2. In UDPglucose, the sugar is attached to the UDP by means of an ester link between the C_1-hydroxyl group and the terminal phosphate. UDPglucose is formed by the action of the enzyme glucose-1-phosphate uridylyltransferase as follows:

$$UTP^{4-} + \text{glucose 1-phosphate}^{2-} + H^+ \longrightarrow UDPglucose^{2-} + PP_i^{3-}$$

This reaction is described in more detail in Chapter 16; Section A.

3. It appears that a single aldolase is responsible for cleaving both fructose 1,6-bisphosphate and fructose 1-phosphate in the liver. It must, however, be a distinct enzyme from that in muscle, since the latter has only a very feeble activity towards fructose 1-phosphate.

4. The decrease in the ATP/ADP concentration ratio also leads to activation of the enzyme AMP deaminase, which catalyses the conversion of AMP to inosine monophosphate (IMP). Under normal conditions this enzyme is inhibited by the high concentrations of ATP and phosphate present in the liver. Since AMP is in equilibrium with ATP and ADP (through the reaction catalysed by adenylate kinase) but AMP deaminase catalyses a non-equilibrium reaction, activity of the latter leads to a reduction in the total concentration of adenine nucleotides, as they are converted to inosine monophosphate. The latter is further metabolized to inosine by 5'-nucleotidase and this inosine is converted to hypoxanthine, xanthine, and finally uric acid (see Figure 12.7) which diffuses into the bloodstream and is excreted via the kidneys. Hence there may be high concentrations of uric acid in blood and urine after fructose infusion. The hyperuricaemia produced by fructose infusion may not, in itself, be harmful but it is an indication of reduced energy status in the liver which is dangerous.

5. A small whisky or a small sherry can (especially on an empty stomach) raise the blood concentration of ethanol to 2–4 mM. In Britain it is an offence to drive if the blood ethanol concentration exceeds 18 mM (80 mg.100 cm^{-3}). In the U.S.A. limits of 22–33 mM (100–150 mg.100 cm^{-3}) apply according to the State.

6. In rat liver, the concentration of ethanol was varied between 10–60 μmol.g^{-1} fresh wt. (approximately 20–120 mM) which produced changes in the acetaldehyde concentration between 0.1–0.25 μmol.g^{-1} (about 0.2–0.5 mM).

7. In the perfused liver of the rat, inhibition is observed at 2–4 mM ethanol (Krebs, 1968) which is the concentration reached in man after drinking a small glass of sherry. The precise biochemical mechanism(s) for the inhibition of gluconeogenesis by ethanol is unclear. The decreased cytosolic NAD$^+$/NADH concentration ratio markedly reduces the pyruvate concentration which will reduce the activity of pyruvate carboxylase (since this enzyme does not catalyse a flux-generating reaction—Krebs, 1968). In addition, acetaldehyde is known to inhibit mitochondrial oxidative phosphorylation

so that the mitochondrial and cytosolic ATP/ADP concentration ratios may be decreased, which would result in further inhibition of gluconeogenesis and activation of glycolysis (see Section B.4.b).

8. Three effects could play a part in causing the increased triacylglycerol deposition in the liver: an increased rate of lipogenesis, a decreased rate of fatty acid oxidation and a reduced rate of triacylglycerol secretion. Ethanol oxidation within the liver raises the acetyl-CoA concentration and this may inhibit fatty acid oxidation (see Chapter 7, Section D.2). Furthermore, with a high intake of ethanol, the rate of acetyl-CoA production could exceed the energy requirement of liver so the extra acetyl-CoA will be converted to fatty acids. The decreased $NAD^+/NADH$ concentration ratio will cause an increase in the concentration of glycerol 3-phosphate which is required for esterification of acyl-CoA to triacylglycerol (see Chapter 17; Section A.7) and this change may stimulate this process. The increase in the lipogenic rate in the liver and increased rate of release of VLDL could cause hypertriglyceridaemia (Janus and Lewis, 1978). Eventually the processes involved in the synthesis and secretion of VLDL could become saturated so that triacylglycerol would accumulate within the hepatocyte. In man, the malnutrition that usually accompanies high ethanol consumption could also contribute to the inability of the liver to synthesize and secrete VLDL; the low intake of the essential amino acid methionine would restrict the formation of choline, an important component of phospholipids (see Chapter 17; Section C.2). Nonetheless, experiments in animals (including primates) that were otherwise well-fed show that excessive ethanol consumption leads to the development of 'fatty liver' (see Lieber *et al.*, 1975).

There is uncertainty concerning the amount of alcohol that can be consumed without any serious liver damage. On the basis of liver biopsy studies on men admitted to an alcohol-detoxification unit, Lelbach (1974) claimed that those with an intake below 140 g.day^{-1} (a bottle of sherry or eight pints of beer) had normal livers, whereas intakes in excess of this caused 'fatty liver', hepatitis, or cirrhosis. Other work suggests that lower daily intakes may be dangerous; a lower limit for men may be 80 g.day^{-1} but may be much lower for women at 40 g.day^{-1} (Saunders *et al.*, 1981).

CHAPTER 12

METABOLISM OF AMMONIA

Ammonia is formed in a number of reactions in the body. Under most conditions, the quantitatively important source of ammonia is the metabolism of amino acids but ammonia is also produced during the degradation of amines, purines, and pyrimidines. Since it is produced in degradation reactions, ammonia is usually considered only as an endproduct of metabolism which must be converted to a non-toxic compound for excretion. In most mammals, including man, this compound is urea which is produced in the liver by a cyclical series of reactions, known as the urea cycle. However, the complexity of amino acid metabolism described in Chapter 10, and in particular the contribution of different tissues, means that provision of ammonia for urea formation is not as straightforward as originally considered (Section B). In addition to providing the nitrogenous substrate for the urea cycle, ammonia can play at least four other roles in metabolism: as a substrate for the glutamate dehydrogenase reaction in the formation of non-essential amino acids; as a substrate for glutamine synthetase in the formation of glutamine in muscle (Chapter 10; Section F.2.b); as an important regulator of the rate of glycolysis in muscle, brain, and possibly other tissues; and in the regulation of acid/base balance by the kidney.

A. FORMATION OF AMMONIA

Although the urea cycle occurs exclusively in the liver, ammonia is produced in several other tissues in a number of different reactions. The production of ammonia in non-hepatic tissues means that it must be transported directly or indirectly to the liver for urea formation (see Section B.1).

The metabolism of amino acids, amines, adenosine (a purine), and cytosine (a pyrimidine) generates the major part of the ammonia which is used to synthesize urea. The reactions involved are described below.

1. Glutamate dehydrogenase reaction

It now seems likely that a substantial part of the ammonia arising from amino acids in the liver does so by way of the glutamate dehydrogenase reaction:

$$\text{glutamate}^- + \text{NAD(P)}^+ + H_2O \longrightarrow$$

$$\text{2-oxoglutarate}^{2-} + \text{NADPH} + H^+ + NH_4^+$$

The important role of the glutamate dehydrogenase reaction as part of the transdeamination system is described in Chapter 10; Section C.2.a. In addition to the liver, this reaction occurs in brain, muscle and kidney.

2. Specific amino acid deamination reactions

A number of amino acids are deaminated in specific reactions that produce ammonia (for example, serine, threonine, cysteine, and glycine—see Appendix 10.1). These reactions occur principally in the liver.

3. Amine oxidase reaction

Many of the monoamines (e.g. noradrenaline, adrenaline, dopamine, 5-hydroxytryptamine, histamine) serve as hormones or neurotransmitters. Both roles imply a rapid turnover and their catabolism generates ammonia. Monoamines also enter the body in the diet and certain foods, such as soft cheese, contain large amounts. Dietary monoamines and the hormones secreted by the adrenal medulla are metabolized by intestinal and hepatic amine oxidase, while neurotransmitters are similarly metabolized by the enzyme within nervous tissue (see Chapter 21; Section B.2.c). Amine oxidase (formerly known as monoamine oxidase) is a flavoprotein attached to the outer membrane of the mitochondria and catalyses the following reaction:

$$R.CH_2NH_2 + \tfrac{1}{2}O_2 \longrightarrow R.CHO + NH_3$$

However, the amount of ammonia from this source is likely to be small. In normal human subjects, in the absence of dietary monoamine, perhaps 1.0 mg of ammonia is produced from these biogenic amines each day.

4. Purine nucleotide metabolism

In terms of their excretory fate, the nitrogen atoms in a purine fall into two classes. The four present in the purine ring itself are excreted in man as the purine, urate (uric acid) (see Appendix 12.1), while the amino groups attached to the ring (in the 6-position in adenine and the 2-position in guanine) are released as ammonia. Most of the ammonia produced from purines is derived from adenine nucleotides. Normal metabolism results in the hydrolysis of ATP to ADP and subsequent re-phosphorylation, but if there is a marked increase in the rate of ATP utilization, the ATP/ADP concentration ratio decreases which increases the concentration of AMP, via the reaction catalysed by adenylate kinase:

$$ATP + AMP \rightleftarrows 2ADP$$

(This equilibrium probably occurs in all tissues and appears to be of significance in the regulation of glycolysis in muscle (see Chapter 7; Section B.1.b).) An

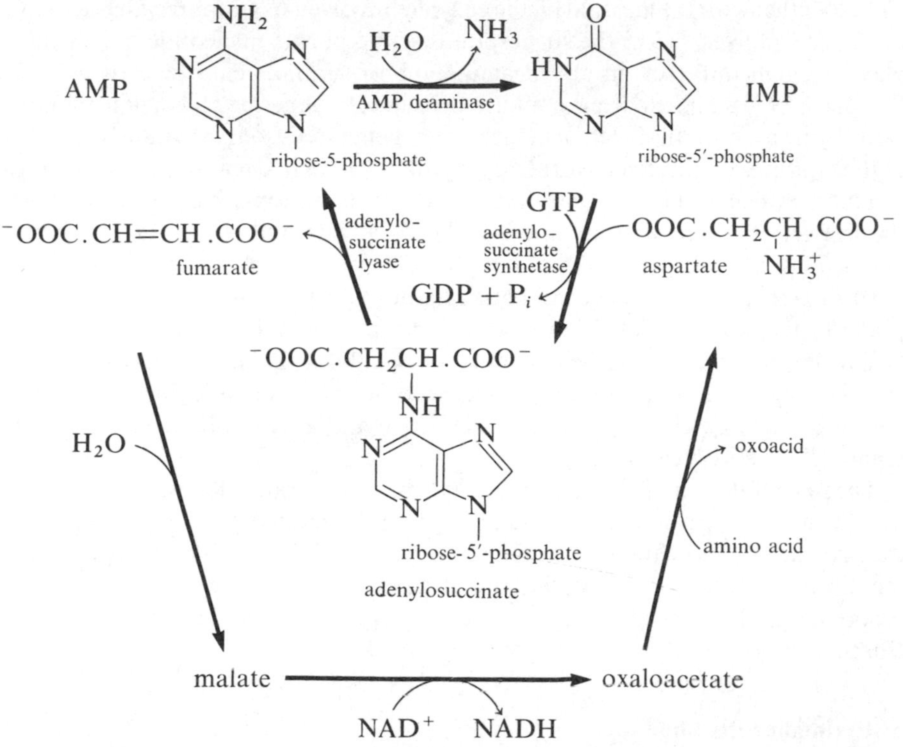

NH$_2$

N N

N

N

AMP

ribose-5-phosphate

H$_2$O   NH$_3$

AMP deaminase

O

HN N

N N

IMP

ribose-5'-phosphate

$^-$OOC.CH=CH.COO$^-$

fumarate

adenylo-
succinate
lyase

GTP

adenylo-
succinate
synthetase

GDP + P$_i$

OOC.CH$_2$CH.COO$^-$

aspartate   NH$_3^+$

$^-$OOC.CH$_2$CH.COO$^-$

NH

N N

N N

ribose- 5'-phosphate

adenylosuccinate

H$_2$O

oxoacid

amino acid

malate $\longrightarrow$ oxaloacetate

NAD$^+$ NADH

Figure 12.1   Reactions of the purine nucleotide cycle. The conversion of fumarate to aspartate involves enzymes of the TCA cycle plus an aminotransferase

increase in the concentration of AMP increases the activity of AMP deaminase, which catalyses the following reaction:

$$AMP^{2-} + H_2O \longrightarrow IMP^{2-} + NH_3$$

An alternative pathway for the deamination of AMP, when its concentration increases, involves the sequential action of the enzymes 5'-nucleotidase and adenosine deaminase:[6]

$$AMP^{2-} + H_2O \longrightarrow adenosine + P_i^{2-}$$
$$adenosine + H_2O \longrightarrow inosine + NH_3$$

The deamination of AMP by these routes was formerly thought to play a minor metabolic role but the recognition that AMP deaminase is present in many tissues and that there is a pathway for the resynthesis of AMP from IMP (see Figure 12.1) led to the proposal that a purine nucleotide cycle operated in some tissues (Lowenstein and Goodman, 1978). The net effect of the cycle is the deamination of aspartate as follows:

$$aspartate^- + GTP^{4-} \longrightarrow fumarate^{2-} + GDP^{3-} + P_i^{2-} + NH_4^+ + H^+$$

This, together with the fact that all the enzymes involved have appreciable activity in extracts of liver, led to the suggestion that the purine nucleotide cycle could play a significant part in the deamination of amino acids (McGivan and Chappell, 1975). The aspartate could, in principle, be generated from other amino acids by transamination with oxaloacetate (regenerated from fumarate by action of the enzymes fumarate dehydratase and malate dehydrogenase). However, the available evidence suggests that the generation of ammonia by this pathway in the liver is quantitatively unimportant (Van der Berghe *et al.*, 1980; Evered, 1981). This raises the question of the physiological role of the reactions of the purine nucleotide cycle in liver and other tissues. It is probably fair to say that, at present, it is unclear but, in muscle and brain, it could be important in the generation of ammonia for the regulation of glycolysis (Chapter 7; Section B.1.b) and it may be important for the control of the total adenine nucleotide concentrations which, in any given tissue, are remarkably similar from animal to animal (Beis and Newsholme, 1975).

The main function of 5'-nucleotidase may be to generate adenosine which acts as a local messenger in many tissues, including the heart, where its release from the myocardial cells causes relaxation of the smooth muscle fibres in the circular muscle layer of the arterioles of the coronary vessels, with consequently increased blood flow and therefore oxygen and fuel supply to the heart. This is discussed in Chapter 7; Section B.1.b.

5. Pyrimidine metabolism

The pyrimidine, cytosine, possesses an amino group which can be released as ammonia in a reaction catalysed by cytosine deaminase in which cytosine is converted to uracil:

$$\text{cytosine} + H_2O \longrightarrow \text{uracil} + NH_3$$

Deamination can also take place at the nucleoside level, catalysed by cytidine deaminase:

$$\text{cytidine} + H_2O \longrightarrow \text{uridine} + NH_3$$

In the complete degradation of the pyrimidine ring one of the nitrogen atoms is converted to ammonia while the other appears in β-alanine (in the case of cytosine and uracil) or β-aminoisobutyrate (in the case of thymine). The degradative pathways of both purine and pyrimidine metabolism are described in Appendix 12.1.

6. Ammonia from the small intestine

It has been known for many years that the intestine produces ammonia but until recently it was assumed that this arose from the action of the enzyme urease, present in intestinal bacteria, on urea which diffused into the intestine. The current view, however, is that the ammonia is derived from the amide nitrogen atom of glutamine and that it represents a quantitatively important source of ammonia for urea synthesis in the liver. The glutamine is taken up both from digested protein in the lumen of the small intestine and from the blood by the absorptive cells of the small intestine where it is deamidated by the enzyme glutaminase:

$$
\begin{array}{c}
CONH_2 \\
| \\
(CH_2)_2 \\
| \\
CHNH_3^+ \\
| \\
COO^-
\end{array}
\quad + H_2O \quad \longrightarrow \quad
\begin{array}{c}
COO^- \\
| \\
(CH_2)_2 \\
| \\
CH\ NH_3^+ \\
| \\
COO^-
\end{array}
\quad + NH_3^+
$$

L-glutamine   L-glutamate

The nitrogen atom from the α-amino group of the glutamine (together with that from dietary glutamate, aspartate, and asparagine—Chapter 10; Section F.4) leaves the small intestine in the form of ammonia, alanine, citrulline and proline, which are transported via the hepatic portal vein to the liver (Hanson and Parsons, 1980). It has been estimated that these nitrogenous compounds, including ammonia, account for about 60% of the nitrogen atoms excreted as urea in the rat (Windmueller and Spaeth, 1980).

B. THE UREA CYCLE

Animals which have internal access to large volumes of water are able to excrete ammonia directly. The majority of fresh water organisms come into this category. They are able to avoid the toxic effects of ammonia (see Section C.1) by preventing its accumulation in the body. Terrestrial organisms are denied the large volumes of water necessary to excrete a very dilute solution of ammonia and so convert the ammonia into less toxic compounds which can be stored for periodic excretion. Marine organisms are generally in a similar position because osmotic factors limit the acquisition of large volumes of water. In mammals and many amphibia and marine fish,[1] the ammonia is converted to urea by reactions of the urea cycle described below. Urea is highly soluble ($2\ mol.l^{-1}$ at 25°C), is non-toxic, has a high nitrogen content (47%) and requires relatively little chemical energy for synthesis (0.5 molecule ATP per atom of nitrogen, Section B.3.a). Normal human subjects excrete about 30 g urea each day on a Western diet but a high protein diet can increase this to at least 100 g. An alternative strategy is employed by birds, insects, and reptiles, all of which have an egg stage in which water availability is severely restricted. Under these circumstances the synthesis of a highly soluble excretory product would have serious osmotic consequences and most of the ammonia is converted to the virtually insoluble purine, uric acid (urate).[7] This product can be

Table 12.1. Concentration of ammonia in blood and other tissues

Tissue	Ammonia concentration (mM)
Arterial blood	0.02
Hepatic portal venous blood	0.26
Hepatic venous blood	0.03
Liver	0.71
Skeletal muscle (thigh)	0.26
Brain	0.34
Heart	0.20
Spleen	0.20
Arterial blood (in ammonia toxicity)	0.50

The concentrations of ammonia refer to the total ammonia, that is NH_3 and NH_4^+, in tissues of the rat; these are in equilibrium as follows:

$$NH_3 + H^+ \rightleftarrows NH_4^+$$

At physiological pH the equilibrium of this reaction is far to the right, so that at pH 7.1 about 99 % of the ammonia is in the ionic form. It is generally accepted that NH_3, but not NH_4^+, is able to diffuse across cell membranes whereas NH_4^+ is transported by a carrier-mediated process (for review see Kleiner, 1981). Hence in the absence of the latter process, movement across the cell membrane will be slow and this might explain the low concentration of ammonia in the blood in comparison to that in the cells.

safely retained in the egg or excreted as a slurry of fine crystals by the adult. Uric acid has a lower nitrogen content than urea (33 %) and is more 'expensive' to synthesize (2.25 mol ATP per atom of nitrogen, see Chapter 18; Section B.1.a). Mammals also produce uric acid but as a product of purine catabolism. Whether this is excreted as such or further degraded depends on the species of mammal (see Appendix 12.1).

1. Transport of ammonia in blood

As indicated in Section A.6 a large proportion of the ammonia used in urea synthesis does not arise from catabolism in the liver but in extra-hepatic tissues. Only a small proportion of this leaves the extra-hepatic tissues as ammonia and, consequently, the concentration of ammonia in blood is low (0.02–0.03 mM). This is consistent with the toxicity of ammonia and is reflected in the low concentrations of ammonia in tissues (Table 12.1).

Most tissues release nitrogen mainly as alanine or glutamine. In the former case, ammonia is not formed as an intermediate but the amino group is transferred by transamination from glutamate to pyruvate (itself arising from the degradation of amino acids or from glycolysis). Glutamine synthesis, on the other hand, utilizes ammonia itself and is catalysed by glutamine synthetase:

$$glutamate^- + ATP^{4-} + NH_3 \longrightarrow glutamine + ADP^{3-} + P_i^{2-}$$

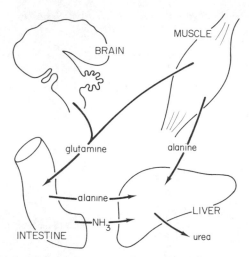

Figure 12.2 The provision of nitrogenous compounds for urea synthesis in the liver. In addition, some amino acids after absorption from the intestine will be degraded by the liver and contribute to urea synthesis (Chapter 10; Section F.1).

Indeed, if the glutamate arises by amination of 2-oxoglutarate, catalysed by glutamate dehydrogenase, glutamine can be seen as the transporter of two molecules of ammonia. The pathways of alanine and glutamine synthesis in muscle have been described in Chapter 10; Section F.2. Those in brain, although quantitatively less important, are probably similar.

The absorptive cells of the small intestine are exceptional in that they release ammonia into the hepatic portal vein, in which the ammonia may attain concentrations ten-fold higher than in other vessels (Table 12.1), and account for 30 % of the nitrogen of the urea synthesized in the liver.

Despite the relatively high concentration of ammonia in the hepatic portal vein, that in the hepatocytes is higher so that ammonia transport into the cell may be an active process (Brosnan, 1976; Kleiner, 1981) or ammonia may be chemically bound within the hepatocyte (Sainsbury, 1980). The quantitatively important tissues providing 'nitrogen' for the liver are shown in Figure 12.2.

2. Reactions of the urea cycle

The reactions by which ammonia is converted into urea were discovered in the 1930s and the novel suggestion of a metabolic cycle of reactions for the synthesis of urea was proposed by Krebs and Henseleit in 1932. At that time, not all the biochemical details of the individual reactions were known but they were clarified by later work. The history of the discovery of the urea cycle (also know as the ornithine cycle) has been recounted by Krebs (1976). Detailed descriptions of the reactions of the pathway are to be found in many texts and for this reason the descriptions given here will be brief.[2] Details of the metabolism of the cycle that are of physiological and clinical importance will be discussed in greater detail.

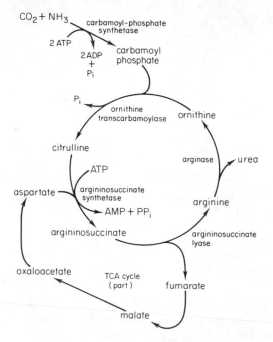

Figure 12.3   Reactions of the urea cycle

The reactions that are usually considered to constitute the cycle are summarized in Figure 12.3 and the physiological pathway and its regulation are discussed in Section B.4. In the first reaction, ammonia reacts with carbon dioxide and ATP to produce carbamoyl phosphate within the mitochondrion. The reaction is catalysed by carbamoyl-phosphate synthetase:

$$2ATP^{4-} + CO_2 + NH_3 + H_2O \longrightarrow$$
$$NH_2CO.OPO_3^{2-} + 2ADP^{3-} + P_i^{2-} + 2H^+$$
$$\text{carbamoyl phosphate}$$

The carbamoyl phosphate then reacts with ornithine to produce citrulline in a reaction catalysed by ornithine transcarbamoylase, which is also localized within mitochondria:

$$
\begin{array}{lll}
\text{(CH}_2)_3\text{NH}_3^+ & \text{NH}_2 & \text{(CH}_2)_3\text{NH.CONH}_2 \\
\quad | & \quad | & \quad | \\
\text{CH.NH}_3^+ \; + & \text{CO} \longrightarrow & \text{CH.NH}_3^+ \qquad + P_i^{2-} + H^+ \\
\quad | & \quad | & \quad | \\
\text{COO}^- & \text{OPO}_3^{2-} & \text{COO}^- \\
\text{L-ornithine} & & \text{L-citrulline}
\end{array}
$$

Citrulline is transported out of the mitochondria by a transporter and, in the cytosol, reacts with aspartate to produce argininosuccinate. In this reaction (catalysed by argininosuccinate synthetase), ATP is hydrolysed to AMP and pyrophosphate:

$$(CH_2)_3NH.CONH_2 \quad\quad +H_3N.CH.COO^-$$
$$| \quad\quad\quad\quad + \quad\quad\quad | \quad\quad + ATP^{4-} \longrightarrow$$
$$CH.NH_3^+ \quad\quad\quad\quad\quad CH_2COO^-$$
$$|$$
$$COO^-$$

$$\quad\quad\quad\quad NH_2^+$$
$$\quad\quad\quad\quad ||$$
$$(CH_2)_3NH.C.NH. \quad CH.COO^- \quad + AMP^{2-} \quad + PP_i^{3-} \quad + H^+$$
$$| \quad\quad\quad\quad\quad\quad |$$
$$CH.NH_3^+ \quad\quad\quad\quad CH_2COO^-$$
$$|$$
$$COO^-$$
$$\quad\quad\quad\quad \text{L-argininosuccinate}$$

Argininosuccinate is hydrolysed to form arginine and fumarate in the cytosol in a reaction catalysed by the enzyme argininosuccinate lyase.

$$\quad NH_2^+ \quad\quad\quad\quad\quad\quad\quad\quad NH$$
$$\quad || \quad\quad\quad\quad\quad\quad\quad\quad\quad\quad ||$$
$$(CH_2)_3NH.C.NH.CH.COO^- \quad (CH_2)_3NH.C.NH_3^+ \quad\quad CH.COO^-$$
$$| \quad\quad\quad\quad\quad\quad | \quad\quad\quad\quad | \quad\quad\quad\quad\quad\quad\quad ||$$
$$CH.NH_3^+ \quad\quad\quad CH_2COO^- \longrightarrow CH.NH_3^+ \quad\quad + {}^-OOC.CH$$
$$| \quad\quad\quad\quad\quad\quad\quad\quad\quad\quad\quad | \quad\quad\quad\quad\quad\quad \text{fumarate}$$
$$COO^- \quad\quad\quad\quad\quad\quad\quad\quad\quad COO^-$$
$$\quad\quad\quad\quad\quad\quad\quad\quad\quad \text{L-arginine}$$

Finally, this arginine is hydrolysed to ornithine and urea by the enzyme arginase:

$$\quad\quad NH$$
$$\quad\quad ||$$
$$(CH_2)_3NH.C.NH_3^+ \quad +H_2O \quad (CH_2)_3NH_3^+ \quad\quad\quad NH_2$$
$$| \quad\quad\quad\quad\quad\quad\quad\quad\quad\quad\quad | \quad\quad\quad\quad\quad\quad |$$
$$CH.NH_3^+ \quad\quad\quad\quad\quad \longrightarrow CH.NH_3^+ \quad + O{=}C.NH_2$$
$$| \quad\quad\quad\quad\quad\quad\quad\quad\quad\quad\quad | \quad\quad\quad\quad\quad\quad \text{urea}$$
$$COO^- \quad\quad\quad\quad\quad\quad\quad\quad COO^-$$
$$\quad\quad\quad\quad\quad\quad\quad \text{L-ornithine}$$

The urea diffuses into the blood which carries it to the kidney where it passes into the glomerular filtrate for excretion in the urine.

In order to continue the cycle, ornithine must traverse the mitochondrial membrane so that it is available for the formation of citrulline. The transport of ornithine into the mitochondrion occurs via the same transporter system that carries citrulline out of the mitochondria.

3. **Metabolically significant aspects of the cycle**

(a) *Energy 'cost'*

The reactions of the cycle can be summarized in the equation:

$$3ATP^{4-} + NH_3 + CO_2 + 2H_2O + aspartate^- \longrightarrow$$
$$fumarate^{2-} + 2ADP^{3-} + 4P_i^{2-} + AMP^{2-} + urea + 5H^+$$

From this, it appears that four molecules of 'energy-rich' phosphate are required for the synthesis of one molecule of urea. However, this is misleading since, to complete the cycle, fumarate must be converted to aspartate. This is achieved by reactions of the TCA cycle and transamination of oxaloacetate to aspartate. This series of reactions involves the conversion of malate to oxaloacetate, with the generation of NADH the reoxidation of which should produce three molecules of ATP. Hence, the energetic cost of the cycle is only one ATP molecule for each molecule of urea synthesized.

(b) *Sources of nitrogen for the cycle*

It is apparent from consideration of the reactions of the urea cycle that only one of the two nitrogen atoms comes directly from ammonia. If insufficient is available

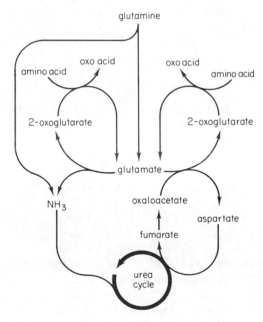

Figure 12.4 Alternative routes of entry for nitrogen into the urea cycle. Note that the operation of the purine nucleotide cycle (Figure 12.1) would provide additional pathways

Table 12.2. Rate of urea production in man on normal and low protein diets

Diet	Time on diet (day)	Rate of urea production (g.day^{-1})
Normal	—	26
Low protein	2	10
	5	7
	7	3
	10	2
	13	2
Normal	1	17

Data from Folin (1905) (see Raijman, 1976). After a low protein diet (starch and cream for 13 days) the subjects returned to a normal diet.

from the intestine (Section A.6) the deficit will be made up by the hepatic transdeamination of amino acids from the diet or released from other tissues (see Figure 12.4). The aspartate required is formed from glutamate by transamination, the latter having arisen from the deamidation of glutamine or by transamination from other amino acids. As the reactions of transamination and transdeamination are near-equilibrium, the concentration of ammonia and aspartate will remain relatively constant in the liver.

4. Regulation of the urea cycle

Work on human subjects by Folin in 1905 showed that the rate of urea excretion could change dramatically according to the dietary intake of protein. A marked decrease in urea production occurred on a very low protein diet but on the day on which the subject returned to a normal diet, the rate of urea excretion increased nearly ten-fold (Table 12.2). The establishment of the mechanism by which urea cycle flux is controlled requires delineation of the physiological pathway.

(a) *Flux-generating step*

There has been no systematic study of the equilibrium status of the reactions of the urea cycle. Except for ornithine transcarbamoylase and arginase, the maximum catalytic activities of all the enzymes in the liver of the rat are similar to the maximum measured flux (Table 12.3), so that the authors consider that most, if not all, the reactions of the cycle are non-equilibrium.

A comparison of the K_m values of the enzymes of the cycle for the pathway-substrates with the concentrations of the substrates in the liver (Table 12.4) suggests that the only reaction to approach saturation is ornithine transcarbamoylase (for which the pathway-substrate is carbamoyl phosphate). The cycle is thus initiated at the ornithine transcarbamoylase reaction which explains earlier work indicating that this is the 'rate-limiting' step for the cycle

Table 12.3. Effect of content of protein in the diet on activities of urea cycle enzymes in the liver of the rat*

Condition	Enzyme activities (μmol.min^{-1}.g^{-1})				
	Carbamoyl-phosphate synthetase	Ornithine transcarbamoylase	Argininosuccinate synthetase	Argininosuccinate lyase	Arginase
Normal diet (15% protein)	5.7	223	2.9	4.5	583.
High protein diet (30% for two weeks)	9.7	447	4.9	8.1	758
High protein diet (60% for two weeks)	16.0	736	7.8	15.3	991
Zero protein† diet	2.6	60.2	0.8	1.2	350

* Data from Raijman (1976). The maximum rate of urea synthesis in isolated hepatocytes is calculated to be about 3 μmol.min^{-1}.g^{-1} at 25 °C. The hepatocytes were isolated from rats fed a normal diet.

† Total starvation of rats increases the activities of all the urea cycle enzymes (Schimke, 1962).

Table 12.4. Hepatic substrate concentrations and K_m values of enzymes of the urea cycle*

Enzyme	Substrate	K_m value (mM)	Hepatic substrate concentration (mM)
Carbamoyl-phosphate synthetase	ammonia	0.5	0.7
Ornithine transcarbamoylase	carbamoyl phosphate	0.02	0.11
	ornithine	1.8	<0.1
Argininosuccinate synthetase	citrulline	0.13	0.05
	aspartate	0.03	0.50
Argininosuccinate lyase	argininosuccinate	—	0.03
Arginase	arginine	10.0	0.04

* Data from Raijman (1976) and Brosnan (1976).

(Krebs *et al.*, 1973). In this respect, the urea cycle is similar to the TCA cycle since, in both, the first reaction of the cycle (citrate synthase in the TCA cycle) is flux-generating. (The high maximum activity *in vitro* is explained by the enzyme being limited *in vivo* by the concentration of its cosubstrate, ornithine—see below.)

(b) *Regulation of the flux through the cycle*[3]

Since the non-equilibrium reactions are not known for certain, any hypothesis of control must be tentative. At least four processes could be of regulatory significance; control of the flux of ammonia to the hepatocyte, regulation of carbamoyl-phosphate synthetase, regulation of ornithine transcarbamoylase (which, as the flux-generating step must play a key role in regulation), and control of the availability of aspartate (for review, see Evered, 1981).

1. The flux of ammonia will be regulated by the rates of several processes occurring outside the liver. These include amino acid release (predominantly alanine and glutamine) from muscle; glutamine metabolism in the absorptive cells of the small intestine and amino acid metabolism in liver (itself dependent on the rates of amino acid transport into the cell, protein synthesis and activities of key enzymes involved in the degradation of essential amino acids).
2. The compound *N*-acetylglutamate is an allosteric activator of carbamoyl-phosphate synthetase. There is some evidence that the hepatic concentration of this compound changes when the rate of urea formation is changed (Cheung and Raijman, 1980) although it is unclear how this comes about.[4] Since this enzyme is not flux-generating, the flux through the cycle cannot normally be regulated by changes in its activity. The importance of this allosteric control is presumably to enable the flux through this reaction to change without marked changes in the hepatic concentration of ammonia. If this reaction was controlled only by changes in the concentration of ammonia (that is, by

internal control) this might have to increase to such an extent (on a high protein diet, for example), that the resulting ammonium ion concentration gradient would be so great that transport into the liver could not occur and the blood ammonia concentration would reach toxic levels.

3. If ornithine transcarbamoylase is flux-generating, changes in the concentration of carbamoyl phosphate will not influence the rate of the cycle. However, the concentration of ornithine is below the K_m value of the enzyme, so that changes in the intramitochondrial concentration of ornithine could be important in the regulation of the activity of this enzyme and hence the flux through the cycle. (This would be an example of external control by cofactor availability in a similar manner to control of citrate synthase by oxaloacetate availability in the TCA cycle—see Chapter 7; Section B.3.a.) It is of interest that the absorptive cells of the intestine release not only ammonia and alanine but also citrulline and proline into the hepatic portal blood. Since these two amino acids are metabolized to ornithine in the liver, an increase in their rate of release by the intestinal cells could result in an increased concentration of ornithine in the liver. Hence, the intestinal cells may provide not only a quantitatively important source of ammonia but also two key precursors for a specific regulator of the flux-generating reaction in the urea cycle. There may, of course, be other mechanisms in the liver for the regulation of the concentration of ornithine (see Evered, 1981).

4. As indicated above, aspartate is required in equimolar amounts with ammonia for the cycle. However, aspartate availability is unlikely to determine the rate of urea synthesis since the near-equilibrium nature of the aspartate aminotransferase and glutamate dehydrogenase reactions ensure that ammonia and aspartate are always available in adequate concentrations (Figure 12.4).

In addition to the acute regulation of the cycle by factors outlined in this section, long-term regulation is achieved by changes in the actual concentration of enzymes. An increase in amount of protein in the diet increases the amount of all the enzymes of the cycle (in from four to eight days) whereas a decrease reduces the enzyme concentrations (Table 12.3). This is a mechanism of control by which the tissue adjusts the capacity of the enzymes to meet the *expected* flux, but how this is achieved is not known. The concentration of the 'flux-generating' enzyme will probably be the major limitation in the maximum rate of ammonia utilization when the diet changes from low protein (or zero protein) to high protein. The acute control mechanisms discussed above will increase the rate but if the rate of ammonia production exceeded this capacity, the ammonia accumulation could be dangerous (Chapter 14; Section C.2).

C. CLINICAL ASPECTS OF AMMONIA METABOLISM

Ammonia is a central metabolite produced in large quantities in the course of normal metabolism; it is also very toxic. These two facts lead directly to the clinical interest in ammonia metabolism since a slight imbalance in the rates of ammonia production and utilization can very rapidly threaten life.

The toxicity of ammonia is well known and perhaps most dramatically demonstrated by the early experiments of J. B. Sumner in which injection of purified urease (an enzyme which hydrolyses urea to ammonia and carbon dioxide) into rabbits rapidly caused their death (Kirk and Sumner, 1931). Despite this, it is important to appreciate that ammonia also performs a number of vital functions in the body. Its biosynthetic role in providing nitrogen atoms for the synthesis of non-essential (and indeed essential) amino acids has been described in Chapter 10; Section E. Via incorporation into the amide group of glutamine, ammonia also contributes to the biosynthesis of purines and the formation of CTP from UTP (see Chapter 18; Section B.1.a). Ammonia also has two important regulatory roles: in muscle and brain, ammonium ions are de-inhibitors of the ATP inhibition of 6-phosphofructokinase, thereby controlling the flux through this important reaction in glycolysis (Chapter 7; Section B.1.a); in the kidney, ammonia formation and excretion may be necessary to enable protons to be excreted to maintain acid/base balance (Chapter 13; Section B.4.c).

The activity of the urea cycle is such as to maintain the peripheral blood concentration of ammonia at about 0.02 mM. Any impairment of the cycle increases this and although there is no precise concentration of ammonia at which toxicity becomes apparent, it is considered that a concentration of 0.2 mM or above in the blood is dangerous. Ammonia toxicity in very young children is usually associated with vomiting and coma and is almost invariably due to the deficiency of an enzyme of the urea cycle. In adults, ammonia toxicity may be responsible for hepatic coma or encephalopathy which usually results from liver damage due to poisons, alcohol, or viral infection.

1. Mechanism of ammonia toxicity

The biochemical basis of ammonia toxicity is still unclear but there is no shortage of hypotheses.

1. The enzyme glutamate dehydrogenase is present in mitochondria of the brain so that, assuming it catalyses a near-equilibrium reaction as in liver, an increase in the ammonia concentration would decrease the concentration of 2-oxoglutarate.

$$H_2O + glutamate^- + NAD^+(P) \rightleftharpoons$$
$$2\text{-oxoglutarate}^{2-} + NAD(P)H + NH_4^+ + H^+$$

Since 2-oxoglutarate is an important intermediate in the TCA cycle, a decrease in its concentration could reduce the flux through the latter half of the cycle and could lead to a serious depletion in the ATP concentration in the cells of the brain, for which there is some evidence (Bessman and Pal, 1976). One difficulty with this suggestion is that, if brain is similar to muscle, oxoglutarate dehydrogenase should be a flux-generating step for the latter part of the TCA cycle so that a moderate decrease in 2-oxoglutarate concentration should make little or no difference to the flux.

2. Through the same equilibrium, an increase in the concentration of ammonia should also lead to an increase in the $NAD^+/NADH$ concentration ratio within the mitochondrion and this could lead to a decrease in the rate of production of ATP in the electron transfer chain. However, since the electron transfer process is close to equilibrium and the rate is regulated by the ATP/ADP concentration ratio, any change in the $NAD^+/NADH$ concentration ratio would be compensated for by a change in that of ATP/ADP, thus maintaining the rate of electron transfer. Any change in flux through the electron transfer chain could only occur if the flux-generating steps within the TCA cycle were affected by the $NAD^+/NADH$ concentration ratio (Chapter 7; Section A.3.c). In view of these criticisms, and the fact that these explanations would not account for the restriction of the effect of ammonia to the brain, other hypotheses have gained credence.

3. Glutamate is an excitatory neurotransmitter in brain. This action of glutamate may be arrested by conversion of glutamate to glutamine via the glutamine synthetase reaction in the glial cells (see Chapter 21 for details of neurotransmitter metabolism). This glutamine is eventually returned to the nerve cell where, by the action of glutaminase, re-synthesis of glutamate occurs. However, glutaminase can be inhibited by concentrations of ammonia greater than 1 mM, which could therefore result in depletion of this excitatory neurotransmitter thus disturbing brain function (Quastel, 1979).

4. In ammonia toxicity there is an increased membrane permeability to potassium and chloride ions which could interfere with electrical activity in the brain. The change in permeability might be brought about by an increase in the proton concentration, due to the increase in ammonia concentration de-inhibiting 6-phosphofructokinase and thus leading to an increased glycolytic flux with consequent formation of lactic acid (Quastel, 1979).

2. Deficiencies of urea cycle enzymes

A number of inborn errors of metabolism in which one of the enzymes of the cycle is defective are known (for review, see Shih, 1978). A major deficiency of any enzyme in the cycle would be lethal, since this would remove the capacity to detoxify the ammonia so that the inborn errors that have been reported are caused either by a partial reduction in the maximum capacity of a given enzyme or an increase in its K_m for substrate.

Deficiencies reported include those of carbamoyl-phosphate synthetase, ornithine transcarbamoylase, argininosuccinate synthetase (which is associated with accumulation and excretion of citrulline) and argininosuccinate lyase (which is associated with accumulation and excretion of argininosuccinate). Finally, there are reports of inhibition of arginase by high levels of lysine in patients suffering from congenital lysine intolerance (Chapter 10; Appendix 10.1).

These deficiencies usually mean that the liver cannot remove sufficient ammonia when the protein content of the diet is high, so that the blood concentration of ammonia increases (hyperammonaemia) and symptoms of ammonia toxicity develop. For those patients with deficiencies of the enzymes, episodes of vomiting and lethargy often follow infections, metabolic stresses (when endogenous protein breakdown increases—Chapter 14; Section D) or an increase in protein intake. These problems can, in some cases, be overcome with restriction of protein intake but many patients progress to irreversible coma and death. If infants survive but remain undiagnosed and untreated for some time, they may become mentally retarded. Treatment of children suffering from a urea cycle enzyme deficiency with oxoacids (or their analogues) of the essential amino acids (see Chapter 10; Section G, for details) has proved to be of value in overcoming the acute episodes of ammonia toxicity that can frequently occur with illness etc. (Close, 1974).

3. Hepatic coma

Since the urea cycle is localized exclusively in the liver, ammonia retention can occur in adults who are suffering from severe liver disease (see Sherlock, 1978). Cirrhosis of the liver in which there is generalized fibrosis (large deposition of collagen) is one of the commonest forms of chronic liver disease. The condition arises from a variety of causes including carbon tetrachloride poisoning, infective hepatitis in which the damage to the liver has been heavy and prolonged, and fatty degeneration of the liver due to malnourishment and excessive consumption of alcohol (see Chapter 11; Section F.5.b). Cirrhosis may be asymptomatic for long periods or the earliest stages may be so subtle as to escape notice, but the patient may suffer irritability, loss of drive and there may be disturbance of sleep rhythm. This condition can rapidly develop into confusion, drowsiness, stupor and finally deep coma. Factors that can cause the chronic patient to develop encephalopathy which leads to coma include excess dietary protein, infection, gastrointestinal haemorrhage, and certain drugs. If none of these causes appear to be the precipitant factor, it is possible that, for unknown reasons, there has been a deterioration in the underlying disease. In alcoholics, who are already suffering from reduced liver function, an excessive bout of drinking can precipitate liver failure, although in such cases hypoglycaemia and acidosis could also contribute to the coma. Measurements of blood pH and concentrations of glucose and ammonia are therefore necessary to identify the major contributory factor. In contrast to chronic liver malfunction, in which the onset of neurological symptoms is usually gradual, and from which the patient usually recovers, in acute hepatic failure (in the absence of any previous liver disease) coma usually occurs very rapidly and the prognosis is then poor. Fulminant viral hepatitis is often the cause but, fortunately, the condition is rare.

Although the toxicity of ammonia is well established, and children with inborn errors of the cycle do suffer from coma, it is still not clear that accumulation of ammonia is the only, or even the major, cause of the encephalopathy in adults. There is, in fact, a lack of correlation between arterial and cerebrospinal fluid concentrations of ammonia and the degree of unconsciousness and, indeed, a deep coma can occur with a normal blood concentration of ammonia. Moreover, ammonia toxicity is known to affect the cortex but not the brain stem and the lower centres that are particularly involved in hepatic coma. Other factors that could be responsible are lactic acidosis or the accumulation of false neurotransmitters.[5] It has been suggested that bacteria in the large intestine could be responsible for production of such neurotransmitters. This will, of course, occur in normal subjects but these compounds are removed by the functioning liver, whereas in hepatic failure, the false neurotransmitter cannot be destroyed. A correlation between the blood concentration of octopamine and the degree of consciousness in hepatic encephalopathy has indeed been observed (Manghani *et al.*, 1975).

The treatment of patients with hepatic coma is designed to reduce the blood concentration of compounds that may be responsible for the coma (Triger, 1981). Since amino acid metabolism, especially by bacteria in the intestine, can result in the formation of both ammonia and false neurotransmitters, restriction of protein intake and administration of an antibiotic (to reduce the number of colonic bacteria) are recommended treatment. Other more subtle metabolic interventions to reduce the ammonia level are still at the experimental stage. However, somewhat surprisingly, oral administration of the synthetic disaccharide, lactulose, have been shown to have beneficial results. The rationale for this treatment is that lactulose is not hydrolysed so can reach the colon where it is fermented by bacteria. The acidic products of fermentation lowers the pH of the colon, favouring the formation of the less diffusible ammonium ion (NH_4^+) or amine cation, so that less ammonia or false neurotransmitter enters the body (Conn and Lieberthal, 1979; Lancet editorial 1983b).

Recently, with an increasing knowledge of metabolism, a different explanation for hepatic coma has been put forward. It has been emphasized in Chapter 10; Section F that, whereas a larger number of amino acids are metabolized by the liver, the branched-chain amino acids are metabolized almost exclusively by skeletal muscle. Hence in liver failure the metabolism of the latter amino acids is unaffected, whereas that of other amino acids, especially the aromatic ones (tyrosine and tryptophan) is seriously reduced. In consequence, the blood concentration of branched-chain amino acids are normal, whereas those of the aromatic amino acids are increased. Since both types of amino acid are transported into the brain via the same carrier (Chapter 10; Section C.1), the change in concentration ratio increases the amount of tyrosine and tryptophan that enters the brain cells. Since these amino acids are the precursors of the important biogenic amines, dopamine, noradrenaline, and 5-hydroxytryptamine and particularly since the pathway for 5-hydroxytryptamine synthesis in brain does not contain a flux-generating step (Chapter 21; Section B.3.b) these changes in the blood amino acid levels increase the concentrations of amines in the brain. There is considerable evidence that 5-hydroxytryptamine promotes sleep, so a very large increase in the concentration of tryptophan and hence 5-hydroxytryptamine in the brain could explain the coma. This hypothesis has led to the idea of treating patients suffering from hepatic coma by infusion of branched-chain amino acids to raise their concentration in the blood so that they compete more effectively with the aromatic amino acids for the transport process (Fisher and Baldessarini, 1976; James *et al.*, 1979).

Finally, if liver failure is due to chronic and extensive cirrhosis so that there is no chance of recovery, liver transplantation offers the only hope for the patient (Calne, 1980; Starzl *et al.*, 1981).

APPENDIX 12.1 PATHWAYS FOR THE DEGRADATION OF PURINE AND PYRIMIDINE NUCLEOTIDES

The pathway for the biosynthesis of nucleotides is described in Chapter 18, Section B.1 but, as their breakdown contributes nitrogenous products to the urine, the catabolic pathways are considered here. In broad terms, nucleotide catabolism has much in common with that of amino acids. Thus nucleotides arise from the hydrolysis of both dietary and endogenous nucleic acids. Although DNA is not normally broken down, except after the death of cells and perhaps during DNA repair, RNA is turned over in much the same way as protein. Nucleic acids are broken down to their constituent nucleotides from which the bases are produced for further degradation, in man, to uric acid (see below). However, not all the bases that are produced will be degraded since a proportion can be reconverted to the nucleotides by routes known as the salvage pathways. (For more detailed information see Watts, 1981; Seegmiller, 1980.)

1. **Purines**

(a) *Degradation of nucleotides to the bases*

Nucleases (ribonucleases and deoxyribonucleases) act on nucleic acids in the lumen of the small intestine. The best known is probably ribonuclease A which is secreted by the pancreas and which catalyses the hydrolysis of ribonucleic acid to yield 3'-nucleotides (that is, cleavage of the phosphodiester link occurs on the side attached to the 5'-hydroxyl group of the ribose). In fact, this particular enzyme acts both as an endonuclease, cleaving ribonucleic acids into oligonucleotides, and as an exonuclease removing nucleotides from the 3'-terminal end. It acts only on links where the ribose contributing the 3'-hydroxyl group to the phosphodiester linkage is attached to a pyrimidine; other intestinal ribonucleases have different specificities. The hydrolysis of oligonucleotides is completed by the

site of action of lysosomal ribonuclease

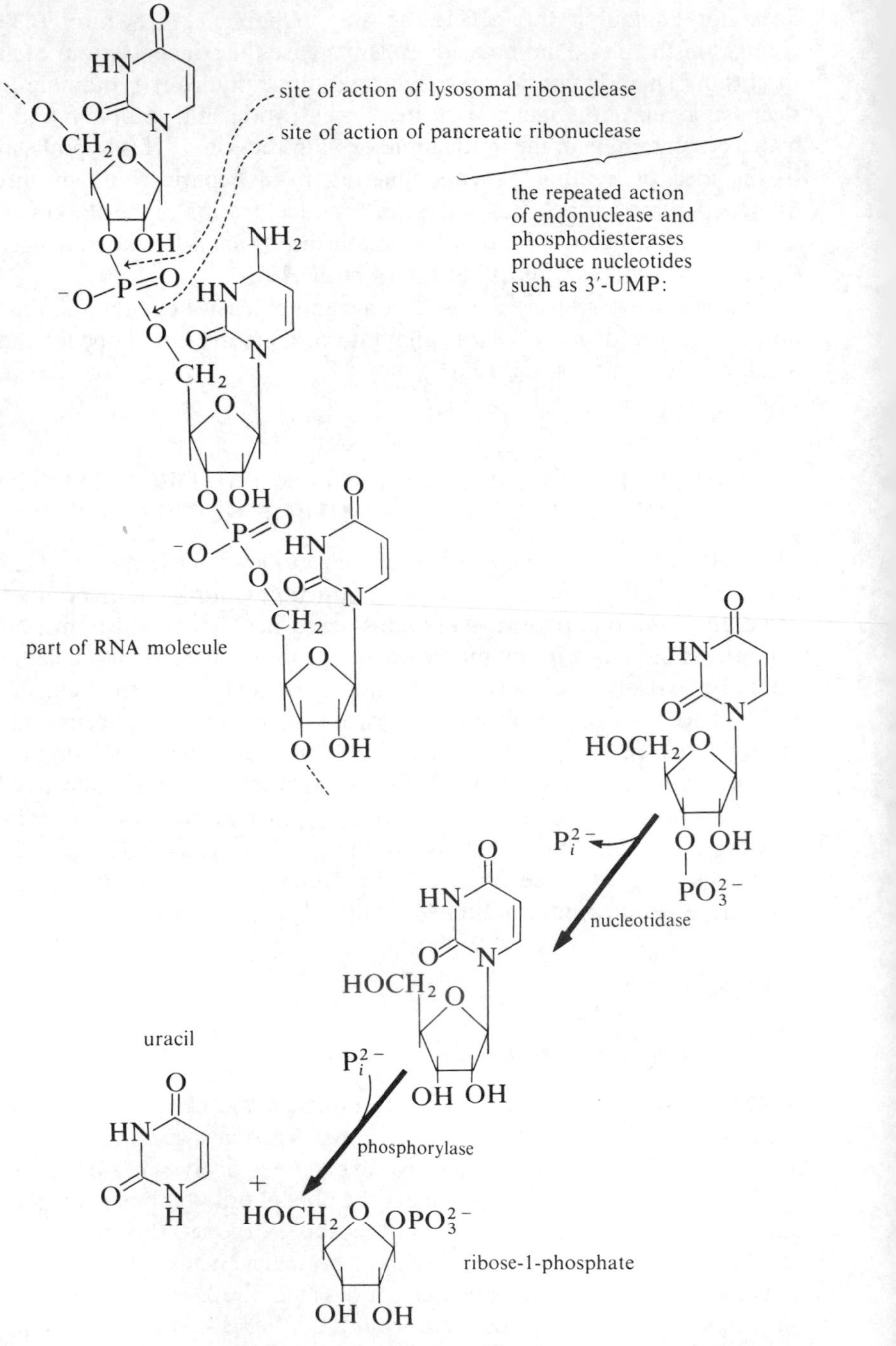

site of action of pancreatic ribonuclease

the repeated action
of endonuclease and
phosphodiesterases
produce nucleotides
such as 3'-UMP:

part of RNA molecule

nucleotidase

uracil

phosphorylase

ribose-1-phosphate

Figure 12.5   Degradation of ribonucleic acid. DNA is broken down in an analogous
manner

action of phosphodiesterases (exonucleases) secreted by the intestinal cells. The intracellular hydrolysis of RNA occurs within the lysosome; a ribonuclease acts as an endonuclease cleaving phosphodiester links to leave the phosphate group attached to the 5'-hydroxyl group of the oligonucleotide (Figure 12.5) and hydrolysis is completed by phosphodiesterases. DNA is degraded by similar enzymes to those involved in the degradation of RNA.

The hydrolytic reactions described above result in the formation of purine mononucleotides (e.g. AMP, GMP, and IMP). The nucleotides are further degraded by a similar mechanism in both the lumen of the intestine and intracellularly. A nucleotidase catalyses the hydrolytic removal of phosphate to produce the nucleoside which is then converted to the free base and either ribose 1-phosphate or deoxyribose 1-phosphate by the action of nucleoside phosphorylases[6] (Figure 12.6). In some tissues, including muscle and liver, the quantitatively important pathway for degradation of AMP involves deamination to IMP, catalysed by AMP deaminase (Section A.4) and, if AMP is not reformed via part of the purine nucleotide cycle, the IMP is hydrolysed to the nucleoside, inosine, by a nucleotidase. The inosine can be converted to hypoxanthine via the purine nucleoside phosphorylase reaction as follows:

$$\text{AMP} \xrightarrow{\text{NH}_3} \text{IMP} \xrightarrow{\text{P}_i} \text{inosine} \longrightarrow \text{hypoxanthine}$$

(b) *Conversion of bases to uric acid*

The purine bases arising as a result of the reactions described above are hypoxanthine, adenine, and guanine. The further catabolism, which occurs primarily in the liver, involves progressive oxidation to uric acid[7] (urate) which in man is excreted via the kidneys. The central intermediate in the pathway is hypoxanthine which is formed from adenine, guanine, or inosine and is oxidized first to xanthine and then to urate in reactions catalysed by xanthine oxidase* (Figure 12.7).

In birds, some reptiles, and insects, ammonia is converted to urate rather than urea for nitrogen excretion and this involves *de novo* purine synthesis, as described in Chapter 18; Section B.1.a. Furthermore, in the mammals, urate is excreted by relatively few species (primates and, apparently, Dalmatian dogs). In other species, the urate is further hydrolysed to a variety of products (Figure 12.8). The enzymes required for the conversion of urate to urea (Figure 12.8) are present in micro-organisms in the intestine of man. Thus the liver of man synthesizes approximately 0.8 g of urate each day but less than this appears in the urine because between 20% and 50% enters the gut (in gastric juice and bile) and is degraded by micro-organisms. In addition, small quantities of purines other than urate appear in the urine having arisen from substituted purines present in transfer- and ribosomal-RNA (Chapter 18; Section B.2).

* The reaction also results in the formation of the superoxide radical which is destroyed by superoxide dismutase (Chapter 4; Section E.5.b).

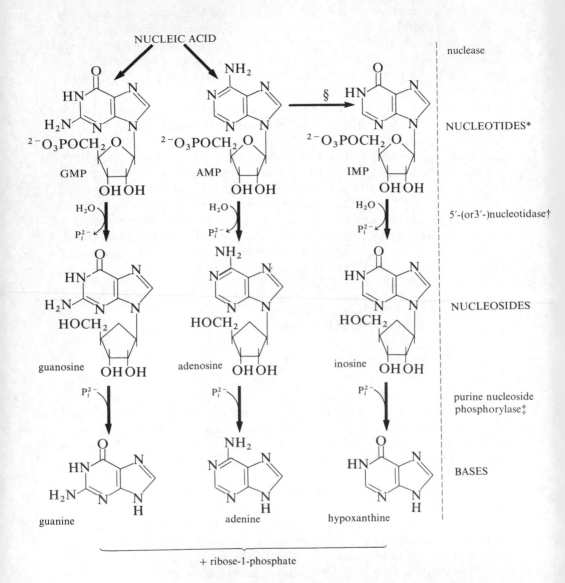

Figure 12.6 Pathway of nucleic acid degradation. For convenience, ribonucleotides are shown but DNA is degraded by similar enzymes. Names of enzymes and intermediates are shown in the right hand column.

* The products of nuclease action include 5'-nucleotides (shown) together with 3'-nucleotides.

† Separate 3 - and 5'-nucleotidases are present but each has broad specificity with respect to the purine component.

‡ A single enzyme acts on all nucleosides.

§ The reaction involved in this conversion is shown in Figure 12.1

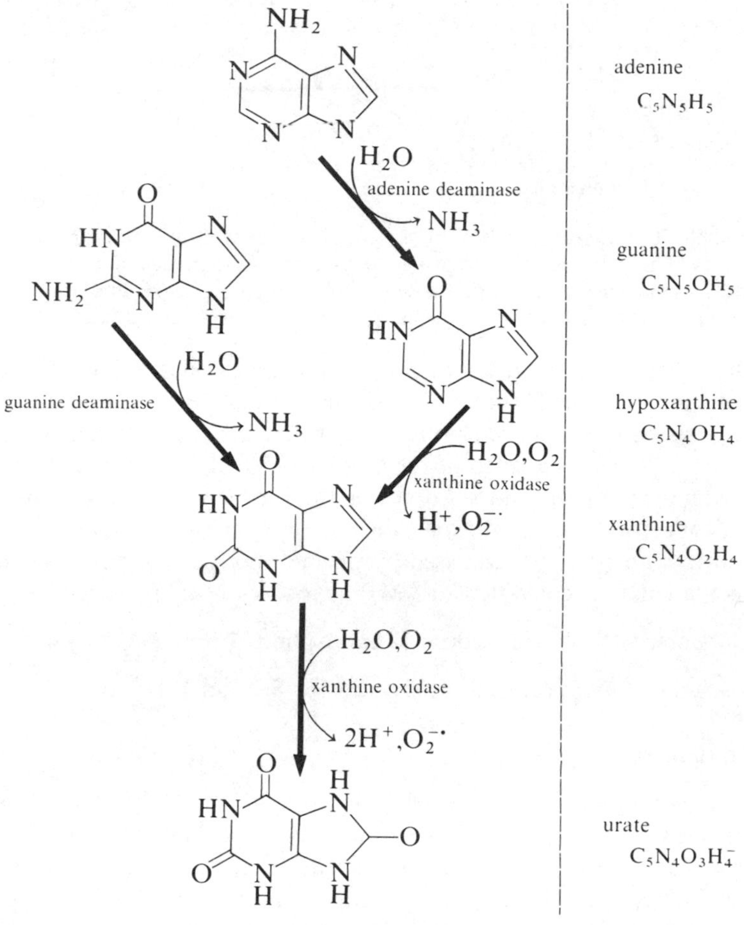

Figure 12.7 Pathway of purine metabolism arranged to emphasize progressive oxidation. Adenosine nucleotides are normally deaminated to IMP which is hydrolysed to hypoxanthine. Xanthine oxidase is a complex flavoprotein containing non-haem iron-centres and molybdenum. The reduced product of its action is the superoxide radical (see Chapter 4; Section E.5). The same enzyme oxidizes both hypoxanthine and xanthine

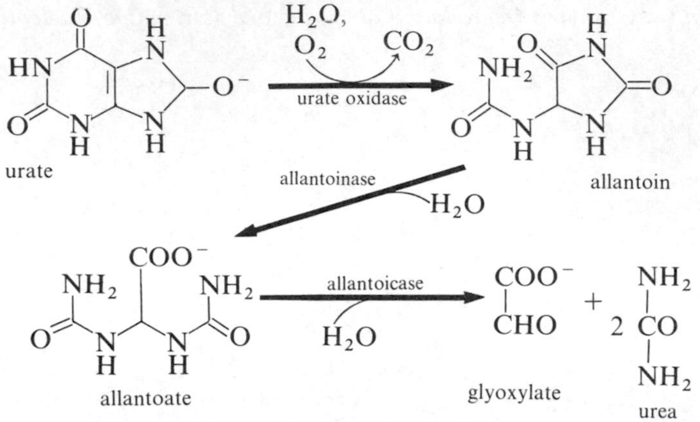

Figure 12.8 The further metabolism of urate. The extent of the metabolism depends on the species. The primates excrete urate; other mammals, turtles, and molluscs excrete allantoin; some fishes excrete allantoate and the remaining fishes and amphibia secrete glyoxylate and urea as endproducts of purine degradation

It should be emphasized that not all the bases that are produced in the reactions indicated in Figure 12.6 are converted to urate. A proportion of the purines arising from the reactions are converted back into nucleotides. This involves a different sequence of reactions to that involved in *de novo* purine nucleotide biosynthesis and is known as the purine salvage pathway. The free base is condensed with 5-phosphoribose 1-diphosphate (phosphoribosylpyrophosphate) in a reaction catalysed by purine phosphoribosyltransferases. For example, adenine is converted to AMP as follows:

$$\text{adenine} + \text{5-phosphoribose 1-diphosphate}^{5-} \longrightarrow \text{AMP}^{2-} + \text{PP}_i^{3-}$$

These reactions are discussed in Chapter 18; Section B.1.

2. Pyrimidines

The pyrimidine nucleotides (predominantly UMP, CMP, and dTMP) are hydrolysed to their respective bases (uracil, cytosine, and thymine) by reactions comparable with those for purine nucleotides. The pathways of degradation of the bases, depicted in Figure 12.9, generate products (ammonia, malonyl-CoA, and methylmalonyl-CoA) which are also produced in amino acid catabolism so that no distinctive nitrogenous end products result. Note that β-alanine, an intermediate in cytosine and uracil degradation, is required for coenzyme A synthesis. The pyrimidine bases can also be converted back to the corresponding nucleotides by the salvage pathway (Chapter 18; Section B.1).

Figure 12.9 Pathway of pyrimidine degradation. It is likely that deamination of cytosine largely occurs at the level of the nucleosides. Further metabolism of the endproducts of this pathway occurs by way of their coenzyme A derivatives. Malonyl-CoA is an intermediate in fatty acid synthesis and methylmalonyl-CoA metabolism has been outlined in Figure 10.4. 2-OG^{2-} and glut^- are abbreviations for 2-oxoglutarate and glutamate.

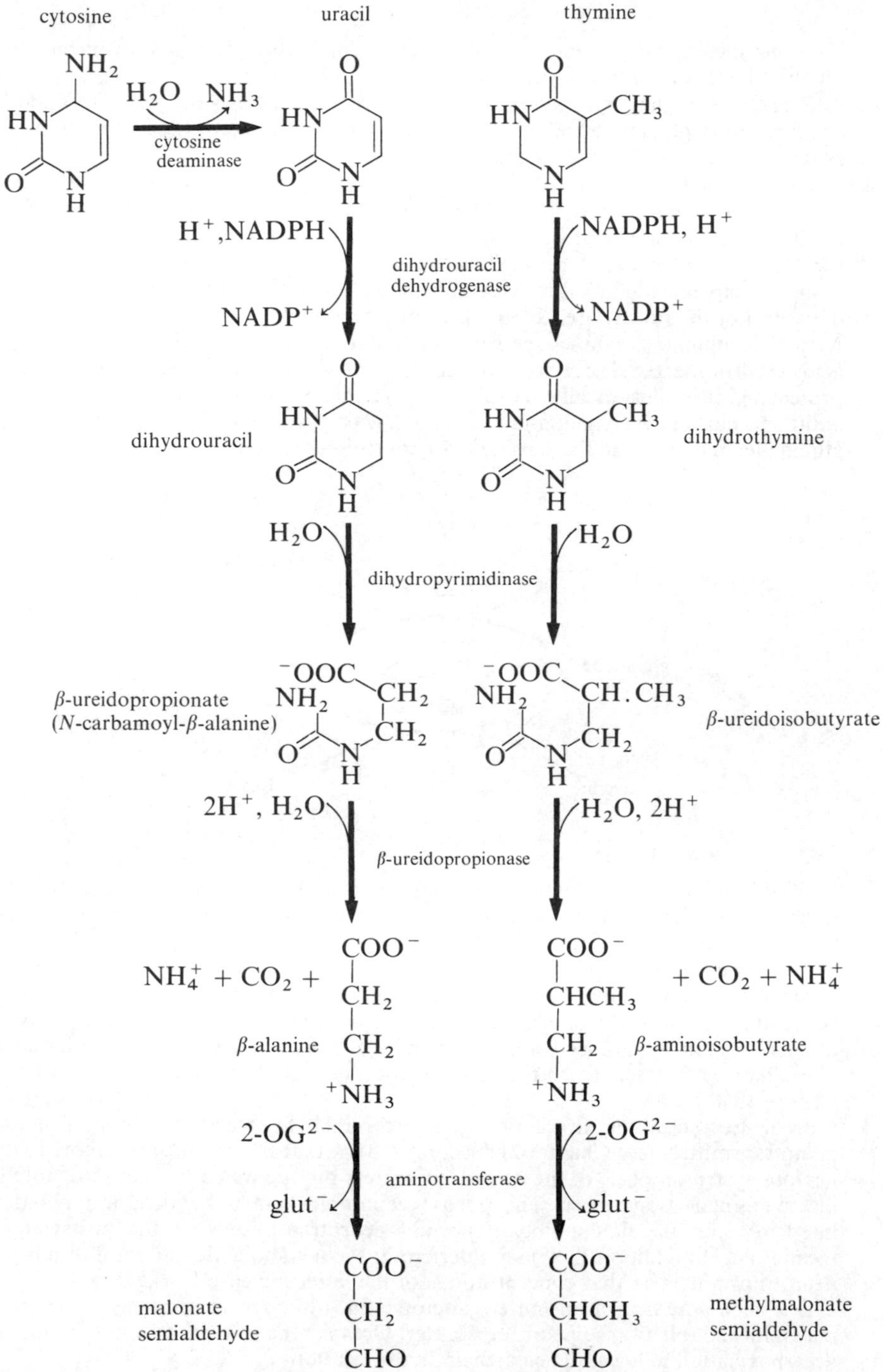

cytosine

uracil

thymine

H_2O NH_3

cytosine deaminase

$H^+, NADPH$

dihydrouracil dehydrogenase

$NADPH, H^+$

$NADP^+$

$NADP^+$

dihydrouracil

dihydrothymine

H_2O

dihydropyrimidinase

H_2O

β-ureidopropionate
(N-carbamoyl-β-alanine)

β-ureidoisobutyrate

$2H^+, H_2O$

β-ureidopropionase

$H_2O, 2H^+$

$NH_4^+ + CO_2 +$

$+ CO_2 + NH_4^+$

β-alanine

β-aminoisobutyrate

2-OG^{2-}

aminotransferase

2-OG^{2-}

$glut^-$

$glut^-$

malonate semialdehyde

methylmalonate semialdehyde

NOTES

1. Some marine fish convert ammonia to trimethylamine oxide $(CH_3)_3NO$ which is also highly soluble and of low toxicity.
2. The reactions of this pathway are not unique to the synthesis of urea; ornithine is produced from glutamate and, via the reactions of the cycle, ornithine can be used to synthesize arginine (Chapter 10; Section E).
3. Cahill (1974) has pointed out that the vampire bat has a capacity for urea synthesis perhaps 1000-fold greater than that of man. The bat eats one half of its weight in blood in 10–15 minutes and begins to urinate water and urea even before it has finished its meal. The control of the flux through the urea cycle must be of considerable importance in the vampire bat and might provide a fruitful system for investigating the mechanism of control of the rate of urea synthesis in other mammals.
4. *N*-Acetylglutamate is synthesized from glutamate in the liver by the enzyme amino-acid acetyltransferase. The concentration of this enzyme rises with the intake of dietary protein and it is allosterically activated by arginine. A regulatory cycle involving the additional enzyme, acetylglutamate deacylase, which hydrolyses *N*-acetylglutamate to glutamate, may increase the sensitivity of control at this stage:

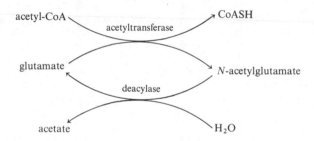

Of clinical interest is the fact that hyperammonemia is associated with elevated blood levels of propionate and methylmalonate, caused by metabolic defects in the metabolism of these compounds. These conditions are associated with decreased hepatic concentrations of acetyl-CoA and coenzyme A probably due to the accumulation of propionyl-CoA and methyl malonyl-CoA in this tissue. The decreased concentration of acetyl-CoA decreases the rate of formation of *N*-acetylglutamate so that its concentration is decreased which, in turn, decreases the activity of carbamoyl-phosphate synthetase. Hence the concentration of ammonia increases (Stewart and Walser, 1980).

5. False neurotransmitters are amines that are similar in structure to normal amine neurotransmitters (see Chapter 21; Section A.3) so that they bind to receptors in the nervous system in place of the neurotransmitters but are much less active or totally inactive as neurotransmitters. They act as receptor blockers so that their accumulation interferes with the functioning of normal neurotransmitters at the postsynaptic membranes. In addition they may interfere in the metabolic degradation of normal neurotransmitters so that concentrations of the latter increase markedly.

   One such false neurotransmitter is octopamine although in some invertebrates it may fulfil the role of true neurotransmitter. Octopamine is formed from tyrosine by decarboxylation followed by side-chain hydroxylation:

$$\text{HO}\langle\bigcirc\rangle\text{CH}_2\text{CH}.\text{COO}^- \xrightarrow{\quad -\text{CO}_2\quad} \text{HO}\langle\bigcirc\rangle\text{CH}_2\text{CH}_2 \xrightarrow[\text{O}_2]{\text{AH}_2,\;\text{A},\;\text{H}_2\text{O}} \text{HO}\langle\bigcirc\rangle\overset{\text{OH}}{\underset{}{\text{CH}}}.\text{CH}_2$$

tyrosine $\;\;\overset{|}{\text{NH}_3^+}\qquad\qquad$ tyramine $\;\;\overset{|}{\text{NH}_3^+}\qquad\qquad$ octopamine $\;\;\overset{|}{\text{NH}_3^+}$

6. Somewhat surprisingly, inherited deficiencies of either purine nucleoside phosphorylase or adenosine deaminase result in severe immunodeficiency. The deficiency of the purine nucleoside phosphorylase is characterized by failure of the T-lymphocytes to respond to an immunological challenge (Gibblet, 1979). These patients usually present within the first few months of life with severe and recurrent infections that cause failure to thrive. In the absence of purine nucleoside phosphorylase there is failure to convert, in particular, guanosine and deoxyguanosine to the bases guanine and deoxyguanine so that these nucleosides accumulate (Figure 12.6). It has been suggested that the accumulation of the nucleosides increases their rate of phosphorylation and this results in a marked increase in the concentrations of dGMP and hence dGTP. The latter changes the affinity of ribonucleotide reductase for the various nucleoside diphosphates, a mechanism which normally causes the production of balanced concentrations of all the deoxynucleoside triphosphates necessary for DNA synthesis (Chapter 18; Section B.2.c). Excess dGTP thus prevents the formation of the correct concentrations of deoxynucleoside triphosphates, which results in inhibition of DNA synthesis. Since an immunological response requires extensive proliferation of T-lymphocytes, and since this is dependent upon DNA replication, the response of the T-lymphocytes to an immunological challenge is poor.

The inherited deficiency of adenosine deaminase prevents the degradation of adenosine and particularly deoxyadenosine and this causes severe immunodeficiency characterized by failure of both T- and B-lymphocytes. A similar explanation to the above is proposed. The accumulation of deoxyadenosine increases the amount of the nucleoside that is phosphorylated by adenosine kinase to produce high concentrations of dATP in both T- and B-lymphocytes. The accumulation of the latter results in the inhibition of ribonucleotide reductase and this causes low concentrations of other deoxynucleoside triphosphatases and hence an inhibition of DNA synthesis (Osborne, 1981).

7. Like all purines, uric acid can undergo tautomerism so that it exists as an equilibrium mixture of the lactam (keto) and lactim (enol) forms (Figure 12.10). Partial lactim forms can also exist. In the lactim form, the 8-hydroxyl group ionizes with a pK of 5.4. This ensures the predominance of the ionized partial lactim form (urate) at physiological hydrogen ion concentration.

Undissociated uric acid has a very low solubility ($6.5\,\text{mg}.100\,\text{cm}^{-3}$, at $37°\text{C}$). Although the solubility of monosodium urate in water is much greater than this, the sodium concentration in serum is such that the solubility product for monosodium urate is exceeded at about this concentration and can crystallize in the body with unfortunate consequences. Occasionally, and especially when the urine is unusually acid, calcium urate stones can form in the kidney and bladder. More commonly, in patients suffering from hyperuricaemia, monosodium urate crystals form in the joints to cause the painful affliction known as gout. The inflammatory response to these crystals in the joint provides the characteristic pathology. Gout is unlikely to develop if the urate concentration remains below $6\,\text{mg}.100\,\text{cm}^{-3}$ (0.4 mM) but any factor which increases urate production or reduces its rate of elimination by the kidney, so that the concentration rises above $9\,\text{mg}.100\,\text{cm}^{-3}$ (0.6 mM), is virtually certain to cause gout. The disease is found predominantly in males and it usually affects the elderly (often in

lactam form $+H^+ \quad -H^+$ lactim form

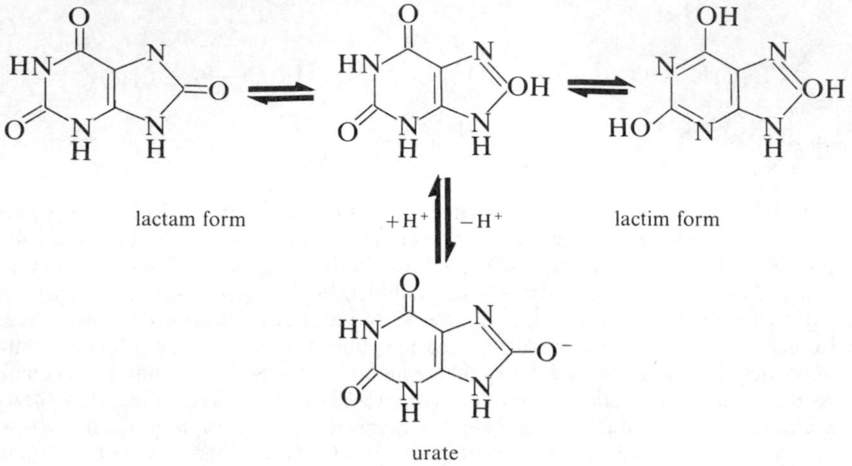

urate

Figure 12.10 Tautomerism and ionization of uric acid. Uric acid exists predominantly as urate at pH 7.0. Other purines exist predominantly in the unionized lactam forms

the early stages of the disease only in the joints of the big toe). It is traditionally associated with the aristocratic families and thier indulgence in excessive eating and drinking. This may be explained by the large amount of wine contaminated with lead that was consumed by members of this class in the 18th and 19th centuries (Nriagu, 1983). The aetiology and treatment are more fully considered in Chapter 18. Suffice it to say that at least some of the sufferers of gout (and there are 800 000 in the USA alone) must ponder the possible significance of the evolutionary quirk that has denied man the ability to convert urate to more soluble products.

CHAPTER 13

METABOLISM AND ACID/BASE BALANCE

The relationship between acid/base balance and respiration has been known for many years but only recently has its relationship to metabolism been explored. This chapter contains a brief discussion of how the acid/base balance of the extracellular fluid is achieved, how it is related to metabolism and how it is impaired in certain pathological states giving rise to lactic acidosis and ketoacidosis (for more detailed discussion see Leaf and Cotran, 1980).

Acidity is determined by the concentration of free hydrogen ions (protons); if the concentration of hydrogen ions is greater than 10^{-7} M (the concentration in pure water) the solution is said to be acid and if less than 10^{-7} M, alkaline. A useful means of representing this concentration is the pH, which is the negative logarithm (base ten) of the free hydrogen ion concentration in molar terms. Examples should make this clearer.

1. A pH of 7.0 represents a hydrogen ion concentration of $10^{-7.0}$ M (or 0.0001 mM)
2. A pH of 7.4 represents a hydrogen ion concentration of $10^{-7.4}$ M (or 0.00004 mM)[1].

Although the pH notation is a convenient way of representing small numbers it has two serious disadvantages in practice. First it can obscure the fact that the absolute concentrations of hydrogen ions in organisms are very small compared with other solutes (for example, sodium ions are present in extracellular fluid at a concentration of 145 mM). Secondly, pH units are difficult to handle because they provide a logarithmic and not a linear scale. Thus a rise in pH of one unit is in fact a ten-fold rise in hydrogen ion concentration. This goes some way to explaining why such apparently small changes in pH (< 1 unit) have fatal consequences.

The pH of extracellular fluid (including plasma) is normally stabilized within the range 7.35–7.45. Values above and below this indicate alkalosis and acidosis respectively. The pH limits of the extracellular fluid compatible with life are from about pH 7.0 to pH 7.8 (i.e. 0.1 to 0.016 μM). The precise biochemical reasons for death if a pH change exceeds this range are unclear but both alkalosis and acidosis inhibit brain function.

A. PRODUCTION OF PROTONS IN THE BODY

Before considering the means by which the body reduces the impact of changes in hydrogen ion concentration, it is important to know the sources of these changes.

509

Protons are produced both in the lumen of the intestine during digestion and within the cells of the body during normal metabolism. The main sources of protons in the intestine are phospholipids and nucleic acids. Both of these are diesters, the hydrolysis of which can be represented as follows:

$$A-O-\overset{\displaystyle O}{\underset{\displaystyle O^-}{\overset{\displaystyle \|}{P}}}-O-B + 2H_2O \longrightarrow A.OH + B.OH + H_2PO_4^-$$

The phosphoric acid produced (or more correctly the $H_2PO_4^-$ ion) immediately dissociates (almost completely) to generate the proton (which in a sense has come from water):

$$H_2PO_4^- \longrightarrow H^+ + HPO_4^{2-}$$

A second major dietary source of protons are the sulphur-containing amino acids (cysteine and methionine) derived from the digestion of protein. Their degradation (in the liver) generates sulphate ions and protons (equivalent to sulphuric acid, H_2SO_4) from uncharged molecules. These dietary contributions are particularly significant in omnivores and carnivores. Man on a normal diet produces 40–80 mmoles of protons daily from dietary sources and on a high-protein diet this can reach 150 mmoles per day. (In the absence of any mechanism for removing the protons the normal diet would cause a fall in blood pH to about 2.0 in a day.)

Another source of protons, estimated at 12.5 moles per day, arises from the carbon dioxide produced in respiration, much of which is hydrated to carbonic acid, which dissociates as follows:

$$CO_2 + H_2O \longrightarrow H_2CO_3 \longrightarrow H^+ + HCO_3^-$$

These protons are, of course, effectively lost when the carbon dioxide is exhaled. Other metabolic products that contribute to the acid load in the blood include lactic acid and the ketone bodies (acetoacetic and 3-hydroxybutyric acids). All of these compounds vary in concentration according to metabolic circumstances and can cause pathological acidosis (see Sections D and E). The rate of ketone body production, for example, can rise to 800 mmoles per day during diabetic ketoacidosis.

Many of the reactions of metabolism involve protons, for example, hydrolysis of ATP and reduction of NAD^+. Such reactions result in the production of a staggering 150 moles (150 g) of protons each day. However, these reactions are reversed by other processes so that no net change in proton concentration occurs.

The total acid load on the body is reduced when weak acids such as citric acid are ingested as their sodium and potassium salts and are metabolized to carbon dioxide. Paradoxically (in view of their taste) fruits are the major source of such alkali in the human diet. Loss of carbon dioxide from the body (e.g. by hyperventilation) can also raise blood pH, as can vomiting, due to the loss of hydrochloric acid from the stomach.

B. MECHANISMS FOR THE ADJUSTMENT OF pH

The body has three lines of defence against increasing acidity: buffers, respiration, and renal acid secretion (for detailed reviews see Pitts, 1971; 1973). Only the latter involves metabolism to any great extent but, in order to provide a comprehensive account, the first two mechanisms will be briefly described. In addition to these processes, the pattern of fuel utilization and metabolism may change in such a direction as to oppose pH changes (see Section B.5).

1. Buffers

The first line of defence to a change in pH is the action of buffers. Buffers are solutions of weak acids or bases containing significant amounts of both undissociated and dissociated forms in equilibrium:

$$\text{e.g.} \qquad HA \rightleftharpoons H^+ + A^-$$

The addition of H^+ to such a system will displace the equilibrium to the left, thereby removing H^+ so that very little change in its concentration occurs. Conversely, removal of H^+ will displace the reaction to the right so generating more H^+ to replace that removed. The ability of such a system to act as a buffer will depend on two factors: the concentration of the buffer and its dissociation constant (that is, the equilibrium constant of the dissociation reaction, usually given as its negative logarithm (pK_a)). In practice, useful buffering is only achieved within one pH unit either side of the pK_a value.

Three major buffers operate in the body (plus a host of minor ones occurring in low concentrations but which happen to have appropriate pK_a values) as follows.

(a) *Protein*

Protein acts as a buffer by virtue of its many ionizable groups (each with different pK_a values) and high concentration. Its action can be represented over-all by:

$$H.Prot \rightleftharpoons H^+ + Prot^-$$

or by

$$H.Prot^+ \rightleftharpoons H^+ + Prot$$

A particularly important protein buffer is haemoglobin present in very large amounts within the erythrocytes. Due to the permeability of the red cell membrane to hydrogen ions this buffering action influences the pH of the plasma.

(b) *Phosphates*

The main phosphate for buffering is inorganic phosphate which has one ionizable group with a pK_a of 6.8:

$$H_2PO_4^- \rightleftharpoons H^+ + HPO_4^{2-}$$

Organic phosphates play some part but the over-all contribution of phosphates to buffering capacity is relatively small.

(c) *Hydrogencarbonate* (*bicarbonate*)

Hydrogencarbonate ions are in equilibrium with carbonic acid:

$$H_2CO_3 \rightleftharpoons H^+ + HCO_3^-$$

This reaction has a pK_a of about 3.0 and for reasons given above should not contribute significantly to the buffering capacity at pH 7.4. However the importance of hydrogencarbonate in acid/base regulation is that the concentrations of both hydrogencarbonate and carbonic acid are regulated, in response to changes in pH, by other physiological processes (see below). Thus the hydrogencarbonate system could be described as a 'dynamic buffer system'.

2. Respiration and changes in hydrogencarbonate concentration

The concentration of carbonic acid depends on the concentration of carbon dioxide (that is, its partial pressure) since carbon dioxide is the anhydride of carbonic acid:

$$H_2CO_3 \rightleftharpoons H_2O + CO_2$$

The rate of this reaction is normally low but is increased enormously by the enzyme carbonate dehydratase (more widely known as carbonic anhydrase), which is present in erythrocytes (and indeed most other tissues, see Carter, 1972) so that the reaction is brought close to equilibrium.

The pH of any weak acid or base can be calculated from the Henderson–Hasselbalch equation:

$$pH = pK_a + \log \frac{\text{concentration of conjugate base (salt)}}{\text{concentration of undissociated acid}}$$

In the case of carbonic acid this becomes:

$$pH = 3.0 + \log \frac{[HCO_3^-]}{[H_2CO_3]}$$

An increase in the rate of lung ventilation will increase the rate of loss of carbon dioxide from the lungs so that its partial pressure in the blood, and hence the concentration of carbonic acid, will fall. This, as can be seen from the equation, will tend to raise the pH; a condition known as respiratory alkalosis. Conversely, decreased ventilation will lead to respiratory acidosis.

In normal situations this relationship exerts a stabilizing effect on blood pH. A rise in hydrogen ion concentration (fall of pH) in the blood will lead to a decrease in the hydrogencarbonate concentration. However, the respiratory centre in the brain stem is affected by the proton concentration, so that a decrease in pH increases the ventilation rate. This lowers the carbonic acid concentration so that both components of the $\log \dfrac{[HCO_3^-]}{[H_2CO_3]}$ term in the Henderson–Hasselbalch equation are lowered and the pH alters only slightly. In addition, the hydrogencarbonate ion concentration is regulated by the kidney (see below) and is one of the ways in which this organ contributes to the control of blood pH.

3. Quantitative importance of physiological buffers handling an acid load

The quantitative importance of the various buffers in the body in response to an acid load has been investigated by infusion of hydrochloric acid into dogs that had had both kidneys removed (Pitts, 1973). Sufficient acid was given to lower the plasma pH from 7.4 to 7.1, the plasma hydrogencarbonate concentration decreased from 24 to 7 mM and plasma carbonic acid from 1.2 to 0.7 mM (the latter would be due to increased respiration). The following processes were involved in buffering.

1. Plasma proteins provided only about 1 % of the total buffering of the acid administered.
2. Slightly more buffering, about 6 %, was achieved by the erythrocytes, probably by penetration of the protons into the cell where they would be buffered by the haemoglobin.
3. Not surprisingly, the extracellular carbonic acid/hydrogencarbonate system (plasma and interstitial fluid) played a very large part in the buffering action with about 42 % of the acid being buffered in this way.
4. Perhaps more surprisingly, the largest contribution to the buffering action (51 %) was provided by intracellular proteins with extracellular protons being exchanged for intracellular sodium ions (36 %) and potassium ions (15 %). It is probable that some of the sodium ions that were exchanged for protons were obtained from bone since sodium ions bind to the apatite crystals of bone and are exchangeable. This may be particularly important in chronic acidosis when actual resorption of bone may occur with liberation of calcium and phosphate ions.

The above description provides a quantitative indication of the acute response to an acid load. As protons are excreted into the urine, hydrogencarbonate will be released into the bloodstream by the kidney. Studies have shown that in the first 24 hr after such a load, 25 % of the acid is excreted in the urine and the pH of the blood returns to 7.4. Over the next few days the acid that was buffered intracellularly is returned to the extracellular fluid, in exchange for sodium and potassium ions, and it is then excreted in the urine or as carbon dioxide.

4. Role of the kidney in controlling acid/base balance

The kidney serves to maintain many of the solutes in the plasma at relatively constant concentrations by adjusting the rate at which they are lost in the urine. For some substances this involves recovering them from the glomerular filtrate (reabsorption); for others it involves discharging them into the urine (secretion).

The glomerular filtrate contains high concentrations of sodium and hydrogencarbonate ions (virtually identical to their concentrations in plasma). Loss of this sodium hydrogencarbonate from the body would cause the pH of the blood to fall so that its recovery can be seen as a means of preventing acidosis. The process can be visualized as starting with the secretion of protons into the lumen from the tubular cells. This is an active process and proceeds against a concentration gradient. (In order to maintain electrical neutrality the protons are exchanged for sodium ions.) Once in the lumen, the protons will tend to associate with the hydrogencarbonate ions present:

$$HCO_3^- + H^+ \longrightarrow H_2CO_3$$

The carbonic acid thus formed decomposes into carbon dioxide and water in a spontaneous reaction that is greatly accelerated by the enzyme carbonate dehydratase present in the brush border:

$$H_2CO_3 \longrightarrow H_2O + CO_2$$

Carbon dioxide diffuses readily through cell membranes and re-enters the tubular cells where the above reactions are reversed to generate hydrogen-

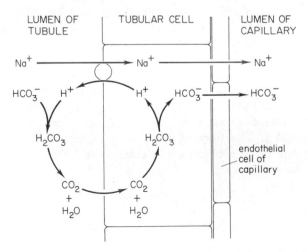

Figure 13.1   Reabsorption of sodium hydrogencarbonate from the kidney tubule

carbonate ions. Both hydrogencarbonate and sodium ions pass into the blood so that the sodium hydrogencarbonate has been recovered. This sequence of events is supported by the observation that inhibitors of carbonate dehydratase reduce the ability of the kidney to remove acid from the blood. Note, however, that this cycle does not achieve net proton excretion (only hydrogencarbonate salvage) because the same number of protons are regenerated in the tubular cells as are exchanged for sodium ions (Figure 13.1). Compared with the alternative of losing this hydrogencarbonate, however, this process will serve to raise the pH of the blood.

The major problem in the net excretion of protons by the kidney cells is buffering the protons in the tubular fluid. Proton secretion (which as indicated above occurs in exchange for sodium ions) ceases when the proton concentration gradient exceeds approximately 1000. This would limit the pH of the urine to approximately 4.5. If there was no mechanism for buffering protons in the tubular fluid, and hence allowing a greater concentration to be excreted, only about 0.1 mmol of protons per day would be excreted into the urine. Fortunately, buffering of the urine is brought about by several systems (see Pitts, 1973).

(a) *Carbonic acid/hydrogencarbonate system*

This system, which has been discussed above, plays some part since a proportion of the carbonic acid and carbon dioxide produced in the lumen will be lost in the urine (see Figure 13.1).

(b) *Phosphate system*

A rather larger part is played by phosphates. At the blood pH of 7.4 a considerable proportion of phosphate exists as the monohydrogenphosphate ion, HPO_4^{2-}. Consequently, this ion is present in the glomerular filtrate. However, as protons are excreted by the renal cells, the pH falls and this favours the formation of the dihydrogenphosphate ion, so buffering the urine (Figure 13.2):

$$HPO_4^{2-} + H^+ \rightleftharpoons H_2PO_4^-$$

If the urine is collected and titrated against alkali, these protons will dissociate and react with the added hydroxyl ions. For this reason, dihydrogenphosphate ions form the major part of what is known as the titratable acid of the urine. Since anions must be accompanied by equivalent amounts of cations, this mechanism leads to a loss of sodium and potassium from the body. Furthermore, the restricted availability of phosphate (from the diet, and for short periods 'borrowed' from bone and extracellular fluid) prevents this mechanism from coping with the massive proton excretion required to correct metabolic acidosis.

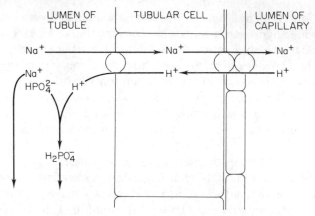

Figure 13.2 The operation of the phosphate buffer system in the kidney tubule

(c) *Ammonia system*

To cope with large acid loads, a third renal system comes into operation. Ammonia is generated from glutamine within the tubule cells and diffuses into the tubule to combine with the protons as follows:

$$NH_3 + H^+ \longrightarrow NH_4^+$$

The glutamine is produced in muscle (although brain and liver may contribute small amounts) and transported to the kidney where the two nitrogen atoms are released as ammonia by the pathway discussed in Chapter 10; Section F.3. Since the glutamine release by muscle may be at the expense of alanine (see Chapter 10; Section F.2), the excretion of the nitrogen from amino acid metabolism in muscle is effectively switched from urea, which is produced from the metabolism of alanine in the liver, to ammonia, which is produced from glutamine metabolism in the kidney. In this way, the endproduct of nitrogen metabolism is put to good use by the body in achieving control of acidosis. The ammonium ions do, of course, have to be excreted with a counter-ion but that is the price paid for acid excretion. The major anion lost is chloride so that ammonium chloride is effectively excreted (see Figure 13.3).

In conditions of extreme acidosis, both nitrogen atoms of glutamine can give rise to ammonia via the sequential reactions of glutaminase and glutamate dehydrogenase as follows:

$$H_2NOC.CH_2CH_2CH.COO^- \xrightarrow{\hspace{2cm}} {}^-OOC.CH_2CH_2CH.COO^-$$

$$\underset{\text{glutamine}}{\overset{|}{NH_3^+}} \quad H_2O \quad \overset{NH_3,}{\underset{H^+}{}} \quad \underset{\text{glutamate}}{\overset{|}{NH_3^+}}$$

$$\xrightarrow{\hspace{2cm}} {}^-OOC.CH_2CH_2CO.COO^-$$

$$\underset{\text{NADH}}{NH_3, 2H^+} \quad \text{2-oxoglutarate}$$

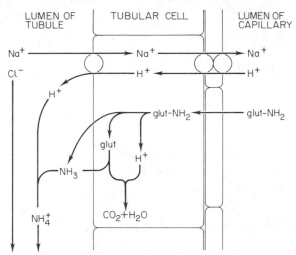

Figure 13.3 Secretion into the kidney tubule.
Abbreviations: glut-NH_2, glutamine; glut, glutamate

The fact that the reactions catalysed by glutaminase and glutamate dehydrogenase both generate protons as well as ammonia have led to much controversy concerning their role in the maintanance of acid/base balance. The view that ammonia production in the kidney should not be seen as a means of excreting hydrogen ions has, perhaps, been put most strongly by R. W. McGilvery in his textbook (1979). He claims that the importance of ammonia production in the kidney with respect to acid/base balance lies in it providing an alternative means of nitrogen excretion since there is one less proton (per nitrogen atom) to handle when nitrogen is excreted as NH_4^+ rather than urea. This would appear to be so.[2] However, the present authors believe that the claim that ammonium ion excretion does in fact achieve a net loss of protons from the body is also valid. A net proton loss is difficult to demonstrate by writing equations for two reasons. First it is necessary, as clearly appreciated by McGilvery, to write all reactants in their correct ionized form at physiological pH. Wherever possible this has been attempted in the present text. Secondly, the need to include all contingent reactions in the analysis can render it very complex. In particular, it is necessary in the accounting to consider the further metabolism of glutamate. In some circumstances at least, the glutamate will be oxidized to 2-oxoglutarate which, through reactions of the TCA cycle plus phosphenol-pyruvate carboxykinase (or the malic enzyme), pyruvate kinase and pyruvate dehydrogenase, will be totally oxidized to carbon dioxide. The over-all equations are presented in Figure 13.4. Analysis is simplified if the not unreasonable assumption is made that the concentrations of all other metabolites involved, e.g. ATP, ADP, acetyl-CoA, remain unchanged, that is, a steady state pertains. From this scheme it can be seen that both protons generated are in fact utilized by subsequent reactions so that ammonia can be used to eliminate protons arising in

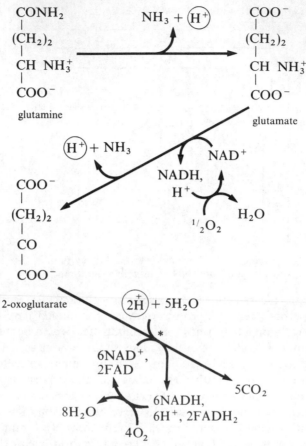

Figure 13.4 Proton balance in glutamine metabolism. Strictly, proton changes occurring in glutamine synthesis must also be included. The formation of glutamate by hydrolysis of protein involves no net proton change; its amination to glutamine actually consumes a proton if the nitrogen atom comes from another amino acid.

* For those who wish to confirm this by drawing out the whole pathway, note that each of the following reactions involve net proton changes:

pyruvate carboxylase; proton used:	isocitrate dehydrogenase; proton used:
pyruvate dehydrogenase; proton used:	citrate synthase; proton consumed

the processes outlined in Section A. Consideration of alternative fates of glutamate, for example conversion to glucose, lead to similar conclusions concerning net proton production.

(d) *The importance of ammonia production in regulation of acid/base balance*

An experiment that has been performed ön man illustrates the importance of ammonia production by the kidney in maintaining acid/base balance and the consequences of proton excretion on the loss of other cations (Pitts, 1971). On a normal mixed diet, man may excrete 40 mmol of titratable acid (that is, protons buffered by phosphate) and 40 mmol of ammonia in a 24 hr period. This is mainly a consequence of the phosphoric and sulphuric acids produced in the digestion and metabolism of meat. During each day for an experimental period of five days, a subject ingested 15 g of ammonium chloride; this produced 280 mmol of hydrochloric acid each day,[3] approximately that which is produced by a patient in moderately severe diabetic ketoacidosis. At first, the hydrochloric acid was buffered by the systems described in Section A, namely plasma proteins, haemoglobin, carbonic acid/hydrogencarbonate and intracellular proteins plus bone. Thus on the first day, most of the chloride of the ammonium chloride was excreted as sodium chloride; the sodium ions having arisen from the body fluid. On the second and third days, potassium ions replaced those of sodium. The potassium had arisen by displacement from intracellular proteins as these played an increasing role in buffering as the acid load increased. However, during the fourth and fifth days, the excretion of ammonium ions increased enormously and towards the end of the period this ion had replaced most of the sodium and potassium in the urine. On restoration of a normal diet, only very small quantities of sodium and potassium were excreted and protons continued to be eliminated in the urine in combination with ammonia. The explanation of this was that, once the dietary acid load was removed, the protons were released from the intracellular proteins and the bone to be replaced by sodium and potassium. Hence, ammonia production by the kidney is important even after the acute acid load has been removed, and six days after ceasing to administer ammonium chloride the deficit in body potassium had not been restored.

These findings are important when considering the care of a patient recovering from diabetic ketoacidosis or lactic acidosis. Of particular importance is the necessity for potassium ion supplementation either in the diet or via infusion (Section B.4). In addition, it emphasizes the importance of renal ammonia production in the preservation of body cation reserves.

5. Metabolic feedback

Although buffers ameliorate sudden pH changes, for example, arising from a massive increase in the glycolytic rate in muscle during sprinting, it would appear that excessive pH changes can be prevented by allosteric effects of protons on several regulatory enzymes in metabolism. Thus, the activities of the two key glycolytic enzymes, glycogen phosphorylase and 6-phosphofructokinase are inhibited by a fall in pH* which reduces the rate of glycolysis and hence that of lactic acid formation. This can be considered as a protective effect since a marked fall in pH would damage the muscle cells (Chapter 9; Section A.4.a). In addition, a

---

* The allosteric inhibition of 6-phosphofructokinase by ATP is markedly increased by a decrease in pH below about 7.0 (Danforth, 1965).

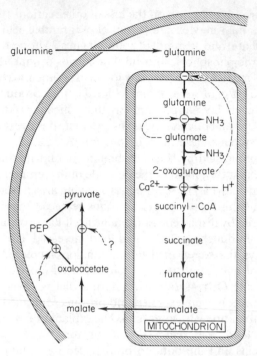

Figure 13.5 Control of glutamine metabolism in the kidney during acidosis

fall in pH increases the activity of pyruvate dehydrogenase so that the rate of pyruvate oxidation and hence proton utilization will be increased. The rate of pyruvate oxidation, however, will be limited by the demand for ATP; if this is low, the effect of even a large decrease in pH on the rate of pyruvate oxidation will be small.

C. CONTROL OF RATE OF GLUTAMINE METABOLISM IN THE KIDNEY

As discussed above, the important metabolic role of glutamine in the kidney is to produce ammonia to enable a sufficient amount of protons to be excreted and to protect the body against acidosis (See Goldstein, 1980). It has been shown that the greater the degree of acidosis, the greater is the rate of renal ammonia production but the question of how the rate of ammonia production is related to the acidosis has puzzled physiologists and biochemists for many years. Although a complete answer is not yet possible, the question is tackled here in relation to the principles developed in Chapter 7; Section A.1.

The pathway of glutamine metabolism in the kidney is described in Figure 10.13 and a possible mechanism of regulation is given in Figure 13.5. This mechanism is based on the assumptions that the transport of glutamine across

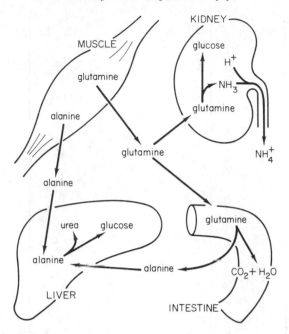

Figure 13.6 Inter-relationship of organs in glutamine metabolism. Other tissues that contain rapidly dividing cells may oxidize glutamine in a similar manner to the intestine (e.g. lymphocytes, see Ardawi and Newsholme, 1983) but they are quantitatively much less important

the cell membrane and the mitochondrial membrane and the reactions catalysed by glutaminase, oxoglutarate dehydrogenase and phosphoenolpyruvate carboxykinase (or the malic enzyme) are non-equilibrium. The other reactions of the pathway are assumed to be near-equilibrium.

The flux-generating step in a pathway is identified by comparison of K_m values of the enzyme and transport processes with intracellular concentrations of the pathway-substrates. These data are given in Table 13.1, from which it can be seen that none of the early reactions in the glutamine pathway is saturated with substrate. This is supported by the fact that the rate of glutamine metabolism (and ammonia production) in the kidney is increased by raising the blood glutamine concentration above the normal (Hems, 1972). Since most of the glutamine present in blood is synthesized in muscle, the flux-generating step for ammonia production is likely to be in this tissue.* The most likely candidates are either glutamine transport out of the muscle or the glutamine synthetase reaction within the muscle (Figure 13.6). Nonetheless, the rate of glutamine utilization by the kidney can be changed independently of the glutamine concentration, that is,

* It can be calculated that at maximal rates of glutamine utilization by the kidney, all the glutamine in the blood would be used in less than 30 min if it were not released from the muscle (Goldstein *et al.*, 1980).

Table 13.1. Properties of some 'non-equilibrium' enzymes and transport systems in the renal glutamine pathway

Non-equilibrium enzymes or transport systems	K_m for pathway substrate	Properties of enzymes		
		Physiological concentration of pathway-substrate	Activators	Inhibitors
Mitochondrial transport of glutamine	2.1 mM (glutamine)	2 mM*	N.K.	2-oxoglutarate
Glutaminase	5 mM (glutamine)	2 mM*	phosphate	glutamate, 2-oxoglutarate
Oxoglutarate dehydrogenase	0.3 mM (2-oxoglutarate)	2.6 mM	H^+, Ca^{2+}	N.K.
Malate transport across mitochondrial membrane	N.K. (malate)	0.6 mM*	N.K.	N.K.
Phosphoenolpyruvate carboxykinase	N.K. (oxaloacetate)	1.0 μM†	N.K.	N.K.

N.K. = not known.

* Assuming an equal distribution between cytosolic and mitochondrial compartments.

† Assuming an equal distribution of malate between cytosolic and mitochondrial compartments and calculation of concentration of oxaloacetate from cytosolic $NAD^+/NADH$ ratio.

there is external control of the glutamine pathway in the kidney. The evidence for this is as follows. First, in the isolated perfused kidney, the rate of glutamine utilization is higher if the kidney has been taken from an acidotic animal[3] (Hems, 1972). Secondly, when the rate of glutamine utilization is increased *in vivo* (by acidosis), the blood concentration of glutamine remains unchanged or decreases, so that the increase in glutamine uptake is not due to an increase in the concentration of the pathway-substrate.

Reactions of the pathway that are subject to external control are indicated by changing the flux and this can most readily be achieved by inducing acidosis or alkalosis; acidosis increases the rate of glutamine utilization and alkalosis decreases it. The changes in the concentrations of intermediates of the pathway are given in Table 13.2 from which it can be seen that the concentrations of all intermediates are either unchanged or decrease when the flux is increased, or increase when the flux is decreased. This suggests that all the non-equilibrium reactions of the pathway are subject to external control. Candidates for the external regulators are obtained from consideration of the properties of the enzymes, which are presented in Table 13.1. However, as with other transport systems it is difficult to study the transport of glutamine across the cell membrane *in vitro*, so that there are no known external regulators of this system.

One important enzyme in the regulation of the pathway in the kidney is oxoglutarate dehydrogenase, the activity of which is increased by acidosis resulting in a reduction in the concentration of 2-oxoglutarate which, via the near-equilibrium of glutamate dehydrogenase, results in a reduction in the concentration of glutamate. The latter is an allosteric inhibitor of glutaminase, and 2-oxoglutarate is also an allosteric inhibitor of mitochondrial transport of glutamine (Figure 13.5). Hence, a decrease in concentration of 2-oxoglutarate, due to activation of oxoglutarate dehydrogenase, should lead to increase in the rate of both processes.

The lack of changes in concentrations of malate and PEP suggest that the activity of phosphoenolpyruvate carboxykinase is increased to a similar extent to that of oxoglutarate dehydrogenase.

The major problems at the present time are how acute changes in blood pH can cause changes in the activities of oxoglutarate dehydrogenase, phosphoenol-pyruvate carboxykinase and the transport of glutamine across the cell membrane in the kidney. A pH change may affect oxoglutarate dehydrogenase directly Table 13.1) or produce an increase in the mitochondrial concentration of Ca^{2+}, which is an activator of oxoglutarate dehydrogenase, but it is unclear how a pH change could affect the mitochondrial concentration of Ca^{2+} in the kidney (Ross and Lowry, 1981).

1. Chronic acidosis

The mechanism of control of the glutamine pathway given in Figure 13.5 applies to acute changes in pH.[3] However, the response to acidosis may need to occur

Table 13.2. Concentrations of intermediates of the glutamine pathway in the rat kidney

Conditions	Concentrations (μmol.g^{-1} wet wt.)						
	Glutamine	Glutamate	Ammonia	2-Oxoglutarate	Malate	PEP	Pyruvate
Normal	1.14	3.04	1.29	0.16	0.34	0.09	0.06
Acute acidotic (1 hr)*	1.19	2.66	2.99	0.05	0.20	0.07	0.02
Acute alkalotic (1 hr)*	1.20	3.68	0.77	0.38	0.34	0.04	0.04
Normal	1.72	3.01	0.29	0.30	0.37	0.03	0.03
Chronic acidotic (1 week)†	1.17	2.48	1.11	0.11	0.14	0.06	0.03

* Acute acidosis and alkalosis were produced by giving ammonium chloride or sodium hydrogencarbonate solutions by stomach tube (Boyd and Goldstein, 1979).

† Chronic acidosis was produced by inclusion of ammonium chloride in the drinking water of rats. (Hems and Brosnan, 1971). Kidneys were removed and perfused prior to freeze-clamping and extraction for assay of intermediates.

Table 13.3. Activities of key enzymes of the glutamine pathways in kidney and liver of normal and chronically acidotic rats*

Tissue	Condition of animal	Enzyme activities (μmol.min.$^{-1}$.g^{-1} wet wt.)			
		Glutaminase	Glutamate dehydrogenase	Oxoglutarate dehydrogenase	Phosphoenol-pyruvate carboxykinase
Kidney	Normal	30.2	112.6	11.2	7.1
	Chronic acidosis	80.9	274.1	11.8	20.9
Liver	Normal	5.6	252.2	5.8	5.7
	Chronic acidosis	6.8	307.0	5.5	5.4

* Rats were made acidotic by addition of ammonium chloride to the drinking water. Data from Newsholme et al., 1982.

over a long period of time (Section B.3). The capacity of the kidney to utilize glutamine, and hence produce ammonia, increases during the first two to three days of acidosis; that is, the rate of ammonia production in an isolated kidney perfused at normal pH and at a normal concentration of glutamine is very much greater than that in a kidney from a non-acidotic animal (Hems, 1972). This adaptation to acidosis is due to increased capacities of the following processes, mitochondrial transport of glutamine, glutaminase, glutamate dehydrogenase and phosphoenolpyruvate carboxykinase but not that of oxoglutarate dehydrogenase (see Table 13.3).* These changes are probably due to an increase in the concentrations of the enzymes (or carriers) but how this is achieved and how it is co-ordinated with the acute mechanism is unclear. Furthermore, after a period of acidosis the kidney continues to excrete protons and ammonia for several days after the blood pH has returned to normal. The protons are obtained from the intracellular fluid (see Section B.3).

2. Integration of glutamine metabolism between muscle, intestine, and kidney

The probability that the flux-generating step for the renal glutamine pathway is in muscle (Figure 13.6) raises several questions; which reaction in muscle is flux-generating; how is the control of this reaction related to the control of glutamine utilization in the kidney; is the control of the flux-generating step in muscle dependent upon other metabolic changes in muscle? These questions cannot be answered at present, since metabolism in muscle has been infrequently studied in relation to provision of glutamine for the kidney. One interesting study, however, does show that, during acidosis, the rate of glutamine release by muscle is increased and that this is proportional to the increased uptake by the kidney (Schröch *et al.*, 1980). This indicates concerted control over the pathway in the two tissues but the mechanism is unknown. The problem may even be more complex than this since, in addition to muscle and kidney, any over-all control mechanism for glutamine metabolism must take into account the utilization of glutamine by the absorptive cells of intestine and other rapidly-dividing cells. A large proportion of the glutamine that is released by the muscle is normally taken up by the absorptive cells of the intestine and either oxidized or converted to alanine (see Chapter 10; Section F.4). Experiments with the isolated perfused intestine indicate that, unlike the kidney, the capacity of the pathway is unchanged in the acidotic animal. However, since the blood concentration of glutamine decreases in chronic acidosis and since the glutamine pathway in the intestine is not saturated with glutamine, the rate of glutamine utilization by the intestine *in vivo*, is probably decreased under this condition. This would make more glutamine available for the kidney (Hanson and Parsons, 1980).

---

* The specificity of this effect is shown by the fact that there is no change in the activities of these enzymes in the liver (Table 13.3).

D. LACTIC ACIDOSIS

The ways in which the body is able to cope with an acid load, sometimes a very high acid load, have been discussed to Section B. Despite this capacity there are pathological conditions in which the pH of the blood does fall to dangerous levels; these conditions are usually characterized by a high concentration of lactate or ketone bodies (or both) in the blood[4] and are considered under the headings of lactic acidosis and ketoacidosis.

Since the early 1960s, the problem of the accumulation of lactic acid in the blood as a serious consequence of a number of diseases has become apparent. Lactate can be measured easily and accurately and its concentration in blood is now part of the standard biochemical information that will be available to the physician, who should be aware of the possible metabolic reasons underlying high concentrations ($>5\,$mM) of lactate.

There is no evidence that a raised lactate concentration *per se* (i.e. hyperlactataemia) has any deleterious effects on organisms. The problems arise when the high concentration of lactate is accompanied by a low pH (hyperlactic acidaemia or lactic acidosis). Lactic acid is a weak acid and in solution dissociates according to the equation:

$$CH_3CH(OH)COOH \rightleftharpoons H^+ + CH_3CH(OH)COO^-$$

Since it is only the protons that cause physiological problems, it is necessary to look at factors affecting their concentration. If only water were present, apart from the lactic acid, the proton concentration would depend solely on the lactic acid concentration. However, the blood contains buffers (Section B). If the buffering capacity is large, the pH will change negligibly on addition of lactic acid so that hyperlactataemia will prevail. If, however, the buffer capacity is reduced, if acid is produced for a prolonged period or if additional protons are added (e.g. due to the production of ketone bodies—see Section E) then the pH will fall, lactic acidosis will pertain and harm could ensue. This discussion emphasizes that factors in addition to the rate of lactate production and utilization may have to be taken into account when considering the aetiology of the lactic acidosis.

Nonetheless, understanding the problems of lactic acidosis requires knowledge of the mechanisms of regulation of both lactate production and pyruvate utilization (see Figure 13.7). These processes have been dealt with previously in Chapter 7; Section C and Chapter 11; Section B.4.b, but they will be briefly reiterated here for convenience. For reviews of lactic acidosis, see Krebs *et al.* (1975); Cohen (1976); Cohen and Woods (1976).

1. Production of lactate

The process of anaerobic glycolysis can be represented as follows:

$$glucose \longrightarrow 2\,lactate^- + 2H^+$$

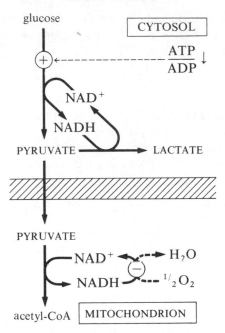

Figure 13.7 The mechanism of increased lactic acid production as a result of hypoxia

So that, in a sense, glycolysis produces lactic acid not lactate.[5] This becomes an important matter when the rate of glycolysis in muscle (or other tissues) exceeds the rate of utilization of lactate (see below).

Lactate is produced continuously from the anaerobic metabolism of glucose in the blood cells, kidney medulla, and tissues of the eye (see Chapter 5; Section F). The turnover of lactate in man is difficult to measure but it has been estimated at 20 mmol.kg^{-1} body wt (that is, about 120 g each day; 40 g from the anaerobic tissues with the remainder most likely arising from muscle, brain, intestine, and skin—Krebs *et al.*, 1975). Lactate is also produced from muscle glycogen during short duration vigorous exercise such as sprinting or climbing stairs. Indeed, complete activation of phosphorylase in all the muscles of man could lead to an increase in the concentration of lactate in the blood from 1 mM to almost 1 M within a few minutes. The fact that, even in the most severe exercise, the blood lactate concentration never exceeds 30 mM (Hermansen, 1981), indicates both the efficacy of the processes that remove lactate from the blood and the degree of control over glycolysis.

The rate of lactate release from tissues into the bloodstream depends upon its rate of formation within the tissue from pyruvate via the lactate dehydrogenase reaction and this depends upon the concentrations of pyruvate and NADH. In hypoxia, (a major cause of lactic acidosis), for example, the pyruvate concentration is raised because the rate of glycolysis is increased and that of pyruvate oxidation is decreased. In addition, the cytosolic $NAD^+/NADH$ concentration ratio will be low. Both of these factors will markedly increase the rate of lactate formation (Figure 13.7).

2. Utilization of lactate

Lactate can be utilized in two ways, oxidation to carbon dioxide and water via the TCA cycle and conversion to glucose via gluconeogenesis. Both these processes utilize protons, so that they can be considered as utilizing lactic acid, just as glycolysis can be considered as producing lactic acid:

$$\text{oxidation:} \qquad \text{lactate}^- + H^+ + 3O_2 \longrightarrow 3CO_2 + 3H_2O$$

$$\text{gluconeogenesis:} \qquad 2\,\text{lactate}^- + 2H^+ \longrightarrow \text{glucose}$$

The main tissues in which oxidation occurs will be liver, kidney cortex, heart and type I muscle fibres. However, the rate of oxidation of lactate will depend upon the metabolic activity of these tissues, oxidation will not increase just because the lactate concentration in the blood has increased. (Remember that both pyruvate dehydrogenase and citrate synthase are flux-generating steps so that an increase in lactate, and hence pyruvate, concentration in these tissues will not of itself increase the rate of oxidation.) Furthermore, it is likely that oxidation of ketone bodies or fatty acids will be preferred to that of lactate, since oxidation of these fuels inhibits pyruvate dehydrogenase (see Chapter 8; Section B.1.b) so that, under some conditions, the rate of lactate oxidation *in vivo* could be very low. Gluconeogenesis, therefore, becomes the quantitatively important process for removing lactate, and hence protons, from the blood.

An increase in the blood lactate concentration may stimulate the rate of gluconeogenesis directly, since it will increase the concentration of hepatic pyruvate and hence increase the activity of pyruvate carboxylase (Chapter 11; Section B.3.a) and , in addition, increase the hepatic concentration of citrate (Start, 1971) which is an allosteric inhibitor of 6-phosphofructokinase, the activity of which reduces the gluconeogenic flux. However, the rate of gluconeogenesis will depend on other factors including the blood concentrations of a number of hormones (e.g. glucagon, insulin, glucocorticoids). Of particular importance in the present context is the oxygen supply which will determine the ATP/ADP concentration ratio in the liver; a decrease in the ATP/ADP ratio will inhibit gluconeogenesis and if severe enough will increase the rate of glycolysis (see Chapter 11; Section B.4.b).

3. Clinical causes of lactic acidosis

In metabolic acidosis, the most common of the acid/base balance disorders, the blood pH, the plasma hydrogencarbonate concentration and the partial pressure of carbon dioxide in the blood and body fluids are all decreased. In addition, in lactic acidosis, the concentration of lactate is increased. Although there is no clearcut dividing line between normal and abnormal plasma concentrations of lactate, it is usually assumed that 1–2 mM is normal, 5 mM is abnormal and above 10 mM is dangerous. Lactate concentrations between 2 mM and 5 mM represent a grey area, which may represent the upper end of normality or may indicate an abnormality.

The symptoms of lactic acidosis include sudden onset of malaise, weakness, anorexia, and vomiting. These symptoms can lead to stupor and eventually coma. The first objective clinical feature may be hyperpnoea, which may be accompanied by a decrease in blood pressure, a rise in pulse rate and a general condition of shock. These signs and symptoms are similar to those observed in patients suffering from the other major metabolic acidosis, diabetic ketoacidosis, except of course, the lactate concentration is unlikely to be so high.

There are various clinical causes of lactic acidosis and the examples selected below represent those in which the integration of metabolism is impaired.

(a) *Tissue hypoxia*

Tissue hypoxia is the most common cause of lactic acidosis. It occurs in shock, especially haemorrhagic shock when the blood volume is decreased. It is also seen in patients with congestive heart failure, when there is inadequate oxygenation of the blood and the circulation to most organs is seriously reduced. Hypoxia in tissues that are normally aerobic will result in an increased rate of lactic acid formation, although in the liver it will first inhibit gluconeogenesis and only if sufficiently severe will the rate of glycolysis increase. Since the liver is the major organ for removal of lactic acid from the blood, transition from gluconeogenesis to glycolysis in this tissue is particularly serious. Under these circumstances, lactate concentration in the blood may exceed 25 mM and the blood pH fall to below 7.0.

(b) *Drugs*

Two drugs are frequently responsible for lactic acidosis; ethanol and the hypoglycaemic biguanides (e.g. phenformin).

(i) *Ethanol*   Ethanol inhibits gluconeogenesis (see Chapter 11; Section C.2.e) which leads to an increased concentration of blood lactate. However, for most individuals, even alcoholics, the accumulation is not sufficiently large to be dangerous (<5 mM). The problem usually arises with the alcoholic patient who suddenly goes on a drinking 'binge' so that the alcohol intake is very high but the patient does not eat. The combination of starvation and inhibited gluconeogenesis produces hypoglycaemia which results in mobilization of fatty acids from the adipose tissue and ketone body formation in the liver (see Chapter 8; Section D.1). These acidic lipid-fuels exacerbate the problem of acid/base balance and, even worse, their oxidation inhibits pyruvate dehydrogenase

(Chapter 8; Section B.1.b) thus reducing the rate of lactic acid oxidation in muscle, kidney and liver. The combination of acidosis and hypoglycaemia can rapidly result in coma.

(ii) *Phenformin*   Phenformin has been used in the treatment of hyperglycaemia, especially in maturity-onset diabetes (see Chapter 19; Section C.4.a). A normal dose of the drug increases the conversion of glucose to lactate in peripheral tissues and decreases glucose absorption from the intestine. Consequently, the blood glucose concentration is decreased whereas that of lactate is increased. Under normal circumstances the increase in the latter is not sufficient for concern but unfortunately the drug, at higher than normal concentrations, inhibits both gluconeogenesis and the oxidation of lactate. These latter effects can cause the blood lactate concentration to increase to 10–25 mM. One obvious reason for an elevated drug concentration is an accidental overdose but more commonly a decrease in the rate of metabolism or excretion is responsible. For example, administration of a normal dose of the drug to patients with impaired renal function, so that the rate of excretion of the drug is decreased, can lead to a dangerously high concentration of lactate.

One problem with many drugs including phenformin is that they are metabolized by an enzyme that also metabolizes ethanol so that ethanol competes with the drug for this enzyme and the rate of inactivation of both is reduced. For this reason, ingestion of ethanol by patients taking phenformin, especially if their liver has a lower than normal functional capacity, can be dangerous. A sufficient number of patients on phenformin treatment developed lactic acidosis in the U.S.A. that its use was banned (Williams and Palmer, 1975).

(iii) *Other chemicals*   There are a number of other chemical agents that can also precipitate lactic acidosis. Those occasionally ingested include methanol, salicylates, and ethylene glycol. These chemicals or their metabolites interfere with hepatic metabolism and decrease the rate of gluconeogenesis and increase that of glycolysis in the liver. For a discussion of the metabolic acidosis caused by ethylene glycol intoxication in a young man, see Levinsky and Robert, (1979).

(c) *Tumours*

The high rate of lactate production by tumour tissue is discussed in Chapter 5; Section G.1. Patients with acute leukaemia may suffer from lactic acidosis due to over-production of lactic acid by the tumour cells (Roth and Porte, 1970). Not surprisingly other tumours can cause lactic acidosis if there is involvement of the liver (i.e. secondary hepatic tumours) and the function of the liver is impaired (Spechler *et al.*, 1978).

(d) *Renal failure*

Since the kidney plays an important role in acid/base balance, it is not surprising that patients with renal failure suffer from acidosis.

4. Treatment of lactic acidosis

Since lactic acidosis is a serious problem that develops in many cases due to a specific clinical problem (e.g. circulatory distress, liver damage, renal failure) treatment of the underlying cause is vital. However, rapid correction of the pH is important and this can be done by intravenous infusion of sodium hydrogencarbonate. Unfortunately, this treatment is not without its problems; it increases total extracellular volume which may be dangerous, especially in the case of cardiac failure. Over-rapid correction of blood pH has a further complication; the increase in pH increases the affinity of haemoglobin for oxygen. In patients with a chronically low blood pH, which decreases the affinity of haemoglobin for oxygen (the Böhr effect), metabolism in the erythrocytes adapts by reducing the intracellular concentration of 2,3-bisphosphoglycerate (see Chapter 5; Section F.3.c) so that the affinity of haemoglobin, despite acidosis, is normal. Unfortunately, the concentration of 2,3-bisphosphoglycerate in the red blood cells does not change rapidly, so that sudden correction of blood pH by hydrogencarbonate administration has the effect of temporarily increasing the affinity of haemoglobin for oxygen. This can result in exacerbation of tissue hypoxia and is a further threat to survival. (This illustrates the fact that serious problems can readily arise when even simple metabolic interventions are attempted and emphasizes the importance to the clinician of a detailed knowledge of metabolic inter-relations.)

A new approach to the treatment of lactic acidosis, based on knowledge of the control of pyruvate dehydrogenase, has been suggested. The drug dichloro-acetate activates pyruvate dehydrogenase (see Chapter 7; Section B.2) so that administration of this drug should stimulate pyruvate oxidation and hence lower the lactic acid concentration in blood (Relman, 1978). Unfortunately, this idea suffers from the problem that even if the activity of pyruvate dehydrogenase is increased, the acetyl-CoA will not be oxidized by the cycle unless the demand for ATP has been increased to stimulate the rate of the cycle. It will lead only to an increase in the concentration of acetyl-CoA in the tissues.

E. KETOACIDOSIS

1. Origin of ketoacidosis

Ketone bodies are acids (acetoacetic and 3-hydroxybutyric) with a lower pK_a value than acetic acid (see Chapter 6; Note 13). Their presence in the blood can therefore impose a considerable acid load on the buffering capacity. Although the concentration of ketone bodies achieved even in prolonged starvation (approximately 8 mM) does not normally cause serious acidosis, the ketoacidosis resulting from uncontrolled diabetes is frequently life-threatening. In severe cases, the total ketone body concentration may reach 25–30 mM, and the acidosis is usually exacerbated by the presence of high concentrations of fatty acid and lactic acid. Indeed, Sestoft *et al.* (1980) have pointed out that lactic acidosis may be an important factor in diabetic ketoacidosis. In the early stages of the development of the acidosis there is a depletion of sodium, potassium and phosphate reserves of the body since these ions are lost in the urine as a consequence of proton excretion by the kidney (see Section B.4). In addition, this can be accompanied by severe dehydration and reduction in blood volume. Hence the blood flow to a number of tissues including the liver may be reduced resulting in generalized hypoxia and, with the very high concentration of glucose, the latter is readily converted to lactic acid. If the blood flow to the liver is seriously reduced, this tissue will also become glycolytic resulting in high levels of lactic acid (> 5 mM) and severe acidosis.

Not only is diabetic ketoacidosis life-threatening; it is also relatively common. (An average of one such case occurs every two weeks in the Radcliffe Infirmary, Oxford, which serves a population of 200 000.) It can occur not only in newly diabetic patients but also in normally well-controlled diabetics who are subjected to an additional stress (e.g. infection, depression, bereavement).

2. Treatment of diabetic ketoacidosis

Once the diagnosis is made, insulin is normally administered as soon as possible. This reduces markedly the rate of fatty acid mobilization from adipose tissue and the rate of ketogenesis in the liver so that the concentrations of the acidic lipid-fuels are lowered. In addition to insulin, isotonic sodium hydrogencarbonate is sometimes infused to correct the acidosis rapidly. Its use in such patients can, however, lead to tissue hypoxia as described in Section D.4.

Although the accepted practice has been administration, either intravenously or subcutaneously, of relatively large amounts of insulin at intervals of 2–4 hr, it has recently been suggested that continuous intravenous infusion of much smaller doses of insulin may be preferable (Alberti and Hockaday, 1977). One advantage of smaller doses is that the changes in concentrations of fuels in the blood and metabolic intermediates within the cell may occur more gradually. It is particularly important that the blood glucose concentration should be reduced gradually and, to this end, infusion of glucose during administration of insulin has been recommended (McGarry and Foster, 1972). This recommendation has been made in an attempt to reduce the incidence of cerebral oedema, which may be responsible for many of the deaths in patients during treatment for coma. Cerebral oedema causes increased intracranial pressure, which can decrease blood flow to the brain and lead to death. A rapid change in the blood glucose concentration may be one factor contributing to cerebral oedema in patients.

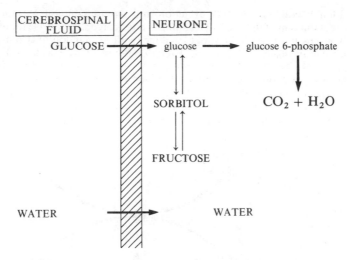

Figure 13.8 Metabolic origin of cerebral oedema. Bold arrows indicate large fluxes. Upper case letters indicate relatively high concentrations

During the hyperglycaemic period preceding the coma (which may have existed for several months), glucose concentration in the cells of the brain is increased and this stimulates the formation of sorbitol and fructose via the polyol pathway (see Chapter 5; Section F.2.b). The slow accumulation of these sugars over several months ensures that the osmotic balance between the concentrations of sugars in the cells and glucose in the cerebrospinal fluid is maintained. However, a rapid fall in the glucose concentration in blood causes the glucose concentration in the cerebrospinal fluid to fall rapidly and in turn reduces the intracellular concentration of glucose. Because sorbitol and fructose are metabolized more slowly by the polyol pathway than is glucose by the glycolytic pathway, their intracellular concentrations do not adjust rapidly to the change in glucose concentration. With a higher concentration of sugar inside the neurone than in the cerebrospinal fluid, water moves into the cells to cause oedema (Figure 13.8).

NOTES

1. The simplest way of converting ten raised to a complex power into decimal form is to multiply the power (index) by the logarithm of 10 (that is, by one) and to take the antilogarithm. In this example:

   $$\log_{10} 10 \times (-7.4) = -7.4$$

   $$\text{antilog}\,(-7.4) = \text{antilog}\,(-8.0 + 0.6)$$

   (necessary since the mantissa must be positive)

   $$= 3.98 \times 10^{-8}\,\text{M}$$

   $$= 3.98 \times 10^{-5}\,\text{mM}$$

   $$= 0.00004\,\text{mM}$$

2. Net proton production in urea excretion and ammonium ion excretion can be compared if the sum of the over-all reactions is considered. In this case the excretion of nitrogen atoms from two molecules of alanine (with production of two molecules of pyruvate) will be taken as an example. Excretion of the nitrogen as urea (Figure 13.9) involves the net production of two protons (at the glutamate dehydrogenase and ornithine transcarbamoylase reactions). Excretion as ammonium ions (Figure 13.10) involves no net proton production, thus establishing McGilvery's point. It is assumed that a steady state pertains so that the concentrations of adenine and nicotinamide nucleotides do not change and these metabolites are thus omitted from the pathways depicted in the figures.

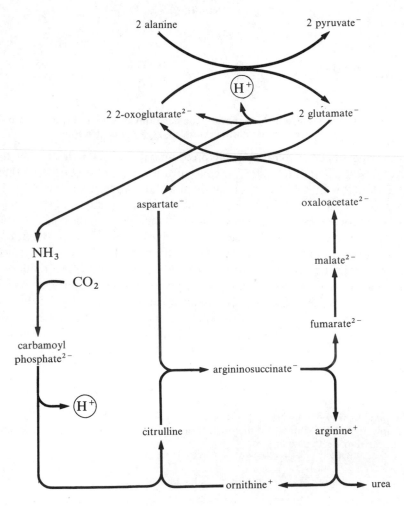

Figure 13.9 Pathway for the synthesis of one molecule of urea from two of alanine, showing sites of proton production. Intermediates are depicted with the net charge born at pH 7.0. The concentration of all other reactants is assumed to remain constant so that no change in their net charge will occur

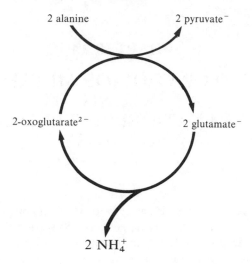

2 NH$_4^+$

Figure 13.10 Pathway for the synthesis of ammonium ions from alanine—to be compared with Figure 13.9

3. Acidosis can be induced rapidly in animals by infusion or oral administration of hydrochloric acid and alkalosis by administration of sodium hydrogencarbonate (Boyd and Goldstein, 1979). Chronic acidosis is readily produced by ingestion of ammonium chloride, which is absorbed into the body and in the blood the ammonium ions dissociate as follows:

$$NH_4^+ \rightarrow NH_3 + H^+$$

The ammonia is rapidly taken up by the liver and converted completely to urea:

$$2NH_3 + CO_2 \rightarrow CO(NH_2)_2 + H_2O$$

leaving the proton to lower the pH of the blood.
4. The other important information that will be available to the physician is the concentrations of electrolytes in the blood, from which the 'anion gap' can be calculated. The anion gap is obtained by subtraction of the sum of the serum chloride and hydrogencarbonate concentrations from the serum sodium concentration, $[Na^+] - ([Cl^-] + [HCO_3^-])$. The usual 'anion gap' is $10–16 \, mE.l^{-1}$. If the gap is greater than 16, this indicates the addition of acid stronger than carbonic acid (e.g. lactic acid) to the body at rates which exceed the rate of loss of the anion (e.g. lactate) from the body (Gabow *et al.*, 1980).
5. The fact that the glycolytic process involves phosphorylated intermediates including, ATP, ADP, and phosphate, complicates the calculation of proton formation in glycolysis. Indeed this is a controversial area and readers who are interested are referred to Gevers, 1977; Wilkie, 1979.

CHAPTER 14

THE INTEGRATION OF METABOLISM DURING STARVATION, REFEEDING, AND INJURY

The role of the glucose/fatty acid/ketone body cycle in control of carbohydrate and fat metabolism has been discussed in Chapter 8; Section B.2. The cycle is now extended to include protein and amino acid metabolism and is discussed in relation to prolonged starvation, refeeding after starvation, and physical injury. Although it is usually assumed that starvation does not afflict man in industrialized societies, this is not necessarily so. In some institutions (e.g. boarding schools, prisons, hospitals) the overnight period of starvation may be longer than 12 hr and illness can extend this period to several days. More prolonged starvation in man occurs in the less well-developed countries of the world, especially when they are affected by natural disaster or war. Not surprisingly, there have been relatively few detailed studies of the metabolic effects of prolonged starvation until those carried out in the latter part of the Second World War, especially in western Netherlands, when a large number of people suffered severe shortage of food in the winter of 1944–1945[1] (see Davidson et al., 1979 for historical accounts of various studies). More recently, it has become clear that knowledge of the control of metabolism during prolonged starvation is of considerable importance in understanding integration of metabolism and providing information about the metabolic problems of patients who have undergone intestinal surgery or suffered severe injury. To this end, studies have been carried out using normal subjects fasting for about a week or very obese subjects fasting for several weeks (Cahill, 1970; Owen et al., 1979). Studies of the controls operating during the opposite changes are also of interest and the metabolic response to refeeding after a period of starvation is considered in Section C.

Since the pioneering work of Sir David Cuthbertson in the 1930s there has been continuous interest in the metabolic response to injury (including burns and infection) which is discussed in this chapter since the response is not unlike that seen in prolonged starvation. Patients undergoing surgery have to fast for more than 12 hr prior to the operation and if the surgery involves any part of the intestine may have to extend their fast for many days. In such cases, intravenous feeding is usually given, thus raising the question of the fuels that should be provided. This problem is discussed in Section E.

536

A. THE METABOLIC RESPONSES TO STARVATION

For convenience, prolonged starvation can be divided, somewhat arbitrarily, into four phases which will be discussed in order; postabsorptive period, early starvation, intermediate starvation, and prolonged starvation. These periods are characterized by different metabolic problems but, of course, the transition from one period to another is very gradual. For the quantitative aspects of the discussion, it is assumed that the subject is adult and resting.

1. Postabsorptive period

The postabsorptive period begins several hours after the last meal when the contents of the small intestine have been absorbed. The precise time will depend upon the quantity and contents of the previous meal. (For example, a large meal containing fat and protein will reduce the rate of gastric emptying and extend the length of the absorptive period as will a meal rich in fibre.) For the following discussion it is assumed that the last meal was a normal mixed meal containing a high proportion of digestible carbohydrate. Under these circumstances, the postabsorptive period will last approximately 3 or 4 hr and will terminate either with another meal or with the onset of the next period of starvation. It is characterized by the mobilization of liver glycogen to provide glucose required by peripheral tissues (Figure 14.1).* Towards the end of the absorptive period, the concentration of glucose, and hence insulin, in the hepatic portal blood will have fallen, thus reducing glucose uptake by the liver and by the peripheral tissues. In the postabsorptive period, the glucose and insulin concentrations will diminish to levels that lead to inhibition of glycogen synthase and activation of phosphorylase in the liver so that hepatic glycogen is broken down to provide blood glucose (see Chapter 11; Section E.1.c for details of the mechanism).

2. Early starvation

Early starvation commences after the postabsorptive period and persists until about 24 hr after the last meal. It is characterized by glucose release from the liver and fatty acid mobilization from adipose tissue. The changes occurring are summarized in Figure 14.2. It must be emphasized that the rate of glucose utilization by some tissues will change throughout the period; in the early stages most tissues will utilize glucose but in the latter part of this period only the brain and the 'anaerobic' tissues will utilize significant amounts of glucose. These changes are largely controlled by small decreases in the concentrations of glucose and insulin (Table 14.1) although nervous stimulation of fatty acid mobilization from adipose tissue may also play an important role. This increase in the plasma

---

* The importance of glycogen in the maintenance of the blood glucose concentration is emphasized by the severe hypoglycaemia in patients who have a glycogen storage disease and cannot convert hepatic glycogen to glucose (Chapter 16; Section E).

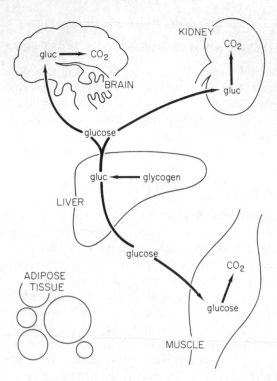

Figure 14.1 Pattern of fuel utilization during the postabsorptive period

fatty acid concentration will increase the rate of fatty acid oxidation by muscle and other tissues (e.g. kidney) so that glucose utilization and oxidation will be reduced through the operation of the glucose/fatty acid cycle (see Chapter 8; Section B.1).

The rates of glucose utilization at the end of this post-absorptive period will be about $4\,g.hr^{-1}$ for brain and about $1.5\,g.hr^{-1}$ for the 'anaerobic' tissues. Direct measurement of liver glycogen levels in biopsy samples obtained from human volunteers shows that the mean rate of hepatic glycogen breakdown between 4 hr and 24 hr is about $3\,g.hr^{-1}$ (Hultman, 1978; Table 14.2). Since the rate is likely to be higher in the earlier part of this period of starvation, glycogenolysis alone will not be able to produce sufficient glucose towards the end of the period and additional glucose must be provided by gluconeogenesis.

If glucose utilization was not inhibited by fatty acid oxidation, it would be oxidized at a rate of $9–10\,g.hr^{-1}$ so that liver glycogen would only last about 10 hr. Even when fatty acids are oxidized, liver glycogen stores only last about 24 hr (Table 14.2).

3. Intermediate starvation

The period of intermediate starvation lasts from about 24 hr until about 24 days of starvation during which time complex changes occur in fuel supply. The high

Table 14.1. Concentrations of some fuels, metabolites, and hormones in man during starvation*

Fuel, metabolites or hormone	Source	Units	Fed	Post-absorptive	Concentration at different stages of starvation					
					1 day	3 day	7 day	8 day	10 day	28–42 day
Glucose	Blood	mM	5.5	4.8	4.4	3.8	3.6	3.5	3.8	3.6
Free fatty acids	Blood	mM	0.3	0.61	0.5	1.18	1.14	1.88	1.36	1.44
Acetoacetate	Blood	mM	0.01	0.02	0.10	0.45	0.95	—	1.00	1.32
3-Hydroxybutyrate	Blood	mM	0.01	0.05	0.16	1.36	3.58	—	4.30	6.00
Total ketone bodies	Blood	mM	0.02	0.07	0.26	1.81	4.53	5.3	5.30	7.32
Lactate	Blood	mM	—	0.71	—	0.74	0.60	—	0.64	0.61
Glycerol	Blood	mM	0.10	0.13	—	0.19	0.16	—	0.17	0.18
Blood urea nitrogen	Blood	mg.100 cm^{-3}	—	10.9	—	16.2	13.0	—	—	5.2
Glutamine	Blood	mM	—	0.59	—	0.51	0.46	—	0.46	0.48
Alanine	Blood	mM	—	0.34	—	0.32	—	—	—	0.14
Leucine	Plasma	mM	—	0.11	—	0.15	—	—	—	0.07
Insulin	Plasma	μU.cm^{-3}	50	15	—	8	—	—	—	6
Glucagon	Plasma	pg.cm^{-3}	—	100	—	150	—	—	—	120

* Data given are taken from various sources (Cahill et al., 1966; Owen, et al., 1967; Williamson and Whitelaw, 1978). For days 3–42, data are from the same source and therefore comparable.

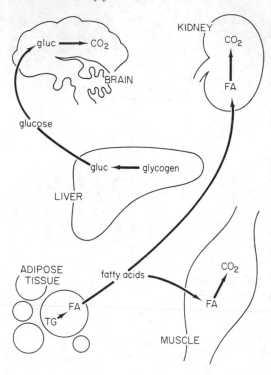

Figure 14.2   Pattern of fuel utilization during early starvation

rate of gluconeogenesis in the early stages decreases in the later stages as the concentration of ketone bodies in the blood rises. Information about metabolic changes occurring during this period comes from measurements of the blood concentrations of various fuels and metabolites and from arterio-venous concentration differences across various tissues in normal subjects fasted for eight days and in obese subjects fasted for five to six weeks or longer (Cahill, 1970; Ruderman, 1975; Ruderman *et al.*, 1976; Owen *et al.*, 1979).

The first few days of the period are dominated by gluconeogenesis which must provide about 105 g of glucose each day for the first two days of starvation, but this is progressively reduced so that towards the end of the period only about 75 g is produced each day. The gluconeogenic precursors are lactate, glycerol, and the amino acids (see Chapter 11; Section B.3). Lactate is derived from glucose metabolism in the 'anaerobic' tissues and is converted back to glucose in the liver for re-use by the anaerobic tissues. Glycerol is produced from lipolysis in adipose tissue and provides about 19 g of glucose each day under resting conditions. Since the brain oxidizes about 100 g of glucose each day the remaining glucose must be derived from amino acids released by protein degradation in muscle.* Arterio-

---

* Degradation of muscle actomyosin can be determined by measuring the rate of release of 3-methylhistidine in the urine (see Chapter 10; Note 7).

Table 14.2 Effect of starvation on liver glycogen in man

Time of starvation (hr)	Glycogen content*	
	$\mu mol.g^{-1}$ fresh liver	g in total liver
0	300	97
2	260	84
4	216	70
24	42	14
64	16	5

*Glycogen was measured in human liver after removal of a small piece of liver by biopsy needle (Hultman, 1978). Values are means from several volunteers.

venous concentration differences show that both the liver and kidney remove amino acids from the circulation to synthesize approximately 60 g of glucose after one day, 48 g after 2–4 days and about 16 g after 3 weeks starvation (Table 14.3). Although it is difficult to be precise about the requirements for glucose and the rates of gluconeogenesis over the first one to two days of starvation (since it is a rapidly changing situation) it is likely that amino acid catabolism cannot provide sufficient glucose so that the brain must use an alternative fuel. This is provided by ketone bodies and in the early part of the period their oxidation may account for 10–20% of the energy requirement of the brain (Figure 14.3).

Some of the amino acids derived from muscle protein during starvation undergo metabolism within the muscle to form alanine and glutamine which are

Table 14.3. Estimation of glucose production from glycerol, lactate (plus pyruvate), and amino acids during starvation in man

Glucose precursor	Approximate amount (g) of glucose produced each day		
	1 day starvation	2–4 day starvation	Several weeks starvation
Lactate	39	39	39
Glycerol	19	19	19
Amino acids	60	48	16
Total glucose produced from above precursors	118	106	74
Maximum glucose available for brain (i.e. glycerol and amino acids as precursors)	79	67	35
Fuel requirements for brain (glucose equivalents)	100	100	100

* Data taken from Cahill *et al.*, 1966, Owen *et al.*, 1969, Ruderman *et al.*, 1976, Owen *et al.*, 1979.

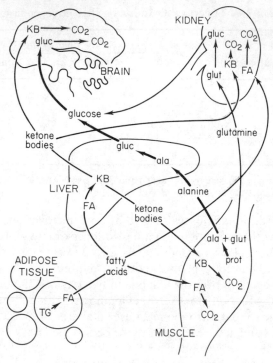

Figure 14.3 Pattern of fuel utilization during intermediate starvation

subsequently released into the bloodstream (see Chapter 10; Section F.2). Some of this glutamine is metabolized by the absorptive cells of the intestine and of this about 50 % is converted to alanine, which is released back into the bloodstream. The net result of these processes is to make a substantial part of the degraded muscle protein available to the liver as alanine, an excellent gluconeogenic precursor. Of course, the other gluconeogenic amino acids released by muscle are also converted to glucose in the liver or the kidney.

It might seem that the metabolic condition described above could be maintained for the whole period of starvation but it cannot. The major problem is the high rate of breakdown of skeletal muscle protein since for every gram of carbohydrate synthesized from amino acids, 1.75 g of protein must be degraded ((Krebs, 1964). Hence to provide even 50 g of glucose each day, almost 90 g of protein would have to be degraded. The body is unable to withstand a loss of more than half of its muscle protein (see below) and at this rate would take only 17 days.*

Since man, in fact, can starve for one or two months, it is obvious that this initial rate of protein degradation must be reduced. Indeed, it is progressively

---

* The average 70 kg man possesses about 30 kg muscle, which contains about 3 kg protein; if 90 g of protein were lost each day, 1.5 kg would be lost in about 17 days.

Table 14.4. Rates of urea and ammonia excretion after prolonged starvation in man*

	Excretion rates (g.day^{-1})	
Excreted compound	Normal	Prolonged starvation
Urea	13	0.5–1.0
Ammonia	1	3
Others	1.5	1.5

* The values given are approximate and are taken from Cahill (1970).

reduced throughout this period* but this can only occur because the brain gradually changes from oxidation of glucose to oxidation of ketone bodies (Table 14.3). During this period of starvation, the blood concentration of ketone bodies gradually rises from low levels to about 8 mM, an increase which is the signal for the brain to increase its rate of ketone body oxidation and hence reduce that of glucose (see Section B.1). The latter is reduced from almost 100 g.day^{-1} at the beginning of the period to about 35 g.day^{-1}, so that the amount of glucose that has to be synthesized from amino acids decreases to about 16 g.day^{-1} (Table 14.3). This is essential for survival during prolonged periods of starvation. The increase in ketone body oxidation by the brain during this period is dependent only upon the increase in the blood ketone body concentrations. There is no increase in the activities of ketone body utilizing enzymes in the brain of experimental animals during prolonged starvation and, in addition, sufficient activities of these enzymes to account for rates of ketone body oxidation in prolonged starvation are present in the brain of normal fed man obtained *post mortem* (Page and Williamson, 1971).

One further point of interest is that, during this period of starvation, there is a change in both the tissue carrying out gluconeogenesis and in the amino acid precursor. In the initial period, gluconeogenesis from alanine and other amino acids occurs in the liver where nitrogen atoms from the amino acids are converted to urea. However, towards the end of this period of starvation there is a problem of acidosis, due to increased concentrations of circulating fatty acids and ketone bodies, and the kidney increases the rate of proton excretion accordingly. In order to buffer the protons in the urine, the kidney produces ammonia from the nitrogen atoms of glutamine, while the carbon atoms are converted to glucose (Chapter 10; Section F.3). Hence this amino acid becomes an important gluconeogenic precursor, and a large proportion of the nitrogen loss appears as ammonia (Table 14.4). Since the demand for glucose is decreased during this period of starvation, the rate of release of amino acids from muscle is also decreased, eventually to about one third of that in the postabsorptive state (Marliss *et al.*, 1971).

*Rates of 3-methylhistidine excretion in the urine indicate that the rate of myofibrillar protein degradation is reduced in man during prolonged starvation (Young *et al.*, 1973).

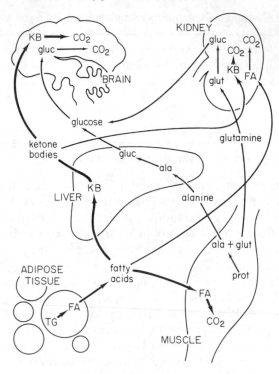

Figure 14.4 Pattern of fuel utilization during prolonged starvation

Throughout this long period of starvation the blood glucose concentration remains remarkably constant (see Table 14.1) despite the following changes: glucose production by hepatic glycogenolysis followed by hepatic gluconeogenesis and finally by both hepatic and renal gluconeogenesis; a progressive decrease in the rate of glucose utilization by the brain accompanied by a decreased rate of gluconeogenesis; a change in the major gluconeogenic precursor from alanine to glutamine. What is known of the control mechanisms that ensure that these metabolic changes take place in a smooth and co-ordinated way is discussed in Section B.

4. Prolonged starvation

After about 24 days starvation (at least in obese subjects) the steady-state rates of carbohydrate, lipid, and protein metabolism appear to be established. This metabolic condition is summarized in Figure 14.4. The blood concentrations of ketone bodies remain constant at about 8 mM, nitrogen excretion reaches a low plateau at about 5 g per day and gluconeogenesis produces about 74 g of glucose each day. Of this glucose, about 39 g is derived from lactate and 19 g from glycerol (mostly in the liver) and 16 g is derived from amino acids, (mostly from glutamine in the kidney). It is not known whether brain has an absolute requirement for the 35 g of glucose used per day at this stage or whether ketone bodies, if their concentration could be increased further, could satisfy an even greater proportion of its fuel requirements.

Prolonged starvation ends either with refeeding or death. Refeeding is discussed in Section C. There are probably several reasons why starvation eventually leads to death and the study of starvation in western Netherlands in 1944–1945 showed a remarkable lack of consistency in factors that caused death in the terminal stages of starvation (Burger *et al.*, 1948). One common cause is pneumonia. Because of loss of myofibrillar protein from the diaphragm and intercostal muscles there may be inadequate removal of fluid from the bronchioles and lungs so that the lung is open to infection (Ruderman, 1975). Furthermore, the immunological response to an infection may be reduced because of a low level of circulating antibodies, a small number of white cells and an impairment of the ability of the lymphocyte system to respond to an infection (Chandra, 1981). Another cause of death can be shock from depletion of blood volume; this may be due to a decreased rate of production of albumin and hence a decrease in its concentration in the blood, which lowers the colloid osmotic pressure so that fluid is lost from the blood to the interstitial space.

B. INTEGRATION OF CARBOHYDRATE, LIPID, AND PROTEIN METABOLISM DURING STARVATION

Brief details of metabolic control mechanisms during the first two periods of starvation have been given above but the mechanisms of control responsible for the changes in fuel supply during the intermediate period of starvation (1–24 days) are not so well established. In the early part of this period (first two days), the increase in the concentrations of glucagon, growth hormone, and glucocorticoids, together with the decrease in that of insulin, will be important in increasing the rates of fatty acid mobilization, ketogenesis, and gluconeogenesis. Furthermore, the decrease in the insulin concentration, together with the increase in glucocorticoid concentration may be important in increasing the rate of breakdown of body protein. However, the mechanisms that increase the concentration of ketone bodies during this period and reduce the rate of protein degradation are still unclear. The concentration of ketone bodies in the blood increases about ten-fold during the period of 2–24 days despite the fact that there is only a very small increase in concentration of fatty acids and the concentration of glucose remains very constant (Table 14.1; Figure 14.5). Somewhat surprisingly, however, this is not due to an increase in hepatic production of ketone bodies as arterio-venous difference concentrations across the liver indicate that the rate of ketone body production is similar at day 3 to that at day 24 (Table 14.5). The rise in the ketone body concentration can be accounted for by a gradual reduction in the rate of utilization by the muscle and an increase in the rate of re-absorption of ketone bodies from the glomerular filtrate by the

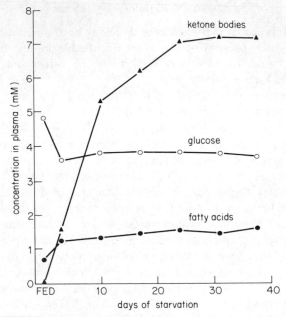

Figure 14.5 Changes in the concentrations of fuels in plasma during starvation of obese human subjects. Data from Owen *et al.*, (1967)

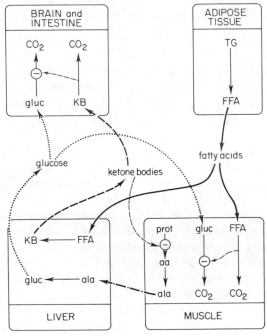

Figure 14.6 Role of ketone bodies in the regulation of protein degradation and glucose utilization; the extended glucose/fatty acid/ketone body cycle

Table 14.5. Rates* of ketogenesis in man and rat during starvation

	Period of starvation (days)	Rate of ketogenesis (μmol.min.$^{-1}$ 100 g^{-1} body weight)
Man	3	0.96
	24	0.96
Rat	0	1.7
	2	4.2

* Rates are calculated from data in Garber *et al.* (1974), Reichard *et al.* (1974), Bates *et al.* (1968).

kidney (Robinson and Williamson, 1980). At day 3 of starvation, sufficient ketone bodies are taken up by muscle to account for 50% of the energy requirement whereas at day 24 they provide only about 10% of the energy requirement (Table 14.6). The mechanism for the decreased uptake of ketone bodies by the muscle during this period of starvation is not known (see Aoki *et al.*, 1978).

1. Regulatory roles of ketone bodies

The increase in the concentration of ketone bodies from day 2 to day 24 suggests that they might play an important regulatory role as well as providing an important fuel. This increase stimulates the rate of ketone body oxidation by a number of tissues (including brain and nervous tissue, kidney cortex, lactating mammary gland, epithelial cells of the small intestine, and possibly certain vital muscles—see Chapter 6; Section E.3.c) and so decreases their rate of glucose utilization.

There is some evidence that a high concentration of ketone bodies inhibits the rate of protein degradation in muscle (see Chapter 10; Section B.3.c). This may be one important factor reducing the rate of protein degradation during the later part of the intermediate period of starvation (and, in addition, after severe injury, infection or burns). If the blood concentration of ketone bodies is artificially raised in normal subjects who have starved for several days, the rate of alanine release from muscle falls (Sherwin *et al.*, 1975). In addition, the rate of protein degradation in severely injured patients is lower if the blood ketone body concentration is elevated. The mechanism for this effect of ketone bodies is not known but the lack of effects of ketone bodies *in vitro* suggest that it is indirect. It is known that one of the metabolic intermediates of leucine oxidation reduces the rate of protein degradation in muscle (Goldberg *et al.*, 1980b), so that it is tempting to speculate that the concentration of this intermediate is increased by ketone bodies or their metabolism in muscle.

The inhibition of protein degradation in muscle and of glucose utilization in brain by ketone bodies could provide a concerted mechanism for maintaining the blood glucose concentration constant during prolonged starvation. The inhibition

Table 14.6. Utilization of ketone bodies by muscle of human forearm during starvation*

Period of starvation	Arterio-venous difference (mM)		Percentage of arterial concentration extracted		Percentage contribution of ketone bodies to oxygen consumption of muscle	Rate of ketone body utilization by muscle (nmol.min^{-1}.g^{-1})
	acetoacetate	3-hydroxybutyrate	acetoacetate	3-hydroxybutyrate		
Overnight	0.043	0.022	40	12	10	2.5
3 days	0.21	0.92	25	4	51	12.4
24 days	0.17	−0.082	11	—	16	2.8

* Data from Owen and Reichard (1971).

of protein degradation effectively reduces the rate of glucose formation but less glucose is required because its utilization by the brain is reduced.

The effects of ketone bodies on the rates of protein degradation in muscle extend the regulatory properties of the glucose/fatty acid/ketone body cycle (described in Chapter 8; Section B.2) and the extended cycle is shown in Figure 14.6. This extension emphasizes the importance of the concentration of alanine in controlling the rate of gluconeogenesis in the liver. As the blood alanine concentration decreases (due to a decreased rate of protein degradation in the muscle) this will reduce the rate of gluconeogenesis, which is matched by the decreased rate of glucose utilization by the brain. However, the capacity of hepatic gluconeogenesis is not affected, so that lactate and glycerol can still be converted to glucose at the same rate throughout this entire period of starvation.

One question, to which an answer cannot yet be given, is why the high concentration of 3-hydroxybutyrate observed in prolonged starvation (6 mM) does not inhibit lipolysis or stimulate insulin secretion as would be expected from knowledge of control mechanisms discussed in Chapter 8, Section B.2.b.* Perhaps the sensitivity of adipose tissue and the pancreas to the regulatory effects of 3-hydroxybutyrate is reduced in prolonged starvation.

2. The role of triiodothyronine in starvation

Although it is likely that ketone bodies play some role in the regulation of protein degradation in prolonged starvation, there is now considerable evidence that the thyroid hormones are also involved. Studies in man have shown that the basal metabolic rate is decreased by as much as 25 % after 2–3 days of starvation (Owen *et al.*, 1979). The hormone thyroxine (3,5,3′,5′-tetraiodothyronine, abbreviated to T_4) has been known to be involved in the control of the basal metabolic rate for many years and it might be expected that its concentration would be reduced in starvation. However, this is not the case. This does not mean that this hormone is not involved since there is evidence that thyroxine is not the active form of the thyroid hormones. The active form is considered to be 3,5,3′-triiodothyronine (abbreviated to T_3) which is produced from thyroxine by deiodination in peripheral tissues, especially the liver and the kidney (Visser, 1980; Frieden, 1981). Support for the importance of T_3 is provided by the fact that the rate of conversion of T_4 to T_3 is decreased in starvation so that the concentration of T_3 is decreased. However this is not the whole story. In the formation of T_3, iodine is removed from the 5′-position of thyroxine, that is, in the tyrosine ring furthest from the α-carbon atom to produce 3,5,3′-triiodothyronine, but iodine can also be removed from the 5-position (that is, from the tyrosine ring nearest to the α-carbon atom) to produce 3,3′,5′-triiodothyronine, which is considerably less active than T_3 and is known as reverse-T_3 (Figure 14.7). In prolonged starvation, not only is the conversion of T_4 to T_3 reduced but conversion of T_4 to reverse-T_3

---

* The concentrations of fatty acids and insulin remain remarkably constant in prolonged starvation—Table 14.1.

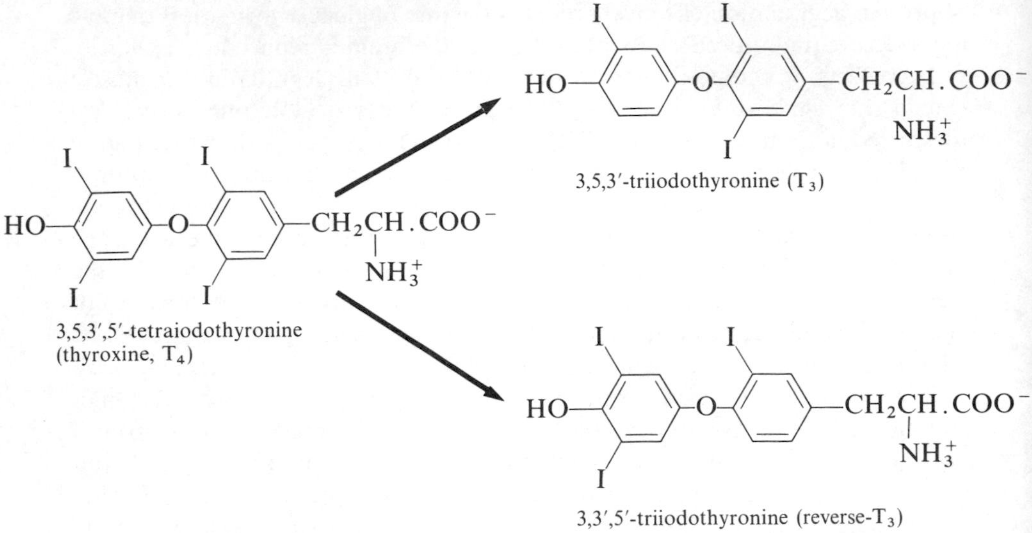

Figure 14.7 Structures of the naturally occurring thyroid hormones

is increased so that while the plasma concentration of T_3 falls, that of reverse-T_3 rises (Table 14.7).* The mechanisms controlling these changes are not known.

These changes in thyroid hormones are important for understanding metabolic integration in starvation, since T_3 is involved in the control of protein degradation in muscle (see Chapter 10; Section B.3.c) and energy expenditure in the whole animal (Chapter 22; Section B.2). Experiments with animals show that whereas starvation increases nitrogen excretion in normal rats there is no effect in thyroidectomized rats (Figure 14.8) (Goldberg *et al.*, 1978). Furthermore, the rate of amino acid release from muscle incubated *in vitro*, which is an index of protein degradation, is less from muscles of thyroidectomized animals than from those of normal animals and, similarly, the loss of body weight in starvation is less in thyroidectomized rats. In addition, thyroidectomy dramatically increased survival during starvation; after seven days of starvation, all normal rats were dead whereas some hypothyroid rats survived for as long as 20 days[2] (Figure 14.9). This work suggests that the hypothyroid state in man during starvation reduces the rate of body protein breakdown and is consequently important for survival. If, however, the concentration of T_3 is maintained (by oral administration) there is an increase in the rate of urea and 3-methylhistidine excretion (Gardner *et al.*, 1979), both indices of protein breakdown.

It is also important that the basal metabolic rate in starvation is decreased to a minimum to conserve fuel, and this too is probably caused by the decreased T_3 concentration, since this will reduce energy expenditure and hence fuel

* It is possible that the concentration of T_3 in the brain is controlled by the rate of deiodination of T_4 within the cells of the cerebral cortex and cerebellum. This suggests a still further extension to the means for controlling the action of the thyroid hormones (Bleech and Moore, 1982).

Table 14.7. Effect of starvation on the serum concentrations of thyroxine, triiodothyronine, and reverse triiodothyronine in man*

Period of starvation (hr)	Concentration of hormone in serum† (ng.100 cm^{-3})	
	Triiodothyronine	Reverse triiodothyronine
0	120	26
16	115	24
24	115	28
32	110	29
40	100	36
48	85	36
64	70	42
72	70	50

* Data from Gardner *et al.* (1979).
† The concentration of thyroxine at all times was 7000 ng.100 cm^{-3}.

utilization. This knowledge is of importance not only in consideration of the metabolic response to starvation but also the response to over-feeding since in this case there is an increase in metabolic rate and T_3 also appears to be involved (see Chapter 19, Section C.3.b).

What remains to be clarified is whether the effects of ketone bodies and T_3 are the only ones involved in the control of protein breakdown. This is an extremely important question for there are several common clinical conditions (see Section

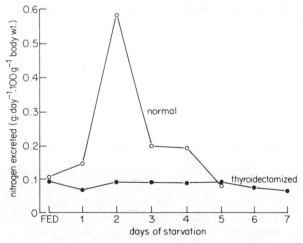

Figure 14.8 Nitrogen excretion by normal and thyroidectomized rats during starvation. Data from Goldberg *et al.* (1978)

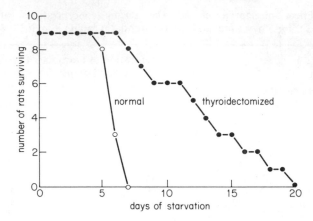

Figure 14.9   Survival of thyroidectomized and control rats during starvation. Data from
Goldberg *et al.* (1978)

D) characterized by a high rate of protein degradation which is considered to be
detrimental to the well-being of the patient.

C. THE METABOLIC RESPONSE TO REFEEDING

In general, man eats intermittently so that refeeding after a period of starvation is
a daily occurrence. It has been estimated that man normally spends about 5 % of
his time eating although absorption will take much longer than this. Two
refeeding situations will be considered here; after the overnight fast and after a
prolonged period of starvation. In both cases the energy stores of the body have
been mobilized during the preceding period of starvation, but, soon after
refeeding, these processes must be strongly inhibited so that the fuels that are
absorbed from the intestine can be utilized and, if necessary, stored to avoid large
changes in the blood concentrations. In other words, during refeeding the
processes that release glucose, fatty acids and amino acids must be strongly
inhibited but those that utilize and store these fuels must be increased to an extent
depending on the quality and quantity of the food consumed.

1. Metabolic response to refeeding after overnight starvation

The day/night cycle usually imposes a diurnal rhythm on the feeding behaviour of
an animal so that sufficient fuel must be stored during the active period (daylight
or darkness) to provide for the metabolic demands during the period of sleep.
This is a particular problem for small warm-blooded animals, especially during
the winter. As indicated in the introduction to this chapter, man may regularly
starve for more than 12 hr during the overnight period and this requires use of
stored fuel.

(a) *Changes in concentrations of fuels and hormones*

In man, overnight starvation will end with breakfast, which in many cases will be a carbohydrate-rich meal. Undoubtedly, changes in the circulating concentration of insulin will play an important role in the smooth metabolic transition from use of endogenous fuel stores to use of exogenous fuels. After a meal that contains a high content of digestible carbohydrate most, if not all, of the tissues will utilize glucose for their energy requirements (so that the respiratory exchange quotient approaches or exceeds unity, see Owen *et al.*, 1979). Despite the rapid ingestion of large quantities* of easily digested carbohydrate, the peripheral blood glucose concentration does not increase more than two fold. There are several reasons for this.

1. There appears to be a reflex release of insulin in response to the sight of the meal (Berthoud *et al.*, 1981). This increase in insulin concentration will antagonize the action of glucagon on the liver, so that rates of glycogenolysis and gluconeogenesis will be reduced. It will also inhibit fatty acid mobilization from adipose tissue. Hence the liver will be 'primed' to take up glucose if and when the concentration in the hepatic portal blood increases.
2. The presence of carbohydrate in the duodenum causes the release of intestinal hormones which markedly increase the sensitivity of the β-cells of the islets of Langerhans to the increased concentration of glucose. Thus a small increase in the concentration of glucose in the hepatic portal blood will markedly increase the rate of insulin secretion (Chapter 22; Section D.2.c).
3. The changes in the concentration of glucose in the hepatic portal vein will be much larger than the peripheral changes and this change will inhibit phosphorylase and activate glycogen synthase so that the liver will take up glucose as described in Chapter 11; Section E.1.c.
4. A reduction in the blood concentration of fatty acids will reduce their inhibitory action on glucose utilization in the peripheral tissues, via the glucose/fatty acid cycle, and so facilitate glucose uptake by these tissues.

Qualitative indications of the amounts of glucose utilized by the various tissues in the absorptive period after a high carbohydrate-containing meal are presented in Figure 14.10.

If the meal contains lipid, the rate of digestion, absorption, and processing of the lipid to produce chylomicrons will be considerably slower than the digestion of carbohydrate and absorption of glucose. Nonetheless, the release of insulin in response to the meal will increase the activity of lipoprotein lipase in adipose tissue and so facilitate uptake of triacylglycerol into this tissue. Finally, since insulin stimulates the rate of amino acid transport and protein synthesis in many tissues and, in addition, inhibits protein degradation, amino acids absorbed from a mixed meal will be taken up by the tissues and converted into protein. A

---

* It has been calculated that man can ingest at least 50% of the daily energy requirement in 10 min (Owen *et al.*, 1979).

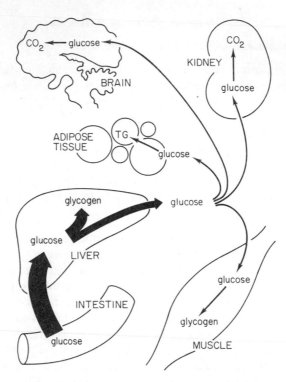

Figure 14.10   Fate of glucose absorbed from the small intestine. The thickness of lines is approximately proportional to the flux through that route

problem could arise after a meal that contains much protein and little carbohydrate. Secretion of insulin under these conditions could result in serious hypoglycaemia, due to inhibition of glucose production and facilitation of glucose utilization, but this is prevented by the additional secretion of glucagon in response to a high protein content. The high concentration of glucagon ensures that gluconeogenesis in the liver is maintained in order to supply blood glucose; the raised insulin concentration ensures that high rates of amino acid and triacylglycerol uptake, together with protein and lipid synthesis, can still occur in the peripheral tissues.

b. Substrate cycles in metabolic control after a meal

Although it is possible to explain how changes in hormone concentrations determine the direction of metabolism after refeeding, what is surprising are the remarkably small changes in the concentrations of hormones and fuels (see Table 14.8).* There is now evidence in experimental animals that substrate cycles play

* The changes indicated in Table 14.8 represent those in peripheral blood. As suggested in Chapter 11; Section E.1 changes in glucose concentration in the hepatic portal blood upon refeeding are much greater than those in the periphery. This is also the case for other fuels (see Hopkirk and Bloxham, 1977).

Table 14.8. Response of blood insulin and fuel concentrations to refeeding*

Hormone or fuel	Factor by which the concentration is increased in response to feeding
Insulin	2–6
Glucose	1.5
Amino acids	1.5
Triacylglycerol	2
Free fatty acids	2

* Data taken from Owen *et al.* (1979).

an important role in the absorptive period after a meal in providing sensitivity to the small changes in fuel and hormone concentrations. The cycles that will be important in the immediate control of the rates of mobilization and synthesis of the reserve fuels will be the glucose/glucose 6-phosphate and glycogen/glucose 1-phosphate cycle in liver (see Chapter 16; Section C), the protein/amino acid cycle in muscle (Chapter 19; Section C.3.6) and the triacylglycerol/fatty acid cycle in adipose tissue (Chapter 8; Section C.2). Evidence has been obtained *in vitro* that the rate of the triacylglycerol/fatty acid cycle is increased in adipose tissue by the catecholamines (Brooks *et al.*, 1982) and in experimental animals *in vivo* by feeding (Brooks *et al.*, 1983; Table 14.9). Furthermore, the blood catecholamine concentration is increased by feeding (Young and Lansberg, 1977), and the increased rate of the triacylglycerol/fatty acid cycle after feeding can be prevented by the prior injection of a β-blocker, a pharmacological agent that prevents the action of catecholamines on the tissues. It is proposed that the increased catecholamine concentration after a meal, caused by increased sympathetic nervous activity, is a mechanism which increases the rate of substrate cycling to

Table 14.9. Effects of refeeding and administration of a β-blocker and a β-agonist on the rate of triacylglycerol/fatty acid cycling in adipose tissue of the mouse *in vivo*†

Conditions	Rates (μmol.hour^{-1}.g^{-1} dry wt. of tisue)		
	Triacyl-glycerol synthesis	Fatty acid synthesis	Cycling*
Starved (24 hr)	91	2.7	270
Fed (after 20 hr starvation)	222	146	520
Fed (injected with β-blocker)	80	8	232
Starved (24 hr (injected with β-agonist)	508	89	1435

* The rate of cycling is given by, (rate of triacylglycerol synthesis × 3) − rate of fatty acid synthesis, since three fatty acid molecules are incorporated into one triacylglycerol.

† Data from Brooks *et al.* (1983).

provide a marked improvement in sensitivity of control mechanisms to the small changes in hormone and fuel concentrations. The effects of a meal on the rates of the other two cycles have not been studied but it is known that the protein/amino acid cycle (that is, the rate of protein turnover) is increased when protein synthesis is increased (Waterlow and Jackson, 1981). It should be pointed out that increased rates of substrate cycling will involve energy expenditure and may be partly responsible for the thermic response of food (Chapter 19; Section C.3.b).

2. Metabolic response to refeeding after prolonged starvation

It is likely that the metabolic response to refeeding after prolonged starvation is similar to that described for short term starvation, with the exception of the metabolism of protein and amino acids. During prolonged starvation there is a marked reduction in the rate of protein degradation in muscle and other tissues and in the maximum activities of enzymes metabolizing amino acids in the liver and perhaps other tissues. This knowledge has led to concern over what should provide sensible nutrition in subjects recovering from prolonged starvation. The low capacity for metabolism of amino acids could lead to a marked increase in the concentration of amino acids in the blood and in the tissues, thus causing the rate of deamination to exceed that of the urea cycle, so that the concentration of ammonia could reach toxic levels. Indeed there are reports of deaths, possibly due to toxic effects of ammonia, in prisoners-of-war who were fed steak after release from concentration camps towards the end of the Second World War (Freedland and Briggs, 1977). Certainly, injection of ammonium acetate into rats fed a low protein diet caused greater mortality than injection into rats fed a high protein diet (Wergedal and Harper, 1964). On the other hand, during the refeeding of the population of the western Netherlands after liberation in 1945, a high protein-high energy diet (in many case, 300 g of protein and 13 400 kJ each day) produced excellent results in restoration of the chronically starved to normal health. Indeed the medical report of this carefully monitored operation expressed surprise at the lack of complications and indicated that even prior hydrolysis of the protein was an unnecessary precaution. Although the degree of starvation was severe (with over 10 000 deaths attributed to this cause), serious vitamin deficiencies were not present and this may have assisted a rapid return to normal. The major problems encountered in this particular rehabilitation programme were psychological (Burger *et al.*, 1948).

D. METABOLISM AFTER INJURY

The metabolic response to injury is not unlike the response to starvation and, since severe injury leads to anorexia, 'voluntary' fasting can accompany the stress of injury. Furthermore, if the injury requires surgery, food will not be given to the patient. The magnitude of the problem of severe injury has increased in recent years due to the number of road traffic accidents and, if surgery is included in the

Table 14.10. Concentrations of fuels and hormones in blood immediately and several days after injury

	Blood concentration (mM)*	
	immediately after injury	several days after injury
Glucose	6.8	5.4
Lactate	1.9	1.1
Total ketone bodies	0.23	0.10
Fatty acids	1.0	0.4
Glycerol	0.13	0.05
Insulin	11.9†	35.1†
Cortisol	0.8‡	0.5‡

* Blood was taken for measurement as soon as possible after injury and follow-up samples were taken 1–7 days after injury (Stoner *et al.*, 1979).
† μU.cm^{-3}.
‡ nmol.cm^{-3}.

definition of injury, most of the population at some time or other will be 'injured'. Knowledge of the metabolic changes associated with this condition is likely to be of considerable importance in helping the physician to improve the treatment offered to such patients.

1. Metabolic response to injury

The similarities between the metabolic response to injury and to starvation are apparent from analyses performed on blood and urine samples from injured patients (Tables 14.10 and 14.11) and from studies on experimental animals (Cuthbertson and Tilstone, 1969). These include an increase in the plasma concentration of fatty acids* and a negative nitrogen balance. However, there are important differences which include an increase in the metabolic rate, a raised blood glucose concentration, which may be explained by high concentrations of the catecholamines and glucocorticoids, and a rapid loss of body protein. The latter is the response that has perhaps received most attention and the following questions have been raised: which tissues lose protein; is protein synthesis or degradation affected; what controls the changes in these processes and is the loss of body protein advantageous in the injured state?

It seems likely that most of the protein is lost from muscle, which is not surprising since that is the site of the highest rate of protein turnover (Chapter 10; Section B.3). Studies on the over-all rate of protein synthesis in the body suggest that in mild injury there is no change in protein degradation but a decrease in the

* Despite the fact that the plasma concentration of fatty acids is usually increased in injury and that of insulin decreased, many injured patients present with normal plasma concentration of ketone bodies (Williamson, 1981).

Table 14.11. Excretion rates for nitrogen, 3-methylhistidine, histidine, and tryptophan in patients suffering orthopaedic surgery or injury*

Condition	Excretion rate				
	nitrogen g.day^{-1}	3-methyl histidine mmol.day^{-1}	histidine mmol.day^{-1}	tryptophan mmol.day^{-1}	nitrogen balance g.day^{-1}
Orthopaedic surgery	10.7	0.31	1.4	0.17	−2.6
Injury with elevated blood ketone body concentration	12.0	0.35	1.2	0.21	−4.9
Injury with normal blood ketone body concentration	17.5	0.65	2.1	0.47	−12.4

* Data are taken from Williamson *et al.* (1977).

rate of protein synthesis. However, in more severe injury the rate of protein degradation is increased.[3] Of considerable interest is the observation that severe injury does not result in an increased rate of degradation if the blood concentration of ketone bodies is increased (Williamson *et al.*, 1977; Munro, 1979).[4] (These observations on injured patients have been used as evidence that ketone bodies can reduce the rate of protein degradation, see Section B.1.)

The metabolic changes observed after injury are likely to be the result of changes in hormone concentrations; the concentrations of insulin and T_3 are decreased whereas those of the catecholamines, glucagon, and glucocorticoids are increased (Stoner *et al.*, 1979; Traynor and Hall, 1981). The elevated concentrations of the catecholamines, glucagon, and cortisol, and the decreased concentration of insulin, could explain the hyperglycaemia since these hormonal changes would lead to increased rates of hepatic glycogenolysis and gluconeogenesis. Furthermore, the increased concentration of blood fatty acids would reduce the rate of glucose utilization in muscle and other tissues (Chapter 8; Section B.1). In addition, the decreased concentration of insulin could explain the decreased rate of protein synthesis. The increased rate of protein degradation may be caused by an increased concentration of cortisol, which, after severe injury, can increase three-fold in man (Munro, 1979).

It seems unlikely that a state of massively-increased protein degradation in the whole body is ideal for the encouragement of repair and healing processes in damaged tissues. Furthermore, if generalized, it could impair the immunological capacity of the body, and so increase the risk of infection, and it could also reduce the concentration of plasma proteins and so interfere with maintenance of the extracellular volume. For these reasons, patients with severe injury, especially burns, are often treated with large amounts of insulin (and glucose to prevent the hypoglycaemia) in order to reduce generalized protein degradation (Woolfson *et al.*, 1979). Although the work of Williamson *et al.* (1977) referred to above indicated that a high blood concentration of ketone bodies may prevent the

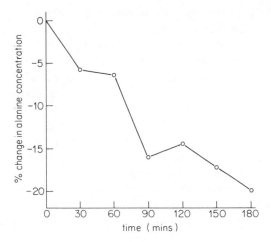

Figure 14.11 Changes in plasma alanine concentration in response to infusion of 3-hydroxybutyrate. Mean concentration of 3-hydroxybutyrate in plasma was 0.94 mM. Data taken from Sherwin *et al.* (1975)

increase in the rate of protein degradation caused by injury, the authors are unaware of any attempt to treat injured patients by infusions of ketone bodies (or short-chain fatty acids that readily give rise to ketone bodies). However, 3-hydroxybutyrate has been used to reduce protein degradation in obese subjects on low energy diets (Pawan and Semple, 1983; Smith and Williamson, 1983).

The final intriguing question is whether there is any physiological significance in the degradation of protein especially muscle protein after injury?* Despite statements made in the previous paragraph, a case can be made for its importance. It may be that since injury to animals in the wild would normally cause starvation (since their ability to forage for food would be impaired) gluconeogenesis would be necessary. Furthermore, since repair processes involve cell growth and division, and since injury would also lead to proliferation of lymphocytes in the immune response, enhanced rates of gluconeogenesis might be essential after severe injury to provide adequate glucose for these processes. In addition, it is possible to speculate that some protein is sacrificed from intact cells to provide the amino acids, especially glutamine, needed for the synthesis of purine and pyrimidine nucleotides for DNA and RNA synthesis during the formation of new cells in the healing process.

2. Atrophy of muscle fibres during immobilization

One of the major problems in sports medicine is the atrophy of muscle fibres which is responsible for the muscle wasting seen during immobilization after injury (Sargeant *et al.*, 1977; Eriksson, 1981). It is also a problem for all patients who are confined to bed for any period of time; the muscle wastage that occurs

---

* The isolation of a peptide from the blood of patients suffering from sepsis or trauma that increases the rate of proteolysis in the isolated incubated muscle of a rat provides evidence of the physiological significance of this process (Clowes *et al.*, 1983).

delays the rehabilitation of the patient after recovery. Although both type I and type II fibres atrophy during immobilization,* a greater proportion of type I fibres is usually affected. There is no satisfactory metabolic basis for the loss of myofibrillar protein that occurs under these conditions but since nervous stimulation is necessary for maintaining protein synthesis and hence protein levels in muscle (Chapter 10; Section B.3.c) this could provide a physiological explanation. (The mechanism by which nervous stimulation affects protein synthesis is not known.) Normally, type I fibres receive a more continuous nervous stimulation than type II fibres. Even in an immobilized muscle there is probably sufficient nervous stimulation to maintain a satisfactory rate of protein synthesis in type II fibres but not in type I fibres. Consequently, any treatment that maintains some nervous activity to muscle should reduce atrophy and it has been shown that if a muscle is immobilized such that it is held under some tension, the atrophy of the type I fibres is indeed less. Electrical stimulation applied to the immobilized muscle may also be beneficial but although continuous electrical stimulation may be feasible for specific muscles and for the top class sports personalities who suffer such injuries, it is probably impractical for the larger number of patients confined to bed for several weeks in hospital or at home. Whether anabolic steroids (Chapter 20; Section E.3.f) could reduce the effects of immobilization is therefore a question of some importance (Johnston and Chenneour, 1962).

E. AMINO ACIDS AND PARENTERAL NUTRITION

A good nutritional state for any patient is of the utmost importance in aiding recovery from injury, illness, or operation. Although many patients can resume oral nutrition soon after an operation or during recovery from an illness, in some cases (for example, with operations involving the alimentary tract) this may not be possible for a considerable period. During this time, stored fuels would be utilized and degradation of body protein could be sufficiently severe to hinder recovery (Hoover *et al.*, 1975). Consequently, parenteral nutrition[5] is of vital importance in such patients and glucose is usually given intravenously (see Chapter 11; Section F.2.b). However, in the early 1970s, it was observed that infusion of an isotonic mixture of amino acids instead of glucose improved the nitrogen balance of many patients (Blackburn *et al.*, 1973). This raised the problem of whether amino acids were providing increased amounts of precursor for protein synthesis or whether the infusion of amino acids provided the metabolic conditions that favoured reduced rates of protein degradation (see Greenberg *et al.*, 1976). The infusion of amino acids could provide a source of energy (oxoacids and glucose via gluconeogenesis) and still allow mobilization of fat stores and hence favour ketosis, which reduces protein breakdown (Hoover *et al.*, 1975). It remains unclear which is the best parenteral nutrient or mixture of

---

* Immobilization appears to reduce the rate of muscle protein synthesis and is without effect on the rate of degradation (based on changes in the rate of 3-methylhistidine release).

nutrients for patients who may have to rely on this form of nutrition for a prolonged period (Blackburn *et al.*, 1977; Fischer, 1977; Lee, 1977; Floch, 1981). Recent studies with injured experimental animals show that amino acid infusion, especially branched-chain amino acids, increases the rate of whole body protein synthesis (Sakamoto *et al.*, 1983). No doubt more basic knowledge of the mechanism of protein degradation in muscle and its control would provide a more rational approach to long-term parenteral nutrition.

NOTES

1. From the start of the occupation until September, 1944, the rationing system had provided the population with near-adequate nutrition. However, from September, 1944, until after the liberation in May, 1945, the population of western Netherlands (more than 4 million) suffered severe malnutrition. On September 17, 1944, the Dutch government in London declared a general railway strike in Holland and in retaliation the occupying forces prohibited all transport of food from the northern and eastern production areas to western Holland. This embargo was removed on November 8, 1944, when food was distributed by water transport. However, this was soon prevented by cold weather and the limited stocks of food were rapidly depleted. The energy content of the average daily food intake during the period was probably less than 2500 kJ. It is estimated that well over 10 000 people died from starvation alone during this period and probably an equal number from diseases directly related to starvation. A detailed study of the problems caused by starvation, the response to refeeding and the best methods for refeeding were carried out by an international team of medical and nutritional experts whose findings were published in 1948 by The Netherlands Government (Burger *et al.*, 1948).

2. This does not mean that the thyroid hormones are normally detrimental to survival in starvation; laboratory animals are protected from such factors as temperature changes and predators, the response to which in the wild would require the presence of these hormones. Thyroxine, or its active agent, triiodothyronine, may increase the sensitivity of processes (such as protein synthesis and degradation, glycogen synthesis and degradation, and triacylglycerol synthesis and degradation) to other control mechanisms. This could involve increased rates of energy expenditure due to increased rates of substrate and interconversion cycles (see Chapter 19; Section C.3.b for detailed discussion). Hypothyroidism would, therefore, decrease energy expenditure and so prolong survival during starvation.

3. Studies with patients undergoing surgery (total hip replacement) have shown that the rate of protein synthesis in muscle is increased but there is no change in the rate of protein degradation. On the other hand, in severe injury (e.g. traffic accidents, severe burns, severe infections) both processes can be affected; protein degradation is increased while synthesis is decreased. Changes in synthesis and degradation can be determined from the changes in the specific activity of infused [^{14}C]leucine (O'Keefe *et al.*, 1974) and changes in degradation rates of myofibrillar protein from the rate of excretion of 3-methylhistidine (Chapter 10; Note 7); (Elia *et al.*, 1981). In severe injury, some of the nitrogen lost in the urine may, in part, be derived from protein degradation due to resorption of damaged tissue.

4. In the absence of ketosis, injured patients have a large negative nitrogen balance and there is loss of 3-methylhistidine as well as other amino acids in the urine, whereas if the injury is accompanied by ketosis, these losses are considerably less (Table 14.11).

5. Parenteral nutrition is the provision of food to the body other than by mouth. This usually involves infusion into a vein.

CHAPTER 15

INSULIN-DEPENDENT DIABETES MELLITUS

One disease that illustrates particularly well the problems that can arise when the integration of metabolism is impaired is diabetes mellitus. The disease involves carbohydrate, lipid, and protein metabolism and, unless controlled, soon leads to death. One type of diabetes mellitus (insulin-dependent—see below) is a disease that can be logically discussed at this point since it is a pathological extension of starvation. One of the problems in reviewing this subject for the medical student is the vast amount of information available so that it is necessary to restrict discussion to those aspects which impinge directly on the integration of metabolism or that are of current clinical interest (for other details, see Renold *et al.*, 1978; Skyler and Cahill, 1981).

The chapter contains a description of the symptoms and clinical features, an explanation of these on the basis of our knowledge of the integration of metabolism, a discussion of the aetiology of the disease, an account of the treatment, and of the long-term problems of the patient. However, since diabetes has been recognized as a disease for almost 2000 years, a brief historical introduction is in order (see Allen *et al.*, 1919) but first it is important to appreciate which disease is being discussed. Diabetes mellitus is not a single

Table 15.1. Distinguishing characteristics for insulin-dependent and non-insulin-dependent diabetes. See text for details

Characteristics	Insulin-dependent diabetes	Non-insulin-dependent diabetes
Usual age of onset (years):	< 30	> 35
Nature of onset	Rapid	Insidious
Symptoms	Polyuria, polydipsia, polyphagia	May be asymptomatic
Signs	Wasting, dehydration, loss of consciousness	Commonly obese
Tendency to ketosis	Very prone	Little
Response to insulin	Sensitive	Commonly resistant
Treatment:		
(i) diet	Insufficient without insulin	May be sufficient
(ii) insulin	Essential	Needed in some cases
(iii) sulphonylureas	Ineffective	May be effective

disease but cases usually fall into one of two classes, which have been known for many years as juvenile-onset and maturity-onset diabetes. However, age is a poor criterion to use for classification of the disease so that more clinically based terms, insulin-dependent (or insulin-deficiency) indicating a requirement for insulin (abbreviated to IDDM), and non-insulin-dependent diabetes (NIDDM) are now recommended. This, unfortunately, is also not very satisfactory since, in both conditions, there is an inability of the β-cell of the islets of Langerhans to produce sufficient insulin to control the blood glucose concentration; in the former conditions there is damage specific to the β-cell (see below for possible mechanisms) and in the latter condition there are changes in the peripheral tissues so that very high concentrations of insulin are required. To overcome this difficulty, the less descriptive terms type I and type II diabetes are sometimes used for IDDM and NIDDM respectively. A discussion of non-insulin-dependent diabetes is delayed until Chapter 19, where obesity is discussed, since there may be a causal relationship between the two conditions. Some of the characteristics of the two types of diabetes are given in Table 15.1.

A. EARLY REFERENCES TO DIABETES

The first known recognition of diabetes appears to have been made by Aulus Cornelius Celsus (c. 30 B.C.–50 A.D.) who wrote as follows:

When urine, even in excess of the drink,* and flowing forth without pain, causes emaciation and danger, if it is thin, exercise, and massage are indicated, especially in the sun or before a fire; the bath should be infrequent, nor should one linger long in it; the food should be constipating, the wine sour and unmixed, in summer cold, in winter lukewarm; but everything in smallest possible quantity. The bowels also should be moved by enema, or purged with milk. If the urine is thick, both exercise and massage should be more vigorous; one should stay longer in the bath; the food should be light, the wine likewise. In each disease, all things should be avoided that are accustomed to increase urine.'

The recognition of the disease was so new that it had not been given a name. Aretaeus of Cappadocia (c. 30–90 A.D.) provided a more detailed description.

'Diabetes is a strange disease, which fortunately is not very frequent. It consists in the flesh and bones running together into urine. It is like dropsy in that the cause of both is moisture and coldness, but in diabetes the moisture escapes through the kidneys and bladder. The patients urinate increasingly; the urine keeps running like a rivulet. The illness develops very slowly. Its final outcome is death. The emaciation increases very rapidly, so that the existence of the patients is a sad and painful one. The patients are tortured by an unquenchable thirst; they never cease drinking and urinating, and the quantity of the urine exceeds that of the liquid imbued. Neither is there any use in trying to prevent the patient from urinating and from drinking; for if he abstains only a short time from drinking his mouth becomes parched, and he feels as if a consuming fire were ranging in his bowels. The patient is tortured in a terrible manner by thirst. If he retains the urine, the hips, loins, and

* Celsus introduced an error into the clinical description, since fluid output is not greater than intake.

testicles begin to swell; the swelling subsides as soon as he passes the urine. When the illness begins, the mouth begins to be parched, and the saliva is white and frothy. A sensation of heat and cold extends down into the bladder as the illness progresses; and as it progresses still more there is a consuming heat in the bowels. The integuments of the abdomen become wrinkled, and the whole body wastes away. The secretion of the urine becomes more copious, and the thirst increases more and more. The disease was called diabetes, as though it were a siphon,* because it converts the human body into a pipe for the transflux of liquid humors. Now, since the patient goes on drinking and urinating, while only the smallest portion of what he drinks is assimilated by the body, life naturally cannot be preserved very long, for a portion of the flesh also is excreted through the urine. The cause of the disease may be that some malignity has been left in the system by some acute malady, which afterwards is developed into this disease.'

With remarkable foresight, Artaeus predicted that the disease may be caused by an acute illness; after almost 2000 years we are beginning to have some biochemical understanding of the aetiology of diabetes and the causal link between an acute viral attack and diabetes mellitus (see Section C).

Prior to the isolation of insulin from the pancreas in 1921 and the first use in the treatment of a patient in January 1922 (Banting *et al.*, 1922; and Sönksen, 1977, Marliss 1982 for reviews), the only treatment was diet. 'Starvation diets', which were low in energy and carbohydrate but high in fat, reduced the osmotic diuresis and the accompanying symptoms (see Section E) but the progressive breakdown of both lipid and protein continued (Allen *et al.*, 1919; Denning, 1982). Patients gradually wasted away.

B. DEFINITION, SYMPTOMS, AND CLINICAL FEATURES OF INSULIN-DEPENDENT DIABETES

Although considerable progress is being made in this complex disease, it is not possible to define it in terms of the primary defect. Indeed there may be several primary defects so that, until further work provides more definite conclusions as to the basic cause, the best definition is 'inappropriate hyperglycaemia' (Renold *et al.*, 1978).

Symptoms of insulin-dependent diabetes include polyuria, polydipsia, polyphagia, loss of body weight with increasing tendency towards weakness, lassitude and drowsiness; clinical characteristics include wasting, dehydration, glycosuria, and possibly ketonuria (Table 15.1). The disease can develop so rapidly and to such an extent that the patient may lose consciousness before diabetes is diagnosed. In other cases, the clouding of consciousness may be the symptom that sufficiently worries the patient, or more especially the family of the patient, that the advice of the physician is sought. The condition of diabetic coma is one of the more common life-threatening situations, but the cause of the loss of consciousness is still not understood. It may be due to the decrease in pH, increase

---

* In Greek, διαβήτης is the word for a siphon.

in glucose concentration or increase in osmolality (see Alberti and Hockaday, 1977). The treatment of diabetic coma has been discussed in Chapter 13; Section E.2.

The symptoms given above are normally observed when the diabetes has become severe, whereas many people may be diabetic for a period without any obvious symptoms. In such cases, diabetes may be suspected when patients are found to have glycosuria during a routine examination or are known to suffer from hyperglycaemia. Alternatively, a family history of diabetes may reveal patients at increased risk. In order to confirm a diagnosis of diabetes, these patients must be further investigated.

1. Diagnostic tests

On the basis of the definition given above, glucose intolerance is the cardinal clinical feature of diabetes mellitus, so that clinical diagnosis depends upon tests to demonstrate such intolerance.

An elevated fasting blood glucose concentration together with an elevated blood ketone body concentration and ketonuria may be sufficient to diagnose diabetes mellitus. In addition, the oral glucose tolerance test (oral GTT), which is designed to test the ability of the patient's beta cells to secrete insulin, may also be used.[1]

As in other situations, there is no such thing as a normal blood glucose concentration or normal concentrations after specified times following an oral glucose load. Nonetheless, for diagnostic purposes, certain glucose concentrations are assumed to be abnormal. The European Association for the Study of Diabetes has recently put forward concentrations at or above which diabetes mellitus is indicated; these are given in Table 15.2. If the fasting blood glucose

Table 15.2. Blood glucose concentration criteria* for diabetes mellitus

Condition	Glucose concentration (mM)		
	Venous whole blood	Capillary whole blood	Venous plasma
Patient fasted overnight	7.0	7.0	8.0
120 min after oral glucose	10.0	11.0	11.0
Any other intervening time interval after glucose	10.0	11.0	11.0

* Criteria recommended by European Association for Study of Diabetes (see Keen *et al.*, 1979). Diabetes mellitus is indicated if the three values are at or above those indicated.
Oral glucose tolerance test performed under standard conditions; 75 g glucose in 200–500 cm^3, or 1.75 g.kg^{-1} body wt for children.

concentration, that at 120 min after oral glucose and one other concentration at an earlier time interval are at or above those given in Table 15.2, diabetes mellitus is diagnosed. If only one or two are at, or above, the given concentration, impaired glucose tolerance is indicated and further tests may need to be carried out to investigate the cause.

On the basis of the complete clinical examination, diabetes is diagnosed under one of the following categories.

(a) *Clinical diabetes*

The patient has a diabetic glucose tolerance response together with the symptoms and clinical features of diabetes.

(b) *Chemical diabetes*

The patient has a diabetic glucose tolerance response but has no clinical abnormalities.

(c) *Latent diabetes*

The patient has a normal glucose tolerance response but has had either a previous abnormal response or a diabetic response to tolbutamide or steroid tests.

(d) *Potential diabetes*

A patient with a normal glucose tolerance response but with higher than normal chance of developing the disease due to family history, or presence of another disease associated with diabetes (e.g. acromegaly).

C. AETIOLOGY OF INSULIN-DEPENDENT DIABETES MELLITUS

It has been known for many years that insulin-dependent diabetes is due to irreversible damage to the β-cells of the islets of Langerhans in the pancreas and that this damage results from an inflammation of the cells (see Christy *et al.*, 1977). Recent research has shown that the interplay of at least three factors is involved in this destructive process, namely genetics, viral infections, and autoimmunity (see Craighead, 1979). Precisely how these factors are involved and the sequence of events that leads to damage of the cells is still a matter for speculation rather than established theory.

1. Genetics

Although a familial tendency for insulin-dependent diabetes has been long recognized, less than 20 % of patients have first degree relations with a diabetic history and studies with identical twins have revealed that this type of diabetes was concordant in only about 50 % of such cases whereas non-insulin-dependent diabetes was concordant in 90 % of cases. This suggests that environmental factors are also involved (see below).

A new explanation of the genetic link in insulin-dependent diabetes was provided by the observation that the disease is associated with certain of the major transplantation (histocompatibility) antigens which were first identified on human lymphocytes and were termed human leucocyte antigens (abbreviated to HLA). These antigens are glycoproteins present on the external surface of all nucleated cells and they are coded for by genes situated at the A, B, C, and D loci in a small segment of the sixth chromosome (the major histocompatibility complex). There are more than 60 genes in the major histocompatibility complex and they control the immunological defence mechanism of the body. The main function of the antigens produced from the A, B, and C loci is to determine the number and specificity of cytotoxic killer lymphocytes (T-lymphocytes). These lymphocytes will only attack and kill cells that have at least one of the HLA-A or B antigens on their surface. The role of the antigens coded by the HLA-D locus is not understood but they may aid the interaction between B and T-lymphocytes and between T-cells and macrophages (for further discussion see Bodmer, 1976, and McDevitt, 1980). Of importance for the aetiology of diabetes is that certain HLA types occur with considerably increased frequency in insulin-dependent but not non-insulin-dependent diabetes[2] (Irvine *et al.*, 1977; Thomsen *et al.*, 1980; Rotter and Rimoin, 1981). This finding supports the view that lymphocytes are involved in the damage of the β-cell and a speculative mechanism of how this might occur is given in Section C.3.

2. Viral infections

There are four lines of evidence that viral infections play an important role in the development of insulin-dependent diabetes, namely clinical reports, epidemiological studies, pathological studies, and induction of diabetes in experimental animals. A seasonal clustering of new cases of diabetes was the first epidemiological evidence to suggest a link with viral infections. There is increased incidence of new cases of diabetes in the winter or late autumn of each year and this coincides with a high incidence of common viral infections in the community. This seasonal variation was found to be similar in three widely separated communities in North America and also in Australia, with a yearly consistency that extended back for more than 15 years (Fleegler *et al.*, 1979). The viruses implicated from both epidemiological and clinical studies include Coxsackievirus B, cytomegalovirus, and those causing mumps, rubella, infectious mononucleosis (glandular fever), and chicken pox (Drash, 1979; Craighead, 1981). The abrupt onset of the disease is also consistent with a viral aetiology. However, perhaps the best evidence comes from animal studies. First, it is known that an encephalomyocarditis virus can infect the β-cells and that it produces diabetes.

Secondly, a virus was isolated from the pancreas of a ten-year-old boy who was admitted to hospital with diabetic ketoacidosis and died of cerebral oedema seven days later (see Chapter 13; Section E.2). Inoculation of mice with the isolated virus caused inflammation of the beta cells and finally necrosis and cell death (Yoon *et al.*, 1979). It is likely that in the ten-year-old boy the virus had also caused damage to the β-cells which precipitated the severe diabetes.

3. Autoimmunity

At least three lines of evidence support the view that damage to the β-cells involves the immune system. First, pathological studies of islet tissue from patients with juvenile diabetes have shown inflammatory lesions characteristic of an immunologically mediated process (Gepts, 1965; Craighead, 1979). Secondly, antibodies to the cells of the islets have been found in diabetic patients at the time of onset of the disease although, in many cases, they disappear as the disease progresses (Irvine *et al.*, 1977). Thirdly, lymphocytes from diabetic patients are cytolytic for the cultured human β-cells (Huang and Maclaren, 1976). The evidence outlined above leads to the conclusion that autoimmunity could be responsible for juvenile diabetes in a significant number of patients (for review, see Nerup and Lernes 1981).

A speculative suggestion as to the cause and development of the disease is as follows. The viral attack on the patient causes damage to one or more tissues of the body including the cell membrane of the β-cell. This damage exposes proteins to the extracellular fluid that are not normally exposed and these are 'seen' as foreign protein by the immune system so that both B- and T-lymphocytes are activated against this protein. This immunological attack produces further damage and more 'foreign' proteins are exposed upon which the immune system can act (see also Chapter 17; Section B.2.d). Susceptible individuals (those with certain lymphocyte antigens—see above) produce a particularly massive immunological response which causes more damage and, eventually, death of most of the β-cells. Some support for this hypothesis is provided by the observation that certain chemicals can damage the membranes of the β-cell and cause insulin-dependent diabetes with the appearance of antibodies to the β-cells. A number of cases of diabetes have been reported after ingestion of a rodent poison[3] which causes damage to the islet cell membranes (Karam *et al.*, 1978).

This speculative hypothesis has considerable clinical implications since it suggests that immunosuppressive agents could be beneficial in the treatment of insulin-dependent diabetes, particularly if they could be administered at the critical stage prior to the irreversible damage to the cell. If such agents could reduce the extent of the immunological attack on the β-cells, some cells might recover so that the inflammatory damage could be repaired and the antigenic site removed thereby removing the cause of the autoimmunity (Rossini, 1983). However, such treatment would depend upon the early detection of the onset of the disease so that treatment could start before extensive damage to the β-cells had occurred. Knowledge of susceptible subjects would facilitate such treatment.

Although the studies outlined above have sufficiently changed our knowledge of the aetiology of the disease so that eventually its development may be prevented, current clinical attention must be directed to the amelioration of the pathological effects of the disease (see Section E).

D. METABOLIC EFFECTS OF INSULIN

The insulin-dependent diabetic fails to produce sufficient insulin to control satisfactorily the many metabolic processes affected by this hormone. In order to understand the pathology of the diabetic patient (see Section E) it is necessary to review these processes, which is done in outline below. (A more detailed account of the metabolic actions of insulin is given in Chapter 22; Section C.)

1. Effects of insulin on carbohydrate metabolism

1. Insulin increases transport of glucose across the cell membrane in adipose tissue and muscle.[4]
2. Insulin stimulates glycogen synthase in a number of tissues including adipose tissue, muscle and liver.[5]
3. Insulin increases glycolysis, indirectly, by stimulating other processes (triacylglycerol, glycogen and protein syntheses) requiring increased rates of ATP formation.
4. Insulin inhibits the rates of glycogenolysis and gluconeogenesis in the liver.[6]
5. Insulin increases the activity of glucokinase in the liver (probably by increasing its concentration) and this apparently plays an important part in decreasing the rate of glucose release and facilitating an increase in the uptake of glucose by the liver after a meal (Chapter 11; Section E.1.a). These two latter effects of insulin—(4) and (5)—may explain why the liver of the diabetic continues to release glucose even in the absorptive period after a carbohydrate-containing meal.
6. Finally, insulin increases the rate of glucose oxidation by the pentose phosphate pathway in liver and adipose tissue but this is secondary to the stimulation of fatty acid synthesis by this hormone (Chapter 17; Section A.8).

It must be emphasized that certain tissues are largely or totally insensitive to the direct action of insulin on carbohydrate metabolism (e.g. kidney, brain, intestine). In order to reduce glucose utilization in these tissues during starvation, hypoglycaemia, etc., it is essential that fatty acids are mobilized from adipose tissue and ketogenesis is increased in the liver so that the oxidation of these fuels will reduce the rate of glucose utilization and oxidation.

2. Effects of insulin on lipid metabolism

1. Insulin inhibits lipolysis in adipose tissue[7] (see Chapter 8; Section C.1).
2. Insulin inhibits ketone body synthesis in the liver (see Chapter 8; Section D.1.b). Both these effects are fundamentally important in the control of glucose utilization in most tissues of the body (the glucose/fatty acid/ketone body cycle—see Chapter 8; Section B.2).
3. Insulin stimulates fatty acid and triacylglycerol synthesis in adipose tissue and liver; the reactions affected by insulin are given in Chapter 17; Section A.9.

3. Effects of insulin on protein metabolism

1. Insulin increases the rate of amino acid transport into muscle, adipose tissue and liver cells.
2. Insulin increases the rate of protein synthesis in muscle, adipose tissue, liver and other tissues.
3. Insulin decreases the rate of protein degradation in muscle (and perhaps other tissues). (See Chapter 10; Section B.3.c.)

These effects of insulin serve to encourage protein synthesis rather than amino acid oxidation, so that the hormone produces positive nitrogen balance in the normal subject.

4. Effects of insulin and glucagon on metabolism in the liver

For reasons that will become apparent in the next section, it is important to bear in mind that there are two important effects of glucagon on carbohydrate metabolism in the liver. Glucagon increases the rate of both glycogenolysis and gluconeogenesis, and so increases the rate of glucose release by the liver (Chapter 11; Section B.4.b). However, this effect can be reduced or abolished by insulin (Cherrington *et al.*, 1978; see Chapter 11; Section B.4).

E. PATHOLOGY OF INTEGRATION OF METABOLISM IN DIABETES

Patients who suffer from a severe deficiency of insulin (as do most insulin-dependent diabetics) will die from their inability to control metabolism if they are not treated. Since the discovery of insulin, the life-threatening situation can be removed by injection of the hormone. Nonetheless the life of the diabetic patient can be endangered prior to proper diagnosis, or when, although stabilized on insulin, the patient suddenly requires an increase in dose because of an acute illness (e.g. gastroenteritis).

Although intolerance to glucose is the clinical feature used in diagnosis of diabetes, the more important symptoms and clinical features of insulin-dependent diabetes mellitus are due to the fact that insulin is one of the major anabolic hormones. Changes in the blood glucose concentration and glucose intolerance are almost incidental consequences despite being given pre-eminence in metabolic discussions of diabetes. After all, the 'starvation diets' recommended for treatment prior to the isolation of insulin prevented the glycosuria but could not prevent the patient losing body protein and wasting away.

As can be seen by reference to the varied effects of insulin described above, a deficiency of insulin will have effects on many processes and lead to a general disturbance of metabolism. Opinions have changed as to which aspect of insulin action is fundamental in the development of insulin-dependent diabetes. Two such theories are described in Sections E.1 and E.2 while a third, more in line with current knowledge of metabolic integration, is presented in Section E.3 (Figure 15.1).

1. Insulin and glucose transport into cells

Early work on diabetic animals and success of insulin in reducing the blood glucose concentration and improving the well-being of the diabetic patient indicated the importance of insulin in facilitating glucose utilization. Consequently, when in 1950 it was discovered that insulin increased the rate of transport of glucose into the muscle cell, it was an obvious logical extension to suggest that glucose intolerance and hyperglycaemia in the insulin-dependent diabetic were due to lack of stimulation of this process. In addition, decreased glucose utilization in adipose tissue would lower the concentration of glycerol 3-phosphate thus reducing the rate of esterification of fatty acids and so leading to excessive rates of fatty acid mobilization. This would account for the markedly elevated plasma fatty acid concentration and the latter would favour ketogenesis in the liver and explain ketosis in the diabetic patient. Although this explanation is widely accepted, its quantitative importance in accounting for any of the clinical features is probably minimal (Espinal *et al.*, 1983).

2. Insulin and the glucose/fatty acid cycle

The advent of the concepts of the glucose/fatty acid and glucose/fatty acid/ketone body regulatory cycles in the 1960s and 1970s (Chapter 8; Section B.2) enabled a different interpretation of the metabolic problem in the diabetic patient to be put forward. Since insulin is a potent antilipolytic hormone, a decreased concentration of this hormone would increase the rate of adipose tissue lipolysis, even in the fed state, so that blood fatty acid (and ketone body) concentrations would remain elevated and this would reduce the rate of glucose uptake by muscle and other tissues.

These effects undoubtedly play a part in reducing glucose tolerance but their quantitative importance *in vivo* is difficult to assess.

3. **The concentration ratio of insulin to glucagon and the effect on liver metabolism**

Both glucagon and insulin are secreted by the cells of the islets of Langerhans, the former from the α-cells and the latter from the β-cells. The endocrine secretions of the pancreas enter the hepatic portal vein, so that the liver is the first tissue to be exposed to these hormones. Since the liver inactivates a high proportion of the insulin that flows through it (60–90%), this tissue is exposed to considerably higher concentrations of insulin (three- to tenfold) than peripheral tissues (Blackard and Nelson, 1970). Hence in the insulin-deficient state, the liver will be exposed to much lower concentrations of insulin than normal but to the usual high concentrations of glucagon. In the absence of the restraining effects of insulin, glycogenolysis, gluconeogenesis, and ketogenesis will be favoured (Chapter 11; Section B.4.b; Chapter 8; Section D.1; Unger, 1971, 1974). The result of this will be that, after a carbohydrate meal, much higher concentrations of glucose in the hepatic portal blood will be necessary to inhibit glucose release and stimulate glucose uptake. (Except at very high concentrations, the regulatory effect of glucose on glycogenolysis, glycogen synthesis and glucose uptake will be overcome by the effects of glucagon (via cyclic AMP) on the phosphorylase system; that is, the high concentration of cyclic AMP will maintain more than 10% of phosphorylase in the active form which will interfere in the 'pull' mechanism for control of glucose uptake by the liver—Chapter 11; Section E.1. In addition, the high concentration of cyclic AMP in the liver will facilitate gluconeogenesis even when the blood glucose concentration is high.)

All three mechanisms described above (summarized in Figure 15.1), namely failure to stimulate glucose transport into muscle (and adipose tissue), decreased rates of glucose utilization and oxidation in many tissues due to fatty acid and ketone body oxidation and increased rates of glucose release (and decreased glucose uptake) by the liver, could all play a role in glucose intolerance characteristic of the insulin-dependent diabetic. However, it is suggested that the malfunctioning of the control mechanism in the liver plays the dominant role in failure to maintain normal glucose tolerance and normal blood glucose concentrations. The evidence to support this statement is derived from two observations. In the first, a detailed study of the concentrations of glucose, fatty acids, and ketone bodies in the blood either of patients in diabetic coma, who have been brought into hospital for treatment, or in diabetic patients from whom insulin was withdrawn for 12 h, demonstrates that the blood glucose and ketone body concentrations are grossly abnormal, whereas that of fatty acids is within the normal range expected during starvation (Table 15.3). This implicates control in the liver because blood glucose and ketone body concentrations are dependent upon the control of metabolism in the liver, whereas the fatty acid concentration is largely dependent upon the rate of fatty acid mobilization from adipose tissue.

The second observation is that the diabetic patient, even when receiving a properly controlled dose of insulin, has much more difficulty in maintaining a

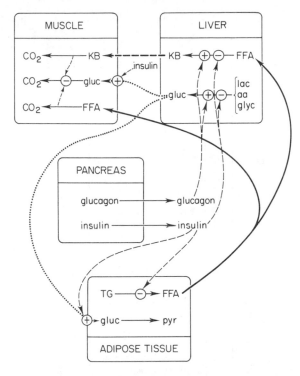

Figure 15.1 Control steps of probable importance in impaired glucose tolerance of insulin-dependent diabetes.
Abbreviations: KB, ketone bodies; gluc, glucose; lac, lactate; aa, amino acids; glyc, glycerol; pyr, pyruvate; TG, triacylglycerol

correct blood glucose concentration than does the normal subject. This may be in part because the balance of the effects of glucagon and insulin on the liver is not restored by insulin injections. Thus the injection of insulin, which is done either subcutaneously or intramuscularly, does not permit a high concentration of the hormone to reach the liver so that even in a well-controlled diabetic, the liver is under-exposed to insulin whereas the peripheral tissues may be over-exposed.

4. Metabolic explanation of symptoms and clinical features of the diabetic

The very marked elevation in the concentrations of glucose and ketone bodies in the blood can explain many of the short-term problems of the uncontrolled diabetic patient. If the blood glucose concentration exceeds about 10 mM, the rate of glomerular filtration of glucose exceeds that of the tubular reabsorption process in the kidney and glucose appears in the urine. This causes osmotic diuresis (the glucose in the urine exerts an osmotic pressure which reduces the

Table 15.3. Effect of insulin withdrawal in diabetic patients with no β-cell function and insulin infusion in diabetic patients in coma on plasma fuel concentrations*

Conditions	Time after insulin withdrawal or administration	Fuel or metabolite concentration (mM)				
		Glucose	Lactate	Fatty acids	Glycerol	3-hydroxy-butyrate
Insulin withdrawal	0	7.2	0.8	0.5	0.1	0.4
	12	17.2	0.8	2.0	0.2	5.5
Insulin administration	0	45.0	2.1	1.6	0.2	11.3
	4	18.0	1.5	0.9	0.2	5.7

* Data for withdrawal (taken from Madsbad *et al.* (1979)) are means for seven patients. Their blood glucose was controlled overnight with insulin infusion and was normal at 8 a.m. when the infusion was stopped. The patients were fasted and confined to bed for the 12-hr period of insulin withdrawal. The effect of intravenous infusion of insulin was investigated in 38 patients admitted to hospital, 31 of whom were ketoacidotic and 28 had a depressed level of consciousness (Page *et al.*, 1974).

rate of water reabsorption in the collecting ducts) which results in polyuria, thirst, and hence polydypsia. Loss of weight and wasting occurs because of the breakdown of body protein and the mobilization of adipose tissue triacylglycerol,[7] processes that are normally inhibited by insulin. The amino acids derived from protein breakdown and the glycerol from triacylglycerol lipolysis provide precursors for hepatic gluconeogenesis but much of the glucose synthesized will be lost in the urine. (This provides the metabolic explanation for the 'observation' of Artaeus, made almost 2000 years ago, that the flesh is excreted through the urine.) Not only is there a loss of fluid resulting in dehydration and a reduction in extracellular fluid volume (hypovolaemia) but also a loss of ions especially phosphate and potassium. This is explained by the acidosis that results from the formation of fatty acids, and ketone bodies; the hypovolaemia which accompanies the loss of water can exacerbate the acidosis due to peripheral hypoxia and resultant lactic acid formation (see Chapter 13; Section E). The kidney controls the pH of the blood by excretion of protons into the urine, which has to be buffered. This is achieved by excretion of phosphate and production of ammonia (see Chapter 13 for detailed description of mechanism). Phosphate buffers the urine as follows:

$$HPO_4^{2-} + H^+ \rightarrow H_2PO_4^-$$

but the dihydrogenphosphate must be excreted along with a counter cation, sodium or potassium, which, together with the phosphate, are lost from the body. The loss of potassium and sodium ions may be considerable (Sestoft *et al.*, 1980). Replacement of fluid and ions in addition to the administration of insulin, is important in the acute treatment of the diabetic ketoacidotic patient (see Alberti and Hockaday, 1977).

F. CURRENT TREATMENT OF DIABETIC PATIENTS

There is no doubt that insulin treatment can be described as a cure for insulin-dependent diabetes. Not surprisingly, since the original use of the hormone, there have been many improvements in the types of insulin used and in the management of the diabetic patient (see Bloom, 1977; Sönksen, 1977).

1. Insulin administration

The first aim in the management of the diabetic patient is to prevent excessive hyperglycaemia, with the *sequelae* indicated above, and hypoglycaemia with its inherent dangers of the loss of consciousness and coma. To this end, treatment usually consists of injections of either a mixture of short- and long-lasting insulin[8] twice a day (before breakfast and after the evening meal) or short-acting insulin before each main meal plus an injection of long-lasting insulin in the evening to provide basal insulin levels through the night (see Oakley *et al.*, 1978). In normal subjects, the maximum concentration of insulin in the blood occurs 30–60 min after a meal which coincides with the period of rapid glucose absorption. However, subcutaneous injection of short-acting insulin results in the maximum blood insulin concentration 2–3 hr after injection, which does not normally coincide with maximum blood glucose concentrations which are, therefore, much higher than normal. Furthermore, hypoglycaemia can be a problem 2–3 hr after a main meal. Thus, although this treatment prevents extreme variations in the blood glucose concentration, it does not maintain the blood glucose concentration within the narrow limits of the normal subject. There is some evidence that minimizing the variations in the blood glucose concentration slows the development of, or even prevents, the secondary complications of the diabetic patient (microangiopathy, macroangiopathy, and neuropathy; see Cahill *et al.*, 1976; Job *et al.*, 1976; Tchobroutsky, 1978 and Section G).

In order to try and reduce these variations in the blood glucose concentration, biomedical engineers have been attempting to develop an artificial pancreas. Two general types have been developed. The most complicated is the closed-loop feedback control system in which insulin is delivered in response to measured changes in the blood glucose concentration. The system comprises a pump for injection of insulin, a sensor for measuring the blood glucose concentration (or the change in concentration) and a computer for calculating the amount of insulin to be injected in response to any change and for providing instructions to the pump. At the present time, the systems are large, non-portable, expensive and extracorporeal, so that they can only be used for short periods. They have, however, been particularly useful, for example, in treatment of ketoacidosis, and in the management of diabetics during surgery and labour (Schwartz *et al.*, 1979; Nattress *et al.*, 1978). The current direction of research in this field is the development of much smaller systems that can be implanted within the body and replaced only when they fail. Such systems have been used in animals but there are major problems to be overcome before they will be 100 % reliable for use in man.

Perhaps more success has been achieved with the open-loop control system,* which is simpler because a glucose measurement is not required. Insulin is infused continuously but the rate of infusion can be varied by the patient according to the conditions; for example, the rate is increased at mealtimes and varies with the size of the meal (Irsigler and Kritz, 1979). This system has had some success in the treatment of unstable ('brittle') diabetes. However, there is controversy over whether it provides better control of the concentrations of blood glucose compared to that provided by multiple daily injections (that is, intensified conventional therapy—Rizza *et al.*, 1980).

An alternative to either of the artificial systems is the transplanted pancreas. Two approaches are being tried, transplantation of islets and transplantation of the tail of the pancreas. A particular advantage is that the transplant could be

---

* These systems use either the intravenous or subcutaneous route for infusion. The former carries with it the dangers of infection and venous thrombosis so that subcutaneous infusion may provide the safest long term mechanism for good control for the majority of diabetic patients (Pickup *et al.*, 1978).

placed so that secretions would enter the hepatic portal vein and the liver could be exposed to more normal concentrations of insulin than is the case with insulin injections. There are, however, the major problems common to all transplantation treatment, inadequate sources of islet tissue and rejection (*Lancet* editorial, 1979a; Garvey *et al.*, 1979).

Two totally different approaches to the ones indicated above involve physiological manipulations that reduce the demand by the patient for insulin so that injections, when they are made, are more effective in control of the blood glucose concentration; these are exercise and inclusion of fibre in the diet.

2. Exercise

It has been shown for many years that physical exercise is beneficial in the treatment of diabetes mellitus (see Vranic and Berger, 1979; Richter *et al.*, 1981). Although the precise mechanisms through which exercise is effective are not understood, several suggestions have been made. Exercise can increase the rate of absorption of insulin into the bloodstream from the subcutaneous site of injection and this would be beneficial after a meal although it could promote hypoglycaemia between meals (Lawrence, 1926). Exercise increases the rate of utilization of all fuels if they are present in the blood, glucose, fatty acids and ketone bodies (Felig and Wahren, 1975), which will obviously improve the hyperglycaemia. In addition, exercise reduces the glycogen content of the liver and muscle (Chapter 9; Section B.2) which will increase the activity of glycogen synthase and reduce the requirement for insulin in the stimulation of glycogen synthesis after a meal (Chapter 16; Section C.1). Finally, in normal subjects, endurance-training markedly increased insulin sensitivity, so that very much

Table 15.4. Plasma glucose and insulin concentrations in response to oral glucose in well trained middle-aged athletes and control middle-aged subjects*

	Concentrations			
	control subjects		athletes	
Time after ingestion of glucose (min)	glucose (mM)	insulin (μU.cm^{-3})	glucose (mM)	insulin (μU.cm^{-3})
0	3.5	10	4.0	22
30	6.7	79	5.9	32
60	6.0	95	4.4	34
90	4.7	96	3.8	24
120	3.8	55	4.2	21
Summation	27.7	337	22.3	112

* Data from Björntorp *et al.* (1972). The study involved 15 well-trained men aged 52–56 and controls, who were similarly aged and from the same region of Sweden. The subjects were given 100 g of glucose solution to drink after overnight starvation.

lower concentrations of insulin are required to control the blood glucose concentration after an oral glucose load (Table 15.4; Bjorntorp *et al.*, 1972). Similar findings have been obtained in experimental animals (Berger *et al.*, 1979). If these findings can be extrapolated to the diabetic patient, much smaller doses of injected insulin should be adequate for control in the endurance-trained patient and, moreover, if any β-cell function remains in the patient, this could play a larger role in control and hence reduce the variations in blood glucose concentration.

3. Dietary fibre

Insulin is secreted after a meal in order to reduce the rate of the mobilization of fuel reserves and to ensure that excess fuel is converted into storage compounds (see Chapter 14; Section C.1). In general, the more rapid the digestion and absorption of the food the greater is the change in the blood fuel concentrations (especially glucose) and consequently the larger is the requirement for insulin. Hence the type of carbohydrate ingested by the diabetic patient is important. The ingestion of sugars such as glucose or sucrose should be avoided (except for hypoglycaemic emergencies) and only carbohydrate that is slowly digested (e.g. starchy foods) should be eaten by diabetic subjects. A logical extension of this is to increase further the time of digestion by use of the natural agent, fibre (see Chapter 5; Section A.3). Consequently, there is considerable interest in the observation that the presence of fibre, especially the gums, in the diet reduces the rate of absorption of glucose and hence the peak concentrations of this sugar in the blood in both normal and diabetic subjects, and reduces urinary excretion of glucose and ketone bodies in diabetic patients (Jenkins *et al.*, 1976, 1978, 1979; Jenkins, 1979; Simpson *et al.*, 1981). Furthermore, dietary fibre improves the control of the blood glucose concentration and reduces the peak concentration of insulin in non-insulin-dependent diabetics.

It is possible that better control of the blood glucose concentration could be obtained in many diabetic patients by a carefully planned exercise regime together with a fibre-enriched diet. This could so increase the effectiveness of endogenous insulin that the dependence on exogenous insulin would be less and good control more readily achieved with only minimal help from insulin injections.

G. LONG-TERM COMPLICATIONS OF INSULIN-DEPENDENT DIABETES MELLITUS

Before the discovery of insulin, almost half of all diabetics died in coma. With the advent of insulin this has decreased to about 1 %. The problems of ketotic coma, and the treatment of these patients by insulin, have been considered previously (Chapter 13; Section E.2). Apart from the possibility of ketotic coma or hypoglycaemia, and the need for daily injections, the major problem for the diabetic patient is the long term complications of the disease, which affect the arteries, capillaries, eyes, kidneys, and nerves. These can be separated into three groups, although all may be causally related: accelerated atherosclerosis, sometimes known as macroangiopathy (see Chapter 20; Section G.1.a); microangiopathy, which is a thickening of the basement membrane and hence an increased size of the endothelial lining of arterioles and capillaries; and a collection of seemingly unrelated problems such as premature senile cataract and neuropathy (Renold *et al.*, 1978; Tchobroutsky, 1978; Brownlee and Cerami, 1981). The importance of these problems is emphasized by the fact that most diabetics die prematurely of kidney failure, heart attack, or a stroke. In addition, there is some evidence that poor control of blood glucose concentration in pregnant women increases the incidence of congenital abnormalities in their babies.[10]

It is possible that the microangiopathy can explain many of the chronic problems of the diabetic patient. At the same time as the thickening of the basement membrane, there is an increase in the permeability of the capillaries so that plasma proteins, especially albumin, escape more rapidly into the interstitial space. It is not clear if the thickening of the membrane causes the increase in permeability or whether the two phenomena occur at the same time. The consequences of the microangiopathy will obviously depend on the location of the affected vessels. In the retina, the increased size of the small blood vessels, increased permeability, increased incidence of aneurysms, and, because of the poor provision of fuel and oxygen, new vessel formation, all interfere with light absorption so that impairment of sight and eventual blindness can ensue. These changes are known as retinopathy. In the kidney, the thickening of the basement membrane interferes with the normal filtration properties of the capillaries in the glomeruli, resulting in proteinuria and a decrease in glomerular filtration rate. As the severity increases, the filtration can become so poor that toxic endproducts accumulate in the blood. Finally, total kidney failure occurs, which can only be resolved by dialysis or kidney transplantation. Microangiopathy of the small vessels in the lower leg and foot gives rise to intermittent claudication (severe pain in the legs on walking which disappears upon rest), rest pain, and necrosis or gangrene, which is usually precipitated by a minor local injury. The problems can be compounded by peripheral neuropathy, so that there is a loss of sensory function and damage to the feet is less easily noticed by the patient (see below). This problem is sometimes known as 'diabetic foot'.

The metabolic reason for the thickening of the basement membranes is not known but there is an increase in the amount of glycoprotein in, for example, glomeruli of the kidneys of diabetic patients and evidence of increased synthetic activity in the membrane. This is probably due to an increased rate of synthesis of the protein moiety of the glycoprotein of the basement membrane. The glycoprotein is collagen-like and has a high proportion of lysine and proline residues, many of which are hydroxylated. This process occurs after the peptide bonds have been synthesized (a post-translational event) and is catalysed by an hydroxylating enzyme(s) which, it has been suggested, increases in activity in the basement membranes of diabetics (see Tchobroutsky, 1978). However, it has also been suggested that there is an increased rate of synthesis of the carbohydrate moiety of the glycoprotein which may increase the overall synthesis of glycoprotein. Spiro and Spiro (1971) have shown that the activity of a glucosyltransferase is increased in the renal capillary basement membrane of diabetic animals and man. This enzyme is involved in the synthesis of a disaccharide of glucose and galactose that is attached to the protein via hydroxylysine. It is possible that an increase in the intracellular concentration of glucose could lead to an increase in the activity of enzymes such as the

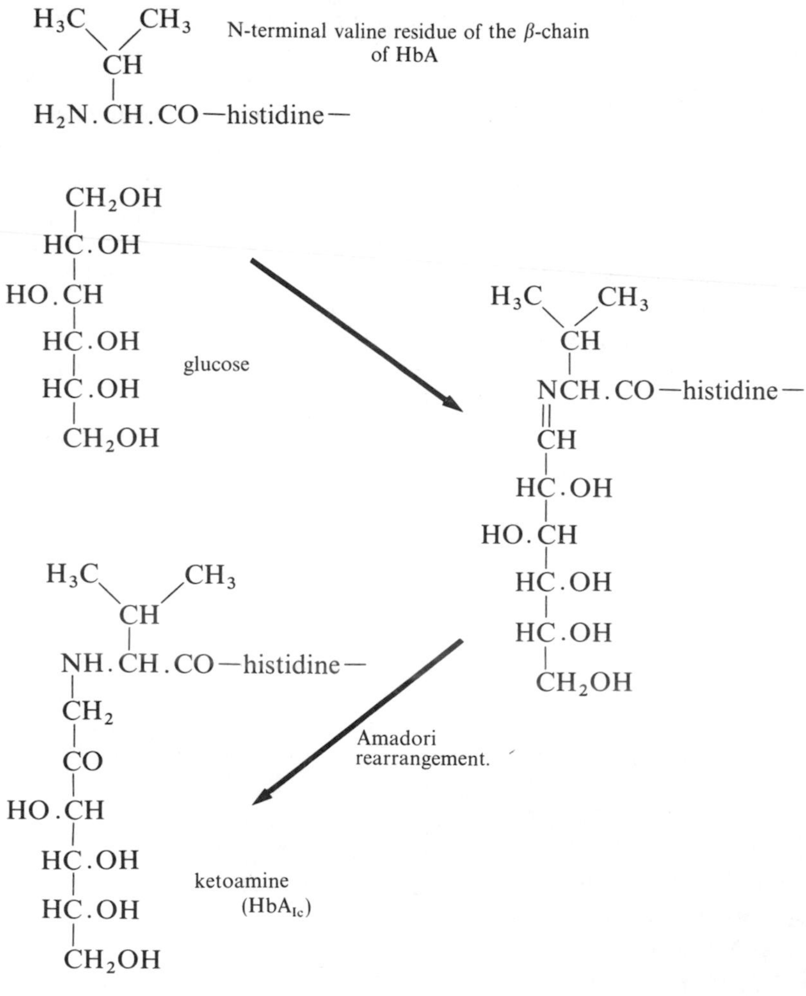

$$H_3C \diagdown \diagup CH_3$$
$$CH$$
$$|$$
$$H_2N.CH.CO{-}histidine{-}$$

N-terminal valine residue of the β-chain of HbA

$$CH_2OH$$
$$|$$
$$HC.OH$$
$$|$$
$$HO.CH$$
$$|$$
$$HC.OH$$
$$|$$
$$HC.OH$$
$$|$$
$$CH_2OH$$

glucose

$$H_3C \diagdown \diagup CH_3$$
$$CH$$
$$|$$
$$NCH.CO{-}histidine{-}$$
$$||$$
$$CH$$
$$|$$
$$HC.OH$$
$$|$$
$$HO.CH$$
$$|$$
$$HC.OH$$
$$|$$
$$HC.OH$$
$$|$$
$$CH_2OH$$

$$H_3C \diagdown \diagup CH_3$$
$$CH$$
$$|$$
$$NH.CH.CO{-}histidine{-}$$
$$|$$
$$CH_2$$
$$|$$
$$CO$$
$$|$$
$$HO.CH$$
$$|$$
$$HC.OH$$
$$|$$
$$HC.OH$$
$$|$$
$$CH_2OH$$

ketoamine (HbA$_{1c}$)

Amadori rearrangement.

Figure 15.2 Reactions involved in the formation of the modified haemoglobin (Hb A$_{1c}$) in diabetic patients

glucosyltransferase of the basement membrane and thus lead to an increase in the synthesis of the disaccharide component of the glycoprotein. This could then result in increased synthesis of the protein moiety of the glycoprotein. Alternatively, the elevated glucose concentration itself might increase the rate of protein synthesis in the basement membrane.

Finally, the diabetic patient also suffers damage to nerves leading to a loss of both sensory and motor function. Since this condition particularly affects the peripheral nerves, especially in the foot, it is known as peripheral neuropathy and results in paraesthesia, pain, and loss of sensation especially to high temperature often resulting in the injury remaining unnoticed. However, the sympathetic system can also be affected leading to nocturnal diarrhoea, postural hypotension, impotence in the male due to erectile dysfunction, and disturbance of bladder function (Levine, 1976; Jackson and Vinik, 1977). The basis of the malfunction of the nerves appear to be demyelination caused by damage to the Schwann cells and may be explained as follows. The hyperglycaemia causes an increase in the intracellular glucose concentration of the Schwann cells, and this increases the rate of synthesis of fructose and sorbitol by the polyol pathway (see Chapter 13; Section E.2). These sugars are not readily transported out of the cell so that the intracellular osmotic pressure increases, which causes water and sodium ions to move into the cells, thus increasing the intra-cellular volume, damaging membranes and intracellular structures and disturbing the ion balance. Although this suggestion is still somewhat speculative for the nerve, there is strong evidence for this as one mechanism[9] producing opacity in the lens of diabetic patients (see Thomas and Ward, 1975; Chapter 5; Section F.2.b).

Some support for the idea that an elevated concentration of glucose can cause changes in the structure of protein has been obtained from an unexpected source, haemoglobin. Diabetic patients have several unusual haemoglobin molecules in their blood, haemoglobins A_{1a}, A_{1b}, and A_{1c}. The structure of haemoglobin A_{1c} has been elucidated; the N-terminal valine of the haemoglobin β-chain has reacted with a glucose molecule to form a stable ketoamine linkage (Figure 15.2; Koenig, *et al.*, 1977). This reaction occurs spontaneously in the erythrocyte after haemoglobin has been synthesized and the rate is increased by an elevated blood glucose concentration. The fact that the reaction is slow and irreversible has given an indirect, but a more functional, basis for assessing the control of the glucose concentration; a low level of haemoglobin A_{1c} is indicative of good control[10] (Boden *et al.*, 1980).

NOTES

1. In the oral glucose tolerance test, 75 g (or $1.75 \, \text{g.kg}^{-1}$ body wt.) of glucose is dissolved in flavoured water (about 400 ml) and taken orally over a period of 5–10 min after an overnight fast. Venous or capillary (e.g. fingertip) blood is drawn before and at time intervals (30, 60, 90, 120 min) after ingestion of glucose. Not surprisingly, many factors can affect the glucose tolerance test, including prior dietary history so that patients are usually asked to take a high carbohydrate diet for at least three days before the test

and no food at all 8–12 hr immediately prior to the test. In addition to the oral test, an intravenous glucose tolerance test is occasionally used: 25 g of glucose (in 50 cm^3) is rapidly injected intravenously over 2–4 min and samples are withdrawn before and at 5 min or 10 min intervals up to an hour after the midpoint of the injection.

2. On the basis of the HLA types associated with the disease, it is suggested that at least two genes are predisposing for insulin-dependent diabetes mellitus (Cahill and McDevitt, 1981).

3. From 1976–1978 approximately 20 cases of insulin-dependent diabetes mellitus have been reported after the ingestion of the rodenticide Vacour (*N*-3-pyridylmethyl-*N'*-*p*-nitrophenylurea) (Karam *et al.*, 1978).

4. The authors consider that this effect on glucose transport is given too much emphasis in discussion on the control of carbohydrate metabolism by insulin. Although it plays some role in the regulation of glucose utilization, the metabolic activity of the tissue and the oxidation of alternative fuels may play a more quantitatively important role in the regulation of glucose transport and glycolysis (see Espinal *et al.*, 1983; Chapter 8; Section B.1).

5. The rate of glycogen synthesis is also controlled by the level of glycogen within the cell which will modify the extent of the response to insulin. When the content of glycogen is high the activity of glycogen synthase is low and insulin will have little effect on the rate of synthesis (see Chapter 16; Section C.1).

6. These effects, apparent from studies with isolated perfused liver (see Exton, 1972) and the intact animal (Cherrington *et al.*, 1978) account for the fact that insulin reduces the rate of glucose release and may facilitate an increase in the rate of glucose uptake by the liver.

7. Insulin appears to be the only hormone that is antilipolytic (apart from local hormones—such as prostaglandins, see Table 8.5). Consequently, there is no other agent that can adapt in the diabetic patient to take over the role of insulin. Similarly, it is perhaps the only hormone that acutely stimulates the rate of protein synthesis.

8. Insulin can be prepared in such a way as to increase the time during which it is active. Normal soluble preparations of the hormone (short-lasting insulin) have a maximal effect for about 2–4 hr after injection. However, combination of insulin with other proteins (protamine or globin), so that hydrolysis is necessary for activity, or insoluble zinc insulin (administered as a suspension), which is only slowly absorbed into the blood, have a maximum effect for up to 12 hr (Bloom 1977).

9. Another proposed mechanism is that non-enzymic glycosylation of the protein of the lens (similar to glycosylation of haemoglobin) leads to a modification of the structure of these proteins which could favour disulphide bond formation between proteins and result in high molecular weight aggregates capable of scattering light (see Brownlee and Cerami, 1981).

10. The attachment of the glucose molecule to haemoglobin interferes in the binding of 2,3-bisphosphoglycerate so that the oxygen affinity of these haemoglobin molecules is increased. Since in diabetic patients the concentration of 2,3-bisphosphoglycerate may also be reduced, the rate of oxygen delivery to peripheral tissues may be seriously reduced leading to hypoxia and hence the formation of lactic acid. Of particular importance is the evidence that the higher incidence of congenital abnormalities in children born of diabetic mothers (compared to the incidence in the general population) may be caused by poor control of the blood glucose concentration. In a study of 115 births of insulin-dependent diabetic mothers, the major congenital abnormalities were correlated with the concentration of haemoglobin A_{1c} in the blood of the mother (Miller *et al.*, 1981). Improvement in the control of the blood glucose concentration may therefore be particularly important for mothers in the early stages of pregnancy (Freinkel, 1981).

CHAPTER 16

THE SYNTHESIS OF GLYCOGEN AND OTHER POLYSACCHARIDES

For most purposes glycogen can be considered as a polymer of glucose and details of its structure will be found in Chapter 5; Section C.1. Although most tissues contain some glycogen, and the pathway of synthesis is the same in all, the quantitatively important stores are found only in muscle and liver (Table 8.1). The role of glycogen in these two tissues has been discussed in Chapters 5, 8, 9, and 11.

A. THE PATHWAY OF GLYCOGEN SYNTHESIS

It is usually considered that the pathway of glycogen synthesis begins with phosphorylation of glucose to form glucose 6-phosphate. In the liver this reaction is catalysed by glucokinase whereas in the muscle and other tissues the reaction is catalysed by hexokinase. The enzyme phosphoglucomutase converts the glucose 6-phosphate to glucose 1-phosphate which then reacts with UTP to form UDPglucose, the immediate precursor for glycogen synthesis. The latter reaction is catalysed by glucose-1-phosphate uridylyltransferase (also known as UDPglucose pyrophosphorylase). As the latter name implies, the products of the reaction are UDPglucose and pyrophosphate (Figure 16.1). The structure of UDPglucose may be more familiar to the reader if it is considered as a dinucleotide (Chapter 4; Section C.1.e) in which the second sugar is glucose but the second base is missing. The nucleotide is the glucose donor for transfer onto the end of an uncompleted chain in the glycogen molecule and this reaction is catalysed by the enzyme glycogen synthase.* In more detail, the synthase catalyses the formation of a 1,4-α-glycosidic linkage by condensing the hydroxyl group on the 1-position of the glucose on the UDPglucose molecule with the hydroxyl group at the 4-position of a glucose on the growing chain of the glycogen molecule. The sequence of reactions is given in Figure 16.1.

If these reactions were the only ones involved in the synthesis of glycogen, they would result in a linear chain of glucose residues, whereas glycogen has a branched tree-like structure (Figure 5.13). This is achieved by the action of the glycogen branching enzyme (1,6-α-glucosyltransferase), which catalyses the

---

* The synthase reaction also produces UDP, which is reconverted back to UTP in a reaction with ATP catalysed by the enzyme nucleosidediphosphate kinase.

$$\text{ATP}^{4-} + \text{UDP}^{3-} \longrightarrow \text{ADP}^{3-} + \text{UTP}^{4-}$$

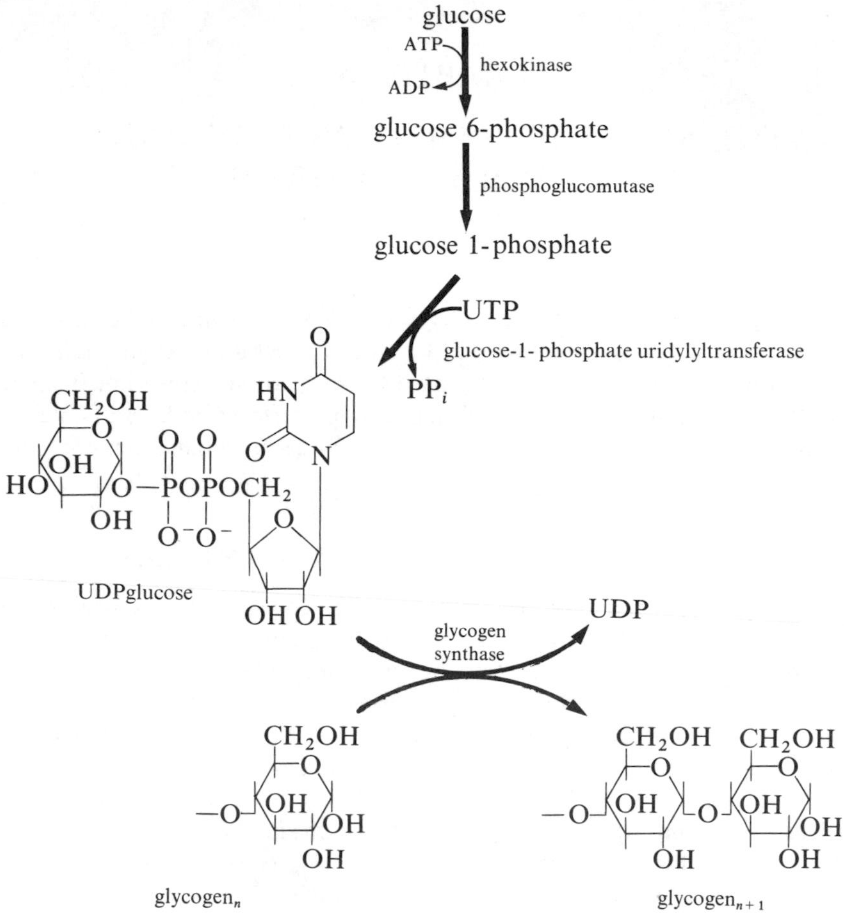

Figure 16.1 Pathway for glycogen synthesis

transfer of a sequence of seven glucose residues from a chain of at least eleven residues (that is, glucose molecules linked by 1,4-α-glycosidic linkages) onto the 6-position of a glucose residue in a neighbouring chain. The new branch is established on a chain so that its nearest branch point neighbour is at least four residues away. The new branch and the 'stub' from which it was obtained can now grow through further activity of glycogen synthase. This ensures the branched nature of glycogen which is probably essential for its function and for the structure of the glycogen particle (Chapter 5; Section C).

1. The precursor for glycogen synthesis

The enzyme system described in Figure 16.1 can only synthesize glycogen if a primer is already present to which glucosyl units can be added. *In vitro* studies

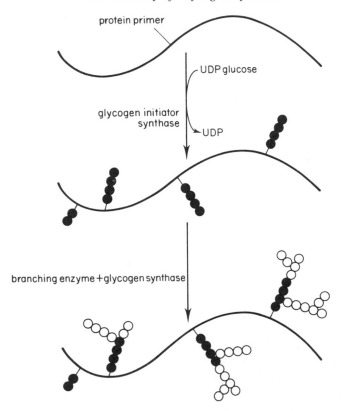

Figure 16.2 Model for synthesis of glycogen on a protein primer (after Matcham *et al.* 1978)

have shown that a branched oligosaccharide which possesses several reducing ends, to which addition of glucosyl units can take place, constitutes an excellent primer. Although the existence of such a primer *in vivo* cannot be ruled out, there is now evidence that the physiological primer is in fact a protein bearing several covalently attached oligosaccharide groups (Krisman and Barengo, 1975; Matcham *et al.*, 1978). This glucosylated primer is synthesized in a reaction that requires a further enzyme, tentatively known as glycogen initiator synthase, which catalyses the successive transfer of several glucosyl units from UDPglucose to the acceptor protein, forming small, linear, 1,4-α-oligosaccharides (Figure 16.2).

In addition to its role as a primer, it is possible that the glucosylated protein may serve as a site of temporary attachment for the glycogen synthase and branching enzymes. Furthermore, since the individual glycogen molecules may remain attached to the primer after their completion, the primer could be involved in the formation of glycogen particles.

Table 16.1. Free energy changes in some of the reactions involved in glycogen synthesis†

Reaction catalysed by	ΔG (kJ.mol^{-1})	Suggested equilibrium nature
Hexokinase (glucokinase)	−31.9	non-equilibrium
Phosphoglucomutase	− 0.4	near-equilibrium
Glucose 1-phosphate uridylyltransferase	3.4	near-equilibrium
Glycogen synthase	—	non-equilibrium*
Branching enzyme	—	non-equilibrium*

* Equilibrium constant not known but considered to be large.

† Calculations are based on concentrations of metabolites reported for the rat liver (Williamson and Brosnan, 1974).

2. The physiological pathway of glycogen synthesis

The free energy changes of some of the reactions involved in glycogen synthesis are given in Table 16.1 from which it can be seen that hexokinase (or glucokinase), glycogen synthase and branching enzyme catalyse non-equilibrium reactions whereas phosphoglucomutase and the uridylyltransferase catalyse near-equilibrium reactions. Of the non-equilibrium reactions, only glycogen synthase approaches saturation with pathway-substrate (UDPglucose). This therefore is the flux-generating reaction so that the physiological pathway for glycogen synthesis starts at UDPglucose and comprises only two reactions, those catalysed by the synthase and the branching enzyme. However, since the concentration of UDPglucose is low in comparison to the rate of glycogen synthesis, its concentration must be maintained whenever glycogen is being synthesized. This is achieved by the reactions catalysed by glucose-1-phosphate uridylyltransferase and phosphoglucomutase, utilizing glucose 6-phosphate produced from the glycolytic or (in liver) gluconeogenic pathways (Figure 16.3).

B. SOURCE OF GLUCOSE RESIDUES FOR GLYCOGEN SYNTHESIS

1. Liver

There is a marked increase in the glycogen content of muscle, and especially the liver, after a meal and the increase is greater if the meal contains a large amount of digestible carbohydrate (Hultman, 1978). Not surprisingly, it is commonly accepted that blood glucose is the precursor for glycogen synthesis in these tissues and for the liver there are at least two lines of evidence that support this view. First, measurement of arterio-venous differences across the liver in human subjects indicates that a large proportion (60%) of an oral glucose load is taken up by the liver and it can be assumed that most of this is converted to glycogen (Felig *et al.*, 1976). Indeed, direct measurement of glycogen in biopsy samples of

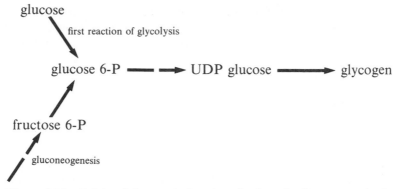

Figure 16.3 Origin of glucose 6-phosphate for hepatic glycogen synthesis

the liver of human volunteers demonstrates that glycogen is synthesized in the liver after a meal (Hultman, 1978). Secondly, liver, unlike muscle, contains a high activity of the enzyme glucokinase, the properties of which well suit it to its specific role of removing glucose from the blood for glycogen synthesis (see Chapter 11; Section E.1.a). On the other hand, liver possesses an alternative source of precursor for glycogen synthesis, namely the pathway of gluconeogenesis by which non-carbohydrate precursors are converted to glucose 6-phosphate. When glycogen is formed from this, the over-all process is sometimes known as glyconeogenesis. There are several lines of evidence to suggest that glyconeogenesis is quantitatively important in the deposition of glycogen in the liver after a meal. First, the rate of glycogen synthesis in *in vitro* preparations (that is isolated perfused rat liver or isolated incubated hepatocytes) is higher when lactate or alanine are supplied as precursors than with glucose (Hems *et al.*, 1972; Clark *et al.*, 1973). Secondly, experiments have been performed with isolated hepatocytes and the perfused liver which attempt to mimic the *in vivo* situation in a starved and re-fed animal. In the starved situation, lactate and alanine were converted into glucose. Addition of a high concentration of glucose did not reduce the rate of gluconeogenesis from lactate or alanine but directed glucose 6-phosphate to glycogen (Hems *et al.*, 1972; Boyd *et al.*, 1981). If these experiments can be extrapolated to the *in vivo* situation, they suggest that the absorption products of a meal lead to glycogen synthesis in the liver, not from the glucose present in the portal vein, but from lactate, alanine, other amino acids, and glycerol, derived from metabolism in the intestine (Chapter 5; Section A.2.c).

The contradiction between the two experimental approaches could be explained if the hormones that are released by the gastrointestinal tract (especially the duodenum) during digestion and absorption of a meal (e.g. gastric inhibitory peptide) facilitate glucose uptake and glycogen deposition in the liver (Morgan, 1980). Evidence supporting this comes from experiments with human subjects in which increasing both the blood glucose and the insulin concentrations by intravenous infusion produced only a small increase in hepatic glucose uptake. However, if the glucose was given orally, the hepatic uptake of glucose increased six-fold (DeFronzo *et al.*, 1978b). Such gastrointestinal factors appear to play an important role in modifying metabolism of the body in accord with the contents of the meal so that when the digested products are absorbed they are rapidly and smoothly assimilated (see also Chapter 14; Section C.1).*

2. Muscle

Glucose is readily converted into glycogen by isolated muscle preparations. The feeding of glucose labelled with carbon-14 at specific carbon atoms to experimental animals has shown that muscles incorporate glucose into glycogen without extensive metabolism and, in particular, without preliminary cleavage into 3-carbon compounds.

The possibility that muscle can also synthesize glycogen from non-carbohydrate precursors has been discussed in the literature for 50 years. Certainly, it is well established that muscle does not carry out gluconeogenesis, that glucose is readily converted into glycogen in muscle, and that lactate produced in muscle can be released into the bloodstream and transported to the liver for glucose or glycogen synthesis (that is, the Cori cycle). However, there is increasing evidence that, provided the glycogen concentration of the muscle is low and that of lactate is high, glycogen synthesis from lactate can occur in muscle and that this could be quantitatively important. Careful experiments by Hermansen and Vaage (1977) have shown conclusively that this pathway operates in quadriceps muscle of man. After volunteers had sprinted at maximum rate for short periods and then rested, measurements of arterio-venous differences across their leg muscles showed that very little lactate escaped into the bloodstream and virtually no glucose was removed from the blood. However, biopsy samples of the muscles showed that during rest the lactate concentration fell and the glycogen concentration rose (Table 16.2). Furthermore, activities of the enzymes catalysing the non-equilibrium reaction of gluconeogenesis (e.g. fructose-bisphosphatase, phosphoenolpyruvate carboxykinase, and pyruvate carboxylase) are present in human muscle at sufficient activities to account for the observed rate of glycogen synthesis (Table 9.4). Glyconeogenesis has also been demonstrated in isolated rat and frog muscle, especially when the concentration of glycogen is low and that of lactate is high (Bendall and Taylor, 1970; McLane and Holloszy, 1979).

The authors believe that the ability of muscles to resynthesize glycogen from lactate reflects the fact that substrate cycles are important in control of the rate of glycolysis both before and after exercise. When the concentration of lactate is high and that for glycogen is low, and cycling is taking place, the activities of the 'glyconeogenic' enzymes (pyruvate carboxylase, phosphoenolpyruvate carboxy-kinase, fructose-bisphosphatase and glycogen synthase) could exceed those of glycolysis, resulting, incidentally, in the conversion of lactate to glycogen.

---

* For review of gastrointestinal hormones see Rayford *et al.*, 1978; Mutt, 1980.

Table 16.2. Concentrations of glycogen, lactate, ATP, and phosphocreatine in quadriceps muscle of man before and immediately after exhaustive sprinting activity*

Time after exercise (min)	Concentration in muscle (μmol.g^{-1} fresh wt.)			
	Glycogen	Lactate	ATP	Phosphocreatine
0	88	1.1	4.6	17.0
0.25	58	30.5	3.4	3.7
5	55	25.0	3.3	13.7
10	61	22.0	3.7	17.9
20	68	11.0	4.1	17.7
30	70	6.5	4.0	18.8

* Data from Hermansen and Vaage (1977).

C. REGULATORY MECHANISMS IN GLYCOGEN SYNTHESIS

The fact that glycogen synthase catalyses a flux-generating step indicates that knowledge of its properties is necessary to formulate a theory of control. Glycogen synthase in both the muscle and liver exists in active and less active forms and, in contrast to phosphorylase, the less active forms are phosphorylated. The rates of interconversion between the phosphorylated and non-phosphorylated forms control the glycogen synthase activity. The properties of the interconverting systems differ in muscle and liver, undoubtedly reflecting the different physiological importance of glycogen storage in these tissues.

1. Regulation of glycogen synthesis in muscle

Glycogen synthase in muscle exists in a non-phosphorylated and in several phosphorylated forms. At physiological concentrations of glucose 6-phosphate (0.1–0.5 mM) the phosphorylated forms are much less active than the non-phosphorylated form. There are at least three protein kinase enzymes that can phosphorylate, and hence lower the activity of, the synthase; these are known as glycogen synthase kinase-1, -2, and -3 (abbreviated to GSK_1, GSK_2, and GSK_3). An additional complexity is that GSK_1 is, in fact, cyclic AMP-dependent protein kinase and GSK_2 is phosphorylase kinase; the third kinase, GSK_3, may be specific for glycogen synthase (Rylatt *et al.*, 1980). This obviously affords some common control between phosphorylase and glycogen synthase so that when the former is activated the latter is inhibited (Figure 16.4). The three kinases phosphorylate different serine residues on the synthase enzyme leading to different forms of the phosphorylated enzyme, all of which possess lower activities than the non-phosphorylated form[1] (Figure 16.5 and Table 16.3).

The dephosphorylation of all the phosphorylated forms is achieved by a single phosphatase enzyme, glycogen synthase phosphatase, also known as protein phosphatase-I since it has hydrolytic activity towards other phosphoproteins.

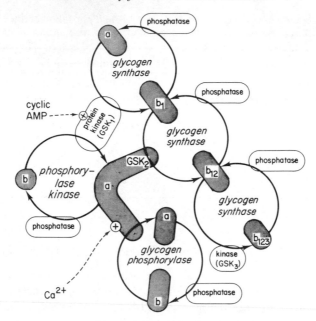

Figure 16.4 Inter-relationships in the control of glycogen synthase and glycogen phosphorylase by interconversion cycles. The three glycogen synthase kinases can act in any order (see Figure 16.5)

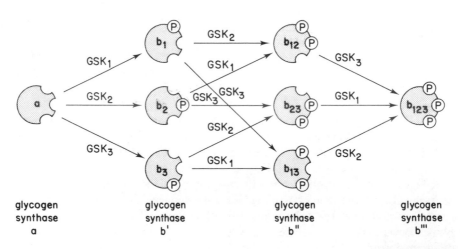

Figure 16.5 Patterns of phosphorylation which reduce the activity of glycogen synthase in muscle. The phosphorylations are catalysed by a series of glycogen synthase kinases (GSK_1, GSK_2, and GSK_3) which can act in any order. The activity of glycogen synthase descreases with increasing phosphorylation (see Table 16.3). There may be other sites that are phosphorylated but the significance of these is not clear (Cohen, 1982)

Table 16.3. Properties and activities of glycogen synthases

Form of glycogen synthase	Moles of phosphate incorporated per mole of enzyme	Concn. of glucose 6-phosphate required for 50% maximum activity (mM)	Relative activity *in vivo*
a	none	0.05	+ + + + + + + +
b_1	one	0.2	+ + + +
b_2	one	0.2	+ + + +
b_3	one	0.7	+ + +
$b_{1,2}$	two	0.3	+ + + +
$b_{1,3}$	two	1.1	+ +
$b_{2,3}$	two	1.2	+ +
$b_{1,2,3}$	three	1.8	+

The catalytic activity of glycogen synthase will depend mainly on the proportion of the enzyme in the various forms since the concentration of glucose 6-phosphate changes very little in muscle. The activity will therefore be affected by the concentration of cyclic AMP, which activates cyclic AMP-dependent protein kinase (GSK_1), and Ca^{2+} which activates phosphorylase kinase (GSK_2). These properties explain how adrenaline, which raises the concentration of cyclic AMP, and muscular contraction which raises the concentration of Ca^{2+}, reduce the activity of glycogen synthase in muscle (Figure 16.4).

Two important factors, circulating insulin and a low glycogen content in the muscle, are of physiological importance in increasing the activity of glycogen synthase but it is unclear how they act at a molecular level. Evidence that glycogen synthase can be regulated by the glycogen content is provided by the negative correlation between the proportion of synthase in the active form and the glycogen content, which suggests that the activity of one or more of the interconverting enzymes can be influenced by the glycogen content.

Activation of glycogen synthase by insulin has been known since 1960 (see Villar-Palasi and Larner, 1968) and there has been intensive research to elucidate the mechanisms involved but, unfortunately, they remain unclear. Since insulin does not change the cyclic AMP concentration in muscle, it is unlikely that insulin inhibits GSK_1 and must, therefore, inhibit GSK_2 or GSK_3 or activate protein phosphatase (Cohen, 1980). However, there is no direct evidence to support any of these possibilities at present.

The complexity of control of glycogen synthesis in muscle reflects the importance of maintaining the muscle glycogen store. The store is important in providing energy for muscle, especially for sprinting activity (Chapter 9; Section A.2). Furthermore the total store (approximately 300 g) is greater than the average daily carbohydrate intake so that, if the glycogen content of muscle was low, a high rate of resynthesis could result in severe hypoglycaemia.

2. Regulation of glycogen synthesis in the liver

In comparison with muscle, there is less information on the enzymology of glycogen synthesis in the liver. Hepatic glycogen synthase exists in at least two forms and at physiological concentrations of ATP and phosphate the *a* form (non-phosphorylated) is fully active whereas the *b* form (phosphorylated) has a very low activity (Stalmans, 1976). Phosphorylation of hepatic synthase is carried out by cyclic AMP-dependent protein kinase, as in muscle, but there is no evidence of further phosphorylation by phosphorylase kinase or other protein kinase. The phosphorylated form is dephosphorylated (and hence activated) by a protein phosphatase which is, in contrast to muscle, a distinct enzyme from that which dephosphorylates hepatic glycogen phosphorylase (Laloux *et al.*, 1978). The properties of the interconverting enzymes provide the basis for a physiologically important mechanism of control which is quite different to that in muscle and which has been described briefly in Chapter 11.

In the liver, the control of synthase activity is intimately related to that of phosphorylase so that knowledge of the control of the latter enzyme is required (Chapter 11; Section E.1.c). As in muscle, hepatic glycogen phosphorylase exists in two forms, phosphorylase *a*, the active form, and phosphorylase *b*, the inactive form. Phosphorylase kinase catalyses the conversion of the inactive form to the active, whereas phosphorylase phosphatase catalyses the reverse reaction. Hepatic phosphorylase *a* possesses a binding site for glucose and, when glucose is

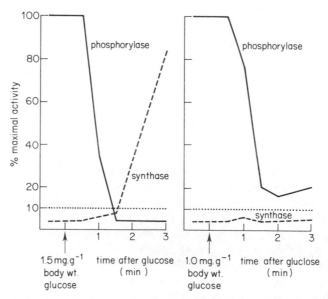

Figure 16.6 Activation of glycogen synthase by changes in the activity of glycogen phosphorylase. Note that if sufficient glucose is injected into the animal the activity of phosphorylase falls below 10 % of its maximal activity which leads to activation of glycogen synthase (After Stallmans, 1976)

bound, phosphorylase *a* is a better substrate for phosphorylase phosphatase. The concentration of glucose that produces a 50% effect is about 20 mM so that an increase in the concentration of glucose in the liver above the physiological level of 5 mM should lead to a decreased proportion of phosphorylase in the *a* form and a reduction in the rate of glycogenolysis. However, of importance for the control of glycogen synthesis is the fact that glycogen synthase phosphatase (the enzyme that activates glycogen synthase) is inhibited by phosphorylase *a*. This is probably due to some interaction between the two proteins. Hence a reduction in the activity of phosphorylase, due to an increase in glucose concentration in the liver, will lead to an activation of glycogen synthase. This provides a very effective and sensitive control mechanism, both for the rate of glycogenolysis and that of glycogen synthesis. Since the glucose transport process in the liver is near-equilibrium, an increase in the blood glucose concentration will rapidly and proportionally increase the concentration of glucose in the cells of the liver. This will reduce the rate of glycogenolysis and, if the increase in glucose concentration is large enough, increase the rate of glycogen synthesis. Some very elegant experiments have been performed by Hers and coworkers (see Hers, 1976, for review) which demonstrate that the proportion of phosporylase in the active form must fall below 10% (of the total phosphorylase) before glycogen synthase is activated (Figure 16.6). This ensures that an increase in the glucose concentration in the hepatic portal vein after a meal will first reduce the rate of glycogenolysis and then, if the concentration of glucose in the blood and hence in the liver remains high, increase the rate of glycogen synthesis (for mechanism, see Figure 11.9). The control of glycogen synthase via this 'glucose mechanism' is related to the regulation of the rate of glucose uptake by the liver as described in Chapter 11; Section E.1.c.

It is interesting to compare the mechanism of control of glycogen synthesis in liver and muscle. The major difference is that the hepatic control mechanism depends primarily on changes in intracellular glucose concentration, whereas in muscle the most important factors are the amount of glycogen already stored in the muscle and the blood concentration of insulin (Figure 16.7). There is some evidence that insulin plays a part in control of glycogen synthesis in the liver but this may be by reducing the proportion of phosphorylase in the active form.

D. SYNTHESIS OF OTHER CARBOHYDRATE COMPOUNDS

From a metabolic point of view, preoccupation with glycogen is excusable but it is not the only large carbohydrate molecule of importance in the body. Because of the greater problems they present in terms of purification and determination of structure, polysaccharides have, until recently, received much less attention than proteins. It is to be expected, however, that our rapidly improving knowledge of their structure and metabolism will lead to a better understanding of the many diseases involving aberrant polysaccharide metabolism (e.g. rheumatism and lysosomal deficiency diseases at least as numerous at the glycogenoses described

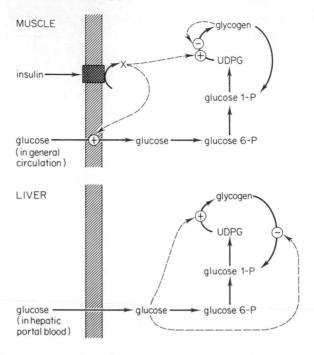

Figure 16.7 Comparison of the regulation of glycogen synthesis in muscle and liver. See text for details

in Section E). A survey of molecules of importance in the body which contain carbohydrates is given in Table 16.4. This section briefly describes the functions and synthesis of a few such carbohydrates; it shows that the same general mechanism is employed as that described above for glycogen synthesis. For further details the reader is referred to Candy (1980).

1. Lactose

Lactose is a disaccharide of glucose and galactose linked 1,4-β (see Figure 5.30). It is the only disaccharide synthesized in large quantities in mammals where its sole function is to provide the soluble sugar in milk. The synthesis occurs by the transfer of a galactosyl residue from UDPgalactose to a glucose molecule in a reaction catalysed by the enzyme lactose synthase:

$$\text{UDPgalactose} + \text{glucose} \longrightarrow \text{lactose} + \text{UDP}$$

The galactose is produced from glucose but the conversion actually occurs while the sugars are attached to UDP. Glucose is converted to UDPglucose in the cell of the mammary gland by glucose-1-phosphate uridylyltransferase (see Section A). Since the structural difference between glucose and galactose lies only in the position of the hydroxyl group on carbon atom 4, and the monosaccharide

Table 16.4. Example of carbohydrate-containing molecules in man

Class	Definition	Example	Function
Disaccharides	condensation of two monosaccharide units	lactose	soluble sugar in milk
Glycosides	condensation of monosaccharide with a non-sugar alcohol	steroid glucuronides (Chapter 20; Section C.2)	mostly detoxification products in man
Polysaccharides glucans	homopolymers of glucose	glycogen	storage of glucose residues
glycosamino-glycans	heteropolymers of a uronate and an amino sugar (plus a small proportion of protein)	hyaluronate	synovial fluid, vitreous body of eye
Carbohydrate/protein complexes proteoglycans (mucopoly-saccharides)	glycosaminoglycans attached to protein	chondroitin sulphate	ground substance in cartilage
glycoproteins	proteins with oligosaccharide chains attached	fibrinogen, immunoglobulin, collagen, glycophorin, mucin	blood-clotting, antibodies, structural protein, membrane protein, lubrication
Carbohydrate/lipid complexes glycolipid	oligosaccharides attached to sphingosine or ceramide (see Appendix 6.1)	cerebrosides, gangliosides	membranes

is attached to the nucleotide at carbon atom 1, the conversion can readily take place. The reaction is catalysed by UDPglucose 4-epimerase which requires NAD^+ (see Chapter 11; Section F.1.a):

UDPglucose   UDPgalactose

It is of interest to speculate on the presence of lactose rather than glucose in milk. One possible reason for the use of an 'unusual' sugar is that glucose metabolism in the mammary gland can be directed towards the synthesis of a specific sugar for the milk. However, this reasoning would lead one to predict the presence of a sugar such as galactose rather than lactose. The significance of a disaccharide may lie in its effective halving of the osmotic concentration (for a given energy content) in mammary gland cells.

2. Glycosaminoglycans

A large number of the polysaccharides that occur in mammalian connective tissue are glycosaminoglycans (see Table 16.4). They form a fairly homogeneous group of carbohydrates and are broadly defined by possession of the following characteristics.

1. They contain a repeating disaccharide unit, consisting of an amino sugar (such as glucosamine, usually N-acetylated) and a uronic acid (except in keratan sulphate).
2. The uronic acid is usually esterified with sulphate (except in hyaluronate).
3. The polysaccharide is usually covalently linked to a protein to form a proteoglycan (also known as a mucopolysaccharide).

Most glycosaminoglycan chains contain fewer than 100 monosaccharide units but there may be as many as 50 chains attached to a single core protein in the proteoglycan. Hyaluronate chains are longer and may contain 5000 monosaccharide residues.

(a) *Functions of glycosaminoglycans*

Glycosaminoglycans possess an exceptionally large negative charge density and therefore assume a random, extended, conformation. This accounts for the highly viscous solutions and gels (in which water molecules are 'trapped' in a loose framework of polysaccharide molecules) which these substances form and also for the lubricant action of, for example, hyaluronate, in synovial fluid. In combination with protein, the sulphates of chondroitin, dermatan and keratan (each with a distinctive monosaccharide composition, see Figure 16.8) form the ground substance of connective tissue. They provide the resistance of such tissues to compression, while the protein fibres of collagen and elastin embedded within them provide, respectively, the resistance to stretch and the elasticity characteristic of particular connective tissues. Heparin, although structurally similar to other glycosaminoglycans, does not occur in connective tissue but is stored in mast cells beneath the epithelium of blood vessels. Here it acts as an anticoagulant and as a structural component of lipoprotein lipase (see Chapter 6; Section C.1.a).

Like all other cell constituents, glycosaminoglycans undergo continual turnover in the tissues. Their degradation is brought about by lysosomal enzymes and a number of diseases are caused by the failure of this system to work smoothly. In rheumatoid arthritis, excessive release of lysosomal enzymes

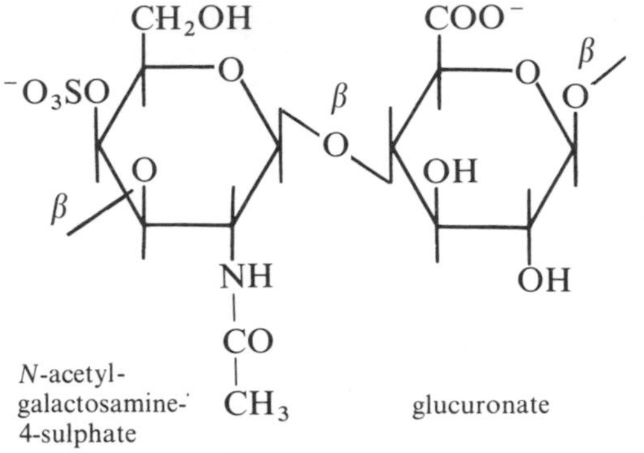

CHONDROITIN 4-SULPHATE

In chondroitin 6-sulphate the sulphate groups are attached to the 6-position of the N-acetylgalactosamine units.

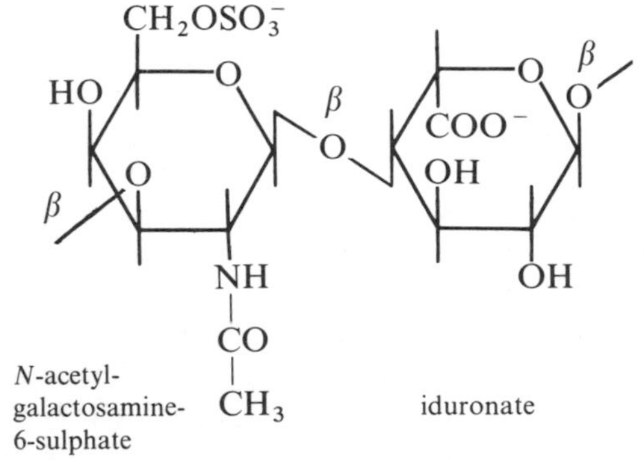

N-acetyl-galactosamine-4-sulphate

glucuronate

DERMATAN SULPHATE

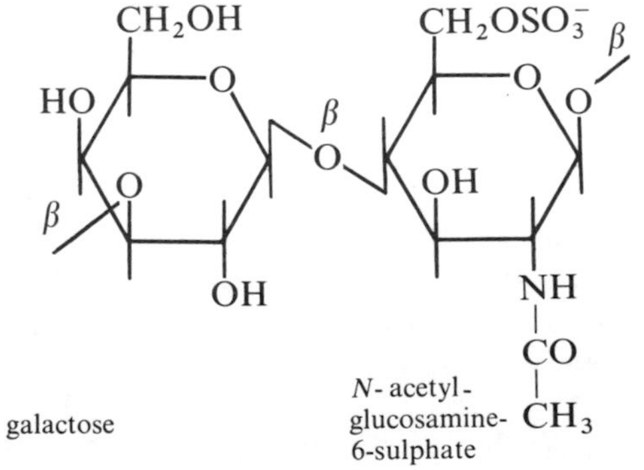

N-acetyl-galactosamine-6-sulphate

iduronate

KERATAN SULPHATE

galactose

N-acetyl-glucosamine-6-sulphate

Figure 16.8 Predominant disaccharide repeating units in the glycosaminoglycans of connective tissue. No attempt has been made to indicate conformation

degrades the articular cartilage covering the surfaces of bone in apposition in the joints. A number of rare inborn errors of metabolism are known in which particular lysosomal enzymes are absent so that partially degraded glycosaminoglycans accumulate and cause problems (often by damaging the lysosomes so that their contents leak into the cytosol).

(b) *Synthesis of glycosaminoglycans*

The glycosaminoglycans are built up by the sequential addition of monosaccharide units from uridine diphosphate sugar precursors to the protein component. The latter is synthesized on ribosomes attached to endoplasmic reticulum and then passes through the cisternae of the endoplasmic reticulum to the Golgi apparatus during which time the polysaccharide chains are assembled and sulphation occurs.

(i) *Synthesis of UDP monosaccharides*   N-acetylglucosamine (2-acetamido-2-deoxy-D-glucose) is a derivative of glucosamine (2-amino-2-deoxy-D-glucose). The osamines are sugars that have one hydroxyl group replaced by an amino group, usually on carbon atom 2. Glucose is converted, via the glycolytic pathway, to fructose 6-phosphate, which reacts with glutamine receiving an amino group in the 2-position and simultaneously isomerizing to the aldose so that the product of the reaction is glucosamine 6-phosphate. The amino group is then acetylated with acetyl-CoA providing the acetyl group. The N-acetylglucosamine 6-phosphate is converted, via a reaction catalysed by a mutase, to N-acetylglucosamine 1-phosphate which, in an analogous reaction to that of glucose 1-phosphate, reacts with UTP to produce UDP-N-acetylglucosamine. UDP-N-acetylgalactosamine is formed by isomerization of UDP-N-acetylglucosamine. These reactions and the enzymes catalysing them, are shown in Figure 16.9.

Glucuronate is formed from glucose by oxidation of the $-CH_2OH$ group at the 6-position to a carboxylic acid group in a reaction catalysed by UDPglucose dehydrogenase. The reaction takes place in two stages while glucose is attached to the nucleotide (Figure 16.10).

(ii) *Polymerization*   Monosaccharide units are transferred from UDP-sugars to growing polysaccharide chains in reactions catalysed by glycosyltransferase enzymes. The specificity of these enzymes ensures that the appropriate amino sugars and uronates are added alternately.

(iii) *Sulphation*   It is characteristic of the glycosaminoglycans that the monosaccharide units, usually the amino sugars, are sulphated, especially in the 4- and 6-positions. Sulphation occurs after polysaccharide synthesis by transfer of sulphate groups from 3'-phosphoadenosine-5'-phosphosulphate in reactions catalysed by sulphotransferases. 3'-Phosphoadenosine-5'-phosphosulphate (also

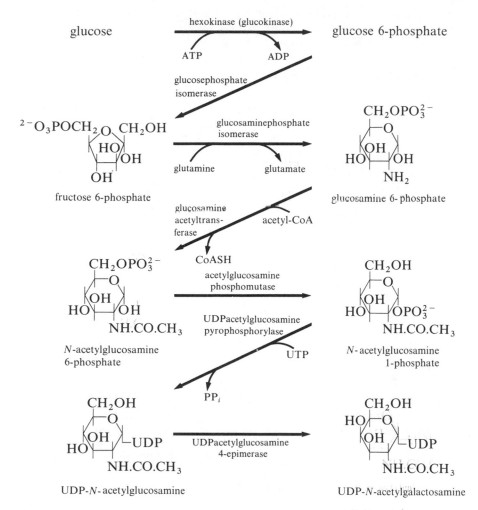

Figure 16.9 Pathways of UDP-*N*-acetylgalactosamine and UDP-*N*-acetylglucosamine synthesis

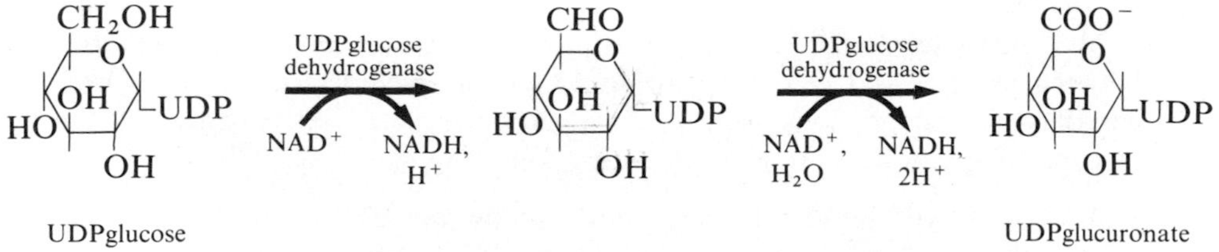

Figure 16.10 Pathway of UDPglucuronate synthesis

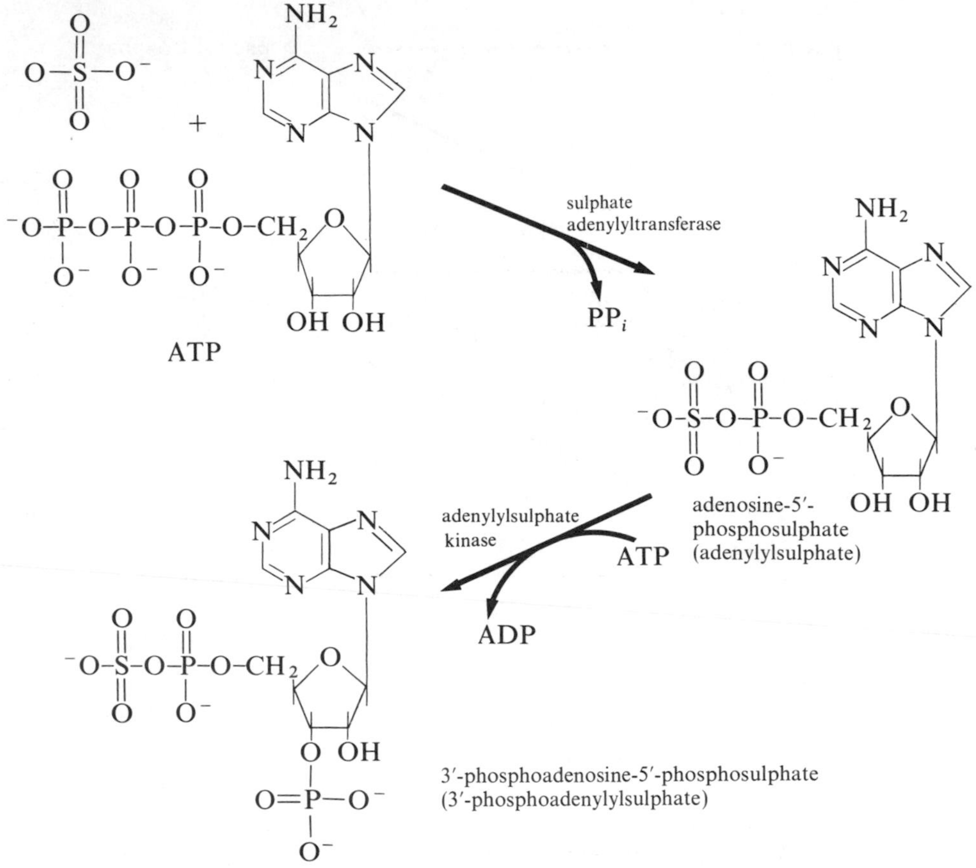

Figure 16.11　Pathway of 3′-phosphoadenosine-5′-phosphosulphate synthesis

known as 3′-phosphoadenylylsulphate) is analogous to ADP but bears a sulphate group in place of the terminal phosphate (as well as an additional phosphate group on the ribose moiety). It is formed from ATP and sulphate as shown in Figure 16.11.

3. Glycoproteins

In the widest sense, the term glycoprotein includes all covalent associations between proteins and carbohydrates and therefore includes the proteoglycans described in the previous section. In a more restricted sense, however, the glycoproteins can be distinguished by the following features.

1. The carbohydrate moiety does not usually possess a simple repeating sequence.
2. Uronic acids (see above) are usually absent.

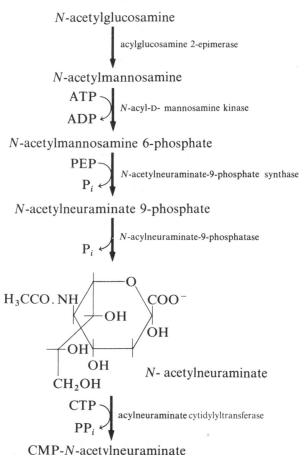

Figure 16.12 Pathway of *N*-acetylneuraminate (sialate) synthesis. Transfer of *N*-acetylneuraminate residues to oligosaccharides is achieved via formation of the CMP-derivative

3. There is a greater diversity of monosaccharide units than in the glycosaminoglycans, and neutral sugars can be present.
4. Sialic acids (e.g. *N*-acetylneuraminate, Figure 16.12) are frequently present.
5. There are, in general, a large number of short, but branched, oligosaccharide chains attached to the peptide chain (Figure 16.13).
6. Finally, the carbohydrate content of glycoprotein can vary from 2% to 90% (by weight).

(a) *Functions of glycoproteins*

Many glycoproteins are secretory proteins that are rendered soluble by virtue of their carbohydrate content. These include fibrinogen (involved in blood clotting),

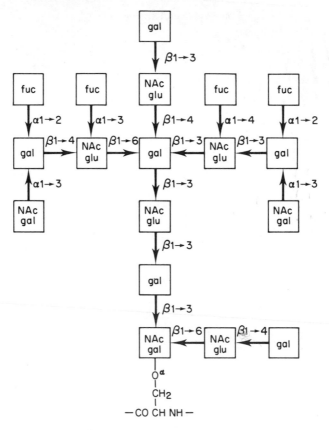

Figure 16.13   An oligosaccharide of a typical glycoprotein (that is responsible for the A antigenic site on red blood cells).
Abbreviations: fuc, fucose; gal, galactose; NAc gal, *N*-acetylgalactosamine; NAc gluc, *N*-acetylglucosamine

immunoglobulins (antibodies), thyroid stimulating hormone, pepsin (a digestive enzyme) and salivary mucin. Other glycoproteins are membrane components and posess discrete hydrophobic and hydrophilic regions. One such glycoprotein is glycophorin, found in the plasma membranes of erythrocytes. In this protein, the oligosaccharide chains are all at one end of the molecule which projects from the surface of the cell. Such features provide antigenic determinants and hence are responsible for the existence of blood groups (Figure 16.13; for detailed review of this subject, see Watkins *et al.*, 1981).

(b) *Synthesis of glycoproteins*

In most glycoproteins, the oligosaccharide is joined to the protein component by *O*-glycosidic linkages involving serine or threonine residues (or, less commonly, hydroxylysine or hydroxyproline residues). Such glycoproteins are assembled in the same way as the proteoglycans described above. The mechanism of synthesis of those glycoproteins in which the carbohydrate is linked by an *N*-glycosidic linkage with an asparagine residue is, however, fundamentally different. In this case the oligosaccharide is assembled on a polyisoprenoid (Chapter 20; Section B.1) compound known as a dolichol and then transferred complete to the protein. Further modification of the oligosaccharide may then occur.

E. GLYCOGEN STORAGE DISEASES

Most of the glycogen storage diseases, fortunately, are very rare so that the average physician will rarely come across such patients. However, the symptoms and clinical features of some of these diseases do illustrate metabolic principles and, for this reason, are briefly discussed below. In these diseases, except in types V and VII, there is an impairment in the ability to degrade liver glycogen when required to maintain the blood glucose concentration near to normal. Hence patients usually suffer from hypoglycaemia which in some cases can be very severe. As a consequence of this hypoglycaemia, the concentration of insulin is reduced and those of catecholamine, growth hormone, and glucagon may be elevated. These hormonal changes result in the mobilization of fatty acids and increased rates of ketone body formation so that the plasma concentrations of these fuels are elevated. The concentrations of glucose, lactate and ketone bodies in several of the glycogenoses are given in Table 16.5.

As in the case of hyperlipoproteinaemias, the various types have been classified with Roman numerals but a number are also known by the name of the person who first recognized the disease. For more detailed discussion the reader is referred to Mahler (1976), Howell (1978), and Felig (1980).

1. Type I glycogenosis

This is also known as von Gierke's disease and is due to a deficiency of the enzyme glucose-6-phosphatase in the liver, kidney, and intestine. The disease is characterized by the child being of short stature with a disproportionately large head. The abdomen is huge and rotund due to enlargement of the liver and spleen. Muscles are poorly developed and flabby, and there may be mental retardation (possibly due to the hypoglycaemia, see below). Xanthomata (Chapter 6; Section F.3.a) may be present, especially on the buttocks and exterior surface of the limbs. The plasma concentration of lactate, pyruvate, fatty acids, triacylglycerol, phospholipid, cholesterol, and ketone bodies are increased (Table 16.5). Not surprisingly, the fasting blood glucose concentration is low, sometimes less than 0.05 mM, but despite this there may be no symptoms of hypoglycaemia.

There are two functional tests that can easily be carried out for diagnosis. The first involves administration of glucagon. In normal subjects, glucagon administration raises the blood glucose concentration markedly but has no effect

Table 16.5. Concentration of fuels and metabolites in the blood* of control children and children suffering from glycogen storage diseases

Type of glycogenosis	Glycogen	Concentrations in blood (mM)				
		Glucose	Lactate	3-Hydroxy-butyrate	Aceto-acetate	Total ketone bodies
Control	Normal	3.3	1.2	0.1	0.05	0.15
Glucose-6-phosphatase deficiency	Increased	2.0	11.0	0.7	0.3	1.0
Amylo-1,6-glucosidase deficiency	Increased	2.7	1.1	2.6	1.0	3.6
Phosphorylase deficiency	Increased	2.7	1.6	4.5	1.4	5.9
Glycogen synthase deficiency	Decreased	1.5	0.7	6.4	2.2	8.2

* Blood was taken after an overnight fast (12–15 hr). Data are taken from Williamson and Whitelaw (1978).

on that of lactate. However, when glucagon is administered to patients with type I glycogenosis, there is only a small increase in the blood glucose concentration but there is a marked increase in that of lactate. Thus glycogenolysis can occur but the glucose residues are converted to lactate. Secondly, infusion of galactose does not raise the blood glucose concentration as it would in normal subjects (see Chapter 11; Section F.1.a) but again raises that of lactate since the pathway from galactose to glucose 6-phosphate is intact. The diagnosis can be confirmed by taking a biopsy sample of the liver and assaying the activity of glucose-6-phosphatase, which is non-detectable.

Most of the clinical features are anticipated from the known role of glucose-6-phosphatase. During starvation, the liver cannot release glucose so that there is hypoglycaemia. Indeed, since glucokinase will be active in the liver and not 'balanced' by the activity of glucose-6-phosphatase, there may be little restraint on the liver to take up glucose so that the blood glucose concentration falls to very low levels. This causes changes in hormone concentrations (a decrease in that of insulin and increase in those of glucagon, adrenaline, growth hormone) which favour lipolysis in adipose tissue so that fatty acids largely replace blood glucose as a fuel. The esterification of fatty acids by the liver will result in the elevation of plasma concentration of triacylglycerol, phospholipids, and cholesterol. Somewhat surprisingly, the ketone body concentration is not very high in these patients (Binkiewicz and Senior, 1973) which supports the view that high levels of glycogen in the liver may be one factor reducing the rate of ketogenesis. This is consistent with the fact that the highest ketone body concentration (about 8 mM) is seen in the glycogenosis in which the enzyme glycogen synthase is deficient with, consequently, a very low level of glycogen in the liver. The possibility is that, in type I glycogenosis, tissues such as the brain adapt to using lactate, since its concentration is very high (Table 16.5).

The ability of the liver to take up glucose and synthesize glycogen, without the usual ability to degrade it, results in excessive levels of glycogen. Furthermore, the liver cannot convert lactate to glucose (via gluconeogenesis) or glycogen (since the level is already very high) so that patients usually suffer from lactic acidosis, which can be chronic and result in osteoporosis (loss of calcium from the bone, see Chapter 13; Section D).

The disease should be suspected in the newborn child who has enlarged liver and kidneys, acidosis, and hypoglycaemia. There is little treatment except to maintain a high carbohydrate diet which is fed frequently. The prognosis is poor. Death usually occurs after about two years but if the patient survives to four, the prognosis is better since the severity of the metabolic abnormalities appears to decrease with increasing age.

2. Type II glycogenosis

This is sometimes known as Pompe's disease and is due to a deficiency of a 1,4-α-glucosidase which has a pH optimum about 4.0 (also known as acid maltase) which is found in the lysosomes of most tissues. This enzyme hydrolyses the 1,4-linkage in glycogen and it may play an important role in the degradation of glycogen in human tissues, especially in the early months of life. In the absence of this enzyme, there is a massive increase in the glycogen content of many tissues including heart, skeletal muscle, liver, and brain (especially in the motor nuclei of the brain stem and anterior horn cells of the spinal cord). There are no abnormalities of carbohydrate or fat metabolism and there is a normal response to glucagon administration. It would appear that, although excess glycogen accumulates in the lysosomes, there is a normal level of glycogen present in the glycogen particles and this can be degraded in the usual manner by glycogen phosphorylase. However, this enzyme cannot attack the glycogen in the lysosomes, due presumably to the lysosomal membrane. (Indeed, even if phosphorylase could enter the lysosome it would be inactive because of the low pH within this organelle.) The affected child has severe muscular weakness which develops, along with the other symptoms of the disease, at about two months of age. Death occurs before nine months of age due to cardio-respiratory failure, possibly because the large amount of glycogen in the heart interferes with contractility. At present there is no known treatment.

The existence of the disease emphasizes our lack of knowledge about the hydrolytic function of the lysosomes in the cell. The fact that large amounts of glycogen are contained in these organelles suggests that, at least in the early stage of life, the lysosomes could play an important role in degrading intracellular glycogen, especially in muscle and liver.

3. Type III glycogenosis

This is also known as Cori's disease or 'limit dextrinosis' and is due to a deficiency of amylo-1,6-glucosidase activity (debranching enzyme, see Chapter 5; Section C.2). In general, the symptoms and clinical features are similar to type I disease except that they are less severe. Thus there is enlargement of the liver, growth retardation, fasting hypoglycaemia, increased blood concentration of lipid-fuels, and enhanced levels of glycogen in the liver. However, the glycogen that accumulates is abnormal in that the outer chains are very short and the structure resembles a 'limit dextrin' produced by the action of glycogen phosphorylase on glycogen (Chapter 5; Section C.2). It can readily be distinguished from type I glycogenosis by the following functional tests. First, administration of glucagon after an overnight fast (or a fast as long as the child can withstand) results in little or no increase in blood glucose or lactate; conversely, administration of glucagon after feeding (when the outer chains will have been synthesized due to action of glycogen synthase) results in a normal increase in the blood glucose concentration. Secondly, the administration of galactose results in a normal increase in the blood glucose concentration.

Patients are treated by frequent feeding throughout the day and night with a high protein diet. Since the gluconeogenic pathway is intact it is possible for the patient to produce glucose from amino acids absorbed from the intestine. If the patient survives longer than four years of age, the prognosis can be fairly good. Failure to maintain the blood glucose concentration during the early stages of life can result in mental retardation.

4. Type IV glycogenosis

This is also known as Anderson's disease and it is due to a deficiency of glycogen branching enzyme. It is a very rare disease. The infants are normal at birth but fail to thrive. The clinical features include enlarged liver, poor weight gain, muscular weakness, and progressive cirrhosis of the liver. Death usually occurs in the second year of life. There may be a moderate hypoglycaemia on fasting and a poor response to glucagon. The concentrations of blood glucose and lipid-fuels are normal and there may not be excessive glycogen levels in the liver. However, the structure of glycogen in liver, muscle and other tissues is abnormal in that there are many fewer branch points and the material produced is less soluble than normal glycogen. The liver cell responds to the abnormal glycogen molecule by treating it as a foreign body which results in the production of a large amount of fibrous material. This sufficiently reduces the functional capacity of the liver cell that cirrhosis results. Treatment is that described in Chapter 12; Section C.3.

5. Type V glycogenosis

This is also known as McArdle's syndrome and it is due to a deficiency of the enzyme glycogen phosphorylase in muscle (the enzyme in liver is present and there is normal response to glucagon). The deficiency has been described previously to illustrate one role of glycogen in muscle (Chapter 5; Section F.1.c). Patients have been diagnosed as suffering from McArdle's syndrome over a wide age range (six to 60) but, in most, severe symptoms develop towards the end of their second decade. There is a moderate increase in the level of glycogen in the muscle and intolerance to exercise; patients suffer severe pain and cramps at the onset of exercise. However, if they can overcome this initial pain and discomfort, they can continue to exercise for longer periods without any apparent difficulty. This has been described as the 'second wind phenomenon' and is probably due to vasodilation making available glucose and fatty acids in the blood (Pernow *et al.*, 1967).

A functional test for the disease is to exercise under ischaemic conditions, when there is no accumulation of lactate in the bloodstream.[2] The usual advice given to such patients is to avoid all forms of extreme exercise but this is questionable, since lack of exercise will reduce the aerobic capacity of the muscle (that is, the number of mitochondria and the activities of mitochondrial enzymes are reduced, see Chapter 9; Section B.5.a). This will place a greater reliance on 'anaerobic exercise' which uses glycogen as substrate (Chapter 5; Section F.1.b). Knowledge of exercise biochemistry suggests that the patients should be encouraged to perform 'aerobic' exercise, provided that they initiate the exercise very gradually. It is possible that treatment of such patients with vasodilator drugs prior to exercise might be beneficial.

6. Type VII glycogenosis

Type VII is described immediately after type V since the symptoms and clinical features are very similar. This disease is due to a deficiency of 6-phosphofructokinase in the muscle[3] so that it is not possible to obtain energy from either glycogen or glucose metabolism.* It has been described in a family in Japan and in one subject in the U.S.A. (Tarui *et al.*, 1965; Layzer *et al.*, 1967). Energy can only be obtained from oxidation of lipid-fuels and the same advice should be given as described for type V.

7. Type VI glycogenosis

This is sometimes known as Hers' disease and it appears to be due to a decreased activity of hepatic glycogen phosphorylase. The disease has similar symptoms and clinical characteristics to those described for type I glycogenosis, although they are usually milder. This suggests that there is some residual activity of the phosphorylase. The main clinical features are enlarged liver, due to the increased glycogen level, and mild hypoglycaemia. The test administration of glucagon results in little or no increase in either blood glucose or lactate concentration, whereas galactose administration raises the blood glucose level.

8. Other glycogen storage diseases

It is possible to extend this series of glycogen storage diseases which are given higher roman numerals. These are very rare and some are probably due to deficiencies of enzymes involved in the control of phosphorylase activity; for example, in type IX there is a deficiency of hepatic phosphorylase kinase and in type X there is probably a deficiency of hepatic cyclic AMP-dependent protein kinase. Since these regulatory enzymes may play other roles in the cell, the clinical features may not be due solely to problems of glycogen metabolism.

One child has been discovered who suffers from glycogen synthase deficiency, so that glycogen cannot be synthesized in the liver (Aynesley-Green *et al.*, 1977). The clinical features of this child illustrate the importance of hepatic glycogen. During the overnight starvation, in the absence of any glycogen, the blood glucose concentration frequently fell to 10 mM or less whereas it increased to 10 mM after breakfast and remained elevated for the remainder of the day. In order to compensate for the hypoglycaemia, the blood concentration of ketone bodies increased to above 8 mM overnight but fell to almost normal values during the day (Figure 16.14). The massive fluctuations of glucose and ketone body concentrations and the hypoglycaemic symptoms were removed by frequent

---

* There is also a reduction in the 6-phosphofructokinase activity of the red cell.

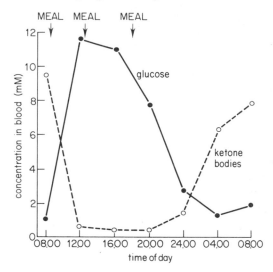

Figure 16.14 Diurnal changes in the blood concentration of glucose and ketone bodies in a child with glycogen synthase deficiency. See text

feeding of a high protein diet (that is, a diet providing sufficient gluconeogenic precursors to maintain the blood glucose concentration).

Of interest is that in this child, the concentration of ketone bodies increased in a matter of hours to levels that were only achieved in obese subjects after about 24 days starvation (Chapter 14; Section A.4). It could be that the control mechanisms for ketogenesis are considerably more sensitive to hormones and other regulators in children.

NOTES

1. Since all three sites on glycogen synthase can be phosphorylated, seven phosphorylated forms of the enzyme exist, GSb_1, GSb_2, GSb_3, GSb_{12}, GSb_{13}, GSb_{23}, GSb_{123}. The lowest activity is achieved when all the sites have been phosphorylated (Table 16.3; Figure 16.5). Since glucose 6-phosphate is less effective as an activator of the enzyme for the phosphorylated forms, an indication of the degree of 'inactivity' is obtained by determining the concentration of glucose 6-phosphate required for half-maximal activity (Table 16.5). The non-phosphorylated form requires 0.05 mM (less than the physiological concentration) whereas the fully phosphorylated forms requires 1.8 mM glucose 6-phosphate (which is well in excess of the physiological concentration; Cohen, 1978, 1979, 1980).

2. Phosphorus nuclear magnetic resonance studies have established that ischaemic exercise of the arm muscle of normal subjects results in a decrease in the intracellular pH to about 6.4 (due to lactic acid accumulation). However, there is no change in the intracellular pH during similar exercise performed by patients suffering from McArdle's syndrome (see Chapter 13; Note 7 and Ross *et al.*, 1981).

3. The concentration of the hexose monophosphates in the muscle of these patients is increased while that of fructose bisphosphate is decreased. The elevated glucose 6-phosphate concentration will reduce the activity of phosphorylase which may explain the increased level of glycogen in the muscle.

CHAPTER 17

BIOSYNTHESIS OF FATTY ACIDS, TRIACYLGLYCEROL, PHOSPHOLIPIDS AND PROSTAGLANDINS

Although the chemical and metabolic relationships between triacylglycerol, phospholipids, and prostaglandins (see Appendix 6.1) justify consideration of their biosynthesis in a single chapter, their very different roles give rise to large differences in the rates of synthesis of these lipids. Triacylglycerols constitute the major stored fuel compound and are thus synthesized and broken down (see Chapter 6) on a massive scale. Phospholipids serve a primarily structural function in membranes and their turnover is accordingly smaller than for triacylglycerols. Prostaglandins are involved in the control of a large number of processes and are present in many tissues but at very low concentrations.

A. BIOSYNTHESIS OF FATTY ACIDS AND TRIACYLGLYCEROL

All three major dietary components (lipid, protein, and carbohydrate) can be used as a source of carbon for the formation of triacylglycerol which occurs during the absorptive and immediate postabsorptive periods. The fate of ingested triacylglycerol, the most obvious source of triacylglycerol for storage, has already been described in Chapter 6. After hydrolysis in the lumen of the intestine and absorption into, and resynthesis within, the epithelial cells of the small intestine, triacylglycerol is transported to the tissues in the form of chylomicrons. Under normal conditions most of this triacylglycerol will be taken up for storage by adipose tissue. Hydrolysis of the triacylglycerol by lipoprotein lipase in the extracellular space of adipose tissue, uptake of the fatty acid and re-esterification of the latter within the cells are necessary stages in the entry of this lipid-fuel (Chapter 6; Section C.1.c). One advantage of this complex process of absorption and assimilation of triacylglycerol has been given in Chapter 8; it effectively 'conceals' fatty acids from all tissues except those that have an active lipoprotein lipase (e.g. adipose tissue in the fed state). Another advantage is that it allows the insertion of newly-synthesized fatty acids or modified dietary fatty acids into triacylglycerol in the storage tissues. Most of these alterations (described in subsequent sections) involve relatively minor changes in chain length and degree of saturation.

1. Synthesis of acetyl-CoA from dietary carbohydrate

The average Western man ingests 100–150 g of triacylglycerol each day which is normally more than enough to replace all the triacylglycerol broken down during

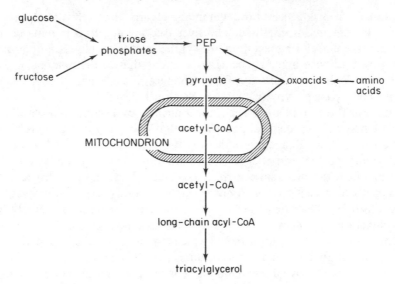

Figure 17.1 Pathways involved in the synthesis of triacylglycerol from carbohydrates and amino acids

the periods of fasting between meals (including the overnight fast) so that formation of lipid from other sources will not be required. Indeed, the absorption of triacylglycerol from the diet inhibits fatty acid synthesis in the lipogenic tissues (see Section A.9). Nonetheless, carbohydrates or amino acids can be used for *de novo* fatty acid and triacylglycerol synthesis and these conversions will be particularly important on a low fat diet or with high carbohydrate or high protein intakes. The pathways involved are summarized in Figure 17.1.

Although both the liver and white adipose tissue carry out significant amounts of triacylglycerol synthesis, it has proved difficult until recently to establish the quantitative importance of each *in vivo*. The development of a technique for labelling NADPH with tritium by the use of tritiated water, (which can be injected into animals and rapidly equilibrates with intracellular water) and following the incorporation of tritium from NADPH into the fatty acid moiety of triacylglycerol, has enabled such data to be obtained (Jungas, 1968). Although these experiments have shown that liver and white adipose tissue are quantitatively important sites of lipogenesis in the intact animal, less than 50% of the total lipogenesis occurs in these tissues in rats and mice. Other tissues which may contribute substantially to lipogenesis include the skin, brown adipose tissue, and the small intestine (Agius and Williamson, 1980; Hollands and Cawthorne, 1981).

Milk contains a high proportion of lipid, which is used as an energy source by the infant. Although some of the triacylglycerol required for its synthesis can be taken up from the bloodstream (via lipoprotein lipase) perhaps 50% is synthesized from glucose in the mammary gland. The pathway appears to be similar to that in adipose tissue and liver (see below). Experiments in the rat demonstrate that during lactation, the synthesis of triacylglycerol in both white and brown adipose tissue is decreased whereas that in the mammary gland is increased (Williamson, 1980). These changes are similar to those seen in lipoprotein lipase activity (Chapter 6; Section E.3.a).

Although the acetyl-CoA destined for triacylglycerol synthesis is synthesized in adipose tissue by an identical pathway to that operating in muscle (Chapter 5; Section B) there are differences in the way the process is controlled. As in muscle, glucose transport into adipocytes is stimulated by insulin but, unlike muscle, this probably provides the main control of glycolytic flux in adipose tissue, with the rates of all other reactions being controlled by changes in their pathway-substrate concentrations. The rate of glycolysis is, therefore, determined largely by the concentration of insulin in the blood which depends on the concentration of glucose in the hepatic portal blood which, in turn, depends upon the rate of absorption of glucose from the small intestine. This latter process is probably flux-generating for glycolysis in adipose tissue (as in muscle and other tissues). The logic of this apparently indirect control system depends on the fact that the concentration of glucose changes much more in hepatic portal blood than it does in peripheral blood after a meal. Since the pancreas releases insulin in response to changes in the hepatic portal blood glucose concentration, larger amounts of the hormone are released after a meal and the change in the concentration of insulin in the peripheral blood is greater. This has a correspondingly larger effect on lipogenesis in adipose tissue (see Figure 17.2).

In adipose tissue, any fructose in the blood which has escaped phosphorylation by the liver is phosphorylated to fructose 6-phosphate by hexokinase and so follows precisely the same pathway of glycolysis as does glucose. This contrasts with fructose metabolism in the liver, where the major pathway for fructose metabolism is phosphorylation to fructose 1-phosphate followed by cleavage to dihydroxyacetone phosphate and glyceraldehyde which enter glycolysis after the reaction catalysed by 6-phosphofructokinase (Chapter 11; Section F.2.a). The metabolic importance of this difference is that the latter enzyme plays an important role in the control of the rate of glycolysis in liver so that the rate of conversion of fructose to pyruvate and hence acetyl-CoA may not be subject to the same metabolic restrictions as that from glucose. Indeed there is evidence that fructose is a better lipogenic substrate than glucose in the liver and that sucrose feeding elevates the plasma concentration of triacylglycerol. It is interesting that sucrose ingestion in western society shows a positive correlation with obesity (Yudkin, 1972), but such relationships may not be causally linked.

Pyruvate is transported into mitochondria by a specific transport process where it is converted to acetyl-CoA in a reaction catalysed by pyruvate dehydrogenase. It is likely that this multienzyme complex is flux-generating for pyruvate conversion to acetyl-CoA and the activity of the enzyme in adipose tissue is known to be regulated by insulin (Section A.9).

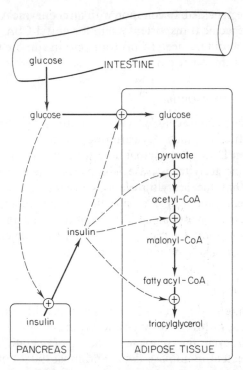

Figure 17.2 Regulation of lipogenesis in adipose tissue by insulin. The figure shows the relationship between glucose absorption into the body, the stimulation of insulin release from the pancreas by this glucose and the points on the pathway for conversion of glucose to triacylglycerol in adipose tissues at which insulin exerts its regulatory role (see Section A.9 for further details). The effects of insulin are assumed to be due to release of a secondary messenger from the cell membrane (Chapter 22; Section C.1)

2. Synthesis of acetyl-CoA from amino acids

The pathways by which amino acids are degraded are considered in Chapter 10 (see Appendix 10.1). All amino acids can be degraded to acetyl-CoA which can be used as a precursor for lipogenesis and, although the liver plays a quantitatively important part in this degradation, there is increasing evidence that a significant rate of conversion of amino acids (including the branched-chain amino acids which are less readily oxidized by liver) to lipid takes place in adipose tissue (Chapter 10; Section F.5).

3. Transport of acetyl-CoA out of mitochondria

Since the final stages in the synthesis of acetyl-CoA from either glucose or the amino acids occur within the mitochondrion but fatty acid synthesis occurs in the cytosol, some means of transport of acetyl-CoA across the mitochondrial membrane must be provided. In common with all coenzyme A derivatives of fatty acids, there is no specific transporter system for acetyl-CoA. Two shuttles exist which could bring about net acetyl-CoA transport in adipose tissue, one utilizing acetylcarnitine and the other citrate.

(a) *Transport via acetylcarnitine*

The transport of *long-chain* acyl-CoA into the mitochondrion prior to oxidation occurs via a shuttle in which acylcarnitine is the transported intermediate (Chapter 6; Section D.1.c). An analogous shuttle, involving acetylcarnitine and the enzyme carnitine acetyltransferase, is known to operate in lipogenic tissues but it differs in that the acetyltransferases are not structurally part of the mitochondrial membrane but are free in the two compartments. In the mammalian systems investigated, this shuttle appears to play only a quantitatively minor role;* the transport of acetyl-CoA via citrate appears to be the major mechanism (Martin and Denton, 1970).

(b) *Transport via citrate*

Within the mitochondrion, acetyl-CoA condenses with oxaloacetate to form citrate in the reaction catalysed by citrate synthase. Although this is the first reaction of the TCA cycle, the complete oxidation of 'acetate' via the cycle occurs only to a limited extent in adipose tissue and most (90%) of the citrate that is produced in the mitochondrion is transported as such into the cytosol via the tricarboxylate carrier. In the cytosol, acetyl-CoA is reformed from citrate in a reaction catalysed by ATP citrate lyase. This reaction involves the hydrolysis of ATP as acetyl-CoA and oxaloacetate are formed from citrate and coenzyme A:

$$\text{citrate}^{3-} + \text{CoASH} + \text{ATP}^{4-} \longrightarrow$$
$$\text{acetyl-CoA} + \text{oxaloacetate}^{2-} + \text{ADP}^{3-} + \text{P}_i^{2-}$$

(The cleavage of citrate by reversal of the citrate synthase reaction has a large positive ΔG^0. In the lyase reaction, the reaction is coupled with ATP hydrolysis so that the reaction can occur at physiological concentrations of reactants—see Chapter 2; Section A.2.d) Although this enzyme generates acetyl-CoA in the correct compartment for fatty acid synthesis, the over-all process of acetyl-CoA transport is not complete until the oxaloacetate has been returned to the mitochondrion. Since oxaloacetate is another of the intermediates not transported directly, it too must be transformed into one which can be. This is achieved by conversion into pyruvate through the activity of two malate dehydrogenases. The first of these is specific for NADH as the reducing agent and is identical to that involved in the TCA cycle:

$$\text{oxaloacetate}^{2-} + \text{NADH} + \text{H}^+ \longrightarrow \text{malate}^{2-} + \text{NAD}^+$$

---

* Another role of the acetylcarnitine transport system is in the regulation of the rate of oxidation of fatty acids in muscle (Chapter 7; Section D.2).

The second is known by the full title of malate dehydrogenase (oxaloacetate-decarboxylating) ($NADP^+$) and formerly as malic enzyme. It catalyses the oxidation of malate to pyruvate:

$$malate^{2-} + NADP^+ \longrightarrow pyruvate^- + CO_2 + NADPH$$

The pyruvate then enters the mitochondrion where the shuttle is completed by conversion of the pyruvate to oxaloacetate by the action of pyruvate carboxylase (Chapter 11; Section B.2.a):

$$pyruvate^- + HCO_3^- + ATP^{4-} \longrightarrow$$
$$oxaloacetate^{2-} + ADP^{3-} + P_i^{2-} + H^+$$

The complete sequence, known as the pyruvate/malate shuttle, is summarized in Figure 17.3. It serves not only to transport most of the acetyl-CoA required for fatty acid synthesis[1] but also to generate some of the NADPH needed for this process.

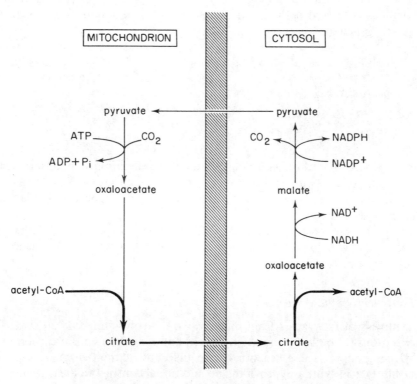

Figure 17.3 Pyruvate/malate shuttle for the transport of acetyl-CoA out of the mitochondria

4. Synthesis of long-chain fatty acids

Once the acetyl-CoA is present in the cytosol it functions as the immediate precursor for fatty acid synthesis. In principle, fatty acids can be synthesized by the progressive addition of two-carbon acetyl units to acetyl-CoA and reduction of the carbonyl groups to methylene groups; that is, by reversal of the β-oxidation pathway. However, for reasons already considered, synthetic reactions rarely, if ever, proceed by reversal of the corresponding degradative reactions and fatty acid synthesis is no exception. There are several important differences. First, the chain is progressively extended, not by addition of acetyl-CoA, but by addition of malonyl-CoA, a three-carbon compound formed by carboxylation of acetyl-CoA. Malonyl-CoA is a better condensing agent than is acetyl-CoA and this favours the biosynthetic process. Secondly, the reducing agent is NADPH rather than NADH. Thirdly, coenzyme A is not involved in the elongation or reductive processes. The reactions of fatty acid synthesis are given in sequence below.

(a) *Acetyl-CoA carboxylase reaction*

Acetyl-CoA carboxylase catalyses the carboxylation of acetyl-CoA in a reaction that requires ATP and hydrogencarbonate (biotin is required as a prosthetic group, see Appendix A).

$$HCO_3^- \qquad \text{biotin} \leftarrow \qquad ^-OOC.CH_2CO.S.CoA \quad \text{malonyl-CoA}$$

$$ATP^{4-}$$

$$ADP^{3-} + P_i^{2-} \leftarrow \quad \varepsilon\text{-carboxybiotin} \quad CH_3CO.SCoA$$
$$+ H^+$$

Since acetyl-CoA carboxylase catalyses the flux-generating step for fatty acid and triacylglycerol synthesis from acetyl-CoA (see Section A.6), it is not surprising that its activity is regulated by factors that control the rate of lipogenesis (see Section A.9).

During the subsequent addition reaction, the carbon dioxide that is incorporated into malonyl-CoA is eliminated, so that the net result is extension of the chain by two carbon atoms.

(b) *Fatty acid synthase reactions*

The synthesis of fatty acids from malonyl-CoA involves an ordered series of condensations (to build up the chain length) and reductions (from carbonyl to methylene groups). These reactions take place on a multienzyme complex (termed fatty acid synthase) providing an additional contrast to the degradative process. This close association of functionally related catalytic activities may well be similar in general organization to, for example, the oxoglutarate dehydrogenase complex, described in detail in Chapter 4; Section B.1.d. It is apparent, however, that the organization of this complex differs from species to species. The complex from *E. coli*, which is most amenable to study, consists of seven separable enzymes plus a small additional protein (acyl-carrier protein) which transfers covalently-attached intermediates from one active site to another. In the mammalian liver, the functions are not readily separated and it is possible that more than one catalytic site is present on a single polypeptide chain. Despite this, it is believed that the sequence of reactions is similar, if not the same, in all species. Although mammalian fatty acid synthase appears to lack a distinct acyl-carrier protein it does possess the same functional unit of that protein, namely a 4'-phosphopantetheine group. This is identical to part of the coenzyme A molecule (Figure 4.2) and consists of a flexible chain of 14 atoms including a terminal sulphydryl group which can esterify with the fatty acids. Linking of the other end of the arm to a serine residue on one of the components of the fatty acid synthase extends its length to 16 atoms. It appears to be able to transfer intermediates from one active site to another as successive cycles of condensations and reductions are carried out. In the first cycle, an acetyl and a malonyl group condense with the elimination of carbon dioxide and the resulting acetoacetyl group is reduced to a butyryl (butanoyl) group. In succeeding cycles (up to a maximum of seven) further malonyl groups condense with the growing fatty acyl chain and are reduced. The seven functional activities of the complex are considered in order.

(i) *Acetyltransferase** This enzyme catalyses the reactions between the sulphydryl group of the 4'-phosphopantetheine (E–PP.SH) arm and acetyl-CoA, thereby introducing the first two carbon atoms:

$$CH_3CO.S.CoA + E-PP.SH \longrightarrow CH_3CO.S.PP-E + CoASH$$

The acetyl group is subsequently transferred to a second sulphydryl group near the active site of 3-oxoacyl synthase (below) leaving the 4'-phosphopantetheine arm free:

$$CH_3CO.S.PP-E + E'-SH \longrightarrow CH_3CO.S-E' + E-PP.SH$$

In mammalian liver and mammary gland, butyryl-CoA can replace acetyl-CoA so that synthesis is primed with a four-carbon compound.

(ii) *Malonyltransferase* This enzyme catalyses the transfer of each successive incoming malonyl group to the 4'-phosphopantetheine arm:

$$^-OOCCH_2CO.S.CoA + E-PP.SH \longrightarrow {}^-OOCCH_2CO.S.PP-E + CoASH$$

malonyl-CoA

(iii) *3-Oxoacyl synthase* All is now set for the first condensation reaction, catalysed by 3-oxoacyl synthase, in which attack by the acetyl group on the

---

* The names given here to the constituent activities of the fatty acid synthase complex are intended to be descriptive rather than definitive names.

malonyl group attached to the 4'-phosphopantetheine arm is followed by de-carboxylation to yield the 3-oxobutyryl (acetoacetyl) derivatives:

$$CH_3CO.S-E' + {}^-OOC.CH_2CO.S.PP-E + H^+ \longrightarrow$$
$$CH_3CO.CH_2CO.S.PP-E + CO_2 + E'-SH$$

(iv) *3-Oxoacyl reductase*   Both reductions of the extended carbon chain involve NADPH. The first of these generates D-hydroxybutyrate in the first cycle of reactions (as opposed to L-hydroxybutyrate, the isomer involved in fatty acid oxidation):

$$CH_3COCH_2CO.S.PP-E + NADPH + H^+ \longrightarrow$$
$$CH_3CH(OH)CH_2CO.S.PP-E + NADP^+$$

(v) *3-Hydroxyacyl dehydratase*   This enzyme catalyses the removal of a molecule of water from the 3-hydroxyacyl derivative:

$$CH_3CH(OH)CH_2CO.S.PP-E \longrightarrow CH_3CH{=}CH.CO.S.PP-E + H_2O$$

(vi) *Enoyl reductase*   Reduction of this enoyl derivative by a second molecule of NADPH finally generates a fatty acid which remains attached to the 4'-phosphopantetheine arm:

$$CH_3CH{=}CH.CO.S.PP-E + NADPH + H^+ \longrightarrow$$
$$CH_3CH_2CH_2CO.S.PP-E + NADP^+$$

This acyl group is then transferred to the sulphydryl group adjacent to the active site of the 3-oxoacyl synthase so that reactions (ii)–(vi) can be repeated. This continues until the 16-carbon palmitoyl group is formed. This group, still attached to the 4'-phosphopantetheine arm, is a substrate for the remaining enzyme of the complex, thioester hydrolase, the activity of which prevents transfer of the palmitoyl group to the synthase.

(vii) *Thioester hydrolase*   This enzyme liberates free palmitate from the 4'-phosphopantetheine arm:

$$CH_3(CH_2)_{14}CO.S.PP-E + H_2O \longrightarrow CH_3(CH_2)_{14}COO^- + E-PP.SH + H^+$$

Figure 17.4 attempts to depict the whole sequence of reactions involved in fatty acid synthesis.

(c) *Chain elongation*

At least 60% of triacylglycerol molecules in human adipose tissue contain fatty acids with 18 carbon atoms, and fatty acids with 20, 22, and 24 carbon atoms also occur. Since the reactions catalysed by fatty acid synthase yield only palmitate,

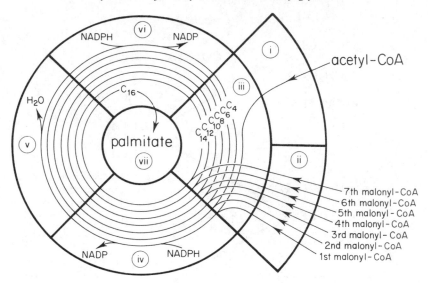

Figure 17.4 Diagrammatic representation of the fatty acid synthase complex. Note that there is no evidence for the actual spatial relationship of the active sites.
(i) Acetyltransferase. (ii) Malonyltransferase. (iii) 3-Oxoacyl synthase. (iv) 3-Oxoacyl reductase. (v) 3-Hydroxyacyl dehydratase. (vi) Enoyl reductase. (vii) Thioester hydrolase

and the longer-chain fatty acids occur in adipose tissue even when none are present in the diet, it is clear that some additional chain elongation must occur. The reactions of elongation appear to be identical to those involved in palmitate synthesis but the enzymes are found in association with the endoplasmic reticulum and are specific for the coenzyme A derivatives of intermediates. The over-all reaction for each stage in chain elongation can be represented as follows:

$$palmitoyl\text{-}CoA + malonyl\text{-}CoA^- + 2NADPH + 3H^+ \longrightarrow$$

$$stearoyl\text{-}CoA + 2NADP^+ + CoASH + CO_2$$

This process occurs mainly in liver and is equally important in the elongation of the polyunsaturated fatty acids, described in Section B.

Chain elongation can also occur within the mitochondrion but this is probably important only for the synthesis of fatty acids for incorporation into mitochondrial membranes. Furthermore, it is achieved by a reversal of β-oxidation so that acetyl-CoA is the source of the additional carbon atoms and reducing power is provided by NADH. Involvement of this reducing agent is not surprising since the $NAD^+/NADH$ redox ratio is considerably more reduced in mitochondria compared to that in the cytosol (see Chapter 5; Section D).

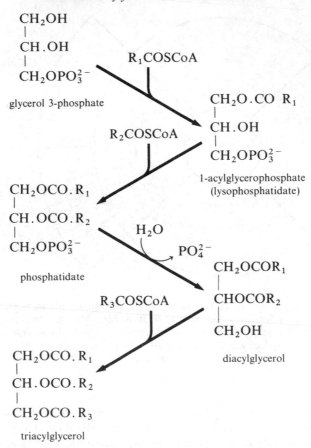

Figure 17.5 Pathway of triacylglycerol synthesis from long-chain acyl-CoA and glycerol 3-phosphate in adipose tissue

5. Esterification

Since long-chain fatty acids, the products of the reactions outlined above, do not accumulate in adipose tissue, they cannot be considered the endproducts of the pathway. The true products are the triacylglycerols which are formed by esterification of glycerol with three molecules of fatty acid. The reactants are actually long-chain acyl-CoA and glycerol 3-phosphate so that the over-all process can be represented:

$$3 \text{ long-chain acyl-CoA} + \text{glycerol 3-phosphate}^{2-} + H_2O \longrightarrow$$
$$\text{triacylglycerol} + 3 \text{ CoASH} + P_i^{2-}$$

The origin of the glycerol 3-phosphate is considered in Section A.7.

The individual reactions involved in esterification are shown in Figure 17.5.

Table 17.1. Nomenclature and structural formulae of some widely occurring unsaturated fatty acids. See Section B for explanation of nomenclature

---

Oleic acid   18:1 (9)

Δ^9 octadecenoic acid
$CH_3(CH_2)_7CH=CH(CH_2)_7COOH$

Linoleic acid   18:2 (9,12)

$\Delta^{9,12}$ octadecadienoic acid
$CH_3(CH_2)_4CH=CHCH_2CH=CH(CH_2)_7COOH$

Linolenic acid   18:3 (9,12,15)

$\Delta^{9,12,15}$ octadecatrienoic acid
$CH_3CH_2CH=CHCH_2CH=CHCH_2CH=CH(CH_2)_7COOH$

γ-Linolenic acid   18:3 (6,9,12)

$\Delta^{6,9,12}$ octadecatrienoic acid
$CH_3(CH_2)_4CH=CHCH_2CH=CHCH_2CH=CH(CH_2)_4COOH$

Arachidonic acid   20:4 (5,8,11,14)

$\Delta^{5,8,11,14}$ eicosatetraenoic acid
$CH_3(CH_2)_4CH=CHCH_2CH=CHCH_2CH=CHCH_2CH=CH(CH_2)_3COOH$

The specificities of the enzymes involved ensure that the fatty acids incorporated into triacylglycerols are not randomly selected. The initial acylation at the 1-position of glycerol 3-phosphate, catalysed by glycerophosphate acyltransferase, almost always involves a saturated fatty acid (e.g. palmitate, stearate). On the other hand, the enzyme catalysing the second acylation, 1-acylglycerophosphate acyltransferase, is specific for an acyl-CoA with either one or two double bonds (e.g. oleate, palmitoleate, linoleate—see Table 17.1). The final acylation, catalysed by diacylglycerol acyltransferase, which occurs after hydrolytic removal of the phosphate group by phosphatidate phosphatase, is less specific and an unsaturated or saturated fatty acid may be inserted. Note that, in the epithelial cells of the intestine, a different metabolic pathway is involved in the resynthesis of triacylglycerol after absorption of fatty acids and monoacylglycerols (see Figure 6.3).

6. The physiological pathway of triacylglycerol formation

In common with many of the other pathways discussed in this book, the subdivisions used to clarify the above description do not correspond to those derived from application of the principles given in Chapter 2. Since an analysis in terms of the physiological pathways is useful for a full understanding of the way in which the rate of lipogenesis is controlled (Section A.9) it is presented here. Despite the paucity of information concerning the equilibrium nature of the individual reactions and the consequent difficulty of identifying flux-generating

Table 17.2. Free energy changes of some of the reactions involved in fatty acid synthesis in the liver*

Reaction catalysed by	ΔG (kJ.mol^{-1})	Suggested equilibrium nature
Glucokinase	-32.1	non-equilibrium
6-Phosphofructokinase	-20.9	non-equilibrium
Pyruvate kinase	-17.8	non-equilibrium
Pyruvate dehydrogenase	-47.3	non-equilibrium
Citrate synthase	-19.6	non-equilibrium
ATP citrate lyase	-31.8	non-equilibrium
Acetyl-CoA carboxylase	$-$	non-equilibrium
NADP$^+$-linked malate dehydrogenase	$+7.4$	near-equilibrium
Glucose-6-phosphate dehydrogenase	-17.6	non-equilibrium
Phosphogluconate dehydrogenase	-0.2	near-equilibrium

* Data from Stanley (1980).

steps, it is considered that the sequence of reactions linking dietary glucose with triacylglycerol, in adipose tissue, can be divided into four physiological pathways (Figure 17.6). The available data, on which these conclusions are based, are presented in Table 17.2. The physiological pathways (with the flux-generating step being given first in each case) are as follows.

1. From the absorption of glucose from the lumen of the small intestine to the pyruvate kinase reaction. (There appears to be no flux-generating step in glycolysis in adipose tissue.)
2. The conversion of pyruvate to acetyl-CoA. (Pyruvate dehydrogenase is flux-generating.)
3. From the citrate synthase reaction to the pyruvate carboxylase reaction. (That is, the shuttle which operates to transport acetyl-CoA out of the mitochondrion.)
4. From the acetyl-CoA carboxylase reaction to the formation of triacylglycerol. (This final pathway includes esterification; the evidence that this process does not contain a flux-generating step is that raising the fatty acid concentration in the tissue increases the rate of esterification.)

7. Origin of glycerol 3-phosphate

Glycerol 3-phosphate, the cosubstrate in esterification, can arise in two ways. The enzyme glycerol kinase catalyses the phosphorylation of glycerol (which most likely will have arisen from lipolysis) as follows:

$$\begin{array}{ccc} \mathrm{CH_2OH} & & \mathrm{CH_2OH} \\ | & & | \\ \mathrm{CH.OH} + \mathrm{ATP}^{4-} \longrightarrow & \mathrm{CH.OH} + \mathrm{ADP}^{3-} + \mathrm{H}^+ \\ | & & | \\ \mathrm{CH_2OH} & & \mathrm{CH_2OPO_3^{2-}} \end{array}$$

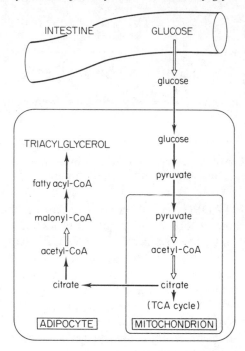

Figure 17.6 The physiological pathways of triacylglycerol synthesis. The flux-generating steps (which initiate each pathway) are indicated by open arrows

A respectable activity of this enzyme is present in liver (about 1.5 μmol.min^{-1}.g^{-1} in rat liver) and this presumably supplies much or all of the glycerol 3-phosphate needed. However, the activity of glycerol kinase in adipose tissue is very low (about 0.004 μmol.min^{-1}.g^{-1}) so that in this tissue an alternative pathway is required to account for glycerol phosphate formation.

This involves the reduction of dihydroxyacetone phosphate[2] catalysed by glycerol 3-phosphate dehydrogenase (see Chapter 5; Section D.2), and occurs in both liver and adipose tissue. Since dihydroxyacetone phosphate is a glycolytic intermediate, its concentration is maintained by glycolysis.[3] In both liver and adipose tissue, the concentration of glycerol 3-phosphate could play a role in regulating the rate of esterification, a possibility considered further in Section A.9.

8. Origin of NADPH: the pentose phosphate pathway

The conversion of several molecules of acetyl-CoA to one of long-chain acyl-CoA involves reductions in which NADPH, rather than NADH, acts as the reducing agent. The reasons for this are explored in Section A.8.c. One source of NADPH has already been described, namely that arising incidentally as a result of the transport of acetyl-CoA into the cytosol as citrate (Figure 17.3). If all the acetyl-CoA for fatty acid synthesis was transported in this way, the pathway should

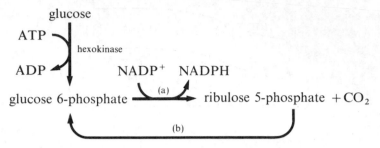

Figure 17.7   An outline of the pentose phosphate pathway. No attempt has been made to indicate stoichiometry. (See Figure 17.8)

provide half of the NADPH needed. (Since it is known that less than half is provided by this route, it follows that other shuttles must play some part in transport of acetyl-CoA.) The remainder of the NADPH required is provided by a pathway variously known as the pentose phosphate, hexose monophosphate, or phosphogluconate pathway, cycle, or shunt. Although the pentose phosphate pathway achieves oxidation of glucose, this is not its function. The distribution of the pathway in different tissues, and the fact that only NADPH is produced, clearly indicate that its function is to generate reducing power for biosynthetic processes. (Although it may also be important in the generation of pentoses for nucleic acid synthesis, see Chapter 18, Section B.)

The over-all pathway is represented by the equation:

$$C_6H_{12}O_6 + 12NADP^+ + 6H_2O \longrightarrow 6CO_2 + 12NADPH + 12H^+$$

This equation erroneously suggests that a single glucose molecule is progressively degraded but this is not, however, the case. The pathway can most clearly be envisaged as consisting of two parts: first, the oxidation of glucose to pentose phosphate plus carbon dioxide; secondly, the resynthesis of hexose from the pentose (Figure 17.7). The first part of the pathway (a) contains the flux-generating step and the formation of hexose in the second portion (b) explains why such terms as 'cycle' and 'shunt' have been used in naming the pathway. All the reactions occur in the cytosol.

(a) *Oxidation of glucose 6-phosphate*

The pentose phosphate pathway begins with glucose 6-phosphate which is oxidized by $NADP^+$ in a reaction catalysed by glucose 6-phosphate dehydrogenase, thereby generating the first of two molecules of NADPH:

A second enzyme, 6-phosphogluconolactonase, catalyses the hydrolysis of this product with consequent ring opening, to form 6-phosphogluconate:

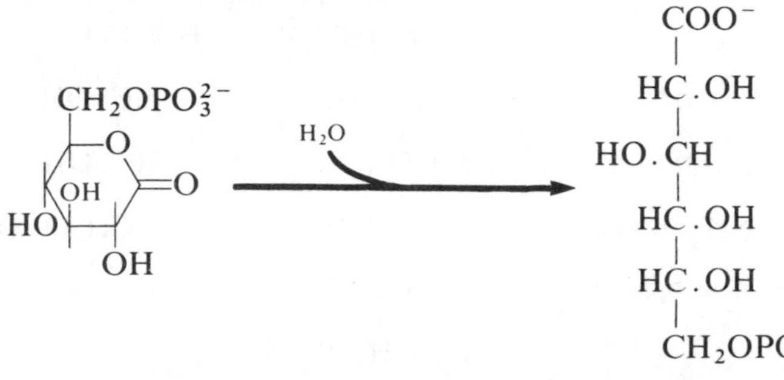

$$
\begin{array}{l}
CH_2OPO_3^{2-} \\
\quad\text{(ring structure with O, OH, HO, OH, =O)} \qquad \xrightarrow{\; H_2O \;} \qquad
\begin{array}{l}
COO^- \\
HC.OH \\
HO.CH \\
HC.OH \\
HC.OH \\
CH_2OPO_3^{2-}
\end{array}
\quad \text{6-phosphogluconate}
\end{array}
$$

The pentose phosphate (namely ribulose 5-phosphate) arises by oxidation of this compound by $NADP^+$, catalysed by phosphogluconate dehydrogenase, generating the second NADPH molecule and eliminating carbon dioxide:

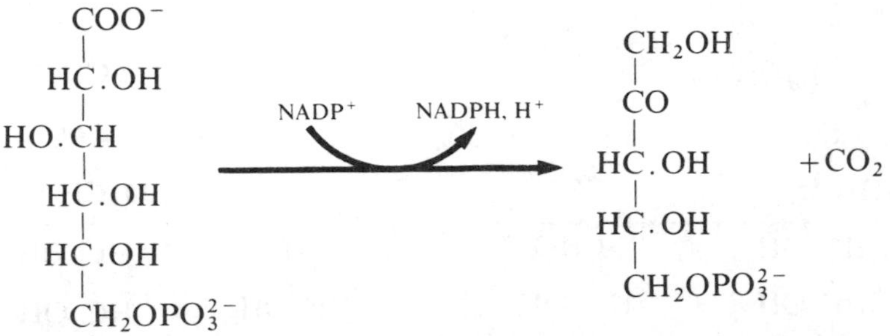

$$
\begin{array}{l}
COO^- \\
HC.OH \\
HO.CH \\
HC.OH \\
HC.OH \\
CH_2OPO_3^{2-}
\end{array}
\qquad \xrightarrow[\text{NADP}^+ \quad \text{NADPH, H}^+]{} \qquad
\begin{array}{l}
CH_2OH \\
CO \\
HC.OH \\
HC.OH \\
CH_2OPO_3^{2-}
\end{array}
\quad + CO_2
$$

Glucose 6-phosphate dehydrogenase catalyses a non-equilibrium reaction (see Table 17.2) and in both liver and adipose tissue, it is the flux-generating step and hence initiates the pentose phosphate pathway.

(b) *Conversion of ribulose 5-phosphate to hexose phosphate*

Although the individual reactions of the second stage of the pathway are straightforward, it is impossible to depict them in a simple linear sequence so that description is difficult. Basically, the atoms in six molecules of pentose phosphate are rearranged to form five molecules of hexose phosphate. The reactions involved are of four main types.

(i) *Transketolases* These enzymes catalyse the transfer of a two-carbon unit from a ketose with five or more carbon atoms to an acceptor sugar (ketose or aldose) with at least three carbon atoms, e.g.:

$$
\begin{array}{c}
CH_2OH \\
| \\
CO \\
\text{-----|-----} \\
HO.CH \\
| \\
HC.OH \\
| \\
HO.CH \\
| \\
HC.OH \\
| \\
HC.OH \\
| \\
CH_2OPO_3^{2-}
\end{array}
\quad + \quad
\begin{array}{c}
CHO \\
| \\
HC.OH \\
| \\
HC.OH \\
| \\
CH_2OPO_3^{2-} \\
\text{erythrose 4-phosphate}
\end{array}
\quad \longrightarrow \quad
\begin{array}{c}
CHO \\
| \\
HC.OH \\
| \\
HO.CH \\
| \\
HC.OH \\
| \\
HC.OH \\
| \\
CH_2OPO_3^{2-}
\end{array}
\quad + \quad
\begin{array}{c}
CH_2OH \\
| \\
CO \\
\text{-----|-----} \\
HO.CH \\
| \\
HC.OH \\
| \\
HC.OH \\
| \\
CH_2OPO_3^{2-} \\
\text{fructose 6-phosphate}
\end{array}
$$

glycero-ido-octulose 8-phosphate   glucose 6-phosphate

(ii) *Transaldolases*   These enzymes catalyse the transfer of a three-carbon unit from a ketose with at least six carbon atoms to an aldose with at least three carbon atoms, e.g.:

$$
\begin{array}{c}
CH_2OH \\
| \\
CO \\
| \\
HO.CH \\
\text{-----|-----} \\
HC.OH \\
| \\
HC.OH \\
| \\
HC.OH \\
| \\
CH_2OPO_3^{2-}
\end{array}
\quad + \quad
\begin{array}{c}
CHO \\
| \\
HC.OH \\
| \\
CH_2OPO_3^{2-}
\end{array}
\quad \longrightarrow \quad
\begin{array}{c}
CHO \\
| \\
HC.OH \\
| \\
HC.OH \\
| \\
CH_2OPO_3^{2-}
\end{array}
\quad + \quad
\begin{array}{c}
CH_2OH \\
| \\
CO \\
| \\
HO.CH \\
\text{-----|-----} \\
HC.OH \\
| \\
HC.OH \\
| \\
CH_2OPO_3^{2-}
\end{array}
$$

sedoheptulose   glyceraldehyde   erythrose   fructose
7-phosphate   3-phosphate   4-phosphate   6-phosphate

(iii) *Epimerases*   These enzymes catalyse a change in configuration at a single carbon atom, e.g.:

$$
\begin{array}{c}
CH_2OH \\
| \\
CO \\
| \\
HC.OH \\
| \\
HC.OH \\
| \\
CH_2OPO_3^{2-}
\end{array}
\quad \longrightarrow \quad
\begin{array}{c}
CH_2OH \\
| \\
CO \\
| \\
HO.CH \\
| \\
HC.OH \\
| \\
CH_2OPO_3^{2-}
\end{array}
$$

ribulose-5-phosphate   xylulose-5-phosphate

(iv) *Isomerases* These enzymes catalyse isomerizations between ketoses and aldoses, e.g.:

$$
\begin{array}{ccc}
\mathrm{CH_2OH} & & \mathrm{CHO} \\
| & & | \\
\mathrm{CO} & & \mathrm{HC.OH} \\
| & & | \\
\mathrm{HC.OH} & \longrightarrow & \mathrm{HC.OH} \\
| & & | \\
\mathrm{HC.OH} & & \mathrm{HC.OH} \\
| & & | \\
\mathrm{CH_2OPO_3^{2-}} & & \mathrm{CH_2OPO_3^{2-}} \\
\text{ribulose-5-phosphate} & & \text{ribose-5-phosphate}
\end{array}
$$

A further complexity is that the sequence of reactions may differ according to the tissue. On the basis of detailed isotopic experiments and enzyme activity measurements, Williams (1980) has suggested that the conversion of pentose to hexose follows one sequence of reactions in adipose tissue (and probably cornea, lens, brain, ovary, testes, and adrenal cortex), the so-called F-type pathway, whereas a slightly different sequence occurs in the liver, the so-called L-type pathway.* These are depicted in Figure 17.8. The stoichiometry of the over-all pathway is complicated by the fact that triose phosphates, which are produced in the pentose pathway, can enter the glycolytic pathway and be oxidized.

The pentose pathway serves to generate NADPH which, in adipose tissue and liver, is used mainly in fatty acid sythesis. In other tissues NADPH is used differently; in adrenal cortex, testis and ovary, it is required primarily in the pathway for formation of steroid hormones (Chapter 20; Section B.3); in the adrenal medulla and in nervous tissue, it is required for hydroxylation reactions involved in dopamine, adrenaline and noradrenaline synthesis (Chapter 21; Section B.2.a) and in some tissue (e.g. lens, cornea, red blood cells, and red muscle) NADPH is required to maintain the concentration of reduced glutathione, which decreases the concentration of the toxic intermediates generated by oxygen radicals (see Chapter 4; Section E.5.b).

(c) *Significance of separate $NAD^+/NADH$ and $NADP^+/NADPH$ redox systems*

Reference to a map of metabolic pathways shows that, in general, the $NAD^+/NADH$ system is involved in oxidative catabolic processes (e.g. glycolysis, TCA cycle, electron transfer) whereas the $NADP^+/NADPH$ system is usually involved in reductive anabolic processes. The explanation for this metabolic 'division of labour' has been elucidated by H. A. Krebs and coworkers (see Krebs, 1973).

* It should, however, be noted that alternative interpretations of the experiments of Williams and colleagues have been proposed.

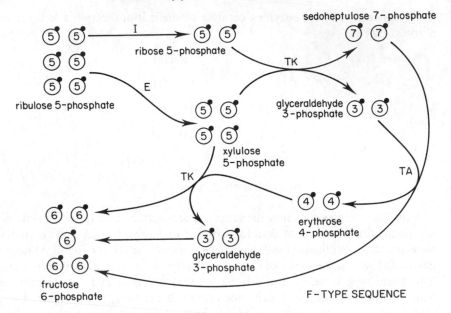

F-TYPE SEQUENCE

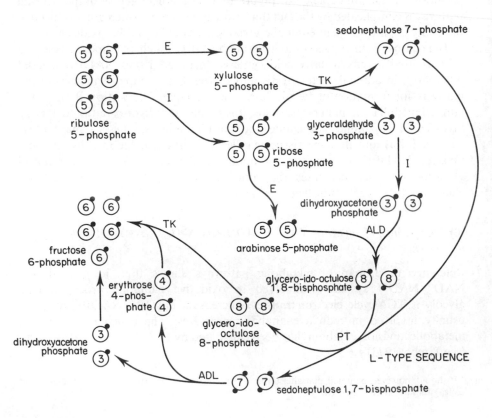

L-TYPE SEQUENCE

In Chapter 7; Note 1, the method for measuring the cytosolic $NAD^+/NADH$ concentration ratio was described. A similar method has been developed for measuring the cytosolic $NADP^+/NADPH$ concentration ratio.[4] When these two ratios are compared a difference of almost five orders of magnitude is apparent between them. In the cytosol of the liver, and presumably other tissues, the $NAD^+/NADH$ concentration ratio is about 1000, whereas that for $NADP^+/NADPH$ is about 0.01. The latter ratio favours reductive reactions by enabling such reactions to take place at concentrations of precursors that are very much lower than would be the case if $NAD^+/NADH$ were involved. On the other hand, a high $NAD^+/NADH$ concentration ratio favours oxidative reactions (e.g. the glyceraldehyde-3-phosphate dehydrogenase reaction of glycolysis).

Since both NAD^+ and $NADP^+$-linked reactions occur in the cytosol of the liver, the maintenance of the difference in ratio requires that the dehydrogenase enzymes exhibit a very high degree of specificity for NAD^+ or $NADP^+$. This has been shown to be the case for the dehydrogenases that have been studied in detail (Dalziel, 1980). Any enzyme capable of using either nucleotide pair will, if active enough, bring the concentration ratios of the pairs to identical values in the compartment in which it is present. This does not occur in the cytosol but may occur in the mitochondrion where the concentration ratio for both pairs is approximately ten; the enzyme glutamate dehydrogenase may be responsible since it can react with either NAD^+ or $NADP^+$.

9. Regulation of fatty acid and triacylglycerol synthesis

Higher animals including man have a large capacity to convert carbohydrate and some amino acids to triacylglycerol in adipose tissue, liver and some other tissues (see Section A.1 and 2). Some of the non-equilibrium reactions are indicated in Table 17.2 but information is not available to calculate ΔG values for the reactions of the fatty acyl-CoA synthase complex or the reactions of esterification. The major external regulators of fatty acid synthesis and esterification are hormones; in liver, insulin increases whereas glucagon decreases lipogenesis; in adipose tissue insulin increases and adrenaline decreases lipogenesis. Studies on intact adipose tissue have shown that insulin stimulates the rate of glucose transport into the cell and the activities of pyruvate dehydrogenase, acetyl-CoA carboxylase, and the esterification process (see Figure 17.2). It is likely that the latter two systems are also affected in liver.

The mechanism by which insulin increases the rate of these reactions is an area of intense investigation. (For reviews of control of lipogenesis by insulin, see

---

Figure 17.8 Comparison of the F-type and L-type pathways for conversion of ribulose 5-phosphate to hexose phosphates.
Abbreviations: TK, transketolase; ALD, aldolase; I, isomerase; E, epimerase; PT, phosphotransferase. The numbers indicate the number of carbon atoms and the small dark circles represent phosphate groups.

Denton *et al.*, 1978, 1981; Hardie, 1980.) The effect on glucose transport is discussed in Chapter 22; Section C.2.b. As in other tissues, pyruvate dehydrogenase is located within the mitochondria and is controlled by an interconversion cycle, the properties of the interconverting enzymes are similar if not identical to those given in Figure 7.9. Insulin increases the proportion of the enzyme in the active form (that is, the dephosphorylated form) and the available evidence suggests that it increases the activity of pyruvate dehydrogenase phosphatase.[5]

Acetyl-CoA carboxylase is also activated by insulin. The enzyme exists as a protomer of molecular weight about 500 000 (which is composed of two subunits), which has a very low activity, and as a polymer which is active. The polymer is sufficiently large that it can be seen as a filamentous structure in the electron microscope. Incubation of adipose tissue with insulin increases the proportion of the enzyme in the polymer form. *In vitro* studies of the enzyme demonstrate that citrate markedly increases the activity of the enzyme and causes polymerization, whereas long-chain fatty acyl-CoA favours dissociation to the less active form. It is, however, unclear whether insulin or other hormones change the concentrations of citrate or fatty acyl-CoA in the direction that would support the above hypothesis. (One of the problems is that it is the cytosolic concentrations of the regulators that will affect these enzymes whereas measurements in experimental studies indicate the total cellular concentration (Chapter 7; Note 1).) In addition to the above mechanism, there is increasing evidence that the activity of acetyl-CoA carboxylase is regulated by an interconversion cycle. The enzyme is phosphorylated by two separate protein kinases, cyclic AMP-dependent protein kinase and a more specific kinase, acetyl-CoA carboxylase kinase-2 at two different sites on the molecule (Figure 17.9; Hardie, 1980; 1981). Phosphorylation leads to inactivation (that is, formation of the *b* form) similarly to other biosynthetic systems, for example glycogen synthase (Chapter 16; Section C.1), HMG-CoA synthase (Chapter 20; Section B.1.a), and initiation factor for protein synthesis (Chapter 18; Section D.2). Dephosphorylation is brought about by a phosphatase (probably protein phosphatase-I) which leads to an increased activity of the enzyme (Figure 17.9). The similarity to the mechanism of control of glycogen synthase in muscle is emphasized by comparing Figures 17.9 and 16.4. Since insulin decreases whereas adrenaline (or glucagon) increases the concentration of cyclic AMP in liver and adipose tissue, it is possible that some of the changes in the activity of acetyl-CoA carboxylase caused by these hormones are produced through cyclic-AMP protein kinase. Alternatively, or in addition, insulin might increase the activity of the phosphatase (see Chapter 22; Section C).

Insulin appears to be particularly important in control of esterification in adipose tissue and probably liver. However, it is unclear which of the enzymes in the esterification process is subject to external regulation. It is possible that glycerol phosphate is an external regulator of this process which requires that its concentration should be similar to, or below, the K_m of glycerophosphate

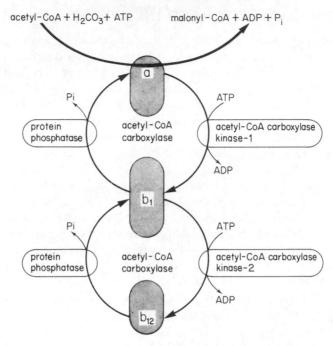

Figure 17.9 Control of activity of acetyl-CoA carboxylase activity by an enzyme interconversion cycle. Acetyl-CoA carboxylase kinase-1 is probably identical to cyclic AMP-dependent protein kinase.

acyltransferase for glycerol phosphate, but this is still uncertain. Insulin increases the concentration of glycerol phosphate in adipose tissue and also increases the activity of glycerophosphate acyltransferase. However, there is some evidence that the latter enzyme may be regulated by an interconversion cycle; cyclic AMP-dependent protein kinase results in inhibition of the enzyme (formation of the *b* form, see Nimmo, 1980). Further studies on the significance of this mechanism in the control of esterification will require purification of the enzyme. Nonetheless, this finding is consistent with insulin increasing the rate of esterification either by decreasing the cyclic AMP concentration or increasing the activity of a phosphatase.

Finally, one interesting feature of the lipogenic process from glucose is that, if the energy requirements are calculated for the over-all pathway indicated in Figure 17.2, more ATP is generated than required (see Appendix 17.1). Of course, this calculation excludes consideration of the ATP expenditure involved in regulation of the process, especially that consumed by the interconversion cycles and the triacylglycerol/fatty acid cycle. Under conditions when the rate of lipogenesis is increased, Brooks *et al.* (1983) have shown the rate of the triacylglycerol/fatty acid cycle is increased (see Figure 8.8, Table 8.6 and Table 14.9) and the rates of the interconversion cycles may also be increased.[5]

B. DESATURATION AND THE ESSENTIAL FATTY ACIDS

A significant proportion ($>50\%$) of the fatty acids in human triacyglycerol are unsaturated (Gellhorn and Marks, 1961). Although desaturation (that is the conversion of $-CH_2CH_2-$ to $-CH=CH-$) can be carried out by enzymes that are present in the liver, certain polyunsaturated fatty acids which are required by the body cannot be made in this way and must be provided in the diet; these are known as the essential fatty acids (EFA).

Although many of the more widely occurring unsaturated fatty acids have acquired trivial names, discussion is facilitated by the use of systematic names which demonstrate structural relationships more clearly. Several systems are in widespread use. In one, the acid is named according to the number of carbon atoms present with the addition of suffixes, -enoic, -dienoic, -trienoic, etc. indicating one, two, three, or more double bonds. The positions of the double bonds are indicated by the symbol Δ with superscript numbers specifying the carbon atoms (numbered from the carboxyl end) involved in double bonds. Examples are given in Table 17.1. In a second, more abbreviated nomenclature, three sets of numbers are used to specify, respectively, the number of carbon atoms, the number of double bonds and their positions. For example, linoleic acid is described as 18:2 (9,12) (see Table 17.1, p. 621).

1. Provision of unsaturated fatty acids

(a) *Enzymes catalysing desaturation*

The liver contains enzymes that can catalyse the insertion of double bonds into long-chain acyl-CoA using molecular oxygen as the oxidizing agent and NADPH or NADH as the co-reductant (for review—see Jeffcoat, 1979). These enzymes, which are known as acyl-CoA desaturases, are attached to the endoplasmic reticulum and contain both a flavoprotein and cytochrome b_5 or cytochrome P-450. The oxygen is partially reduced by the NAD(P)H to produce an enzyme-bound superoxide radical (Chapter 4; Section E.5a) which oxidizes the acyl-CoA. The over-all reaction can be written as follows:

$$CH_3(CH_2)_xCH_2CH_2(CH_2)_yCO.S.CoA \xrightarrow[\substack{NAD(P)H + H^+ \qquad NAD(P)^+}]{\substack{O_2 \qquad 2H_2O}}$$

$$CH_3(CH_2)_xCH=CH(CH_2)_yCO.S.CoA$$

There are, however, important restrictions concerning the desaturations that the mammalian liver can carry out. First, only four desaturases are present (known as Δ^9-, Δ^6-, Δ^5-, and Δ^4-desaturases) and these insert double bonds at positions 9, 6, 5, and 4, respectively. Secondly, if the substrate is fully saturated, the first double bond is always inserted at position 9. When the substrate is already unsaturated, double bonds are inserted between the carboxyl group and the double bond nearest to the carboxyl group. Thirdly, desaturation occurs in such a way as to maintain a methylene-interrupted distribution of double bonds (that is, $-CH=CH.CH_2CH=CH.CH_2-$). Fourthly, desaturation usually alternates with chain elongation and, finally, all double bonds produced by the activity of desaturases of either plant or animal origin are *cis* rather than *trans*.

(b) *Essential fatty acids*

Although some of the unsaturated fatty acids required for the functions described in Section B.2 can be synthesized in the mammalian liver, there are several that cannot, so that these must be present in the diet. From the description of the properties of desaturases given above, it should be clear that the essential fatty acids contain at least one double bond in a position that is greater than nine carbon atoms from the carboxyl end of the fatty acid. It was originally proposed that three unsaturated fatty acids were essential, linoleic acid, linolenic acid and arachidonic acid (Table 17.1). However, since these fatty acids together with γ-linolenic acid can be synthesized in the liver from linoleic acid by the desaturation and elongation systems described above, only linoleic acid is essential if sufficient is provided for the synthesis of the other polyunsaturated fatty acids (see Figure 17.10). These essential fatty acids are found in plant material, and the triacylglycerols in vegetable seed oils (e.g. sunflower seed soil) are particularly rich in the essential polyunsaturated fatty acids (for review see Rivers and Frankel, 1981).

Although the chemistry of the essential fatty acids is known, details of their biochemical and physiological effects are only just being clarified. There has even been some debate concerning whether they are in fact essential in man. In 1929,

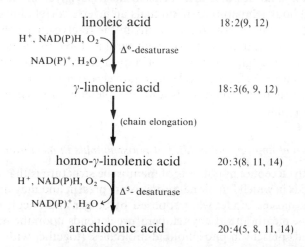

Figure 17.10 Pathway of 'essential' fatty acid synthesis from linoleic acid in the liver

Burr and Burr noted that a deficiency of certain fatty acids in the diet of the rat resulted in scaliness of skin, loss of hair, decreased reproductive capacity and decreased rate of growth despite an increased energy intake and that death occurred early in the life of the animal. Surprisingly, early nutritional studies in man failed to demonstrate a deficiency disease when subjects were maintained on fat-free diets for prolonged periods (Brown *et al.*, 1938; Bryant *et al.*, 1952). However, these experiments were performed on adults and it is now known that the requirement for essential fatty acids is higher in children and, more importantly, that in adults the adipose tissue can provide a store of about 2 kg of essential fatty acids. Signs of deficiency (including scaly and thickened skin, decreased utilization of ingested food and a decreased growth rate) have been demonstrated in patients given fat-free parenteral nutrition. In this condition, the carbohydrate supplied stimulated insulin secretion and thus prevented mobilization of the adipose tissue triacylglycerol (Connor, 1975; Steginsk *et al.*, 1980). All of the deficiency symptoms and signs disappear on feeding linoleic or arachidonic acids.

2. Functions of unsaturated fatty acids

The demonstration that certain unsaturated fatty acids are essential implies that they serve a vital function in the organism. At least four such functions are known.

(a) *Adjustment of melting point of triacylglycerol reserves*

To prevent damage to membranes it is important that triacylglycerol droplets stored in cells remain liquid at the temperature of the organism. However, tripalmitin (tripalmitoylglycerol) has a melting point of approximately 65 °C. The inclusion of unsaturated fatty acids in the triacylglycerols ensures that they are liquid at body temperature. In animals an inverse correlation exists between the proportion of unsaturated fatty acids in the reserve triacylglycerol and the body temperature and in plants triacylglycerol is stored in the form of oils (low melting-point triacylglycerols rich in unsaturated fatty acids) rather than fats. Note that unsaturated fatty acids, being relatively oxidized, have a lower enthalpy for oxidation than saturated acids and are, therefore, less efficient as energy-storage compounds.

(b) *Provision of appropriate fluidity of phospholipids in membranes*

The currently accepted hypothesis of membrane structure is that of a bilayer of phospholipids in which functional proteins are present and may diffuse laterally. Such a fluid mosaic model was proposed by Singer and Nicolson (1972). The fluidity of the membrane at any temperature depends upon the unsaturation of the fatty acids present in phospholipid molecules, (together with the content of cholesterol—see Chapter 20; Section D.1). That fluidity is important is shown by the fact that the properties of membranes change very markedly at temperatures which correspond to the melting temperature of the constituent phospholipids (see Raison, 1973; Martin *et al.*, 1976).

(c) *Provision of precursors for prostaglandin synthesis*

Prostaglandins and related compounds are synthesized from essential polyunsaturated fatty acids. Since these compounds have important regulatory roles in the body they are discussed separately below (Section D).

(d) *Treatment of autoimmune diseases*

For the immune system to act effectively against invading organisms there must exist a mechanism whereby the system can distinguish the individual's own potential antigens (mostly proteins) from those which are foreign, that is, an ability to distinguish 'self' from 'non-self'. It has been suggested that during the perinatal period the immune system in some way 'learns' the body constituents to which it is exposed at that time thereby creating a permanent tolerance, so that it does not react to these proteins. If this is so, it is likely that any proteins (and other components) present but not normally exposed to the immune system (for example, those locked within the cell by an impermeable membrane) will not be learned as 'self'. If subsequent cell damage, for example as a result of viral attack, exposes such proteins, the immune system could become activated to attack them. This response might not only prevent repair but could, if large, lead to further tissue damage with exposure of more antigenic sites, so that the eventual extent of tissue damage could be large and severely detrimental to health and well-being. This problem is termed autoimmunity or, sometimes, autoaggression. Autoimmunity is now considered to be the basis of a wide variety of diseases including multiple sclerosis, rheumatoid arthritis, polyneuritis, and possibly insulin-dependent diabetes mellitus (Chapter 15; Section C.3). Current clinical treatment of autoimmune diseases involves use of steroids and a variety of immunosuppressive drugs, some of which are highly toxic (Salaman, 1981). However, although it is controversial, the suggestion has been made that treatment of these diseases with a diet high in polyunsaturated fatty acids may be beneficial (Meade and Mertin, 1978; Bower and Newsholme, 1978; Crawford and Stevens, 1981). The polyunsaturated fatty acids are conveniently administered as the oils derived from certain plant seeds, for example, those of the sunflower and the evening primrose. One of the authors (E.A.N.) has suggested that the presence of polyunsaturated fatty acids in the normal diet could have a natural immunosuppressive role. He has pointed out that such fatty acids probably reduce immunological activity by effects on lymphocytes. A considerable number of lymphocytes are released into the bloodstream from mesenteric lymph nodes and these in particular will be exposed to high concentrations of any polyunsaturated fatty acids present in the diet, since the latter are transported (in esterified form in chylomicrons) through the lymphatic system before entering the general circulation. It is argued that this natural immunosuppression would be automatically removed during infection by pathogens (when it would be highly undesirable) because such infections rapidly cause anorexia and so reduce the concentration of polyunsaturated fatty acids in the lymph (Newsholme, 1977). If this hypothesis has any validity, then administration of polyunsaturated fatty acids to patients suffering from autoimmune diseases can be seen as an extension of a natural mechanism and not as a drug therapy, thus justifying its inclusion under the heading of functions of unsaturated fatty acids.

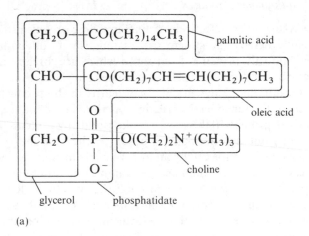

(a)

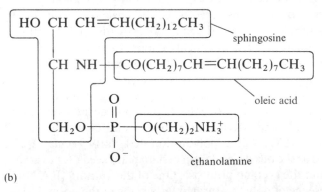

(b)

Figure 17.11 Representative structures of phospholipids.

(a) A phosphatidylcholine (lecithin). Other phosphatidylcholines may have different fatty acids.

(b) A sphingomyelin. Other sphingomyelins may differ in the structures of the sphingosine moiety and the fatty acid; in addition, it may have choline in place of ethanolamine

C. PHOSPHOLIPIDS

1. Structure

Phospholipids may be defined as substances in which both fatty acids and phosphoric acid are esterified to an alcohol. They are thus compound lipids (see Appendix 6.1) and can be divided into two classes depending on whether the alcohol is glycerol (giving derivatives of phosphatidic acid) or sphingosine (giving sphingophospholipids) (Figure 17.11). Groups such as choline, ethanolamine, serine, and inositol may esterify with the second acid group on the phosphate. Further diversity is provided by the variation in the fatty acids present, with palmitic, stearic, oleic, linoleic, and arachidonic being most common. In general, the hydroxyl group at position-1 of the glycerol moiety of a phosphatidic acid is esterified with a saturated fatty acid and that at position-2 is esterified with an unsaturated fatty acid reflecting the specificity of the enzymes involved in the synthesis. If only one of these positions is esterified, the compound is known as a lysophospholipid, so named because of its strong detergent properties so that it readily lyses red blood cells. For further information concerning the structure of phospholipids, see Gurr and James (1971).

Regardless of structure, all phospholipids are markedly polar, possessing both a hydrophobic tail (the two hydrocarbon chains which come to lie close together) and a hydrophilic head (containing the phosphate group and the second alcohol which rotates away from the tail). They are thus well-suited for their roles in membranes and as surfactants.

2. Biosynthesis

The principles of phospholipid synthesis are illustrated by reference to the synthesis of representative examples, namely phosphatidylcholine (lecithin) and diphosphatidylglycerol (cardiolipin).

(a) *Phosphatidylcholine (glycerophosphocholine)*

Cursory examination of the structure of phosphatidylcholine might suggest that its biosynthesis would involve the esterification of choline with phosphatidate formed as an intermediate in the process of triacylglycerol synthesis (see Section A.5). This is not the case. Rather the phosphatidate is hydrolysed to a diacylglycerol and the phosphocholine group transferred from cytidine diphosphocholine (CDPcholine) to the diacylglycerol in a reaction catalysed by cholinephosphotransferase:

$$\text{CDPcholine} + \text{1,2-diacylglycerol} \longrightarrow \text{phosphatidylcholine*} + \text{CMP}$$

* A similar pathway is involved in the synthesis of phosphatidylethanolamine but phosphatidylserine is synthesized by exchange of ethanolamine from phosphatidylethanolamine for serine.

In this reaction, CDPcholine is acting as a donor of the phosphorylated base in a similar way to that of UDPglucose acting as a donor of a glucosyl residue in polysaccharide synthesis. The CDPcholine is formed in two reactions. First, choline is phosphorylated, by choline kinase:

$$\text{choline} + \text{ATP}^{4-} \longrightarrow \text{phosphocholine}^- + \text{ADP}^{3-}$$

Secondly, phosphocholine reacts with the nucleoside triphosphate in a reaction catalysed by cholinephosphate cytidylyltransferase:

$$\text{phosphocholine}^- + \text{CTP}^{4-} \longrightarrow \text{CDPcholine}^{2-} + \text{PP}_i^{3-}$$

Note that conversion of the diacylglycerol to the phospholipid involves the use of three energy-rich phosphate bonds.

Both phospholipids and triacylglycerols are synthesized from phosphatidates, so that the nature of the fatty acid at position-2 should be determined by the specificity of 1-acylglycerophosphate acyltransferase. However, phospholipids frequently possess, at position-2, fatty acids such as arachidonic acid which have a greater degree of unsaturation than do those in triacylglycerols. This is explained by the existence of a pair of enzymes catalysing the deacylation and re-acylation of phospholipids at position-2, so that the fatty acid originally present is replaced by a more unsaturated one (Jeffcoat, 1979). The first of these enzymes, phospholipase A_2, catalyses the hydrolysis of the unsaturated fatty acid from position-2:

$$\text{phosphatidylcholine} + \text{H}_2\text{O} \longrightarrow$$
$$\text{lysophosphatidylcholine} + \text{unsaturated fatty acid}$$

While the second enzyme, 1-acylglycerophosphocholine acyltransferase (1-AGCAT) catalyses esterification of the lysophosphatidylcholine with arachidonoyl-CoA to form the more unsaturated phospholipid:

$$\text{lysophosphatidylcholine} + \text{arachidonoyl-CoA} \longrightarrow$$
$$\text{phosphatidylcholine} + \text{CoASH}$$

A similar acylation/deacylation cycle is responsible for the synthesis of dipalmitoylphosphatidylcholine which is produced in large quantities in the epithelial cells of the lung alveoli where, together with sphingomyelins and protein, it forms pulmonary surfactant (King, 1974). This is secreted from the cells onto the exterior surface of the alveoli where it is essential for the reduction of surface tension. The high surface tension otherwise present would tend to oppose the opening of the alveoli as the lungs expanded (in much the same way as it is difficult to open a wet polythene bag). Furthermore, even if the alveoli opened, in the absence of the surfactant, pressures would be such that fluid would move from the blood into the alveoli thus decreasing the rate of gas diffusion.

The deacylation/reacylation cycle involves, in this case, the replacement of an unsaturated fatty acid by a saturated one (palmitate) and of particular clinical significance is the fact that this occurs only in the late stages of foetal development. The respiratory distress syndrome frequently exhibited by premature babies is a consequence of the failure of the lungs of these babies to synthesize pulmonary surfactant before birth so that the lungs will not fully inflate.[6] The metabolic disturbances are as expected from generalized hypoxia; lactic acidosis and leakage of potassium from the cells which raises the plasma potassium concentration and can result in cardiac arrest. Treatment consists of administration of air enriched with oxygen which, however, is not without its hazards (see Chapter 4; Section E.5), and correction of the metabolic acidosis (Chapter 13: Section D.4).

(b) *Diphosphatidylglycerol*

Diphosphatidylglycerols (cardiolipins) are 'double' phospholipids in which a glycerol molecule acts as a bridge between the phosphate groups of two phosphatidic acid molecules. They occur, in particular, in the inner mitochondrial membrane and are suspected of playing an important role in the organization and operation of electron transfer and oxidative phosphorylation. In the synthesis of cardiolipins the role of CDP is 'reversed' in that it is used to transfer phosphatidyl units onto glycerol (as the phosphate). The pathway is summarized in Figure 17.12. A similar pathway is involved in the synthesis of phosphatidylinositol in animal tissues.

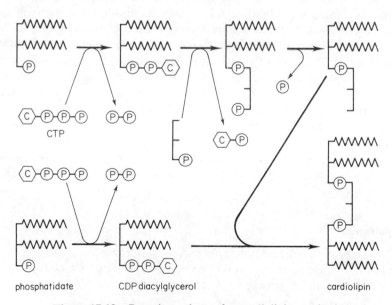

Figure 17.12 Reaction scheme for cardiolipin synthesis

D. PROSTAGLANDINS

The therapeutic effects of semen have been known for centuries; in ancient China it was used to treat gastric ulceration and by North African tribes to initiate labour in pregnant women. von Euler and Goldblatt independently took the first steps towards elucidating the physiological mechanisms involved when, in the 1930s, they demonstrated that the contractile activity of smooth muscle could be increased by extracts of human seminal fluid and the blood pressure could be lowered (Goldblatt, 1935; von Euler, 1934). In the belief that the active agent originated from the prostate gland, it was termed prostaglandin. The discovery that it was actually secreted by the seminal vesicles and is, in fact, produced in very many other tissues, came too late to influence its name. In the early 1960s, as a result of work by Bergström and coworkers, the structure of prostaglandins as cyclic unsaturated fatty acids was elucidated (Bergström *et al.*, 1962). During the next ten years, there was an enormous increase in the understanding of the chemistry, biochemistry, and role of prostaglandins in the regulation of a very wide range of physiological functions. During the last ten years this has continued with increasing appreciation of their clinical importance and therapeutic value (for full details of prostaglandins and their chemical relations, the prostacyclins and thromboxanes—see Berti and Velo, 1981).

1. Structure and nomenclature

The prostaglandins form a large family of related compounds derived from prostanoic acid (Figure 17.13). All prostaglandins possess 20 carbon atoms, a cyclopentane ring, two aliphatic side-chains, a double bond between carbon atoms 13 and 14, a terminal carboxyl group and a hydroxyl group at carbon atom 15. An abbreviated system has evolved for the description of individual prostaglandins. The letters PG are followed by a third (A, B, C, D, E, or F)* to indicate the nature of the cyclopentane ring (which may bear hydroxyl or keto groups and be unsaturated). This is followed by a numerical subscript to indicate the number of double bonds in the side chain. For further details on the nomenclature of prostaglandins the reader is referred to Flower (1979).

Two other groups of compounds, thromboxanes and prostacyclins are now known to be metabolically and functionally related to prostagandins but do not possess the prostanoic acid structure (see Flower, 1979, for review). Examples are shown in Figure 17.13.

2. Synthesis

Prostaglandins are sometimes described as local hormones. Unlike conventional hormones, they are normally synthesized in the cells upon which they act, or in cells of the organ in which they act on other cells. All are derived from

---

* Other letters have been used for intermediates of prostaglandin synthesis.

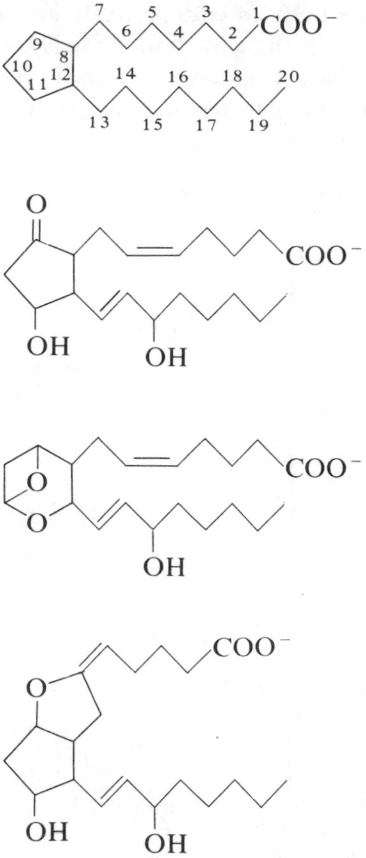

prostanoate

prostaglandin PGE$_2$

thromboxane A$_2$

prostacyclin PGI$_2$

Figure 17.13 Structural formulae of a prostaglandin and related compounds

polyunsaturated fatty acids that have arisen from the essential fatty acids. Using the terminology introduced in Section B, prostaglandins with one double bond arise from the 20:3 (8,11,14) acid (homo-γ-linolenic), prostaglandins with two double bonds from the 20:4 (5,8,11,14) acid (arachidonic), and prostaglandins with three double bonds from the 20:5 (5,8,11,14,17) acid (eicosapentaenoic acid). The reaction sequence for the synthesis of prostaglandin PGE$_2$ from arachidonic acid is outlined below and in Figure 17.14.

The concentration of free arachidonate is very low in all tissues ($< 10^{-6}$ M) but much higher concentrations are present in esterified form in the 2-position of membrane phospholipids. The acid is made available for prostaglandin synthesis by the action of phospholipase A$_2$.[7] The free arachidonate is then oxidized by molecular oxygen to PGG$_2$ in a complex reaction catalysed by fatty acid cyclo-oxygenase, part of the prostaglandin synthase multienzyme complex, which is present in the microsomal fraction of every mammalian tissue so far investigated

(Flower, 1979). PGG_2 possesses both an endoperoxide and a hydroperoxide group and the next reaction involves the reduction of the latter by a peroxidase also in the prostaglandin synthase complex to form PGH_2. By means of isomerization reactions, this intermediate can give rise to PGE_2, thromboxane A_2 or prostacyclin (see Figure 17.14).

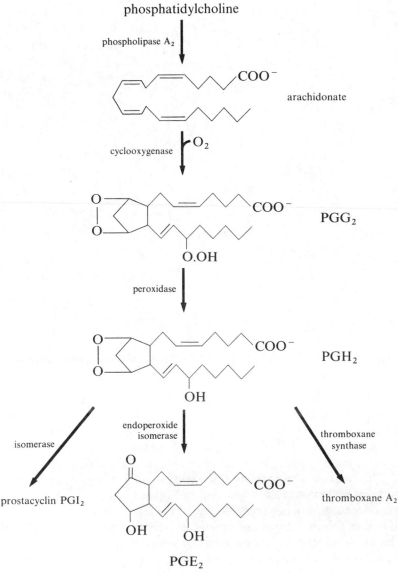

Figure 17.14 Pathway for the synthesis of prostaglandin PGE_2 and related compounds. The cyclo-oxygenase, peroxidase and endoperoxide isomerase are all constituents of prostaglandin synthase

The mechanism of control of the formation of prostaglandins, thromboxanes or prostacyclin is not understood but the flux-generating enzyme is probably the phospholipase (Samuelsson, 1970) and the synthase probably catalyses a non-equilibrium reaction that is controlled by the concentration of the unsaturated fatty acid (that is, by internal control). Many phospholipases require Ca^{2+} as a cofactor so that changes in the intracellular concentration of this ion could be important in control of prostaglandin synthesis.

Arachidonic acid can be degraded by another pathway to produce leukotrienes, so named because they were discovered in the white cells and possess at least three alternating double bonds. Arachidonic acid is converted to monohydroxyeicosatetraenoic acid (5-HPETE) catalysed by the enzyme 5-lipoxygenase, and 5-HPETE is further converted into the various leukotrienes e.g. leukotriene B_4 (5,12-dihydroxy-6,8,10,14-icosatetraenoic acid). The pathway is known as the lipoxygenase pathway. Leukotrienes influence the contraction of smooth muscle and have been strongly implicated as one component responsible for such contractions during asthma (Weissmann, 1983). They, like prostaglandins (see below), may have a large number of regulatory functions.

3. Metabolism

To achieve control of the concentration of a hormone or a messenger, a mechanism for degradation as well as for synthesis must exist (see Chapter 22). Degradation usually involves metabolism to a less active metabolite and in the case of prostaglandins this involves two reactions: the C_{15}-hydroxyl group is oxidized to a keto group and the double bond at C_{13} is reduced in reactions catalysed by 15-hydroxyprostaglandin dehydrogenase and Δ^{13}-prostaglandin reductase, respectively. The carboxylate side chain of the resultant metabolite suffers β-oxidation and the other side-chain is hydroxylated and oxidized to form a second carboxylate group which results in a water soluble compound that is excreted in the urine. The metabolism of thromboxanes and prostacyclins has not yet been elucidated.

4. Physiological functions

Limitations of space prevent a full account of prostaglandin function (for which the review by Hillier and Karim (1979) is recommended) and attention is focussed on those with clinical implications, some of which are considered in more detail in the next section.

Despite the extensive cataloguing of the effects of prostaglandins that has been going on since their characterization, surprisingly little information concerning their molecular mechanism of action has emerged. Undoubtedly, some actions of prostaglandins are mediated by intracellular changes in cyclic AMP concentration and hence change in protein kinase activity (Chapter 7; Section C.1.b; Chapter 22; Section A.4.b).

(a) *Female reproductive system*

Although prostaglandins were first discovered in association with the male reproductive system, their role in the male remains obscure. In contrast, several important roles have been established in the female (see Karim and Hillier, 1979). Some of these may be better appreciated with a knowledge of the hormonal control of the menstrual cycle which is outlined in Chapter 20; Section E.4.

(i) *Steroidogenesis*   Prostaglandin (PGE_2) stimulates steroidogenesis, especially the formation of progesterone in the ovary, and this may be due to an increased activity of adenylate cyclase raising the cyclic AMP concentration.

(ii) *Ovulation*   Prostaglandins PGE and PGF may be involved in the process of ovulation (see Chapter 20; Section E.4.a).

(iii) *Regression of the corpus luteum*   Towards the end of the menstrual cycle, if pregnancy has not occurred, the corpus luteum degenerates. This degeneration appears to be initiated by prostaglandins but the biochemical basis of the degeneration process is unclear. The prostaglandins are produced, not in the ovary, but in the endometrial cells of the uterus in response to the decreasing blood concentration of luteinizing hormone that occurs towards the end of the cycle. The prostaglandins, via the circulatory link that exists between the horn of the uterus and the ovary, provide an informational link between the two tissues.

(iv) *Menstruation*   Prostaglandins appear to be involved in the initiation of menstruation by increasing the contraction of smooth muscle of the myometrium, resulting in ischaemia. Demonstration of prostaglandins in menstrual fluid provided the first evidence of a prostaglandin involvement (Pickles *et al.*, 1965; see also Abel, 1979). This knowledge has led to the use of prostaglandins as abortifacients (see Section G). A study of disorders associated with menstruation has been carried out by Richards (1979) who has shown that dysmenorrhoea (painful menstruation), which is common between 15 and 24 years of age, and menorrhagia (excessive or prolonged menstruation) can, in some cases, be successfully treated with prostaglandin synthesis inhibitors (see also Dingfelder, 1982).

The intra-uterine contraceptive device may be effective because it stimulates the local production of prostaglandins (Chapter 20; Section C.2.b). It is possible that the continuous increased rate of production of prostaglandins causes increased contractility of the myometrium preventing implantation (Karim and Hillier, 1979).

(v) *Parturition*   The concentration of prostaglandins in the amniotic fluid increases more than ten-fold at the onset of labour and this probably reflects the concentration change in the myometrium. Prostaglandins are considered to play a major role in increasing and perhaps initiating the contractile activity of the myometrium at this time (Mitchell, 1979; Anderson, 1980). In contrast, prostaglandins cause relaxation of the muscles of the cervix during labour, a phenomenon vital to normal parturition (Ellwood, 1979).

(b) *Gastric secretion*

Injections of prostaglandins PGE and PGA decrease the secretion of protons and pepsin by the stomach and may, therefore, be involved in the natural regulation of the secretory response of the stomach to food, gastrin and histamine.

(c) *Control of blood pressure*

The effects of prostaglandins on the kidney are to increase the blood flow, increase the rate of renin release and increase sodium excretion. All of these effects could be achieved simply by a vasodilatory effect on the kidney. Furthermore, since these effects would be expected to lower the blood pressure it is possible to suggest that one of the causes of essential hypertension is the inability of the kidney to produce prostaglandins at a sufficient rate (McGiff *et al.*, 1981; *Lancet* editorial, 1981).

(d) *Pain and inflammation*

Prostaglandins make several contributions in the inflammatory response to injury or infection, including vasodilation and increased vascular permeability. Unphysiologically large concentrations of prostaglandin probably cause pain by direct action on pain receptors but the physiological role of prostaglandins in pain stimulation is to sensitize the receptors to the algesic effects of other stimuli. This hyperalgesic action of prostaglandins persists for some time after the concentration has fallen to normal levels.

(e) *Fever*

The temperature regulatory centre is in the hypothalamus and there is evidence implicating prostaglandins in the mechanism by which this centre maintains the body temperature. Thus, raising the concentration of prostaglandins in this centre results in fever; injection of PGE_1 into the third ventricle also induces fever and pyrogens raise the concentration of prostaglandins in the cerebrospinal fluid (Feldberg, 1974). Furthermore, drugs that reduce fever inhibit prostaglandin synthesis (e.g. aspirin, paracetamol; the latter specifically inhibits brain cyclooxygenase and is known to be very effective at reducing body temperature— Vane, 1974).

(f) *Blood clotting*

One of the initial reactions in the complex process of blood clotting is the aggregation of platelets. It has been suggested that prostaglandins are involved in this process but recent evidence suggests that thromboxanes and prostacyclins play a more important part (see McGiff, 1978; Moncada and Vane, 1979a).

Thromboxanes are produced by platelets and are potent inducers of their aggregation. In contrast, prostacyclins, formed and released into the blood by the endothelial cells of artery wall and also produced in the lungs, inhibit platelet aggregation. Under normal conditions the production of prostacyclin is sufficient to override the effect of thromboxanes so that there is no significant aggregation. However, damage to the arterial wall reduces the rate of prostacyclin synthesis and release, so that the effect of thromboxane predominates causing platelet aggregation and initiation of clotting. The presence of an atherosclerotic plaque could similarly reduce the rate of production and release of prostacyclin from the artery thereby increasing the chance of thrombus formation in an otherwise undamaged artery, with possibly fatal consequences.

5. Clinical applications

The increasing knowledge of the ways in which prostaglandins are involved in cellular control processes has been used clinically in two main ways.

1. A number of drugs are in use which reduce the concentration of prostaglandins by inhibiting their synthesis (see below). In fact, most of these were introduced before their mechanism of action was known. Perhaps the best known is acetylsalicylic acid (aspirin). This drug has been used as an analgesic but, although there have been many hypotheses on its mechanism of action (Collier, 1969) there was little or no evidence to support them, until the discovery by Vane (1971) that aspirin inhibits the cyclooxygenase in the prostaglandin synthetic pathway. This can explain all of the known effects of aspirin which, in addition to effecting analgesia, include the ability to reduce fever and inflammation and increase gastric secretion. The increased acidity, which results, explains the damage to the gastric mucosa caused by chronic aspirin intake (Mocada and Vane, 1979b). Furthermore, chronic ingestion of aspirin delays the onset of labour (Rudolph, 1981). In addition, the possible benefit of low doses of aspirin in reducing mortality after myocardial infarction may have some basis if the rate of thromboxane synthesis in platelets is reduced more than that of prostacyclin synthesis in the arterial wall (Preston *et al.*, 1981; Weksler, 1983).

2. The concentration of prostaglandins, prostacyclins or thromboxanes can be increased by administration or by local application. However, the effects of administering the naturally occurring compounds are short-lived since these compounds are rapidly inactivated by the metabolising systems described above. Synthetic prostaglandin analogues which are metabolised more slowly are being produced by the pharmaceutical industry and may be important in clinical practice.

(a) *Drugs reducing prostaglandin concentrations*

Apart from the analgesic effect of aspirin, the main use of drugs that inhibit prostaglandin synthesis has been in the treatment of the inflammatory diseases, such as rheumatoid arthritis and psoriasis, and inflammation caused by injury (Vane, 1974). Two classes of drug, steroid and non-steroid, are currently in use for the treatment of these conditions and they act in different ways. The anti-inflammatory steroids (e.g. prednisolone) inhibit phospholipase A_2 and thus reduce the supply of arachidonate for prostaglandin synthesis. The non-steroidal anti-inflammatory drugs (e.g. acetylsalicylic acid, indomethacin) inhibit the cyclooxygenase.

One further interesting use of prostaglandin synthesis inhibitors (particularly indomethacin) is in the treatment of pregnant women who have a history of premature labour which usually results in loss of the baby (Zuckerman and Harpaz-Kepiel, 1979).

(b) *Administration of prostaglandins or their analogues*

The main clinical use of prostaglandins or their analogues concerns the induction of abortions. They have been used in three different ways to achieve this end (see below).

(i) *Menstrual regulation* The termination of very early pregnancy (known somewhat euphemistically as menstrual regulation) is normally achieved by vacuum aspiration which is a skilled procedure. A pharmacological means of achieving the same end would present some advantages, and knowledge of the role of prostaglandins in causing regression of the corpus luteum and initiating menstruation has been used with advantage. Satisfactory abortions have been achieved by vaginal administration of prostaglandin PGE_2 and $PGF_{2\alpha}$ or their analogues, provided this is done within three weeks of the first missed menstrual period (Bygdeman, 1979). After this period, prostaglandin administration may not result in a complete abortion so that surgery is necessary. Unfortunately, prostaglandin-induced abortions are often accompanied by complications. These include nausea, low abdominal pain, and prolonged bleeding. Understandably, the popular press has followed developments in this field with enthusiasm, if a little uncritically, and has coined the name 'morning-after-pill' to describe this use of the prostaglandins.

(ii) *Abortion during the second trimester* Abortions at this late stage are normally only performed when the pregnancy is abnormal or there is evidence of malformation of the foetus (Karim, 1979; Karim *et al.*, 1979). Intra-amniotic or intravaginal administration of prostaglandin $PGF_{2\alpha}$ or its analogues have been used in such cases, in combination with other methods such as oxytocin infusion or the intra-amniotic infusion of saline.

(iii) *Preoperative cervical dilation* Prostaglandin administration has been used as an aid to surgical abortion. Prostaglandin PGE_2 and its analogues (administered intravaginally or intramuscularly) cause dilation of the cervix, a necessary pre-operative requirement for vacuum aspiration. The use of prostaglandins in this way is considerably less damaging to the cervix than mechanical dilation (Karim and Prasad, 1979).

With increasing knowledge of the mechanism of control of prostaglandin concentrations in tissues and of their mechanism of action, the development of more specific analogues and the increasing acceptability of abortions, it seems likely that prostaglandins will be used more frequently to produce abortions with minimal medical attendance and a high degree of safety.

(iv) *Prostacyclin as an antithrombotic agent* In a number of clinical manipulations, anticoagulation treatment is necessary. For example, during haemodialysis it is necessary to prevent the blood clotting on contact with the dialysis membrane. Heparin is the usual anticoagulant but it has the disadvantage that its action is on the clotting mechanism so that generalized anticoagulation results, with the risk of severe haemorrhage in some patients. Prostacyclin, however, inhibits platelet aggregation and not the clotting mechanism *per se* (Zussman *et al.*, 1981) and it has been suggested that prostacyclin (PGI_2) could replace heparin with advantage to patients undergoing haemodialysis.

APPENDIX 17.1   ATP FORMATION DURING LIPOGENESIS

An analysis of the requirements of acetyl CoA and NADPH for fatty acid synthesis and the stoichiometry of the reactions producing these precursors demonstrate that during synthesis from glucose a net generation of ATP should occur within the adipocyte (Flatt, 1970). The number of molecules of ATP produced per fatty acid molecule synthesized depends upon the relative contributions of the pyruvate/malate cycle (Figure 17.3) and the pentose phosphate pathway to the supply of NADPH. An example of a stoichiometry of the pathway illustrating ATP production is given in Table 17.3. In this example, 57% of the NADPH required for fatty acid synthesis is provided by the pyruvate/malate cycle whereas *in vivo*, it may contribute about 40%. Consequently more than five molecules of ATP could be produced per molecule of palmitate synthesized.

NOTES

1. The importance of the pyruvate/malate shuttle in lipogenesis is shown by the fact that administration of *threo*-hydroxycitric acid, which is a specific inhibitor of ATP citrate lyase, markedly reduces fatty acid and triacylglycerol synthesis *in vivo*. Such specific inhibitors of lipogenesis could be important in the treatment of obesity, and hydroxycitric acid does have an anorexic effect (Sullivan and Triscari, 1977).

Table 17.3. Stoichiometric production of palmitate and ATP from glucose*

Reaction number	Reaction
1	$4\frac{1}{2}$ glucose + $4\frac{1}{2}$ ATP → $4\frac{1}{2}$ glucose 6-phosphate + $4\frac{1}{2}$ADP
2	3 glucose 6-phosphate + 6NADP$^+$ → glyceraldehyde 3-phosphate + 2F6P + $3CO_2$ + 6NADPH
3	$1\frac{1}{2}$ glucose 6-phosphate → $1\frac{1}{2}$ fructose 6-phosphate
4	$3\frac{1}{2}$ fructase 6-phosphate + $3\frac{1}{2}$ ATP → 7 glyceraldehyde 3-phosphate + $3\frac{1}{2}$ADP
5	8 glyceraldehyde 3-phosphate + 8NAD$^+$ + 16ADP + $8P_i$ → 8 pyruvate + 8NADH + 16ATP
6	8 pyruvate + $4O_2$ + 24ADP + $24P_i$ + 8CoA → 8 acetyl-CoA$_{(mito)}$ + $8CO_2$ + 24ATP
7	8 acetyl-CoA$_{(mito)}$ + 8NADH + 8NADP + 20ATP → 8 acetyl-CoA$_{(cytosol)}$ + 8NAD + 8NADPH + 20ADP + $20P_i$
8	8 acetyl-CoA$_{(cytosol)}$ + 7ATP + 14NADPH → palmitate + 14NADP + 8CoA + 7ADP + $7P_i$
Sum:	$4\frac{1}{2}$ glucose + $4O_2$ + 5ADP + $5P_i$ → palmitate + $11CO_2$ + 5ATP

* Data from McGilvery (1979). Of the $4\frac{1}{2}$ molecules of glucose required to produce one molecule of palmitate, three enter the pentose phosphate pathway (reaction 2) and generate six molecules of NADPH (reaction 2). The other $1\frac{1}{2}$ glucose 6-phosphate molecules, together with the two molecules of fructose 6-phosphate produced by the pentose phosphate pathway enter the glycolytic pathway to give seven molecules of glyceraldehyde 3-phosphate (reaction 4). Since one molecule of the latter is produced in the pentose phosphate pathway (reaction 2) a total of eight continue in the glycolytic pathway to form eight molecules of pyruvate and eight molecules of NADH (reaction 5). After entering the mitochondrion, the pyruvate is oxidized to acetyl-CoA (reaction 6). Transfer of the latter out of the mitochondrion is achieved via citrate which in the cytosol produces acetyl-CoA and oxaloacetate which is converted to pyruvate producing NADPH (reaction 7 and Figure 17.3). Finally, the cytosolic acetyl-CoA is converted to palmitate, using the NADPH produced in reactions 2 and 7.

2. It has been suggested that dihydroxyacetone itself could act as a substrate for acyl transfer in esterification. However, since its concentration is at least ten-times lower than that of glycerol 3-phosphate it is unlikely to be quantitatively important in this role.

3. Since both pyruvate carboxylase and phosphoenolpyruvate carboxykinase are present in adipose tissue, glycerol 3-phosphate could be synthesized from lactate, pyruvate or alanine. However, the major role of pyruvate carboxylase in this tissue is probably in the pyruvate/malate shuttle as described in Section A.3 and that of phosphoenolpyruvate carboxykinase is in amino acid metabolism as described in Chapter 10; Section C.4.a.

4. The reaction catalysed by the $NADP^+$-linked malate dehydrogenase (see A.3.b) is considered to be near-equilibrium so that the concentration ratio, $NADP^+/NADPH$, is given by:

$$[NADP^+]/[NADPH] = \frac{[pyruvate^-][CO_2]}{[malate^-]} \cdot \frac{1}{K_{eq}}$$

From the measurement of concentrations of pyruvate, carbon dioxide and malate in the rat liver, the cytosolic ratio is calculated to be about 0.01 (Krebs, 1973).

5. The activity of pyruvate dehydrogenase kinase can be measured in intact mitochondria isolated from adipose tissue by measurement of incorporation of radioactivity from $[^{32}P]ATP$ into pyruvate dehydrogenase. Mitochondria prepared from adipose tissue incubated with insulin actually show an increase in the activity of the kinase which is probably due to an increase in the concentration of the substrate, dephosphorylated pyruvate dehydrogenase. The activation by insulin, therefore, does not involve inhibition of kinase but probably an activation of the phosphatase. Unfortunately, the activity of the latter cannot be measured in intact mitochondria and has to be assayed in mitochondrial extracts. This enzyme shows no change in activity in extracts of mitochondria from adipose tissue incubated with insulin. It is probable that the activity of the phosphatase is increased in the intact mitochondria by a change in the concentration of a regulator (perhaps Ca^{2+} or Mg^{2+}) which may be affected by insulin. (For a review of insulin effects on pyruvate dehydrogenase activity, see Denton *et al.*, 1978; Denton *et al.*, 1981).

6. The condition is characterized by a rapid, grunting respiration with indrawing of both ribs and sternum. Dyspnoea and cyanosis progressively increase. The concentration of saturated phosphatidylcholine in amniotic fluid samples has been measured in an attempt to predict respiratory distress syndrome in abnormal pregnancies (Torday *et al.*, 1979).

7. Phospholipase enzymes are classified according to the bond in the phospholipid that is hydrolyzed (Figure 17.15). Phospholipase B attacks lysophospholipids (Section C1) from which it releases a fatty acid.

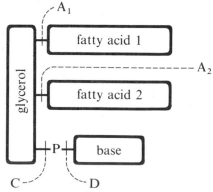

Figure 17.15 Sites of action of phospholipases.

CHAPTER 18

BIOSYNTHESIS OF NUCLEIC ACIDS AND PROTEINS

The synthesis of the macromolecules, protein and nucleic acid, is sometimes omitted from discussions of metabolism and considered under the separate heading of molecular biology. Macromolecule synthesis too, however, depends on the formation of covalent bonds by enzyme-catalysed reactions so that the process and its regulation should be as amenable to analysis along the lines laid down in Chapters 2, 3, and 7 as is the synthesis of smaller molecules. Since so much is now known of the mechanisms of protein and nucleic acid synthesis, only a summary of the processes is given here but this should be sufficient to understand the current views on the mechanisms of the control of both rate and specificity in protein synthesis. For comprehensive reviews on protein and nucleic acid synthesis in higher animals, see Hoagland, 1978; Weissman, 1980.

Until the early 1950s very little was known of protein or nucleic acid synthesis. The development of techniques to study these processes and the challenges posed by the problem of information transfer in living organisms led to the enormous and rapid growth of these subjects in the following decades. Much of the earlier work was carried out on the experimentally more amenable prokaryotes (that is, bacteria) and viruses. Studies on eukaryotes[1] have shown that the basic mechanisms in all organisms are similar but there are important differences in detail. Emphasis in this chapter is given to mechanisms in eukaryotes (for reviews, see Campbell, 1977 and Pain, 1978).

A. AN OVERVIEW OF STORAGE AND TRANSFER OF INFORMATION IN PROTEIN SYNTHESIS

The central dogma of information transfer in protein synthesis is that the specific information contained in the deoxyribonucleic acid (DNA) molecules is transcribed into ribonucleic acid (RNA) molecules and the latter are responsible for the synthesis of the many thousands of proteins in a given organism (see Section C). The portion of DNA that carries information for synthesis of a single polypeptide chain is termed a structural gene.[2] The question arises how can a molecule contain information? The answer is through its covalent structure. An analogy may make this clearer. A book contains information encoded in about 30 symbols (the letters, the space, and a few punctuation marks). Given a few conventions (such as starting in the top left-hand corner and reading sequentially

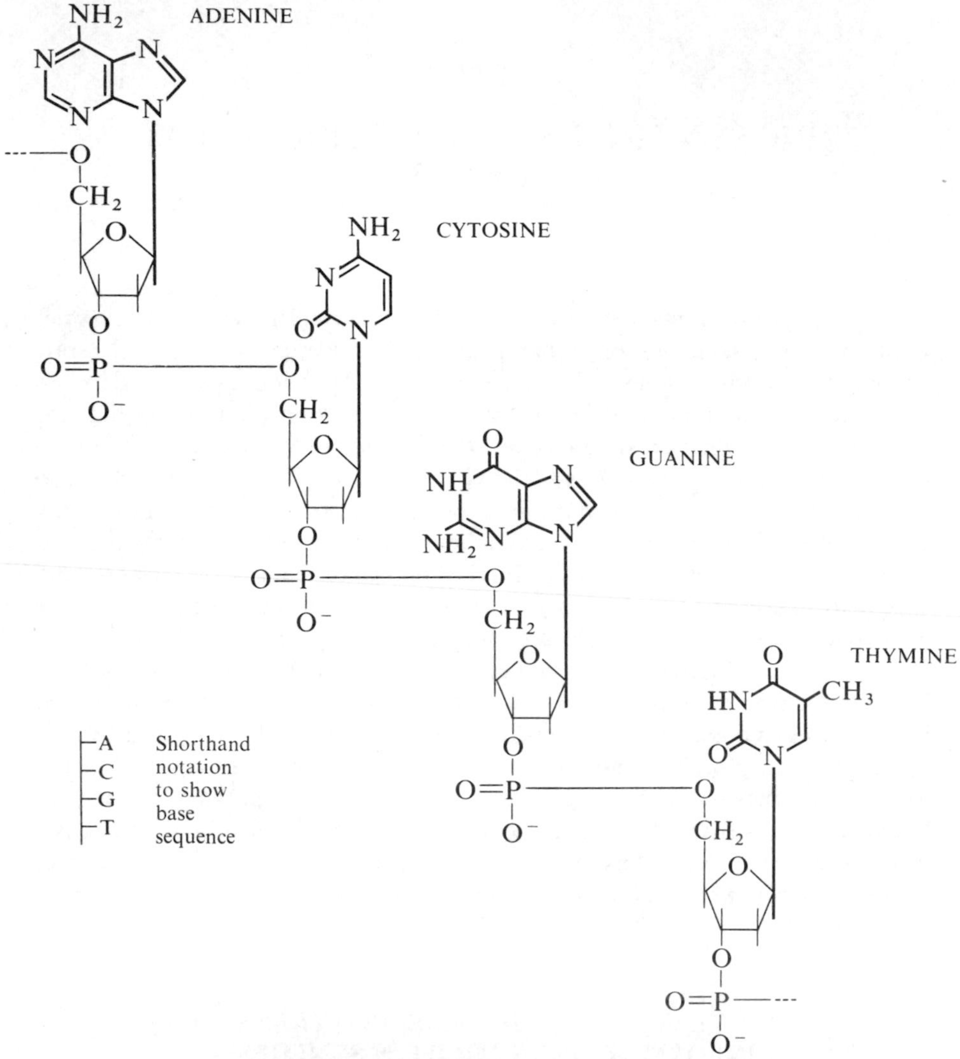

Figure 18.1 Covalent structure of section of a single strand of DNA. Note that interatomic distances are considerably distorted in this representation

from left to right in successive horizontal lines) vast amounts of information can be contained in a relatively small volume. The same message could be represented by fewer symbols but the message would then occupy more space (if, for example, the book were written in Morse code). In DNA there are only four 'symbols', namely adenine, cytosine, guanine, and thymine (the purine and pyrimidine bases), differing in covalent structure and conveniently known by their initial letters A, C, G, and T (Figure 18.1). The linear sequence of these bases in the DNA molecules (see Section B) determines the primary structures of all proteins

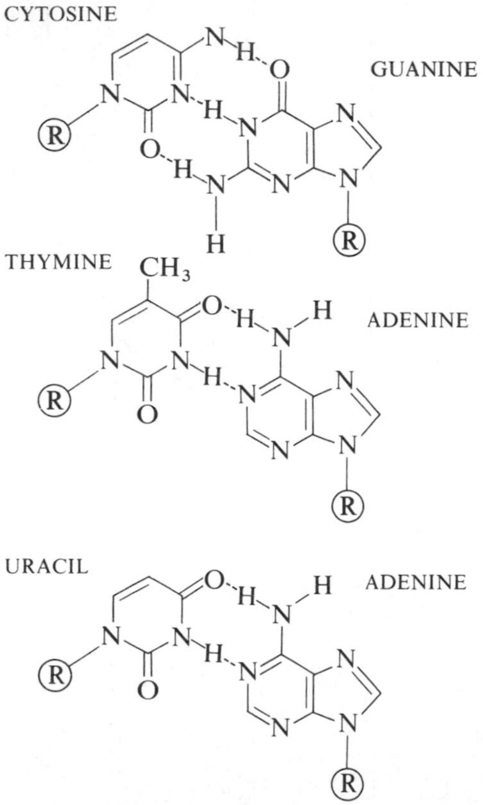

Figure 18.2 Hydrogen bonding (indicated by a broken line) between complementary base pairs. Although the interatomic distances are not drawn to scale it should be noted that the total width of all base pairs is virtually identical. The pairs C–G and T–A are found in DNA; the pairs C–G and U–A are found in RNA. Strands of DNA and RNA can associate through these pairs

produced by a cell. In prokaryotes, the total information is contained in a single DNA molecule; in eukaryotic cells the DNA is present in several separate molecules each forming a separate chromosome* (like the volumes of a complete work). Precise replication of DNA (that is, synthesis of an exact copy) and chromosome movements during cell division ensure that each daughter cell (and, ultimately, the offspring) has an exact copy[3] of the genetic information possessed by its parent.

The key to understanding the mechanism of information transfer in living organisms lies in the phenomenon of base-pairing exhibited by the bases in the

---

* The DNA of a human gamete is contained in 23 different chromosomes and in a normal cell there are 23 pairs of chromosomes. These chromosomes become visible in the light microscope during cell division. They contain not only DNA but also protein molecules which presumably play important structural and regulatory roles.

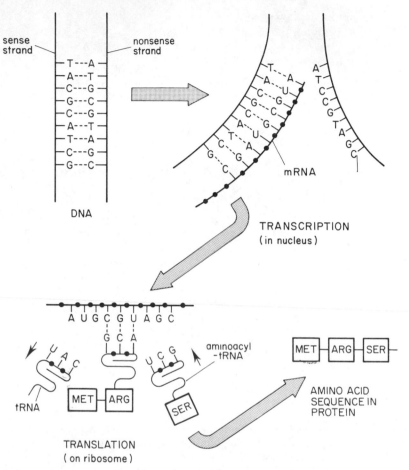

Figure 18.3 Information transfer in protein synthesis. See text for explanation

nucleic acids. Although each base is capable of forming hydrogen bonds[4] with any other base, relatively strong hydrogen bonds are only formed between specific pairs of bases, that is, between adenine and thymine (or uracil in RNA) and between guanine and cytosine[5] (Figure 18.2). The significance of the hydrogen bonds is that they are strong enough to keep the bases together, especially when a sequence of bases in one chain is bonded to a complementary sequence in a second chain (double-stranded DNA molecules) but are weak enough to be broken during replication (see below).

The process of the transfer of this information to the structure of proteins is best illustrated by reference to a specific example. Let us assume that the base sequence TACGCATCG represents a (small) part of the DNA in the gene determining the structure of a protein, P, which consists of a single polypeptide chain (Figure 18.3). In the chromosome, this sequence will exist in association with a complementary length of DNA possessing the base sequence ATGCGTAGC. Since the strand containing this latter sequence does not contain information that is used to determine protein structure, it is known as the nonsense strand.

The first step in information transfer is the formation of a molecule of RNA known as messenger RNA (mRNA). By the catalytic action of an RNA polymerase enzyme, ribonucleotides are polymerized in a specific sequence determined by the base sequence in the aligned DNA-sense strand. This sequence is achieved by base-pairing between the DNA and the ribonucleotides so that the sequence in the RNA is completementary to that of the DNA. Indeed, the same base-pairing 'rules' that determine DNA pairing determine that involved in mRNA synthesis, except that the base uracil (U) is substituted for thymine (T). Thus the newly-formed mRNA contains the base sequence AUGCGUAGC. This process is called transcription (since the nucleotide language is transcribed from one form to another) and takes place in the nucleus. The RNA remains single-stranded and is transported across the nuclear membrane into the cytosol. Here it associates with ribosomes where translation of the base sequence into an amino acid sequence occurs.* A sequence of three bases, in a given mRNA molecule, corresponds to a specific amino acid and these triplet sequences are known as codons.[6] The amino acids are assembled along the mRNA according to the base sequence (or codon sequence) and are polymerized so that a polypeptide is formed. The successive assembly of amino acids along the mRNA molecule requires a series of smaller RNA molecules, known as transfer RNAs (tRNA). These act as adaptors since one particular kind of tRNA molecule binds a specific amino acid to one end and possesses the complementary base sequence of the codon for this amino acid, that is, the anticodon, at the other. It is therefore responsible for the correct positioning of the amino acid onto the mRNA during translation. The sequence of mRNA used in the above example comprises three codons; the first (AUG) associates with the tRNA that possesses the anticodon UAC. Since this tRNA binds with methionine, this amino acid is incorporated into the growing peptide chain. This is followed by incorporation of the amino acids arginine (codon, GGU) and serine (codon, AGC). The process is outlined in Figure 18.3 and certain aspects are described in more detail in subsequent sections.

B. NUCLEOTIDES AND NUCLEIC ACIDS

Since nucleic acids and their structural components, the nucleotides, play a fundamental role in protein synthesis, their structure and biosynthesis are described briefly in this section. Nucleotides are polymerized to form nucleic acids but, in addition, they have a wide variety of metabolic roles† which are discussed in other chapters.

---

* The process is called translation because the language changes from that of nucleotides to that of amino acids.

† For example, as ATP, ADP, cyclic AMP, FAD, coenzyme A, NAD^+, $NADP^+$, UDPglucose

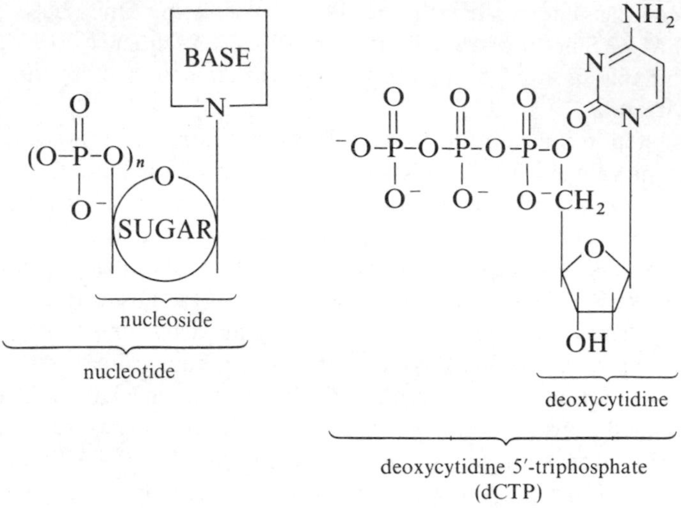

Figure 18.4 General structure of a nucleotide with representative example. For structural formulae of other nucleotides see Figures 18.1 and 18.2

1. Nucleotides

Nucleotides consist of three parts: a nitrogen-containing ring system (known as the base which, in nucleic acids, is either a purine or pyrimidine); a sugar (ribose or deoxyribose in nucleic acids) and one or more phosphate groups. A base linked to the carbonyl carbon atom of a sugar but without a phosphate group is known as a nucleoside. The structures of a nucleotide and a nucleoside are depicted in Figure 18.4 (see also Figure 2.2 for structure of ATP). Those contributing to nucleic acids are listed in Table 18.1.

All of the common nucleotides can be synthesized from simple intermediates in most tissues of man and a brief description of the pathways involved are given below and summarized in Figure 18.5. However, in the degradation of nucleic acids or nucleotides, free purines and pyrimidines are produced and these can be converted back to nucleotides via a synthetic process known as the 'salvage' pathway (see below). A more complete, albeit a more complex, picture is given in Figure 18.6 and details are provided by Seegmiller (1980). Since purines and pyrimidines are necessary for synthesis of nucleic acids which are essential for cell division and growth, there is considerable interest in antimetabolites that interfere in the biosynthetic pathways as antitumour drugs (see Montgomery *et al.*, 1979).

(a) *Purine ribonucleotide synthesis*

Both pathways for purine nucleotide synthesis, that is, the *de novo* pathway and the salvage pathway are outlined in this section.

Table 18.1. Bases commonly occurring in nucleic acids

Base*	Abbreviation	Type	Nucleoside†	DNA or RNA
Adenine	A	purine	adenosine	DNA plus RNA
Cytosine	C	pyrimidine	cytidine	DNA plus RNA
Guanine	G	purine	guanosine	DNA plus RNA
Thymine	T	pyrimidine	thymidine	DNA only
Uracil	U	pyrimidine	uridine	RNA only

* A large number of other bases (mostly methylated derivatives of the above) occur in small amounts in ribonucleic acids (especially tRNA).

† Nucleotides are named by adding a term denoting the number of phosphate groups (and their position) to the name of the nucleoside, e.g. adenosine 5′-triphosphate (ATP). In general, ribonucleotides are assumed unless the prefix deoxy (or d- in abbreviations) is used. See Figures 18.1 and 18.2 for structural formulae.

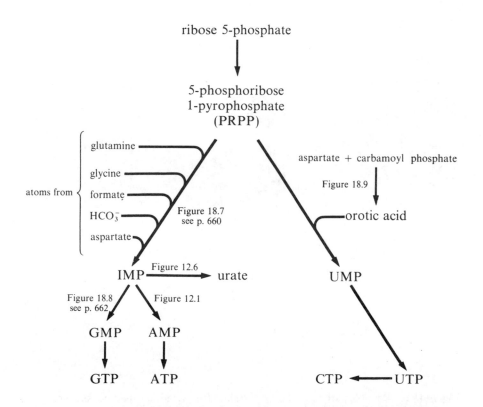

Figure 18.5 Outline of purine and pyrimidine ribonucleotide biosynthesis

658

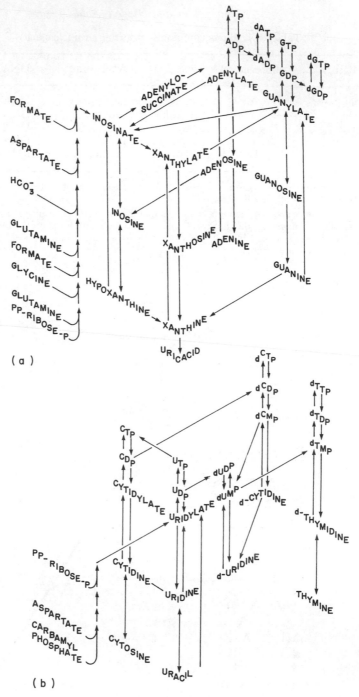

Figure 18.6 Possible pathways of (a) purine and (b) pyrimidine metabolism. No attempt is made to indicate quantitative importance. Taken from Henderson *et al.* (1977) with kind permission

(i) *De novo pathway*   The basic starting point for the formation of a purine nucleotide is a ribose 5-phosphate molecule (provided by 5-phosphoribosyl-pyrophosphate now more correctly known as 5-phosphoribose 1-diphosphate) upon which carbon and nitrogen atoms are added until the purine nucleotide is formed (see Figure 18.5 for overview and Figure 18.7 for details of the individual reactions). Two of the nitrogen atoms are provided by the amide groups of glutamine, one by aspartate and one by glycine which also provides two carbon atoms. Two more carbon atoms are transferred from folate derivatives (see Appendix A.6) and one enters as the hydrogencarbonate ion. The first nucleotide formed is inosine monophosphate (IMP, known also as inosinic acid or inosinate) from which both GMP and AMP can be produced (see Figure 18.5 and also Figures 12.1 and 18.8). To convert both purine and pyrimidine nucleotides to nucleoside triphosphate, two additional kinases are required, neither of which is specific for the base involved:

nucleosidemonophosphate kinase
ATP + nucleoside monophosphate $\longrightarrow$ ADP + nucleoside diphosphate

nucleosidediphosphate kinase
ATP + nucleoside diphosphate $\longrightarrow$ ADP + nucleoside triphosphate

There have been no systematic studies on the equilibrium nature of the reactions of the purine nucleotide synthesis pathway so that it is not possible to identify flux-generating steps. Based on the *in vitro* properties of the enzyme, it is assumed that the inhibition of amidophosphoribosyltransferase by IMP is a feedback inhibitory mechanism *in vivo*[7] (see Figure 18.7). Similarly the conversion of IMP to GMP or AMP may be regulated by a feedback inhibitory mechanism; the *in vitro* properties of the enzymes suggest that AMP is an allosteric inhibitor of adenylosuccinate synthetase (Figure 12.1) and GMP is an allosteric inhibitor of IMP dehydrogenase (see Figure 18.8).

(ii) *Salvage pathway*   In the pathway for the degradation of the purine nucleotides, described in detail in Figure 12.6, the enzyme 5'-nucleotidase converts AMP, GMP, and IMP to their respective nucleosides. The bases adenine, guanine and hypoxanthine are released from these nucleosides by the action of purine nucleoside phosphorylase. These bases can be converted to urate for excretion but can also be converted back into their respective nucleotides by 'salvage' reactions in which a 5-phosphoribosyl group is transferred from 5-phosphoribosylpyrophosphate (PRPP) to the base in reactions catalysed by phosphoribosyltransferases:

base + $PRPP^{5-} \longrightarrow$ mononucleotide^{2-} + PP_i^{3-}

Two enzymes are involved. Adenine phosphoribosyltransferase and hypoxanthine phosphoribosyltransferase. The former is specific for adenine but the

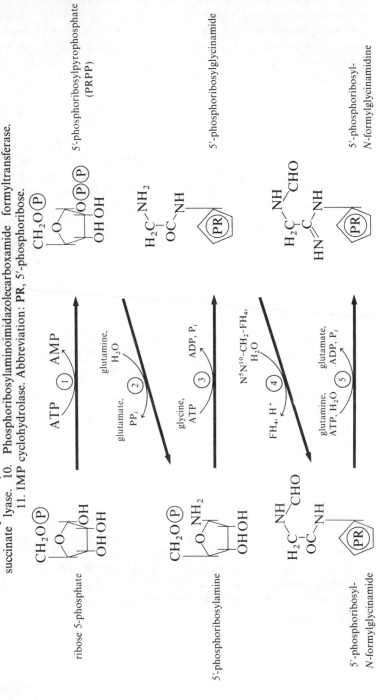

Figure 18.7 Pathways of purine nucleotide synthesis. The following enzymes are involved: 1. Ribosephosphate pyrophosphokinase. 2. Amidophosphoribosyltransferase. 3. Phosphoribosylglycinamide synthetase. 4. Phosphoribosylglycinamide formyltransferase. 5. Phosphoribosylformylglycinamide synthetase. 6. Phosphoribosylaminoimidazole synthetase. 7. Phosphoribosylaminoimidazole carboxylase. 8. Phosphoribosylaminoimidazole-succinocarboxamide synthetase. 9. Adenylosuccinate lyase. 10. Phosphoribosylaminoimidazolecarboxamide formyltransferase. 11. IMP cyclohydrolase. Abbreviation: **PR**, 5'-phosphoribose.

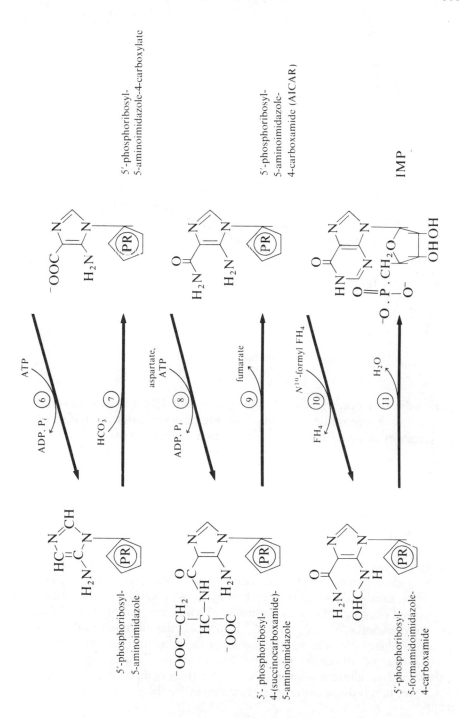

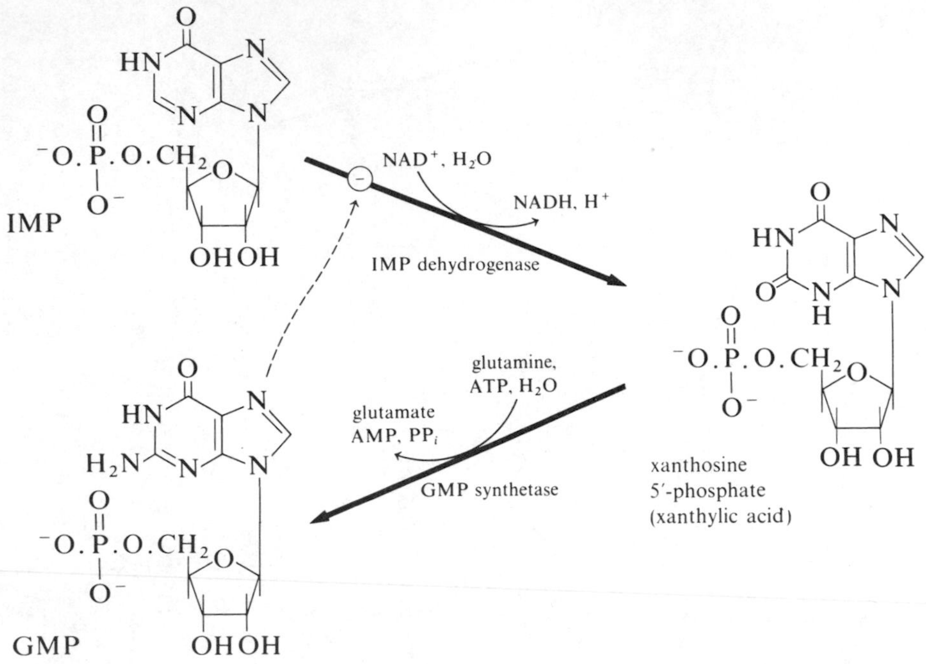

Figure 18.8 Pathway from IMP to GMP

latter will utilize either hypoxanthine or guanine (and for this reason is sometimes known as hypoxanthine-guanine phosphoribosyltransferase (HGPRT). A deficiency of this enzyme results in the Lesch–Nyhan syndrome.[8]

(b) *Pyrimidine ribonucleotide synthesis*

(i) *De novo* pathway A fundamentally different pathway is involved in the biosynthesis of the pyrimidine nucleotides (UTP, CTP and TTP) since the pyrimidine ring (in the form of orotate) is synthesized before reaction with phosphoribosylpyrophosphate (Figure 18.9). All the atoms of the pyrimidine ring are brought together by the initial condensation of carbamoyl phosphate with aspartate, catalysed by aspartate carbamoyltransferase. Unlike the formation of carbamoyl phosphate for urea synthesis, which utilizes ammonia and occurs in the mitochondria (Chapter 12; Section B.2), the formation of carbamoyl phosphate for pyrimidine synthesis occurs in the cytosol and utilizes glutamine as the source of nitrogen. On the basis of *in vitro* properties of the carbamoyltransferase, it is assumed that the activity of this enzyme, and hence the rate of the biosynthetic pathway, is regulated by feedback inhibition by CTP and UTP.

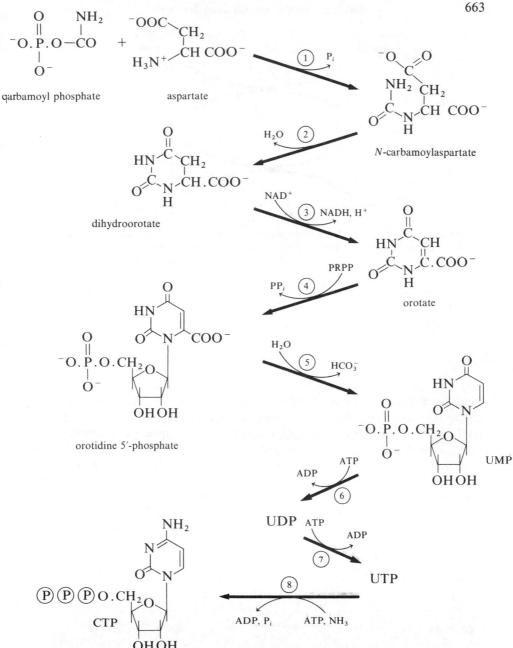

Figure 18.9 Pathway for the synthesis of pyrimidine nucleotides. The following enzymes are involved: 1. aspartate carbamoyltransferase; 2. dihydro-orotase; 3. orotate reductase; 4. orotate phosphoribosyltransferase; 5. orotidine-5′-phosphate decarboxylase; 6. nucleosidemonophosphate kinase; 7. nucleosidediphosphate kinase; 8. CTP synthetase

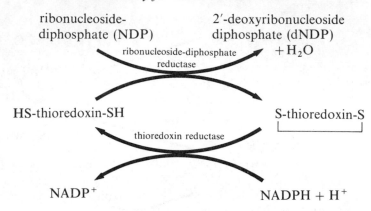

ribonucleoside-diphosphate (NDP) → 2'-deoxyribonucleoside diphosphate (dNDP) + H₂O

ribonucleoside-diphosphate reductase

HS-thioredoxin-SH — S-thioredoxin-S

thioredoxin reductase

NADP⁺ — NADPH + H⁺

Figure 18.10 Pathway for deoxyribonucleotide synthesis. ATP and Mg^{2+} are required for the ribonucleoside-diphosphate reductase system, so named because it probably comprises two enzymes

(ii) *Salvage pathway* In a similar manner to the purines, the pyrimidine bases can be converted to the mononucleotides by reaction with 5-phosphoribosylpyrophosphate.

(c) Deoxyribonucleotide synthesis

The nucleotides involved in DNA formation are deoxyribonucleotides rather than ribonucleotides, that is, the hydroxyl group in the 2'-position of the ribose is replaced by hydrogen. The formation of deoxyribose from ribose occurs at the level of the nucleoside diphosphates, rather than triphosphates, despite the fact that the latter are precursors in nucleic acid synthesis. ADP, GDP, CDP, and UDP are converted to their deoxy-derivatives in a reduction reaction catalysed by ribonucleoside-diphosphate reductase which is part of a multienzyme system which includes thioredoxin, a small protein containing two cysteine groups which are cyclically oxidized by the ribonucleotide and reduced by NADPH (see Figure 18.10). However, although deoxyuridine diphosphate is produced by this

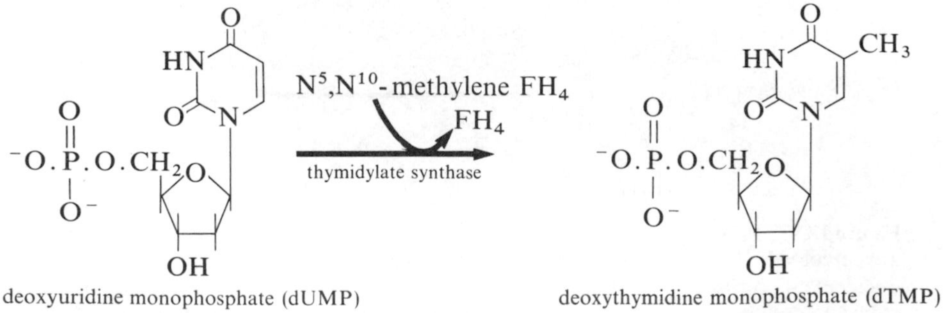

N^5,N^{10}-methylene FH₄

FH₄

thymidylate synthase

deoxyuridine monophosphate (dUMP)

deoxythymidine monophosphate (dTMP)

Figure 18.11 The synthesis of deoxythymidine monophosphate

reaction, deoxythymidine, required for DNA synthesis, cannot be. In fact, deoxythymidine monophosphate (from which dTDP and dTTP are formed) arises from deoxyuridine monophosphate in a reaction involving N^5N^{10}-methylene-tetrahydrofolate catalysed by thymidylate synthase (Figure 18.11).

In order for DNA to be synthesized from the deoxynucleotides, all must be available at approximately similar concentrations. Although there appears to be only one enzyme (ribonucleoside-diphosphate reductase) for reducing all the nucleosides to deoxynucleotides, specificity is controlled by allosteric regulation in which changes in the concentrations of the various deoxynucleoside triphosphates either increase or decrease the affinities for different ribonucleoside diphosphates (Thelander and Reichard, 1979). (This has important clinical implications in patients deficient in adenosine deaminase or purine nucleoside phosphorylase, see Chapter 12; Note 6.)

2. Nucleic acids

(a) *General structure*

Nucleic acids are polymers of nucleotides in which phosphate groups form diester links between adjacent nucleotides. In naturally occurring nucleic acids the link is between the 3′-position of one nucleotide and the 5′-position of the next. This results in the formation of very long chains of alternating sugar and phosphate groups to which the bases are attached in an all-important sequence (Figure 18.1).

The importance of complementary base-pairing in the function of nucleic acids in the transfer of information has been described in Section A. Base-pairing in nucleic acid can occur between different regions of the same molecule (e.g. in tRNA), between complementary molecules to form a double molecule or duplex (e.g. in DNA) or between complementary regions of quite different molecules (e.g. between DNA and mRNA). Despite the importance of base-pairing in nucleic acid structures, other non-covalent interactions are also important. For example, repulsion between the negatively-charged phosphate groups favours an extended conformation and, in addition, the phosphate groups form ionic links with divalent metal ions (especially Mg^{2+}), polyamines (such as spermine) and basic proteins (e.g. histones).

(b) *Biosynthesis of nucleic acids*

The precursors for the formation of nucleic acid are the nucleoside triphosphates. Since deoxyribonucleotides are produced from ribonucleoside diphosphates, it is necessary to convert dexoyribonucleoside diphosphates into triphosphates, by the action of nucleosidediphosphate kinase, prior to nucleic acid synthesis.

The conversion of the triphosphate to nucleic acid is catalysed by enzymes widely known as nucleic acid polymerases but classified as nucleotidyl-transferases. The basic reaction can be represented as follows:

$$n\text{NTP} \rightarrow \text{nucleic acid} + n\text{PP}_i$$

A large number of such polymerases have now been isolated. They vary according to the specificity for ribo- or deoxyribonucleotides, the direction of chain extension ($3' \rightarrow 5'$ or $5' \rightarrow 3'$), the requirement for a primer (an extended polynucleotide, similar to the requirement of a small oligosaccharide for glycogen synthase in glycogen synthesis) and the requirement for a template (a polynucleotide which determines, through complementary base-pairing, the nucleotide sequence). An enzyme catalysing DNA synthesis during replication is known as a DNA-dependent DNA polymerase while one catalysing mRNA synthesis during transcription is a DNA-dependent RNA polymerase. (The dependency indicates the requirement for a template.) In addition to these polymerases, a number of enzymes which can join together two nucleotide chains are widely distributed. These are sometimes known as ligases or repair enzymes. Since ATP is used, but not incorporated, they are classified as polynucleotide synthetases. They serve not only to repair broken strands in duplexes but also play an essential part of DNA·replication which proceeds in short lengths to produce fragments of DNA which are subsequently joined. They are also involved in the exchange of sections of DNA molecules that occurs during crossing-over.

(c) *Deoxyribonucleic acid (DNA)*

DNA appears to have a single function, that of storing the information required for protein synthesis. In all cells, except for some viruses, DNA molecules exist as a pair of complementary strands which assume the now classical double helical conformation. The association between strands is so intimate that the duplex is often referred to as a single molecule although there is no covalent link between the two strands. This produces a very stable structure so important for the storage of information. It also permits semi-conservative replication, that is, separation of strands and synthesis of new complementary strands along each of the separated ones to give two identical duplex molecules. The process of replication is, however, far from simple (for detailed description, see Kornberg, 1978).

(d) *Messenger RNA (mRNA)*

Messenger RNAs are polymers of ribonucleotides that transfer information for protein synthesis from the nucleus to the ribosomes in the cytosol of the cell. The polymers are single-stranded which allows interaction, by complementary base-pairing, with tRNA, and probably ribosomal RNA, during protein synthesis. Since three bases are required to determine the incorporation of each amino acid, a typical mRNA molecule might contain about 450–1200 bases. Some may be substantially larger due to the fact that they may carry information for several polypeptides produced from adjacent genes (forming a polycistronic messenger, especially in prokaryotes). Finally, in most eukaryotes, additional polynucleotide sequences are added to both ends of the mRNA molecules after transcription: at the 3'-end a sequence of up to 200 adenylate units is added which may be important in transport of the mRNA out of the nucleus; at the 5'-end a small sequence (known as a 'cap'), beginning with the base 7-methylguanine, is added which may serve to regulate the stability of mRNA or facilitate its interaction with ribosomes.

In most organisms the half-life of messenger RNA varies from a few minutes to several days. For those proteins, which are produced from mRNA that has a very short half-life, the control of the rate of mRNA synthesis can regulate the rate of protein synthesis and hence the concentration of protein can be regulated at the level of transcription (see Chapter 22; Section B.1).

(e) *Ribosomal RNA (rRNA)*

Ribosomes are complex assemblies of RNA molecules* and proteins, and they are present in the cytosol. At least 70 different proteins are found in the ribosome, some of which are structural while others have a more direct role in translation. All ribosomes are composed of two subunits, one larger than the other, which are characterized by their sedimentation coefficients in the ultracentrifuge so that, in eukaryotes, the subunits are termed 60S and 40S. The larger subunit contains three different RNA molecules and the smaller subunit contains only one. They are single-stranded and contain a high proportion of methylated bases.

(f) *Transfer RNA (tRNA)*

Transfer RNAs are relatively small, containing around 70 bases. There is at least one specific tRNA molecule for each amino acid, which is attached covalently to the tRNA. All tRNA molecules possess the base sequence-CCA at the 3'-terminal end and it is to this that the aminocyl group is esterified (see Section C.1). Although tRNAs are single-stranded, there is extensive internal base-pairing between complementary regions resulting in a complex tertiary structure. In one of the unpaired regions, a sequence of three bases is responsible for the specific association with a codon for the specific amino acid on mRNA. This region on the tRNA is known as the anticodon.

A characteristic of tRNA molecules is the high proportion of 'unusual' bases (that is, other than A, C, G, or U). In yeast alanine tRNA, for example, nine out of 77 bases are unusual. Most of these bases are methylated derivatives of the regular bases and this modification occurs after transcription. The functions of these unusual bases may be to increase the stability of the molecule by their ability to resist degradation by ribonucleases. Alternatively, they may modify the

* In eukaryotes, RNA for the ribosomes is synthesized in the nucleolus, which contains the ribosomal genes.

conformation of the tRNA to provide specificity for interaction with the aminoacylating enzyme (see Section C.1).

Transfer RNA is synthesized in the nucleus as a much larger precursor molecule which is then cleaved by specific ribonucleases in the nucleus.

C. THE TRANSLATION PROCESS

The translation process has been outlined in Section A. Although the principles were established in the 1960s, current work is revealing the complexity of the processes. Each stage in the process requires one or more enzymes but the precise roles of many have not yet been fully elucidated. Those which are bound to the ribosome for only part of the translation process (so that they are not normally considered as components of the ribosomes) are generally referred to as 'factors'.

The process of translation of the message contained in the RNA molecule into a peptide chain, and thence into protein, is the aspect of biosynthesis in this chapter most relevant to the theme of this text. For this reason, several aspects are discussed in more detail.

1. Formation of aminoacyl-tRNA

The first reaction in the pathway of protein synthesis (probably a flux-generating step) is the formation of the tRNA-amino acid complex, catalysed by aminoacyl-tRNA synthetase enzymes. The reaction can be considered to take place in two stages. In the first stage, ATP reacts with the amino acid to form an aminoacyladenylate intermediate which remains attached to the enzyme. In the second, the aminoacyl group is transferred to the tRNA:

The controversy over whether the amino acid is esterified to the 2'- or 3'-hydroxyl group of the terminal adenosine group has not been fully resolved but it seems likely that interconversion of the two forms can occur. As indicated above, there is at least one separate tRNA for each amino acid and a given synthetase must bind not only the correct amino acid but also the correct tRNA; each synthetase has specific recognition sites for both. The importance of the enzyme in relation to fidelity of translating information in mRNA is indicated by the fact that once an amino acid is bound to tRNA, its identity as an amino acid for protein synthesis is dictated by the anticodon site on the transfer RNA and not by the amino acid itself. (The enzyme can be considered as a dictionary, since it provides a cross-reference between the nucleic acid and amino acid codes.)

2. Initiation

Initiation of the process of translation is defined as the positioning of the first amino acid into the correct site on the ribosome. The first amino acid is always methionine for which the codon is AUG. Although the same codon is used to place methionine in an internal position in the polypeptide, a different transfer RNA molecule is involved.* The sequence of events in eukaryotes is as follows. Methionyl-tRNA interacts with GTP and an initiation factor known as eIF-2 to produce a complex which associates with the smaller ribosomal subunit (the 40S subunit) to form a larger complex. At this stage, a molecule of mRNA is bound to the ribosome subunit (with the first AUG codon nearest the 5'-end of the mRNA molecule in register with the anticodon of the charged $tRNA_{met}$). This latter reaction appears to involve other initiation factors (e.g. eIF-3 and eIF-5) and results in the hydrolysis of a molecule of ATP (to form ADP). Finally in the initiation process, a 60S subunit joins the complex to produce the 80S ribosomal complex that is involved in translation; the original GTP molecule is hydrolysed to GDP. The over-all process is summarized in Figure 18.12 and can be compared to a short sequence of reactions in a metabolic pathway. The available evidence suggests that the first reaction in this sequence is flux-generating and is controlled by the activity of the initiation factor eIF-2 (see Section D.2).[9]

3. Elongation

The ribosomal complex possesses two binding sites for aminoacyl-tRNA complexes: the aminoacyl site (or A-site) and the peptidyl site (or P-site). After initiation, the peptidyl site is occupied by methionyl-$tRNA_{met}$, so that the next aminoacyl-tRNA complex must bind to the aminoacyl site. Which aminoacyl-tRNA is bound is specified by the next codon on the mRNA. The binding involves a protein known as an elongation factor-I (EF-I) in eukaryotes and hydrolysis of

---

* In prokaryotes, the methionyl residue attached to $tRNA_{met}$ that is involved in initiation is N-formylated but this is not the case for the process in eukaryotes. In eukaryotes, there are two species of $tRNA_{met}$ and one of them functions as initiation tRNA.

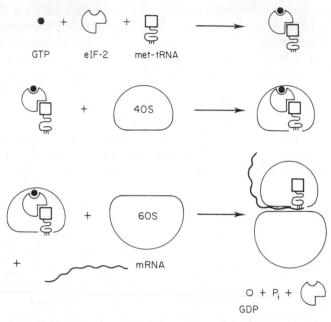

Figure 18.12   Steps in the formation of the polypeptide chain-initiation complex. 40*S* and
60*S* represent the two ribosomal components and details are given in the text

GTP to GDP. Prior to the binding of a further aminoacyl-tRNA complex at the
aminoacyl site, two reactions must occur, as follows.

1. Formation of a peptide bond between the second amino acid and methionine.
   This is catalysed by a peptide synthase (also known as peptidyltransferase)
   which is a component of the ribosome. Its activity results in the transfer of
   methionine from its tRNA to the second amino acid which remains attached to
   its tRNA at the aminoacyl site. The deacylated tRNA$_{met}$ is now released from
   the complex.
2. Translocation now occurs. The dipeptidyl-tRNA is transferred from the
   aminoacyl site to the peptidyl site by a second elongation factor (EF-2), with
   hydrolysis of a further GTP molecule. The mRNA is moved to bring the third
   codon into register with the aminoacyl site. The third aminoacyl-tRNA
   complex is bound to this site and the sequence of reactions is repeated (see
   Figure 18.13).

4. Termination

The series of reactions that constitutes elongation continues until a termination
codon is encountered. Three such codons, namely UAA, UAG, and UGA occur.
No tRNA molecule has an anticodon corresponding to these so that

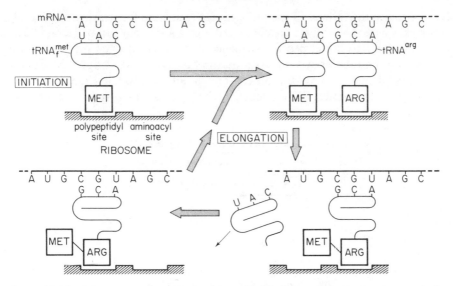

Figure 18.13 Diagrammatic representation of initiation and elongation in translation. See text for further explanation

translocation ceases and the components dissociate (that is, peptide chain, ribosomal subunits and mRNA molecule). In eukaryotes, at least, the termination codons are not at the physical end of the mRNA molecule. Probably two proteins (termination factors) are involved in termination and the process probably involves GTP hydrolysis. One factor binds to the aminoacyl site instead of a charged tRNA and this leads to cleavage of the last tRNA from the polypeptide chain.

5. Post-translational modification

Although the primary structure of a protein is established as described above, various covalent modifications may be required before a fully functional protein is produced. Such modifications fall into several classes including hydrolysis of part of protein molecule, addition of non-protein groups and modification of specific amino acids.

(i) *Hydrolytic removal of part of the protein* One function of this modification is to enable protein to be stored in a non-functional form and be activated only when required by the action of a specific proteinase.* Alternatively, some proteins (for export) have an additional sequence at one end of the molecule to facilitate

---

* For example, digestive enzymes (e.g. trypsin stored as trypsinogen), hormones (e.g. insulin stored as proinsulin—Chapter 22; Section D.2) and the blood-clotting protein, fibrin (stored in the blood as fibrinogen).

transport into the tubules of the endoplasmic reticulum for transfer to the Golgi complex, where the proteins are converted into a form that can be secreted from the cells. The additional peptide sequence is hydrolysed once the protein has passed into the tubular system; parathyroid hormone (see Chapter 20; Section F.3) provides an example of this process and details are given in Figure 18.14. See also Chapter 22; Section D.2 for details of the processes involved in the secretion of insulin.

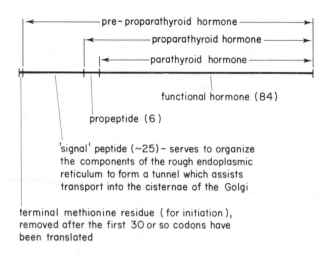

Figure 18.14 Postranslational hydrolysis of parathyroid hormone precursor. The numbers in parentheses indicate the number of amino acid residues

(ii) *Addition of non-amino acid groups* (*especially carbohydrate*)   The addition of carbohydrate to proteins to produce glycoproteins is described in Chapter 16; Section D.3.

(iii) *Modification of individual amino acids*   A large number of modifications are possible. In many proteins, pairs of sulphydryl groups become oxidized (apparently spontaneously) to produce disulphide links either within one polypeptide chain, between chains in the same protein or between protein molecules. In some proteins, for example collagen, proline residues are hydroxylated to form hydroxyproline, and in others the nitrogen atom of the imidazole ring of histidine is methylated to produce 3-methylhistidine (for example, in actin and myosin, see Chapter 10; Note 13).

6. **Summary of energy requirements in translation**

The process of peptide formation during translation involves the hydrolysis of a considerable amount of ATP. At least, five molecules of ATP (or equivalent) are converted to ADP for each amino acid incorporated, as summarized in Table

Table 18.2. Requirement of ATP and GTP for incorporation of each amino acid into a polypeptide chain

Process	Initiation or elongation	Nucleotide	No. of molecules
Formation of initiation complex	I	ATP	1†
Joining ribosomal subunits	I	GTP	1
Charging of tRNA	I plus E	ATP	2*
Binding of aminoacyl-tRNA to aminoacyl binding site	E	GTP	1
Translocation	I plus E	GTP	1

* Equivalent to 2 since the product is AMP.
†Only one used per complete peptide chain.

18.2. In this respect alone, protein synthesis is more expensive (in terms of the number of high energy molecules utilized per monomer incorporated) than the synthesis of any other polymer. In addition, the synthesis of all the complex components involved in protein synthesis (e.g. various RNA molecules) requires a considerable expenditure of energy, but is necessary to ensure fidelity of protein synthesis. In addition to this expenditure, which is required solely for the biosynthetic processes and is necessary for structural information transfer, energy may be required to control the rate of protein synthesis. In order to achieve sensitivity in metabolic regulation, synthesis and degradation of protein may take place at the same time (that is, a substrate cycle between protein and amino acids may occur in most tissues). The summation of these various factors may explain for the fact that synthesis of protein accounts for 15–20% of the basal metabolic rate (Chapter 10; Section B.3).

D. REGULATION OF PROTEIN SYNTHESIS

It might be expected that in order to increase the concentration of a protein in a cell its rate of synthesis would be increased. However, there are at least four ways in which the concentration could be changed:

1. The rate of synthesis of the mRNA that codes for the particular protein(s) could be increased (known as transcriptional control).
2. The rate of synthesis of the polypeptide chain by the ribosomal-mRNA complex could be increased (known as translational control).
3. The rate of degradation of the mRNA could be decreased (this would also be translational control).
4. The rate of degradation of the protein could be decreased. (The mechanism of protein degradation and its control is discussed in Chapter 10; Section B.3.b.)

It is likely that the process involved will depend upon whether the increase in the concentration of one or a few specific proteins is to take place or whether the

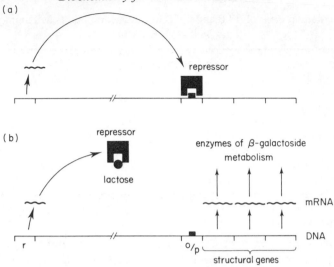

Figure 18.15 Transcriptional control at the lac operon. In (a), lactose, the inducer, is absent so that the repressor can bind at the operator-promoter site and prevent transcription of the structural genes. In (b) lactose is present and binds to repressor which is unable to bind at the operator–promoter

concentrations of all proteins in the cell are to change. The latter will probably be achieved by an over-all increase in the rate of translation, due probably to an increase in the rate of initiation (that is, a more efficient use of already existing mRNA). However, for an increase in the concentration of one or a few selected proteins, an increase in the rate of synthesis of the necessary mRNA molecules (or an increase in their stability) must occur. Both these forms of control are described below but it should be appreciated that in order to obtain satisfactory growth of a cell or an organism, the rate of synthesis of all the macromolecular components must be co-ordinated. In higher animals, this is likely to involve the complex interplay of hormones and is discussed in Chapter 19; Section B.

Unfortunately there is no information concerning the equilibrium nature, or possible flux-generating steps, of the various reactions in the over-all process of protein synthesis. Hence the approach to rate-control discussed in Chapter 7 cannot, unfortunately, be applied with any commitment to protein synthesis.

1. Transcriptional control

If the availability of mRNA is limiting for protein synthesis, then the rate of the latter can be regulated by changes in the rate of mRNA synthesis (or degradation). However, this will provide a flexible means of control only if the mRNA has a relatively short lifetime. Control of the synthesis of protein via mRNA synthesis was first established in prokaryotes from studies on the synthesis of three enzymes involved in galactose metabolism, β-galactosidase

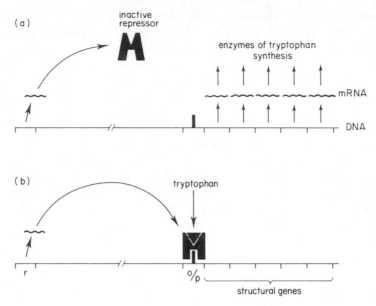

Figure 18.16 Transcriptional at the trp operon. In (a) the repressor is unable to bind to the operator–promotor site so that enzyme synthesis occurs. In the presence of tryptophan (b) the repressor binds tryptophan and then undergoes a conformational change which allows it to bind to the operator–promoter site

(which hydrolyses lactose), β-galactoside permease and galactoside transacetylase (both involved in transport of lactose) in *E. coli* (Jacob and Monod, 1961). It was shown that the genes coding for the three enzymes are contiguous on the genome and that the genes were controlled by a operator-promoter site (Figure 18.15). A regulator gene produces an mRNA molecule which codes for a protein that binds to the operator-promoter site so that, when this protein is bound, DNA-dependent RNA-polymerase cannot initiate the process of transcription; the protein is known as a repressor. The repressor is synthesized continually so that under normal circumstances the rate of transcription of these genes is very low and hence the concentrations of the enzymes are very low. However, if substrates for the β-galactosidase (e.g. lactose) become available in the environment, a small quantity enters the cell, binds to the repressor protein and inhibits its ability to bind to the operator-promoter site. Hence the genes can now be transcribed, mRNA* is formed, the three enzymes are synthesized and lactose can be metabolized and used as an energy and carbon source by the bacteria. The enzymes are said to have been induced by the substrate which is thus termed the inducer. (Certain non-metabolizable analogues of lactose can also act as inducers.) The advantage of this mechanism to the bacteria is that the enzymes are only produced when the substrate is available.

*In the absence of the inducer, mRNA is rapidly degraded; its half-life is about 1 min.

In a similar manner, the rate of synthesis of enzymes that are normally present and functioning in a given pathway in the cell could be decreased by the presence of a repressor-protein that is active as a repressor only when the end-product of the pathway accumulates. For example, in *E. coli* the enzymes of the pathway for the synthesis of tryptophan are coded by five contiguous genes together with promoter and operator sites. A regulator gene produces an mRNA molecule that codes for an inactive repressor which, however, becomes active upon binding tryptophan. In this way, the accumulation of tryptophan in the cell inhibits the rate of synthesis of the enzymes specifically involved in the tryptophan synthetic pathway (see Walker, 1977) (Figure 18.16). The regulatory unit, comprising regulator, promoter-operator and structural genes is known as an operon.

Although the operon model could provide a basis for control of transcription in animal cells, this is no evidence to support such a mechanism. Indeed, it may be that these models apply only to bacterial cells since there are several important differences in the process of transcription in cells of higher animals.

1. There is considerably more genetic information in eukaryotic cells and it is organized into a number of discrete chromosomes. This requires the involvement of structural proteins and it is possible that these proteins play an additional role in the regulation of the process of transcription.
2. The genetic material is physically separated from the rest of the cell by the nuclear membrane so that mRNA, after synthesis within the nucleus, has to traverse this membrane before it can function in protein synthesis (Figure 18.17). This barrier introduces further possibilities for control. For example, the mRNA has to be modified by polyadenylation (see above) or methylation before it can traverse the nuclear membrane and these processes could be regulated. This possibility is supported by the fact that there is normally a high rate of mRNA synthesis in the nucleus, much of which is degraded and does not enter the cytosol; the role of this nuclear turnover of RNA is unclear but it could indicate that transport into the cytosol is a key regulatory process.
3. The role of changes in the concentration of enzymes and proteins in the life of the bacterial cell and in that of the cells of higher animals is quite different. In the bacterial cell, the rate of protein synthesis must be carefully controlled and co-ordinated according to the availability of carbon and nitrogen sources so that the cell can grow and divide with the most efficient use of these resources. In the cells of higher animals, the control of protein synthesis is of the utmost importance in differentiation. Despite the presence of complete genetic information, cells can become part of muscle, nerve, liver, etc., so that many of the genes must remain totally inactive for the life of the cell. This requirement for total repression of some genetic activity may have caused the mechanism of control to be completely different from that in the bacterial cell.

Evidence that transcriptional control of enzyme concentration does occur in cells of adult mammals has been obtained by measuring the changes in the concentration of specific mRNA molecules, particularly in the liver. For example,

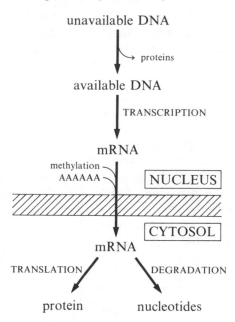

unavailable DNA

→ proteins

available DNA

TRANSCRIPTION

mRNA

methylation
AAAAAA

NUCLEUS

CYTOSOL

mRNA

TRANSLATION   DEGRADATION

protein   nucleotides

Figure 18.17   Some processes involved in control of protein synthesis in eukaryotic cells

the concentrations of mRNA molecules that code for tryptophan2,3-dioxygenase, tyrosine aminotransferase, and phosphoenolpyruvate carboxy-kinase are increased by administration of a glucocorticoid to an animal. The glucorticoid is known to exert this effect by entering the nucleus, to combine with a specific receptor and bind to the genome. This presumably reduces the effect of a repressor system and allows increased access of the genome to DNA-dependent RNA polymerase (Chapter 22; Section B.1).

2. Translational control

It seems likely that the over-all rate of protein synthesis in the eukaryotic cell is regulated by the rate of translation, probably at the level of initiation. Most of the studies in this area have been carried out with reticulocytes (premature erythrocytes) in which globin is the major protein synthesized and the rate is controlled by the concentration of the prosthetic group haem. In homogenates of reticulocytes the rate of globin synthesis is decreased if haemin is not available. Evidence has been obtained that one of the initiation factors (eIF-2) limits the rate of translation and that the activity of this factor is controlled by an interconversion cycle; phosphorylation of eIF-2 inhibits its initiating activity whereas dephosphorylation increases it.* The phosphorylation is catalysed by a

* One view is that in order to form the first complex in the initiation process, eIF-2, possibly in the form of an eIF$_2$-GTP complex, must complex with a further protein and that phosphorylation of eIF-2 prevents association with this additional protein (Ochoa, 1979).

cyclic AMP-independent protein kinase (eIF-2 kinase)[10] and the activity of this latter enzyme is inhibited by haemin. An additional protein inhibitor of eIF-2 has been isolated from homogenates of reticulocytes and this appears to act on the eIF-2 kinase; it is possibly cyclic AMP-dependent protein kinase which phosphorylates eIF$_2$ kinase and leads to activation of the latter. Hence a double interconversion cycle (that is, an enzyme-cascade, see Chapter 7; Section C.1.b) may be involved in the control of the activity of the key initiation factor in protein synthesis (Ochoa *et al.*, 1981; Figure 18.18). In addition to the effect on initiation, at least one of the ribosomal proteins is known to undergo phosphorylation and dephosphorylation (ribosomal protein S-6) and when this protein is dephosphorylated, the ribosomes are less active in protein synthesis (see Hunt, 1980; Leader, 1980). One important question is how far the mechanism of control of protein synthesis in the reticulocyte is similar to that in other cells. The available evidence suggests that it is similar, so that the variety of factors known to be involved in the control of protein synthesis in tissues such as muscle (see Table 18.3), probably all act at the level of translation and may well affect the rate of initiation through the phosphorylation of eIF-2 (Jackson and Hunt, 1980). These factors could change the activity of eIF-2 kinase, eIF-2 phosphatase, cyclic AMP-dependent protein kinase or the phosphatase that inactivates eIF-2 kinase (see Figure 18.18).

One further question is whether hormones could change the rate of synthesis of specific enzymes or other proteins at the translational level rather than that of transcription as discussed in Section D.1. For example, insulin causes an increase in the hepatic concentrations of glucokinase and pyruvate kinase whereas glucagon increases that of phosphoenolpyruvate carboxykinase. These

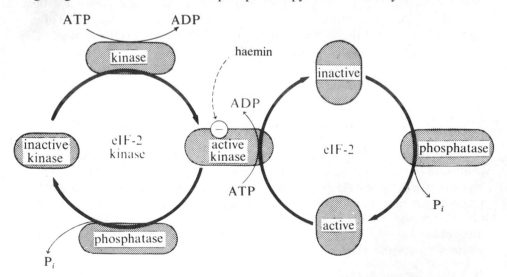

Figure 18.18   Inhibition of initiation by cascade phosphorylation

Table 18.3. A summary* of conditions or factors affecting protein synthesis

Conditions or factors	Effect on rate of protein synthesis	Cross reference to other chapters
Decreased intake of protein	decreased	Chapter 10; Section B.3.c
Decreased energy intake	decreased	Chapter 10; Section B.3.c
Increased intake of leucine in presence of sufficiency of other amino acids	increased	Chapter 10; Section B.3.c
Lack of nervous stimulation	decreased	Chapter 14; Section D.2
Muscle stretch, or exercise	increased	Chapter 10; Section B.3.c
		Chapter 14; Section D.2
Insulin	increased	Chapter 15; Section D.3
Growth hormone	increased	Chapter 19; Section B.2
Somatomedins	increased	Chapter 19; Section B.6
Thyroxine	increased	Chapter 10; Section B.3.c
Glucocorticoids	decreased	Chapter 14; Section D.1
Physical trauma, infection	decreased	Chapter 14; Section D.1

* For further details see Young and Pluskal (1977), Goldberg (1980).

hormones change the concentration of cyclic AMP and hence the activity of cyclic AMP-dependent protein kinase in the liver (Chapter 11; Section B.4.6). How could this influence the process of translation of specific enzymes? Perhaps phosphorylation of the nuclear membrane increases the rate of transport of specific mRNA molecules into the cytosol or perhaps specific mRNA molecules are phosphorylated leading to increased stability (Wicks, 1974).

E. MUTATIONS

The relative stability of the information carried in the DNA of an organism, and its exact transmission to daughter cells, is essential to life as we know it. However, it is not absolutely stable and mutations can, and do, occur. Mutations are changes in the covalent structure (that is, the base sequence) of the DNA which are perpetuated by replication and thus inherited. They lead to changes in protein structure and without mutation there would be no inherited differences between individuals of the same species and, indeed, no evolution. However, the living state is only metastable and even quite small changes in the processes occurring within an organism are likely to affect its well-being. Consequently most mutations are deleterious. Indeed a number of diseases (most of them very rare) are caused by an individual being unable to produce a particular protein (usually an enzyme) in an amount or form adequate for normal functioning. These diseases are inherited and are variously known as genetic disorders, inborn errors of metabolism, or deficiency diseases. Examples are discussed in several parts of this text (see Chapter 3; Section D.1.d, Chapter 10; Appendix 1; Chapter 16; Section E.). For further discussion see Rosenberg (1980).

1. Nature of mutations

Mutations can be classified according to the nature of the changes occurring in the DNA. Their origin is discussed in Section E.2.

1. Replacement of one base by another. The consequences depend upon the amino acid change and the role of the affected amino acid in the protein. In sickle-cell anaemia, for example, a glutamate residue in position 6 of the β-chain of haemoglobin is replaced by a valine residue. The consequences for the homozygous patient are very serious.
2. Insertion or deletion of bases cause more extensive changes since all codons from the mutation to the end of the polypeptide will be out of register and 'read' incorrectly. This is known as a frameshift mutation.
3. Chromosome mutations in which base sequences are reversed (inverted) or moved to new positions (translocations) are likely to be more serious but this depends on the extent of the changes. Sometimes a section of a chromosome is duplicated so that additional copies of a gene may be present. In such cases the 'spare' copies of the gene have few constraints on their viable mutations and genes with new functions are thought to arise in this way.
4. The most profound mutations are those in which whole chromosomes are missing as a result of faulty cell division (nondisjunction). These are almost invariably lethal (unless involving a sex chromosome) but more surprising, is the serious, often lethal, consequence of possessing a third copy of a chromosome (trisomy). One of the few trisomies compatible with life (with the exception of those involving the sex chromosomes) is trisomy-21, an extra chromosome 21, which causes Down's syndrome (mongolism).

2. Origin of mutations

The occurrence of a mutation in any one gene is normally a rare event but its frequency can be greatly increased by a variety of chemical and physical agents termed mutagens. Some mutagens are base analogues which become incorporated into new DNA molecules but exhibit faulty base-pairing. Examples include 5-bromouracil (which can be incorporated in place of thymine but can base-pair with guanine) and 2-aminopurine (incorporated in place of adenine but sometimes base-pairing with cytosine). Other chemical mutagens cause covalent modification (often cross-linkage) of bases already incorporated in DNA but with similar consequences. Particularly potent are alkylating agents, such as ethylene oxide, the nitrogen and sulphur mustards, and nitrosamines such as N-methyl-N'-nitro-N-nitrosoguanidine. Physical agents such as ultraviolet and γ-

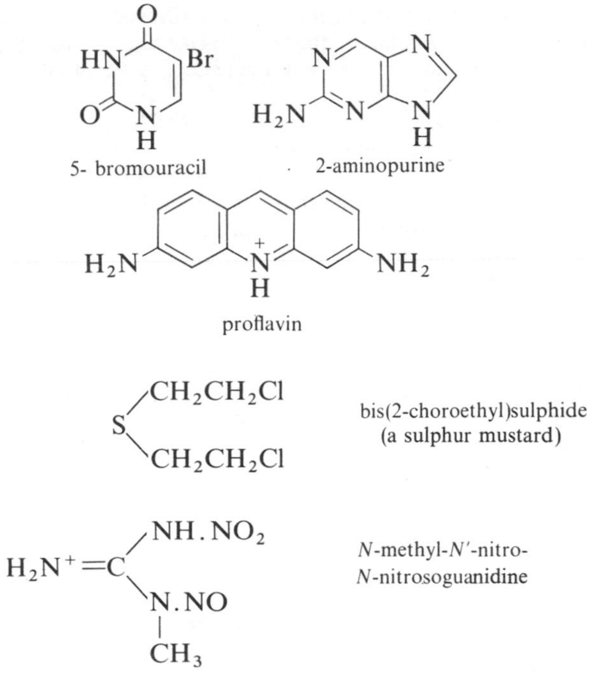

Figure 18.19 Molecular structure of some chemical mutagens

radiation also cause covalent changes either directly or via free radicals induced by the radiation. Acridine dyes such as proflavin exert their mutagenic effect by becoming intercalated between the strands of the DNA duplex and interfering with base-pairing in this way. Finally, a number of drugs such as colchicine (which has a therapeutic use in the treatment of gout) interfere with meiosis and increase the frequency of nondisjunction. The structures of some chemical mutagens are shown in Figure 18.19.

A number of mutagens occur naturally and are present in the environment[11] and it is now known that cells have some ability to repair mutationally altered DNA. However, man has the ability to create many more mutagens than occur naturally and vigilance is continually required to prevent the unintentional distribution of such agents in the guise of useful chemical products.

NOTES

1. With the exception of viruses, all living organisms are either prokaryotes or eukaryotes. The prokaryotes (bacteria and blue-green algae, now considered to comprise the kingdom Monera) are simpler and lack much of the internal structure characteristics of eukaryotic cells. They lack nuclei and chromosomes (although they do possess nucleic acids) and so do not undergo mitosis and meiosis. In addition to the apparatus associated with these processes, prokaryotes also lack mitochondria and an

endoplasmic reticulum (although they possess ribosomes). For these reasons they have been considered acellular but the term 'bacterial cell' is now widely accepted. All organisms belonging to the kingdoms Protista, Plantae, Fungi, and Animalia are eukaryotes.

2. It is now known that less than 5% of the DNA of higher organisms carries information for the synthesis of structural proteins. The remainder is involved in the regulation of protein synthesis, in the synthesis of non-messenger ribonucleic acids (see below) and, in ways yet to be established, in maintaining the correct conformation of the DNA. It has recently been shown that genes of eukaryotes are 'split', that is to say that the information for a single polypeptide chain is encoded in a series of non-contiguous stretches of DNA (as if the information in this book were contained in some of the words only) (see O'Malley *et al.*, 1981).

3. Onto this basis of constancy are superimposed mechanisms that allow slight variations (mutations, see Section E) to appear in the DNA and mechanisms which bring together variations arising in different cells (meiosis and fertilization).

4. Hydrogen bonds involve a partial sharing of electrons between atoms which are already covalently bonded. Some of the most frequently encountered hydrogen bonds in biology are represented by dotted lines in the following structures:

$$\diagdown \atop \diagup O \text{---} H \text{---} O \diagup$$

$$= O \text{---} H \text{---} N \diagup_{\diagdown}$$

$$N \text{---} H \text{--} N \diagup_{\diagdown}$$

Individual hydrogen bonds are invariably weak but when many occur in parallel they can become a very significant determinant of conformation. In the first example above, the O---H link has approximately 4% of the bond strength of the O—H bond at 37 °C but in order for the hydrogen bond to have any significant strength the atoms involved must lie in a straight line at precise interatomic distances (0.177 nm for O---H compared with 0.097 nm for O—H). Hydrogen bonds are readily disrupted by heat and a number of agents, including urea, which form good alternative hydrogen bonds.

5. Note that each of the base pairs involves a purine and a pyrimidine and that in each case the association results in the same distance between the carbon atom of position-1 of the pentose unit and the nitrogen of the base to which it is attached (see Figure 18.2). This means that when two nucleic acid strands are linked together by such hydrogen bonds the distance between the two pentose phosphate backbones is constant.

6. The relationship between the base sequence of the codon and the amino acid specified is known as the genetic code and appears to be universal. Four bases, taken in any order, can form 64 different triplet codons. Since only 20 different amino acids are normally incorporated into proteins, the code is degenerate, that is, some amino acids are specified by more than one codon.

7. The purine synthesis pathway as described in Figure 18.7 is also important for nitrogen excretion in those animals in which urate is the major nitrogenous excretory product (e.g. insects, reptiles, and birds). In such animals the pathway occurs in the liver (see Appendix 12.1) which possesses high activity of the 5'-nucleotidase so that the concentration of IMP is maintained low thus preventing significant inhibition of

the amidophosphoribosyltransferase so that the rate of synthesis of urate remains high (Krebs, 1978).

8. The low solubility of urate can lead to deposition of sodium urate crystals in tissues if its concentration becomes raised for any reason (see Chapter 12; Appendix 1). Urate kidney stones can form but the most widespread affliction resulting from urate deposition is gout. In this condition urate cyrstals form in the cartilage of joints (especially in the foot) rendering movement painful. Blood urate concentrations may become elevated because of poor control of purine synthesis, poor excretion or because of an increased availability of purine nucleotides in the blood. The latter is a consequence of a number of diseases that result in a high rate of cell destruction so that hyperuricaemia is frequently secondary. A high rate of urate synthesis leading to gout is associated with type I-glycogen storage disease, in patients with an increased activity of ribosephosphate pyrophosphokinase (possibly due to lack of control of this enzyme) and deficiency of guanine-hypoxanthine phosphoribosyltransferase (Section A.1.a). Since this latter enzyme is involved in conversion of purine bases back to nucleotides, deficiency increases the rate of degradation of nucleotides to uric acid. In addition, the deficiency leads to an increased concentration of 5-phosphoribosylpyro-phosphate which interferes with the regulation of *de novo* purine synthesis so that excessive amounts of IMP and urate are produced. An almost complete deficiency of this enzyme results in a five- to sixfold increase in the rate of excretion of uric acid in the urine and presents clinically as a severe neurological disorder (consisting of mental retardation, spasticity, choreoathetosis, and a compulsive form of self-mutilation*) known as the Lesch–Nyhan syndrome. This may be accompanied by gout. In patients with a less severe deficiency of the enzyme, the neurological disturbances may be mild but gout may well be present. The link between a deficiency of this one enzyme and the neurological disturbances may be that high concentrations of the purine bases, especially hypoxanthine, in the brain interfere with the binding of ATP (or adenosine) to its postsynaptic receptors in the central nervous system (Chapter 21; Section B.9; see Kopin, 1981).

Several drugs are used in the treatment of gout; for example allopurinol inhibits xanthine oxidase and hence the production of uric acid (see Seegmiller, 1980 for details).

9. Once a ribosome has moved part way along the mRNA during translation, a second ribosome may attach. This process can be repeated so that many polypeptide chains can be synthesized simultaneously. Aggregates of one mRNA molecule with many ribosomes are termed polysomes. The proportion of ribosomes in this state can be used as index of the rate of initiation.

10. Initiation factor, eIF2, consists of three subunits, α, β, and γ of approximate molecular weight, 38 000, 52 000, and 55 000 respectively. The eIF2 kinase phosphorylates the α-subunit which leads to a loss of initiation activity (see Ochoa *et al.*, 1981). It is tempting to speculate that phosphorylation of one of the other subunits could further decrease the loss of initiation activity (this would be another example of multisite phosphorylation—see Chapter 7; Section C.1.6 for other examples of multisite phosphorylation).

11. Mutagenicity and carcinogenicity are closely related and nitrosamines are amongst the most potent carcinogens known. Of particular concern is that they arise in reactions between secondary amines and nitrous acid. Secondary amines cannot be excluded from the diet and nitrites are used to cure certain meats. The latter will form nitrous acid under the acidic conditions of the stomach making nitrosamine synthesis a possibility.

---

* Patients bite their own lips and hands so that partial amputation of the fingers can occur.

CHAPTER 19

INTEGRATION OF ANABOLIC PROCESSES

In principle, an understanding of the growth process, what happens when growth ceases and the way in which adult size is maintained should emerge from a full knowledge of anabolic processes and how they are integrated. However, our knowledge of the integration of anabolic processes lags behind that of the integration of catabolic processes and is still largely at the descriptive stage. It is not therefore possible to apply the principles employed earlier in this text (e.g. in Chapters 8, 9, and 14) although it is presumed that they apply. Nevertheless, an attempt is made in the present chapter to tackle two questions. How are the rates of synthesis of polysaccharides, proteins, lipids, and nucleic acids integrated to permit gradual and coordinated growth (or to maintain size despite turnover)? What controls the size of a particular tissue or organ within the body?

As in other areas, the instance where regulation fails can shed considerable light on the mechanisms involved. This, together with the desire to alleviate suffering caused by abnormalities of growth, has directed research into the conditions described in Section C. They include, in addition to the mercifully rare congenital disorders of gigantism and dwarfism, the much more widespread conditions of obesity, non-insulin dependent diabetes mellitus, and anorexia nervosa.

Growth is difficult to define precisely. It implies an increase in size but this can be measured in a number of ways. Almost always it involves an increase in cell number and control of cell division must play a central role in regulating growth. Because of cell death, cell division must continue even when no net growth is occurring. The process of cell division is outlined in Section A.

In practice, dimensional or mass measurements are usually used to quantify growth. Difficulties arise, however, because growth is often accompanied by a change in shape and, particularly in the early stages of life, an increase in complexity. The term development is often used to encompass growth, cell differentiation and morphological change. While it is clear that the course of development is basically determined by the genetic information specific to an individual organism, environmental factors play some part in the interpretation of this information, particularly with respect to its rate. It seems likely that the broad mechanisms which control this rate (and which must involve the integration of anabolic processes) will be similar in related organisms and they are investigated in Section B.

A. THE CELL CYCLE ·

The fundamental feature of growth is an increase in cell number and the sequence of events that occurs in the reproduction of individual cells is known as the cell cycle (Shall, 1981). Cell division can be divided into two processes, the replication of DNA (and separation of chromosomes) and the approximate doubling in amount of the other components of the cell. The former process, constituting the reproductive part of the cycle, is a periodic event while the increase in cell mass during the non-reproductive part of the cycle may be continuous. During the reproductive phase of the cycle, DNA replication is initiated and completed (S phase), chromatin is organized in preparation for mitosis (G_2 phase), mitosis occurs (M phase) and the interphase nucleus reforms (Figure 19.1). Once this process has begun, the sequence of events is almost always completed in a time which is approximately constant for the given cell. The non-reproductive phase of the cell cycle (G_1 phase) can, however, be long or short depending upon the reproductive activity in the tissue. Quiescent cells have a very long G_1 phase and are considered to be in a G_0 phase by some authors (Baserga, 1981). Finally, some non-dividing cells leave the cycle altogether although it may be a long time before they die.

The rate of cell division, therefore, would appear to be the rate of entry into the reproductive phase and, in animal cells, the first identifiable event is the initiation of DNA synthesis. At the present time, there is little information on the identity of

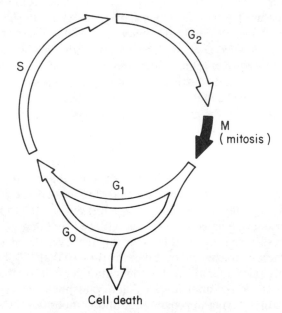

Figure 19.1   The cell cycle. Initiation of DNA replication occurs at the beginning of the S phase. Cells undergoing a lengthy non-reproductive period enter the G_0 phase

the factor that initiates DNA replication in animal cells but it is known that hormones such as the somatomedins and growth factors stimulate cell division and must therefore cause this initiation.

At least some aspects of the cell-cycle appear to be controlled by phosphorylation of histone H1, a key component of chromatin (Matthews, 1980). The evidence is best for phosphorylation (by a non-cyclic-AMP dependent kinase) causing chromosome condensation during early prophase of mitosis but phosphorylation could also initiate DNA replication and so be involved in the transition from the G_1 to the S phase. It is not difficult to imagine secondary messengers, released in response to the binding of somatomedins, activating a protein kinase within the nucleus.

B. HORMONAL INTEGRATION OF GROWTH

The consequence of growth is an increase in size. To accommodate any increase in the size of the organs the structural elements of the body too must enlarge. These are composed of connective tissues which share a number of structural and chemical features.

All connective tissues consist of protein fibres (collagen or elastin) embedded in a matrix of proteoglycan (see Chapter 16; Section D.2). The mechanical properties of the tissue depend on the proportion of these different components present and whether or not mineralization has occurred. Collagen molecules contain a high proportion of the amino acids glycine, proline and hydroxyproline and are organized into fibrils which aggregate to form the fibres visible under the light microscope (Woodhead-Galloway, 1980). These fibres are virtually inextensible and have a very high tensile strength. Elastin is a similar protein but contains less hydroxyproline and its conformation is less ordered so that it is readily extensible and contributes elasticity to tissues containing it.

In loose (areolar) connective tissue, bundles of collagen and elastin fibres are widely separated by the matrix. In a more specialized form, known as adipose tissue, masses of triacylglycerol-containing adipocytes cluster between the protein strands. In dense connective tissue, as is found, for example, in tendon and the aortic wall, collagen fibres predominate and there is little matrix.

In skeletal tissues, cartilage and bone, the connective tissue is firmer. In the case of cartilage, this is largely a consequence of matrix composition. The vertebrate skeleton is originally formed in cartilage but during development (in all species except sharks and related fish) this is replaced by bone to leave cartilage only where its resilience is a valuable property (e.g. over the articulating surfaces in joints). Bone is characterized by mineralization so that 75 % by weight of a typical bone is inorganic material, most of the remainder being collagen. Mineralization involves deposition of amorphous calcium phosphate and crystals of the mineral hydroxyapatite (having approximately the composition $3Ca_3(PO_4)_2.Ca(OH)_2$) in the organic matrix.

In all these tissues, both protein fibrils and proteoglycan matrix are secreted by cells which remain in the connective tissue. These cells are known by different names according to the connective tissue which they form. Thus fibroblasts form connective tissue, chondroblasts form cartilage and osteoblasts form bone. As mineralization proceeds, the osteoblasts become imprisoned in cavities of the calcified matrix and in this state are known as osteocytes. A second type of cell, the osteoclast, is also found in bone. Its function is to resorb bone, that is, to remove the minerals and hydrolyse some of the matrix, so that the growth of bone depends on a balance between osteoblast and osteoclast activities. The simultaneous activity of both types of cell leads to bone remodelling which occurs even in adult man and especially during periods of rapid skeletal growth.

Hormones have long been known to play a major role in the co-ordination of the anabolic processes involved in growth although many of the mechanisms involved have remained obscure. Renewed interest in this area has come from the discovery of a group of peptide growth factors known as somatomedins. Their role, together with that of other hormones involved in the integration of anabolism, is described in detail in this Section.

1. Parathyroid hormone, vitamin D and calcitonin

These three hormones interact to regulate the plasma calcium concentration (see Chapter 20; Section F.3). One way in which this is achieved is by altering the balance between calcium deposition in bone and resorption of bone but whether these hormones play a role in normal growth is not clear (Raisz and Kream, 1981).

Parathyroid hormone increases the activity of both osteocytes and osteoclasts so increasing the rate of calcium turnover in bone. This may increase the sensitivity to other factors affecting calcium deposition or release. Calcitonin inhibits the resorptive activity of osteoclasts and may therefore play a role in skeletal growth.

2. Growth hormone

(a) *Structure of hormone and control of its secretion*

Growth hormone is secreted by acidophilic cells of the anterior pituitary and, even in adult man, these cells can contain up to 10% of their dry weight as the hormone. Human growth hormone (abbreviated to hGH) is usually considered to be a single peptide chain of 191 amino acid residues with two disulphide bridges. However, recent studies suggest that the hormone exists in the gland and is presumably secreted from it as a mixture of peptides which differ in size, amino acid sequence and post-translational modifications. For example, the 191 amino acid peptide is readily cleaved to lose amino acid residues between 134 and 150 so that a two-chain form of the hormone is formed in which the two chains are linked by the disulphide bridge between cysteine residues at 53 and 165 (Figure 19.2). A

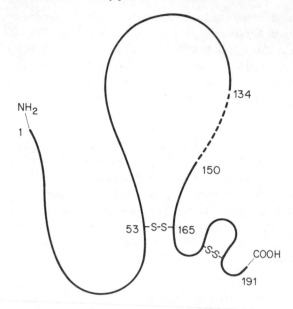

Figure 19.2 Diagrammatic representation of the human growth hormone molecule. Limited hydrolysis removes small peptides from the region between amino acids 134 and 150 to produce a two-chain form of the hormone. After Lewis *et al.* (1980)

further variant, identical to the 191-residue growth hormone except that there are 15–20 fewer amino acids between 31 and 60, is known as the 20K peptide because its molecular weight is about 20 000. Acetylation of the N-terminal amino acid further increases the number of forms present so that, in the absence of unequivocal evidence indicating any of the peptides to be dominant in growth promotion, growth hormone should be considered, at present, as a family of peptides (Lewis *et al.*, 1980).

The rate of secretion of growth hormone(s) from the anterior pituitary is controlled by the balance of concentrations of a growth hormone-releasing and growth hormone release-inhibiting hormones which are secreted by (different) cells in the hypothalamus and carried to the anterior pituitary in the hypothalamic-pituitary portal system (see Chapter 20; Section E.4.c). Factors, or conditions that increase the rate of secretion, and hence the blood concentration of growth hormone, include stress, exercise, sleep, hypoglycaemia, and some amino acids such as arginine. Hyperglycaemia and glucocorticoids decrease the rate of secretion (Sönksen and West, 1979). Since the secretion of growth hormone is episodic, it is difficult to interpret a single measurement and a number of measurements must be made, over a period of time, to determine the level of the hormone in the blood for clinical purposes. Nonetheless, none of these conditions are obviously related to growth-promotion.

(b) *Role in growth promotion*

The major growth-producing role of growth hormone is its ability to stimulate the release of the somatomedin peptides from the liver (see Section B.6). Indeed it may be that growth hormone should be considered as a trophic hormone (equivalent, for example, to the gonadotrophins or ACTH) which regulates the secretion of growth promoting peptides by the liver. Consistent with this suggestion is the observation that a high blood concentration of somatomedins inhibits growth hormone secretion. Such a feedback inhibitory mechanism is typical of the trophic hormones, see Phillips and Vassilopoulou-Sellin, 1980.

(c) *Role in metabolism*

Growth hormone has a number of metabolic functions apparently unrelated to its role in growth. The most important of these is its stimulation of lipolysis in adipose tissue to provide adequate supplies of fatty acids during prolonged starvation, exercise, and chronic hypoglycaemia. For this effect, glucocorticoids must be present (see Chapter 8; Section C.1). If the concentration of growth hormone remains elevated for some time it leads to a decreased sensitivity to insulin. This insulin antagonism may be a consequence of raised fatty acid concentrations which, through the operation of the glucose/fatty acid cycle, will tend to inhibit glucose utilization.

In considering the diversity of effects attributed to growth hormone, it must be remembered that it is a family of polypeptides and that the different functions could be those of different members of the family. Indeed other functions may remain to be discovered.

3. Insulin

Since insulin is the one hormone which increases the rate of synthesis of almost all macromolecules its association with growth is inevitable. Indeed it could provide a link between good nutrition (insulin is secreted after meals) and satisfactory growth (see Section B.6). Details of possible mechanisms by which insulin acts to stimulate anabolic processes will be found in Chapter 16; Section C (glycogen); Chapter 17; Section A.9 (triacylglycerol); and Chapter 18; Section D.2 (proteins). Summaries of other effects of insulin will be found in Chapter 15; Section D and Chapter 22; Section C.2. In addition to these direct effects of insulin, the hormone may also influence growth processes by stimulating somatomedin secretion (see Section B.6.e).

4. Thyroxine

The involvement of thyroxine in growth and development is well known because congenital hypothyroidism produces cretins, dwarfs who have infantile features and are mentally retarded. Despite this, the precise role of thyroxine in growth is unclear. Although administration of growth hormone to hypophysectomized animals increases the rate of growth towards normal, the effect of this hormone is greater if administered with thyroxine, which if given alone has no effect. In hyperthyroid children, linear growth is accelerated but this may not be due to a direct effect of thyroid hormones since they increase the rate of secretion of growth hormone in the normal animal. However, they increase directly the resorption of bone by increasing the activity of the osteoclasts, so that they probably increase the rate of bone remodelling (Mundy *et al.*, 1976). It may be a general feature of thyroid hormone function that they increase the rate of substrate cycles not only in catabolic processes but also in anabolism, which involves the macromolecules so that small increases in the concentration of growth-promoting hormones (e.g. insulin, somatomedins) can cause larger increases in the rates of synthesis of the macromolecules. Indeed Waterlow and Jackson (1981) point out that growth is associated with an increased rate of protein turnover. This increased rate of cycling in a number of processes may play a role in weight control (see Section C.3.b).

5. Androgens

Androgens are largely responsible for the growth-spurt of puberty in both males and females although it occurs at different times (see Brook, 1981). In girls, the height spurt occurs early in the sequence of pubertal changes (the first change is probably the enlargement of the breasts) whereas in boys it occurs up to two years later. This probably reflects a different source of the androgen; the adrenal cortex secretes most of the androgens in the female whereas the testes are responsible in the male. The growth spurt may need to occur early in the female since menarche occurs soon after the growth spurt is complete. The growth-promoting action of the androgens does not involve somatomedins and it would appear to be a direct stimulation of growth in certain parts of the body. The pubertal growth spurt involves primarily an increase in length of the trunk rather than growth of the legs. Thus the androgens appear to control growth of the vertebral column and also the width of the shoulders and hips; in contrast, growth hormone influences growth of the long bones in the legs.[1] The effect of androgens on secondary sex characteristics and the mechanism of action of steroids in general are described in Chapter 20; Section E.3.d and Chapter 22; Section B.1 respectively.

6. Somatomedins

It is possible that much of the confusion and ignorance concerning the control of growth by hormones will be clarified by the studies on the role of a group of peptide hormones (or factors) known as the somatomedins. For reviews, see Chochinov and Daughaday, 1976; Hall and Fryklund, 1979; Philips and Vassilopoulou-Sellin, 1980. Some of the evidence indicating their importance in growth is as follows.

1. There is a strong correlation between the plasma concentration of somatomedins and the rate of growth in children (Hall and Filipsson, 1975).
2. In children with hypopituitarism but treated with growth hormone, the rate of growth is proportional to the increase in the biological activity of the somatomedins in the blood.
3. The rate of release of somatomedins by the liver is modified by many of the hormones or conditions known to play a role in growth.
4. The biochemical effects of somatomedins are entirely consistent with a role in growth; that is, stimulation of anabolic processes (insulin-like activity) and increasing the rate of mitosis (mitogenic activity).

Somatomedins were discovered (under the name of 'sulphation-factor') by Salmon and Daughaday (1957) as a factor which restored the rate of sulphation of chondroitin in the cartilage that had been removed from hypophysectomized rats. Growth hormone was known to restore this activity when administered to hypophysectomized rats but was without effect *in vitro*. 'Sulphation-factor' (somatomedin) was obtained from the plasma of hypophysectomized rats which had received growth hormone. It thus became apparent that growth hormone caused release of somatomedin and, in time, the liver was identified as the major site of release.

(a) *Nature of the somatomedins*

Somatomedins are a family of peptides which includes somatomedins A, B, and C, insulin-like growth factors (IGF-I and IGF-II)[2] and multiplication stimulating activity (MSA). The peptides have a molecular weight of about 7000 and are similar in general structure; indeed, somatomedin-C may be identical to IGF-I. One reason for the uncertainty surrounding the identity of these peptides is the difficulty in purification of large amounts since, until recently, the only source was serum, where they are present at very low concentrations and are attached to much larger transport proteins from which they must be separated before purification. In addition, bioassays for measurement of the concentration were non-specific. However, the development of protein-binding assays (Chapter 22; Section E.3) provides more precision (see Merimee *et al.*, 1981).

(b) *Effects of somatomedins*

The *in vivo* effects of somatomedins on growth have been demonstrated unequivocally in dwarf mice in which administration resulted in increased weight, length of the animals and rate of synthesis of cartilage. These effects are similar to those observed on administration of growth hormone. *In vitro* studies have led to a greater understanding of the role played by somatomedins and these are summarized as follows.

(i) *Effects on cartilage*   Somatomedins increase the rate of cartilage formation by stimulating the following processes in chondrocytes (derived from chondroblasts); transport of amino acids across the cell membrane, synthesis of DNA and RNA, incorporation of sulphate into proteoglycans and of proline into collagen. Cartilage from young animals is considerably more sensitive to these effects than is cartilage from older animals. Some of these effects can also be produced by insulin but much higher concentrations are required.

It is possible that all these effects stem from a more fundamental effect on cell division, since the rate of mitosis of both fibroblasts in culture and chondrocytes in cartilage is stimulated by somatomedins. The peptide known as multiplication stimulating activity (MSA) is particularly active and may act by initiating DNA synthesis. This effect of somatomedins may account for the requirement for serum shown by fibroblasts in culture.

(ii) *Insulin-like activity*   It is not known whether the similar, if not identical, effects of somatomedins to those of insulin on liver and muscle simply reflect a structural similarity between the hormones or whether it has some other functional significance.

(c) *Somatomedin inhibitors*

The growth-promoting effects of somatomedins on cartilage can be antagonized by somatomedin inhibitors present in serum, especially from starved, hypophysectomized or diabetic rats or from malnourished man. These inhibitors have not been purified and their identity is uncertain although they are likely to be proteins with a larger molecular weight than that of the somatomedins. Their source has not been unequivocally identified although it is likely to be the liver. Both somatomedins and their inhibitors are present in normal serum so that the growth-promoting effect probably depends on their relative concentrations.

(d) *Site of origin of the somatomedins*

The evidence that somatomedins arise in the liver is strong. Release has been detected by arterio-venous difference measurements and their serum concentration decreases after partial hepatectomy. Somatomedins are released from the perfused liver *in vitro* at a rate dependent upon the hormonal or nutritional state of the animal from which the liver was obtained.*

The central position of the liver in receiving absorbed nutrients, its key role in control of the blood concentrations of glucose, amino acids, and ketone bodies and its important contribution to the biosynthesis of carbohydrate and lipid emphasize the ability of this organ to monitor the nutritional state of the animal and to release somatomedins (or its inhibitor) accordingly (Figure 19.3).

---

* The rate of release is decreased by inhibitors of protein synthesis.

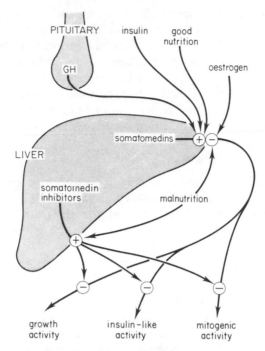

Figure 19.3 Factors affecting the release of somatomedins from the liver and their peripheral activity

(e) *Factors controlling the release of somatomedins from the liver*

As with steroids (Chapter 22; Section B.1) the rate of release of somatomedins from the liver is probably controlled by their rate of synthesis although how this is controlled is not known. As anticipated from the early studies of Salmon and Daughaday, release of somatomedin is stimulated by growth hormone. More recently it has been shown that deficiency of the latter, due to hypophysectomy, results in very low serum concentrations of somatomedins. In patients who have low concentrations of growth hormone, administration of the hormone results in an increase in the concentration of circulating somatomedins within hours. Higher than normal concentrations of growth hormone are associated with raised somatomedin concentrations.

Both insulin and good nutrition increase somatomedin release and may be able to maintain the concentration of somatomedins and normal growth even if the concentration of growth hormone is low.* Conversely, poor nutrition and low insulin concentrations can delay growth; for example, children with kwashiorkor (Chapter 10; Section B.1.a) suffer from poor growth despite normal or elevated

---

* This effect of good nutrition may explain the progressive increase in size of the population in western societies over the last few generations (see Rona, 1981).

concentrations of growth hormone. Studies with animals indicate that both the protein and the energy content of the diet is important in control of somatomedin release and that, in addition, low protein diets may cause a reduction in the sensitivity of cartilage to the effects of somatomedins. Glucocorticoids and oestrogens decrease the rate of release of somatomedins from the liver and this may explain, in part, how these hormones decrease growth (Loeb, 1976). In addition, high concentrations of glucocorticoids decrease the sensitivity of cartilage to the action of somatomedins (Figure 19.3).

7. Other tissue growth factors

Factors which stimulate growth in specific tissues have been extracted from various sources. These include fibroblast growth factor, nerve growth factor, epidermal growth factor and bone-derived growth factor. The nature of these substances, and the ways in which their actions are integrated with those of somatomedins, remain to be elucidated.

C. CLINICAL CONDITIONS

The two most obvious growth abnormalities are gigantism and dwarfism. In both cases, the growth of most parts of the body has been affected similarly. However, poor integration of anabolic processes can lead to abnormal growth of some tissues only. In particular, the normal proportion of adipose tissue (containing triacylglycerol) to body size (or protein content) may be disturbed. This control of normal body weight is impaired in insulin-dependent diabetes mellitus (Chapter 15), when wasting occurs, and in obesity. The conditions known as non-insulin-dependent diabetes mellitus and anorexia nervosa are also discussed in this section since the former may be a pathological extension of obesity and the latter indicates the complexity of control mechanisms involved in the maintenance of the body's triacylglycerol store, especially in the female.

1. Gigantism and acromegaly

A tumour of the acidophilic cells of the anterior pituitary can result in secretion of large amounts of growth hormone. Since this hormone is involved particularly in growth of the long bones an excess before closure of the epiphyseal plates at puberty results in disproportionate extension of the long bones and hence gigantism. After closure of the epiphyseal plates, growth of the long bones can no longer occur and excess of the hormone results in growth of other bones, for example those in the hands and feet and the lower jaw, to produce the condition known as acromegaly.[3]

2. Dwarfism

Dwarfism, or short stature, has a large number of causes (Parkin, 1981). However, dwarfs who are of normal body proportion and sexually mature (known as sexual ateliotic dwarfs) probably have either a defect in the secretion of growth hormone, a resistance to growth hormone, a defect in the production or secretion of one or more of the somatomedins or a resistance to these peptides.[4] Provided that treatment is started soon enough, children suffering from simple growth hormone deficiency can be restored to normal by administration of this hormone, which increases the blood concentration of somatomedins (Hall and Olin, 1972). In Laron dwarfs, there is an elevated plasma growth hormone concentrations and no response to exogenous hormone. However, such patients have low concentrations of somatomedins, so that their liver may be unable to respond to growth hormone (a possible receptor disorder; see Chochinov and Daughaday, 1976).

African pygmies are a race characterized by their short stature and studies have been carried out to establish the mechanism causing this. They have normal serum concentrations of growth hormone which are increased by stimuli that normally increase the rate of secretion and no apparent decrease in the biological activity of somatomedins. However, a detailed study of the concentrations of the individual somatomedins using specific immunoassays has demonstrated that in some pygmies from the Central African Republic the concentration of one of the somatomedins, insulin-like growth factor-I, is very low, although that of insulin-like growth factor-II is normal* (Merimee *et al.*, 1981). This study indicates the importance of one specific somatomedin in growth and suggests that each peptide of the group may have a specific role in this complex process.

Finally, short stature can result from chronic malnutrition, insulin deficiency or chronic diseases of the liver or kidney (Chochinov and Daughaday, 1976; Vassilopoulou-Sellin, 1980). Low rates of growth in these conditions are probably due to low concentrations of circulating somatomedins.

3. Obesity

Obesity can be defined as a condition caused by an excessive amount of adipose tissue (Garrow, 1978). However, as usual it is difficult to define a norm and know what is meant by excessive. It is further complicated by the difficulty of measuring the amount of adipose tissue. It is usually done by relating the body weight to tables of standard weights or measurement of skinfold thickness† (Womersley and Durnin, 1977). Other less practical methods involve density measurements (weighing the subject in air and immersed in water), measurement of total body fat using fat-soluble gases (e.g. krypton), measurement of lean body mass using distribution of labelled water or potassium and measurement of fat cell size and number after removal of a sample of adipose tissue by biopsy needle (Garrow,

---

* Previous studies on pygmies had shown that the biological activity of somatomedins in stimulation of [^{35}S]sulphate incorporation into cartilage ('sulphation-factor' activity) was normal, which could now be explained by the normal concentration of somatomedins other than IGF-I.

† The skinfold thickness should be measured in four positions, triceps, subscapular, biceps and suprailiac (see Davidson *et al.*, 1979).

1978; Garrow and Blaza, 1982). The normal content of triacylglycerol in adipose tissue is about 12% for adult men and 26% for adult women (see Table 8.1). If the proportion is more than 20% for man and more than 30% for women, they are considered obese. In the United Kingdom, the record for obesity is held by William Campbell who died aged 22 in 1878. He weighed 340 kg, (Garrow and Blaza, 1982).

There is increasing evidence that obesity shortens the lifespan with increased incidence of atherosclerosis, hypertension, gall bladder disease, non-insulin-dependent diabetes mellitus, and psychological disturbances. This, together with the large number of individuals in western societies who suffer from this condition and the changing attitudes of society to obesity, have led to a renewed interest in the causes of the problem. In simple terms, obesity is caused by the ingestion of too much energy and utilization of too little. This is a consequence of the first law of thermodynamics (Chapter 2; Section A.1) and the metabolic mechanisms for energy storage; energy consumed in excess of requirements will be retained as chemical energy and in the non-growing adult the only major storage form for this energy is triacylglycerol in the adipose tissue. However, there is evidence that a mechanism exists for the maintenance of the normal proportion of adipose tissue (12% for men, 26% for women) so that obesity is caused by an impairment of this mechanism. The fascinating and controversial question concerns the metabolic basis of this mechanism, knowledge of which might explain the aetiology of obesity. The regulation of body weight probably involves both control of energy intake and control of energy expenditure. Possible metabolic mechanisms of control of these two processes are discussed below (Cawthorne, 1982).

(a) *Control of energy intake*

Evidence that an animal can control its food intake is obtained from studies on the effects of stimulation or damage to specific areas of the hypothalamus. Lesions in the lateral hypothalamus lead to anorexia (which can be sufficiently serious that the animal starves to death) whereas stimulation leads to feeding. This led to the idea of a hunger or feeding centre in this part of the brain. On the other hand, lesions in the ventromedial hypothalamus leads to hyperphagia (over-eating) and obesity (if sufficient food is available) whereas stimulation leads to cessation of feeding, thus leading to the idea of a satiety centre. A balance between the two centres would therefore result in normal feeding behaviour and recent neuropharmacological manipulations[5] and electrical recordings support this. It appears that there is negative interaction between the two centres so that stimulation of one leads to inhibition of the other (for reviews, see Rolls, 1981; Booth, 1981; Morley and Levine, 1983).

It seems likely that these centres must recieve a number of sensory inputs, some external but others metabolic in nature. For example, both glucose and insulin (which will rise in concentration after a meal) stimulate the satiety centre and there is evidence that this stimulation depends on the utilization of glucose by cells in this part of the hypothalamus.[6] Much information is received, via nerves, from the mouth and facial area concerning odour, taste, texture, and appearance of food. Gastric distension may also initiate neural signals to the hypothalamus. The hormones released from the duodenum in response to the entry of food, including cholecystokinin and bombesin, probably provide further sensory input to the centres. Finally, it is known that the presence of high concentrations of glucose in the hepatic portal vein cause a sensory nerve output from the liver which induces a feeling of satiety. Whether this is produced as a direct response to the binding of glucose to a receptor or whether it requires metabolism of glucose within the liver is not known.

The possibility that different neurotransmitters are responsible for the different behavioural effects controlled by the two centres has generated much pharmacological interest. The feeding centre appears to involve adrenaline and dopamine whereas the satiety centres involve noradrenaline and 5-hydroxytryptamine. Compounds that produce changes in the concentrations of these neurotransmitters, or interfere with their action, could thus result in anorexia or over-eating. Indeed, obesity can be produced by administration of *p*-chlorophenylalanine which inhibits the formation of 5-hydroxytryptamine (Chapter 21; Section B.3.a). Of possible relevance is that frequently administered small pain stimuli (e.g. tail pinching), which may increase the activity of dopamine-containing neurones, does result in obesity.

(b) *Control of energy expenditure*

There is considerable evidence that body weight can be maintained despite marked variations in the food intake. Perhaps the first recorded investigation was carried out by Neuman (1902) who performed a 725-day experiment on himself, in which he varied his intake of energy over a wide range and despite this he maintained a constant weight. Similar studies have been carried out by Gulick (1923) and Passmore *et al.* (1955). Normal subjects fattened by over-eating require more energy in relation to their body surface for maintenance of the obese state than they require at their natural weight and, also, more than spontaneously obese individuals (Sims *et al.*, 1973). This suggests that these subjects have a natural tendency to 'burn-off' excess food rather than store it. Similarly, feeding rats a high energy diet, in which their normal laboratory diet was supplemented with food palatable to man (so-called 'cafeteria-feeding'), causes a marked increase in heat production after feeding* (Rothwell and Stock, 1979; 1981). Experiments in non-ruminant animals and in man have shown that the efficiency of utilization of carbohydrate is considerably higher when the food intake is below maintenance requirements than when the intake is above this requirement (Blaxter, 1970). These observations, together with the first law of thermodynamics, suggest that one or more metabolic processes exist in which excess energy is dissipated in the form of heat.

* Rothwell and Stock (1981) claim that this heat production was sufficient to 'burn off' all the extra energy consumed but this has been questioned by Hervey and Tobin (1981).

The question therefore arises as to the nature of the processes that can convert excess chemical energy into heat. At least two mechanisms have been suggested: uncoupling of oxidative phosphorylation from electron transfer in brown adipose tissue and an increase in the activity of substrate cycles.

(i) *Brown adipose tissue thermogenesis in weight control*   The mechanism of uncoupling of mitochondrial oxidative phosphorylation in this tissue is described in Chapter 4; Section E.4. Brown adipose tissue is present in many animals including man (Heaton, 1972). In experimental animals the amount of this tissue increases with overfeeding (e.g. 'cafeteria feeding', see James and Trayhurn, 1981). Furthermore, the increase in the amount of brown adipose tissue is paralleled by an increase in the concentration of the proton-conductance protein* in the mitochondria, which is responsible for transporting protons into the mitochondria without concomitant ATP formation hence dissipating the high-energy state (Chapter 4; Section E.4). Thermogenesis in brown adipose tissue is increased by noradrenaline and it is suggested that the increased concentration of this hormone after feeding (see Chapter 14; Section C.1.a) increases thermogenesis in this tissue which thus accounts for the thermic response to food.

The suggestion that a defect in brown adipose tissue mitochondria or the response of this tissue to noradrenaline could play a part in development of obesity in man receives some support from work with the genetically obese mouse. This animal has a reduced capacity, or an impaired control, of thermogenesis since it rapidly dies of hypothermia when exposed to temperatures of 4 °C. Furthermore, mitochondria from brown adipose tissue of obese mice have a low concentration of the proton-conductance protein which does not increase to the same extent as that in non-obese mice when exposed to a low temperature (Himms-Hagen and Desautels, 1978; Himms-Hagen, 1979).

Although brown adipose tissue does occur in man (Heaton, 1972) the proportion may be much less than in small rodents in which thermogenesis in response to cold may be of enormous survival value. It is also unclear if this tissue in man responds to overfeeding as in the rodents and whether it is defective in obesity (Hervey and Tobin, 1983; Rothwell and Stock, 1983).

(ii) *Substrate cycles and weight control*   The role of substrate cycles in improving sensitivity in metabolic control of flux through pathways is discussed in Chapter 7; Section A.2.b. The improvement is achieved by an increase in the rate of cycling (in relation to the flux through the pathway) which involves an increase in the rate of hydrolysis of ATP to ADP and phosphate. Since cycling does not produce a net metabolic change, except for the hydrolysis of ATP, it results in the conversion of chemical energy into heat. The amount of heat produced will depend upon two

---

* This protein binds GDP (and other nucleotides) as part of the control of its transporting activity and the binding of radiolabelled GDP to mitochondria is used as an assay for this protein. The latter has been given the name thermogenin.

Table 19.1. List of possible substrate cycles that might be involved in thermogenesis

Cycle	Chapter reference
Glucose/glucose 6-phosphate	Chapter 11; Section E.1
Glycogen/glucose 1-phosphate	Chapter 7; Section C.2
Fructose 6-phosphate/fructose bisphosphate	Chapter 7; Section C.3
Pyruvate/phosphoenolpyruvate	Chapter 9; Section A.3.b
Triacylglycerol/fatty acid	Chapter 8; Section C.2
Protein/amino acid	Chapter 10; Section B.3.a
Cholesterol/cholesterol ester	Chapter 20; Section B.2

factors, the rate of cycling and the number of ATP molecules hydrolysed per turn of the cycle. For the fructose 6-phosphate/fructose bisphosphate cycle one molecule of ATP is hydrolysed for each turn and on the basis of the maximum activity of fructose bisphosphatase in the muscle of man ($2.0\,\mu\text{mol.min}^{-1}.\text{g}^{-1}$ at $37\,°C$) it can be calculated that, if this cycle were fully active in all muscles for 24 hr, almost 7000 kJ of energy could be liberated as heat. This is more than half of the average daily energy intake. Although it is very unlikely that this cycle would ever be maximally active for a prolonged period, this calculation demonstrates the large capacity for heat production. Furthermore since there are many substrate cycles in metabolism* (Table 19.1) a small stimulation of a large number could result in a considerable rate of conversion of chemical energy into heat.

The rate of these cycles might be increased *in vivo* in at least three different conditions; stress, after exercise and after feeding. It is likely that both anxiety-stress and aggression-stress increase the activity of many cycles involved in the mobilization and utilization of fuels, especially glucose and fatty acids. For example, adrenaline and glucagon increase the rate of the triacylglycerol/fatty acid cycle in adipose tissue (Chapter 14; Section C.1.b).

In Chapter 9 (Section A.3.b) it was suggested that the rates of substrate cycling are increased prior to exercise but of more importance in weight control is the possibility that such cycles remain stimulated for a prolonged period after exercise. It is well established that in the post-exercise period, oxygen is consumed in excess of that required to support resting metabolism; this is known as the oxygen debt (Hill *et al.*, 1925). It is possible to divide the debt into three phases, the rapid phase, the slow phase and the ultra-slow phase (see Figure 19.4). The rapid phase may be explained by replenishment of phosphocreatine stores and re-oxygenation of myoglobin and haemoglobin; the slow phase may be partially

* These cycles can also include translocation cycles in which ions (e.g. sodium ions) leak into a cell and are extruded by a sodium-pump (i.e. an ATP-requiring reaction, see Crabtree and Newsholme, 1976). There is some evidence that such translocation cycles may be less active in obese man (De Luise *et al.*, 1980).

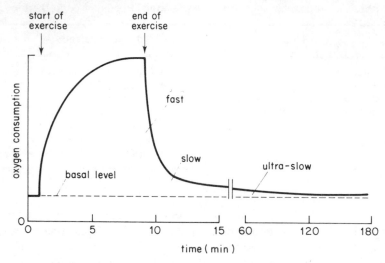

Figure 19.4 Stylized plot of oxygen consumption against time, during and after exercise to show the components of the oxygen debt. Note the change of time scale

explained by reconversion of lactate to glucose and glycogen, but the extra oxygen consumed in this phase is usually considerably larger than can be explained by carbohydrate synthesis (see Newsholme, 1978). It is proposed that some of the oxygen consumed in the slow phase and all of the oxygen consumed in the ultra-slow phase is due to the stimulation of substrate cycles. Since the small increase in oxygen consumption associated with the ultra-slow phase may persist for several hours after exercise,* it could make a significant contribution to energy expenditure.

The increase in heat production after a meal is known as the thermic response (previously known as the specific dynamic action). It has been pointed out that many of the properties of the thermic response are consistent with a stimulation of cycling rates after a meal (Crabtree and Newsholme, 1976) and this could be caused by the increased concentrations of noradrenaline (Chapter 14; Section C.1.a).

In order for any mechanism to control the amount of triacylglycerol in the adipose tissue, information must be conveyed from the stores to those systems controlling food intake and energy dissipation (presumably via the feeding and satiety centres in the hypothalamus) (Figure 19.5). Candidates for the role of messenger include insulin and triiodothyronine. Increases in the concentrations of both of these hormones could increase any or all of the following; the amount of brown adipose tissue, the rate of substrate cycling and the sensitivity of either of these to noradrenaline (Figure 19.6).

*Recent work has established that after sustained exercise, the rate of oxygen consumption was increased for at least 24 hours (L. Hermansen, 1984, Medice and Sport Science, Karger, Basel, New York, Vol. 17, pp. 119–129).

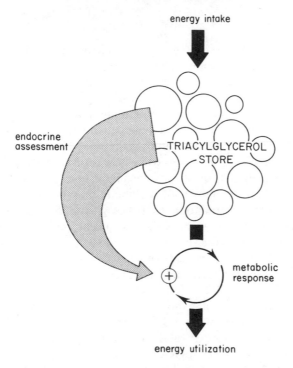

Figure 19.5 Speculative system for feedback control of triacylglycerol stores in adipose tissue

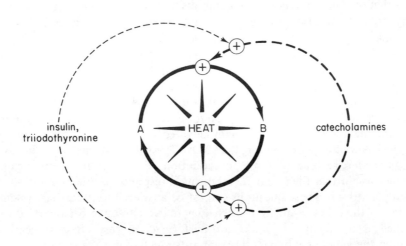

Figure 19.6 Possible role of insulin or triiodothyronine in modifying the control of thermogenic cycles by noradrenaline

(c) *Treatment of obesity*

One of the commonest instructions given by general practitioners to their patients is to lose weight but despite a plethora of treatments, success is rarely achieved. The forms of treatment have been well-reviewed by Garrow (1978) and fall into the following classes: dietary restriction, drugs, exercise, and surgery. Although dietary restriction may be adequate to control weight in some individuals, in many it is not. If the defect in these patients is a decreased ability to stimulate thermogenic mechanisms, a lowering of the energy intake may have little effect on body weight unless it is so low as to be near total starvation. Such diets lead to a number of metabolic problems including loss of lean body mass (Chapter 14; Section A.3). A number of metabolic manipulations have been devised to decrease protein degradation, such as high protein (ketotic) diets in which the high concentrations of ketone bodies and amino acids decrease the rate of body protein degradation. Total dependence on such diets can however be dangerous (Frank *et al.*, 1981). Oxoacids of some of the essential amino acids have been used to decrease protein degradation in patients undergoing otherwise total starvation (see Chapter 14; Section D.1).

Pharmacological solutions involve the use of anorexic or thermogenic drugs. Amphetamine and its derivatives are well known as 'slimming pills' but their effect is probably totally anorexic and they are central nervous system stimulators (they can produce severe psychotic reactions, tachycardia, insomnia) and patients rapidly become habituated to them. Other anorexic drugs include fenfluramine and phenmetrazine but they are only of value over short periods since their efficacy decreases rapidly. At the present there is no satisfactory thermogenic drug although triiodothyronine has occasionally been used, but doses that produce an increase in metabolism also cause tremor, diarrhoea, and tachycardia. In addition, the dangers of uncontrolled thermogenesis have been exemplified by 2,4-dinitrophenol (Chapter 4; Section D.3.c).

There is contradictory evidence on the benefits of exercise in treatment of obesity (see Garrow, 1978); at least two important questions remain to be answered. First, what is the level of exercise that is required to have an effect on energy consumption? Simple calculations based on the rate of energy expenditure during a given form of exercise (see Table 2.2) may be misleading since the metabolic rate may remain elevated for a considerable period after exercise has ceased (possibly more than 24 h). This will greatly improve the efficacy of regular exercise as an aid to weight reduction. Secondly, could regular exercise also increase further the rate of metabolism in response to other conditions (e.g. feeding or stress)? These questions cannot be answered until further systematic studies on the effect of different intensities of exercise have been carried out.

Finally some forms of surgery are performed on very obese subjects. For example, the jejunoileal bypass short-circuits much of the absorptive area of the small intestine. Patients must then reduce their intake of food to prevent diarrhoea, flatulence, and abdominal discomfort. Alternatively, the jaws of the patient can be wired together with a dental splint so that drinking but not eating is possible (Garrow, 1978).

(d) *Endocrine disturbances in obesity*

The major endocrine disturbance in obesity is insulin resistance, which is discussed in detail below but, in addition, there is evidence of a diminished response to catecholamines (Jung *et al.*, 1979). This could minimize the stimulation of thermogenesis in brown adipose tissue or reduce the substrate cycling rate so that less energy is dissipated as heat in response to feeding, stress or exercise. How this is related to the other major endocrine problem, insulin resistance, is unclear. It has been suggested that the normal response of thyroid hormones to overfeeding (that is, an increased conversion of thyroxine to T_3 and decreased formation of reverse-T_3) does not occur in obese subjects. However, this is not the case.

(e) *Insulin resistance in obesity*

Insulin resistance is a state in which higher-than-normal concentrations of the hormone are required to produce a given biological effect. This could be caused by the presence of antibodies to the hormone, the production of an abnormal, less effective, insulin or a decreased response of the target tissue to the hormones. In obesity and most cases of non-insulin-dependent diabetes mellitus, the latter problem is present. Studies in man and experimental animals demonstrate that in obesity the stimulation of glucose utilization in muscle and adipose tissue, the inhibition of lipolysis in adipose tissue and inhibition of glucose release by liver are resistant to insulin.[7] This resistance results in much higher concentrations of insulin being required to maintain normal metabolic functions so that patients are hyperinsulinaemic. Since this state can lead to non-insulin-dependent diabetes it is important to establish its metabolic basis. Although a complete answer cannot yet be given there are two areas of interest. First, in mild hyperinsulinaemia, the number of insulin receptors is reduced. This means that in order to obtain the same biological response from a tissue, the concentration of the hormone must be increased (this is described as decreased tissue sensitivity to the hormone—see Chapter 22; Section A.3.a). Secondly, in more severe cases of hyperinsulinaemia, there is not only a decreased number of receptors but also a further post-receptor defect. Thus, a very high concentration of insulin (sufficient to saturate all the receptors) does not elicit the maximal biological response observed in the non-resistant state (see Figure 19.7; for reviews see Kahn, 1978; Blecher and Bar, 1980; Olefsky and Kolterman, 1981). The nature of the post-receptor defect is not known. It could be that the coupling process that links the insulin receptor to the production of the intracellular signal (secondary messenger) is defective (an example of such a receptor-effector coupling defect for parathyroid hormone is

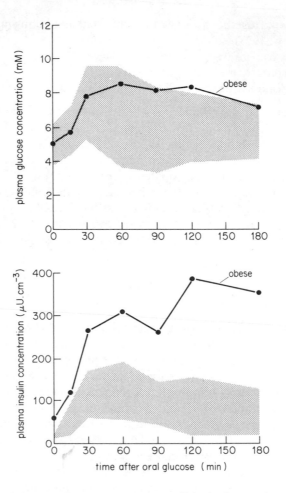

Figure 19.7 Oral glucose tolerance tests on a non-diabetic obese patient. The shaded area indicates the mean ± 2 standard deviations for a control group of non-diabetic thin subjects. Note that although 'normal' glucose concentrations are just maintained in the obese subject, this requires a massive production of insulin. After Kahn *et al.* (1977)

described in Chapter 22; Section A.3.b). Alternatively, the sensitivities of the control mechanisms that regulate the key enzymes in the tissues could be reduced so that a given concentration of the hormone would be less effective (e.g. a decreased rate of substrate cycling).

The degree of insulin resistance observed in obese patients is of pathological interest since it is associated with hyperglycaemia and can lead to diabetes mellitus (see Section C.4). Nonetheless, a decreased receptor number may have physiological importance in normal man; it might provide an informational link between the important anabolic and storage tissues (e.g. adipose tissue and liver) and the satiety and hunger centres in the hypothalamus. For example, if when the triacylglycerol stores in a given adipocyte approach a normal repletion state, this resulted in a decrease in the rate of receptor-protein synthesis so decreasing the number of receptors then, in order to produce the same biological effect and control of the blood glucose concentration, a higher circulating concentration of insulin would be required (presumably produced by an increased rate of secretion). However, this higher concentration of insulin would facilitate glucose utilization in the insulin-sensitive cells of the hypothalamus which would stimulate the satiety centre (see Section C.3.a). Consequently, the insulin resistance seen in the obese condition may be an exaggeration of a normal mechanism due to a defect in the response of the hypothalamic centres to an increased insulin concentration. Hence, despite repleted stores of triacylglycerol there would be little or no restriction on the amount of food ingested which, if thermogenic mechanisms were also deficient, would lead to obesity.

4. Non-insulin-dependent diabetes mellitus

Diabetes mellitus was discussed in Chapter 15. Discussion of the non-insulin-dependent form (also known as type II) has been deferred to the present chapter since the great majority of patients with this disease are obese and suffer from insulin resistance. Monocytes and adipocytes isolated from patients with diabetes have many fewer insulin receptors[8] but these have the same affinity as those from normal cells. Patients with mild carbohydrate intolerance suffer only from decreased receptor number but with more severe carbohydrate intolerance and fasting hyperglycaemia there is both a decrease in receptor number and a post-receptor defect (as in the severe forms of obesity). Between these two extremes there is probably a complete spectrum as the contribution of the two defects varies.

Non-insulin-dependent diabetes could be explained as a further pathological development of obesity. Consider the changes in the blood concentrations of glucose and insulin in response to an oral glucose load in the obese and diabetic patients in comparison to normals (Figure 19.7). In the obese, changes in glucose concentration are within the normal range whereas those of insulin are considerably in excess of the normal. In other words, despite the resistance to insulin, the beta-cells of the pancrease can secrete sufficient insulin to control the blood glucose concentration within the normal range (the concentrations are below those considered to indicate diabetes mellitus—see Table 15.2). However, if the resistance is too severe and the pancreas is unable to secrete sufficient insulin to overcome the resistance, the blood glucose concentration will not be maintained within the normal limits, despite the high levels of circulating insulin, and diabetes results. Hence, provided there is sufficient 'insulin secretory reserve' in the β-cells, the patient will remain obese whereas, with a small 'reserve', the patient could become diabetic.

(a) *Treatment of non-insulin-dependent diabetes mellitus*

Non-insulin-dependent diabetic patients are usually treated with diet or drugs. In patients suffering from mild diabetes, dietary restriction, particularly reduction of the digestible carbohydrate content of the diet, is sometimes sufficient to decrease the blood glucose concentration so that it is within the normal range; replacing digestible carbohydrate with fibre may be particularly beneficial (Chapter 15; Section F.3). If, however, this is not sufficient, hypoglycaemic drugs are recommended. The sulphonylurea drugs* are widely used to lower the blood glucose concentration. They are known to increase the secretion of insulin from the pancreas which would explain their ability to increase peripheral glucose utilization. However, it has been found that the sulphonylureas increase glucose utilization without changing the blood concentration of insulin. Of considerable interest is that this effect may be the result of an increase in the number of insulin receptors. How they change the number of receptors is unknown, but if this is a specific and long-lasting effect it should be a very safe mechanism by which to lower the blood glucose concentration (Olefsky and Kolterman, 1981). The biguanides, phenformin and metformin, are also used but they have a higher incidence of adverse side effects, including the danger of lactic acidosis (Chapter 13; Section D.3.b) and their use is banned in the U.S.A.

5. **Anorexia nervosa and amenorrhoea**

Anorexia nervosa is a disease that mainly affects young women, usually aged 12 to 18 years. It is characterized by a marked reluctance to eat or a disturbed eating pattern, weight loss which may be severe (body weight can be 35 kg or less), an intense fear of gaining weight, cold extremities (hands and feet), low blood pressure, slow pulse, and amenorrhoea. It is unclear whether the patients are truly anorexic or repress the sensation of hunger. The latter would appear to be the case in some patients since they occasionally engage in an eating orgy, after which they may induce vomiting. Indeed many show an active interest in food, for example, preparing it for others but not eating. The term anorexia is, therefore, probably a misnomer. Many patients deny that they are suffering from an illness and seek medical attention reluctantly and only after pressure from family or friends when weight loss is profound. Anorexia is common in patients with other psychiatric disorders such as depression and schizophrenia but in such cases the signs and symptoms are different.

Somewhat surprisingly, there appear to be few metabolic studies on anorexia nervosa. Since it is essentially a condition of starvation, the metabolite and endocrinological changes would be expected to be similar to those described in Chapter 14; Section A. It is known that the concentrations of insulin and T_3 are decreased whereas those of growth hormone and reverse-T_3 are increased (see Halmi, 1978). However, the metabolic interest in anorexia nervosa, and the reason

---

* Sulphonylurea drugs include tolbutamide, chlorpropamide, glibenclamide, and glipizide.

it is discussed in this chapter, does not lie in its similarity to starvation but in the fact that the decrease in body weight, and particularly the decrease in the amount of adipose tissue, results in amenorrhoea. The normal proportion of body fat in the female is 26 % and there is considerable evidence[9] that if it falls much below this value the normal menstrual cycle stops (amenorrhea) resulting in infertility. For the average non-obese woman a loss of as little as 1 kg of triacylglycerol may be sufficient to cause amenorrhoea.

The biological advantage for this relationship between adipose tissue stores of triacylglycerol and fertility is provision of fuels to satisfy the enormous requirements of pregnancy and especially lactation (Pond, 1978). This would explain why fertility can be maintained in men despite much lower proportion of body weight. Indeed in some elite long-distance runners it may be 6 % or lower (Costill, 1979).

The intriguing question is how the amount of adipose tissue in the female is assessed and how the information is conveyed to the systems that control the reproductive cycle. Knowledge of the menstrual cycle (Chapter 20; Section E.4) suggests that control of ovulation is the most likely process to be influenced by a general endocrine imbalance, and that the effects of oestradiol-17β and gonadotrophin-releasing hormone might be influenced by other factors. For example, it is suggested that the concentration of blood insulin may be one factor that is influenced by the amount of adipose tissue and this information is transferred to the cells of the satiety or hunger centres (Section C.3.a); if these cells of the hypothalamus also communicate with the cells that secrete gonadotrophin-releasing hormone, a connection between the amount of adipose tissue and the control of the menstrual cycle could be established. Alternatively, adipose tissue cells are known to be capable of converting androgens to oestrogens[10] (androgens produced from the adrenal cortex in the female), so that the larger the amount of adipose tissue the greater may be the rate of peripheral oestrogen formation which could be important as part of the mechanism controlling the surge of LH secretion by the pituitary[11] (see Chapter 20; Section E.4.c).

One psychiatric view of anorexia nervosa is that the patient is able to control her own development. Any unwillingness to accept the responsibilities of adulthood or sexual maturity, at what may now be a very early chronological age, could lead to anorexia nervosa since this arrests part of the development in the female (Crisp, 1977; 1980). The complex interplay of metabolism, endocrinology, and psychology provides a fascinating probe for investigating control of body weight and its physiological and indeed psychological importance, especially in the female.

NOTES

1. In the long bones, growth occurs at each end by mineralization of a plate of proliferating cartilage known as the epiphyseal plate. The width of this epiphyseal plate governs the rate of growth of the long bone, the wider the plate the greater is the rate of cartilage formation. Growth hormone increases the rate of cartilage formation at the epiphyseal plate but androgens appear to accelerate closure of the plate, preventing further growth. Thus a partial deficiency of growth hormone leads to a relatively long trunk with short legs whereas a partial deficiency of androgens will produce a short trunk and long legs.

2. Insulin-like growth factors are present in serum but are not inactivated (or suppressed) by antibodies to insulin (that is, the molecules are not identical to insulin; they are also known as non-suppressible insulin-like activity or NSILA).

3. Acromegaly is also associated with increased body hair, gynaecomastia, cardiac enlargement, hypertension, atherosclerosis, and ischaemic heart disease.

4. In addition to the phenotypic abnormalities, patients with growth hormone deficiency also suffer metabolic abnormalities detected by various treatments. For example, administration of exogenous insulin produces exaggerated and prolonged hypoglycaemia and there is a reduced rate of insulin secretion after ingestion of glucose or arginine.

5. Injections of gold-thioglucose cause permanent damage to cells in the ventromedial hypothalamus and induce obesity.

6. Systemic administration of 2-deoxyglucose increases feeding. This is explained by its phosphorylation (by hexokinase) to 2-deoxyglucose 6-phosphate which is not further metabolized and accumulates to inhibit hexokinase and so prevent glucose utilization in cells of the satiety or hunger centres.

7. In experimental animals, insulin resistance can be readily demonstrated with isolated tissue preparations. In man, a number of experiments have been performed to demonstrate insulin resistance in different tissues *in vivo*.

   (a) Injection of a standard dose of insulin produces a small decrease in the blood glucose concentration in obese patients compared to normal control subjects. This suggests that glucose utilization by muscle is resistant to insulin.

   (b) Similarly, the blood concentrations of fatty acids and glycerol (increased by infusion of noradrenaline) decrease less in response to insulin in obese subjects, suggesting that the antilipolytic response is also resistant (Ratzmann, 1977).

   (c) Glucose production by the liver, as measured by arterio-venous concentration differences, is also resistant to insulin in obese man (Felig *et al.*, 1974).

   (d) Insulin resistance has been demonstrated by use of an elegant technique known as the 'euglycaemic insulin clamp' (Sherwin *et al.*, 1974). In this method, insulin is infused at whatever rate is required to raise and maintain the plasma insulin concentration by a standard amount. At the same time, glucose is infused to maintain the normal blood glucose concentration. The rate of glucose infusion required to maintain a constant blood glucose concentration is an index of the subject's response to insulin. Use of this technique has demonstrated conclusively that the obese are resistant to insulin (DeFronzo *et al.*, 1978b).

8. Insulin resistance has been demonstrated *in vitro* using adipocytes isolated from small pieces of adipose tissue removed by biopsy needle. Insulin binding to the receptor in man is usually measured on isolated monocytes from blood since these possess specific receptors for insulin. The number of binding sites is reduced in obesity and non-insulin dependent diabetes mellitus (Olefsky and Kolterman, 1981).

9. Evidence for the importance of body fat in control of menstrual cycling activity in woman is as follows:

(a) Epidemiological studies demonstrate a decreasing age at menarche in successive, better nourished generations of women in USA (the average age of menarche in the USA is currently 12.8 years), and it usually occurs at a set weight/height ratio. A loss of 10% to 15% of normal weight for height produces amenorrhoea or delays menarche (Frisch and McArthur, 1974).

(b) The secretion of gonadotrophins in the adult female occurs during both the day and night, but in prepubertal females and patients with anorexia nervosa the secretion occurs only during sleep (Boyar, 1978) (i.e. the anorexic patient assumes a prepubertal secretion pattern).

(c) Young girls attending professional schools of ballet in the USA have been studied since they restrict their food intake to maintain a low body weight and are highly physically active. Such dancers had delayed menarche, amenorrhoea, or a high incidence of irregular cycles that correlated with excessive thinness. Furthermore, menarche occurred in dancers after injury which prevented physical activity and allowed a weight-gain (Frisch *et al.*, 1980).

(d) There is a strong correlation between the physical activity of female runners (miles per week) and the incidence of amenorrhoea (Feicht *et al.*, 1978).

10. Obesity in human beings is associated with increased conversion of androgens to oestrogens (Grodin *et al.*, 1973).

11. Factors other than the amount of adipose tissue must play a role in control of the LH surge since it is known that some patients who have suffered from anorexia nervosa but regained their normal weight do not always regain normal menstrual cycle activity without pharmacological intervention (administration of gonadotrophin-releasing hormone or an active analogue—see Chapter 20; Section E.4.c; Nillius and Wide, 1975).

CHAPTER 20

BIOSYNTHESIS AND METABOLISM OF CHOLESTEROL AND THE STEROID HORMONES

The steroids include cholesterol, the bile acids, vitamin D, and hormones of the adrenal cortex, ovary, and testes. They form a discrete chemical class with affinities with the lipids. The chapter on steroids has been delayed until now since their biosynthesis follows logically from that of triacylglycerol and phospholipids. However, in addition to synthesis, the chapter contains a description of the pathways of degradation and an account of the physiological and clinical importance of steroids, emphasis being given to those aspects which fall within the scope of this book. In this way it provides a link between the biosynthesis chapters and the two final chapters which discuss control and chemical communication.

A. STRUCTURE OF THE STEROIDS

The basic structure of all steroids is that of the fully reduced phenanthrene ring (perhydrophenanthrene) to which is fused a five-membered ring producing the cyclopentanoperhydrophenanthrene nucleus, the four rings of which are known as A, B, C, and D (Figure 20.1). The parent compound from which cholesterol is derived, cholestane, possesses a side-chain of eight carbon atoms attached to the D-ring at carbon-17 and two 'angular' methyl groups attached to the rings at carbon-10 and carbon-13. Because of normal bond angles, the steroid nucleus has a corrugated structure that is conventionally represented on paper as planar. The upper side is defined as that to which the side-chain (or, if absent, 'angular' methyl groups) is attached. Substituents attached to this side are drawn with full lines

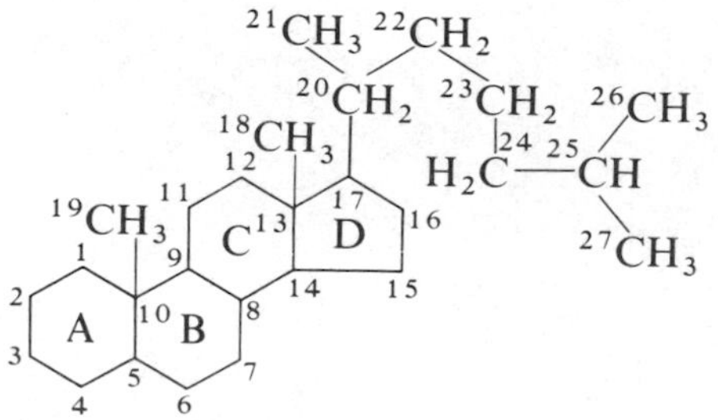

Figure 20.1 Structural formula of cholestane. The four rings (without side-chains) form the cyclopentanoperhydrophenanthrene nucleus on which all steroid structures are based

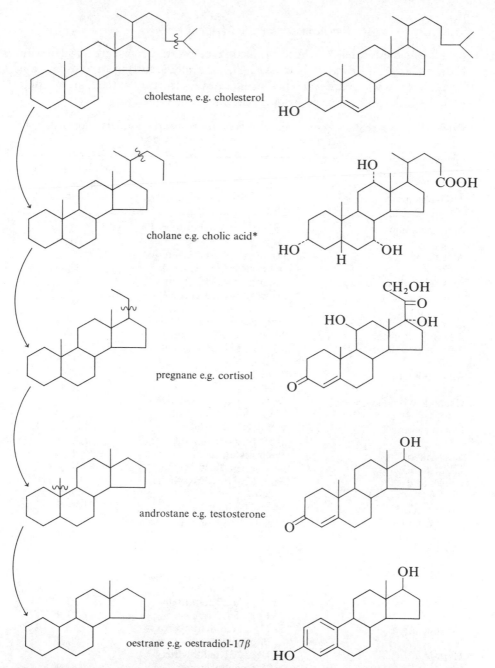

Figure 20.2 Structural relationships between the major groups of steroids with examples of biologically important derivatives. From cholestane to oestrane the number of carbon atoms is decreased and the sign ∿ indicates which carbon-carbon bond is 'broken' and hence which carbon atoms are lost (see text).

* The three hydroxyl groups are drawn with dotted lines to show that they are α-groups, that is pointing in the opposite direction to the methyl groups at C_{10} and C_{13}

(——) and designated β; those projecting below the rings are drawn with broken lines (————) and designated α. The naturally occurring steroids fall into several groups which are named according to the structure of the 'parent' compound which can be derived from cholestane as described on p. 714.

Table 20.1. Systematic names and references to structural formulae for naturally occurring steroids referred to in the text

Trivial name	Systematic name	Figure depicting structural formula
Aetiocholanolone	3α-hydroxy-5β-androstan-17-one	20.36
Aldosterone	11β,21-dihydroxy-3,20-dioxo-4-pregnen-18-ol	20.36
Androstanediol	5α-androstane-3β,17β-diol	20.36
Androstanedione	5α(or 5β)-androstane-3,17-dione	20.36
5-Androstenediol	5(6)-androstene-3β,17β-diol	20.36
4-Androstenedione	4-androstene-3,17-dione	20.9
4-Androstenedione-19-al	3,17-dioxo-4-androstene-19-al	20.36
Androsterone	3α-hydroxy-5α-androstan-17-one	20.36
Chenodeoxycholic acid	3α,7α-dihydroxycholan-24-oic acid	20.14
Cholecalciferol	9(10)seco-cholesta-5,7,10(9)-trien-3β-ol	20.29
Cholestanol	5α-cholestan-3β-ol	20.36
Cholesterol	5-cholesten-3β-ol	20.2
Cholic acid	3α,7α,12α-trihydroxycholan-24-oic acid	20.2
Coprostanol	5β-cholestan-3β-ol	20.36
Corticosterone	11β,21-dihydroxy-4-pregnene-3,20-dione	20.36
Cortisol	11β,17α,21-trihydroxy-4-pregnene-3,20-dione	20.2
Cortisone	17α,21-dihydroxy-4-pregnene-3,11,20-trione	
Cortol*	3α,11β,17α,20,21-pentahydroxy-5β-pregnane	20.36
Cortolic acid*	3α,11β,17α,20-tetrahydroxy-5β-pregnan-21-oic acid	20.36
Cortolone*	3α,17α,20,21-tetrahydroxy-5β-pregnan-11-one	20.18
Cortolonic acid*	3α,17α,20-trihydroxy-11-oxo-5β-pregnan-21-oic acid	20.36
7-Dehydrocholesterol	5,7-cholestadien-3β-ol	20.4

* These steroids occur in both 20α- and 20β-forms.

Dehydroepiandrosterone (DHEA)	3α-hydroxy-5-androsten-17-one	20.18
Deoxycholic acid	3α,12α-dihydroxycholan-24-oic acid	20.36
11-Deoxycorticosterone	21-hydroxy-4-pregnene-3,20-dione	20.36
11-Deoxycortisol	17α,21-dihydroxy-4-pregnene-3,20-dione	20.36
Dihydrotestosterone	17β-hydroxy-5α-androstan-3-one	20.23
Ergosterol	24-methyl-5,7,22-cholesta-trien-3β-ol	20.36
19-Hydroxy-4-androstene-dione	19-hydroxy-4-androstene-3,17-dione	20.36
25-Hydroxycholecalciferol	25-hydroxy-9(10)-seco-cholesta-5,7,10(19)-trien-3β-ol	20.29
7-Hydroxycholesterol	5-cholestene-3β,7α-diol	20.14
22-Hydroxycholesterol	5-cholestene-3β,22α-diol	20.7
18-Hydroxycorticosterone	18,21-dihydroxy-4-pregnene-3,20-dione	20.36
17-Hydroxypregnenolone	3β,17α-dihydroxy-5-pregnen-20-one	20.36
17-Hydroxyprogesterone	17α-hydroxy-4-pregnen-3,20-dione	20.9
19-Hydroxytestosterone	17β,19-dihydroxy-4-andro-sten-3-one	20.36
Lanosterol	4,4-dimethyl-8,24-cholesta-dien-3β-ol	20.36
Lithocholic acid	3α-hydroxycholan-24-oic acid	20.36
19-Nortestosterone	17β-hydroxy-4-oestren-3-one	20.36
Oestradiol-17β	1,3,5(10)-oestratriene-3,17β-diol	20.2
Oestriol	1,3,5(10)-oestratriene-3,16α,17β-triol	20.36
Oestrone	17-oxo-1,3,5(10)-oestratrien-3-ol	20.36
Pregnanediol	5β-pregnane-3α,20α-diol	20.36
Pregnenolone (Δ⁵pregnenolone)	3β-hydroxy-5-pregnen-20-one	20.7
Progesterone	4-pregnene-3,20-dione	20.9
Testosterone	17β-hydroxy-4-androsten-3-one	20.2
Testosterone-19-al	17β-hydroxy-3-oxo-4-andro-sten-19-al	20.36
Tetrahydrocortisol	3α,11β,17α,21-tetrahydroxy-5β-pregnan-20-one	20.36
Tetrahydrocortisone	3α,17α,21-trihydroxy-5β-preg-nane-11,20-dione	20.36

1. Cholane (24 carbon atoms). Removal of a three-carbon fragment from the side-chain of cholestane by cleavage between carbon-24 and carbon-25 produces cholane. This group contains the bile acids such as cholic acid (Figure 20.2).
2. Pregnane (21 carbon atoms). Removal of the side-chain of cholane by cleavage between carbon-20 and carbon-22 produces pregnane. This group contains progesterone, the corticosteroids, and aldosterone (Figure 20.2).
3. Androstane (19 carbon atoms). Removal of all the side-chain of pregnane by cleavage between carbon-17 and carbon-20 produces androstane. This group contains the androgens, testosterone, and androstenedione (Figure 20.2).
4. Oestrane (18 carbon atoms). Removal of the methyl group from androstane at carbon-10 produces oestrane. This group contains the oestrogens (Figure 20.2).

All of the important steroids have a trivial as well as a systematic name. Only trivial names are used in this text, but the systematic names are given in Table 20.1, which can also be used as an index of molecular formulae. For further details of steroid structure and basis for the systematic nomenclature, the reader is referred to Kellie (1975) and Gower (1979).

B. FORMATION OF CHOLESTEROL, CHOLESTEROL ESTER, AND STEROID HORMONES

1. **Cholesterol**

Cholesterol is usually obtained from the diet but, if necessary, sufficient for normal requirements can be synthesized in the liver, intestine, and other tissues. Indeed, virtually all nucleated cells have the capacity to synthesize this compound, but the quantitatively important tissue is the liver.

Evidence for the precursor of cholesterol was obtained when isotopic experiments demonstrated that all the carbons were derived from acetate and, moreover, they were incorporated alternately so that the carbon–carbon bond in the acetate was not broken during the synthesis. The elucidation of the metabolic pathway for cholesterol synthesis, through the studies of J. W. Cornforth, G. Popjak, K. Bloch, and F. Lynen, proved that acetyl-CoA was indeed the precursor for cholesterol biosynthesis (see Packter, 1973, for references). (However, whether this is the beginning of the physiological pathway, which starts with the substrate for the flux-generating step, is discussed in Section B.1a.) The pathway of cholesterol synthesis from acetyl-CoA is depicted in Figures 20.3 and 20.4.

Acetyl-CoA, in the cytosol of the liver, is converted to 3-hydroxy-3-methylglutaryl-CoA[1] (HMG-CoA) which is reduced to mevalonate (a six-carbon compound) by HMG-CoA reductase in a reaction that requires NADPH. This reaction plays a key role in the regulation of the rate of cholesterol synthesis (Section B.1.a). Mevalonate undergoes three separate phosphorylations to

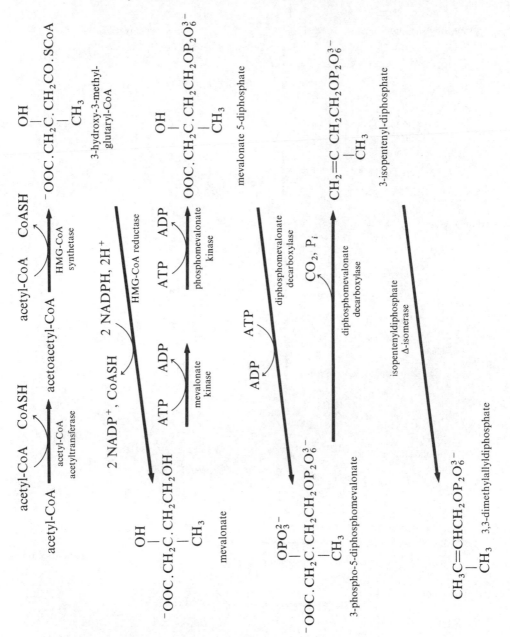

Figure 20.3 Pathway of prenyl (isoprene) unit biosynthesis. Note that mevalonate 5-diphosphate and related compounds have been known as mevalonate 5-pyrophosphate and 5-diphosphomevalonate, etc. 3-Isopentenyldiphosphate has also been known as $^{3}\Delta$-isopentenylpyrophosphate

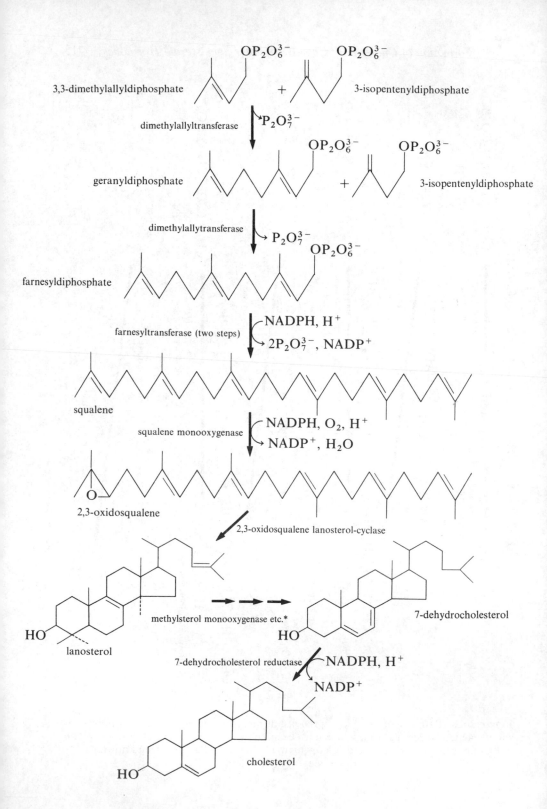

3,3-dimethylallyldiphosphate $+$ 3-isopentenyldiphosphate

dimethylallyltransferase $\rightarrow P_2O_7^{3-}$

geranyldiphosphate $+$ 3-isopentenyldiphosphate

dimethylallyltransferase $\rightarrow P_2O_7^{3-}$

farnesyldiphosphate

farnesyltransferase (two steps) $\begin{array}{l} \text{NADPH, H}^+ \\ \rightarrow 2P_2O_7^{3-}, \text{NADP}^+ \end{array}$

squalene

squalene monooxygenase $\begin{array}{l} \text{NADPH, O}_2, \text{H}^+ \\ \rightarrow \text{NADP}^+, \text{H}_2\text{O} \end{array}$

2,3-oxidosqualene

2,3-oxidosqualene lanosterol-cyclase

lanosterol

methylsterol monooxygenase etc.*

7-dehydrocholesterol

7-dehydrocholesterol reductase $\begin{array}{l} \text{NADPH, H}^+ \\ \rightarrow \text{NADP}^+ \end{array}$

cholesterol

Figure 20.4 Pathway of cholesterol synthesis from prenyl units.
*A large number of steps, involving other enzymes, are involved in the oxidative
removal of these three methyl groups

produce 3-phospho-5-diphosphomevalonate which is then decarboxylated in a
reaction that also hydrolyses the phosphate from the 3-position to produce 3-
isopentenyldiphosphate. (This compound contains the five-carbon prenyl group,
which is an important building block for many compounds, e.g. side chains of
vitamins A, E, and K, ubiquinone, carotenes—see Packter, 1973.) The
isopentenyldiphosphate is the donor of the prenyl group in the succeeding
reactions (Figure 20.3). It isomerises to form 3,3-dimethylallyldiphosphate and
these two compounds condense to produce geranyldiphosphate, which then
contains two isoprenyl units. Another molecule of isopentenyldiphosphate
condenses with geranyldiphosphate to produce farnesyldiphosphate, a 15-
carbon compound. Two molecules of the farnesyldiphosphate condense head-to-
head to form squalene[2] in a reaction that requires NADPH and releases
pyrophosphate. Squalene contains 30 carbon atoms but has no rings. These are
produced in a series of reactions initiated by the conversion of squalene to the
epoxide, 2,3-oxidosqualene, catalysed by a mixed function oxidase. This
undergoes a concerted internal condensation to form lanosterol which is the
parent steroid. Lanosterol is converted to cholesterol by the loss of three methyl
groups (two from carbon-4 and one from carbon-14 so that cholesterol has 27
carbon atoms), by saturation of the double bond in the side-chain and relocation
of the double bond from position 8–9 to position 5–6 in ring B. Methyl groups are
removed as carbon dioxide by oxidation, and relocation of double bonds occurs
by appropriate reduction and oxidations. For a detailed description of these
reactions the reader is referred to Packter (1973) and Gower (1979).

Quantitatively the most important fate of cholesterol is its incorporation into
membranes; this will be particularly important in rapidly growing and dividing
cells (see Section D.1). It is also a precursor for the synthesis of bile acids (Section
C.1). However, of particular relevance to this chapter is its role as precursor of the
steroid hormones described in Section B.3.

(a) *Control of the rate of cholesterol synthesis*

There have been no studies of the equilibrium nature of the reactions of the
cholesterol biosynthetic pathway or determination of which reactions may be
flux-generating. Nonetheless, HMG-CoA reductase,* the first enzyme specific to
the cholesterol biosynthetic pathway, is likely to catalyse a non-equilibrium
reaction and so may be an important regulatory reaction although it is unclear

* There has been considerable interest in the mechanism of regulation of this enzyme since it is one
means by which pharmacological intervention could reduce the concentration of cholesterol in cells
and blood particularly in patients suffering from familial hypercholesterolaemia (see Rodwell *et al.*,
1976; Endo, 1981; Brown and Goldstein, 1981). Drugs that inhibit the activity of HMG-CoA reductase
plus cholestyramine have been particularly effective in lowering the concentration of blood cholesterol
(Mabuchi, 1983).

whether it is flux-generating. It is possible that the first reaction from cytosolic acetyl-CoA, acetyl-CoA acetyltransferase, is flux-generating and, if this is the case, there must be some communication between the reductase reaction and the acetyltransferase to prevent a massive accumulation of HMG-CoA; one possibility is indicated as follows:

acetyl-CoA ── ⊖ ──→ acetoacetyl-CoA ──→ HMG-CoA ──→ mevalonate

The evidence implicating the reductase in the regulation of the pathway is that changes in its activity, which probably reflects changes in its concentration, correlate closely with rates of cholesterol synthesis (see Rodwell *et al.*, 1976). In rodents, in which the effect has been studied, the reductase activity as well as the rate of cholesterol synthesis in the liver exhibits a diurnal rhythm; the highest rates of synthesis occur in the middle of the feeding period (for the rat about midnight—assuming that darkness occurs at 6 p.m) and the lowest rate occurs at about 12 noon. This appears to be due to changes in the rate of synthesis* of the reductase which may be controlled by the hormones, insulin, glucagon, and glucocorticoids. In addition, there is evidence that cholesterol or one of its metabolites regulates the activity of the reductase by direct feedback, but the precise mechanism is not known. Cholesterol enters cells in the form of LDL-cholesterol (see Section D.2) and this cholesterol (or a metabolite) regulates the rate of cholesterol synthesis by controlling the rate of synthesis of the reductase and perhaps the activity via a feedback inhibitory mechanism.

In addition to any regulation through protein synthesis, HMG-CoA reductase is also regulated through an interconversion cycle (Figure 20.5). Almost all the work on the interconversion cycle has been carried out on enzymes isolated from the liver but it is probable that it also applies to other tissues (for details of control mechanism, see Ingebritsen and Gibson, 1980). HMG-CoA reductase is phosphorylated by a protein kinase that is *not* cyclic AMP-dependent; phosphorylation leads to inactivation (i.e. it produces the *b* form of the enzyme— similar to glycogen synthase, see Chapter 16; Section C.1). As expected from the general principles of interconversion cycles, the reductase is activated by dephosphorylation. This is catalysed by a protein phosphatase which is similar, if not identical, to hepatic glycogen phosphorylase phosphatase. However, the system is more complex than this and resembles that of glycogen phosphorylase in that the HMG-CoA reductase kinase is also regulated by an interconversion cycle. The kinase can be phosphorylated by a further protein kinase (also *not* cyclic AMP-dependent and termed HMG-CoA reductase kinase kinase) to form the active (*a*) form of the reductase kinase. The same phosphatase that dephosphorylates reductase *b* also dephosphorylates reductase kinase *a* (Figure 20.5). Consequently, control of the activity of the protein phosphatase in the liver could cause large changes in the activity of HMG-CoA reductase. Current

---

* Changes in rates of synthesis of the enzyme can have a marked effect on the enzyme activity since it has a short half-life, 2–4 hr (Rodwell *et al.*, 1976; Endo, 1981).

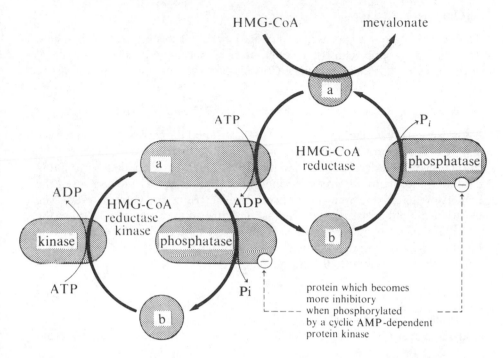

Figure 20.5   Regulation of HMG-CoA reductase activity by enzyme interconversions

evidence suggests that the phosphatase may be controlled by a specific protein which can inhibit its activity (and hence decrease that of the reductase) but, to indicate the complexity of control, this protein-inhibitor is also regulated by a further interconversion cycle. Phosphorylation of the inhibitor-protein by cyclic AMP-dependent protein kinase increases its inhibitory effect on the phosphatase. Hence a decrease in the cyclic AMP concentration in the cell, decreases the activity of cyclic-AMP dependent protein kinase and this reduces the degree of phosphorylation of the phosphatase inhibitor-protein, so that it is less inhibitory. Therefore, the phosphatase activity increases, leading to a marked increase in the activity of HMG-CoA reductase, as explained above.

This effect of cyclic AMP-dependent protein kinase could account for the effects of glucagon and insulin on HMG-CoA reductase activity; the former hormone lowers the activity whereas insulin increases it. In addition, it is possible that insulin modifies the activity of the phosphatase inhibitor-protein by a cyclic AMP independent mechanism (Chapter 22; Section C). How far the control of HMG-CoA reductase activity by cholesterol (or other intermediates) involves these interconversion cycles remains to be established. In addition, the link, if any, between control via changes in enzyme concentration and the interconversion cycles is not known.

2. Cholesterol ester

The hydroxyl group at position-3 of cholesterol can be esterified by long-chain acyl-CoA to produce cholesterol esters in a reaction catalysed by cholesterol acyltransferase (also known as acyl-CoA cholesterol: acyltransferase and hence ACAT):

$$\text{acyl-CoA} + \text{cholesterol} \longrightarrow \text{cholesterol ester} + \text{CoASH}$$

This reaction takes place in the cytosol of the cell and the cholesterol ester acts as a store of cholesterol in this compartment. It can be hydrolysed to cholesterol by the action of cholesterol esterase and the importance of this reaction in providing cholesterol for steroid hormone formation is emphasized by the control of the activity of this enzyme by an interconversion cycle (Boyd and Gorban, 1980). The existence of both transferase and esterase activities in the same cell also raises the possibility of a substrate cycle between cholesterol and cholesterol ester, the role of which may be to provide a precise control mechanism for the intracellular concentration of cholesterol (Figure 20.6).

3. Steroid hormones

The major tissues involved in steroid hormone formation are the adrenal cortex, testes, and ovary. There are three sources of cholesterol for the synthesis of steroid

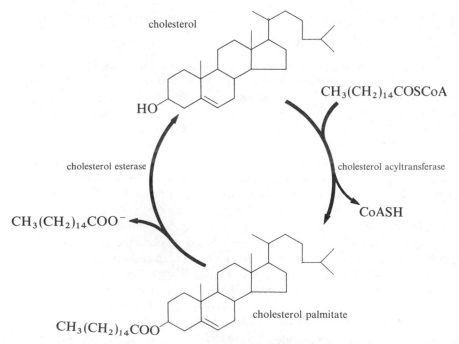

Figure 20.6 Substrate cycle between free and esterified cholesterol. Unsaturated fatty acyl-CoA can also act as substrate for ACAT.

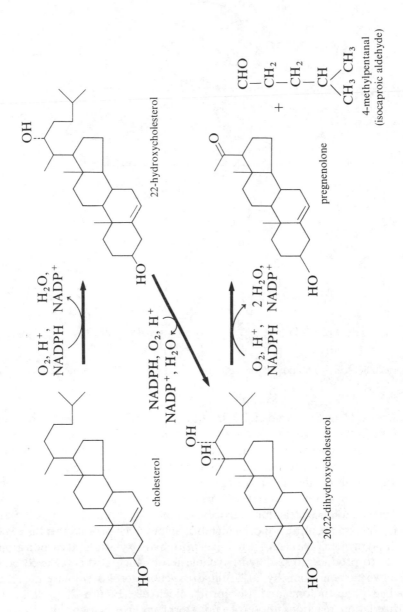

Figure 20.7   Pathway for conversion of cholesterol to pregnenolone. The three successive monooxygenase reactions and final lyase reaction occur in a mitochondrial complex containing the cytochrome P-450 system

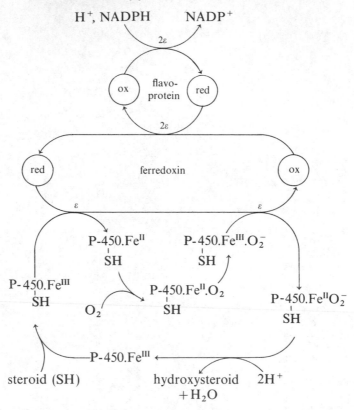

Figure 20.8   Likely reaction sequence for monooxygenases containing cytochrome P-450

hormones: from the bloodstream in the form of LDL-cholesterol (see Section D.2), hydrolysis of stored cholesterol ester in the cell catalysed by cholesterol esterase, and biosynthesis from acetyl-CoA as described above. The relative importance of each process may vary from one tissue to another but has been investigated in detail only in the adrenal cortex (Brown, *et al.*, 1979). The first reaction in the synthesis of steroid hormones is the cleavage of the side-chain to form pregnenolone, a reaction that occurs in the mitochondria so that cholesterol has to traverse the mitochondrial membrane for which process there is a specific transport protein. The cleavage requires prior hydroxylations at carbon atoms 20 and 22 to produce 20,22-dihydroxycholesterol which is then cleaved between these two carbon atoms by 20,22-dihydroxycholesterol desmolase (or lyase) to produce pregnenolone and isocaproic aldehyde (Figure 20.7). Like many hydroxylation reactions those of cholesterol involve cytochrome P-450 as a prosthetic group of the mono-oxygenase. A likely sequence of events is depicted in Figure 20.8.

Through chemical transformations of relatively few basic kinds (see Figure

20.9), pregnenolone can be converted into all the steroid hormones. The reactions leading from pregnenolone to the other steroid hormones are summarized in Figures 20.10–20.13. These summaries identify the intermediates of the pathways and the type of reactions involved. Further details of the individual reactions are given in Appendix 20.1 and the structures of some of the intermediates in Table 20.1.

HYD   *Hydroxylation.* Involves simultaneous oxidation of a steroid and NADPH by oxygen. Catalysed by monooxygenase (or hydroxylase). See Figure 20.8

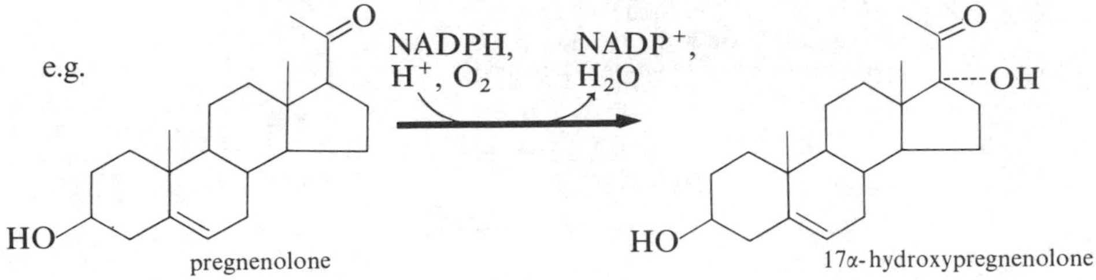

RED   *Reduction of keto (oxo) group to an hydroxyl group*

OX   *Oxidation of an hydroxyl group to a keto (oxo) group*

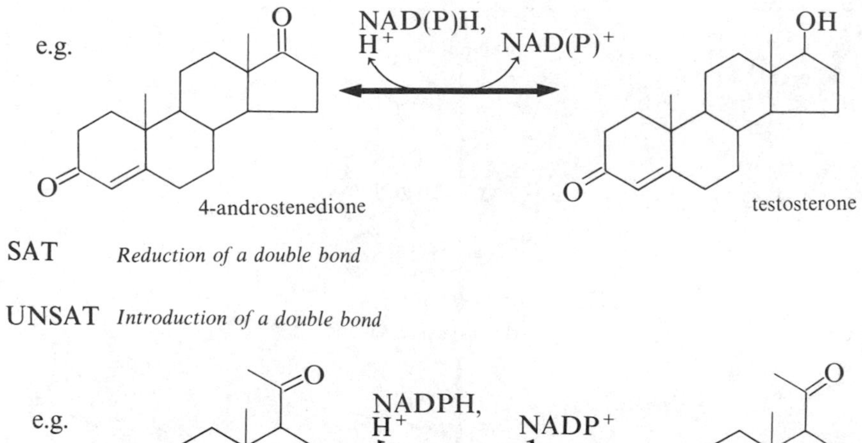

SAT   *Reduction of a double bond*

UNSAT   *Introduction of a double bond*

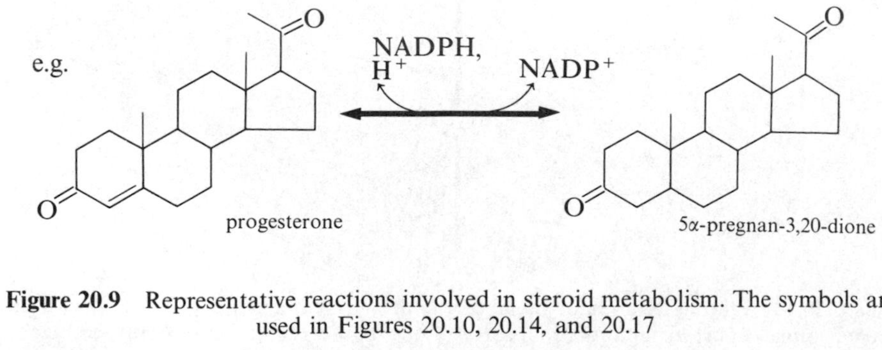

Figure 20.9   Representative reactions involved in steroid metabolism. The symbols are used in Figures 20.10, 20.14, and 20.17

Transposition of double bonds. Catalysed by steroid isomerase

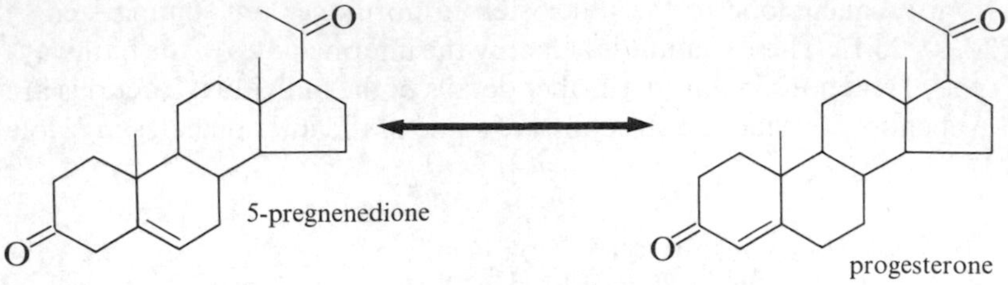

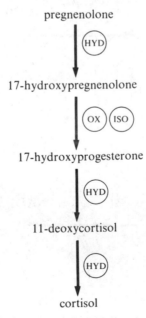

5-pregnenedione

progesterone

LY *Side-chain cleavage* Catalysed by lyases. Frequently associated with oxidation and catalysed by multienzyme complexes known as desmolases

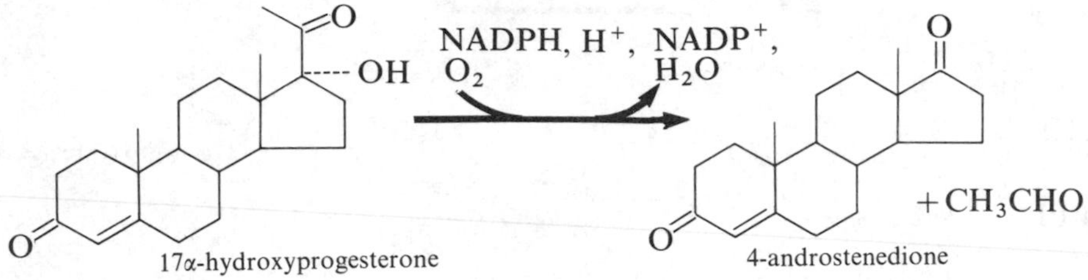

NADPH, H⁺, NADP⁺,
O₂   H₂O

17α-hydroxyprogesterone

4-androstenedione

$+ CH_3CHO$

Figure 20.9 Continued

pregnenolone

(HYD)

17-hydroxypregnenolone

(OX) (ISO)

17-hydroxyprogesterone

(HYD)

11-deoxycortisol

(HYD)

cortisol

Figure 20.10 Outline of a major metabolic pathway for the synthesis of cortisol from pregnenolone occurring in the adrenal cortex. Other pathways may also occur. See Figure 20.9 for explanation of symbols and Table 20.1 for steroid structures and alternative names

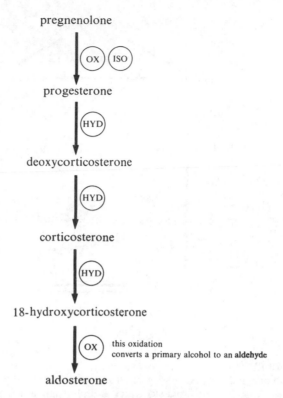

Figure 20.11 Outline of a major metabolic pathway for the synthesis of aldosterone from pregnenolone occurring in the adrenal cortex. Other pathways may also occur. See Figure 20.9 for explanation of symbols and Table 20.1 for steroid structures and alternative names

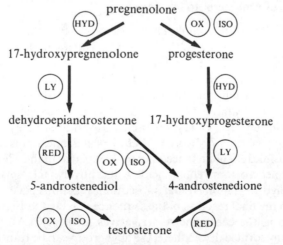

Figure 20.12 Outlines of major pathways for synthesis of testosterone from pregnenolone in the testis. See Figure 20.9 for explanation of symbols and Table 20.1 for steroid structures and alternative names

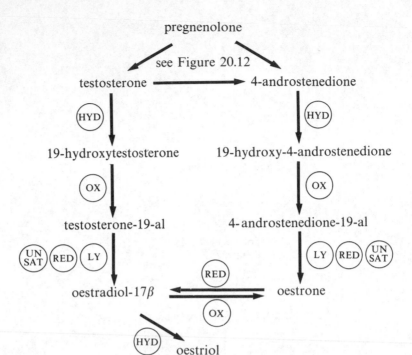

Figure 20.13 Outlines of major pathways for the synthesis of oestrogens from pregnenolone in the ovary. See Figure 20.9 for explanation of symbols and Table 20.1 for steroid structures and alternative names

C. METABOLISM OF STEROIDS

1. Conversion of cholesterol to bile acids

The major metabolic fate of cholesterol, and indeed the major route of excretion from the body, is conversion to the bile acids (cholate and chenodeoxycholate) which occurs only in the liver. Full details of the conversion are given in Percy-Robb (1975) and summarized in Figures 20.14 and 20.15. The first reaction involves hydroxylation of cholesterol in the 7-position to give 7α-hydroxycholesterol, catalysed by cholesterol 7α-monooxygenase, a cytochrome P-450 containing enzyme. Oxidation and transposition of the double bond from ring B to ring A generates 7α-hydroxy-4-cholesten-3-one. If this remains attached to the endoplasmic reticulum it becomes hydroxylated in the 12α-position and so becomes destined to form cholic acid. Alternatively, release into the cytosol obviates this hydroxylation so that subsequent reactions generate chenodeoxycholic acid. In the final reaction of the sequence depicted in Figure 20.15, or one subsequent to it, the coenzyme A derivatives, choloyl-CoA and chenodeoxycholoyl-CoA are formed. From there, the acyl group can be transferred to glycine

Figure 20.14 Pathway for conversion of cholesterol to main bile acids. See Figure 20.9 for explanation of symbols

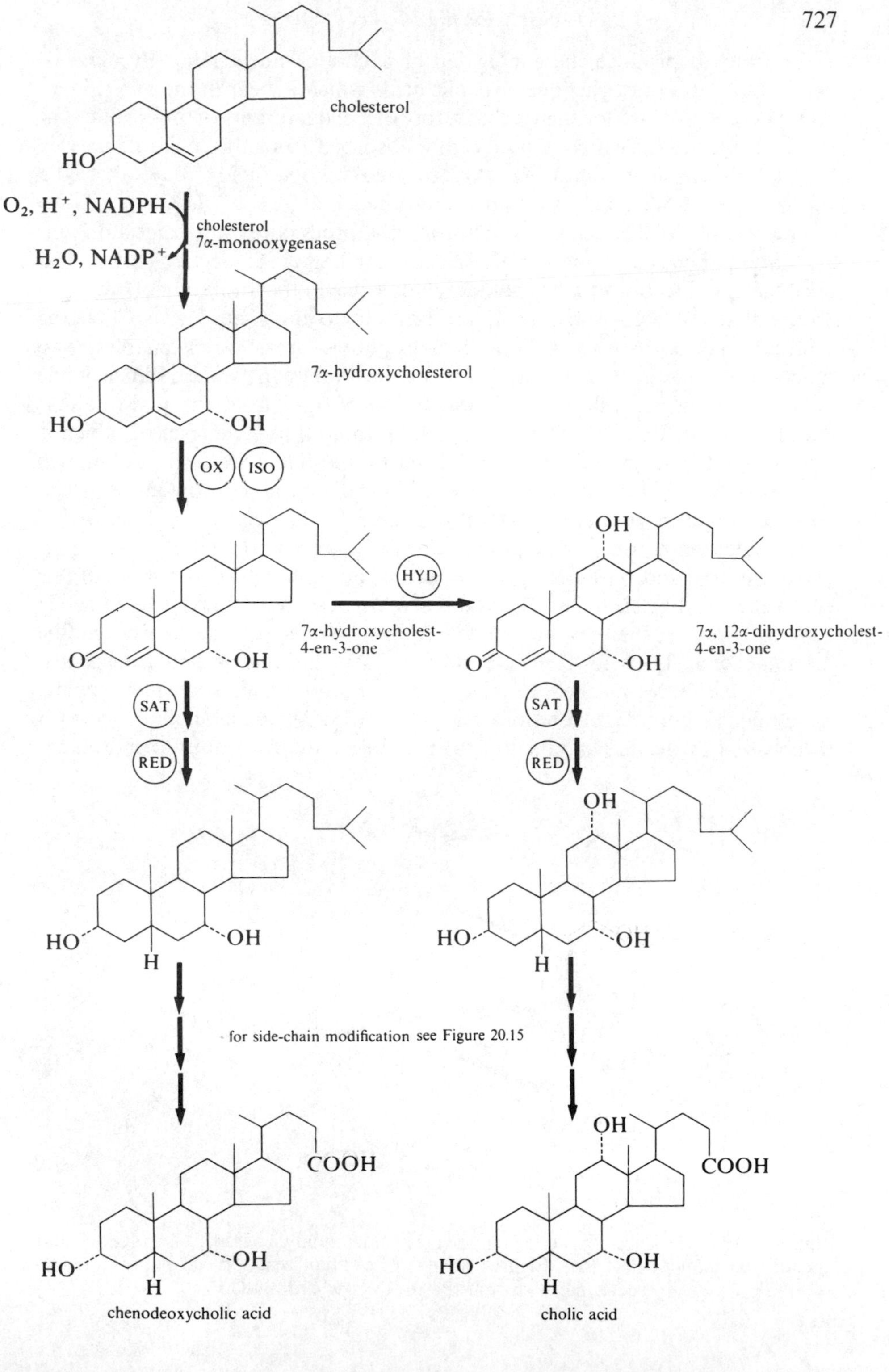
cholesterol
HO
O_2, H^+, NADPH
H_2O, $NADP^+$
cholesterol 7α-monooxygenase
7α-hydroxycholesterol
HO
OH
OX
ISO
7α-hydroxycholest-4-en-3-one
O
OH
HYD
7α, 12α-dihydroxycholest-4-en-3-one
OH
O
OH
SAT
RED
SAT
RED
HO
OH
H
OH
HO
OH
H
for side-chain modification see Figure 20.15
COOH
HO
OH
H
chenodeoxycholic acid
OH
COOH
HO
OH
H
cholic acid

or taurine[3] to produce the conjugated glycocholic, taurocholic, glycocheno-deoxycholic and taurochenodeoxycholic acids which appear in the bile (Figure 6.1). There is evidence for the synthesis, too, of small amounts of lithocholic acid (Percy-Robb, 1975) which is not readily absorbed from the lower ileum (see below). Bacterial modification (e.g. to deoxycholic acid) may alter the composition of bile acids in the small intestine.

The role of the bile acids in solubilizing the products of lipid digestion and facilitating their absorption is described in Chapter 6; Section B.1.a. The chemical modifications to the cholesterol molecule in the formation of bile acids ensure that the polar substituents are brought to the same face, so that the molecule has both hydrophilic and hydrophobic areas which enables it to function as a detergent in the absorption of lipid. Failure to produce sufficient bile acids by the liver results in poor absorption of lipid from the intestine and consequent steatorrhoea (Donaldson, 1965). In addition, cholesterol, which is also found in bile and which is solubilized by the bile acids and lecithin, will precipitate if the concentration of bile acids and lecithin are low and this can result in the formation of cholesterol gallstones[4] (Vlahcevic *et al.*, 1970).

The total amount of bile acids in the body (largely in the biliary tract, liver, small intestine and portal blood) is about 3 g but considerably more than this enters the intestine each day. This is possible because more than 90% of the bile acids are re-absorbed by an energy-requiring transport mechanism in the terminal ileum. The acids then enter the hepatic portal vein and are transported back to the liver attached to albumin. This constitutes the enterohepatic circulation (Figure 20.16). The importance of this system in metabolic economy is emphasized by the fact that the total pool of bile acids cycles four to eight times

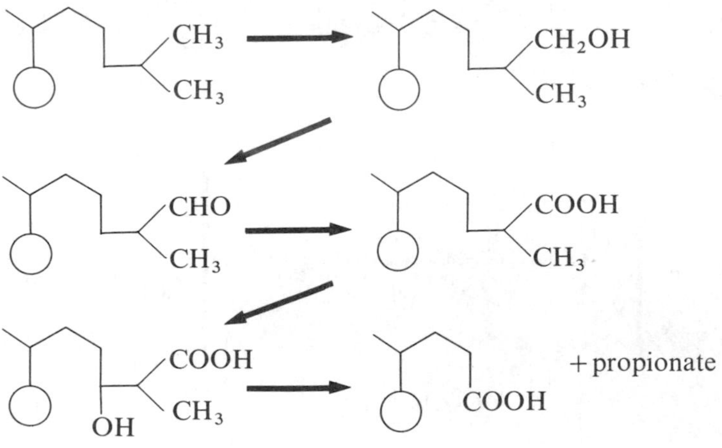

Figure 20.15 Sequence of reactions involved in side-chain cleavage to form cholic and chenodeoxycholic acids (see Figure 20.14). The final reaction probably involves coenzyme A so that the product is choloyl-CoA

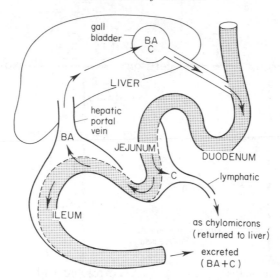

Figure 20.16   The enterohepatic circulation of bile acids (BA) and cholesterol (C)

every 24 hr yet only 10 % is lost in the faeces. Of the cholesterol secreted into the bile (about 1 g.day^{-1}), about half is taken up by the jejunum and is used in the formation of chylomicrons: the remainder (about 500 mg.day^{-1}) is lost in the faeces.[5] An additional 100 mg is lost each day due to loss of cells from the surface of skin. Thus the total steroids (including bile acids) lost each day from the body amounts to about 900 mg, which could be readily synthesized or obtained from the diet.

The loss of bile acids in the faeces can be increased markedly by the ingestion of the ion-exchange compound cholestyramine (Chapter 6; Note 17) which binds the bile acids and renders them unavailable for reabsorption in the ileum. However, under these circumstances the total amount of bile acid in the body changes very little since the loss is made up by an increase in the rate of bile acid synthesis due to an increase in the activity of cholesterol 7α-monooxygenase in the liver. Under normal conditions, the activity of this enzyme is inhibited allosterically by the bile acids (possibly chenodeoxycholate) but if the rate of synthesis has to increase markedly, the concentration of the enzyme itself increases. The ingestion of cholestyramine can sufficiently increase the rate of conversion of cholesterol to bile acids that hepatic and blood concentrations of cholesterol decrease. However, the effect of this will in part be offset by the increase in rate of synthesis of cholesterol due to increases in the activity and concentration of hepatic HMG-CoA reductase (see Section B.1.a).

2. *Catabolism of the steroid hormones*

Knowledge of the endproducts of steroid hormone metabolism is clinically important since measurement of their concentration in the urine provides an

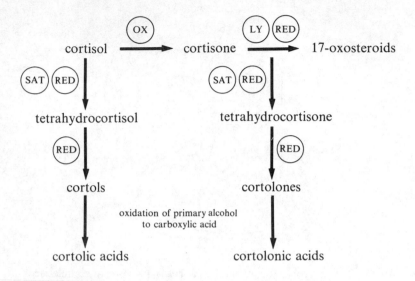

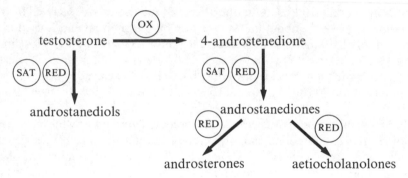

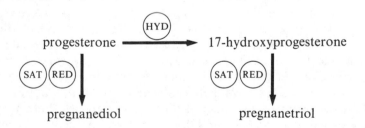

Figure 20.17 Outline of some of the major pathways involved in the degradation of steroid hormones. Oestrogens are mostly excreted as oestrone and oestriol (see Figure 20.13). See Figure 20.9 for explanation of symbols and Table 20.1 for steroid structures and alternative names

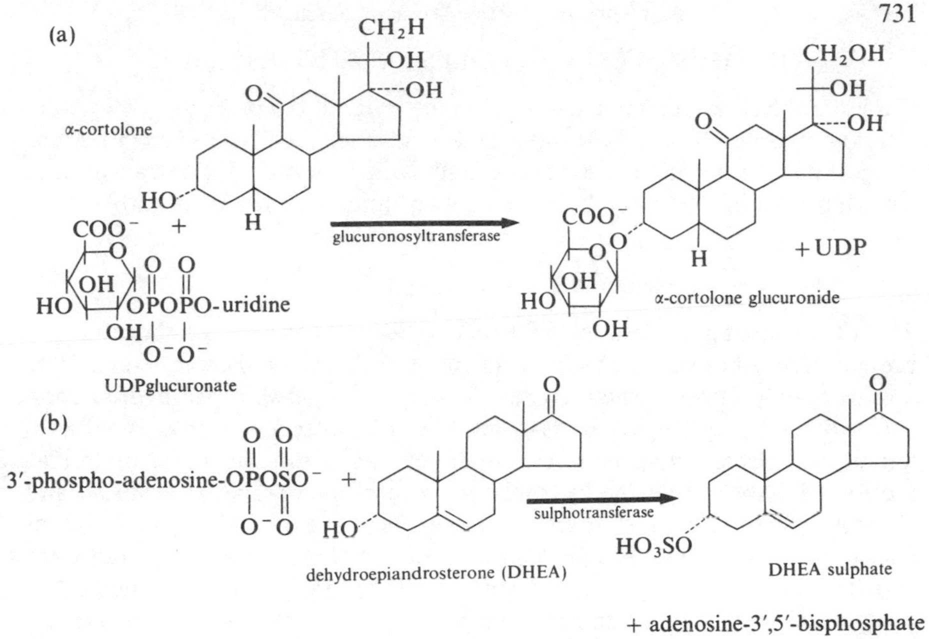

(a)

α-cortolone

UDPglucuronate

glucuronosyltransferase →

α-cortolone glucuronide

+UDP

(b)

3'-phospho-adenosine-OPOSO⁻ +

dehydroepiandrosterone (DHEA)

sulphotransferase →

DHEA sulphate

+ adenosine-3',5'-bisphosphate

Figure 20.18 Conjugation pathways for steroid excretion.
(a) Glucuronide formation.
(b) Sulphation.
For origin of 3'-phosphoadenosine-5'-phosphosulphate see Figure 16.11

indication of the rates of turnover of the steroid hormones.[6] The liver and kidney carry out reactions which render the molecules more polar and this polarity is further increased by formation of sulphate esters or glucuronides, which are water soluble conjugates that can be excreted in the urine (see below).

Reactions of the kinds shown in Figure 20.9 produce characteristic endproducts derived from each of the steroid hormones (Figure 20.17): cortisol produces endproducts that include cortolic acids and cortolonic acids; progesterone produces pregnanediol and pregnanetriol; testosterone loses 80% of its androgenic property when the 17β-hydroxyl group is converted to an oxo-group to produce 4-androstenedione which can be converted into other compounds (e.g. androsterone, aetiocholanolone); conversion of oestradiol-17β to oestrone reduces the oestrogenic activity and the molecule may be hydroxylated in a large number of positions so that at least 20 metabolites of oestrogens can be found in human urine.

Most of the metabolites are converted to sulphate or glucuronide derivatives, probably in the liver, before excretion via the kidney. Glucuronides are glycosides formed by the transfer of a glucuronosyl group from UDPglucuronic acid (Chapter 16; Section D.2) to the steroid in a reaction catalysed by a glucuronosyltransferase (Figure 20.18). Steroid sulphates are formed in a reaction between one of the hydroxyl groups on the steroid and 3'-phosphoadenosine-5'-phosphosulphate (see Chapter 16; Section D.2.b) catalysed by a sulphotransferase (Figure 20.18).

D. PHYSIOLOGICAL ROLES OF CHOLESTEROL

Cholesterol has several important roles in the body: it is a precursor of both the steroid hormones (Section B.3) and the bile acids (Section C.1), it is an important structural component of membranes, and it is involved in the transport of triacylglycerol in the blood. These latter two functions are discussed below.

1. Cholesterol in cell membranes

The total content of cholesterol in the body is about 140 g of which about 120 g is present in membranes. The highly hydrophobic character of cholesterol and its approximately planar structure enables it to fit in between the hydrocarbon chains of the phospholipids in the membrane. This interdigitation of cholesterol among the phospholipid molecules is important in modifying the molecular motion (fluidity) within the hydrophobic core of membranes (see Shattil and Cooper, 1978). Hence changes in molecular motion, which may be of considerable importance in control of the function of membranes (e.g. transport, activities of membrane-bound enzymes) can be produced by changing the cholesterol/phospholipid ratio in the membrane. How this is controlled is not known but there is evidence that an increase in the blood concentration of cholesterol can increase this ratio in the cell membrane.*

2. Transport of lipoproteins in the blood

A summary of lipoprotein metabolism is given in Chapter 6 but the aspects that are particularly related to cholesterol, namely the metabolism of LDL in peripheral tissues and the function of HDL, are considered in this chapter.

(a) *Pathway of LDL metabolism*

As well as transporting triacylglycerols in the plasma, the lipoprotein particles also transport cholesterol. LDL, which consists mainly of cholesterol ester and protein (Table 6.1) arises from the progressive removal of triacylglycerol and phospholipid from VLDL and intermediate density lipoprotein as they pass through the adipose tissue (Figure 6.6). LDL is metabolized in most, if not all, tissues by a specific LDL pathway[7] that results not only in the degradation of LDL but in the provision of cholesterol for membrane synthesis in that cell. LDL binds with high affinity to a specific receptor on the surface of a cell (the apolipoprotein-B is probably the component of the LDL particle that binds to

---

* There is an increase in the cholesterol/phospholipid ratio in blood platelets incubated in a medium containing a high concentration of cholesterol. In addition, membrane fluidity changes and this affects the activity of the cyclooxygenases so that the rate of formation of thromboxane-A_2 (an agent that favours platelet aggregation—see Chapter 17; Section D.4.f) increases. This is one factor that could account for the correlation between elevated cholesterol concentrations and incidence of coronary heart disease—see also Section G.1 (Stuart *et al.*, 1980; Cooper and Shattil, 1980).

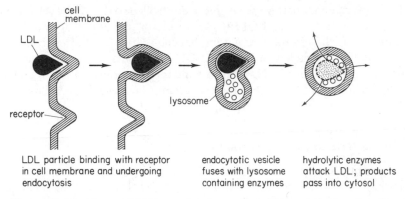

LDL particle binding with receptor
in cell membrane and undergoing
endocytosis

endocytotic vesicle
fuses with lysosome
containing enzymes

hydrolytic enzymes
attack LDL; products
pass into cytosol

Figure 20.19   Entry of LDL particle into cell and its metabolism therein

the receptor) and together with part of the cell membrane is then taken into the cell by endocytosis to produce an intracellular vesicle (an endocytotic vesicle) which fuses with the membrane of a lysosome. The components of the LDL receptor-complex are then degraded by the lysosomal enzymes (Chapter 10; Section B.3.b). The protein is hydrolysed to amino acids and the cholesterol ester is hydrolysed by a lipase to fatty acid and free cholesterol (Figure 20.19). The latter traverses the lysosomal membrane to enter the cytosol of the cell where it is available for incorporation into cell membranes. However, if the rate of production of cholesterol by this process exceeds the rate of utilization, the cytosolic concentration of free cholesterol increases above normal and it is then esterified as described in Section B.2. This cholesterol ester is then stored in the cell for subsequent use. Within the cell, there are two regulatory mechanisms that prevent excessive accumulation of cholesterol or cholesterol ester. First, cholesterol, or one of its metabolites, inhibits the activity of HMG-CoA reductase (Section B.1.a). Secondly, the increased intracellular cholesterol (or one of its metabolites) causes a decrease in the number of LDL-receptors on the cell membrane (probably by inhibiting their synthesis), thus reducing the rate of LDL uptake in that particular cell. This control mechanism is demonstrated by the fact that in rapidly dividing cells, which require a large amount of cholesterol for membrane synthesis (e.g. small intestine), the number of receptors for LDL is high, whereas in cells that are slowly-dividing or non-dividing (e.g. muscle) the number of receptors is low.

This pathway for metabolism of LDL has been established for a number of human tissues including fibroblasts, lymphocytes, and smooth muscle cells and calculations suggest that, in the tissues of normal man, about 70% of the metabolism of LDL occurs in this way. The remaining 30% is metabolized by a pathway that does not require a specific receptor so that the LDL enters the cell by non-specific endocytosis. It is, however, unclear which cells are normally involved in this 'scavenger pathway' but the liver, which metabolizes most of the LDL in man, does not utilise the 'scavenger pathway'. The rate of entry of LDL into a cell by the scavenger pathway may not be controlled so that, if the cell is exposed to high concentrations of LDL, it can result in a marked accumulation of cholesterol and cholesterol ester within that cell. Patients suffering from familial hypercholesterolaemia have a defective LDL receptor; heterozygotes have about half the normal number of LDL-receptors and there is compensation by increasing the concentration of LDL in the blood which may increase the metabolism of the LDL through the scavenger pathway (Havel *et al.*, 1980).

(b) *Role of HDL in cholesterol transport*

HDL accepts apolipoproteins from degraded lipoproteins; it also accepts cholesterol from tissues and transfers cholesterolester to the products of chylomicron and VLDL metabolism during the hydrolytic removal of triglyceride from these particles (Chapter 6; Section C.1.c and Figure 6.6).

HDL particles arise in two ways. Some are produced in the liver and intestine and are secreted into the blood or lymph. Additional HDL particles are formed in the bloodstream from the surface of the chylomicrons (and possibly VLDL) as the triacylglycerol is removed by lipoprotein lipase activity.* The HDL that is produced from either of these two sources is known as 'nascent' HDL and has a sheet or discoidal shape comprising phospholipid, protein, and cholesterol in a bilayer-lamellar structure which, like the cell membrane, is stable. However, this nascent HDL is modified in the bloodstream to produce HDL. The plasma enzyme, lecithin-cholesterol acyltransferase (LCAT) transfers fatty acid from the phospholipid onto the cholesterol of the nascent HDL to form cholesterol ester which then moves towards the interior of the lamellar structure. Consequently, the HDL particle takes up more cholesterol, probably from the cell membranes of many tissues and from arterial walls, and it too is esterified by LCAT. With sufficient formation of cholesterol ester to produce a hydrophobic 'core', the discoidal HDL becomes spherical (Figure 20.20). In this form, the HDL particle acts as a large reserve of cholesterol for transfer to the metabolic products of chylomicrons and VLDL during formation of the end products, remnant and intermediate-density lipoproteins (see Tall and Small, 1978; Havel *et al.*, 1980). Since the intermediate-density lipoprotein is the precursor of LDL, the cholesterol ester transferred from HDL can, via the degradation of LDL, find its way into the cells of many tissues as described above. It is estimated that the turnover of cholesterol in this way amounts to about $1100 \, mg.day^{-1}$ and about 500 mg is taken by the liver from metabolism of intermediate-density lipoproteines. The HDL, therefore, plays a very important role in transfer of cholesterol from tissues to the lipoprotein particles thus maintaining a turnover not only of triacylglycerol but also of cholesterol and a low concentration of HDL may lower the ability of the body to remove cholesterol from the tissues.[8] The clinical importance of high levels

---

* As the triacylglycerol is lost from the core of the chylomicron or VLDL, the particle shrinks and some of the stabilizing 'coat' of lipoprotein, phospholipid and cholesterol is split off to produce HDL (Tall and Small, 1978).

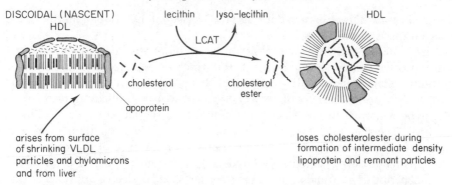

DISCOIDAL (NASCENT) HDL lecithin lyso-lecithin HDL

LCAT

cholesterol cholesterol ester

apoprotein

arises from surface of shrinking VLDL particles and chylomicrons and from liver

loses cholesterolester during formation of intermediate density lipoprotein and remnant particles

Figure 20.20   Formation of HDL particles. After Tall and Small, (1978)

of HDL is discussed in Section G.1.b. The metabolism of the protein and phospholipid of HDL occurs in both the liver and extrahepatic tissues. The HDL is probably taken up by an endocytotic mechanism, transported to the lysosomes and the components subsequently hydrolysed in a similar way to that of LDL.

E. PHYSIOLOGICAL ROLES OF STEROID HORMONES

The steroid hormones have fundamentally important physiological effects which are described below together with a brief account of the control of the rate of synthesis of each hormone. The mechanism of action of all steroid hormones is probably similar, stimulation of specific mRNA synthesis, thus increasing the rate of synthesis of specific proteins in the cell; this is discussed in Chapter 22; Section B.1.

1. **Glucocorticoids**

Glucocorticoids are so named because they increase the blood glucose concentration due to an increased rate of production of glucose by the liver and probably a decreased rate of utilization by other tissues. The human adrenal cortex secretes 16 mg cortisol each day and lesser amounts of corticosterone (4.4 mg), 11-deoxycorticosterone (0.2 mg) and 11-deoxycortisol (0.4 mg) (Bondy, 1980).

(a) *Control of the secretion of cortisol*

The steroid hormones are not stored to any large extent in the secretory tissues so that the rate of synthesis from cholesterol controls the rate of secretion. The biochemical mechanisms for the control of the rate of steroidogenesis in any endocrine tissue are still far from clear but most information is available for the adrenal cortex; this is discussed in Chapter 22; Section D.1. but an outline of the mechanism is given below.

The rate of cortisol synthesis and secretion is controlled by the adrenocorticotrophic hormone (ACTH) which is secreted by the anterior pituitary gland. The secretory activity of this gland is controlled both by the central nervous system, via the secretion of corticotropin-releasing factor from the hypothalamus, and by feedback inhibition via the blood concentration of cortisol. The fact that cortisol secretion follows a daily rhythm,* which is characteristic for each individual, indicates the importance of central control. Perhaps the most physiologically important variations occur just prior to the normal period of sleep and upon waking. As the period of sleep is approached, the plasma cortisol concentration decreases and it reaches a minimum during the first few hours of sleep then increases again as the normal awakening time approaches and reaches its maximum at the time of awakening or soon afterwards. These changes in blood concentrations of cortisol reflect changes in rates of steroidogenesis that are controlled by variations in the level of ACTH, which in turn are controlled by the hypothalamus. Since the physiological role of cortisol is to prepare the body for metabolic and physical activity (e.g. increased production of blood glucose, increased mobilization of fatty acids, increased sensitivity to hormones such as adrenaline and thyroxine) this diurnal pattern can be understood. ACTH controls the rate of steroidogenesis in the adrenal cortex by binding to a receptor on the cell membrane, increasing the activity of adenylate cyclase and thus raising the intracellular concentration of cyclic AMP. This activates cyclic AMP-dependent protein kinase and leads to activation of the LDL-uptake mechanism, cholesterol esterase and HMG-CoA reductase all of which increase the cytosolic concentration of cholesterol. In addition, phosphorylation of one or more of the proteins involved in the initiation of protein synthesis increases the rate of production of a carrier protein that transports cholesterol into the mitochondrion. Finally, the activity of cholesterol side-chain cleavage system may also be activated (see Figure 20.21).

(b) *Physiological effects of cortisol*

Cortisol (and the glucocorticoids in general) has a diverse range of metabolic and physiological effects but, since few if any can be observed with isolated tissues *in vitro*, the view has developed that its effects are permissive, that is, it changes the response to other hormones or regulators. These effects are outlined below.

1. To achieve stimulation of gluconeogenesis, the hormones glucagon and adrenaline (see Chapter 11; Section B.4.b) require the presence of glucocorticoids. A deficiency of the latter, therefore, in Addison's disease (see Section E.1.c) results in hypoglycaemia. Glucocorticoids increase the capacity of the liver to produce glucose and glycogen due, at least in part, to an increase

---

* From seven to 13 peak concentrations can be observed in any one individual during a 24 hr period.

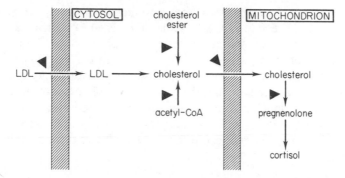

Figure 20.21 Potential control sites in cortisol synthesis. ▶ Indicates control sites

in the concentration of the key gluconeogenic enzymes (e.g. phosphoenol-pyruvate carboxykinase, fructose bisphosphatase, glycogen synthase). However, to produce a steady state increase in the rate of gluconeogenesis, the rate of production of precursors must also increase. This explains the increases seen in activities of enzymes involved in degradation of amino acids in the liver (e.g. tyrosine aminotransferase, tryptophan 2,3-dioxygenase).

2. Glucocorticoids increase the rate of protein degradation and hence amino acid release in muscle. The molecular basis of this effect is unclear but it may involve an increased number of lysosomes (Goldberg, 1980).

3. Glucocorticoids increase the sensitivity of adipose tissue to lipolytic hormones (e.g. growth hormone, catecholamines), so that for a given concentration of a lipolytic hormone, the rate of lipolysis will be higher in the presence of glucocorticoids (Fernandez and Saggerson, 1978). This may explain why high concentrations of glucocorticoids increase the blood concentration of fatty acids. The mechanism of this 'permissive' effect is not known.

4. Glucocorticoids potentiate the β-adrenergic effect of catecholamines and may increase the rate of synthesis of adrenaline. Since a deficiency of cortisol can lead to depression whereas excess can lead to a state of euphoria, it is tempting to suggest that these effects are due to changes in sensitivity to catecholamines in the central nervous system (Chapter 21; Section E.1).

5. Glucocorticoids have an anti-inflammatory action and for this reason are used in the treatment of autoimmunity, for immunosuppression after transplant operation,[9] and also in treatment of septic shock in which the inflammatory response causes massive vasodilation producing hypovolaemia and hypotension (see Sheagren, 1981). In brief, the inflammatory response consists of two components; local vascular changes, such as arteriolar dilation and increased capillary permeability, and mobilization and activation of the lymphocytes, marcophages, polymorphonuclear leucocytes, and complement. Since the vascular changes and the activation of the lymphocytes etc. may be caused by prostaglandins, glucocorticoids may be effective because of their ability to inhibit phospholipase A_2 and hence the production of the

unsaturated fatty acid precursors for prostaglandin synthesis (see Sheagren, 1981; Chapter 17; Section D.2). In some animals, glucocorticoids reduce the number of lymphocytes and the amount of lymphoid tissue. This is probably due to inhibition of protein synthesis in lymphoid tissue so that net protein breakdown increases. However, in man, glucocorticoids may not influence lymphoid tissue but may reduce the access of lymphocytes to the location of the infection or damage, decrease the rate of formation of antibodies and decrease the conversion of T-lymphocytes to killer lymphocytes.

6. Glucocorticoids increase the appetite and hence increase food intake and this can result in obesity.

(c) *Adrenal insufficiency* (Addison's disease)

Inadequate secretion of glucocorticoids and mineralocorticoids occurs in Addison's disease. Anorexia is one of the first symptoms if onset is acute and is followed rapidly by nausea, vomiting, diarrhoea, abdominal cramping pain, and fever which may be severe. The blood pressure is low and there is dehydration. If there has been insufficiency for some time, the skin may be abnormally pigmented. The reduction in blood glucose and sodium concentrations explains some of the symptoms described above, and there may be lactic acidosis (Chapter 13; Section D). Treatment involves hormone and fluid replacement.

(d) *Adrenocortical hyperactivity* (Cushing's syndrome)

This can be caused by a tumour of the adrenal cortex or by the massive intake of corticosteroids for therapeutic purposes. Weight gain is common (producing 'moon face') and obesity can result. Lassitude, emotional disturbances, loss of muscle strength, and bone pain (due to osteoporosis) are common. In women, a disturbance in the menstrual cycle is usually the first symptom and this often progresses to amenorrhea. The blood glucose may be elevated with a 'diabetic' glucose tolerance curve. The white cell count can be reduced. The disease is usually confirmed by the elevated plasma concentration of cortisol and the high urinary concentration of 17-hydroxycorticosteroids.

2. Mineralocorticoids

The main mineralocorticoid is aldosterone and it is synthesized in the zona glomerulosa cells of the adrenal cortex (see Figure 20.11). The conversion of cholesterol to pregnenolone and corticosterone to aldosterone occur within the mitochondria and are probably important regulatory steps. Three factors are important in the control of the rate of secretion of aldosterone, angiotensin II,[10] potassium ions, and ACTH. The biochemical mechanism of control is unclear but it probably involves activation of adenylate cyclase, increasing the cyclic AMP concentration, activation of protein kinase and phosphorylation of key enzymes in the pathway as described in Chapter 22; Section A.4. (see above).

The major biological role of aldosterone is to enhance the rate of reabsorption of sodium ions and reduce that of potassium ions in the kidney. The hormone acts on the cells of the distal tubules and the collecting duct to increase sodium ion uptake probably in exchange for potassium ions and protons (see Chapter 13; Section B.4). It may also increase the concentration of potassium and reduce that of sodium in muscle and brain cells. It is likely that the mineralocorticoids increase the production of a specific mRNA which determines the synthesis of a protein involved in sodium ion transport in the kidney and other tissues.

In adrenal insufficiency (i.e. Addison's disease), the acute problems are due to cortisol deficiency but the loss of sodium and the reduction in plasma volume and hypotension may be caused, in part, by lack of aldosterone (hypoaldosteronism). The addisonian patient is maintained on cortisol (or prednisolone), a liberal salt diet, and a mineralocorticoid. Aldosterone is not active by mouth, so that a synthetic mineralocorticoid (e.g. 9α-fluorohydrocortisone) is given (Tan and Mulrow, 1980). Primary aldosteronism is usually due to a secreting tumour in the adrenal cortex. The clinical characteristics are hypertension, hypokalaemia, and alkalosis; the symptoms are polyuria, polydipsia, nocturia, weakness, intermittent paraesthesia, and tetany. Usual treatment is surgery.

3. Androgens

In order to understand the role of the androgens, it is necessary to have some knowledge of the structure of the testes and the male reproductive tract. (For reviews of androgenic steroids, see Lincoln, 1979; Griffin and Wilson, 1980.)

(a) *Structure of the testis*

The testis consists of two components, loops of convoluted seminiferous tubules plus Sertoli cells, in which primordial germ cells mature into spermatozoa (a process which takes more than 70 days), and the interstitial cells of Leydig which produce and secrete androgenic steroids. The seminiferous tubules drain into a network of ducts known as the epididymis. From here, sperm enter the vas deferens and pass through the ejaculatory duct into the urethra mainly at the time of ejaculation.

(b) *Role of gonadotrophins in the male*

The anterior pituitary secretes two gonadotrophins, luteinizing hormone and follicle stimulating hormone (abbreviated to LH and FSH) which have distinct roles in both the testis and the ovary. In the male, LH controls the rate of secretion of testosterone by the Leydig cells (see below), whereas FSH governs the secretory activity of the Sertoli cells. The latter produce proteins necessary for maturation of the spermatozoa, the steroid oestradiol (produced from testosterone) and a protein known as inhibin. Oestradiol, testosterone, and

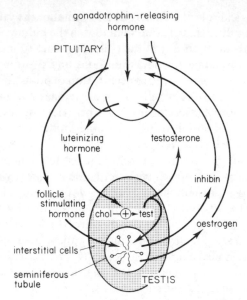

Figure 20.22   Feedback relationships between testis and pituitary gland

inhibin all regulate the rate of secretion of LH and FSH by the pituitary (by feedback control—see Figure 20.22).

The control of the rate of secretion of LH and FSH in the male appears to be similar to that in the female (see Section E.4.c). As well as the feedback control mentioned above there is central control via the release of gonadotrophin-releasing hormone (GnRH) which is transported from the secretory cells of the hypothalamus to the anterior pituitary via the portal blood system.

(c) *Production of androgens*

The major steroid products of the Leydig cells are testosterone, dihydro-testosterone (and possibly oestradiol) but the production of testosterone[11] is predominant. The pathway from cholesterol to testosterone develops at maturity in the Leydig cells and is shown in Figure 20.12. It is controlled by LH which binds to a cell membrane receptor thus activating adenylate cyclase resulting in an increase in the intracellular concentration of cyclic AMP and hence activation of protein kinase. The consequences of this are probably identical to those of protein kinase activation by ACTH in the adrenal cortex (see Section E.1.a and Figure 20.21).

(d) *Physiological actions of testosterone*

After secretion from the Leydig cells, testosterone is transported in the blood, bound both to albumin and testosterone-oestradiol binding globulin (a specific transport protein), to the target tissues. Testosterone is involved in a number of important changes at puberty and maintenance of such changes in the adult. The effects of testosterone are complicated by the fact that it can be converted to oestrogens in other tissues (e.g. Sertoli cells, adrenal cortex) and these may antagonize its action, and it can also be converted to another active androgen, dihydrotestosterone (see Section E.3.e). Testosterone or dihydrotestosterone are responsible for the increase in size of the external genitalia, the increased secretory capacity of the seminal vesicles, prostate and bulbourethral glands, the deepening of the voice, changes in the distribution and amount of body hair, increased secretions of the sebaceous gland, an increase in the amount of muscle (anabolic effect) and more aggressive behaviour. Testosterone plus FSH have effects on the Sertoli cells that are important in the maturation of spermatozoa.

Two of the more metabolic aspects of androgens are singled out for detailed discussion below.

(e) *Steroid 5α-reductase deficiency and male pseudohermaphroditism*

The work reported in this section indicates that both the androgens, testosterone and dihydrotestosterone, are involved in sexual development in the male but that each hormone has specific roles. This has emerged particularly from studies on male pseudohermaphroditism in a number of families in the village of Salinas in the Dominican Republic (Imperato-McGinley *et al.*, 1974). The affected males are born with a marked ambiguity of the external genitalia such that they were raised by their families as girls (until the inhabitants became aware of the condition). However, at puberty, their voices deepen and they develop male phenotype characteristics, with substantial increase in muscle mass, and the phallus enlarges to become a functional penis. The Leydig cells, concentration of plasma testosterone, the epididymis, and vas deferens are all normal, there is an ejaculate, and the psychosexual orientation of the individuals is undoubtedly male. Nonetheless, the prostate remains small, there is little beard growth and there is no anterolateral recession of the hairline.

Studies showed that, although plasma testosterone concentrations were normal, those of dihydrotestosterone were low. The latter is derived from testosterone by a simple metabolic reaction, saturation of the double bond

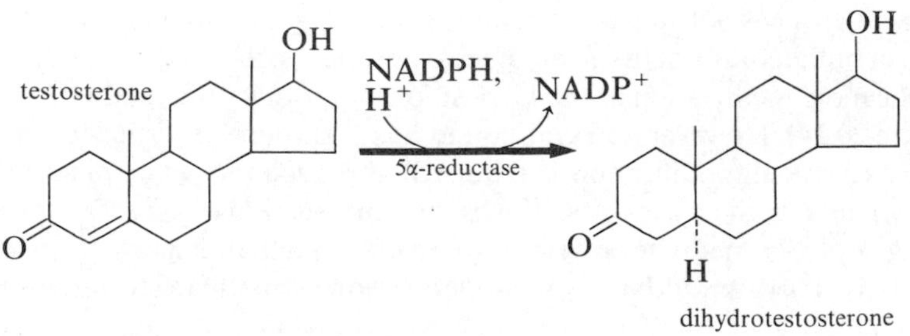

Figure 20.23   Conversion of testosterone to dihydrotestosterone

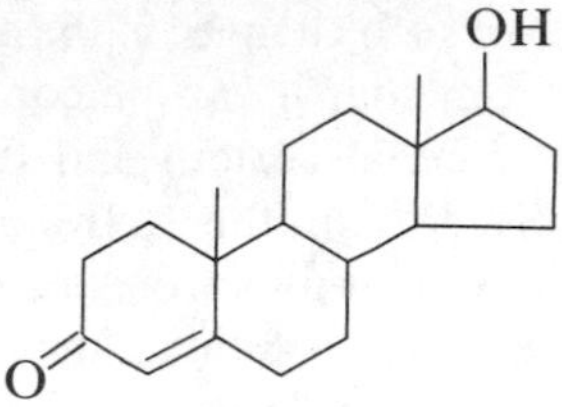

testosterone

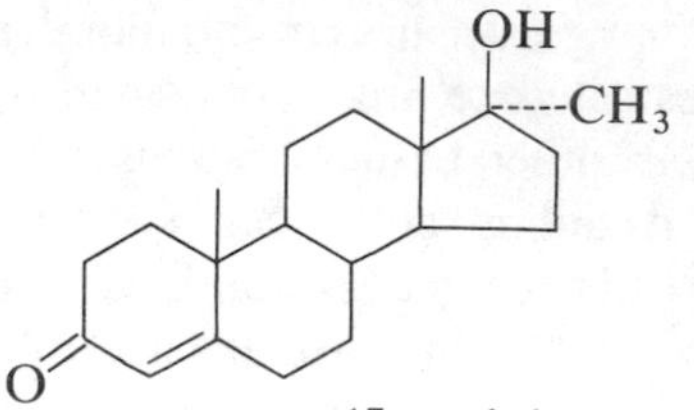

17α-methyltestosterone

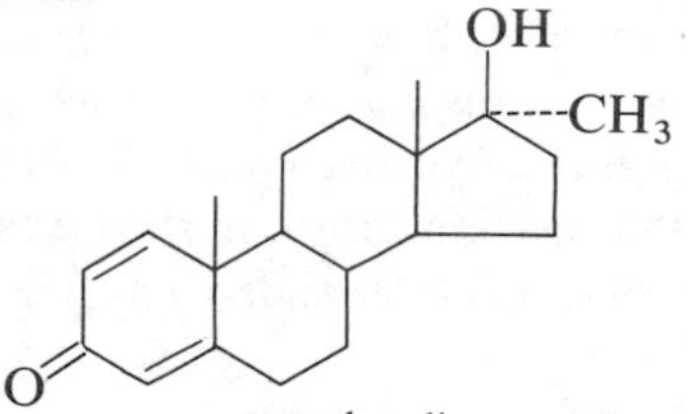

methandienone (dianabol)

Figure 20.24 Structural formulae of synthetic anabolic steroids (plus testosterone for comparison)

between atoms 5 and 6 which is brought about by the enzyme, steroid 5α-reductase (Figure 20.23). It was observed that the activity of this enzyme in various tissues of these subjects was very low, thus explaining the low concentration of dihydrotestosterone. This strongly suggests a specific role for this hormone in male sexual development and, in particular, that dihydrotestosterone is important in producing complete differentiation of the male external genitalia *in utero*, in beard growth, recession of hair line, and development of the prostate at puberty.

(f) *Anabolic steroids and the athlete*

The anabolic (in this context, 'body-building') property of androgens has been used in a number of ways. For example, in the treatment of emaciation (which may be due to prolonged illness or be self-inflicted), administration of a synthetic androgen, plus a high protein-high energy food intake, produces a marked positive nitrogen balance. In addition, the use of anabolic steroids by some athletes to increase muscle mass is common knowledge (Lucking, 1982).

Testosterone is not very active if administered by mouth due to metabolism in the liver but an orally active form of testosterone is produced by simple chemical modification, methylation in the 17-position to produce 17α-methyltestosterone* (Figure 20.24). However, this compound has considerable virilizing activity and further chemical modification is required to reduce this while maintaining the anabolic action. Such a modification is the introduction of a further double bond in ring A of 17α-methyltestosterone to produce methandienone (also known as dianabol). This steroid has the anabolic action of methyltestosterone but with

* This prevents conjugation of the 17-hydroxyl group in the liver which results in inactivation.

little of the virilizing effects.* Anecdotal evidence suggests that many male and female athletes take more than 100 mg of methandienone each day to increase muscle bulk, muscle strength, and performance, especially in the explosive sports such as shot, discus, and javelin throwing (Hervey, 1975). Control studies with volunteers have shown that muscular strength increases with anabolic steroids only if regular training is performed (Freed and Banks, 1975). Indeed, athletes have reported that anabolic steroids increase aggressiveness and decrease fatigue, allowing training to be carried out for longer and more intensively (Lucking, 1981, 1982). It would be of interest to know if specific biochemical changes occurred to provide a greater concentration of the important 'instant' fuel, phosphocreatine and whether there were increases in the activity of creatine kinase or perhaps the key glycolytic enzymes, phosphorylase and 6-phosphofructokinase.

Monitoring of the urine of athletes for the degradation products of these androgens is one means by which the officials of the sport hope to prevent this form of misuse of drugs. Perhaps more success would be achieved if, additionally, the side-effects of ingesting such large amounts of steroids were to be advertised more widely; these include sterility, impotence, and increased risk of diabetes, coronary heart disease, and cancer (Wynn, 1975; Sutton, 1981).

4. Oestrogens and progesterone

The reproductive system of the female mammal shows regular cyclic changes that represent periodic preparations for fertilization and pregnancy. In primates, the cycle is characterized by periodic vaginal bleeding when the uterine mucosa is shed (a process known as menstruation). By common usage, the cycle starts with the first day of menstruation and in woman has an average length of 28 days. The changes in the reproductive system during the cycle and also during pregnancy depend largely upon changes in the plasma concentrations of oestrogen and progesterone. These hormones are produced by the ovary (and during pregnancy by the placenta and the adrenal cortex of the foetus) and this production is controlled by the secretions of gonadotrophic hormones from the anterior pituitary. Before this control is described a brief account of the structure of the ovary is necessary.

(a) *The ovary and follicular development*

The ovary at birth contains many primordial follicles (perhaps 2 million) each of which contains a primary oocyte (immature ovum). A follicle consists of an external membrane known as the theca externa enclosing cells of the theca interna which are external to the granulosa cells that also surround the ovum.

*Anecdotal evidence suggests that athletes may be taking large quantities of testosterone because of the ease with which synthetic steroids can be detected during routine monitoring of urine. Large quantities are needed to overcome the metabolism in the liver.

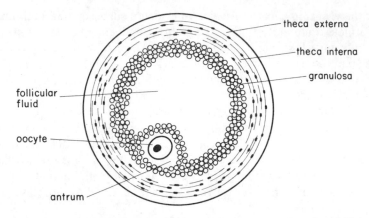

theca externa
theca interna
granulosa
follicular fluid
oocyte
antrum

Figure 20.25 Diagrammatic section through mature Graafian follicle

The space in the centre of the follicle is filled with follicular fluid (Figure 20.25). At the start of each cycle several follicles enlarge but, in woman, on about day six of the cycle, one of the follicles grows more rapidly while the others undergo atresia. The growth of the follicle is under control of the pituitary gonadotrophins. This maturing follicle, known as a Graafian follicle, contains the ovum that will be shed in the middle of the cycle. Up to the shedding of the ovum, the cells of the theca interna and granulosa proliferate and between them increase the rate of secretion of oestradiol-17β (see below). At about the 14th day, the follicle ruptures and the ovum is extruded into the abdominal cavity—a process known as ovulation. The ovum is picked up by the mobile fronds at the end of the funnel-shaped ends of the uterine tubes and transported by ciliary action down the fallopian tube to the uterus. After ovulation, the cells of both the theca interna and, particularly, the granulosa hypertrophy and there is a marked increase in vascularization of the granulosa. The follicle is now known as the corpus luteum, which secretes both oestrogen (theca interna) and progesterone (granulosa cells). If fertilization of the ovum occurs, the corpus luteum persists and maintains the correct endocrine balance to maintain pregnancy during the initial period. If pregnancy does not occur, the corpus luteum begins to regress on about day 24 of the cycle (probably due to the action of prostaglandins—see Chapter 17; Section D.4.a) and this results in menstruation.

Only about 400 of the 2 million follicles available at birth actually ovulate; the remainder (99.9%) become atretic. At the end of the reproductive life of the female there are very few immature follicles remaining. As this stage is approached, the frequency of cycles is reduced and they become irregular. This state, known as the climacteric, may last for several years until, at the menopause, the cycles, and hence ovulation, cease completely. The decrease in plasma oestrogen concentration that occurs during the climacteric and during the post-menopausal state can cause a number of problems which have been reduced by oestrogen replacement therapy (see Section G.3).

(b) *Secretion of oestrogen and progesterone during the menstrual cycle*

In the first part of the menstrual cycle, there is a gradual increase in the blood concentration of the gonadotrophins, especially FSH. These stimulate the growth and development of a small number of follicles, in which the cells of the theca interna and the granulosa replicate. At one time it was thought that only the theca interna cells were involved in oestrogen synthesis but it is likely that both the theca interna and the granulosa cells are involved. In the early part of the cycle, the granulosa cells cannot cleave the side-chain of cholesterol so that they must be provided with a precursor steroid hormone. This is done by the cells of the theca interna, which convert cholesterol to testosterone and androstenedione. However, the thecal cells cannot metabolize these androgens further as the activities of the enzymes catalysing reduction of the keto group, introduction of a double bond and of the 10,19-lyase, which leads to aromatization of ring A, are very low (see Figure 20.13). They are therefore released, traverse the basement membrane and enter the granulosa cells, which contains the 10,19-lyase, so that oestradiol-17β is produced and is secreted into the ovarian vein (Figure 20.26). This idea of co-operation between the two types of cells in the ovary to produce the oestrogen is known as the 'two-cell' concept and is supported by the fact that, in the follicular phase, the theca interna is well vascularized so that cholesterol and fuels are freely available for hormone synthesis. Although no studies have been carried out, it is likely that the flux-generating step is either the provision of free cholesterol in the cytosol of the thecal cells or the transport of cholesterol into the mitochondria. Receptors for LH are present in the thecal cells and it is suggested that this hormone controls both the rate of synthesis of the enzymes involved in steroidogenesis and the activities of the processes involved. These include cholesterol uptake, cholesterol esterase, mitochondrial transport and the

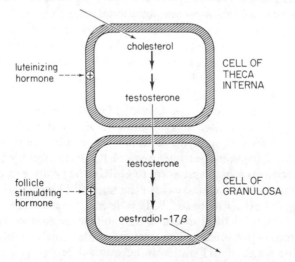

Figure 20.26   The 'two-cell' concept of oestradiol-17β synthesis in the Graafian follicle

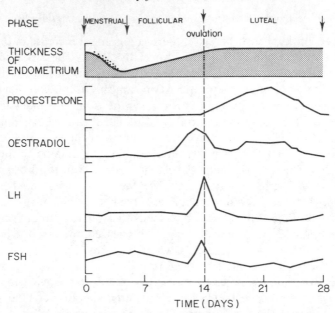

Figure 20.27 Changes in plasma concentration of hormones during a normal menstrual cycle

mitochondrial side-chain cleavage system (probably via activation of adenylate cyclase, raising the concentration of cyclic AMP and activation of protein kinase, see Chapter 22; Section A.4). On the other hand, the concentration and the activity of the 10,19-lyase in the granulosa cells is under the control of FSH (Figure 20.26).

The division of labour between the thecal and granulosa cells in production of oestradiol-17β may have an important functional role; that of determining which follicles undergo atresia. Atresia occurs if high concentrations of androgens accumulate in the granulosa cells. The most likely follicle to mature is therefore the one that has most FSH receptors on the granulosa cells and hence the greatest 10,19-lyase activity so that androgens are maintained at a low concentration (Kase and Speroff, 1980).

As the blood concentrations of oestradiol-17β increase towards the end of the pre-ovulatory period, the effect of the hormone on the rate of secretion of LH changes from negative to positive feedback. In combination with increased secretion of gonadotrophin releasing hormone (GnRH) from the hypothalamus and increased sensitivity of the pituitary to GnRH, the positive feedback effect of oestradiol results in a marked increase in the rate of LH secretion, known as the mid-cycle surge of LH[12] (Figure 20.27). The latter initiates a series of events in the follicle which results in the rupture of the follicle, release of the ovum and formation of the corpus luteum. The rupture of the follicle involves controlled proteolysis. The surge of LH probably activates a proteolytic enzyme in the follicular fluid and may also increase the rate of production of prostaglandins E and F. The precursor proteolytic enzyme, plasminogen, is present in follicular fluid and may be converted to plasmin (see Chapter 3; D.5) directly by LH or, more likely, by its intracellular messenger, that is by the increased cyclic AMP concentration and consequent activation of protein kinase. The action of this proteolytic enzyme weakens the follicle wall so that contraction of the theca externa cells (which have properties similar to smooth muscle) caused by the increased concentration of prostaglandins (see Chapter 17; Section D.4.a) finally results in rupture of the follicle.

The surge of LH also results in changes in the pattern of steroidogenesis by the follicle. A reduction in the number of LH receptors on thecal cells leads to a decrease in the activities of the enzymes responsible for androgen synthesis. However, the vascularization of granulosa cells increases the availability of cholesterol and, in addition, there is an increase in the activity of the enzymes responsible for side-chain cleavage of cholesterol and decreases in the activities of the enzymes responsible for 17-hydroxylation and further side-chain cleavage so that cholesterol is now converted to progesterone rather than oestradiol-17β[13] (reference to Figure 20.28 may help to follow these changes). After ovulation, the

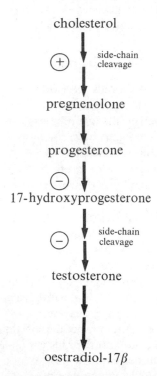

Figure 20.28 Simplified pathway of oestradiol and progesterone synthesis in the follicle. Signs refer to enzyme activity changes at ovulation which lead to increased progesterone formation in granulosa cells.

concentrations of LH and FSH decrease due to feedback inhibition by the progesterone and oestrogens secreted by the corpus luteum.

There is little doubt that a prostaglandin (probably $PGF_{2\alpha}$) is involved in regression of the corpus luteum (Chapter 17; Section D.4.a). The production of the prostaglandin occurs in the uterus and the rate of production is probably increased by the low concentration of LH.

Although the steroid hormones secreted by the ovary, in combination with the pituitary and hypothalamus, provide the cyclical changes in this tissue, they also control the development of the secretory activity of the endometrium necessary for fertilization of the ovum and the changes necessary for successful implantation. In the follicular phase of the cycle, oestradiol-17β causes growth of the epithelial cells and increases vascularization of the endometrium and development of the uterine glands. After ovulation, the endometrium is subjected to the effects of both oestrogen and progesterone and there is an increase in vascularization of the stroma and the endometrial glands secrete fluid into the uterus. The endometrium is now in a state in which it is capable of accepting the fertilized ovum. When the corpus luteum regresses, the concentrations of oestrogen and progesterone are rapidly reduced and this causes an increased rate of production of prostaglandins in the endometrium which are involved in the initiation of menstruation (Chapter 17; Section D.4.a).

(c) *Regulation of secretion of gonadotrophins*

The importance of the gonadotrophins, LH and FSH, in controlling the secretion of steroids by the ovary is apparent from the account presented above. However, the question arises, what controls the rate of release of the gonadotrophins by the pituitary? The anatomical arrangement in the pituitary provides at least part of the answer. The hypothalamus is connected to the anterior pituitary by a portal blood system, and hypothalamic cells secrete gonadotrophin-releasing hormone (GnRH) which is carried directly to the anterior pituitary. The mechanism of GnRH secretion from the hypothalamus cells involves the biogenic amine, dopamine (Eddy *et al.*, 1971). GnRH binds to a receptor on the relevant cells of the pituitary and the current evidence suggests that the mechanism of control of secretion involves activation of adenylate cyclase, increase in cyclic AMP concentration and increase in the activity of protein kinase. Presumably phosphorylation of a key protein(s) leads to an increased rate of secretion of FSH and LH (see Chapter 22; Section A.4 for general discussion of the problem). It is, however, unclear how differential rates of secretion of LH and FSH are achieved.

A further point of considerable importance is that the rate of secretion of GnRH and hence that of the gonadotrophins is not steady but is pulsatile. The full significance of this pulsatile release may not yet be appreciated but it is known that in prepubertal females pulsatile release of GnRH occurs only in the night whereas, in the mature female, release is pulsatile during the day and night; indeed, this change in secretory pattern is probably important in initiating puberty (see

Chapter 19; Note 9). One advantage of pulsatile release of hormones is that an increased rate of secretion will not produce a prolonged increase in blood concentration of the hormone so that, although acute interaction with the receptor can occur, the more chronic effect of a decreased number of receptors and hence a decreased sensitivity to the hormone may not result. In other words, pulsatile secretion protects against the phenomenon of 'down regulation' of receptors (Chapter 22; Section A.3.a; see Knobil, 1981). Since GnRH is a decapeptide, structural analogues can be synthesized[14] and the effect of long acting agonists provides evidence for this suggestion. Administration of such agonists decreases the sensitivity of the pituitary to GnRH and results in a decrease in the rate of secretion of FSH and LH. This could be of importance in providing a safe means of fertility control in women (Fink, 1979; see Section G.2.c) and in the treatment of precocious puberty (Comite *et al.*, 1981). Such analogues have also been used to treat infertility in both men and women when the infertility is due to a failure of secretion of GnRH (e.g. due to removal of a tumour). Furthermore, the failure of normal menstrual cycles that accompanies anorexia nervosa (see Chapter 19; Section C.5) may not be cured when the patient is restored to a normal food intake and normal body weight but treatment of such patients with analogues of GnRH can result in normal cycles and fertility.

(d) *Oestrogen and progesterone in pregnancy*

If fertilization occurs, the corpus luteum continues to secrete progesterone and oestradiol. Regression of the corpus luteum is prevented by the implanting blastocyst[15] which secretes a hormone, known as human chorionic gonadotrophin (hCG).[16] This binds to the LH receptor on the granulosa cells and maintains the production of cyclic AMP and hence the rates of oestrogen and progesterone synthesis. It may also bind to prostaglandin receptors on these cells, preventing the binding of $PGF_{2\alpha}$ which normally initiates regression (that is, it acts as a physiological receptor blocker). The corpus luteum is important in the production of progesterone and oestradiol for the first 9–10 weeks of pregnancy. After this time, the placenta and also the adrenal cortex of the foetus gradually take over the production of these hormones and the corpus luteum becomes unimportant. The interaction between the adrenal cortex of the foetus and the placenta in production of steroid hormones is complex. In outline, the placenta produces progesterone from cholesterol (which is available from the maternal blood) whereas the foetal adrenal cortex produces corticosteroids and androgens from progesterone produced in the placenta. The placenta also produces some oestrogens from the androgens produced by the foetal adrenal cortex. The interplay between the placenta and the foetus is underlined by the term 'foeto-placental unit' used in discussing steroidogenesis during pregnancy. Especially during the second half of pregnancy, very large quantities of the steroids are secreted by the foeto-placental unit and this is reflected in the excretion of large amounts of pregnanediol (from progesterone) and oestradiol in the urine. The rate of excretion of the latter may be 1000-fold greater than during the follicular stage of the menstrual cycle. Indeed, the plasma concentration or the excretion of oestradiol by the mother can be used to indicate foeto-placental well-being; low levels of oestradiol indicating possible toxaemia or foetal death.

The above discussion implies the importance of progesterone during pregnancy. Relevant effects of this hormone include, proliferation of cells of endometrium, a decrease in the contractile activity of the myometrium (which is necessary for implantation and maintenance of the developing placenta and foetus and is achieved by inhibition of prostaglandin formation), development of the endometrium to form the placenta, prevention of follicular development in the ovary and immunosuppressive activity which may be important in preventing rejection of the blastocyst in the early stages of pregnancy.

F. VITAMIN D

The classification of vitamin D as a vitamin (that is, an essential dietary component required only in trace amounts, see Appendix A) is largely an historical accident. To a much greater extent vitamin D conforms to the definition of a hormone, that is a substance produced in one part of an organism which has an effect on another part of that organism. Vitamin D plays an important role in the control of the concentration of calcium in the plasma. Since it is a steroid its effects are discussed in this chapter (but see also Appendix A).

Vitamin D_3 or cholecalciferol[17] can be produced in sufficient quantities in man provided that the subject is adequately exposed to sunlight. In primitive man, living with very few clothes and exposed to sunlight for considerable periods of time, sufficient cholecalciferol would have been synthesized in the skin. Only when man began to live in colder climes and covered most of his body with clothes would he become dependent upon a dietary source of cholecalciferol.

Whether as a hormone produced in the skin, or as a vitamin provided in the diet, vitamin D has to be chemically modified by two hydroxylations, which occur sequentially in the liver and the kidney, before it can play an active part in control of calcium distribution. The discovery of the necessity for hepatic and renal hydroxylations of cholecalciferol has led to explanations for a number of clinical problems and their more satisfactory treatment. However, a complete understanding of the actions of cholecalciferol and its usefulness in clinical conditions is still lacking (see Section F.4). The following reviews on the hormone are recommended DeLuca (1974, 1980); Norman and Henry (1979); Fraser (1980, 1981).

1. Synthesis of the active hormone

Cholecalciferol (vitamin D_3) is a seco-steroid, that is a steroid in which one of the rings has been broken by fission of a carbon–carbon bond. In this case, fission of the bond between atoms 9 and 10 in ring B of 7-dehydrocholesterol produces cholecalciferol.

Cholesterol is converted to 7-dehydrocholesterol by desaturation of the 9,10-carbon bond and, in the epidermis of the skin, ultraviolet light breaks this bond to produce cholecalciferol (Figure 20.29).[18] The cholecalciferol is transported via the bloodstream to the liver where the first step in the activation of the hormone occurs, namely hydroxylation by cholecalciferol 25-monooxygenase (hydroxylase) to produce 25-hydroxycholecalciferol. The hydroxylation reaction occurs on the endoplasmic reticulum and requires NADPH and oxygen but does not appear to involve the cytochrome P-450 system. The 25-hydroxycholecalciferol, which can be stored in the liver and possibly in the muscle, is transported from the liver to the kidney in the bloodstream. In the kidney, a further hydroxylation takes place at the 1-position to produce 1α,25-dihydroxycholecalciferol, which is the active form of the hormone (Figure 20.30). The hydroxylase (25-hydroxycholecalciferol 1-monooxygenase) is a mixed function oxidase present in the renal mitochondria which requires molecular oxygen and NADPH and does involve the cytochrome P-450 system. The 1,25-dihydroxycholecalciferol is not stored in the kidney but passes into the bloodstream where, along with the other derivatives of cholecalciferol, it is transported on a specific plasma protein. (As with other steroid hormones, the rate of synthesis, in this case 1α-hydroxylation, controls the rate of secretion.) Another hydroxylation of 25-hydroxycholecalciferol can also occur in the kidney; a 24-hydroxylation which produces 24,25-dihydroxycholecalciferol.* Although it is generally accepted that 1α,25-dihydroxycholecalciferol is the active form of the hormone (DeLuca, 1980), the role of 24,25-dihydroxycholecalciferol is unclear. Ornoy *et al.* (1978) suggest that it plays an important role in mineralization of bone of rachitic chicks, whereas studies with patients with rickets suggests that is unimportant in the recovery from this disease (Papapoulos *et al.*, 1980; see also Smith, 1979).

2. Physiological effects of 1α,25-dihydroxycholecalciferol

Dihydroxycholecalciferol has several physiological effects described below, at least two of which provide the basis for control of the plasma calcium concentration.[19]

(a) *Calcium uptake in the intestine*

The hormone increases the concentration of a specific calcium-transporting protein in the absorptive cells of the small intestine and thus increases the absorption of calcium into the body. The hormone also increases phosphate uptake by the intestine but this may be due to effects on calcium uptake since the

---

* In a normal subject, the relative concentrations of the three hydroxylated forms, 25-hydroxy-, 24,25-dihydroxy-, and 1α,25-dihydroxycholecalciferol are 100:10:1.

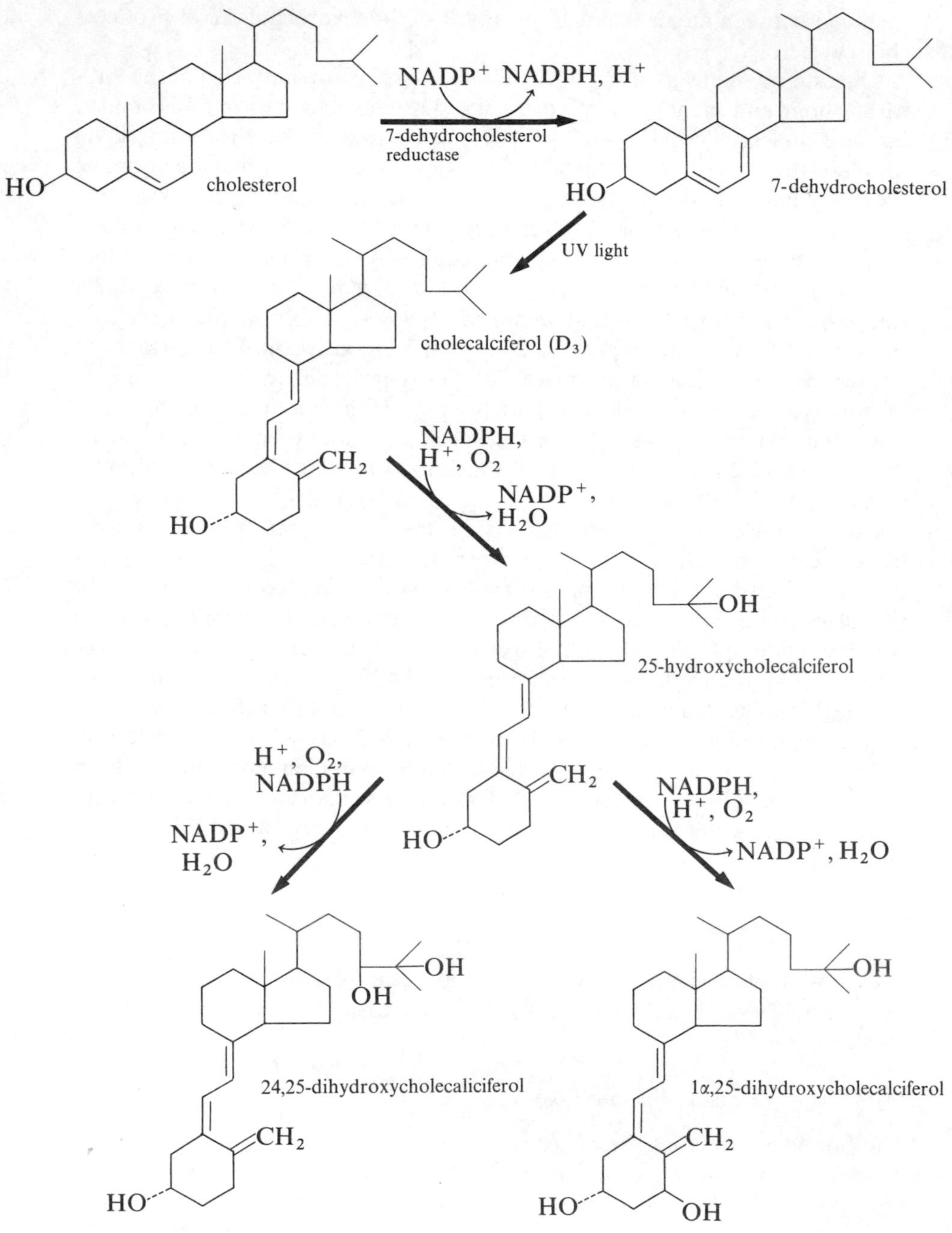

Figure 20.29 Synthesis and chemical conversions of vitamin D

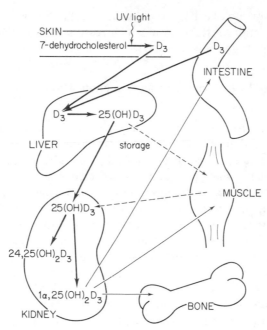

Figure 20.30 Role of different tissues in vitamin D metabolism. Abbreviations: D₃, cholecalciferol; 25(OH)D₃, 25-hydroxycholecalciferol; 1α,25(OH)₂D₃ and 24,25(OH)₂D₃, dihydroxycholecalciferols

two processes are linked (to achieve electroneutrality). The hormone may also increase calcium uptake from glomerular filtrate in the kidney (DeLuca, 1980).

(b) *Muscle*

Deficiency of vitamin D results in myopathy with particular weakness of the proximal muscles of the limbs. This is not surprising since calcium is involved in several important facets of muscle function including control of contractility and control of the rates of glycogenolysis and the TCA cycle (see Chapter 7; Sections B.3.b and C.1.c) but it is not known which, if any, of these processes is affected.

(c) *Calcium mobilization from bone*

Although it has been a controversial area it is now generally accepted that dihydroxycholecalciferol affects bone by increasing the rate of release of calcium and phosphate. This is probably due to activation of the osteoclasts, which cause resorption of the bone.[20] This effect of dihydroxycholecalciferol on bone is somewhat surprising in view of the fact that low levels of this hormone (that is vitamin D deficiency) results in osteomalacia (deficiency of mineral in the bone) which is also known as rickets in children (see Section 4.a). Probably the most important effect of cholecalciferol on bone mineralization is via the increase in calcium concentration in the plasma which results from the increased rate of uptake of calcium from the intestine.

3. Role of 1α,25-dihydroxycholecalciferol in control of plasma calcium concentration

Several factors control the activity of the enzyme 25-hydroxycholecalciferol 1-monooxygenase and hence the rate of production of 1α,25-dihydroxychole-calciferol by the kidney. These include parathyroid hormone, growth hormone, cortisol, oestrogens, and prolactin. However, under normal conditions, parathyroid hormone probably plays the most important role. A low plasma calcium concentration increases the rate of secretion of parathyroid hormone which increases the activity of the monooxygenase[21] and hence the production of 1α,25-dihydroxycholecalciferol. This results in an increase in the rate of absorption of calcium from the intestine and raises the plasma calcium concentration which, in turn, reduces the rate of secretion of parathyroid hormone and hence the rate of production of 1α,25-dihydroxycholecalciferol. If the plasma calcium concentration increases above the normal, the C-cells of the thyroid increase the rate of secretion of calcitonin which inhibits mobilization of calcium from bone and may increase calcium and phosphate excretion via the kidney. In this way parathyroid hormone, 1α,25-dihyroxycholecalciferol and calcitonin are all involved in the long-term control of plasma calcium concentration. Short-term control is achieved by the action of parathyroid hormone which together with a normal concentration of 1,25-dihydroxychole-calciferol causes mobilization of calcium from bone.

4. Vitamin D deficiency

Because of the complexity of the metabolism of cholecalciferol, deficiency can occur in several ways; the usual deficiency state arises due to poor diet and minimum exposure to sunlight.

(a) *Dietary or sunlight deficiency*

The impaired control of calcium distribution in the vitamin D deficient state results in rickets in growing children and osteomalacia in adults. In the latter there is defective mineralization of bone that results in deformity of the skeleton and bone pain (especially backache). In rickets the bones may continue to grow but without mineralization so that on the growing ends of the long bones there is general enlargement with widening of the epiphysial plate and increase in osteoid tissue. There may be bending of the bones and greenstick fractures of the long bones. The plasma concentrations of calcium and phosphate are decreased but the activity of alkaline phosphatase (released from the osteoblasts) is increased.

Osteomalacia and rickets were prevalent in the large cities of Britain during the nineteenth century owing to poor nutrition and lack of sunlight. Improved nutrition, and fortification of foods with vitamin D_2, have improved the situation enormously, but, in recent years, there has been an increase in this disease in the elderly and also in children and pregnant women of immigrant Asian families. Fraser (1981) considers that in both groups there is a lack of exposure to sunshine and, furthermore, in the Asian immigrants some factor in the high cereal diet may interfere with absorption of vitamin D or with its formation in the skin. He also suggests that endogenously produced vitamin D may be more effective than that in the diet; if this is true, the approach of treating vitamin D deficiency by application of precursors of the active form of the hormone to the skin (e.g. 1,25-dihydroxy-7-dehydrocholesterol), which are then absorbed and treated as a normal product of skin metabolism, may have much to commend it (Holick *et al.*, 1980). The importance of ultraviolet light in increasing the concentration of 1,25-dihydroxycholecalciperol in the blood of vitamin D-deficient patients has been established (Adams *et al.*, 1982).

There is a greater requirement for vitamin D during pregnancy and lactation when the plasma concentration of 1,25-dihydroxycholecalciferol increases—no doubt to ensure adequate rates of calcium absorption for the developing foetus and for milk formation. A dietary deficiency of vitamin D can, therefore, more readily occur in these conditions. The increased plasma concentration of 1,25-dihydroxycholecalciferol may be brought about by stimulation of the renal 1-monooxygenase by steroid hormones and placental human chorionic gonadotrophin during pregnancy and by prolactin during lactation (Kumar *et al.*, 1980).

(b) *Other conditions*

There are several other conditions in which there is a lack of vitamin D due to causes other than a primary deficiency.

1. In patients suffering from impaired bile production or secretion, coeliac disease or short bowel syndrome there may be malabsorption of the vitamin. In such cases larger oral doses of the vitamin may be beneficial or, alternatively, it may be administered by injection.
2. Patients suffering from glomerular failure have a low capacity of the 1-monooxygenase so that the active form of the hormone cannot be produced. Deformity of bones and bone pain are common symptoms. In this, and subsequently described conditions, patients usually respond well to administration of 1α,25-dihydroxycholecalciferol (or 1α-hydroxycholecalciferol, which is then further hydroxylated in the liver).
3. Hypoparathyroidism, (e.g. due to surgical removal of the parathyroid glands) leads, as predicted, to inadequate concentration of 1α,25-dihydroxycholecalciferol and consequent low plasma concentration of calcium but,

interestingly, not to osteomalacia or rickets. In the disease known as vitamin D resistant rickets the renal 1-monooxygenase system is unable to respond to normal levels of the parathyroid hormone (see Chapter 22; Section A.3.b).

4. Patients who are on chronic anti-convulsant therapy (Chapter 21; Section E.4) often suffer from symptoms of vitamin D deficiency because the anti-convulsants compete with 25-hydroxycholecalciferol as substrates for the renal hydroxylation systems with a consequent low rate of formation of the active hormone.

G. CLINICAL INTEREST IN THE STEROIDS

A large number of clinical problems relate to overproduction, underproduction or imbalance of the steroid hormones. Few of these can be considered adequately here but space is taken to discuss areas in which biochemistry and metabolic principles have been brought to bear on clinically important problems. These include fertility control in women, hormone replacement therapy in climacteric and post-menopausal women and the use of anti-oestrogens in treatment of breast cancer. First, however, the controversial involvement of cholesterol in atherosclerosis and coronary heart disease will be discussed.

1. **Cholesterol, atherosclerosis, and coronary heart disease**

That coronary heart disease is one of the most intensively investigated subjects in medical science is not surprising considering the vulnerability of members of modern industrialized societies to fatal or debilitating heart attacks (Stamler, 1974). Despite this there is still considerable controversy over the causes of this disease and, although there is no doubt that cholesterol and cholesterol ester are involved in the development of atherosclerosis, whether this is the major cause of the disease or simply a manifestation of it remains unclear.

Evidence implicating lipoproteins, and especially cholesterol, in the development of atheroma (the fatty deposits which reduce the lumen diameter of arteries) and coronary heart disease is obtained from pathological, clinical and epidemiological[22] studies on man and studies with experimental animals (for review, see Havel *et al.*, 1980). Some results of such studies are illustrated by the following.

1. Atheromatous plaques contain large quantities of cholesterol and cholesterol ester.
2. Patients who suffer from familial hypercholesterolaemia (Chapter 6; Section F.3.a) suffer from premature atherosclerosis and coronary heart disease. In the heterozygotes, the disease is severe in the third and fourth decades whereas in the homozygotes, in which the cholesterol concentrations are very high, the disease can be present as early as the second decade (Slack, 1975; Havel *et al.*, 1980).

3. The incidence of coronary heart disease shows a positive correlation with plasma cholesterol concentration and the dietary intake of saturated fat[22] (see Oliver, 1981, for review).
4. Monkeys who were fed an average North American diet (high in saturated fat and cholesterol) had more marked atherosclerosis than monkeys fed a 'prudent' low-cholesterol diet. Furthermore, changing from the average to the prudent diet resulted in less marked atherosclerosis and a reduction in cholesterol deposition (Wissler and Vesselinovitch, 1977).

(a) *Development of atherosclerotic plaque*

The development of atheroma involves the smooth muscle cells of the tunica media of the artery wall[23] which proliferate and migrate to involve the tunica intima and produce the plaque (for review see Havel *et al.*, 1980).

There are small depositions of lipid in arterial walls even in the first decade of life and these are known as fatty streaks. They may have no pathological significance except as an indication of the ability of the arterial wall to accumulate lipid. In the third decade of life, probably in most individuals in industrialized societies, the amount of lipid, especially cholesterol ester, accumulating in the artery wall increases markedly and this is assumed to be indicative of the development of atheroma. How does this come about? The primary lesion is considered to be damage to the endothelial cells that line the intima; This damage may be caused by various factors including hypertension, cigarette smoking, viral attack and high lipid concentration (including cholesterol or long-chain fatty acids—Chapter 6; Section F.2.a). This damage increases the permeability of the endothelial lining so that plasma and blood platelets can enter the tunica media. This has two effects: within the tunica media blood platelets aggregate and release mitogenic agents that cause the smooth muscle cells to proliferate; and the smooth muscle cells become exposed to much higher concentrations of LDL particles than normal. Probably by a non-specific endocytotic uptake process, the LDL particles enter the smooth muscle cells and the contents are degraded by the enzymes of the lysosomes. Most of the amino acids and fatty acids leave the cell but cholesterol remains and becomes esterified to form cholesterol ester with which the cells become overloaded.* As the proliferating cells migrate into the tunica intima, they synthesize collagen, glycosaminoglycans, and elastin that may help to strengthen the arterial wall at the site of the original damage so initiating plaque formation. In this way the atheromatous thickening comes to consist of smooth muscle cells loaded with cholesterol ester and surrounded by fibrous material which also contains cholesterol ester probably produced from dead and dying smooth muscle cells.

Possible relationships between atherosclerosis and the coronary heart attack are discussed in Chapter 5; Section G.2. In brief, the reduction in diameter of the

---

* Many of these cells are so swollen with droplets of cholesterol ester that their cytoplasm appears to be vacuolated and they are termed 'foam cells'.

coronary arteries causes a reduction in blood flow and an increase in the likelihood of aggregation of platelets and thrombus formation. Alternatively, or in addition, the atheroma may reduce prostacyclin formation (an anti-aggregation agent) by the endothelial cells of the arterial wall so that the aggregation-effect of thromboxane, which is produced by the platelets, dominates (Chapter 17; Section D.4.f). Indeed, the high plasma concentration of cholesterol may have led to an increase in the cholesterol/phospholipid ratio in the cell membrane of the platelets, which increases membrane fluidity and this could result in a greater rate of thromboxane synthesis than normal (Stuart *et al.*, 1980).

(b) *Normal and abnormal concentrations of plasma cholesterol, LDL, and HDL*

It is generally assumed that $200 \, mg.100 \, cm^{-3}$ (5.3 mM) or below is a normal serum cholesterol concentration, whereas $230 \, mg.100 \, cm^{-3}$ is abnormal and dangerous. Dietary or drug intervention is aimed at reducing the concentration below $220 \, mg.100 \, cm^{-3}$. The diets that are advised usually have a lower percentage of total lipid and a higher proportion of polyunsaturated to saturated fatty acids than normal diets.

The specific LDL-receptor mechanism for uptake of LDL by various cells (Section D.2) provides a means of predicting a normal concentration of LDL. The affinity of the receptor for LDL can be measured and assuming that the concentration of LDL will normally be below the affinity constant, it is suggested that the 'normal' interstitial LDL concentration* should be about $2.5 \, mg.100 \, cm^{-3}$. Since the concentration of LDL in the plasma is ten-fold greater than in the interstitial fluid the 'normal' plasma concentration should be around $25 \, mg.100 \, cm^{-3}$. Even in normal man in the industrialized societies, it is about $120 \, mg.100 \, cm^{-3}$ or four- to five-fold higher than the predicted norm. It is interesting to note that the plasma concentration of LDL in mammals other than man and in newborn infants, prior to exposure to the normal diet, is considerably lower than $120 \, mg.100 \, cm^{-3}$.

Drugs or conditions that can lead to a decrease in the concentration of LDL-cholesterol may be beneficial in subjects with an increased risk of atherosclerosis (e.g. patients with familial hypercholesterolaemia). The concentration of LDL-cholesterol could be decreased by raising the number of LDL-receptors on the cell membrane of various tissues, since this would increase the capacity of one specific process in the lipoprotein metabolic pathway, and hence lower the concentration of substrate LDL-cholesterol. Since production of endogenous cholesterol by the biosynthetic pathway described in Section B.1 causes a decrease in the number of LDL-receptors, inhibition of this pathway should lead to an increase in the number of LDL-receptors. For this reason, there is considerable interest in a drug (Compactin) that is a potent inhibitor of hepatic HMG-CoA reductase (see Brown and Goldstein, 1981).

---

* These concentrations refer to LDL-cholesterol. The LDL can be measured according to protein content or cholesterol content; 1 μg of LDL-protein is equivalent to 1.67 μg of LDL-cholesterol.

The role of HDL in removal of cholesterol from tissues and in transfer of cholesterol ester to intermediate-density lipoprotein particles that may be metabolized, and hence transfer their cholesterol to tissues other than the smooth muscle cells of the arterial wall, has increased the interest in this lipoprotein. This was strengthened in the 1970s when a strong inverse relationship was shown between plasma concentrations of HDL-cholesterol and death from coronary heart disease (Miller and Miller, 1975) and it was suggested that increased concentrations of HDL-cholesterol may actually protect against atherosclerosis (Rhoads *et al.*, 1976; Lees and Lees 1982). This could account for the established fact that, up to the menopause, women have a much lower incidence of coronary heart disease than men; their HDL concentration is higher. Furthermore, frequent sustained exercise (e.g. aerobic training—see Chapter 9; Section B.5.c) increases the level of HDL-cholesterol and lowers that of LDL-cholesterol in comparison to non-exercising controls (Wood *et al.*, 1977; Hartung *et al.*, 1980; see Table 20.2); such exercise is considered to reduce the incidence of atherosclerosis and coronary heart disease (Froelicher, 1978; Paffenbarger and Hyde, 1980; Milvy and Siegel, 1981; Kramsch *et al.*, 1981). Furthermore, moderate ethanol intake, which is inversely related to the incidence of coronary heart disease, also increases the concentration of HDL-cholesterol. The effects of sustained exercise and a moderate ethanol intake are, furthermore, additive (Willett *et al.*, 1980).

Table 20.2. Plasma lipid and lipoprotein concentrations of marathon runners, joggers and inactive subjects*

Condition	Plasma concentrations (mg.100 cm^{-3})			
	Total cholesterol	Triacylglycerol	LDL	HDL
Inactive subjects	211.7	154.2	136.5	43.3
Joggers	204.2	105.7	125.0	58.0
Marathon runners	187.2	77.1	107.0	64.8

* Data from Hartung *et al.* (1980).

2. Fertility control

Knowledge of the mechanism of control of the periodic changes in the menstrual cycle and of the maintenance of the early stages of pregnancy provides a basis for the search for improved contraceptives. Two modern methods based on this knowledge are the oral ovulation-inhibiting steroid, known as the 'pill', and the intrauterine device. These are discussed below, followed by an outline of new developments in these and other areas of contraceptive research.

(a) *Oral steroid contraceptives*

The history of the development of the oral contraceptives has been reviewed by Petrow (1966). In 1921, Häberlandt showed that if ovaries of pregnant rabbits were transplanted into non-pregnant animals, temporary sterility resulted. He suggested that this 'sterilization' method might be applied to fertile women. In 1936, it was shown that daily injections of progesterone inhibited the oestrus cycle of the rat, and that the inhibition of ovulation after mating in the rabbit and guinea pig was due to progesterone. In 1938, Kurzrok noted that, during treatment of dysmenorrhoea with oestrogen, the normal menstrual rhythm was upset and ovulation was delayed; thus oestrogen could inhibit ovulation and this offered an approach to fertility control. The idea of attempting to mimic the normal cycle pattern of steroid hormone secretion by oral dosing of oestrogen followed by oestrogen plus progesterone was proposed by Albright in 1945. In the early 1950s it was shown that the daily administration of progesterone to women from the fifth to the 25th day of the cycle caused a large reduction in the frequency of ovulation. Partially synthetic progestins (steroids having a progesterone-like action) became commercially available in the mid-1950s and the work of Rock *et al.* (1957) showed that they could be used as oral contraceptives with few untoward effects provided that the correct dosage was given. The troublesome occurrence of 'breakthrough bleeding' was overcome if oestrogens were given with the progestins.

The oral contraceptive steroids in general use at the present time* include the synthetic oestrogens, ethinyl oestradiol, and mestranol (Figure 20.31); and either substituted progesterones (e.g. megestrol acetate, chlormadinone) or derivatives of 19-nortestosterone (e.g. norethindrone acetate, norethynodrel, norgestrel). These are administered in either of two ways: the combined pill contains a synthetic progestin plus an oestrogen and one pill is taken each day from day five to day 25 of the cycle; the sequential regime requires one pill to be taken each day containing only oestrogen from day five of the cycle for 15 days which is followed for five days by pills that contain a progestin plus an oestrogen.

The interesting question is how the oral contraceptive steroids provide fertility control. It is likely that several effects are involved. Probably the most important is interference with the normal balance between the negative and positive feedback effects of oestradiol that produces the LH surge (Section E.4.c). Possibly the high plasma concentrations of oestrogen from the early part of the cycle reduces the number of receptors for steroid hormone on the pituitary (down-regulation) so that the positive feedback response to the normal cyclical increase in oestradiol may not occur and the LH surge fails. Furthermore, there is probably inhibition of FSH secretion so that the follicle fails to develop. In addition, the progestins may change the composition of the secretions of the endometrium and the cervix so that sperm motility is decreased and implantation of a fertilized ovum made more difficult (Figure 20.32).

* It is estimated that more than 50 million women are using these contraceptives.

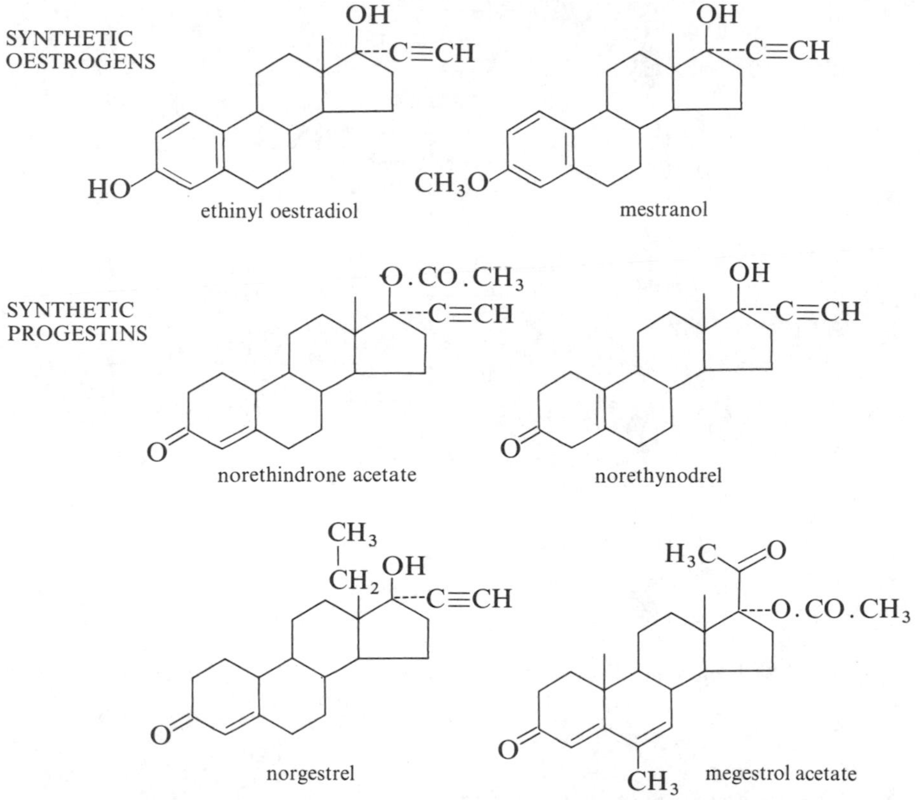

Figure 20.31 Synthetic steroids used in contraceptive pills

Unfortunately, there are a number of side-effects of the oral contraceptive steroids including deep vein thrombosis of the leg, stroke, myocardial infarction, insulin resistance, gall bladder disease, and hypertension. There has been a considerable amount of work into these side-effects and the following reviews are recommended to the interested reader (Andrews, 1979; Guilleband, 1980; Vessey, 1980; Stadel, 1981).

(b) *Intrauterine device*

Insertion of a spiral, ring, loop, or bow into the uterus reduces the chance of pregnancy. It is likely that the presence of this device stimulates a higher than normal rate of prostaglandin production by the uterus although it is unclear how this prevents pregnancy. One possibility is that prostaglandins increase motility of the myometrium throughout the cycle and this interferes in the critical process of blastocyst implantation after fertilization. Possible side-effects include pain, menorrhagia, reproductive tract infection, and perforation of the uterus (Kase and Speroff, 1980).

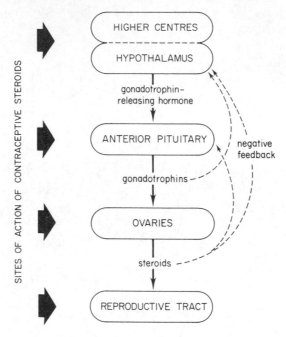

Figure 20.32 Hierarchical representation of the female reproductive system showing sites of action of steroid contraceptives

(c) *New developments in contraception*

The secretion of chorionic gonadotrophin by the blastocyst at the time of implantation in order to maintain a functional corpus luteum is critical in the early stages of pregnancy. If an immunogen could be designed that would elicit the formation of antibodies against chorionic gonadotrophin the action of this hormone might be prevented and the pregnancy would fail. This would be an exmple of selective autoimmunity in which the immunological system would be turned against a specific protein essential for the maintenance of pregnancy. Similarly, drugs that could block the chorionic gonadotrophin receptor on the granulosa cells of the corpus luteum and which would have no biological activity could also result in failure of pregnancy. Such effects should be very specific and hence should have few if any side-effects.

Analogues of GnRH that are either agonists or antagonists could have contraceptive activity. The agonists might increase the concentration of LH in the blood which would decrease the number of ovarian receptors for LH (down-regulation) so that the ovary would fail to respond to the LH surge. Antagonists, on the other hand, could prevent the LH surge from the pituitary (Schally *et al.*, 1980). One attraction is that these analogues are simple peptides[14] so that they could be administered daily by a nasal spray and absorbed from the nasal membranes. A small clinical trial has shown that such agonists could be used in this way successfully as contraceptives (Bergquist *et al.*, 1979).

The possible use of prostaglandin analogues in contraception is discussed in Chapter 17; Section D.5.b.

3. Use of oestrogens in climacteric and post-menopausal women

Women between 45 and 55 gradually lose the ability to menstruate since at this age the ovary becomes depleted of follicles. The change occurs gradually over a period of perhaps several years during which cycles become less regular and more infrequent. This period is known as the climacteric and menopause begins after the last menstrual period. Once cycles have ceased, or become very infrequent, the plasma concentration of oestrogen is reduced and this may lead to a number of symptoms and clinical problems. These include hot flushes, sweating, depression, psychological disturbances, decreased libido, atherosclerosis, and osteoporosis.[24] The latter results in increased incidence of hip, radial, and vertebral fractures (Quigley and Hammond, 1979).

Some or all of these problems may be due to severely decreased concentrations of oestrogens. Not surprisingly, therefore, oestrogen-replacement therapy has been recommended for climacteric and post-menopausal women. The treatment has reduced symptoms and improved the problem of osteoporosis in such women (Weiss *et al.*, 1980; Eskin, 1980). Unfortunately, it has led to an increased incidence of endometrial cancer and other problems such as gall bladder disease (Weinstein, 1980). Since progesterone appears to protect against endometrial cancer, combined treatment with oestrogen plus progesterone might be perferable[25] (see *Lancet* editorial, 1979b; Hirvonen *et al.*, 1981; Nisker and Woods, 1982).

4. Anti-steroids

The initial action of most, if not all, hormones is an interaction with a receptor, a reaction which depends upon structural complementarity between the hormone and the receptor (Chapter 22; Section A.3). This has been exploited by the pharmaceutical industry for a number of years to produce compounds that either mimic (agonists) or inhibit (antagonists) the action of hormones and other agents by binding to the specific receptors. There is considerable interest in the therapeutic use of 'steroid blockers' or anti-steroids which could bind to the cytosolic steroid receptor and elicit either no biological response or an erroneous response. The successful treatment of a proportion of mammary gland tumours in both experimental animals and women with various anti-oestrogens, especially tamoxifen (*trans*-1-*p*-β-dimethylaminoethoxyphenyl-1,2-diphenylbut-1-ene— see Figure 20.33) has raised the question of the mechanism of action of such drugs. At a biochemical level, it competes with oestradiol-17β for binding to the specific cytosolic receptor but, although it appears to have anti-oestrogenic

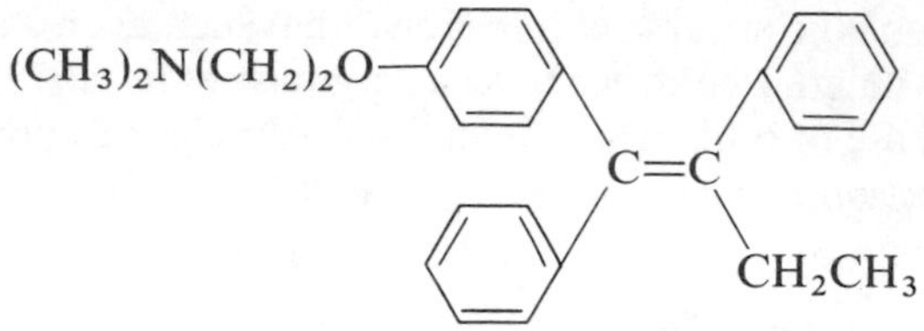

$$(CH_3)_2N(CH_2)_2O-\underset{\overset{|}{\underset{\bigcirc}{}}}{\bigcirc}\overset{\displaystyle\bigcirc}{\underset{\displaystyle C=C}{}}\overset{\displaystyle\bigcirc}{\underset{\displaystyle CH_2CH_3}{}}$$

Figure 20.33　Structural formula of tamoxifen

properties, it does enter the nucleus (as a complex with the receptor) and initiate transcriptional activity. It is unclear if the initiation is inefficient or whether the mRNA product is abnormal. The efficacy of the drug supports the view that some breast tumours are oestrogen-dependent and that this hormone plays a role in initiation and maintenance of the cancerous state (McGuire, 1975). Since this applies not only to primary breast tumours but also to metastases, anti-oestrogen therapy is important even after mastectomy, provided that there is evidence of oestrogen dependence as indicated by measurement of oestrogen-binding sites in the tumours removed from the breast (Fisher *et al.*, 1981; Henderson, 1981).

APPENDIX 20.1　BIOSYNTHESIS OF THE STEROID HORMONES

Despite the relatively small number of types of reactions involved (see Figure 20.9) the pathways for the biosynthesis of steroids are dauntingly complex. In this appendix a few metabolic details are added to the summary in Section B.3. The enzymes are referred to by the names by which they are widely known rather than Enzyme Commission names which have not always been assigned. The structural formulae and systematic names of the steroids involved may be determined by reference to Table 20.1.

1. **Corticosteroids**

In man, the major pathway for the formation of cortisol involves the conversion of pregnenolone to 17-hydroxypregnenolone by hydroxylation at carbon-17, followed by conversion to 17α-hydroxyprogesterone by means of reactions catalysed by 5-ene-3β-hydroxysteroid dehydrogenase and an isomerase (Figure 20.10). This is followed by hydroxylation at carbon-21 to produce 17,21-hydroxyprogesterone (11-deoxycortisol) which occurs on the endoplasmic reticulum. However, the enzyme catalysing the final reaction producing cortisol, the 11β-hydroxylase which requires NADPH and molecular oxygen, is bound to the inner mitochondrial membrane. Cortisone is produced from cortisol in the endoplasmic reticulum by the activity of 11β-hydroxysteroid dehydrogenase.

　　Corticosterone and aldosterone are produced via progesterone. Pregnenolone is converted to progesterone in reactions catalysed by 5-ene-3β-hydroxysteroid dehydrogenase and an isomerase (Figure 20.11). Progesterone is hydroxylated at carbon-21 to produce deoxycorticosterone, which is hydroxylated at carbon-11 to produce corticosterone (as indicated above the 11-β-hydroxylase is bound to the inner mitochondrial membrane). Aldosterone is produced from corticosterone by hydroxylation of the methyl group (carbon-18) to produce the $-CH_2OH$ group which is converted to the aldehyde by the enzyme, 18-dehydroxysteroid dehydrogenase. Both the latter enzymes are present inside the mitochondria. This latter pathway is quantitatively much less important than the pathway for formation of cortisol; the rate of cortisol secretion in man is about $20\,mg.day^{-1}$ whereas that for corticosterone is $3\,mg.day^{-1}$ and aldosterone is $0.15\,mg.day^{-1}$.

The adrenal cortex consists of three histologically defined zones, zona glomerulosa (ZG), the zona fasciculata (ZF), and the zona reticularis (ZR). The largest zone, the ZF, which contains large amounts of cholesterol ester, and the ZR are both responsible for the synthesis of corticosteroids (and the androgens and oestrogens produced in this tissue). The cells of the ZG produce aldosterone (and the small amounts of corticosterone) since the 18-hydroxylase and 18-hydroxysteroid dehydrogenase enzyme are specifically localized in these cells.

2. Androgens

There appear to be two major routes for the synthesis of testosterone in the human testes (Figure 20.12). First, pregnenolone is converted to 17-hydroxypregnenolone, as in the adrenal cortex, but is then converted to dehydroepiandrosterone (sometimes abbreviated to DHEA) by side-chain cleavage catalysed by a 17,20-desmolase, which is bound to the endoplasmic reticulum and requires NADPH and molecular oxygen for activity. The reduction of the oxo-group at carbon-17 of DHEA, transfer of the double bond from 5,6 to 4,5 and oxidation of the hydroxyl at carbon-3 produces testosterone. These reactions are catalysed by 17β-hydroxysteroid dehydrogenase, an isomerase and 5-ene-3-β-hydroxysteroid dehydrogenase respectively.

In the second pathway, pregnenolone is converted to progesterone (as in the adrenal cortex) which is then hydroxylated at carbon-17 prior to side-chain cleavage to produce 4-androstenedione. The latter is converted to testosterone by reduction of the 17-oxo-group by 17-β-hydroxysteroid dehydrogenase.

DHEA sulphate (see Figure 20.18) and 5-androstenediol sulphate may act as a local store of androgens since sulphatases exist which can produce free steroids from which testosterone can be synthesized.

3. Oestrogens

In pre-menopausal women, the ovaries are important in the production of oestrogens. These are produced by way of the androgens, 4-androstenedione and testosterone, which are synthesized as described above. Oestrogen synthesis involves removal of the methyl group at carbon-10; this is achieved first by hydroxylation of carbon-19, followed by dehydrogenation (catalysed by 19-hydroxysteroid dehydrogenase) to produce the aldehyde and finally removal of the carbon as methanol by a 10,19 lyase. Since spontaneous rearrangement in the A ring results in the formation of an aromatic ring, the desmolase enzyme is sometimes referred to as an aromatase. The products are oestradiol-17β from testosterone and oestrone from 4-androstenedione. Finally a 16α-hydroxylase catalyses the formation of oestriol from oestradiol-17β (Figure 20.13).

NOTES

1. Similar enzymes are involved in the formation of this compound in the pathway of ketone body synthesis which occurs within the mitochondria of the liver (Chapter 6; Section D.2). Since the enzyme HMG-CoA lyase does not occur in the cytosol, ketone bodies cannot be formed in this compartment.
2. Squalene received its name from its presence in large quantities in the liver of sharks, especially of the genus *Squalus*. Its role in these fish is not fully understood, although it may increase buoyancy.
3. Taurine is formed by decarboxylation of cysteic acid, which in turn is formed from the oxidation of the amino acid, cysteine (see Chapter 21; Section B.8).
4. Cholesterol gallstones can also be formed as a result of excessive cholesterol secretion into the bile (Bennion and Grundy, 1975).
5. Cholesterol reaching the lower intestinal tract is modified by bacteria in the colon, mainly by saturation of ring B, producing coprostanol and also cholestanol, which are found in the faeces.
6. For example, in Cushing's syndrome there is an overproduction of cortisol that can be detected by the elevated concentration of 17-hydroxysteroids in the urine (Section C.2). These are often known as the 17-oxogenic steroids since their estimation involves side-chain cleavage to yield 17-oxosteroids which can be estimated with alkaline *m*-dinitrobenzene (Zimmerman reaction).
7. The details of this pathway have been established in human cells by the elegant work of Drs J. L. Goldstein and M. S. Brown using human fibroblasts obtained from biopsy samples of skin. It is likely that the pathway is similar in other cells (Goldstein and Brown, 1976).
8. Patients with Tangier disease, a rare autosomal recessive disorder, have a very low level of plasma HDL. Cells of the reticuloendothelial system (which are found in tonsils, lymph nodes, bone marrow, thymus, rectal mucosa) accumulate large amounts of cholester ester. The concentration of LDL is low and there are abnormal lipoprotein particles that probably represent incompletely formed remnants of intermediate-density lipoprotein. It is suggested that these particles are taken up by endocytosis into the reticuloendothelial cells which become so loaded with cholesterol ester that the tonsils become enlarged and orange-coloured. Patients have a low plasma cholesterol concentration and suffer from peripheral neuropathy (Havel *et al.*, 1980).
9. Prednisone (or its hydroxy-form, prednisolone) is the synthetic glucocorticoid usually chosen for such treatment. A combination of cyclosporin A and prednisone has been used with success in both liver and kidney transplants (Starzl *et al.*, 1981).
10. The enzyme renin, secreted by the kidney, hydrolyses a peptide bond in angiotensinogen in the blood to produce angiotensin I (a decapeptide) which is then converted to angiotensin II by a converting enzyme present in the blood. The main functions of angiotensin II are to increase the peripheral resistance and hence raise blood pressure and to stimulate aldosterone synthesis.

11. About 5 mg of testosterone is secreted each day in a normal young man but very little is stored in the Leydig cell. Hence synthesis and secretion occur simultaneously and the control of the rate of steroidogenesis by LH is thus very important.

12. The control of the LH surge is dependent upon the complex interplay of changing hormone concentrations. However, the progressive increase in the concentration of oestradiol-17β is of the utmost importance, particularly the peak concentration and the time during which the high concentration is maintained. It is not surprising therefore that oral contraceptive steroids interfere with this delicate control mechanism and prevent the LH surge.

13. These changes in enzyme activities may result from hypertrophy of the granulosa cells which results in an increase in the number of LH receptors. The increased binding of this hormone activates adenyl cyclase and this raises the cyclic AMP concentration which activates protein kinase. Phosphorylation of key proteins may result in increased rates of synthesis of the enzymes necessary to convert cholesterol to progesterone and decreased rates of synthesis of the other enzymes.

14. The structure of human GnRH is:

$$\text{(Pyro)Glu-His-Trp-Ser-Tyr-Gly-Leu-Arg-Pro-Gly-CONH}_2$$
$$1 \quad\; 2 \quad 3 \quad 4 \quad 5 \quad 6 \quad 7 \quad 8 \quad 9 \quad 10$$

The N-terminal glutamate is in the pyroform in which the terminal amino group forms a peptide bond with the carboxyl group at position 5. The C-terminal glycine is in the amide form. (Schally, 1978). If, for example, glycine in position 6 is replaced by D-leucine (rather than the natural L-leucine) and ethylamide replaces glycinamide in position 10, the potency for releasing LH from the pituitary is increased markedly; replacement of amino acids in positions 2, 3, and 6 by D-phenylalanine, D-tryptophan and D-phenylalanine respectively produces a strong inhibitor of LH secretion (see Schally, 1978).

15. Cell division begins as soon as the ovum is fertilized and the developing embryo is known as the blastocyst.

16. Human chorionic gonadotrophin has a similar structure to FSH and LH; it is a glycoprotein with two chains, α and β, linked together non-covalently. The α-subunits of all these hormones are very similar but the structure of the β-subunit is different. The only known function for hCG is to take over the role of LH in the maintenance of the corpus luteum about eight days after fertilization. The presence of hCG in urine is used in testing for pregnancy; it is detected in the urine by an immunological agglutination test which distinguishes it from FSH and LH (Chapter 22; Section E.3.b).

17. The most abundant natural form of vitamin D is D_3 which is derived from cholesterol, hence its name cholecalciferol. It is found in fish liver oils (especially cod and halibut) and in butter and egg yolk. Vitamin D_2 (Appendix A; Section B.2) is produced synthetically by ultraviolet irradiation of ergosterol (hence its name, ergocalciferol) and is added to some foods (e.g. milk, margarine) in industrial societies. The nomenclature is such that there is no vitamin D_1.

18. 7-Dehydrocholesterol is stored in the epidermis and some of this is converted first to previtamin-D_3 by ultraviolet radiation (in the spectral range 290–320 nm). Vitamin D_3 (cholecalciferol) is formed only slowly from previtamin D_3 by a temperature dependent thermal isomerization as follows:

$$\text{7-dehydrocholesterol} \xrightarrow{\text{UV light}} \text{previtamin } D_3 \xrightarrow{\text{heat}} \text{vitamin } D_3$$

This slow rate of isomerization permits a continuous synthesis of vitamin D for up to four days following a single exposure to ultraviolet radiation. It may also be one factor preventing too high a rate of synthesis of the vitamin during prolonged exposure to ultraviolet radiation (high levels of active hormone can be dangerous) (Adams *et al.*, 1982).

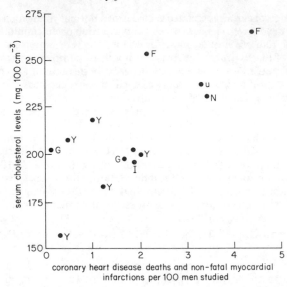

Figure 20.34 Relationship between serum cholesterol concentrations and coronary heart disease. Each dot represents the mean value for different populations of men aged 40–59 years in seven countries, Yugoslavia (5 areas), Greece (2 areas), Finland (2 areas), Italy (2 areas), Netherlands, USA (railroad workers). Data from Stamler (1974).

19. The molecular basis of action of 1α,25-dihydroxycholecalciferol is probably identical to other steroid hormones; it binds to a cytosolic receptor in the target tissue, the complex enters the nucleus and increases the rate of formation of specific mRNA molecules (see Chapter 22) (Norman and Henry, 1979).

20. Bone consists of a protein and proteoglycan matrix which is impregnated with mineral salts, especially phosphates of calcium. Adequate amounts of both protein and mineral must be available for maintenance of normal structure. The mineral is mainly hydroxyapatite, $3Ca_3(PO_4)_2.Ca(OH)_2$. Activation of the osteoclasts causes resorption of the whole bone (collagen plus mineral).

21. Parathyroid hormone binds to a receptor on the kidney cell, activates adenylate cyclase and raises the intracellular cyclic AMP concentration. This suggests that cyclic AMP-dependent protein kinase phosphorylates the 1-monooxygenese enzyme system, which causes activation (see also Chapter 22; Section A.3.b).

22. In an epidemiological investigation, a disease frequency is compared with other characteristics of the population in a search for an association. The well-known seven countries study by Keys (1970), which studied over 11 000 men in seven countries (Greece, Holland, Yugoslavia, Italy, Japan, the U.S.A. and Finland), showed a strong association between the incidence of coronary heart disease and the serum total cholesterol (i.e. cholesterol plus cholesterol ester) concentration (Figure 20.34). This association has been confirmed after a further ten years experience with the study (Keys, 1980).

   Further epidemiological studies have indicated a number of other risk factors in coronary heart diseases. The major ones are, hypertension, heavy cigarette smoking, and diabetes mellitus; minor risk factors include lack of exercise, obesity, hypertriglyceridaemia, drinking soft water, and stress.

   The suggestion from a number of such studies is that the plasma cholesterol concentration should be reduced to below 220 mg.100 cm^{-3}. However, the total plasma cholesterol concentration is not the only important factor, since one of the highest rates of coronary heart disease occurs in Scotland, where the average total cholesterol concentration is below 220 mg.100 cm^{-3} (Logan *et al.*, 1978). Although many factors could contribute to the disease in Scotland, a low intake of essential fatty acids compared to saturated fats is one that is gaining credence.*

Essential fatty acids have at least three roles that might explain their beneficial effect: they are involved in the transport of cholesterol as cholesterol ester in HDL (Section D.2.b), in the structure of membranes and in the formation of prostacyclin and thromboxanes (Chapter 17; Section D.2). These compounds are of interest since they provide one explanation for the remarkably low rate of myocardial infarction in Greenland Eskimos despite their high lipid intake. Although Eskimos have a low intake of arachidonic and linoleic acids they have a high intake of eicosapentaenoic acid. This latter polyunsaturated fatty acids leads to the production of a thomboxane A_3, which has little platelet-aggregating activity, and also to prostacyclin I_3, which has potent anti-aggregating activity (see Oliver, 1981; Thorngren and Gustafson, 1981).

A clinical trial of clofibrate, a hypocholesterolaemic drug (see Figure 6.16) (involving approximately 30 000 male volunteers over a period of four to six years) showed a 20 % fall in the incidence of cardiovascular disease and a 9 % fall in serum cholesterol concentration in comparision with control subjects with a similar initial high cholesterol level. Unfortunately, deaths due to non-cardiovascular diseases were increased in the clofibrate group (see Oliver, 1978, 1981).

Despite the evidence implicating nutritional factors in the development of atherosclerosis, there is no doubt that other factors play a role. It is generally accepted that stress can accelerate the development of this disease and the simple but important observation that the plasma level of cholesterol increases markedly in tax accountants as the end of the financial year approaches and the work load increases, suggests that plasma cholesterol levels may be involved. Rosenman and Friedman (1974) have suggested that psychosocial factors play an important role. Subjects who are described as exhibiting type A personality behaviour, 'ambition, competitive drive, aggressiveness, a strong sense of time urgency', are much more likely to suffer from atherosclerosis and coronary heart disease than subjects with a type B personality whose behaviour is more relaxed, unhurried, and mellow. A correlation between A type behaviour and extent of coronary heart disease has, however, not always been observed (Bass and Wade, 1982). The interaction between the psychosocial make-up of the individual, the endocrine balance and metabolic indicators of stress (e.g. plasma fatty acid concentration) has not been investigated (see Jenkins, 1978).

23. The artery wall consists of three layers: the tunica intima or endothelium, which is the innermost layer and consists of endothelial cells and the basement membrane: the tunica media, which is the middle layer composed of smooth muscle cells and elastic and collagen fibres, and the tunica adventitia, which is the outermost layer consisting of collagen and elastic fibres without smooth muscle cells (Figure 20.35).

24. The osteoporosis may be due to insufficient synthesis of 1α,25-dihydroxychole-calciferol (Section F.3 and Wheddon, 1981).

25. It is estimated that there are over 40 million women over 50 years of age in the U.S.A. alone. Since this number is likely to increase enormously over the next 20 years the economic cost of such treatment cannot be easily dismissed (Weinstein, 1980).

---

* The suggestion was made as early as 1956 by H. Sinclair (1980) but no major prevention clinical trial using a diet low in saturated fat has been conducted (Oliver, 1981). However, two large primary prevention trials have demonstrated that, when the proportion of polyunsaturated to saturated fat is increased, the serum cholesterol concentration is decreased (by 10–15 %) and the incidence of cardiovascular diseases and non-fatal myocardial infarctions also decreases.

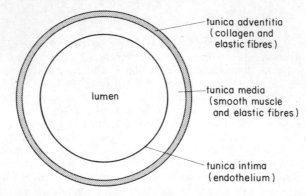

Figure 20.35 Diagrammatic cross-section of an artery to show structure of wall

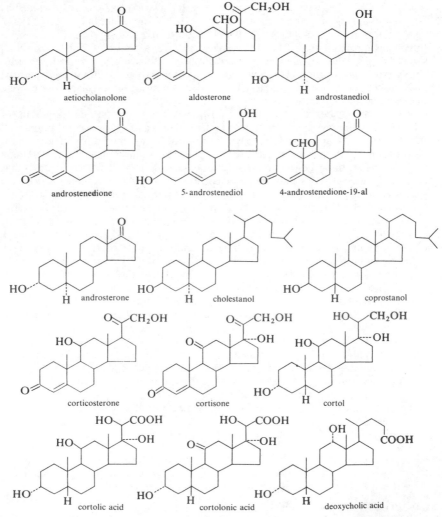

Figure 20.36 Structural formulae of steroids not depicted elsewhere

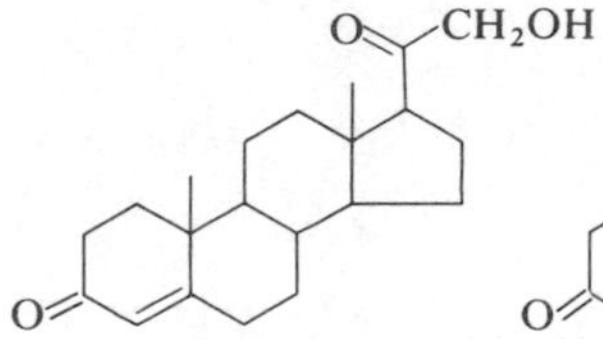

11-deoxycorticosterone

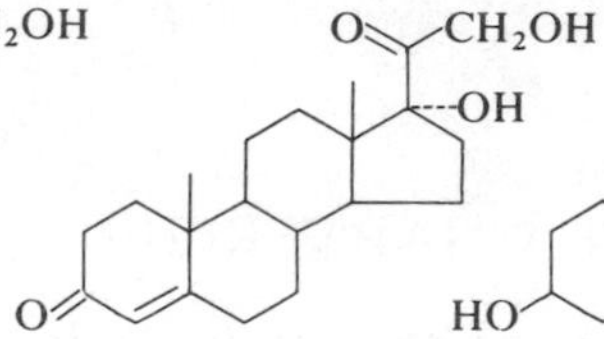

11-deoxycortisol

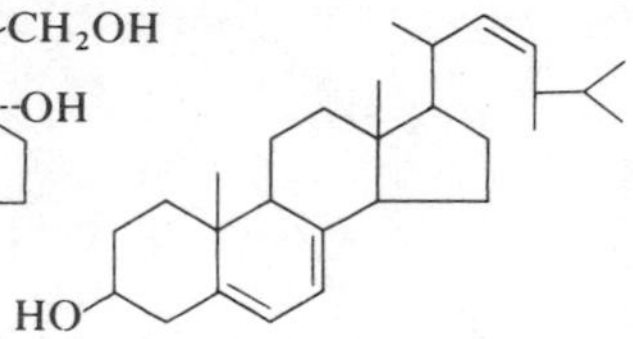

ergosterol

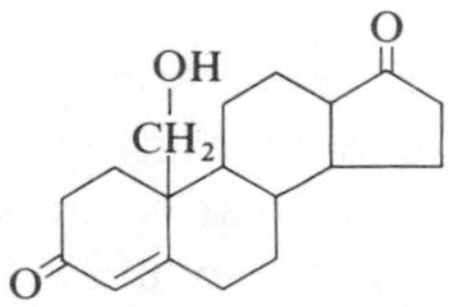

19-hydroxy-4-andiostenedione

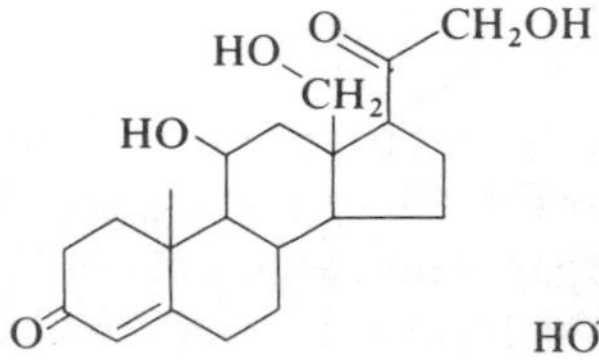

18-hydroxycorticosterone

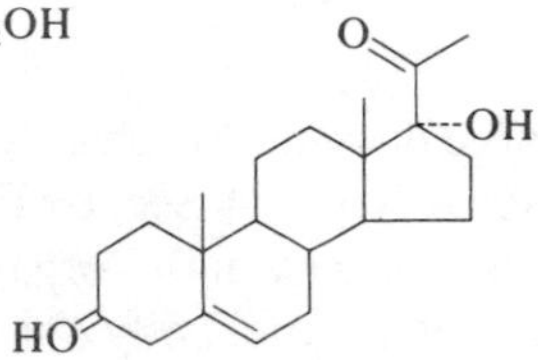

17-hydroxypregnenolone

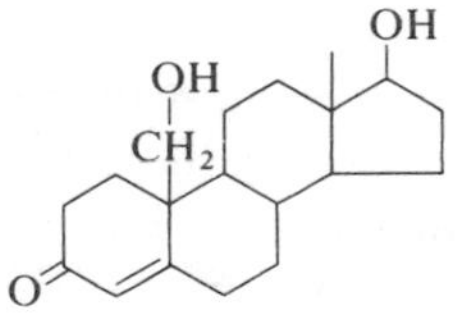

19-hydroxytestosterone

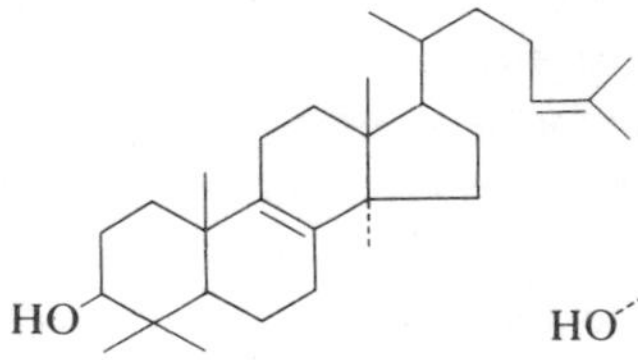

lanosterol

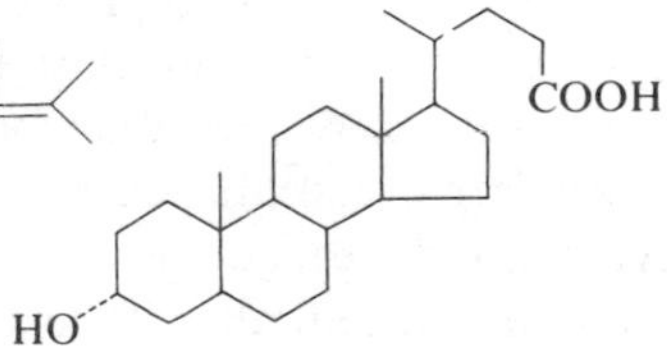

lithocholic acid

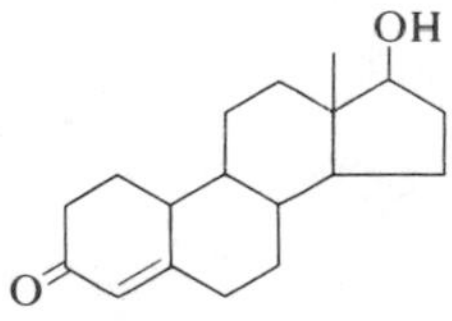

19-nortestosterone

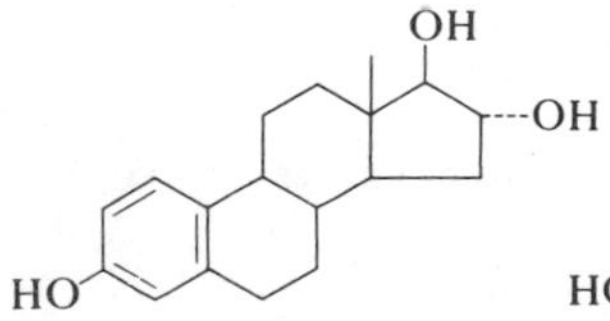

oestriol

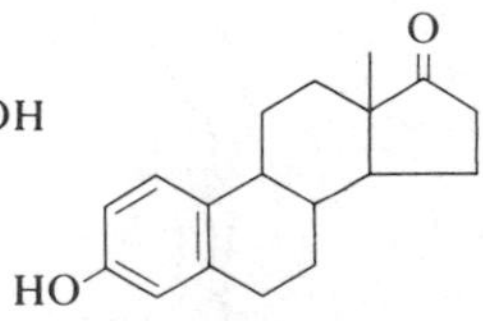

oestrone

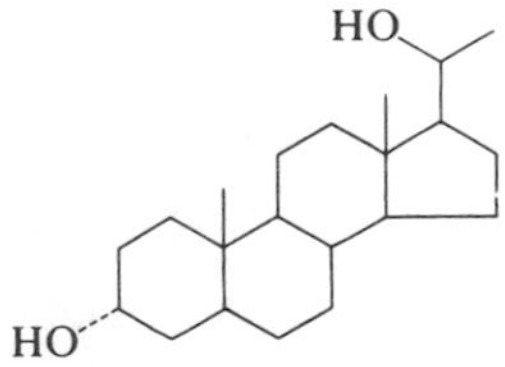

pregnanediol

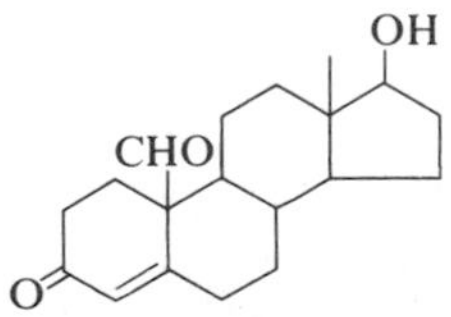

testosterone-19-al

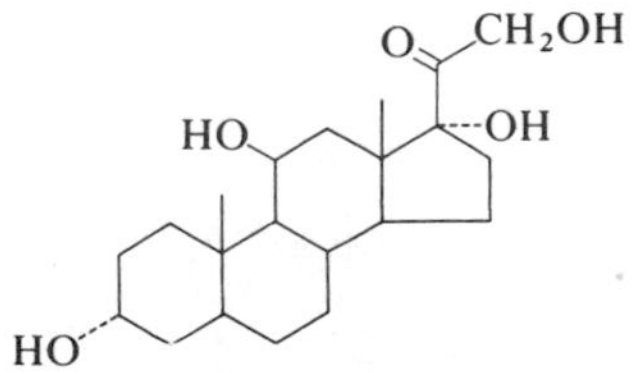

tetrahydrocortisol

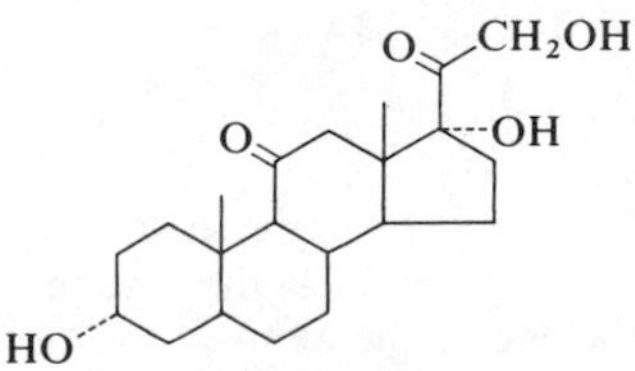

tetrahydrocortisone

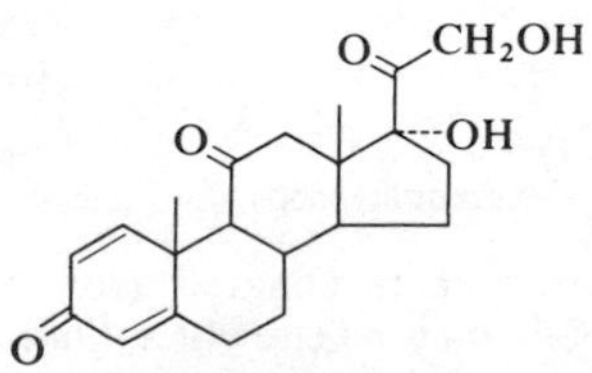

prednisone

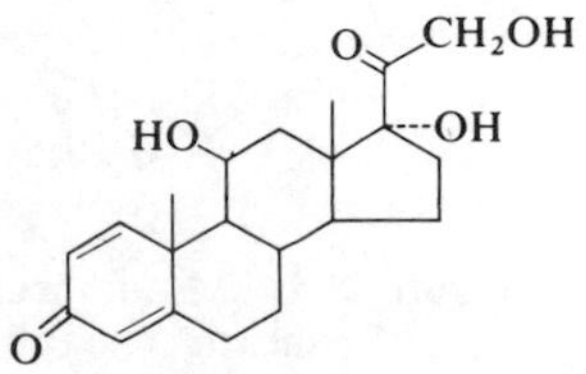

prednisolone

CHAPTER 21

NEUROTRANSMITTER METABOLISM

The large demand for glucose and oxygen by the brain is emphasized in earlier chapters (Chapter 4, Section E.1; Chapter 8, Table 8.3). This results from the high metabolic activity of this organ. Thus, although the brain of adult man comprises only 2 % of the total body weight, its metabolic activity accounts for about 25 % of the basal metabolic rate (Sokoloff *et al.*, 1977). This is explained by the constant electrical activity of the brain which requires communication between the large number of neurones (approximately 10^{13}) and it is the processes involved in this interneurone communication that demands most of the energy.

The neurones comprising the nervous system of an animal are among the most specialized of cells. In general, they consist of a cell body (containing the nucleus, cytosol, and the machinery for protein synthesis), an elongated portion known as the axon and a number of shorter processes known as dendrites (Figure 21.1). The

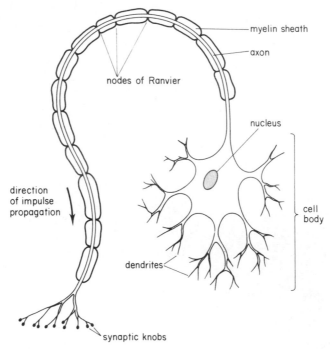

Figure 21.1 Motor neurone with myelinated axon. Some neurones lack the myelin sheath. The cell body may be elsewhere than at the end of the axon

importance of the axon is that it transmits, along its length, electrical signals. These are transient changes in electrical potential that propagate along the axon. It is the temporal pattern of these impulses that conveys information and, by linking together several such cells, a conduction pathway or 'circuit' is established. Although, along the axon, information is transmitted electrically, between the different neurones in such a circuit it is transmitted chemically. The chemical is released when the impulse reaches the terminus of the axon (of one cell); it then diffuses across the gap between the two cells and may initiate an impulse in the second cell. Junctions of this type are known as synapses and the chemicals fulfilling this role as neurotransmitters. Within the central nervous system a single neurone may connect with a very large number (perhaps more than 1000) of other neurones via such junctions. Not only does the maintenance of basic life functions (respiration, heart beat, movement, etc.) depend on the pattern of these connections but also higher functions such as intelligence and memory, thus establishing the importance of the biochemistry of the neurones. Discussion in this chapter is limited to the nature of the neurotransmitters, their metabolism, their mode of action and how the malfunctioning of neurotransmitter metabolism might explain some neurological disorders (see Section E).

A. INTRODUCTION TO NEUROTRANSMITTERS

In order to appreciate the role of neurotransmitters as a chemical link in an electrical circuit it is important to understand the basis of the nerve impulse.

1. The nerve impulse

The axon of a neurone is, in effect, a tube of cytosol bounded by a membrane. When no impulses are passing along the axon it possesses a resting potential difference across the cell membrane so that the inside of the neurone is some 70 mV negative with respect to the outside. In this resting state it is said to be polarized. The potential difference arises because the membrane is permeable to potassium ions, the concentration of which is higher inside the cell than outside. As potassium ions move through the membrane, down their concentration gradient, they carry with them positive charge. As this accumulates on the outer surface of the membrane it opposes further ion movement, so that an equilibrium potential is established which depends upon the concentration difference of potassium ions across the membrane. The relationship between this difference and the potential is given by the Nernst equation.[1]

An impulse is a transient reversal of this membrane potential so that the inside of the axon becomes positive relative to the outside. This depolarization results in an action potential of over 100 mV (Figure 21.2). This depolarization is caused by a marked increase in the permeability of the membrane to sodium ions. Since the concentration of the latter is high outside the axon, sodium ions tend to move

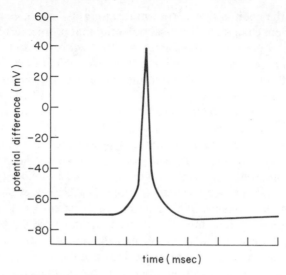

Figure 21.2   An action potential recorded with one electrode within the neurone

inwards. The depolarization is transmitted along the axon since the flow of current, which occurs as a result of the depolarization, causes the increase in sodium permeability. The change in sodium permeability is, however, short-lived, and potassium ions rapidly move out to restore the resting potential (repolarization). For a short time, after repolarization (the refractory period), the axon cannot be depolarized and this limits the frequency at which impulses can travel along the axon.

It should be clear from this description that the passage of each impulse results in the flow of sodium ions into and potassium ions out of the axon. Although the ion fluxes are extremely small, this would eventually lead to abolition of the concentration differences and an inability to transmit impulses. This is prevented by an energy-requiring translocation mechanism in the membrane which transports sodium ions out of, and potassium ions into, the axon. This process is known as the 'sodium pump' or, since ATP hydrolysis accompanies the translocation, the 'Na^+/K^+ ATPase'. However, the quantitative importance of this process in ATP utilization is probably very small; most of the ATP utilization is involved in neurotransmitter metabolism (see Sections B and C).

2. The synapse

During the propagation of an impulse along the axon, depolarization is caused by an electrical current flowing through the membrane. However, depolarization can also be caused by chemicals and the transmission of impulses from cell to cell within a neuronal circuit is usually achieved in this way. The junction between two neurones is known as a synapse and is represented diagrammatically in

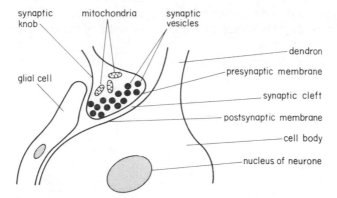

Figure 21.3 A schematic representation of a synapse

Figure 21.3. The neuromuscular junction is essentially similar. When the presynaptic membrane becomes depolarized due to the arrival of an impulse, the neurotransmitter is discharged into the synaptic cleft. The neurotransmitter diffuses across the cleft (rapidly since it is only 20 nm across) and causes depolarization of the postsynaptic membrane (by changing its permeability to sodium ions), thus initiating an impulse in the second cell. A neurotransmitter acting in this way is said to be excitatory; other neurotransmitters decrease the ease with which the postsynaptic membrane can be depolarized and are known as inhibitory neurotransmitters. In all cases, it is important that the neurotransmitter is removed from the synaptic cleft as soon as the signal has been transmitted. This is achieved by uptake of the neurotransmitter back into the presynaptic nerve terminal, into the postsynaptic neurone, into an adjacent neuroglial* cell, or by conversion of the neurotransmitter to an inactive metabolite within the synapse.

3. Neurotransmitters

The most widespread excitatory neurotransmitters outside the central nervous system are acetylcholine and the catecholamines, adrenaline and noradrenaline.† Neurones secreting the former are known as cholinergic, and the latter as adrenergic. However, a larger (and ever increasing) number of compounds are considered to function as neurotransmitters in the central nervous system (the number now exceeds 40—Iversen, 1982). These include the amino acids, glutamate, aspartate, glycine, and taurine, and the amines, dopamine, 5-

---

* In the central nervous system neuroglial cells are found in close association with neurones and synapses. Their precise function is not clear but there is evidence that they play a role in neurotransmitter metabolism.

† Adrenaline and noradrenaline are also known as epinephrine and norepinephrine respectively. Together with dopamine they are termed catecholamines.

hydroxytryptamine, 4-aminobutyric acid, and histamine, as well as ATP and a number of peptides. The structural formulae of some of these neurotransmitters are given in Figure 21.4.

For reviews, see Agranoff, (1975); Cooper *et al.*, (1978); Quastel, (1979), and Snyder, (1980).

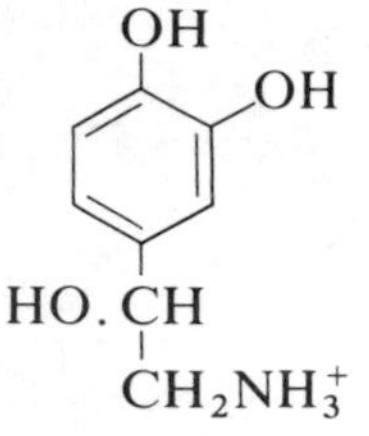

noradrenaline

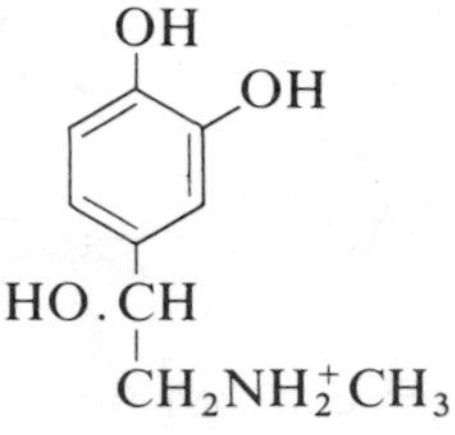

adrenaline

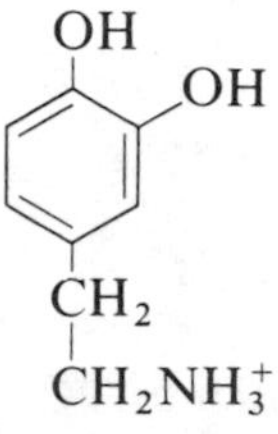

dopamine

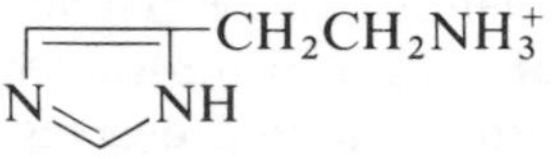

histamine

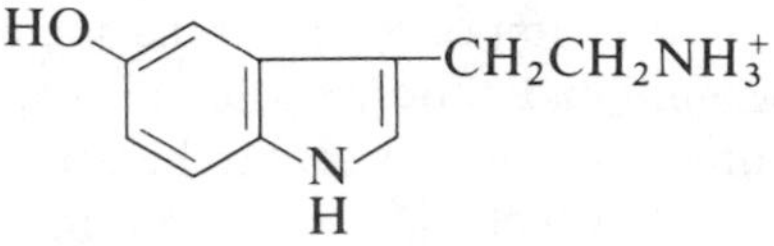

5-hydroxytryptamine

$(CH_3)_3N^+CH_2CH_2O.CO.CH_3$

acetylcholine

$H_3^+NCH_2CH_2CH_2COO^-$

4-aminobutyrate

$^-OOC.CH_2CH_2CH.COO^-$
$\quad\quad\quad\quad |$
$\quad\quad\quad NH_3^+$

glutamate

CH_2COO^-
$|$
NH_3^+

glycine

$^-OOC.CH_2CH.COO^-$
$\quad\quad\quad |$
$\quad\quad NH_3^+$

aspartate

$H_2^+NCH_2CH_2SO_3^-$

taurine

Figure 21.4 Structural formulae of some established neurotransmitters. See Figure 21.18 for structures of neuropeptides

(a) *Criteria to establish identity as a neurotransmitter*

A consideration of criteria needed to establish that a given substance is indeed a neurotransmitter helps to focus attention on their essential properties. Examples of the application of these criteria are to be found in Section B.

1. The presence of neurotransmitter in the presynaptic axon terminal must be demonstrated.[2]
2. The enzymes necessary for the synthesis of the neurotransmitter must be shown to be present in the presynaptic neurone.
3. Physiological stimulation of the presynaptic neurone should result in release of the neurotransmitter.
4. A mechanism must exist for the rapid termination of the action of the neurotransmitter.
5. Direct application of the neurotransmitter to the postsynaptic terminal should mimic the action of nervous stimulation.
6. Drugs that modify metabolism of the neurotransmitter should result in physiological or behavioural effects *in vivo*.

(b) *General aspects of neurotransmitter metabolism*

In general, the metabolism of neurotransmitters can be represented as a series of processes that include synthesis, storage, release, action and termination of action. The latter may involve the conversion of a neurotransmitter to an inactive product that is either a common intermediate of metabolism or a readily excreted endproduct. Such a sequence of processes has similarities to a metabolic pathway and, consequently, it can be analysed and the mechanism of regulation studied according to the principles described in Chapters 2, 3, and 7. This involves establishing the equilibrium nature of the reactions and identifying the flux-generating steps and positions of external regulation. Unfortunately, sufficient information is not available to carry out a complete analysis of any pathway of neurotransmitter metabolism so that the theories of control given below are only tentative. Despite the lack of information on the precise details of the mechanism of regulation, there is no doubt that several of these processes involved in neurotransmitter metabolism require energy (that is, ATP hydrolysis). These include transport of amino acids into the nerve terminal, storage and uptake of the neurotransmitter from the synaptic cleft. Together with the energy requirement of regulation, these processes contribute to the high energy demand of the brain.

B. METABOLISM OF NEUROTRANSMITTERS

An important feature of neurotransmitter metabolism is that many are synthesized from amino acids; indeed several amino acids are neurotransmitters in their own right.

1. Acetylcholine

As a consequence of O. Loewi's classic experiments, carried out in 1920, demonstrating chemical transmission from the vagus nerve to the heart, acetylcholine was the first neurotransmitter to be identified. Its metabolism is usefully considered first here as it illustrates a number of metabolic principles relating to neurotransmitter function.

(a) *Synthesis of acetylcholine*

Acetylcholine is synthesized by the acetylation of choline by acetyl-CoA, catalysed by choline acetyltransferase in the presynaptic nerve terminal:

$$(CH_3)_3N^+CH_2CH_2OH + CH_3CO.SCoA \longrightarrow$$

choline $(CH_3)_3N^+CH_2CH_2OCO.CH_3 + CoASH$

acetylcholine

The choline required for this synthesis may be obtained from the diet,[3] be recycled from acetylcholine degraded in the synapse or synthesized from ethanolamine,[4] which is derived from serine (a non-essential amino acid, see Appendix A). Since choline acetyltransferase is present in the cytosol of the presynaptic neurone but acetyl-CoA is produced from pyruvate in the mitochondria, the problem of transport of acetyl-CoA across the mitochondrial membrane arises. This is comparable with the situation for fatty acid synthesis considered in Chapter 17; Section A.3. The same two mechanisms, namely transport via citrate or via acetylcarnitine, probably operate but it is unclear which is the more important in nervous tissue.

(b) *Catabolism of acetylcholine*

The action of acetylcholine on the postsynaptic membrane is terminated by its hydrolysis to acetate and choline, catalysed by acetylcholinesterase:

$$acetylcholine^+ + H_2O \longrightarrow acetate^- + choline^+ + H^+$$

The enzyme is bound to the postsynaptic membrane but neither product is able to bind to the postsynaptic receptor. The choline is taken up into the presynaptic nerve terminal (apparently by a specific transport system) where it is used for the resynthesis of acetylcholine (Figure 21.5). Since choline acetyltransferase is not saturated with choline, the rate of acetylcholine synthesis will be decreased if this choline is not reabsorbed (see Anderson, 1981). The fate of the acetate, however, is less clear. [^{14}C]Acetate is a poor precursor for acetylcholine in the brain, probably due to the high K_m for acetate exhibited by brain acetyl-CoA synthetase. Hence, most of the acetate produced in the brain may be lost and metabolized in other tissues. Even if all the acetate were lost to the brain the metabolic effect would be negligible: the brain is able to supply acetyl-CoA

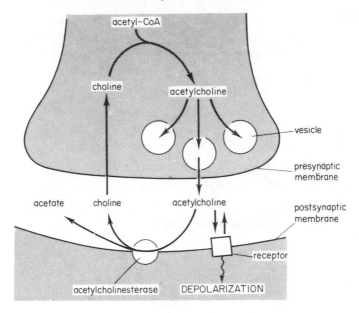

Figure 21.5 Reactions occurring at a cholinergic synapse

(from glycolysis) at a rate one-hundred times greater than that needed for acetylcholine synthesis (as indicated by the maximal activity of key 'indicator' enzymes, see Sugden and Newsholme, 1977).

2. Catecholamines

Catechol is the dihydroxyphenol with two adjacent hydroxyl groups. Derivatives with a side chain bearing an amino group are known as catecholamines. Those of importance as neurotransmitters are 3,4-dihydroxyphenylethylamine (abbreviated to dopamine), noradrenaline and adrenaline (Figure 21.4). Catecholamines are found in the central nervous system, peripheral nerves and the adrenal medulla (from which tissue, adrenaline and noradrenaline can be released into the blood as hormones).

(a) *Synthesis of catecholamines*

The pathway is given in Figure 21.6. The precursor for the formation of the catecholamines in the nervous tissue is extracellular tyrosine,* which is transported into the presynaptic nerve terminal by an energy-dependent transport process as described in Chapter 10; Section C.1. The first reaction in the cell is conversion of tyrosine to 3,4-dihydroxyphenylalanine (abbreviated to

---

* The conversion of phenylalanine to tyrosine (a reaction catalysed by phenylalanine 4-monooxygenase) occurs in liver but not in brain (see Chapter 10; Section E).

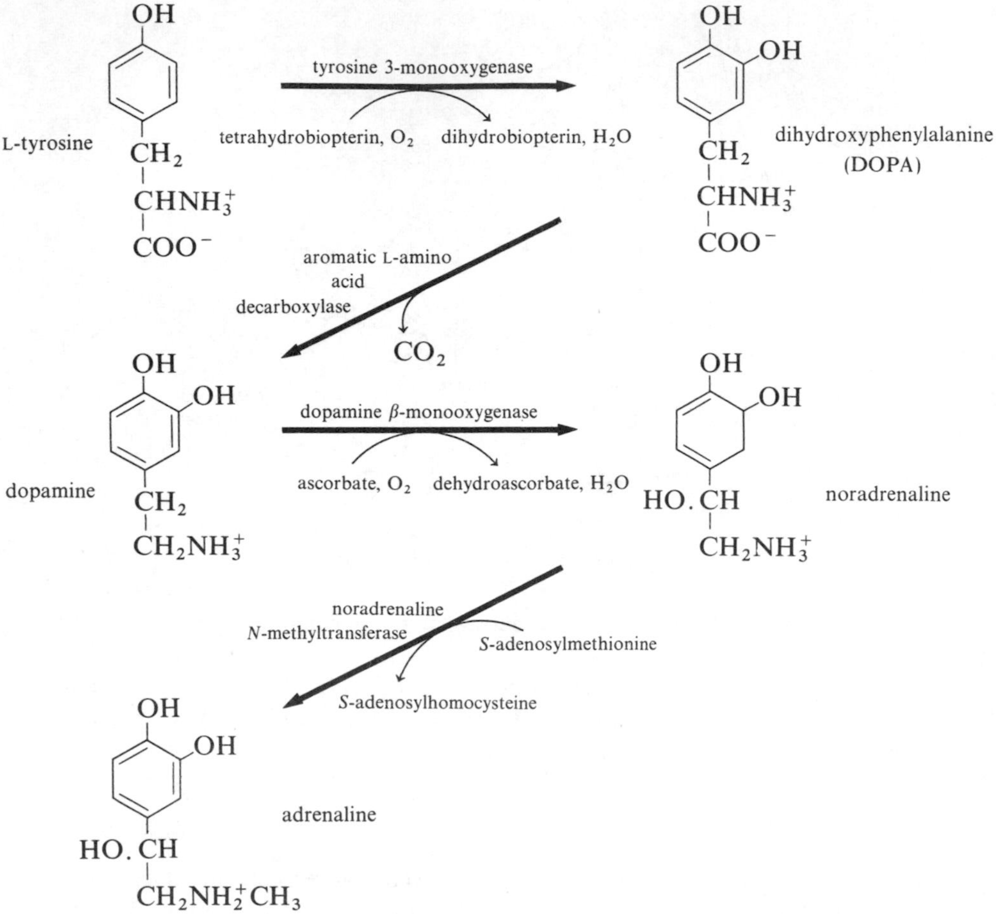

Figure 21.6 Pathway of dopamine, noradrenaline and adrenaline synthesis

DOPA) catalysed by tyrosine 3-monooxygenase (tyrosine hydroxylase[5]). The enzyme is a mixed-function oxidase and requires molecular oxygen and NADPH. However, the latter does not provide the hydrogen directly; this is achieved via a further co-substrate, tetrahydrobiopterin which is oxidised to dihydrobiopterin in this reaction (Figure 21.7). The NADPH then reduces the dihydrobiopterin. Tyrosine 3-monooxygenase is unique to the catecholamine-producing pathway and probably catalyses the flux-generating step in this pathway (Section B.2.d). Decarboxylation of DOPA produces dopamine, 2-(3,4-dihydroxyphenyl)-ethylamine, in a reaction catalysed by aromatic L-amino acid decarboxylase, a widely distributed enzyme which requires pyridoxal phosphate (vitamin B_6) as a prosthetic group. It is not specific for DOPA and is involved in the formation of other amines including 5-hydroxytryptamine (see below). In the parts of the nervous system that release dopamine as a neurotransmitter, (that is, dopaminergic neurones) no further metabolism occurs and the dopamine is stored in vesicles in the presynaptic nerve terminal. However, in noradrenergic neurones, the side-chain of dopamine is further hydroxylated (on the second, or β, carbon) to produce noradrenaline, 1-(3,4-dihydroxyphenyl)-2-aminoethanol. This hydroxylation is catalysed by dopamine β-monooxygenase (β-hydroxylase) and requires molecular oxygen and ascorbate or tetrahydrobiopterin. In the adrenergic neurones (and the adrenal medulla), the terminal amino group of noradrenaline is methylated, using *S*-adenosylmethionine as the methyl donor, to produce adrenaline, 1-(3,4-dihydroxyphenyl)-2-methylaminoethanol. This reaction is catalysed by noradrenaline N-methyltransferase.[6] The pathway appears to be similar in all tissues including brain, peripheral nerves and the adrenal medulla.

(b) *Control of rate of catecholamine synthesis*

Although no systematic studies on the equilibrium nature of the reactions of the catecholamine synthetic pathway have been carried out, it is likely that all are non-equilibrium. The evidence that tyrosine 3-monooxygenase catalyses the flux-generating step in the pathway is that an increase in the intracellular

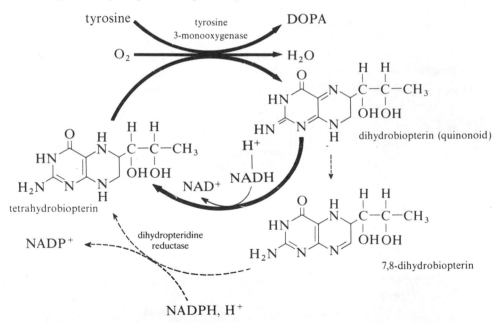

Figure 21.7 Details of the reactions catalysed by tyrosine 3-monooxygenase and related enzymes. The formation of tetrahydrobiopterin via dihydrofolate reductase is of minor importance. Note that the reaction catalysed by phenylalanine 4-monooxygenase (Chapter 10, Appendix 1) proceeds by a similar mechanism.

concentration of tyrosine does not lead to an increase in the concentration of the catecholamines in nervous tissue and that the K_m of the enzyme for tyrosine is about 30 μm whereas the tyrosine concentration is about 190 μM (Vacarro *et al.*, 1980). (However, there is a suggestion that this enzyme is not saturated with tyrosine and that changes in the blood concentration of this amino acid can change the concentration of catecholamines in the brain, see Anderson, 1981.) Since an increase in flux through the pathway is associated with a decrease in the concentration of tyrosine,[7] it is suggested that control of the monooxygenase can be achieved by external factors. The nature of these external regulators is not clear but several possibilities exist. First, since the enzyme is not saturated with its cosubstrate, tetrahydrobiopterin, this could function as an external regulator. However, it is not known whether the concentration of the cosubstrate varies sufficiently to play a regulatory role. Secondly, dopamine and noradrenaline are allosteric effectors which are known to decrease the activity of monooxygenase *in vitro*. Thirdly, cyclic AMP-dependent protein kinase phosphorylates the mono-oxygenase and this leads to activation.[8] Since stimulation of nervous tissue converts the enzyme to a form which has similar properties to the phosphorylated enzyme, nervous activity probably activates adenylate cyclase, increases the concentration of cyclic AMP and so activates the monooxygenase (Figure 21.8) (Ames *et al.*, 1978; Vulliet *et al.*, 1980). Finally, in addition to these mechanisms of acute control, the concentration of tyrosine 3-monooxygenase and other enzymes in the pathway in the neurone can also be changed to achieve chronic control. Increased nervous activity over a period of time leads to increased concentrations of these enzymes, whereas a decrease in nervous activity lowers the concentration. The mechanism by which nervous activity changes the extent

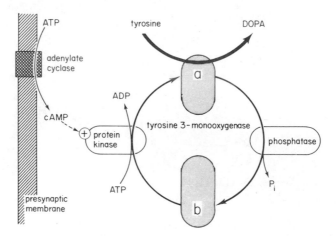

Figure 21.8 Regulation of tyrosine 3-monooxygenase by an enzyme interconversion cycle. The adenylate cyclase is activated by depolarization of the presynaptic membrane

of phosphorylation and the concentration of the monooxygenase may be the same. Receptors for the catecholamines are present on the presynaptic membrane (as well as the postsynaptic membrane) where they probably act as regulatory receptors. The extent of binding to these receptors will be proportional to the rate of release of catecholamine from the presynaptic terminal and will activate adenylate cyclase and thus raise the cyclic AMP concentration accordingly. This should increase the extent of phosphorylation of monooxygenase but, if at the same time it modified the mRNAs for the synthesis of these enzymes in the presynaptic neurone, this could also lead to increased rates of synthesis of these enzymes.

(c) *Catabolism of catecholamines*

The amines are degraded by a series of reactions which require the enzymes amine oxidase (formerly monoamine oxidase[9]), catechol methyltransferase and aldehyde dehydrogenase or alcohol dehydrogenase* (Figure 21.9). These enzymes have been found in all tissues investigated including nervous tissue, liver, intestine, and kidney. Catechol methyltransferase occurs in the cytosolic compartment of the cell and catalyses the transfer of the labile methyl group from S-adenosylmethionine to one of the phenolic groups of a catechol to form a methoxyphenol derivative (Figure 21.9). Whether amines are first attacked by amine oxidase or catechol methyltransferase is unclear, but the second possibility in supported by the lower K_m for the amine and its cytosolic location (Axelrod, 1975).

The products of catecholamine degradation are of some clinical importance because they appear in the cerebrospinal fluid, blood, and urine where measurement of their concentrations can indicate changes in turnover rates of the individual neurotransmitters.[10] Consequently, it is important to be aware that in the brain the reduction of the aldehyde derived from noradrenaline (or adrenaline) takes preference over oxidation so that the endproduct is mainly 3-methoxy-4-hydroxyphenylglycol (MOPEG) whereas the opposite is the case in peripheral tissues and 3-methoxy-4-hydroxymandelic acid ('VMA') is the major endproduct. The endproduct of dopamine metabolism is homovanillic acid (HVA).

3. **5-Hydroxytryptamine**

5-Hydroxytryptamine (also known as serotonin) is the best known of a group of indolealkylamines which also includes *N*-acetyl-5-methoxytryptamine (melatonin) which is produced from 5-hydroxytryptamine in the pineal gland (Section B.3.c). 5-Hydroxytryptamine is found in the brain, the enterochromaffin granules

---

* Readers may recall that one of the explanations for the neurological effects of alcohol is competitive inhibition of one or both of these enzymes by acetaldehyde, which is produced from alcohol in the liver—Chapter 11; Section F.5.b.

of the intestine and blood platelets but only in the brain does it have a neurotransmitter role.

(a) *Synthesis of 5-hydroxytryptamine*

Tryptophan is transported into the presynaptic nerve terminal by an active transport mechanism of the type described in Chapter 10; Section C.1. It is then hydroxylated by tryptophan 5-monooxygenase[11] in the cytosol to produce 5-hydroxytryptophan in a reaction that requires molecular oxygen and tetrahydrobiopterin. The decarboxylation to produce 5-hydroxytryptamine is catalysed by aromatic L-amino acid decarboxylase, a similar if not identical enzyme to the one involved in the catecholamine pathway so that whether a cell synthesizes 5-hydroxytryptamine depends on whether it possesses tryptophan 5-mono-oxygenase. The pathway for 5-hydroxytryptamine synthesis is depicted in Figure 21.10.

(b) *Control of rate of 5-hydroxytryptamine synthesis*

As with the catecholamine pathway, there have been no systematic studies on the equilibrium nature of the reactions but it is assumed that they are non-equilibrium. However, in the 5-hydroxytryptamine pathway there does *not*

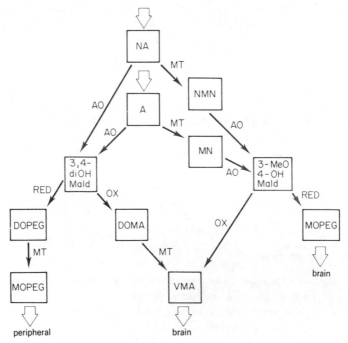

Figure 21.9   Pathways of catecholamine degradation. Dopamine degradation follows similar pathways to produce homovanillic acid (oxidized product) or 3-methoxy-4-hydroxyphenylethanol (reduced product). See opposite page for abbreviations.

Abbreviations and formulae of compounds cited in Figure 21.9

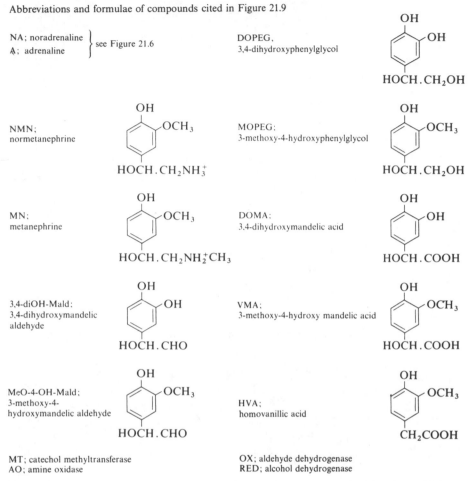

NA; noradrenaline	} see Figure 21.6	
A; adrenaline		

DOPEG, 3,4-dihydroxyphenylglycol

NMN; normetanephrine

MOPEG; 3-methoxy-4-hydroxyphenylglycol

MN; metanephrine

DOMA; 3,4-dihydroxymandelic acid

3,4-diOH-Mald; 3,4-dihydroxymandelic aldehyde

VMA; 3-methoxy-4-hydroxy mandelic acid

MeO-4-OH-Mald; 3-methoxy-4-hydroxymandelic aldehyde

HVA; homovanillic acid

MT; catechol methyltransferase
AO; amine oxidase

OX; aldehyde dehydrogenase
RED; alcohol dehydrogenase

Figure 21.9 (*cont.*)   Abbreviations and formulae of compounds cited in Figure 21.9

appear to be a flux-generating step; K_m of the monooxygenase for tryptophan is about $50\,\mu M$ and the concentration of tryptophan is about $25\,\mu M$. In addition, there is little evidence for a simple feedback control mechanism (that is, inhibition of the monooxygenase by the amine). This suggests that the rate of 5-hydroxytryptamine formation in the brain will depend on the concentration of tryptophan in the blood. Although this may remain constant under many circumstances, it can change. Some of the consequences of changes due to diet and other conditions are explored in Section D.

Although there is no evidence of a direct feedback mechanism in control of 5-hydroxytryptamine production, an increased rate of neurotransmitter release (caused by stimulation of the neurones) increases the flux through the synthetic pathway and *vice versa*. The explanation for this is the existence of regulatory

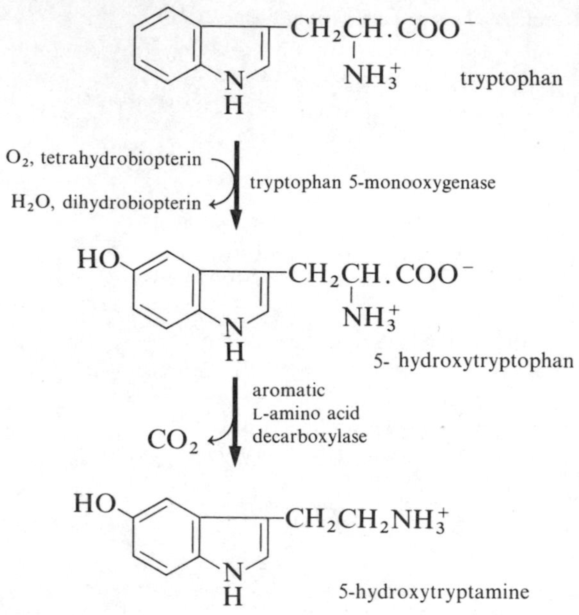

Figure 21.10   Pathway of 5-hydroxytryptamine (serotonin) synthesis from tryptophan

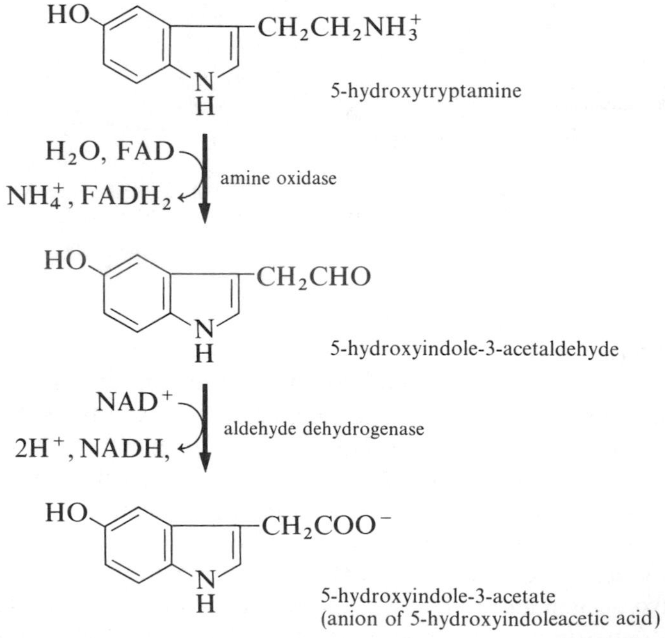

Figure 21.11   Pathway of 5-hydroxytryptamine degradation

receptors on the presynaptic membranes in the synapse; these receptors bind the neurotransmitter in the synapse and in this way assess the rate of neurotransmitter release.[12] Thus increased binding to the presynaptic membrane could increase the activity of tryptophan 5-monooxygenase, possibly through cyclic AMP and cyclic AMP-dependent protein kinase (as indicated for catecholamine synthesis, above).

(c) *Catabolism of 5-hydroxytryptamine*

5-Hydroxytryptamine is degraded to 5-hydroxyindoleacetic acid (5-HIAA) by the sequential action of amine oxidase and aldehyde dehydrogenase (Figure 21.11). Although this is probably the major route for catabolism of 5-hydroxytryptamine, methylation of the amine, or its derivatives, can occur as well as conversion to 5-hydroxytryptamine-*O*-sulphate in a reaction that requires 3'-phosphoadenosine-5'-phosphosulphate and a sulphotransferase enzyme (see Chapter 16; Section D.2). Sulphation of the amine causes loss of biological activity and promotes its excretion. In addition to the brain, three other tissues have high activities of the metabolizing enzymes, blood platelets, the liver and endothelial cells of the capillaries in the lung.

In the pineal gland, 5-hydroxytryptamine is converted to melatonin in two reactions catalysed by 5-hydroxytryptamine *N*-acetyltransferase and hydroxy-indole methyltransferase respectively (Figure 21.12). Melatonin is a hormone which, in rats at least, is involved in the regulation of the breeding cycle in accordance with environmental conditions.

4. 4-aminobutyrate (GABA)

4-Aminobutyrate, formerly known as γ-aminobutyrate (and hence as GABA) is an important neurotransmitter in nervous tissue in many different animals and is widely distributed in mammalian brain with high concentrations in the substantia nigra, globus pallidus and hypothalamus. Indeed, it is probably the major inhibitory neurotransmitter in mammalian brain. Of particular clinical interest is the suggestion that most tranquillizing drugs, including alcohol, barbiturates and benzodiazepines, may act by increasing the effectiveness of 4-aminobutyrate at postsynaptic receptors (Cowen and Nutt, 1982).

(a) *Synthesis of 4-aminobutyrate*

4-Aminobutyrate is produced by decarboxylation of glutamate in a reaction catalysed by glutamate decarboxylase:

$$H_2O + {}^-OOCCH_2CH_2\underset{\underset{NH_3^+}{|}}{CH} COO^- \longrightarrow {}^-OOCCH_2CH_2CH_2NH_3^+ + HCO_3^-$$

L-glutamate   4-aminobutyrate

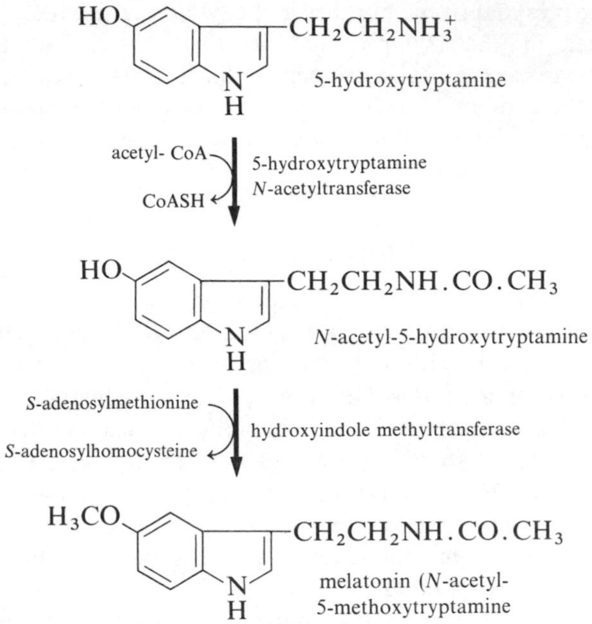

Figure 21.12   Pathway of melatonin synthesis from 5-hydroxytryptamine

However, in contrast to the synthesis of other amines discussed above, most of the glutamate required for the formation of 4-aminobutyrate is not taken up from the bloodstream but is produced within the neurone from 2-oxoglutarate, either by transamination or by amination via glutamate dehydrogenase. The source of the 2-oxoglutarate for either reaction is most likely the TCA cycle and the original precursor is probably glucose since it can give rise to both acetyl-CoA and oxaloacetate from pyruvate, via pyruvate dehydrogenase and pyruvate carboxylase, respectively. This means that the cycle can continue even if some of the intermediates are removed. This synthetic pathway is, therefore, identical to that described in Chapter 10; Section E. The suggestion has also been made that the glutamate arises from glutamine. Glutamate decarboxylase is cytosolic so that glutamate (which is synthesized with the mitochondria) must be transported out of the mitochondria prior to decarboxylation.

(b) *Catabolism of 4-aminobutyrate*

Aminobutyrate is degraded to succinate in two sequential reactions, catalysed by aminobutyrate aminotransferase (sometimes abbreviated to GABA-AT) and succinate-semialdehyde dehydrogenase as follows:

$$^-OOC.CH_2CH_2CH_2NH_3^+ \longrightarrow {}^-OOC.CH_2CH_2CHO$$

4-aminobutyrate

2-oxoglutarate   glutamate   succinate semialdehyde

$$\longrightarrow {}^-OOC.CH_2CH_2COO^-$$

NAD^+,H_2O $NADH,2H^+$ succinate

The combination of glutamate formation from 2-oxoglutarate, conversion to aminobutyrate and further metabolism of this to succinate can be seen to bypass some of the reactions of the cycle (oxoglutarate dehydrogenase and succinyl-CoA synthetase) and for this reason is sometimes known as the 4-aminobutyrate shunt (or more commonly the GABA-shunt—Figure 21.13). An interesting metabolic question is whether the shunt normally diverts a quantitatively important proportion of the TCA cycle flux in order to produce the neurotransmitter; estimates as large as 40% have been suggested. Comparison of the maximum activity of glutamate decarboxylase of human brain with the calculated TCA cycle flux suggests that maximal flux through the shunt could divert up to 30% of the TCA cycle flux.[13]

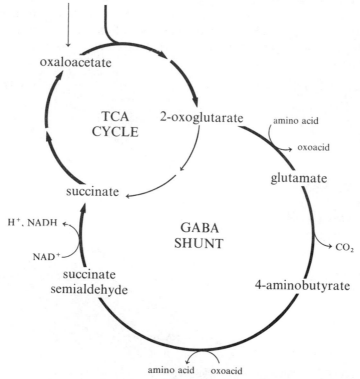

Figure 21.13   Metabolism of 4-aminobutyrate via the 'GABA shunt'

5. Histamine

Histamine, the amine of the amino acid histidine, is found in the mast cells, which are present in most tissues, especially connective tissue, and liberate the amine during an inflammatory response. In addition, there is a high content of histamine in the anterior and posterior lobes of the pituitary and in some areas of the brain where it is considered that it functions as a neurotransmitter. Histamine is produced from histidine by decarboxylation catalysed by a specific histidine decarboxylase. Histamine is stored in granules in the presynaptic neurones of the brain and its effect is probably terminated by a specific uptake mechanism and subsequent metabolism to N-methylimidazole acetic acid via reactions catalysed by histamine methyltransferase, amine oxidase, and aldehyde dehydrogenase (Figure 21.14).

Two types of histamine receptor, H_1, and H_2, have been identified by differential effects of agonists and antagonists (antihistamines). The effect of histamine on the smooth muscles of the bronchioles, which possess H_1 receptors, is bronchoconstriction which occurs during allergic asthma and hay fever. H_1-specific blockers (e.g. promethazine and diphenylhydramine) are of value in treating these conditions. On the other hand, the effect of histamine in stimulating secretion of acid in the stomach involves H_2 receptors. This can be inhibited by H_2-specific blockers such as cimetidine which are important in the treatment of gastric ulceration. Both H_1 and H_2 receptors have been found in the central nervous system which provides further evidence of the involvement of histamine as a central neurotransmitter.

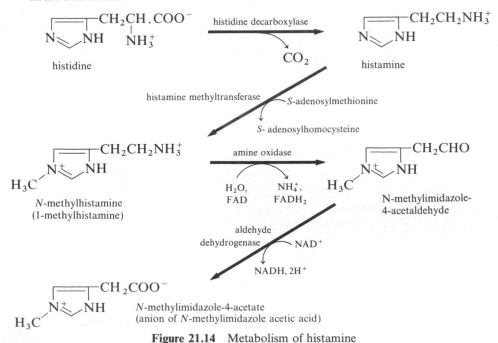

Figure 21.14 Metabolism of histamine

6. Glutamate and aspartate

Both glutamate and aspartate are excitatory neurotransmitters in the central nervous system but because of the similarity of the two molecules it is difficult to distinguish between their metabolic effects and neurotransmitter roles. Glutamate is probably synthesized by the reactions described in Section B.4 and aspartate from the transamination of oxaloacetate as described in Chapter 10; Section E. However, the immediate precursor of glutamate in both 'glutamate' and '4-aminobutyrate' neurones may be glutamine, which is accumulated by the glial cells, transported into the neurones and converted into glutamate by glutaminase (Quastel, 1979).

The termination of the action of these amino acids is almost certainly achieved by uptake into the presynaptic neurone or the glial cells. However, it is unclear if they are metabolized separately from the normal metabolic processes. Undoubtedly, excess glutamate can be converted to glutamine but it could also be converted to 2-oxoglutarate, by transamination, and then oxidized. In view of the involvement of these amino acids in metabolic processes, localization in vesicles would appear to be a necessary prerequisite for their neurotransmitter role. However, an interesting alternative is chemical compartmentation since brain contains high concentrations (5–10 mM) of N-acetylaspartate and N-acetylglutamate and hydrolysis of these components could produce the neurotransmitters in the axon terminals when required:

$$^-OOC.CH_2CH.COO^- \qquad\qquad ^-OOC.CH_2CH.COO^-$$
$$\underset{\substack{|\\ NH.CO.CH_3}}{} \xrightarrow{\quad H_2O \quad CH_3COO^-\quad} \underset{\substack{|\\ NH_3^+}}{}$$

N-acetylaspartate — aspartate

$$^-OOC.CH_2CH_2CH.COO^- \qquad\qquad ^-OOC.CH_2CH_2CH.COO^-$$
$$\underset{\substack{|\\ NH\ CO\ CH_3}}{} \xrightarrow{\quad H_2O \quad CH_3COO^-\quad} \underset{\substack{|\\ NH_3^+}}{}$$

N-acetylglutamate — glutamate

7. Glycine

Glycine is an inhibitory neurotransmitter in the central nervous system including the spinal cord. The metabolism of this amino acid is poorly understood due to the diversity of pathways involved (see Chapter 10; Appendix 1). However, glycine is probably produced from serine by the action of serine hydroxymethyltransferase and metabolised to carbon dioxide, ammonia and N^5N^{10}-methylenetetrahydrofolate by glycine cleavage enzyme (glycine synthase) (Figure 21.15). Glycine is likely to be stored in vesicles in the presynaptic neurone and its postsynaptic action terminated by a specific high-affinity uptake process into the presynaptic nerve terminal.

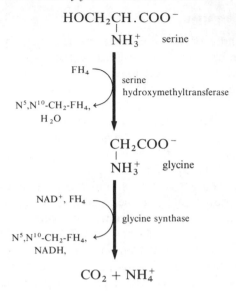

$$HOCH_2CH.COO^-$$
$$|$$
$$NH_3^+ \quad \text{serine}$$

FH$_4$

serine
hydroxymethyltransferase

N^5,N^{10}-CH$_2$-FH$_4$,
H$_2$O

$$CH_2COO^-$$
$$|$$
$$NH_3^+ \quad \text{glycine}$$

NAD$^+$, FH$_4$

glycine synthase

N^5,N^{10}-CH$_2$-FH$_4$,
NADH,

$$CO_2 + NH_4^+$$

Figure 21.15   Metabolism of glycine

8. Taurine

Taurine is a sulphur-containing amino acid that possesses a sulphonyl rather than a carboxyl group. It is present in the diet but can also be produced from cysteine by oxidation to cysteine sulphinate followed by decarboxylation to hypotaurine and finally oxidation to form taurine (Figure 21.16). Taurine may act as an inhibitory neurotransmitter in the brain stem, spinal cord and retina. The high concentrations present in these areas suggest that it is localized in vesicles in the presynaptic nerve terminals. A high affinity uptake process has been identified in the brain and retina, which supports this mechanism of the neurological action of taurine.

9. ATP

Perhaps surprisingly, some nerves use ATP as a neurotransmitter, and have thus been termed the purinergic nerves (Burnstock, 1977). It seems likely that ATP is stored in vesicles, is released into the synapse in a similar manner to other neurotransmitters and binds to specific ATP receptors on the postsynaptic membrane. Of metabolic interest is the mechanism by which the neuro-transmitter action of ATP is terminated. This could be achieved by a specific uptake process but, because of the large negative charge on ATP, this would cause problems in maintaining electrical neutrality. On the other hand, ATP could be readily removed by hydrolysis within the synaptic cleft and there is some evidence that in nervous tissue ATP can be degraded to adenosine and inosine (see Figure 21.17). The fate of the adenosine or inosine is unclear but they may be re-incorporated into nucleotides by 'salvage' pathways—see Chapter 18; Section B.1.

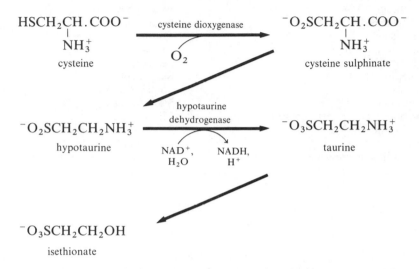

$$HSCH_2CH.COO^- \xrightarrow[\quad O_2 \quad]{cysteine\ dioxygenase} {}^-O_2SCH_2CH.COO^-$$

cysteine

cysteine sulphinate

hypotaurine dehydrogenase

$$^-O_2SCH_2CH_2NH_3^+ \xrightarrow[\substack{NAD^+,\\ H_2O} \quad \substack{NADH,\\ H^+}]{} {}^-O_3SCH_2CH_2NH_3^+$$

hypotaurine

taurine

$$^-O_3SCH_2CH_2OH$$

isethionate

Figure 21.16   Metabolism of taurine

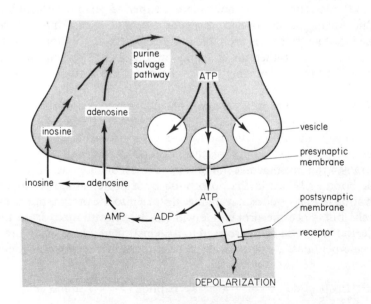

Figure 21.17   Reactions occurring at a purinergic synapse

Table 21.1. Some peptides present in the brain*

Adrenocorticotropic hormone (ACTH)	α-Melanocyte-stimulating
Angiotensin II	hormone (α-MSH)
Bombesin	Neurophysin(s)
Bradykinin	Neurotensin
Carnosine	Oxytocin
Cholecystokinin	Prolactin
β-Endorphin	Secretin
Enkephalins	Sleep peptide(s)
Gastrin	Somatostatin
Glucagon	Substance-P
Gonadotropin releasing hormone	Thyrotropin
Growth hormone	Thyrotropin-releasing
Insulin	hormone (TRH)
Luteinizing hormone	Vasoactive intestinal peptide

* Data from Kreiger and Martin (1981).

10. Peptides

Since about 1970, it has become clear that a number of peptides play informational roles in the central nervous system. Some are listed in Table 21.1 (Kreiger and Martin, 1981, Snyder 1982). It is not always clear whether they play a direct neurotransmitter role or whether they influence the effectiveness of other neurotransmitters (*Lancet* editorial, 1982b). Most, if not all of these peptides, also occur in the intestine, where they function either as general or local hormones. This dual location does not necessarily indicate a functional link between brain and the intestine (since it is probable that these peptides do not cross the blood brain barrier) but rather an economy in the use of specific informational molecules (in the same manner that cyclic AMP has various roles in different cells).

It is likely that the peptides are synthesized from the constituent amino acids via a messenger RNA-ribosomal system (Chapter 18; Section C) and that they are produced in the form of an inactive precursor (propeptide) which is stored in vesicles in the nerve endings of the neurone. Release of the peptide as a neurotransmitter must involve not only exocytosis but also cleavage of a specific peptide bond or bonds in the propeptide by a proteinase(s) (endopeptidase). Inactivation of the peptides may involve uptake into the presynaptic neurones or glial cells but, since the propeptide will not be resynthesized from the active peptide, the latter will be hydrolysed to its constituent amino acids, presumably by specific peptidases. The over-all process can be represented as follows:

$$\text{Propeptide} \xrightarrow{\text{proteinase}} \text{active peptide} \xrightarrow{\text{peptidases}} \text{amino acids}$$
$$\searrow$$
$$\text{amino acid(s)}$$

In view of the probable role of peptides in a number of neural activities including the ability to learn, memory, control of feeding, temperature regulation, the sensation of pain and possibly in the aetiology of certain neurological disturbances (e.g. epilepsy, schizophrenia, Huntington's disease, Alzheimer's disease—Edwardson and McDermott, 1982), there is considerable interest in establishing not only the precise action and localization of the neuro-peptides but the nature of the specific proteinases involved in their synthesis and peptidases involved in their inactivation. The clinical use of peptides or their analogues in neurological disorders is hampered by their low permeability across the blood–brain barrier, so that there is considerable pharmaceutical interest in the control of the specific proteinases or peptidases.

The group of peptides, known as the endorphins, have already aroused considerable clinical interest. These are known as the opiate (or opioid) peptides since they appear to mimic the action of morphine or its analogues[14] and control the response to impulses in nerve fibres that relay painful sensations.[15] Electrical stimulation of discrete areas of the brain, especially the ventrolateral periaqueductal grey matter of the midbrain, produces analgesia, and high concentrations of receptors for endorphins (opioid receptors) are found in this area of the central nervous system. A large number of naturally occurring endorphins have been described. They may all belong to three main chemical families; the enkephalins (which comprise two pentapeptides, Leu-enkephalin and Met-enkephalin, that differ only in their C-terminal amino acid; that in Leu-enkephalin is leucine whereas that in Met-enkephalin is methionine—Figure 21.18); extended versions of the Met-enkephalins up to the 31 amino acid, β-endorphin; and extended versions of Leu-enkephalin up to the 17 amino acid, dynorphin. All endorphins are produced from larger propeptides (pro-opiocortins). For example, β-endorphin is produced by proteolysis of β-lipotropin which itself is part of a larger precursor molecule that also contains ACTH (see Figure 21.19).

There are at least two different opiate receptors in the central nervous system

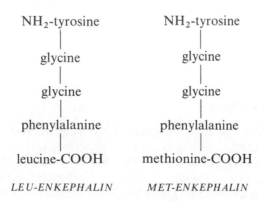

Figure 21.18 Amino acid sequences of enkephalins

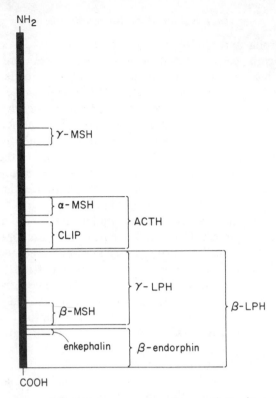

Figure 21.19 Structure of the protein pro-opiomelanocortin to show sequences of other peptide hormones contained within (Nakanishi *et al.*, 1979).
Abbreviations: MSH, melanocyte-stimulating hormone; ACTH, adrenocorticotropin; CLIP, corticotropin-like intermediate-lobe peptide; LPH, lipotropin

which are referred to as μ and δ. The larger endorphins bind to both types of receptors but the pentapeptide enkephalins bind only to the δ receptors. The μ receptors, to which morphine predominantly binds, are known to be involved in the analgesic response whereas δ receptors are involved in seizure activity and perhaps other behavioural responses (Paterson *et al.*, 1983). So far there is only one antagonist (receptor blocker) to the opiate receptors, naloxone, and this is used to investigate whether physiological or biochemical actions are opiate mediated (Sawynok *et al.*, 1979).

11. Other neurotransmitters

There is no doubt that cyclic nucleotides (cyclic AMP and cyclic GMP) and prostaglandins are involved in neurotransmission but it is unclear if they are true transmitters (as defined in Section A.3.a) or if they modify the effectiveness of other neurotransmitters. The action of the cyclic nucleotides may be to increase the activities of specific protein kinase enzymes which cause changes in the state of phosphorylation of key proteins, especially transport proteins, in the neurones.

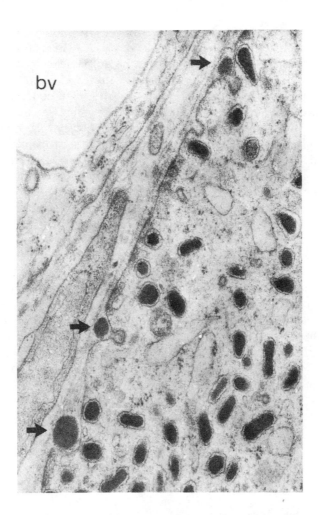

Figure 21.20 Electronmicrograph (magnification 50000) of the apical part of an adrenal chromaffin cell from the golden hamster. Note the membrane-limited chromaffin granules, which show electron-dense cores. This is an adrenaline-storing cell and the dense cores represent proteins (mainly chromagranins and dopamine β-hydroxylase) because the adrenaline is lost during fixation. Some of the dense cores can be seen lying under the basement membrane in the extracellular space (arrows) following exocytosis. Following exocytosis, the membrane of the chromaffin granule gains a bristle coat on its cytoplasmic face; it is then withdrawn into the cell: b.v., lumen of a blood vessel. Micrograph provided by Drs. I. Benedeczky and A. D. Smith Department of Pharmacology, University of Oxford.

C. ORGANIZATION AND ACTION OF NEUROTRANSMITTERS

1. Compartmentation

Since nervous activity requires the immediate release of neurotransmitter, most neurotransmitters are stored in membrane bound vesicles in the presynaptic nerve terminal. This provides the immediate source of neurotransmitter for release into the synapse without the necessity of a high concentration in the cytosol, which could lead to side reactions including neurotransmitter degradation. However, such compartmentation need not be physical; some neurotransmitters are present in a precursor form (e.g. propeptides).

Much of our knowledge of physical compartmentation has been gained from the study of the adrenal medulla which stores adrenaline (and some noradrenaline) and releases it in an apparently similar manner to the way in which neurotransmitters are released (Njus *et al.*, 1981). The catecholamines are stored in vesicles known as chromaffin granules (Figure 21.20). The membrane of the vesicle is similar to other membranes in that it is composed of phospholipid, cholesterol, and protein, and neurotransmitters are transported into the vesicles by an energy-dependent transport process.[16] The high concentration of an electrically-charged neurotransmitter within the vesicle poses the problem of neutralization of the charge. This is achieved, for the positively charged neurotransmitters (e.g. amines, acetylcholine), by the presence of ATP (and ADP) in the vesicle. Studies on the chromaffin granule have established a

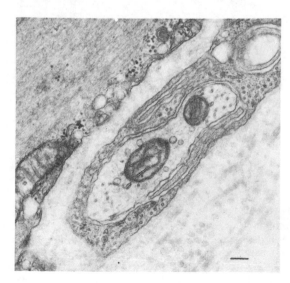

Figure 21.21   Axon terminal enclosing a Schwann cell showing vesicles, neurotubules and two mitochondria. The line is 0.1 μm in length. The electronmicrograph was provided by Dr. M. Fillenz, Laboratory of Physiology, University of Oxford

catecholamine/ATP concentration ratio of 4.5 (which agrees reasonably well with that predicted from the four negative charges on the ATP molecule neutralizing the single charge on each amine molecule). In addition to these molecules, vesicles contain protein[17] plus calcium and magnesium ions which may help to stabilize the amine-nucleotide complexes, perhaps in the form of high molecular weight aggregates.

The neurotransmitter that is stored in the vesicles arises either from synthesis within the nerve terminal or from that taken up from the synaptic cleft. However, what proportion of the neurotransmitter that is transported back into a presynaptic terminal is re-incorporated into vesicles and how much is metabolized is unclear. The fact that degradation products of the neuro-transmitters appear in the cerebrospinal fluid and that their concentration increases with electrical activity suggests that a proportion must be metabolized. Finally, it should be noted that both the transport of neurotransmitters and ATP into the vesicle and the uptake of neurotransmitter from the synapse require ATP, accounting, in part, for the mitochondria present in nerve terminals (Figure 21.21). Under most conditions, the fuel for this ATP production will be glucose and it is noteworthy that the carrier which transports glucose into the nerve terminals of the brain has an unusually high affinity for glucose (0.1 mM compared to the usual value of about 10 mM—see Bachelard, 1978). This indicates that glucose transport into the nerve terminal will be a flux-generating process so that the rate of glycolysis in these cells will be independent of small changes in the blood glucose concentration.

2. Neurotransmitter release

The mechanism of neurotransmitter release in the central nervous system is considered to be similar to that of catecholamine release from the adrenal medulla, which has been more intensively studied. Here, the process involved is exocytosis, in which fusion of the vesicle membrane with the presynaptic membrane results in the discharge of the entire contents of the vesicle[18] (Figure 21.22). The precise details of this process are the province of the membrane

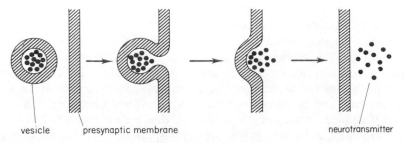

vesicle   presynaptic membrane   neurotransmitter

Figure 21.22 Schematic representation of neurotransmitter release by exocytosis from a presynaptic vesicle

biochemist and the biophysicist (see Borgese *et al.*, 1979). An important physiological question, however, is the nature of the link between stimulation of the presynaptic nerve and release of the transmitter, that is, the stimulus-secretion coupling. As with many other secretory processes, the coupling mechanism involves calcium ions and the sequence of events may be as follows (see Douglas, 1974; Rasmussen *et al.*, 1979). When the action potential reaches the nerve terminal it increases the rate of entry of calcium ions from the extracellular space as well as that from the endoplasmic reticulum. The increase in concentration of calcium ions in the nerve terminal increases the contractile activity of the microtubules to which neurotransmitter vesicles are attached and this causes the movement of vesicles towards the membrane. When a vesicle reaches the cell membrane it fuses with it and the contents of the vesicle are released completely into the synaptic cleft.

A similar mechanism is involved in the control of the rate of insulin release from the pancreas (Chapter 22; Section D.2.b) and it may be involved in the control of the secretion rates in many secretory tissues both exocrine and endocrine.

3. **Neurotransmitter action on the postsynaptic membrane.**

The release of the neurotransmitter into the synaptic cleft is the first part of the process of chemical communication between nerves. The neurotransmitter then diffuses across the synaptic cleft in order to transfer information to the postsynaptic neurone. How is this information received? Knowledge of other communication and regulatory systems provides the answer that interaction of the postsynaptic membrane with the neurotransmitter is achieved through a specific receptor. The basis of the interaction is probably that of equilibrium-binding so that, at any given time, the proportion of neurotransmitter molecules bound to the receptor is controlled solely by the concentration of neurotransmitter in the synaptic cleft. The receptor-neurotransmitter complex must communicate with the postsynaptic neurone via an effector process that produces a functional change in this cell. Two such effector systems have been proposed. First, the effector could be an ion-transport protein in the postsynaptic cell membrane, the activity of which would be increased as a result of transmitter-receptor interaction, so that the permeability to the given ion rises. This permeability change would either increase the likelihood of action potential generation (e.g. by increasing the permeability to sodium ions) or decrease the likelihood (e.g. by increasing the permeability to chloride ions). A second possibility is that the effector could be the enzyme adenylate cyclase, the activity of which could be increased by transmitter-receptor interaction. This would elevate the concentration of cyclic AMP and thus increase the activity of protein kinase and cause an increase in the rate of phosphorylation of key proteins in the postsynaptic cell (see for example, Creese, 1981). Details of adenylate cyclase effector systems are given in Chapter 22; Section A.4.

4. Termination of neurotransmitter action

The role of the transmitter is to convey information quickly and transiently across the synapse and to amplify the electrical signal. To this end, the effect of the transmitter must be rapidly terminated. The process of termination means removal of the neurotransmitter from the synapse but, to appreciate how this is achieved, it is first necessary to consider the process of neurotransmission in metabolic terms. It is likely that there is always a low rate of neurotransmitter release into the synaptic cleft but the concentration within the synapse is maintained very low by the activity of the terminating mechanisms. The marked increase in the rate of transmitter release during nervous activity is such that it exceeds the rate of termination and the neurotransmitter concentration in the synapse will increase so that binding to the postsynaptic receptor increases. However, the increased rate of release is only transient and, as this decreases, the rate of termination of the neurotransmitter will exceed the rate of release so that the concentration of neurotransmitter in the synapse rapidly decreases. This causes dissociation from the receptors and the effect of the neurotransmitter ceases. (The process is, in principle, similar to control of the changes in the concentration of a secondary messenger of hormones, see Chapter 22; Section A.4.)

At least two mechanisms exist for the removal of the neurotransmitter from the synapse. First, enzymatic metabolism of the neurotransmitter occurs in the synaptic cleft; for example, acetylcholine is broken down to acetate and choline by acetylcholinesterase which is bound to the postsynaptic membrane. Secondly, the neurotransmitter can be taken up from the synapse by the presynaptic nerve terminal, the glial cell or the postsynaptic neurone. Such uptake processes are usually specific, high-capacity, high-affinity and energy-requiring. In both the postsynaptic and glial cells, the neurotransmitter must be metabolized to endproducts that can be excreted.

D. EFFECT OF DIET ON NEUROTRANSMITTER CONCENTRATIONS

Since the reactions involved in the formation of some neurotransmitters in the brain may not be saturated with pathway-substrate, the flux-generating step for their formation must exist in other tissues. This step may be either the transport of amino acids (or other precursors) across the absorptive cells of the intestine or the provision of precursor in the diet. In the latter case, dietary manipulations may be capable of modifying the concentration of neurotransmitter in the central nervous system and thus influencing behaviour. The phenomenon is illustrated by reference to tryptophan and the formation of 5-hydroxytryptamine in serotoninergic neurones (Wurtman, 1983).

There is evidence that an increase in the amount of tryptophan in the diet can increase the concentration of 5-hydroxytryptamine in the brain (see Anderson,

1981). Diets low in tryptophan can do the opposite and, in those for whom corn, which is particularly low in tryptophan, is the main source of dietary protein, the 5-hydroxytryptamine concentration in the brain may be below normal (Fernstrom and Wurtman, 1971). Since 5-hydroxytryptamine is probably involved in the control of sleep, food intake, mood, pain sensitivity, and pituitary hormone release, there is considerable interest in the possibility that simple dietary manipulations can influence these centrally controlled activities. For example, tryptophan administration could be of value in treatment of insomnia or depression but the problem with this is that tryptophan is metabolized by the liver. This can be avoided if an inhibitor of tryptophan 2,3-dioxygenase (e.g. nicotinic acid) is administered together with the additional tryptophan. Indeed success has been claimed for this latter approach in the treatment of depression (Chouinard *et al.*, 1977) and chronic pain (Hosobouchi, 1979; Bender, 1982).

The control of the tryptophan concentration in the presynaptic nerve terminals is likely to involve factors other than variations in the blood concentration of this amino acid. For example, a considerable proportion of tryptophan in the blood is bound to albumin but it is the free concentration that governs the rate of entry into the neurone. A number of compounds (e.g. other amino acids, certain drugs, long-chain fatty acids) can displace tryptophan from its binding site on the albumin and hence increase its free concentration. Secondly, branched-chain amino acids are transported into the neurone by the same carrier that transports the aromatic amino acids (including tryptophan—see Table 10.5). Hence a decrease in the concentration of branched-chain amino acids relative to that of the aromatic amino acids will favour entry of the latter into brain neurones and could result in an increase in the concentration of 5-hydroxytryptamine in the presynaptic neurone. Such an explanation has been put forward to account for the coma that characterizes malfunction of the liver (hepatic coma) since the rate of metabolism of aromatic amino acids, which occurs in the liver, is decreased whereas that of the branched-chain amino acids, which occurs in the muscle, remains unchanged (see Chapter 12; Section C.3).

It is tempting to speculate that other conditions that are known to facilitate sleep or produce the feeling of fatigue may be caused by changes in this amino acid concentration ratio; e.g. fatigue that follows sustained physical activity. Such activity increases the rate of utilization of the branched-chain amino acids by muscle and hence decreases their concentrations in the plasma in relation to those of the aromatic amino acids. Such changes have been shown after a 100 km race (Décombaz *et al.*, 1979). Similarly, insulin increases the uptake of branched-chain amino acids by muscle but has no direct effect on that of tryptophan by the liver so that the acute effect of this hormone would be to decrease the branched-chain amino acid/aromatic amino acid concentration ratio. If this raised the 5-hydroxytryptamine concentration in the brain, it could account for the phenomenon of postprandial sleepiness (see Munro and Crim, 1980).[19]

E. NEUROLOGICAL DISEASE AND NEUROTRANSMITTERS

One of the most important developments from our knowledge of neurotransmitter biochemistry has been a better understanding of some neurological disorders, which has led to improvements in their treatment. Moskowitz and Wurtman (1975) have put forward the lines of evidence which indicate that a given disorder could be due to a disturbance in one or more facets of neurotransmitter metabolism described above. They are summarized as follows.

1. The concentration of the neurotransmitter, one of its metabolites or the activity of one or more enzymes involved in metabolism of the neurotransmitter in samples of brain taken (post mortem) from patients suffering from the given disorder is different from those of normal subjects. Perhaps the best example is the decrease in concentration of dopamine in specific areas of the brain of patients suffering from Parkinson's disease (see below).
2. The concentration of a metabolite of the neurotransmitter in the cerebrospinal fluid of the patient is different from that of normal subjects. Similar differences may also be observed in the blood or urine but these are more difficult to interpret since they may be due to changes in peripheral metabolism of the neurotransmitter.
3. The condition of the patient improves in response to a drug that is known to change the concentration, or the effectiveness, of a known neurotransmitter. Drugs are now available that can mimic or antagonize the action of a neurotransmitter on a postsynaptic receptor, interfere with the release of the neurotransmitter into the synapse, inhibit its uptake from the synapse or interfere with its synthesis or metabolism in the presynaptic neurone.
4. More direct experiments can be performed on experimental animals. This is particularly so if an animal model of the disorder can be produced; changes in neurotransmitter concentration in the brain can be measured and the effect of drugs can be investigated directly.

1. Depression and mania

Depression and mania are the two major affective disorders (disorders which affect the feeling-state in contrast to the processes involved in thought). The symptoms of depression usually include feelings of profound sadness, diminished appetite, general fatigue, insomnia, and loss of sexual desire; when the disorder becomes very severe, patients may become abnormally pessimistic, become convinced that they have an incurable physical illness and may make plans for suicide, (see Watts, 1973; Snyder, 1980). Depressive disorders may follow a single unhappy event, for example, bereavement, or a series of minor misfortunes. On the other hand, there may be no obvious reason for the depression, in which case it is known as endogenous depression. Mania is in most respects the opposite of depression: the patient is abnormally cheerful, overactive, over confident, lacks self criticism, and may embark on grandiose plans. However, they share some common symptoms, for example, both kinds of patients sleep badly, (Snyder, 1980). Some patients can suffer from marked swings in mood (manic-depressive illness), for example, from a period of depression to one of mania but the period between the different moods is usually fairly long (several months or years).

The possibility that the affective disorders could be due to a decrease in the concentration of catecholamines or of 5-hydroxytryptamine in the brain was raised in the 1950s when hypertensive patients who were treated with reserpine became severely depressed. It was known that this drug decreased the brain concentrations of monoamines in experimental animals. Consequently, the amine-hypothesis was proposed which stated that depression is due to a decreased concentration, or decreased effectiveness, of catecholamines or 5-hydroxytryptamine in specific areas of the brain.[20] In this case, it was logical to consider use of drugs that reduce the rate of degradation of the monoamines and not surprisingly, amine oxidase inhibitors[21] (known as monoamine oxidase or MAO inhibitors) were the first to be employed (Fig. 21.23). However, although of some benefit in treatment of depression, the use of these drugs can be dangerous. Amine oxidase activity is not restricted to brain; it is present in the cells of the small intestine, stomach, liver, and kidney where it plays an important role in degrading amines present in food or produced by bacterial action in the large intestine. For example, cheese, chocolate, citrus fruits, marmite, and red wine contain large amounts of tyramine (the decarboxylation product of tyrosine). If peripheral amine oxidases are inhibited, these dietary amines enter the general

$$CO.NH.NH.CH(CH_3)_2$$

iproniazid

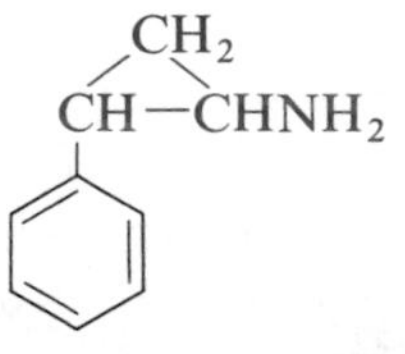

$$CH_2$$
$$CH-CHNH_2$$

tranylcypromine

$$CH_3$$
$$CH_2NCH_2C\equiv CH$$

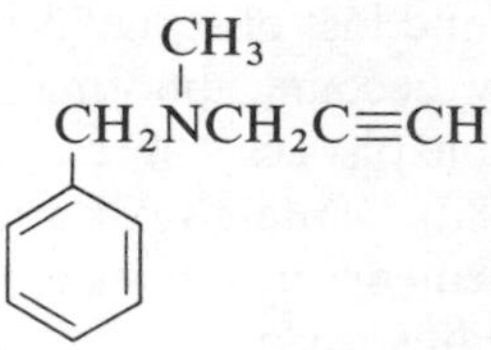

pargyline

Figure 21.23   Inhibitors of amine oxidase (MAO inhibitors) which are used, or have been used, clinically

circulation and cause release of noradrenaline from sympathetic nerve endings and adrenaline from the adrenal medulla. This results in peripheral vasoconstriction and increased cardiac output which causes severe hypertension and can lead to headache, palpitations, subdural haemorrhage, stroke, or myocardial infarction. Patients on monoamine oxidase inhibitors must, therefore, be advised not to eat such foods.

Another means of increasing the effectiveness of the monoamines is to increase their concentration in the synapse by inhibiting their rate of uptake from the synaptic cleft. A class of antidepressant drugs known as tricyclic anti-depressants[22] may act in this manner[23] (Figure 21.24). They have been much more successful in the treatment of depression than amine oxidase inhibitors (for review, see Van Praag, 1982).

In cases of depression that are unresponsive to drugs, patients may be given electroconvulsive therapy (ECT). There is some evidence that such treatment increases the effectiveness of amine neurotransmitters in the brain, possibly by increasing the affinity of the postsynaptic receptor (*British Medical Journal* editorial, 1977; Costain *et al.*, 1982).

A remarkable finding is that administration of lithium salts (lithium carbonate or lithium aspartate) has been extremely successful in treating mania. Unfortunately, there is still little evidence to support a mechanism of action for lithium but one possibility is that it facilitates the uptake of monoamines from the synapse. This would suggest that mania is due to hyperactivity of the monoamine-neurotransmitter system but there is no evidence for this.

2. Parkinson's disease

Patients with Parkinson's disease suffer from muscular rigidity, rhythmic tremor, slowness of movement and stiffness of the trunk. *Post mortem* studies on the brain of patients suffering from this disease show degenerative changes in specific areas of the brain, particularly the nigrostriatal system, (that is, the substantia nigra, the

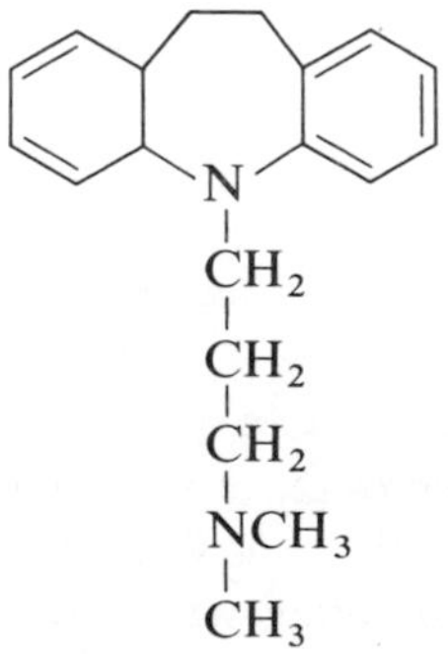

imipramine 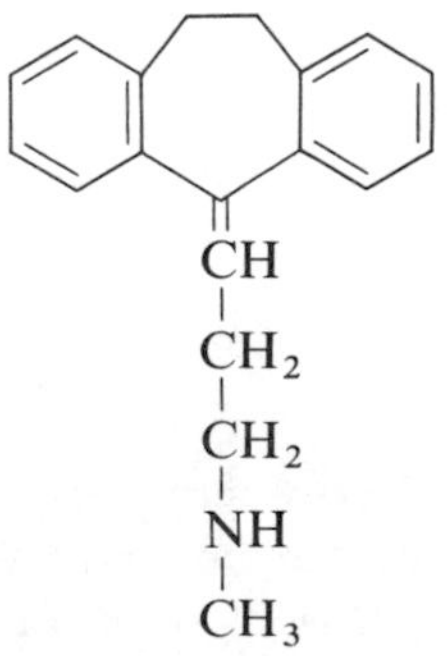   nortriptyline

Figure 21.24   Representative tricyclic antidepressants in clinical use

striatum and the pathway linking the two), that could account for most of the clinical manifestations. The major neurotransmitter in this area of the brain is dopamine and *post mortem* studies in the early 1960s showed that the concentration of dopamine and its degradation product homovanillic acid, together with the activities of the enzymes of the dopamine synthetic pathway, were severely decreased in patients with Parkinson's disease compared to normal subjects (Hornykiewicz, 1973; Marsden, 1982). It is, however, unclear what leads to the degeneration of neurones in this specific area of the brain. The question was raised as to whether administration of this neurotransmitter to the patients would be beneficial. Since dopamine does not enter the central nervous system (probably because it does not traverse the endothelial cells of the capillaries in the brain—the blood/brain barrier) it was necessary to use the precursor L-DOPA.[24] When very large quantities of DOPA were given to such patients (6–8 g each day) there was a remarkable improvement in their condition. Large quantities were necessary because L-DOPA is decarboxylated to dopamine in the peripheral tissues.[25] This problem has been partially overcome by administration of α-methyldopahydrazine, a decarboxylase inhibitor that does not enter the central nervous system (Papavasiliou *et cl.* 1972).

3. Huntington's disease

Huntington's disease or chorea is a degenerative disorder of the brain characterized by progressive dementia and involuntary movements (chorea) from early to middle adult life, terminating in death 12 to 15 years after the appearance of the early symptoms. Huntington's disease is one of several that can be classified under the general term of dementia (Rossor, 1982). There is loss of cells mainly in the putamen and caudate nucleus of the cerebral cortex (the same part of the brain as that of the nigrostriatal system) where the concentrations of 4-aminobutyrate and acetylcholine, together with the activities of the enzymes glutamate decarboxylase and choline acetyltransferase, are decreased. In addition, there may be increased activity of dopamine in the striatal system and changes in the concentrations of a number of neuro-peptides (see Kreiger and Martin, 1981). The cause of these changes is not known. A review of the changes in neurotransmitter concentrations and function in dementia, in general, is given by Rossor (1982).

4. Epilepsy

Epilepsy is a collective term describing a group of chronic convulsive disorders which can be divided into grand mal and petit mal. Grand mal epilepsy is characterized by periods of unconsciousness together with motor convulsions, whereas petit mal epilepsy is characterized by loss of consciousness, usually for short periods, without loss of motor control (and is usually confined to children). Epilepsy may be caused by a sudden spread of electrical activity in the brain, and since 4-aminobutyrate is a major inhibitory neurotransmitter in brain, a defect in its production or release may be the cause of this disease. Indeed one of the more successful anticonvulsive drugs, sodium valproate, inhibits the enzymes involved in the degradation of 4-aminobutyrate (both aminobutyrate aminotransferase and succinate-semialdehyde dehydrogenase). However, this *in vitro* inhibition is observed only at high concentrations of the drug, so that, *in vivo*, it is more likely to interfere with the uptake of 4-aminobutyrate from the synapse (for reviews, see Browne, 1980; Spero, 1982).

5. Schizophrenia

Perhaps the most fascinating possibility in this field is that schizophrenia too may be caused by impairment of neurotransmitter metabolism. Schizophrenia is not an easy disease to define but there is no doubt that schizophrenics have difficulty with thought processes rather than experience problems of mood. The disorders of thought are usually seen as poor association, and the connections between statements are confused or may be non-existent. In addition, the patient may suffer from delusions; persecution delusions occur in the paranoid schizophrenic. Auditory hallucinations are common; the patient hears voices which are usually derogatory and critical. The emotional responses of the patient are either blunted or inappropriate. Unlike the affective disorders, which are usually episodic, schizophrenia is usually chronic.

One hypothesis to explain this condition is hyperactivity of the dopamine transmitter system in the limbic areas of the forebrain, including the central amygdaloid nucleus, nucleus accumbens, olfactory tubercle and parts of the frontal cerebral cortex (Snyder, 1982). Agents that cause dopamine release (e.g. amphetamines) can induce schizophrenia-like psychosis in normal subjects and can exacerbate the symptoms in schizophrenic subjects. Conversely, drugs that ameliorate the symptoms of schizophrenia are known to be dopamine antagonists. Hyperactivity could be caused by overproduction of dopamine, a decreased rate of uptake from the synapse or an increased number or affinity of postsynaptic dopamine receptors. Since there does not appear to be an increased concentration of dopamine or its metabolites in the cerebrospinal fluid of schizophrenic patients, and since drugs which are used to treat such patients (the antipsychotic or neuroleptic drugs, Figure 21.25) probably bind to dopamine receptors without eliciting a response, it seems likely that the dopamine receptors are abnormal. Binding studies suggest that there is an increase in the number of receptors (see Creese, 1981). Since the antipsychotic drugs will probably reduce dopamine binding in all parts of the brain it is not surprising that they give rise to symptoms of Parkinson's disease (due to blocking of dopamine receptors in the nigrostriatal system) and may result in infertility and amenorrhea. This is explained by dopamine receptor blockade in the anterior pituitary where this neurotransmitter is involved in the gonadotrophin-stimulated release of LH and FSH (Chapter 20; Section E.4.c).

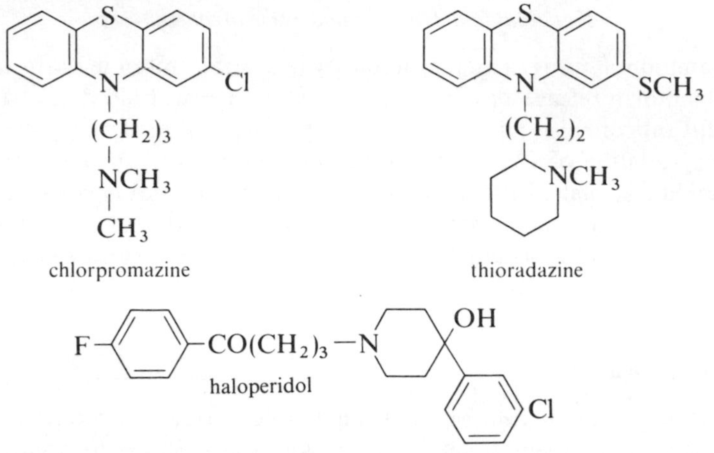

chlorpromazine

thioradazine

haloperidol

Figure 21.25 Three drugs used in the treatment of schizophrenia

6. Migraine

Migraine is defined as a recurrent headache with which are associated at least two of the following: unilateral headache, nausea, visual or other neurological disturbance, family history of migraine and a history of travel sickness, asthma, eczema, and hay fever. Although many of the neurotransmitters described in this chapter have been implicated as the cause of migraine (see Crook, 1981), it seems likely that the initial cause is a decrease in blood flow to specific parts of the brain, due to release of 5-hydroxytryptamine by blood platelets (Hanington, *et al.*, 1981). It is suggested that some stimulus causes aggregation of platelets which release 5-hydroxytryptamine. This results in local vasoconstriction of the intracerebral arteries (since this neurotransmitter causes contraction of the arterial smooth muscle) and the resulting cerebral hypoxia causes disturbances which are responsible for the aura or prodromal phase. This vasoconstriction results in the production of local vasodilators (e.g. adenosine, see Chapter 7; Section B.1.b) which, although normalizing the blood flow in the intracerebral arteries, causes marked vasodilation in the extracerebral vessels. These changes, in concert with changes in concentrations of peptides or prostaglandins that influence pain receptors in the brain, result in the headache and other symptoms.

The evidence that certain foods can trigger a migraine attack could be explained by effects on neurotransmitter release or metabolism in the brain. For example, a number of foods contain tyramine (see Section E.1) which is known to cause release of other amines (dopamine, noradrenaline and 5-hydroxytryptamine) from both nerve terminals and platelets. This release could initiate the sequence of events that results in the migraine attack. Elimination of such foods from the diet can decrease the number of headaches and compounds that discourage platelet aggregation (e.g. aspirin) may prevent such attacks (*Lancet* editorial, 1982c).

NOTES

1. The Nernst equation relates the potential difference at equilibrium (ΔE) across a membrane, which is permeable to an ion, to the concentration of that ion:

$$\Delta E = \frac{RT}{nF} \ln \frac{[I^+]_o}{[I^+]_i}$$

   where R is the gas constant, T the absolute temperature, F the Faraday constant, and $[I^+]_o$ and $[I^+]_i$ the concentration of an ion bearing a positive charge n outside and inside the cell respectively. For a negatively charged ion the ratio $[I^-]_i/[I^-]_o$ is substituted.

2. For the monoamines, localization involves exposure of slices of nervous tissue to formaldehyde (as a gas) to form a fluorescent complex. The complexes with different monoamines can be readily distinguished.

3. An increase in the amount of choline in the diet could result in an increased concentration in the blood and availability for acetylcholine synthesis in the brain. This could be of therapeutic value in some degenerative conditions. Since fish is rich in choline, the belief that 'fish is food for the brain' may have some validity.

4. This pathway involves, first, the formation of phosphatidylethanolamine (Chapter 17; Section C). The phosphatidylethanolamine then undergoes three successive methylations (with S-adenosylmethionine acting as methyl donor) to form phosphatidylcholine. This is mostly used as a phospholipid but could be hydrolysed to produce choline for acetylcholine synthesis.

5. Tyrosine 3-monooxygenase can be inhibited by a structural analogue of the substrate α-methyl-β-tyrosine, which is used experimentally to deplete stores of catecholamines. Dopamine β-monooxygenase, which requires copper ions for activity, is inhibited by copper-chelating compounds such as disulphiram (antabuse).

6. The tissue with the highest activity of noradrenaline N-methyltransferase is, not surprisingly, the adrenal medulla. In the brain, the highest activities are found in the brain stem and hypothalamus.

7. A two-fold stimulation of the flux through the pathway of dopamine synthesis decreases the tyrosine content of phaeochromocytoma cells by 50% (Vacarro *et al.*, 1980).

8. This activation probably results from an increase in the K_i for dopamine or noradrenaline so that higher concentrations of pathway-product are required to inhibit the enzyme. At normal tissue concentrations of the monoamines, phosphorylation would remove virtually all the feedback inhibition. Phosphorylation may also activate the enzyme by increasing the V_{max} or decreasing the K_m for the cosubstrate, tetrahydrobiopterin.

9. Because of its intrinsic importance in the brain, amine oxidase has been studied in some detail (for review, see Tipton, 1975). It contains covalently bound FAD and at least 90% of the enzyme is localized on the outer membrane of the mitochondria. More recently it has been discovered that there are two forms of this enzyme, termed type A and type B, which can be distinguished by their substrate specificity and sensitivity to inhibitors. Type A oxidizes 5-hydroxytryptamine, noradrenaline, octopamine, tyramine, and dopamine, and is inhibited by the drug clorgyline. Type B oxidizes phenylenthylamine, benzylamine, tryptamine, tyramine, and dopamine, and is inhibited by the drug deprenyl. Most tissues contain both types of enzymes but the small intestine contains mainly type A and the liver mainly type B.

10. More accurate turnover rates can be obtained in experimental animals from measurement of the increased amine concentration in the brain after an amine oxidase inhibitor has been injected to prevent breakdown of the amine. Alternatively, the

increase in the concentration of the degradation product in the brain can be measured after the injection of the drug probenecid, which prevents release of the degradation product.

11. Tryptophan 5-mono-oxygenase is inhibited both *in vitro* and *in vivo* by *p*-chlorophenylalanine which can be used in experimental animals to deplete neurones of 5-hydroxytryptamine.
12. There is indirect evidence for presynaptic receptor regulation of 5-hydroxytryptamine synthesis in that D-lysergic acid diethylamide (LSD) which is a receptor-blocker for this neurotransmitter, reduces the flux through this pathway (Haigler, 1981).
13. The maximal activity of glutamate decarboxylase in the cortex of human brain is 0.17 μmol.min^{-1}.g^{-1} fresh wt. (E. Philips and E. A. Newsholme, unpublished work). The calculated flux through the TCA cycle based on oxygen consumption (Table 4.7) is about 0.6 μmol.min^{-1}.g^{-1}.
14. Morphine is an alkaloid obtained from the seed capsule of the poppy *Papaver somniferum*. Its structure, together with those of related compounds, is given in Figure 21.26. The clinical interest in morphine stems from its potent analgesic properties

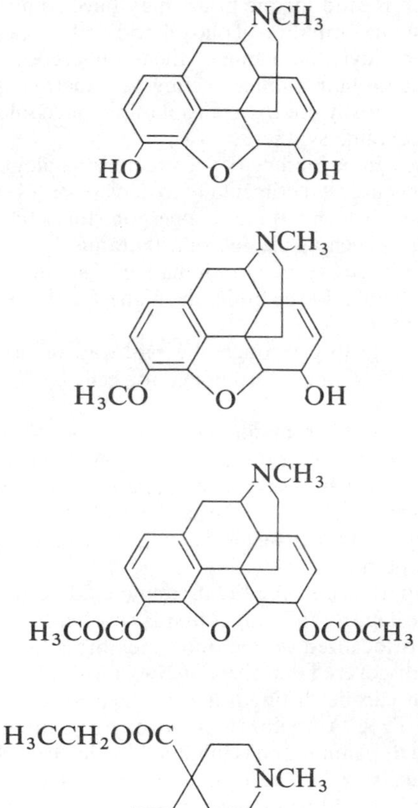

morphine

codeine

heroin

pethidine

Figure 21.26 Structural formulae of some narcotic analgesics (unfortunately accompanied by addiction) but other effects include respiratory depression, cough suppression, vomiting, constipation, and sedation. The depressive effects of morphine on the central nervous system differ from those of other agents, such as barbiturates, in that they are selective. This suggests that binding to specific receptors is involved. The prediction that morphine mimicked the action of endogenous compounds which interacted with these receptors in the central nervous system to modify pain led to the discovery of endorphins and enkephalins (see Hughes, 1975; Hughes and Kosterlitz, 1983).

15. Pain is transmitted by sensory nerves to pain centres in the central nervous system where the integration and interpretation of the impulses take place. Hence pain can be relieved peripherally by interfering with transmissions of the impulses or centrally by interfering with integration and interpretation. Morphine acts centrally not only in altering the pain threshold but also in altering the attitude to pain. Although pain can be still perceived by the patient it is less effective, maybe because morphine has an euphoric effect. The highest concentration of binding sites for opiates is found in the periaqueductal grey area but it is possible that other sites are also involved in the analgesic action (e.g. limbic system, hypothalamus, basal ganglia) (Hosobuchi *et al.*, 1978; Parkhouse *et al.*, 1979).

16. This would explain how metabolic poisons readily interfere with neurotransmitter function, since if the compound cannot enter the vesicle it will be metabolized. The mechanism of the energy-linking is that extragranular ATP is hydrolysed by an ATPase that pumps protons into the granule and these are exchanged for catecholamine and ATP which are transported into the granule (Njus *et al.*, 1981). The drug reserpine inhibits the energy-linked transport of amines into vesicles.

17. The major protein found in the acetylcholine-containing vesicles is termed vesiculin and in the chromaffin granule, chromogranin-A. In addition, the chromaffin granules contain dopamine β-mono-oxygenase which suggests that some of the final stages of synthesis of noradrenaline may occur within the vesicles.

18. The main evidence implicating exocytosis in the secretion of adrenaline from the chromaffin granule is that it occurs together with the release of ATP, chromogranin-A and the enzyme dopamine β-mono-oxygenase, all of which are present in the granules. Nonetheless, there is an alternative view that, for some neurotransmitters, the vesicles may act only as a store and, close to the cell membrane, the contents of the vesicle are released so that the high transient concentration of neurotransmitter favours rapid diffusion into the synaptic cleft.

19. The consumption of carbohydrate-containing drinks late in the evening will cause an increase in the plasma concentration of insulin and could change the amino acid ratio to produce high neuronal concentrations of 5-hydroxytryptamine and encourage sleep. A combination of tryptophan and carbohydrate-containing drinks could be particularly beneficial for patients suffering from mild insomnia (Anderson, 1981).

20. Indirect support for this hypothesis is provided from the observation that chronic abuse of depressant drugs (e.g. alcohol, barbiturates, other sedatives) which lower the concentration of monoamines is associated with depressive illness.

21. Iproniazid was one of the first drugs used, since treatment of tuberculosis patients with this drug was observed to produce euphoria.

22. Tricyclic antidepressants include imipramine, desipramine, amitriptyline, and nortriptyline.

23. This suggested mode of action and also that of the monoamine oxidase inhibitors is based on *in vitro* experiments in which the drugs act very rapidly. Unfortunately, in practice, the antidepressants have no obvious benefit for approximately 4–10 days after treatment is started, which does raise doubts about the suggested mechanism of action.

24. The effectiveness of L-DOPA therapy depends upon the fact that it enters the pathway after the flux-generating step, tyrosine 3-mono-oxygenase. If it entered the pathway prior to the step, no matter how large the increase in substate concentration it would not influence the rate of endproduct formation (Chapter 7; Section A.1).

25. One consequence of this large dose, and of peripheral metabolism, is the increased rate of use of S-adenosylmethionine which provides the methyl group required in dopamine degradation (see Section B.2.c). This in turn leads to an increased rate of synthesis of methionine from homocysteine in a reaction catalysed by tetrahydro-pteroyltriglutamate methyltransferase, a vitamin B_{12} requiring enzyme (see Appendix A). Patients on L-DOPA, therefore, frequently benefit from a supplement of vitamin B_{12} in their diet.

CHAPTER 22

HORMONES AND METABOLISM

Hormones are chemical signals that are synthesized in and secreted from endocrine glands into the circulation to elicit specific responses from one or more target tissues. They play a fundamental role in the regulation and integration of metabolism and their effects on metabolism have already been described where appropriate. This chapter outlines the principles of hormone action, which are illustrated by reference to hormones that have major effects on metabolism (catecholamines, thyroxine, steroids, and insulin). Because insulin has such a diverse range of metabolic effects and is so important in the integration of metabolism, a section summarizing its major metabolic effects is also included. Although many hormones modify metabolism, the mechanism of action of some is at the level of protein synthesis and any effects on metabolism are achieved through changes in the concentration of key proteins in the cell (see Section B.1).

Hormones, in common with neurotransmitters, initiate their effects on a target tissue by binding to a receptor and they can be classified into those hormones that bind to cell-surface receptors and those that bind to intracellular receptors. Two examples of each are given; catecholamines and insulin bind at the cell surface whereas steroids and thyroxine bind intracellularly. The control of secretion of the steroid hormones, which is synonymous with the control of their synthesis since there is no storage of the steroid hormones in the secretory cells, was briefly described in Chapter 20, and is discussed in detail in this chapter. Insulin, adrenaline, and thyroxine, however, are stored after synthesis and the control of the rates of secretion of these hormones from the stored form will also be discussed in this chapter. Finally, a brief description of the methods for measuring the concentrations of hormones, so very important in current clinical practice, will be described.

A. GENERAL PRINCIPLES OF HORMONE ACTION

1. Terminology

The terms, effect, action and function, are often used indiscriminately in the endocrine field and to avoid confusion they are defined as follows (see Levine and Goldstein, 1955; Wool, 1964). An effect of a hormone is an experimental observation made either *in vitro* or *in vivo*. In initial studies, all observed effects will be of equal importance. However, as more effects are discovered it should be possible to organize them into a hierarchical system in which one effect is

813

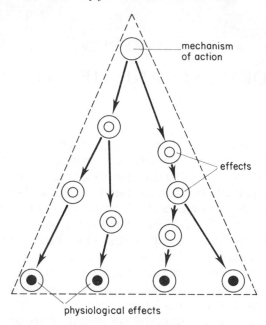

mechanism
of action

effects

physiological effects

Figure 22.1   Hierarchical representation of hormone action

dependent upon another. Eventually, such an approach should lead to the elucidation of the primary effect, that is, the effect from which *all* other effects can be causally related. This primary effect is known as the mechanism of action of the hormone. Recent studies have established that this primary effect is the binding of the hormone to a specific receptor either on the cell membrane or within the cell. Since an effect higher in the hierarchical system may give rise to more than one effect lower in the system, the hormone profile can be considered as a pyramid, with the mechanism of action at the tip of the pyramid and the physiological effects on the base (Figure 22.1). Even if the hormone has more than one mechanism of action (that is, two separate receptors), the principle still holds since it is possible to organize the effects into two separate hierarchical systems.

The function of the hormone is the inference made according to its physiological effect(s). Unfortunately, the description of the function at any time will be limited by the system under study. For example, when a factor isolated from the anterior pituitary was observed to promote growth, especially in animals from which the pituitary had been removed, it was given the name growth hormone, implying a major role in growth. If, however, the plasma fatty acid concentration, rather than growth, had been measured, the hormone might have been termed lipid-mobilizing hormone. It must be realized that the function of a hormone is an integrative theory based on various observed effects and must always be open to revision in the light of new effects or reinterpretation of known effects.

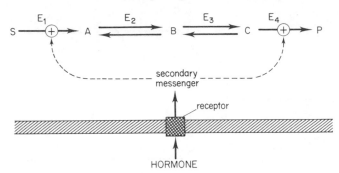

Figure 22.2 Generalized mechanism of action of a hormone on a metabolic pathway. E_1 is flux-generating; E_4 is also non-equilibrium; E_2 and E_3 are near-equilibrium. Note that some hormones can traverse cell membranes

2. Locus of hormone effects in metabolic pathways

If the role of a hormone is to change the flux through a metabolic pathway, which reactions should the hormone (or its intracellular messenger) affect? This question is answered by reference to the principles of metabolic control described in Chapter 7; Section A.1. If the physiological pathway is linear (without branch points), the hormone must change the activity of the enzyme that catalyses the flux-generating step. Then the flux through the remaining reaction of the pathway could be changed simply by internal regulation. However, the latter may not be sufficient to transmit the flux through some or all of the non-equilibrium reactions so that the hormone may additionally modify the activities of enzymes catalysing these reactions (Figure 22.2).

Several examples of hormones affecting flux-generating steps have been given in this text; adrenaline, glucagon and vasopressin modify hepatic phosphorylase activity (see Chapter 11; Section D); a large number of hormones may modify the activity of triacylglycerol lipase in adipose tissue (see Chapter 8; Section C.1); insulin, the important anabolic hormone, increases the activity of pyruvate dehydrogenase in adipose tissue (an important flux-generating step in the lipogenic pathway—Chapter 17; Section A.9); the activity of glycogen synthase in muscle and probably liver (the important flux-generating enzyme in glycogen synthesis[1]) and the initiating process in protein synthesis (Chapter 18; Section D.2).

It is possible for a hormone to change the flux through a series of reactions without affecting the flux-generating step if it acts at a branch point. If a single pathway branches into two and the only flux-generating step is in the single pathway, a change in maximum activity or the K_m of one of the enzymes at the branch point will change the distribution of the flux at that branch (Figure 22.3). For example, there is an important branch point in the metabolism of fatty acyl-CoA in liver, leading either to oxidation or esterification, and insulin plays an

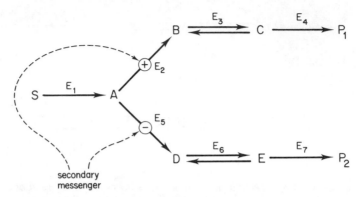

Figure 22.3 Action of a hormone at a branch-point to determine the product formed. E_1
is the flux-generating reaction and both E_2 and E_5 are non-equilibrium

important role in controlling the fate of the fatty acyl-CoA at this branch point
(Chapter 8; Section D.1.b).

Since a hormone, or more likely its secondary messenger (see below), influences
enzyme activities in which it does not participate as a substrate or products, the
effect on the enzyme must be allosteric (Chapter 3; Section C.5). This limits direct
effects of the hormone to non-equilibrium reactions (Chapter 7; Section A.2). The
allosteric effect could occur directly upon the pathway-enzyme but more likely it
will be an effect on an interconverting enzyme as part of an interconversion cycle
that controls the activity of the pathway-enzyme (see below).

3. Hormones and receptors

How can a hormone influence the metabolism in a cell? The hormone binds to
some component of the cell and this complex must then directly or indirectly
influence metabolism. The component is known as a receptor, which can be found
either on the outer surface of the cell membrane, so that the hormone does not
have to enter the cell, or within the cell, in which case the hormone (e.g. steroids,
thyroxine) must be lipid-soluble to traverse the cell membrane.

(a) *Properties of receptors*

Several of these properties can be used to identify a specific hormone receptor.

(i) *Affinity of the receptor for the hormone*   The concentrations of hormones in
the blood are normally very low (10^{-9} to 10^{-12} M)* so that the binding constant
of the receptor for the hormone must be large (in fact, dissociation rather than
binding constants[2] are usually given and are frequently in the range 10^{-9} to
10^{-10} M).

---

* The concentration range of most metabolic intermediates is 10^{-3} to 10^{-4} M.

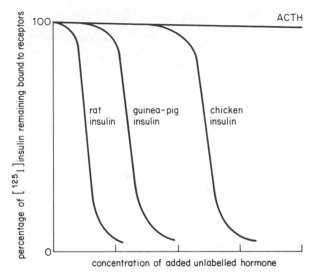

Figure 22.4 The displacement of ^{125}I-labelled rat insulin from rat cell membranes by other hormones. The more similar the structure of the unlabelled hormone is to rat insulin, the more readily it will displace the rat insulin

(ii) *Specificity* The specificity of a receptor for a hormone is usually high although it is not necessarily absolute. Specificity is usually indicated by binding a radiolabelled hormone to the purified receptor (or to membranes bearing the receptor) and following its displacement from the receptor by increasing concentrations of other non-labelled hormones (Figure 22.4).

(iii) *Receptor binding and biological activity* If the binding of the hormone to the receptor is the mechanism of action of that hormone, the extent of the biological response elicited should be directly proportional to the amount of hormone bound. Although this relationship is often found, it is not always so since post-receptor events can influence the information transfer from hormone-receptor complex to the target enzyme (see below).

(iv) *Number of receptors* Binding studies enable not only the dissociation constant for the receptor to be measured but also the number of receptor sites.[3] The number of receptor sites is usually large and considerably greater than required to elicit the maximum biological response of the hormone. In other words, the maximum response is achieved when only a small proportion of receptors (perhaps 10%) are occupied. This has led to the idea of 'spare' receptors but this is a misleading description since it suggests that these receptors have no role. A greater number of receptors are present to ensure a satisfactory response to a change in hormone concentration. If the binding of hormone to receptor follows a hyperbolic curve, greatest *sensitivity* is achieved on the linear part of the curve, that is, when the effect (and hence the number of receptors occupied) is less than half maximal (Chapter 7; Section A.2).

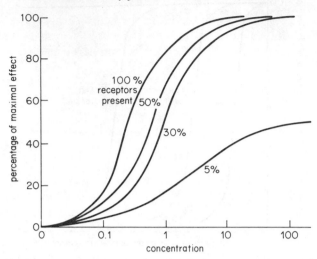

Figure 22.5   Effect of receiver number on the response to different concentrations of the hormone. As the number of receptors decreases, there is a progressive loss in sensitivity but the maximal effect remains attainable until the number of receptors falls to a very low level (e.g. 5%). Based on predictions by Olefsky and Kolterman (1981)

(v) *Regulation of receptor number*   One way by which a tissue can vary its response to a hormone is to change the number of receptors. For example, a decrease in the number of receptors will increase the concentration of the hormone required to produce the same biological response (Figure 22.5). For some hormones, the number of receptors can be varied quite quickly. What controls the number of receptors? Although there are probably several factors that can do this, the concentration of the hormone itself plays an important role. Exposure of a tissue to a high concentration of a hormone for even a short period (e.g. 1–2 hr) can reduce the number of receptors. This phenomenon has been termed 'down-regulation' and its role may be to prevent too large a response for a prolonged period of time by a tissue to a given hormone. It has been demonstrated for a number of hormones including insulin, growth hormone, thyrotrophin-releasing hormone, catecholamines, gonadotrophins, oestrogens and glucagon. This phenomenon has been described in earlier chapters; it may be partially responsible for reducing tissue sensitivity to insulin in obesity and in non-insulin-dependent diabetes (Chapter 19; Section C.4) and for reducing the sensitivity of the anterior pituitary to oestradiol in women taking oral steroid contraceptive pills (Chapter 20; Section G.2).

(vi) *Two types of receptor*   The use of specific agents that bind to a receptor and either elicit a response (agonists) or prevent the response of the natural hormone (antagonists or blockers) has been very important in identifying more than one type of receptor and, therefore, more than one point or mechanism of action of a hormone. The approach was first used for the neurotransmitter acetylcholine; some effects of acetylcholine can be mimicked by the drug nicotine and not muscarine, whereas others can be mimicked by muscarine and not nicotine. For hormones, it is well established that two types of receptor exist for the catecholamines and are known as the α- and β-adrenergic receptors. Several neurotransmitters can interact with more than one receptor, for example, dopamine, histamine (Michell, 1979). It is possible that receptors for a given hormone in one tissue may be slightly different for the same hormone in another tissue. Such tissue differences are usually analysed by the use of agonists and antagonists.[4] This is a field in which pharmacology has been, and is, making a major contribution to clinical practice since it enables drugs to be developed that have tissue specificity (e.g. cardioselective β-blockers, have their major effect by binding to the β-adrenergic receptors on the heart and not other tissues).

(b) *Receptor-effector interaction*

The next question is how the binding of hormone to the receptor can elicit a metabolic response? The receptor-hormone complex may, in fact, be the effector system which produces an effect on the target enzyme or system. This appears to be the case for steroid and thyroid hormones, which have intracellular receptors (see Section B). However, for hormones that bind to an extracellular receptor, the receptor-hormone complex must produce a change in a further membrane component (the effector) which results in an intracellular event. For many hormones the effector is the enzyme adenylate cyclase, and the intracellular event is cyclic AMP production. Here, the β-adrenergic receptor (for catecholamines) is taken as an example to illustrate the details of the interaction.

Binding of hormones to the β-adrenergic receptor activates the enzyme adenylate cyclase, which is located on the inner side of the cell membrane, and this catalyses the intracellular conversion of ATP to cyclic AMP:

$$ATP + H_2O \rightarrow \text{cyclic AMP} + PP_i$$

In the uncomplexed state (that is, with no hormone bound) the receptor has little or no effective contact with the cyclase enzyme and its activity is low. When the hormone-receptor complex is formed, however, this has a high affinity for adenylate cyclase, so that a hormone-receptor-adenylate cyclase complex is formed, which increases the enzyme activity and leads to a rise in the intracellular concentration of cyclic AMP. However, the process of activation of adenylate cyclase by the hormone-receptor complex is not as straightforward as this. The adenylate cyclase system comprises, in addition to a catalytic subunit, a coupling subunit that interacts with the receptor (see Figure 22.6). (In the other words, the receptor-effector-complex comprises three distinct types of protein component— receptor, adenylate cyclase, and a coupling protein, Rodbell, 1980). But in addition to coupling, this subunit has regulatory properties due to its ability to bind the nucleotide GTP so that it is known as the nucleotide-binding

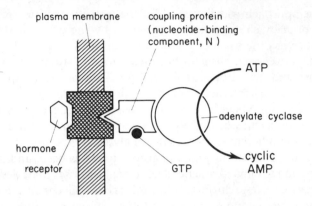

Figure 22.6   Schematic representation of the adenylate cyclase system

component, N). It can also catalyse the hydrolysis of GTP to GDP and P_i (that is, it has GTPase activity). The presence of GTP bound to this component is necessary to ensure maximal activity of adenylate cyclase when the hormone receptor-coupling protein-adenylate cyclase complex is formed. However, association of the adenylate cyclase complex with the hormone-receptor complex not only increases cyclase activity but also increases the GTPase activity of the coupling subunit. This lowers the concentration of bound GTP and so, in turn, decreases the adenylate cyclase activity. This is known as desensitization. Full activity of the cyclase can only be restored by dissociation of the coupling protein from the receptor, exchange of GDP for GTP and reassociation with the hormone-receptor complex. The importance of this GTPase activity may lie in providing a *cellular* mechanism for regulating the activity of adenylate cyclase, so that the cell can modify its response to the external hormone. For example, if the GTPase activity could be increased, the adenylate cyclase activity, once it has been activated by association with the hormone-receptor complex, would rapidly be decreased, so that the increase in the concentration of cyclic AMP within the cell would be small and hence the metabolic response to the hormone should be minimal. On the other hand, if the GTPase activity were to be decreased, the increase in adenylate cyclase activity would be maintained and the increase in the concentration of cyclic AMP should be large so that the effect of the hormone could be very marked. Indeed this is precisely the mechanism of action of certain toxins. Cholera toxin, for example, increases and maintains a high activity of adenylate cyclase and a high concentration of cyclic AMP in cells of the small intestine and this inhibits their ability to absorb water, resulting in severe diarrhoea and hence dehydration, which can be sufficiently severe to cause shock and even death. The toxin consists of two peptide chains, A and B; peptide B binds to the cell surface and A enters the cell where it catalyses the transfer of ADP-ribose from the nucleotide, NAD^+, to the GTP-coupling protein as follows:

GTP-coupling protein + NAD$^+$ $\longrightarrow$

ADP-ribosyl-GTP-coupling protein + nicotinamide + H$_2$O

This is known as ADP-ribosylation* (see Chapter 7; Note 6; Ueda *et al.*, 1982) of the protein and inhibits the GTPase activity of the complex and so maintains a high activity of adenylate cyclase (Gill and Meren, 1978).

In Chapter 20, the effect of parathyroid hormone on the activity of the renal 1α-hydroxylating enzyme system that converts 25-hydroxycholecalciferol into 1α,25-dihydroxycholecalciferol is described. Parathyroid hormone has this effect by binding to a receptor on the kidney cell membrane, activating adenylate cyclase and thereby raising the cyclic AMP concentration within the cell. However, it has been shown that in some patients suffering from pseudohypoparathyroidism (in which there is only a small increase in cyclic AMP concentration in response to parathyroid hormone) the GTP-coupling protein is defective† (Farfel *et al.*, 1980; Spiegel *et al.*, 1982). How far other hormone defects may be due to poor receptor-effector coupling remains to be determined.

(c) *Fate of the receptor-hormone complex*

The effect of a hormone in the body is achieved by an increase in the plasma concentration of hormone which increases the concentration of the hormone-receptor complex which activates the effector system. For the termination of hormone action, at least three mechanisms are known.

1. A decrease in the rate of secretion of hormone decreases the plasma concentration which allows the hormone to dissociate from the receptor.
2. After a short period of interaction, the hormone-receptor complex is shed from the cell membrane either into the bloodstream or into the cell. In the former case, the complex is degraded in the tissues such as liver, lung and kidney and in the latter case by the lysosomes of the cell. In addition to terminating the hormone action, this mechanism can regulate the number of receptors when the hormone concentration is high (that is 'down-regulation', see above). The migration of the receptor-hormone complex into the cell (known as 'internalization') and the degradation of the complex within the cell may be the most important means of terminating the action of the peptide hormones (King and Cuatrecasas, 1981).
3. The hormone is inactivated while bound to the receptor and the inactive molecule released into the bloodstream for complete degradation by other tissues.

*Diphtheria toxin similarly consists of two polypeptides, A and B, and A enters the cell where it catalyses ADP-ribosylation of elongation factor-2. The latter is involved in translocation of the growing peptide chain on the ribosome which requires the hydrolysis of GTP (Chapter 18; Section C.3). ADP-Ribosylation of this protein inhibits elongation and hence protein synthesis (Gill, 1978; see Ueda *et al.*, 1982). Modification of such toxic peptides so that they prevent protein synthesis in selected cells (e.g. tumour cells) is of current interest (Edwards and Thorpe, 1981; Lancet editorial, 1983c).

†It is likely that the ability to bind GTP is reduced, although the same effect could be achieved if the activity of the GTPase was increased.

It is likely that all three mechanisms play some part in the termination of hormone-mediated processes, but the importance of each probably varies from hormone to hormone. For those hormones that have a short half-life in the blood, variations in the hormone concentration can occur quickly so that the mechanism (1) could be a satisfactory means of control. Most of the peptide hormones (e.g. insulin, glucagon, growth hormone, somatomedins, pituitary gonadotrophins, TSH, ACTH) have half-lives of about 20 min. However, other hormones have much longer half-lives; about 60 min for cortisol and about seven days for thyroxine. For these hormones, changes in response due to dissocation of the hormone will occur only gradually and degradation of the hormone-receptor complex is probably the major mechanism for terminating hormone action.

4. Secondary messengers and hormone action

The identification of cyclic AMP as the intracellular compound which was produced in the tissue after incubation with adrenaline and which could cause activation of glycogen phosphorylase provided the first insight into the molecular mechanism of the action of hormones. It led to the concept of the secondary messenger in hormone action discussed by Sutherland *et al.* (1965). It is very likely that there are additional secondary messengers which include cyclic GMP, calcium ions, adenosine and prostaglandins although the evidence for their involvement is not as strong as that for cyclic AMP. Nonetheless, the principles that govern the secondary messenger system are as valid for these compounds as they are for cyclic AMP. To appreciate these principles, a hypothetical situation is first considered.

A secondary messenger (X) can be considered as an intermediate in a metabolic pathway as follows.

$$S \xrightarrow{E_1} X \xrightarrow{E_2} B \longrightarrow$$

where E_1 is the reaction producing the messenger and E_2 is the utilizing reaction which converts the message into an inactive metabolite, which itself is further metabolized. Several important points arise from this.

1. The rate of conversion of S to B and further will normally be in a steady state, even in the absence of the hormone, due to the flux-generating nature of E_1; this results in a steady-state concentration of the secondary messenger X.
2. Enzyme E_2 catalyses a non-equilibrium reaction; if it were near-equilibrium, changes in the concentration of metabolite B could influence the concentration of the messenger.
3. It is likely that the concentration of X will be below the K_m of E_2 so that changes in the concentration of X will regulate the activity of E_2 (that is, a first order process). This provides not only a very simple system for the regulation of the concentration X but a very precise system (see below).

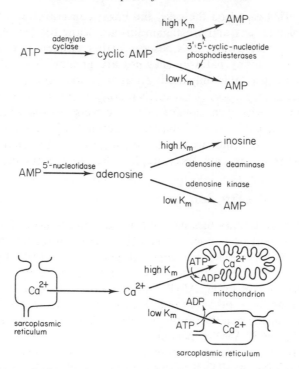

Figure 22.7 Removal of secondary messengers by alternative pathways with high and low K_m

4. The steady state concentration of X can be changed by modifying the V_{max} of E_1 and the K_m or V_{max} of E_2.

The importance of the kinetics of the utilizing-enzyme in the regulation of the concentration of the secondary messenger is illustrated by the fact that in the case of cyclic AMP, where phosphodiesterase corresponds to E_2, there appear to be two forms of the enzyme that have different values of the K_m for cyclic AMP. The combination of two enzymes provides a large concentration range of secondary messenger over which the rate of degradation is approximately linear (Arch and Newsholme, 1976). This property may also apply to other possible secondary messengers (see Figure 22.7).

(a) *Concentrations of secondary messengers*

Secondary messengers are produced within the cell for the specific purpose of regulation but the properties outlined above lead to a steady-state rate of production of the messenger even when there is no hormonal stimulation and this must represent an energy cost to the cell.[5] For maximum 'economy' the rate of secondary messenger turnover must be low and hence the concentration must also be low. This is indeed the case, the basal concentration of cyclic AMP, adenosine, calcium ions, and prostaglandins are below 10^{-6} M and may be even lower than 10^{-7} M. However, this poses a further problem in control since the concentrations of many pathway-enzymes are greater than this, so that the secondary messenger could only modify a small proportion of this enzyme.* For this reason, secondary messengers usually modify not the activity of the pathway-enzyme but the activity of enzymes of an interconversion cycle which controls the activity of the pathway-enzyme (Chapter 7; Section C.1.b). The concentrations of the interconverting enzymes are usually much lower (perhaps ten-fold less) than the pathway-enzyme and are, therefore, less than that of the messenger.[6]

(b) *Action of cyclic AMP*

Although other secondary messengers exist, the mechanism of action of cyclic AMP is undoubtedly best understood and is used in this section to exemplify the mode of action of secondary messengers in general. This discussion will extend the accounts of the action of cyclic AMP given in other chapters. The properties and characteristics of cyclic AMP and the principles concerning its action probably also apply to other secondary messengers such as calcium ions.

An increase in the concentration of cyclic AMP will increase the extent of binding to the cyclic AMP-dependent protein kinase, which will increase the activity of this enzyme. This enzyme exists as a complex of two catalytic (C) and two regulatory subunits (R) which upon binding cyclic AMP dissociates into its components as follows:

$$C_2R_2 + 2\,\text{cyclic AMP} \rightleftarrows 2\,\text{R-cyclic AMP} + 2C$$

The C_2R_2 complex has a very low activity but this is markedly increased upon dissociation since C is the active form of the enzyme that will catalyse the phosphorylation of a number of intracellular proteins. Phosphorylation of these proteins results in a marked change in their activity. Enzymes phosphorylated in this way include glycogen phosphorylase kinase (Chapter 7; Section C.1.b), triacylglycerol lipase (Chater 8; Section C.1), pyruvate kinase (Chapter 11; Section B.4.b), glycogen synthase (Chapter 16; Section C.1), acetyl-CoA carboxylase (Chapter 17; Section A.9), initiation factor (Chapter 18; Section D.2), HMG-CoA reductase and cholesterol esterase (Chapter 20; Section B.1.a).

In addition to the phosphorylation of these enzymes, it has also been suggested that phosphorylation of membrane proteins plays an important role in the control of transport of both metabolites and ions. This, however, illustrates a particular problem in assessing the physiological importance of protein phosphorylation by cyclic AMP-dependent protein kinase and indeed by other protein kinases; the apparent lack of specificity of such kinases. In other words, many proteins can be

---

* The highest concentration of cyclic AMP is about 10^{-6} M whereas that of phosphorylase in muscle is probably about 10^{-5} M.

phosphorylated and the question arises whether the phosphorylation is part of a control mechanism or is it simply non-specific. Consequently, a series of criteria, which should be satisfied before physiological importance is assumed, have been put forward[7] (Cohen, 1980). Very few of the phosphorylation systems indicated above can be said to satisfy all these criteria.

In addition to cyclic AMP-dependent protein kinase, there are several other protein kinases that are not responsive to this particular secondary messenger; the secondary messengers, if there are any, responsible for controlling the activity of these kinases are not known. Some of these protein kinases have already been mentioned (e.g. pyruvate dehydrogenase kinase; cholesterol kinase; cholesterol esterase kinase, glycogen synthase kinase-3).

5. Hormones and substrate cycling

It is usually considered that hormones change flux through a pathway by modifying the activity of an enzyme or a transport system which forms part of that pathway. However, rather than modifying the flux directly, some hormones may increase the sensitivity of metabolic control so that a given change in a metabolic regulator (or another secondary messenger) will have a greater (or perhaps smaller) effect on the flux. One way in which a hormone could change sensitivity is to change the rate of substrate cycling since the sensitivity to a given change in concentration of a regulator depends upon the ratio, cycling rate/pathway flux. The role of the catecholamines in increasing the rate of cycling between glycogen and glucose 1-phosphate, fructose 6-phosphate and fructose bisphosphate, and PEP and pyruvate is discussed in Chapter 9; Section A.3.b and between triacylglycerol and fatty acid in adipose tissue in response to feeding in Chapter 14; Section C.2. (See also Chapter 19; Section C.3.b).

B. ACTION OF STEROID HORMONES AND THYROXINE

Although steroid hormones and thyroxine interact with a receptor, the mechanism is somewhat different to that described above since the receptors for these hormones exist within the cell and much less is known about the receptor-effector interaction. Furthermore, the major effect of these hormones is to control the rate of synthesis of specific proteins by influencing the expression of particular genes. Since it is unclear how gene expression is regulated (see Chapter 18; Section D.1) the effect of these hormones cannot yet be fully understood at the molecular level. What is known is briefly described for each hormone below.

1. Steroid hormones

The metabolic and physiological effects of steroid hormones are discussed in Chapter 20. Their physiological effects arise as a result of their influence on the

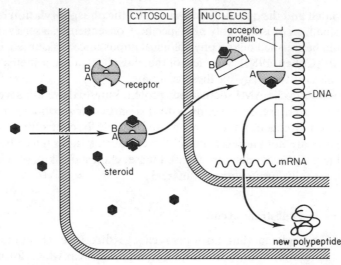

Figure 22.8   Schematic representation of the mechanism of steroid hormone action

rate of protein synthesis or the type of protein that is synthesized in the target tissue. For reviews of steroid action readers are referred to King (1976), O'Malley (1978) and O'Malley *et al.* (1981).

The concentration of a steroid hormone in the blood is typically about 10^{-10} M but, since it is transported by specific protein in the plasma, the free concentration is probably less than this. Since the hormone must traverse the cell membrane (probably by simple diffusion) and bind to a specific receptor in the cytosol of the cell, it is likely that the concentration of the hormone in the cytosolic compartment of the cell is much less than 10^{-10} M. It follows that the affinity of the receptor for the hormone must be high (dissociation constants are known to be about 10^{-9} M). The receptors exhibit very high specificity for particular steroids despite the chemical similarity between steroid hormones. Two important structural features of the steroid hormones are the conformation of the ring system and the nature of the substituents and, not surprisingly, changes in either of these can seriously interfere with binding.

The steroid hormone receptors so far identified are proteins, which consists of two different subunits, A and B, each of which probably binds one steroid molecule.[8] It seems likely that the conformations of the A and B subunits of the receptor protein change upon binding the hormone and this may facilitate transport of the complex into the nucleus (which may require ATP). However the major importance of the hormone-induced conformational change in the receptor may be in unmasking sites that can interact with acceptor sites on the chromatin, probably on the non-histone proteins. It is suggested that the B subunit possesses specific binding sites for the acceptor(s) on the chromatin, which identifies the position on the genome for the effect of the hormone. Upon binding to this acceptor, the A subunit dissociates and attaches itself to the DNA which causes some change that allows access of DNA-dependent RNA polymerase to the DNA so that this gene (or group of genes) is then transcribed (Figure 22.8). It is possible that binding to the DNA actually causes a change in the state of the protein that is associated with the DNA (e.g. phosphorylation, ADP-ribosylation, acetylation) which is responsible for permitting access to the polymerase enzyme (Johnson *et al.*, 1978). In this way the steroid could elicit the synthesis of new protein rather than increasing the rate of synthesis of proteins already being synthesized.

One important difference between the hormone receptors described in Section A and the steroid hormone receptors is that the latter appear to possess both receptor sites and effector activity; that is, there is no secondary messenger analogous to cyclic AMP. However, the interaction between the receptor and the steroid hormone has characteristics similar to that of the interaction between other hormones and their receptors: receptors are present in responsive cells but not in unresponsive ones; there is a good correlation between the steroid specificity of a receptor and the biological effectiveness of the steroid; *in vitro* effects of the steroid require the presence of the receptor and formation of the receptor-steroid complex; there is an excess of receptors in the cytoplasm and it is likely that the receptor number is subject to control (see Section A.3.a).

It should be emphasized, finally, that perhaps not all effects of steroids occur through this mechanism. The steroid-receptor complex may have direct effects in the cytosol independent of their action on the nucleus (e.g. on the lysosome—see Chapter 10; Section B.3.b). Alternatively, binding sites for the steroid hormone may be present in the nucleus. For example, these hormones have actions on the chicken liver, yet no receptors have been found in the cytosol. Since binding sites for the hormone can be found in the nucleus, it is possible that some steroid receptors may already be attached to the target site in the nucleus and the steroid hormone itself must traverse the nuclear membrane to elicit a response.

2. Thyroxine

The fact that thyroxine (or its active form, triiodothyronine—see Chapter 14; Section B.2) is small and lipid soluble, that receptors exist in the cytosol but not on the cell membrane and that it stimulates mRNA and protein synthesis in responsive cells, suggests a similar mechanism of action to that of steroids (see Sterling, 1980, for review). However, receptors for triiodothyronine, rather than the receptor-hormone complex, are found in the non-histone protein fraction of the nucleus suggesting a direct action of hormone within this organelle (see Oppenheimer and Dillmann, 1978; Figure 22.9). This is supported by the good agreement between the potency of binding of the thyroid hormones and their analogues to the nuclear receptors and their biological activity.[*] In addition, the nuclear receptors exhibit the phenomenon of down-regulation.

[*] The receptors for the thyroid hormones have a much higher affinity for triiodothyronine (T_3) than thyroxine which may explain why the former is the active form of the hormone.

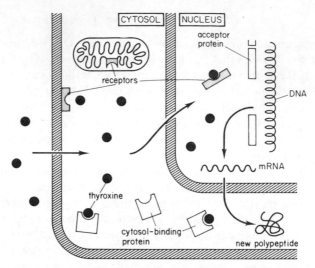

Figure 22.9 Schematic representation of the mechanism of thyroxine action. Although thyroxine is indicated, the physiologically active form may be triiodothyronine. The cytosol-binding protein may facilitate entry of thyroxine into the cell by possessing a high-affinity binding site.

In addition to those in the nucleus and the cytosol, high affinity and high specificity binding sites for thyroid hormones have been found on the cell membranes and the inner mitochondrial membrane. These locations are consistent with the fact that thyroid hormones may increase the mitochondrial capacity for energy generation and may also increase the cellular capacity for energy dissipation via substrate cycles (Chapter 19; Section C.3.b).

C. ACTION OF INSULIN

One of the problems in discussing the hormone insulin is that it has a large number of metabolic effects but, since the secondary messenger has not yet been identified, it is not possible to organize these into a satisfactory 'hierarchy' in order to aid understanding and memory. This section is divided into two parts; a discussion of current views on the secondary messenger and mechanism of action of the hormone followed by a catalogue of the effect of the hormone in various tissues.

1. Secondary messenger for insulin

For many years biochemists have been trying to discover the nature of the secondary messenger for insulin. There has been no shortage of candidates but a paucity of evidence in support. One of the technical problems in this quest has been that an intact tissue preparation was necessary to demonstrate the effect of insulin. The lack of success in identification of a secondary messenger led some workers to suggest that a conventional messenger, as described above, did not exist and the effects of insulin were due to subtle conformational changes in the structure of the internal skeleton of the cell (Hechter and Calek, 1974). However,

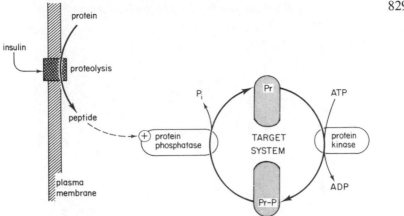

Figure 22.10 Speculative scheme for the mechanism of action of insulin

demonstration that insulin reacts with a cell membrane receptor in an analogous manner to the catecholamines does support the existence of a 'classical' messenger. Consequently, a search for compounds that are released from the cell membrane in response to insulin has been initiated in several laboratories and has provided evidence that the secondary messenger may be a peptide or glycopeptide and that the effector is a proteolytic enzyme (Larner *et al.*, 1982; Seals and Czech, 1982). In other words, the insulin-receptor complex activates a membrane proteolytic enzyme and this produces a peptide or similar compound* which leads to the known effects of insulin. The isolation, characterization and study of the physiological properties of the peptide is likely to be of considerable interest.

Assuming that the peptide is confirmed as the secondary messenger, the next question is its intracellular site of action. Since insulin has such a large number of effects in various tissues (Section C.2) it is possible that the peptide will not affect directly each of these target systems. In an analogous manner to the cyclic AMP activation of protein kinase, current work suggests that the secondary messenger for insulin may increase the activity of a protein phosphatase. In this way, it would influence target systems by reducing the degree of phosphorylation of a given protein through interconversion cycles (Cohen, 1980). A speculative proposal for the action and effects of insulin is given in Figure 22.10 and a summary of target systems that could be affected by changing the activity of a protein phosphatase is given in Table 22.1.

2. Metabolic effects of insulin

Although some of the effects of insulin have already been referred to in various chapters (especially in Chapter 15) they are brought together here, in brief, for ease of reference.

* A suggestion has been made that the compound produced from the membrane is a protein kinase which might cause phosphorylation of protein phosphatase or a protein that regulates the activity of protein phosphatase (Denton *et al.*, 1981).

Table 22.1. Effect of insulin on phosphorylation of key regulatory catabolic or anabolic enzymes and suggested mode of action of the hormone*

Reaction or process	Chapter and Section	Regulatory enzyme	Effect of phosphorylation on enzyme	Effect of insulin	Suggested mechanism of action of insulin
Glycogen degradation	7; C.1.b	phosphorylase (phosphorylase kinase)	activation	? decreased activity in liver	(i) ? activation of protein phosphatase (ii) decrease in [cyclic AMP]
Triacylglycerol hydrolysis	8; C.1	triacylglycerol lipase	activation	decreased activity in adipose tissue	activation of protein phosphatase
Gluconeogenesis in liver	11; B.4.b	pyruvate kinase	inhibition	increased activity	(i) ? activation of protein phosphatase (ii) decrease in [cyclic AMP]
Glycogen synthesis	16; C.1	glycogen synthase	inhibition	increased activity in muscle (? liver)	activation of protein phosphatase

Pyruvate conversion to acetyl-CoA in adipose tissue	17; A.9	pyruvate dehydro-genase	inhibition	increased activity in adipose tissue	activation of pyruvate dehydrogenase phosphatase
Acetyl-CoA carboxylation	17; A.9	acetyl-CoA carboxylase	inhibition	increased activity in adipose tissue	(i) activation of protein phosphatase (ii) decrease in [cyclic AMP]
Esterification	17; A.9	glycerophosphate acyltransferase	inhibition	increased activity	?
Cholesterol synthesis	20; B.1.a	HMG-CoA reductase	inhibition	increased activity	(i) activation of protein phosphatase (ii) decrease in [cyclic AMP]
Initiation of protein synthesis (translation)	18; D.2	initiation factor-2 (eIF-2)	inhibition	increased activity	?

* See Cohen (1980) and Figure 22.10.

(a) *Muscle*

Insulin affects both glucose and protein metabolism in muscle.

(i) *Glucose metabolism*   Insulin increases the rate of glucose utilization by skeletal and cardiac muscle and this can be divided into effects on membrane transport, glycolysis and glycogen synthesis.

The effect on membrane transport is due to an increase in the V_{max} of the transport process with little, if any, effect on the K_m which is explained by an increase in the number of functional carrier molecules in the cell membrane (Cushman *et al.*, 1981). This increase in the rate of glucose transport probably stimulates the process of glycolysis by internal regulation. Of the extra glucose taken up by muscle in response to insulin, a large proportion is converted to glycogen, due to an increase in the activity of glycogen synthase (Villar-Palasi and Larner, 1968). Insulin increases the proportion of the enzyme in the dephosphorylated form and this could be achieved by activation of protein phosphatase-I (see Section C.1 and Chapter 16; Section C.1).

(ii) *Protein metabolism*   Insulin increases the transport of a large number of amino acids into muscle (Manchester, 1970; Munro, 1974; Morgan *et al.*, 1979). It also increases the rate of protein synthesis and decreases the rate of protein degradation (Morgan *et al.*, 1979; Munro and Crim, 1980). In addition to these effects, insulin may decrease the rate of endogenous lipolysis in heart and type I and IIA muscle fibres but the mechanism and physiological importance are unclear.

(b) *White and brown adipose tissue*

Insulin increases the rates of glucose utilization and synthesis of triacylglycerol and decreases the rate of lipolysis in white adipose tissue. There is some evidence that insulin exerts similar effects in brown adipose tissue and that this may be quantitatively important for lipogenesis in the animal.

(i) *Glucose metabolism*   As in muscle, the rate of membrane transport is increased due to an increase in V_{max}, probably due to an increase in the number of carriers (Olefsky, 1978). The rate of glycolysis and the activity of glycogen synthase are also increased by similar mechanisms to those described for muscle.

(ii) *Lipid metabolism*   Insulin increases the rate of fatty acid synthesis in adipose tissue due to stimulation of a number of processes including glucose transport, pyruvate dehydrogenase and acetyl-CoA carboxylase (Denton and Halestrap, 1979). The mechanism of control of the latter enzymes are discussed in Chapter 17; Section A.9.

The substrates for triacylglycerol synthesis in adipose tissue are fatty acyl-CoA and glycerol phosphate (Chapter 17; Section A.9) and the sources of the acyl-CoA are endogenous fatty acid synthesis, endogenous lipolysis or fatty acids derived from the blood. The latter will arise primarily from the extracellular hydrolysis of triacylglycerol from VLDL or chylomicrons, catalysed by lipoprotein lipase (Chapter 6; Section C.1.c). This enzyme probably catalyses the flux-generating step for uptake of extracellular triacylglycerol and, in accord with this, the activity is increased by insulin (Garfinkel *et al.*, 1976).

Insulin decreases the rate of lipolysis in adipose tissue (see Green and Newsholme, 1979) and the mechanism and physiological importance of this effect are described in Chapter 8; Section B.

(iii) *Protein metabolism*   Insulin increases the rate of uptake of amino acids into the cell and the rate of protein synthesis (Minemura *et al.*, 1970).

(c) *Liver*

(i) *Carbohydrate metabolism*   The glucose transport system in liver is near-equilibrium so that it is not affected by insulin. Indeed insulin probably does not influence the glucose uptake by liver directly, which is controlled mainly by changes in the portal vein glucose concentration (see Chapter 11; Section E). However, insulin increases the activity of glucokinase, an effect which probably involves synthesis of new enzyme-protein, rather than activation of existing enzyme molecules (Weinhouse, 1976). Whether insulin has any direct effect on glycogen synthesis or breakdown in the liver is still unclear (see Chapter 16; Section C.2). However, it inhibits adrenaline- or glucagon-stimulated gluconeogenesis but has no effect on the basal rate of gluconeogenesis (Chapter 11; Section B.4.b).

(ii) *Lipid metabolism*   Insulin increases the rate of fatty acid synthesis in the liver, an effect likely to involve activation of pyruvate dehydrogenase and acetyl-CoA carboxylase (Geelen *et al.*, 1978). In addition, the rate of secretion of VLDL is also increased by insulin (Topping and Mayes, 1976) but it is not known if insulin affects the rate of endogenous lipolysis in this tissue. Insulin also increases the rate of cholesterol synthesis in the liver.

(iii) *Protein metabolism*   Insulin increases the rates of uptake of amino acids and of protein synthesis (Kletzien *et al.*, 1976) and decreases proteolysis in the liver (Mortimore and Mondon, 1970).

D. CONTROL OF HORMONE SECRETION

Since there are a large number of hormones, there are a large number of mechanisms for the control of their release into the bloodstream. However, it is possible to classify these mechanisms into three general types which depend upon how the hormone is stored in the tissue. For steroid hormones there is little or no storage so that the rate of secretion is controlled by the rate of synthesis from steroid hormone precursors; many other hormones are stored intracellularly in granules or vesicles and are released by exocytosis; finally, thyroxine is stored extracellularly and is released by partial hydrolysis of the storage thyroglobulin. Some examples of these mechanisms are described below.

1. Control of cortisol secretion from the adrenal cortex

Steroid hormone production occurs in the ovary, the testes and the adrenal cortex (and also in the placenta during pregnancy—see Chapter 20; Section B). There has been no systematic study of the equilibrium nature of the reactions involved in conversion of cholesterol to the steroid hormones and there is little or no information on the flux-generating step or steps in these pathways. Most of the work on regulation of the rate of this process has been carried out on the adrenal cortex but the available evidence suggests that it is similar for both the testes and ovary. Individual differences in control of the activities or concentrations of specific enzymes during development of the Graafian follicle are discussed in Chapter 20; Section E.4.b.

The action of ACTH on the adrenal cortex (see Chapter 20; Section E.1.a) and the gonadotrophic hormones on the ovary or testes involve binding to the specific cell membrane receptor which activates adenylate cyclase leading to an increase in the cyclic AMP concentration and hence activation of protein kinase as described in Section A.3 above; it is also possible that changes in the intracellular concentrations of calcium ions and cyclic GMP maybe involved (Sala *et al.*, 1979).

Since the amount of cortisol in the adrenal cortex is so low ($25\,\mu g$) in comparision with the amount secreted each day, the trophic hormone controls secretion by controlling the rate of synthesis. There are at least three important reactions in the control of the rate of cortisol formation; provision of cholesterol in the cytosol, transport of cholesterol into the mitochondria and the enzyme system involved in cleavage of the side-chain of cholesterol to form pregnenolone within the mitochondria.

(a) *Provision of cytosolic cholesterol*

There are three sources of cholesterol for the steroid hormone-producing endocrine glands, synthesis from acetyl-CoA within the gland, hydrolysis of stored cholesterol ester and uptake of cholesterol in the form of LDL. These three processes are discussed in Chapter 20, and the quantitative importance of each has been studied for the adrenal cortex by Brown *et al.* (1979).

The sequence of events in the control of provision of cholesterol in the adrenal cortex in response to ACTH is probably as follows. The raised intracellular concentration of cyclic AMP activates both HMG-CoA reductase and cholesterol esterase so that the rates of both biosynthesis and of cholesterol ester hydrolysis are rapidly increased. However, if the stimulus is maintained for a period, the capacity of the LDL-cholesterol uptake process by the adrenal cells is increased (probably via protein synthesis although a direct role of protein kinase cannot be ruled out) and this then becomes the dominant process for provision of cholesterol (indeed, this process will lead to an increase in the cytosolic concentration of cholesterol which will inhibit HMG-CoA reductase and favour cholesterol ester synthesis—see Chapter 20; Section B.1.a).

(b) *Mitochondrial transport of cholesterol*

The activation of protein kinase may increase the phosphorylation of a ribosomal protein or of a mRNA molecule which results in the increased rate of synthesis of a protein involved in the transport of cholesterol across the mitochondrial membrane. The increased rate of synthesis of this protein can have a marked effect on cholesterol metabolism in the mitochondria because this transport-protein has a very short half-life (8 min; compare with half-lives given in Table 10.3). A low rate of synthesis means that there is little protein available to transport cholesterol into the mitochondria.

(c) *Cholesterol side-chain cleavage*

Once inside the mitochondria, the important regulatory process for cholesterol metabolism appears to be the cleavage of the side-chain. However, it is unclear how the activity of this enzyme system is controlled. It is possible that control is achieved simply via the availability of cholesterol (internal control), in which case, the mitochondrial cholesterol transport process would play a very important role in regulation. However, it is more probable that the enzyme is regulated externally by an unknown mechanism.

In addition to these acute effects of the ACTH and other hormones, it is probable that they can increase the concentration of most of the enzymes involved in the pathways of steroid hormone formation but the mechanism is not known.

2. Control of insulin secretion from the islets of Langerhans

Although any of a number of hormones could have been chosen to exemplify the control of secretion of a hormone stored intracellularly, insulin is discussed here since it is an important metabolic hormone and the control of the rate of its secretion appears to involve the metabolism of glucose (for reviews, see, Ashcroft, 1980; Hedeskov, 1980).

Insulin is secreted by the β-cells of the islets of Langerhans in the pancreas. Biochemical studies on the mechanism of control of secretion have been possible since the introduction of methods for isolation and incubation of these islets from the pancreas of experimental animals (Lacy and Kostianovsky, 1967).

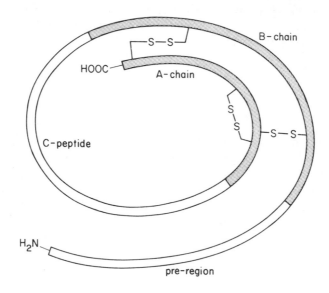

Figure 22.11   Diagrammatic representation of the preproinsulin molecule

(a) *Formation and packaging of insulin in granules*

Protein synthesis in the β-cell produces a single-chain precursor of insulin, termed preproinsulin, which consists of proinsulin plus a peptide of 23 amino acids (the pre-region) which occurs at the amino end of the molecule (Permutt *et al.*, 1976). The pre-region contains a large number of hydrophobic amino acids which enable the protein to traverse the membrane of the endoplasmic reticulum and enter the cisternal space. The pre-region is then cleaved by a specific endoproteolytic enzyme (known as signal peptidase) associated with the reticulum (Blobel and Dobberstein, 1975). This forms proinsulin, a single-chain polypeptide in which the carboxyl-terminal residue of the insulin B-chain is joined to the amino-terminal residue of the insulin A-chain by a connecting peptide (the C-peptide) of about 30 amino acids (Figure 22.11). Proinsulin is transported through the cisternae of the endoplasmic reticulum to the Golgi complex for package into secretory granules. The conversion of proinsulin to insulin (by proteolytic removal of the C-peptide) takes place in the secretory granules soon after their formation and insulin is stored as a hexamer which is stabilized by a zinc ion. The entire contents of the granules (insulin, C-peptide, and some basic amino acids) are released into the extracellular fluid during insulin secretion, so that the concentration of the C-peptide in the blood can be used as an index of endogenous insulin secretion in diabetic patients receiving insulin.

(b) *Secretion of insulin*

Insulin is probably released from the β-cell directly from the granules by exocytosis. Movement of the granules to the cell membrane in response to stimulation probably involves microfilaments and microtubules. Microfilaments are 5–7 nm in diameter and in many cells they can be seen as separate single filaments in the cytosol. In the β-cells they are organized in a subcytosolic web with individual microfilaments inserting in the cell membrane and the microtubules (Malaisse *et al.*, 1975). The latter have an outer diameter of 25 nm and a hollow core 17 nm in diameter and are composed of the polymerized protein tubulin. Depolymerization of tubulin, which results in breakdown of the microtubules, is caused by the agents colchicine and vincristine (Dustin, 1978). These changes inhibit the secretion of insulin by the β-cells and, indeed, the secretion of many other glands. The most important stimulus for insulin secretion is an increase in the extracellular concentration of glucose; an increased rate of secretion occurs within 1 min of the addition of glucose to the tissue. It is of interest that the response of insulin secretion to the change in glucose concentration is sigmoid so that there is little response below 5 mM, and a 50% response at about 8 mM (Ashcroft, 1976). This provides an improvement in sensitivity of the secretory process to changes in glucose above the normal blood concentration and it is a similar response to that provided by the glucose/glucose 6-phosphate substrate cycle in the liver for the uptake of glucose (Chapter 11; Section E.1). Indeed, the existence of both glucokinase and glucose-6-phosphatase in the β-cell (see Randle *et al.*, 1968) provides support for the role of this cycle in the regulation of glucose metabolism and raises the possibility that changes in the rate of glycolysis in the β-cell provide the stimulus for insulin secretion. If this is so, what is the means by which the secretory process assesses the rate of the glycolytic process? Since the concentrations of intermediates of a pathway are quantitatively related to the flux through it (see Chapter 7; Section A.1) any of the intermediates could act as the link with the secretory process. However, the fact that addition of glyceraldehyde to isolated islets can elicit insulin release suggests that an intermediate towards the end of the pathway is responsible; PEP has been suggested as the most likely candidate (Sugden and Aschroft, 1977).

The next question is the nature of the link between the glycolytic intermediate and secretion. This link, sometimes known as a 'coupling factor' is considered to involve calcium ions;[9] both the rate of entry of extracellular calcium ions into the cell and the release of calcium ions from intracellular stores are considered to be important factors controlling the rate of insulin secretion. The intracellular protein that detects the change in concentration of calcium ions is probably calmodulin, a heat stable protein of about 17 000 molecular weight, which can bind up to four ions of calcium. This changes the three dimensional structure of the protein which somehow controls the contractile activity of the microtubules. The following hypothesis is therefore proposed. An increase in the extracellular glucose concentration above 5 mM increases proportionally the rate of glycolysis (through operation of the glucose/glucose 6-phosphate cycle) and this raises the concentration of PEP which, by an unknown mechanism, increases the rate of uptake of calcium ions and probably increases the rate of release from intracellular stores. An increased cytosolic concentration of calcium ions, via calmodulin, causes contraction of the microfilaments or microtubules, which may possess actin and myosin and hence contract in a similar manner to skeletal muscle. Somehow this contraction results in increased rate of exocytosis and insulin secretion.

(c) *The role of cyclic AMP in insulin secretion*

The above hypothesis attempts to link the rate of glycolysis, calcium ions, and rate of secretion, but it fails to take into account the effect of cyclic AMP, which also increases the rate of insulin secretion. However, although certain compounds that increase the concentration of cyclic AMP in the β-cell (e.g. glucagon, caffeine) stimulate the rate of insulin release, they only do this if glucose is present. Since these compounds alone cannot maintain an increased rate of insulin secretion, they are consequently known as potentiators of insulin secretion. Furthermore, there is no correlation between the concentration of cyclic AMP in islet tissue and the rate of the insulin secretion. Thus it appears that cyclic AMP does not act as a coupling factor nor does it control the rate of secretion directly; rather it can increase the magnitude of the secretory response to glucose or other sugars which can enter the glycolytic pathway. It is likely that this effect is achieved by activation of cyclic AMP-dependent protein kinase and phosphorylation of a protein involved in control of calcium transport, so that for a given concentration of glucose more calcium enters the cytosolic compartment of the cell; or phosphorylation of microtubules could increase their responsiveness to a given change in the calcium ion concentrations (that is, to the calmodulin–calcium ion complex).

3. Control of thyroxine secretion from the thyroid gland

The thyroid secretes thyroxine (and triiodothyronine) from the follicular cells and calcitonin from the C cells. Only the former is discussed here. In the thyroid, a large number of follicles are surrounded by the thyroid follicular cells and this structural relationship is important in the formation and secretion of thyroid hormones (Figure 22.12). The lumen of the follicle contains a protein known as thyroglobulin, which is synthesized in the thyroid cell and can be considered as a prohormone. Thus it is a glycoprotein with a molecular weight of about 660 000 and consists of several peptide chains held together by disulphide bridges. It contains about 3 % tyrosine (about 140 tyrosyl residues) of which about 20 % are iodinated. Iodination is a post-translational process that occurs probably at the surface of the thyroid cell

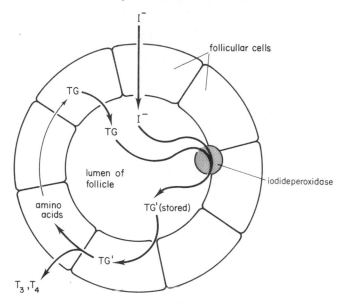

Figure 22.12 Schematic representation of thyroid hormone synthesis in the thyroid follicle. See text for details. TG represents thyroglobulin and TG′ represents iodinated thyroglobulin

but within the lumen of the follicle. There are several sequential steps in the production and eventual release of the hormones and these are as follows. (For review, see Van Herle *et al.*, 1979).

1. Iodide is actively transported from the extracellular fluid across the basal membrane of the thyroid cell from where it is concentrated in the follicular lumen.
2. The iodide reacts with hydrogen peroxide at the apical membranes, catalysed by an iodide peroxidase which is sometimes known as thyroid peroxidase. *In vitro*, this catalyses the following reaction

$$2H^+ + 2I^- + H_2O_2 \longrightarrow I_2 + 2H_2O$$

but, *in vivo*, the enzyme probably forms an oxidized species of iodine which is enzyme bound, the iodinium ion $(E - I^+)$. These 'activated' forms iodinate tyrosine residues in the thyroglobulin molecule producing both mono-iodotyrosine and diiodotyrosine residues (Figure 22.13).
3. The iodothyronine hormones, thyroxine, and triiodothyronine, are synthesized by coupling together two iodotyrosine residues within the thyroglobulin molecule. The process may proceed through a free radical mechanism which requires oxygen and produces serine as the side product and may be catalysed by the same thyroid peroxidase. Approximately five or six iodothyronyl residues per molecule of thyroglobulin can be produced.

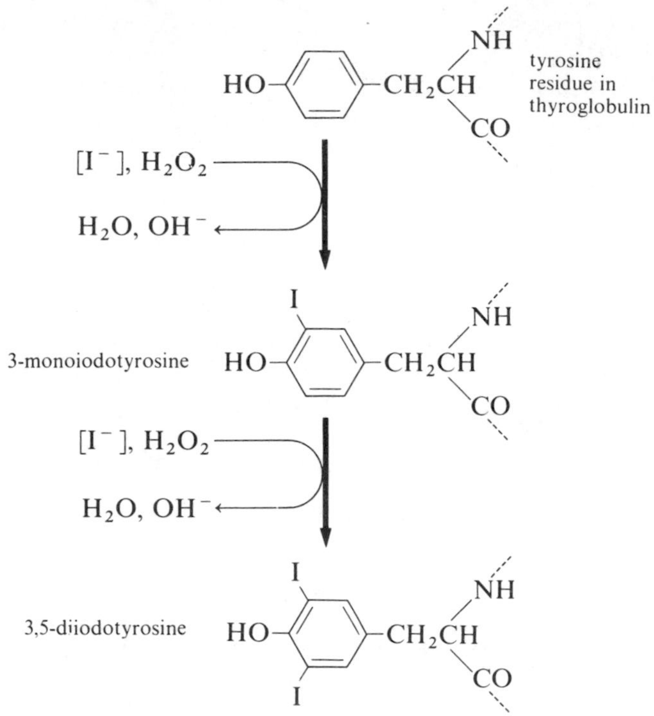

Figure 22.13   Reactions catalysed by iodide peroxidase

In response to thyrotropin[10] (thyroid-stimulating hormone, TSH), the thyroid cells take up some of the thyroglobulin from the follicle by endocytosis and degrade it by proteolytic enzymes and oligosaccharidases in the lysosomes. This degradation will produce the constituent amino acids, sugars, thyroxine (and triiodothyronine), as well as iodotyrosines. The iodothyronines diffuse from the thyroid cell into the extracellular fluid and bloodstream whereas the iodotyrosines are deiodinated by an iodotyrosine deiodinase in the thyroid cell. This enzyme is attached to the endoplasmic reticulum and requires NADPH.

In the absence of thyrotropin, very little thyroxine is secreted by the thyroid gland. Thyrotropin binds to the cell membrane of the thyroid cells which leads to activation of adenylate cyclase, an increase in the intracellular concentration of cyclic AMP and activation of protein kinase. (It is, however, unclear if all the effects of this hormone can be explaind by this increase in cyclic AMP concentration.) One of the first effects (presumably of protein phosphorylation) is an increase in the endocytotic activity of the apical membrane so that the uptake of thyroglobulin into the cell is increased, together with an increase in both the number of lysosomes and their activity. Since both these activities involve membranes, it is not surprising that thyrotropin increases phospholipid synthesis. There is also an increase in the rate of glucose uptake and oxidation via the pentose phosphate pathway; the latter provides NADPH necessary for reactions involved in iodination of thyroglobulin and other reactions (e.g. iodotyrosine deiodinase); the pathway may also be involved in provision of ribose for nucleotide synthesis, since TSH also results in growth and proliferation of the thyroid cells.

When the intake of iodine is low over a prolonged period, the increased thyrotropin concentrations cause such a proliferation of thyroid cells that a goitre (simple goitre) results. This ensures that the extraction of the iodide from the blood by the thyroid cells is very efficient. Goitre can also result from agents that will compete with iodide uptake in the thyroid, so effectively reducing the iodine concentration of the gland (e.g. SCN^-, ClO_4^-), or agents that inhibit iodination (e.g. propylthiouracil, which inhibits iodide peroxidase, or thiourea which reduces the oxidized forms of iodine involved in the iodination).

The storage capacity of the gland for iodinated thyroglobulin is such as to allow a normal thyroid to maintain normal rates of thyroxine secretion for several weeks in the absence of iodination or thyroglobulin synthesis.

(a) *Peripheral activation of thyroxine*

The thyroid gland secretes mainly thyroxine although a small amount of triiodothyronine is released. Thyroxine is converted to 3,5,3'-triiodothyronine (T_3) in peripheral tissues by the action of a deiodinase attached to the endoplasmic reticulum (liver and kidney are probably the most important tissues). Since T_3 is considered to be the active form of the hormone, this enzyme is important in controlling the biological activity of the hormone. However, thyroxine can also be converted to the inactive 3,3',5'-triiodothyronine (reverse-T_3) in these tissues and a different deiodinase appears to be involved. The activities of the two deiodinases appear to be co-ordinately controlled since conditions that decrease the circulating concentration of T_3 increase that of reverse-T_3 (e.g. starvation, severe illness), and *vice versa* (e.g. carbohydrate or over-feeding) (see Robbins and Rall 1979; and Chapter 14; Section B.2).

E. MEASUREMENT OF HORMONE CONCENTRATION

Many endocrine disturbances are satisfactorily diagnosed only when the concentration of one or more hormones has been measured in a sample of plasma taken from the patient. The problems of hormone assays arise from the fact that their concentrations in blood (or other body fluids) are very low (e.g. 10^{-10} M or below) and similar substances are often present. How can they be measured accurately, reproducibly and precisely? In general, three methods are available; bioassays using intact tissues or intact animals, use of a particular chemical or physical property of the hormone and binding assays using specific binding proteins. For review of methods for measuring hormone concentrations, see Ekins (1974), Landon *et al.*, (1979).

1. Bioassay

Hormones produce a biological response in a tissue or animal and this response can be used to provide an assay. Since the responses are wide-ranging so are the bioassays for hormones (from, for example, the production of hypoglycaemic convulsions in the mouse for insulin to production of a specific mRNA for a given steroid hormone). Perhaps one of the better bioassays for insulin, which illustrates the procedure, is the measurement of the increase in ^{14}C-carbon dioxide production from isolated rat adipose tissue (e.g. epididymal fat pads) incubated *in vitro* with glucose labelled with ^{14}C specifically in the 1-position. This method depends upon the fact that insulin markedly increases the rate of the pentose phosphate pathway in adipose tissue and, in this pathway, carbon in the 1-position of the glucose is oxidized to carbon dioxide (Chapter 17; Section A.8).

Most, if not all, bioassays suffer from lack of sensitivity and specificity and in order to overcome these problems it may be necessary to extract and partially purify the hormone, thus precluding use of the procedure for routine diagnostic purposes. In principle, however, bioassays have the advantage of measuring the concentration of *functional* hormone.

2. Physical or chemical techniques

Hormones can be assayed by specific chemical or physical techniques which depend upon a particular chemical group in the molecule that either reacts with an added reagent to produce a product that is capable of being measured or has a particular physical characteristic that enables it to be detected quantitatively; for example, prostaglandins have been estimated by mass spectroscopy. However, the problem with most of these assays is that the hormones have to be purified prior to their assay and this prevents such techniques from being used routinely.

3. Binding assays

The binding assay could, in principle, be described as a 'bioassay' since a biological molecule, the binding protein, is used to interact with the hormone so that an equilibrium is established:

$$\text{hormone} + \text{binding protein} \rightleftarrows [\text{binding protein—hormone}]$$

However, the process takes place *in vitro* and a range of chemical procedures may be involved in the assay. Three classes of binding protein have been used, receptor proteins for the hormone, antibodies to the hormone (or to the hapten if the hormone is not normally antigenic) and plasma binding proteins for the hormone. The first quantitative binding-assays were reported in 1960: Yalow and Bersen developed an assay for insulin that depended upon the binding to insulin antibody and Ekins developed an assay for thyroxine that depended upon binding to the plasma thyroxine-binding globulin. The antibody technique has proved to be the most versatile tool and subsequent discussion will be limited to such examples. The binding assays can be further subdivided into two classes, involving the use of excess binding protein or the use of limited binding protein.

(a) *Binding assays involving excess binding protein*

In this type of assay, excess binding protein is present so that almost all the hormone added is converted to the complex; this is sometimes known as a stoichiometric assay and is similar, in principle, to many other chemically based assays or titrations. It depends upon a change in property of the binding protein in the form of the complex or separation of labelled complex from the reactant. Two examples are used to illustrate the range of techniques.

(i) *Immunoradiometric assay for insulin* Antibodies to insulin are raised, purified and radiolabelled (e.g. by iodination with ^{125}I-iodine). Excess antibody is allowed to react with insulin as follows:

$$^{125}\text{I-antibody} + \text{insulin} \rightleftarrows [^{125}\text{I-antibody—insulin}]$$

The uncomplexed antibody is removed (e.g. by binding to insulin firmly attached to a solid support, that is, using an affinity column) and the radioactivity remaining is measured. This will be proportional to the hormone concentration.

(ii) *Sandwich technique* This technique is suitable for larger protein hormones with more than one antigenic determinant. An antibody to one determinant (A) is attached to a solid support (e.g. small beads or granules) and interacted with hormone. A second antibody (B) which is labelled with radioisotope (e.g. ^{125}I-iodine) is added which interacts with another group on the hormone and forms the complex [antibody A—hormone—antibody B]. The labelled antibody that does not interact is easily separated from the complex (e.g. by washing) and the radioactivity bound to the support is proportional to the hormone concentration.

(b) *Binding assays involving a limited amount of binding protein*

In these assays, a limited amount of binding protein is present so that only a proportion of the hormone is present in the complex. Advantage is taken of the fact that at equilibrium between, for example, labelled hormone, antibody and complex, addition of further unlabelled hormone (either standard or unknown) will displace some of the labelled hormone from the complex. For $[^{125}\text{I}]$insulin, equilibrium is as follows:

$$[^{125}\text{I}]\text{insulin} + \text{antibody} \rightleftarrows [^{125}\text{I-insulin—antibody}]$$

The addition of unlabelled hormone to this equilibrium will cause some $[^{125}\text{I}]$insulin to be displaced from the complex and the decrease in the

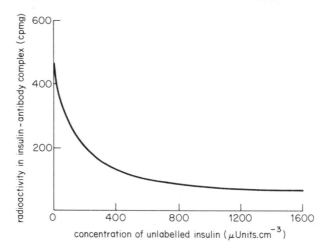

Figure 22.14 Radioactivity remaining in the [^{125}I] insulin–antibody complex in the presence of different concentrations of unlabelled insulin

radioactivity is proportional to the concentration of unlabelled hormone (Figure 22.14). The principle is sometimes known as the substoichiometric isotope dilution principle (Ruzicka and Stary, 1964) and the relationship between the decrease in label in the complex and the concentration of hormone is derived in Appendix 22.1.

Although the principle is straightforward, the assays depend upon the ability to separate labelled complex from labelled hormone (or labelled antibody) and a large number of ingenious methods for separation have been devised. For example, in the assay of insulin at least three separation procedures have been used; electrophoresis, precipitation by salt, and precipitation by addition of antibody against the antibody. Solid supports have provided a simple means of separation, since an antibody chemically linked to a support is easily separated from the hormone by washing, centrifugation, filtration, etc. An extension of this is to use particles coated with antibodies that will interact with several antigenic sites on the hormone. Addition of the hormone will cause the particles to agglutinate, which can be detected by simple visual observation. Such assays are, of course, only qualitative but are useful in a number of cases, for example, the pregnancy test that indicates the presence of chorionic gonadotropin (see Chapter 20; Section E.4.d).

(c) *Extensions of the method*

The substoichiometric isotopic principle, described above, was so named because all the early methods for measuring the interaction between binding-protein and hormone used radiolabels. However, later developments have utilized other means of 'labelling' the hormone so that measurement of the change in the amount of 'labelled' hormone bound to the antibody is simpler. (For this reason, the general term 'saturation analysis' to describe the principle is preferable.) For example, if a hormone is chemically attached to an enzyme such that enzyme activity is retained but is inhibited when the hormone binds to the antibody, the decrease in enzyme activity will be proportional to the amount of hormone bound to the antibody. Addition of unlabelled hormone will displace some of the enzyme-labelled hormone from the complex and increase the enzyme activity. A standard curve can be constructed using known amounts of purified hormone.

APPENDIX 22.1. THEORY OF SUBSTOICHIOMETRIC ISOTOPE DILUTION

In order to appreciate the theory, a specific example is taken. Insulin is assayed by binding $[^{125}I]$insulin to its antibody and the antibody-insulin complex is separated from insulin as described above.

$$[^{125}I]\text{insulin} + \text{antibody} \rightleftarrows [^{125}I\text{-insulin—antibody}]$$

Let C_0 represent the total radioactivity in the antibody-insulin complex prior to addition of unlabelled hormone; C_i represents the total radioactivity in the antibody-insulin complex after addition of unlabelled hormone; X_0 represents the specific radioactivity of insulin when no unlabelled insulin is added; X_i represents the specific radioactivity after addition of unlabelled insulin; P_0 represents the amount of insulin bound to antibody prior to the addition of unlabelled insulin and P_i represents the amount of insulin bound to antibody after addition of unlabelled insulin, so that

$$C_0 = P_0 X_0 \tag{1}$$

$$C_i = P_i X_i \tag{2}$$

If S_0 is the concentration of labelled insulin and S_i is the concentration of unlabelled insulin, then according to the isotope dilution principle:

$$X_i = X_0 \cdot \frac{S_0}{S_0 + S_i} \tag{3}$$

Substituting for X_i in equation (2)

$$C_i = P_i X_0 \left(\frac{S_0}{S_0 + S_i} \right) \tag{4}$$

Dividing equation (1) by equation (4)

$$\frac{C_0}{C_i} = \frac{P_0}{P_i} \cdot \frac{S_0 + S_i}{S_0}$$

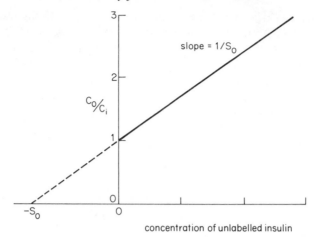

Figure 22.15   Relationship between the ratio of radioactivies in the insulin–antibody complex in the absence and presence of unlabelled insulin, and the concentration of unlabelled insulin

If the same amount of insulin is bound by the antibody in the absence and in the presence of unlabelled insulin, so that $P_0 = P_i$, then

$$\frac{C_0}{C_i} = \frac{S_0 + S_i}{S_0} = 1 + \frac{S_i}{S_0}$$

In this case, a plot of C_0/C_i against S_i will be linear with a slope of $1/S_0$ and the value of C_0/C_i will be unity when S is zero. This has in fact been shown to be the case for insulin (see Figure 22.15 and Hales and Randle, 1963).

Of course, even if different amounts of insulin are bound by the antibody ($P_0 \neq P_i$) the concentration of insulin can still be obtained from the standard curve, despite the fact that the relationship is not as elegant (Hales, 1963).

NOTES

1. Stimulation of glycogen synthase alone would not lead to an increase in the steady-state rate of glycogen synthesis since the concentrations of UDPglucose and hexose monophosphates would be decreased unless the rate of glucose transport was increased. Hence insulin stimulates glucose transport. This could be achieved, in the absence of any effect of insulin, by a large increase in the blood glucose concentration since the process is not saturated with substrate but such large changes are known to cause, in the long term, damage to tissues (see Chapter 15; Section G). The concentration of insulin in the blood is, of course, controlled by the rate of absorption of carbohydrate into the hepatic portal vein so that an increased rate of glucose transport into tissues and increased glycogen synthesis only occurs when glucose is available for storage. This rationale explains, in part, why injections of insulin can be dangerous, especially in the absence of glucose absorption. Stimulation of glucose transport, a non-saturated reaction, will cause marked hypoglycaemia which can be

sufficiently severe to be fatal. This explains the effectiveness of the insulin syringe as a murder weapon when used by medically knowledgeable personnel.
2. Dissociation constants are the reciprocal of binding or affinity constants.
3. Values for the dissociation constant and the number of binding sites are usually obtained from a Scatchard plot. If the concentration of the ligand (e.g. hormone, neurotransmitter) that is bound to the receptor is B and that which is free is F, and a plot of B against B/F produces a straight line, the slope is the dissociation constant and the intercept on the B axis is the total number of binding sites (B_{max}). The plot is described by the equation

$$B = B_{max} - K_d \frac{B}{F}$$

This is identical to the Eadie-Hofstee manipulation of the Michaelis–Menten equation

$$V = V_{max} - K_m \left[\frac{V}{S} \right]$$

which is derived in Chapter 3; Section C.2.f.
4. The difficulty of studying the β-adrenergic receptor has been overcome by the discovery of agonists or antagonists that would bind specifically to the receptor and which could be labelled with suitable radioactive isotopes so that binding could be readily detected. Several β-blockers have now been used for such studies including tritiated propranolol and alprenolol, and radioiodinated iodohydroxybenzylpindolol (I-HYP). They show a high degree of specificity for the β-receptor and saturation-binding kinetics.
The use of these radioactive-labelled analogues has shown the following properties of the β-receptor.
   (a) The binding of antagonists to the receptor and the inhibition of adenylate cyclase are closely linked.
   (b) Similarly, binding of agonists and activation of the cyclase are closely linked, provided the GTP-binding subunit does not inhibit the cyclase (this is avoided by use of an analogue of GTP that binds but is not hydrolysed.
5. In rat heart, the basal rate of cyclic AMP formation is about 30 $nmol.min^{-1}.g^{-1}$ (Arch and Newsholme, 1976) whereas the normal rate of ATP turnover is about 78 $\mu mol.min^{-1}.g^{-1}$ (Table 5.6). Hence only 0.05 % of the energy production in rat heart is used for the synthesis of cyclic AMP.
6. The concentrations of the enzymes involved in the interconversion cycles controlling phosphorylase activity are known: the proportions of the concentrations of the three enzymes, protein kinase, phosphorylase kinase, and phosphorylase in white skeletal muscle are 1:10:240, respectively (Cohen, 1980).
7. The criteria needed to establish the physiological importance of protein phosphorylation are as follows.
   (a) A protein that is phosphorylated by protein kinase should have a regulatory role in a process considered to be affected by cyclic AMP.
   (b) The function or property of the protein should change upon phosphorylation and this should be reversed by dephosphorylation, catalysed by a protein phosphatase.
   (c) The change in function or property of this protein should occur *in vivo* in response to cyclic AMP (or an analogue such as dibutyryl cyclic AMP which traverses membranes more readily).
   (d) The protein should be phosphorylated *in vivo* in response to a hormone that raises the concentration of cyclic AMP and phosphorylation should occur at the same position in the protein *in vivo* as occurs *in vitro*.

8. Both A and B subunits are single polypeptide chains of molecular weight about 72 000 and 117 000 respectively (O'Malley, 1978).

9. The evidence that implicates calcium ions in the coupling between glycolysis and secretion is as follows: for insulin to be secreted by islets, calcium ions have to be present in the incubation medium; an increase in the extracellular concentration of calcium stimulates insulin secretion in the absence of glucose; glucose increases the rate of uptake of calcium by the islets and this increase is proportional to the increase in glycolytic rate.

10. Thyrotropin is a glycoprotein, of molecular weight about 29 000, secreted by cells of the anterior pituitary. The rate of release of the hormone is controlled by thyrotropin-releasing hormone (TRH) secreted into the portal system by cells in the hypothalamus (see Chapter 20; Section E.4.c). TRH is a tripeptide, with the structure pyroglutamyl–histidyl–prolinamide. It binds to receptors on the cell membrane, increases the activity of adenylate cyclase and raises the intracellular concentration of cyclic AMP. This leads to activation of protein kinase which may increase the intracellular concentration of calcium ions leading to increased secretion of thyrotropin (probably in a similar manner to that of insulin—see Section D.2.b).

APPENDIX A: VITAMINS

Vitamins are complex organic molecules required in relatively small quantities for metabolism but which cannot be synthesized either at all or in sufficient amounts. Hence they must be provided in the diet and this implies that they must be synthesized by some organisms (particularly plants and micro-organisms). Species differ in their vitamin requirements but, with the exception of vitamin C which, outside the primates is only required by the guinea-pig, those described below are required by all mammals as far as is known. In some cases a proportion of the requirement for these substances may be synthesized but the situation is complicated by the contributions made by intestinal bacteria. By definition rather than rational decision, the essential fatty acids (Chapter 17; Section B.1.b) are not considered to be vitamins.

Vitamins were discovered, as dietary factors required for healthy growth, in the first half of the twentieth century. Their discovery makes exciting reading but has been chronicled elsewhere (Davidson *et al.*, 1979). Gradually their chemical natures were elucidated and diverse they proved to be. In many cases a vitamin was found to be small group of chemically related substances. It has proved useful to divide the vitamins into those that are water-soluble and those that are fat-soluble. This classification has some clinical utility since it indicates the likely method of absorption and transport and has been retained in this appendix.

In most cases, the absence of adequate amounts of a vitamin in the diet or poor absorption produces a deficiency disease with characteristic symptoms and, since many of the vitamins provide cofactors or prosthetic groups, in some cases these symptoms can be traced to a deficient catalytic activity of a specific enzyme. In such deficiency diseases, the cure affected by therapy with the appropriate vitamin is often dramatic. In recent years, at least in Western society, emphasis has shifted to preventive therapy and literally tons of purified vitamins are added to foods or taken (mostly by healthy individuals) in pill form. With this in mind it is well to note that an excessive intake of some vitamins is harmful. Although most are only toxic at very high doses, the hypervitaminoses A and D can occur clinically. The former (causing a number of deviations from good health) has occurred in food faddists who have over-dosed themselves and, reputedly, in Arctic explorers who have consumed large quantities of polar bear liver, a particularly rich source. Vitamin D toxicity (causing hypercalcaemia) can occur in some children from a relatively normal intake of vitamin D-fortified foods.

Readers requiring a more detailed account of the vitamins are recommended to consult the relevant chapters in Goodhart and Shils (1980) (see also Scrimshaw and Young, 1976).

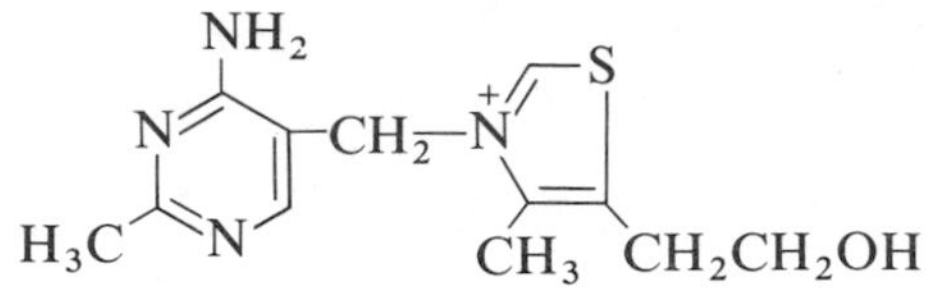

Figure A.1 Structural formula of thiamine (3-(2'-methyl-4' amino-5 pyrimidylmethyl)-5-(2-hydroxyethyl)-4-methylthiazole). The solid is prepared as the double chloride (with both the thiazole nitrogen atom and the amino group bearing a positive charge) known as thiamine chloride hydrochloride

A. WATER SOLUBLE VITAMINS

1. Vitamin B_1 (thiamine)

Good souces of thiamine are fresh vegetables and various meats including liver. It is also present in the husk of cereal grains but this is often removed before consumption. The structure of thiamine is given in Figure A.1.

Thiamine pyrophosphate (or, more correctly now, diphosphate), the active form of the vitamin, is formed by phosphorylation of thiamine by ATP in a reaction catalysed by thiamine pyrophosphokinase. Thiamine pyrophosphate is required as a cofactor in a variety of reactions but especially in oxidative decarboxylation reactions, such as those catalysed by oxoglutarate dehydrogenase and pyruvate dehydrogenase (see Chapter 4; Section B.1.d and Chapter 5; Section B.13.b). Since these reactions are of central importance in ATP generation, a deficiency of this vitamin will particularly affect tissues with a high energy requirement such as nerves, heart, and kidney.

Deficiency in man gives rise to beriberi, the symptoms and clinical features of which are as follows: paresthesia, muscular weakness and atrophy, foot and wrist drop, enlargement of the heart, and digestive disturbances including anorexia and constipation. Cardiac failure may be precipitated by physiological stress which requires an increase in the rate of cardiac metabolism. In the Western world, beriberi is most frequently encountered in alcoholics who may have eaten little or nothing for several weeks; the deficiency is caused not only from a lack of food but to low stores of thiamine due to poor absorption when food was being eaten.

2. Vitamin B_2 (riboflavin)

Riboflavin is present in most foods but especially in milk and meat products. Riboflavin must be present in the diet for cellular growth to occur so that it is needed in proportionally higher amounts in infants and children and in wound-healing (e.g. after injury or surgery). The structure is given in Figure A.2. Riboflavin is a constituent of flavin mononucleotide and flavin adenine dinucleotide, which are involved in many oxido-reduction reactions (see Chapter

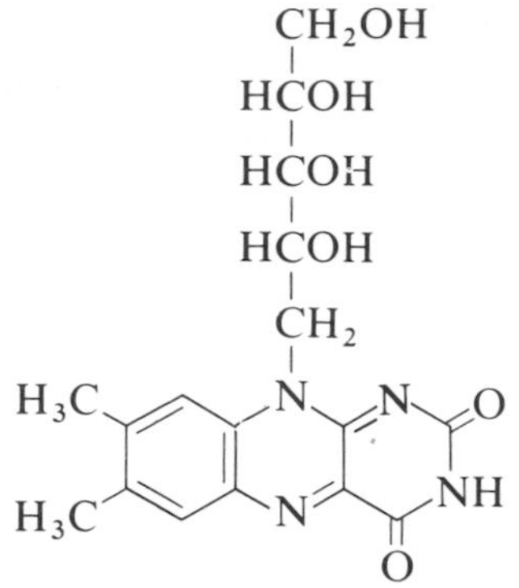

$$CH_2OH$$
$$|$$
$$HCOH$$
$$|$$
$$HCOH$$
$$|$$
$$HCOH$$
$$|$$
$$CH_2$$

Figure A.2 Structural formula of riboflavin (6,7-dimethyl-9-ribitylisoalloxazine). See Figure 4.10 for structures of FMN and FAD

4; Section C.1.b; Chapter 10; Section C.2.a). Riboflavin deficiency in man (ariboflavinosis) produces corneal vascularization, lesions of the skin at the corners of the mouth and on the face, nose, and scrotum.

3. Nicotinic acid (niacin)

Nicotinic acid is present in liver, lean meats, cereals and legumes and, in addition, is formed (as NAD^+) in a minor pathway of tryptophan catabolism. (It is estimated that 1 mg is produced per 60 mg of tryptophan metabolized.) The structures of nicotinic acid and nicotinamide are given in Figure A.3. Nicotinamide forms part of NAD^+ and $NADP^+$ (see Chapter 4; Section C.1.e) and consequences of its deficiency are diverse. In man, a deficiency of nicotinic acid (or tryptophan) causes pellagra, the early signs and symptoms of which include weakness, lassitude, anorexia, and indigestion, and these are followed by dermatitis, diarrhoea, and nervous symptoms. The dermatitis characteristically appears on those parts of the body exposed to sunlight, heat or mild trauma. The nervous symptoms are irritability, headaches, sleeplessness, loss of memory, and other signs of emotional instability.

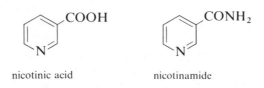

nicotinic acid   nicotinamide

Figure A.3 Structural formulae of nicotinic acid and nicotinamide

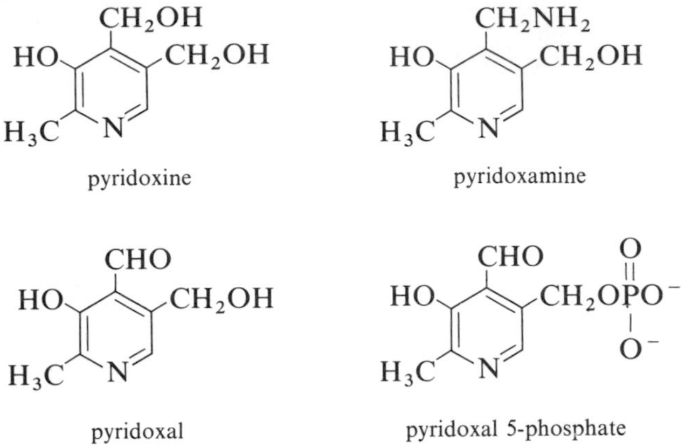

Figure A.4   Forms of vitamin B_6

4. Vitamin B_6 (pyridoxine)

Vitamin B_6 is present in many foods, including meats, cereals, lentils, nuts, some fruits and vegetables. In animal products it occurs largely as pyridoxal and pyridoxamine, while in vegetable products it occurs as pyridoxine (Figure A.4).

Pyridoxal 5-phosphate is the coenzyme form of vitamin B_6, and its synthesis from pyridoxal and ATP is catalysed by pyridoxal kinase. The number of enzymes known to require pyridoxal 5-phosphate is more than 60 and almost all are involved in amino acid metabolism. The reactions include transamination (see Figure 10.3), racemization, decarboxylation, cleavage, synthesis, dehydration, and desulfhydration. Aromatic L-amino acid decarboxylase and glutamic acid decarboxylase also require pyridoxal phosphate (Chapter 21; Section B.2.a). In addition, pyridoxal 5-phosphate plays an important structural role in the enzyme glycogen phosphorylase.

Pyridoxine was first reported to be an essential nutrient when patients on poor diets were observed to have an ill-defined syndrome characterized by weakness, irritability, nervous disorder, insomnia, and difficulty in walking. The widespread occurrence of convulsive seizures and nervous irritability in infants fed an autoclaved commercial liquid milk diet, which was low in vitamin B_6, confirmed its essentiality to man.

Early experiments on volunteers fed diets low in vitamin B_6 produced no clearcut symptoms, though mental depression was noted. However, derangement of tryptophan metabolism, manifested by the excretion of xanthuronic acid, was observed within three weeks. Vitamin B_6 deficiency induced by a pyridoxine antagonist (4-deoxypyridoxine) produced lesions over the face similar to those produced by nicotinic acid or riboflavin deficiency. Hyperpigmented scaly pellagra-like dermatitis also developed occasionally and, in addition, weight loss, apathy, somnolence, and increased irritability were noted.

Vitamin B_6 deficiency may occur in patients with kidney failure and in patients with various forms of liver disease (e.g. in chronic alcoholics). Indications of deficiency have also been reported from patients on a high leucine diet and women taking steroid hormones for contraceptive purposes.

5. Pantothenic acid

Panothenic acid occurs in most foods especially in liver, meat, cereals, milk, egg yolk, and fresh vegetables. It is part of the coenzyme A molecule (see Figure 4.2 and Chapter 4; Section A.3), and hence is involved in many of the major metabolic pathways. It is also present in fatty acid synthase (Chapter 17; Section A.4.b).

Pantothenic acid is of such widespread distribution in foods that a deficiency under normal conditions is exceedingly rare. The following symptoms were recorded from volunteers on a pantothenic acid deficient diet; vomiting, malaise, abdominal distress, burning cramps and, at a later stage, tenderness in the heels, fatigue, and insomnia. Pantothenic acid has been reported to improve the ability of well-nourished subjects to withstand stress. In experimental animals, failure to grow was the most obvious feature of the deficiency.

6. Folates

(a) *Structure and function*

All folates can be considered as being derived from pteroylmonoglutamic acid (folic acid) in which a pteridine* moiety is linked by a methylene bridge to *para*-aminobenzoic acid which is joined by a peptide bond to glutamic acid. However, pteroylmonoglutamic acid is not normally found, as such, in significant amounts in foods or in the body. The structure is modified in several ways.

1. The folates are reduced to give 5,6,7,8-tetrahydrofolates (Figure A.5). For many purposes this can be abbreviated to FH_4 but for more detailed descriptions the abbreviation $H_4PteGlu$ is more useful. Other reduced forms can be specified, for example as $7,8\text{-}H_2PteGlu$.
2. The number of glutamate residues may vary from one to eight, with each one being linked to the next by a peptide bonds between the amino group of one glutamate and the γ-carboxyl group on the previous glutamate (Figure A.5). This glutamate 'chain' is the part of the vitamin that is attached to enzymes. The number of glutamyl residues is indicated by a subscript in the abbreviation; hence the pentaglutamate becomes $H_4PteGlu_5$.
3. Various one-carbon groups (e.g. $-CH_3$, $-CHO$, $-CH=NH$ and $-CH_2-$) may be attached to nitrogen atoms in positions 5 or 10 or may form a bridge between them (Figure A.5). These structures are fundamental to the role of

---

* Tetrahydrobiopterin, which is involved as a cofactor in some hydroxylation reactions (see Chapter 21; Section B.2.a. and Figure 21.7) is a derivative of pteridine.

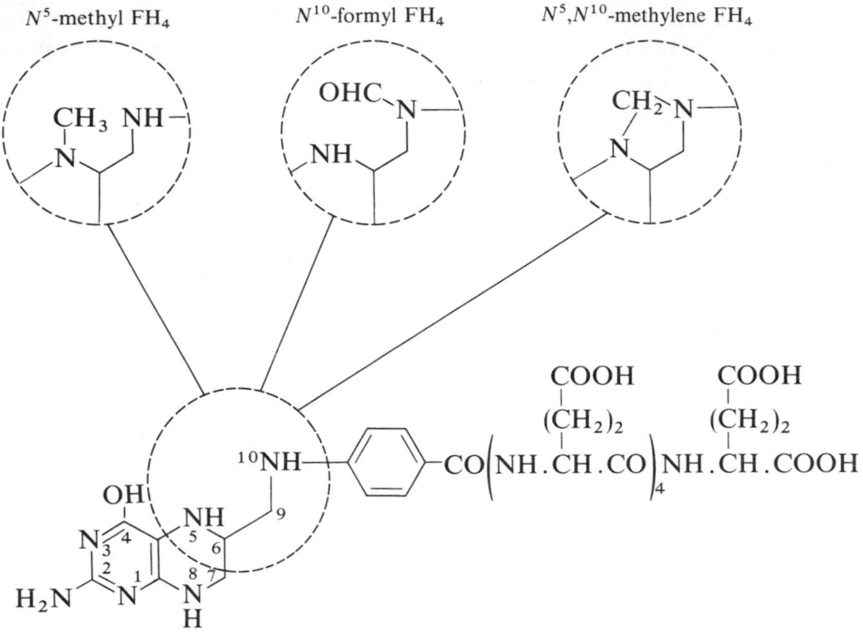

5,6,7,8-tetrahydropteroylpentaglutamate (FH$_4$)

Figure A.5 Structural formula of tetrahydrofolate and representative derivatives. The pentaglutamate shown is the predominant form in man but tetrahydrofolate with from one to eight glutamate residues occurs. The parent compound, pteroylglutamate (folate), lacks four hydrogen atoms, one each from carbon atoms 5, 6, 7, and 8

folates. They are interconvertible and are used in different reactions, for details of which see Colman and Jacob (1980). The nature of the one carbon group and its position on the vitamin can be indicated in a number of ways, e.g. 5,10-CH$_2$—H$_4$PteGlu$_5$ or N^5N^{10}-methylene FH$_4$ for 5,10-methylenetetra-hydropteroylpentaglutamate.

Conversion of dietary folate in the intestinal lumen and absorptive cells is such that only 5-CH$_3$—H$_4$PteGlu$_1$ enters the portal blood and is transported to tissues. Folate is transported, by an energy-dependent process, into bone marrow cells, reticulocytes, hepatocytes, renal tubular cells, and cerebrospinal fluid. In man, the pentaglutamate form is most common intracellularly.

Folates form prosthetic groups of many enzymes that are involved in transfer of one-carbon units. These one-carbon units arise mainly from the β-carbon of serine, the α-carbon of glycine and carbon-2 of the imidazole ring of histidine (see Chapter 10; Appendix 1) and also from the utilization of formate. The transfer of one-carbon units from folates occurs in synthetic reactions which include:

1. *De novo* purine nucleotide synthesis (to provide carbon atoms 2 and 8 of the purine ring—see Figure 18.7).

2. Pyrimidine nucleotide synthesis (methylation of dUMP to produce dTMP).
3. Conversion of homocysteine to methionine, which also requires vitamin B_{12}.
4. Methylation of transfer RNA.

(b) *Deficiency*

Folates are present in nearly all natural foods. Those with the highest folate content include yeast, liver and other organ meats, fresh green vegetables, and some fresh fruits. However, folates are highly susceptible to oxidative destruction and 50% to 90% of the folate content of foods may be destroyed by protracted cooking or other processing such as canning. Because of this widespread occurrence, folate deficiency is more likely to arise from inadequate usage (e.g. inadequate absorption from the intestine) than from a deficiency of folate in the diet. Although folate deficiency may induce personality changes such as hostility and paranoid behaviour, anaemia is the most easily recognizable sign of folate deficiency. The megaloblastic anaemia arising from folate deficiency is characterized by weakness, tiredness, dyspnoea, sore tongue, diarrhoea, irritability, anorexia, headache, and palpitations. Even when adequate amounts of folate are absorbed, deficiency symptoms may occur if insufficient vitamin B_{12} or methionine is available in tissues. It has been suggested that the functional relationship between folate and vitamin B_{12} in the tetrahydropteroylglutamate methyltransferase reaction (Figure A.6) may cause a deficiency of intracellular folate resulting in megaloblastic anaemia. In vitamin B_{12} deficiency, homocysteine cannot be converted to methionine so that $5\text{-}CH_3\text{-}H_4PteGlu$ cannot be converted to $H_4PteGlu$, in which form it is retained within the cell by conversion to the polyglutamate form (pentaglutamate in man): only $H_4PteGlu$ can be converted to polyglutamate forms. Hence a deficiency occurs in the bone marrow cells. This has

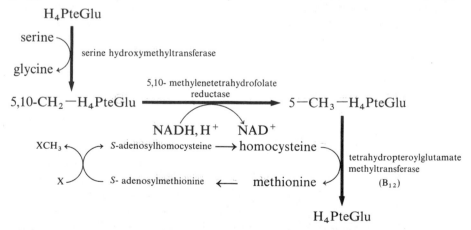

Figure A.6 Pathway of methyl group transfer. The folate in these reactions is most probably in the pentaglutamyl form. Note that $5,10\text{-}CH_2\text{-}H_4$ PteGlu arises by reactions in addition to that shown

been termed the 'methylfolate trap' and prevents the synthesis of purines and pyrimidines so inhibiting cell division an l particularly erythrocyte formation. Similarly, it has been suggested that in protein deficiency states, methionine concentrations will be too low to provide sufficient methyl groups for methylation, especially in the brain and nervous tissue where methylation is essential for the phospholipids and sphingomyelins in myelin. To compensate, conversion of homocysteine to methionine must be enhanced which is done by increasing the concentration of $5\text{-}CH_3\text{-}H_4PteGlu$, again at the expense of the intracellular form, $H_4PteGlu$. This may explain why patients with kwashiorkor can suffer from anaemia, despite normal blood concentrations of vitamin B_{12} and folate (Scott and Weir, 1981).

7. Vitamin B_{12} (cobalamin)

Vitamin B_{12} is not found in plants and the sole source in nature is provided by micro-organisms. The usual dietary sources for man are meat and meat products (including shell-fish, fish, and poultry) and to a lesser extent milk and milk products.

The structure of vitamin B_{12} is complex (Figure A.7); it contains four linked pyrrole rings known as the corrin nucleus, to which are attached three acetamide, four propionamido, and eight methyl groups. The corrin nucleus provides four of the ligands co-ordinating a cobalt atom in its centre. The fifth ligand is the ribonucleotide of 5,6-dimethylbenzimidazole which is also linked to the nucleus through 1-amino-2-propanol. A variety of ligands can occupy the sixth co-ordination position of cobalt (see legend to Figure A.7) but the deoxyadenosyl and methyl forms predominate in animal tissues.

There are several steps in the absorption of vitamin B_{12}. In the stomach and lumen of the small intestine, it is hydrolysed from its (peptide) links with proteins and attaches itself to the gastric intrinsic factor, a glycoprotein of molecular weight about 50 000, to form a complex of two molecules of intrinsic factor and two molecules of vitamin B_{12}. In this form it is carried into the ileum and binds to the surface of the brush border of the absorptive cells. The B_{12} is released from the intrinsic factor probably at the surface of the brush border but possibly within the absorptive cell. The vitamin now finds its way into the portal venous blood, where it is bound to the vitamin B_{12}-binding proteins (of which there are three types, transcobalamin I, II, III). Methylcobalamin appears to be the main serum transport form, whereas deoxyadenosylcobalamin appears to be the main storage form in the liver and kidney.

Vitamin B_{12} provides prosthetic groups to two classes of enzyme. The first, exemplified by methylmalonly-CoA mutase, is involved in intramolecular group transfers (isomerizations), while the second consists of methyltransferase reactions of the type illustrated in Figure A.6. Methylmalonyl-CoA mutase (Chapter 6; Section D.1.e; Chapter 10; Section C.3.b) requires deoxyadenosylcobalamin while tetrahydropteroylglutamate methyltransferase, not surprisingly, employs methyl-cobalamin.

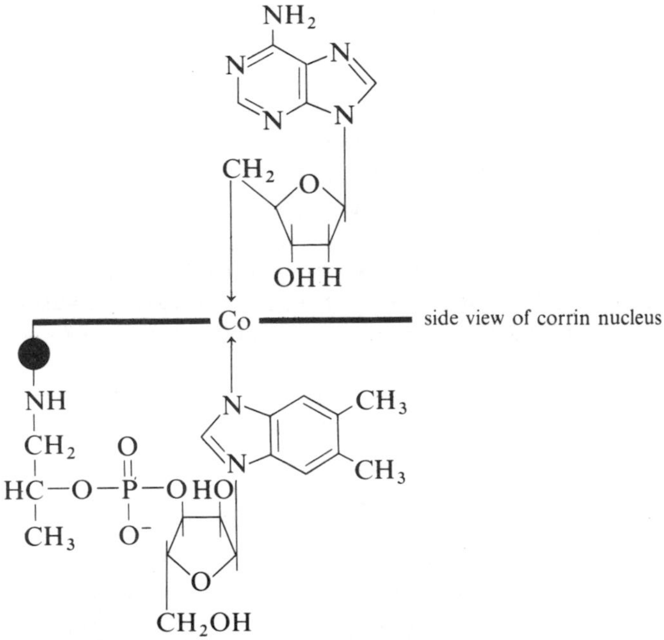

Figure A.7 Structural formula of deoxyadenosylcobalamin (coenzyme B_{12}).

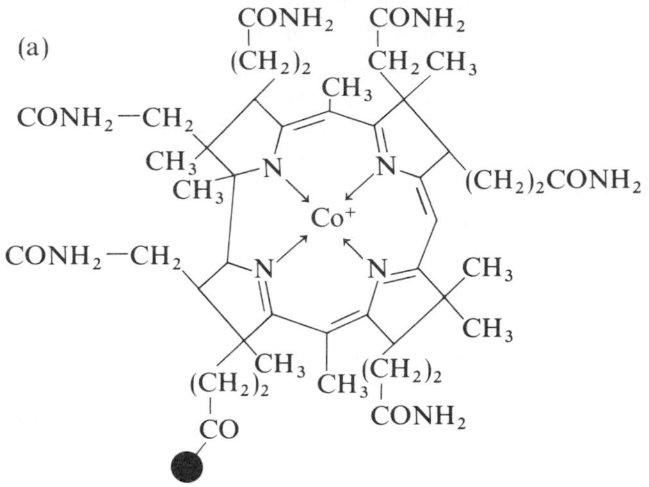

(a) shows a 'plan' view of the corrin nucleus with substituents.

(b) shows the position of the remaining two ligands of the cobalt atom. No attempt is made to show correct stereochemical relationships. Related compounds have different groups in place of the 5′-deoxyadenosyl group: cyanocobalamin, vitamin B_{12}, $-CN$; hydroxycobalamin, vitamin B_{12a}, $-OH$; methylcobalamin, methyl B_{12}, $-CH_3$

Vitamin B_{12} deficiency in man results in two serious problems, megaloblastic anaemia and neurological disorders. However, vitamin B_{12} deficiency is rare, because of the very low concentrations required. A normal healthy adult may survive more than a decade without dietary vitamin B_{12} with no obvious deficiency symptoms. However, anaemia occurs fairly rapidly in patients who have a defective vitamin B_{12} absorption system. Poor absorption may be due to deficiency of intrinsic factor or to a primary intestinal disease such as sprue. Anaemias resulting from vitamin B_{12} inadequacies have been termed pernicious anaemias because they do not respond to treatment with iron. Symptoms include tiredness, dyspnoea, sore tongue, paresthesia, constipation, anorexia, headache, and palpitations. Megaloblastic anaemia is characterized by a low concentration of haemoglobin, abnormally large red cells and abnormal erythroblasts in the bone marrow. These are probably caused by an inability to produce sufficient S-adenosylmethionine in the marrow for methyl group transfer in formation of nucleotides for DNA synthesis (see below and Figure A.6). The neuropathy, which affects peripheral nerves as well as the central nervous system, is probably due to a lack of methionine for methyl transfer to form choline for the phospholipids and sphingomyelins needed in the formation of myelin.

8. Biotin

Biotin occurs in a wide variety of foods (including yeast, organ meats, muscle meats, diary products, grains, fruits, and vegetables) and it is also produced by micro-organisms in the large intestine. Biotin is a coenzyme for a number of enzymes including acetyl-CoA carboxylase, propionyl-CoA carboxylase, pyruvate carboxylase, β-methyl-crotonoyl carboxylase, and methylmalonyl-CoA carboxyltransferase. In each of these enzyme systems, ATP hydrolysis is coupled to the carboxylation of biotin, to form N-1'-carboxybiotin (see Figure A.8) which then carboxylates the substrate of the enzyme.

Biotin deficiency is rare in man since intestinal micro-organisms can provide sufficient for the normal requirements.* However, biotin deficiency can be induced in subjects by feeding raw egg-whites which contain a protein, avidin, which binds

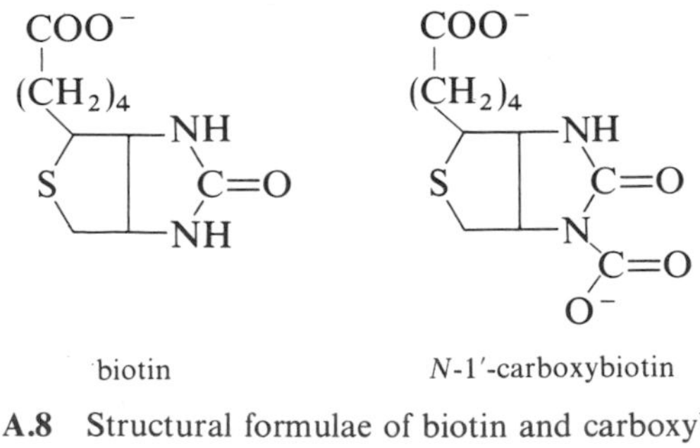

biotin N-1'-carboxybiotin

Figure A.8   Structural formulae of biotin and carboxybiotin

* Deficiency can occur in man due to poor intestinal absorption (see Thoene *et al.*, 1983).

biotin very strongly and so prevents it binding to carboxylases. Such deficiency produces the following signs and symptoms: a scaly dermatitis, atrophy of lingual papillae, graying of mucous membranes, increasing skin dryness, depression, lassitude, muscle pains, paresthesia, anorexia, nausea, and electrocardiogram abnormalities. Low concentrations of biotin in the liver have been found in patients suffering from fatty livers as a result of alcoholism, possibly indicating a decreased storage capacity for biotin in the fat-infiltrated liver.

9. Choline

Choline (see Figure 17.10 for structure) can be synthesized in man but it is not clear if the rate of synthesis is sufficient for all the physiological purposes so that it might be a vitamin. There have been no reports of experimental studies designed to produce choline deficiency in man. It is found in many foods especially organ meats, muscle meats, egg yolk, legumes, and cereals. Choline is synthesized by the transfer of methyl groups form S-adenosylmethionine to the phospholipid, phosphatidylethanolamine with the formation of lecithin; complete hydrolysis of the lecithin releases free choline. As well as its structural role in phospholipids (see Chapter 17; Section C.1) choline is a precursor in the synthesis of the neurotransmitter, acetylcholine (Chapter 21; Section B.1.a). A deficiency of choline in experimental animals results in development of a 'fatty liver' (see Chapter 11; Section F.5.b), due to a lowered capacity to remove the triacylglycerol, and damage to the blood vessels in the liver.

10. Vitamin C (ascorbic acid)

Vitamin C is obtained primarily from plants, especially fresh, rapidly growing fruits and vegetables. In rats, and presumably other mammals that do not require a dietary source of ascorbic acid, it is synthesized from glucose. Ascorbic acid functions as a cosubstrate in a number of oxido-reduction reactions including that catalysed by dopamine β-monooxygenase (Chapter 21; Section B.2.a) and the post-translational hydroxylation of proline in the formation of collagen, so that it is important in growth and healing of both wounds and bone fractures. In these processes, ascorbate is reversibly oxidized to dehydroascorbate (Figure A.9). Ascorbic acid is a water-soluble antioxidant and can protect other

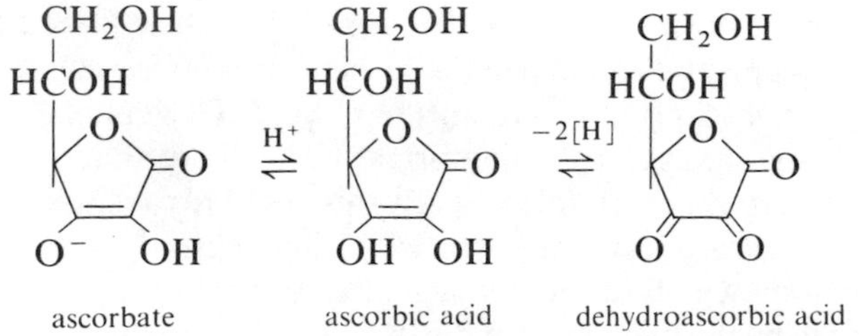

Figure A.9 Structural formulae of forms of ascorbic acid

antioxidants (e.g. vitamin E, vitamin A, and the essential fatty acids). This property may explain its role in the maintenance of epithelia and mucous secretory cells. It is also a chelating agents and facilitates the absorption of iron from the intestine. Vitamin C is essential for normal protein metabolism and deficiency leads to lowering of the serum albumin concentration and a high rate of nitrogen loss in the urine. Finally, ingestion of large amounts of vitamin C have been recommended for protection against the common cold but clinical trials have shown only minor improvements.

Vitamin C deficiency produces a disease known as scurvy. Symptoms and clinical features include fatigue, lassitude, bleeding gums, depression, dry skin, xerophthalmia, xerostomia, follicular hyperkeratosis, impaired iron absorption, and impaired wound healing.

B. FAT-SOLUBLE VITAMINS

1. Vitamin A

The term vitamin A now includes several biologically active compounds including all-*trans*-retinol (vitamin A_1; Figure A.10) and 3-dehydroretinol (vitamin A_2; having an additional double bond between carbon atoms 3 and 4). In animal tissues, and especially the liver, vitamin A is stored as retinyl esters (with long-chain fatty acids) so that these are the major dietary source. Hydrolysis to retinol occurs in the lumen of the small intestine. The other major sources of this vitamin are the carotenoid pigments present in green plants. The carotenoids are known as provitamins A, and conversion to vitamin A involves oxidative fission catalysed by β-carotene 15,15′-dioxygenase in the intestine (Figure A.10). This forms retinal, the corresponding aldehyde which is the active form of the vitamin in the visual cycle, but which is reduced to the alcohol by retinol dehydrogenase for transport and storage.

Vitamin A is absorbed from the small intestine with the aid of bile salts and incorporated into the chylomicrons which transport it to the liver for storage until required.

The most readily apparent role of vitamin A is its participation in the detection of light by the retinal cells. The retina contains two distinct photoreceptor systems, cones and rods (see Figure 5.24). In both, the photoreceptor pigments consist of proteins, opsin in rods and iodopsin in cones, linked covalently to retinal. In these pigments, retinal is in the 11-*cis* form which is converted to the 11-*trans* form by the absorption of a quantum of light. This change in shape of the retinal produces a change in the conformation of the protein which probably leads to an increase in calcium ion permeability. This results in a change in potential that initiates the action potential sending nerve impulses, via the optic nerves, to the brain which result in visual sensations. In the rods, these changes are accompanied by the hydrolysis of the link between retinal and opsin to release all-*trans*-retinal which is then isomerized to the 11-*cis* isomer which recombines

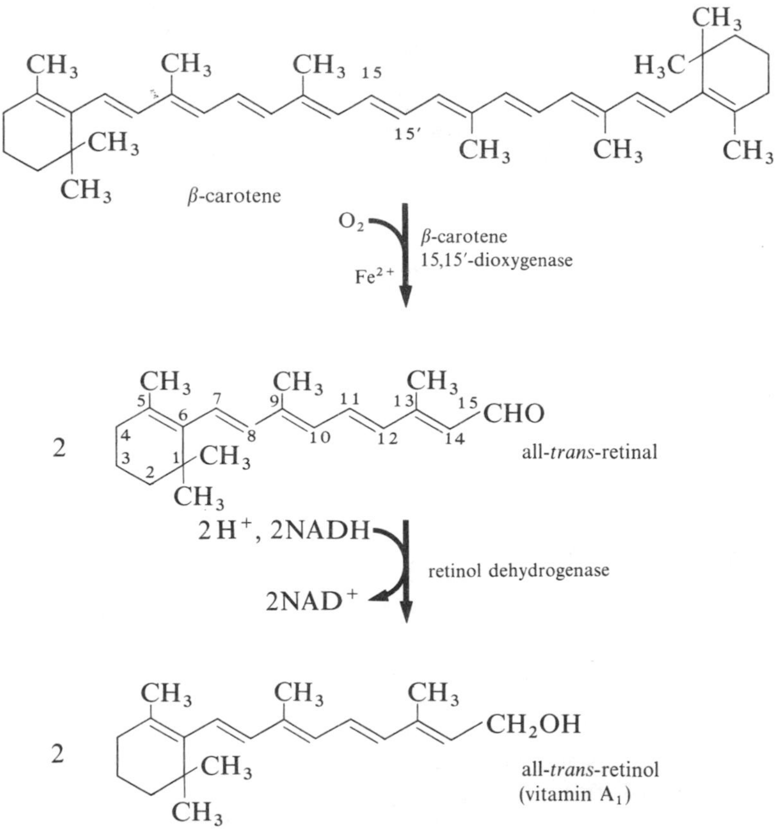

Figure A.10 Formation of vitamin A from β-carotene. Alcohol dehydrogenase is also able to catalyse the reduction

with opsin to regenerate rhodopsin (Figure A.11). A similar mechanism operates in cones.

Although the first symptom is night-blindness, deficiency of vitamin A causes many lesions indicating that it has roles in the body other than those associated with photoreception. A thickening of membranes (keratinization) occurs in a number of sites in the body, particularly the cornea where it gives rise initially to xerophthalmia which, if untreated, can lead to blindness through perforation of the cornea and loss of the lens.

Patients with kwashiorkor have very low reserves of vitamin A and since their requirement for this vitamin increases when they are supplied with protein, they can quickly become vitamin A deficient unless it is administered with the protein. A deficiency of vitamin A can also occur in fat absorption diseases, for example, if there is a deficency of bile (Chapter 6; Section B.1.b) or an increase in its loss from the body.

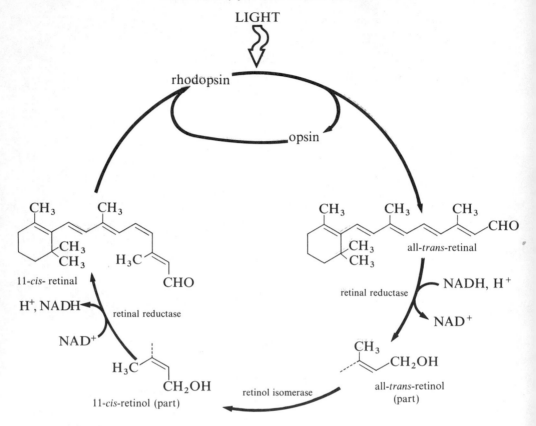

Figure A.11   Role played by vitamin A derivatives in the detection of light by rod cells in the retina

2. Vitamin D

Since Vitamin D can be considered as a steroid hormone it is discussed in Chapter 20; Section F. The naturally occurring form of the vitamin is cholecalciferol (D_3) but large amounts of ergocalciferol (D_2) are manufactured from ergosterol and used as a food additive. The structures of both are shown in Figure A.12. Vitamin D_2 is as active as vitamin D_3 in the rat but does not appear to be converted to it.

3. Vitamin E (tocopherol)

Vitamin E is a group of eight naturally occurring compounds called tocopherols (Figure A.13). The most important source of vitamin E is vegetable seed oils (corn oil, sunflower seed oil, wheat germ oil). Other sources include milk, eggs, meat, leafy vegetables.

    There is considerable debate concerning the role of vitamin E although it seems

cholecalciferol (D$_3$)

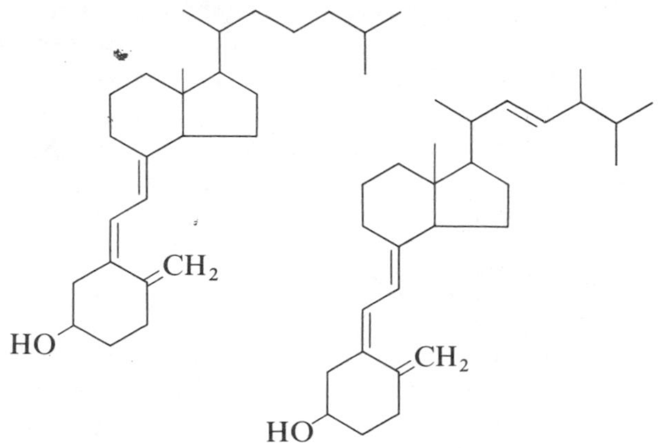

ergocalciferol (D$_2$)

Figure A.12   Structural formulae of two forms of vitamin D

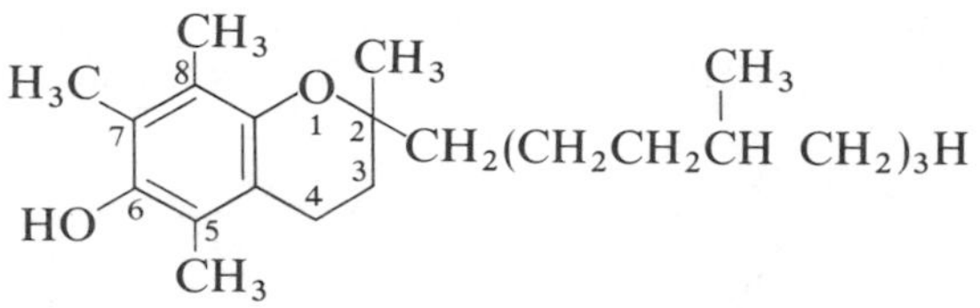

$$CH_2(CH_2CH_2CH\ CH_2)_3H$$

Figure A.13   Structural formula of α-tocopherol. In β-tocopherol, the 7-methyl group is absent, in γ-tocopherol, the 5-methyl group is absent, and in δ-tocopherol, both methyl groups are absent. In the tocotrienols, the side-chain at position-2 is replaced by:

$$CH_2(CH_2CH{=}CCH_2)_3H$$
$$CH_3$$

likely that it functions as an antioxidant since it is a scavenger for oxygen free radicals (Chapter 4; Section E.5.b). However, the large number of different physiological abnormalities produced by a deficiency of the vitamin (see below) has led to the suggestion that it functions as a component of a number of enzymes or transport systems (see Horwitt, 1980).

   In man, deficiency of vitamin E is rare, although this can occur due to abnormalities in lipid absorption. Deficiency causes premature destruction of erythrocytes and a high intake of this vitamin may reduce the abnormally high rate of destruction in patients with glucose-6-phosphate dehydrogenase deficiency (Chapter 4; Section E.5.b). Premature infants can suffer from vitamin E deficiency which results in anaemia and may also be responsible for oedema, skin lesions, and an elevated number of platelets. There is some evidence to suggest that vitamin E decreases the rate of ageing (possibly by decreasing the concentration of free radicals). Studies with animals show that deficiency affects the reproductive system; in the male, there is inhibition of spermatogenesis and eventually testicular atrophy; in the female, poor vascularization of the placenta can lead to fuel and oxygen starvation of the foetus and its eventual resorption. Deficiency also affects the muscular system causing dystrophy and paralysis and, if the heart is affected, death by myocardial failure may occur. Deficiency can also result in extensive lesions in some parts of the brain and an increased susceptibility to haemorrhage.

4. Vitamin K

The vitamin K requirement of mammals is met by a combination of dietary intake (vitamin K_1) and microbiological synthesis in the large intestine (vitamin K_2). Each provides about half of the daily requirement in man. The richest dietary source of vitamin K is provided by green leafy vegetables but it is also obtained from meats and dairy produce. The structural formulae of vitamin K_1 (phylloquinone) and vitamin K_2 (menaquinone) are given in Figure A.14.

Vitamin K is needed for the post-translational carboxylation of certain glutamate residues in proteins to γ-carboxyglutamate. This process is important in the activation of four blood-clotting factors (Factors II (prothrombin), VII, IX,

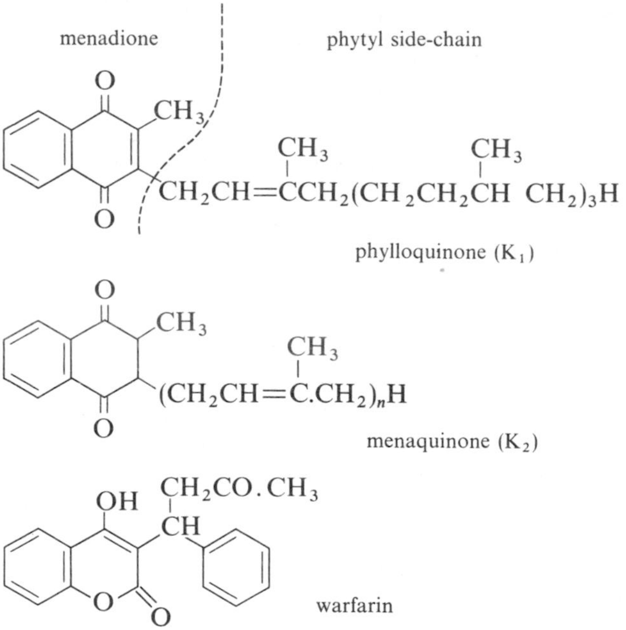

Figure A.14 Forms of vitamin K and the structure of a vitamin K antagonist. Menadione (2-methyl-1,4-napthoquinone) arises in the intestine due to bacterial action on phylloquinone and can be converted into the active vitamin by animal tissues

and X). An antagonist of vitamin K, warfarin, has been developed as a rat poison and as an anti-coagulant in the treatment of thrombosis (Figure A.14).

Primary vitamin K deficiency is uncommon in man not only because of the presence of vitamin K in many plant and animal tissues but especially due to the production of this vitamin by micro-organisms in the intestine. However, newborn infants can readily become deficient since transport across the placenta is low (so that there is little or no store of the vitamin in the newborn) and the intestine is sterile during the first few days of life. Adults on a low vitamin K diet can show signs of deficiency (a marked decrease in the plasma prothrombin concentration) if they are given broad spectrum antibiotics such as neomycin.

REFERENCES

Adams, J. S., Clemens, T. L., Parrish, J. A. *et al.* (1982). Vitamin-D synthesis and metabolism after ultraviolet irradiation of normal and vitamin-D deficient subjects. *New Engl. J. Med.* **306**, 722–725.

Abel, M. H. (1979). Production of prostaglandins by the human uterus; are they involved in menstruation. *Research Clinical Forums*, **1**, 33–37.

Adair, G. S., Barcroft, J., and Bock, A. V. (1921). The identity of haemoglobin in human beings. *J. Physiol.* (*London*), **55**, 332–338.

Adibi, S. A., Morse, E. L., Masilamani, S. S. *et al.* (1975). Evidence for two different modes of tripeptide disappearance in human intestine; uptake by peptide carrier systems and hydrolysis by peptide hydrolases. *J. Clin. Invest.*, **56**, 1355–1363.

Ager, J. Y., and Young, R. T. (1974). Non-hypoglycaemia is an epidemic condition. *New Engl. J. Med.*, **291**, 907–908.

Agius, L., and Williamson, D. H. (1980). Lipogenesis in interscapular brown adipose tissue of virgin, pregnant, and lactating rats. *Biochem. J.*, **190**, 477–480.

Agranoff, B. W. (1975). Neurotransmitters. *Fed. Proc.*, **34**, 1911–1914.

Ahlborg, B., Bergström, J., Ekelund, L.-G. *et al.* (1967). Muscle glycogen and muscle electrolytes during prolonged physical exercise. *Acta Physiol. Scand.*, **70**, 129–142.

Ahlborg, G., Felig, P., Hagenfeldt, L. *et al.* (1974). Substrate turnover during prolonged exercise in man. *J. Clin. Invest.*, **53**, 1080–1090.

Alberti, K. G. M. M., and Hockaday, T. D. R. (1977). Diabetic coma: an appraisal after five years. *Clin. Endocrinol. Metab.*, **6**, 421–456.

Alberti, K. G. M. M., Johnston, D. G., Burrin, J. *et al.* (1981). Ketogenesis: regulatory factors *in vivo*. *Biochem. Soc. Trans.*, **9**, 8–9.

Albright, F. (1945). Disorders of the gonads. In: *Internal Medicine—its Theory and Practice*. (Ed. Musser, J. H.) Lea and Febiger, Philadelphia. pp. 951–980.

Alfin-Slater, R. B., and Aftergood, L. (1980). Lipids In: *Modern Nutrition in Health and Disease*, (Ed. by Goodhart, R. S., and Shils, M. E.) Lea and Febiger, Philadelphia. pp. 113–141.

Allen, F. M., Stillman, E., and Fitz, R. (1919). Total dietary regulation in the treatment of diabetes. *Monograph of the Rockefeller Institute for Medical Research*, Rockefeller Institute, New York, vol. 11, 1–646.

Alleyne, G. A. O., Hay, R. W., Picon, D. I., *et al.* (1977), *Protein Energy Malnutrition*, Arnold, London.

Almy, T. P., and Howell, D. A. (1980). Diverticular disease of the colon. *New Engl. J. Med.*, **302**, 324–331.

Alp, P. R., Newsholme, E. A., and Zammit, V. A. (1976). Activities of citrate synthase and NAD$^+$-linked and NADP$^+$-linked isocitrate dehydrogenase in muscle from vertebrates and invertebrates. *Biochem. J.* **154**, 689–700.

Ames, M. M., Lerner, P., and Lovenberg, W. (1978). Tyrosine hydroxylase: activation by protein phosphorylation and end-product inhibition. *J. Biol. Chem.*, **253**, 27–31.

Anderson, A. (1980). The genital system. In: *Clinical Physiology in Obstetrics*, (Ed. Hytten, F., and Chamberlain, G.) Blackwell Scientific Publications, Oxford. pp. 328–382.

Anderson, G. H. (1981). Diet, neurotransmitters and brain function. *Brit. Med. Bull.*, **37**, 95–100.

Anderson, M. P., Bechsgaard, P., Frederiksen, J., *et al.* (1979). Effect of alprenolol on mortality among patients with definite or suspected acute myocardial infarction. *Lancet*, **ii**, 865–867.

Andrews, W. C. (1979). Oral contraception. *Clinic. Obstetrics and Gynaecol.*, **6**, 3–26.

Annison, E. F., and Lewis, D. (1959). *Metabolism in the Rumen*, London, Methuen.

Aoki, T. T., Finley, R. J., and Cahill, G. F. (1978). The redox state and regulation of amino acid metabolism in man. *Biochem. Soc. Symp.*, **43**, 17–30.

Arch, J. R. S., and Newsholme, E. A. (1976). Activities and properties of adenylate cyclase and phosphodiesterase in muscle, liver and nervous tissue in relation to control of concentration of adenosine 3′,5-cyclic monophosphate. *Biochem. J.*, **158**, 603–622.

Arch, J. R. S., and Newsholme, E. A. (1978). The control of the metabolism and the hormonal role of adenosine. *Essays in Biochem.*, **14**, 82–123.

Ardawi, M. S. M., and Newsholme, E. A. (1982). Maximal activities of some enzymes of glycolysis, the tricarboxylic acid cycle and ketone body and glutamine utilization pathways in lymphocytes of the rat. *Biochem., J*, **208**, 743–748.

Ardawi, M. S. M., and Newsholme, E. A. (1983). Glutamine metabolism in lymphocytes of the rat. *Biochem. J.* **212**, 835–842.

Arnone, E., and Perutz, M. F. (1974). Structure of inositol hexaphosphate—human deoxyhaemoglobin complex. *Nature*, **249**, 34–36.

Ashcroft, S. J. H. (1976). The control of insulin release by sugars. *CIBA Found. Symp.*, **41**, 117–139.

Ashcroft, S. J. H. (1980). Glucoreceptor mechanisms and the control of insulin release and biosynthesis. *Diabetologia*, **18**, 5–15.

Asmussen, E. (1979). Muscle fatigue. *Med. Sci. Sports*, **11**, 313–321.

Atkins, E. Fever. *New Engl. J. Med.* **308**, 958–960.

Atkinson, D. E. (1977). *Cellular Energy Metabolism and its Regulation.* Academic Press, New York, London.

Axelrod, J. (1975). Catechol-*O*-methyltransferase and other *O*-methyltransferases. In: *Handbook of Physiology, Endocrinology* II, *Adrenal gland*, (Ed. Blaschko, H., Sayers, G., and Smith, A. D.) American Physiological Society, Washington, D. C. pp. 669–676.

Aynesley-Green, A., Williamson, D. H., and Gitzelmann, R. (1977). Hepatic glycogen synthetase deficiency. *Arch. Dis. Child.*, **52**, 573–579.

Babior, B. M. (1978). Oxygen dependent microbial killing by phagocytes. *New Engl. J. Med.*, **298**, 659–668; 721–725.

Bachelard, H. S. (1978). Glucose as a fuel for the brain. *Biochem. Soc. Trans.*, **6**, 520–524.

Badawy, A. A. B. (1978). The metabolism of alcohol. *Clinics in Endocrinol. Metabol.*, **7**, 247–272.

Badwey, J. A., Curnutte, J. T., and Karnovsky, M. L. (1979). The enzymes of granulocytes that produce superoxide and peroxide. *New Engl. J. Med.*, **300**, 1157–1160.

Bailey, I. A., Williams, S. R., Radda, G. K., *et al.* (1981). Activity of phosphorylase in total global ischaemia in the rat heart. A phosphorus-31 nuclear-magnetic-resonance study. *Biochem. J.*, **196**, 171–178.

Baldwin, K. M., Winder, W. W., Terjung, R. L., *et al.* (1973). Glycolytic enzymes in different types of skeletal muscle in adaptation to exercise. *Am. J. Physiol.*, **225**, 962–966.

Baldwin, J. E., and Krebs, H. A. (1981). The evolution of metabolic cycles. *Nature*, **291**, 381–382.

Ballard, F. J. (1977). Intracellular protein degradation. *Essays in Biochem.*, **13**, 1–38.

Balasse, E. O., and Neef, M. A. (1973). Influence of nicotinic acid on the rates of turnover and oxidation of plasma glucose in man. *Metabolism*, **22**, 1193–1204.

Balasse, E. O., and Neef, M. A. (1974). Operation of the 'Glucose-fatty-acid cycle' during experimental elevations of plasma free fatty acid levels in man. *Europ. J. Clin. Invest.,* **4,** 247–252.

Bank, W. J., Dimauro, S., Bonilla, E., *et al.* (1975). The disorder of muscle lipid metabolism and myoglobinuria; absence of carnitine palmitoyl transferase. *New Engl. J. Med.,* **292,** 443–449.

Banks, B. E. C., and Vernon, C. A. (1970). Reassessment of the role of ATP *in vivo. J. Theor. Biol.,* **29,** 301–326.

Bannister, J. V., and Hill, H. A. O. (1982). Chemical reactivity of oxygen derived radicals with reference to biological systems. *Biochem. Soc. Trans.,* **10,** 68–71.

Banting, F. G., Best, C. H., Collip, J. B., *et al.* (1922). Pancreatic extracts of diabetes mellitus. *Canadian Med. Ass. J.,* **12,** 141–146.

Barclay, S. (1969). *Air Crash Detective.* Hamish Hamilton, London.

Baserga, R. (1981). The cell cycle. *New. Engl. J. Med.,* **304,** 453–459.

Bass, C., and Wade, C. (1982). Type A behaviour, not specifically pathogenic. *Lancet,* **ii,** 1147–1149.

Bates, M. W., Krebs, H. A., and Williamson, D. H. (1968). Turnover rates of ketone bodies in normal, starved and alloxan-diabetic rats. *Biochem. J.,* **110,** 655–661.

Baum, M., Brinkley, D. M., Dossett, J. A. *et al.* (1983). Controlled trial of tamoxifen as adjuvent agent in management of early breast cancer, *Lancet,* **6,** 257–260.

Bayless, T. M., and Rosensweig, N. S. (1966). A racial difference in incidence of lactase deficiency. *J. Am. Med. Ass.,* **197,** 968–972.

Beaumont, J. L. (1970). Classification of hyperlipidaemias and hyperlipoproteinaemias. *Bull. Wld. Hlth. Org.,* **43,** 891–915.

Beis, A., Zammit, V. A., and Newsholme, E. A. (1980). Activities of 3-oxoacid CoA-transferase and acetoacetyl-CoA thiolase in relation to ketone body utilization in muscles from vertebrates and invertebrates. *Eur. J. Biochem.* **104,** 209–215.

Beis, I., and Newsholme, E. A. (1975). The contents of adenine nucleotides, phosphagens and some glycolytic intermediates in resting muscles from vertebrates and invertebrates. *Biochem. J.,* **152,** 23–32.

Belfrage, P., Berg, B., Hägerstrand, I., *et al.* (1977). Alterations of lipid metabolism in healthy volunteers during long-term ethanol intake. *Eur. J. Clin. Invest.,* **7,** 127–131.

Bendall, J. R., and Taylor, A. A. (1970). The Meyerhof quotient and the synthesis of glycogen from lactate in frog and rabbit muscle. *Biochem. J.,* **118,** 887–893.

Bender, A. E. (1975). Nutritional aspects of amino acids. In: *Amino Acid Metabolism* (Ed. Bender, D. A.), John Wiley, London, New York, pp. 192–202.

Bender, D. A. (1982) Biochemistry of tryptophan in health and disease. *Molec. Aspects Med.,* **6,** 101–197.

Benesch, R., and Benesch, R. E. (1967). The effect of organic phosphates from the human erythrocyte on the allosteric properties of haemoglobin. *Biochem. Biophys. Res. Commun.,* **26,** 162–167.

Bennion, L. J., and Grundy, S. M. (1975). Effects of obesity and caloric intake on biliary lipid metabolism in man. *J. Clin. Invest.,* **56,** 996–1011.

Berger, M., Kemmer, F. W., Becker, K., *et al.* (1979). Effect of physical training on glucose tolerance and on glucose metabolism of skeletal muscle in anaesthetized normal rats. *Diabetologia,* **16,** 179–184.

Bergquist, C., Nillius, S. J., and Wide, L. (1979). Intranasal gonadotrophin releasing hormone agonist as a contraceptive agent. *Lancet,* **ii,** 215–216.

Bergmeyer, H.-U. (1974). *Methods of Enzymatic Analysis,* 2nd Edn. Verlag Chemie/Academic Press. New York and London, Vols. 1–4.

Bergmeyer, H.-U. (1978). Determination of the catalytic activity of enzymes. In: *Principles of Enzymatic Analysis,* (Ed. Bergmeyer, H.-U.) Verlag Chemie, New York, pp. 56–78.

Bergström, J. (1962). Muscle electrolytes in man. *Scand. J. Clin. Invest.,* **14,**

Supplementum 68, 1–110.

Bergström, J., Hermansen, L., Hultman, E., *et al.* (1967). Diet, muscle glycogen and physical performance. *Acta Physiol. Scand.,* **71,** 140–150.

Bergström, J. P., Fürst, P., and Noree, L. O. (1975). Treatment of chronic uremia patients with protein poor diet and oral supply of essential amino acids. *Clin. Nephrol.,* **3,** 187.

Bergström, S., Ryhage, R., Samuelsson, B., *et al.* (1962). The structure of prostaglandins E_1, F_1, and F_2. *Acta Chem. Scand.,* **16,** 501–502.

Berthoud, H. R., Bereiter, D. A., Trimble, E. R., *et al.* (1981). Cephalic phase, reflex insulin secretion. *Diabetologia,* **20,** 393–401.

Berti, F., and Velo, G. P. (1981). *The Prostaglandin System,* Plenum Press, New York.

Bessman, S. P., and Pal, N. (1976). The Krebs cycle depletion theory of hepatic coma. In: *The Urea Cycle,* (Ed. Grisolia, S., Baguena, R., and Mayor, F.) John Wiley, New York, London, pp. 83–89.

Beutler, E. (1972). Disorders due to enzyme defects in the red blood cell. *Adv. Metab. Disord.,* **6,** 131–160.

Beutler, E. (1981). Enzyme replacement therapy. *Trends Biochem. Sci.,* **6,** 95–97.

Bigland-Ritchie, B. (1981). EMG and fatigue of human voluntary and stimulated contractions. *CIBA Found. Symp.,* **82,** 130–156.

Binkiewicz, A., and Senior, B. (1973). Decreased ketogenesis in von Gierke's disease. *J. Pediatrics,* **83,** 973–978.

Bjorntorp, P., Fahlén, M., Grimby, G., *et al.* (1972). Carbohydrate and lipid metabolism in middle-aged physically well-trained men. *Metabolism,* **21,** 1037–1044.

Blackard, W. G., and Nelson, N. C. (1970). Portal and peripheral vein immunoreactive insulin concentrations before and after glucose infusion. *Diabetes,* **19,** 302–306.

Blackburn, G. L., Flatt, J. P., Clowes, G. H. A., *et al.* (1973). Protein sparing therapy during periods of starvation with sepsis or trauma. *Ann. Surg.,* **177,** 588–594.

Blackburn, G. L., Rienhoff, H., Miller, J. D. B., *et al.* (1977). Amino acid infusion after surgical injury. *Current Concepts in Parenteral Nutrition.* (Ed. Greep, J. M., Soeters, P. B., Wesdorp, *et al.*) Martinus Nizhoff Medical Division, The Hague, pp. 299–312.

Blass, J. P. (1978). Pyruvate dehydrogenase deficiencies. *Inherited Disorders of Carbohydrate Metabolism.* (Ed. Burman, D., Holton, J. B., and Pennock, C.A.) M.T.P. Press, Lancaster, U.K., pp. 239–267.

Blaxter, K. L. (1970). Methods of measuring the energy metabolism of animals and interpretation of results obtained. *Fed. Proc.,* **30,** 1436–1443.

Blecher, M., and Bar, R. S. (1980). *Receptors and Human Disease.* Williams and Wilkins, Baltimore.

Bleech, H. L., and Moore, M. J. (1982). Feedback regulation of thyrotropin secretion by thyroid hormones. *New Engl. J. Med.,* **306,** 23–32.

Blobel, G., and Dobberstein, B. (1975). Transfer of proteins across membranes. *J. Cell. Biol.,* **67,** 835–862.

Block, A. J. (1982). Dangerous sleep; oxygen therapy for nocturnal hypoxemia. *New Engl. J. Med.,* **306,** 166–167.

Bloom, A. (1977). Some practical aspects of the management of diabetes. *Clin. Endocrinol. Metabol.,* **6,** 499–517.

Boden, G., Master, R. W., and Gordon, S. S. (1980). Monitoring metabolic control in diabetic outpatients with glycosylated haemoglobin. *Ann. Intern. Med.,* **92,** 357–360.

Bodmer, W. F. (1976). HLA: A Super Supergene. *Harvey Lectures,* **72,** 91–138.

Bohr, C., Hasselback, M. A., and Krogh, A. S. (1904). *Scand. Arch. Physiol.,* **16,** 402–409.

Bondy, P. K. (1980). The adrenal cortex. In: *Metabolic Control and Disease,* (Ed. Bondy, P. K., and Rosenberg, L. E.) Saunders Co. Philadelphia, pp. 1427–1500.

Bone, Q. (1966). The function of the two types of myotomal muscle fibre in elasmobranch fish. *J. Mar. Biol. Ass. U.K.,* **46,** 321–349.

Booth, D. A. (1981). The physiology of appetite. *Brit. Med. Bull.*, **37**, 135–140.

Borgese, N., De Camilli, P., Tanaka, Y., *et al.* (1979). Membrane interactions in secreting cell systems. *Symp. Soc. Expt. Biol.*, **33**, 117–144.

Bougneres, P. F., Saudabray, J. M., Marsac, C., *et al.* (1980). Decreased ketogenesis due to deficiency of hepatic carnitine acyltransferase. *New Engl. J. Med.*, **302**, 123–124.

Bower, B., and Newsholme, E. A. (1978). Treatment of idiopathic polyneuritis by a polyunsaturated fatty acid diet. *Lancet*, **i**, 583–585.

Boyar, R. M. (1978). Control of the onset of puberty. *Annu. Rev. Med.*, **29**, 509–520.

Boyd, G. S., and Gorban, A. M. S. (1980). Protein phosphorylation and steroidogenesis. In: *Recently Discovered Systems of Enzyme Regulation by Reversible Phosphorylation*, (Ed. Cohen, P.) Elsevier/North Holland, Amsterdam, pp. 95–133.

Boyd, M. E., Albright, E. B., and Foster, D. W., *et al.* (1981). *In vitro* reversal of the fasting state of liver metabolism in the rat. Re-evaluation of the roles of insulin and glucose. *J. Clin. Invest.*, **68**, 142–152.

Boyd, T. A. and Goldstein, L. (1979). Kidney metabolite levels and ammonia production in acute acid-base alterations in the rat. *Am. J. Physiol.*, **236**, E289–E295.

Boyer, P. D. (1964). Carboxyl activation as a possible common reaction in substrate level and oxidative phosphorylation and in muscle contraction. *Oxidases and Related Redox Systems*. (Ed. King, T. E.). John Wiley and Sons, New York, pp. 994–1031.

Brady, R. O., Pentchey, P. G., and Gal, A. E. (1975). Investigations in enzyme replacement therapy in lipid storage diseases. *Fed. Proc.*, **34**, 1310–1315.

Braunstein, A. E., and Bychkov, S. M. (1939). *Nature*, **144**, 751–752.

Braunwald, E. (1978a). Protection of the ischemic myocardium. *Harvey Lectures*, **71**, 247–282.

Braunwald, E. (1978b). Coronary spasm and acute myocardial infarction. New possibility for treatment and prevention. *New Engl. J. Med.*, **299**, 1301–1303.

Braunwald, E. (1980). Treatment of the patient after myocardial infarction. *New Engl. J. Med.*, **302**, 290–293.

Brentzel, H. J., and Thurman, R. G. (1977). In: *Alcohol and Aldehyde Metabolising Systems*. (Ed. Thurman, R. G.) Academic Press, New York. Vol. II, pp. 373–380.

Briggs, G. E., and Haldane, J. B. S. (1925). A note on the kinetics of enzyme action. *Biochem. J.*, **19**, 338–339.

Brindley, D. N. (1974). *Biomembranes*, **4B**, 621–671.

British Medical Journal editorial (1977). Treatment of depression. *Brit. Med. J.*, **ii**, 1105.

Brooks, B. J., Arch, J. R. S., and Newsholme, E. A. (1982). Effect of some hormones on the rate of the triacylglycerol/fatty acid substrate cycle in adipocytes and epididymal fat pads. *FEBS Lettr.* **146**, 327–330.

Brooks, B. J., Arch, J. R. S., and Newsholme, E. A. (1983). Effect of some hormones on the rate of the triacylglycerol/fatty acid substrate cycle in adipose tissue of the mouse *in vivo*. Bioscience Reports, **3**, 263–267.

Brook, C. G. D. (1981). Endocrinological control of growth at puberty. *Brit. Med. Bull.*, **37**, 281–285.

Broquist, H. P. (1982). Carnitine biosynthesis and function. *Fed. Proc.*, **41**, 2840–2842.

Brosnan, J. T. (1976). Factors affecting intracellular ammonia concentration in liver. In: *The Urea Cycle*, (Ed. Grisolia, S., Baguena, R., and Mayor, F.) John Wiley & Sons, New York, pp. 443–458.

Brown, E. M., Spiegal, A. M., and Gardner, J. D. (1978). Direct identification of β-adrenergic receptors and functional relationship of adenyl cyclase. In: *Receptors and Hormone Action*. (Ed. Birnbayner, L., and O'Malley, B. W.) Academic Press, New York. Vol. III, pp. 102–132.

Brown, M. S., and Goldstein, J. L. (1981). Lowering plasma cholesterol by raising LDL receptors. *New Engl. J. Med.*, **305**, 515–517.

Brown, M. S., Kovanen, P. T., and Goldstein, J. L. (1979). Receptor mediated uptake of lipoprotein-cholesterol and its utilization for steroid synthesis in the adrenal cortex. *Rec. Progr. Horn. Res.*, **35**, 215–258.

Brown, W. R., Hansen, A. E., Barr, G. O., *et al.* (1938). *J. Nutr.*, **16**, 511–524.

Browne, T. R. (1980). Drug therapy: valproic acid. *New Engl. J. Med.*, **302**, 661–665.

Brownlee, M., and Cerami, A. (1981). Biochemistry of the complications of diabetes mellitus. *Annu. Rev. Biochem.*, **50**, 385–432.

Bryant, H. H., Griffiths, J. J., and Smith, D. W. (1952). *Jackson Mem. Hosp. Bull.*, **6**, 19–24.

Bryant, M. P. (1978). Cellulose digesting bacteria from human faeces. *Am. J. Clin. Nutr.*, **31**, S.113–115.

Bullough, J. (1958). Protracted foetal and neonatal asphyxia. *Lancet*, **i**, 999–1000.

Bunn, H. F. (1981). Non-enzymatic glycosylation of protein: relevance to diabetes. *Am. J. Med.*, **70**, 325–330.

Burger, G. C. E., Drummond, J. C., and Sandstead, H. R. (1948). *Malnutrition and Starvation in Western Netherlands. September, 1944-July 1945.* General State Printing Office, The Hague, Netherlands, Pt. I and II.

Burnstock, G. (1977). The purinergic nerve hypothesis. *Ciba Found. Symp.*, **48**, 295–314.

Burr, G. O., and Burr, M. M. (1929). A new deficiency disease produced by the rigid exclusion of fat from the diet. *J. Biol. Chem.*, **82**, 345–367.

Burston, D., Marrs, T. C., Sleisenger, M. H., *et al.* (1977). Mechanism of peptide transport. *Ciba Found. Symp.*, **50**, 79–98.

Bygdeman, M. (1979). Menstrual regulation with prostaglandins. In: *Practical Applications of Prostaglandins and Their Synthesis Inhibitors*, (Ed. Karim, S. M. M.) M.T.P. Press, Lancaster, U.K., pp. 267–282.

Cahill, G. F. (1970). Starvation in man. *New Engl. J. Med.*, **282**, 668–675.

Cahill, G. F. (1974). Nitrogen versatility in bats, bears and man. *New Engl. J. Med.*, **290**, 686–687.

Cahill, G. F., and McDevitt, H. O. (1981). Insulin-dependent diabetes mellitus: the initial lesion. *New Engl. J. Med.*, **304**, 1454–1465.

Cahill, G. F., and Soeldner, J. S. (1974). A non-editorial on non-hypoglycaemia. *New Engl. J. Med.*, **291**, 905–906.

Cahill, G. F., Herrera, M. G., Morgan, A. P., *et al.* (1966). Hormone-fuel interrelationships during fasting. *J. Clin. Invest.*, **45**, 1751–1769.

Cahill, G. F., Etzwiler, D. D., and Freinkel, N. (1976) Blood glucose control in diabetes. *Diabetes*, **25**, 237–

Calne, R. Y. (1980). Transplant surgery; current status. *Br. J. Surg.*, **67**, 765–771.

Campbell, P. N. (1977). Recent advances in eucaryotic protein synthesis. In: *Symp. Protein Metabolism and Nutrition. The Netherlands May 2–6, 1977.* Proceedings of Europ. Associat. for Animal Production. (EAAP, Ser. No. 20). Unipubl., pp. 12–14.

Campbell, P. N. (1979). Post-translational events associated with biosynthesis of secretory proteins. In: *Companion to Biochemistry* (Ed. Bull, A. T., et al.) Longman Group, London, vol. 2, pp. 229–244.

Candy, D. J. (1980). *Biological Functions of Carbohydrates.* Blackie, Glasgow.

Carlson, L. A. (1970). Mobilization and utilization of lipids after trauma: relation to caloric homeostasis. In: *Ciba Found. Symp.; Energy Metabolism in Trauma.* (Ed. Porter, R., and Knight, J. J. A.) Churchill, London, pp. 155–172.

Carlson, L. A., Havel, R. J., and Ekelund, L.-G. (1963). Effect of nicotinic acid on the turnover rate and oxidation of the free fatty acids of plasma in man during exercise. *Metabolism*, **12**, 837–845.

Carruthers, M. E. (1969). Aggression and atheroma. *Lancet*, **ii**, 1170–1171.

Carter, M. J. (1972). Carbonic anhydrase: isoenzymes, properties, distribution and functional significance. *Biol. Rev.*, **47**, 465–513.

Caspary, W. F. (1978). Sucrose malabsorption in man after ingestion of glucoside-hydrolase inhibitor. *Lancet*, **i**, 1231–1233.

Cawthorne, M. A. (1982). Metabolic aspects of obesity. *Molec. Aspects Med.* **5**, 293–400.

Chakraborty, J. (1978). Alcohol and its metabolic interactions with other drugs. *Clinics in Endocrinol. Metabol.*, **7**, 273–296.

Chance, B. (1943). The kinetics of the enzyme-substrate compound of peroxidase. *J. Biol. Chem.*, **151**, 553–577.

Chance, B. (1981). The cycling of oxygen through intermediates in the cytochrome oxidase-oxygen reaction. *Curr. Topics in Cellul. Regul.*, **18**, 343–360.

Chance, B., and Williams, G. R. (1956). The respiratory chain and oxidative phosphorylation. *Adv. Enzymol.*, **17**, 65–134.

Chance, B., Boveris, A., and Oshino, N. (1977). Peroxide generation in mitochondria and utilization by catalase. In: *Alcohol and Aldehyde Metabolizing Systems*, (Ed. Thurman, R. G., Williamson, J. R., Drott, H. R., *et al.*) Academic Press, New York, pp. 261–274.

Chance, B., Sies, H., and Boveris, A. (1979). Hydroperoxide metabolism in mammalian organs. *Physiol. Rev.*, **59**, 527–605.

Chandra, R. K. (1981). Immunocompetence as a functional index of nutritional status. *Brit. Med. Bull.*, **37**, 89–94.

Chanutin, A., and Curnish, R. R. (1967). Effect of organic and inorganic phosphates on the oxygen equilibrium of human erythrocytes. *Arch. Biochem. Biophys.*, **121**, 96–102.

Chapoy, P. R., Angelini, C., Brown, J., *et al.* (1980). Systemic carnitine deficiency—a treatable inherited lipid-storage disease presenting as Reye's syndrome. *New Engl. J. Med.*, **303**, 1389–1394.

Cherrington, A. D., Chiasson, J. L., Liljenquist, J. E., *et al.* (1978). Control of hepatic glucose output by glucagon and insulin in the intact dog. *Biochem. Soc. Symp.*, **43**, 31–46.

Cheung, C.-W., and Raijman, L. (1980). The regulation of carbamyl phosphate synthetase (ammonia) in rat liver mitochondria. *J. Biol. Chem.*, **255**, 5051–5057.

Chiba, H., and Sasaki, R. (1978). Functions of 2,3-bisphosphoglycerate and its metabolism. *Curr. Topics Cellul. Regul.*, **14**, 75–116.

Chochinov, R. H., and Daughaday, W. H. (1976). Current concepts of somatomedins and other biologically related growth factors. *Diabetes*, **25**, 994–1004.

Chouinard, G., Young, S. N., Annable, L. *et al.* (1977). Tryptophan-nicotinamide combination in depression. *Lancet*, **i**, 249.

Chowdhury, F., and Bleicher, S. J. (1973). Studies of tumour hypoglycaemia. *Metabolism*, **22**, 663–674.

Christensen, E. H., and Hansen, O. (1939). *Skand. Arch. Physiol.*, **81**, 160–175.

Christy, M., Deckert, T., and Nerup, J. (1977). Immunity and autoimmunity in diabetes mellitus. *Clin. Endocrinol. Metab.*, **6**, 305–332.

Chua, B., Kao, R., Rannels, D. E., *et al.* (1978). Hormonal and metabolic control of proteolysis. *Biochem. Soc. Symp.*, **43**, 1–16.

Clark, D. G., Rognstad, R., and Katz, J. (1973). Isotopic evidence for futile cycles in liver cells. *Biochem. Biophys. Res. Commun.*, **54**, 1141–1148.

Clark, J. M., and Lambertsen, C. J. (1971). Pulmonary oxygen toxicity: a review. *Pharmacol. Rev.*, **23**, 37–133.

Cleave, T. L. (1974). *The Saccharine Disease*. John Wright & Sons, Bristol.

Close, J. H. (1974). The use of amino acid precursors in nitrogen-accumulation diseases. *New Engl. J. Med.*, **290**, 663–667.

Clowes, G. H. A., George, B. C., Villee, C. A., *et al.* (1983). Muscle proteolysis induced by a circulating peptide in patients with sepsis or trauma. *New Engl. J. Med.*, **308**, 545–552.

Cohen, G., and Collins, M. (1970). Alkaloids from catecholamines in adrenal tissues: a possible role in alcoholism. *Science*, **167**, 1749–1751.

Cohen, R. D. (1976). Disorders of lactic acid metabolism. *Clinics in Endocrinol. Metab.*, **5**, 613–625.

Cohen, R. D., and Woods, H. F. (1976). *Clinical and Biochemical Aspects of Lactic Acidosis*. Blackwell Scientific Publications, Oxford.

Cohen, P. (1978). The role of cyclic-AMP dependent protein kinase in the regulation of glycogen metabolism in mammalian skeletal muscle. *Curr. Top. Cellul. Regulat.*, **14**, 117–196.

Cohen, P. (1979). The hormonal control of glycogen metabolism in mammalian muscle by multivalent phosphorylation. *Biochem. Soc. Trans.*, **7**, 459–480.

Cohen, P. (1980). Well established systems of enzyme regulation by reversible phosphorylation. *Molec. Aspects of Cellul. Regulat.*, **I**, 1–10; 255–268.

Cohen, P. (1982). The role of protein phosphorylation in neural and hormonal control of cellular activity. *Nature*, **296**, 613–620.

Cole, R. P. (1982). Myoglobin function in exercising skeletal muscle. *Science*, **216**, 523–525.

Coleman, N., and Jacob, E. (1980). Folic acid and vitamin B_{12}. In: *Modern Nutrition in Health and Disease*, (Eds. Goodhart, R. S., and Shils, M. E.) Lea & Febiger, Philadelphia, pp. 229–259.

Collier, H. O. J. (1969). A pharmacological analysis of aspirin. *Adv. Pharmacol. Chemother.*, **7**, 333–405.

Comite, F., Cutler, G. B., Rivier, J., *et al.* (1981). Short-term treatment of idiopathic precocious puberty with a long-acting analogue of luteinizing hormone-releasing hormone. *N. Engl. J. Med.* **305**, 1546–1550.

Conn, H. O., and Lieberthal, M. M. (1979). *The Hepatic Coma Syndromes and Lactulose*. Williams & Wilkins Co., Baltimore.

Connell, A. M. (1978). The effects of dietary fibre on gastrointestinal motor function. *Am. J. Clin. Nutr.*, **31**, S.152–156.

Connell, E. B. (1979). Future methods of fertility regulation. *Clinics Obstetrics and Gynaecol.*, **6**, 171–184.

Connor, W. E. (1975). Pathogenesis and frequency of essential fatty acid deficiency during total parenteral nutrition. *Ann. Intern. Med.*, **83**, 895–896.

Conyers, R. A. J., and Goldsmith, L. E. (1971). A case of organophosphorus-induced psychosis. *Med. J. of Australia*, **1**, 27–29.

Cooney, G. J., Taegtmeyer, H., and Newsholme, E. A. (1981). Tricarboxylic acid cycle flux and enzyme activities in the isolated working rat heart. *Biochem. J.*, **200**, 701–703.

Cooper, J. R., Bloom, F. E., and Roth, R. H. (1978). *The Biochemical Basis of Neuropharmacology*, 3rd Edn. Oxford University Press, London.

Cooper, R. A., and Shattil, S. J. (1980). Membrane cholesterol—is enough too much? *New Engl. J. Med.*, **302**, 49–50.

Corasch, L., Spielberg, S., Bartsocas, C. *et al.*, (1980). Reduced chronic haemolysis during high-dose vitamin E administration in mediterranean-type glucose 6-phosphate dehydrogenase deficiency. *New Engl. J. Med.*, **303**, 416–420.

Cornblath, R. (1968). Hypoglycaemia. In: *Carbohydrate Metabolism and Its Disorders*. (Eds. Dickens, F., Randle, P. J., and Whelan, W. J.) Academic Press, London. Vol. 2, pp. 51–86.

Costain, D. W., Cowen, P. J., Gelder, N. G., *et al.* (1982). Electroconvulsive therapy and the brain: evidence for increased dopamine mediated responses. *Lancet*, **ii**, 400–404.

Costill, D. L. (1979). *A Scientific Approach to Distance Running*. Track & Field News, Los Altos, California.

Costill, D. L., and Fox, E. L. (1969). Energetics of marathon running. *Medicine and Science in Sports*, **1**, 81–86.

Costill, D. L., and Miller, J. M. (1980). Nutrition for endurance sport; carbohydrate and fluid balance. *Int. J. Sports Med.*, **1**, 2–14.

Costill, D. L., Branam, G., Eddy, D., *et al.* (1971). Determinants of marathon running success. *Int. Z. Angew. Physiol.*, **29**, 249–254.

Costill, D. L., Daniels, J., Evans, W., *et al.* (1976). Skeletal muscle enzymes and fibre composition in male and female track athletes. *J. Applied Physiol.*, **40**, 149–154.

Costill, D. L., Coyle, E., and Dalsky, G. (1977). Effects of elevated plasma FFA and insulin on muscle glycogen usage during exercise. *J. Appl. Physiol.*, **43**, 695–699.

Cowan, J. C., and Vaughan-Williams, E. N. (1977). The effects of palmitate on intracellular potentials recorded from Langendorff-perfused guinea pig hearts in normoxia and hypoxia and during perfusion at a reduced rate of flow. *J. Molec. Cellul. Cardiology*, **9**, 327–342.

Cowen, P. J., and Nutt, D. J. (1982). Abstinence symptoms after withdrawal of tranquilizer drugs: is there a common neurochemical metabolism? *Lancet*, **ii**, 360–362.

Coward, W. A., and Lunn, P. G. (1981). The biochemistry and physiology of kwashiorkor and marasmus. *Brit. Med. Bull.*, **37**, 19–24.

Crabtree, B. (1976). Reversible (near-equilibrium) reactions and substrate cycles. *Biochem. Soc. Trans.*, **4**, 1046–1048.

Crabtree, B., and Newsholme, E. A. (1972a). The activities of phosphorylase, hexokinase, phosphofructokinase, lactate dehydrogenase and glycerol 3-phosphate dehydrogenase in muscles from vertebrates and invertebrates. *Biochem J.*, **126**, 49–58.

Crabtree, B., and Newsholme, E. A. (1972b). The activities of lipases and carnitine palmitoyltransferase in muscles from vertebrates and invertebrates. *Biochem. J.*, **130**, 697–705.

Crabtree, B., and Newsholme, E. A. (1976). Substrate cycles in metabolic regulation and heat generation. *Biochem. Soc. Symp.*, **41**, 61–109.

Crabtree, B., and Newsholme, E. A. (1978). Sensitivity of a near-equilibrium reaction in a metabolic pathway to changes in substrate concentration. *Eur J. Biochem.*, **89**, 19–22.

Crabtree, B., and Taylor, D. J. (1979). Thermodynamics and metabolism. In: *Biochemical Thermodynamics*, (Ed. Jones, N. M.) Elsevier, Amsterdam, pp. 333–378.

Crabtree, B., Higgins, S. J., and Newsholme, E. A. (1972). The activities of pyruvate carboxylase, phosphoenolpyruvate carboxykinase and fructose diphosphatase in muscles from vertebrates and invertebrates. *Biochem. J.*, **130**, 391–396.

Crabtree, B., Leech, A. R., and Newsholme, E. A. (1979). Measurement of enzyme activities in crude extracts of tissues. In: *Techniques in Metabolic Research*, (Ed. Pogson, C.) Elsevier/North Holland, B211, pp. 1–37.

Craighead, J. E. (1979). Current views on the etiology of insulin dependent diabetes mellitus. *New Engl. J. Med.*, **299**, 1439–1445.

Craighead, J. E. (1981). Viral diabetes mellitus in man and experimental animals. *Am. J. Med.*, **70**, 127–134.

Crane, R. K. (1977). The gradient hypothesis and other models of carrier-mediated active transport. *Rev. Physiol. Biochem. Pharmacol.*, **78**, 99–159.

Crawford, M. A., and Stevens, P. (1981). A study on essential fatty acids and multiple sclerosis. *Progr. Lipid Res.*, **20**, 255–258.

Creese, I. (1981). Dopamine receptors. In: *Neurotransmitter Receptors*, (Ed. Yamamura, H. I., and Enna, S. J.) Chapman & Hall, London, pp. 129–184.

Cresshull, I. P., Dawson, M. J., Edwards, R. H. T., *et al.* (1981). Human muscle analysed by ^{31}P nuclear magnetic resonance in intact subjects. *J. Physiol.*, **317**, 18P.

Crisp, A. H. (1977). Anorexia nervosa. *Proc. Roy. Soc. Med.*, **70**, 464–470.

Crisp, A. H. (1980). *Anorexia Nervosa: Let Me Be.* Academic Press, London.

Crook, M. (1981). Migraine: a biochemical headache. *Biochem. Soc. Trans.*, **9**, 351–357.

Cummings, J. H. (1981). Dietary fibre. *Brit. Med. Bull.*, **37**, 65–70.

Cushman, S. W., Hissin, P. J., Wardzala, L. J., *et al.* (1981). Mechanism of insulin-resistance in the adipose cell in the ageing rate, model of obesity. *Biochem. Soc. Trans.*, **9**, 518–522.

Cuthbertson, D. P., and Tilstone, W. J. (1969). Metabolism during the post injury period. *Adv. Clin. Chem.*, **12**, 1–55.

Cutting, W. C., Mehrtens, H. G., and Tainter, M. L. (1933). Actions and uses of dinitrophenol. *J. Am. Med. Ass.*, **101**, 193–195.

Dalziel, K. (1980). Isocitrate dehydrogenases and related oxidative decarboxylases. *FEBS Lettr.*, **117**, K45–55.

Danforth, W. H. (1965). Activation of glycolytic pathway in muscle. In: *Control of energy metabolism*, (Ed. Chance, B., and Eastabrook, R. W.) Academic Press, New York, pp. 287–297.

Davidson, S., Passmore, R., Bronk, J. F., *et al.* (1979). *Human Nutrition and Dietetics.* 7th Edn. Churchill Livingstone, Edinburgh.

Davies, C. T. M., and Thompson, M. W. (1979). Aerobic performance of female marathon and male ultramarathon athletes. *European Journal of Applied Physiology*, **41**, 233–245.

Dawes, G. S., and Shelley, H. J. (1968). In: *Carbohydrate Metabolism and its Disorders*, (Ed. Dickens, F., Randle, P. J., and Whelan, W. J.) Academic Press, London, New York, pp. 87–121.

Dawson, A. G. (1979). Oxidation of cytosolic NADH formed during aerobic metabolism in mammalian cells. *Trends Biochem. Sci.*, **4**, 171–176.

Dawson, A. P., and Selwyn, M. J. (1974). Mitochondrial oxidative phosphorylation. *Companion to Biochemistry*, (Ed. Bull, A. T., Lagnado, J. R., Thomas, J. D., *et al.*) Longman, London. Vol. 1, pp. 553–586.

Dawson, M. J., Gadian, D. G., and Wilkie, D. R. (1978). Muscular fatigue investigated by phosphorus nuclear magnetic resonance. *Nature*, **274**, 861–866.

Dean, R. T. (1980). Lysosomes and protein degradation. *Ciba Found. Symp.*, **75**, 139–149.

Dean, R. T., and Barrett, A. J. (1976). Lysosomes. *Essays in Biochem.*, **12**, 1–40.

Décombaz, J., Reinhardt, P., Anantharaman, K., *et al.* (1979). Biochemical changes in a 100 km run: free amino acids, urea and creatinine. *Eur. J. Appl. Physiol.*, **41**, 61–72.

DeFronzo, R. A., Soman, V., Sherwin, R. S., *et al.* (1978a). Insulin binding to monocytes and insulin action in human obesity, starvation and refeeding. *J. Clin. Invest.*, **62**, 204–213.

DeFronzo, R. A., Ferrannini, E., Hendler, R., *et al.* (1978b). Influence of hyperinsulinaemia, hyperglycaemia and route of glucose administration on splanchnic glucose exchange. *Proc. Nat. Acad. Sci.*, **75**, 5173–5177.

Delivoria-Papadopoulos, M., Oski, F. A., and Gottlieb, A. J. (1969). Oxygen haemoglobin dissociation curves: Effect of inherited enzyme defects of the red cell. *Science*, **165**, 601–602.

DeLuca, H. F. (1974). Vitamin D: the vitamin and the hormone. *Fed. Proc.*, **33**, 2211–2219.

DeLuca, H. F. (1980). Some new concepts emanating from a study of the metabolism and function of vitamin D. *Nutrition Rev.*, **38**, 169–182.

DeLuise, M., Blackburn, G. L., and Flier, J. S. (1980). Reduced activity of the red-cell sodium-potassium pump in human obesity. *New Engl. J. Med.*, **303**, 1017–1022.

Deneke, S. M., and Fanburg, B. L. (1980). Normobaric oxygen toxicity of the lung. *New Engl. J. Med.*, **303**, 76–86.

Denning, D. W. (1982). A case of diabetes mellitus. *New Engl. J. Med.*, **307**, 128.

Denny-Brown, D. E. (1929). The histological features of striped muscle in relation to its functional activity. *Proc. Roy. Soc. B.*, **104**, 371–411.

Denton, R. M., and Halestrap, A. P. (1979). Regulation of pyruvate metabolism in mammalian tissues. *Essays in Biochem.*, **15**, 37–77.

Denton, R. M., and McCormack, J. G. (1980). Role of calcium ions in the regulation of intramitochondrial metabolism. *FEBS Letters*, **119**, 1–8.

Denton, R. M., Bridges, B. J., and Brownsey, R. W., *et al.* (1978). Acute hormonal regulation of fatty acid synthesis in mammalian tissues. In: *FEBS 11th Meeting, Copenhagen (1977)*, (Ed. Dils, R., and Knudsen, J.) Pergamon Press, Oxford. Vol. 46, pp. 21–30.

Denton R. M., Brownsey, R. W., and Belsham, G. J. (1981). A partial view of the mechanism of insulin action. *Diabetologia*, **21**, 347–362.

De Wulf, H., and Carton, H. (1981). Neural control of glycogen metabolism. *Short-term Regulation of Liver Metabolism*. (Ed. Hue, L., and Van de Werve, G.) Elsevier/North Holland, Amsterdam, pp. 63–76.

Dice, J. F., and Goldberg, A. L. (1975). A statistical analysis of the relationship between degradative rates and molecular weights of proteins. *Arch. Biochem. Biophys.*, **170**, 213–219.

DiMauro, S., Bonilla, E., Lee, C. P., *et al.* (1976). Luft's disease. Further biochemical and ultrastructural studies of skeletal muscle in the second case. *J. Neurol. Sci.*, **27**, 217–232.

Dingfelder, J. R. (1982). Prostaglandin inhibitors: New Treatment for an Old Nemesis. *New Engl. J. Med.*, **307**, 746–747.

di Prampero, P. E., and Magnoni, P. (1981). Maximum anaerobic power in man. *Physiological Chemistry of Exercise and Training*, (Ed. di Prampero, P. E., and Poortmans, J. R.) Karger, Basel, pp. 38–44.

Dixon, M., and Webb, E. C. (1979). *Enzymes*, 3rd Edn. Longman, London.

Dodge, A. D. (1982). Oxygen radicals and herbicide action. *Biochem. Soc. Trans.*, **10**, 73–75.

Donaldson, R. M. (1965). Studies on the pathogenesis of steatorrhea in the blind loop syndrome. *J. Clin. Invest.*, **44**, 1815–1825.

Donaldson, S. K. B., Kerrick, W., and Hermansen, L. (1978). Differential, direct effects of H^+ on Ca^{2+}-activated form of skinned fibres from soleus, cardiac and adductor magnus muscles of rabbits. *Pflügers Arch.*, **376**, 55–65.

Donnellan, J. F., Barker, M. D., and Wood, J., *et al.* (1970). Specificity and locale of the L-3-glycerophosphate-flavoprotein oxidoreductase of mitochondria isolated from the flight muscle of *Sarcophaga barbata*. *Biochem. J.*, **120**, 467–478.

Dormandy, T. L. (1978). Free-radical oxidation and antioxidants. *Lancet*, i, 647–650.

Douglas, W. W. (1974). Involvement of calcium in exocytosis and the exocytosis-vesiculation sequence. *Biochem. Soc. Symp.*, **39**, 1–28.

Drash, A. L. (1979). The etiology of diabetes mellitus. *New Engl. J. Med.*, **300**, 1211–1213.

Drochmans, P., and Dantan, E. (1968). Size distribution of liver glycogen particles. In: *Control of Glycogen Metabolism*, (Ed. Whelan, W. J.) Academic Press, New York, pp. 187–292.

Dufaux, B. (1981). Plasma lipoproteins and physical activity. *Int. J. Sports Medicine*, **2**, 195–196.

Durnin, J. G. V. A., and Passmore, R. (1967). *Energy, Work and Leisure*. Heinemann, London.

Dustin, P. (1978). *Microtubules*, Springer-Verlag, Berlin.

Eddy, R. L., Jones, A. L., Chakmakjian, Z. H., *et al.* (1971). Effect of levodopa (L-DOPA) on human hypophyseal trophic hormone release. *J. Clin. Endocrinol. Metab.*, **33**, 709.

Edwards, A. V. (1972). The sensitivity of the hepatic glycogenolytic mechanism to stimulation of the splanchnic nerves. *J. Physiol.*, **220**, 315–334.

Edwards, D. C., and Thorpe, P. E. (1981). Targeting toxins—the retaliation approach to chemotherapy. *Trends Biochem. Sci.*, **6**, 313–316.

Edwards, R. H. T. (1981). Human muscle function and fatigue. *Ciba Found. Symp.*, **82**, 1–18.

Edwards, R. H. T., Young, A., and Wiles, M. (1980). Needle biopsy of skeletal muscle in diagnosis of myopathy and the clinical study of muscle function and repair. *New Engl. J. Med.*, **320**, 261–271.

Edwards, R. H. T., Harris, R. C., Hultman, E., *et al.* (1972). Effect of temperature on muscle energy metabolism and endurance during successive isometric contractions sustained to fatigue on the quadriceps muscle in man. *J. Physiol.*, **220**, 335–352.

Edwardson, J. A., McDermott, J. R. (1982). Neurochemical pathology of brain peptides. *Brit. Med. Bull.* **38**, 259–264.

Ehrenkranz, R. A., Ablow, R. C., and Warshaw, J. B. (1979). Prevention of bronchopulmonary dysplasia with vitamin E administration during the acute stages of respiratory distress syndrome. *J. Pediatr.*, **95**, 873–878.

Eisenberg, S., and Levy, R. I. (1975). Lipoprotein metabolism. *Adv. Lipid Res.*, **13**, 1–89.

Ekelund, L. G. (1969). Exercise including weightlessness. *Annu. Rev. Physiol.*, **31**, 85–116.

Ekins, R. P. (1960). *Clin. Chim. Acta.*, **5**, 453–459.

Ekins, R. P. (1974). Basic principles and theory of radioimmunoassay and saturation analysis. *Brit. Med. Bull.*, **30**, 3–11.

Elders, M. J., Garland, J. T., and Daughaday, W. A. (1973). Laron's dwarfism: studies on the nature of the defect. *J. Pediatr.*, **83**, 253–263.

Elia, M., Carter, A., Bacon, S., *et al.* (1981). The clinical usefulness of urinary 3-methylhistidine excretion in indicating muscle protein breakdown. *Brit. Med. J.*, **282**, 351–354.

Ellingboe, J., and Varanelli, C. C. (1979). Ethanol inhibits testosterone biosynthesis by direct action on Leydig cells. *Res. Commun. Chem. Pathol. Pharmacol.*, **24**, 87–102.

Elliot, K. R. F. (1979). The preparation, characterization and use of isolated cells for metabolic studies. *Techniques in Metabolic Research*, **B.204**, 1–20.

Ellwood, D. A. (1979). The hormonal control of connective tissue changes in the uterine cervix in pregnancy and at parturition. *Biochem. Soc. Trans.*, **8**, 662–667.

Elwood, T. C., and Sweetnam, P. M. (1979). Aspirin and secondary mortality after myocardial infarction. *Lancet*, **ii**, 1313–1315.

Endo, A. (1981). Biological and pharmacological activity of inhibitors of 3-hydroxy-3-methylglutaryl coenzyme A reductase. *Trends in Biochem. Sci.*, **6**, 10–13.

Engel, P. C. (1977). *Enzyme Kinetics*. Chapman and Hall Ltd., London.

Endo, A. (1981). Biological and pharmacological activity of inhibitors of 3-hydroxy-3-methylglutaryl coenzyme-A reductase. *Trends Biochem. Sci.*, **6**, 10–13.

Enzyme Nomenclature (1979). *Recommendation of the Nomenclature Committee of the International Union of Biochemistry on the Nomenclature and Classification of Enzymes.* Academic Press, New York.

Erecińska, M., and Wilson, D. F. (1978). Homeostatic regulation of cellular energy metabolism. *Trends Biochem. Sci.*, **3**, 219–223.

Eriksson, E. (1981). Rehabilitation of muscle function after sport injury—major problem in sports medicine. *Int. J. Sports Med.*, **2**, 1–6.

Eriksson, E. (1982). Role of sports medicine today and in the future. *Int. J. Sports Med.*, **3**, 2–3.

Ernst, N., and Levy, R. I. (1980). Diet, hyperlipidemia and atherosclerosis. In: *Modern Nutrition in Health and Disease*, 6th Edn. (Ed. Goodhart, R. S., and Shils, M. E.) Lea & Febiger, Philadelphia, pp. 1045–1070.

Eskin, B. A. (1980). *The Menopause: Comprehensive Management.* (Ed. Eskin, B. A.) Masson Publishing, New York, U.S.A.

Espinal, J., Challiss, R. A. J., Newsholme, E. A. (1983). Effect of adenosine deaminase and an adenosine analogue on insulin sensitivity in soleus muscle of the rat. *FEBS Lett.*, in press.

Essén, B. (1978). Studies on the regulation of metabolism in human skeletal muscle using intermittent exercise as an experimental model. *Acta Physiol. Scand.*, **Suppl. 454**.

Essen, B., and Kaijser, L. (1978). Regulation of glycolysis in intermittent exercise in man. *J. Physiol.*, **281**, 499–511.

Evered, D. F. (1981). Advances in amino acid metabolism in mammals. *Biochem. Soc. Trans.*, **9**, 159–169.

Exton, J. H. (1972). Gluconeogenesis. *Metabolism*, **21**, 945–990.

Exton, J. H. (1979). Hormonal control of gluconeogenesis. *Adv. Expt. Med. Biol.*, **111**, 125–167.

Farfel, Z., Brickman, A. S., Kaslow, H. R., et al. (1980). Defect in receptor-cyclase coupling protein in pseudohypoparathyroidism. *New Engl. J. Med.*, **303**, 237–242.

Fatania, H. R., and Dalziel, K. (1980). Intracellular distribution of NADP-linked isocitrate dehydrogenase, fumarase and citrate synthase in bovine heart muscle. *Biochim. Biophys. Acta*, **631**, 11–19.

Feicht, C. B., Johnson, T. S., Martin, B. J., et al. (1978). Secondary amenorrhea in athletes. *Lancet*, **ii**, 1145–1146.

Feldberg, W. (1974). Fever, prostaglandins and antipyretics. In: *Prostaglandin Synthetase Inhibitors*, (Ed. Robinson, H. J., and Vane, J. R.) Raven Press, New York, pp. 197–204.

Felig, P. (1975). Amino acid metabolism in man. *Ann. Rev. Biochem.*, **44**, 993–955.

Felig, P. (1980). Disorders of carbohydrate metabolism. In: *Metabolic Control and Disease* 8th edn. (Ed. Bondy, P. K., and Rosenberg, L. E.) Saunders Co., Philadelphia, pp. 276–392.

Felig, P., and Lynch, V. (1970). Starvation in human pregnancy: hypoglycemia, hypoinsulinemia and hyperketonemia. *Science*, **170**, 990–992.

Felig, P., and Wahren, J. (1975). Fuel homeostasis in exercise. New Engl. J. Med., **293**, 1078–1084.

Felig, P., Kim, Y. J., Lynch, V., et al. (1972). Amino acid metabolism during starvation in human pregnancy. *J. Clin. Invest.*, **51**, 1195–1202.

Felig, P., Wahren, J., Hendler, R., et al. (1974). Splanchnic glucose and amino acid metabolism in obesity. *J. Clin. Invest.*, **53**, 582–590.

Felig, P., Wahren, J., Sherwin, R., et al. (1976). Insulin, glucagon and somatostatin in normal physiology and diabetes mellitus. *Diabetes*, **25**, 1091–1099.

Feller, D. D. (1965). In: *Handbook of Physiology, Section 5, Adipose Tissue*, (Ed. Renold, A. E., and Cahill, G. F.) American Physiological Society, Washington, D.C., pp. 363–374.

Ferdinand, W. (1976). *The Enzyme Molecule*. John Wiley & Sons, London, New York.

Fernandes, J., and Pikaar, N. A. (1972). Ketosis in hepatic glycogenesis. *Archiv. Dis. Child.*, **47**, 41–46.

Fernandez, B. M., and Saggerson, E. D. (1978). Alterations in response of rat white adipocytes to insulin, noradrenaline, corticotropin and glucagon after adrenalectomy. *Biochem. J.*, **174**, 111–118.

Fernstrom, J. D., and Wurtman, R. J. (1971). Effect of chronic corn consumption on serotonin content of rat brain. *Nature*, **234**, N.B.62–64.

Fersht, A. (1977). *Enzyme Structure and Mechanism*. W. H. Freeman & Co., Reading and San Francisco, U.S.A.

Fink, G. (1979). Neuroendocrine control of gonadotrophin secretion. *Brit. Med. Bull.*, **35**, 155–160.

Fink, W. J., Costill, D. L., and Pollock, M. L. (1977). Submaximal and maximal working capacity of elite distance runners: muscle fiber composition and enzyme activities. *Annals N.Y. Acad. Sci.*, **301**, 323–327.

Fischer, J. E. (1977). Complications of parenteral nutrition. *Current Concepts in Parenteral Nutrition*, (Ed. Greep, J. M., Soeters, P. B., Westorp, R. I. C., et al.) Martinus Nizhoff Medical Division, The Hague, pp. 171–178.

Fischer, J. E., and Baldessarini, R. J. (1976). Pathogenesis and therapy of hepatic coma. *Progress in Liver Diseases*, **5**, 363–397.

Fisher, B., Redmond, C., Brown, A., *et al.* (1981). Treatment of primary breast cancer with chemotherapy and tamoxifen. *New Engl. J. Med.*, **305**, 1–6.

Fishman, J. (1980). Fatness, puberty and ovulation. *New Engl. J. Med.*, **303**, 42–43.

Flatt, J. P. (1970). Conversion of carbohydrate to fat in adipose tissue: an energy yielding and, therefore, self-limiting process. *J. Lipid Res.*, **11**, 131–143.

Fleegler, F. M., Rogers, K. D., Drash, A. L., *et al.* (1979). Age, sex and season of onset of juvenile diabetes in different geographic areas. *Paediatrics*, **63**, 374–379.

Floch, M. H. (1981). *Nutrition and Diet Therapy in Gastrointestinal Disease.* Plenum Publishing Co., New York.

Flower, R. J. (1979). Biosynthesis of prostaglandins. *Ciba Found. Symp.*, **65**, 123–142.

Folin, O. (1905). *Am. J. Physiol.*, **13**, 66–115.

Food and Agriculture Organization of the United Nations (1949). Food Composition Tables for International Use. FAO Nutritional Studies No. 3, Rome.

Forssner, G. (1909). *Skand. Arch. Physiol.*, **22**, 393–405.

Frank, A., Graham, C., and Frank, S. (1981). Fatalities on the liquid-protein diet: an analysis of possible causes. *Int. J. Obesity*, **5**, 243–248.

Fraser, D. R. (1980). Regulation of the metabolism of vitamin D. *Physiol. Rev.*, **60**, 551–613.

Fraser, D. R. (1981). Biochemical and clinical aspects of vitamin D function. *Brit. Med. Bull.*, **37**, 37–42.

Fratantoni, J. C., Ness, P., and Simon, T. L. (1975). Thrombolytic therapy. Current status. *New Engl. J. Med.*, **293**, 1073–1078.

Fredrickson, D. S., and Gordon, R. S. (1958). Transport of fatty acids. *Physiol. Rev.*, **38**, 585–630.

Fredrickson, D. S., Levy, R. I., and Lees, R. S. (1967). Fat transport in lipoproteins—an integrated approach to mechanisms and disorders. *New Engl. J. Med.*, **276**, 148–156; 215–225; 273–281.

Freed, D. L. J., and Banks, A. J. (1975). A double-blind cross-over trial of Methanedienone ('Dianabol') in moderate dosage on highly trained experienced athletes. *Brit. J. Sports Medicine*, **9**, 78–81.

Freedland, R. A., and Briggs, S. (1977). *A Biochemical Approach to Nutrition.* Chapman & Hall, London.

Frenkel, R. A., and McGarry, J. D. (1980). *Carnitine Biosynthesis, Metabolism, and Functions.* Academic Press, New York.

Freiden, E. (1981). Iodine and the thyroid hormones. *Trends in Biochem. Sci.*, **6**, 50–53.

Freinkel, N. (1981). Pregnant thoughts about metabolic control and diabetes. *New Engl. J. Med.*, **304**, 1357–1359.

Fridovich, I. (1978). The biology of oxygen radicals. *Science*, **201**, 875–880.

Fridovich, I., and Hassan, H. M. (1979). Paraquat and the exacerbation of oxygen toxicity. *Trends Biochem. Sci.*, **4**, 113–115.

Frieden, C. (1976). The regulation of glutamate dehydrogenase. In: *The Urea Cycle*, (Ed. Grisolia, S., Baguena, R., and Mayor, F.) John Wiley & Sons, Ltd., London & New York, pp. 59–72.

Frisch, R. E., and McArthur, J. W. (1974). Menstrual cycles: fatness as a determinant of minimum weight for height necessary for their maintenance or onset. *Science*, **185**, 949–951.

Frisch, R. E., Wyshak, G., and Vincent, L. (1980). Delayed menarche and amenorrhea in ballet dancers. *New Engl. J. Med.*, **303**, 17–19.

Froelicher, V. F. (1978). Exercise and the prevention of coronary atherosclerotic heart disease. *Cardiovasc. Clin.*, **9**, (3) 12–23.

Froesch, E. R. (1978). Essential fructosuria, hereditary fructose intolerance and fructose 1,6-diphosphatase deficiency. In: *The Metabolic Basis of Inherited Disease*. 4th edn. (Ed. Stanbury, J. B., Wyngarden, J. B., and Fredrickson, D. S.) McGraw Hill Books Co., New York, pp. 121–136.

Froesch, E. R., Zapt, J., and Widmer, U. (1982). Hypoglycaemia associated with non-islet-cell tumour and insulin-like growth factors. *New Engl. J. Med.*, **306**, 1178–1179.

Gabow, P. A., Kaehny, W. D., Fennessey, P. V., et al. (1980). Diagnostic importance of an increased serum anion gap. *New Engl. J. Med.*, **303**, 854–858.

Gadian, D. G., and Radda, G. K. (1981). NMR studies of tissue metabolism. *Annu. Rev. Biochem.*, **50**, 69–84.

Galbo, H., Holst, J. J., and Christensen, N. J. (1975). Glucagon and plasma catecholamine response to graded and prolonged exercise in man. *J. Applied Physiol.*, **38**, 70–76.

Galbo, H., Holst, J. J., and Christensen, N. J. (1979). The effect of different diets and of insulin on the hormonal response to prolonged exercise. *Acta Physiol. Scand.*, **107**, 19–32.

Garber, A. J., Menzel, P. H., Boden, G., et al. (1974). Hepatic ketogenesis and gluconeogenesis in humans. *J. Clin. Invest.*, **54**, 981–989.

Gardner, D. F., Kaplan, M. M., and Stanley, C. A. (1979). Effect of triiodothyronine replacement on the metabolic and pituitary response to starvation. *New Engl. J. Med.*, **300**, 579–584.

Garfinkel, A. S., Nilsson-Ehle, P., and Schotz, M. C. (1976). Regulation of lipoprotein lipase. Induction by insulin. *Biochim. Biophys. Acta*, **414**, 264–273.

Garrow, J. S. (1970). The effects of severe protein deficiency on the human infant. In: *Protein Metabolism and Biological Function*, (Ed. Bianchi, C. P., and Hilt, R.) Rutgers University Press, pp. 28–47.

Garrow, J. S. (1978). *Energy Balance and Obesity in Man*. Elsevier/North Holland, Amsterdam.

Garrow, J. S., and Blaza, S. (1982). Energy requirements in human beings. *Human Nutrition, Current Issues and Controversies*, (Ed. Neuberger, A., and Jukes, T. H.) M.T.P. Press, Lancaster, England, pp. 1–22.

Garvey, J. F. W., Morris, P. J., and Finch, D. R. A., et al. (1979). Experimental pancreas transplantation. *Lancet*, **i**, 971–972.

Geelen, M. J. H., Beynen, A. C., Christiansen, R. Z., et al. (1978). Short-term effects of insulin and glucagon on lipid synthesis in isolated rat hepatocytes. *FEBS Lett.*, **95**, 326–330.

Gellhorn, A., and Marks, P. A. (1961). The composition and biosynthesis of lipids in human adipose tissue. *J. Clin. Invest.*, **40**, 925–932.

Gepts, W. (1965). Pathological anatomy of the pancreas in juvenile diabetes mellitus. *Diabetes*, **14**, 619–633.

Gevers, W. (1977). Generation of protons by metabolic processes in heart cells. *J. Mol. Cell. Cardiol.*, **9**, 867–874.

Giblett, E. R. (1979). Adenosine deaminase and purine nucleoside phosphorylase deficiency: how they were discovered and what they mean. *Ciba Foundation Symp.*, **68**, 3–18.

Gill, D. M. (1978). *Bacterial Toxins and Cell Membranes*. (Ed. Jeljarwicz, J., and Wodström, T.) Academic Press, New York, pp. 291–332.

Gill, D. M., and Meren, R. (1978). ADP-ribosylation of membrane proteins catalysed by cholera toxin: basis of activation of adenylate cyclase. *Proc. Natl. Acad. Sci. U.S.A.*, **75**, 3050–3054.

Giovannetti, S., and Maggiore, O. (1964). A low nitrogen diet with proteins of high biological value for severe chronic uraemia. *Lancet*, **i**, 1000–1003.

Gjone, E. (1974). Familial lecithin: cholesterol acyltransferase deficiency. A clinical survey.

J. Clin. Lab. Invest, **33, suppl. 137,** 73–82.

Glaymore, C. N. (1970). In: *Biochemistry of the Eye* (Ed. Glaymore, C. N.) Academic Press, London, New York, pp. 645–736.

Gloster, J., and Harris, P. (1977). Fatty acid binding to cytoplasmic proteins of myocardium and red and white skeletal muscle in the rat. A possible new role for myoglobin. *Biochem. Biophys. Res. Commun.,* **74,** 506–513.

Goldberg, A. L. (1980). Regulation of protein turnover by endocrine and nutritional factors. In: *Plasticity of Muscle*, (Ed. Pette, P.) Walter de Gruyter, Berlin, pp. 469–492.

Goldberg, A. L., and St. John, A. C. (1976). Intracellular protein degradation in mammalian and bacterial cells. *Ann. Rev. Biochem.,* **45,** 747–803.

Goldberg, A. L., Demartino, G., and Chang, T. W. (1978). Release of gluconeogenic precursors from skeletal muscle. *FEBS 11th Meeting, Copenhagen,* **42,** 347–358.

Goldberg, A. L., Strand, N. P., and Swamy, K. H. S. (1980a). Studies of the ATP dependence of protein degradation in cells and cell extracts. *Ciba Found. Symp.,* **75,** 227–252.

Goldberg, A. L., Tischler, M, and Libby, P. (1980b). Regulation of protein degradation in skeletal muscle. *Biochem. Soc. Trans.,* **8,** 497.

Goldblatt, M. W. (1935). Properties of human seminal plasma. *J. Physiol.,* **84,** 208–218.

Goldstein, J. L., and Brown, M. S. (1976). The LDL-pathway in human fibroblasts: a receptor-mediated mechanism for the regulation of cholesterol metabolism. *Curr. Top. Cell. Reg.,* **11,** 147–181.

Goldstein, L. (1980). Adaptation of renal ammonia production to metabolic acidosis. *The Physiologist,* **23,** 19–25.

Goldstein, L., Schröck, H., and Cha, C.-J. M. (1980). Relationship of muscle glutamine production to renal ammonia metabolism. *Biochem. Soc. Trans.,* **8,** 509–510.

Gollnick, P. D., and Hermansen, L. (1973). Biochemical adaptations to exercise: anaerobic metabolism. *Exercise and Sports Sciences Reviews,* **1,** 1–43.

Gonen, B., and Rubenstein, A. H. (1978). Haemoglobin A_1 and diabetes mellitus. *Diabetologia,* **15,** 1–8.

Goodhart, R. S., and Shils, M. E. (1980). *Modern Nutrition in Health and Disease.* Lea and Febiger, Philadelphia.

Goodman, L. S., and Gilman, A. (1955). *Pharmacological Basis of Therapeutics.* MacMillan Co., New York, pp. 1542–1543.

Gordon, R. S., and Cherkes, A. (1956). Unesterified fatty acid in human blood plasma. *J. Clin. Invest.,* **35,** 206–212.

Gower, D. B. (1979). *Steroid Hormones.* Croom Helm, London.

Gray, G. M. (1975). Carbohydrate digestion and absorption. Role of the small intestine. *New Engl. J. Med.,* **292,** 1225–1230.

Green, A., and Newsholme, E. A. (1979). Sensitivity of glucose uptake and lipolysis of white adipocytes of the rat to insulin and effects of some metabolites. *Biochem. J.,* **180,** 365–370.

Greenberg, G. R., Marliss, E. B., Anderson, G. H., *et al.* (1976). Protein sparing therapy in post-operative patients. *New Engl. J. Med.,* **294,** 1411–1416.

Greville, G. D., and Tubbs, P. K. (1968). The catabolism of long chain fatty acids in mammalian tissues. *Essays in Biochem.,* **4,** 155–212.

Griffin, J. E., and Wilson, J. D. (1980). The testis. In: *Metabolic Control and Disease.* (Ed. Bondy, P. K., and Rosenberg, L. E.) Saunders Co., Philadelphia, U.S.A., pp. 1535–1578.

Grodin, J. M., Siitteri, P. K., and MacDonald, P. C. (1973). Source of estrogen production in post-menopausal women. *J. Clin. Endocrinol. Metab.,* **36,** 207–214.

Guidotti, G. G., Borghetti, A. F., and Gazzola, G. C. (1978). The regulation of amino acid transport in animal cells. *Biochim. Biophys. Acta,* **515,** 329–366.

Guilleband, J. (1980). *The Pill*. O.U.P., Oxford, U.K.

Gulick, A. (1923). A study of weight regulation in the adult human body during overnutrition. *Am. J. Physiol.*, **60**, 371–395.

Gurr, M. I., and James, A. T. (1971). *Lipid Biochemistry*. Chapman and Hall Ltd., London.

Guy, P. S., and Snow, D. H. (1977). A preliminary survey of skeletal muscle fibre types in equine and canine species. *J. Anat.*, **124**, 499–500.

Häberlandt, L. (1921). Uber hormonale Sterilisierung des weiblichen Tierkörpers. *Münch. Med. Wschn.*, **68**, 1577–1578.

Hagenfeldt, L. (1979). Metabolism of free fatty acids and ketone bodies during exercise in normal and diabetic man. *Diabetes*, **28, Suppl. 1**, 66–70.

Haigler, H. J. (1981). Serotonergic receptors in the central nervous system. In *Neurotransmitter Receptors*, (Ed. Yamamura, H., and Enna, S. J.) Chapman & Hall, London, pp. 1–70.

Hales, C. N. (1963). The Metabolic Action of Insulin. Ph.D. Thesis, Cambridge University.

Hales, C. N., and Randle, P. J. (1963). Immunoassay of insulin with insulin antibody precipitate. *Lancet*, **i**, 200.

Hales, E. N., Luzio, J. P., and Siddle, K. (1978). Hormonal control of adipose tissue lipolysis. *Biochem. Soc. Symp.*, **43**, 97–135.

Halestrap, A. P. (1975). The mitochondrial pyruvate carrier. *Biochem. J.*, **148**, 85–96.

Hall, G. M., Young, C., Holdcroft, A., *et al.* (1978). Substrate mobilization during surgery. *Anaesthesia*, **33**, 924–930.

Hall, G. M., Lucke, J. N., Masheter, K., *et al.* (1981). Metabolic and hormonal changes during prolonged exercise in the horse. In: *Biochemistry of Exercise IV, A.* (Ed. Poortmans, J., and Niset, G.) University Park Press, Baltimore, pp. 88–92.

Hall, K. (1971). Effect of intravenous administration of human growth hormone on sulphation factor activity in serum of hypopituitary subjects. *Acta Endocrinol.*, **66**, 491–497.

Hall, K., and Filipsson, R. (1975). Correlation between somatomedin A in serum and body height development in healthy children and children with certain growth disturbances. *Acta Endocrinol. Copenh.*, **78**, 239–250.

Hall, K., and Fryklund, L. (1979). Somatomedins. In: *Hormones in Blood*, (Ed. Gray, C. H., and James, V. H. T.) Academic Press, New York, Vol. I, pp. 255–278.

Hall, K., and Olin, P. (1972). Sulphation factor activity and growth rate during long-term treatment of patients with pituitary dwarfism with human growth hormone. *Acta Endocrinol. Copenh.*, **69**, 417–433.

Halmi, K. A. (1978). Anorexia nervosa. *Annu. Rev. Med.*, **29**, 137–148.

Hamosh, M., Clary, T. R., Chernick, S. S., *et al.* (1970). Lipoprotein lipase activity of adipose and mammary tissue and plasma triglyceride in pregnant and lactating rats. *Biochim. Biophys. Acta*, **210**, 473–482.

Hanington, E., Jones, R. J., Amess, J. A. L., *et al.* (1981). Migraine: a platelet disorder. *Lancet*, **ii**, 720–723.

Hanson, P. J., and Parsons, D. S. (1976). The utilization of glucose and production of lactate by *in vitro* preparations of rat small intestine: effects of vascular perfusion. *J. Physiol.*, **255**, 775–795.

Hanson, P. J., and Parsons, D. S. (1980). The interrelationship between alanine and glutamine in the intestine. *Biochem. Soc. Trans.*, **8**, 506–509.

Hardie, D. G. (1980). Regulation of fatty acid synthesis by reversible phosphorylation of acetyl-CoA carboxylase. *Molecular Aspects of Cellular Regulation*, **1**, 33–62.

Hardie, F. (1981). The role of covalent enzyme modification in lipid synthesis. *Trends Biochem. Sci.*, **6**, 75–77.

Harris, D. A., (1982). How ATP is made. *Biochemical Education*, **10**, 50–55.

Hartung, G. H., Foreyt, J. P., Mitchell, R. E., *et al.* (1980). Relationship of diet to high-density-lipoprotein cholesterol in middle-aged marathon runners, joggers and inactive men. *New Engl. J. Med.*, **302**, 357–361.

Harvengt, C., and Desager, J. P. (1976). Colestipol in familial type II hyperlipoproteinemia: a three year trial. *Clin. Pharmacol. Ther.*, **20**, 310–314.

Haskell, W. L. (1982). Sudden cardiac death during vigorous exercise. *Int. J. Sprts. Med.*, **3**, 45–48.

Hatefi, T., Haavik, A. G., Fowler, L. R., *et al.* (1962). Studies on the electron transfer system. *J. Biol. Chem.*, **237**, 2661–2669.

Havel, R. J., and Fredrickson, D. S. (1956). The metabolism of chylomicra 1. The removal of palmitic acid $1\text{-}C^{14}$ labelled chylomicra from dog plasma. *J. Clin. Invest.*, **35**, 1025–1032.

Havel, R. J., Carlson, L. A., Ekelund, L.-G., *et al.* (1964). Turnover rate and oxidation of different free fatty acids in man during exercise. *J. Appl. Physiol.*, **19**, 613–618.

Havel, R. J., Goldstein, J. L., and Brown, M. S. (1980). Lipoproteins and lipid transport. In: *Metabolic Control and Disease*, 8th edn. (Ed. Bondy, P. K., and Rosenberg, L. E.) Saunders & Co., Philadelphia, U.S.A., pp. 393–493.

Haymond, M. W., Karl, I. E., and Pagliara, A. S. (1974). Increased gluconeogenic substrates in the small-for-gestational-age infant. *New Engl. J. Med.*, **291**, 322–328.

Hearse, D. J. (1980). In: *Lactate, Physiologie, Methodologie and Pathologie Approach*, (Ed. Moret, P. R., Weber, J., Haissly, J.-Cl., and Denolin, H.) Springer-Verlag, Berlin, pp. 230–246.

Hearse, D. J., Braimbridge, M. V., and Jynge, P. (1981). *Protection of the Ischemic Myocardiumi Cardioplegia*. Raven Press, New York.

Heaton, K. W. (1976). *Bile Salts in Health and Disease*. Churchill Livingstone, Edinburgh, London.

Heaton, J. M. (1972). The distribution of brown adipose tissue in the human. *J. Anat.*, **112**, 35–41.

Hechter, O., and Calek, A. (1974). Hormone action considered as an informational transaction: the problem of molecular linguistics. *Acta Endocrinol.*, **77**, **Suppl. 191**, 39–66.

Hedeskov, C. J. (1980). Mechanism of glucose-induced insulin secretion. *Physiol. Rev.*, **60**, 442–509.

Heidland, A., Kult, J., Röckel, A., *et al.* (1978). Evaluation of essential amino acids and ketoacids in uremic patients on low-protein diet. *Am. J. Clin. Nutrit.*, **31**, 1784–1792.

Heldt, H. W., Klingenberg, M., and Milovancev, M. (1972). Differences between the ATP/ADP ratios in the mitochondrial matrix and in the extramitochondrial space. *Eur. J. Biochem.*, **30**, 434–440.

Helmreich, E., Danforth, W. H., Karpatkin, S. *et al.* (1965). The response of the glycolytic system of anaerobic frog sartorius muscle to electrical stimulation. In: *Control of Energy Metabolism.* (Ed. Chance, B., Estabrook, R. W., and Williamson, J. R.) Academic Press, New York, pp. 299–312.

Hems, D. A. (1972). Metabolism of glutamine and glutamic acid by isolated perfused kidneys of normal and acidotic rats. *Biochem. J.*, **130**, 671–680.

Hems, D. A., and Brosnan, J. T. (1971). Effects of metabolic acidosis and starvation on the content of intermediary metabolites in rat kidney. *Biochem. J.*, **123**, 391–397.

Hems, D. A., and Whitton, P. D. (1980). Control of hepatic glycogenolysis. *Physiol. Rev.*, **60**, 2–50.

Hems, D. A., Whitton, P. D., and Taylor, E. A. (1972). Glycogen synthesis in perfused liver of the starved rat. *Biochem. J.*, **129**, 529–538.

Henderson, I. C. (1981). Less toxic treatment for advanced breast cancer. *New Engl. J. Med.*, **305**, 575–577.

Henderson, J. F., Lower, J. K., and Barankiewicz, J. (1977). Purine and pyrimidine metabolism: pathways, pitfalls and perturbations. *Ciba Found. Symp.*, **48**, 3–22.

Henderson, K. M. (1979). Gonadotrophic regulation of ovarian activity. *Brit. Med. Bull.*, **35**, 161–166.

Hendrix, T. R. (1975). In: *Intestinal Absorption and Malabsorption.* (Ed. Czaky, T. Z.) Raven Press, New York, pp. 253–271.

Hennekens, C. H., Willett, W., Rosner, B., *et al.* (1979). Effects of beer, wine and liquor in coronary deaths. *J. Am. Med. Ass.*, **242**, 1973–1974.

Hermansen, L. (1979). Effect of acidosis on skeletal muscle performance during maximal exercise in man. *Bulletin European de Physiopathologie Respiratoire*, **15**, 229–238.

Hermansen, L. (1981). Effect of metabolic changes on force generation in skeletal muscle during maximal exercise. *Ciba Found. Symp.*, **8**, 75–88.

Hermansen, L., and Vaage, O. (1977). Lactate disappearance and glycogen synthesis in human muscle after maximal exercise. *Amer. J. Physiol.*, **233**, E422–E429.

Hermansen, L., Hultman, E., and Saltin, B. (1967). Muscle glycogen during prolonged severe exercise. *Acta Physiol. Scand.*, **71**, 129–139.

Hermansen, L., *et al.* (1983) (In preparation).

Hers, H. G. (1976). The control of glycogen metabolism in the liver. *Annu. Rev. Biochem.*, **45**, 167–190.

Hers, H. G. (1980). Carbohydrate metabolism and its regulation. In: *Inherited Disorders of Carbohydrate Metabolism,* (Ed. Burman, D., Holton, J. B., and Pennock, C. A.) M.T.P. Press Ltd., Lancaster, U.K., pp. 3–18.

Hers, H. G., and Van Schaftingen, E. (1982). Fructose 2,6-bisphosphate 2 years after its discovery. *Biochem. J.*, **206**, 2–12.

Hervey, G. R. (1975). Are athletes wrong about anabolic steroids. *Brit. J. Sports Medicine*, **9**, 74–77.

Hervey, G. R., and Tobin, G. (1981). Brown adipose tissue and diet-induced thermogenesis. *Nature*, **289**, 699.

Hervey, G. R., and Tobin, G. (1983). Luxuskonsumption diet-induced thermogenesis and brown fat: a critical review. *Clin. Science*, **64**, 7–18.

Herzfeld, A., and Raper, S. M. (1976). Enzymes of ornithine metabolism in adult and developing rat intestine. *Biochim. Biophys. Acta*, **428**, 600–610.

Hess, B., and Brand, K. (1974). Cell and tissue disintegration. In: *Methods of Enzymatic Analysis*, 2nd edn., (Ed. Bergmeyer, H. U.) Verlag Chemie/Academic Press, New York, London. Vol. 1, pp. 399–409.

Hickson, R. C., Rennie, M. J., Conlee, R. K., *et al.* (1977). Effects of increased plasma fatty acids on glycogen utilization and endurance. *J. Appl. Physiol.*, **43**, 829–833.

Hill, A. V., Long, C. N. A., and Lupton, H. (1925). Muscular exercise, lactic acid and the supply and utilization of oxygen v. the recovery process after exercise in man. *Proc. Roy. Soc. (B)*, **97**, 96–138.

Hill, H. A. O. (1979). The chemistry of dioxygen and its reduction products. *Ciba Found. Symp.*, **65**, 5–18.

Hillier, M., and Karim, S. M. M. (1979). General introduction and practical implications of some pharmacological action of prostaglandins, thromboxanes and their synthesis inhibitors. *Practical Applications of Prostaglandins and Their Synthesis Inhibitors.* (Ed. Karim, S. M. M.) M.T.P. Press, Lancaster, U.K., pp. 1–24.

Hilz, H., and Stone, P. R. (1976). Poly (ADP-ribose) and ADP-ribosylation of proteins. *Rev. Physiol. Biochem. Pharmacol.*, **76**, 1–58.

Himms-Hagen, J. (1979). Obesity may be due to a malfunctioning of brown fat. *Can. Med. Ass. J.*, **121**, 1361–1364.

Himms-Hagen, J., and Desautels, M. (1978). A mitochondrial defect in brown adipose tissue of the obese (ob/ob) mouse. *Biochem. Biophys. Res. Commun.*, **83**, 628–634.

Hirvonen, E., Mälkönen, M., and Manninen, V. (1981). Effects of different progestogens on lipoproteins during post-menopausal replacement therapy. *New Engl. J. Med.,* **304**, 560–563.

Hoagland, M. B. (1978). Coding, information transfer and protein synthesis. In: *The Metabolic Basis of Inherited Disease*, 4th edn. (Ed. Stanbury, J. B., Wyngaarden, J. B., and Fredrickson, D. B.) McGraw-Hill, New York, pp. 33–50.

Hoffman, B. B., and Lefkowitz, R. J. (1982). Adrenergic receptors in the heart. *Annu. Rev. Physiol.,* **44**, 474–484.

Holbrook, J. J., Liljas, A., and Steindel, S. J. (1975). Lactate dehydrogenase. *The Enzymes,* 3rd edn. (Ed. Boyer, P. D.) Academic Press, New York, pp. 191–293.

Holick, M. F., Uskokovic, M., Henley, J. W., *et al.* (1980). The photoproduction of 1α,25-dihydroxy vitamin D_3 in skin. *New Engl. J. Med.,* **303**, 349–384.

Hollands, M. A., and Cawthorne, M. A. (1981). Important sites of lipogenesis in the mouse other than liver and white adipose tissue. *Biochem. J.,* **196**, 645–647.

Hommes, F. A., Polman, H. A., and Reerink, J. D. (1968). Leigh's encephalomyelopathy: an inborn error of gluconeogenesis. *Arch. Dis. Child.,* **43**, 423–426.

Hood, W. B. (1982). More on sulfinpyrazone after myocardial infarction. *New Engl. J. Med.,* **306**, 988–989.

Hoover, H. C., Grant, J. P., Gorschboth, C., *et al.* (1975). Nitrogen sparing intravenous fluids in post-operative patients. *New Engl. J. Med.,* **293**, 172–175.

Hopkirk, T. J., and Bloxham, D. (1977). Fatty acid synthesis and metabolite fluctuations in meal-fed rats. *Biochem Soc. Trans.,* **6**, 1294–1297.

Hornykiewicz, O. (1973). Dopamine in the basal ganglia: its role and therapeutic implications. *Br. Med. Bull.,* **29**, 172–178.

Horwitt, M. K. (1980). In: *Modern Nutrition in Health and Disease*, 6th edn. (Ed. Goodhart, R. S., and Shils, M. E.) Lea and Febiger, Philadelphia, pp. 181–190.

Hosobuchi, Y. (1978). In: *Endogenous and Exogenous Opiate Agonists and Antagonists.* (Ed. Leong-way, E.) Pergamon Press, Toronto, pp. 375–378.

Hosobuchi, Y., Rossier, J., Bloom, F. E., *et al.* (1979). Stimulation of human periaqueductal gray for pain relief increases immunoreactive β-endorphin in ventricular fluid. *Science,* **203**, 279–281.

Houghton, C. R. S. (1971). Studies on the metabolism of exercise with special reference to perfused isolated skeletal muscle. D. Phil. thesis. Oxford University.

Hoult, D. I., Busby, S. J. W., Gadian, D., *et al.* (1974). Observation of tissue metabolites using ^{31}P nuclear magnetic resonance. *Nature,* **252**, 285–287.

Howell, R. R. (1978). The glycogen storage diseases. In: *The Metabolic Basis of Inherited Disease.* (Ed. Stanbury, J. B., Wyngaarden, J. B., and Fredrickson, D. S.) McGraw-Hill, New York. 4th edn., pp. 137–159.

Huang, S.-W., and MacLaren, N. K. (1976). Insulin dependent diabetes: a disease of autoaggression. *Science,* **192**, 64–66.

Huber, W., Saifer, M. F. P., and Williams, L. D. (1980). Superoxide dismutase pharmacology and orgotein efficacy: new perspectives. *Biological and Clinical Aspects of Superoxide and Superoxide Dismutase.* (Ed. Bannister, W. H., and Bannister, J. V.) Elsevier/North Holland, New York, pp. 395–407.

Hübscher, G. (1970). In: *Lipid Metabolism*, (Ed. Wakil, S. J.) Academic Press, New York, London, pp. 280–370.

Hue, L. (1979). Short term control of liver carbohydrate metabolism. *Biochem. Soc. Trans.,* **7**, 850–854.

Hughes, J. (1975). Isolation of an endogenous compound from the brain with pharmacological properties similar to morphine. *Brain Res.,* **88**, 295–308.

Hughes, J., and Kosterlitz, H. W. (1983). Introduction. *Brit. Med. Bull.,* **39**, 1–3.

Huijing, F. (1975). Glycogen metabolism and glycogen storage diseases. *Physiol. Rev.,* **55**, 609–658.

Hultman, E. (1978). Regulation of carbohydrate metabolism in the liver during rest and exercise with special reference to diet. *3rd International Symposium on Biochemistry of Exercise*. (Ed. Landry, P., and Orban, W. A. R.) Symposia Specialist Inc., Miami, Florida, pp. 99–126.

Hume, D. A., and Weidemann, M. J. (1979). Role and regulation of glucose metabolism in proliferating cells. *J. Natl. Cancer Inst.*, **62**, 3–8.

Hume, D. A., Radik, J. L., Ferber, E., *et al.* (1978). Aerobic glycolysis and lymphocyte transformation. *Biochem. J.*, **174**, 703–709.

Hunt, T. (1980). Phosphorylation and control of protein synthesis in reticulocytes. *Molecular Aspects of Cellular Regulation*, **1**, 175–202.

Imperato-McGinley, J., Guerrero, L., Gautier, T., *et al.* (1974). Steroid 5-reductase deficiency in man: an inherited form of male pseudohermaphroditism. *Science*, **186**, 1213–1215.

Ingebritsen, T. S., and Gibson, D. M. (1980). Reversible phosphorylation of hydroxymethylglutamyl-CoA reductase. *Molecul. Aspects Cellul. Regul.*, **1**, 63–94.

Irving, L. (1966). Adaptations to cold. *Scientific American*, **214**, 94–101.

Irsigler, K., and Kritz, H. (1979). Long term continuous intravenous insulin therapy with a portable insulin dosage-regulating apparatus. *Diabetes*, **28**, 196–203.

Irvine, W. J., McCallum, C. J., Gray, R. S., *et al.* (1977). Pancreatic islet-cell antibodies in diabetes mellitus correlated with the duration and type of diabetes, coexistent autoimmune disease and HLA type. *Diabetes*, **26**, 138–147.

Iversen, L. L. (1982). Neurotransmitter and CNS disease, *Lancet*, **ii**, 914–918.

Jackson, R. J., and Hunt, T. (1980). Mechanism of control of polypeptide-chain initiation in reticulocytes. *Biochem. Soc. Trans.*, **8**, 457–458.

Jackson, W. P., and Vinik, A. I. (1977). *Diabetes Mellitus, Clinical and Metabolic*. Edward Arnold, London.

Jacob, F., and Monod, J. (1961). Genetic regulatory mechanisms in the synthesis of proteins. *J. Mol. Biol.*, **3**, 318–356.

Jagendorf, A. T., and Uribe, E. (1966). ATP formation caused by acid-base transition of spinach chloroplasts. *Proc. Natl. Acad. Sci. U.S.A.*, **55**, 170–177.

James, J. H., Zapero, V., Jeppsson, B., *et al.* (1979). Hyperammonaemia, plasma amino acid imbalance and blood-brain amino acid transport: a unified theory of portal-systemic encephalopathy. *Lancet*, **ii**, 772–775.

James, W. P. T., and Trayhurn, P. (1981). Thermogenesis and obesity. *Brit. Med. Bull.*, **37**, 43–48.

Jansson, E. (1980). Diet and muscle metabolism in man. *Acta Physiol. Scand.*, **Suppl. 487**.

Jansson, E., Sjödin, B., and Tesch, P. (1978). Changes in muscle fibre type distribution in man after physical training. *Acta Physiol. Scand.*, **104**, 235–237.

Janus, E. D., and Lewis, B. (1978). Alcohol and abnormalities of lipid metabolism. *Clinics. Endocrinol. Metab.*, **7**, 321–332.

Jeffcoat, R. (1979). The biosynthesis of unsaturated fatty acids and its control in mammalian liver. *Essays in Biochem.*, **15**, 1–36.

Jenkins, C. D. (1978). Behavioural risk factors in coronary artery disease. *Ann. Rev. Med.*, **29**, 543–562.

Jenkins, D. J. A. (1979). Dietary fibre, diabetes and hyperlipidaemia. Progress and prospects. *Lancet*, **ii**, 1287–1290.

Jenkins, D. J. A., Leeds, A. R., and Wolever, T. M. S. (1976). Unabsorbable carbohydrates and diabetes: decreased post-prandial hyperglycaemia. *Lancet*, **ii**, 172–174.

Jenkins, D. J. A., Wolever, T. M. S., and Nineham, R. (1978). Guar crispbread in the diabetic diet. *Brit. Med. J.*, **ii**, 1744–1746.

Jenkins, D. J. A., Hockaday, T. D. R., and Wolever, T. M. S. (1979). Dietary fibre and ketone bodies; reduced urinary 3-hydroxy-butyrate excretion in diabetics on guar. *Brit. Med. J.*, **ii**, 1555.

Jessell, T. M. (1982). Pain. *Lancet*, **ii,** 1084–1087.

Job, D., Eschwege, E., Guyot-Argenton, C., *et al.* (1976). Effect of multiple daily insulin injections on the course of diabetic retinopathy. *Diabetes*, **25,** 463–470.

Joel, C. D. (1965). The physiological role of brown adipose tissue. In: *Handbook of Physiology: Adipose Tissue*, (Ed. Renold, A. E., and Cahill, G. F.) American Physiological Society, Washington, D.C., pp. 59–85.

Johnson, L. K., Baxter, J. D., and Rousseau, G. G. (1978). Mechanisms of glucocorticoid receptor function: an analysis of selected models. In: *Glucocorticoid Hormone Action.* (Ed. Baxter, J. D., and Rousseau, G. G.) Springer-Verlag, Heidelberg, pp. 305–326.

Johnson, M. K. (1975). The delayed neuropathy caused by some organophosphorus esters: mechanisms and challenge. *C.R.C. Crit. Rev. Toxicol.,* **3,** 289–316.

Johnson, R. A., Baker, S. S., Fallon, J. T., *et al.* (1981). An occidental case of cardiomyopathy and selenium deficiency. *New Engl. J. Med.,* **304,** 1210–1212.

Johnston, I. O. A., and Chenneour, R. (1962). The effect of methandienone on the metabolic response to surgical operations. *Br. J. Surg.,* **50,** 924–928.

Johnston, J. M. (1978). In: *Disturbances in Lipid and Lipoprotein Metabolism*, (Ed. Dietsky, J. M., Gotto, A. M., and Ontko, J. A.) American Physiological Society, Bethesda, pp. 57–68.

Joplin, G. F., and Wright, A. P. (1968). The detection of diabetes in man. In *Carbohydrate Metabolism and its Disorders*, (Ed. Dickens, F., Randle, P. J., and Whelan, W. J.) Academic Press. London, New York, pp. 1–24.

Jung, R. T., Shetly, P. S., and James, W. P. T., *et al.* (1979). Reduced thermogenesis in obesity. *Nature*, **279,** 322–323.

Jungas, R. L. (1968). Fatty acid synthesis in adipose tissue incubated in tritiated water. *Biochemistry*, **7,** 3708–3717.

Kagawa, Y., and Racker, E. (1966). Partial resolution of the enzymes catalysing oxidative phosphorylation. *J. Biol. Chem.,* **241,** 2461–2482.

Kahn, C. R. (1978). Insulin resistance, insulin in-sensitivity and insulin unresponsiveness: a necessary distinction. *Metabolism*, **27,** 1893–1902.

Kahn, C. R., Megyesi, K., Bar, R. S., *et al.* (1977). Receptors for peptide hormones: new insights into pathophysiology of disease states in man. *Ann. Intern. Med.,* **86,** 205–219.

Kakkar, V., and Scully, M. F. (1978). Thrombolytic therapy. *Brit. Med. Bull.,* **34,** 191–200.

Kaplan, N. O., and Everse, J. (1972). Regulatory characteristics of lactate dehydrogenases. *Adv. Enz. Regul.,* **10,** 323–336.

Karam, J. H., Prosser, P. R., and Lewitt, P. A. (1978). Islet cell surface antibodies in a patient with diabetes mellitus after rodenticide ingestion. *New Engl. J. Med.,* **299,** 1191.

Karim, S. M. M. (1979). Termination of second trimester pregnancy with prostaglandins. In: *Practical Applications of Prostaglandins and Their Synthesis Inhibitors*, (Ed. Karim, S. M. M.) M.T.P. Press, Lancaster, U.K., pp. 375–410.

Karim, S. M. M., Hillier, K. (1979). Prostaglandins in the control of animal and human reproduction. *Brit. Med. Bull.,* **35,** 173–180.

Karim, S. M. M., and Prasad, R. N. V. (1979). Pre-operative cervical dilation with prostaglandins. In: *Practical Applications of Prostaglandins and Their Synthesis Inhibitors*, (Ed. Karim, S. M. M.) M.T.P. Press, Lancaster, U.K., pp. 283–300.

Karim, S. M. M., Ny, S. G., and Ratmans, S. (1979). Termination of abnormal intrauterine pregnancy with prostaglandins. In: *Practical Applications of Prostaglandins and Their Synthesis Inhibitors*, (Ed. Karim, S. M. M.) M.T.P. Press, Lancaster, U.K., pp. 319–374.

Kase, N. G., and Speroff, L. (1980). The ovary. In: *Metabolic Control and Disease*, 8th edn. (Ed. Bondy, P. K., and Rosenberg, L. E.) Saunders & Co., Philadelphia, U.S.A., pp. 1579–1620.

Katz, J. (1979). Use of isotopes for the study of glucose metabolism in vivo. In: *Techniques in Metabolic Research* **B,207,** (Ed. Kornberg, H. L., Metcalfe, J. C., Northcote, D. H., *et al.*) Elsevier/North Holland, Amsterdam, pp. 1–22.

Katz, J., and Rognstad, R. (1976). Futile cycles in the metabolism of glucose. *Current Topics in Cellul. Regul.,* **10,** 237–289.

Katz, J., Rognstad, R. (1978). Futile cycling in glucose metabolism. *Trends Biochem. Sci.,* **3,** 171–174.

Kauzmann, W. (1959). Some factors in the interpretation of protein denaturation. *Adv. Prot. Chem.,* **14,** 1–63.

Kealey, T. (1983). The metabolism and hormonal response of human eccrine sweat glands isolated by collagence digestion. *Biochem. J.,* **212,** 143–148.

Keen, H., Jarrett, R. J., and Alberti, K. G. M. M. (1979). Diabetes mellitus: a new look at diagnostic criteria. *Diabetologie,* **16,** 283–285.

Keilin, D. (1925). *Proc. Roy. Soc. (London), B,* **98,** 312.

Kellie, A. E. (1975). Structure and nomenclature. In: *Biochemistry of the Steroid Hormones,* (Ed. Makin, H. L. J.) Blackwell Scientific Publications, Oxford, U.K., pp. 1–16.

Keys, A. (1970). Coronary heart disease in seven countries. *Circulation 41,* **suppl. I,** 1–211

Keys, A. (1980). *Seven Countries: a Multivariate Analysis of Death and Coronary Heart Disease.* Harvard Univ. Press, Cambridge, Mass, U.S.A.

Kikucki, G. (1973). The glycine cleavage system: composition, reaction mechanism and physiological significance. *Mol. Cell Biochem.,* **1,** 169–187.

Kilberg, M. S., Handlogten, M. E., and Christensen, H. N. (1980). Characteristics of an amino acid transport system in rat liver for glutamine, asparagine, histidine and closely related analogs. *J. Biol. Chem.,* **255,** 4011–4019.

King, A. C., and Cuatrecasas, P. (1981). Peptide hormone induced receptor mobility, aggregation and internalization. *New Engl. J. Med.,* **305,** 77–88.

King, R. J. (1974). The surfactant system of the lung. *Fed. Proc.,* **33,** 2238–2247.

King, R. J. B. (1976). Intracellular receptors of steroid hormones. *Essays in Biochem.* **12,** 41–76.

Kirk, J. S., and Sumner, J. B. (1931). *J. Biol. Chem.,* **94,** 21.

Kleiner, D. (1981). The transport of NH_3 and NH_4^+ across biological membranes. *Biochem. Biophys. Acta,* **639,** 41–52.

Kletzien, R. F., Pariza, M. W., Becker, J. E., *et al.* (1976). Induction of amino acid transport in primary cultures of adult rat liver parenchymal cells by insulin. *J. Biol. Chem.,* **251,** 3014–3020.

Klingenberg, M. (1964). Reversibility of energy transformations in the respiratory chain. *Angewandte Chemie International,* **3,** 54–61.

Klingenberg, M. (1976). The adenine nucleotide transport of mitochondria. *Mitochondria: Bioenergetics, Biogenesis and Membrane Structure,* (Ed. Packer, L., and Gomez-Puyou, A.) Academic Press Inc, London, pp. 127–149.

Klingenberg, M., and Bücher, T. (1960). Biological oxidations. *Annu. Rev. Biochem.,* **29,** 669–708.

Kluger, M. J. (1979). Phylogeny of fever. *Fed. Proc.,* **38,** 30–34.

Knobil, E. (1981). Patterns of hormonal signals and hormone action. *New Engl. J. Med.,* **305,** 1582–1583.

Koenig, R. J., Blobstein, S. A., and Cerami, A. (1977). Structure of carbohydrate of haemoglobin A_{1c}. *J. Biol. Chem.,* **252,** 2992–2997.

Koeslag, J. H., Noakes, T. D., and Sloan, A. W. (1980). Post-exercise ketosis. *J. Physiol.,* **301,** 79–90.

Koike, M., Koike, K., and Hiraoka, T. (1980). Structure, function and an inherited defect of 2-oxo acid dehydrogenase multienzyme complexes. *Biochem. Soc. Trans.,* **8,** 6–7.

Kopin, I. J. (1981). Neurotransmitters and the Lesch-Nyhan syndrome. *New Engl. J. Med.,* **305,** 1148–1149.

Kornberg, A. (1978). Aspects of DNA replication. *Cold Spring Harbor Symp.*, **43**, 1–10.

Koshland, D. E. (1958). Application of a theory of enzyme specificity to protein synthesis. *Proc. Nat. Acad. Sci. U.S.A.*, **44**, 98–104.

Kramsch, D. M., Aspen, A. J., Abramowitz, B. M., *et al.* (1981). Reduction of coronary atherosclerosis by moderate conditioning exercise in monkeys on atherogenic diet. *New Engl. J. Med.*, **305**, 1483–1489.

Krebs, H. A. (1964). The metabolic fate of amino acids. *Mammalian Protein Metabolism*, (Ed. Munro, H. N., and Allison, J. B.) Academic Press, New York. Vol. 1, pp. 125–177.

Krebs, H. A. (1968). The effects of ethanol on the metabolic activities of the liver. *Adv. Enzym. Regul.*, **6**, 467–480.

Krebs, H. A. (1972). The Pasteur effect and the relations between respiration and fermentation. *Essays in Biochem.*, **8**, 1–34.

Krebs, H. A. (1973). Pyridine nucleotides and rate control. *Symp. Soc. Expt. Biol.*, **27**, 299–318.

Krebs, H. A. (1976). Discovery of the ornithine cycle. In: *The Urea Cycle*, (Ed. Grisolia, S., Baguena, R., and Mayor, F.) John Wiley & Sons, New York, U.S.A., pp. 1–12.

Krebs, H. A. (1978). Regulatory mechanism in purine biosynthesis. *Adv. Enzym. Regulat.*, **16**, 409–422.

Krebs, H. A. (1980). Glutamine metabolism in the animal body. *Glutamine: Metabolism, Enzymology and Regulation*, (Ed. Mora, J., and Palacios, R.) Academic Press, New York, pp. 319–329.

Krebs, H. A. (1981). *Reminiscences and Reflections*. Oxford University Press.

Krebs, H. A., and Henseleit, K. (1932). *Z. Physiol. Chem.*, **210**, 33–66.

Krebs, H. A., and Johnson, W. A. (1937). The role of citric acid in intermediate metabolism in animal tissues. *Enzymologia*, **48**, 148–156.

Krebs, H. A., and Kornberg, H. L. (1957). A survey of the energy transformations in living matter. *Ergeb. Physiol. Biol. Chem. und Experim. Pharmak.*, **49**, 212–198.

Krebs, H. A., and Lund, P. (1977). Aspects of the regulation of the metabolism of branched-chain amino acids. *Adv. Enz. Regul.*, **15**, 375–394.

Krebs, H. A., Williamson, D. H., and Bates, M. W., *et al.* (1971). The role of ketone bodies in caloric homeostasis. *Adv. Enz. Regul.*, **9**, 387–409.

Krebs, H. A., Hems, R., and Lund, P. (1973). Some regulatory mechanisms in the synthesis of urea in the mammalian liver. *Adv. Enzyme Regul.*, **11**, 361–377.

Krebs, H. A., Woods, H. F., and Alberti, K. G. M. M. (1975). Hyperlactataemia and lactic acidosis. *Essays Med. Biochem.*, **1**, 81–104.

Kricka, L. J., and Clark, P. M. S. (1979). *Biochemistry of Alcohol and Alcoholism*, Ellis Horwood Ltd., Chichester, U.K.

Krieger, D. T., and Martin, J. B. (1981). Brain peptides. *New Engl. J. Med.*, **304**, 876–885; 944–951.

Krisman, C. R., and Barengo, R. (1975). A precursor of glycogen biosynthesis: α-1:4-glucan-protein. *Eur. J. Biochem.*, **52**, 117–123.

Kromhout, D., Bosschieter, E. M., and de Lezenne Coulander, C. (1982). Dietary fibre and 10 year mortality from coronary heart disease, cancer and all causes. *Lancet*, **ii**, 518–521.

Kuck, J. F. R. (1970). In: *Biochemistry of the Eye*. (Ed. Glaymore, C. N.) Academic Press, London, New York, pp. 261–318.

Kumar, R., Cohen, W. R., and Epstein, F. H. (1980). Vitamin D and calcium hormones in pregnancy. *New Engl. J. Med.*, **302**, 1143–1146.

Kurzrok, R. (1937). The prospects for hormonal sterilisation. *J. Contracept.*, **2**, 27–29.

Lacy, P. E., and Kostianovsky, M. (1967). Method for the isolation of intact islets of Langerhans from the rat pancreas. *Diabetes*, **16**, 35–39.

Laloux, M., Stalmans, W., and Hers, H. G. (1978). Native and latent forms of liver phosphorylase phosphatase. The non-identity of native phosphorylase phosphatase and synthase phosphatase. *Eur. J. Biochem.*, **92**, 14–24.

Lancet editorial (1979a). New insulin-delivery systems for diabetics. *Lancet*, **i**, 1275–1277.

Lancet editorial (1979b). Oestrogen therapy and endometrial cancer. *Lancet* **i**, 1121–1122.

Lancet editorial (1981). Prostaglandins in the kidney. *Lancet*, **ii**, 343–345.

Lancet editorial (1982). Adequate dialysis, *Lancet*, **i**, 147–148.

Lancet editorial (1982a). Salsolinol and alcoholism, *Lancet*, **ii**, 80–81.

Lancet editorial (1982b). Endogenous opiates and their actions. *Lancet*, **ii**, 305–307.

Lancet editorial (1982c). Treatment of migraine, *Lancet*, **ii**, 1338–1340.

Lancet editorial (1983a). Selenium perspective, *Lancet*, **i**, 685.

Lancet editorial (1983b). Diet and hepatic encephalopathy, *Lancet*, **i**, 625–626.

Lancet editorial (1983c). Drug targeting in cancer, *Lancet*, **i**, 512.

Lancet editorial (1983d). The starch-blocker idea, *Lancet*, **i**, 569–570.

Land, J. M., and Clark, J. B., (1978). Mitochondrial myopathies. *Biochem. Soc. Trans*, **7**, 231–245.

Landon, J., Hassan, M., Pourferzaneth, M., et al. (1979). Non-isotopic immunoassay of hormones in blood. In: *Hormones in Blood*, 3rd Edn. (Ed. Gray, C. H., and James, V. H. T.) Academic Press, New York. Vol. 3, pp. 1–40.

Lanman, J. T., Guy, L. P., and Dancis, J. (1954). Retrolental fibroplasia and oxygen therapy. *J. Am. Med. Asso.*, **155**, 223–226.

LaNoue, K. F., and Schoolwerth, A. C. (1979). Metabolite transport in mitochondria. *Annu. Rev. Biochem.*, **48**, 871–922.

Lantigua, R., Amatruda, J. M., Biddle, T. L., et al. (1980). Cardiac arrhythmias associated with a liquid protein diet for the treatment of obesity. *New Engl. J. Med.*, **303**, 735–737.

Larner, J., Cheng, K., Schwartz, C., et al. (1982). A proteolytic mechanism for the action of insulin via oligopeptide mediator formation. *Fed. Proc.*, **41**, 2724–2729.

Lassers, B. W., Wahlqvist, M. L., Kaijser, L., et al. (1971). Relationship in man between plasma free fatty acids and myocardial metabolism of carbohydrate substrates. *Lancet*, **ii**, 448–450.

Lawrence, J. C., and Larner, J. (1978). Activation of glycogen synthase in rat adipocytes by insulin and glucose involves increased glucose transport and phosphorylation. *J. Biol. Chem.*, **253**, 2104–2113.

Lawrence, R. A. (1926). The effects of exercise on insulin action in diabetes. *Br. Med. J.*, **i**, 648–652.

Layzer, R. B., Rowland, L. P., and Ranney, H. M. (1967). Muscle phosphofructokinase deficiency. *Arch. Neurol.*, **17**, 512–523.

Leader, D. P. (1980). The control of phosphorylation of ribosomal proteins. *Molecular Aspects of Cellular Regulation*, **1**, 203–234.

Leaf, A., and Cotran, R. S. (1980). *Renal Pathophysiology*. 2nd edn. O.U.P. Oxford.

Lee, H. A. (1977) The application of parenteral nutrition to renal failure patients. *Current Concepts in Parenteral Nutrition*, (Ed. Greep, J. M., Soeters, P. B., and Wesdorp, R. I. C.) Martinus Nijhoff Medical Division, The Hague, pp. 217–226.

Lees, R. S., and Lees, A. M. (1982). High-density lipoproteins and the risk of atherosclerosis. *New Engl. J. Med.*, **306**, 1546–1548.

Lehninger, A. (1973). *Biochemistry*, Worth Inc. New York, U.S.A.

Lehninger, A. L. (1975). *Biochemistry*, 2nd edn. Worth Inc. New York, U.S.A.

Lelbach, W. K. (1974). Organic pathology related to volume and pattern of alcohol use. In: *Research Advances in Alcohol and Drug Problems*, (Ed. Gibbins, R. J., Israel, Y., Kalant, H., et al.) John Wiley, New York, U.S.A., pp. 93–198.

Lerner, J. (1978). *A Review of Amino Acid Transport Processes in Animal Cells and Tissues*. University of Maine at Orono Press, Orono, Maine, U.S.A.

Levine, R., and Goldstein, M. S. (1955). On the mechanism of action of insulin. *Rec. Progr. Hormone Res.*, **11**, 343–380.

Levine, S. B. (1976). Marital sexual dysfunction; erectile dysfunction. *Ann. Intern. Med.,* **85,** 342–350.

Levinsky, N. G., and Robert, N. J. (1979). Severe metabolic acidosis in a young man. *New Engl. J. Med.,* **301,** 650–657.

Lev-Ran, A., and Anderson, R. W. (1981). The diagnosis of postprandial hypoglycaemia. *Diabetes,* **30,** 996–999.

Lewis, B. (1976). *The Hyperlipaemias.* Blackwell Scientific Publications. Oxford, London.

Lewis, B., Katan, M., Merkz, I., *et al.* (1981). Towards an improved lipid lowering diet: additive effects of changes in nutrient intake. *Lancet,* **ii,** 1310–1313.

Lewis, U. J., Singh, R. N. P., and Tutwiler, G. F. (1980). Human growth hormone: a complex of proteins, *Rec. Progr. Horm. Res.,* **36,** 477–504.

Lieber, C. S. (1976). Metabolism and metabolic actions of ethanol. In: *The Year in Metabolism,* (Ed. Freinkel, N.) Plenum Press, New York, pp. 317–342.

Lieber, C. S. (1977). Metabolism and metabolic action of ethanol. In: *The Year in Metabolism,* (Ed. Freinkel, N.) Plenum Press, New York, pp. 411–434.

Lieber, C. S., DeCarli, L. M., and Rubin, E. (1975). Sequential production of fatty liver, hepatitis and cirrhosis in sub-human primates fed ethanol with adequate diets. *Proc. Natl. Acad. Sci., U.S.A.,* **72,** 437–441.

Lienhard, G. E. (1973). Enzymatic catalysis and transition-state theory. *Science,* **180,** 149–154.

Lincoln, G. A. (1979). Pituitary control of testicular activity. *Brit. Med. Bull.,* **35,** 167–172.

Linford, J. H. (1966). *An Introduction to Energetics.* Butterworths, London, U.K.

Lipmann, F. (1941). Metabolic generation and utilization of phosphate bond energy. *Adv. in Enzymol.,* **1,** 99–160.

Lithell, H., Hellsing, K., Lundqvist, G., *et al.* (1979). Lipoprotein-lipase activity of human skeletal-muscle and adipose tissue after intensive physical exercise. *Acta Physiol. Scand.,* **105,** 312–315.

Littleton, J. (1978). Alcohol and neurotransmitters. *Clinics in Endocrinol. Metabol.,* **7,** 369–384.

Littleton, J. (1983). Cell-membrane lipids in ethanol tolerance and dependence. *Biochem. Soc. Trans.,* **11,** 61–62.

Lloyd, D., and Coakley, W. T. (1979). Techniques for the breakage of cells and tissues. In: *Techniques in Metabolic Research,* (Ed. Kornberg, H. L., Metcalfe, J. C., Northcote, D. H., *et al.*) Elsevier/North Holland, Amsterdam. Vol. B.201, pp. 1–18.

Lloyd, D., and Poole, R. K. (1979). Subcellular fractionation: isolation and characterization of organelles. In: *Techniques in Metabolic Research,* (Ed. Kornberg, H. L., Metcalfe, J. C., Northcote, D. H., *et al.*) Elsevier/North Holland, Amsterdam. Vol. B.202, pp. 1–27.

Loeb, J. N. (1976). Corticosteroids and growth. *New Engl. J. Med.,* **295,** 547.

Logan, R. L., Riemersma, R. A., Thomson, M., *et al.* (1978). Risk factors for ischaemic heart disease in normal men aged 40. *Lancet,* **i,** 949–955.

Lowenstein, J. M., and Goodman, M. N. (1978). The purine nucleotide cycle. *Fed. Proc.,* **37,** 2308–2312.

Lowry, M., and Ross, B. D. (1980). Activation of oxoglutarate dehydrogenase in the kidney in response to acidosis. *Biochem. J.,* **140,** 771–780.

Lucking, M. T. (1981). Obligatory random spot testing for administration of sex hormones and their derivatives to influence international sports performance. *Int. J. Sports Medicine,* **2,** 196–199.

Lucking, M. T. (1982). Steroid hormones in sports. *Int. J. Sports Medicine,* **3,** 65–68.

Luft, R., Ikkos, D., Palmieri, G., *et al.* (1962). A case of severe hypermetabolism of non-thyroid origin with a defect in the maintenance of mitochondrial respiratory control. A correlated clinical, biochemical and morphological study. *J. Clin. Invest.,* **41,** 1776–1804.

Lund-Andersen, H. (1979). Transport of glucose from blood to brain. *Physiol Rev.*, **59**, 305–352.

Mabuchi, H., Sakai, T., Sakai, Y., *et al.* (1983). Reduction of serum cholesterol in heterozygous patients with familial hypercholesterolemia. *New Engl. J. Med.*, **308**, 609–613.

MacKay, E. M. (1943). The significance of ketosis, *J. Clin. Endocrinol.*, **3**, 101–110.

Madsbad, S., Alberti, K. G. M. M., and Binder, C. (1979). Role of residual insulin secretion in protecting against ketoacidosis in insulin-dependent diabetes. *Brit. Med. J.*, **ii**, 1257–1259.

Mahler, R. F. (1976). Disorders of glycogen metabolism. *Clin. Endocrinol. Metabol.*, **5**, 579–598.

Malaisse W. J., Malaisse-Lagae, F., and Van Obberghen, E., *et al.* (1975). *Ann. N.Y. Acad. Sci.*, **253**, 630–652.

Manchester, K. L. (1970). Sites of hormonal regulation of protein metabolism. In: *Mammalian Protein Metabolism.* (Ed. Munro, H. N.) Academic Press Inc., New York, Vol. 4, pp. 229–298.

Manghani, K. K., Lunzer, M. R. and Billing, B. H. (1975). Urinary and serum octopamine in patients with portal-systemic encephalopathy. *Lancet*, **ii**, 943–946.

Margaria, R. (1976). *Biomechanics and Energetics of Muscular Exercise.* Clarendon Press, Oxford, U.K.

Margaria, R., Aghemo, P., and Rovelli, E. (1966). Measurement of muscular power (anaerobic) in man. *J. Applied Physiol.*, **21**, 1662–1664.

Marks, V. (1978). Alcohol and carbohydrate metabolism. *Clin. Endocrinol. Metab.*, **7**, 333–349.

Marliss, E. B. (1982). Insulin: Sixty years of use. *New Engl. J. Med.*, **306**, 362–364.

Marliss, E. B., Aoki, T. T., Pozefsky, T., *et al.* (1971). Muscle and splanchnic glutamine and glutamate metabolism in post-absorptive and starved man. *J. Clin. Invest.*, **50**, 814–817.

Marsden, C. D. (1982). Basal ganglia disease. *Lancet*, **ii**, 1141–1146.

Martin, B. R., and Denton, R. M. (1970). The intracellular localization of enzymes in white-adipose tissue fat-cells and permeability of fat cell mitochondria. *Biochem. J.*, **117**, 861–877.

Martin, C. E., Hiramitsu, K., Kitajima, Y., *et al.* (1976). Molecular control of membrane properties during temperature acclimation: fatty acid desaturase regulation of membrane fluidity in acclimating Tetrahymena cells. *Biochemistry*, **15**, 5218–5227.

Maseri, A., L'Abbate, A., and Baroldi, G. (1978). Coronary vasospasm as a possible cause of myocardial infarction. *New Engl. J. Med.*, **299**, 1271–1277.

Mason, E. E. (1981). *Surgical treatment of obesity.* W. B. Saunders Co., Philadelphia.

Masters, W. H., and Johnson, V. E. (1970). *Human Sexual Inadequacy.* Little, Brown & Co., Boston.

Matcham, G. W. J., Patil, N. B., and Smith, E. E., *et al.* (1978). The glycoprotein nature of liver glycogen. *FEBS 11th Meeting Copenhagen. Symp. A1*, (Ed. Esmann, V.) Pergamon Press, Oxford, pp. 305–316.

Matthews, H. R. (1980). Modification of histone H1 by reversible phosphorylation and its relation to chromosome condensation and mitosis. *Molecular aspects of cellular regulation*, **1**, 235–254.

Maurice, D. M., and Riley, M. V. (1970). In: *Biochemistry of the Eye*, (Ed. Glaymore, C. N.) Academic Press, London, New York, pp. 1–104.

McArdle, B. (1951). Myopathy due to a defect in muscle glycogen breakdown. *Clin. Sci.*, **10**, 13–35.

McCance, R. A., and Widdowson, E. M. (1960). *The Composition of Foods. Spec. Rep. Med. Res. Council*, No. 297, H.M.S.O., London, U.K.

McCann, S. M. (1977). Luteinizing hormone releasing hormone. *New Engl. J. Med.*, **296,** 797–802.

McCord, J. M., and Fridovich, I. (1968). The reduction of cytochrome c by milk xanthine oxidase. *J. Biol. Chem.*, **247,** 5753–5760.

McCord, J. M., and Fridovich, I. (1978). The biology and pathology of oxygen radicals. *Annals Intern. Med.*, **89,** 122–127.

McCormack, J. G., and Denton, R. M. (1980). Role of calcium ions in the regulation of intramitochondrial metabolism. *Biochem. J.*, **190,** 95–105.

McDevitt, H. O. (1980). Regulation of the immune response by the major histocompatibility system. *New Engl. J. Med.*, **303,** 1514–1517.

McGarry, J. D. (1979). New perspectives in regulation of ketogenesis. *Diabetes*, **28,** 517–523.

McGarry, J. D., and Foster, D. W. (1972). Regulation of ketogenesis and clinical aspects of the ketotic state. *Metabolism*, **21,** 471–489.

McGarry, J. D., and Foster, D. W. (1979). In support of the role of malonyl-CoA and carnitine acyltransferase I in the regulation of hepatic fatty acid oxidation and ketogenesis. *J. Biol. Chem.*, **254,** 8163–8168.

McGarry, J. D., and Foster, D. W. (1980). Regulation of hepatic fatty acid oxidation and ketone body production. *Annu. Rev. Biochem.*, **49,** 395–420.

McGiff, J. C. (1978). Thromboxane and prostacyclin: implications for function and disease of the vasculature. *Adv. Intern. Med.*, **25,** 199–216.

McGiff, J. C., Spokas, E. G., and Wong, P. Y.-K. (1981). Prostaglandin mechanism in blood regulation and hypertension. In: *The Prostaglandin System*, (Ed. Berti, F., and Velo, G. P.) Plenum Press, New York, pp. 235–246.

McGilvery, R. W. (1979). *Biochemistry, A Functional Approach*. W. B. Saunders Co., Philadelphia, U.S.A.

McGivan, J. D., and Chappell, J. B. (1975). On the metabolic function of glutamate dehydrogenase in rat liver. *FEBS Lettr.*, **52,** 1–7.

McGuire, W. L. (1975). Endocrine therapy of breast cancer. *Ann. Rev. Med.*, **26,** 353–363.

McLane, J. A., and Holloszy, J. O. (1979). Glycogen synthesis from lactate in the three types of skeletal muscle. *J. Biol. Chem.*, **254,** 6548–6553.

Meade, C. J., and Merton, J. (1978). Fatty acids and immunity. *Adv. Lipid Research*, **16,** 127–166.

Medical Research Council (1968). Potassium, glucose and insulin treatment for acute myocardial infarction. *Lancet*, **ii,** 1355–1360.

Meister, A. (1981). On the cycles of glutathione metabolism and transport. *Curr. Top. Cellul. Regul.*, **18,** 21–58.

Menander-Haber, K. B., and Haber, W. (1977). In: *Superoxide and Superoxide Dismutase*, (Eds. Michelsen, A. M., McCord, J. M., and Fridovich, I.) Academic Press, London & New York, pp. 537–549.

Mendelson, J. H., and Mello, N. K. (1979). Biologic concomitants of alcoholism. *New Engl. J. Med.*, **301,** 912–921.

Menzies, I. (1980). In: *Inborn Errors of Metabolism*. (Ed. by R. Ellis) Redwood Burn Ltd., Trowbridge & Esher, pp. 45–57.

Merimee, T. J., Dimoin, D. L., Hall, J. D., *et al.* (1969). A metabolic and hormonal basis for classifying ateliotic dwarfs. *Lancet*, **i,** 963–965.

Merimee, T. J., Zapf, J., and Froesch, E. R. (1981). Dwarfism in the pigmy. *New Engl. J. Med.*, **305,** 965–968.

Michaelis, L., and Menten, M. L. (1913). Die Kinetik der Invertinwirkung. *Biochem. Z.*, **49,** 333–369.

Michell, R. H. (1979). Mechanisms of cell surface receptors for hormones and neurotransmitters. *Companion to Biochemistry*, **2,** 205–228.

Mier, P. D., and Cotton, D. W. K. (1976). *The Molecular Biology of Skin.* Blackwell Scientific Publications, Oxford, U.K.

Miller, E., Hare, J. W., Cloherty, P. G., *et al.* (1981). Elevated maternal hemoglobin A_{1c} in early pregnancy and major congenital abnormalities in infants of diabetic mothers. *New Engl. J. Med.,* **304,** 1331–1334.

Miller, G. J., and Miller, N. E. (1975). Plasma high-density-lipoprotein concentration and development of ischaemic heart disease. *Lancet,* **i,** 16–19.

Miller, L. L. (1962). The role of the liver and the non-hepatic tissues in the regulation of free amino acid levels in the blood. In: *Amino Acid Pools,* (Ed. Holden, J. T.) Elsevier, Amsterdam, pp. 708–722.

Milvy, P. (1977). *Annals N.Y. Acad. Sci.,* **301,** 1–2.

Milvy, P., and Siegel, A. J. (1981). Physical activity levels and altered mortality from coronary heart disease with an emphasis on marathon running: a critical review. *Cardiovasc. Rev. Rep.,* **2,** 233–231.

Minemura, T., Lacy, W. W., and Crofford, O. B. (1970). Regulation of transport and metabolism of amino acids in isolated fat cells. *J. Biol. Chem.,* **245,** 3872–3881.

Minneman, K. P. (1981). Adrenergic receptor molecules. *Neurotransmitter Receptors,* (Ed. Yamamura, H. I., and Enna, S. J.) Chapman & Hall, London, New York, pp. 185–268.

Mitchell, M. D. (1979). Prostaglandins at parturition and in the neonatal infant. *Biochem. Soc. Trans.,* **8,** 659–662.

Mitchell, P. (1961). Coupling of phosphorylation to electron and hydrogen transfer by a chemi-osmotic type of mechanism. *Nature,* **191,** 144–148.

Molla, A. M., Sarker, S. A., Hossain, M., *et al.* (1982). Rice-powder electrolyte solutions as oral in diarrhoea. *Lancet,* **i,** 1317–1319.

Moncada, S., and Vane, J. R. (1979a). Arachidonic acid metabolites and the interactions between platelets and blood vessel walls. *New. Engl. J. Med.,* **300,** 1142–1147.

Moncada, S., and Vane, J. R. (1979b). Mode of action of aspirin like drugs. *Adv. Intern. Med.,* **24,** 1–22.

Montgomery, J. A., Elliott, R. D., Allan, P. W., *et al.* (1979). Metabolism and mechanisms of action of some new purine antimetabolites. *Adv. Enz. Regulat.* **17,** 419–436.

Morgan, H. E., Henderson, M. J., Regen, D. M., *et al.* (1961). Regulation of glucose uptake in muscle. *J. Biol. Chem.,* **236,** 253–261.

Morgan, H. E., Rannels, D. E., and McKee, E. E. (1979). Protein metabolism of the heart. In: *Handbook of Physiology: The Cardiovascular System.* (Ed. Berne, R. M.) American Physiological Society, Washington, D.C., U.S.A. Vol. 1, pp. 845–871.

Morgan, L. M. (1980). The entero-insular axis. *Biochem. Soc. Abstracts,* **8,** 17–19.

Morgan-Hughes, J. A. (1978). *Medicine, (second series)* **12,** 151–160.

Morley, J. E., and Levine, A. S. (1983). The central control of appetite. *Lancet,* **i,** 398–461.

Mortimer, C. H., McNeilly, A. J., Fisher, R. A., *et al.* (1974). Gonadotrophin-releasing hormone therapy in hypogonadal males with hypothalamic or pituitary dysfunction. *Brit. Med. J.,* **iv,** 617–621.

Mortimore, G. E., and Mondon, C. E. (1970). Inhibition by insulin of valine turnover in liver. *J. Biol. Chem.,* **245,** 2375–2383.

Mortimore, G. E., and Schworer, C. M. (1980). Application of liver perfusion as an *in vitro* model in studies of intracellular protein degradation. *Ciba Found. Symp.,* **75,** 281–306.

Moskowitz, M. A., and Wurtman, R. J. (1975). Catecholamines and neurological diseases. *New Engl. J. Med.,* **293,** 274–280; 332–338.

Motulsky, H. J., and Insel, P. A. (1982). Adrenergic receptors in man. Direct identification, physiologic regulation and clinical alterations. *New Engl. J. Med.,* **307,** 18–29.

Muller, J. E., Stone, P. H., Markis, J. E., *et al.* (1981). Let's not let the genie escape from the bottle—again. *New Engl. J. Med.,* **304,** 1294–1296.

Multicentre International Study (1977). Reduction in mortality after myocardial infarction with long-term beta-adrenoceptor blockade. *Br. Med. J.,* **ii,** 419–421.

Mundy, G. R., Shapiro, J. L., Bandelin, J. G., *et al.* (1976). Direct stimulation of bone resorption by thyroid hormones. *J. Clin. Invest.*, **58**, 529–534.

Munro, H. N. (1969). A general survey of techniques used in studying protein metabolism in whole animals and intact cells. In: *Mammalian Protein Metabolism*, (Ed. Munro, H. N.) Academic Press, New York. Vol. 3, pp. 237–262.

Munro, H. N. (1974). Control of plasma amino acid concentrations. *Ciba Found. Symp.*, **22**, 5–24.

Munro, H. N. (1979). Hormones and the metabolic response to injury. *New Engl. J. Med.*, **300**, 41–42.

Munro, H. N., and Crim, M. C. (1980). The proteins and amino acids. In: *Modern Nutrition in Health and Disease* 6th Edn. (Ed. Goodhart, R. S., and Shils, M. E.) Lea and Febiger, Philadelphia, U.S.A., pp. 51–98.

Mutt, V. (1980). Chemistry, isolation and purification of gastrointestinal hormones. *Biochem. Soc. Trans.*, **8**, 11–14.

Nakamura, Y., and Schwartz, A. (1972). The influence of hydrogen ion concentrations on calcium binding and release by skeletal muscle sarcoplasmic reticulum. *J. Gen. Physiol.*, **59**, 22–32.

Nakanishi, S., Inoue, A., Kita, T., *et al.* (1979). Nucleotide sequence of cloned cDNA for bovine corticotrophin—lipotropin precursor. *Nature*, **278**, 423–427.

Nattress, M., Alberti, K. G. M. M., Dennis, K. J., *et al.* (1978). A glucose controlled insulin infusion system for diabetic women during labour. *Br. Med. J.*, **2**, 599–601.

Needham, D. M. (1926). Red and white muscle. *Physiol. Rev.*, **6**, 1.

Neely, J. R., and Morgan, H. E. (1974). Relationship between carbohydrate and lipid metabolism and energy balance of the heart muscle. *Ann. Rev. Physiol.*, **36**, 413–459.

Nelson, O. E. (1969). Genetic modification of protein quality in plants. *Advances in Agronomy*, **21**, 171–194.

Nelson, R. A. (1973). Winter sleep in the black bear: a physiologic and metabolic marvel. *Mayo Clin. Proc.*, **48**, 733–737.

Nerup, J., and Lern, (1981). Autoimmunity in insulin-dependent diabetes mellitus. *Am. J. Med.*, **70**, 135–141.

Neuman, R. O. (1902). *Arch. Hyg.*, **45**, 1–87.

Newsholme, E. A. (1976a). The role of the fructose 6-phosphate/fructose 16-diphosphate cycle in metabolic regulation and heat generation. *Biochem. Soc. Trans.*, **4**, 978–984.

Newsholme, E. A. (1976b). Carbohydrate metabolism *in vivo*: regulation of the blood glucose level. *Clinics in Endocrin. Metabolism*, **5**, 543–578.

Newsholme, E. A. (1977). Mechanism for starvation suppression and refeeding activation of infection. *Lancet*, **i**, 654.

Newsholme, E. A. (1978a). Substrate cycles: their metabolic, energetic and thermic consequences in man. *Biochem. Soc. Symp.*, **43**, 183–205.

Newsholme, E. A. (1978b). Control of energy provision and utilization in muscle in relation to sustained exercise. In: *3rd Internation Symposium on Biochemistry of Exercise*, (Ed. Landry, F., and Orban, W. A. R.) Symposia Specialists, Florida, U.S.A. Vol. 3, pp. 1–28.

Newsholme, E. A. (1980). A possible metabolic basis for the control of body weight. *New Engl. J. Med.*, **302**, 400–405.

Newsholme, E. A., and Crabtree, B. (1973). Metabolic aspects of enzyme activity regulation. *Symp. Soc. Exp. Biol.*, **27**, 429–460.

Newsholme, E. A., and Crabtree, B. (1976). Substrate cycles in metabolic regulation and heat generation. *Biochem. Soc. Symp.*, **41**, 61–110.

Newsholme, E. A., and Crabtree, B. (1979). Theoretical principles in the approaches to control of metabolic pathways and their application to glycolysis in muscle. *J. Mol. Cellul. Cardiol.*, **11**, 839–856.

Newsholme, E. A., and Crabtree, B. (1981a). Flux-generating and regulatory steps in metabolic control. *Trends in Biochem. Sci.,* **6,** 53–55.

Newsholme, E. A., and Crabtree, B. (1981b). Control of flux through metabolic pathways. In: *Short Term Control in the Liver,* (Ed. Hue, L., and Van der Werve, G.) Elsevier/North Holland, Amsterdam, pp. 3–18.

Newsholme, E. A., and Leech, A. R. (1983). *The Runner: Energy and Endurance,* Walter Meagher, P.O. Box 382, Roosevelt, New Jersey.

Newsholme, E. A., and Start, C. (1973). *Regulation in Metabolism.* John Wiley & Sons, London, U.K.

Newsholme, E. A., and Williams, T. (1978). The role of phosphoenolpyruvate carboxykinase in amino acid metabolism in muscle. *Biochem. J.* **176,** 623–626.

Newsholme, E. A., Sugden, P. H., and Williams, T. (1977). Effect of citrate on the activities of 6-phosphofructokinase from nervous and muscle tissue from different animals and its relationship to the regulation of glycolysis. *Biochem. J.,* **166,** 123–129.

Newsholme, E. A., Zammit, V. A., and Crabtree, B. (1978). The roles of glucose and glycogen as fuels for muscle. *Biochem. Soc. Trans.,* **6,** 512–520.

Newsholme, E. A., Crabtree, B., and Zammit, V. A. (1980). Use of enzyme activities as indices of maximum rates of fuel utilization. *Ciba Found. Symp.,* **73,** 245–258.

Newsholme, E. A., Lang, J., and Relman, A. S. (1982). Control of rate of glutamine metabolism in the kidney. *Contrib. Nephr.,* **31,** 1–4.

Nicholls, D. G. (1976). The bioenergetics of brown adipose tissue mitochondria. *FEBS Lett.,* **61,** 103–110.

Nicholls, D. G. (1979). Brown adipose tissue mitochondria. *Biochim. Biophys. Acta,* **549,** 1–29.

Nicholls, D. G. (1982). *Bioenergetics: An Introduction to the Chemiosmotic Theory.* Academic Press, London.

Nichols, C. W., and Lambertsen, C. J. (1969). Effect of high oxygen pressures on the eye. *New Engl. J. Med.,* **281,** 25–30.

Nicholson, M. R., and Somerville, K. W. (1978). Heat stroke in a 'run for fun'. *Br. Med. J.,* **i,** 1525–1526.

Nicholson, R. I. (1979). Biochemistry of tamoxifen therapy in breast cancer. *Biochem. Soc. Trans.,* **7,** 569–572.

Nicoll, A., Miller, N. E., and Lewis, B. (1980). High density lipoprotein metabolism. *Adv. Lipid Res.,* **17,** 54–106.

Nieto, A., and Castano, J. G. (1980). Control *in vivo* of rat liver phosphofructokinase by glucagon and nutritional changes. *Biochem. J.,* **186,** 953–957.

Nillius, S. J., and Wide, L. (1975). Gonadotrophin-releasing hormone treatment for induction of follicular maturation and ovulation in amenorrhoric women with anorexia nervosa. *Br. Med. J.,* **iii,** 405–408.

Nimmo, H. G. (1980). The hormonal control of triacylglycerol synthesis. *Molecular Aspects of Cellular Regulation,* **1,** 135–152.

Nisker, J. A., and Woods, B. T. (1982). Female hormones and the postmenopausal endometrium. *New Engl. J. Med.,* **306,** 1424–1425.

Njus, D., Knoth, J., and Zallakian, M. (1981). Proton linked transport in chromaffin granules. *Curr. Topics Bioenergetics,* **11,** 108–149.

Norman, A. W., and Henry, H. L. (1979). Vitamin D to 1,25-dihydroxycholecalciferol: evolution of a steroid hormone. *Trends Biochem. Sci.,* **4,** 14–18.

Northway, W. H. (1978). Bronchopulmonary dysplasia and vitamin E. *New Engl. J. Med.,* **299,** 599–601.

Norum, K. R., and Gjone, E. (1967). Familial plasma lecithin: cholesterol acyltransferase deficiency. Biochemical study of a new inborn error of metabolism. *Scand. J. Clin. Lab. Invest.,* **20,** 231–243.

Nowacki, P., Schmid, E., and Weist, F. (1969). The turnover of sympathicoadrenal hormones of sportsmen in training, anticipation and during competition judged by measurement of the urinary excretion of 3-methoxy-4-hydroxy-mandelic acid. *Biochemistry of Exercise: Medicine and Sport*, (Ed. Poortmans, J. R.) Karger, Basel, New York, Vol. 3, pp. 205–208.

Nriagu, J. O. (1983). Saturnine gout among Roman aristocrats. *New Engl. J. Med.*, **308**, 660–663.

Oakley, W. G., Pyke, D. A., and Taylor, K. W. (1978). *Diabetes and its Management* 3rd Edn. Blackwell Scientific Publications, Oxford.

Ockner, R. K., and Manning, J. M. (1974). Fatty acid binding protein in small intestine. Identification, isolation and evidence for its role in cellular fatty acid transport. *J. Clin. Invest.*, **54**, 326–338.

Ochoa, S. (1979). Regulation of protein synthesis. *Eur. J. Cell Biol.*, **19**, 95–101.

Ochoa, S., de Haro, C., Siekierka, J., *et al.* (1981). Role of phosphorylation-dephosphorylation cycles in the control of protein synthesis in eukaryotes. *Curr. Topics Cellul. Regul.*, **18**, 421–436.

Odum, E. P. (1965). In: *Handbook of Physiology, Section 5, Adipose Tissue*, (Ed. Renold, A. E., and Cahill, G. F.) American Physiological Society, Washington, D.C., pp. 37–44.

Ogston, A. G. (1948). Interpretation of experiments on metabolic processes using isotopic tracer elements. *Nature*, **162**, 963.

Ohniski, K., and Lieber, C. S. (1977). In: *Alcohol and Aldehyde Metabolizing Systems*, (Ed. Thurman, R. G.) Academic Press, New York. Vol. II, pp. 341–350.

O'Keefe, S. J. D., Sender, P. M., and James, W. P. T. (1974). 'Catabolic' loss of nitrogen in response to surgery. *Lancet*, **ii**, 1035–1038.

Oldham, K. (1968). *Radiochemical Methods of Enzyme Assay*. Radiochemical Centre, Amersham, England.

Olefsky, J. M. (1978). Mechanisms of the ability of insulin to activate the glucose-transport system in rat adipocytes. *Biochem. J.*, **172**, 137–145.

Olefsky, J. M., and Kolterman, O. G. (1981). Mechanisms of insulin resistance in obesity and non-insulin dependent (type II) diabetes. *Am. J. Media.*, **70**, 151–168.

Oliver, M. F. (1978). In: *Developments in Cardiovascular Medicine*, (Ed. Dickinson, C. J., and Marks, J.) M.T.P. Press, Lancaster, England, pp. 145–164.

Oliver, M. F. (1980). Primary prevention of coronary heart disease: an appraisal of clinical trials of reducing raised plasma cholesterol. *Prog. Cardiol.*, **9**, 1–24.

Oliver, M. F. (1981). Diet and coronary heart disease. *Brit. Med. Bull.*, **37**, 49–58.

Oliver, M. F., Kurien, V. A., and Greenwood, T. W. (1968). Relation between serum free-fatty acids and arrhythmias and death after acute myocardial infarction. *Lancet*, **i**, 710–715.

O'Malley, B. W. (1978). Studies on the molecular mechanism of steroid hormone action. *Harvey Lectures*, **72**, 53–90.

O'Malley, B. W., and Mears, A. R. (1974). Female steroid hormones and target cell nuclei. *Science*, **183**, 610–620.

O'Malley, B. W., Woo, S. L. C., and Tsai, M.-J. (1981). Structure and hormonal control of the ovalbumin gene cluster. *Curr. Topics. Cellul. Regul.*, **18**, 437–453.

Opie, L. H. (1970). The glucose hypothesis: relation to acute myocardial ischaemia. *J. Molec. Cell Cardiol.*, **1**, 107–115.

Opie, L. H. (1975). Sudden death and sport. *Lancet*, **i**, 263–266.

Opie, L. H. (1980). Drugs and the heart. III-Calcium antagonists. *Lancet*, **i**, 806–810.

Opie, L. H., and Owen, P. (1975). Effects of increased mechanical work by isolated perfused rat heart during production or uptake of ketone bodies. *Biochem. J.*, **148**, 403–415.

Opie, L. H., and Owen, P. (1976). Effect of glucose-insulin-potassium infusions on arteriovenous differences of glucose and of free fatty acids and on tissue metabolic changes in dogs with developing myocardial infarction. *Am. J. Cardiol.*, **38**, 310–321.

Opie, L. H., and Stubbs, W. A. (1976). Carbohydrate metabolism in cardiovascular disease. *Clin. Endocrinol. Metabol.*, **5**, 703–729.

Oppenheimer, J. H., and Dillman, W. H. (1978). Nuclear receptors for triiodothionine: a physiological perspective. In: *Receptors and Hormone Action*. (Ed. Birnbaamer, L., and O'Malley, B. W.) Academic Press, New York. Vol. III, pp. 1–33.

Ornoy, A., Goodwin, D., Noff, D., et al. (1978). 24,25-dihydroxyvitamin D is a metabolite of vitamin D essential for bone formation. *Nature*, **276**, 517–519.

Osborne, J. C., and Brewer, H. B. (1977). The plasma lipoproteins. *Adv. Prot. Chem.*, **31**, 253–337.

Osborne, W. R. A. (1981). Inherited absence of purine recycling enzymes associated with defects of immunity. *Trends in Biochem. Sciences*, **6**, 80–83.

Oski, F. A. (1980). Vitamin E. A radical defence. *New Engl. J. Med.*, **303**, 454–455.

Otway, S., and Robinson, D. S. (1968). The significance of changes in tissue clearing factor lipase activity in relation to the lipaemia of pregnancy. *Biochem. J.*, **106**, 677–682.

Owen, J. S., and McIntyre, N. (1982). Plasma lipoprotein metabolism and lipid transport. *Trends Biochem. Sci.*, **7**, 95–98.

Owen, O. E., and Reichard, G. A. (1971). Human forearm metabolism during prolonged starvation. *J. Clin. Invest.*, **50**, 1536–1545.

Owen, O. E., Morgan, A. P., Kemp, H. G., et al. (1967). Brain metabolism during fasting. *J. Clin. Invest.*, **46**, 1589–1595.

Owen, O. E., Felig, P., Morgan, A. P., et al. (1969). Liver and kidney metabolism during prolonged starvation. *J. Clin. Invest.*, **48**, 574–583.

Owen, O. E., Reichard, G. A., Patel, M. S., et al. (1979). Energy metabolism in feasting and fasting. *Adv. Expt. Med. Biol.*, **111**, 119–188.

Packter, N. M. (1973). *Biosynthesis of Acetate-Derived Compounds*. John Wiley & Sons Ltd., London.

Paffenbarger, R. S., and Hyde, R. T. (1980). Exercise as protection against heart attacks. *New Engl. J. Med.*, **302**, 1026–1027.

Page, M. A., and Williamson, D. H. (1971). Enzymes of ketone body utilization in human brain. *Lancet*, **ii**, 66–68.

Page, M. McB., Alberti, K. G. M. M., Greenwood, R., et al. (1974). Treatment of diabetic coma with continuous low-dose infusion of insulin. *Brit. Med. J.*, **ii**, 687–690.

Pagliara, A. S., Karl, I. E., De Vivo, D. C., et al. (1972). Hypoalaninaemia: a concomitant of ketotic hypoglycaemia. *J. Clin. Invest.*, **51**, 1440–1449.

Pagliara, A. S., Karl, I. E., Haymond, M., et al. (1973). Hypoglycemia in infancy and childhood. *J. Pediatr.*, **82**, 365–379.

Pain, V. M. (1978). Protein synthesis and its regulation. In: *Protein Turnover in Mammalian Tissues and in The Whole Body*. (Ed. Waterlow, J. C., Garlic, P. J., and Millward, D. J.) Elsevier, Amsterdam, pp. 16–54.

Papapoulos, S. E., Clemens, T. L., Fraher, L. J., et al. (1980). Metabolites of vitamin D in human vitamin D deficiency: effect of vitamin D_3 or 1,25-dihydroxycholecalciferol. *Lancet*, **ii**, 612–615.

Papavasiliou, P. S., Cotzias, G. C., Düby, S., et al. (1972). Levodopa in Parkinsonism: potentiation of central effects with a peripheral inhibitor. *New Engl. J. Med.*, **286**, 8–14.

Parkhouse, J., Pleuvry, B. J., and Rees, J. M. H. (1979). *Analgesic Drugs*. Blackwell Scientific Publications. Oxford.

Parkin, J. M. (1981). Dysmorphology and short stature. *Brit. Med. Bull.*, **37**, 297–302.

Parsons, D. S. (1979). Fuels of the small intestinal mucosa. *Topics in Gastroenterology*, **7**, 253–271.

Passmore, R., Meiklejohn, A. P., Dewar, A. D., et al. (1955). An analysis of the gain in weight of overfed thin young men. *Br. J. Nutr.*, **9**, 27–33.

Paterson, S. J., Robson, L. E., and Kosterlitz, H. W. (1983). Classification of opioid receptors. *Brit. Med. Bull.*, **39**, 31–36.

Paul, A. A., and Southgate, D. A. T. (1978). *The Composition of Foods*. MRC Special Report No. 297. McCance and Widdowson's, H.M.S.O. London.

Paul, J. M. (1979). Comparative Studies on the Citric Acid Cycle in Muscle. D. Phil. Thesis, Oxford University.

Paul, P., and Holmes, W. L. (1975). Free fatty acids and glucose metabolism during increased energy expenditure and after training. *Medicine and Science in Sports*, **7**, 176–184.

Pawan, G. L. S., and Semple, S. J. G. (1983). Effect of 3-hydroxybutyrate in obese subjects on very low energy diets and during therapeutic starvation. *Lancet*, **i**, 15–17.

Pearson, S. B., Banks, D. C., and Patrick, J. M. (1979). The effect of β-adrenoceptor blockade on factors affecting exercise tolerance in normal man. *Br. J. Clin. Pharmacol.*, **8**, 143–148.

Percy-Robb, L. W. (1975). Bile acid synthesis. *Essays in Medical Biochemistry*, **1**, 59–80.

Permutt, M. A., Biesbroeck, J., Chyn, R., *et al.* (1976). Isolation of a biologically active messenger RNA: preparation from fish pancreatic islets by oligo (2'-deoxythymidylic acid) affinity chromatography. *Ciba Found. Symp.*, **41**, 97–116.

Pernow, B. B., Havel, R. J., and Jennings, D. B. (1967). The second wind phenomenon in McArdle's syndrome. *Acta Med. Scand.*, **Suppl. 472**, 294–307.

Perutz, M. F. (1970). Steroechemistry of co-operative effects in haemoglobin. Part I—Haem-haem interaction and the problem of allostery. *Nature*, **228**, 726–734.

Peter, J. B., Barnard, R. J., Edgerton, V. R., *et al.* (1972). Metabolic profiles of three fibre types of skeletal muscle in guinea-pigs and rabbits. *Biochemistry*, **11**, 2627–2633.

Peters, R. A. (1951). Lethal synthesis. *Proc. Roy. Soc. B.*, **139**, 143–170.

Peters, T. J. (1982). Ethanol metabolism. *Brit. Med. Bull.*, **38**, 17–20.

Petrow, V. (1966). Steroidal oral contraceptive agents. *Essays in Biochem.*, **2**, 117–145.

Phillips, L. S., and Vassilopoulou-Sellin, R. (1980). Somatomedins. *New Engl. J. Med.*, **302**, 371–380; 438–446.

Pickles, V. R., Hall, W. J., Best, F. A., *et al.* (1965). Prostaglandins in endometrium and menstrual fluid from normal and dysmenorrhoeic subjects. *J. Obstet. Gynaecol. Brit. Comm.*, **72**, 185–192.

Pickup, J. C., Keen, H., Parsons, J. A., *et al.* (1978). Continuous subcutaneous insulin infusion: an approach to achieving normoglycaemia. *Br. Med. J.*, **i**, 204–207.

Pilkis, S. J., Park, C. R., and Claus, T. H. (1978). Hormonal control of hepatic gluconeogenesis. *Vitamins and Hormones*, **36**, 383–460.

Pitts, R. F. (1971). The role of ammonia production and excretion in regulation of acid-base balance. *New Engl. J. Med.*, **284**, 32–38.

Pitts, R. F. (1973). Production and excretion of ammonia in relation to acid-base regulation. In: *Handbook of Physiology: Section 8: Renal Physiology*, American Physiological Society, Washington, D.C., pp. 455–496.

Pond, C. M. (1978). Morphological aspects and the ecological and mechanical consequences of fat deposition in wild vertebrates. *Ann. Rev. Ecol. Syst.*, **9**, 519–570.

Porte, D., Graf, R. J., Halter, J. B., *et al.* (1981). Diabetic neuropathy and plasma glucose control. *Am. J. Med.*, **70**, 195–200.

Porteous, J. W. (1978). Glucose as a fuel for the small intestine. *Biochem. Soc. Trans.*, **6**, 534–539.

Porter, K. R., and Tucker, J. B. (1981). The ground substance of the living cell. *Scientific American*, **44**, 40–51.

Preslock, J. P. (1980). A review of *in vitro* testicular steroidogenesis in rodents, monkeys and humans. *J. Steroid Biochem.*, **13**, 965–975.

Preston, F. E., Whipps, S., Jackson, C. A., *et al.* (1981). Inhibition of prostacyclin and platelet thromboxane A2 after low-dose aspirin. *New Engl. J. Med.*, **304**, 76–80.

Purnell, M. R., Stone, P. R., and Whish, W. J. D. (1980). ADP-ribosylation of nuclear proteins. *Biochem. Soc. Trans.*, **8**, 215–228.

Quastel, J. H. (1979). The role of amino acids in the brain. *Essays in Medical Biochem.*, **4**, 1–48.

Quigley, M. M., and Hammond, C. B. (1979). Estrogen-replacement therapy—help or hazard? *New Engl. J. Med.*, **301**, 646–648.

Racker, E. (1976). *A New Look at Mechanisms in Bioenergetics*, Academic Press, New York.

Raijman, L. (1976). Enzyme and reactant concentrations and the regulation of urea synthesis. In: *The Urea Cycle*, (Ed. Grisolia, S., Baguena, R., and Major, F.) John Wiley & Sons, New York, pp. 243–259.

Raison, J. K. (1973). Temperature-induced phase changes in membrane lipids and their influence on metabolic regulation. *Symp. Soc. Expt. Biol.*, **27**, 485–512.

Raisz, L. G., and Kream, B. E. (1981). Hormonal control of skeletal growth. *Ann. Rev. Physiol.*, **43**, 225–238.

Randle, P. J. (1981a). Phosphorylation-dephosphorylation cycles and regulation of fuel selection in mammals. *Curr. Topics Cellul. Reg.*, **18**, 107–130.

Randle, P. J. (1981b). Molecular mechanisms regulating fuel selection in muscle. In: *Biochemistry of Exercise*, (Ed. Poortmans, J., and Niset, G.) University Park Press, Baltimore. Vol. IV-A, pp. 13–32.

Randle, P. J., and Tubbs, P. K. (1979). Carbohydrate and fatty acid metabolism. In: *Handbook of Physiology: The Cardiovascular System*, (Ed. Berne, R. M.) American Physiological Society, Bethesda. Vol. 1, pp. 805–844.

Randle, P. J., Garland, P. B., Hales, C. N., *et al.* (1963). The glucose fatty acid cycle. Its role in insulin sensitivity and the metabolic disturbances of diabetes mellitus. *Lancet*, **i**, 785–789.

Randle, P. J., Newsholme, E. A., and Garland, P. B. (1964). Effects of fatty acids, ketone bodies and pyruvate and of alloxan-diabetes and starvation on the uptake and metabolic fate of glucose in rat heart and diaphragm muscles. *Biochem. J.*, **93**, 652–665.

Randle, P. J., Ashcroft, S. J. H., and Gill, J. R. (1968). Carbohydrate metabolism and release of hormones. In: *Carbohydrate Metabolism and its Disorders*. (Ed. Dickens, F., Randle, P. J., and Whelan, W. J.) Academic Press, London. Vol. 1, pp. 427–448.

Ranvier, L. (1873). C.R. Acad, Sci. (Paris), **77**, 1030.

Rasmussen, H., Clayberger, C., and Gustin, M. C. (1979). The messenger function of calcium in cell activation. *Symp. Soc. Expt. Biol.*, **33**, 161–198.

Ratzmann, K. P. (1977). *Endokrinologie*, **70**, 199–211.

Rayford, P. L., Miller, T. A., and Thompson, J. C. (1978). Secretion of cholecystokinin and newer gastrointestinal hormones. *New Engl. J. Med.*, **294**, 1157–1164.

Redhead, I. H. (1968). Poisoning on the farm. *Lancet*, **i**, 686–688.

Redinger, R. N., and Small, D. M. (1972). Bile composition, bile salt metabolism and gall-stones. *Arch. Int. Med.*, **130**, 618–631.

Reichard, G. A., Owen, O. E., Haff, A. C., *et al.* (1974). Ketone-body production and oxidation in fasting obese humans. *J. Clin. Invest.*, **53**, 508–515.

Reimuth, O. M., Scheinberg, P., and Bourne, B. (1965). Total cerebral blood flow and metabolism. *Archiv. Neurol.*, **12**, 49–66.

Relman, A. S. (1978). Lactic acidosis and a possible new treatment. *New Engl. J. Med.*, **298**, 564–565.

Rennie, M. J., and Edwards, R. H. T. (1981). Carbohydrate metabolism of skeletal muscle and its disorders. In: *Carbohydrate Metabolism and its Disorders*. (Ed. Randle, P. J., Steiner, D. F., and Whelan, W. J.) Academic Press, London. Vol. 3, pp. 1–120.

Rennie, M. J., and Holloszy, J. O. (1977). Inhibition of glucose uptake and glycogenolysis by availability of oleate in well oxygenated perfused skeletal muscle. *Biochem. J.*, **168**, 161–170.

Rennie, M. J., Jennett, S., and Johnson, R. H. (1974). The metabolic effects of strenuous exercise: A comparison between untrained subjects and racing cyclists. *Quart. J. Expt. Physiol.*, **59**, 201–212.

Renold, A. E., Muller, W. A., Mintz, P. H., *et al.* (1978). Diabetes mellitus. In: *The Metabolic Basis of Inherited Disease*, (Ed. Stanbury, J. B., Wyngaarden, J. B., and Fredrickson, D. S.) McGraw-Hill, New York, pp. 80–109.

Rhoads, G. G., Gulbrandsen, C. L., and Kagan, A. (1976). Serum lipoproteins and coronary heart disease in a population study of Hawaii Japanese men. *New Engl. J. Med.*, **294**, 293–298.

Richards, D. H. (1979). A general practice view of functional disorders associated with menstruation. *Research Clinical Forums*, **1**, 39–45.

Richards, P. (1978). The metabolism and clinical relevance of the keto acid analogues of essential amino acids. *Clin. Sci. Mol. Med.*, **54**, 589–593.

Richter, E. A., Ruderman, N. B., and Schneider, S. H. (1981). Diabetes and exercise. *Am. J. Med.*, **70**, 201–209.

Rivers, J. P. W., and Frankel, T. L. (1981). Essential fatty acid deficiency. *Brit. Med. Bull.*, **37**, 59–64.

Rizza, R. A., Gerich, J. E., Haymond, M. W., *et al.* (1980). Control of the blood sugar in insulin-dependent diabetes: comparison of an artificial endocrine pancreas, continuous subcutaneous insulin infusion and intensified conventional insulin therapy. *New Engl. J. Med.*, **303**, 1313–1318.

Robbins, J., and Rall, J. C. (1979). The iodine-containing hormones. In: *Hormones in Blood*, (Ed. Gray, C. H., and James, V. H. T.) Academic Press, London, Vol. I, pp. 576–688.

Robinson, A. M., and Williamson, D. H. (1980). Physiological roles of ketone bodies as substrates and signals in mammalian tissues. *Physiol. Rev.*, **60**, 143–187.

Rock, J., Garcia, C. R., and Pincus, F. (1957). Synthetic progestins in the normal human menstrual cycle. *Rec. Progr. Horm. Res.*, **13**, 323–346.

Rodbell, M. (1980). The role of hormone receptors and GTP-regulatory proteins in membrane transduction. *Nature*, **284**, 17–22.

Rodwell, V. W., Nordstrom, J. L., and Mitschelen, J. J. (1976). Regulation of HMG-CoA reductase. *Adv. Lipid Res.*, **14**, 2–74.

Rona, R. J. (1981). Genetic and environmental factors in growth in childhood. *Brit. Med. Bull.*, **37**, 265–272.

Rolls, E. T. (1981). Central nervous mechanisms related to feeding and appetite. *Brit. Med. Bull.*, **37**, 131–134.

Roos, D., and Loos, J. A. (1973). Changes in the carbohydrate metabolism of mitogenically stimulated human peripheral lymphocytes. *Expt. Cell Research*, **77**, 127–135.

Rose, W. C. (1938). The nutritive significance of the amino acids. *Physiol. Rev.*, **18**, 109–136.

Rosenberg, L. E., and Scriver, C. R. (1980). Disorders of amino acid metabolism. In: *Metabolic Control and Disease*, (Ed. Bondy, P. K., and Rosenberg, L. E.) W. B. Saunders & Co., Philadelphia, pp. 73–103.

Rosenberg, L. E., and Tanaka, K. (1978). Metabolism of amino acids and organic acids. In: *The Year in Metabolism (1977)*, (Ed. Freinkel, N.) Plenum Press, New York, London, pp. 219–251.

Rosenberg, R. N. (1981). Biochemical genetics of neurologic disease. *New Engl. J. Med.*, **305**, 1181–1193.

Rosenman, R. H., and Friedman, M. (1974). The central nervous system and coronary heart disease. In: *The Myocardium: Failure and Infarction*, (Ed. Braunwald, E.) H.P. Publishing Co., New York, pp. 237–246.

Ross, B. D. (1972). *Perfusion Techniques in Biochemistry*. Clarendon Press, Oxford.

Ross, B. D. (1979). Techniques for investigation of cell metabolism. In: *Techniques in metabolic research*, (Ed. Kornberg, H., Metcalfe, J. C., Northcote, D. H., *et al.*) Elsevier/North Holland, New York. B.203, pp. 1–22.

Ross, B. D., and Lowry, M. (1981). Recent developments in renal handling of glutamine and ammonia. In: *Renal Transport of Organic Substances*, (Ed. Silbernagl, S., Lang, F. C., and Greger, R. S.) Springer-Verlag, Heidelberg, pp. 78–82.

Ross, B. D., Radda, G. K., Gadian, D. G., *et al.* (1981). Examination of a case of suspected McArdle's syndrome by ^{31}P nuclear magnetic resonance. *New Engl. J. Med.*, **304**, 1338–1342.

Rossini, A. A. Immunotherapy for insulin-dependent diabetics? *New Engl. J. Med.*, **308**, 333–335.

Rossor, N. M. (1982). Dementia. *Lancet*, **ii**, 1200–1204.

Roth, G. J., and Porte, D. (1970). Chronic lactic acidosis and acute leukaemia. *Arch. Intern. Med.*, **125**, 317–321.

Rothwell, N. J., and Stock, M. J. (1981). Regulation of energy balance. *Ann. Rev. Nutr.*, **1**, 235–256.

Rothwell, N. J., and Stock, M. J. (1983). Luxuskonsumption, diet-induced thermogenesis and brown fat: the case in favour. *Clin. Science*, **64**, 19–23.

Rotter, J. I., and Rimoin, D. L. (1981). The genetics of the glucose intolerance disorders. *Am. J. Med.*, **70**, 116–126.

Roussel, B., and Bittel, J. (1979). Thermogenesis and thermolysis during sleeping and waking in the rat. *Pflügers Arch.*, **382**, 225–231.

Rowan, A. N., and Newsholme, E. A. (1979). Changes in contents of adenine nucleotides and intermediates of glycolysis and the citric acid cycle in flight muscle of the locust upon flight and their relationship to the control of the cycle. *Biochem. J.*, **178**, 209–216.

Rowan, A. N., Newsholme, E. A., and Scrutton, M. C. (1978). Partial purification and some properties of pyruvate carboxylase from the flight muscle of the locust (*Schistocerca gregaria*). *Biochim. Biophys. Acta*, **522**, 270–275.

Rowell, L. B. (1974). Human cardiovascular adjustments to exercise and thermal stress. *Physiol. Rev.*, **54**, 75–119.

Ruderman, N. B. (1975). Muscle amino acid metabolism and gluconeogenesis. *Ann. Rev. Med.*, **26**, 245–258.

Ruderman, N. B., Ross, P. S., Berger, M., *et al.* (1974). Regulation of glucose and ketone-body metabolism in brain of anaesthetized rats. *Biochem. J.*, **138**, 1–10.

Ruderman, N. B., Aoki, T. T., and Cahill, G. F. (1976). Gluconeogenesis and its disorders in man. In: *Gluconeogenesis; Its Regulation in Mammalian Species*, (Ed. Hanson, R. W., and Mehlman, M. A.) John Wiley, London, New York, pp. 515–532.

Rudman, D., Kutner, M. H., Blackston, R. D. *et al.* (1981). Children with normal variant short stature: treatment with human growth hormone for six months. *New Engl. J. Med.*, **305**, 123–131.

Rudolph, A. M. (1981). Effect of aspirin and acetaminophen in pregnancy and in the new born. *Arch. Intern. Med.*, **141**, 358–363.

Ruzicka, J., and Stary, J. (1964). *Atomic Energy Rev., Vienna* **2**, 3–15.

Rylatt, D. B., Aitken, A., Bilham, T., *et al.* (1980). Glycogen synthase from rabbit skeletal muscle. *Eur. J. Biochem.* **107**, 529–537.

Ryman, B. E., and Tyrrell, P. A. (1980). Liposomes—bags of potential. *Essays in Biochem.*, **16**, 49–98.

Sahlin, K. (1978). Intracellular pH and energy metabolism in skeletal muscle of man, with special reference to exercise. *Acta Physiol. Scand.*, **Suppl. 455**, 1–56.

Sahlin, K., Palmskog, G., and Hultman, E. (1978). Adenine nucleotide and IMP contents of the quadriceps muscle in man after exercise. *Pflügers Archiv.*, **374**, 193–198.

Sainsbury, G. M. (1980). Distribution of ammonia between hepatocytes and extracellular fluid. *Biochim. Biophys. Acta*, **631**, 305–316.

Sakamoto, A., Moldawer, L. L., and Palombo, J. D., *et al.* Alterations in tyrosine and protein kinetics produced by injury and branched chain amino acid administration in rats. *Clin. Sci.*, **64**, 321–331.

Sala, G. B., Hayashi, K., Catt, K. J., *et al.* (1979). Adrenocorticotropin action in isolated adrenal cells. *J. Biol. Chem.*, **254**, 3861–3865.

Salaman, J. R. (1981). Pharmacological immunosuppressive agents. In: *Immunosuppressive therapy*. (Ed. Salaman, J. R.) M.T.P. Press Ltd., Lancaster, U.K., pp. 3–18.

Salmon, W. D., and Daughaday, W. H. (1957). A hormonally controlled serum factor which stimulates sulphate incorporation by cartilage *in vitro*. *J. Lab. Clin. Med.*, **49**, 825–836.

Saltin, B., Henriksson, J., Nygaard, E., *et al.* (1977). Fibre types and metabolic potentials of skeletal muscles in sedentary man and endurance runners. *Annal. N.Y. Acad. Sci.*, **301**, 3–29.

Samuelsson, B. (1970). Biosynthesis of prostaglandins. In: *Proc. 4th Int. Conr. Pharmacol.* Schwabe, Basle, pp. 12–31.

Santosham, M., Daum, R. S., Dillman, L., *et al.* (1982). Oral rehydration therapy of infantile diarrhoea. *New Engl. J. Med.*, **306**, 1070–1076.

Sapir, D. G., Owen, O. E., Pozefsky, T., *et al.* (1974). Nitrogen sparing induced by a mixture of essential amino acids given chiefly as their keto-analogues during prolonged starvation in obese subjects. *J. Clin. Invest.*, **54**, 974–980.

Sapir, D. G., and Walser, M. (1977). Nitrogen sparing induced early in starvation by infusion of branched-chain keto acids. *Metabolism*, **26**, 301–308.

Sargeant, A. J., Davies, C. T. M., Edwards, R. H. T., *et al.* (1977). Functional and structural changes after disuse of human muscle. *Clin. Sci. Mol. Med.*, **52**, 337–342.

Saunders, B. C. (1957). *Some Aspects of the Chemistry and Toxic Action of Organic Compounds Containing Phosphorus and Fluorine.* Cambridge University Press, London.

Saunders, J. B., Davies, M., and Williams, R. (1981). Do women develop alcoholic liver disease more readily than men? *Brit. Med. J.*, **282**, 1140–1143.

Sawynok, J., Pinsky, C., and La Bella, F. S. (1979). Mini review on the specificity of naloxone as an opiate antagonist. *Life Sci.*, **25**, 1621–1632.

Scanu, A. M. (1978). Plasma lipoproteins: Structure, function and regulation. *Trends in Biochem. Sci.*, **3**, 202–205.

Schalch, D. S., and Kipnis, D. M. (1964). The impairment of carbohydrate tolerance by elevated plasma free fatty acids. *J. Clin. Invest.*, **43**, 1283–1284.

Schally, A. V. (1978). Aspects of hypothalamic regulation of the pituitary gland. *Science*, **202**, 18–28.

Schally, A. V., Arimura, A., and Coy, D. H. (1980). Recent approaches to fertility control based on derivatives of LH-RH. *Vitamin and Hormone*, **38**, 257–323.

Scheuer, P. J. (1982). The morphology of alcoholic liver disease. *Brit. Med. Bull.*, **38**, 63–65.

Schimke, R. T. (1962). Differential effects of fasting and protein free diets on levels of urea cycle enzymes in rat liver. *J. Biol. Chem.*, **237**, 1921–1924.

Schmidt, E., and Schmidt, F. W. (1974). Diagnosis, control of progress and therapy. In: *Methods of Enzymatic Analysis*, (2nd. Edn.) (Ed. Bergmeyer, H.-U.) Verlag Chemie/Academic Press, New York, London. Vol. 1, pp. 14–30.

Schröch, H., Cha, C.-J. M., and Goldstein, L. (1980). Glutamine release from the hindlimb and uptake by kidney in acutely acidotic rat. *Biochem. J.*, **188**, 557–560.

Schuler, P. (1978). Potentiometry and polarometry. In: *Principles of Enzymatic Analysis*, (Ed. Bergmeyer, H.-U.) Verlag Chemie, New York, pp. 155–162.

Schwartz, S. S., Horwitz, D. L., Zehfus, B., *et al.* (1979). Use of glucose controlled insulin infusion system (artificial beta cell) to control diabetes during surgery. *Diabetologia*, **16**, 157–164.

Scott, J. M., and Weir, D. G. (1981). The methyl folate trap. *Lancet*, **ii**, 337–340.

Scrimshaw, N. S., and Young, V. R. (1976). The requirements of human nutrition. *Scientific American*, **235, No. 3,** 50–73.

Seals, J. R., Czech, M. P. (1982). Production by plasma membranes of a chemical mediator of insulin action. *Fed. Proc.*, **41,** 2730–2735.

Seegmiller, J. E. (1980). Diseases of purine and pyrimidine metabolism. In: *Metabolic Control and Disease*, (Ed. Bondy, P. K., and Rosenberg, L. E.) W. B. Saunders & Co., Philadelphia, pp. 777–938.

Segal, S. (1978). Disorders of galactose metabolism. In: *The Metabolic Basis of Inherited disease*, 4th Edn. (Ed. Stanbury, J. B., Wyngaarden, J. B., and Fredrickson, D. S.) McGraw-Hill Book Co., New York, pp. 160–181.

Seiss, E. A., and Wieland, O. H. (1976). Phosphorylation state of cytosolic and mitochondrial adenine nucleotides and of pyruvate dehydrogenase in isolated rat liver cells. *Biochem. J.*, **156,** 91–102.

Sestoft, L., Folke, M., Gammeltoft, S., *et al.* (1980). Development of diabetic ketoacidosis: some observations on and deductions about the source of acid. *Acta Med. Scand. Supl.*, **639,** 7–16.

Shall, S. (1981). Control of cell reproduction. *Brit. Med. Bull.*, **37,** 209–214.

Shapiro, A., and Shapiro, B. (1979). Role of the liver in intestinal glucose absorption. *Biochim. Biophys. Acta*, **586,** 123–127.

Shapot, V. S. (1972). Some biochemical aspects of the relationship between the tumour and the host. *Adv. Cancer Res.*, **15,** 253–386.

Sharma, G. V. R. K., Cella, G., Parisi, A. F., *et al.* (1982). Thrombolytic therapy. *New Engl. J. Med.*, **306,** 1268–1276.

Shattil, S. J., and Cooper, R. A. (1978). Role of membrane lipid composition, organization and fluidity in human platelet function. *Prog. Hemost. Thromb.*, **4,** 59–86.

Shaw, G. K. (1982). Alcohol dependence and withdrawal. *Brit. Med. Bull.*, **38,** 99–102.

Sheagren, J. N. (1981). Septic shock and corticosteroids. *New Engl. J. Med.*, **305,** 456–458.

Shephard, R. J. (1978). *The Fit Athlete*. Oxford University Press, Oxford, U.K.

Sherlock, S. (1978). The portal vein and portal venous hypertension. *Advanced Medicine*, **14,** 41–58.

Sherlock, S. (1982). Patterns of hepatic injury in man. *Lancet*, **i,** 782–786.

Sherwin, R. S., Kramer, K. J., Tobin, J. D., *et al.* (1974). A model of the kinetics of insulin in man. *J. Clin. Invest.*, **53,** 1481–1492.

Sherwin, R. S., Hendler, R. G., and Felig, P. (1975). Effect of ketone infusions on amino acid and nitrogen metabolism in man. *J. Clin. Invest.*, **55,** 1382–1390.

Shih, V. E. (1978). In: *The Metabolic Basis of Inherited Disease*, 4th Edn. (Ed. Stanbury, J. B., Wyngaarden, J. B., and Fredrickson, D. S.) McGraw-Hill, New York, pp. 362–286.

Shulman, R. G. (1982). NMR Spectroscopy of living cells *Scientific Amer.*, **248, No. 1,** 76–83.

Siegmiller, J. E. (1980). Diseases of purine and pyrimidine metabolism. In: *Metabolic Control and Disease*, (Ed. Bondy, P. K., and Rosenberg, L. E.) Saunders & Co., Philadelphia, pp. 777–938.

Simons, L. A., and Gibson, J. C. (1980). *Lipids: A Clinician's Guide*, M.T.P. Press Ltd., Lancaster, U.K.

Simpson, H. C. R., Simpson, R. W., and Lousley, S. (1981). A high carbohydrate leguminous fibre diet improves all aspects of diabetic control. *Lancet*, **i,** 1–5.

Simpson, W. T. (1977). Nature and incidence of unwanted effects with atenolol. *Postgrad. Med. J.*, **53, Suppl. 3,** 162–167.

Sims, E. A. H., Danforth, E., Horton, E. S., *et al.* (1973). Endocrine and metabolic effects of experimental obesity in man. *Recent Prog. Horm. Res.*, **29,** 457–496.

Sinclair, H. M. (1956). Deficiency of essential fatty acids and atherosclerosis etcetera. *Lancet*, **i,** 381–383.

Sinclair, H. M. (1980). Prevention of coronary heart disease: the role of essential fatty acids. *Postgraduate Med. J.*, **56,** 579–584.

Singer, S. J. (1981). Current concepts of molecular organization in cell membranes. *Biochem. Soc. Trans.,* **9,** 203–206.

Singer, S. J., and Nicolson, G. L. (1972). The fluid mosaic model of the structure of cell membranes. *Science,* **175,** 720–731.

Sizonenko, P. C., Paunier, L., and Vallotton, M. B. (1973). Response to 2-deoxy-D-glucose and to glucagon in 'ketotic hypoglycaemia' of childhood; evidence of epinephrine deficiency and altered alanine availability. *Pediat. Res.,* **7,** 983–993.

Skyler, J. S., and Cahill, G. F. (1981). Diabetes mellitus: progress and directions. *Am. J. Med.,* **70,** 101–104.

Slack, J. (1975). The genetic contribution to coronary heart disease through lipoprotein concentrations. *Postgraduate Med. J.,* **51, Suppl. 8,** 27–32.

Slater, E. C. (1953). Mechanism of phosphorylation in the respiratory chain. *Nature,* **172,** 975–978.

Slater, T. F. (1982). Lipid peroxidation. *Biochem Soc. Trans.,* **10,** 70–71.

Slater, T. F., and Benedetto, C. (1981). Free radical reactions in relation to lipid peroxidation, inflammation and prostaglandin metabolism. In: *The Prostaglandin System,* (Ed. Berti, F., and Velo, G. P.) Plenum Press, New York, pp. 109–126.

Sleight, P. (1981). Beta-adrenergic blockage after myocardial infarction. *New Engl. J. Med.,* **304,** 837–838.

Sleisenger, M. H., and Young, S. K. (1979). Protein digestion and absorption. *New Engl. J. Med.,* **300,** 659–663.

Smith, R. (1979). *Biochemical Disorders of the Skeleton.* Butterworths, London.

Smith, R. and Williamson, D. H. (1983). Biochemical effects of human injury. *Trends Biochem. Sci.,* 142–146.

Smith, R. A. G., Dupe, R. J., English, P. D., *et al.* (1981). Fibrinolysis with acyl-enzymes: a new approach to thrombolytic therapy. *Nature,* **290,** 505–508.

Smith, R. E., and Hock, R. J. (1963). Brown fat: thermogenic effector of arousal in hibernators. *Science,* **140,** 199–200.

Smith, R. E., and Horwitz, B. A. (1969). Brown fat and thermogenesis. *Physiol. Rev.,* **49,** 330–425.

Snell, K. (1980). Muscle alanine synthesis and hepatic gluconeogenesis. *Biochem. Soc. Trans.,* **8,** 205–213.

Snow, D. H., and Guy, P. S. (1980). Muscle fibre type composition of a number of limb muscles in different types of horse. *Res. Vet. Sci.,* **28,** 137–144.

Snyder, S. H. (1980). *Biological Aspects of Mental Disorders.* O.U.P., New York.

Snyder, S. H. (1982). Schizophrenia. *Lancet,* **ii,** 970–973.

Sobel, B. E. (1974). Biochemical and morphologic changes in infarcting myocardium. In: *The Myocardium: Failure and Infarction.* (Ed. Braunwald, E.) N.P. Publishing Co., Inc., New York, pp. 247–260.

Soderling, T. R., Corbin, J., and Park, C. R. (1975). Regulation of adenosine 3,5-monophosphate dependent-protein kinase. *J. Biol. Chem.,* **248,** 1813–1821.

Sodi-Pallares, D., Gautam, H. P., and Medrano, G. A. (1969). Improved cardiac performance with potassium, glucose and insulin. *Lancet,* **i,** 1315–1316.

Sokoloff, L., Fitzgerald, G. G., and Kaufman, E. E. (1977). In: *Nutrition of the Brain.* (Ed. Wurtman, R. J., and Wurtman, J. J.) Raven Press, New York. Vol. 1, pp. 87–139.

Sönksen, P. H. (1977). The evolution of insulin treatment. *Clin. Endocrinol. Metab.,* **6,** 481–498.

Sönksen, P. H., and West, T. E. T. (1979). Growth hormone. In: *Hormones in blood,* (Ed. Gray, C. H., and James, V. H. T.) Academic Press, London. Vol. 1, pp. 225–254.

Southgate, D. A. T. (1978). Dietary fibre: analysis and food sources. *Am. J. Clin. Nutr.,* **31,** S107–S110.

Southgate, D. A. T., and Durnin, J. V. G. A. (1970). Caloric conversion factors and experimental reassessment of the factors used in calculations of the energy value of human diets. *Brit. J. Nutr.*, **24**, 517–535.

Spanner, D. C. (1964). *Introduction to Thermodynamics*. Academic Press, London, New York.

Spechler, S. J., Esposito, A. L., Koff, R. S., *et al.* (1978). Lactic acidosis in cat cell carcinoma with extensive metastases. *Arch. Intern. Med.*, **138**, 1663–1664.

Spector, A. A., and Fletcher, J. E. (1978). Transport of fatty acid in the circulation. In: *Disturbances in Lipid and Lipoprotein Metabolism*, (Ed. Dietschy, J. M., Gotto, A. M., and Ontko, J. A.) American Physiological Society, Bethesda, pp. 229–250.

Spero, L. (1982). Epilepsy. *Lancet*, **ii**, 1319–1322.

Spiegel, A. M., Levine, M. A., Aurbach, G. D., *et al.* (1982). Deficiency of hormone receptor-adenylate cyclase coupling protein: basis for hormone resistance in pseudohypoparathyroidism. *Am. J. Physiol.*, **243**, E37–E42.

Spiro, R. G., and Spiro, M. J. (1971). Effect of diabetes on the biosynthesis of renal glomerular basement membrane. *Diabetes*, **20**, 641–648.

Stadel, B. V. (1981). Oral contraceptives and cardiovascular disease. *New Engl. J. Med.*, **305**, 612–618.

Stadtman, E. R., and Chock, P. B. (1978). Interconvertible enzyme cascades in metabolic regulation. *Curr. Topics Cellul. Reg.*, **13**, 53–95.

Stalmans, W. (1976). The role of the liver in the homeostasis of blood glucose. *Curr. Top. Cellul. Reg.*, **11**, 51–97.

Stamler, J. (1974). The primary prevention of coronary heart disease. In: *The Myocardium: Failure and Infarction*, (Ed. Braunwell, E.) H.P. Publishers Co., New York, pp. 219–236.

Stanley, J. C. (1980). Effect of dietary fibre on lipogenic enzymes in liver. D. Phil. Thesis, University of Oxford.

Start, C. (1971). The physiological significance of hepatic citrate. D. Phil. Thesis, Oxford University.

Starzl, T. E., Klintmalm, G. B. G., and Porter, K. A. (1981). Liver transplantation with use of cyclosporin A and prednisone. *New Engl. J. Med.*, **305**, 266–269.

Stave, U. (1974). Metabolism of the small-for-gestational age infant. *New Engl. J. Med.*, **291**, 359–360.

Stegink, L. D., Freeman, J. B., and Wispe, J., *et al.* (1980). *Am. J. Clin. Nutr.*, **33**, 2490–2494.

Steinberg, D. (1963). Fatty acid mobilization—mechanisms of regulation and metabolic consequences. *Biochem. Soc. Symp.*, **24**, 111–144.

Steinberg, D. (1976). Interconvertible enzymes in adipose tissue regulated by cyclic AMP-dependent protein kinase. *Adv. Cyclic Nucleotide Research*, **7**, 158–197.

Steincrohn, P. J. (1972). *Low Blood Sugar*. Allison & Busby, London.

Sterling, K. (1980). Thyroid hormone action at the cell level. *New Engl. J. Med.*, **300**, 117–122.

Stewart, P. M., and Walser, M. (1980). Proposed mechanism for the hyperammonaemia of propionic and methylmalonic acidaemia. *J. Clin. Invest.*, **66**, 484–492.

Stoner, H. B., Frayn, K. N., Barton, R. N., *et al.* (1979). The relationships between plasma substrates and hormones and severity of injury in 277 recently injured patients. *Clin. Science*, **56**, 563–573.

Storm, D. R., and Koshland, D. E. (1972). An indication of the magnitude of orientation factors in esterification. *J. Am. Chem. Soc.*, **94**, 5805–5814.

Stuart, M. J., Gerrard, J. M., and White, J. G. (1980). Effect of cholesterol on production of thromboxane B_2 by platelets *in vitro*. *New Engl. J. Med.*, **302**, 6–10.

Stubbs, M. (1979). Inhibitors of the adenine nucleotide translocase. *Pharmac. Ther.*, **7**, 329–349.

Sugden, M. C., and Ashcroft, S. J. H. (1977). Phosphoenolpyruvate in rat pancreatic islets: a possible intracellular trigger of insulin release? *Diabetologia*, **13**, 481–486.

Sugden, P. H., and Newsholme, E. A. (1975). Effects of ammonium, inorganic phosphate and potassium ions on the activity of phosphofructokinases from muscle and nervous tissue of vertebrates and invertebrates. *Biochem. J.*, **150**, 113–122.

Sugden, P. H., and Newsholme, E. A. (1977). Activities of choline acetyltransferase acetylcholine esterase, glutamate decarboxylase, 4-aminobutyrate aminotransferase and carnitine acetyltransferase in nervous tissue from some vertebrates and invertebrates. *Comp. Biochem. Physiol.*, **560**, 89–94.

Sullivan, A. C., and Triscari, J. (1977). Influence of hydroxycitrate on experimentally induced obesity in the rodent. *Am. J. Clin. Nutr.*, **30**, 767–776.

Sutherland, E. W., Øye, I., and Butcher, R. W. (1965). The action of epinephrine and the role of adenyl cyclase system in hormone action. *Rec. Progr. Horm. Res.*, **21**, 623–642.

Sutton, J. R. (1981). Drugs used in metabolic disorders. *Medicine and Science in Sports and Exercise*, **13**, 266–271.

Taggart, P., and Carruthers, M. (1971). Endogenous hyperlipidaemia induced by emotional stress of racing driving. *Lancet*, **i**, 363–366.

Tall, A. R., and Small, D. M. (1978). Plasma high-density lipoproteins. *New Engl. J. Med.*, **299**, 1232–1236.

Tan, S. Y., and Mulrow, P. J. (1980). Aldosterone in hypertension and oedema. In: *Metabolic Control and Disease*. (Ed. Bondy, P. K., and Rosenberg, L. E.) Saunders Co., Philadelphia, pp. 1501–1534.

Tarlow, M. J., Hadorn, B., Arthurton, M. W., *et al.* (1970). Intestinal enterokinase deficiency: a newly-recognized disorder of protein digestion. *Arch. Dis. Childhood*, **45**, 651–655.

Tarui, S., Okuno, G., Ikura, Y., *et al.* (1965). Phosphofructokinase deficiency in skeletal muscle. *Biochem. Biophys. Res. Commun.*, **19**, 517–523.

Tate, S. S., and Meister, A. (1973). *The Enzymes of Glutamine Metabolism* (Ed. Prusiner, S. and Stadtmann, E. R.) Academic Press, New York. pp. 77–125.

Taylor, S. H., Silke, B., and Lee, P. S. (1982). Intravenous beta-blockade in coronary heart disease. *New Engl. J. Med.*, **306**, 631–635.

Tchobroutsky, G. (1978). Relation of diabetic control to development of microvascular complications. *Diabetologia*, **15**, 143–152.

Thelander, L., and Reichard, P. (1979). Reducation of ribonucleotides. *Annu. Rev. Biochem.*, **48**, 133–158.

Thoene, J. G., Lemons, R., and Baker, H. (1983). Impaired intestinal absorption of biotin in juvenile multiple carboxylase deficiency. *New Engl. J. Med.*, **308**, 639–642.

Thom, M. H., White, P. J., Williams, R. M., *et al.* (1979). Prevention and treatment of endometrial disease in climacteric women receiving oestrogen therapy. *Lancet*, **ii**, 455–457.

Thomas, P. K., and Ward, J. P. (1975). Diabetic neuropathy. In: *Complications of Diabetes*, (Ed. Keen, H., and Jarret, J.) Arnold, London, pp. 151–177.

Thomsen, M., Nerup, J., Christy, M., *et al.* (1980). HLA antigens and diabetes. In: *Inherited Disorders of Carbohydrate Metabolism*, (Ed: Barman, D., Holton, J. B., and Pennock, C. A.) M.T.P. Press, Lancaster, U.K., pp. 401–409.

Thomson, A. B. R. (1978). Intestinal absorption of lipids: influence of the unstirred water layer and bile acid micelle. In: *Disturbances in Lipid and Lipoprotein Metabolism*. (Ed. Dietschy, J. M., Gotto, A. M., and Ontko, J. A.) American Physiological Society, Bethesda, pp. 29–56.

Thorngren, M., and Gustafson, A. (1981). Effects of 11-week increase in dietary eicosapentaenoic acid on bleeding time, lipids and platelet aggregation. *Lancet*, **ii**, 1190–1193.

Tipton, K. F. (1975). Monoamine oxidase. In: *Handbook of Physiology, Endocrinology VI, Adrenal gland*, (Ed. Blaschko, J., Sayers, G., and Smith, A. D.) American Physiological Society, Washington, D. C., pp. 677–697.

Tischler, M. E., and Goldberg, A. L. (1980). Leucine degradation and release of glutamine and alanine by adipose tissue. *J. Biol. Chem.*, **255**, 8074–8081.

Topping, D. L., and Mayes, P. A. (1976). Regulation of lipogenesis by insulin and free fatty acids in the perfused rat liver. *Biochem. Soc. Trans.*, **4**, 717.

Torday, J., Carson, L., and Lawson, E. E. (1979). Saturated phosphatidylcholine in amniotic fluid and prediction of the respiratory distress syndrome. *New Engl. J. Med.*, **301**, 1013–1018.

Traynor, C., and Hall, G. M. (1981). Endocrine and metabolic changes during surgery: anaesthetic implications. *Brit. J. Anaesthesia*, **53**, 153–160.

Trichopoulos, D., Klatsouyanni, K., Zavitsanos, X. (1983). Psychological stress and fatal heart attack: the Athens (1981) Earthquake natural experiment. *Lancet*, **i**, 441–444.

Triger, D. R. (1981). *Practical Management of Liver Disease*. Blackwell Scientific Publ., Oxford.

Trowell, H. C. (1973). Kwashiorkor. In: *Food: Readings from Scientific American*, Freeman & Co., San Francisco, pp. 52–56.

Trowell, H. C., and Burkett, D. P., (1981). *Western Diseases: Their Emergence and Prevention*. Edward Arnold, London.

Trowell, H. C., Southgate, D. A. T., Wolever, T. M. S., *et al.* (1976). Dietary fibre redefined. *Lancet*, **i**, 967.

Tunstall-Pedoe, D. (1979). Exercise and sudden death. *Brit. J. Sports Med.*, **12**, 215–219.

Ueda, K., Ogata, N., Kawaichi, M., *et al.* (1982). ADP-Ribosylation reactions. *Curr. Topic. Cellul. Regul.*, **21**, 175–186.

Umbreit, W. W., Burris, R. H., and Stauffer, J. F. (1964). *Manometric Techniques*. Burgers Publishing Co., Minneapolis, U.S.A.

Unger, R. H. (1971). Glucagon and the insulin:glucagon ratio in diabetes and other catabolic illnesses. *Diabetes*, **20**, 834–838.

Unger, R. H. (1974). Alpha- and beta-cell interrelationships in health and disease. *Metabolism*, **23**, 581–593.

Vacarro, K. K., Liang, B. T., Perelle, B. A., *et al.* (1980). Tyrosine 3-mono-oxygenase regulates catecholamine synthesis in pheochromocytoma cells. *J. Biol. Chem.*, **255**, 6539–6541.

Van Biervliet, J. P. G. M., Bruinvis, L., and Ketting, D. (1977). Hereditary mitochondrial myopathy with lactic acidemia: Defective respiratory chain in voluntary striated muscles. *Paediatric Res.*, **11**, 1088–1093.

Van De Werve, G. (1977). The effect of insulin on glycogenolytic cascade and on the activity of glycogen synthase in the liver of anaesthetized rabbits. *Biochem. J.*, **162**, 143–146.

Van Den Berghe, G., Bontemps, F., and Hers, H.-G. (1980). Purine catabolism in isolated rat hepatocytes. *Biochem. J.*, **188**, 913–920.

Van Herle, A. J., Vassat, G., and Dumont, J. E. (1979). Control of thyroglobulin synthesis and secretion. *New Engl. J. Med.*, **301**, 239–248; 307–314.

Van Heyningen, S. (1976). The molecular mechanism of cholera. *New Scientist* **25 March 1976**, 692–694.

Van Praag, H. M. (1982). Depression. *Lancet*, **ii**, 1259–1264.

Van Schaftingen, E., Hue, L., and Hers, H.-G. (1980). Study of the fructose 6-phosphate/fructose 1,6-bisphosphate cycle in liver *in vivo*. *Biochem. J.*, **192**, 263–271.

Van Thiel, D. H., and Lester, R. (1974). Sex and alcohol. *New Engl. J. Med.*, **291**, 251–253.

Van Weemen, B. K., and Schuurs, A. H. M. M. (1978). Principles of enzyme-immunoassays. In: *Principles of Enzymatic Analysis*, (Ed. Bergmeyer, H.-U.) Verlag Chemie, New York, pp. 93–98.

Van Wyk, J. J., and Underwood, L. E. (1978). The somatomedins and their actions. In: *Biochemical Actions of Hormones*, (Ed. Litwack, G.) Academic Press, New York. Vol. 5, pp. 102–148.

Vane, J. R. (1971). Inhibition of prostaglandin synthesis as a mechanism of action for aspirin-like drugs. *Nature New Biology*, **231**, 232–235.

Vane, J. R. (1974). Mode of action of aspirin and similar compounds. In: *Prostaglandin Synthetase Inhibitors*, (Ed. Robinson, H. J., and Vane, J. R.) Raven Press, New York, pp. 155–163.

Vary, T. C., Reibel, D. K., and Neely, J. R. (1981). Control of energy metabolism of heart muscle. *Ann. Rev. Physiol.*, **43**, 419–430.

Veech, R. L., Lawson, J. W. R., Cornell, N. W., et al. (1979). Cytosolic phosphorylation potential. *J. Biol. Chem.*, **254**, 6538–6547.

Vessey, M. P. (1980). Female hormones and vascular disease—an epidemiological overview. *Br. J. Family Plann.*, **6, Suppl.** 1–12.

Vignais, P. V., and Lauquin, G. J. M. (1979). Mitochondrial adenine nucleotide transport and its role in the economy of the cell. *Trends Biochem. Sci.*, **4**, 90–92.

Villar-Palasi, L., and Larner, J. (1968). The hormonal regulation of glycogen metabolism in muscle. *Vitamin and Hormones*, **26**, 65–118.

Visser, T. J. (1980). Deiodination of thyroid hormone and the role of glutathione. *Trends Biochem. Sci.*, **5**, 222–224.

Vlahcevic, Z. R., Bell, C. C., Buhac, I., et al. (1970). Diminished bile acid pool size in patients with gall stones. *Gastroenterology*, **59**, 165–173.

Voaden, M. J. (1979). Vision: the biochemistry of the retina. In: *Companion to Biochemistry*, (Ed. Bull, A. T., et al.) Longman Group Ltd., London. Vol. 2, pp. 451–473.

Von Euler, U. S. (1934). *Nounyn-Schmiedeberg's Arch. exp. Path. Pharmak.*, **175**, 78–84.

Vranic, M., and Berger, M. (1979). Exercise and diabetes mellitus. *Diabetes*, **28**, 147–167.

Vulliet, P. R., Langon, T. A., and Weiner, N. (1980). Tyrosine hydroxylase: Substrate of cyclic AMP-dependent protein kinase. *Proc. Natl. Acad. Sci., U.S.A.*, **77**, 92–96.

Wahren, J. (1979). Glucose turnover during exercise in healthy man and in patients with diabetes mellitus. *Diabetes*, **28**, 82–88.

Walker, P. R. (1977). Regulation of enzyme synthesis in animal cells. *Essays in Biochem.*, **13**, 39–69.

Walker, W. A., and Isselbacher, K. J. (1974). Uptake and transport of macromolecules by the intestine: possible role in clinical disorders. *Gastroenterology*, **67**, 531–550.

Walli, R. A. (1978). Interrelation of aerobic glycolysis and lipogenesis in isolated perfused liver of well fed rats. *Biochim. Biophys. Acta*, **539**, 62–80.

Walser, M. M. D. (1978). Principles of ketoacid therapy in uremia. *Am. J. Clin. Nutrition*, **31**, 1756–1760.

Walser, M. M. D. (1983). Nutritional support in renal failure. *Lancet*, **i**, 340–342.

Walsh, M. J. (1971). Role of acetaldehyde in the interactions of ethanol with neuroamines. In: *Biological Aspects of Alcohol*, (Ed. Roach, M. K., McIsaac, W. M., and Creaven, P. J.) University of Texas Press, Austin, London, p. 233–266.

Warburg, O. (1956a). The origin of cancer cells. *Science*, **123**, 309–314.

Warburg, O. (1956b). *Science*, **124**, 269–270.

Warshaw, J. B. (1975). An energy crisis in muscle. *New Engl. J. Med.*, **292**, 476–477.

Waterlow, J. C., and Jackson, A. A. (1981). Nutrition and protein turnover in man. *British Medical Bulletin*, **37**, 5–10.

Waterlow, J. C., Garlick, P. J., and Millward, D. J. (1978). *Protein Turnover in Mammalian Tissues and in the Whole Body.*, Elsevier, Amsterdam.

Watkins, W. M., Yates, A. D., and Greenwell, P. (1981). Blood group antigens and the enzymes involved in their synthesis: past and present. *Biochem. Soc. Trans.*, **9**, 186–190.

Watt, B. K., and Merrill, A. L. (1963). *Composition of Foods*. U.S. Dept. Agric. Handbook No. 8, Washington D.C.

Watts, C. A. H. (1973). *Depression: the Blue Plague*. Priory Press Ltd., London.

Watts, R. W. E. (1981). Purine metabolism. *Biochem. Soc. Trans.*, 3–5.

Weeds, A. (1980). Myosin light chains, polymorphism and fibre types in skeletal muscle.

In: *Plasticity of Muscle*, (Ed. Pette, D.) Walter de Grayter & Co., Berlin, New York, pp. 55–68.

Weinberg, E. D. (1980). Roles of iron in fever and infectious disease. In: *Thermoregulatory Mechanisms and Their Therapeutic Implications*, (Ed. Cox, B., Lomax, P., Milton, A. S., *et al.*) S. Karger, Basel, pp. 105–110.

Weinhouse, S. (1976). Regulation of glucokinase in liver. *Curr. Topics Cellul. Regul.*, **11**, 1–50.

Weinstein, M. C. (1980). Estrogen use in postmenopausal women—costs, risks and benefits. *New Engl. J. Med.*, **303**, 308–316.

Weis-Fogh, T. (1952). Fat combustion and metabolic rate of flying locusts (*Schistocerca gregaria*). *Phil. Trans. Roy. Soc. B (London)*. **237**, 1–36.

Weis-Fogh, T. (1961). Power in flapping flight. In: *The Cell and the Organism*, (Ed. Ramsay, J. A., and Wigglesworth, V. B.) Cambridge University Press, London, pp. 283–300.

Weiss, N. S., Ure, C. L., Ballard, J. H., *et al.* (1980). Decreased risk of fractures of the hip and lower forearm with postmenopausal use of oestrogen. *New Engl. J. Med.*, **303**, 1195–1198.

Weissmann, G. (1983). The eicosanoids of asthma. *New Engl. J. Med.*, **308**, 454–455.

Weissman, S. M. (1980). Gene structure and function. In: *Metabolic Control and Disease*. (Ed. Bondy, P. K., and Rosenberg, L. E.) W. B. Saunders and Co., Philadelphia. 8th Edn., pp. 1–26.

Weiter, J. J. (1981). Retrolental fibroplasia: an unsolved problem. *New Engl. J. Med.*, **305**, 1404–1406.

Weksler, B. B., Pett, S. B., Alonso, D., ·*et al.* (1983). Differential inhibition by aspirin of vascular and platelet prostaglandin synthesis in atherosclerotic patients. *New Engl. J. Med.*, **308**, 800–805.

Wellner, D., and Meister, A. (1981). A survey of inborn errors of amino acid metabolism and transport in man. *Annu. Rev. Biochem.*, **50**, 911–968.

Wergedal, J. E., and Harper, A. E. (1964). Effect of high protein intake on amino nitrogen catabolism *in vivo*. *J. Biol. Chem.*, **239**, 1156–1163.

Whedon, G. D. (1981). Osteoporosis. *New Engl. J. Med.*, **305**, 397–399.

WHO/FAO Report (1973). *Energy and Protein Requirements*. WHO Technical Report Series No. 522. Geneva.

Wicks, W. D. (1974). Regulation of protein synthesis by cyclic AMP. *Adv. Cyclic Nucleotide Res.*, **4**, 335–438.

Wilkie, D. R. (1970). *Chem. Brit.*, **6**, 475–479.

Wilkie, D. R. (1979). Generation of protons by metabolic processes other than glycolysis in muscle cells. *J. Mol. Cellul. Cardiol.*, **11**, 325–330.

Willett, W., Hennekens, C. H., Siegel, A. T., *et al.* (1980). Alcohol consumption and high density lipoprotein cholesterol in marathon runners. *New Engl. J. Med.*, **303**, 1159–1161.

Williams, C. D. (1933). A nutritional disease of children associated with a maize diet. *Archiv. Dis. Childhood*, **8**, 423–433.

Williams, J. F. (1980). A critical examination of the evidence for the reactions of the pentose pathway in animal tissues. *Trends Biochem. Sci.*, **5**, 315–320.

Williams, R. H., and Palmer, J. P. (1975). Farewell to phenformin for treating diabetes mellitus. *Ann. Intern. Med.*, **83**, 567–568.

Williams, R. S. (1978). The management of liver failure. *Advanced Medicine* **14**, 19–29.

Williamson, D. H. (1973). Tissue-specific direction of blood metabolites. *Symp. Soc. Expt. Biol.*, **27**, 283–298.

Williamson, D. H. (1980). Integration of metabolism in the tissues of the lactating rat. *FEBS Lettr.*, **117**, K93–K105.

Williamson, D. H. (1981). Regulation of ketone body metabolism and the effects of injury.

Acta Chirurgica Scandinavica, **Suppl. 507**, 22–29.

Williamson, D. H., and Brosnan, J. T. (1974). Concentrations of metabolites in animal tissues. In: *Methods of Enzymatic Analysis*, (Ed. Bergmeyer, H.-U.) Academic Press/Verlag Chemie, New York. Vol. 4, pp. 2266–2302.

Williamson, D. H., and Whitelaw, E. (1978). Physiological aspects of the regulation of ketogenesis. *Biochem. Soc. Symp.*, **43**, 137–161.

Williamson, D. H., Farrell, R., Kerr, A., *et al.* (1977). Muscle-protein catabolism after injury in man, as measured by urinary excretion of 3-methylhistidine. *Clin. Sci. Mol. Med.*, **52**, 527–533.

Williamson, J. R., and Cooper, R. H. (1980). Regulation of the citric acid cycle in mammalian systems. *FEBS Lettr.*, **117**, K73–K85.

Williamson, J. R., Safer, R., Lanoue, K. F., *et al.* (1973). Mitochondrial cytosolic interactions in cardiac tissue: role of the malate-aspartate cycle in the removal of glycolytic NADH from the cytosol. *Symp. Soc. Exp. Biol.*, **27**, 241–282.

Wilmore, D. W. (1979). Metabolic changes in burns. In: *Burns*, (Ed. Artz, C. P., Moncrief, J. A., and Pruitt, B. A.) W. B. Saunders & Co., Philadelphia, U.S.A., pp. 130–131.

Wilson, D. H., Nishiki, K., and Erecinska, M. (1981). Energy metabolism in muscle and its regulation during individual contraction relaxation cycles. *Trends Biochem. Sci.*, **6**, 16–19.

Windmueller, H. C., and Spaeth, A. E. (1974). Uptake and metabolism of plasma glutamine by the small intestine. *J. Biol. Chem.*, **249**, 5070–5079.

Windmueller, H. G., and Spaeth, A. E. (1980). Respiratory fuels and nitrogen metabolism *in vivo* in small intestine of fed rats. *J. Biol. Chem.*, **255**, 107–122.

Wissler, R. W., and Vesselinovitch, S. (1977). *Med. Concepts Cardiovascular Dis.*, **46**, 27–32.

Wollenberger, A., Ristau, O., and Schoffa, G. (1960). *Pflügers Archiv. ges. Physiol.*, **270**, 399.

Womersley, J., and Durnin, J. V. G. A. (1977). A comparision of skinfold method with extent of overweight and various weight-height relationships in the assessment of obesity. *Brit. J. Nutr.*, **38**, 271–284.

Wood, P. D., Haskell, W. L., and Stern, M. P. (1977). Plasma lipoprotein distributions in male and female runners. *Annals New York Acad. Sci.*, **301**, 748–763.

Woodhead-Galloway, J. (1980). *Collagen: The Anatomy of a Protein*. Edward Arnold, London.

Woods, H. F. (1974). Some metabolic aspects of parenteral feeding. *Topics in Gastroenterology*, **2**,

Woods, H. F. (1980). Pathogenic mechanisms of disorders in fructose metabolism. In: *Inherited Disorders of Carbohydrate Metabolism*, (Ed. Burman, D., Holton, J. B., and Pennock, C. A.) M.T.P. Press Ltd., Lancaster, U.K., pp. 191–203.

Woods, H. F., and Alberti, K. G. M. M. (1972). Dangers of intravenous fructose. *Lancet*, **ii**, 1354–1357.

Woods, H. F., Eggleston, L. V., and Krebs, H. A. (1970). The cause of hepatic accumulation of fructose-1-phosphate on fructose loading. *Biochem. J.*, **119**, 501–510.

Wool, I. G. (1964). Theoretical endocrinology. In: *Bone Biodynamics*, Little Brown & Co., pp. 285–314.

Woolfson, A. M. J., Heatley, R. V., and Allison, S. P. (1979). Insulin to inhibit protein catabolism after injury. *New Engl. J. Med.*, **300**, 14–17.

Wright, J. (1978). Endocrine effects of alcohol. *Clinics in Endocrinol. Metab.*, **7**, 351–367.

Wurtman, R. J. (1983). Behavioural effects of nutrients. *Lancet*, **i**, 1145–1147.

Wyatt, P. A. H. (1967). *Energy and Entropy in Chemistry*, London, Macmillan.

Wyndham, C. H. (1977). Heat stroke and hyperthermia in marathon runners. *Annals New York Acad, Sci.*, **301**, 128–138.

Wynn, V. (1975). Metabolic effects of anabolic steroids. *Brit. J. Sports Medicine*, **9**, 60–64.

Xenophon, *Anabasis III*, (Trans. Carleton, L.) Heinemann, London. Books I–VII, p. 45.

Yager, J., and Young, R. T. (1974). Non-hypoglycaemia is an epidemic condition. *New Engl. J. Med.*, **291**, 907–908.

Yalow, R. S., and Berson, S. A. (1960). Immunoassay of endogenous plasma insulin in man. *J. Clin. Invest.*, **39**, 1157–1175.

Yoon, J.-W., Austin, M., Onodera, T., *et al.* (1979). Virus induced diabetes mellitus. *New Engl. J. Med.*, **300**, 1173–1179.

Youmans, J. B. (1964). The changing face of nutritional disease in America. *J. Am. Med. Ass.*, **189**, 672–676.

Young, J. B., and Landsberg, L. (1977). Stimulation of the sympathetic nervous system during sucrose feeding. *Nature*, **269**, 615–617.

Young, V. R. (1981). Selenium: a case for its essentiality in man. *New Engl. J. Med.*, **304**, 1228–1230.

Young, V. R., and Munro, H. N. (1978). N^t—methylhistidine (3-methylhistidine) and muscle protein turnover: an overview. *Fed. Proc.*, **37**, 2291–2300.

Young, V. R., and Pluskal, M. G. (1977). Mode of action of anabolic agents with special reference to steroids and skeletal muscle: a summary review. In: *Int. Symp. Protein Metabolism and Nutrition. The Netherlands, May 2–6 1977*. Proc. Europ. Asse. Animal Production. Unpubl., pp. 17–28.

Young, V. R., Haverberg, L. N., Bilmazes, C., *et al.* (1973). Potential use of 3-methylhistidine excretion as an index of progressive reduction in muscle protein catabolism during starvation. *Metabolism*, **22**, 1429–1436.

Yudkin, J. (1972). *Pure, White and Deadly*. Davis-Poynter Ltd., London.

Zammit, V. A. (1981). Intrahepatic regulation of ketogenesis. *Trends Biochem. Sci.*, **6**, 46–49.

Zammit, V. A., and Newsholme, E. A. (1976). The maximum activities of hexokinase, phosphorylase, phosphofructokinase, glycerol phosphate dehydrogenase, lactate dehydrogenase, octopine dehydrogenase, phosphoenolpyruvate carboxykinase, nucleoside diphosphatekinase, glutamate oxaloacetate transaminase and arginine kinase in relation to carbohydrate utilization in muscles from marine invertebrates. *Biochem. J.*, **160**, 447–462.

Zammit, V. A., and Newsholme, E. A. (1979). Activities of enzymes of fat and ketone body metabolism and effects of starvation on blood concentrations of glucose and fat fuels in teleost and elasmobranch fish. *Biochem. J.*, **184**, 313–322.

Zammit, V. A., Beis, I., and Newsholme, E. A. (1978). Maximum activities and effects of fructose bisphosphate on pyruvate kinase from muscles of vertebrates and invertebrates in relation to control of glycolysis. *Biochem. J.*, **174**, 989–998.

Zammit, V. A., Beis, A., and Newsholme, E. A. (1979). The role of 3-oxo-acid-CoA transferase in the regulation of ketogenesis in the liver. *FEBS Lettr.*, **103**, 212–215.

Zipes, D. P. (1981). New approaches to antiarrhythmic therapy. *New Engl. J. Med.*, **304**, 475–476.

Zuckerman, H., and Harpaz-Kepiel, S. (1979). Prostaglandins and their inhibitors in premature labour. In: *Practical Applications of Prostaglandins and Their Synthesis Inhibitors*, (Ed. Karim, S. M. M.) M.T.P. Press Ltd., Lancaster, U.K., pp. 411–436.

Zussman, R. M., Rubin, R. H., Cato, A. E., *et al.* (1981). Hemodialysis using prostacyclin instead of heparin as the sole antithrombotic agent. *New Engl. J. Med.*, **304**, 934–939.

INDEX